Satellite Communications: The First Quarter Century of Service
David W. E. Rees

Fundamentals of Telecommunication Networks
Tarek N. Saadawi, Mostafa Ammar, with Ahmed El Hakeem

Meteor Burst Communications: Theory and Practice
Donald L. Schilling, Editor

Signaling in Telecommunication Networks
John G. van Bosse

Telecommunication Circuit Design
Patrick D. van der Puije

Worldwide Telecommunications Guide for the Business Manager
Walter H. Vignault

DIGITAL COMMUNICATION RECEIVERS

Digital Communication Receivers

Synchronization, Channel Estimation, And Signal Processing

Heinrich Meyr
Marc Moeneclaey
Stefan A. Fechtel

A Wiley-Interscience Publication
JOHN WILEY & SONS, INC.
New York • Chichester • Weinheim • Brisbane • Singapore • Toronto

This book is printed on acid-free paper ♾.

Published simultaneously in Canada.

Library of Congress Cataloging-in-Publication Data:

Meyr, Heinrich.
Digital communication receivers : synchronization, channel estimation and signal processing / Heinrich Meyr, Marc Moeneclaey, and Stefan A. Fechtel.
p. cm. — (Wiley series in telecommunications and signal processing)
Includes bibliographical references.
ISBN 0-471-50275-8 (alk. paper)
1. Digital communications. 2. Radio—Receivers and reception.
3. Signal Processing—Digital techniques. I. Moeneclaey, Marc.
II. Fechtel, Stefan A. III. Title. IV. Series.
TK5103.7.M486 1997
621.382—dc21 97-17108
CIP

Printed in the United States of America.

10 9 8 7 6 5

To Nicole, Claudia, and Tarek
and to Mieke, Sylvia, and Bart

Contents

PART C Baseband Communications

Preface

This book is about receivers for digital communications. The word *digital* carries a double meaning. It refers to the fact that information is transported in digital form. It also refers to the property of the receiver to retrieve the transmitted information, apart from the analog front end, entirely by means of digital signal processing.

The ever-increasing demand for mobile and portable communication ultimately calls for optimally utilizing the available bandwidth. This goal is only attainable by digital communication systems capable of operating close to the information theoretic limits. The implementation of such systems has been made possible by the enormous progress in semiconductor technology which allows the communication engineer to economically implement "systems on silicon" which execute:

1. Advanced compression algorithms to drastically reduce the bit rate required to represent a voice or video signal with high quality.
2. Sophisticated algorithms in the receiver for power control, channel estimation, synchronization, equalization, and decoding.
3. Complex protocols to manage traffic in networks.
4. User-friendly graphical man–machine interfaces.

Seen from a different perspective we argue that the communication engineer today can trade the familiar physical performance measures *bandwidth* and *power efficiency* for signal processing *complexity*. As a consequence, the design process is characterized by a close interaction of architecture and algorithm design, as opposed to a separation between theory and implementation in the traditional way.

This book differs substantially from other texts in communication engineering in the selection and treatment of topics. We focus on channel estimation, synchronization, and digital signal processing. In most books on digital communications, synchronization and channel estimation are addressed only superficially, if at all. This must give the reader the impression that these tasks are trivial and that the error performance is always close to the limiting case of perfect channel knowledge and synchronization. However, this is a most unfortunate misconception for the following reasons:

1. Error performance: Synchronization and channel estimation are critical to error performance;
2. Design effort: A large amount of design time is spent in solving these problems;

3. Implementation: A very large portion of the receiver hardware and software is dedicated to synchronization and channel estimation.

In our endeavour to write this book we have set the following goals. First, we have made an effort to systematically structure the information on these topics to make it accessible to the communication engineer, and second, we hope to convince the reader that channel estimation and synchronization algorithms can be developed very elegantly using mathematical estimation theory. By this we hope to supply the reader not only with a collection of results but to provide an elegant mathematical tool to solve design problems.

The book falls naturally into four major parts. Following the introduction (Part A), Part B is devoted to mathematical background material. Part C is concerned with baseband communication. Part D deals with passband communication over time variant channels. Part E addresses transmission over fading channels. The structuring of the material according to types of channels is motivated by the fact that the receiver is always designed for a given channel. For example, the engineer designing a clock recovery circuit for a fiber-optical channel receiver does not need to worry about channel estimation algorithms for fading channels.

BASIC MATERIAL

The purpose of this part of the book is to provide a concise treatment of those topics in random process theory, signal analysis, and estimation theory that are used later on in the book. It is not intended as a replacement for formal courses in these topics but might serve to fill in a limited number of gaps in the reader's knowledge. For the mathematically versed reader it serves as a quick summary of results and introduction to the notation used.

BASEBAND COMMUNICATION

There are four sections in this part. Section 2.1 provides an introduction to baseband communication with emphasis on high data rate applications. We discuss the practically important topic of line coding which is often ignored in more theoretically inclined books on communication theory. Section 2.2 provides a classification of clock recovery circuits and the disturbances which impair the performance of such circuits. The next two sections (2.3 and 2.4) are devoted to an in-depth treatment of the two main classes of clock recovery circuits. We discuss the differences and commonalities of error feedback synchronizers (Section 2.3) and spectral line generating synchronizers (Section 2.4). We also study the effect of jitter accumulation in a chain of repeaters. In section 2.5 we analyze in detail the performance of three practically important synchronizers. The influence of different disturbances such as self-noise and additive noise are discussed. Self-noise reducing techniques, which are crucial when clock recovery circuits are cascaded, are described.

PASSBAND COMMUNICATION OVER TIME INVARIANT CHANNELS

Chapter 3 briefly reviews the fundamentals of passband transmission over a time invariant channel. In Chapter 4 the optimum maximum-likelihood receiver is rigorously derived. We differ from the usual treatment of the subject by deriving a fully digital receiver structure, comprising digital matched filter, interpolator, variable rate decimator, and carrier phase rotator. While the optimal receiver is clearly nonrealizable, it forms the basis for suitable approximations leading to practical receiver structures, the key issue being the principle of synchronized detection. In Chapter 5 estimation theory is used to systematically derive algorithms for timing and carrier phase synchronization. Chapter 6 is concerned with performance analysis of synchronizers. We derive lower bounds on the variance of synchronization parameter estimates. These bounds serve as a benchmark for practical synchronizers. The tracking performance of carrier and symbol synchronizers is analyzed in Section 6.3. Nonlinear phenomena, such as acquisition and cycle slips, are discussed in Section 6.4. Chapter 7 derives the bit error rate performance degradation as function of the carrier phase and timing error variance for practically important modulation formats. Chapter 8 is concerned with the derivation of digital algorithms for frequency estimation and the performance analysis of these algorithms. Chapter 9 discusses timing recovery by interpolation and controlled decimation. Chapter 10 is devoted to implementation. We address the design methodology and CAD tools. A recently completed DVB (Digital Video Broadcasting) ASIC chip design serves as a case study to illustrate the close interaction between algorithm and architecture design.

COMMUNICATION OVER FADING CHANNELS

Chapter 11 gives an introduction to fading channels. Starting with time-continuous and discrete-equivalent flat and selective fading channel models, the statistical characterization of fading channels as well as techniques of modeling and simulation of discrete-equivalent fading channels are discussed. Chapter 12 is concerned with the fundamentals of detection and parameter synchronization on fading channels. Based on mathematical transmission models, optimal estimation-detection receiver structures for joint detection and synchronization are derived. Chapter 13 presents realizable receiver structures for synchronized detection on flat and selective fading channels. The concept of outer and inner receivers is reviewed, with emphasis on the role of diversity. Inner receiver structures, particularly suited for serial or TDMA-like channel access, are discussed for both flat and selective fading channels, detailing on decision metric computation and preprocessing of the received signal. A brief extension to CDMA systems concludes the chapter. Parameter synchronization for flat fading channels is the topic of Chapter 14. Algorithms for non-data-aided (NDA), decision-directed (DD), and data-aided (DA) flat fading channel estimation and their performance are discussed in detail. In Chapter 15 adaptive algorithms for NDA/DD selective fading channel estimation, as well as methods of

DA snapshot acquisition and channel interpolation, are presented and discussed. Bit error results, both for uncoded and coded transmission, and a comparison between NDA/DD and DA performance conclude this final chapter.

We invite comments or questions regarding the text to be sent to us at the following email addresses: meyr@ert.rwth-aachen.de, fechtel@hl.siemens.de, mm@lci.rug.ac.be.

POST SCRIPT

This book was intended as a companion volume to *Synchronization In Digital Communications*, by H. Meyr and G. Ascheid, published by Wiley in 1990. In writing this book we increasingly felt that the 1990 title did not adequately describe the content, and so the title of this volume was changed. Nevertheless, when we speak of Volume 1, we refer to the 1990 publication.

HEINRICH MEYR
MARC MOENECLAEY
STEFAN A. FECHTEL

Aachen, Germany
August 1997

Acknowledgments

We wish to express our gratitude to anyone who has helped to write this book. We (S. Fechtel, H. Meyr) are deeply indebted to our former and present colleagues at the Institute for Integrated Systems in Signal Processing (ISS) at the Aachen University of Technology (RWTH), who provided research results, shared their knowledge with us, reviewed portions of the manuscript, contributed to the text, and engaged in many exciting discussions on the subjects. In particular we are thankful for:

CONTRIBUTIONS TO THE TEXT

The Chapter of frequency estimation is based on the research of Dr. F. Classen who also contributed much to the writing and reviewing of Chapter 8. Dr. V. Zivojnovic provided the results on digital interpolation and carefully reviewed Chapter 9. The case study reported in Chapter 10 greatly profited from the expert knowledge of ASIC designers, Dr. H. Dawid, Dr. O. Joeressen, Dr. U. Lambrette, and M. Vaupel who also contributed to writing the chapter. S. Bitterlich's expertise in Viterbi decoder design was instrumental in the bibliography on conventional decoders. He also reviewed parts of the manuscript.

VALUABLE DISCUSSIONS AND COMMENTS

The following colleagues have given freely of their time to read the manuscript, in part or full, Dr. H. Dawid, Dr. G. Fettweis, Dr. O. Joeressen, Dr. O. Mauss, Dr. F. Mehlan, Dr. M. Oerder, Dr. M. Pankert, Dr. S. Ritz, and Dr. P. Zepter. We are also grateful to our graduate students, J. Baltersee, G. Fock, T. Grötker, F. Munsche, G. Post, R. Schoenen, C. Schotten, M. Speth, and M. Willems for reading parts of the manuscript and making helpful suggestions to improve the book.

We would like to thank Dr. Floyd Gardner for many valuable inputs during the course of writing this book. The DVB chip which serves as a case study in Chapter 10 was a collaboration of Siemens and ISS. We gratefully acknowledge the leadership of the management by R. Schwendt and Dr. C. von Reventlow, who supported the project, and the technical contribution of Dr. K. Müller and F. Frieling. Our sincered thanks also go to our colleagues at Synopsys, Dr. M. Antweiler, Dr. G. Ascheid, J. Kunkel, Dr. J. Stahl, and Dr. R. Subramanian for many inputs and stimulating discussions. We are also grateful to Dr. P. Höher of DLR, for providing valuable input with respect to the parts on fading channel estimation.

One of us (M. Moeneclaey) wants to express his gratitude to Katrien Bucket, Geert De Jonghe, and Mark Van Bladel for their assistance during the preparation of certain parts of the manuscript.

PREPARATION OF THE MANUSCRIPT

The book would have never been completed without the diligent effort of Birgit Merx and Gustl Stüsser. Gustl Stüsser did an outstanding job creating almost all the illustrations. Birgit Merx created and edited the entire camera-ready manuscript. Besides many other duties, she admirably handled the input of three authors, a publisher in New York, and a software package stressed to its limits.

H. M.
M. M.
S. A. F.

About the Authors

Heinrich Meyr has worked extensively in the areas of communication theory, synchronization, and digital signal processing for the last 30 years. His research has been applied to the design of many industrial products. He received his M.S. and Ph.D. from ETH Zürich, Switzerland. Since 1977 he has been a professor of Electrical Engineering at Aachen University of Technology (RWTH, Aachen), where he heads the Institute for Integrated Systems in Signal Processing involved in the analysis and design of complex signal processing systems for communication applications. Dr. Meyr co-founded the company CADIS GmbH which in 1993 was acquired by SYNOPSYS of Mountain View, California. CADIS commercialized the tool suite COSSAP, now used extensively by industry worldwide. Dr. Meyr is a member of the Board of Directors of two communication companies. He has published numerous IEEE papers and is co-author with Dr. Gerd Ascheid of *Synchronization in Digital Communications, Volume 1,* published by Wiley in 1990. He is a Fellow of the IEEE.

Marc Moeneclaey is Research Director for the Flemish Fund for Scientific Research and a part-time professor at the Communications Engineering Laboratory, University of Gent, Belgium. His main research topics are statistical communication theory, carrier and symbol synchronization, bandwidth-efficient modulation and coding, spread-spectrum, and satellite and mobile communication. Dr. Moeneclaey received his Diploma and Ph.D. in Electrical Engineering from the University of Gent and has authored more than 150 scientific papers published in international journals and conference proceedings. From 1992 to 1994, he edited Synchronization topics for the *IEEE Transactions on Communications.*

Stefan A. Fechtel has served since 1993 as a project manager and senior research engineer at the Institute for Integrated Systems in Signal Processing, Aachen University of Technology, where his main topics are communication and estimation theory, channel fading, synchronization for single- and multicarrier modulation, and channel coding. Dr. Fechtel received his Diploma from Aachen University of Technology, Aachen, Germany and M.S. degree from the University of Kansas at Lawrence in 1987 and earned his Ph.D. from Aachen University in 1993. He has written 20 scientific papers which have been published in international journals and conference proceedings. Since 1992, he has been co-editor and reviewer for several journals, including the *IEEE Transactions on Communications.*

DIGITAL COMMUNICATION RECEIVERS

PART A
Introduction

Introduction and Preview

It is a certainty that communication will be essentially fully digital by the end of the century. The reason for this development is the progress made in microelectronics which allows to implement complex algorithms economically to achieve bit rates close to the information theoretic limits.

The communication model studied in information theory is shown in Figure 1. Information theory answers two fundamental questions:

1. What is the ultimate data compression? (answer: the entropy H)
2. What is the ultimate transmission rate of communications? (answer: the channel capacity C)

In information theory we are concerned with sequences. Source symbols from some alphabet are mapped onto sequences of channel symbols $\mathbf{x} = (x_1, ..., x_n, ...)$ which then produce the output sequence $\mathbf{y} = (y_1, ..., y_n, ...)$ of the channel. The output sequence is random but has a distribution that depends on the input sequence. From the output sequence we attempt to recover the transmitted message.

In any physical communication system a time-continuous waveform $s(t, \mathbf{x})$ corresponding to the sequence $\mathbf{x}$ is transmitted, and not the sequence itself. The assignment of the channel symbol sequence to the waveform is done by the modulator. In addition to the sequence $\mathbf{x}$ the waveform depends on a set of parameters $\boldsymbol{\theta} = \{\boldsymbol{\theta}_T, \boldsymbol{\theta}_C\}$. The subset $\boldsymbol{\theta}_T$ is related to the transmitter and $\boldsymbol{\theta}_C$ are parameters of the channel. These parameters are unknown to the receiver. In order to be able to retrieve the symbol sequence $\mathbf{x}$ the receiver must estimate these unwanted parameters from the received signal. The estimates are then used as

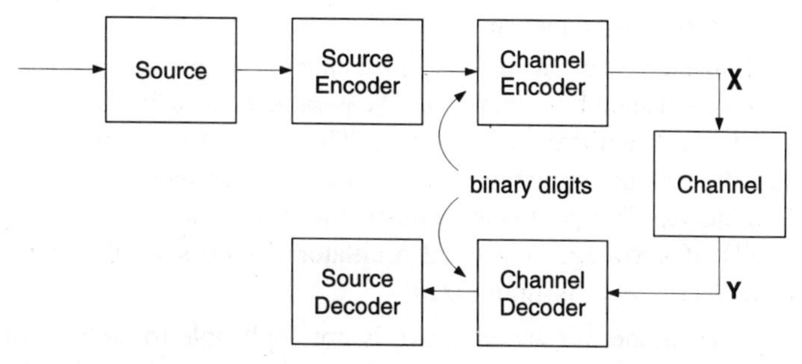

Figure 1 Idealized Communication Model Studied in Information Theory

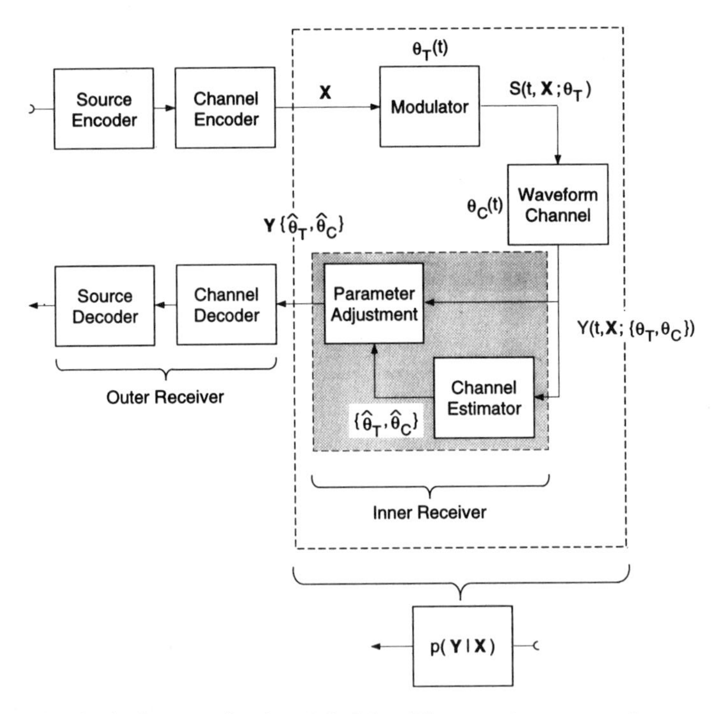

Figure 2 Physical Communication Model. The receiver comprises an **inner receiver** to estimate the channel parameters and an **outer receiver** for the decoding of the data.

if they were true values. Even if the parameters are not strictly related to units of time we speak of "synchronized detection" in abuse of the precise meaning of synchronization.

In our physical communication model we are thus concerned with an *inner* and an *outer* receiver, as shown in Figure 2. The sole task of the *inner* receiver is to produce an (output) sequence $\mathbf{y}(\hat{\boldsymbol{\theta}}_T, \hat{\boldsymbol{\theta}}_C)$ such that the "synchronized channel" has a capacity close to that of the information theoretic channel – which assumes there are no unknown parameters. The *outer* receiver has the task to optimally decode the transmitted sequence.

In the simplest case of an additive noise channel the parameters $\boldsymbol{\theta} = \{\boldsymbol{\theta}_T, \boldsymbol{\theta}_C\}$ may be assumed constant but unknown. The parameter set $\boldsymbol{\theta}$ includes, for example, the phase θ or the fractional time delay ε. The *channel estimator* of Figure 2 in this case has the task of estimating a set of unknown parameters in a known signal corrupted by noise. The parameter adjustment is done, for example, by shifting the phase $\theta(t)$ of a voltage-controlled oscillator (VCO) such that the estimation error $\phi(t) = \theta(t) - \hat{\theta}(t)$ is minimized.

The channel model discussed above is not applicable to mobile communications. These channels are time variant. The task of the *channel estimator* then consists of estimating a time-variant set of parameters. Mathematically, we have to solve the problem of estimating a random signal – the channel impulse response

– in a noisy environment. The *parameter adjustment* block performs tasks such as the computation of channel-dependent parameters (the taps of a matched filter or an equalizer, to mention examples).

We said earlier that the sole task of the inner receiver is to produce a good channel for the decoder. To achieve this goal, it is necessary to fully utilize the "information" content of the received signal. Since we are interested in digital implementations of the receiver, this implies that we must find conditions such that the entire information of the continuous time signal is contained in the samples $\{y(t = kT_s, \mathbf{x}; \boldsymbol{\theta})\}$ of the received signal ($1/T_s$: sampling rate). It will be shown that conditions exist such that *time discretization* causes no loss of information. We mathematically say that the samples $y(kT_s, \mathbf{x}; \boldsymbol{\theta})$ represent *sufficient statistics*.

It is obvious that the *amplitude quantization* must be selected in a way that the resulting distortion is negligible. This selection requires a careful balance between the two conflicting goals: the performance-related issues – which require high resolution – and an implementation complexity, which strongly increases with the increasing number of quantization levels.

The picture of Figure 2 deceivingly suggests that it is (readily) possible to separate the estimation of unwanted parameters $\boldsymbol{\theta}$ and the estimation of the useful data sequence $\mathbf{x}$. Unfortunately, this is not the case.

In the general case the received signal $y(t, \mathbf{x}; \boldsymbol{\theta})$ contains two random sequences $\mathbf{x}$ (useful data) and $\boldsymbol{\theta}$ – the randomly varying channel parameters which cannot be separated and must in principle be jointly estimated. For complexity reasons, this is not feasible.

An elegant way out of this dilemma is to impose a suitable *frame structure* on the physical layer. (At the higher level of the OSI model information is always structured in frames.) A frame structure allows to separate the task of estimating the random channel parameters from the detection of the useful data sequence. Because of this separation of the two random sources, the complexity of the algorithm is drastically reduced. In many practical cases it is only this separation which makes algorithms realizable.

Figure 3 illustrates the *principle of separation*. The bit stream is composed of frame symbols (shown shaded) and useful data information. Using the *known* frame

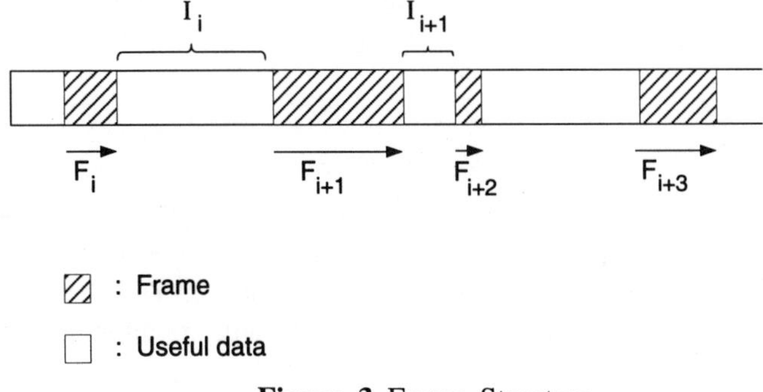

Figure 3 Frame Structure

symbols, the channel parameters are estimated in the shaded bit stream segments. Conversely, during the transmission of the data symbols, the channel parameters are assumed to be known and the information bits are decoded. It is important to notice that the frame structure can be matched to the channel characteristics to meet two conflicting requirements. On the one hand the number of known frame symbols should be minimized since the efficiency of the channel use

$$\eta = \frac{\overline{I}}{\overline{I}+\overline{F}} \qquad \left.\begin{matrix}\overline{I}\\ \overline{F}\end{matrix}\right\} \begin{matrix}\text{average length of}\\ \text{frame/data segments}\end{matrix} \tag{1}$$

decreases with increasing $\overline{F}$. On the other hand, the accuracy of the estimate of the channel parameters increases with increasing number of channel symbols $\overline{F}$.

Example 1: Time-Invariant Channel

In this case we need no frame structure for the estimation of the channel parameters such as phase θ or the fractional time delay ε ($\overline{F} = 0$). The data dependence of the received signal can be eliminated by various means.

Example 2: Time-Variant (Fading) Channel

Since the channel impulse is time-variant, a frame structure is required. The exact form depends on the channel characteristics.

We can summarize the discussions on models and bounds as follows. The separation principle defines two different tasks which are performed by an inner and an outer receiver, respectively. The *outer receiver* is concerned with an idealized model. The main design task includes the design of codes and decoders possibly using information about the channel state from the inner receiver. The channel is assumed to be known and is defined by the probability $p(\mathbf{y}|\mathbf{x})$. The *inner receiver* has the task to produce an output such that the performance of the outer receiver is as close as possible to the ideal of perfect channel knowledge. The optimal inner receiver assumes perfect knowledge of the transmitted symbol sequence in the frame segments, see Figure 3. (An unknown random sequence is considered as unwanted parameter.) In this volume we are concerned with the design of the inner receiver.

The outer receiver is the domain of information theory. Information theory promises error-free transmission at a bit rate of at most equal to the channel capacity C,

$$R_b \leq C \tag{2}$$

In a practical system we must accept a nonzero transmission error probability. The acceptable error probability for a bit rate R_b measured in bits per second is achieved for a bandwidth B and the signal-to-noise ratio γ. The numerical value of B and γ depend on the system characteristic. In order to allow a meaningful comparison among competing transmission techniques, we must define properly normalized performance measures. For example, it would not be very meaningful to compare the error probability as a function of the signal-to-noise ratio (SNR) unless this comparison was made on the basis of a fixed bandwidth, or equivalently a fixed data rate. The most compact comparison is based on the normalized bit rate

$$\nu = \frac{R_b}{B} \qquad [\text{bits/s/Hz}] \tag{3}$$

and the signal-to-noise ratio γ.

The parameter ν is a measure for the *spectral efficiency* as it relates the transmission rate R_b measured in bits per second to the required bandwidth measured in Hz. Spectral efficiency has the unit bits per second per hertz.

The parameter γ is a measure for the *power efficiency* since it specifies the relative power required to achieve the specified bit error rate. Both measures are affected by coding and the inner receiver.

The performance measure of the inner receiver is the variance of the unbiased estimate. The variance of any estimate can be shown to be greater than a fundamental bound known as the Cramér-Rao bound.

The inner receiver causes a loss Δ of spectral and power efficiency due to the allocation of bandwidth to the frame structure

$$\Delta = \frac{\overline{F}}{\overline{F} + \overline{I}} \tag{4}$$

and a *detection loss* Δ_D due to imperfect synchronization. Detection loss Δ_D is defined as the required increase in SNR γ, associated with an imperfect reconstruction of the parameter set $\boldsymbol{\theta}$ (see Figure 2) at the receiver relative to complete knowledge of these parameters (perfect sync) for maintaining a given error probability.

Traditionally the communication engineer is concerned with a design space comprising the dimensions power and bandwidth. In this design space the communication engineer can trade spectral efficiency versus power efficiency.

State-of-the-art communication systems operate close to the theoretical limits. These systems have become feasible because of the enormous progress made in microelectronics which makes it possible to implement complex digital signal processing algorithms economically. As a consequence a third dimension has been added to the design space: signal processing complexity (see Figure 4).

The communication engineer today has the option to trade physical performance measures – power and bandwidth – for signal processing complexity, thus approaching the information theoretic limits. This very fact has revolutionized the entire area of communications. We have tried to take this aspect into account when writing this book by giving implementations of digital algorithms the proper weight.

While we have clearly defined performance measures for bandwidth and

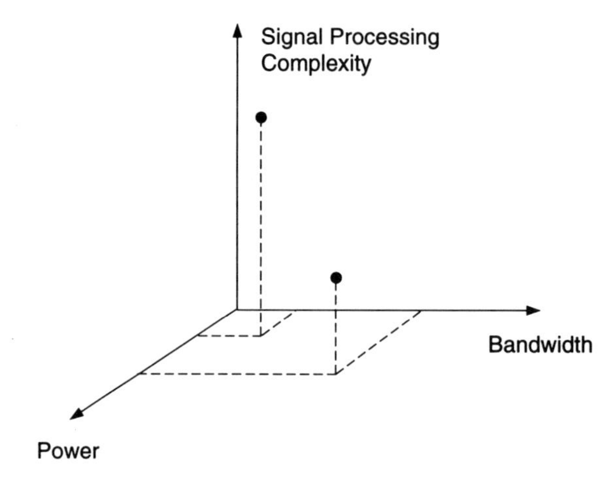

Figure 4 The 3-Dimensional Design Space

power, unfortunately, there exist no such universally applicable measures of complexity. If the algorithms run on a programmable processor, then the number of operations per unit of time T (where $1/T$ is the rate at which the value of the algorithm is computed) is a meaningful indicator of complexity. If an ASIC (application-specific integrated circuit) is employed, a possible measure is the so-called AT product, which is the ratio of silicon area A divided by the processing power (proportional to $1/T$).

PART B
Basic Material

Chapter 1 Basic Material

1.1 Stationary and Cyclostationary Processes

In telecommunications, both the information-bearing transmitted signal and the disturbances introduced by the channel are modeled as random processes. This reflects the fact that the receiver knows only some statistical properties of these signals, rather than the signals themselves. From these known statistical properties and the observation of the received signal, the receiver computes an estimate of the transmitted information.

The simplest random processes to deal with are *stationary* random processes. Roughly speaking, their statistical properties (such as the variance) do not change with time (but, of course, the instantaneous value of the process does change with time). Several channel impairments, such as additive thermal noise and the attenuation of a non-frequency-selective channel, can be considered as stationary processes. However, the transmitted signal, and also interfering signals (such as adjacent channel interference) with similar properties as the transmitted signal, cannot be considered as stationary processes. For example, the variance of an amplitude-modulated sinusoidal carrier is not independent of time: it is zero at the zero-crossing instants of the carrier, and it is maximum halfway between zero crossings. The transmitted signal (and interfering signals with similar properties) can be modeled in terms of *cyclostationary* processes; these are processes with statistical properties that vary periodically with time.

In the following, we briefly review some basic properties of stationary and cyclostationary processes which will be used in subsequent chapters; both continuous-time and discrete-time processes are considered. For a more complete coverage of random processes on an engineering level, the reader is referred to the Bibliography at the end of of this section[1–3].

1.1.1 Continuous-Time Stationary Processes

A complex-valued continuous-time random process $x(t)$ is *strict-sense stationary* (or *stationary* for short) when its statistical properties are invariant under an arbitrary time shift t_0. In other words, $x(t)$ and $x(t-t_0)$ have identical statistical properties. The processes $x_1(t), x_2(t), ..., x_N(t)$ are jointly stationary when they have jointly the same statistical properties as $x_1(t - t_0)$, $x_2(t - t_0)$, ..., $x_N(t - t_0)$.

The statistical expectation $E[x(t)]$ of a stationary process does not depend on the time variable t. Indeed,

$$E[x(t)] = E[x(t - t_0)] \Big|_{t_0 = t} = E[x(0)] \tag{1-1}$$

We will use the notation $m_x = E[x(t)]$ for the statistical expectation value of the stationary process $x(t)$. Also, the autocorrelation function $E[x^*(t)\, x(t + u)]$ of a stationary process does not depend on the time variable t. Indeed,

$$E[x^*(t)\, x(t + u)] = E[x^*(t - t_0)\, x(t + u - t_0)] \Big|_{t_0 = t} = E[x^*(0)\, x(u)] \tag{1-2}$$

The autocorrelation function will be denoted as $R_x(u) := E[x^*(t)\, x(t + u)]$. The power spectral density $S_x(\omega)$ of a stationary process is the Fourier transform of the autocorrelation function $R_x(u)$:

$$S_x(\omega) = \int_{-\infty}^{+\infty} R_x(u) \exp(-j\omega u)\, du \tag{1-3}$$

Note that the autocorrelation function $R_x(u)$ is complex conjugate symmetric about $u = 0$:

$$\begin{aligned} R_x^*(-u) &= E[x(t)\, x^*(t - u)] \\ &= E[x^*(t)\, x(t + u)] \\ &= R_x(u) \end{aligned} \tag{1-4}$$

Consequently, the power spectral density is real-valued:

$$S_x^*(\omega) = \int_{-\infty}^{+\infty} R_x^*(-u) \exp(j\omega u)\, du = \int_{\infty}^{-\infty} R_x^*(-u) \exp(-j\omega u)\, du = S_x(\omega) \tag{1-5}$$

Also, it can be shown that $S_x(\omega) \geq 0$ [3]. The power of a stationary process $x(t)$ is defined as $E\left[|x(t)|^2\right]$. As $R_x(u)$ and $S_x(\omega)$ are a Fourier transform pair, we obtain

$$E\left[|x(t)|^2\right] = R_x(0) = \int_{-\infty}^{+\infty} S_x(\omega) \frac{d\omega}{2\pi} \tag{1-6}$$

In a similar way, it is easily verified that the cross-correlation function $E[x(t)\, x(t + u)]$ of $x^*(t)$ and $x(t)$ does not depend on t; this cross-correlation function will be denoted as $R_{x^*x}(u)$. The cross spectrum $S_{x^*x}(\omega)$ of $x^*(t)$ and $x(t)$ is the Fourier transform of $R_{x^*x}(u)$. Both $R_{x^*x}(u)$ and $S_{x^*x}(\omega)$ are even functions:

$$R_{x^*x}(-u) = E[x(t)\,x(t-u)] = E[x(t)\,x(t+u)] = R_{x^*x}(u) \tag{1-7}$$

$$\begin{aligned} S_{x^*x}(-\omega) &= \int_{-\infty}^{+\infty} R_{x^*x}(u)\,\exp(j\omega u)\,du \\ &= \int_{-\infty}^{+\infty} R_{x^*x}(-u)\,\exp(-j\omega u)\,du = S_{x^*x}(\omega) \end{aligned} \tag{1-8}$$

When $x(t)$ is a real-valued stationary process, then $R_{x^*x}(u) = R_x(u)$ and $S_{x^*x}(\omega) = S_x(\omega)$; in this case, both $R_x(u)$ and $S_x(\omega)$ are real-valued even functions.

A process $x(t)$ is said to be *wide-sense stationary* when its statistical expectation $E[x(t)]$ and the auto- and cross-correlation functions $E[x^*(t)\,x(t+u)]$ and $E[x(t)\,x(t+u)]$ do not depend on the time variable t. The processes $x_1(t)$, $x_2(t)$, ..., $x_N(t)$ are jointly wide-sense stationary when their statistical expectations $E[x_i(t)]$ and the correlation functions $E[x_i^*(t)\,x_j(t+u)]$ and $E[x_i(t)\,x_j(t+u)]$ do not depend on t.

A time-invariant (not necessarily linear, not necessarily memoryless) operation on one or more jointly stationary processes yields another stationary process. Indeed, because of the time-invariant nature of the operation, applying to all input processes the same time shift gives rise to an output process with an identical time shift; because of the jointly stationary nature of the input processes, their statistics are invariant under this time shift, so that also the statistical properties of the output process do not depend on the time shift. A special case of a time-invariant operation is a linear time-invariant filter, characterized by its impulse response $h(t)$. The output $y(t)$ is related to the input $x(t)$ by

$$y(t) = \int_{-\infty}^{+\infty} h(t-u)\,x(u)\,du \tag{1-9}$$

When $x(t)$ is (wide-sense) stationary, it is easily verified that $y(t)$ is also (wide-sense) stationary. The statistical expectation $m_y = E[y(t)]$, the power spectral density $S_y(\omega)$, and the cross spectrum $S_{y^*y}(\omega)$ are given by

$$m_y = H(0)\,m_x \tag{1-10}$$

$$S_y(\omega) = |H(\omega)|^2\,S_x(\omega) \tag{1-11}$$

$$S_{y^*y}(\omega) = H(\omega)\,H(-\omega)\,S_{x^*x}(\omega) \tag{1-12}$$

where $H(\omega)$ is the filter transfer function, i.e., the Fourier transform of the impulse response $h(t)$.

1.1.2 Continuous-Time Cyclostationary Processes

A complex-valued continuous-time random process $x(t)$ is *strict-sense cyclostationary with period* T (or *cyclostationary with period* T for short) when its statistical properties are invariant under a time shift kT, where k is an arbitrary positive or negative integer. In other words, $x(t)$ and $x(t-kT)$ have identical statistical properties. The processes $x_1(t)$, $x_2(t)$, ..., $x_N(t)$ are jointly cyclostationary with period T when they have jointly the same statistical properties as $x_1(t-kT)$, $x_2(t-kT)$, ..., $x_N(t-kT)$.

The statistical expectation $m_x(t) = E[x(t)]$ of a cyclostationary process is periodic in the time variable t with period T. Indeed,

$$m_x(t) = E[x(t)] = E[x(t-kT)] = m_x(t-kT) \tag{1-13}$$

Hence, the expectation can be expanded into a Fourier series:

$$m_x(t) = \sum_{k=-\infty}^{+\infty} m_{x,k} \exp\left(j2\pi \frac{kt}{T}\right) \tag{1-14}$$

where the Fourier coefficients $\{m_{x,k}\}$ are given by

$$m_{x,k} = \frac{1}{T} \int_{-T/2}^{T/2} m_x(t) \exp\left(-j2\pi \frac{kt}{T}\right) dt \tag{1-15}$$

Also, the auto- and cross-correlation functions $R_x(t, t+u) = E[x^*(t)\, x(t+u)]$ and $R_{x^*x}(t, t+u) = E[x(t)\, x(t+u)]$ of a cyclostationary process are periodic in t with period T. Indeed,

$$\begin{aligned} R_x(t, t+u) &= E[x^*(t)\, x(t+u)] \\ &= E[x^*(t-kT)\, x(t-kT+u)] \\ &= R_x(t-kT, t-kT+u) \end{aligned} \tag{1-16}$$

$$\begin{aligned} R_{x^*x}(t, t+u) &= E[x(t)\, x(t+u)] \\ &= E[x(t-kT)\, x(t-kT+u)] \\ &= R_{x^*x}(t-kT, t-kT+u) \end{aligned} \tag{1-17}$$

This implies that these correlation functions can be expanded into a Fourier series:

$$R_x(t, t+u) = \sum_{k=-\infty}^{+\infty} r_{x,k}(u) \exp\left(j2\pi \frac{kt}{T}\right) \tag{1-18}$$

$$R_{x^*x}(t, t+u) = \sum_{k=-\infty}^{+\infty} r_{x^*x,k}(u) \exp\left(j2\pi \frac{kt}{T}\right) \tag{1-19}$$

where the Fourier coefficients $\{r_{x,k}(u)\}$ and $\{r_{x^*x,k}(u)\}$ are given by

$$r_{x,k}(u) = \frac{1}{T} \int_{-T/2}^{T/2} R_x(t, t+u) \exp\left(-j2\pi \frac{kt}{T}\right) dt \tag{1-20}$$

$$r_{x^*x,k}(u) = \frac{1}{T} \int_{-T/2}^{T/2} R_{x^*x}(t, t+u) \exp\left(-j2\pi \frac{kt}{T}\right) dt \tag{1-21}$$

The Fourier transforms of $r_{x,k}(u)$ and $r_{x^*x,k}(u)$ will be denoted as $S_{x,k}(\omega)$ and $S_{x^*x,k}(\omega)$, respectively.

A process $x(t)$ is said to be *wide-sense cyclostationary with period* T when its statistical expectation $m_x(t)$ and the auto- and cross-correlation functions $R_x(t, t+u)$ and $R_{x^*x}(t, t+u)$ are periodic in t with period T. The processes $x_1(t)$, $x_2(t)$, ..., $x_N(t)$ are jointly wide-sense cyclostationary with period T when their statistical expectations $E[x_i(t)]$ and the correlation functions $E[x_i^*(t)\, x_j(t+u)]$ and $E[x_i(t)\, x_j(t+u)]$ are periodic in t with period T.

From a (wide-sense) cyclostationary process $x(t)$ one can derive a (wide-sense) stationary process $\chi(t)$ by applying a random shift to $x(t)$:

$$\chi(t) = x(t - \tau) \tag{1-22}$$

where τ is uniformly distributed in the interval $[0, T]$, and independent of $x(t)$. The statistical expectation m_χ and the correlation functions $R_\chi(u)$ and $R_{\chi^*\chi}(u)$ of this (wide-sense) stationary process $\chi(t)$ can be expressed in terms of the corresponding moments of the (wide-sense) cyclostationary process $x(t)$:

$$m_\chi = \langle m_x(t) \rangle_t = m_{x,0} \tag{1-23}$$

$$R_\chi(u) = \langle R_x(t, t+u) \rangle_t = r_{x,0}(u) \tag{1-24}$$

$$R_{\chi^*\chi}(u) = \langle R_{x^*x}(t, t+u) \rangle_t = r_{x^*x,0}(u) \tag{1-25}$$

where $\langle ... \rangle_t$ denotes averaging over the time variable t; for instance,

$$\langle R_x(t, t+u) \rangle_t = \frac{1}{T} \int_{-T/2}^{T/2} R_x(t, t+u)\, dt \tag{1-26}$$

Hence, the power spectral density $S_\chi(\omega)$ and the cross-spectral density $S_{\chi^*\chi}(\omega)$ are given by $S_{x,0}(\omega)$ and $S_{x^*x,0}(\omega)$, the Fourier transforms of $\langle R_x(t, t+u) \rangle_t$ and $\langle R_{x^*x}(t, t+u) \rangle_t$, respectively. Consequently, the power of the randomly shifted process $\chi(t)$ is given by

$$E\left[|\chi^2(t)|\right] = R_\chi(0) = \int_{-\infty}^{+\infty} S_\chi(\omega)\,\frac{d\omega}{2\pi}$$
$$= \langle R_x(t,t)\rangle_t = r_{x,0}(0) = \left\langle E\left[|x^2(t)|\right]\right\rangle_t = \int_{-\infty}^{+\infty} S_{x,0}(u)\,\frac{d\omega}{2\pi} \tag{1-27}$$

Sometimes, $S_{x,0}(\omega)$ and $r_{x,0}(0)$ are also referred to as the power spectral density and the power of the (wide-sense) cyclostationary process $x(t)$, respectively. Although $S_{x,0}(\omega)$ is *not* the Fourier transform of $R_x(t, t+u)$, the interpretation of $S_x(\omega)$ as the spectrum of $x(t)$ is reasonable, because, as will be shown in Section 1.1.4, $S_{x,0}(\omega)$ is the quantity measured by a spectrum analyzer operating on the (wide-sense) cyclostationary process $x(t)$.

A similar reasoning as for stationary processes reveals that a time-invariant (not necessarily linear, not necessarily memoryless) operation on one or more jointly cyclostationary processes with period T yields another cyclostationary process with the same period. Taking a time-invariant linear filter with impulse response $h(t)$ as a special case of a time-invariant operation, the output $y(t)$ is related to the input $x(t)$ by

$$y(t) = \int_{-\infty}^{+\infty} h(t-u)\; x(u)\; du \tag{1-28}$$

When $x(t)$ is (wide-sense) cyclostationary with period T, then $y(t)$ is also (wide-sense) cyclostationary with period T. Its statistical expectation $m_y(t)$ and the auto- and cross-correlation functions $R_y(t, t+u)$ and $R_{y^*y}(t, t+u)$ can be expressed in terms of the corresponding moments of the process $x(t)$:

$$m_y(t) = \int_{-\infty}^{+\infty} h(t-v)\; m_x(v)\; dv \tag{1-29}$$

$$R_y(t, t+u) = \iint_{-\infty}^{+\infty} h^*(t-v)\; h(t+u-v-w)\; R_x(v, v+w)\; dv\; dw \tag{1-30}$$

$$R_{y^*y}(t, t+u) = \iint_{-\infty}^{+\infty} h(t-v)\; h(t+u-v-w)\; R_{x^*x}(v, v+w)\; dv\; dw \tag{1-31}$$

Note that (1-29)–(1-31) are valid for any process $x(t)$ at the input of the filter with impulse response $h(t)$. The above functions are periodic in t with period

T, and can be expanded into the following Fourier series that are analogous to (1-14), (1-18), and (1-19):

$$m_y(t) = \sum_{k=-\infty}^{+\infty} m_{y,k} \exp\left(j2\pi \frac{kt}{T}\right) \tag{1-32}$$

$$R_y(t, t+u) = \sum_{k=-\infty}^{+\infty} r_{y,k}(u) \exp\left(j2\pi \frac{kt}{T}\right) \tag{1-33}$$

$$R_{y^*y}(t, t+u) = \sum_{k=-\infty}^{+\infty} r_{y^*y,k}(u) \exp\left(j2\pi \frac{kt}{T}\right) \tag{1-34}$$

The Fourier coefficients in the above series are related to the corresponding Fourier coefficients of the process $x(t)$ by

$$m_{y,k} = H\left(\frac{2\pi k}{T}\right) m_{x,k} \tag{1-35}$$

$$S_{y,k}(\omega) = H^*\left(\omega - \frac{2\pi k}{T}\right) H(\omega)\, S_{x,k}(\omega) \tag{1-36}$$

$$S_{y^*y,k}(\omega) = H\left(-\omega + \frac{2\pi k}{T}\right) H(\omega)\, S_{x^*x,k}(\omega) \tag{1-37}$$

where $H(\omega)$ is the transfer function of the filter, while $S_{y,k}(\omega)$ and $S_{y^*y,k}(\omega)$ are the Fourier transforms of the coefficients $r_{y,k}(u)$ and $r_{y^*y,k}(u)$. The power spectral density $S_{y,0}(\omega)$ of the randomly shifted process $y(t-\tau)$ is given by

$$S_{y,0}(\omega) = |H(\omega)|^2\, S_{x,0}(\omega)$$

where $S_{x,0}(\omega)$ is the power spectral density of the randomly shifted process $x(t-\tau)$. Now, in the special case where $H(\omega) = 0$ for $|\omega| > \pi/T$, we obtain

$$\begin{aligned} H\left(\frac{2\pi k}{T}\right) &= H^*\left(\omega - \frac{2\pi k}{T}\right) H(\omega) \\ &= H\left(-\omega + \frac{2\pi k}{T}\right) H(\omega) = 0 \quad \text{for} \quad k \neq 0 \end{aligned} \tag{1-38}$$

so that the Fourier coefficients with $k \neq 0$ in (1-32)–(1-34) are identically zero.

Consequently, (1-32)–(1-34) reduce to

$$m_y(t) = H(0)\, m_{x,0} \tag{1-39}$$

$$R_y(t, t+u) = r_{y,0}(u), \text{ with } S_{y,0}(\omega) = |H(\omega)|^2\, S_{x,0}(\omega) \tag{1-40}$$

$$R_{y^*y}(t, t+u) = r_{y^*y,0}(u), \text{ with } S_{y^*y,0}(\omega) = H(\omega)\, H(-\omega)\, S_{x^*x,0}(\omega) \tag{1-41}$$

It is important to note that $m_y(t)$, $R_y(t, t+u)$ and $R_{y^*y}(t, t+u)$ in (1-39)–(1-41) do not depend on t. Hence, passing a (wide-sense) *cyclostationary* process $x(t)$ through a lowpass filter, whose bandwidth (in hertz) does not exceed $1/2T$, yields a wide-sense stationary process $y(t)$. The corresponding power spectral density $S_y(\omega)$ and cross-spectral density $S_{y^*y}(\omega)$ are given by $S_{y,0}(\omega)$ and $S_{y^*y,0}(\omega)$, respectively, and are the same as for the randomly shifted process $y(t-\tau)$.

1.1.3 Discrete-Time Stationary Processes

A discrete-time process is a sequence $\{x_k\}$ of random variables; we can think of these variables as being produced at a rate of one random variable per interval of duration T. Many properties of continuous-time random processes can be extended to discrete-time processes.

A discrete-time complex-valued process $\{x_k\}$ is *strict-sense stationary* (or *stationary* for short) when it has the same statistical properties as $\{x_{k+K}\}$, where K is an arbitrary integer. Following a similar reasoning as for continuous-time stationary processes, it can easily be shown that the statistical expectation $E[x_k]$ of x_k, the autocorrelation function $E[x_k^*\, x_{k+m}]$ of $\{x_k\}$, and the cross-correlation function $E[x_k\, x_{k+m}]$ between $\{x_k^*\}$ and $\{x_k\}$ do not depend on the time index k; these moments will be denoted as m_x, $R_x(m)$, and $R_{x^*x}(m)$, respectively. The power spectral density $S_x(\exp(j\omega T))$ of $\{x_k\}$ and the cross-power spectral density $S_{x^*x}(\exp(j\omega T))$ between $\{x_k^*\}$ and $\{x_k\}$ are defined as the discrete-time Fourier transforms of the corresponding correlation functions

$$S_x(\exp(j\omega T)) = \sum_m R_x(m)\, \exp(-j\omega m T) \tag{1-42}$$

$$S_{x^*x}(\exp(j\omega T)) = \sum_m R_{x^*x}(m)\, \exp(-j\omega m T) \tag{1-43}$$

Note that the above spectra are periodic in ω with period $2\pi/T$. As for continuous-time stationary processes, it can be shown that $R_x(m)$ has complex conjugate even symmetry, $S_x(\exp(j\omega T))$ is real-valued and nonnegative, and both $R_{x^*x}(m)$ and $S_{x^*x}(\exp(j\omega T))$ are even functions. The sequence $\{y_k\}$, resulting from a time-invariant (not necessarily linear, not necessarily memoryless) operation on a stationary sequence $\{x_k\}$, is also stationary.

A complex-valued random sequence $\{x_k\}$ is wide-sense stationary when its statistical expectation $E[x_k]$, its autocorrelation function $E[x_k^* \, x_{k+m}]$, and the cross-correlation $E[x_k \, x_{k+m}]$ between $\{x_k^*\}$ and $\{x_k\}$ do not depend on the time index k.

Applying a (wide-sense) stationary sequence $\{x_k\}$ to a linear filter with impulse response $\{h_k\}$ yields a (wide-sense) stationary sequence $\{y_k\}$, given by

$$y_k = \sum_m h_{k-m} \, x_m$$

As for continuous-time (wide-sense) stationary processes, it can be shown that

$$m_y = H\left(e^{j0}\right) m_x \tag{1-44}$$

$$S_y(\exp(j\omega T)) = |H(\exp(j\omega T))|^2 \, S_x(\exp(j\omega T)) \tag{1-45}$$

$$S_{y^*y}(\exp(j\omega T)) = H(\exp(j\omega T)) \, H(\exp(-j\omega T)) \, S_{x^*x}(\exp(j\omega T)) \tag{1-46}$$

where $H(\exp(j\omega T))$ is the transfer function of the discrete-time filter:

$$H(\exp(j\omega T)) = \sum_m h_m \, \exp(-j\omega m T) \tag{1-47}$$

Note that (1-44)–(1-46) are the discrete-time counterpart of (1-10)–(1-12). When no confusion with continuous-time processes is possible, occasionally the notation $S_y(\omega)$, $S_{y^*y}(\omega)$, and $H(\omega)$ will be used instead of $S_y(\exp(j\omega T))$, $S_{y^*y}(\exp(j\omega T))$, and $H(\exp(j\omega T))$, respectively.

A discrete-time (wide-sense) stationary process $\{x_k\}$ can be obtained by sampling a continuous-time (wide-sense) stationary process $s(t)$ at a fixed rate $1/T$. Defining $x_k = s(kT + t_0)$, where t_0 denotes the sampling phase, it is easily verified that

$$m_x = m_s \tag{1-48}$$

$$R_x(m) = R_s(mT) \tag{1-49}$$

$$R_{x^*x}(m) = R_{s^*s}(mT) \tag{1-50}$$

where m_s, $R_s(u)$, and $R_{s^*s}(u)$ are the expectation of $s(t)$, the autocorrelation function of $s^*(t)$, and the cross-correlation function of $s^*(t)$ and $s(t)$, respectively. Translating (1-49) and (1-50) into the frequency domain yields

$$S_x(\exp(j\omega T)) = \frac{1}{T} \sum_m S_s\left(\omega - \frac{2\pi m}{T}\right) \tag{1-51}$$

$$S_{x^*x}(\exp(j\omega T)) = \frac{1}{T} \sum_m S_{s^*s}\left(\omega - \frac{2\pi m}{T}\right) \tag{1-52}$$

where $S_s(\omega)$ and $S_{s^*s}(\omega)$ are the spectrum of $s(t)$ and the cross spectrum of $s^*(t)$ and $s(t)$, respectively. Note that the statistical expectation m_x and the correlation functions $R_x(m)$ and $R_{x^*x}(m)$ do not depend on the sampling phase t_0, because of the (wide-sense) stationarity of $s(t)$.

Taking samples, at a rate $1/T$, of a (wide-sense) cyclostationary process $s(t)$ with period T yields a (wide-sense) stationary sequence $\{x_k\} = \{s(kT + t_0)\}$. One easily obtains

$$m_x = m_s(t_0) = \sum_k m_{s,k} \exp\left(j \frac{2\pi k}{T} t_0\right) \tag{1-53}$$

$$R_x(n) = R_s(t_0, t_0 + nT) = \sum_k r_{s,k}(nT) \exp\left(j \frac{2\pi k}{T} t_0\right) \tag{1-54}$$

$$R_{x^*x}(n) = R_{s^*s}(t_0, t_0 + nT) = \sum_k r_{s^*s,k}(nT) \exp\left(j \frac{2\pi k}{T} t_0\right) \tag{1-55}$$

In the above, $m_s(t)$, $R_s(t, t+u)$, and $R_{s^*s}(t, t+u)$ are the expectation of $s(t)$, the autocorrelation function of $s(t)$, and the cross-correlation function of $s^*(t)$ and $s(t)$, respectively; $m_{s,k}$, $r_{s,k}(u)$, and $r_{s^*s,k}(u)$ are the coefficients in the Fourier series expansion of these functions. Converting (1-54) and (1-55) into the frequency domain yields

$$S_x(\exp(j\omega T)) = \sum_k \exp\left(j \frac{2\pi k}{T} t_0\right) \left[\frac{1}{T} \sum_m S_{s,k}\left(\omega - \frac{2\pi m}{T}\right)\right] \tag{1-56}$$

$$S_{x^*x}(\exp(j\omega T)) = \sum_k \exp\left(j \frac{2\pi k}{T} t_0\right) \left[\frac{1}{T} \sum_m S_{s^*s,k}\left(\omega - \frac{2\pi m}{T}\right)\right] \tag{1-57}$$

where $S_{s,k}(\omega)$ and $S_{s^*s,k}(\omega)$ are the Fourier transforms of $r_{s,k}(u)$ and $r_{s^*s,k}(u)$, respectively. As the (wide-sense) cyclostationary process $s(t)$ is in general not wide-sense stationary, the statistical expectation m_x and the correlation functions $R_x(m)$ and $R_{x^*x}(m)$ corresponding to the (wide-sense) stationary sequence $\{x_k\}$ depend on the sampling phase t_0.

1.1.4 Examples

Additive Thermal Noise

Additive thermal noise $n(t)$ corrupting the received signal is modeled as a real-valued stationary random process with zero mean and power spectral density $S_n(\omega)$. When $S_n(\omega)$ takes on a constant value of $N_0/2$ within the bandwidth of the useful signal, $n(t)$ is referred to as being white within the signal bandwidth. Truly white noise has a power spectral density which equals $N_0/2$ for *all* ω; its corresponding autocorrelation function equals $(N_0/2)\ \delta(u)$, where $\delta(u)$ denotes the Dirac impulse. As in practice the receiver contains a filter which rejects the frequency components of the received signal that are not contained within the bandwidth of the useful signal, additive noise which is white within the signal bandwidth can be replaced by truly white noise.

Data Symbol Sequence

At the transmitter, the digital information to be sent to the receiver is translated into a sequence $\{a_k\}$ of data symbols, which take values from an alphabet of size M; usually M is an integer power of 2. When the data symbol sequence is uncoded, $\{a_k\}$ is modeled as a stationary sequence of statistically independent equiprobable symbols. When coding for error detection or error correction is used, redundancy is added to the data symbol sequence; in this case, the data symbols are no longer statistically independent.

Linear Modulation

At the transmitter, the data symbol sequence $\{a_k\}$ is converted by means of a modulator into a real-valued signal, which is sent over the communications channel. Often used is linear modulation, yielding a transmitted signal which is a linear function of the data symbol sequence.

In the case of baseband transmission, linear modulation gives rise to a transmitted signal $s(t)$, given by

$$s(t) = \sum_{m} a_m \, g(t - mT) \tag{1-58}$$

where $\{a_m\}$ is a sequence of real-valued data symbols, $1/T$ is the channel symbol rate, and $g(t)$ is the real-valued baseband pulse-amplitude modulation (PAM) pulse. In the case of passband transmission with center frequency ω_0 in radians per second, linear modulation yields the transmitted signal $\sqrt{2}\,\mathrm{Re}[s(t)\,\exp(j\omega_0 t)]$, where $s(t)$ is again given by (1-58), but the data symbols a_k and/or the baseband pulse $g(t)$ can now be complex-valued.

When the data symbols a_k are real-valued, the modulation is called one-dimensional. Often used is M-ary pulse-amplitude modulation (M-PAM), where the data symbols take values from an alphabet of size M, given by $\{\pm 1, \pm 3, \ldots, \pm(M-1)\}$ with M an integer power of 2. In baseband transmission, also ternary signaling corresponding to the alphabet $\{-1, 0, 1\}$ is encountered. Figure 1-1 shows an M-PAM alphabet for $M = 4$.

When the data symbols a_k are complex-valued, the modulation is called two-dimensional. Often encountered are M-ary phase-shift keying (M-PSK) and M-ary quadrature amplitude modulation (M-QAM). For M-PSK, the symbol alphabet is the set $\{\exp(j2\pi m/M) \mid m = 0, 1, \ldots, M-1\}$, where M is usually a power of 2; the cases $M = 2$ and $M = 4$ are also called binary phase-shift keying (BPSK) and quaternary phase-shift keying (QPSK), respectively. For M-QAM, M is usually an even power of 2; in this case, the real and imaginary part of a_k

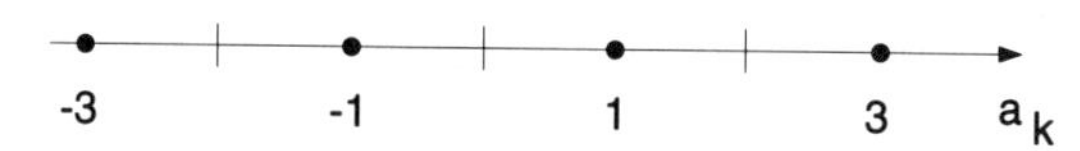

Figure 1-1 4–PAM Constellation

both belong to the alphabet $\left\{\pm1;\ \pm3,\ \ldots,\pm\left(\sqrt{M}-1\right)\right\}$. Figures 1-2 and 1-3 show the M-PSK alphabet for $M = 8$ and the M-QAM alphabet for $M = 16$, respectively.

When the data sequence $\{a_k\}$ is stationary, $s(t)$ from (1-58) is cyclostationary with period T, in both cases of baseband and passband transmission. The cyclostationarity can be verified by showing that $s(t)$ and $s(t-kT)$ have identical statistical properties. Indeed, as the channel symbol sequence is stationary, $\{a_m\}$ has the same statistical properties as $\{a_{m-k}\}$. Hence, the process $s(t-kT)$, given by

$$
\begin{aligned}
s(t-kT) &= \sum_m a_m\ g(t-kT-mT) \\
&= \sum_m a_{m-k}\ g(t-mT)
\end{aligned}
$$

has the same statistical properties as $s(t)$. Let us assume that $E[a_k]=0$, so that also $E[s(t)]=0$. The correlation functions $R_s(t,t+u)$ and $R_{s^*s}(t,t+u)$ are given by

$$
R_s(t,t+u) = \sum_{m,n} R_a(n)\ g(t-mT)\ g(t+u-mT-nT) \tag{1-59}
$$

$$
R_{s^*s}(t,t+u) = \sum_{m,n} R_{a^*a}(n)\ g(t-mT)\ g(t+u-mT-nT) \tag{1-60}
$$

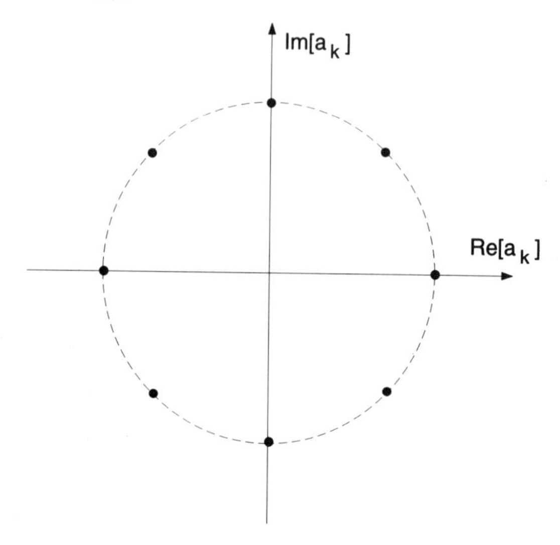

Figure 1-2 8–PSK Constellation

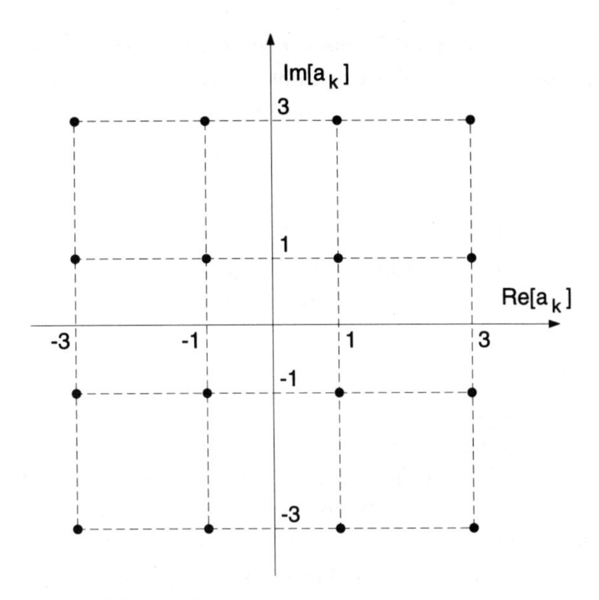

Figure 1-3 16–QAM Constellation

where $R_a(n)$ and $R_{a^*a}(n)$ are the autocorrelation function of $\{a_k\}$ and the cross-correlation function of $\{a_k^*\}$ and $\{a_k\}$, respectively. As $s(t)$ is cyclostationary, $R_s(t, t+u)$ and $R_{s^*s}(t, t+u)$ can be expanded into the Fourier series (1-18) and (1-19), with x replaced by s. Denoting the Fourier transforms of $r_{s,k}(u)$ and $r_{s^*s,k}(u)$ by $S_{s,k}(\omega)$ and $S_{s^*s,k}(\omega)$, respectively, it can be verified from (1-59) and (1-60) that

$$S_{s,k}(\omega) = \frac{1}{T}\, S_a(\exp(j\omega T))\, G(\omega)\, G^*\left(\omega - \frac{2\pi k}{T}\right) \tag{1-61}$$

$$S_{s^*s,k}(\omega) = \frac{1}{T}\, S_{a^*a}(\exp(j\omega T))\, G(\omega)\, G\left(-\omega + \frac{2\pi k}{T}\right) \tag{1-62}$$

In the above, $S_a(\exp(j\omega T))$ and $S_{a^*a}(\exp(j\omega T))$ are the spectrum of $\{a_k\}$ and the cross spectrum of $\{a_k^*\}$ and $\{a_k\}$, respectively. The spectrum of the randomly shifted process $s(t-\tau)$, where τ is uniformly distributed in the interval $(0,\, T)$, and the cross spectrum of $s^*(t-\tau)$ and $s(t-\tau)$ are given by

$$S_{s,0}(\omega) = \frac{1}{T}\, S_a(\exp(j\omega T))\, |G(\omega)|^2 \tag{1-63}$$

$$S_{s^*s,0}(\omega) = \frac{1}{T}\, S_{a^*a}(\exp(j\omega T))\, G(\omega)\, G(-\omega) \tag{1-64}$$

When $G(\omega) = 0$ for $|\omega| > \pi/T$, it follows from (1-61) and (1-62) that $S_{s,k}(\omega) = S_{s^*s,k}(\omega) = 0$ for $k \neq 0$. Hence, when the bandwidth (in hertz) of the baseband pulse $g(t)$ does not exceed $1/2T$, the signal $s(t)$ is wide-sense stationary. In the case of independent data symbols, we obtain $R_a(n) = R_{a^*a}(n) = 0$ for $n \neq 0$, yielding $S_a(\exp(j\omega T)) = R_a(0) = E\left[|a_k|^2\right]$ and $S_{a^*a}(\exp(j\omega T)) = R_{a^*a}(0) = E\left[a_k^2\right]$, which simplifies (1-59)–(1-60) and (1-61)–(1-62).

The Spectrum Analyzer

The operation of a spectrum analyzer is illustrated in Figure 1-4. In order to obtain an estimate of the power spectral density of a signal $s(t)$ at $\omega = \omega_0$, the following operations are performed:

- The signal $s(t)$ is multiplied by $\exp(-j\omega_0 t)$, in order to shift the frequency component at $\omega = \omega_0$ to $\omega = 0$.
- The resulting signal $x(t) = s(t)\ \exp(-j\omega_0 t)$ is applied to a narrowband lowpass filter with transfer function $H(\omega)$, which yields the signal $y(t)$, containing only the frequency components of $s(t)$ near $\omega = \omega_0$.
- The squared magnitude of $y(t)$ is obtained and applied to an averaging filter with transfer function $H_{\text{av}}(\omega)$, to produce an estimate $z(t)$ of the direct current (DC) component of the squared magnitude of $y(t)$.

In the following we will investigate the statistical expectation $E[z(t)]$, for the cases of a wide-sense stationary and a wide-sense cyclostationary zero-mean input signal $s(t)$.

In the case of a zero-mean wide-sense stationary process $s(t)$, it is easily verified that the autocorrelation function $E[x^*(t)\ x(t+u)]$ of $x(t)$ does not depend on t; denoting this autocorrelation function by $R_x(u)$, one obtains

$$R_x(u) = R_s(u)\ \exp(-j\omega_0 u) \tag{1-65}$$

from which the power spectral density of $x(t)$ is found to be

$$S_x(\omega) = S_s(\omega + \omega_0) \tag{1-66}$$

However, note that $x(t)$ is *not* wide-sense stationary, because $R_{x^*x}(t, t+u)$ depends on t:

$$R_{x^*x}(t, t+u) = R_{s^*s}(u)\ \exp[-j\omega_0\ (2t+u)] \tag{1-67}$$

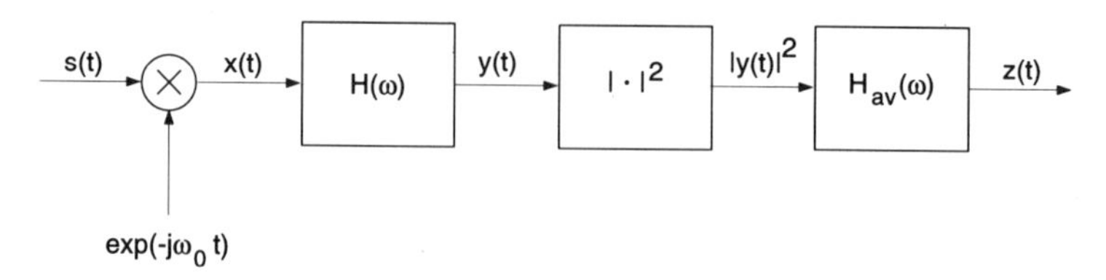

Figure 1-4 Block Diagram of Spectrum Analyzer

As the autocorrelation function of $x(t)$ does not depend on t, it follows from (1-30) that also the autocorrelation function of $y(t)$ is not dependent on t. In this case, (1-11) is valid, yielding

$$S_y(\omega) = |H(\omega)|^2 \, S_s(\omega+\omega_0) \tag{1-68}$$

Hence, the expectation of the squared magnitude of $y(t)$ is given by

$$E\left[|y(t)|^2\right] = \int_{-\infty}^{+\infty} |H(\omega)|^2 \, S_s(\omega+\omega_0) \, \frac{d\omega}{2\pi} \tag{1-69}$$

which is independent of t. Consequently,

$$E[z(t)] = \int_{-\infty}^{+\infty} h_{\rm av}(t-u) \, E\left[|y(u)|^2\right] \, du = H_{\rm av}(0) \int_{-\infty}^{+\infty} |H(\omega)|^2 \, S_s(\omega+\omega_0) \, \frac{d\omega}{2\pi} \tag{1-70}$$

where $h_{\rm av}(t)$ is the impulse response of the averaging filter. When the bandwidth of the lowpass transfer function $H(\omega)$ is so small that $S_s(\omega+\omega_0)$ is essentially constant within this bandwidth, then $E[z(t)]$ is well approximated by

$$E[z(t)] \cong \left[H_{\rm av}(0) \int_{-\infty}^{+\infty} |H(\omega)|^2 \, \frac{d\omega}{2\pi} \right] S_s(\omega_0) \tag{1-71}$$

which is proportional to the power spectral density $S_s(\omega)$ of the wide-sense stationary process $s(t)$, evaluated at $\omega = \omega_0$.

In the case of a wide-sense cyclostationary process $s(t)$, the autocorrelation function $R_x(t, t+u)$ is given by

$$R_x(t,t+u) = R_s(t,t+u) \, \exp(-j\omega_0 u) \tag{1-72}$$

which is periodic in t with period T. Both $R_x(t,t+u)$ and $R_s(t,t+u)$ can be expanded into a Fourier series like (1-18); it is easily verified that their Fourier series coefficients are related by $r_{x,k}(u) = r_{s,k}(u) \, \exp(-j\omega_0 u)$, which yields $S_{x,k}(\omega) = S_{s,k}(\omega+\omega_0)$. However, note that $s(t)$ is in general not wide-sense cyclostationary, because the cross-correlation function $R_{x^*x}(t,t+u)$ is not periodic in t with period T (unless ω_0 happens to be a multiple of π/T):

$$R_{x^*x}(t,t+u) = R_{s^*s}(t,t+u) \, \exp[(-j\omega_0(2t+u))] \tag{1-73}$$

As $R_x(t,t+u)$ is periodic in t with period T, it can be derived from (1-30) that also $R_y(t,t+u)$ is periodic in t with period T. Hence, both $R_y(t,t+u)$ and $R_x(t,t+u)$ can be expanded into a Fourier series [see (1-18) and (1-33)]; their Fourier series coefficients $r_{y,k}(u)$ and $r_{x,k}(u)$ are related by (1-36). Assuming that $H(\omega) = 0$ for $|\omega > \pi/T|$, the Fourier series coefficients $r_{y,k}(u)$ with $k \neq 0$ in (1-33) are identically zero, so that (1-40) holds: $R_y(t, t+u)$ does not depend on t. Consequently,

$$E\left[|y(t)|^2\right] = \int_{-\infty}^{+\infty} |H(\omega)|^2 \, S_{x,0}(\omega) \, \frac{d\omega}{2\pi} = \int_{-\infty}^{+\infty} |H(\omega)|^2 \, S_{s,0}(\omega + \omega_0) \, \frac{d\omega}{2\pi} \quad (1\text{-}74)$$

$$E[z(t)] = \int_{-\infty}^{+\infty} h_{\text{av}}(t-u) \, E\left[|y(u)|^2\right] du = H_{\text{av}}(0) \int_{-\infty}^{+\infty} |H(\omega)|^2 \, S_{s,0}(\omega + \omega_0) \, \frac{d\omega}{2\pi} \quad (1\text{-}75)$$

Assuming that the bandwidth of the lowpass transfer function $H(\omega)$ is so small that $S_{s,0}(\omega + \omega_0)$ is essentially constant within this bandwidth, then $E[z(t)]$ is well approximated by

$$E[z(t)] \cong \left[H_{\text{av}}(0) \int_{-\infty}^{+\infty} |H(\omega)|^2 \, \frac{d\omega}{2\pi} \right] S_{s,0}(\omega_0) \quad (1\text{-}76)$$

which is proportional to the power spectral density $S_{s,0}(\omega)$ of the *wide-sense stationary randomly shifted* process $s(t-\tau)$, evaluated at $\omega = \omega_0$.

The spectrum analyzer output $z(t)$ can be decomposed into its useful component $E[z(t)]$, evaluated above, and a zero-mean self-noise component $z(t) - E[z(t)]$, caused by the statistical fluctuation of the input signal $s(t)$. The self-noise variance can be reduced by decreasing the bandwidth of the averaging filter.

1.1.5 Main Points

A complex-valued process $x(t)$ is stationary (in the strict sense) when it has the same statistical properties as the time-shifted process $x(t - t_0)$, for any time shift t_0. Consequently, the statistical expectation $E[x(t)]$, the autocorrelation function $E[x^*(t)\, x(t+u)]$ and the cross-correlation function $E[x(t)\, x(t+u)]$ do not depend on the time variable t. Stationarity is preserved under time-invariant operations. A process $x(t)$ is wide-sense stationary when $E[x(t)]$, $E[x^*(t)\, x(t+u)]$, and $E[x(t)\, x(t+u)]$ do not depend on the time variable t. The (wide-sense) stationarity is preserved by time-invariant linear filtering.

A complex-valued process $x(t)$ is cyclostationary (in the strict sense) with period T when it has the same statistical properties as the time-shifted process $x(t - kT)$, for any integer k. Consequently, the statistical expectation $E[x(t)]$, the autocorrelation function $E[x^*(t)\, x(t+u)]$ and the cross-correlation function $E[x(t)\, x(t+u)]$ are periodic in t with period T. Cyclostationarity is preserved under time-invariant operations. A process $x(t)$ is wide-sense cyclostationary with period T when $E[x(t)]$, $E[x^*(t)\, x(t+u)]$, and $E[x(t)\, x(t+u)]$ are periodic in t

with period T. The (wide-sense) cyclostationarity is preserved by time-invariant linear filtering; when the filter bandwidth (in hertz) is less than $1/(2T)$, the filter output is wide-sense stationary.

Discrete-time (wide-sense) stationary processes are defined in a similar way as their continuous-time counterpart and have similar properties. Sampling a (wide-sense) stationary process at a fixed rate or a (wide-sense) *cyclostationary* process with period T at a rate $1/T$ yields a discrete-time (wide-sense) stationary process.

Bibliography

[1] L. E. Franks, *Signal Theory*. Englewood Cliffs, NJ: Prentice-Hall, 1969.

[2] W. A. Gardner, *Introduction to Random Processes, with Applications to Signals and Systems*. New York: Macmillan, 1986.

[3] A. Papoulis, *Probability, Random Variables and Stochastic Processes*. Auckland: McGraw-Hill, 1991.

1.2 Complex Envelope Representation

In circuit theory, it is common practice to represent a sinusoidal signal $\sqrt{2}\,A\ \cos(\omega_0 t + \theta)$ by its phasor $A\ \exp(j\theta)$, where A denotes the root-mean-square (rms) amplitude and θ the phase of the sinusoid. The relation

$$\sqrt{2}\,A\ \cos(\omega_0 t + \theta) = \sqrt{2}\,\mathrm{Re}[A\ \exp(j\theta)\ \exp(j\omega_0 t)] \qquad (1\text{-}77)$$

indicates how the sinusoidal signal is to be derived from its representing phasor $A\ \exp(j\theta)$.

In telecommunications, information is often conveyed by means of a bandpass signal, resulting from modulating a sinusoidal carrier; such a signal can be viewed as a sinusoid whose amplitude and phase are fluctuating with time. In a way similar to the phasor representation from circuit theory, a bandpass signal $x(t)$ with center frequency ω_0 can be represented by its *complex envelope* $x_L(t)$. The bandpass signal $x(t)$ is given in terms of its representing complex envelope $x_L(t)$ by

$$x(t) = \sqrt{2}\,\mathrm{Re}[x_L(t)\ \exp(j\omega_0 t)] \qquad (1\text{-}78)$$

This is called the complex baseband representation of the bandpass signal $x(t)$. The real and imaginary parts of $x_L(t)$ are the lowpass signals resulting from demodulating (i.e., translating to baseband) the bandpass signal $x(t)$. The above representation is also valid when $x(t)$ is not bandpass and ω_0 chosen arbitrarily; in this general case, the resulting complex envelope $x_L(t)$ is not necessarily lowpass.

In this section, we will define the complex envelope $x_L(t)$ of a signal $x(t)$ with respect to some frequency ω_0, in both cases of deterministic signals and random processes. Also, some statistical properties of the complex envelope of wide-sense stationary and cyclostationary processes will be derived.

The notion of complex envelope has been used in many textbooks on random processes and on telecommunications, e.g., [1–6]. However, the reader should be aware that some authors drop the scaling factor of $\sqrt{2}$ in (1-78); therefore, some care should be taken when interpreting results from different authors.

1.2.1 Complex Envelope of Deterministic Signals

Any real-valued deterministic signal $x(t)$, whose Fourier transform exists, can be represented by its complex envelope $x_L(t)$ with respect to an unmodulated carrier $\sqrt{2}\,\cos(\omega_0 t)$ at some (angular) frequency ω_0, with $\omega_0 \geq 0$. When $x(t)$ is a carrier-modulated signal, the center frequency is usually (but not necessarily) taken as the frequency of the unmodulated carrier. The signal $x(t)$ is related to its complex envelope $x_L(t)$ by

$$x(t) = \sqrt{2}\,\mathrm{Re}[x_L(t)\;\exp(j\omega_0 t)] \tag{1-79}$$

Expressing $x_L(t)$ in polar form, i.e., $x_L(t) = |x_L(t)|\;\exp[j\;\arg(x_L(t))]$, the above equation reduces to

$$x(t) = \sqrt{2}\,|x_L(t)|\;\cos[\omega_0 t + \arg(x_L(t))] \tag{1-80}$$

where $|x_L(t)|$ and $\arg(x_L(t))$ denote the *instantaneous amplitude and phase*, respectively, of the signal $x(t)$. Expressing $x_L(t)$ in rectangular form, i.e., $x_L(t) = x_C(t) + j\;x_S(t)$ where $x_C(t)$ and $x_S(t)$ are the real and imaginary part of $x_L(t)$, yields

$$x(t) = \sqrt{2}\,x_C(t)\;\cos(\omega_0 t) - \sqrt{2}\,x_S(t)\;\sin(\omega_0 t) \tag{1-81}$$

where $x_C(t)$ and $x_S(t)$ are the *in-phase and quadrature components*, respectively, with respect to an unmodulated carrier $\sqrt{2}\;\cos(\omega_0 t)$.

Given a signal $x(t)$, eq. (1-79) does not define a unique $x_L(t)$; for example, adding in (1-79) a signal $jv(t)\;\exp(-j\omega_0 t)$ to $x_L(t)$ yields the same $x(t)$, for any real-valued $v(t)$. A unique definition of $x_L(t)$ in terms of $x(t)$ follows easily from the frequency-domain interpretation of (1-79):

$$X(\omega) = \frac{\sqrt{2}}{2}\,(X_L(\omega - \omega_0) + X_L^*(-\omega - \omega_0)) \tag{1-82}$$

where $X(\omega)$ and $X_L(\omega)$ are the Fourier transforms of $x(t)$ and $x_L(t)$. Denoting by $u(\omega)$ the unit step function in the frequency domain, we define

$$\frac{\sqrt{2}}{2}\,X_L(\omega - \omega_0) = u(\omega)\;X(\omega) \tag{1-83}$$

This definition implies that $(\sqrt{2}/2)\ X_L(\omega - \omega_0)$ consists of the positive-frequency content of $X(\omega)$. As $x(t)$ is a real-valued signal, $X^*(-\omega) = X(\omega)$, so that

$$\frac{\sqrt{2}}{2}\ X_L^*(-\omega - \omega_0) = u(-\omega)\ X(\omega) \tag{1-84}$$

represents the negative-frequency content of $X(\omega)$. Hence, the Fourier transform $X_L(\omega)$ of the complex envelope $x_L(t)$ is given by

$$X_L(\omega) = \sqrt{2}\ u(\omega + \omega_0)\ X(\omega + \omega_0) \tag{1-85}$$

The construction of $X_L(\omega)$ from $X(\omega)$ is illustrated in Figure 1-5. When the support of $X(\omega)$ is the interval $(-\Omega, \Omega)$, then the support of $X_L(\omega)$ is the interval $(-\omega_0, \Omega - \omega_0)$. Usually, but not necessarily, ω_0 belongs to the support of $X(\omega)$.

The corresponding time-domain definition of $x_L(t)$ is

$$x_L(t) = \frac{\sqrt{2}}{2}\ (x(t) + j\hat{x}(t))\ \exp(-j\omega_0 t) \tag{1-86}$$

where the real-valued signal $\hat{x}(t)$ is the *Hilbert transform* of $x(t)$. Converting

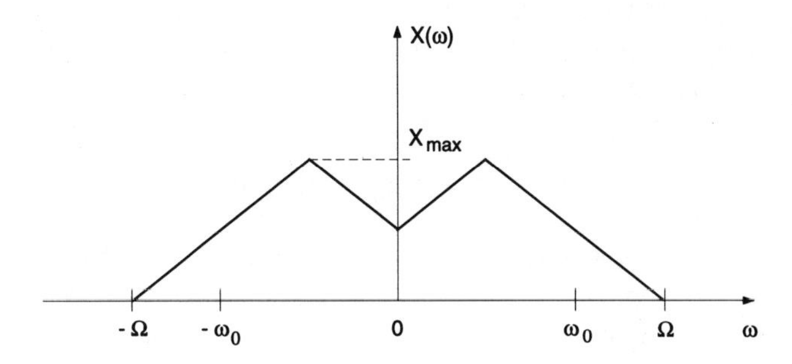

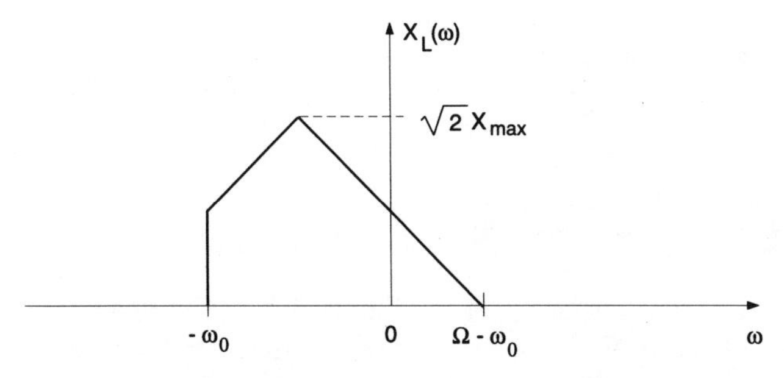

Figure 1-5 Illustration of $X(\omega)$, $X_L(\omega)$

(1-86) into the frequency domain yields

$$X_L(\omega) = \frac{\sqrt{2}}{2}\left(X(\omega+\omega_0) + j\hat{X}(\omega+\omega_0)\right) \tag{1-87}$$

where $\hat{X}(\omega)$ is the Fourier transform of $\hat{x}(t)$. Taking into account that

$$u(\omega) = \frac{1}{2} + \frac{j}{2}\left(-j\ \mathrm{sgn}(\omega)\right) \tag{1-88}$$

where sgn$(\cdot)$ denotes the sign of its argument, identification of the right-hand sides of (1-85) and (1-87) yields

$$\hat{X}(\omega) = -j\ \mathrm{sgn}(\omega)\ X(\omega) \tag{1-89}$$

This indicates that the Hilbert transform of $x(t)$ is obtained by passing $x(t)$ through a linear time-invariant filter with transfer function $-j\ \mathrm{sgn}(\omega)$. Such a filter is a phase shifter of -90 degrees for all positive frequencies.

Figures 1-6 and 1-7 show two equivalent block diagrams, in terms of complex-valued signals, for obtaining the complex envelope $x_L(t)$ from $x(t)$. The corresponding block diagrams in terms of real-valued signals are shown in Figures 1-8 and 1-9. As the transfer functions $\sqrt{2}\ u(\omega)$ and $\sqrt{2}\ u(\omega+\omega_0)$ do not possess complex conjugate symmetry about $\omega = 0$, their corresponding impulse responses are complex-valued. The complex-valued signal $x_A(t)$ in Figure 1-6 has frequency components only in the interval $\omega > 0$; $x_A(t)$ is called the *analytic signal* corresponding to $x(t)$. As

$$x_A(t) = \frac{\sqrt{2}}{2}\left[x(t) + j\hat{x}(t)\right] \tag{1-90}$$

and $x(t)$ and $\hat{x}(t)$ are 90° out of phase, the filter with transfer function $\sqrt{2}\ u(\omega)$ is called a phase splitter.

Let us determine the energy of the signals $x(t)$, $x_A(t)$, and $x_L(t)$. Considering

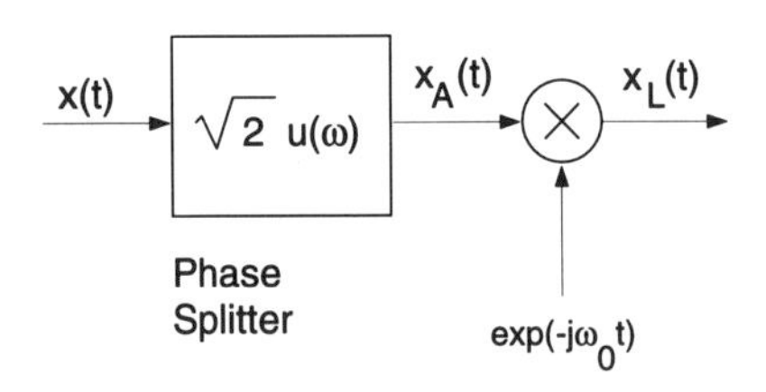

Figure 1-6 Determination of $x_L(t)$ from $x(t)$ by Means of Filtering Followed by Frequency Translation

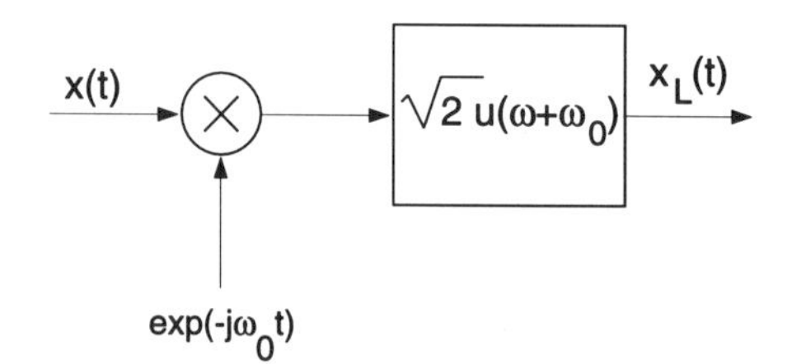

Figure 1-7 Determination of $x_L(t)$ from $x(t)$ by Means of Frequency Translation Followed by Filtering

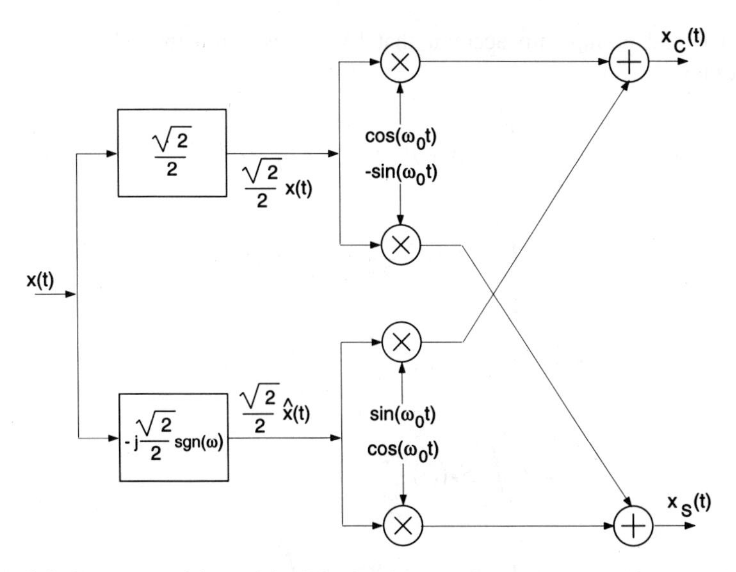

Figure 1-8 Decomposition of the Block Diagram from Figure 1-6 into Real-Valued Operations on Real-Valued Signals

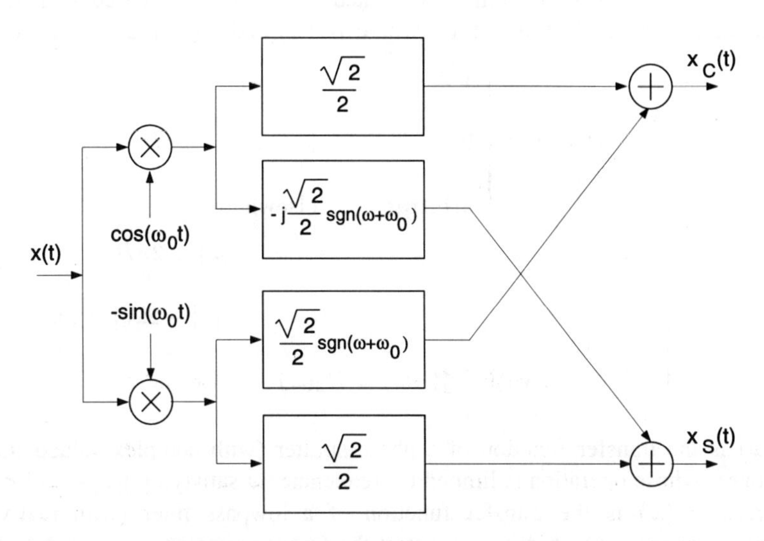

Figure 1-9 Decomposition of the Block Diagram from Figure 1-7 into Real-Valued Operations on Real-Valued Signals

Figure 1-6 and taking into account that $|X(\omega)|$ is symmetric about $\omega = 0$, we easily obtain

$$\int_{-\infty}^{+\infty} |x_A(t)|^2\, dt = \int_{-\infty}^{+\infty} |X_A(\omega)|^2 \frac{d\omega}{2\pi} = 2\int_{0}^{+\infty} |X(\omega)|^2 \frac{d\omega}{2\pi}$$

$$= \int_{-\infty}^{+\infty} |X(\omega)|^2 \frac{d\omega}{2\pi} = \int_{-\infty}^{+\infty} x^2(t)\, dt$$

$$\int_{-\infty}^{+\infty} |x_L(t)|^2\, dt = \int_{-\infty}^{+\infty} |X_L(\omega)|^2 \frac{d\omega}{2\pi} = \int_{-\infty}^{+\infty} |X_A(\omega+\omega_0)|^2 \frac{d\omega}{2\pi} \tag{1-91}$$

$$= 2\int_{0}^{+\infty} S_x(\omega)\frac{d\omega}{2\pi}$$

$$= \int_{-\infty}^{+\infty} |X_A(\omega)|^2 \frac{d\omega}{2\pi} = \int_{-\infty}^{+\infty} x^2(t)\, dt$$

Hence, the analytic signal $x_A(t)$ and the complex envelope $x_L(t)$ have the same energy as the signal $x(t)$.

Let us consider a signal $x(t)$ which is strictly bandpass about ω_0, i.e., $X(\omega)$ is nonzero only for $||\omega| - \omega_0| < 2\pi B$, with $2\pi B < \omega_0$; consequently, $X_L(\omega)$ is nonzero only in the interval $|\omega| < 2\pi B$. In this case, $x_L(t)$ can be obtained from $x(t)$ by performing the operations indicated by one of the two equivalent block diagrams in Figures 1-10 and 1-11, where $H_1(\omega)$ and $H_2(\omega)$ are given by

$$H_1(\omega) = \begin{cases} \sqrt{2} & |\omega - \omega_0| < 2\pi B \\ 0 & |\omega + \omega_0| < 2\pi B \\ \text{arbitrary} & \text{elsewhere} \end{cases} \tag{1-92}$$

$$H_2(\omega) = \begin{cases} \sqrt{2} & |\omega| < 2\pi B \\ 0 & ||\omega| - 2\omega_0| < 2\pi B \\ \text{arbitrary, with } H_2^*(-\omega) = H_2(\omega) & \text{elsewhere} \end{cases} \tag{1-93}$$

$H_1(\omega)$ is the transfer function of a phase splitter (with complex-valued impulse response) whose operation is limited to frequencies ω satisfying $||\omega| - \omega_0| < 2\pi B$ whereas $H_2(\omega)$ is the transfer function of a lowpass filter (with real-valued impulse response) which does not distort the frequency components in the interval $|\omega| < 2\pi B$. The relevant Fourier transforms are also shown in Figures 1-10 and 1-11. Note that the operations indicated in Figures 1-10 and 1-11 correspond to

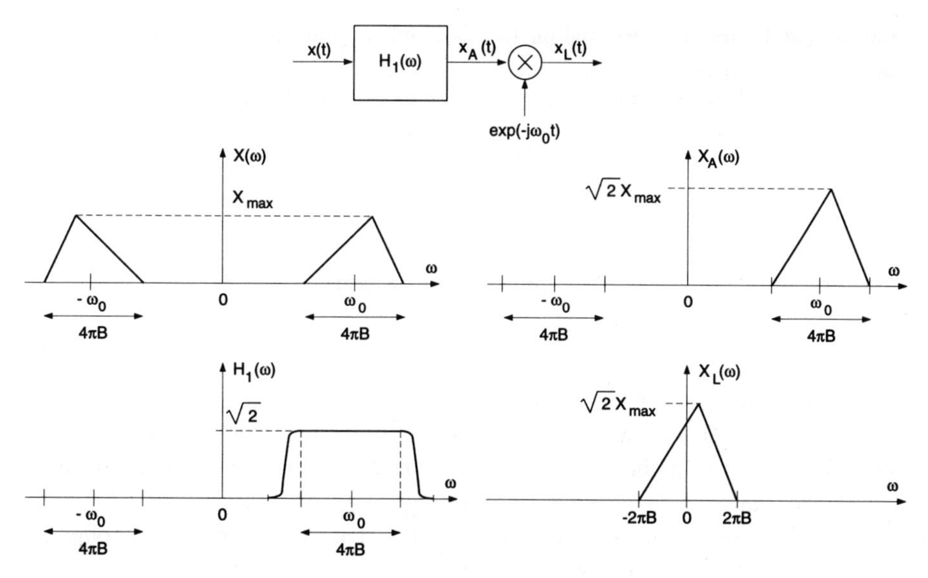

Figure 1-10 Determination of $x_L(t)$ from $x(t)$ by Means of Practical Bandpass Filtering Followed by Frequency Translation

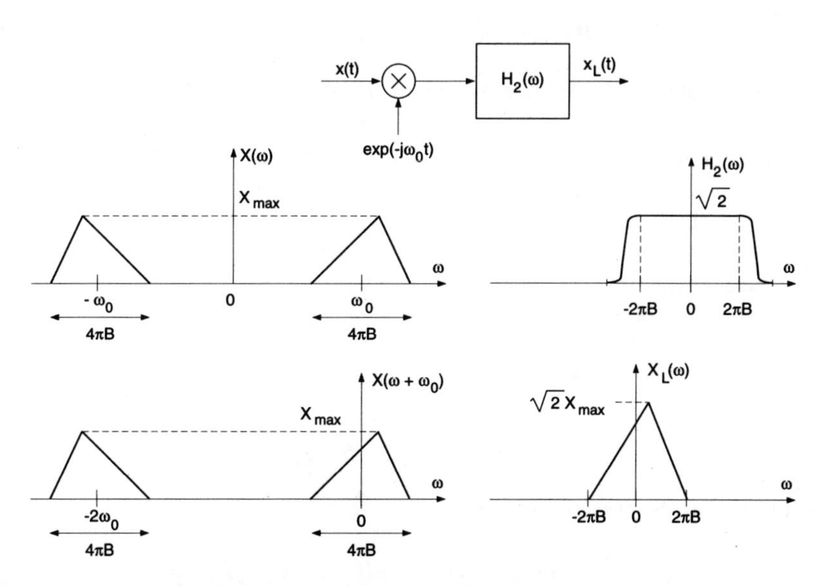

Figure 1-11 Determination of $x_L(t)$ from $x(t)$ by Means of Frequency Translation Followed by Practical Lowpass Filtering

the demodulation (i.e., translation to baseband) of the bandpass signal $x(t)$, as performed in most receivers.

Now suppose that some real-valued signal $x(t)$ is given by

$$x(t) = \sqrt{2}\,\mathrm{Re}[s(t)\,\exp(j\omega_0 t)] \tag{1-94}$$

Let us investigate under which conditions $s(t)$ equals the complex envelope $x_L(t)$ with respect to an unmodulated carrier $\sqrt{2}\,\cos(\omega_0 t)$. Denoting the Fourier transform of $s(t)$ by $S(\omega)$, it follows from (1-85) and (1-94) that

$$X_L(\omega) = u(\omega + \omega_0)\,[S(\omega) + S^*(-\omega - 2\omega_0)] \tag{1-95}$$

It is easily verified that (1-95) reduces to $X_L(\omega) = S(\omega)$, if and only if $S(\omega) = 0$ for $\omega < -\omega_0$. Hence, when a signal $x(t)$ is given by (1-94) and $s(t)$ has no frequency components below $-\omega_0$, then $s(t)$ is the complex envelope of $x(t)$.

Let us consider the signal $y(t)$ at the output of a filter with transfer function $H(\omega)$, driven by the input signal $x(t)$. In the frequency domain, the complex envelope $y_L(t)$ of $y(t)$, with respect to $\sqrt{2}\,\cos(\omega_0 t)$, is determined by

$$\begin{aligned} Y_L(\omega) &= \sqrt{2}\,u(\omega + \omega_0)\,Y(\omega + \omega_0) \\ &= (u(\omega + \omega_0)\,H(\omega + \omega_0))\left(\sqrt{2}\,u(\omega + \omega_0)\,X(\omega + \omega_0)\right) \\ &= H_E(\omega)\,X_L(\omega) \end{aligned} \tag{1-96}$$

where $H_E(\omega)$ is defined by

$$H_E(\omega) = u(\omega + \omega_0)\,H(\omega + \omega_0) \tag{1-97}$$

This indicates that the complex envelope $y_L(t)$ of the signal $y(t)$ at the output of a filter with transfer function $H(\omega)$ is obtained by passing the complex envelope $x_L(t)$ of the input signal $x(t)$ through a filter with transfer function $H_E(\omega)$, given by (1-94); note that $H_E(\omega) = 0$ for $\omega < -\omega_0$. The corresponding impulse response $h_E(t)$, which is the inverse Fourier transform of $H_E(\omega)$, is in general complex-valued. Figure 1-12 shows three equivalent block diagrams for obtaining the complex envelope $y_L(t)$ from $x(t)$.

Let us represent $x_L(t)$, $y_L(t)$, and $h_E(t)$ in rectangular form:

$$\begin{aligned} x_L(t) &= x_C(t) + j x_S(t) \\ y_L(t) &= y_C(t) + j y_S(t) \\ h_E(t) &= h_{E,C}(t) + j h_{E,S}(t) \end{aligned} \tag{1-98}$$

Then (1-96) can be transformed into:

$$\begin{bmatrix} Y_C(\omega) \\ Y_S(\omega) \end{bmatrix} = \begin{bmatrix} H_{E,C}(\omega) & -H_{E,S}(\omega) \\ H_{E,S}(\omega) & H_{E,C}(\omega) \end{bmatrix} \begin{bmatrix} X_C(\omega) \\ X_S(\omega) \end{bmatrix} \tag{1-99}$$

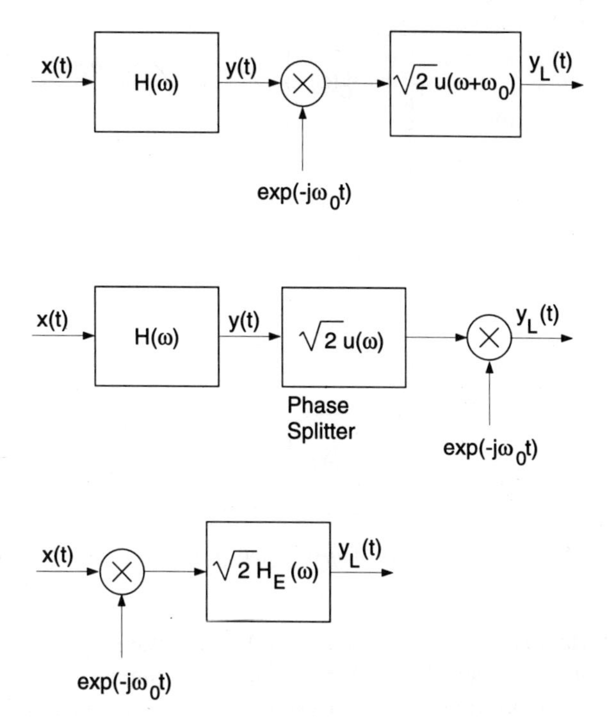

Figure 1-12 Three Ways of Obtaining $y_L(t)$ from $x(t)$

where $H_{E,C}(\omega)$ and $H_{E,S}(\omega)$ are the Fourier transforms of $h_{E,C}(t)$ and $h_{E,S}(t)$, respectively:

$$\begin{aligned} H_{E,C}(\omega) &= \frac{1}{2}\left(H_E(\omega) + H_E^*(-\omega)\right) \\ H_{E,S}(\omega) &= \frac{1}{2j}\left(H_E(\omega) - H_E^*(-\omega)\right) \end{aligned} \tag{1-100}$$

A similar relation holds among $X_C(\omega)$, $X_S(\omega)$, and $X(\omega)$ and among $Y_C(\omega)$, $Y_S(\omega)$, and $Y(\omega)$. The relation (1-99) is illustrated in Figure 1-13. It follows that, in general, the in-phase (or quadrature) component at the output of a filter is determined by both the in-phase and quadrature components at its input.

Let us consider the case where $H(\omega)$ is the transfer function of a bandpass filter, centered at $\omega = \omega_0$: $H(\omega) = 0$ for $||\omega| - \omega_0| > 2\pi B$ with $2\pi B < \omega_0$. In this case, $H_E(\omega)$ in (1-97) satisfies $H_E(\omega) = 0$ for $|\omega| > 2\pi B$; $H_E(\omega)$ is called the *equivalent lowpass transfer function* of the bandpass filter with transfer function $H(\omega)$. The bandpass filter can be specified by means of two lowpass filters, with real-valued impulse responses $h_{E,C}(t)$ and $h_{E,S}(t)$, and corresponding transfer functions $H_{E,C}(\omega)$ and $H_{E,S}(\omega)$. For the considered $H(\omega)$, the transfer functions $\sqrt{2}\,u(\omega)$ and $\sqrt{2}\,u(\omega+\omega_0)$ in Figure 1-12 can be replaced by $H_1(\omega)$ and $H_2(\omega)$, given by (1-92) and (1-93).

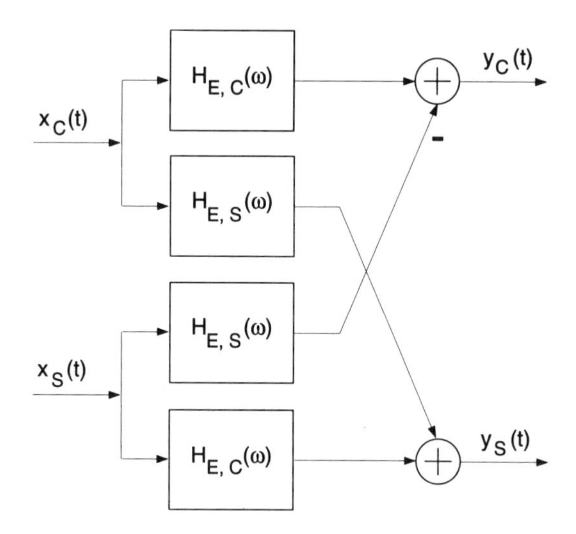

Figure 1-13 Determination of $y_C(t)$ and $y_S(t)$ from $x_C(t)$ and $x_S(t)$

An important special case occurs when the impulse response $h_E(t)$ happens to be real-valued, so that $H_E^*(-\omega) = H_E(\omega)$; this implies that $H_E(\omega)$ is zero for $|\omega| > \omega_0$. For positive ω, the corresponding transfer function $H(\omega)$ exhibits complex conjugate symmetry about ω_0:

$$u(\omega_0 + \omega)\ H(\omega_0 + \omega) = u(\omega_0 - \omega)\ H^*(\omega_0 - \omega) \tag{1-101}$$

A filter with a transfer function $H(\omega)$ satisfying (1-101) is called a *symmetric bandpass filter*. It follows from (1-101) that, for positive ω, the magnitude $|H(\omega)|$ of a symmetric bandpass filter exhibits even symmetry about $\omega = \omega_0$, whereas its argument $\arg(H(\omega))$ is odd-symmetric about $\omega = \omega_0$; this situation is depicted in Figure 1-14. The impulse response $h(t)$ of a symmetric bandpass filter is given by

$$h(t) = 2h_E(t)\ \cos(\omega_0 t) \tag{1-102}$$

As the impulse response $h_E(t)$ is real-valued, it follows that $H_{E,S}(\omega) = 0$, so that $H_E(\omega) = H_{E,C}(\omega)$. Hence, one obtains from (1-99):

$$\begin{aligned} Y_C(\omega) &= H_{E,C}(\omega)\ X_C(\omega) \\ Y_S(\omega) &= H_{E,C}(\omega)\ X_S(\omega) \end{aligned} \tag{1-103}$$

This shows that the in-phase component at the output of a symmetric bandpass filter is determined only by the in-phase component at the input. A similar observation holds for the quadrature component at the output of a symmetric bandpass filter.

In some cases, we need the complex envelope $x_{L,\theta}(t)$ of the signal $x(t)$, with respect to an unmodulated carrier $\sqrt{2}\ \cos(\omega_0 t + \theta)$ rather than $\sqrt{2}\ \cos(\omega_0 t)$; $x_{L,\theta}(t)$ is related to $x(t)$ by

$$x(t) = \sqrt{2}\ \mathrm{Re}[x_{L,\theta}(t)\ \exp(j\ (\omega_0 t+\theta))] \tag{1-104}$$

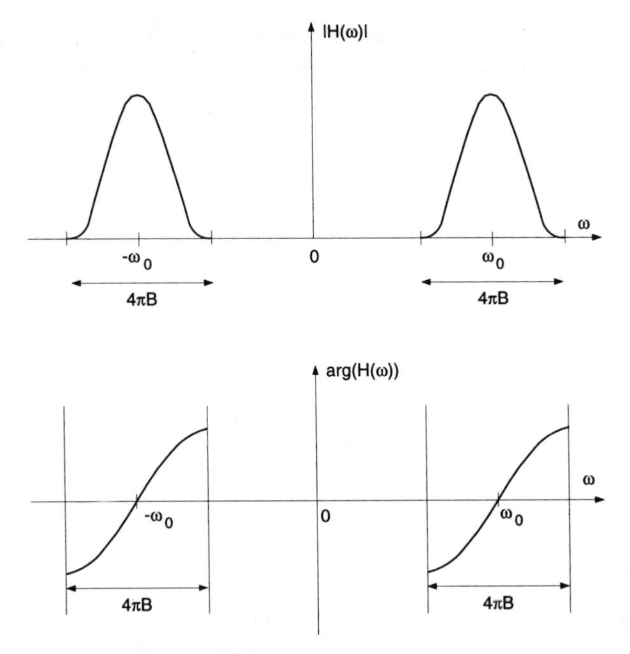

Figure 1-14 Amplitude and Phase Characteristics of a Symmetrical Bandpass Filter

Denoting by $x_L(t)$ the complex envelope of $x(t)$ with respect to $\sqrt{2}\ \cos(\omega_0 t)$, we obtain

$$\begin{aligned} x(t) &= \sqrt{2}\ \mathrm{Re}[x_L(t)\ \exp(j\omega_0 t)] \\ &= \sqrt{2}\ \mathrm{Re}[x_L(t)\ \exp(-j\theta)\ \exp(j\ (\omega_0 t+\theta))] \end{aligned} \tag{1-105}$$

from which we derive

$$x_{L,\theta}(t) = x_L(t)\ \exp(-j\theta) \tag{1-106}$$

Hence, $x_{L,\theta}(t)$ is obtained by simply rotating $x_L(t)$; consequently, $x_{L,\theta}(t)$ and $x_L(t)$ occupy the same bandwidth and have the same energy. The complex envelope $x_{L,\theta}(t)$ can be derived from $x(t)$ by applying $x(t)$ to either of the two equivalent structures from Figures 1-6 and 1-7, with $\exp(-j\omega_0 t)$ replaced by $\exp(-j\ (\omega_0 t + \theta))$. The real and imaginary parts of $x_{L,\theta}(t)$ are the in-phase and quadrature components of $x(t)$ with respect to an unmodulated carrier $\sqrt{2}\ \cos(\omega_0 t + \theta)$.

1.2.2 Complex Envelope of Random Processes

In Section 1.2.1, eq. (1-85) defines the Fourier transform $X_L(\omega)$ of the complex envelope $x_L(t)$ of the deterministic signal $x(t)$ in terms of the Fourier transform $X(\omega)$ of $x(t)$. From eq. (1-85) we derived $x_L(t)$ from $x(t)$ by performing on $x(t)$ either the filtering and the frequency translation indicated in Figure 1-6, or, equivalently, the frequency translation and filtering indicated in Figure 1-7.

For random processes $x(t)$ with finite power, the Fourier transform $X(\omega)$ does not exist. However, it is still possible to define the complex envelope $x_L(t)$ of a random process $x(t)$ with respect to an unmodulated carrier $\sqrt{2}\ \cos(\omega_0 t)$ at some frequency ω_0 as the output of one of the two equivalent structures from Figures 1-6 and 1-7, when driven by $x(t)$; this is because the operations involved (filtering, frequency translation) are well defined also for random processes. Many properties, valid for deterministic signals, can be extended to random processes:

1. When the random process $x(t)$ has frequency components only in the interval $||\omega| - \omega_0| < 2\pi B$, the transfer functions $\sqrt{2}\ u(\omega)$ and $\sqrt{2}\ u(\omega - \omega_0)$ in Figures 1-6 and 1-7 can be replaced by $H_1(\omega)$ and $H_2(\omega)$, given by (1-92) and (1-93).
2. When the random process $x(t)$ can be written as (1-94), then $s(t)$ equals the complex envelope $x_L(t)$ if and only if $s(t)$ has frequency components only in the region $\omega > -\omega_0$.
3. The complex envelope $y_L(t)$, of the random process $y(t)$ resulting from applying the random process $x(t)$ to a filter with transfer function $H(\omega)$, equals the output of a filter with transfer function $H_E(\omega)$, given by (1-97), when driven by the complex envelope $x_L(t)$ of $x(t)$.
4. The complex envelope $x_{L,\theta}(t)$ of a random process $x(t)$, with respect to an unmodulated carrier $\sqrt{2}\ \cos(\omega_0 t + \theta)$, is given by $x_L(t)\ \exp(-j\theta)$, where $x_L(t)$ denotes the complex envelope of $x(t)$ with respect to $\sqrt{2}\ \cos(\omega_0 t)$.

A few remarks about the equality of random processes and the frequency content of random processes are in order. Equality of random processes is meant here in the mean-square sense: two random processes $v(t)$ and $w(t)$ are equal in the mean-square sense when

$$E\left[|v(t) - w(t)|^2\right] = 0 \tag{1-107}$$

A random process $v(t)$ has its frequency content only in the region $\omega \in \Omega$, when $v(t)$ is equal in the mean-square sense to the random process $w(t)$, which is defined by

$$w(t) = \int_{-\infty}^{+\infty} h_\Omega(t-u)\ v(u)\ du \tag{1-108}$$

where $h_\Omega(t)$ is the impulse response of a filter with transfer function $H_\Omega(\omega)$, given by

$$H_\Omega(\omega) = \begin{cases} 1 & \omega \in \Omega \\ 0 & \omega \notin \Omega \end{cases} \tag{1-109}$$

For a wide-sense stationary process $v(t)$, it can be shown that $v(t)$ has its frequency content only in the region $\omega \in \Omega$ when its power spectral density $S_v(\omega)$ is zero for $\omega \notin \Omega$.

In telecommunications, a carrier-modulated signal can be represented by $\sqrt{2}\ \mathrm{Re}[s(t)\ \exp(j\omega_0 t)]$, where $s(t)$ has no (or at least negligible) frequency components in the interval $|\omega| > 2\pi B$ with $2\pi B < \omega_0$; for instance, this condition is fulfilled in the case of linear modulation, where $s(t)$ is given by (1-58), and the Fourier transform $G(\omega)$ of the baseband pulse $g(t)$ is essentially zero for $|\omega| > 2\pi B$. Hence, the complex envelope of the carrier-modulated signal is simply given by $s(t)$. The communications channel adds stationary noise to the carrier-modulated signal. The received signal (carrier-modulated signal plus noise) is applied to the demodulator structure from Figures 1-10 and 1-11. As the additive noise contribution to the demodulator output is a filtered version of the complex envelope of the input noise, we concentrate in the next section on some statistical properties of the complex envelope of wide-sense stationary processes.

In Section 2.4, we will be faced with the complex envelope of a cyclostationary disturbance with period T, with respect to a sinusoid at frequency $\omega_0 = 2\pi/T$. The statistical properties of this complex envelope, needed in Section 2.4, are derived in the section on the complex envelope of wide-sense cyclostationary processes which can be skipped at first reading.

Complex Envelope of Wide-Sense Stationary Processes

Let us consider a real-valued zero-mean wide-sense stationary process $x(t)$, having a complex envelope $x_L(t)$ with respect to $\sqrt{2}\ \cos(\omega_0 t)$:

$$x(t) = \sqrt{2}\ \mathrm{Re}[x_L(t)\ \exp(j\omega_0 t)] \tag{1-110}$$

It follows from Figure 1-6 that $x_L(t)$ can be written as

$$x_L(t) = x_A(t)\ \exp(-j\omega_0 t) \tag{1-111}$$

where the analytic signal $x_A(t)$ is obtained by applying $x(t)$ to a filter with transfer function $\sqrt{2}\ u(\omega)$. Being a filtered version of the zero-mean wide-sense stationary process $x(t)$, $x_A(t)$ is also zero-mean wide-sense stationary; hence, the statistical expectation $E[x_L(t)]$ is zero.

Let us calculate the auto- and cross-correlation functions $E[x_L^*(t)\ x_L(t+u)]$ and $E[x_L(t)\ x_L(t+u)]$. It follows from (1-111) that

$$E[x_L^*(t)\ x_L(t+u)] = R_{x_A}(u)\ \exp(-j\omega_0 u) \tag{1-112}$$

$$E[x_L(t)\ x_L(t+u)] = R_{x_A^* x_A}(u)\ \exp(-j\omega_0 u)\ \exp(-j2\omega_0 t) \tag{1-113}$$

where $R_{x_A}(u)$ and $R_{x_A^* x_A}(u)$ are the autocorrelation function of $x_A(t)$ and the cross-correlation function of $x_A^*(t)$ and $x_A(t)$, respectively. The power spectral density $S_{x_A}(\omega)$ of $x_A(t)$ and the cross-spectral density $S_{x_A^* x_A}(\omega)$ of $x_A^*(t)$ and

$x_A(t)$, which are the Fourier transforms of $R_A(u)$ and $R_{A^*A}(u)$, are easily derived from (1-11) and (1-12):

$$S_{x_A}(\omega) = 2\,|u(\omega)|^2\, S_x(\omega) = 2u(\omega)\, S_x(\omega) \tag{1-114}$$

$$S_{x_A^* x_A}(\omega) = 2u(\omega)\, u(-\omega)\, S_x(\omega) = 0 \tag{1-115}$$

where $S_x(\omega)$ is the power spectral density of $x(t)$. It follows from (1-112) and (1-113) that both the auto- and cross-correlation functions $E[x_L^*(t)\, x_L(t+u)]$ and $E[x_L(t)\, x_L(t+u)]$ do not depend on the time variable t [the latter being identically zero because of (1-115)]; these correlation functions will be denoted as $R_{x_L}(u)$ and $R_{x_L^* x_L}(u)$, respectively. Hence, the complex envelope $x_L(t)$ of a zero-mean wide-sense stationary process $x(t)$ is also zero-mean and wide-sense stationary; the power spectral density $S_{x_L}(\omega)$ of $x_L(t)$ and the cross-spectral density $S_{x_L^* x_L}(\omega)$ of $x_L^*(t)$ and $x_L(t)$ are given by

$$S_{x_L}(\omega) = 2u(\omega+\omega_0)\, S_x(\omega+\omega_0) \tag{1-116}$$

$$S_{x_L^* x_L}(\omega) = 0 \quad \Rightarrow \quad R_{x_L^* x_L}(u) = 0 \tag{1-117}$$

As $S_x(\omega)$ is an even function of ω, we easily obtain

$$\begin{aligned} E\left[|x_A(t)|^2\right] &= \int_{-\infty}^{+\infty} S_{x_A}(\omega)\,\frac{d\omega}{2\pi} = 2\int_{0}^{+\infty} S_x(\omega)\,\frac{d\omega}{2\pi} \\ &= \int_{-\infty}^{+\infty} S_x(\omega)\,\frac{d\omega}{2\pi} = E\left[x^2(t)\right] \\ E\left[|x_L(t)|^2\right] &= \int_{-\infty}^{+\infty} S_{x_L}(\omega)\,\frac{d\omega}{2\pi} = 2\int_{-\omega_0}^{+\infty} S_x(\omega+\omega_0)\,\frac{d\omega}{2\pi} \\ &= \int_{-\infty}^{+\infty} S_x(\omega)\,\frac{d\omega}{2\pi} = E\left[x^2(t)\right] \end{aligned} \tag{1-118}$$

Hence, the random process $x(t)$, its corresponding analytic signal $x_A(t)$ and its complex envelope $x_L(t)$ have the same power. Also, it follows from (1-117) that $x_L^*(t)$ and $x_L(t+u)$ are uncorrelated for any value of u.

The in-phase and quadrature components $x_C(t)$ and $x_S(t)$ of $x(t)$ with respect to $\cos(\omega_0 t)$ are also zero-mean and wide-sense stationary. Their auto- and cross-correlation functions can easily be derived by taking into account that $x_C(t)$ and $x_S(t)$ are the real and imaginary parts of $x_L(t)$. Making use of (1-117) one obtains

$$R_{x_C}(u) = R_{x_S}(u) = 1/2\ \mathrm{Re}[R_{x_L}(u)] \tag{1-119}$$

$$R_{x_C x_S}(u) = -R_{x_C x_S}(-u) = 1/2\ \mathrm{Im}[R_{x_L}(u)] \tag{1-120}$$

This shows that the in-phase and quadrature components of a wide-sense stationary process $x(t)$ have the same autocorrelation function. Their cross-correlation function is an odd function of u, indicating that $x_C(t)$ and $x_S(t)$ are uncorrelated, when taken at the *same* instant of time. From (1-119) we derive

$$\begin{aligned} E\left[x_C^2(t)\right] &= R_{x_C}(0) = 1/2\ R_{x_L}(0) = 1/2\ E\left[|x_L(t)|^2\right] = 1/2\ E\left[x^2(t)\right] \\ E\left[x_S^2(t)\right] &= R_{x_S}(0) = 1/2\ R_{x_L}(0) = 1/2\ E\left[|x_L(t)|^2\right] = 1/2\ E\left[x^2(t)\right] \end{aligned} \tag{1-121}$$

Hence, the in-phase and quadrature components each contain half of the power of $x_L(t)$ or $x(t)$. Denoting the Fourier transforms of $R_{x_C}(u)$, $R_{x_S}(u)$, and $R_{x_C x_S}(u)$ by $S_{x_C}(\omega)$, $S_{x_S}(\omega)$, and $S_{x_C x_S}(\omega)$, respectively, the following frequency-domain relations are derived from (1-119) and (1-120):

$$S_{x_C}(\omega) = S_{x_S}(\omega) = 1/4\ (S_{x_L}(\omega) + S_{x_L}(-\omega)) \tag{1-122}$$

$$S_{x_C x_S}(\omega) = -S_{x_C x_S}(-\omega) = \frac{1}{4j}\ (S_{x_L}(\omega) - S_{x_L}(-\omega)) \tag{1-123}$$

The spectra $S_{x_C}(\omega)$ and $S_{x_S}(\omega)$ are real, even, and nonnegative, whereas $S_{x_C x_S}(\omega)$ is imaginary and odd. The equations (1-116), (1-122), and (1-123) are illustrated in Figure 1-15.

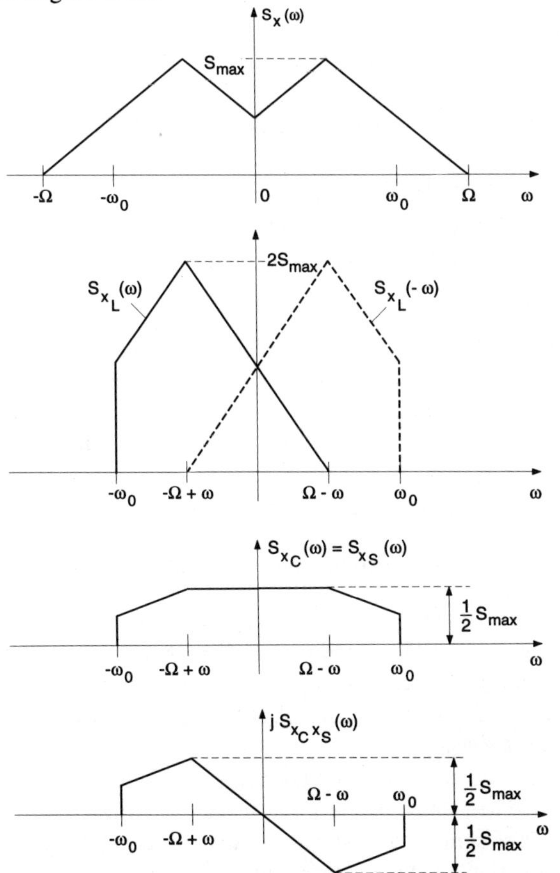

Figure 1-15 Illustration of the Spectra $S_x(\omega)$, $S_{x_L}(\omega)$, $S_{x_C}(\omega)$, $S_{x_S}(\omega)$, and $S_{x_C x_S}(\omega)$

When the positive-frequency content of the spectrum $S_x(\omega)$ is symmetric about the center frequency ω_0, i.e.,

$$u(\omega_0+\omega)\, S_x(\omega_0+\omega) = u(\omega_0-\omega)\, S_x(\omega_0-\omega) \tag{1-124}$$

then $x(t)$ is called a wide-sense stationary *symmetric bandpass process.* In this case $S_{x_L}(\omega)$ is an even function of ω [see (1-116)], so that $R_{x_L}(u)$ is real-valued. Hence, (1-122) and (1-123) reduce to

$$S_{x_C}(\omega) = S_{x_S}(\omega) = 1/2\, S_{x_L}(\omega) = u(\omega+\omega_0) S_x(\omega+\omega_0) \tag{1-125}$$

$$S_{x_C x_S}(\omega) = 0 \quad \Rightarrow \quad R_{x_C x_S}(u) = 0 \tag{1-126}$$

It follows from (1-126) that $x_C(t)$ and $x_S(t+u)$ are uncorrelated for any value of u.

The complex envelope $x_{L,\theta}(t)$ of $x(t)$, with respect to $\sqrt{2}\,\cos(\omega_0 t+\theta)$, is given by $x_L(t)\,\exp(-j\theta)$, where $x_L(t)$ denotes the complex envelope of $x(t)$ with respect to $\sqrt{2}\,\cos(\omega_0 t)$. Hence, when $x(t)$ is wide-sense stationary we obtain

$$\begin{aligned} R_{x_{L,\theta}}(u) &= R_{x_L}(u) \\ R_{x^*_{L,\theta} x_{L,\theta}}(u) &= R_{x^*_L x_L}(u)\, \exp(-2j\theta) = 0 \end{aligned} \tag{1-127}$$

We conclude that the autocorrelation function of $x_{L,\theta}(t)$ and the cross-correlation function of $x^*_{L,\theta}(t)$ and $x_{L,\theta}(t)$ do not depend on the value of θ, and, hence, are the same as for the complex envelope $x_L(t)$.

Let us consider the case where $x(t)$ is a real-valued zero-mean wide-sense stationary *Gaussian* random process. For Gaussian processes, the following general properties hold (see [3]):

1. A wide-sense stationary Gaussian process is also stationary in the strict sense.
2. Any linear operation on a Gaussian process yields another Gaussian process.
3. The statistical properties of a zero-mean complex-valued Gaussian process $v(t)$ are determined completely by the correlation functions $E[v^*(t)\, v(t+u)]$ and $E[v(t)\, v(t+u)]$.
4. Uncorrelated real-valued Gaussian processes are statistically independent.

From the above properties, the following conclusions can be drawn:

1. The signal $x(t)$, the corresponding analytic signal $x_A(t)$, the complex envelope $x_L(t)$, and the in-phase and quadrature components $x_C(t)$ and $x_S(t)$ are strict-sense stationary zero-mean Gaussian processes.
2. The in-phase and quadrature components $x_C(t)$ and $x_S(t)$ have the same statistical properties.
3. The in-phase and quadrature components $x_C(t)$ and $x_S(t)$ are statistically independent, when taken at the same instant of time.
4. When $x(t)$ is a symmetric bandpass process [see (1-124)], then $x_C(t)$ and $x_S(t+u)$ are statistically independent for any u.

5. The complex envelope $x_{L,\theta}(t)$ of $x(t)$, with respect to $\sqrt{2}\ \cos(\omega_0 t + \theta)$, has the same statistical properties as the complex envelope $x_L(t)$ with respect to $\sqrt{2}\ \cos(\omega_0 t)$.

The case of a Gaussian stationary process $x(t)$ is important, because the additive noise at the input of the receiver can be modeled as such.

Complex Envelope of Wide-Sense Cyclostationary Processes

Let us consider a real-valued wide-sense cyclostationary process $x(t)$ with period T, having a complex envelope $x_L(t)$ with respect to $\sqrt{2}\ \cos((2\pi t/T))$. The statistical expectation $E[x(t)]$ and the autocorrelation function $R_x(t,t+u) = E[x^*(t)\ x(t+u)]$ are both periodic in t with period T, and can be expanded into a Fourier series with coefficients $m_{x,k}$ and $r_{x,k}(u)$, respectively [see (1-14) and (1-18)]. The complex envelope $x_L(t)$ is related to $x(t)$ by

$$x_L(t) = x_A(t)\ \exp\left(-j\ \frac{2\pi t}{T}\right) \tag{1-128}$$

where the analytic signal $x_A(t)$ is obtained by applying $x(t)$ to a filter with transfer function $\sqrt{2}\ u(\omega)$. Being a filtered version of the wide-sense cyclostationary process $x(t)$, $x_A(t)$ is also wide-sense cyclostationary with period T. Hence, the statistical expectation $E[x_A(t)]$ and the correlation functions $R_{x_A}(t,t+u)$ and $R_{x_A^* x_A}(t,t+u)$ are periodic in t with period T. The statistical expectation $E[x_A(t)]$ and the correlation functions $R_{x_L}(t,t+u)$ and $R_{x_L^* x_L}(t,t+u)$ are also periodic in t with period T:

$$E[x_L(t)] = E[x_A(t)]\ \exp\left(-j\ \frac{2\pi t}{T}\right) \tag{1-129}$$

$$R_{x_L}(t,t+u) = R_{x_A}(t,t+u)\ \exp\left(-j\ \frac{2\pi u}{T}\right) \tag{1-130}$$

$$R_{x_L^* x_L}(t,t+u) = R_{x_A^* x_A}(t,t+u)\ \exp\left(-j\ \frac{2\pi u}{T}\right)\ \exp\left(-j\ \frac{4\pi t}{T}\right) \tag{1-131}$$

Hence, the complex envelope $x_L(t)$ is also wide-sense cyclostationary with period T.

Let us concentrate on the Fourier series expansions (1-14)–(1-19) of $E[x(t)]$, $R_x(t,t+u)$, and $R_{x^* x}(t,t+u)$, and on the Fourier series expansions of $E[x_A(t)]$, $R_{x_A}(t,t+u)$, and $R_{x_A^* x_A}(t,t+u)$, and of $E[x_L(t)]$, $R_{x_L}(t,t+u)$, and $R_{x_L^* x_L}(t,t+u)$; the expansions related to $x_A(t)$ and $x_L(t)$ are obtained from (1-14)–(1-19) by replacing the subscript x by x_A and x_L, respectively. From

(1-129), (1-130), and (1-131) we obtain

$$m_{x_L,k} = m_{x_A,k+1} \tag{1-132}$$

$$r_{x_L,k}(u) = r_{x_A,k}(u)\ \exp\left(-j\ \frac{2\pi u}{T}\right) \tag{1-133}$$

$$r_{x_L^* x_L,k}(u) = r_{x_A^* x_A,k+2}(u)\ \exp\left(-j\ \frac{2\pi u}{T}\right) \tag{1-134}$$

The frequency-domain relations corresponding to (1-133) and (1-134) are

$$S_{x_L,k}(\omega) = S_{x_A,k}\left(\omega + \frac{2\pi}{T}\right) \tag{1-135}$$

$$S_{x_L^* x_L,k}(\omega) = S_{x_A^* x_A,k+2}\left(\omega + \frac{2\pi}{T}\right) \tag{1-136}$$

where $S_{x_L,k}(\omega)$, $S_{x_A,k}(\omega)$, $S_{x_L^* x_L,k}(\omega)$, and $S_{x_A^* x_A,k}(\omega)$ are the Fourier transforms of $r_{x_L,k}(u)$, $r_{x_A,k}(u)$, $r_{x_L^* x_L,k}(u)$, and $r_{x_A^* x_A,k}(u)$. Using (1-35)–(1-37), with $H(\omega) = \sqrt{2}\ u(\omega)$ and the subscript y replaced by x_L, one obtains

$$m_{x_A,k} = \sqrt{2}\ u\left(\frac{2\pi k}{T}\right)\ m_{x,k} \tag{1-137}$$

$$S_{x_A,k}(\omega) = 2\ u\left(\omega - \frac{2\pi k}{T}\right)\ u(\omega)\ S_{x,k}(\omega) \tag{1-138}$$

$$S_{x_A^* x_A,k}(\omega) = 2\ u\left(-\omega + \frac{2\pi k}{T}\right)\ u(\omega)\ S_{x,k}(\omega) \tag{1-139}$$

where $S_{x.k}(\omega)$ is the Fourier transform of the coefficient $r_{x,k}(u)$ in the expansion (1-18). Hence,

$$m_{x_L,k} = \sqrt{2}\ u\left(\frac{2\pi}{T}\ (k+1)\right)\ m_{x,k+1} \tag{1-140}$$

$$S_{x_L,k}(\omega) = 2\ u\left(\omega - \frac{2\pi}{T}\ (k-1)\right)\ u\left(\omega + \frac{2\pi}{T}\right)\ S_{x,k}\left(\omega + \frac{2\pi}{T}\right) \tag{1-141}$$

$$S_{x_L^* x_L,k}(\omega) = 2\ u\left(-\omega + \frac{2\pi}{T}(k+1)\right)\ u\left(\omega + \frac{2\pi}{T}\right)\ S_{x,k+2}\left(\omega + \frac{2\pi}{T}\right) \tag{1-142}$$

which follows from substituting (1-137), (1-138), and (1-139) into (1-132), (1-135), and (1-136).

Now we consider the process $y_L(t)$, which results from passing the wide-sense cyclostationary process $x_L(t)$ through a (possibly complex-valued) lowpass filter with transfer function $H_{\mathrm{LP}}(\omega)$. Equivalently, $y_L(t)$ is the complex envelope of the process $y(t)$, which results from passing $x(t)$ through a bandpass filter with transfer function $H_{\text{BP}}(\omega) = H_{\text{LP}}(\omega - \omega_0) + H^*_{\text{LP}}(-\omega - \omega_0)$, which corresponds to a real-valued impulse response $h_{\text{BP}}(t)$. We restrict our attention to the case where $H_{\text{LP}}(\omega) = 0$ for $|\omega| > \pi/T$. It has been shown in Section 1.1.2 that the resulting process $y_L(t)$ is *wide-sense stationary* [see (1-39)–(1-41) with $H(\omega) = H_{\text{LP}}(\omega)$ and the subscripts x and y replaced by x_L and y_L, respectively]

$$E[y_L(t)] = H_{\text{LP}}(0)\, m_{x_L,0} = H_{\text{LP}}(0)\, \sqrt{2}\, m_{x,1} \tag{1-143}$$

$$\begin{aligned} S_{y_L}(\omega) &= |H_{\text{LP}}(\omega)|^2\, S_{x_L,0}(\omega) \\ &= |H_{\text{LP}}(\omega)|^2\, 2\, S_{x,0}\left(\omega + \frac{2\pi}{T}\right) \end{aligned} \tag{1-144}$$

$$\begin{aligned} S_{y_L^* y_L}(\omega) &= H_{\text{LP}}(\omega)\, H_{\text{LP}}(-\omega)\, S_{x_L^* x_L,0}(\omega) \\ &= H_{\text{LP}}(\omega)\, H_{\text{LP}}(-\omega)\, 2\, S_{x,2}\left(\omega + \frac{2\pi}{T}\right) \end{aligned} \tag{1-145}$$

It is important to note that, although the complex envelope $y_L(t)$ is wide-sense stationary, the corresponding bandpass signal $y(t) = \sqrt{2}\,\text{Re}[y_L(t)\, \exp(j\omega_0 t)]$ is in general not wide-sense stationary, but only wide-sense cyclostationary. Indeed, let us consider the autocorrelation function of $y(t)$, which is given by

$$\begin{aligned} R_y(t, t+u) =& \text{Re}\left[R_{y_L}(u)\, \exp\left(j\, \frac{2\pi u}{T}\right)\right] \\ &+ \text{Re}\left[R_{y_L^* y_L}(u)\, \exp\left(j\, \frac{2\pi u}{T}\right)\, \exp\left(j\, \frac{4\pi t}{T}\right)\right] \end{aligned} \tag{1-146}$$

Clearly, $R_y(t, t+u)$ becomes independent of t only when

$$R_{y_L^* y_L}(u) = 0 \quad \Rightarrow \quad S_{y_L^* y_L}(\omega) = 0 \tag{1-147}$$

which is usually not fulfilled [see (1-145)] when the wide-sense cyclostationary complex envelope $x_L(t)$ has frequency components within the bandwidth of $H_{\text{LP}}(\omega)$.

1.2.3 Example

Let us consider the demodulator structure shown in Figure 1-16, where the random process $x(t)$ is given by

$$x(t) = \sqrt{2}\,\text{Re}[s(t)\, \exp(j\omega_0 t)] + n(t) \tag{1-148}$$

In (1-148), $s(t)$ has frequency components only in the interval $|\omega| < 2\pi B$, with $2\pi B < \omega_0$; hence, $s(t)$ is the complex envelope of the first term of (1-148). The second term in (1-148) is stationary real-valued zero-mean Gaussian noise $n(t)$, with power spectral density $S_n(\omega)$. The transfer function $H_{\text{LP}}(\omega)$ in Figure 1-16 satisfies $H_{\text{LP}}(\omega) = 0$ for $|\omega| > 2\pi B$. The purpose of the demodulator is to suppress the frequency components of the noise $n(t)$ that are located in the interval $||\omega| - \omega_0| > 2\pi B$, and to provide a signal $y(t)$ consisting of a filtered version of

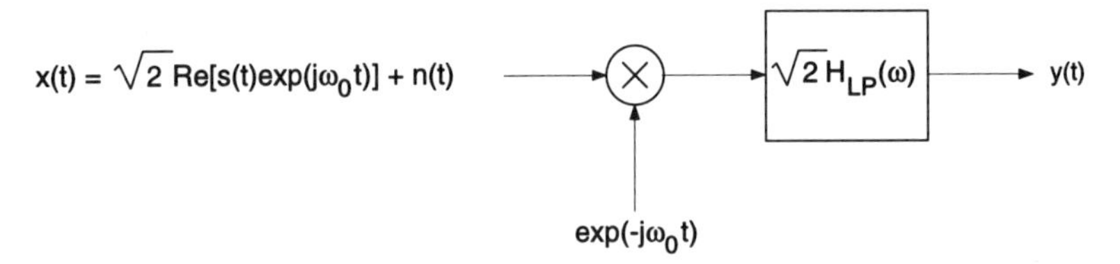

Figure 1-16 Demodulation of a Noisy Bandpass Signal

$s(t)$, disturbed by additive noise caused by the frequency components of the noise $n(t)$ that are located in the interval $||\omega| - \omega_0| < 2\pi B$.

The demodulator from Figure 1-16 is equivalent to the structure shown in Figure 1-17, where $n_L(t)$ is the complex envelope of $n(t)$ with respect to $\sqrt{2}\,\cos(\omega_0 t)$. The contribution from $n(t)$ to the output signal $y(t)$ is a zero-mean stationary Gaussian process, whose power spectral density is given by

$$\begin{aligned} |H_{\mathrm{LP}}(\omega)|^2 S_{n_L}(\omega) &= |H_{\mathrm{LP}}(\omega)|^2 2u(\omega+\omega_0) S_n(\omega+\omega_0) \\ &= 2\,|H_{\mathrm{LP}}(\omega)|^2 S_n(\omega+\omega_0) \end{aligned} \tag{1-149}$$

Hence, as far as the statistics of $y(t)$ are concerned, only the values of the power spectral density $S_n(\omega)$ in the interval $||\omega| - \omega_0| < 2\pi B$ are important; hence, without loss of generality we can assume $S_n(\omega) = 0$ for $||\omega| - \omega_0| > 2\pi B$. In many cases of practical interest, the variation of $S_n(\omega)$ within the interval $||\omega| - \omega_0| < 2\pi B$ can be neglected with respect to the value $S_n(\omega_0)$, so that the following approximation is appropriate

$$S_n(\omega) = \begin{cases} N_0/2 & ||\omega| - \omega_0| < 2\pi B \\ 0 & \text{elsewhere} \end{cases} \tag{1-150}$$

with $N_0/2 = S_n(\omega_0)$. The corresponding noise $n(t)$ is referred to as being "white within the signal bandwidth," with power spectral density $N_0/2$. Figure 1-18 shows the power spectral densities $S_n(\omega)$, $S_{n_L}(\omega)$, $S_{n_C}(\omega)$, and $S_{n_S}(\omega)$ of the noise $n(t)$, its complex envelope $n_L(t)$, and the in-phase and quadrature components $n_C(t)$ and $n_S(t)$. As $S_n(\omega)$ is symmetric about $\omega = \omega_0$ for $\omega > 0$, $n_C(t)$ and $n_S(t)$ are uncorrelated for any value of u. Hence, $n_C(t)$ and $n_S(t)$ are statistically independent identically distributed zero-mean Gaussian random processes, with a power spectral density that equals $N_0/2$ for $|\omega| < 2\pi B$ and zero otherwise. However, as the value of their power spectral density for $|\omega| > 2\pi B$ does not influence the statistical properties of $y(t)$, it is common practice to replace $n_C(t)$

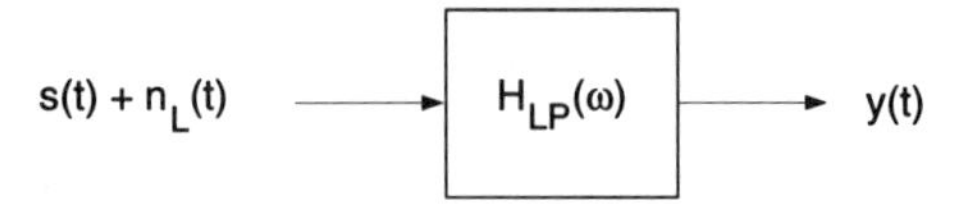

Figure 1-17 Equivalent Demodulator Structure Operating on Noisy Complex Envelope

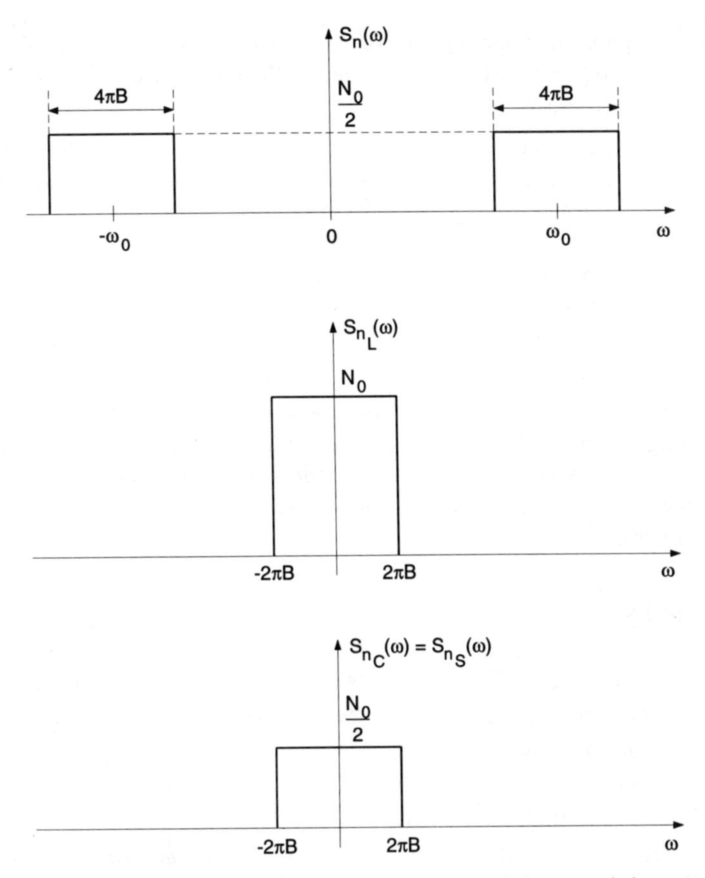

Figure 1-18 Illustration of the Spectra $S_n(\omega)$, $S_{n_L}(\omega)$, $S_{n_C}(\omega)$, and $S_{n_S}(\omega)$ When $n(t)$ Is White Noise

and $n_S(t)$ by truly white processes, whose power spectral density equals $N_0/2$ for *all* ω.

1.2.4 Main Points

The complex envelope $x_L(t)$ of a deterministic or random signal $x(t)$, with respect to an unmodulated carrier $\sqrt{2}\ \cos(\omega_0 t)$, is obtained by performing on $x(t)$ the operations indicated in Figures 1-6 and 1-7; when $x(t)$ is strictly bandpass and centered at $\omega = \omega_0$, $x_L(t)$ is lowpass, and can also be obtained by performing the demodulation indicated in Figures 1-10 and 1-11.

Let $y(t)$ be the output of a filter, with transfer function $H(\omega)$ and real-valued impulse response $h(t)$, when driven by $x(t)$. Then the complex envelope $y_L(t)$ of $y(t)$ is obtained by applying the complex envelope $x_L(t)$ of $x(t)$ to a filter with transfer function $H_E(\omega) = u(\omega + \omega_0)\ H(\omega + \omega_0)$. Unless $H(\omega)$ is the transfer function of a symmetric bandpass filter [see (1-101)], the impulse response $h_E(t)$ is complex-valued.

The complex envelope $x_{L,\theta}(t)$ of $x(t)$, with respect to $\sqrt{2}\ \cos(\omega_0 t + \theta)$, equals $x_L(t)\ \exp(-j\theta)$, where $x_L(t)$ denotes the complex envelope of $x(t)$ with respect to $\sqrt{2}\ \cos(\omega_0 t)$.

The complex envelope $x_L(t)$ of a zero-mean wide-sense stationary process $x(t)$ is itself zero-mean wide-sense stationary. The in-phase and quadrature components $x_C(t)$ and $x_S(t)$ have the same power spectral density; their cross-spectral density is an odd imaginary function of ω, which becomes identically zero when $x(t)$ is a symmetric bandpass process [see (1-124)]. When $x(t)$ is wide-sense stationary and Gaussian, then $x(t)$ is also strict-sense stationary; the corresponding $x_L(t)$, $x_C(t)$, and $x_S(t)$ are also Gaussian strict-sense stationary processes.

When a wide-sense cyclostationary process $x(t)$ with period T is applied to a bandpass filter $H_{\mathrm{BP}}(\omega)$ with $H_{\mathrm{BP}}(\omega) = 0$ for $||\omega| - 2\pi/T| > \pi/T$, then the resulting bandpass process $y(t)$ has a complex envelope $y_L(t)$ with respect to $\sqrt{2}\ \cos(\omega_0 t)$ which is wide-sense stationary. The bandpass process $y(t)$ itself, however, is in general not wide-sense stationary, but only wide-sense cyclostationary.

Bibliography

[1] L. E. Franks, *Signal Theory*. Englewood Cliffs, NJ: Prentice-Hall, 1969.

[2] W. A. Gardner, *Introduction to Random Processes, with Applications to Signals and Systems*. New York: Macmillan, 1986.

[3] A. Papoulis, *Probability, Random Variables and Stochastic Processes*. Auckland: McGraw-Hill, 1991.

[4] S. Benedetto and E. Biglieri and V. Castellani , *Digital Transmission Theory*. Englewood Cliffs, NJ: Prentice-Hall, 1987.

[5] E. A. Lee and D. G. Messerschmitt, *Digital Communications*. Boston: Kluwer Academic, 1988.

[6] J. G. Proakis, *Digital Communications*. Auckland: McGraw-Hill, 1989.

1.3 Band-Limited Signals

1.3.1 Correlation Function and Power Spectrum

Let $y(t)$ be the output of an ideal lowpass filter driven by $x(t)$

$$y(t) = \int_{-\infty}^{\infty} h_{\mathrm{LP}}(t-u)\ x(u)\ du \tag{1-151}$$

with

$$H_{\mathrm{LP}}(\omega) = \begin{cases} 1 & \text{if} \quad |\omega/2\pi| < B \\ 0 & \text{else} \end{cases} \tag{1-152}$$

(see Figure 1-19).

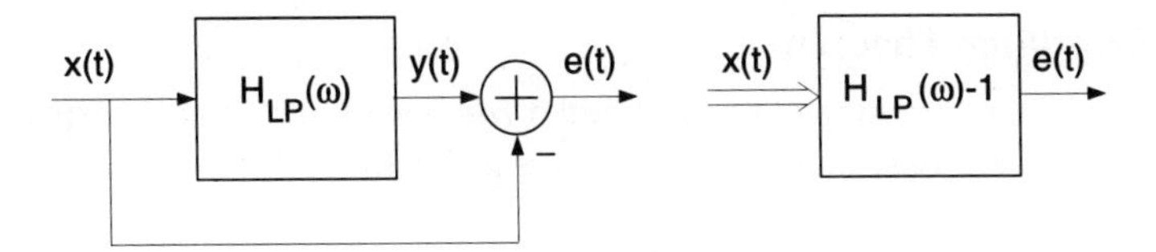

Figure 1-19 Ideal Lowpass Filter

The process $x(t)$ is said to be band-limited to frequency B if it is not altered by the filtering operation, i.e.,

$$E\left[|x(t) - y(t)|^2\right] = 0 \tag{1-153}$$

It is easily shown that this implies that the power spectrum is zero outside $|\omega| > 2\pi B$. Indeed, the difference signal $e(t) = x(t) - y(t)$ can be thought to be generated by a filter $[1 - H_{\mathrm{LP}}(\omega)]$ driven by $x(t)$. The power spectrum of this signal is given by

$$\begin{aligned} Se(\omega) &= S_x(\omega)|1 - H_{\mathrm{LP}}(\omega)|^2 \\ &= 0 \quad \text{for all } \omega \end{aligned} \tag{1-154}$$

if

$$S_x(\omega) = 0 \text{ for } |\omega| > 2\pi B \tag{1-155}$$

For a cyclostationary process $S_x(\omega)$ must be replaced by $S_{x,0}(\omega)$.
Since the power spectrum is deterministic and band-limited the correlation function has a series representation with the base functions.

$$\begin{aligned} \phi_k(t) &= \frac{\sin\left(2\pi B\left(t - \frac{k}{2B}\right)\right)}{2\pi B\left(t - \frac{k}{2B}\right)} \\ &= \mathrm{si}\left(2\pi B\left(t - \frac{k}{2B}\right)\right) \end{aligned} \tag{1-156}$$

For $1/T_s \geq 2B$ we obtain

$$R_x(u) = \sum_k R_x(kT_s)\phi_k(u) \tag{1-157}$$

and for the cyclostationary process [eq. (1-14)]

$$\begin{aligned} R_x(t, t+u) &= \sum_n r_{x,n}(u) e^{(j2\pi/T)nT} \\ &= \sum_n \left[\sum_k r_{x,n}(kT_s)\phi_k(u)\right] e^{(j2\pi/T)nT} \\ &= \sum_k R_x(t, t+kT_s)\phi_k(u) \end{aligned} \tag{1-158}$$

Notice that the series expansion in equation (1-158) is with respect to one variable only.

1.3.2 Sampling Theorem

We will next prove that a band-limited *random* process can be expanded into a series where the base functions $\phi_n(t)$ are given by (1-156).

$$\begin{aligned} x(t+\tau) &= \sum_{n=-\infty}^{\infty} x(\tau + nT_s)\, \phi_n(t) \\ &= \sum_{n=-\infty}^{\infty} x(\tau + nT_s)\, \mathrm{si}\left[\frac{\pi}{T_s}\,(t - nT_s)\right] \end{aligned} \tag{1-159}$$

In the usual form of the theorem it is assumed that $\tau = 0$. If $x(t)$ is a deterministic waveform then (1-159) is the celebrated sampling theorem. In the stochastic version of this theorem the equivalence is to be understood in the mean-square sense.

$$E\left[\left| x(t+\tau) - \sum_{n=-\infty}^{\infty} x(\tau + nT_s)\, \mathrm{si}\left[\frac{\pi}{T_s}\,(t - nT_s)\right]\right|^2\right] = 0$$

Proof: We expand

$$\begin{aligned} &E\left[\left| x(t+\tau) - \sum_{n=-\infty}^{\infty} x(\tau + nT_s)\, \phi_n(t)\right|^2\right] \\ &= R_x(t+\tau; t+\tau) - \sum_{l} R_x(t+\tau; \tau + lT_s)\, \phi_l(t) \\ &\quad - \sum_{n} R_x^*(t+\tau; \tau + nT_s)\, \phi_n(t) \\ &\quad + \sum_{n,l=-\infty}^{\infty}\sum R_x(nT_s + \tau; lT_s + \tau)\, \phi_n(t)\, \phi_l(t) \end{aligned} \tag{1-160}$$

Using the series expansion of $R_x(t, t+u)$ of eq. (1-158) it readily follows that the right-hand side of (1-160) equals zero.

1.3.3 Sampling of Bandpass Signals

We consider a passband signal with a complex envelope $x_L(t)$

$$x(t) = \sqrt{2}\,\mathrm{Re}\left[x_L(t) e^{j\,\omega_0 t}\right] \tag{1-161}$$

When $x(t)$ is strictly bandpass and centered at ω_0 (Figure 1-20)

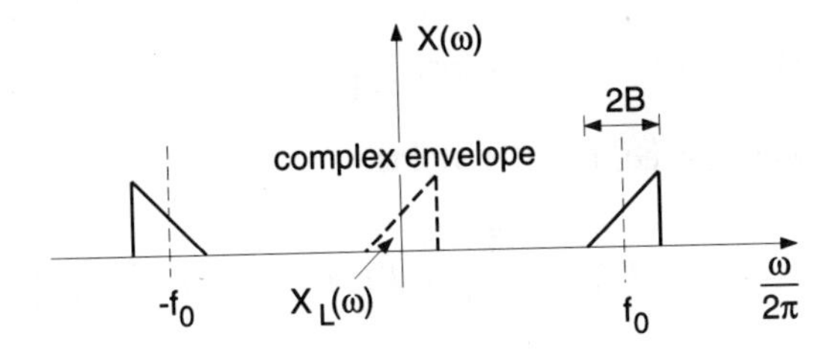

Figure 1-20 Bandpass Signal

then $x_L(t)$ is a complex lowpass process. It can be obtained from $x(t)$ by performing the demodulation

$$\begin{aligned} x(t)e^{-j\,\omega_0 t} &= \sqrt{2}\,\mathrm{Re}\big[x_L(t)e^{j\,\omega_0 t}\big]\,e^{-j\,\omega_0 t} \\ &= \frac{\sqrt{2}}{2}x_L(t) + \frac{\sqrt{2}}{2}x_L^*(t)e^{-j2\omega_0 t} \end{aligned} \tag{1-162}$$

The double-frequency term in eq. (1-162) is removed by a lowpass filter as shown in Figure 1-21.

The analog-to-digital conversion can be moved closer to the input $x(t)$ by simultaneously sampling the signal at carrier frequency ω_0 and frequency translating it to baseband. We denote by $f_s = 1/T_s$ the sampling frequency. Then

$$x(t)e^{-j\,\omega_0 t}\big|_{t=kT_s} = \sqrt{2}\,\mathrm{Re}\big[x_L(t)e^{j\,\omega_0 t}\big]\,e^{-j\,\omega_0 t}\big|_{t=kT_s} \tag{1-163}$$

A very simple realization which avoids complex multiplication is obtained if we sample the signal at four times the carrier frequency f_0:

$$f_s = 4f_0 \tag{1-164}$$

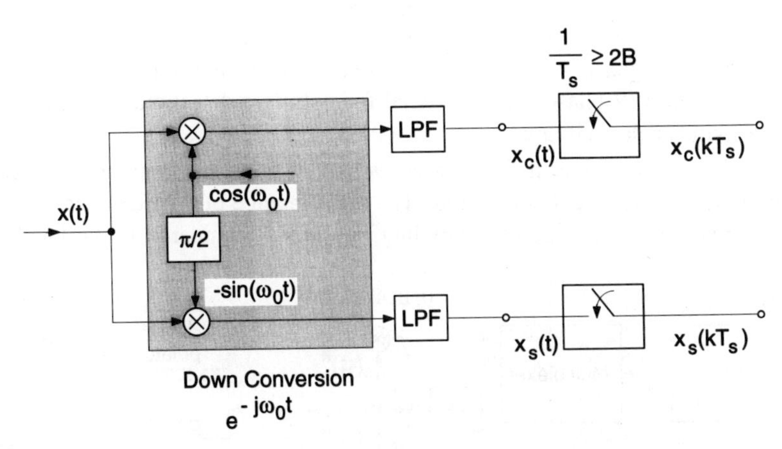

Figure 1-21 Down Conversion Followed by Sampling

Since

$$e^{jk\omega_0 T_s} = (j)^k \tag{1-165}$$

inserting this result into eq. (1-163) yields

$$(-j)^k x(kT_s) = \frac{\sqrt{2}}{2}\left[x_L(kT_s) + x_L^*(kT_s)(-1)^k\right] \tag{1-166}$$

We obtain the in-phase component at the even sampling instants $2kT_s$ and the quadrature component at the odd instants $(2k+1)T_s$:

$$\begin{aligned} (-1)^k x(2kT_s) &= \sqrt{2}\,\mathrm{Re}[x_L(2kT_s)] = \sqrt{2}x_c(2kT_s) \\ (-1)^k x[(2k+1)T_s] &= -\sqrt{2}\,\mathrm{Im}[x_L(2k+1)T_s] = -\sqrt{2}x_s[(2k+1)T_s] \end{aligned} \tag{1-167}$$

For further processing we need the quadrature components at the same sampling instant and not T_s apart. This can be achieved by digital interpolation of the sample values. For symmetry reasons both components are interpolated

$$\begin{aligned} x_c'(2kT_s) &= x_c[2kT_s - T_s/2] \\ x_s'(2kT_s) &= x_s[(2k-1)T_s + T_s/2] \end{aligned} \tag{1-168}$$

The functional block diagram of the combined samples and down converter is depicted in Figure 1-22.

The real-valued signal $x(t)$ is sampled at four times the carrier frequency f_0. Subsequently the samples are sorted by a multiplexer, provided with alternating signs, and finally interpolated to obtain samples at the same time instant. Notice that the multiplexer decimates the sample rate by two.

This digital equivalent of the down converter (Figure 1-22) is very attractive for implementation reasons since it avoids the costly analog mixers and lowpass filters at the expense of digital signal processing circuitry in the interpolator. Functionally, the digital down converter provides the two quadrature components exactly since the phase increment between two consecutive samples equals $\omega_0 T_s = \pi/2$. An exact $\pi/2$ phase shift is crucial for optimum detection. It is difficult to achieve with analog circuits.

If the signal $x(t)$ is first down converted by an analog mixer, a minimum sampling rate of $f_s \geq 2B$ is required. The disadvantage of the digital sampler and down converter is the large oversampling rate of $4f_0$ compared to the bandwidth

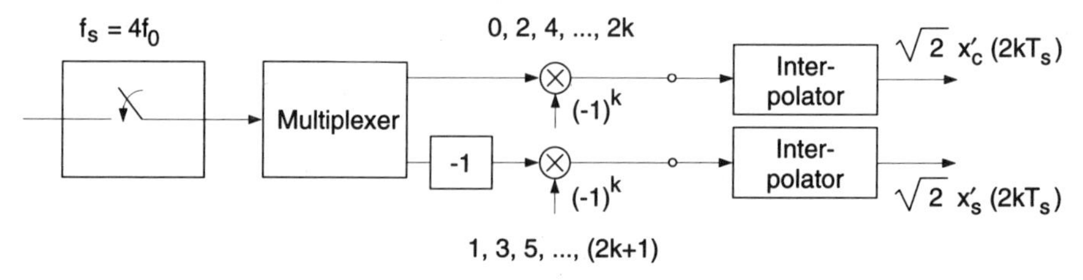

Figure 1-22 Digital Sampling and Down Conversion

of the information-carrying envelope $x_L(t)$. A generalization of the basic concept of the digital sampler and down converter, which maintains the simplicity of the digital circuitry but uses a smaller sampling rate, is discussed next.

The idea behind this is quite simple. We are in search of a sampling rate which produces phase increments of $\pi/2$.

$$e^{j\,k\,2\pi\,f_0\,T_s} = e^{j(\pi/2)k} \tag{1-169}$$

The first solution to the above relation is the one discussed before, $f_0 T_s = 1/4$.

The general solution, however, is

$$f_0 T_s = \pm\frac{1}{4} + N \ > 0 \tag{1-170}$$

where the negative sign yields negative phase increments of $\pi/2$. Hence

$$f_s = \frac{1}{T_s} = \frac{4f_0}{4N \pm 1} \tag{1-171}$$

which can be combined to

$$f_s = \frac{4f_0}{2n+1};\ n \geq 0 \tag{1-172}$$

Basically eq. (1-171) states that we skip N cycles of the phasor $e^{j\omega_0 t}$ before taking the next sample. There is a maximum N determined by the sampling theorem. We obtain samples of the quadrature components every $2T_s$ units of time [see eq. (1-167)].

Hence

$$1/(2T_s) \geq 2B \tag{1-173}$$

or

$$f_s \geq 4B \tag{1-174}$$

A smaller sampling rate simplifies the A/D converter but requires more elaborate interpolation circuitry. This is a trade-off which the design engineer can exploit.

Up to this point, it has been the desire to simplify digital circuitry by assuming certain relationships between carrier frequency and sampling rate. In some cases, it may be impossible to fulfill such relationships exactly. In this case, it may be desirable to choose f_s according to a more general rule. The most general requirement for the sampling rate is given by

$$\begin{gathered}\frac{2f_0 + 2B}{N} \leq f_s \leq \frac{2f_0 - 2B}{N-1} \\ \text{for} \quad N = 1, \ldots, N_{\max}, \qquad N_{\max} = \text{int}\left(\frac{\text{f}_0 + \text{B}}{2\text{B}}\right)\end{gathered} \tag{1-175}$$

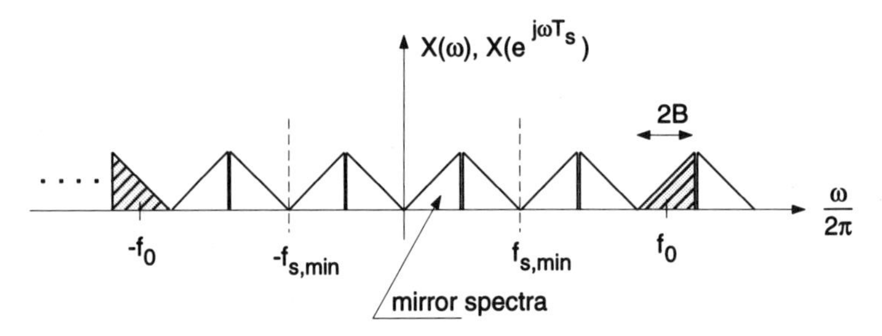

Figure 1-23 Spectrum of Undersampled Signal

(Note that the equality in the right equation only holds for $N > 1$.) If eq. (1-175) is satisfied, the mirror spectra of the sampled signal $x(kT_s)$ will occupy the frequency space between the two components of the original spectrum as shown in Figure 1-23. The minimum required sampling rate $f_{s,\min}$ (for $N = N_{\max}$) is approximately twice the bandwidth $2B$ of the real bandpass signal. If $(f_0 + B)/2B$ is an integer value $f_{s,\min} = 4B$ is exactly fulfilled. In this case, the minimum sampling rates resulting from conditions (1-172) and (1-175) coincide.

The negative mirror images can be eliminated by means of a Hilbert transformer. A discrete-time Hilbert transform filter with a frequency response of

$$H_d\left(e^{j2\pi fT_s}\right) = \begin{cases} -j & \text{for} \quad 0 \le fT_s < \frac{1}{2} \\ j & \text{for} \quad \frac{1}{2} \le fT_s < 1 \end{cases} \tag{1-176}$$

will compute the quadrature component samples $x_s(kT_s)$ from $x(kT_s)$ so that $x(kT_s)$ and $x_s(kT_s)$ form a Hilbert transform pair. Figure 1-24 illustrates the

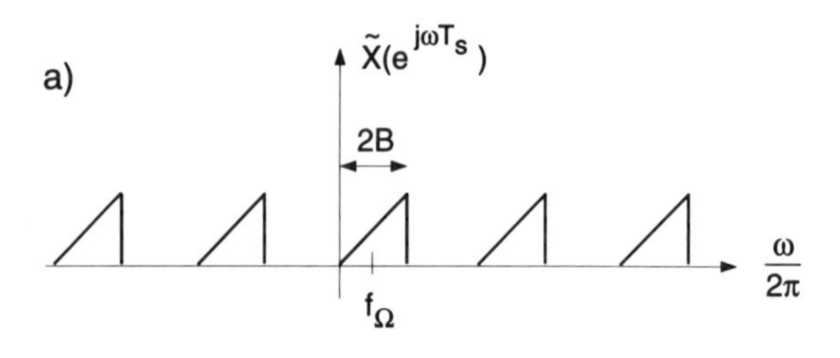

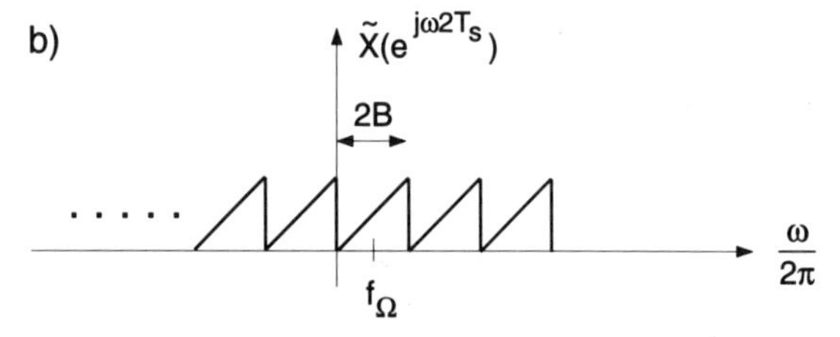

Figure 1-24 (a) Spectrum of Analytic Signal, (b) Spectrum after Decimation

effect of the digital Hilbert transform operation on the spectrum. The undesired negative sideband portion is eliminated leaving behind the spectrum of a sampled complex signal $\widetilde{x}(kT_s)$ which equals the envelope $x_L(kT_s)$ shifted by a frequency offset f_Ω from the origin.

Since the resulting complex signal spectrum consumes only half of the bandwidth of the original real signal, in-phase and quadrature sample streams may be decimated by a factor of 2. Equation (1-176) can efficiently be approximated by a finite-impulse response (FIR) filter with an odd number of coefficients N_{DHT} and whose frequency response $\widetilde{H}\left(e^{j2\pi fT_s}\right)$ fulfills

$$\widetilde{H}_d\left(e^{j2\pi fT_s}\right) = \widetilde{H}_d\left(e^{j(\pi-2\pi fT_s)}\right) \tag{1-177}$$

In this case, the impulse response $\widetilde{h}_d(n),\ n \geq 0$ satisfies

$$\widetilde{h}_d(n) = 0 \quad \text{for} \quad n \quad \text{odd} \tag{1-178}$$

Figure 1-25 shows the functional block diagram of the down converter. Due to the property (1-178), decimation may be performed at the input of the unit by sorting the samples alternatively into in-phase and quadrature branches. The quadrature components are obtained by filtering with $\widetilde{H}_d\left(e^{j2\pi f2T_s}\right)$. To compensate the FIR filter causality, the in-phase samples must be delayed by $N' = (N_{DHT}-1)/2$. Finally, the known frequency offset f_Ω is removed by phase rotation. This may be performed in conjunction with correction of the unknown carrier frequency error.

1.3.4 Main Points

Any band-limited signal can be represented by a series

$$x(t+\tau) = \sum_{n=-\infty}^{\infty} x(\tau + nT_s)\ \mathrm{si}\left[\frac{\pi}{T_s}(t-nT_s)\right] \tag{1-179}$$

with $1/T_s \geq 2B$. If $x(t)$ is a deterministic waveform, then (1-179) is the sampling theorem. In the stochastic version equivalence is understood in the mean-square sense.

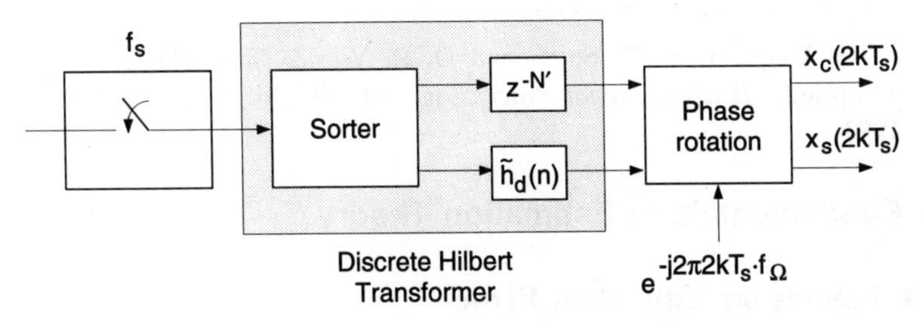

Figure 1-25 Digital Down Converter Using Hilbert Transform

By sampling a bandpass signal, one can simultaneously perform the operation of down conversion, provided the sampling rate fulfills the condition of eq. (1-172). The samples $x(kT_s)$ for even k is equal to the in-phase component of the envelope while for odd k we obtain the quadrature component. Bandpass sampling is very attractive for implementation reasons since it avoids the costly analog mixer. In some cases it may be impossible to fulfill the exact relationship between carrier frequency and sampling rate of eq. (1-172). In this case a generalized relation, eq. (1-175), may be applicable. The digital signal processing then demands a discrete Hilbert transform and a digital down conversion by the residual frequency.

Bibliography

[1] J. K. Cavers and S. P. Stapleton, "A DSP-Based Alternative to Direct Conversion Receivers for Digital Mobile Communications," *Proceedings IEEE Global Telecommunications Conference, GLOBECOM*, p. 2024, Dec. 1990.

[2] L. R. Rabiner and R. W. Schafer, "On the Behavior of Minimax FIR Digital Hilbert Transformers," *The Bell System Technical Journal*, vol. 53, No. 2, pp. 363–390, Feb. 1974.

[3] W. Rosenkranz, "Quadrature Sampling of FM-Bandpass Signals," *Proceedings of the International Conf. Digital Signal Processing, Florence, Italy*, pp. 377–381, 1987.

[4] H. Samueli and B. C. Wong, "A VLSI Arcitecture for High-Speed All-Digital Quadrature Modulator and Demodulator for Digital Applications," *IEEE J. Selected Areas Commun.*, vol. 8, No. 8, pp. 1512–1519, Oct. 1990.

[5] G. J. Saulnier, C. M. Puckette, R. C. Gaus, R. J. Dunki-Jacobs, and T. E. Thiel, "A VLSI Demulator for Digital RF Network Applications: Theory and Results," *IEEE J. Selected Areas Commun.*, vol. 8, No. 8, pp. 1500–1511, Oct. 1990.

[6] J. Sevenhans, A. Vanwelsenaers, J. Wenin, J. Baro, "An Integrated Si Bipolar RF Transceiver for Zero IF 900 MHz GSM Mobile Radio Frontend of a Hand Portable Phone," *Proceedings IEEE Custom Integrated Circuits Conf., San Diego, CA*, May 1991.

[7] N. R. Sollenberger and J. C.-I. Chuang, "Low-Overhead Symbol Timing and Carrier Recovery for TDMA Portable Radio Systems," *IEEE Trans. Commun.*, vol. 38, pp. 1886–1892, Oct. 1990.

[8] R. G. Vaughan, N. L. Scott, and D. R. White, "The Theory of Bandpass Sampling," *IEEE J. Signal Processing*, vol. 39, No. 9, pp. 1973–1984, Sept. 1991.

1.4 Fundamentals of Estimation Theory

1.4.1 Bounds on Estimation Errors

The variance of the estimation error of suboptimal estimators can be obtained analytically only in a few isolated cases. In general it must be obtained by computer simulation. While we thus obtain accurate results, including nonlinear effects such as quantization, for any given estimate we do not know how far from the optimum estimator we are.

In this section we discuss bounds on the variance of the estimation errors for nonrandom but unknown parameters. Such bounds are useful since they allow to compare the variance of a suboptimal (but realizable) estimator to that of an optimal (but nonrealizable) estimator in order to assess the implementation loss of the suboptimal estimator.

This section is based on the book by van Trees [1] where the reader can find a thorough and detailed discussion as well as the mathematical proofs.

Let $\mathbf{r}$ be the received signal vector and $\boldsymbol{\theta} = (\theta_1, \theta_2, ..., \theta_K)$ a set of K nonrandom parameters. The estimator for the ith parameter is denoted by $\hat{\theta}_i(\mathbf{r})$ Then the following theorem can be proved [1, p. 79].

Theorem: Consider any unbiased estimates of θ_i

$$E\left[\hat{\theta}_i(\mathbf{r})\right] = \theta_i \tag{1-180}$$

Then

$$\sigma_{\theta_i}^2 = \mathrm{var}\left[\theta_i - \hat{\theta}_i(\mathbf{r})\right] \geq J^{ii} \tag{1-181}$$

where J^{ii} is the iith element of the $K \times K$ square matrix $\mathbf{J}^{-1}$. The elements in $\mathbf{J}$ are given by

$$\begin{aligned} J_{ij} &= E\left[\frac{\partial \ln p(\mathbf{r}|\boldsymbol{\theta})}{\partial \theta_i} \; \frac{\partial \ln p(\mathbf{r}|\boldsymbol{\theta})}{\partial \theta_j}\right] \\ &= -E\left[\frac{\partial^2 \ln p(\mathbf{r}|\boldsymbol{\theta})}{\partial \theta_i \, \partial \theta_j}\right] \end{aligned} \tag{1-182}$$

The $\mathbf{J}$ matrix is commonly called Fisher's Information Matrix. The equality in (1-181) holds if and only if

$$\hat{\theta}_i(\mathbf{r}) - \theta_i = \sum_{j=1}^{K} K_{i,j}(\mathbf{r}) \, \frac{\partial \ln p(\mathbf{r}|\boldsymbol{\theta})}{\partial \theta_j} \tag{1-183}$$

for all values of θ_i and $\mathbf{r}$. For a single parameter θ the bound is commonly called the Cramér-Rao bound. To understand the significance of the theorem, it is constructive to consider a few simple examples.

Example 1: Maximum-Likelihood (ML) Estimation of a Single Parameter θ
Let

$$r_i = \theta + n_i \qquad i = 1, \ldots, N \tag{1-184}$$

We assume that the n_i are each independent, zero-mean Gaussian variables with variance σ_n^2. Then from

$$p(\mathbf{r}|\theta) = \prod_{i=1}^{N} \frac{1}{\sqrt{2\pi}\,\sigma_n} \exp\left\{-\frac{(r_i - \theta)^2}{2\,\sigma_n^2}\right\} \tag{1-185}$$

and

$$\begin{aligned}\frac{\partial \ln p(\mathbf{r}|\theta)}{\partial \theta} &= \frac{1}{\sigma_n^2}\left[\sum_{i=1}^{N}(r_i - \theta)\right] \\ &= \frac{N}{\sigma_n^2}\left[\left(\frac{1}{N}\sum_{i=1}^{N} r_i\right) - \theta\right]\end{aligned} \tag{1-186}$$

the ML estimate is found by setting

$$\left.\frac{\partial \ln p(\mathbf{r}|\theta)}{\partial \theta}\right|_{\hat{\theta}(\mathbf{r})} = 0 \tag{1-187}$$

From (1-187) it follows

$$\hat{\theta}(\mathbf{r}) = \frac{1}{N}\sum_{i=1}^{N} r_i \tag{1-188}$$

Using (1-184) the expected value of $\hat{\theta}(\mathbf{r})$ is found to be

$$E\left[\hat{\theta}(\mathbf{r})\right] = \frac{1}{N}\sum_{i=1}^{N} E[\theta + n_i] = \theta$$

which shows that the estimate is unbiased. Because the expression (1-186) has exactly the form

$$\frac{\partial \ln p(\mathbf{r}|\theta)}{\partial \theta} = \frac{N}{\sigma_n^2}\left[\hat{\theta}(\mathbf{r}) - \theta\right] \tag{1-189}$$

the estimate reaches the bound with equality. (Any estimate that satisfies the bound with equality is called an efficient estimate.)

The bound is readily found by differentiating (1-189) once more with respect to θ and taking the expected value.

$$E\left[\frac{\partial^2 \ln p(\mathbf{r}|\theta)}{\partial^2 \theta}\right] = -\frac{N}{\sigma_n^2} \tag{1-190}$$

Thus

$$\text{var}\left[\theta - \hat{\theta}(\mathbf{r})\right] = \frac{\sigma_n^2}{N} \tag{1-191}$$

Example 2: Nonlinear ML Estimation of a Single Parameter.
We assume the same conditions as in Example 1 with the exception that the variable θ appears in a nonlinear manner. We denote its dependency by $s(\theta)$

$$r_i = s(\theta) + n_i \tag{1-192}$$

Using

$$p(\mathbf{r}|\theta) = \prod_{i=1}^{N} \frac{1}{\sqrt{2\pi}\,\sigma_n} \exp\left\{-\frac{(r_i - s(\theta))^2}{2\,\sigma_n^2}\right\} \tag{1-193}$$

we obtain

$$\frac{\partial \ln p(\mathbf{r}|\theta)}{\partial\theta} = \frac{N}{\sigma_n^2}\left[\frac{1}{N}\sum_{i=1}^{N}(r_i - s(\theta))\right]\frac{\partial s(\theta)}{\partial\theta} \tag{1-194}$$

In general the right-hand side cannot be written in the form required by (1-183) and therefore an unbiased efficient estimate does not exist.

The likelihood equation is the expectation of

$$\left[\frac{\partial s(\theta)}{\partial\theta}\,\frac{1}{\sigma_n^2}\right]\left[\frac{1}{N}\sum_{i=1}^{N}(r_i - s(\theta))\right]\Bigg|_{\theta=\hat{\theta}(\mathbf{r})} = 0 \tag{1-195}$$

If the range of $s(\theta)$ includes $1/N \sum_{i=1}^{N} r_i$ a solution exists

$$s\left(\hat{\theta}(\mathbf{r})\right) = \frac{1}{N}\sum_{i=1}^{N} r_i \tag{1-196}$$

If (1-196) can be satisfied, then

$$\hat{\theta}(\mathbf{r}) = s^{-1}\left(\frac{1}{N}\sum_{i=1}^{N} r_i\right) \tag{1-197}$$

Observe that we tacitly assume that the inverse function s^{-1} exists. If it does not, even in the absence of noise we shall be unable to determine θ unambiguously. Taking the second derivative of $\ln p(\mathbf{r}|\theta)$ with respect to θ we find

$$\frac{\partial^2 \ln p(\mathbf{r}|\theta)}{\partial^2\theta} = \frac{N}{\sigma_n^2}\left\{-\left[\frac{\partial s(\theta)}{\partial\theta}\right]^2 + \left[\frac{1}{N}\sum_{i=1}^{N}(r_i - s(\theta))\right]\frac{\partial^2 s(\theta)}{\partial^2\theta}\right\} \tag{1-198}$$

Observing that for an unbiased estimate we must have

$$E\left[\frac{1}{N}\sum_{i=1}^{N}(r_i - s(\theta))\right] = 0 \tag{1-199}$$

we obtain the following bound for *any* unbiased estimate

$$\text{var}\left[\theta - \hat{\theta}(\mathbf{r})\right] \geq \frac{\sigma_n^2}{N}\,\frac{1}{[\partial s(\theta)/\partial\theta]^2} \tag{1-200}$$

We see that the bound is exactly the same as in Example 1 except for a factor $[\partial s(\theta)/\partial\theta]^2$. The intuitive reason for this factor and also some feeling for the

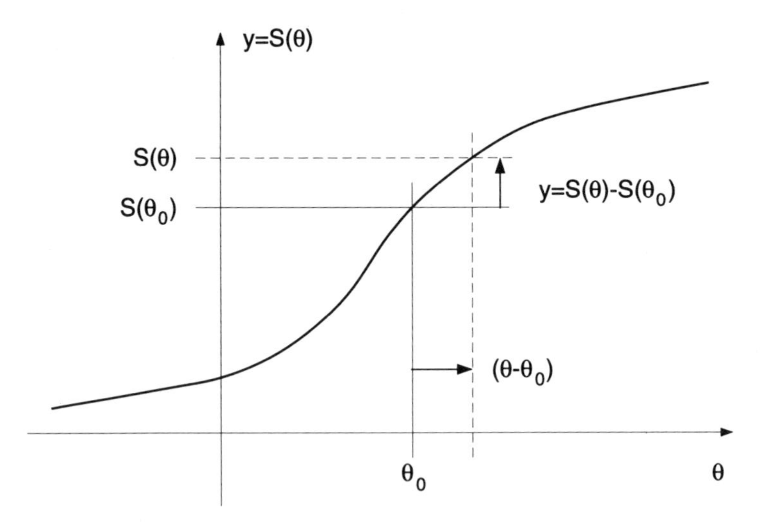

Figure 1-26 Nonlinear Estimation of Single Parameter

conditions under which the bounds will be useful may be obtained by inspection of Figure 1-26.

Define $y = s(\theta)$, then $r_i = y + n_i$. The variance in estimating y is just σ_n^2/N. However, if the error in estimating y is small enough so that the slope is constant, then

$$(\theta - \theta_0) \simeq \frac{y - s(\theta_0)}{\partial s(\theta)/\partial\theta \;\big|_{\theta_0}} \tag{1-201}$$

and

$$\operatorname{var}\left[\theta - \hat{\theta}(\mathbf{r})\right] \simeq \frac{\operatorname{var}(y - s(\theta))}{\left[\partial s(\theta)/\partial\theta\right]^2\big|_{\theta_0}} \tag{1-202}$$

We observe that, if the noise is weak, almost all values of y cluster around the actual value θ_0 where linearization of $s(\theta)$ applies. We can thus expect that the Cramér-Rao (CR) bound gives an accurate answer in the case that the parameter appears in a nonlinear manner. The performance of the nonlinear estimator approaches that of a *linear estimator* which provides an *efficient* estimate.

The properties of the ML estimate, which are valid when the error is small, are generally referred to as *asymptotic*. This occurs if the variance of the noise becomes small, e.g., when the number of independent observations becomes very large.

Nonlinear estimation is generally plagued by threshold effects. At low SNRs there is a range in which the variance rises very rapidly as the SNR decreases. As an example consider the PLL behavior discussed in Chapter 6, Volume 1. Below the threshold, the CR is no longer useful.

Bibliography

[1] H. L. van Trees, *Detection, Estimation, and Modulation Theory*. New York: Wiley, 1968.

PART C
Baseband Communications

Chapter 2 Baseband Communications

2.1 Introduction to Digital Baseband Communication

In baseband communication, digital information is conveyed by means of a pulse train. Digital baseband communication is used in many applications, such as

- Transmission at a few megabits per second (Mb/s) of multiplexed digitized voice channels over repeatered twisted-pair cables
- Transmission of basic rate ISDN (160kb/s) over twisted-pair digital subscriber lines
- Local area networks (LANs) and metropolitan area networks (MANs) operating at 10–100 Mb/s using coaxial cable or optical fiber
- Long-haul high-speed data transmission over repeatered optical fiber
- Digital magnetic recording systems for data storage

This chapter serves as a short introduction to digital baseband communication. We briefly consider important topics such as line coding and equalization, but without striving for completeness. The reader who wants a more detailed treatment of these subjects is referred to the abundant open literature, a selection of which is presented in Section 2.1.5.

2.1.1 The Baseband PAM Communication System

Baseband communication refers to the case where the spectrum of the transmitted signal extends from zero frequency direct current (DC) to some maximum frequency. The transmitted signal is a pulse-amplitude-modulated (PAM) signal: it consists of a sequence of time translates of a baseband pulse which is amplitude-modulated by a sequence of data symbols conveying the digital information to be transmitted.

A basic communication system for baseband PAM is shown in Figure 2-1.

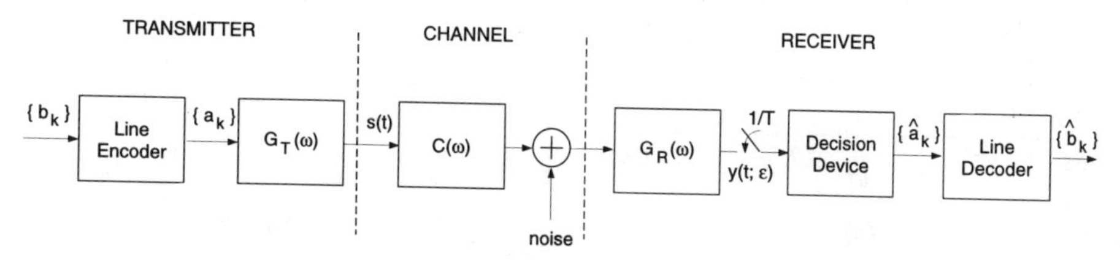

Figure 2-1 Basic Communication System for Baseband PAM

At the transmitter, the sequence of information bits $\{b_k\}$ is applied to an encoder, which converts $\{b_k\}$ into a sequence $\{a_k\}$ of data symbols. This conversion is called *line coding*, and will be considered in more detail in Section 2.1.3. The information bits assume the values binary zero ("0") or binary one ("1"), whereas the data symbols take values from an alphabet of a size L which can be larger than 2. When L is even, the alphabet is the set $\{\pm 1, \pm 3, \ ..., \pm(L-1)\}$, for an odd L the alphabet is the set $\{0, \pm 2, \pm 4, \ ..., \pm(L-1)\}$.

The data symbols enter the transmit filter with impulse response $g_T(t)$, whose Fourier transform is denoted by $G_T(\omega)$. The resulting transmit signal is given by

$$s(t) = \sum_m a_m \, g_T \left(t - mT - \varepsilon T\right) \tag{2-1}$$

where 1/T is the symbol rate, i.e., the rate at which the data symbols are applied to the transmit filter. The impulse response $g_T(t)$ is called the baseband pulse of the transmit signal. The quantity εT is a fractional unknown time delay between the transmitter and the receiver $\left(|\varepsilon| \le \frac{1}{2}\right)$. The instants $\{kT\}$ can be viewed as produced by a hypothetical reference clock at the receiver. At the transmitter, the kth channel symbol a_k is applied to the transmit filter at the instant $kT + \varepsilon T$, which is unknown to the receiver. Figure 2-2 shows a baseband pulse $g_T(t)$ and a corresponding PAM signal $s(t)$, assuming that $L = 2$.

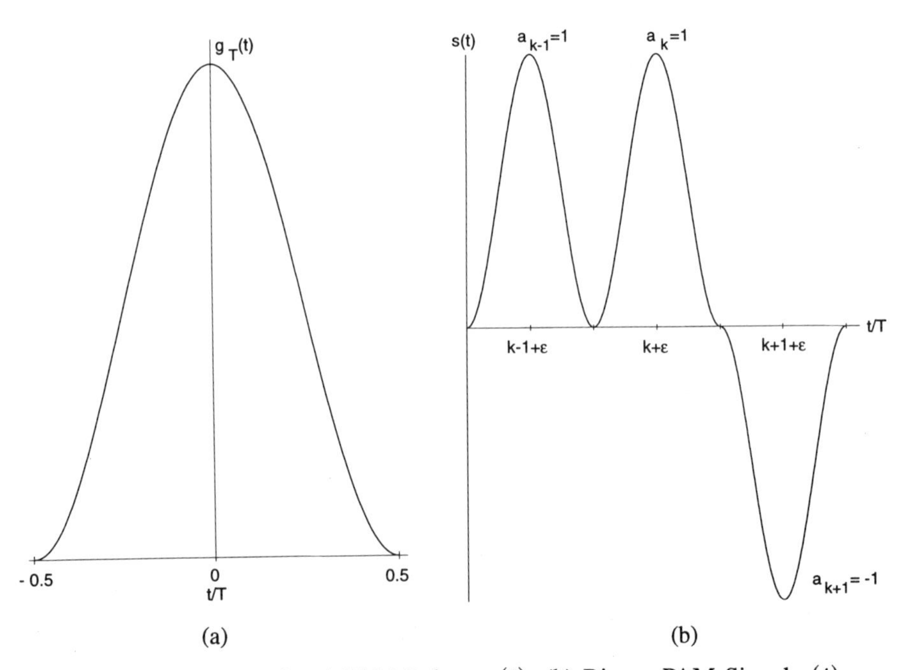

Figure 2-2 (a) Baseband PAM Pulse $g_T(t)$, (b) Binary PAM Signal $s(t)$

The channel is assumed to be linear. It introduces linear distortion and adds noise. The linear distortion (amplitude distortion and delay distortion) is characterized by the channel frequency response $C(\omega)$. It causes a broadening of the transmitted pulses. The "noise" is the sum of various disturbances, such as thermal noise, electronics noise, cross talk, and interference from other communication systems.

The received noisy PAM signal is applied to a receive filter (which is also called data filter) with frequency response $G_R(\omega)$. The role of this filter is to reject the noise components outside the signal bandwidth, and, as we will explain in Section 2.1.2, to shape the signal. The receive filter output signal $y(t;\varepsilon)$ is sampled at symbol rate $1/T$. From the resulting samples, a decision $\{\hat{a}_k\}$ is made about the data symbol sequence $\{a_k\}$. The sequence $\{\hat{a}_k\}$ is applied to a decoder, which produces a decision $\left\{\hat{b}_k\right\}$ on the information bit sequence $\{b_k\}$.

The signal at the output of the receive filter is given by

$$y(t;\epsilon) = \sum_m a_m g(t - mT - \varepsilon T) + n(t) \tag{2-2}$$

where $g(t)$ and $n(t)$ are the baseband pulse and the noise at the receive filter output. The Fourier transform $G(\omega)$ of the baseband pulse $g(t)$ is given by

$$G(\omega) = G_T(\omega)C(\omega)G_R(\omega) \tag{2-3}$$

Let us denote by $\{kT + \hat{\varepsilon}T\}$ the sequence of instants at which the sampler at the receive filter output is activated. These sampling instants are shifted by an amount $\hat{\varepsilon}T$ with respect to the instants $\{kT\}$ produced by the hypothetical reference clock of the receiver. Then the kth sample is given by

$$y_k(e) = a_k\, g_0(e) + \sum_{m \neq 0} a_{k-m}\, g_m(e) + n_k \tag{2-4}$$

where $y_k(e)$, $g_m(e)$, and n_k are short-hand notations for $y(kT + \hat{\varepsilon}T;\varepsilon)$, $g(mT - eT)$, and $n(kT + \hat{\varepsilon}T)$, while $e = \varepsilon - \hat{\varepsilon}$ denotes the difference, normalized by the symbol duration T, between the instant where the kth symbol a_k is applied to the transmit filter and the kth sampling instant at the receiver.

In order to keep the decision device simple, receivers in many applications perform symbol-by-symbol decisions: the decision $\hat{a}_k$ is based only on the sample $y_k(e)$. Hence, only the first term of the right-hand side of (2-4) is a useful one, because it is the only one that depends on a_k. The second term is an *intersymbol interference* (ISI) term depending on a_{k-m} with $m \neq 0$, while the third term is a noise term. When the noise $n(t)$ at the receive filter output is stationary, the statistics of the noise sample n_k do not depend on the sampling instant. On the other hand, the statistics of the useful term and the ISI in (2-4) do depend on the sampling instant, because the PAM signal is cyclostationary rather than stationary.

Let us consider given sampling instants $\{kT + \hat{\varepsilon}T\}$ at the receive filter output.

The symbol-by-symbol decision rule based on the samples $y_k(e)$ from (2-4) is

$$\hat{a}_k = \begin{cases} m & m-1 < y_k(e)/g_0(e) \leq m+1 \qquad m \neq \pm(L-1) \\ L-1 & L-2 < y_k(e)/g_0(e) \\ -L+1 & y_k(e)/g_0(e) \leq -L+2 \end{cases} \tag{2-5}$$

where the integer m takes on only even (odd) values when L is odd (even). This decision rule implies that the decision device is a slicer which determines the symbol value $\hat{a}_k$ which is closest to $y_k(e)/g_0(e)$. In the absence of noise, the baseband communication system should produce no decision errors. A necessary and sufficient condition for this to be true is that the largest magnitude of the ISI over all possible data sequences is smaller than $\mid g_0(e) \mid$, i.e., $M(e) > 0$ where $M(e)$ is given by

$$M(e) = \mid g_0(e) \mid - \max_{\{a_m\}} \left| \sum_{m \neq 0} a_m \; g_{-m}(e) \right| \tag{2-6}$$

When $M(e) < 0$, the data symbol sequence yielding maximum ISI will surely give rise to decision errors in the absence of noise, because the corresponding $y_k(e)$ is outside the correct decision region. When $M(e) > 0$, a decision error can occur only when the noise sample n_k from (2-4) has a magnitude exceeding $M(e)$; $M(e)$ is called the *noise margin* of the baseband PAM system. The noise margin can be visualized by means of an *eye diagram*, which is obtained in the following way. Let us denote by $y_0(t;\varepsilon)$ the receive filter output signal in the absence of noise, i.e.,

$$y_0(t;\varepsilon) = \sum_m a_m \; g(t - mT - \varepsilon T) \tag{2-7}$$

The PAM signal $y_0(t;\varepsilon)$ is sliced in segments $y_{0,i}(t;\varepsilon)$, having a duration equal to the symbol interval T:

$$y_{0,i}(t;\varepsilon) = \begin{cases} y_0(t;\varepsilon) & iT \leq t < (i+1)T \\ 0 & \text{otherwise} \end{cases} \tag{2-8}$$

The eye diagram is a display of the periodic extension of the segments $y_{0,i}(t;\varepsilon)$. An example corresponding to binary PAM $(L = 2)$ is shown in Figure 2-3. As $g(t)$ has a duration of three symbols, the eye diagram for binary PAM consists of $2^3 = 8$ trajectories per symbol interval. Because of the rather large value of $g(\tau)$, much ISI is present when sampling the eye at $t = 0$. The noise margin $M(e)$ for a specific sampling instant is positive (negative) when the eye is open (closed) at the considered instant; when the eye is open, the corresponding noise margin equals half the vertical eye opening.

The noise margin $M(e)$ depends on the sampling instants $kT + \hat{\varepsilon}T$ and the unknown time delay εT through the variable $e = \varepsilon - \hat{\varepsilon}$. The optimum sampling instants, in the sense of minimizing the decision error probability when the worst-

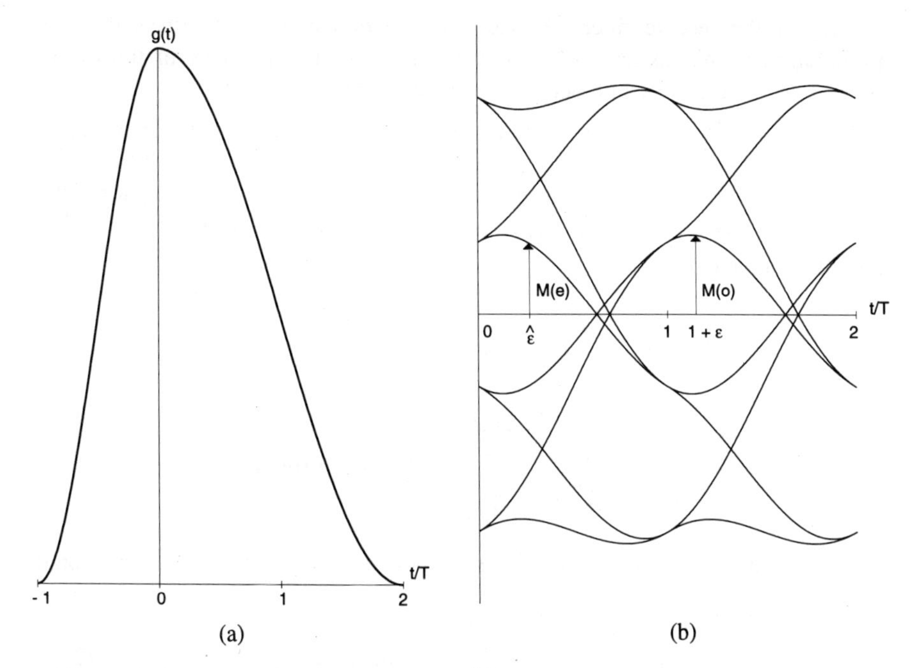

Figure 2-3 (a) Baseband PAM Pulse $g(t)$, (b) Eye Diagram for Binary PAM

case ISI is present, are those for which $M(e)$ is maximum. Using the appropriate time origin for defining the baseband pulse $g(t)$ at the receive filter output, we can assume without loss of generality that $M(e)$ becomes maximum for $e = 0$. Hence, the optimum sampling instants are $kT + \varepsilon T$, and $e = \varepsilon - \hat{\varepsilon}$ denotes the timing error normalized by the symbol interval. The sensitivity of the noise margin $M(e)$ to the normalized timing error e can be derived qualitatively from the eye diagram: when the horizontal eye opening is much smaller than the symbol interval T, the noise margin and the corresponding decision error probability are very sensitive to timing errors.

Because of the unknown delay εT between the receiver and the transmitter, the optimum sampling instants $\{kT + \varepsilon T\}$ are not known a priori to the receiver. Therefore, the receiver must be equipped with a structure that estimates the value of ε from the received signal. A structure like this is called *timing recovery circuit* or *symbol synchronizer*. The resulting estimate $\hat{\varepsilon}$ is then used to activate the sampler at the instants $\{kT + \hat{\varepsilon}T\}$. The normalized timing error $e = \varepsilon - \hat{\varepsilon}$ should be kept small, in order to avoid the increase of the decision error probability, associated with a reduction of the noise margin $M(e)$.

2.1.2 The Nyquist Criterion for Eliminating ISI

It is obvious that the shape of the baseband pulse $g(t)$ and the statistics of the noise $n(t)$ at the output of the receive filter depend on the frequency response

$G_R(\omega)$ of the receive filter. Hence, the selection of $G_R(\omega)$ affects the error probability when making symbol-by-symbol decisions. The task of the receive filter is to reduce the combined effect of noise and ISI.

Let us investigate the possibility of selecting the receive filter such that all ISI is eliminated when sampling at the instants $\{kT + \varepsilon T\}$. It follows from (2-4) that ISI vanishes when the baseband pulse $g(t)$ at the receive filter output satisfies $g(mT) = 0$ for $m \neq 0$. As $G(\omega)$ is the Fourier transform of $g(t)$, $g(mT)$ is for all m given by

$$\begin{aligned} g(mT) &= \int_{-\infty}^{+\infty} G(\omega)\, \exp\left(jm\omega T\right) \frac{d\omega}{2\pi} \\ &= \sum_{n=-\infty}^{+\infty} \int_{(2n-1)\pi/T}^{(2n+1)\pi/T} G(\omega)\, \exp\left(jm\omega T\right) \frac{d\omega}{2\pi} \end{aligned} \tag{2-9}$$

Taking into account that $\exp(jm\omega T)$ is periodic in ω with period $2\pi/T$, we obtain

$$g(mT) = \int_{-\pi/T}^{\pi/T} G_{\text{fld}}(\omega)\, \exp\left(jm\omega T\right) \frac{d\omega}{2\pi} \tag{2-10}$$

where $G_{\text{fld}}(\omega)$ is obtained by folding $G(\omega)$:

$$G_{\text{fld}}(\omega) = \sum_{n=-\infty}^{+\infty} G\left(\omega - \frac{2\pi n}{T}\right) \tag{2-11}$$

Note that $G_{\text{fld}}(\omega)$ is periodic in ω with period $2\pi/T$. It follows from (2-9) that $Tg(-mT)$ can be viewed as the mth coefficient in the Fourier-series expansion of $G_{\text{fld}}(\omega)$:

$$G_{\text{fld}}(\omega) = T \sum_{m=-\infty}^{+\infty} g(-mT)\, \exp\left(jm\omega T\right) \tag{2-12}$$

Using (2-12) and the fact that $G_{\text{fld}}(\omega)$ is periodic in ω, we obtain

$$g(mT) = 0 \quad \text{for} \quad m \neq 0 \quad \Leftrightarrow \quad G_{\text{fld}}(\omega) \quad \text{is constant for} \quad |\omega| < \pi/T \tag{2-13}$$

This yields the well-known *Nyquist criterion for zero ISI*: a necessary and sufficient condition for zero ISI at the receive filter output is that the folded Fourier transform $G_{\text{fld}}(\omega)$ is a constant for $|\omega| < \pi/T$.

The Nyquist criterion for zero ISI is sometimes referred to as the *first Nyquist criterion.* A pulse satisfying this criterion is called an *interpolation pulse* or a *Nyquist-I pulse.* Let us consider the case where $G(\omega)$ is band-limited to some frequency B; i.e., $G(\omega) = 0$ for $|\omega| > 2\pi B$.

- When $B < 1/(2T)$, $G_{\text{fld}}(\omega) = G(\omega)$ for $|\omega| < \pi/T$. As $G(\omega) = 0$ for $2\pi B < |\omega| < \pi/T$, $G_{\text{fld}}(\omega)$ cannot be constant for $|\omega| < \pi/T$. Taking (2-3) into account, it follows that when the bandwidth of the transmit filter, of the channel or of the receive filter is smaller than $1/2T$, it is impossible to find a receive filter that eliminates ISI.
- When $B = 1/(2T)$, $G_{\text{fld}}(\omega)$ is constant only when, within an irrelevant constant of proportionality, $G(\omega)$ is given by

$$G(\omega) = \begin{cases} T & |\omega| < \pi/T \\ 0 & \text{otherwise} \end{cases} \tag{2-14}$$

  The corresponding baseband pulse $g(t)$ equals

$$g(t) = \frac{\sin(\pi t/T)}{\pi t/T} \tag{2-15}$$

  Using (a close approximation of) the pulse $(\sin(\pi t/T))/(\pi t/T)$ is not practical: not only would a complicated filter be required to approximate the abrupt transition in $G(\omega)$ from (2-14), but also the performance is very sensitive to timing errors: as the tails of the pulse $(\sin(\pi t/T))/(\pi t/T)$ decay as $1/t$, the magnitude of the worst-case ISI tends to infinity for any nonzero timing error e, yielding a horizontal eye opening of zero width.
- When $B > 1/(2T)$, the baseband pulse $g(t)$ which eliminates ISI is no longer unique. Evidently, all pulses that satisfy $g(t) = 0$ for $|t| \geq T$ eliminate ISI. Because of their time-limited nature, these pulses have a large (theoretically infinite) bandwidth, so that they find application only on channels having a bandwidth B which is considerably larger than $1/(2T)$; an example is optical fiber communication with on-off keying of the light source. When bandwidth is scarce, one would like to operate at a symbol rate $1/T$ which is only slightly less than $2B$. This is referred to as *narrowband communication.*

When $1/(2T) < B < 1/T$, the Nyquist criterion (2-13) is equivalent to imposing that $G(\omega)$ has a symmetry point at $\omega = \pi/T$:

$$G\left(\frac{\pi}{T} + \omega\right) + G^*\left(\frac{\pi}{T} - \omega\right) = G(0) \qquad \text{for} \quad |\omega| < \pi/T \tag{2-16}$$

A widely used class of pulses with $1/(2T) < B < 1/T$ that satisfy (2-16) are the cosine rolloff pulses (also called raised cosine pulses), determined by

$$g(t) = \frac{\sin(\pi t/T)}{\pi t/T} \, \frac{\cos(\alpha \pi t/T)}{1 - 4\alpha^2 t^2/T^2} \tag{2-17}$$

with $0 \leq \alpha \leq 1$. For $\alpha = 0$, (2-17) reduces to (2-15). The Fourier transform $G(\omega)$ of the pulse $g(t)$ from (2-17) is given by

$$G(\omega) = \begin{cases} T & 0 \leq |\omega T| < \pi(1-\alpha) \\ T/2 \left[1 - \sin\left(\frac{|\omega T| - \pi}{2\alpha}\right)\right] & \pi(1-\alpha) \leq |\omega T| < \pi(1+\alpha) \\ 0 & |\omega T| \geq \pi(1+\alpha) \end{cases} \quad (2\text{-}18)$$

Hence, the bandwidth *B* equals $(1+\alpha)/(2T)$, and $\alpha/(2T)$ denotes the excess bandwidth [in excess of the minimum bandwidth $1/(2T)$]. α is called the rolloff factor. Some examples of cosine rolloff pulses, their Fourier transform and the corresponding eye diagram are shown in Figure 2-4. Note that ISI is absent when sampling at the instants kT. The horizontal eye opening decreases (and, hence, the sensitivity to timing error increases) with a decreasing rolloff factor.

From the above discussion we conclude that a baseband PAM pulse $g(t)$ that

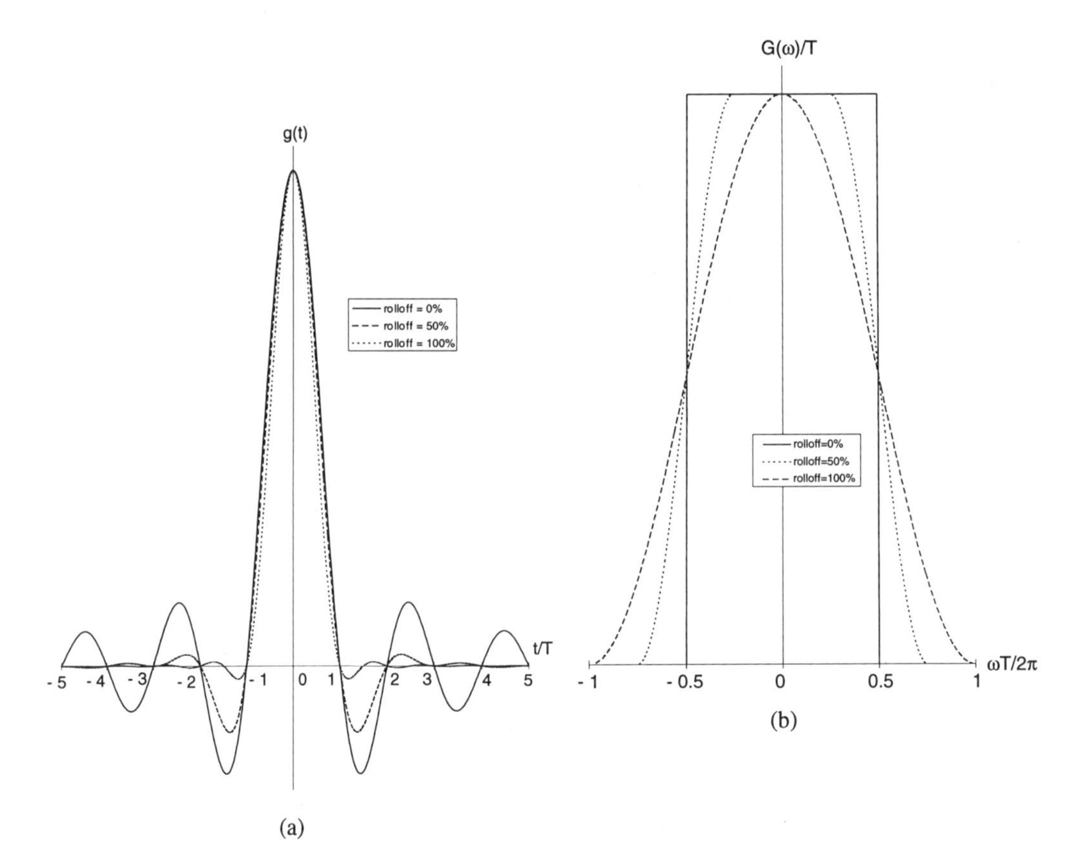

Figure 2-4 Cosine Rolloff Pulses: (a) Baseband Pulse $g_T(t)$, (b) Fourier Transform $G(\omega)$, (c) Eye Diagram for Binary PAM (25% Rolloff), (d) Eye Diagram for Binary PAM (50% Rolloff), (e) Eye Diagram for Binary PAM (100% Rolloff)

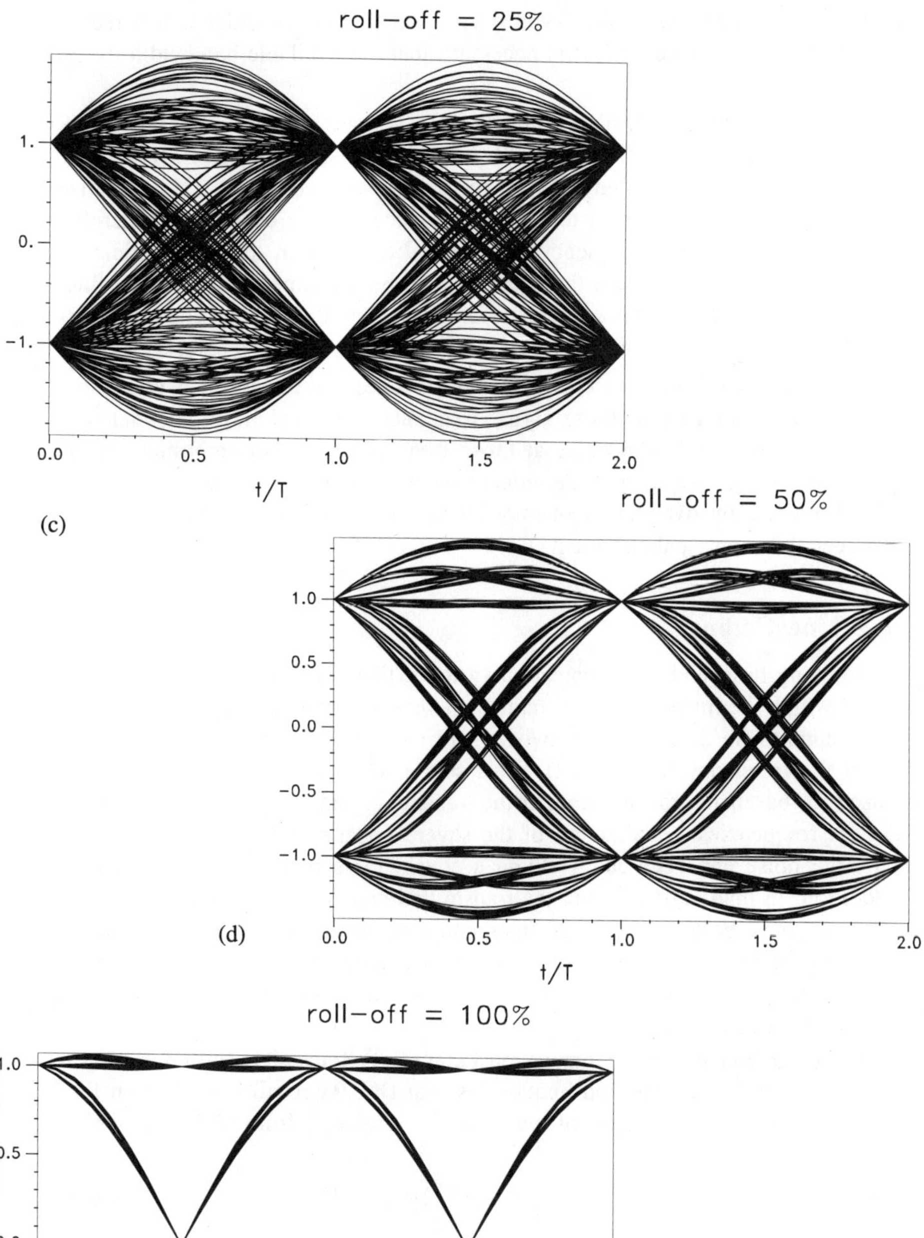
roll−off = 25%
1.
0.
−1.
0.0
0.5
1.0
1.5
2.0
t/T
roll−off = 50%
1.0
0.5
0.0
−0.5
−1.0
0.0
0.5
1.0
1.5
2.0
t/T
roll−off = 100%
1.0
0.5
0.0
0.5
1.0
0.0
0.5
1.0
1.5
2.0
t/T

(c)

(d)

(e)

eliminates ISI must satisfy the Nyquist criterion (2-13). In order that a receive filter exist that eliminates ISI, it is necessary that the available bandwidth exceed half the symbol rate.

As the receive filter which eliminates (or at least substantially reduces) ISI yields a baseband pulse $g(t)$ whose folded Fourier transform $G_{\mathrm{fld}}(\omega)$ is essentially flat, the receive filter is often referred to as a (linear) *equalizer*. The equalizer must compensate for the linear distortion introduced by the channel and therefore depends on the channel frequency response $C(\omega)$. When there is sufficient a priori knowledge about $C(\omega)$, the equalizer can be implemented as a fixed filter. However, when there is a rather large uncertainty about the channel characteristics, the equalizer must be made adaptive.

Generally speaking, the equalizer compensates for the channel attenuation by having a larger gain at those signal frequencies that are more attenuated by the channel. However, the noise at those frequencies is also amplified, so that equalization gives rise to a *noise enhancement*. The larger the variation of the channel attenuation over the frequency interval occupied by the transmit signal, the larger this noise enhancement.

2.1.3 Line Coding

In many baseband communication systems, there is some frequency value B beyond which the linear distortion rapidly increases with frequency. For example, the attenuation (in decibels) of a twisted-pair cable is proportional to the square-root of the frequency (skin effect). This sets a limit on the bandwidth that can be used for baseband transmission: if the transmit signal contained components at too high frequencies, equalization of the severe linear distortion would give rise to a large noise enhancement, yielding a considerable performance degradation. In addition, in many applications the transmitted signal is severely distorted near zero frequency, because of transformer coupling or capacitive coupling, which reject DC. In these cases, the transmit signal should have low spectral content near zero frequency, in order to avoid excessive linear distortion and a resulting performance degradation.

It is clear from the above that we must control the transmit spectrum in order to avoid both high frequencies and frequencies near DC. We recall from Section 1.1.4 that the power spectrum $S_s(\omega)$ of the transmit signal $s(t)$ from (2-1) is given by

$$S_s(\omega) = \frac{1}{T} S_a\left(e^{j\omega T}\right) |G_T(\omega)|^2 \tag{2-19}$$

where $S_a\left(e^{j\omega T}\right)$ is the power spectrum of the data symbol sequence. Hence, the control of the spectrum is achieved by acting on the transmit pulse $g(t)$ and/or on the spectrum of the data symbol sequence. The latter is achieved by means of line coding: line coding involves the specification of the encoding rule that converts the information bit sequence $\{b_k\}$ into the sequence $\{a_k\}$ of data symbols, and therefore affects the spectrum of the data symbol sequence.

In addition to spectrum control, line coding should also provide timing information. Most symbol synchronizers can extract a reliable timing estimate from the received noisy PAM signal, only when there are sufficient data symbol transitions. When there are no data symbol transitions (suppose $a_k = 1$ for all k), the receive filter output signal in the absence of noise is given by

$$\begin{aligned} y_0(t;\varepsilon) &= \sum_{k=-\infty}^{+\infty} g(t-kT-\varepsilon T) \\ &= \frac{1}{T} \sum_{k=-\infty}^{+\infty} G\left(\frac{2\pi k}{T}\right) \exp\left[j\frac{2\pi k}{T}(t-\varepsilon T)\right] \end{aligned} \tag{2-20}$$

Note that $y_0(t;\varepsilon)$ is periodic in t with period T, so that one could be tempted to conclude that timing information can easily be extracted from $y_0(t;\varepsilon)$. However, when $G(\omega) = 0$ for $|\omega| > 2\pi B$ with $B < 1/T$ (as is the case for narrowband communication), the terms with $k \neq 0$ in (2-20) are zero, so that $y_0(t;\varepsilon)$ contains only a DC component, from which obviously no timing information can be derived. The same is true for wideband pulses with $G(2\pi m/T) = 0$ for $m \neq 0$ (as is the case for rectangular pulses with duration T). Hence, in order to guarantee sufficient timing information, the encoding rule should be such that irrespective of the bit sequence $\{b_k\}$, the number of successive identical data symbols in the sequence $\{a_k\}$ is limited to some small value.

As the spectrum $S_a(e^{j\omega T})$ is periodic in ω with period $2\pi/T$, it follows from (2-19) that the condition $S_s(\omega) = 0$ for $|\omega| > 2\pi B$ implies $G_T(\omega) = 0$ for $|\omega| > 2\pi B$. Hence, frequency components above $\omega = 2\pi B$ can be avoided only by using a band-limited transmit pulse. According to the Nyquist criterion for zero ISI, this band-limitation on $G_T(\omega)$ restricts the symbol rate $1/T$ to $1/T < 2B$.

A low spectral content near DC is achieved when $S_s(0) = 0$. This condition is fulfilled when either $G_T(0) = 0$ or $S_a(e^{j0}) = 0$.

- When $G_T(0) = 0$ and $1/T$ is required to be only slightly less than $2B$ (i.e., narrowband communication), the folded Fourier transform $G_{\text{fld}}(\omega)$ of the baseband pulse at the receive filter output is zero at $\omega = 0$. This indicates that equalization can be performed only at the expense of a rather large noise enhancement. The zero in the folded Fourier transform $G_{\text{fld}}(\omega)$ and the resulting noise enhancement can be avoided only by reducing the symbol rate $1/T$ below the bandwidth B. Hence, using a transmit pulse $g_T(t)$ with $G_T(0) = 0$ is advisable only when the available bandwidth is sufficiently large, but not for narrowband applications.
- In the case of narrowband communication, the recommended solution to obtain $S_s(0) = 0$ is to take $S_a(e^{j0}) = 0$. Noting that $S_a(e^{j0})$ can be expressed as

$$S_a(e^{j0}) = \lim_{N\to\infty} E\left[\frac{1}{2N+1}\left|\sum_{n=-N}^{N} a_n\right|^2\right] \tag{2-21}$$

it follows that $S_a\left(e^{j0}\right) = 0$ when the encoding rule is such that the magnitude of the *running digital sum*, RDS(n), given by

$$\mathrm{RDS}(n) = \sum_{k=-\infty}^{n} a_k \tag{2-22}$$

is limited for all N, and for any binary sequence $\{b_k\}$ at the encoder input.

For some encoders, there exist binary input strings (such as all zeroes, all ones, or alternating zeroes and ones) which cause long strings of identical data symbols at the encoder output. In order that the transmitted signal contains sufficient timing information, the probability of occurrence of such binary strings at the encoder input should be made very small. This can be accomplished by means of a scrambler. Basically, a scrambler "randomizes" the binary input sequence by modulo-2 addition of a pseudo-random binary sequence. At the receiver, the original binary sequence is recovered by adding (modulo-2) the same pseudo-random sequence to the detected bits.

Binary Antipodal Signaling

In the case of binary antipodal signaling, the channel symbol a_k equals $+1$ or -1 when the corresponding information bit b_k is a binary one or binary zero, respectively. Unless the Fourier transform $G(\omega)$ of the baseband pulse at the receive filter output satisfies $G(2\pi m/T) \neq 0$ for at least one nonzero integer m, this type of line coding does not provide sufficient timing information when the binary information sequence contains long strings of zeroes or ones. The occurrence of such strings can be made very improbable by using scrambling. Also, in order to obtain a zero transmit spectrum at DC, one needs $G_T(0) = 0$. Consequently, the magnitude of the running digital sum is limited, so that the codes yield no DC.

Quaternary Line Codes

In the case of quaternary line codes, the data symbol alphabet is the set $\{\pm 1, \pm 3\}$. An example is *2B1Q*, where the binary information sequence is subdivided in blocks of 2 bits, and each block is translated into one of the four levels ± 1 or ± 3. The 2B1Q line code is used for the basic rate ISDN (data rate of 160 kb/s) on digital subscriber lines.

2.1.4 Main Points

In a baseband communication system the digital information is conveyed by means of a pulse train, which is amplitude-modulated by the data. The channel is assumed to introduce linear distortion and to add noise. This linear distortion broadens the transmitted pulses; the resulting unwanted pulse overlap gives rise to ISI.

The receiver consists of a receive filter, which rejects out-of-band noise. Data detection is based upon receive filter output samples, which are taken once per symbol. These samples are fed to a slicer, which makes symbol-by-symbol decisions. The decisions are impaired by ISI and noise that occurs at the sampling instants.

The receive filter output should be sampled at the instants of maximum noise margin. These optimum sampling instants are not a priori known to the receiver. A timing recovery circuit or symbol synchronizer is needed to estimate the optimum sampling instants from the received noisy PAM signal.

The receive filter must combat both noise and ISI. According to the Nyquist criterion, the receive filter should produce a pulse whose folded Fourier transform is essentially flat, in order to substantially reduce the ISI. Such a filter is called an equalizer. When the system bandwidth is smaller than half the symbol rate, equalization cannot be accomplished.

In many applications, the channel attenuation is large near DC and above some frequency B. In order to avoid large distortion, the transmit signal should have negligible power in these regions. This is accomplished by selecting a transmit pulse with a bandwidth not exceeding B, and by means of proper line coding to create a spectral zero at DC. Besides spectrum control, the line coding must also provide a sufficient number of data transitions in order that the receiver is able to recover the timing.

2.1.5 Bibliographical Notes

Baseband communication is well covered in many textbooks, such as [1]–[6]. These books treat equalization and line coding in much more detail than we have done. Some interesting topics we did not consider are mentioned below.

Ternary Line Codes

In the case of ternary line codes, the data symbols take values from the set $\{-2, 0, +2\}$. In the following, we will adopt the short-hand notation $\{-, 0, +\}$ for the ternary alphabet.

A simple ternary line code is the alternate mark inversion (AMI) code, which is also called bipolar. The AMI encoder translates a binary zero into a channel symbol 0, and a binary one into a channel symbol + or – in such a way that polarities alternate. Because of these alternating polarities, it is easily verified that the running digital sum is limited in magnitude, so that the transmit spectrum is zero at DC. Long strings of identical data symbols at the encoder output can occur only when a long string of binary zeroes is applied to the encoder. This yields a long string of identical channel symbols 0. The occurrence of long strings of binary zeroes can be avoided by using a scrambler. The AMI decoder at the

Table 2-1 *4B3T* Line Code [1]

Binary input block	Ternary output block	
	Mode A	Mode B
0000	+0−	+0−
0001	−+0	−+0
0010	0−+	0−+
0011	+−0	+−0
0100	++0	−−0
0101	0++	0−−
0110	+0+	−0−
0111	+++	−−−
1000	++−	−−+
1001	−++	+−−
1010	+−+	−+−
1011	+00	−00
1100	0+0	0−0
1101	00+	00−
1110	0+−	0+−
1111	−0+	−0+

receiver converts the detected ternary symbols into binary symbols. 0 is interpreted as binary zero, whereas + and – are interpreted as binary one. The AMI code uses one ternary symbol to transmit one bit of information. Hence the efficiency of the AMI code, as compared to transmitting statistically independent ternary symbols, equals $1/\log_2 3 \simeq 0.63$. AMI is widely used for transmission of multiplexed digitized voice channels over repeatered twisted pair cables at rates of a few megabits per second.

A higher efficiency and more timing information than (unscrambled) AMI are obtained when using ternary block codes, which map blocks of k bits to blocks of n ternary symbols; such codes are denoted as *kBnT*. As an example, we consider the *4B3T*code, which maps 4 bits to 3 ternary symbols according to Table 2-1.

Note that there are two "modes": when the running digital sum is negative (nonnegative), the entries from the first (second) column are used. In this way, the magnitude of the running digital sum is limited, which guarantees a spectral zero at DC. Also, it is not possible to have more than five successive identical ternary symbols, so that timing information is always available. The efficiency of the 4B3T code equals $4/(3\log_2 3) \simeq 0.84$. 4B3T achieves a higher efficiency than AMI, but shows a higher encoding/decoding complexity.

Binary Line Codes

In many applications involving optical fiber communication, the light source is keyed on/off. In this case, the channel symbols are binary, and the symbol alphabet is denoted $\{+, -\}$.

Binary block codes converting a block of m information bits into a block of n binary channel symbols are denoted *mBnB*; their efficiency as compared to uncoded binary transmission equals m/n.

In the case of zero-disparity codes, each block of n binary channel symbols contains the same number $(n/2)$ of $+'$s and $-'$s. The number of possible code words equals $N = (n!)/((n/2)!)^2$, but the number of code words actually used in a mBnB zero-disparity code equals 2^m, which is the largest power of 2 not exceeding N. As the running digital sum is limited to the interval $(-n/2, n/2)$, zero disparity codes yield a spectral null at $\omega = 0$. Also, there are sufficient data transitions for providing timing information. The *1B2B* zero-disparity code is called *Manchester code* (also called biphase) whose two code words are $+-$ and $-+$. The Manchester code is used in coaxial-cable-based Ethernet local area networks in the token ring and in fiber optical communication systems, but its efficiency is only 0.5. The efficiency of the mBnB zero-disparity code increases with increasing n, but at the cost of increased encoding/decoding complexity.

A higher efficiency than for zero-disparity codes is obtained when using *bimode* mBnB *codes*, where the code words do not have the same number of symbols $+$ and $-$. As for the 4B3T ternary code described earlier, each block of m information bits can be represented by one of two code words; the code word reducing the magnitude of the running digital sum is selected by the encoder.

Equalization

- When only a single data symbol is transmitted, the receiver's decision is affected only by additive noise, because ISI is absent. The signal-to-noise ratio at the input of the slicer is maximized when the receive filter is the *matched filter*. The frequency response of the matched filter is $G_T^*(\omega)C^*(\omega)/S_N(\omega)$, where $G_T(\omega)$ and $C(\omega)$ are the Fourier transforms of the transmit pulse and the channel impulse response and $S_N(\omega)$ is the spectrum of the noise at the receiver input. When the additive noise is Gaussian and the values that can be assumed by the data symbols are equiprobable, it can be shown that the optimum receiver, in the sense of minimizing the decision error probability, is the matched filter followed by the slicer. The corresponding decision error probability is called the *matched filter bound*: it is a lower bound on the decision error probability of a baseband PAM communication system that is affected by Gaussian noise and ISI.

- The optimum linear equalizer, in the sense of minimizing the mean-square error between the sample at the input of the slicer and the corresponding data symbol, turns out to be the cascade of a matched filter and a transversal filter operating at a rate $1/T$. Because of the noise enhancement caused by the equalizer, the decision error probability is larger than the matched filter bound [7, 8].
- When symbol-by-symbol decisions are made, but the receiver is allowed to use decisions about previous data symbols, the ISI caused by previous data symbols (i.e., the postcursor ISI) can be subtracted from the samples at the input of the slicer. This yields the decision-feedback equalizer (DFE), consisting of a forward equalizer which combats noise and precursor ISI (i.e., ISI caused by future data symbols) and a feedback filter which generates from the receiver's decisions the postcursor ISI to be subtracted. The optimum forward equalizer (in the sense of minimizing noise and precursor ISI) is the cascade of the matched filter and a noise-whitening transversal filter operating at a rate $1/T$. The DFE yields less noise enhancement than the linear equalizer, assuming that the decisions that are fed back are correct. However, because of the decision-feedback, the DFE can give rise to error propagation [7][8].
- The optimum receiver (in the sense of minimizing sequence error probability) in the presence of ISI is the Viterbi equalizer (VE). It does not make symbol-by-symbol decisions, but exploits the correlation between successive receive filter samples for making a decision about the entire sequence by using a dynamic programming algorithm, operating on matched filter output samples taken at the symbol rate [9–11]. When the eye at the matched filter output is open, the performance of the VE is very close to the matched filter bound [11].
- Fractionally spaced equalizers (i.e., transversal filters operating at a rate exceeding $1/T$) are able to compensate for timing errors [12].
- Various algorithms exist for updating adaptive equalizers [13].

Line Coding

- Other ternary line codes than AMI and 4B3T from Section 2.1.3 are bipolar *n* zero substitution (*BnZS*), high-density bipolar *n* (*HDBn*), pair-selected ternary (*PST*), and *MS43* [14].
- Other binary line codes than those considered in Section 2.1.3 are the Miller code (also called delay modulation), coded mark inversion (CMI), and bit insertion codes such as *mB1C* and *DmB1M* [15, 16].
- The fiber-distributed data interface (FDDI) is a standard for a 100 Mb/s fiber-optic token ring. It uses a *4B5B* line code with ample timing information, but with nonzero DC content.

- In the case of partial response coding (which is also called correlative level encoding), a controlled amount of ISI is introduced deliberately by passing the data symbols to a transversal filter operating at the symbol rate, in order to perform spectral shaping. This transversal filter and the transmit filter can be combined into a single filter. Denoting the frequency response in the z-domain of the transversal filter by $T(z)$, typical frequency responses are

$$\begin{aligned} T(z) &= 1 - z^{-1} && \text{(dicode)} \\ T(z) &= 1 + z^{-1} && \text{(duobinary class 1)} \\ T(z) &= (1 + z^{-1})(1 - z^{-1}) = 1 - z^{-2} && \text{(modified duobinary class 4)} \end{aligned} \tag{2-23}$$

Frequency responses having a factor $1 - z^{-1}$ yield a spectral null at $\omega = 0$, whereas a factor $1 + z^{-1}$ gives rise to a spectral null at the Nyquist frequency $\omega = \pi/T$. A spectral null at $\omega = \pi/T$ allows transmission with the minimum bandwidth $B = 1/(2T)$. The transmit signal is given by (2-1), where $\{a_k\}$ denotes the sequence of filtered symbols at the output of the transversal filter $T(z)$. A simple receiver detects the filtered data symbols by performing symbol-by-symbol decisions (the optimum receiver would use the Viterbi algorithm in order to exploit the correlation between filtered data symbols). As the filtered symbols have more levels than the unfiltered symbols, a degradation of the decision error probability occurs. Decisions about the unfiltered data symbols could be obtained by applying the decisions $\{a_k\}$ about the filtered data symbols to a filter with frequency response $1/T(z)$, as shown in Figure 2-5. As $T(z)$ is a polynomial in z, $1/T(z)$ is the frequency response of an all-pole filter. An all-pole filter can be realized only by means of feedback, so that the performance of the system shown in Figure 2-5 is affected by error propagation. It can be shown that the problem of error propagation is circumvented by precoding the unfiltered data symbols at the transmitter, before entering the transversal filter $T(z)$. Because of the precoding, the unfiltered data symbols can be recovered by means of

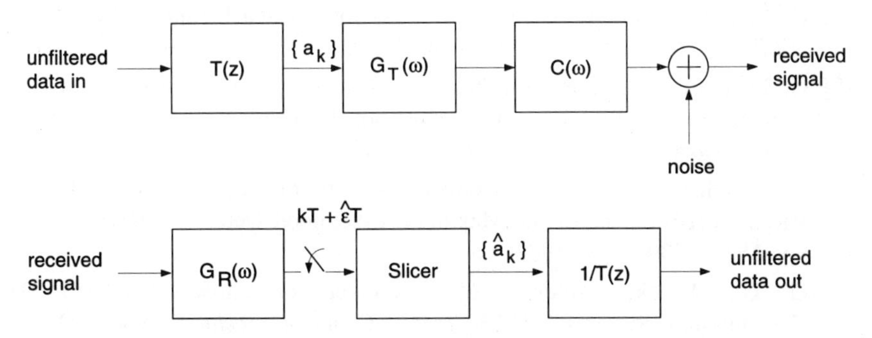

Figure 2-5 Partial Response System with Error Propagation

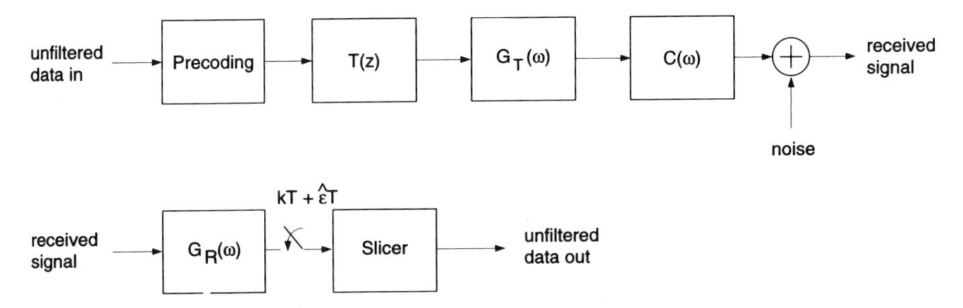

Figure 2-6 Partial Response System with Precoding to Avoid Error Propagation

memoryless symbol-by-symbol decisions (as indicated in Figure 2-6), so that error propagation does not occur [17]–[21].

Bibliography

[1] E. A. Lee and D. G. Messerschmitt, *Digital Communications*. Boston: Kluwer Academic, 1988.

[2] J. Bellamy, *Digital Telephony*. New York: Wiley, 2nd ed., 1991.

[3] S. Benedetto, E. Biglieri, and V. Castellani, *Digital Transmission Theory*. Englewood Cliffs, NJ: Prentice-Hall, 1987.

[4] R. D. Gitlin, J. F. Hayes, and S. B. Weinstein, "Data Communications Principles," *New York: Plenum*, 1992.

[5] R. Lucky, J. Salz, and E. Weldon, "Principles of Data Communication," *New York: McGraw Hill*, 1968.

[6] J. G. Proakis, *Digital Communications*. Auckland: McGraw-Hill, 1989.

[7] D. G. Messerschmitt, "A Geometric Theory of Intersymbol Interference. Part I: Zero-Forcing and Decision Feedback Equalization," *BSTJ*, vol. 52, pp. 1483–1519, Nov. 1973.

[8] C. A. Belfiore and J. H. Park, Jr., "Decision Feedback Equalization," *Proc. IEEE*, vol. 67, pp. 1143–1156, Aug. 1979.

[9] G. D. Forney, "Maximum-Likelihood Sequence Estimation of Digital Sequences in the Presence of Intersymbol Interference," *IEEE Trans. Inform. Theory*, vol. IT-18, pp. 363–378, May 1972.

[10] D. G. Messerschmitt, "A Geometric Theory of Intersymbol Interference. Part II: Performance of the Maximum Likelihood Detector," *BSTJ*, vol. 52, pp. 1521–1539, Nov. 1973.

[11] G. Ungerboeck, "Adaptive ML Receiver for Carrier-Modulated Dat-Transmission Systems," *IEEE Trans. Commun. Technol.*, vol. COM-22, pp. 624–636, May 1974.

[12] G. Ungerboeck, "Fractional Tap-spacing Equalizer and Consequences for Clock Recovery in Data Modems," *IEEE Trans. Commun.*, vol. COM-24, pp. 856–864, Aug. 1976.

[13] S. U. H. Qureshi, "Adaptive Equalization," *Proc. IEEE*, vol. 73, pp. 1349–1387, Sept. 1985.

[14] P. A. Franaszek, "Sequence-State Coding for Digital Transmission," *BSTJ*, vol. 47, pp. 134–157, Jan. 1968.

[15] S. Kawanishi, N. Yoshikai, J. Yamada, and K. Nakagawa, "DmB1M Code and Its Performance in a Very High-Speed Optical Transmission System," *IEEE Trans. Commun.*, vol. COM-36, pp. 951–956, Aug. 1988.

[16] N. Yoshikai, S. Nishi, and J. Yamada, "Line Code and Terminal Configuration for Very Large-Capacity Optical Transmission Systems," *IEEE J. Sel. Areas Commun.*, vol. SAC-4, pp. 1432–1437, Dec. 1986.

[17] P. Kabal and S. Pasupathy, "Partial-Response Signaling," *IEEE Trans. Commun.*, vol. COM-23, pp. 921–934, Sept. 1975.

[18] H. Kobayashi, "Correlative Level Coding and maximum-Likelihood Decoding," *IEEE Trans. Inform. Theory*, vol. IT-17, pp. 586–594, Sept. 1971.

[19] H. Kobayashi, "A Survey of Coding Schemes for Transmission or Recording of Digital Data," *IEEE Trans. Commun. Technol.*, vol. COM-19, pp. 1087–1100, Dec. 1971.

[20] E. R. Kretzmer, "Generalization of a Technique for Binary Data Communication," *IEEE Trans. Commun. Technol.*, vol. COM-14, pp. 67–68, Feb. 1966.

[21] A. Lender, "The Duobinary Technique for High-Speed Data Transmission," *IEEE Trans. Commun. Electon,*, vol. 82, pp. 214–218, May 1963.

2.2 Clock Synchronizers

2.2.1 Introduction

The digital information, embedded in the transmitted PAM signal, is recovered at the receiver by means of a decision device. This decision device operates on samples of the noisy PAM signal $y(t;\varepsilon)$, taken at symbol rate $1/T$ at the receive filter output, which is given by

$$y(t;\varepsilon) = \sum_{m} a_m \, g(t - mT - \varepsilon T) + n(t) \tag{2-24}$$

In (2-24) $\{a_m\}$ is a sequence of zero-mean data symbols, $g(t)$ is the baseband PAM pulse at the receive filter output, εT is an unknown fractional time delay ($-1/2 \le \varepsilon \le 1/2$), and $n(t)$ represents zero-mean additive noise. For maximum noise immunity, the samples upon which the receiver's decision is based should be taken at the instants of maximum eye opening. As the decision instants are a

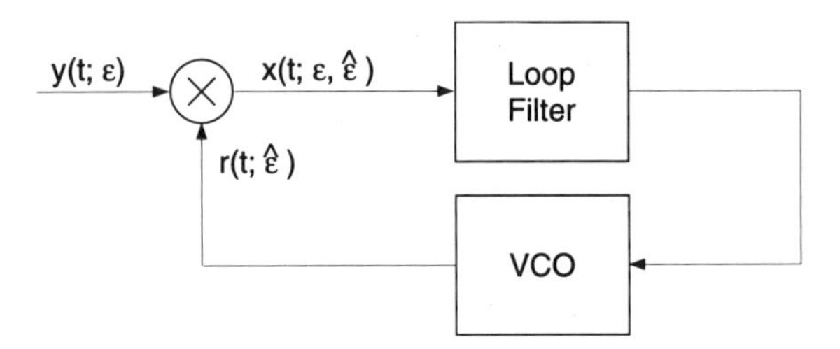

Figure 2-7 Ordinary PLL Operating on PAM Signal

priori unknown (because of the unknown delay εT) the receiver must contain a device which makes an estimate of the normalized delay. Such a device is called a *clock synchronizer* or *symbol synchronizer*. The timing estimate $\hat{\varepsilon}$ is used to bring the sampling clock, which activates the sampler at the receive filter output, in close synchronism with the received PAM signal. This is achieved by adjusting the phase of this sampling clock according to the value of the estimate $\hat{\varepsilon}$.

The received noisy PAM signal contains no periodic components, because the channel symbols $\{a_m\}$ have zero mean. Therefore, an ordinary PLL (see Chapter 2 of Volume I) operating on the filtered received signal $y(t;\varepsilon)$ cannot be used to generate a clock signal which is in synchronism with the received PAM signal. Let us illustrate this fact by considering a PLL with multiplying timing error detector: the local reference signal $r(t;\hat{\varepsilon})$ given by

$$r(t;\hat{\varepsilon}) = \sqrt{2}\,K_r\ \sin\left(\frac{2\pi}{T}\,(t-\hat{\varepsilon}T)\right) \tag{2-25}$$

and is multiplied with the noisy PAM signal $y(t;\varepsilon)$, as shown in Figure 2-7. Taking into account (2-24), the timing error detector output signal equals

$$x(t;\varepsilon,\hat{\varepsilon}) = \left[\sum_m a_m\ g(t-mT-\varepsilon T)+n(t)\right]\sqrt{2}\,K_r\ \sin\left(\frac{2\pi}{T}\,(t-\hat{\varepsilon}T)\right) \tag{2-26}$$

For any values of ε and $\hat{\varepsilon}$, the statistical average of the timing error detector output is identically zero, because the channel symbols $\{a_m\}$ and the additive noise $n(t)$ have zero mean. As the average timing error detector output is zero irrespective of ε and $\hat{\varepsilon}$, there is no deterministic force that makes the PLL lock onto the received PAM signal.

2.2.2 Categorization of Clock Synchronizers

From the operating principle point of view, two categories of synchronizers are distinguished, i.e., *error-tracking* (or feedback, or closed loop) synchronizers and *feedforward* (or open loop) synchronizers.

A general error-tracking synchronizer is shown in Figure 2-8. The noisy PAM signal $y(t;\varepsilon)$ and a locally generated reference signal $r(t;\hat{\varepsilon})$ are "compared" by

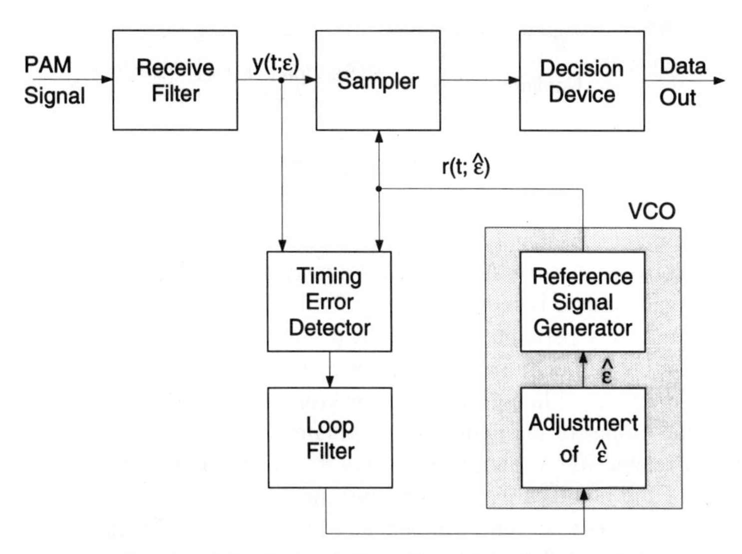

Figure 2-8 General Error-Tracking Synchronizer

means of a *timing error detector*, whose output gives an indication of the magnitude and the sign of the *timing error* $e = \varepsilon - \hat{\varepsilon}$. The filtered timing error detector output signal adjusts the timing estimate $\hat{\varepsilon}$ in order to reduce the timing error e. The timing estimate $\hat{\varepsilon}$ is the normalized delay of the reference signal $r(t;\hat{\varepsilon})$ which activates the sampler operating on $y(t;\varepsilon)$. Hence, error-tracking synchronizers use the principle of the PLL to extract a sampling clock which is in close synchronism with the received PAM signal. Properties of error-tracking synchronizers will be studied in detail in Section 2.3.

Figure 2-9 shows a general feedforward synchronizer. The noisy PAM receive signal $y(t;\varepsilon)$ enters a *timing detector*, which "measures" the instantaneous value of ε (or a function thereof). The noisy measurements at the timing detector output are averaged to yield the timing estimate $\hat{\varepsilon}$ (or a function thereof).

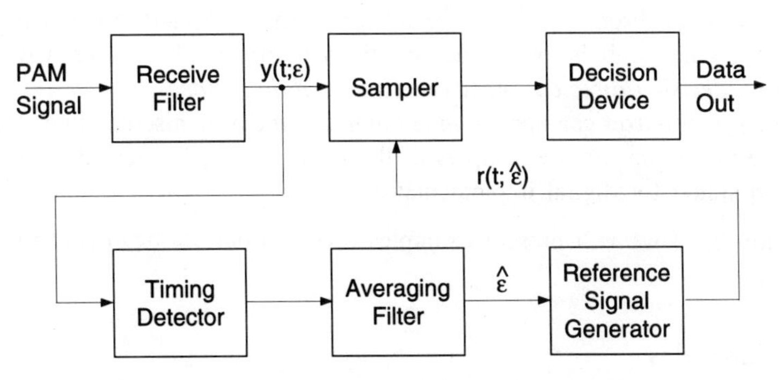

Figure 2-9 General Feedforward Synchronizer

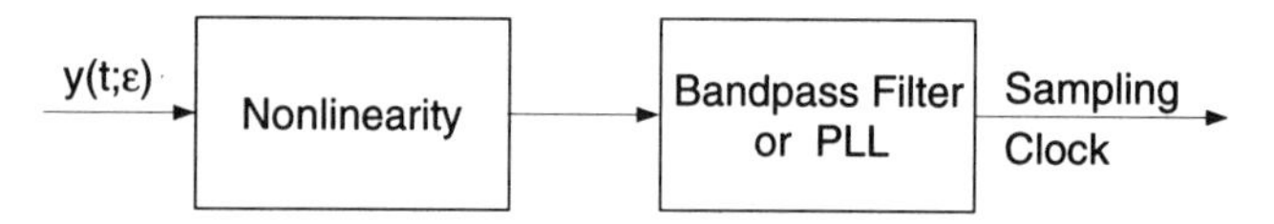

Figure 2-10 General Spectral Line Generating Synchronizer

An important subcategory of both error-tracking and feedforward synchronizers is formed by the *spectral line generating synchronizers.* A general spectral line generating synchronizer is shown in Figure 2-10. The received PAM signal, which does not contain any periodic components, passes through a suitable nonlinearity. In most cases, the output of the nonlinearity is an even function of its input. In Section 2.4, where spectral line generating synchronizers are treated in detail, we will show that the output of the nonlinearity contains periodic components, the phase of which is related to the normalized delay ε. Viewed in the frequency domain, the output of the nonlinearity contains spectral lines at multiples of the channel symbol rate $1/T$. The periodic component with frequency $1/T$ can be tracked by means of an ordinary PLL, whose VCO output signal then controls the sampler at the output of the receive filter. When the periodic component at frequency $1/T$ is extracted by means of a PLL, the spectral line generating synchronizer belongs to the category of error-tracking synchronizers, because the VCO of the PLL is driven by an error signal depending on the normalized delay difference $e = \varepsilon - \hat{\varepsilon}$. Alternatively, the periodic component with frequency $1/T$ can be selected by means of a narrowband bandpass filter, tuned to the channel symbol rate $1/T$, and serves as a clock signal which activates the sampler at the output of the receive filter. When the periodic component at frequency $1/T$ is extracted by means of a bandpass filter, the spectral line generating synchronizer belongs to the category of feedforward synchronizers. Note that the feedforward spectral line generating synchronizer makes no explicit estimate of ε (or a function thereof); instead, the cascade of the nonlinearity and the narrowband bandpass filter gives rise directly to the reference signal $r(t; \hat{\varepsilon})$.

Besides the above categorization into error-tracking and feedforward synchronizers, other categorizations can be made:

- When a synchronizer makes use of the receiver's decisions about the transmitted data symbols for producing a timing estimate, the synchronizer is said to be *decision-directed*; otherwise, it is *non-data-aided.*
- The synchronizer can operate in continuous time or in discrete time. Discrete-time synchronizers use samples of the PAM signal $y(t; \varepsilon)$, and are therefore well-suited for digital implementation.

In Section 2.2.3 we will present examples of synchronizers belonging to various categories.

2.2.3 Examples

In this section we give a few specific examples of synchronizers, belonging to the various categories considered in Section 2.2.2. Their operation will be

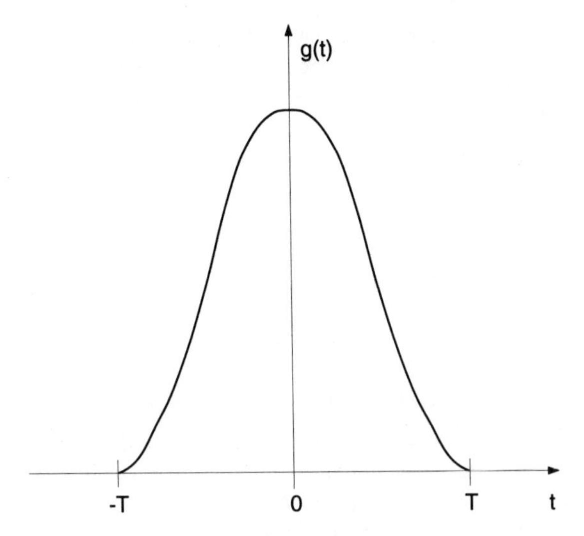

Figure 2-11 Baseband PAM Pulse $g(t)$ Used in Examples

explained mainly in a qualitative way. A detailed quantitative treatment of these synchronizers will be presented in Section 2.5.

For all cases, it will be assumed that the synchronizers operate on a noiseless PAM waveform $y(t;\varepsilon)$, with a baseband pulse $g(t)$, given by

$$g(t) = \begin{cases} \frac{1}{2}\left(1+\cos\left(\frac{\pi t}{T}\right)\right) & |t| < T \\ 0 & \text{otherwise} \end{cases} \tag{2-27}$$

The baseband PAM pulse $g(t)$ is shown in Figure 2-11; note that its duration equals two channel symbol intervals, so that successive pulses overlap. The channel symbols $\{a_k\}$ independently take on the values -1 and 1 with a probability of $1/2$ (binary antipodal signaling).

The Squaring Synchronizer

The squaring synchronizer is a spectral line generating synchronizer. The block diagram of the squaring synchronizer is shown in Figure 2-12. The PAM signal $y(t;\varepsilon)$ enters a prefilter, which in our example is a differentiator. The squared differentiated signal contains spectral lines at DC and at multiples of the channel symbol rate $1/T$; the spectral line at the channel symbol rate is selected by means of a bandpass filter (feedforward synchronizer), or is tracked by means of a PLL (error-tracking synchronizer).

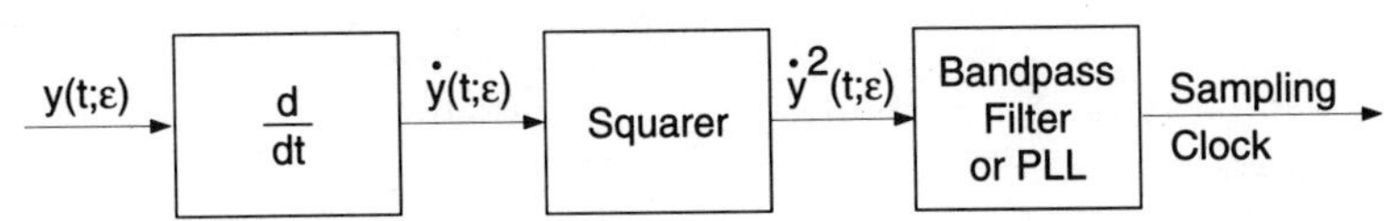

Figure 2-12 Block Diagram of Squaring Synchronizer

Figure 2-13 shows the PAM signal $y(t;\varepsilon)$, its time derivative $\dot{y}(t;\varepsilon)$ and the squarer output $\dot{y}^2(t;\varepsilon)$. The squarer output consists of a sequence of identical pulses, with randomly missing elements; more precisely, the interval $(mT+\varepsilon T, mT+\varepsilon T+T)$ contains a pulse if and only if $a_m \neq a_{m+1}$, i.e., when a channel symbol transition occurs. The squarer output $\dot{y}^2(t;\varepsilon)$ can be decomposed as the sum of two terms. The first term is the statistical expectation $E\left[\dot{y}^2(t;\varepsilon)\right]$, which is periodic in t with period T; this is the useful term, consisting of spectral lines. The second term is a zero-mean disturbance $\dot{y}^2(t;\varepsilon) - E\left[\dot{y}^2(t;\varepsilon)\right]$, which is caused by the random nature of the channel symbols. This term is a *self-noise* term, which does not contain spectral lines and, hence, disturbs the synchronizer operation.

The Synchronizer with Zero-Crossing Timing Error Detector

The synchronizer with zero-crossing timing error detector (ZCTED) is an error-tracking synchronizer. The block diagram of this synchronizer is shown in Figure 2-14 and its operation is illustrated in Figure 2-15.

The VCO output signal $r(t;\hat{\varepsilon})$ is a square wave with a frequency equal to

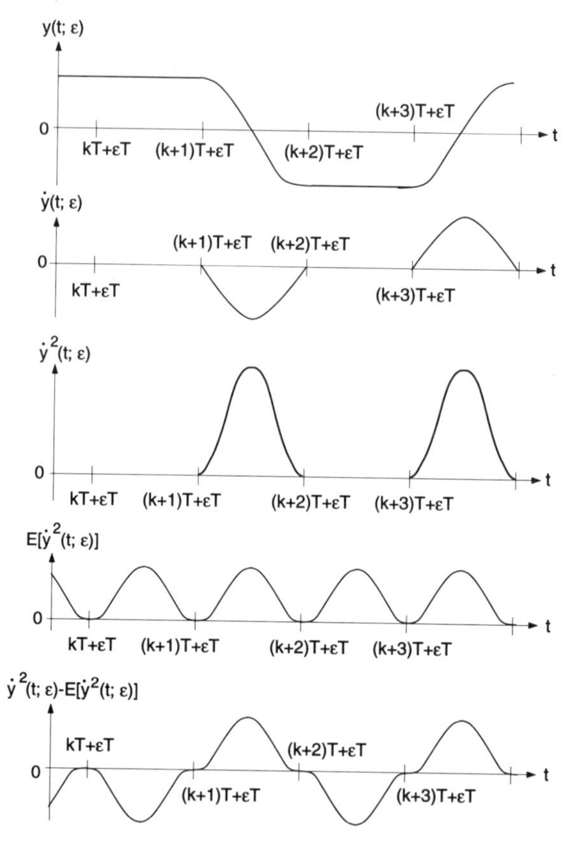

Figure 2-13 Illustration of Squaring Synchronizer Operation

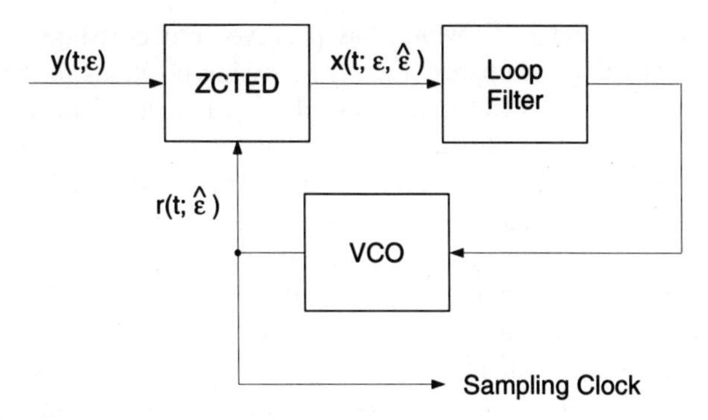

Figure 2-14 Block Diagram of Synchronizer with ZCTED

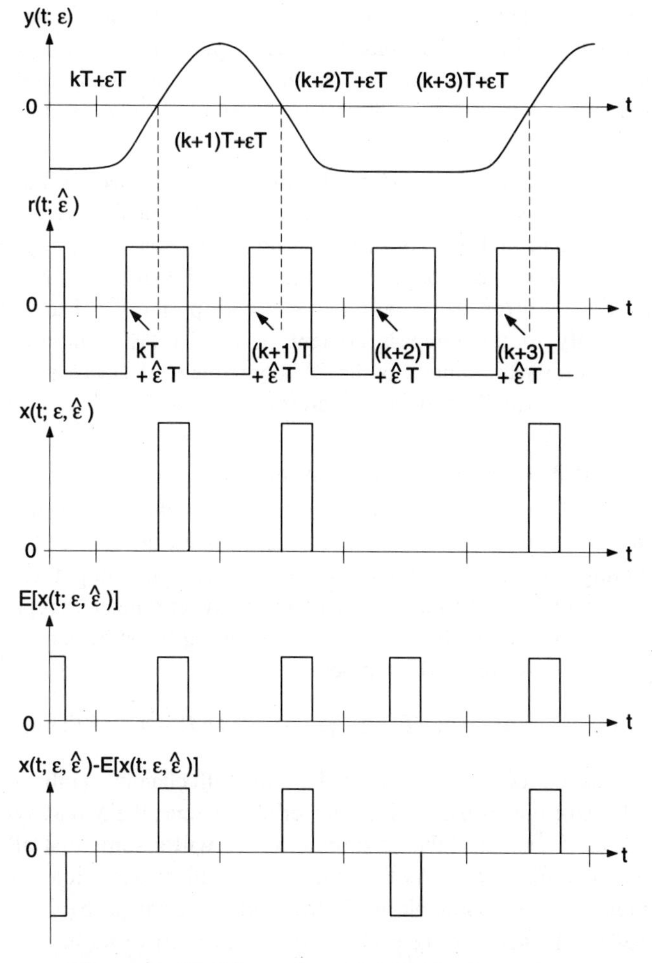

Figure 2-15 Illustration of ZCTED Operation

the channel symbol rate $1/T$, which has positive zero crossings at the instants $kT + \hat{\varepsilon}T$. The ZCTED "compares" the zero-crossing positions of the PAM signal $y(t;\varepsilon)$ with those of the VCO output signal $r(t;\hat{\varepsilon})$ in the following way. The ZCTED looks for the possible (positive or negative) zero crossing of the PAM signal, located between two successive positive zero crossings of the VCO output signal. When there is no such zero crossing of $y(t;\varepsilon)$, the ZCTED output $x(t;\varepsilon,\hat{\varepsilon})$ is identically zero over the considered interval. When $y(t;\varepsilon)$ does exhibit a zero crossing in the considered interval, a rectangular pulse is generated at the ZCTED output. When the zero crossing of $y(t;\varepsilon)$ leads the negative zero crossing of $r(t;\hat{\varepsilon})$ in the considered interval, this rectangular pulse is positive and extends from the zero crossing of $y(t;\varepsilon)$ to the negative zero crossing of $r(t;\hat{\varepsilon})$. On the other hand, when the negative zero crossing of $r(t;\hat{\varepsilon})$ leads the zero crossing of $y(t;\varepsilon)$, the rectangular pulse is negative, and extends from the negative zero crossing of $r(t;\hat{\varepsilon})$ to the zero crossing of $y(t;\varepsilon)$. Summarizing, the ZCTED output signal consists of a sequence of rectangular pulses, with missing elements each time there is no channel symbol transition. The polarity of each pulse gives information about leading or lagging of the zero crossings of the clock with respect to those of the received PAM signal, while the width of each pulse equals the amount of leading or lagging.

The ZCTED output signal $x(t;\varepsilon,\hat{\varepsilon})$ can be decomposed as the sum of a deterministic term $E[x(t;\varepsilon,\hat{\varepsilon})]$, which equals the statistical expectation of the ZCTED output signal, and a zero-mean self-noise term $x(t;\varepsilon,\hat{\varepsilon}) - E[x(t;\varepsilon,\hat{\varepsilon})]$, which is caused by the randomness of the channel symbols. Note that the deterministic term $E[x(t;\varepsilon,\hat{\varepsilon})]$ is periodic in t with period T. The synchronizer loop responds only to the lowpass content of the ZCTED output signal; the useful part of this lowpass content is the DC component of $E[x(t;\varepsilon,\hat{\varepsilon})]$, which is proportional to the normalized delay difference $e = \varepsilon - \hat{\varepsilon}$.

The Mueller and Müller (M&M) Synchronizer

The M&M synchronizer is a discrete-time error-tracking synchronizer, which derives an indication about the delay difference between the received PAM signal and the sampling clock from samples of the receive filter output signal $y(t;\varepsilon)$ taken at the channel symbol rate $1/T$. The M&M synchronizer is represented by the block diagram shown in Figure 2-16. The timing error detector produces a sequence $\{x_k(\varepsilon,\hat{\varepsilon})\}$, which is determined by

$$x_k(\varepsilon,\hat{\varepsilon}) = \hat{a}_{k-1}\, y(kT + \hat{\varepsilon}T;\varepsilon) - \hat{a}_k\, y(kT - T + \hat{\varepsilon}T;\varepsilon) \tag{2-28}$$

where $\hat{a}_m$ denotes the receiver's decision about the mth channel symbol a_m. As the timing error detector makes use of the receiver's decisions, the M&M synchronizer is *decision-directed.* For the sake of simplicity, we will assume that all decisions $\hat{a}_m$ are correct, i.e., $\hat{a}_m = a_m$. As far as the synchronizer operation is concerned, this assumption is most reasonable when the decision error probability is smaller than about 10^{-2}. The actual error probability on most terrestrial baseband links is smaller than 10^{-2} by several orders of magnitude.

The operation of the M&M synchronizer is illustrated in Figure 2-17. The

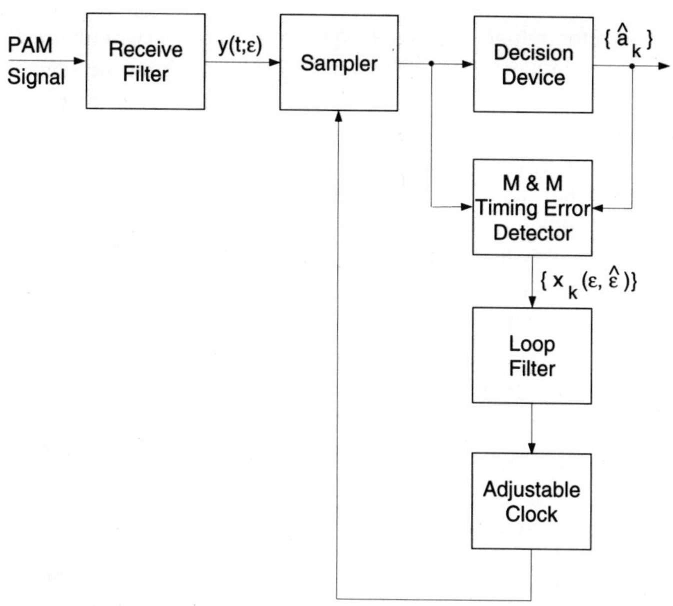

Figure 2-16 Block Diagram of Receiver with M&M Synchronizer

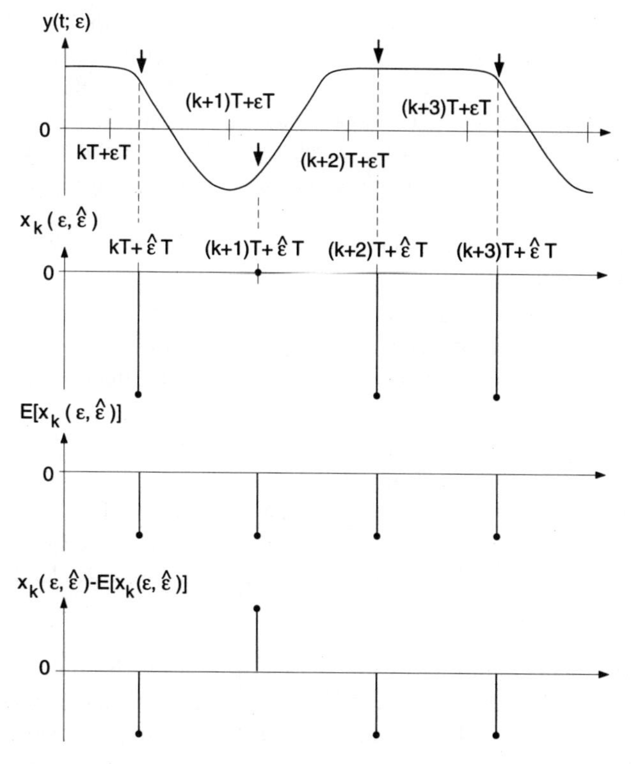

Figure 2-17 Illustration of M&M Synchronzer Operation

timing error detector output sequence $\{x_k(\varepsilon, \hat{\varepsilon})\}$ is a sequence of identical numbers, with randomly missing elements. The nonzero elements have the same sign as the normalized timing error $e = \varepsilon - \hat{\varepsilon}$, while their magnitude is an increasing function of $|e|$. It depends on the channel symbol sequence $\{a_m\}$ which of the elements in the timing error detector output sequence are missing. For $e < 0$, the kth element is zero when $a_{k-1} = a_{k+1}$; for $e > 0$, the kth element is zero when $a_k = a_{k-2}$. The timing error detector output sequence can be decomposed as the sum of two terms. The first term is the statistical average $E[x_k(\varepsilon, \hat{\varepsilon})]$, consisting of identical elements which give an indication about the sign and the magnitude of the normalized delay difference between the PAM signal and the sampling clock. The second term is the zero-mean self-noise term $x_k(\varepsilon, \hat{\varepsilon}) - E[x_k(\varepsilon, \hat{\varepsilon})]$, which disturbs the operation of the synchronizer.

2.2.4 Disturbance in Clock Synchronizers

When the received baseband PAM pulse $g(t)$ extends over more than one channel symbol interval, the successively transmitted pulses partially overlap. In this case, the received PAM signal during the kth symbol interval does not depend on the kth channel symbol a_k only, but also on other channel symbols. This effect is called intersymbol interference (ISI). The ISI affects the operation of the synchronizer, because it gives rise to a random disturbance at the output of the timing error detector of an error-tracking synchronizer, or at the output of the nonlinearity of a spectral line generating synchronizer. This disturbance is called *self-noise*; its statistics depend on the baseband PAM pulse and on the statistics of the channel symbols. In the simple examples considered in Section 2.2.3, the presence of self-noise is obvious from the missing elements in the squarer output (Figure 2-13), in the ZCTED output (Figure 2-15), and in the M&M timing error detector output (Figure 2-17).

Besides self-noise, also additive noise at the input of the receiver affects the operation of the clock synchronizer. However, in the case of terrestrial baseband communication, self-noise is the dominating disturbance for the following reasons:

- Additive noise levels are small (signal-to-noise ratios for terrestrial baseband communication are in the order of 30 dB).
- The bandwidth of the received PAM pulse is often smaller than the symbol rate $1/T$ (i.e., twice the Nyquist frequency), so that the PAM pulse has a duration of several symbol intervals. This gives rise to considerable ISI and much self-noise.

Therefore, the emphasis in the design of clock synchronizers is on the reduction of the self-noise. As we shall demonstrate in Section 2.5, this can be accomplished by prefiltering the received PAM signal, so that the resulting baseband pulse has a suitable shape.

2.3 Error-Tracking Synchronizers

2.3.1 The General Structure of Error-Tracking Synchronizers

Error-tracking synchronizers generate a periodic *reference signal* which must be brought in synchronism with the received waveform. A nonlinear circuit, called the *timing error detector*, "compares" the received waveform with the locally generated reference signal and produces an *error signal* which gives an indication about the sign and the magnitude of their relative misalignment. This error signal is *filtered* and *fed back* to the oscillator which generates the reference signal, so as to reduce the timing error between the received signal and the local reference signal.

The general structure of an error-tracking synchronizer is shown in Figure 2-18. The input signal $y(t;\varepsilon)$ is a noisy PAM signal, given by

$$y(t;\varepsilon) = \sum_{m} a_m \; g(t - mT - \varepsilon T) + n(t) \tag{2-29}$$

where $1/T$ is the channel symbol rate, $\{a_m\}$ is a stationary sequence of zero-mean channel symbols, $g(t)$ is the baseband PAM pulse, εT is an unknown fractional time delay to be estimated by the synchronizer and $n(t)$ is zero-mean stationary noise, which is statistically independent of the channel symbol sequence. It has been shown in Section 1.1.4 that $y(t;\varepsilon)$ is cyclostationary with period T. For more information about cyclostationary processes the reader is referred to Section 1.1.2. The local reference $r(t;\hat{\varepsilon})$ at the output of the VCO is a periodic waveform whose instantaneous frequency is determined by the signal $u(t)$ at the input of the VCO:

$$r(t;\hat{\varepsilon}) = s\left[\frac{2\pi}{T}\,(t - \hat{\varepsilon}T)\right] \tag{2-30}$$

where $s(w)$ is a periodic function of w with period 2π, e.g., $s(w) = \sin\,(w)$, and $\hat{\varepsilon}$ is the estimate of ε. The effect of $u(t)$ on the instantaneous frequency of the local reference is described by

$$\frac{d\hat{\varepsilon}}{dt} = -\left[\frac{1}{T_0} - \frac{1}{T}\right] + K_0\;u(t) \tag{2-31}$$

where $1/T_0$ is the quiescent frequency of the reference signal, corresponding to a control signal $u(t)$ which is identically zero, and the constant K_0 is the "gain" of the VCO. Indeed, when $u(t)$ is identically zero, $\hat{\varepsilon}$ is a linear function of time:

$$\hat{\varepsilon}(t) = -\left[\frac{1}{T_0} - \frac{1}{T}\right] t + \hat{\varepsilon}(0) \tag{2-32}$$

Hence, the resulting reference signal is periodic in t with period T_0:

$$r(t;\hat{\varepsilon})\big|_{\,u(t)=0} = s\left[\frac{2\pi}{T}\,(t - \hat{\varepsilon}T)\right]_{\big|\,u(t)=0} = s\left[\frac{2\pi}{T_0}\,(t - \hat{\varepsilon}(0)T_0)\right] \tag{2-33}$$

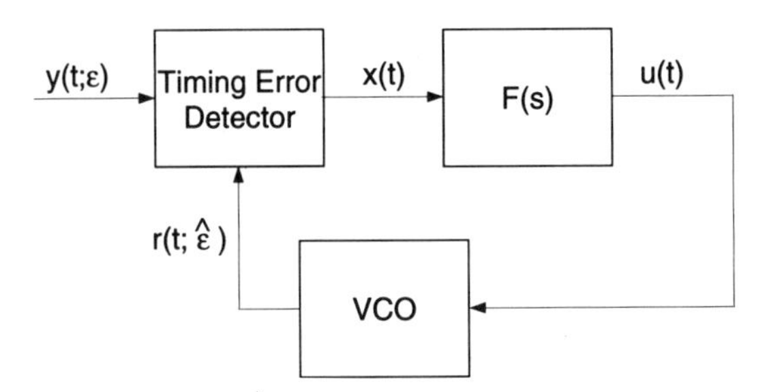

Figure 2-18 Error-Tracking Synchronizer Structure

The timing error detector performs a suitable time-invariant nonlinear operation on the input signal and the local reference, so that its output signal $x(t)$ gives an indication of the instantaneous timing error $e = \varepsilon - \hat{\varepsilon}$. The timing error detector output signal enters a linear time-invariant loop filter. Its frequency response in the Laplace domain is denoted by $F(s)$; the frequency response in the frequency domain is $F(\omega)$. The loop filter output signal $u(t)$ determines the instantaneous frequency of the VCO according to (2-31) such that the timing error e is reduced.

As the input signal $y(t;\varepsilon)$ has zero mean, it does not contain a deterministic periodic component, so that a conventional PLL cannot be used to estimate ε. However, there is still some periodicity embedded in the input signal, because of its cyclostationarity with period T. This cyclostationarity is exploited by the synchronizer in order to make an estimate of ε.

There exists a large variety of timing error detector circuits; some examples are considered in Section 2.2.3. In spite of the many different types of timing error detector circuits encountered in practice, it will be shown that all error-tracking synchronizers can be represented by the same equivalent model, shown in Figure 2-19, so that they can be analyzed in a unified way. Comparing the equivalent model with the synchronizer structure shown in Figure 2-18, we see that the equivalent model replaces the timing error detector output signal by the sum $K_D g(e) + N(t)$, where $g(e)$ and $N(t)$ are called the *timing error detector characteristic* and the *loop noise*, respectively. The useful component $K_D g(e)$ of the timing error detector output signal is periodic in e with period 1; also, it will be shown in Section 2.3.3 that $g(0) = 0$, and the timing error detector gain K_D is usually chosen such that $g'(0)$, the timing error detector slope at the origin, is normalized to 1. The loop noise $N(t)$ represents the statistical fluctuations of the timing error detector output signal. These fluctuations are caused by the additive noise $n(t)$ and the random nature of the channel symbols $\{a_m\}$ in the input signal $y(t;\varepsilon)$. The loop noise is a zero-mean wide-sense stationary process whose power spectral density $S(\omega; e)$ is periodic in the timing error e with period 1. The equivalent model and the actual synchronizer structure have the same loop filter. The remaining part of the equivalent model corresponds to (2-31), with ΔF representing the *frequency detuning* of the VCO:

$$\Delta F = \frac{1}{T_0} - \frac{1}{T} \tag{2-34}$$

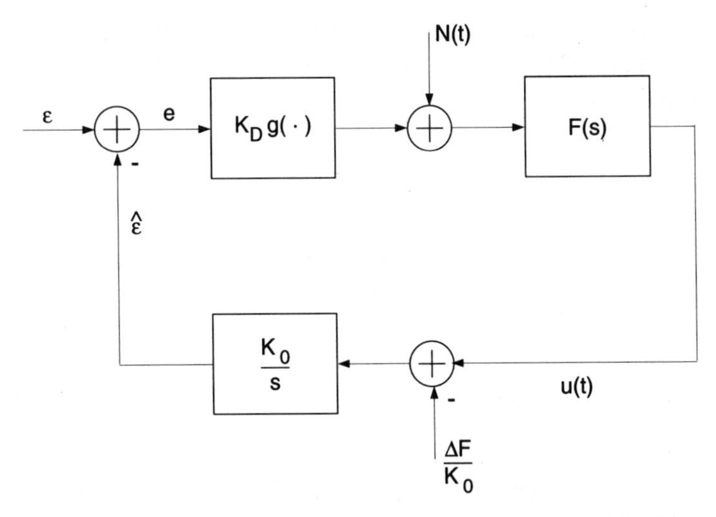

Figure 2-19 Equivalent Model of Error-Tracking Synchronizer

K_0/s is the Laplace domain frequency response of an integrator with gain K_0.

The equivalent model is a powerful means for studying error-tracking synchronizers. Indeed, the results obtained from investigating the equivalent model are valid for the whole class of error-tracking synchronizers. In order to relate a specific synchronizer to the equivalent model, one must be able to derive both the useful component $K_D g(e)$ of the timing error detector output and the loop noise spectrum $S(\omega; e)$ from the description of the synchronizer. In the following, these quantities will be obtained for the general synchronizer structure shown in Figure 2-18.

Actually, $K_D g(e)$ and $S(\omega; e)$ will be derived under "open-loop" conditions. This involves opening the feedback loop (say, at the input of the loop filter) and applying a periodic reference signal $r(t; \hat{\varepsilon})$ with period T and constant value of $\hat{\varepsilon}$. From the resulting timing error detector output signal under open-loop conditions, the useful component $K_D g(e)$ and the loop noise spectrum $S(\omega; e)$ will be determined. These quantities are assumed to remain the same when the loop is closed. This approximation is valid when the loop bandwidth (which will be defined in Section 2.3.4) is small with respect to the channel symbol rate.

2.3.2 Open-Loop Statistics of the Timing Error Detector Output Signal

In order to determine statistical properties of the timing error detector output signal under open-loop conditions, we consider an input signal $y(t; \varepsilon)$ given by (2-29) with a constant value of ε, and a periodic reference signal $r(t; \hat{\varepsilon})$ with period T and a constant value of $\hat{\varepsilon}$. The timing error detector output signal is denoted by $x(t;\varepsilon,\hat{\varepsilon})$ in order to indicate explicitly the dependence on the constant parameters ε and $\hat{\varepsilon}$. The case where both ε and $\hat{\varepsilon}$ are fluctuating in time will be briefly considered also.

Taking into account the cyclostationarity of $y(t;\varepsilon)$ with period T, the periodicity of $r(t;\hat{\varepsilon})$ with period T, and the time-invariant (but not necessarily memoryless) nature of the timing error detector circuit, it follows from Section 1.1.2 that the timing error detector output signal under open-loop conditions is also cyclostationary with period T. In all cases of practical interest, the synchronizer bandwidth (to be defined in Section 2.3.4) is a small fraction (about one percent or even much less) of the channel symbol rate $1/T$, so that the synchronizer operation in closed loop is determined entirely by the lowpass content of the timing error detector output signal. Taking into account that the statistics of the timing error detector output signal in closed loop are essentially the same as in open loop, we can replace $x(t;\varepsilon,\hat{\varepsilon})$ in open loop by its lowpass content $x_{\mathrm{LP}}(t;\varepsilon,\hat{\varepsilon})$, without affecting the operation of the synchronizer when the loop will be closed. The signal $x_{\mathrm{LP}}(t;\varepsilon,\hat{\varepsilon})$ is defined as

$$x_{\mathrm{LP}}(t;\varepsilon,\hat{\varepsilon}) = \int_{-\infty}^{+\infty} h_{\mathrm{LP}}(t-u)\, x(u;\varepsilon,\hat{\varepsilon})\, du \tag{2-35}$$

where $h_{\mathrm{LP}}(t)$ is the impulse response of an ideal lowpass filter with frequency response $H_{\mathrm{LP}}(\omega)$, given by

$$H_{\mathrm{LP}}(\omega) = \begin{cases} 1 & |\omega| < \pi/T \\ 0 & |\omega| > \pi/T \end{cases} \tag{2-36}$$

It has been shown in Section 1.1.2 that passing a cyclostationary process through a lowpass filter with a bandwidth (in hertz) not exceeding $1/(2T)$ yields a wide-sense stationary process. Hence, the lowpass content $x_{\mathrm{LP}}(t;\varepsilon,\hat{\varepsilon})$ of the timing error detector output signal is wide-sense stationary. This implies that the statistical expectation $E[x_{\mathrm{LP}}(t;\varepsilon,\hat{\varepsilon})]$ and the statistical autocorrelation function $E[x_{\mathrm{LP}}(t;\varepsilon,\hat{\varepsilon})\, x_{LP}(t+u;\varepsilon,\hat{\varepsilon})]$ do not depend on t. As εT and $\hat{\varepsilon}T$ are time delays, it is easily verified that the joint statistical properties of $y(t;\varepsilon)$ and $r(t;\hat{\varepsilon})$ are the same as for $y(t-\hat{\varepsilon}T;\varepsilon-\hat{\varepsilon})$ and $r(t-\hat{\varepsilon}T;0)$. Hence, $x_{\mathrm{LP}}(t;\varepsilon,\hat{\varepsilon})$ and $x_{\mathrm{LP}}(t-\hat{\varepsilon}T;\varepsilon-\hat{\varepsilon},0)$ have identical statistical properties.

The lowpass content $x_{\mathrm{LP}}(t;\varepsilon,\hat{\varepsilon})$ can be uniquely decomposed into a useful component $E[x_{\mathrm{LP}}(t;\varepsilon,\hat{\varepsilon})]$ and a zero-mean disturbance $N(t;\varepsilon,\hat{\varepsilon})$ called the loop noise:

$$x_{\mathrm{LP}}(t;\varepsilon,\hat{\varepsilon}) = E[x_{\mathrm{LP}}(t;\varepsilon,\hat{\varepsilon})] + N(t;\varepsilon,\hat{\varepsilon}) \tag{2-37}$$

As $x_{\mathrm{LP}}(t;\varepsilon,\hat{\varepsilon})$ and $x_{\mathrm{LP}}(t-\hat{\varepsilon}T;\varepsilon-\hat{\varepsilon},0)$ have the same statistical properties and are both wide-sense stationary, one obtains

$$\begin{aligned} E[x_{\mathrm{LP}}(t;\varepsilon,\hat{\varepsilon})] &= E[x_{\mathrm{LP}}(t-\hat{\varepsilon}T;\varepsilon-\hat{\varepsilon},0)] \\ &= E[x_{\mathrm{LP}}(0;\varepsilon-\hat{\varepsilon},0)] \\ &= K_D\, g(\varepsilon-\hat{\varepsilon}) \end{aligned} \tag{2-38}$$

which indicates that the useful component at the timing error detector output depends on the delays εT and $\hat{\varepsilon}T$ only through the timing error $e = \varepsilon - \hat{\varepsilon}$. Similarly, it follows that

$$\begin{aligned} E[N(t;\varepsilon,\hat{\varepsilon})\; N(t+u;\varepsilon,\hat{\varepsilon})] &= E[N(t-\hat{\varepsilon}T;\varepsilon-\hat{\varepsilon},0)\; N(t+u-\hat{\varepsilon}T;\varepsilon-\hat{\varepsilon},0)] \\ &= E[N(0;\varepsilon-\hat{\varepsilon},0)\; N(u;\varepsilon-\hat{\varepsilon},0)] \\ &= R_N(u;\varepsilon-\hat{\varepsilon}) \end{aligned} \tag{2-39}$$

so that also the autocorrelation function of the loop noise depends only on the timing error. As the reference waveform $r(t;\hat{\varepsilon})$ is periodic in t with period T, it satisfies $r(t;\hat{\varepsilon}) = r(t;\hat{\varepsilon}+1)$ so that $x_{\mathrm{LP}}(t;\varepsilon,\hat{\varepsilon}) = x_{\mathrm{LP}}(t;\varepsilon,\hat{\varepsilon}+1)$. Hence, the useful component $K_D g(e)$ at the timing error detector output and the autocorrelation function $R_N(u;e)$ of the loop noise are both periodic in the timing error e with period 1. The loop noise spectrum $S_N(\omega;e)$ is the Fourier transform of the autocorrelation function $R_N(u;e)$.

The useful component $K_D g(e)$ and the loop noise spectrum $S_N(\omega;e)$ can easily be derived from the statistical expectation and the autocorrelation function of the timing error detector output signal $x(t;\varepsilon,\hat{\varepsilon})$, because $x_{\mathrm{LP}}(t;\varepsilon,\hat{\varepsilon})$ results from a linear filtering operation on $x(t;\varepsilon,\hat{\varepsilon})$. Taking into account (1-39) and (1-40), one immediately obtains

$$K_D g(\varepsilon-\hat{\varepsilon}) = \langle E[x(t;\varepsilon,\hat{\varepsilon})]\rangle_t \tag{2-40}$$

$$R_N(u;\varepsilon-\hat{\varepsilon}) = [\langle K_x(t,t+u;\varepsilon,\hat{\varepsilon})\rangle_t]_{\mathrm{LP}} \tag{2-41}$$

where $\langle ... \rangle_t$ denotes time-averaging with respect to the variable t, $[...]_{\mathrm{LP}}$ indicates that the quantity between brackets is passed through the ideal lowpass filter, whose frequency response $H_{\mathrm{LP}}(\omega)$ is given by (2-36), and $K_x(t,t+u;\varepsilon,\hat{\varepsilon})$ denotes the autocovariance function of $x(t;\varepsilon,\hat{\varepsilon})$:

$$K_x(t,t+u;\varepsilon,\hat{\varepsilon}) = E[x(t;\varepsilon,\hat{\varepsilon})\; x(t+u;\varepsilon,\hat{\varepsilon})] - E[x(t;\varepsilon,\hat{\varepsilon})]\; E[x(t+u;\varepsilon,\hat{\varepsilon})] \tag{2-42}$$

Equivalently, $K_x(t,t+u;\varepsilon,\hat{\varepsilon})$ is the autocorrelation function of the cyclostationary disturbance $x(t;\varepsilon,\hat{\varepsilon}) - E[x(t;\varepsilon,\hat{\varepsilon})]$ at the timing error detector output.

The important conclusion is that, as far as the operation of the synchronizer is concerned, the timing error detector output under open-loop conditions can be represented as the sum of:

- A useful component $K_D g(e)$, which is a periodic function of the timing error e and
- The zero-mean loop noise $N(t)$, whose spectrum $S(\omega;e)$ is a periodic function of the timing error e [for notational convenience, $N(t;\varepsilon,\hat{\varepsilon})$ will be denoted by $N(t)$ from now on].

In the case where both ε and $\hat{\varepsilon}$ are fluctuating in time, the above representation of the timing error detector output signal under open-loop conditions is still valid, provided that ε and $\hat{\varepsilon}$ are so slowly varying that they are essentially constant over the correlation time of the loop noise (which is in the order of the channel symbol interval T) and over the possibly nonzero memory time of the timing error detector circuit (which is, at most, also in the order of the channel symbol interval T).

2.3.3 The Equivalent Model and Its Implications

Assuming that the loop bandwidth of the synchronizer (which will be defined in Section 2.3.4) is much smaller than the channel symbol rate $1/T$, the statistical properties of the timing error detector output signal under open-loop conditions remain valid when the loop is closed. This yields the equivalent model shown in Figure 2-19. Note that this equivalent model is the same as for a conventional PLL operating on a sinusoid disturbed by additive noise (see Chapter 3 of Volume 1); hence, an error tracking clock synchronizer and a conventional PLL have very similar properties.

In the absence of loop noise $N(t)$ and for zero frequency detuning ΔF, the *stable equilibrium points* of the equivalent model are determined by the positive going zero crossings of $g(e)$, assuming a positive open-loop gain. Because of the periodicity of $g(e)$, there exists an infinite number of stable equilibrium points. By introducing a suitable delay in the definition of the local reference $r(t;\hat{\varepsilon})$, the stable equilibrium points can be made coincident with $e = ... - 2, -1, 0, 1, 2, ...$ Usually, the constant K_D is determined such that the slope $g'(0)$ at the stable equilibrium point is normalized to 1. The negative zero crossings of $g(e)$ correspond to the *unstable equilibrium points* of the synchronizer.

In the absence of loop noise and for nonzero frequency detuning, the stable and unstable equilibrium points correspond to the positive and negative zero crossings, respectively, of $g(e) - \gamma$, where $\gamma = \Delta F/(K_0 K_D F(0))$ is the normalized frequency detuning. Hence, frequency detuning gives rise to a steady-state error, unless the DC gain $F(0)$ of the loop filter becomes infinite. This can be achieved by using a perfect integrator in the loop filter.

We now show that, in the presence of loop noise, the stable equilibrium points coincide no longer with those for zero loop noise, because the loop noise spectrum $S_N(\omega; e)$ depends on the timing error e. This means that the loop noise contributes to the steady-state error; this phenomenon is called *noise-induced drift*. Let us assume that $e = e_s$ is a stable equilibrium point in the presence of loop noise, and that the timing error $e(t)$ takes on the value e_s at $t = 0$. For $t > 0$, the timing error exhibits random fluctuations due to the loop noise $N(t)$. On the average, the rate of change $\Delta e/\Delta t = (e(\Delta t) - e(0))/\Delta t$ over a small time increment must be zero, because $e(0) = e_s$ is a stable equilibrium point. Indeed, a nonzero average rate of change would imply that some deterministic force pushes $e(t)$ away from its initial position e_s, which contradicts our assumption that e_s is a stable equilibrium point. Hence, e_s is such that

$$\lim_{\Delta t \to 0} E\left[\frac{\Delta e}{\Delta t} \;\middle|\; e(0) = e_s\right] = 0 \tag{2-43}$$

which expresses that, at a stable equilibrium point, the input signal and the reference signal have the same average frequency. The left-hand side of the above equation is the definition of the *intensity coefficient* $K_1(e)$, evaluated at the stable equilibrium point (see Volume 1, Section 9.3.1). For a first-order loop [i.e., $F(s) = 1$], the timing error satisfies the system equation

$$\frac{de}{dt} = -K_0\, K_D g(e) + \Delta F - K_0 N(t) \tag{2-44}$$

When the synchronizer bandwidth becomes much smaller than the bandwidth of the loop noise, $e(t)$ converges to the solution of the stochastic differential equation

$$de = \left[-K_0\, K_D g(e) + \Delta F + 1/4 K_0^2\, \frac{dS_N(0;e)}{de} \right] dt + K_0\, S_N(0;e)\, dW \tag{2-45}$$

where $W(t)$ is a Wiener process with unit variance parameter, $S_N(0;e)$ is the loop noise power spectral density at $\omega = 0$, and the term involving the derivative of $S_N(0;e)$ is the Itô correction term (see Volume 1, Section 9.3.2). For the above stochastic differential equation, the intensity coefficient $K_1(e)$ is given by

$$K_1(e) = -K_0\, K_D g(e) + \Delta F + 1/4 K_0^2\, \frac{dS_N(0;e)}{de} \tag{2-46}$$

The stable equilibrium points coincide with the negative zero crossings of $K_1(e)$. When the loop noise spectrum $S_N(\omega;e)$ does not depend on e, the Itô correction term is identically zero, and the stable equilibrium points are the same as for zero loop noise. When the loop noise spectrum does depend on the timing error, the stable equilibrium points in the presence of noise are displaced in the direction of the increasing loop noise spectral density. This situation is depicted in Figure 2-20, where e_1 and e_s denote the stable equilibrium points in the absence and

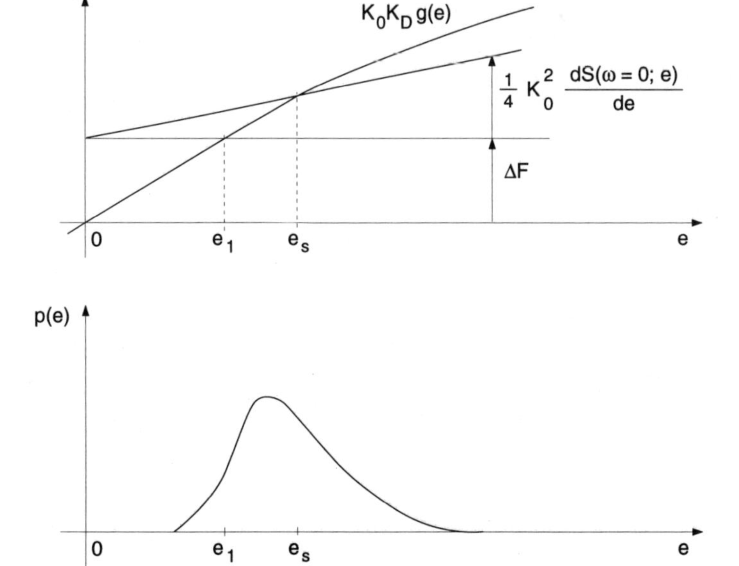

Figure 2-20 Illustration of Noise-Induced Drift

presence of loop noise, respectively. Also shown is a sketch of the probability density function of the timing error e in the presence of loop noise. As there is more noise in the region $e > e_1$ than in the region $e < e_1$, the timing error spends more time in the region $e > e_1$. This explains the strong asymmetry of the probability density function, which is decidedly non-Gaussian. In a similar way, it can be shown that a timing-error-dependent loop noise spectrum $S_N(\omega; e)$ affects also the position of the unstable equilibrium points, which coincide with the positive zero crossings of the intensity coefficient $K_1(e)$.

Most of the time, the timing error $e(t)$ fluctuates in the close vicinity of a stable equilibrium point. These small fluctuations, caused by the loop noise, are called *jitter*. This mode of operation can be described by means of the linearized equivalent model, to be considered in Section 2.3.4. Occasionally, the random fluctuations of the error $e(t)$ are so large that the error moves into the domain of attraction of a neighboring equilibrium point: a *cycle slip* occurs. Cycle slips are nonlinear phenomena and, hence, cannot be investigated by means of a linear model. An in-depth treatment of cycle slips is provided in Volume 1, Chapters 6 and 11.

Let us consider the case where the initial timing error $e(0)$ is very close to an unstable equilibrium point e_u, located between the stable equilibrium points e_s and e_s+1. For the sake of simplicity, let us assume that the synchronizer is a first-order loop, so that the timing error satisfies the stochastic differential equation (2-45). Note that e_u corresponds to a positive zero crossing of the intensity coefficient $K_1(e)$, given by (2-46). This is illustrated in Figure 2-21. When $e(0)$ is slightly smaller than e_u, the intensity coefficient $K_1(e)$ is negative, so that the drift term $K_1(e)\ dt$ in (2-45) has the tendency to drive the timing error toward the stable equilibrium point e_s. However, in the vicinity of the unstable equilibrium point e_u, the intensity coefficient $K_1(e)$ is very small. This can cause the timing error $e(t)$ to spend a prolonged time in the vicinity of the unstable equilibrium point, before it finally reaches the neighborhood of the stable equilibrium point e_s. This phenomenon is called *hang-up*. The occurrence of hang-ups gives rise to long acquisition times. Acquisition is investigated in Volume 1, Chapters 4 and 5.

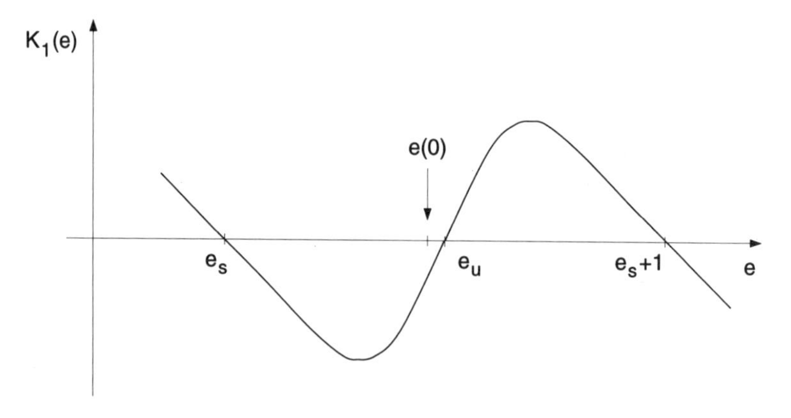

Figure 2-21 Intensity Coefficient $K_1(e)$

It has been shown in Volume 1, Section 3.2, that for some sequential logic timing error detectors, which make use only of the zero crossings of the input signal and the reference signal, the timing error detector characteristic has a sawtooth shape. At the unstable equilibrium points, the sawtooth characteristic jumps from its largest postive value to the opposite value, so that the restoring force near an unstable equilibrium point cannot be considered small. Hence, one might wonder whether hang-ups can occur in this case. The answer is that hang-ups are possible when additive noise is present at the input of the timing error detector. Indeed, Figure 3.2-5 in Volume 1 shows that the timing error detector characteristic has a sawtooth shape only when additive noise is absent. In the presence of additive noise, the slope of the characteristic at the unstable equilibrium point is finite (and decreases with increasing noise level); hence, the restoring force near the unstable equilibrium point is very small, so that hang-ups can occur.

2.3.4 The Linearized Equivalent Model

Under normal operating conditions, the timing error $e(t)$, for most of the time, exhibits small fluctuations about a stable equilibrium point e_s. Suitable measures of performance for this mode of operation are the steady-state error e_s and the variance of the timing error. It is most convenient to analyze these small fluctuations by linearizing the equivalent model about the stable equilibrium point and applying standard linear filter theory.

The nonlinear nature of the equivalent model is caused by:

- The nonlinear timing error detector characteristic $g(e)$
- The dependence of the loop noise spectrum $S_N(\omega; e)$ on the timing error e

Linearization of the equivalent model involves the following approximations:

$$g(e) = g(e_s) + (e - e_s)\, g'(e_s) \tag{2-47}$$

$$S_N(\omega; e) = S_N(\omega; e_s) \tag{2-48}$$

where $g'(e_s)$ is the timing error detector slope at the stable equilibrium point for a small steady-state error e_s and $g'(e_s)$ is well approximated by $g'(0) = 1$. This yields the linearized equivalent model, shown in Figure 2-22. The steady-state

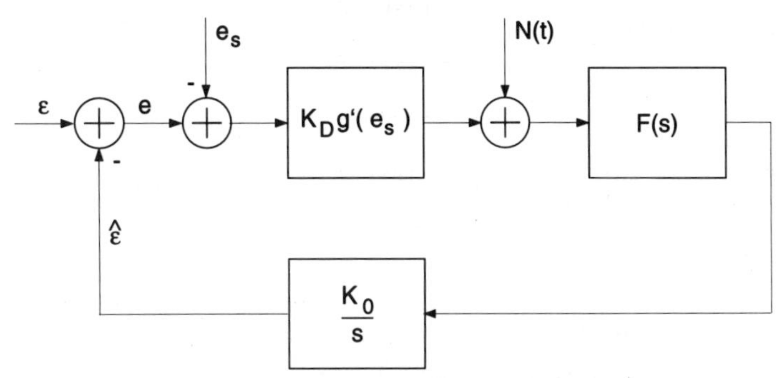

Figure 2-22 Linearized Equivalent Model

error e_s to be used in the linearized equivalent model is the steady-state error for zero loop noise, i.e., a positive zero crossing of $g(e) - \Delta F/(K_0 K_D F(0))$. This is motivated by the fact that, for a low loop noise level and/or a small synchronizer bandwidth, the Itô correction term has only a minor influence on the position of the stable equilibrium point (see Figure 2-20).

The linearized equivalent model can be transformed into the model shown in Figure 2-23, where $H(s)$, given by

$$H(s) = \frac{K_0 K_D g'(e_s) F(s)}{s + K_0 K_D g'(e_s) F(s)} \tag{2-49}$$

which is called the *closed-loop frequency response* (in the Laplace domain) of the synchronizer. Note that $H(s)$ is the frequency response of a lowpass filter with unit gain at DC; consequently, $1 - H(s)$ is the frequency response of a highpass filter. It is evident from Figure 2-23 that the timing error $e(t)$ consists of three terms:

- The steady-state error e_s
- The highpass content of $\varepsilon(t)$. This reflects that the synchronizer can track with a negligible error only a slowly fluctuating delay εT.
- The lowpass content of the disturbance $N(t)/K_D g'(e_s)$

An important parameter is the *one-sided loop bandwidth* B_L (in hertz) of the synchronizer, which is a measure of the bandwidth of the closed-loop frequency response:

$$B_L = \int_0^{\infty} |H(\omega)|^2 \frac{d\omega}{2\pi} \tag{2-50}$$

In many cases of practical interest, $\varepsilon(t)$ is fluctuating very slowly with respect to the channel symbol rate $1/T$. Therefore, it is possible to use a small loop

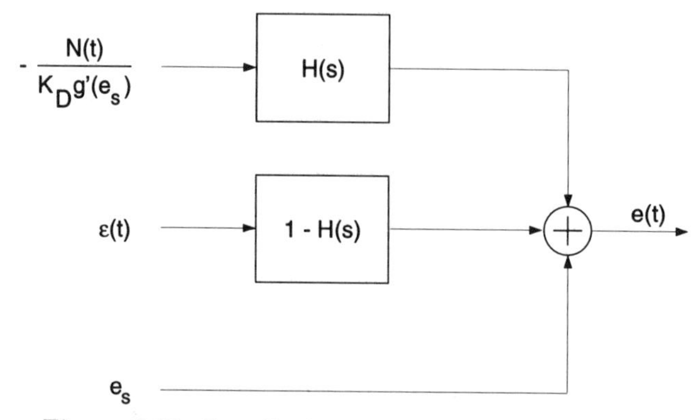

Figure 2-23 Contributions to the Timing Error $e(t)$

bandwidth for reducing the effect of loop noise on the timing error $e(t)$, while still having a negligible contribution to the timing error from the fluctuations of $\varepsilon(t)$. Typical values of $B_L T$ are in the order of one percent, and, in some cases, even much smaller.

2.3.5 The Linearized Timing Error Variance

Assuming that $\varepsilon(t)$ is essentially constant over many channel symbol intervals, so that its contribution to the timing error is negligible, and taking into account the wide-sense stationarity of the loop noise $N(t)$, application of the linearized equivalent model yields a wide-sense stationary timing error $e(t)$. Its statistical expectation and variance do not depend on time, and are given by

$$E[e(t)] = e_s \tag{2-51}$$

$$\mathrm{var}[e(t)] = \frac{1}{[K_D g'(e_s)]^2} \int_{-\infty}^{+\infty} |H(\omega)|^2 \, S(\omega; e_s) \, \frac{d\omega}{2\,\pi} \tag{2-52}$$

Because of the small loop bandwidth B_L, we expect that the timing error variance is determined mainly by the behavior of the loop noise spectrum $S_N(\omega; e_s)$ in the vicinity of $\omega = 0$. When the variation of the loop noise spectrum within the loop bandwidth is small, i.e.,

$$\max_{|\omega| < 2\pi B_L} |S_N(\omega; e_s) - S_N(0; e_s)| \ll S_N(0; e_s) \tag{2-53}$$

the loop noise spectrum is well approximated by its value at $\omega = 0$, as far as the timing error variance eq. (2-52) is concerned. Taking into account the definition eq. (2-50) of the loop bandwidth, this approximation yields

$$\mathrm{var}[e(t)] \cong (2 B_L T) \, \frac{S_N(0; e_s)/\,T}{[K_D g'(e_s)]^2} \tag{2-54}$$

However, for many well-designed synchronizers the loop noise spectrum $S_N(\omega; e_s)$ reaches its minimum value at $\omega = 0$; this minimum value can be so small (and in some cases even be equal to zero) that the variations of the loop noise spectrum within the loop bandwidth cannot be ignored, and the approximation eq. (2-54) is no longer accurate.

When the variation of $S_N(\omega; e_s)$ within the loop bandwidth cannot be neglected, we use the Parseval relation to obtain

$$\int_{-\infty}^{+\infty} |H(\omega)|^2 \, S_N(\omega; e_s) \, \frac{d\omega}{2\pi} = \int_{-\infty}^{+\infty} h_2(t) \, R_N(t; e_s) \, dt \tag{2-55}$$

where $R_N(t; e_s)$ is the autocorrelation function of the loop noise [i.e., the inverse Fourier transform of $S_N(\omega; e_s)$], and $h_2(t)$ is the inverse Fourier transform of $|H(\omega)|^2$. Denoting the inverse Fourier transform of $H(\omega)$ by $h(t)$, it follows that $h_2(t)$ is the autocorrelation function of the pulse $h(t)$. As $h(t)$ is causal [i.e.,

$h(t) = 0$ for $t < 0$], $h_2(t)$ is given by

$$h_2(t) = \int_0^{+\infty} h(u)\, h(t+u)\, du \tag{2-56}$$

We manipulate (2-56) further into

$$\begin{aligned} &\int_{-\infty}^{+\infty} h_2(t) R_N(t; e_s) dt \\ &= h_2(0) \int_{-\infty}^{+\infty} R_N(t; e_s) dt + 2 \int_0^{+\infty} (h_2(0) - h_2(t))(-R_N(t; e_s)) dt \end{aligned} \tag{2-57}$$

Noting that

$$h_2(0) = \int_0^{+\infty} h^2(u)\, du = \int_{-\infty}^{+\infty} |H(\omega)|^2 \frac{d\omega}{2\pi} = 2\, B_L \tag{2-58}$$

and

$$2 \int_0^{+\infty} R_N(t; e_s)\, dt = S_N(0; e_s) \tag{2-59}$$

the first term in eq. (2-57) corresponds to the approximation eq. (2-54), which neglects the variation of $S_N(\omega; e_s)$ within the loop bandwidth. Hence, the second term of eq. (2-57) incorporates the effect of the variation of $S_N(\omega; e_s)$, and can be written as

$$\begin{aligned} &2 \int_0^{+\infty} (h_2(0) - h_2(t))\, (-R_N(t; e_s))\, dt \\ &= \int_{-\infty}^{+\infty} |H(\omega)|^2\, [S_N(\omega; e_s) - S_N(0; e_s)]\, \frac{d\omega}{2\pi} \end{aligned} \tag{2-60}$$

Obviously, this term should not be neglected when the inequality (2-53) does not hold.

For $B_L T \ll 1$, $h_2(t)$ decays much slower than $R_N(t; e_s)$ with increasing $|t|$. For those values of t yielding a nonnegligible contribution of $R_N(t; e_s)$ to the second term of eq. (2-57), $h_2(t)$ is still close to $h_2(0)$, so that $h_2(0) - h_2(t)$ can be approximated by the first term in its Taylor series expansion about $t = 0$. This approximation is carried out in Section 2.3.10; here we simply state the main results.

Case 1:

For large ω, $|H(\omega)|$ decays like $1/\omega$.
The second term in eq. (2-57) is well approximated by

$$\begin{aligned} &2 \int_0^{+\infty} (h_2(0) - h_2(t)) \, (-R_N(t; e_s)) \, dt \\ &\cong h^2(0) \int_0^{+\infty} (-t R_N(t; e_s)) \, dt \end{aligned} \tag{2-61}$$

The right-hand side of (2-61) is proportional to $(B_L T)^2$, because of the factor $h^2(0)$.

Case 2:

For large ω, $|H(\omega)|$ decays faster than $1/\omega$.
The second term in (2-57) is well approximated by

$$2 \int_0^{+\infty} (h_2(0) - h_2(t)) \, (-R_N(t; e_s)) \, dt \cong \frac{1}{2} \, S_N^{(2)}(0; e_s) \int_{-\infty}^{+\infty} \omega^2 \, |H(\omega)|^2 \, \frac{d\omega}{2\pi} \tag{2-62}$$

where

$$S_N^{(2)}(0; e_s) = \frac{d^2}{d\omega^2} S_N(\omega; e_s) \bigg|_{\omega=0} \tag{2-63}$$

The right-hand side of (2-62) is proportional to $(B_L T)^3$, because of the integral over ω. The result (2-62) can also be obtained by simply approximating in (2-60) the variation of $S_N(\omega; e_s)$ by the first term of its Taylor series expansion:

$$S_N(\omega; e_s) - S_N(0; e_s) \cong 1/2 \, S_N^{(2)}(0; e_s) \, \omega^2 \tag{2-64}$$

Note that the approximation (2-62) is not valid for case 1, because the integral over ω does not converge when $|H(\omega)|$ decays like $1/\omega$ for large ω.

From the above results, we conclude that the timing error variance consists of two terms:

- The first term ignores the variation of $S_N(\omega; e_s)$ within the loop bandwidth, and is proportional to $S_N(0; e_s)$ and $B_L T$.

- The second term is caused by the variation of $S_N(\omega; e_s)$ within the loop bandwidth. For small $B_L T$, this term is essentially proportional to the square (when $|H(\omega)|$ decays like $1/\omega$ for large ω) or the third power [when $|H(\omega)|$ decays faster then $1/\omega$ for large ω] of $B_L T$.

When $S_N(\omega; e_s)$ is essentially flat within the loop bandwidth, the first term dominates for small $B_L T$. However, when the variation of $S_N(\omega; e_s)$ within the loop bandwidth is large with respect to $S(0; e_s)$, i.e., when $S_N(\omega; e_s) = 0$ or $S_N(\omega; e_s)$ has a sharp minimum at $\omega = 0$, the second term dominates; in this case it is advantageous that $|H(\omega)|$ decays faster than $1/\omega$, so that the resulting timing error variance is proportional to the third power instead of the square of $B_L T$.

Loop filters with the following frequency response (in the Laplace domain) are commonly used in phase-locked loops:

- First-order loop: $F(s) = 1$
- Perfect second-order loop: $F(s) = 1 + a/s$
- Imperfect second-order loop: $F(s) = (1 + s\tau_1)/(1 + s\tau_2)$

These loop filters all have in common that their frequency response $F(\omega)$ becomes constant for large frequencies. Hence, the magnitude of the corresponding closed-loop frequency response $H(\omega)$ behaves asymptotically like $1/\omega$. This implies that the contribution from the loop noise spectrum variation to the timing error variance is proportional to $(B_L T)^2$. When the loop noise spectrum variations cannot be neglected with respect to the loop noise spectrum at $\omega = 0$, the following numerical example will show the benefit of adding an extra pole to the loop filter, in which case the contribution from the loop noise spectrum variation to the timing error variance is proportional to $(B_L T)^3$.

Numerical Example

We compute the timing error variance (2-52) at $e_s = 0$, as a function of $B_L T$, for two different loop noise power spectral densities $S_1(\omega; 0)$ and $S_2(\omega; 0)$, and two different closed-loop frequency responses $H_1(s)$and $H_2(s)$. It is assumed that the timing error detector slope equals 1. We denote by var(i, j) the timing error variance corresponding to the power spectral density $S_i(\omega; 0)$ and the closed-loop frequency response $H_j(s)$, with $i = 1, 2$ and $j = 1, 2$.

- Power spectral densities:
  The considered power spectral densities $S_1(\omega; 0)$ and $S_2(\omega; 0)$ are given by

$$S_1(\omega; 0) = \begin{cases} T\cos^2(\omega T/2) & |\omega T| < \pi \\ 0 & \text{otherwise} \end{cases} \tag{2-65}$$

$$S_2(\omega; 0) = \begin{cases} T\sin^2(\omega T/2) & |\omega T| < 2\pi \\ 0 & \text{otherwise} \end{cases} \tag{2-66}$$

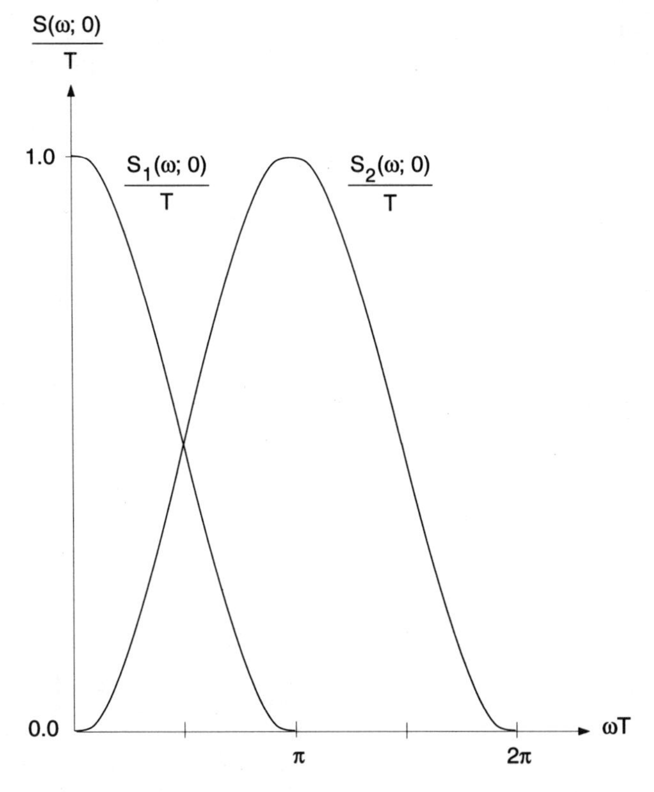

Figure 2-24 Loop Noise Power Spectral Densities $S_1(\omega;\ 0)$ and $S_2(\omega;\ 0)$

and are shown in Figure 2-24. For small $B_L T$, $S_1(\omega; 0)$ is essentially constant within the loop bandwidth. As $S_2(0; 0) = 0$, the variation of $S_2(\omega; 0)$ within the loop bandwidth cannot be ignored, even for very small $B_L T$.

- Closed-loop frequency responses and corresponding bandwidths:

$$H_1(s) = \frac{1}{1 + s\tau} \qquad B_{L,1} = \frac{1}{4\tau} \tag{2-67}$$

$$H_2(s) = \frac{1}{(1 + s\tau)(1 + s\tau/10)} \qquad B_{L,2} = \frac{1}{4.4\,\tau} \tag{2-68}$$

For large ω, $|H_1(\omega)|$ and $|H_2(\omega)|$ decay like $1/\omega$ and $1/\omega^2$, respectively.

- Timing error variances:
The timing error variances $\text{var}(i, j)$ are shown in Figure 2-25, for $i \doteq 1, 2$ and $j = 1, 2$. The following observations can be made:
    - For the considered range of $B_L T$, the curves for $\text{var}(1, 1)$and $\text{var}(1, 2)$ practically coincide. Indeed, as $S_1(\omega; 0)$ is nearly flat within the con-

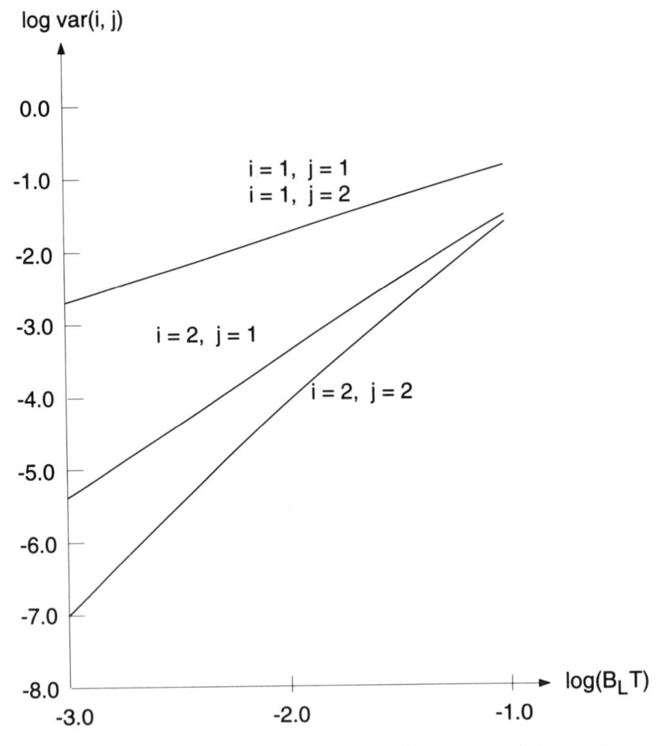

Figure 2-25 Timing Error Variance as a Function of Loop Bandwidth

sidered loop bandwidth, the timing error variance is essentially determined by (and proportional to) $B_L T$, irrespective of the specific shape of $|H(\omega)|$.

- As $S_2(0; 0) = 0$, $\mathrm{var}(2, 1)$ and $\mathrm{var}(2, 2)$ are essentially proportional to the square and the third power of $B_L T$, respectively, when $B_L T$ is small. It is seen from Figure 2-25 that this approximation is very accurate for $B_L T < 0.01$. For small $B_L T$, the additional pole of $H_2(s)$ yields a considerable reduction of the timing error variance.

2.3.6 Discrete-Time Hybrid Synchronizers

Until now we have restricted our attention to continuous-time synchronizers: the timing error detector output signal and the locally generated reference signal are continuous-time waveforms, and the loop filter and VCO operate in continuous time.

There exist many synchronizers which operate in *discrete time.* The general structure of a discrete-time synchronizer is shown in Figure 2-26. The input signal $y(t; \varepsilon)$, given by (2-29), is cyclostationary with period T. The timing error detector involves nonlinear operations and sampling at instants t_k, which are determined

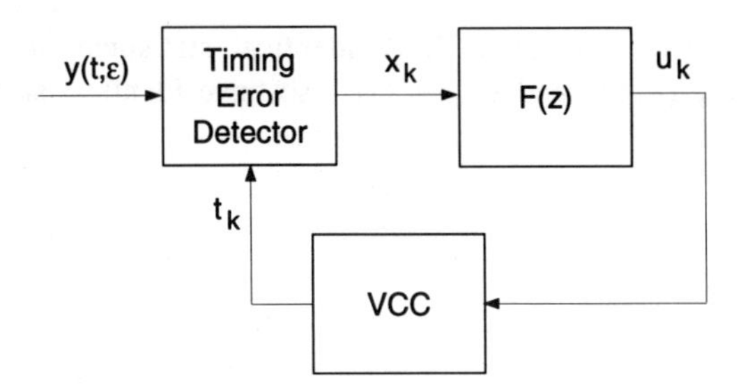

Figure 2-26 Discrete-Time Error-Tracking Synchronizer Structure

by a voltage-controlled clock (VCC). These sampling instants are estimates of the instants $kT + \varepsilon T$. At the output of the timing error detector appears a sequence $\{x_k\}$ of real numbers, produced at the channel symbol rate $1/T$. This sequence enters a discrete-time loop filter, whose frequency response in the *z*-domain is denoted by $F(z)$. The corresponding frequency response in the frequency domain is $F(\exp(\omega T))$. The sequence $\{u_k\}$ at the loop filter output enters the VCC, which generates sampling instants t_k according to

$$t_{k+1} = t_k + T_0 + K_0' u_k \tag{2-69}$$

When u_k is identically zero, the interval between sampling instants is constant and equal to T_0; hence, $1/T_0$ is the quiescent frequency of the VCC.

Note on Terminology

If the error detector output is used to control the sampling instant by means of a VCC we speak of a *hybrid* synchronizer. The term *digital synchronizer* is used for a device which operates entirely in the discrete-time domain. Such devices are discussed in Part D.

As for the continuous-time synchronizer, the key to the equivalent model for the discrete-time synchronizer is the investigation of some statistical properties of the timing error detector output sequence under open-loop conditions. These open-loop conditions are:

- The fractional delay εT of the input signal is constant,
- The instants t_k are given by $t_k = kT + \hat{\varepsilon}T$, where $\hat{\varepsilon}$ is constant.

The timing error detector output at the instant $t_k = kT + \hat{\varepsilon}T$ is denoted by $x_k(\varepsilon, \hat{\varepsilon})$, in order to indicate explicitly the dependence on the constant parameters ε and $\hat{\varepsilon}$.

Under open-loop conditions, the following can be verified:

- An input signal $y(t + mT; \varepsilon)$ and a sampling instant sequence $\{t_k\} = \{kT + \hat{\varepsilon}T\}$ give rise to a timing error detector output sequence $\{x_{k+m}(\varepsilon, \hat{\varepsilon})\}$.

As $y(t+mT;\varepsilon)$ and $y(t;\varepsilon)$ have identical statistical properties, the sequences $\{x_{k+m}(\varepsilon,\hat{\varepsilon})\}$ and $\{x_k(\varepsilon,\hat{\varepsilon})\}$ also have identical statistics. In other words, $\{x_k(\varepsilon,\hat{\varepsilon})\}$ is a stationary sequence.

- As $x_{k+m}(\varepsilon,\hat{\varepsilon}) = x_k(\varepsilon,\hat{\varepsilon}+m)$ and $\{x_k(\varepsilon,\hat{\varepsilon})\}$ is a stationary sequence, $\{x_k(\varepsilon,\hat{\varepsilon})\}$ and $\{x_k(\varepsilon,\hat{\varepsilon}+m)\}$ have identical statistical properties.
- An input signal $y(t-\hat{\varepsilon}T;\varepsilon-\hat{\varepsilon})$ and a sampling instant sequence $\{t_k\} = \{kT+\hat{\varepsilon}T\}$ give rise to a timing error detector output sequence $\{x_k(\varepsilon-\hat{\varepsilon};0)\}$. As $y(t-\hat{\varepsilon}T;\varepsilon-\hat{\varepsilon})$ and $y(t;\hat{\varepsilon})$ have identical statistical properties, the sequences $\{x_k(\varepsilon-\hat{\varepsilon};0)\}$ and $\{x_k(\varepsilon,\hat{\varepsilon})\}$ also have identical statistics.

Hence, $\{x_k(\varepsilon,\hat{\varepsilon})\}$ is a stationary sequence; its statistical properties depend on ε and $\hat{\varepsilon}$ only through the timing error $e=\varepsilon-\hat{\varepsilon}$ and are periodic in e with period 1.

The timing error detector output $\{x_k(\varepsilon,\hat{\varepsilon})\}$ is decomposed as the sum of its statistical expectation and a zero-mean disturbance $N_k(\varepsilon,\hat{\varepsilon})$, which is called the loop noise:

$$x_k(\varepsilon,\hat{\varepsilon}) = E[x_k(\varepsilon,\hat{\varepsilon})] + N_k(\varepsilon,\hat{\varepsilon}) \tag{2-70}$$

Taking into account the statistical properties of the timing error detector output sequence under open-loop conditions, one obtains

$$E[x_k(\varepsilon,\hat{\varepsilon})] = E[x_0(\varepsilon-\hat{\varepsilon},0)] = K_D\, g(\varepsilon-\hat{\varepsilon}) \tag{2-71}$$

$$E[N_k(\varepsilon,\hat{\varepsilon})\, N_{k+m}(\varepsilon,\hat{\varepsilon})] = E[N_0(\varepsilon-\hat{\varepsilon},0)\, N_m(\varepsilon-\hat{\varepsilon},0)] = R_m(\varepsilon-\hat{\varepsilon}) \tag{2-72}$$

The timing error detector characteristic $g(\cdot)$ and the autocorrelation sequence $\{R_m(\cdot)\}$ of the loop noise are periodic functions, with period 1, of the timing error $e=\varepsilon-\hat{\varepsilon}$. The loop noise spectrum $S(\omega;e)$ is defined as the discrete Fourier transform of the autocorrelation sequence $\{R_m(e)\}$:

$$S_N\left(e^{\delta\omega T};e\right) = \sum_{m=-\infty}^{+\infty} R_m(e)\,\exp\left(-j\omega mT\right) \tag{2-73}$$

Note that $S(\omega;e)$ is periodic in ω with period $2\pi/T$.

Assuming that the statistical properties of the timing error detector output sequence do not change when the loop is closed (this approximation is valid when the loop bandwidth is small with respect to the channel symbol rate $1/T$), the equivalent model of Figure 2-27 is obtained. This model takes into account that

$$\hat{\varepsilon}_{k+1} = \hat{\varepsilon}_k - \frac{T-T_0}{T} + K_0\, u_k \tag{2-74}$$

which follows from substituting $t_m = mT+\hat{\varepsilon}_m T$ and $K_0^{'} = K_0 T$ in (2-69). Because of the close similarity between the equivalent models for discrete-time and continuous-time synchronizers, the discussion pertaining to continuous-time synchronizers is also valid for discrete-time synchronizers.

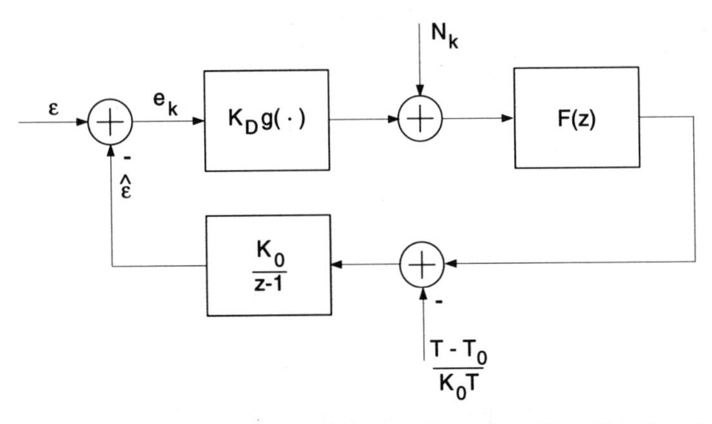

Figure 2-27 Equivalent Model of Discrete-Time Error-Tracking Synchronizer

The linearized model for discrete-time synchronizers is shown in Figure 2-28, where e_s denotes the steady-state timing error, which corresponds to a positive zero crossing of $g(e) - (T - T_0)/(TK_0 K_D F(1))$. An equivalent structure is shown in Figure 2-29, where the closed-loop frequency response $H(z)$ is given (in the z-domain) by

$$H(z) = \frac{K_0 K_D g'(e_s)\, F(z)}{z - 1 + K_0 K_D g'(e_s)\, F(z)} \tag{2-75}$$

The closed loop frequency response in the frequency domain is obtained by replacing z by $\exp\,(j\omega T)$. The one-sided loop bandwidth B_L (in hertz) is given by

$$B_L = \int_0^{\pi/T} |H(\exp\,(\omega T))|^2 \, \frac{d\omega}{2\pi} \tag{2-76}$$

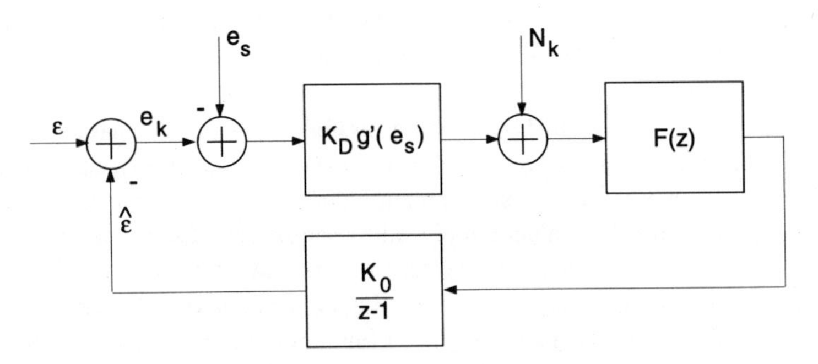

Figure 2-28 Linear Equivalent Model of Discrete-Time Error-Tracking Synchronizer

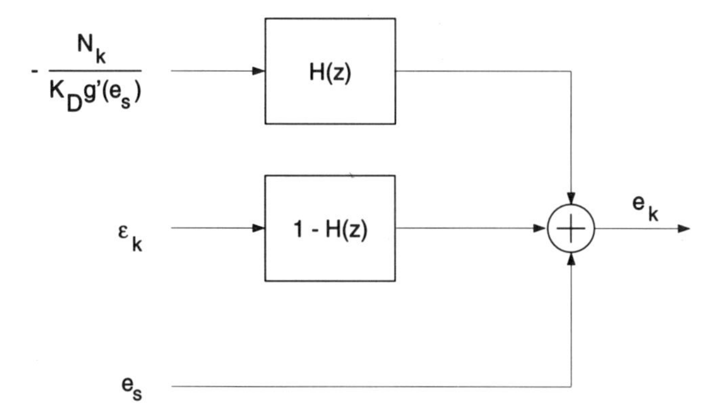

Figure 2-29 Contributions to the Timing Error e_k

while the timing error variance due to the loop noise equals

$$\mathrm{var}[e_k] = \frac{1}{\left[K_D g'(e_s)\right]^2} T \int_{-\pi/T}^{+\pi/T} \left|H(\exp(\omega T))\right|^2 S_N\left(e^{\delta\omega T}; e_s\right) \frac{d\omega}{2\pi} \tag{2-77}$$

The discussion about the influence of the closed-loop frequency response on the timing error variance of continuous-time synchronizers, is also valid for discrete-time synchronizers (see section 2.3.10).

2.3.7 Simulation of Error-Tracking Synchronizers

In Sections 2.3.2, 2.3.5, and 2.3.6 we have pointed out that the timing error detector characteristic, the loop noise power spectral density, and the linearized timing error variance can be evaluated analytically from the appropriate moments of the timing error detector output signal. However, depending on the baseband pulse of the received signal and the type of timing error detector, these computations might be quite involved; in this case, obtaining numerical results by means of computer simulation is much more appropriate than performing an analytical evaluation.

In the following, we indicate how the timing error detector characteristic (and its slope), the loop noise power spectral density, and the linearized timing error variance can be obtained from computer simulation. When using a computer simulation, continuous-time signals are replaced by sequences of samples; the sampling frequency in the simulation should be taken large enough so that frequency aliasing is negligibly small. Also, when simulating continuous-time filters by means of discrete-time finite impulse response (FIR) filters, the impulse response of the FIR filter should be taken long enough so that the continuous-time impulse response is closely approximated.

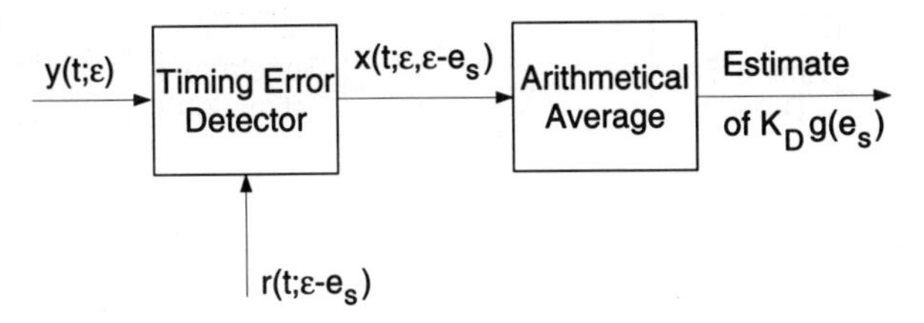

Figure 2-30 Configuration for Simulation of Timing Error Detector Characteristic

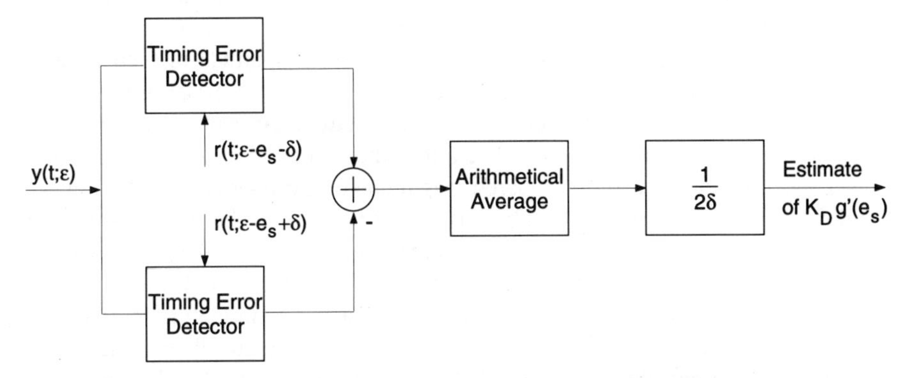

Figure 2-31 Configuration for Simulation of Timing Error Detector Slope

The configuration to be simulated for obtaining estimates of the timing error detector characteristic and its slope, corresponding to a particular value e_s of the timing error, are shown in Figures 2-30 and 2-31. The "arithmetical average" module measures the DC component of its input signal. Assuming that the loop noise power spectral density $S_N(\omega; e_s)$ is essentially flat in the vicinity of $\omega = 0$, the variance of the estimates is inversely proportional to the duration over which the arithmetical average is taken. In Figure 2-31, the expectation of the slope estimate is given by

$$K_D \frac{g(e_s + \delta) - g(e_s - \delta)}{2\delta} \cong K_D \left[g'(e_s) + \frac{\delta^2}{6} \, g^{(3)}(e_s) \right] \tag{2-78}$$

where $g^{(3)}(e_s)$ denotes the third derivative of $g(e)$, evaluated at $e = e_s$. Hence, δ should be taken small enough in order to limit the bias of the slope estimate, caused by the nonlinearity of the timing error detector characteristic. On the other hand, δ should not be taken too small, otherwise the difference of the timing error detector outputs is affected by the limited numerical precision of the computer (rounding errors).

An estimate of the loop noise power spectral density is obtained from the configuration shown in Figure 2-32. Subtracting the DC component $K_D \; g(e_s)$ from the timing error detector output is not strictly necessary, but avoids the

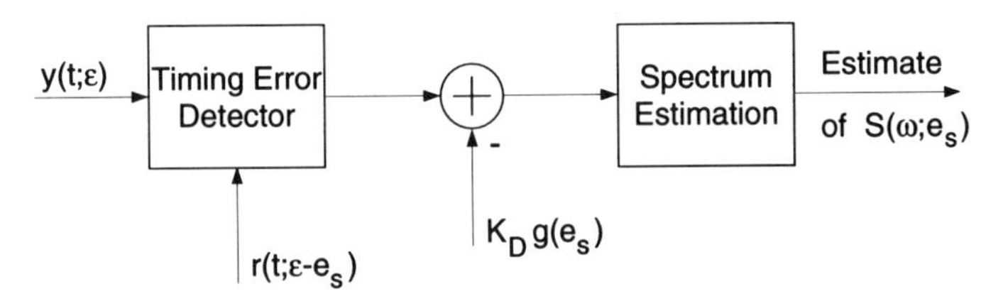

Figure 2-32 Configuration for Simulation of Loop Noise Power Spectral Density

occurrence of a spectral line at $\omega = 0$ in the spectrum estimate. Basically, the "spectrum estimation" module takes windowed segments of the input signal and computes the arithmetical average of the squared magnitude of their Fourier transforms. The variance of the spectrum estimate is inversely proportional to the number of segments. The expectation of the spectrum estimate is not the loop noise power spectral density itself, but rather the convolution of the loop noise power spectral density with the squared magnitude of the Fourier transform of the window. Hence, the spectrum estimate is biased because of the finite duration of the window; this bias can be reduced by increasing the window length. For more information about spectrum estimation, the reader is referred to [1–3].

The timing error variance var[e] can be estimated by simulating the configuration from Figure 2-33. The DC component $K_D g(e_s)$ is removed from the timing error detector output; the result is scaled by the slope $K_D g'(e_s)$ and then passed through a filter with frequency response $H(\omega)$, i.e., the closed-loop frequency response of the synchronizer. The signal at the output of this filter is the fluctuation $e - e_s$ of the linearized timing error with respect to the steady-state error e_s. The "variance estimation" module computes the arithmetical average of the square of its input signal. Because of the small value of $B_L T$, the signal at the output of the filter $H(\omega)$ has a long correlation time, of about $1/B_L T$ symbol intervals. Hence, in order to obtain an estimate of var[e] which has a small variance, the averaging interval in the variance estimation module should be several orders of magnitude larger than $1/(B_L T)$ symbol intervals.

The simulations described above pertain to the error-tracking synchronizer operating under open-loop conditions and are well-suited for obtaining the timing error detector characteristic, the loop noise power spectral density, and the linearized timing error variance. However, when investigating phenomena such as acquisition, hang-ups, and cycle slips, a simulation of the synchronizer operating in closed loop must be performed, in order to obtain the proper timing error

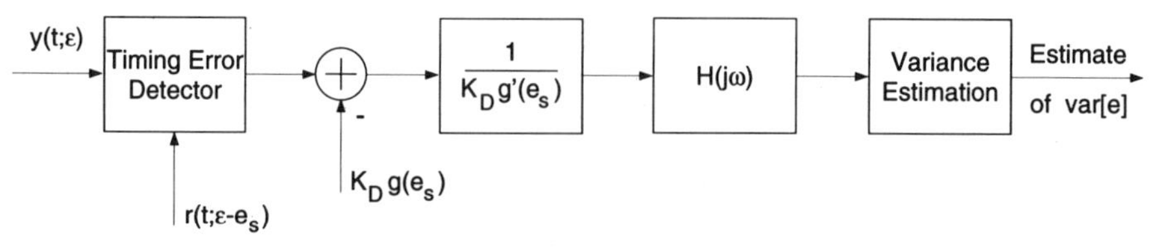

Figure 2-33 Configuration for Simulation of Timing Error Variance

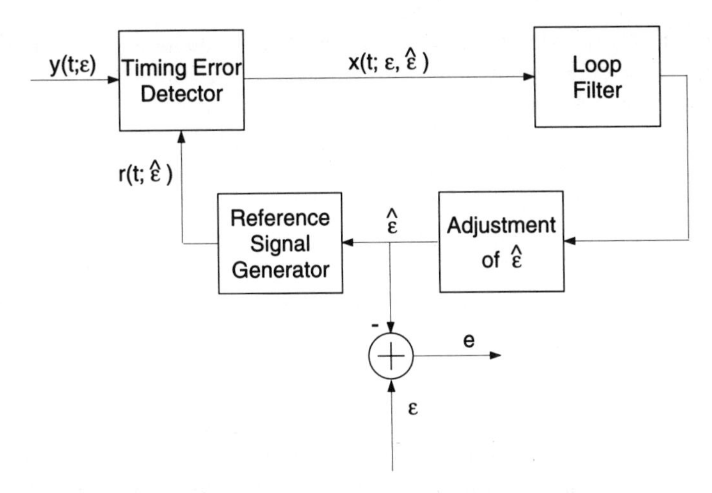

Figure 2-34 Configuration for Closed-Loop Synchronizer Simulation

trajectories. The configuration for the closed-loop simulation is shown in Figure 2-34. Although this closed-loop configuration has the same bandwidth as the open-loop configuration of Figure 2-33, the closed-loop simulation might be more time-consuming than the open-loop simulation. Indeed, because of the feedback in the closed-loop configuration, each module inside the loop necessarily operates sample-by-sample: after each input sample, the module produces one output sample. In the open-loop simulation, the different modules can operate on blocks of input samples, and produce a block of output samples for each block of input samples. In the latter case, the size of the blocks can be selected to minimize the computational load; such an optimization is not possible when a feedback loop is present.

2.3.8 Main Points

- Both continuous-time and discrete-time hybrid synchronizers can be represented by an equivalent model in which the timing error detector output is replaced by the sum of two terms: the first term is the useful term $K_D g(e)$, where the *timing error detector characteristic* $g(e)$ is a periodic function of the timing error $e = \varepsilon - \hat{\varepsilon}$. The second term is wide-sense stationary zero-mean *loop noise*, whose power spectral density $S_N(\omega; e)$ is a periodic function of the timing error e. For a specific synchronizer, $K_D g(e)$ and $S_N(\omega; e)$ can be obtained from an open-loop analysis of the timing error detector output, where both ε and $\hat{\varepsilon}$ are considered as constant parameters; as only the difference $e = \varepsilon - \hat{\varepsilon}$ matters, ε can be set equal to zero without loss of generality, when calculating $K_D g(e)$ and $S_N(\omega; e)$.
- Because of the periodicity of the timing error detector characteristic $g(e)$, error tracking synchronizers have an infinite number of stable and unstable equilibrium points. Under normal operating conditions, the loop noise causes

the error $e(t)$ to exhibit small random fluctuations about a stable equilibrium point. These small fluctuations are called *jitter.* Occasionally, the loop noise is strong enough during a sufficiently long time interval to push the error e from the close vicinity of a stable equilibrium point into the domain of attraction of a neighboring equilibrium point. This phenomenon is called a *cycle slip.* Cycle slips are very harmful to the reliability of the receiver's decisions, because a slip of the receiver clock corresponds to the repetition or omission of a channel symbol. When the initial value of the timing error is close to an unstable equilibrium point, *hang-ups* can occur: because the restoring force is very small near the unstable equilibrium point, the timing error spends a prolonged time in the vicinity of its initial value. Hang-ups give rise to long acquisition times.

- In most applications, the timing error variance caused by the loop noise is very small, so that it can be evaluated by means of the linearized equivalent model. When the loop noise spectrum variation within the loop bandwidth B_L is small, the timing error variance is essentially proportional to B_LT. However, in many cases of practical interest, the loop noise spectrum variation cannot be neglected. In this case, the timing error variance contains an extra term, which is proportional either to $(B_LT)^2$, when the magnitude of the closed-loop frequency response behaves asymptotically like $1/\omega$, or to $(B_LT)^3$, when the magnitude of the closed-loop frequency response decays faster than $1/\omega$ for large ω.
- Obtaining the timing error detector characteristic and the loop noise power spectral density by analytical means might be quite complicated. In this case, a simulation of the synchronizer under open-loop conditions is a valid alternative for obtaining numerical results about the timing error detector characteristic, the loop noise power spectral density, and the linearized timing error variance. For studying acquisition, hang-ups, and cycle slips, a simulation of the synchronizer in closed loop is needed.

2.3.9 Bibliographical Notes

The open-loop analysis in Section 2.3.2 of a general error-tracking synchronizer is an extension of [4] where the open-loop statistics have been derived for a conventional PLL operating on a periodic waveform, corrupted by cyclostationary additive noise (as is the case for a spectral line generating synchronizer).

In [5] hang-ups in a conventional PLL with sinusoidal phase error detector characteristic have been investigated, and an anti-hang-up circuit has been proposed. The theoretical analysis of this anti-hang-up circuit has been carried out in [6], and can also be found in Section 5.1 of Volume 1. For a sinusoidal (or any other continuous) characteristic, it is clear that hang-ups are caused by the small restoring force near the unstable equilibrium point. For a sequential logic phase error detector whose characteristic (in the absence of additive noise at the input of the PLL !) has a sawtooth shape (or, stated more generally, has a discontinuity at the unstable equilibrium point), the restoring force is largest near the unstable equilibrium point; in this case, it has been found in [7] that hang-ups still occur, because additive noise causes equivocation of the zero crossings of the signal at the input of the PLL. Using the general method outlined in Section 3.2.2 of Volume 1, it follows that this equivocation yields an average restoring force (averaging over noise at PLL input) which is continuous at the unstable equilibrium point for a finite signal-to-noise ratio. Hence, hang-up is again caused by the small average restoring force near the unstable equilibrium point.

An extension of the result from Section 2.3.5 has been obtained in [8]: when the magnitude of the closed-loop frequency response decays like ω^{-m} for large ω, and the variation $S_N(\omega; e_s) - S_N(0; e_s)$ of the loop noise spectrum about $\omega = 0$ behaves like ω^{2k}, its contribution to the timing error variance is proportional either to $(B_L T)^{2m}$ when $m \le k$, or to $(B_L T)^{2k+1}$ when $m > k$. For $k = 1$, this general result reduces to the case considered in Section 2.3.5.

2.3.10 Appendix

Continuous-Time Synchronizers

Using a Taylor series expansion of $h_2(0) - h_2(t)$ about $t = 0$, the second term in (2-57) can be transformed into

$$\begin{aligned} &2 \int_0^{+\infty} (h_2(0) - h_2(t)) \left(-R_N(t; e_s)\right) dt \\ &= 2 \sum_{m=1}^{\infty} \left[\frac{-h_2^{(m)}(0)}{m!} \right] \int_0^{+\infty} \left(-t^m \, R_N(t; e_s)\right) dt \end{aligned} \tag{2-79}$$

where

$$h_2^{(m)}(0) = \frac{d^m}{dt^m} h_2(t) \Big|_{t=0} \tag{2-80}$$

In order to investigate the dependence of (2-79) on $B_L T$, we put

$$|H(\omega)|^2 = \left| \tilde{H}\left(\frac{\omega}{B_L T} \right) \right|^2 \tag{2-81}$$

where $\tilde{H}(\omega)$ is a frequency response with a one-sided bandwidth equal to $1/T$; the closed-loop frequency response $H(\omega)$ is obtained by scaling the frequency, as indicated in (2-81). Inverse Fourier transforming (2-81) yields

$$h_2(t) = B_L T \, \tilde{h} \, (B_L T \, t) \tag{2-82}$$

where $\tilde{h}(t)$ is the inverse Fourier transform of $\left|\tilde{H}(\omega)\right|^2$. Using (2-82) in (2-80), we obtain

$$h_2^{(m)}(0) = (B_L T)^{m+1} \, \tilde{h}^{(m)}(0) \tag{2-83}$$

where

$$\tilde{h}_2^{(m)}(0) = \frac{d^m}{dt^m}\, \tilde{h}_2(t)\,\Big|_{t=0} \tag{2-84}$$

It follows that the mth term of the summation in (2-79) is proportional to $(B_L T)^{m+1}$. Hence, for $B_L T \ll 1$, the summation in (2-79) is dominated by the nonzero term with the smallest value of m.

Case of m=1

Using (2-56) yields

$$h_2^{(m)}(0) = \int\limits_0^{+\infty} h(u)\, h^{(m)}(u)\, du \tag{2-85}$$

where

$$h^{(m)}(0) = \frac{d^m}{dt^m}\, h(t)\,\Big|_{t=0} \tag{2-86}$$

For $m = 1$, we obtain from (2-85)

$$h_2^{(1)}(0) = \int\limits_0^{+\infty} h(u)\, h^{(1)}(u)\, du = -1/2h^2(0) \tag{2-87}$$

Hence, when $h(0) \neq 0$, the summation in (2-79) is dominated by the term with $m = 1$, which is proportional to $(B_L T)^2$:

$$2\int\limits_0^{+\infty} (h_2(0) - h_2(t))\,(-R_N(t; e_s))\, dt \cong h^2(0) \int\limits_0^{+\infty} (-t\, R_N(t; e_s))\, dt \tag{2-88}$$

According to the final value theorem,

$$h(0) = \lim_{s \to \infty} s\, H(s) \tag{2-89}$$

We obtain $h(0) \neq 0$ when $|H(\omega)|$ decays like $1/\omega$ for large ω, and $h(0) = 0$ when $|H(\omega)|$ decays faster than $1/\omega$ for large ω.

Case of m=2

Now we assune that $|H(\omega)|$ decays faster than $1/\omega$ for large ω, so that the term with $m = 1$ in (2-79) is identically zero. Let us consider the term with $m = 2$. Taking into account that $h(0) = 0$, we obtain from (2-85)

$$h_2^{(2)}(0) = \int\limits_0^{+\infty} h(u)\, h^{(2)}(u)\, du = -\int\limits_0^{+\infty} \left[h^{(1)}(u)\right]^2 du \tag{2-90}$$

Hence, the summation in (2-79) is for small $B_L T$ dominated by the term with $m = 2$, which is proportional to $(B_L T)^3$:

$$2\int_0^{+\infty} (h_2(0) - h_2(t))\,(-R_N(t;e_s))\,dt \cong \int_0^{+\infty} \left[h^{(1)}(u)\right]^2 du \int_0^{+\infty} \left(-t^2\,R_N(t;e_s)\right) dt \tag{2-91}$$

Taking into account that

$$\int_0^{+\infty} \left[h^{(1)}(u)\right]^2 du = \int_{-\infty}^{+\infty} \omega^2\,|H(\omega)|^2\,\frac{d\omega}{2\pi} \tag{2-92}$$

and

$$\int_0^{+\infty} \left(-t^2\,R_N(t;e_s)\right) dt = \frac{1}{2}\,S_N^{(2)}(0;e_s) \tag{2-93}$$

with

$$S_N^{(2)}(0;e_s) = \frac{d^2}{d\omega^2}\,S_N(\omega;e_s)\,\bigg|_{\omega=0} \tag{2-94}$$

we obtain

$$2\int_0^{+\infty} (h_2(0) - h_2(t))\,(-R_N(t;e_s))\,dt \cong \frac{1}{2}\,S_N^{(2)}(0;e_s) \int_{-\infty}^{+\infty} \omega^2\,|H(\omega)|^2\,\frac{d\omega}{2\pi} \tag{2-95}$$

Discrete-Time Synchronizers

Denoting the inverse z-transform of the closed-loop frequency response $H(z)$ by $\{h_k\}$, the linearized timing error e_k satisfies the following equation:

$$e_k - e_s = \frac{1}{K_D g'(e_s)} \sum_m h_{k-m}\,N_m(e_s) \tag{2-96}$$

where $N_m(e_s)$ is a short-hand notation for $N_m(\varepsilon, \varepsilon - e_s)$, and e_s denotes the steady-state timing error. When $H(z)$ is a rational function of z, it is possible to define a frequency response $H_{\text{eq}}(s)$ in the Laplace domain, which is rational in s, and such that its inverse Laplace transform $h_{\text{eq}}(t)$ satisfies

$$h_{\text{eq}}(kT - MT) = \frac{1}{T}\,h_k \tag{2-97}$$

where M is a suitable nonnegative integer. This is illustrated in the following example.

Example

For $H(z) = (1-a)/(z-a)$ with $0 \le a < 1$, h_k is given by

$$h_k = \begin{cases} (1-a)\, a^{k-1} & k > 0 \\ 0 & k \le 0 \end{cases} \tag{2-98}$$

Now defining

$$h_{\text{eq}}(t) = \begin{cases} \dfrac{1-a}{T} \exp\left[-\frac{t}{T} \ln\left(\frac{1}{a}\right)\right] & t \ge 0 \\ 0 & t = 0 \end{cases} \tag{2-99}$$

so that (2-97) holds with $M = 1$, we obtain

$$H_{\text{eq}}(s) = \frac{1-a}{T} \frac{1}{s + 1/T \ln 1/a} \tag{2-100}$$

which is a rational function of s. Because of (2-97), the Fourier transforms $H(\exp(\omega T))$ and $H_{\text{eq}}(\omega)$ are related by

$$H(\exp(\omega T)) = \sum_k H_{\text{eq}}\left(\omega + j\frac{2\pi k}{T}\right) \exp(-\omega MT - j2\pi Mk) \tag{2-101}$$

When the normalized loop bandwidth $B_L T$ satisfies $B_L T \ll 1$, $H_{\text{eq}}(\omega)$ is essentially zero for $|\omega| > \pi/T$; hence, it follows from (2-101) that

$$H_{\text{eq}}(\omega) \exp(-\omega MT) \cong \begin{cases} H(\exp(\omega T)) & \text{for} \quad |\omega| \le \pi/T \\ 0 & \text{for} \quad |\omega| > \pi/T \end{cases} \tag{2-102}$$

is a very accurate approximation.

Now the summation in (2-96) can be transformed into

$$\sum_m h_{k-m}\, N_m(e_s) = \int_{-\infty}^{+\infty} h_{\text{eq}}(t-u)\, N_{eq}(u; e_s)\, du \bigg|_{t=kT} \tag{2-103}$$

where

$$N_{\text{eq}}(t; e_s) = T \sum_m N_m(e_s)\, \delta(t - mT - MT) \tag{2-104}$$

Hence, the discrete-time synchronizer with closed-loop frequency response $H(z)$ and loop noise $\{N_m(e_s)\}$ yields the same timing error variance as a continuous-time synchronizer with closed-loop frequency response $H_{\text{eq}}(s)$ and loop noise $N_{\text{eq}}(t; e_s)$. The corresponding loop noise autocorrelation function $R_{\text{eq}}(u; e_s)$ for the continuous-time synchronizer is given by

$$\begin{aligned} R_{\text{eq}}(u; e_s) &= \langle E[N_{\text{eq}}(t; e_s)\, N_{\text{eq}}(t+u; e_s)] \rangle_t \\ &= T \sum_m R_m(e_s)\, \delta(t - mT) \end{aligned} \tag{2-105}$$

where $R_m(e_s) = E[N_n(e_s)\, N_{m+n}(e_s)]$. Using (2-102) and (2-105) in the timing error variance expression for continuous-time synchronizers, the following results are obtained for $B_L T \ll 1$.

Case 1: For large ω, $|H_{\text{eq}}(\omega)|$ decays like $1/\omega$; equivalently, $h_M \neq 0$:

$$\text{var}[e] \cong \frac{(2B_L T)\, S_N(0; e_s) + h_M^2 \sum\limits_{k=1}^{\infty} (-k R_k(e_s))}{\left[K_D\, g'(e_s)\right]^2} \tag{2-106}$$

The second term in the numerator is proportional to $(B_L T)^2$.

Case 2: For large ω, $|H_{\text{eq}}(\omega)|$ decays faster than $1/\omega$; equivalently, $h_M = 0$:

$$\text{var}[e] \cong \frac{(2B_L T)\, S_N(0; e_s) + \frac{1}{2}\, S_N^{(2)}(0; e_s) \int\limits_{-\pi}^{+\pi} x^2\, |H(\exp(x))|^2\, 1/(2\pi)\, dx}{\left[K_D\, g'(e_s)\right]^2} \tag{2-107}$$

where

$$\begin{aligned} S_N^{(2)}(0; e_s) &= \frac{d^2}{d(\omega T)^2}\, S_N(\omega; e_s) \Big|_{\omega=0} \\ &= \sum_{k=-\infty}^{+\infty} (-k^2\, R_k(e_s)) \end{aligned} \tag{2-108}$$

The second term in the numerator is proportional to $(B_L T)^3$.

Hence, for $B_L T \ll 1$, the timing error variance resulting from a discrete-time synchronizer with closed-loop frequency response $H(z)$ behaves like the timing error variance resulting from a continuous-time synchronizer with closed-loop frequency response $H_{\text{eq}}(s)$.

Bibliography

[1] L. R. Rabiner and B. G. Gold, *Theory and Application of Digital Signal Processing*. Englewood Cliffs, NJ: Prentice-Hall, 1975.

[2] A. V. Oppenheim and R. W. Shafer, *Digital Signal Processing*. Englewood Cliffs, NJ: Prentice-Hall, 1975.

[3] F. J. Harris, "On the Use of Windows for Harmonic Analysis with Discrete Fourier Transform," *Proc. IEEE*, vol. 66, pp. 51–83, Jan. 1978.

[4] M. Moeneclaey, "Linear Phase-Locked Loop Theory for Cyclostationary Input Disturbances," *IEEE Trans. Commun.*, vol. COM-30, pp. 2253–2259, Oct. 1982.

[5] F. M. Gardner, "Hangup in Phase-Lock Loops," *IEEE Trans. Commun.*, vol. COM-25, pp. 1210–1214, Oct. 1977.

[6] H. Meyr and L. Popken, "Phase Acquisition Statistics for Phase-Locked Loops," *IEEE Trans. Commun.* , vol. COM-28, pp. 1365–1372, Aug. 1980.

[7] F. M. Gardner, "Equivocation as a Cause of PLL Hangup," *IEEE Trans. Commun.*, vol. COM-30, pp. 2242–2243, Oct. 1980.

[8] M. Moeneclaey, "The Optimum Closed-Loop Transfer Function of a Phase-Locked Loop for Synchronization Purposes," *IEEE Trans. Commun.*, vol. COM-31, pp. 549–553, Apr. 1983.

2.4 Spectral Line Generating Clock Synchronizers

This section deals with clock synchronizers which make use of a *nonlinear* operation on the received noisy PAM waveform in order to generate *spectral lines* at the channel symbol rate $1/T$ and at multiples thereof. The spectral line at the channel symbol rate can be isolated by means of a PLL with a multiplying timing error detector and a sinusoidal reference signal or by means of a narrowband bandpass filter tuned to the channel symbol rate. Both cases yield a nearly sinusoidal clock signal at the symbol rate, which is used to control the sampler in the decision branch of the receiver. When the local reference signal is not sinusoidal (but, for instance, a square wave), then also the harmonics at multiples of $1/T$, present at the output of the nonlinearity, contribute to the clock signal.

2.4.1 Nonlinearity Followed by a PLL with Multiplying Timing Error Detector

Description of the Synchronizer

The general structure of a spectral line generating synchronizer making use of a PLL is shown in Figure 2-35. The input signal $y(T;\varepsilon)$ is a noisy PAM signal, given by

$$y(t;\varepsilon) = \sum_m a_m \, g(t - mT - \varepsilon T) + n(t) \tag{2-109}$$

where $\{a_m\}$ is a stationary sequence of zero-mean channel symbols, $g(t)$ is the baseband PAM pulse, $1/T$ is the channel symbol rate, εT is a fractional delay to be estimated by the synchronizer, and $n(t)$ is stationary additive noise. The input signal enters a time-invariant (not necessarily memoryless) nonlinearity, whose output is denoted by $v(t;\varepsilon)$. When the nonlinearity has memory, it is assumed

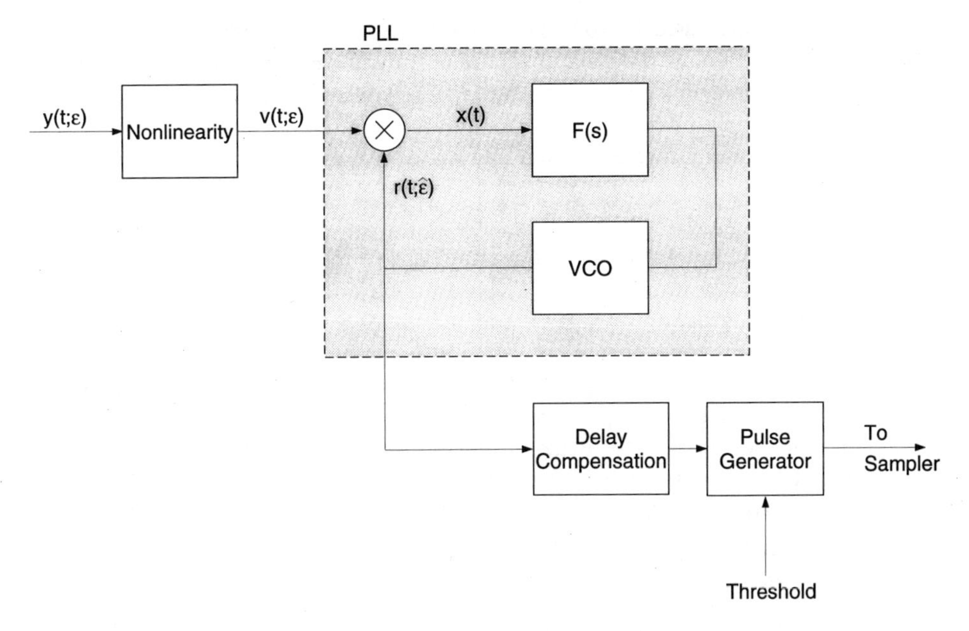

Figure 2-35 General Spectral Line Generating Synchronizer Using a PLL

that the possibly time-varying delay εT is nearly constant over the memory time of the nonlinearity, so that $v(t;\varepsilon)$ is determined essentially by the instantaneous value of ε, and not by past values of ε. The nonlinearity output signal $v(t;\varepsilon)$ and the local reference signal $r(t;\hat{\varepsilon})$ of the PLL are multiplied. This yields the timing error detector output signal $x(t)$, given by

$$x(t) = K_m v(t;\varepsilon)\, r(t;\hat{\varepsilon}) \tag{2-110}$$

where K_m denotes the multiplier gain. The timing error detector output signal is filtered by the loop filter and fed back to the VCO. The VCO modifies the estimate $\hat{\varepsilon}$ so as to reduce the timing error $e = \varepsilon - \hat{\varepsilon}$. The reference signal $r(t;\hat{\varepsilon})$ acts as a clock signal which controls the sampler in the decision branch of the receiver.

In most cases, the sampler in the decision branch of the receiver is activated by a narrow clock pulse. Therefore, the local reference signal enters a pulse generator, which generates a clock pulse each time the reference signal crosses a given threshold level. Ideally, the threshold level should be zero, so that the clock pulses coincide with the positive zero crossings of the local reference signal; however, due to DC offsets, the actual clock pulses correspond to the positive crossings of a nonzero threshold level.

In general, the nominal zero crossings of the local reference signal are biased with respect to the ideal sampling instants in the decision branch; this bias depends on characteristics which are known to the receiver (such as the baseband PAM pulse and the type of nonlinearity), and not on the unknown delay εT. Therefore, the bias can be compensated for, by passing the reference signal, before entering the pulse generator, through a delay circuit which shifts the nominal zero crossings to the correct position; in the case of a sinusoidal reference signal, this compensation network is a simple phase shifter. Alternatively, the delay compensation can be applied to the pulse generator output signal.

Analysis of the Synchronizer

The nonlinearity plus PLL in Figure 2-35 can be redrawn as an error-tracking synchronizer, shown in Figure 2-18, whose timing error detector is the cascade of the nonlinearity acting on $y(t;\varepsilon)$, and the multiplier of the PLL. Hence, the spectral line generating synchronizer with a PLL can be analyzed by means of the equivalent model, shown in Figure 2-19. The useful timing error detector output $K_D g(e)$ and the loop noise spectrum $S_N(\omega; e)$ are determined from the statistical properties of the timing error detector output signal under open-loop conditions (implying that ε and $\hat{\varepsilon}$ are held constant); more specifically, (2-40) and (2-41) must be applied.

Substituting (2-110) into (2-40) and (2-41) yields the following expressions for $K_D g(e)$ and for the loop noise autocorrelation function $R_N(u;e)$:

$$K_D g(e) = K_m \left\langle E[v(t;\varepsilon)]\, r(t;\hat{\varepsilon})\right\rangle_t \tag{2-111}$$

$$R_N(u;e) = K_m^2 \left[\left\langle K_v(t,t+u;\varepsilon)\, r(t;\hat{\varepsilon})\, r(t+u;\hat{\varepsilon})\right\rangle_t\right]_{\text{LP}} \tag{2-112}$$

where $K_v(t,t+u;\varepsilon)$ is the autocovariance function of $v(t;\varepsilon)$:

$$K_v(t,t+u;\varepsilon) = E[v(t;\varepsilon)\, v(t+u;\varepsilon)] - E[v(t;\varepsilon)]\, E[v(t+u;\varepsilon)]$$

and $e = \varepsilon - \hat{\varepsilon}$ denotes the timing error. In the above, $\langle ... \rangle_t$ indicates averaging over t, and $[...]_{\text{LP}}$ denotes filtering by means of an ideal lowpass filter with a bandwidth (in hertz) of $1/(2T)$. Because of the cyclostationarity of $y(t;\varepsilon)$ and the time invariance of the nonlinearity operating on $y(t;\varepsilon)$, the nonlinearity output signal $v(t;\varepsilon)$ is also cyclostationary with period T. Hence, the statistical expectation $E[v(t;\varepsilon)]$ and the autocovariance function $K_v(t,t+u;\varepsilon)$ are both periodic in t with period T, and can be expanded into a Fourier series:

$$E[v(t;\varepsilon)] = \sum_{n=-\infty}^{+\infty} m_{v,n} \exp\left(j\,\frac{2\pi n}{T}\,(t-\varepsilon T)\right) \tag{2-113}$$

$$K_v(t,t+u;\varepsilon) = \sum_{n=-\infty}^{+\infty} k_{v,n}(u) \exp\left(j\,\frac{2\pi n}{T}\,(t-\varepsilon T)\right) \tag{2-114}$$

It is important to note that the Fourier coefficients $m_{v,n}$ and $k_{v,n}(u)$ do not depend on ε; this is because $y(t;\varepsilon)$ and $y(t-\varepsilon T;0)$, and, hence, also $v(t;\varepsilon)$ and $v(t-\varepsilon T;0)$, have identical statistical properties. The periodicity of $E[v(t;\varepsilon)]$ indicates the presence of *spectral lines* at the output of the nonlinearity. Under open-loop conditions, $r(t;\hat{\varepsilon})$ is also periodic in t with period T, and can be expanded as

$$r(t;\hat{\varepsilon}) = \sum_{n=-\infty}^{+\infty} c_n \exp\left(j\,\frac{2\pi n}{T}\,(t-\hat{\varepsilon}T)\right) \tag{2-115}$$

where c_n does not depend on $\hat{\varepsilon}$, because $r(t;\hat{\varepsilon}) = r(t-\hat{\varepsilon}T;0)$. Substituting (2-113) – (2-115) into (2-111) and (2-112) yields

$$K_D g(e) = K_m \sum_n c_n\, m_{v,-n}\, \exp(j2\pi n e) \tag{2-116}$$

$$R_N(u;e) = K_m^2 \left[\sum_{m,n} k_{v,-n}(u)\, c_m\, c_{n-m}\, \exp\left(j\,\frac{2\pi m}{T}\,u\right)\, \exp(j2\pi n e)\right]_{\mathrm{LP}}$$

Taking into account that the loop noise spectrum $S_N(\omega;e)$ is the Fourier transform of $R_N(u;e)$, we obtain

$$S_N(\omega;e) = K_m^2\, |H_{\mathrm{LP}}(\omega)|^2 \sum_{m,n} S_{v,-n}\left(\omega - \frac{2\pi m}{T}\right) c_m\, c_{n-m}\, \exp(j2\pi n e) \tag{2-117}$$

where $S_{v,n}(\omega)$ is the Fourier transform of $k_{v,n}(u)$, and $H_{\mathrm{LP}}(\omega)$ is the frequency response of an ideal lowpass filter with one-sided bandwidth (in hertz) equal to $1/(2T)$ [see (2-36)]. It follows from (2-116) that there is a contribution to the useful component $K_D g(e)$ of the timing error detector output only from those harmonics which are present in both $v(t;\varepsilon)$ and $r(t;\hat{\varepsilon})$. This is according to our intuition: the multiplication of two sinusoids gives rise to a DC component only when both sinusoids have the same frequency.

Let us consider the special case where the local reference signal $r(t;\hat{\varepsilon})$ is a sinusoid with frequency $1/T$. Then only the terms from (2-113) with $n=1$ and $n=-1$ contribute to (2-116). Expressing $m_{v,1}$ in polar form, i.e., $m_{v,1} = \left(A/\sqrt{2}\right)\,\exp\,(j\psi)$, (2-113) reduces to

$$E[v(t;\varepsilon)] = \sqrt{2}\,A\,\cos\left(\frac{2\pi}{T}\,(t-\varepsilon T)+\psi\right) + \text{other terms} \tag{2-118}$$

where "other terms" consist of a DC component and components at multiples of the channel symbol rate $1/T$, which can all be neglected as far as the operation of the synchronizer is concerned. The local reference signal is defined as

$$r(t;\hat{\varepsilon}) = \sqrt{2}\,K_r\,\sin\left(\frac{2\pi}{T}\,(t-\hat{\varepsilon}T)+\psi\right) \tag{2-119}$$

Identification of (2-119) with (2-115) yields

$$c_1 = c_{-1}^* = \frac{\sqrt{2}\,K_r}{2j}\,\exp(j\psi)$$
$$c_n = 0 \quad \text{for} \quad |n| \neq 1$$

Hence, for a sinusoidal reference signal, (2-116) and (2-117) reduce to

$$K_D g(e) = K_m\ K_r\ A\ \sin(2\pi e) \tag{2-120}$$

$$\begin{aligned} S_N(\omega;e) = {} & \frac{K_m^2\ K_r^2}{2}\ |H_{\mathrm{LP}}(\omega)|^2 \\ & \times \left[S_{v,0}\left(\omega - \frac{2\pi}{T}\right) + S_{v,0}\left(\omega + \frac{2\pi}{T}\right) \right. \\ & \quad - S_{v,-2}\left(\omega - \frac{2\pi}{T}\right)\ \exp(2j\psi)\ \exp(j4\pi e) \\ & \quad \left. - S_{v,2}\left(\omega + \frac{2\pi}{T}\right)\ \exp(-2j\psi)\ \exp(-j4\pi e) \right] \end{aligned} \tag{2-121}$$

From (2-120), we observe that the timing error detector characteristic has a sinusoidal shape. Note that the introduction of the phase ψ in the definition (2-119) of the local reference signal guarantees that $g(0) = 0$. It follows from (2-121) that the loop noise spectrum $S_N(\omega; e)$ is periodic in the timing error e with period $1/2$, whereas the general theory from Section 2.3.2 predicts a periodicity with period 1. The periodicity with period $1/2$ comes from the specific nature of the local reference signal (2-119). Indeed, as

$$r(t;\hat{\varepsilon}) = -r(t;\hat{\varepsilon} + 1/2)$$

it follows that increasing $\hat{\varepsilon}$ by $1/2$ changes only the sign of the loop noise, so that the loop noise spectrum is not affected. In general, this observation holds for any local reference signal with period T, containing harmonics only at odd multiples of $1/T$.

In the case of a sinusoidal reference signal, and assuming a steady-state timing error equal to e_s, the linearized equivalent model (see Figure 2-23) yields the following timing error variance, due to the loop noise:

$$\begin{aligned} \mathrm{var}[e(t)] = {} & \frac{1}{K_s^2} \int_{-\infty}^{+\infty} |H(\omega)|^2 \\ & \times \left[S_{v,0}\left(\omega - \frac{2\pi}{T}\right) + S_{v,0}\left(\omega + \frac{2\pi}{T}\right) - S_{v,-2}\left(\omega - \frac{2\pi}{T}\right)\ \exp(2j\psi)\ \exp(j4\pi e_s) \right. \\ & \left. - S_{v,2}\left(\omega + \frac{2\pi}{T}\right)\ \exp(-2j\psi)\ \exp(-j4\pi e_s) \right] \frac{d\omega}{2\pi} \end{aligned} \tag{2-122}$$

where

$$K_s = 2\sqrt{2}\ \pi\ A \cos(2\pi e_s)$$

and

$$H(s) = \frac{(2\pi K_m K_r A \cos(2\pi e_s))\ K_0 F(s)}{s + (2\pi K_m K_r A \cos(2\pi e_s))\ K_0 F(s)}$$

is the PLL's closed-loop frequency response in the Laplace domain. Note from (2-120) that the timing error detector slope at $e = e_s$ equals $2\pi K_m K_r A \cos(2\pi e_s)$.

It is most instructive to reconsider the special case where the local reference signal is a sinusoid with period T, by using complex envelope representation. The nonlinearity output signal is decomposed as

$$\begin{aligned} v(t;\varepsilon) =& E[v(t;\varepsilon)] + w(t;\varepsilon) \\ =& \sqrt{2}\, A \cos\left(\frac{2\pi}{T}(t-\varepsilon T)+\psi\right) \\ &+ \sqrt{2}\,\mathrm{Re}\left[w_L(t;\varepsilon)\exp(j\psi)\exp\left(j\frac{2\pi}{T}(t-\varepsilon T)\right)\right] \\ &+ \text{other terms} \end{aligned} \tag{2-123}$$

where "other terms" are the same as in (2-118), and $w_L(t;\varepsilon)$ denotes the *complex envelope* of the zero-mean cyclostationary disturbance $w(t;\varepsilon)$ with respect to the sinusoid with period T at the output of the nonlinearity [which is the first term of (2-123)]. It follows from Section 1.2.2 that $w_L(t;\varepsilon)$ is also cyclostationary with period T. Under open-loop conditions, the loop noise is given by

$$\begin{aligned} N(t;\varepsilon,\hat{\varepsilon}) &= K_m[w(t;\varepsilon)\, r(t;\hat{\varepsilon})]_{\mathrm{LP}} \\ &= -K_m K_r \mathrm{Im}[[w_L(t;\varepsilon)]_{\mathrm{LP}}\exp(-j2\pi e)] \\ &= -K_m K_r\left([w_S(t;\varepsilon)]_{\mathrm{LP}}\cos(2\pi e)\right. \\ &\qquad \left. - [w_C(t;\varepsilon)]_{\mathrm{LP}}\sin(2\pi e)\right) \end{aligned} \tag{2-124}$$

where $[...]_{\mathrm{LP}}$ indicates that the signal between brackets is passed through an ideal lowpass filter with one-sided bandwidth (in hertz) equal to $1/(2T)$. The signals $w_C(t;\varepsilon)$ and $w_S(t;\varepsilon)$ denote the in-phase and quadrature components of the disturbance $w(t;\varepsilon)$ with respect to the useful periodic component at the output of the nonlinearity:

$$w_L(t;\varepsilon) = w_C(t;\varepsilon) + jw_S(t;\varepsilon)$$

Note that $[w_L(t;\varepsilon)]_{\mathrm{LP}}$, $[w_C(t;\varepsilon)]_{\mathrm{LP}}$, and $[w_S(t;\varepsilon)]_{\mathrm{LP}}$ are *wide-sense stationary* processes [because their bandwidth (in hertz) does not exceed $1/(2T)$], with the same statistical properties as $[w_L(t-\varepsilon T;0)]_{\mathrm{LP}}$, $[w_C(t-\varepsilon T;0)]_{\mathrm{LP}}$ and $[w_S(t-\varepsilon T;0)]_{\mathrm{LP}}$, respectively [because $v(t;\varepsilon)$ and $v(t-\varepsilon T;0)$ have identical statistical properties]. Therefore, the spectra and cross-spectra of the in-phase and quadrature disturbances, which are given by

$$\begin{aligned} S_C(\omega) &= \text{ Fourier transform of } \langle E[w_C(t;\varepsilon)\, w_C(t+u;\varepsilon)]\rangle_t \\ S_S(\omega) &= \text{ Fourier transform of } \langle E[w_S(t;\varepsilon)\, w_S(t+u;\varepsilon)]\rangle_t \\ S_{CS}(\omega) &= \text{ Fourier transform of } \langle E[w_C(t;\varepsilon)\, w_S(t+u;\varepsilon)]\rangle_t \\ S_{SC}(\omega) &= \text{ Fourier transform of } \langle E[w_S(t;\varepsilon)\, w_C(t+u;\varepsilon)]\rangle_t \\ &= S_{CS}^*(\omega) \end{aligned} \tag{2-125}$$

do not depend on ε. The power spectral density $S_N(\omega; e)$ of the loop noise is easily derived from (2-124). The result is

$$\begin{aligned} S_N(\omega; e) = K_m^2 K_r^2 \left|H_{\mathrm{LP}}(\omega)\right|^2 \frac{1}{2} \Big(S_C(\omega) + S_S(\omega) \\ - \left(S_C(\omega) - S_S(\omega)\right) \cos(4\pi e) \\ - \left(S_{SC}(\omega) + S_{CS}(\omega)\right) \sin(4\pi e) \Big) \end{aligned} \tag{2-126}$$

When using the linearized equivalent model, $S_N(\omega; e)$ is replaced by $S_N(\omega; e_s)$, where e_s denotes the steady-state timing error. Let us consider the cases $e_s = 0$ and $e_s = 1/4$:

$$S_N(\omega; e_s = 0) = K_m^2 K_r^2 \left|H_{\mathrm{LP}}(\omega)\right|^2 S_S(\omega) \tag{2-127}$$

$$S_N(\omega; e_s = 1/4) = K_m^2 K_r^2 \left|H_{\mathrm{LP}}(\omega)\right|^2 S_C(\omega) \tag{2-128}$$

Let us assume for a moment that the delay εT to be estimated is constant, so that only the loop noise gives rise to jitter. Then (2-127) and (2-128) indicate that the random fluctuations of the timing error about its steady-state value e_s are caused by the *quadrature* disturbance $w_S(t;\varepsilon)$ when $e_s = 0$, or by the *in-phase* disturbance $w_C(t;\varepsilon)$ when $e_s = 1/4$. This is consistent with (2-124):

$$N(t; \varepsilon, \hat{\varepsilon}) = \begin{cases} -K_m K_r \left[w_S(t;\varepsilon)\right]_{\mathrm{LP}} & e = 0 \\ -K_m K_r \left[w_C(t;\varepsilon)\right]_{\mathrm{LP}} & e = 1/4 \end{cases}$$

For arbitrary values of e_s, the random fluctuations of the timing error about its steady-state value are caused by a linear combination of the in-phase and quadrature disturbance, as indicated by (2-124). The corresponding timing error variance in terms of the in-phase and quadrature spectra and cross-spectra is obtained from the general expression (2-126). When the delay εT is not constant, the timing error e contains an extra term, which is obtained by passing $\varepsilon(t)$ through a highpass filter whose frequency response in the Laplace domain equals $1 - H(s)$, as indicated in Figure 2-23.

Until now, we have considered the timing error e at the PLL. However, what really matters are the sampling instants in the decision branch of the receiver. Taking DC offsets into account, the sampling instants in the decision branch correspond to the positive crossings of a nonzero threshold level by the local reference signal. Assuming a sinusoidal reference signal given by (2-119), the threshold level is denoted by $\sqrt{2}\eta K_r$. The positive level crossing instants are determined by

$$\exp\left[j\left(\frac{2\pi}{T}\left(t_{\eta,k} - \varepsilon T\right) + \psi + 2\pi e(t_{\eta,k})\right)\right] = \exp\left(j\psi_\eta\right) \tag{2-129}$$

where $t_{\eta,k}$ denotes the kth *crossing instant*, and

$$\exp(j\psi_\eta) = \sqrt{1-\eta^2} + j\eta \tag{2-130}$$

The level crossing instants $t_{\eta,k}$ satisfying (2-129) and (2-130) are not a set of equidistant points on the time axis, because of the fluctuations of the timing error e and, possibly, of ε. Let us consider the following decomposition:

$$t_{\eta,k} = T_{\eta,k} + \Delta T_{\eta,k}$$

where $T_{\eta,k}$ would be the kth level crossing instant, if $\varepsilon(t)$ and $e(t)$ were equal to their mean values ε_0 and e_s, while $\Delta T_{\eta,k}$ denotes the variation of the actual kth crossing instant, caused by the fluctuations $\Delta\varepsilon(t)$ and $\Delta e(t)$ of $\varepsilon(t)$ and $e(t)$ about their mean values. It follows from (2-129) and (2-130) that

$$T_{\eta,k} = kT + \varepsilon_0 T - e_s T - \frac{\psi T}{2\pi} + \frac{\psi_\eta T}{2\pi} \tag{2-131}$$

$$\Delta T_{\eta,k} = T(\Delta\varepsilon - \Delta e) \tag{2-132}$$

In (2-132), $\Delta\varepsilon$ and Δe are evaluated at the actual level crossing instant $t_{\eta,k}$, which in turn depends on the fluctuation $\Delta T_{\eta,k}$, so that (2-132) is only an implicit equation for $\Delta T_{\eta,k}$. However, when the fluctuations $\Delta\varepsilon(t)$ and $\Delta e(t)$ are small, the actual level crossing instant $t_{\eta,k}$ is close to the instant $T_{\eta,k}$. Therefore, an accurate approximation is obtained by evaluating in (2-132) the quantities $\Delta\varepsilon$ and Δe at the instant $T_{\eta,k}$ instead of the actual crossing instant $t_{\eta,k}$; this yields an explicit equation for $\Delta T_{\eta,k}$.

The actual zero crossing instants $t_{\eta,k}$ can be decomposed as

$$t_{\eta,k} = T_k - \Delta t_{\eta,k}$$

where

$$T_k = kT + \varepsilon T - \frac{\psi T}{2\pi}$$

$$\Delta t_{\eta,k} = -\frac{\psi_\eta T}{2\pi} + eT$$

In the above, $\varepsilon(t)$ and $e(t)$ are evaluated at $t = T_{\eta,k}$. The *ideal* crossing instant T_k corresponds to the kth positive zero crossing of the sinusoidal reference signal, when the timing error $e(t)$ is identically zero, while the *deviations* $\Delta t_{\eta,k}$ incorporate the effect of the loop noise, of the steady-state error e_s, of the nonzero threshold parameter η, and of the possibly time-varying delay εT. Assuming that the time origin has been chosen such that the ideal sampling instants in the decision branch are given by $\{kT + \varepsilon T\}$, it follows that the instants $\{T_k\}$ contain a bias term $-\psi T/2\pi$. This bias can be compensated for by applying a phase shift ψ to the local reference signal before entering the pulse generator. Taking into account (2-124), and making use of the linearized equivalent model, Figure 2-36 shows how the deviations $\Delta t_{\eta,k}$ from the ideal crossing instants T_k are influenced by the

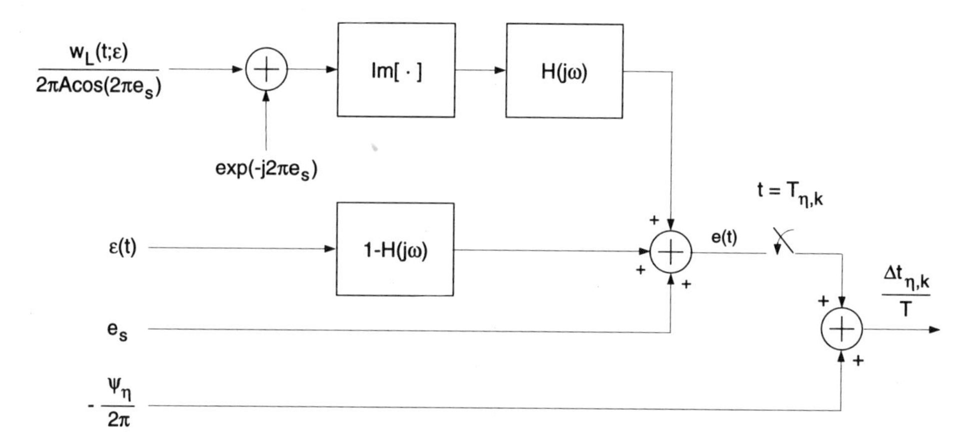

Figure 2-36 Identification of Contributions to Deviation $\Delta t_{\eta,k}$ (Spectral Line Generating Synchronizer Using a PLL)

disturbance at the output of the nonlinearity, by the steady-state error e_s, by the nonzero threshold parameter η, and by the possibly time-varying delay εT. Note that the nonzero threshold level gives rise to a bias term but has no effect on the random fluctuations of the level crossing instants (this is true also when the local reference signal is not sinusoidal). Therefore, $\text{var}[\Delta t_{\eta,k}/T]$ equals $\text{var}[e(T_{\eta,k})]$. As the lowpass content of $w_L(t;\varepsilon)$ is wide-sense stationary, its contribution to this variance does not depend on the position of the instants $\{T_{\eta,k}\}$ and is given by (2-122).

An Erroneous Method of Analysis

As a warning for the readers, we now outline a method of analysis which gives rise to *erroneous* results but which nevertheless has been used in a number of research papers. The method consists of replacing the cyclostationary disturbance $w(t;\varepsilon) = v(t;\varepsilon) - E[v(t;\varepsilon)]$ at the input of the PLL, whose autocorrelation function is the autocovariance function $K_v(t, t+u;\varepsilon)$ of $v(t;\varepsilon)$, by an "equivalent" stationary disturbance with autocorrelation function $\langle K_v(t, t+u;\varepsilon)\rangle_t$. The stationary process is equivalent with the cyclostationary process in the sense that both processes have the same power spectral density. [Remember that the power spectral density of a cyclostationary process is defined as the Fourier transform of the time-average of the autocorrelation function (see Section 1.1.2)]. Also, a spectrum analyzer in the laboratory cannot distinguish between a cyclostationary process and the corresponding spectrally equivalent stationary process. Hence, it might be tempting to replace the cyclostationary disturbance at the input of the PLL by a spectrally equivalent stationary disturbance.

Let us denote by $R_{\text{eq}}(u;e)$ and $S_{\text{eq}}(\omega;e)$ the autocorrelation function and the power spectral density of the loop noise, which result from applying the method of the spectrally equivalent stationary input disturbance. Replacing in (2-112) the autocovariance function $K_v(t, t+u;\varepsilon)$ by its time-average $\langle K_v(t, t+u;\varepsilon)\rangle_t$

yields the following expression for the loop noise autocorrelation function:

$$R_{\text{eq}}(u;e) = K_m^2 [\langle K_v(t,t+u;\varepsilon)\rangle_t \; \langle r(t;\hat{\varepsilon})\, r(t+u;\hat{\varepsilon})\rangle_t]_{\text{LP}}$$

Substituting (2-114) and (2-115) in the above expression yields

$$R_{\text{eq}}(u;e) = K_m^2 \left[k_{v,0}(u) \sum_n |c_n|^2 \, \exp\left(j\, \frac{2\pi n u}{T}\right)\right]_{\text{LP}}$$

or, equivalently,

$$S_{\text{eq}}(\omega;e) = K_m^2 \, |H_{\text{LP}}(\omega)|^2 \sum_n c_n^2 \, S_{v,0}\left(\omega - \frac{2\pi n}{T}\right)$$

It is important to note that the method of the spectrally equivalent disturbance at the input of the PLL gives rise to an erroneous loop noise spectrum which does not depend on the timing error e, because the terms with $n \neq 0$ in the correct expression (2-117) are lost. In the case where the local reference signal is a sinusoid with frequency $1/T$, the method of the spectrally equivalent stationary input disturbance yields

$$S_{\text{eq}}(\omega;e) = \frac{K_m^2 K_r^2}{2} \, |H_{\text{LP}}(\omega)|^2 \left(S_{v,0}\left(\omega - \frac{2\pi n}{T}\right) + S_{v,0}\left(\omega + \frac{2\pi n}{T}\right)\right) \tag{2-133}$$

Taking into account (2-121), (2-127), and (2-128), one obtains

$$\begin{aligned} S_{\text{eq}}(\omega;e) &= 1/2\, (S(\omega;e=0) + S(\omega;e=1/4)) \\ &= K_m^2 \, K_r^2 \, |H_{\text{LP}}(\omega)|^2 \, \frac{S_C(\omega) + S_S(\omega)}{2} \end{aligned} \tag{2-134}$$

The method of the spectrally equivalent stationary input disturbance *ignores* that the actual loop noise (2-124) is a *specific* linear combination (depending on the value e of the timing error) of the in-phase and the quadrature disturbances. This erroneous method gives rise to a loop noise spectrum (2-134) which, unlike the correct loop noise spectrum (2-126), is *independent* of the timing error, and which is proportional to the arithmetical average of the in-phase and quadrature spectra.

The error introduced by using the method of the spectrally equivalent stationary input disturbance can be considerable. When the steady-state error e_s is zero, the true timing error variance is determined by the behavior of the quadrature spectrum $S_S(\omega)$ within the small loop bandwidth of the synchronizer, whereas the method of the spectrally equivalent stationary input disturbance yields an error variance which is determined by the behavior of $(S_C(\omega) + S_S(\omega))/2$ within the loop bandwidth. For well-designed synchronizers, it is very often the case that $S_S(0) \ll S_C(0)$. Hence, the true error variance is determined by the small quadrature spectrum, whereas the incorrect error variance is dominated by the much larger in-phase spectrum. The conclusion is that the method of the spectrally equivalent stationary input disturbance can give rise to a timing error variance which is wrong by several orders of magnitude when the in-phase and quadrature spectra are substantially different near $\omega = 0$.

2.4.2 Nonlinearity Followed by Narrowband Bandpass Filter

The general structure of a spectral line generating synchronizer making use of a narrowband bandpass filter, tuned to the channel symbol rate $1/T$, is shown in Figure 2-37; this structure is very similar to the one shown in Figure 2-35, the only difference being that the PLL is now replaced by a bandpass filter with frequency response $H_{\mathrm{BP}}(\omega)$. It is assumed that the unknown delay εT of the noisy PAM waveform $y(t;\varepsilon)$ is essentially constant over the possibly nonzero memory time of the nonlinearity (which is in the order of the channel symbol interval T), so that the nonlinearity output signal $v(t;\varepsilon)$ depends only on the instantaneous value of ε, and not on past values of ε. Usually, the memory time of the bandpass filter extends over a few hundred channel symbol intervals, so that the fluctuation of εT over the bandpass filter memory time must be taken into account when considering the bandpass filter output signal $v_0(t)$. The sampler in the decision branch of the receiver is activated by the threshold level crossing of the (phase shifted) nearly sinusoidal signal $v_0(t)$.

The signal $v(t;\varepsilon)$ at the input of the bandpass filter can be decomposed as [see (2-123)]

$$\begin{aligned} v(t;\varepsilon) =& \sqrt{2}\,A\,\cos\left(\frac{2\pi}{T}\,(t-\varepsilon T)+\psi\right) \\ &+\sqrt{2}\,\mathrm{Re}\left[w_{0,L}(t;\varepsilon)\,\exp\left(j\psi\right)\,\exp\left(j\,\frac{2\pi}{T}\,(t-\varepsilon T)\right)\right] \end{aligned} \tag{2-135}$$

where we have dropped the "other terms," which are suppressed by the narrowband bandpass filter anyway. The first term in (2-135) is the useful sinusoid at the channel symbol rate $1/T$; the second term represents a zero-mean disturbance, having $w_L(t;\varepsilon)$ as complex envelope with respect to the sinusoid at the channel symbol rate. Let us assume that ε is slowly time varying: $\varepsilon(t)$ exhibits small fluctuations $\Delta\varepsilon(t)$ about its mean value ε_0. Using the approximation

$$\exp\left(-j2\pi\varepsilon\right) = \left(1-j2\pi\Delta\varepsilon\right)\,\exp\left(-j2\pi\varepsilon_0\right)$$

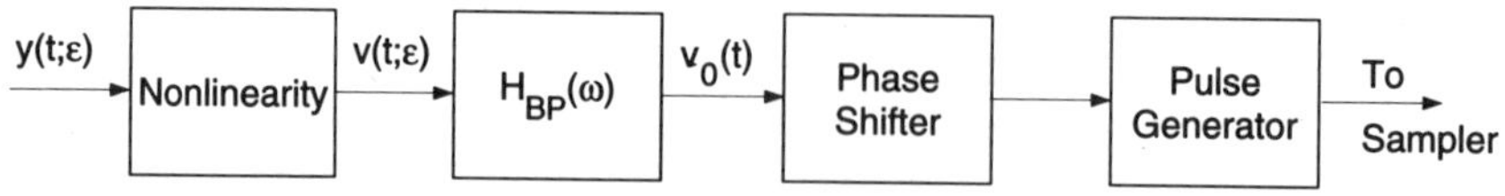

Figure 2-37 General Spectral Line Generating Synchronizer Using a Narrowband Bandpass Filter

and keeping only first-order terms in $w_L(t;\varepsilon)$ and $\Delta\varepsilon(t)$, (2-135) can be transformed into

$$\begin{aligned} v(t;\varepsilon) &= \sqrt{2}\, A \cos\left(\frac{2\pi}{T}(t-\varepsilon_0 T)+\psi\right) \\ &\quad + \sqrt{2}\,\mathrm{Re}\left[(w_L(t;\varepsilon) - j2\pi A\Delta\varepsilon)\exp(j\psi)\exp\left(j\,\frac{2\pi}{T}(t-\varepsilon_0 T)\right)\right] \end{aligned} \tag{2-136}$$

The first term in (2-136) is a sinusoid with constant phase. Note that the fluctuations $\Delta\varepsilon(t)$ give rise to a quadrature component with respect to the first term in (2-136).

The signal at the output of the bandpass filter is given by

$$\begin{aligned} v_0(t) &= \sqrt{2}\, A \left|H_{\mathrm{BP}}\left(\frac{2\pi}{T}\right)\right| \cos\left(\frac{2\pi}{T}(t-\varepsilon_0 T)+\psi+\psi_H\right) \\ &\quad + \sqrt{2}\,\mathrm{Re}\left[w_{0,L}(t)\exp\left(j\,(\psi+\psi_H)\right)\exp\left(j\,\frac{2\pi}{T}(t-\varepsilon_0 T)\right)\right] \end{aligned} \tag{2-137}$$

The first term in (2-137) is the response of the bandpass filter to the first term in (2-136): the bandpass filter introduces a gain $|H_{\mathrm{BP}}(2\pi/T)|$ and a phase shift ψ_H, which are the magnitude and the argument of the bandpass filter frequency response, evaluated at $\omega = 2\pi/T$:

$$H_{\mathrm{BP}}\left(\frac{2\pi}{T}\right) = \left|H_{\mathrm{BP}}\left(\frac{2\pi}{T}\right)\right| \exp(j\psi_H)$$

In order to eliminate the phase shift introduced by the bandpass filter, one could use a bandpass filter which is symmetric about the channel symbol rate $1/T$, i.e.

$$u\left(\frac{2\pi}{T}+\omega\right) H_{\mathrm{BP}}\left(\frac{2\pi}{T}+\omega\right) = u\left(\frac{2\pi}{T}-\omega\right) H_{\mathrm{BP}}^{*}\left(\frac{2\pi}{T}-\omega\right)$$

where $u(.)$ denotes the unit step function. This yields $\psi_H = 0$. Due to frequency detuning and drift of filter component values, the symmetry of the bandpass filter about the channel symbol rate $1/T$ is not perfect, so that one should always take into account the possibility of a nonzero phase shift ψ_H. For a given frequency detuning, this phase shift increases with decreasing filter bandwidth.

The second term in (2-137) is a zero-mean disturbance, having $w_{0,L}(t)$ as complex envelope with respect to the first term in (2-137). This complex envelope is given by

$$w_{0,L}(t) = \exp(-j\psi_H) \int_{-\infty}^{+\infty} h_E(t-u)\,(w_L(u;\varepsilon) - j2\pi A\Delta\varepsilon(u))\,du \tag{2-138}$$

where $h_E(t)$ is the inverse Fourier transform of $H_E(\omega)$, the equivalent lowpass frequency response of the bandpass filter:

$$H_E(\omega) = u\left(\omega + \frac{2\pi}{T}\right) H_{\mathrm{BP}}\left(\omega + \frac{2\pi}{T}\right)$$

The sampler in the decision branch of the receiver is activated each time the bandpass filter output signal $v_0(t)$ crosses the threshold level $\sqrt{2}A\eta|H_{\mathrm{BP}}(2\pi/T)|$ in the positive direction. Ideally, η should be zero, so that the sampler is activated by the zero crossings of $v_0(t)$. However, due to DC offsets, the presence of a nonzero value of η must be taken into account. If the complex envelope $w_{0,L}(t)$ at the bandpass filter output were zero, the level crossing instants would be equidistant points on the time axis, with spacing equal to T. The kth such level crossing instant is denoted by $T_{\eta,k}$ and satisfies the following equation:

$$\exp\left(j\left(\frac{2\pi}{T}(T_{\eta,k} - \varepsilon_0 T) + \psi + \psi_H\right)\right) = \exp\left(j\left(\psi_\eta - \frac{\pi}{2}\right)\right) \tag{2-139}$$

where

$$\exp(j\psi_\eta) = \sqrt{1-\eta^2} + j\eta \tag{2-140}$$

The solution is given by

$$T_{\eta,k} = \left(kT + \varepsilon_0 T - \frac{T}{2\pi}\left(\psi + \frac{\pi}{2}\right)\right) + \frac{T}{2\pi}(\psi_\eta + \psi_H)$$

The complex envelope $w_{0,L}(t)$ at the output of the bandpass filter gives rise to actual level crossing instants $t_{\eta,k}$, which exhibit fluctuations $\Delta T_{\eta,k}$ about the nominal level crossing instants $T_{\eta,k}$:

$$t_{\eta,k} = T_{\eta,k} + \Delta T_{\eta,k}$$

The actual level crossing instants $t_{\eta,k}$ satisfy the following equation:

$$\begin{aligned}
&\sqrt{2}\,A\left|H_{\mathrm{BP}}\left(\frac{2\pi}{T}\right)\right| \cos\left(\frac{2\pi}{T}(t_{\eta,k} - \varepsilon_0 T) + \psi + \psi_H\right) \\
&+\sqrt{2}\,\mathrm{Re}\left[w_{0,L}(t_{\eta,k}) \exp(j(\psi + \psi_H)) \exp\left(j\,\frac{2\pi}{T}(t_{\eta,k} - \varepsilon_0 T)\right)\right] \\
&=\sqrt{2}\,A\left|H_{\mathrm{BP}}\left(\frac{2\pi}{T}\right)\right| \eta
\end{aligned} \tag{2-141}$$

Assuming that the disturbance at the output of the bandpass filter is weak, the zero-mean fluctuations $\Delta T_{\eta,k}$ are small, and are well approximated by the solution of (2-141), linearized about the nominal level crossing instants $T_{\eta,k}$. This linearization procedure is illustrated in Figure 2-38. Taking into account (2-139) and (2-140),

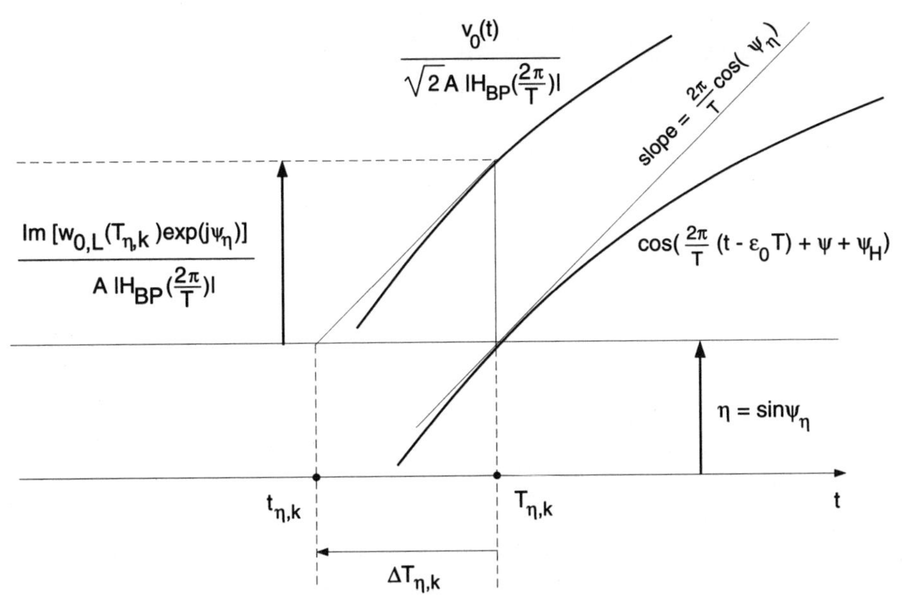

Figure 2-38 Linearization About Nominal Level-Crossing Instant $T_{\eta,k}$

one obtains

$$\frac{\Delta T_{\eta,k}}{T} = \frac{-\text{Im}[w_{0,L}(T_{\eta,k})\ \exp(j\psi_\eta)]}{2\pi A\ |H_{\text{BP}}(2\pi/T)|\ \cos(\psi_\eta)} \tag{2-142}$$

Let us decompose the actual level crossing instants as

$$t_{\eta,k} = T_k - \Delta t_{\eta,k}$$

where

$$T_k = kT + \varepsilon T - \frac{T}{2\pi}\left(\psi + \frac{\pi}{2}\right) \tag{2-143}$$

denotes the kth positive zero crossing of the first term in (2-135) at the input of the bandpass filter. Assuming that the origin of the time axis has been chosen such that $\{kT + \varepsilon T\}$ are the ideal sampling instants in the decision branch, the last term in (2-143) is a bias term, which can be compensated for by introducing a phase shift of $\psi + \pi/2$ at the output of the bandpass filter. The deviations $\Delta t_{\eta,k}$ from the ideal level crossing instants T_k in (2-143) incorporate the effect of the disturbance $w_L(t;\varepsilon)$ at the nonlinearity output, of the bandpass filter asymmetry, of the nonzero threshold level, and of the fluctuations of the delay εT. Making use of (2-142), (2-138), and $|H_{\text{BP}}(2\pi/T)| = |H_E(0)|$, it is easily shown that the random deviations $\Delta t_{\eta,k}$ with respect to the ideal level crossing instants T_k are obtained by performing the operations shown in Figure 2-39. As the lowpass content of $w_L(t;\varepsilon)$ is wide-sense stationary, its contribution to the variance of $\Delta t_{\eta,k}$ does not depend on the position of the instants $T_{\eta,k}$. Taking into account (1-144) and

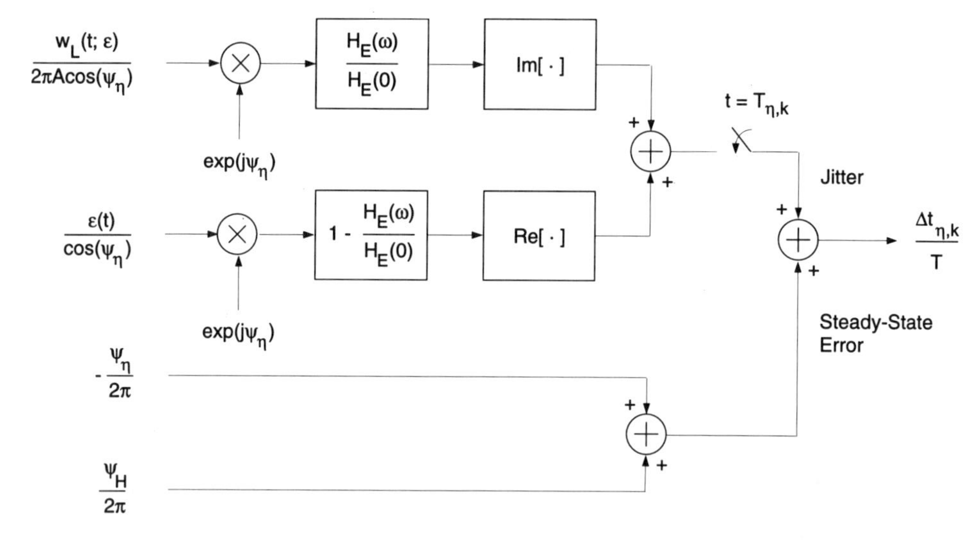

Figure 2-39 Identification of Contributions to Deviation $\Delta t_{\eta,k}$ (Spectral Line Generating Synchronizer Using a Narrowband Bandpass Filter)

(1-145), it follows from (2-142) that this contribution is given by

$$\begin{aligned}
&\operatorname{var}\left[\frac{\Delta t_{\eta,k}}{T}\right]\Bigg|_{\varepsilon=\varepsilon_0} \\
&= \frac{1}{K_s^2} \int_{-\infty}^{+\infty} \left(|H_{\text{eq}}(\omega)|^2 \, S_{v,0}\left(\omega + \frac{2\pi}{T}\right) + |H_{\text{eq}}(-\omega)|^2 \, S_{v,0}\left(\omega - \frac{2\pi}{T}\right) \right. \\
&\quad + H_{\text{eq}}(\omega) \, H_{\text{eq}}(-\omega) \, S_{v,2}\left(\omega + \frac{2\pi}{T}\right) \, \exp\left(2j\,(\psi_\eta - \psi)\right) \\
&\quad \left. + H^*_{\text{eq}}(\omega) \, H^*_{\text{eq}}(-\omega) \, S_{v,-2}\left(\omega - \frac{2\pi}{T}\right) \, \exp\left(-2j\,(\psi_\eta - \psi)\right) \right) \frac{d\omega}{2\pi}
\end{aligned} \tag{2-144}$$

where $K_s = 2\sqrt{2}A\pi\cos(\psi_\eta)$ and $H_{\text{eq}}(\omega) = H_E(\omega)/H_E(0)$. Note that $H_{\text{eq}}(\omega)$ does not possess conjugate symmetry about $\omega = 0$, unless the bandpass filter is symmetric about the channel symbol rate $1/T$.

Finally, using the erroneous method of the spectrally equivalent stationary disturbance at the input of the bandpass filter yields a variance of $\Delta t_{\eta,k}$ which contains only the first two terms of (2-144); this is because $S_{v,2}(\omega)$ and $S_{v,-2}(\omega)$ are identically zero when the disturbance at the output of the nonlinearity [i.e., the second term in (2-135)] is stationary.

2.4.3 Phase-Locked Loops versus Bandpass Filters

In Sections 2.4.1 and 2.4.2 it has been shown that spectral line generating synchronizers consisting of a nonlinearity followed by a PLL (with multiplying timing error detector and sinusoidal reference signal) and those consisting of a nonlinearity followed by a narrowband bandpass filter both yield a nearly sinusoidal clock signal, which is used to control the sampler in the decision branch of the receiver. In this section, a comparison is made between using a PLL and using a bandpass filter for extracting the useful sinusoid at the output of the nonlinearity.

As far as the operation of the PLL or the bandpass filter is concerned, the signal at the output of the nonlinearity is represented as

$$\begin{aligned} v(t;\varepsilon) &= \sqrt{2}\,A\,\cos\left(\frac{2\pi}{T}(t-\varepsilon T)+\psi\right) \\ &\quad + \sqrt{2}\,\mathrm{Re}\left[w_L(t;\varepsilon)\,\exp(j\psi)\,\exp\left(j\,\frac{2\pi}{T}(t-\varepsilon T)\right)\right] \\ &= \sqrt{2}\,A\,\cos\left(\frac{2\pi}{T}(t-\varepsilon T)+\psi\right) + \sqrt{2}\,w_C(t;\varepsilon)\,\cos\left(\frac{2\pi}{T}(t-\varepsilon T)+\psi\right) \\ &\quad - \sqrt{2}\,w_S(t;\varepsilon)\,\sin\left(\frac{2\pi}{T}(t-\varepsilon T)+\psi\right) \end{aligned}$$

where $w_L(t;\varepsilon)$, $w_C(t;\varepsilon)$, and $w_S(t;\varepsilon)$ are the complex envelope, the in-phase component and the quadrature component (with respect to the sinusoid to be tracked), of the disturbance at the output of the nonlinearity. When ε is time varying, its fluctuations should be tracked by the synchronizer. Defining

$$B(t;\varepsilon)\,\exp(j\psi_w(t;\varepsilon)) = A + w_L(t;\varepsilon)$$

or, equivalently,

$$\begin{aligned} B(t;\varepsilon) &= \sqrt{(A+w_C(t;\varepsilon))^2 + (w_S(t;\varepsilon))^2} \\ \psi_w(t;\varepsilon) &= \arctan\left(\frac{w_S(t;\varepsilon)}{A+w_C(t;\varepsilon)}\right) \end{aligned}$$

the signal at the output of the nonlinearity can be transformed into

$$v(t;\varepsilon) = \sqrt{2}\,B(t;\varepsilon)\,\cos\left(\frac{2\pi}{T}(t-\varepsilon T)+\psi+\psi_w(t;\varepsilon)\right)$$

which represents a sinusoid, whose amplitude and phase exhibit unwanted random fluctuations, caused by the disturbance at the output of the nonlinearity. It is important to note that $v(t;\varepsilon)$ would have no unwanted phase fluctuations, if the quadrature component $w_S(t;\varepsilon)$ were identically zero. Under weak noise conditions, the random amplitude and phase are well approximated by

$$\begin{aligned} B(t;\varepsilon) &= A + w_C(t;\varepsilon) \\ \psi_w(t;\varepsilon) &= \frac{w_S(t;\varepsilon)}{A} \end{aligned} \tag{2-145}$$

which indicates that the in-phase component causes amplitude fluctuations and the quadrature component gives rise to phase fluctuations at the output of the nonlinearity.

When the nonlinearity is followed by a PLL, Figure 2-36 shows the influence of various imperfections on the deviations $\Delta t_{\eta,k}$ from the ideal level crossing instants (which are lagging by $T/4$ behind the positive zero crossings of the sinusoidal component at the nonlinearity output). These deviations consist of a static timing offset and a random zero-mean timing jitter. The static timing offset is caused by the steady-state error e_s at the PLL and by the nonzero threshold level parameter η [which yields the nonzero angle ψ_η, determined by (2-130)]. The timing jitter is caused by:

- The highpass content of $\varepsilon(t)$, which falls outside the loop bandwidth of the PLL and, hence, cannot be tracked;
- The lowpass content of $w_L(t;\varepsilon)$, which falls inside the loop bandwidth and, hence, disturbs the operation of the PLL.

Note that the steady-state error e_s has an influence on the jitter contribution from $w_L(t;\varepsilon)$. Let us consider this contribution in more detail. When $e_s = 0$, the timing jitter is caused only by the quadrature component $\text{Im}[w_L(t;\varepsilon)] = w_S(t;\varepsilon)$ of the disturbance at the nonlinearity output. More specifically, the timing jitter consists of samples of the lowpass filtered quadrature component at the nonlinearity output, which, under weak noise conditions, corresponds to samples of the lowpass filtered phase fluctuations at the nonlinearity output [see (2-145)]. The situation becomes more complicated when the steady-state error e_s is nonzero. In this case, Figure 2-36 shows that the timing jitter consists of samples of a lowpass filtered version of the signal $\text{Im}[w_L(t;\varepsilon)\,\exp(-j2\pi e_s)]$. As

$$\text{Im}[w_L(t;\varepsilon)\,\exp(-j2\pi e_s)] = w_C(t;\varepsilon)\,\cos(2\pi e_s) - w_S(t;\varepsilon)\,\sin(2\pi e_s) \tag{2-146}$$

it follows that the timing jitter contains not only a term which is proportional to the lowpass filtered phase fluctuations at the nonlinearity output, but also a term which is proportional to the lowpass filtered amplitude fluctuations [assuming weak noise conditions, so that (2-145) is valid]. Note that (2-146) gives the component of the disturbance at the output of the nonlinearity, which is in phase with the local reference signal of the PLL, and, therefore, is causing the jitter. The conclusion is that a steady-state error at the PLL gives rise to amplitude-to-phase conversion (amplitude fluctuations at the input of the PLL are converted into phase fluctuations of the local reference signal). Decomposing in Figure 2-36 the disturbance $w_L(t;\varepsilon)$ into its real and imaginary parts $w_C(t;\varepsilon)$ and $w_S(t;\varepsilon)$, the jitter is obtained as shown in Figure 2-40.

When the nonlinearity is followed by a bandpass filter, Figure 2-39 shows how the various imperfections give rise to the deviations $\Delta t_{\eta,k}$ from the ideal level crossing instants (which coincide with the positive zero crossings of the sinusoid at the nonlinearity output). These deviations consist of a static timing offset and a

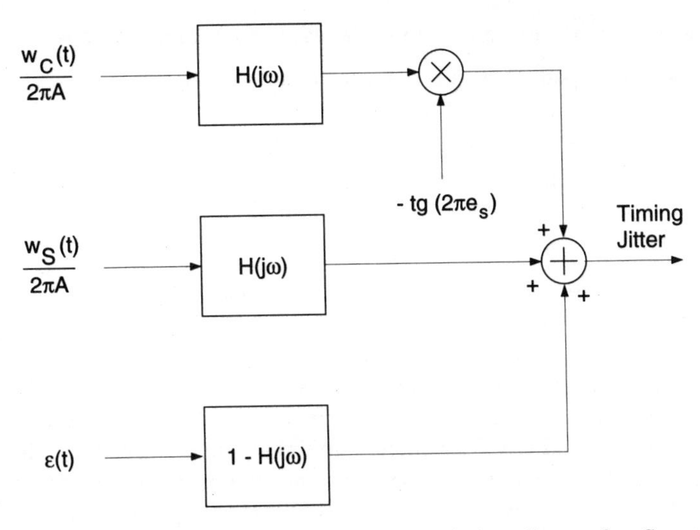

Figure 2-40 Identification of Contributions to Timing Jitter of a Spectral Line Generating Synchronizer with PLL

zero-mean random jitter. The static timing offset is caused by a nonzero threshold level parameter η, and by the phase shift ψ_H introduced by the bandpass filter. The timing jitter is caused by the lowpass content of $w_L(t;\varepsilon)$ and the highpass content of $\varepsilon(t)$. Let us concentrate on the timing jitter.

When the bandpass filter is symmetric about the channel symbol rate $1/T$ [meaning that $H_E^*(\omega) = H_E(-\omega)$, the equivalences indicated in Figure 2-41 can be applied to the structure in Figure 2-39]. Noting further that

$$\text{Im}\left[\frac{-w_L(t;\varepsilon)}{\cos(\psi_\eta)} \exp(j\psi_\eta)\right] = -w_S(t;\varepsilon) - w_C(t;\varepsilon)\tan(\psi_\eta)$$

$$\text{Re}\left[\frac{-\varepsilon(t)}{\cos(\psi_\eta)} \exp(j\psi_\eta)\right] = -\varepsilon(t)$$

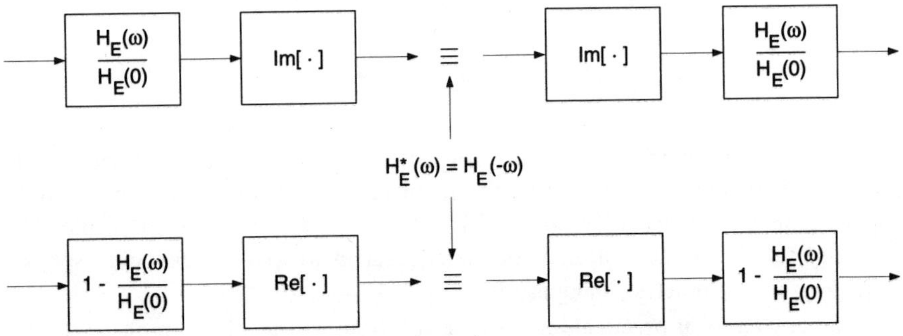

Figure 2-41 Equivalences That Hold When the Bandpass Filter Is Symmetric

the following observations are valid in the case of a symmetric bandpass filter:

- The jitter contribution from $\varepsilon(t)$ and $w_S(t;\varepsilon)$ is not influenced by the threshold level parameter η.
- The jitter contribution from $w_C(t;\varepsilon)$ is influenced by the threshold level parameter: a nonzero threshold level gives rise to amplitude-to-phase conversion (amplitude fluctuations at the output of the nonlinearity are converted into timing jitter).

This is according to our intuition. Indeed, $\varepsilon(t)$ and $w_S(t;\varepsilon)$ give rise to phase fluctuations at the output of the nonlinearity, while $w_C(t;\varepsilon)$ gives rise to amplitude fluctuations (assuming low noise conditions). Hence, because of the symmetry of the bandpass filter, the amplitude fluctuations at the bandpass filter output are caused only by $w_C(t;\varepsilon)$. As a nonzero threshold level converts amplitude fluctuations at the bandpass filter output into timing jitter (while phase fluctuations at the bandpass filter output are converted into timing jitter, irrespective of the threshold level value), the threshold level parameter η affects only the jitter contribution from $w_C(t;\varepsilon)$. It is important to note that the case of a symmetric bandpass filter and a nonzero angle ψ_η yields the same jitter as the case of a PLL with steady-state error $e_s = -\psi_\eta/2\pi$ and closed-loop frequency response $H_E(\omega)/H_E(0)$.

When the bandpass filter is not symmetric about the channel symbol rate $1/T$, it introduces amplitude-to-phase conversion (phase fluctuations at the output of the bandpass filter depend not only on $w_S(t;\varepsilon)$ and $\varepsilon(t)$, but also on $w_C(t;\varepsilon)$) and phase-to-amplitude conversion [amplitude fluctuations at the output of the bandpass filter depend not only on $w_C(t;\varepsilon)$, but also on $w_S(t;\varepsilon)$ and $\varepsilon(t)$]. If the threshold level were zero, then the timing jitter would be proportional with the phase fluctuations at the output of the bandpass filter. When the threshold level is nonzero, the jitter contains an extra term, proportional to the amplitude fluctuations at the bandpass filter output and depending on the threshold level parameter η. This situation is illustrated in Figure 2-42, where $H_C(\omega)$ and $H_S(\omega)$ are the even and odd conjugate symmetric parts of $H_E(\omega)/H_E(0)$:

$$H_C(\omega) = \frac{1}{2}\left(\frac{H_E(\omega)}{H_E(0)} + \frac{H_E^*(-\omega)}{H_E^*(0)}\right)$$

$$H_S(\omega) = \frac{1}{2j}\left(\frac{H_E(\omega)}{H_E(0)} - \frac{H_E^*(-\omega)}{H_E^*(0)}\right)$$

It is important to note that the case of an asymmetric bandpass filter is not equivalent with the case of a PLL; the main reason for this is that the closed-loop frequency response $H(s)$ of a PLL satisfies $H^*(\omega) = H(-\omega)$, while this symmetry relation is not fulfilled for the equivalent lowpass frequency response $H_E(\omega)$ of an asymmetric bandpass filter.

In many cases of practical interest, the spectrum $S_C(\omega)$ of the in-phase disturbance $w_C(t;\varepsilon)$ and the spectrum $S_S(\omega)$ of the quadrature disturbance $w_S(t;\varepsilon)$ be-

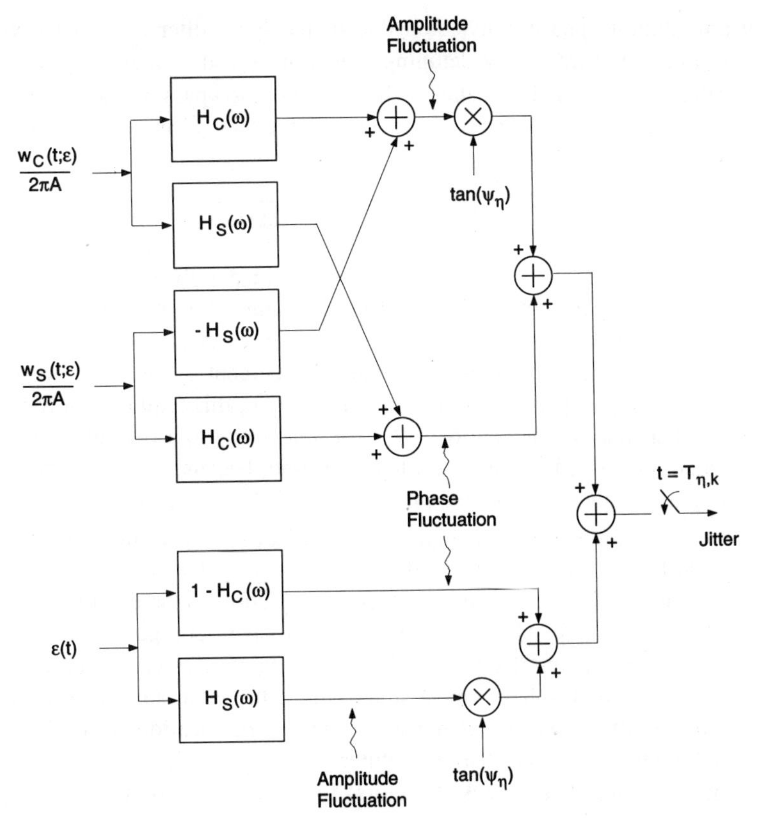

Figure 2-42 Identification of Contributions to Timing Jitter of a Spectral Line Generating Synchronizer with Narrowband Bandpass Filter

have quite differently in the vicinity of $\omega = 0$; more specifically, $S_S(0) \ll S_C(0)$. Therefore, the influence of the in-phase disturbance on the synchronizer performance should be kept as small as possible. When using a PLL, this is accomplished when the steady-state error e_s is close to zero; when using a bandpass filter, the threshold level parameter η should be close to zero, and the bandpass filter should be symmetric. Also, in order to reduce the influence of the quadrature component, the loop bandwidth of the PLL or the bandwidth of the bandpass filter should be small.

Let us consider a bandpass filter, which is symmetric about a frequency $f_0 = \omega_0/2\pi$, where f_0 is slightly different from the channel symbol rate $1/T$. The degree of filter asymmetry with respect to the channel symbol rate $1/T$, due to the frequency detuning $\Delta F = 1/T - f_0$, is closely related to the filter bandwidth: for a given frequency detuning, the degree of asymmetry becomes larger with decreasing filter bandwidth. Hence, when a small filter bandwidth has been selected for a considerable suppression of the disturbance at the output of the nonlinearity, the sensitivity to a frequency detuning is high. In order to keep the phase shift ψ_H

and the amplitude-to-phase conversion due to bandpass filter asymmetry within reasonable limits, the frequency detuning must not exceed a small fraction of the bandpass filter bandwidth. For instance, the equivalent lowpass frequency response $H_E(\omega)$ of a single tuned resonator with center frequency f_0 is determined by

$$H_E(\omega) = \frac{1}{1 + j\left((\omega/\omega_3) + (\Delta\Omega/\omega_3)\right)}$$

where $\Delta\Omega = 2\pi\Delta F$, $\omega_3 = 2\pi f_3$, and f_3 is the one-sided 3 dB bandwidth (in hertz) of the bandpass filter (attenuation of 3 dB at the angular frequencies $\omega_0 + \omega_3$ and $\omega_0 - \omega_3$). If the phase shift $\psi_H = \arg\left(H_E(0)\right)$ is restricted to be less than 12°, the frequency detuning ΔF should not exceed 20 percent of the 3 dB bandwidth f_3; when f_3 equals one percent of the symbol rate $1/T$, this requirement translates into ΔF smaller than 0.2 percent of $1/T$. The sensitivity to a frequency detuning is larger when higher-order bandpass filters are used, because of the steeper slope of the bandpass frequency response phase at the center frequency.

In the case of a (second- or higher-order) PLL, the open-loop gain and the loop bandwidth B_L can be controlled independently. This means that a small loop bandwidth can be selected for considerable rejection of the disturbance at the output of the nonlinearity, while simultaneously the open-loop gain is so large that a frequency detuning gives rise to a negligible steady-state error e_s and, hence, to virtually no amplitude-to-phase conversion. This is the main advantage of a PLL over a bandpass filter. On the other hand, it should be mentioned that PLLs are more complex structures than bandpass filters, as they require additional circuits (such as acquisition aids and lock detectors) for proper operation.

2.4.4 Jitter Accumulation in a Repeater Chain

When digital information is to be transmitted over a long distance that corresponds to a large signal attenuation, a chain of repeaters is often used. The operation of a repeater is illustrated in Figure 2-43. Each repeater contains a receiver and a transmitter. The receiver recovers the digital data symbols that have been sent by the upstream repeater, and the transmitter sends to the downstream

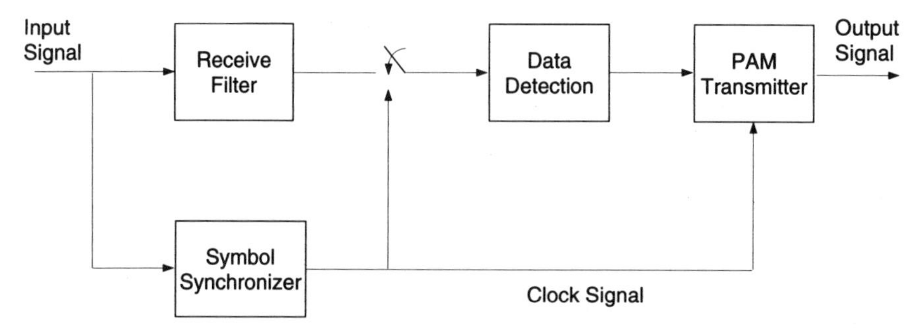

Figure 2-43 Block Diagram of Repeater

repeater a PAM signal containing the recovered data symbols. Each repeater extracts a clock signal from the signal received from its upstream repeater. This clock signal determines the instants at which the receive filter output signal is sampled; from these samples, the digital information is recovered. The same clock signal also provides the timing to the PAM transmitter that sends the recovered data symbols to the downstream repeater.

Let us assume that the PAM signal transmitted to the first repeater of the chain is generated by means of a jitter-free clock: the consecutively transmitted pulses are equidistant in time. Because of additive noise and self-noise, the clock signals extracted at the repeaters exhibit zero-mean random jitter. Let us denote by $\varepsilon_n(t)$ the *total jitter* at the output of the *n*th repeater, i.e., the jitter of the clock extracted by the *n*th repeater, measured with respect to a jitter-free clock. The corresponding clock signals can be represented as follows:

$$\begin{aligned} &\text{jitter} - \text{free clock}: && \sqrt{2}\cos\left(2\pi t/T + \psi\right) \\ &\text{clock at } n\text{th repeater}: && \sqrt{2}\cos\left(2\pi (t - \varepsilon_n(t)T)/T + \psi\right) \end{aligned} \tag{2-147}$$

The *alignment jitter* at the *n*th repeater is given by $\varepsilon_{n-1}(t) - \varepsilon_n(t)$, and represents the random misalignment between the clocks at the *n*th repeater and its upstream repeater. The alignment jitter is caused not only by additive noise and self-noise that occurs on the trnsmission path between the *(n-1)*th and the *n*th repeater, but also by the components of $\varepsilon_{n-1}(t)$ that cannot be tracked at the *n*th repeater (because they fall outside the synchronizer bandwidth). In order not to degrade the bit error rate performance, the alignment jitter should be kept small.

Assuming small jitter, the system equations describing the operation of the synchronizer can be linearized; note that this holds not only for spectral line generating synchronizers but for any type of synchronizer. In this case, a linear relation between $\varepsilon_{n-1}(t)$ and $\varepsilon_n(t)$ exists. This results in the *Chapman jitter model* for the *n*th repeater, shown in Figure 2-44. The quantity $H_\varepsilon(\omega)$ is the *jitter frequency response* of the *n*th repeater. The Chapman model indicates that the jitter $\varepsilon_n(t)$ at the output of the *n*th repeater consists of two terms:

- The first term is a filtered version of the input jitter $\varepsilon_{n-1}(t)$ and represents the contribution from the upstream part of the repeater chain.
- The second term $\Delta_n(t)$ represents the contribution from the *n*th repeater: it is the jitter that would occur at the output of the *n*th repeater if the input jitter $\varepsilon_{n-1}(t)$ were identically zero.

In the case of a spectral line generating synchronizer, the parameters of the Chapman model can be derived from Figure 2-40 (nonlinearity followed by PLL) or Figure 2-42 (nonlinearity followed by narrowband bandpass filter). Noting that the signal at the output of the block diagrams in Figures 2-40 and 2-42 actually represents the alignment jitter, one obtains

$$H_\varepsilon(\omega) = \begin{cases} H(j\omega) & \text{nonlinearity plus PLL} \\ H_C(\omega) - H_S(\omega)\tan(\psi_\eta) & \text{nonlinearity plus bandpass filter} \end{cases} \tag{2-148}$$

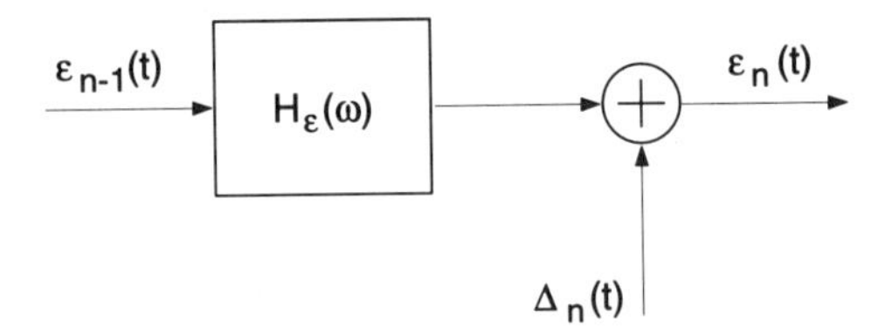

Figure 2-44 Chapman Jitter Model for Repeater

from which it follows that $H_\varepsilon(0) = 1$. Note that the threshold value η affects the jitter frequency response only when the nonlinearity is followed by a non-symmetrical bandpass filter. It follows from Figures 2-40 and 2-42 that the jitter term $\Delta_n(t)$ is a linear combination of filtered versions of the in-phase and quadrature components of the complex envelope of the disturbance at the output of the nonlinearity. For any error-tracking synchronizer, it can be shown that the jitter frequency response $H_\varepsilon(\omega)$ equals the closed-loop frequency response $H(j\omega)$.

In order to investigate the jitter in a chain of N repeaters, we replace each repeater by its Chapman model. Assuming identical jitter frequency responses for all N repeaters, the repeater chain model of Figure 2-45 is obtained, with $\varepsilon_0(t) = 0$ denoting the zero jitter at the input of the first repeater. When the jitter frequency response exhibits peaking [i.e., $|H_\varepsilon(\omega)| > 1$ for some frequency range], the jitter components within that frequency range are amplified by the jitter frequency response. Denoting the maximum value of $|H_\varepsilon(\omega)|$ by $H_{\max}$, the situation $H_{\max} > 1$ results in a total jitter variance var$[\varepsilon_N(t)]$ and an alignment jitter variance var$[\varepsilon_{N-1}(t) - \varepsilon_N(t)]$ that both are essentially proportional to $H_{\max}^{2N}$: when the jitter frequency response exhibits peaking, both variances increase exponentially with the number N of repeaters. Hence, by means of careful circuit design, peaking of the jitter frequency response should be avoided, or at least the amount of peaking, defined as 20 log $H_{\max}$ (in decibels), should be kept very small (in the order of 0.1 dB).

Examples

Figures 2-46 and 2-47 show the magnitudes of the frequency responses $H_C(\omega)$ and $H_S(\omega)$, corresponding to a single tuned resonator. Note that $H_S(\omega) = 0$ for zero detuning. When the detuning $\Delta\Omega/\omega_3$ exceeds $1/\sqrt{2}$ (which yields a phase shift ψ_H of about 35°, or, equivalently, a normalized static timing error of about 10 percent), $|H_C(\omega)|$ exhibits peaking. This peaking gives rise to exponential jitter accumulation along a repeater chain. For large ω, both $|H_C(\omega)|$ and $|H_S(\omega)|$ decay at a rate of 20 dB per decade of ω.

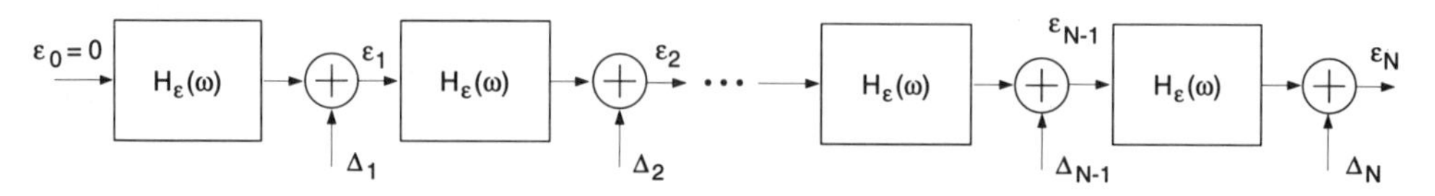

Figure 2-45 Jitter Model for Repeater Chain

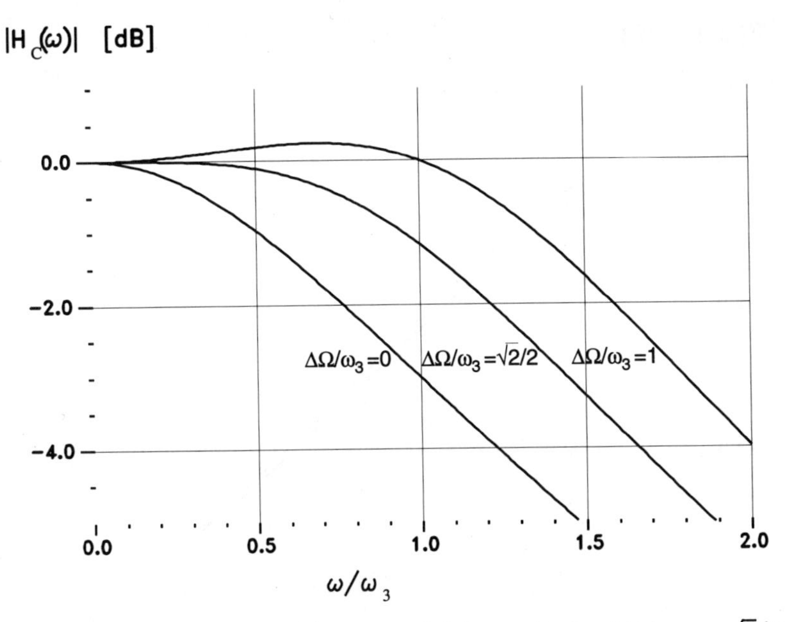

Figure 2-46 Magnitude $|H_C(\omega)|$ for: (a) $\Delta\Omega/\omega_3 = 0$, (b) $\Delta\Omega/\omega_3 = \sqrt{2}/2$, (c) $\Delta\Omega/\omega_3 = 1$

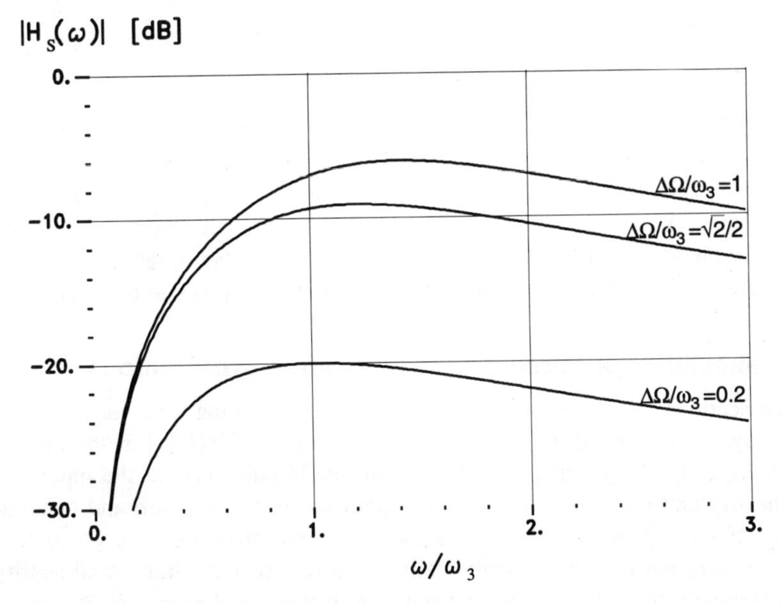

Figure 2-47 Magnitude $|H_S(\omega)|$ for: (a) $\Delta\Omega/\omega_3 = 0.2$, (b) $\Delta\Omega/\omega_3 = \sqrt{2}/2$, (c) $\Delta\Omega/\omega_3 = 1$

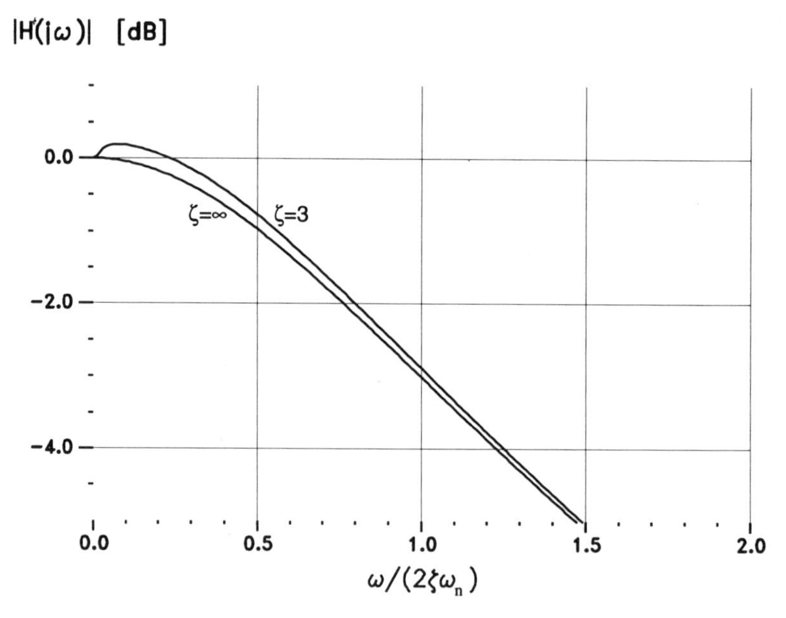

Figure 2-48 Magnitude $|H(j\omega)|$ for: (a) $\zeta = \infty$, (b) $\zeta = 3$

For a perfect second order PLL, the closed-loop frequency response in the Laplace domain is given by

$$H(s) = \frac{2\zeta\omega_n s + \omega_n^2}{s^2 + 2\zeta\omega_n s + \omega_n^2}$$

Keeping $\zeta\omega_n$ fixed, the corresponding magnitude $|H(j\omega)|$ is shown in Figure 2-48, for $\zeta = 3$ and $\zeta = \infty$; in the latter case, $H(s)$ converges to the frequency response of a first-order loop. For any finite value of ζ, $|H(j\omega)|$ exhibits peaking, which increases with decreasing ζ. Hence, exponential jitter accumulation along a repeater chain cannot be avoided. As the jitter accumulation rate decreases with decreasing peak value, large values of ζ are selected in practice ($\zeta > 3$).

2.4.5 Simulation of Spectral Line Generating Synchronizers

In Sections 2.4.1 and 2.4.2 we have evaluated the timing error variance caused by the cyclostationary disturbance $w(t;\varepsilon) = v(t;\varepsilon) - E[v(t;\varepsilon)]$ at the output of the nonlinearity. Depending on the baseband PAM pulse $g(t)$ at the input of the nonlinearity and on the specific type of nonlinearity, the in-phase and quadrature spectra of $w(t;\varepsilon)$, which are needed for the timing error variance computation, might be very hard to obtain analytically; in fact, only a quadratic nonlinearity is mathematically tractable for narrowband transmission, and even that case is quite complicated. A faster way to obtain numerical results is by means of computer simulations.

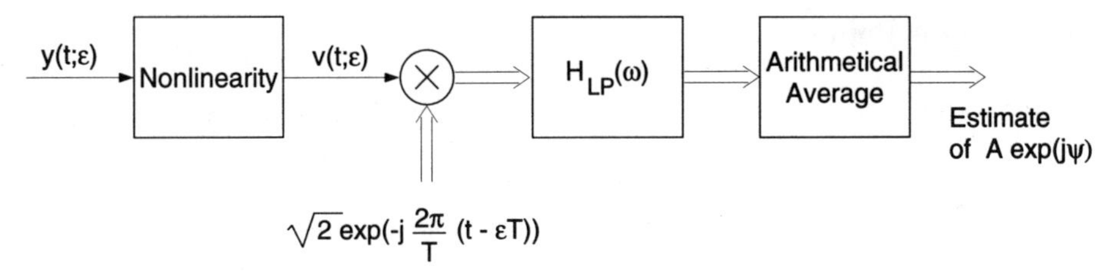

Figure 2-49 Estimation of Quantity $A \ \exp(j\psi)$

Figures 2-40 and 2-42 show how the timing jitter depends on the in-phase and quadrature components $w_C(t;\varepsilon)$ and $w_S(t;\varepsilon)$ and on the normalized delay $\varepsilon(t)$, in the cases where the nonlinearity is followed by a PLL or a bandpass filter, respectively. In order to obtain the timing jitter, the computer simulation must generate $w_C(t;\varepsilon)/A$, $w_S(t;\varepsilon)/A$ and $\varepsilon(t)$, and process them according to Figures 2-40 or 2-42. In the following we concentrate on the generation of $w_C(t;\varepsilon)/A$ and $w_S(t;\varepsilon)/A$.

As far as the synchronizer operation is concerned, the nonlinearity output $v(t;\varepsilon)$ can be represented as [see (2-135)]

$$v(t;\varepsilon) = \sqrt{2}\,\text{Re}\left[[A + w_L(t;\varepsilon)]\ \exp(j\psi)\ \exp\left(j\,\frac{2\pi}{T}\,(t-\varepsilon T)\right)\right]$$

where

$$w_L(t;\varepsilon) = w_C(t;\varepsilon) + j w_S(t;\varepsilon)$$

has zero mean. Hence, the quantity $A \ \exp(j\psi)$ can be estimated by means of the configuration shown in Figure 2-49. The "arithmetical average" module measures the DC component of its input signal; $H_{\text{LP}}(\omega)$ is the frequency response of a lowpass filter, with $H_{\text{LP}}(\omega) = 1$ for $|\omega|$ within the synchronizer bandwidth and $H_{\text{LP}}(\omega) = 0$ for $|\omega| \geq \pi/T$. The variance of the measured DC component is inversely proportional to the duration over which the arithmetical average is taken. Once (an estimate of) $A \ \exp(j\psi)$ is available, it can be used in the configuration of Figure 2-50 to generate $w_C(t;\varepsilon)/A$ and $w_S(t;\varepsilon)/A$.

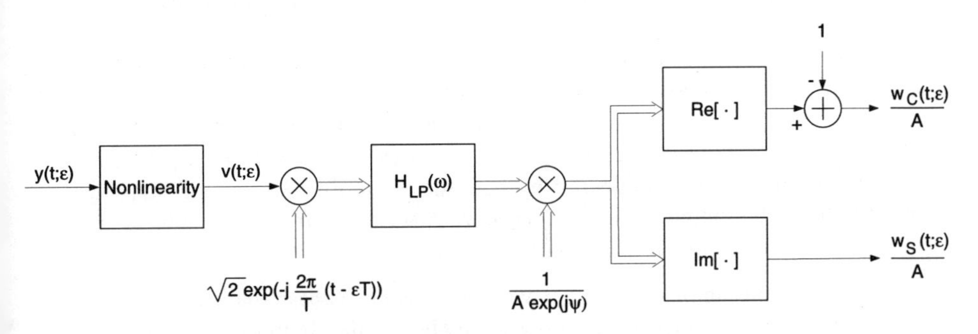

Figure 2-50 Generation of $w_C(t;\varepsilon)/A$ and $w_S(t;\varepsilon)/A$

2.4.6 Main Points

Spectral line generating clock synchronizers perform a nonlinear operation on the received noisy PAM waveform. The resulting signal at the output of the nonlinearity is the sum of a periodic signal with period equal to the channel symbol interval T (this signal contains timing information about the PAM signal), and a zero-mean disturbance. This periodic signal can be decomposed as the sum of sinusoidal components, with frequencies $1/T$ and multiples thereof; this corresponds to spectral lines at the same frequencies. The signal at the output of the nonlinearity enters either a PLL, or a narrowband bandpass filter. In the case of a PLL, a local reference signal is generated, which tracks the periodic signal; when using a PLL with multiplying timing error detector and sinusoidal reference signal with period T, the periodic signal's sinusoidal component at frequency $1/T$ is tracked. In the case of a narrowband bandpass filter, tuned to the channel symbol rate $1/T$, the sinusoidal component at frequency $1/T$, present at the output of the nonlinearity, is isolated. The sampler in the decision branch of the receiver is activated each time the reference signal of the PLL or the signal at the output of the bandpass filter crosses a threshold level; because of DC offsets, this threshold level can be different from zero. The resulting sampling instants do not coincide with the optimum sampling instants; the timing error consists of a static timing offset and a random zero-mean timing jitter.

When using a PLL with zero steady-state error, the timing jitter is caused by the phase fluctuations (or, equivalently, by the quadrature component) of the signal at the output of the nonlinearity. A nonzero threshold level gives rise to a static timing offset only, because the reference signal of the PLL does not exhibit amplitude fluctuations. When the steady-state error is nonzero, it introduces an additional timing offset and gives rise to amplitude-to-phase conversion: the timing jitter is caused not only by the phase fluctuations at the nonlinearity output, but also by the amplitude fluctuations (or, equivalently, by the in-phase component).

In the case of a symmetric bandpass filter and zero threshold level, the timing jitter is caused by the phase fluctuations at the output of the nonlinearity, and there is no static timing offset. When the bandpass filter is still symmetric, but the threshold level is nonzero, both a static timing offset and amplitude-to-phase conversion occur: the timing jitter is caused not only by the phase fluctuations, but also by the amplitude fluctuations, at the output of the nonlinearity. The situation becomes even more complicated when the bandpass filter is not symmetric: this situation gives rise to amplitude-to-phase and phase-to-amplitude conversions from the input to the output of the bandpass filter, so that even with a zero threshold level, the timing jitter is influenced by both the phase fluctuations and amplitude fluctuations at the output of the nonlinearity. Also, an asymmetric bandpass filter gives rise to an additional static timing offset.

For a symmetric bandpass filter with equivalent lowpass frequency response $H_E(\omega)$, and a threshold level yielding an angle ψ_η, the timing jitter is the same as for a PLL with closed loop frequency response $H_E(\omega)/H_E(0)$ and steady-state error $e_s = -\psi_\eta/2\pi$. When the bandpass filter is not symmetric, there exists no equivalent structure with a PLL, giving rise to the same timing jitter.

In many cases, the disturbance at the output of the nonlinearity has an in-phase spectrum which is much larger than the quadrature spectrum, in the vicinity of $\omega = 0$. Therefore, it is advantageous to combine a small bandwidth of the PLL or the bandpass filter with a design which avoids amplitude-to-phase conversion. A bandpass filter should be symmetric about the channel symbol rate $1/T$ in order to eliminate amplitude-to-phase conversion. The degree of asymmetry caused by frequency detuning is closely related to the bandpass filter bandwidth: for a given frequency detuning, the asymmetry and, hence, the amplitude-to-phase conversion, increases with decreasing filter bandwidth. Therefore, when using a narrowband bandpass filter, the jitter variance is very sensitive to frequency detuning. In the case of a PLL, amplitude-to-phase conversion is eliminated when the steady-state error is zero. The steady-state error (due to frequency detuning) can be kept very small by selecting a large open-loop gain. As the open-loop gain and the loop bandwidth can be controlled independently, it is possible to realize simultaneously a small amplitude-to-phase conversion and a small loop bandwidth. From this point of view, a PLL is more attractive than a bandpass filter.

The disturbance at the output of the nonlinearity is a cyclostationary process with period T. It might be tempting to replace this cyclostationary disturbance by a spectrally equivalent stationary process and evaluate the resulting jitter variance. This method yields an erroneous result, because it does not take into account that the true in-phase and quadrature disturbances usually have drastically different spectra near $\omega = 0$. In many cases, the method of the spectrally equivalent stationary disturbance overestimates the true timing jitter variance by several orders of magnitude.

A repeater chain is used when data is to be transmitted over long distances. The linearized timing jitter at the output of a repeater equals the sum of two terms:

- The first term is a filtered version of the jitter at the input of the repeater; the frequency response of the filter is called the jitter frequency response.
- The second term is a contribution from the repeater itself.

When the jitter frequency response exhibits peaking, the jitter variance at the end of the chain increases exponentially with the number of repeaters. For keeping the jitter accumulation rate within reasonable limits, the amount of peaking should be small.

In most cases the analytical evaluation of the timing error variance resulting from a spectral line generating synchronizer is extremely complicated. In these cases, it is advantageous to use computer simulations for obtaining numerical performance results.

2.4.7 Bibliographical Notes

Spectral line generating synchronizers have been used in a large variety of applications, ranging from the early PCM telephone network to present-day fiber optical networks; as a result, there exists a vast literature on the subject. In this section, we consider a few papers which have contributed to a better understanding of the operation of the broad category of spectral line generating synchronizers as a whole. Papers dealing with synchronizers employing specific nonlinearities are covered in Section 6.

The general analysis of spectral line generating synchronizers consisting of a nonlinearity followed by a PLL (Section 2.4.1) is inspired by [1], which investigates the tracking performance of a PLL operating on a periodic signal corrupted by cyclostationary additive noise. The general analysis of Section 2.4.2 follows the same lines as [2, 3], which evaluates the tracking performance of spectral line generating synchronizers consisting of a nonlinearity followed by a bandpass filter.

Both [4] and [5] point out that replacing the cyclostationary disturbance at the output of the nonlinearity by a "spectrally equivalent" stationary disturbance obliterates the distinction between the in-phase and quadrature components of the cyclostationary disturbance. In [5], several practical cases are mentioned where this erroneous approach may overestimate the true timing error variance by orders of magnitude.

The computer simulation approach, outlined in Section 2.4.5, is an extension of [6], where only the quadrature component $w_S(t;\varepsilon)$ was needed to determine the timing error variance resulting from various spectral line generating synchronizers with a symmetrical bandpass filter and zero threshold level.

The use of a surface-acoustic-wave (SAW) bandpass filter for extracting the spectral line at the symbol rate is investigated in [7–10], for high-speed (0.1 to 2 Gbit/s) fiber-optic transmission systems. Both resonator SAW filters and transversal SAW filters are considered. Single-resonator and maximally flat (Butterworth) double-resonator SAW filters have no passband ripple, so that exponential jitter growth along a repeater chain is avoided. Transversal SAW filters exhibit passband ripple, caused by reflections between the transducers; this ripple must be limited by careful circuit design in order to avoid a too rapid jitter increase along the repeater chain.

In Section 2.4.4, the problem of jitter accumulation in a repeater chain has been considered very briefly, although a vast literature exists on that topic. The interested reader is referred to the textbook [11] and the reference list therein (which includes [3, 7, 8]). Cycle slipping in a chain of repeaters with error-tracking symbol synchronizers has been investigated in [12, 13]: it is shown that the bandwidth of the loop noise at the end of a long repeater chain is much smaller than the loop bandwidth; this yields a cycle slip rate that can be orders of magnitude larger than in the usual case where the bandwidth of the loop noise is much larger than the loop bandwidth.

Bibliography

[1] M. Moeneclaey, "Linear Phase-Locked Loop Theory for Cyclostationary Input Disturbances," *IEEE Trans. Commun.*, vol. COM-30, pp. 2253–2259, Oct. 1982.

[2] L. E. Franks and J. P. Bubrouski, "Statistical Properties of Timing Jitter in a PAM Timing Recovery Scheme," *IEEE Trans. Commun.*, vol. COM-22, pp. 913–920, July 1974.

[3] U. Mengali and G. Pirani, "Jitter Accumulation in PAM Systems," *IEEE Trans. Commun.*, vol. COM-28, pp. 1172–1183, Aug. 1980.

[4] L. E. Franks, "Carrier and Bit Synchronization in Data Communications–A Tutorial Review," *IEEE Trans. Commun.*, vol. COM-28, pp. 1107–1121, Aug. 1980.

[5] F. M. Gardner, "Self-Noise in Synchronizers," *IEEE Trans. Commun.*, vol. COM-28, pp. 1159–1163, Aug. 1980.

[6] A. N. D'Andrea and U. Mengali, "A Simulation Study of Clock Recovery in QPSK and 9 QPRS Systems," *IEEE Trans Commun.*, vol. COM-33, pp. 1139–1142, Oct. 1985.

[7] P. D. R. L. Rosenberg, D. G. Ross and C. Armitage, "Optical Fibre Repeatered Transmission System Utilizing SAW Filters," *IEEE Trans. Sonics Ultrason*, vol. SU-30, pp. 119–126, May 1983.

[8] R. D. A. Fishman, "Timing Recovery with SAW Transversal Filters in the Regenerators of Undersea Long-Haul Fiber Transmission Systems," *IEEE J. Selected Areas Commun.*, vol. SAC-2, pp. 957–965, Nov. 1984.

[9] C. Chamzas, "Accumulation of Jitter: A Stochastical Model," *AT&T Tech. J.*, vol. 64, pp. 43–76, Jan. 1985.

[10] D. A. Fishman, R. L. Rosenberg, and C. Chamzas, "Analysis of Jitter Peaking Effects in Digital Long-Haul Transmission Systems Using SAW-Filter Retiming," *IEEE Trans. Commun.*, vol. COM-33, pp. 654–664, July 1985.

[11] P. R. Trischitta and E. L. Varma, *Jitter in Digital Transmission Systems.* 1989.

[12] H. Meyr, L. Popken, and H. R. Mueller, "Synchronization Failures in a Chain of PLL Synchronizers," *IEEE Trans. Commun.*, vol. COM-34, pp. 436–445, May 1986.

[13] M. Moeneclaey, S. Starzak and H. Meyr, "Cycle Slips in Synchronizers with Smooth Narrowband Loop Noise," *IEEE Trans. Commun.*, vol. COM-36, pp. 867–874, July 1988.

2.5 Examples

2.5.1 Example 1: The Squaring Synchronizer

The squaring synchronizer is a spectral line generating clock synchronizer. Its block diagram is shown in Figure 2-10, the nonlinearity being a squarer. In Section 2.4, the tracking performance of spectral line generating clock synchronizers with arbitrary nonlinearity has been derived from the statistical properties of the signal at the output of the nonlinearity. In this section, we determine these statistical properties in the case of a squarer operating on a noisy PAM signal, and derive conditions on the baseband pulse at the squarer input for reducing the self-noise. In addition, we present an alternative version of the squaring synchronizer, where the squarer is followed by Fourier transform computation; this version is well suited for fully digital implementation.

Throughout this section, it will be assumed that the spectral line at the symbol rate is to be extracted; when this is accomplished by means of a PLL, we restrict our attention to a PLL with multiplying timing error detector and sinusoidal reference signal.

Statistical Properties of the Squarer Output

The noisy PAM signal $y(t;\varepsilon)$ at the squarer input is given by

$$y(t;\varepsilon) = \sum_m a_m \, g(t - mT - \varepsilon T) + n(t)$$

where $\{a_m\}$ is a stationary sequence of independent (not necessarily binary) data symbols, $g(t)$ is the baseband pulse, and $n(t)$ is stationary Gaussian noise. Squaring the noisy PAM signal yields

$$\begin{aligned} v(t;\varepsilon) &= y^2(t;\varepsilon) \\ &= \sum_{m,n} a_m \, a_{m+n} \, g(t - mT - \varepsilon T) \, g(t - mT - nT - \varepsilon T) \\ &\quad + 2\, n(t) \sum_m a_m g(t - mT - \varepsilon T) + n^2(t) \end{aligned} \tag{2-149}$$

The three terms in (2-149) are uncorrelated. The first term in (2-149) is a signal $\times$ signal $(S \times S)$ term; this term contains not only spectral lines (at DC and multiples of the symbol rate), but also self-noise, which is caused by the random nature of the data sequence. The second term in (2-149) is a zero-mean signal $\times$ noise $(S \times N)$ term which acts as a disturbance. The third term in (2-149) is a noise $\times$ noise $(N \times N)$ term; it contains a spectral line at DC (which does not influence the synchronizer operation) and a zero-mean random fluctuation which acts as a disturbance.

Let us consider the useful sinusoid at the symbol rate, appearing at the squarer output. Suppose the Fourier transform $G(\omega)$ of the baseband pulse $g(t)$

is zero for $|\omega| \geq \pi W$. In this case, the $(S \times S)$ term in (2-149) is band-limited to $|\omega| < 2\pi W$. Consequently, irrespective of the statistical properties of the stationary data sequence, the $(S \times S)$ term can contain spectral lines at $\omega = 2\pi k/T$ only for $k < WT$. Hence, the bandwidth (in hertz) of the baseband pulse $g(t)$ should exceed $1/(2T)$, otherwise the squarer output contains a spectral line at DC only, which of course, carries no timing information. Indeed, when $G(\omega) = 0$ for $|\omega| > \pi/T$, the PAM signal $y(t;\varepsilon)$ is wide-sense stationary (see Section 1.1), so that the statistical average of its square does not depend on time; a nonlinearity of higher order would be needed to generate spectral lines. In most cases of practical interest, the transmission bandwidth (in hertz) is larger than $1/(2T)$ in order to avoid large ISI at the decision instants; in these cases a squarer can be used for generating spectral lines. For narrowband communication, the bandwidth of $g(t)$ is between $1/(2T)$ and $1/T$; in this case the squarer contains spectral lines only at DC and at the symbol rate. For independent data symbols, the average squarer output signal is given by

$$
\begin{aligned}
E[v(t;\varepsilon)] = &\left(A_2 - A_1^2\right) \sum_m g^2(t - mT - \varepsilon T) \\
&+ A_1^2 \left[\sum_m g(t - mT - \varepsilon T)\right]^2 + R_n(0)
\end{aligned}
\tag{2-150}
$$

where $A_i = E\left[a_k^i\right]$ denotes the ith order moment of the data symbols, and $R_n(u) = E[n(t)\, n(t+u)]$ is the autocorrelation function of the additive noise $n(t)$ at the squarer input. Noting that $A_2 - A_1^2$ equals the variance of the data symbols, the first term in (2-150) is caused by the data symbol transitions; this term would be zero if any given symbol were transmitted continuously. The second term in (2-150) is simply the square of the average $E[y(t;\varepsilon)]$ of the squarer input signal; this term is zero for zero-mean data symbols (i.e., $A_1 = 0$). When the data symbols are not zero-mean, $E[y(t;\varepsilon)]$ is periodic in T with period T, because of the cyclostationarity of the noisy PAM signal $y(t;\varepsilon)$:

$$
\begin{aligned}
E[y(t;\varepsilon)] &= A_1 \sum_m g(t - mT - \varepsilon T) \\
&= \frac{A_1}{T} \sum_m G\left(\frac{m}{T}\right) \exp\left[j\, \frac{2\pi m}{T}\, (t - \varepsilon T)\right]
\end{aligned}
$$

When $G(m/T) = 0$ for $m \neq 0$ (this condition is fulfilled, for example, in the case of narrowband communication), $E[y(t;\varepsilon)]$ reduces to a constant $A_1\, G(0)/T$, independent of time. The third term in (2-150) is independent of time, and equals the additive noise power at the squarer input. Hence, in the case of narrowband communication, only the first term in (2-150) gives rise to a spectral line at the symbol rate; as this term is caused by the transitions of the data symbols, the transmission of long strings of identical symbols should be avoided, for instance, by using a scrambler at the transmitter. Assuming narrowband transmission and/or zero-mean data symbols, the squarer output contains a sinusoid at the symbol rate, given by

$$\sqrt{2}\,A\,\cos\left[\frac{2\pi}{T}\,(t-\varepsilon T)+\psi\right]=2\,\mathrm{R}\left\{\mathrm{em}_{\mathrm{v},1}\,\exp\left[\mathrm{j}\,\frac{2\pi}{\mathrm{T}}\,(\mathrm{t}-\varepsilon\mathrm{T})\right]\right\} \tag{2-151}$$

where

$$\begin{aligned} m_{\mathrm{v},1} &= \frac{1}{T}\left(A_2-A_1^2\right)\int_{-\infty}^{+\infty} g^2(t)\,\exp\left(-j\,\frac{2\pi t}{T}\right)dt \\ &= \frac{1}{T}\left(A_2-A_1^2\right)\int_{-\infty}^{+\infty} G\left(\frac{2\pi}{T}-\omega\right)G(\omega)\,\frac{d\omega}{2\pi} \end{aligned} \tag{2-152}$$

Hence, the phase ψ of the sinusoid at the symbol rate is given by $\psi=\arg\left(m_{v,1}\right)$. The magnitude of the sinusoid at the symbol rate decreases with decreasing overlap between $G(2\pi/T-\omega)$ and $G(\omega)$; hence, this magnitude is very small when the bandwidth in excess of $1/(2T)$ is small.

It has been shown in Section 2.4 that the random part of the timing error resulting from a spectral line generating synchronizer can be viewed as a linear combination of the narrowband lowpass filtered in-phase and quadrature components $w_C(t;\varepsilon)$ and $w_S(t;\varepsilon)$ with respect to the useful sinusoid at the output of the nonlinearity. When using a PLL for extracting the useful sinusoid, the coefficients of the linear combination are a function of the steady-state error e_S, while the filtering depends on the closed-loop frequency response (see Figure 2-36). When using a narrowband bandpass filter for extracting the useful sinusoid, the coefficients of the linear combination are a function of the threshold level η, while the filtering depends on the frequency response of the bandpass filter (see Figure 2-39). In either case (PLL or bandpass filter), the random part of the timing error can be expressed as

$$e(t)=\int_{-\infty}^{+\infty}\left[h_1(t-u)\,w_C(u;\varepsilon)+h_2(t-u)\,w_S(u;\varepsilon)\right]du$$

where $h_1(t)$ and $h_2(t)$ are the impulse responses depending on the synchronizer properties. This yields

$$\begin{aligned} \mathrm{var}[e]=\int_{-\infty}^{+\infty}\Big[&|H_1(\omega)|^2\,S_C(\omega)+|H_2(\omega)|^2\,S_S(\omega) \\ &+H_1(\omega)\,H_2^*(\omega)\,S_{SC}(\omega)+H_2(\omega)\,H_1^*(\omega)\,S_{CS}(-\omega)\Big]\,\frac{d\omega}{2\pi} \end{aligned}$$

where $H_1(\omega)$ and $H_2(\omega)$ are the Fourier transforms of $h_1(t)$ and $h_2(t)$, $S_C(\omega)$ and $S_S(\omega)$ are the spectra of $w_C(t;\varepsilon)$ and $w_S(t;\varepsilon)$, and $S_{CS}(\omega)$ is the cross-spectrum between $w_C(t;\varepsilon)$ and $w_S(t;\varepsilon)$.

Analytical expressions for the spectra $S_C(\omega)$, $S_S(\omega)$ and $S_{CS}(\omega)$ at the squarer output are derived in the appendix at the end of this section, based upon the method described in Section 2.4.1. These computations are straightforward but quite tedious, and the resulting expressions provide only limited insight. The simulation method from Section 2.4.4 is a valid alternative to obtain numerical results.

Self-Noise Reduction

In the following, we derive a condition on the baseband pulse $g(t)$ at the squarer input, such that, in the absence of noise, the quadrature spectrum $S_S(\omega)$ and the in-phase/quadrature cross-spectrum $S_{CS}(\omega)$ at the squarer output are both identically zero within the synchronizer bandwidth, irrespective of the data symbol statistics. Also, less restrictive conditions on $g(t)$ will be presented, for which, in the absence of additive noise, these spectra are zero at $\omega = 0$, irrespective of the data symbol statistics. When the squarer is followed either by a PLL with zero steady-state error, or by a symmetrical bandpass filter centered at the symbol rate and the zero crossings of the bandpass filter output signal are used for timing, then the former condition on $g(t)$ eliminates the self-noise, whereas the less restrictive conditions on $g(t)$ yield a self-noise contribition to the timing error variance which is proportional to the square (or even a higher power) of the normalized synchronizer bandwidth $B_L T$.

Condition yielding $S_S(\omega) = S_{CS}(\omega) = 0$ ***for*** $|\omega| < 2\pi B$

Let us first deal with the condition on $g(t)$ such that, in the absence of noise, $S_S(\omega)$ and $S_{CS}(\omega)$ are zero within the synchronizer bandwidth. We denote by $s(t;\varepsilon)$ the useful signal component of the noisy PAM signal at the squarer input:

$$s(t;\varepsilon) = \sum_m a_m \, g(t - mT - \varepsilon T) \tag{2-153}$$

Now suppose we are able to select the baseband pulse $g(t)$ in such a way that $s(t;\varepsilon)$ is a bandpass signal, centered at $\omega = \pi/T$, whose complex envelope $s_L(t;\varepsilon)$ happens to be real-valued:

$$s(t;\varepsilon) = \sqrt{2}\, s_L(t;\varepsilon) \, \cos\left[\frac{\pi}{T}(t - \varepsilon T)\right] \tag{2-154}$$

Note that $s(t;\varepsilon)$ has no quadrature component with respect to $\cos\left[\pi(t - \varepsilon T)/T\right]$; equivalently, $s(t;\varepsilon)$ is a purely amplitude-modulated signal. In addition, we assume that $s_L(t;\varepsilon)$ is band-limited to the interval $|\omega| \le \pi/T - \pi B$, with $0 < B < 1/T$. When additive noise is absent, the squarer output is given by

$$s^2(t;\varepsilon) = s_L^2(t;\varepsilon) + \, s_L^2(t;\varepsilon) \, \cos\left[\frac{2\pi}{T}(t - \varepsilon T)\right] \tag{2-155}$$

The first term in (2-155) is a lowpass signal band-limited to the interval $|\omega| \le 2\pi/T - 2\pi B$. The second term in (2-155) is a bandpass signal, centered at $\omega =$

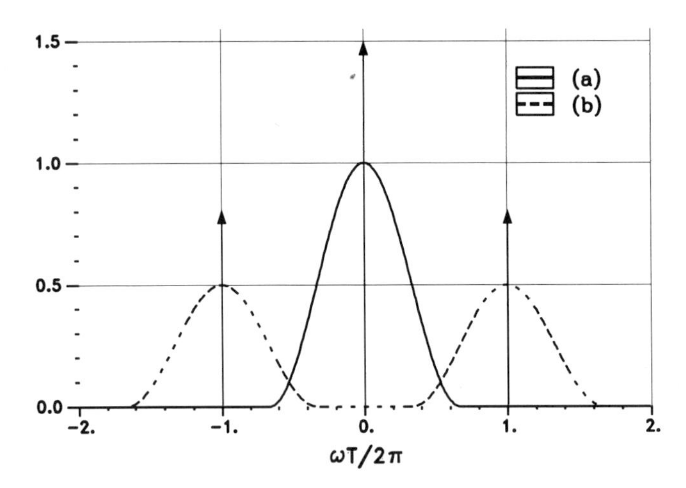

Figure 2-51 (a) Spectrum of $s_L^2(t;\varepsilon)$, (b) Spectrum of $s_L^2(t;\varepsilon)\ \cos(2\pi(t-\varepsilon T)/T)$

$2\pi/T$, and having no quadrature component with respect to $\cos(2\pi(t-\varepsilon T)/T)$; the frequency content of this second term is restricted to the interval $2\pi B \le |\omega| \le 4\pi/T - 2\pi B$. The spectra of the first and second term in (2-155) are sketched in Figure 2-51; because of the band-limited nature of these terms, spectral lines appear only at $\omega = 0$ and $\omega = \pm 2\pi/T$. When the squarer is followed either by a PLL whose closed-loop frequency response is zero for $|\omega| > 2\pi B$ or by a bandpass filter whose equivalent lowpass frequency response is zero for $|\omega| > 2\pi B$, then only the frequency content of $s^2(t;\varepsilon)$ in the interval $||\omega| - 2\pi/T| < 2\pi B$ affects the synchronizer operation. As the spectrum of the first term in (2-155) is zero within this interval, this term can be ignored. As the second term has no quadrature component with respect to $\cos(2\pi(t-\varepsilon T)/T)$, it follows that:

- The useful sinusoid at the squarer output is, within a multiplicative constant, equal to $\cos(2\pi(t-\varepsilon T)/T)$.
- There is no quadrature component with respect to the useful sinusoid at the squarer output that affects the synchronizer operation.

Hence, $S_S(\omega)$ and $S_{CS}(\omega)$ are both zero for $|\omega| < 2\pi B$.

Our remaining task is to find a baseband pulse $g(t)$ for which (2-153) reduces to (2-154), with $s_L(t;\varepsilon)$ band-limited as specified above. We select a pulse $g(t)$ whose complex envelope $g_L(t)$ with respeet to $\cos(\pi t/T)$ is real-valued and band-limited:

$$g(t) = \sqrt{2}\, g_L(t)\ \cos\left[\frac{\pi t}{T}\right] \tag{2-156}$$

where $g_L(t)$ is a real-valued pulse whose Fourier transform $G_L(\omega)$ is zero for

$|\omega| > \pi/T - \pi B$. Substituting (2-156) in (2-153) yields

$$\begin{aligned} s(t;\varepsilon) &= \sqrt{2} \sum_m a_m g_L(t - mT - \varepsilon T) \cos\left[\frac{\pi}{T}(t - mT - \varepsilon T)\right] \\ &= \sqrt{2}\, s_L(t;\varepsilon) \cos\left[\frac{\pi}{T}(t - \varepsilon T)\right] \end{aligned}$$

with

$$s_L(t;\varepsilon) = \sum_m (-1)^m\, a_m\, g_L(t - mT - \varepsilon T)$$

As $g_L(t)$ is real-valued and band-limited to $|\omega| \le \pi/T - \pi B$, the same is true for $s_L(t;\varepsilon)$; consequently, $S_S(\omega) = S_{CS}(\omega) = 0$ for $|\omega| < 2\pi B$. Denoting the Fourier transform of $g(t)$ by $G(\omega)$, it follows from Section 1.2.2 that $G_L(\omega)$ and $G(\omega)$ are related by

$$G_L(\omega) = \sqrt{2}\, u\left(\omega + \frac{\pi}{T}\right) G\left(\omega + \frac{\pi}{T}\right)$$

where $u(\cdot)$ denotes the unit step function. Expressing that $g_L(t)$ is real-valued and band-limited yields the following conditions on $G(\omega)$:

$$u\left(\omega + \frac{\pi}{T}\right) G\left(\omega + \frac{\pi}{T}\right) = u\left(-\omega + \frac{\pi}{T}\right) G^*\left(-\omega + \frac{\pi}{T}\right) \tag{2-157}$$

$$u\left(\omega + \frac{\pi}{T}\right) G\left(\omega + \frac{\pi}{T}\right) = 0 \quad \text{for } |\omega| > \frac{\pi}{T} - \pi B \tag{2-158}$$

The condition (2-157) indicates that $u(\omega)\, G(\omega)$ exhibits complex conjugate symmetry with respect to $\omega = 2\pi/T$, while condition (2-158) imposes the bandwith limitation on $g(t)$.

Conditions yielding $S_S(\omega) = S_{CS}(\omega) = 0$ ***for*** $\omega = 0$

Now we derive less restrictive conditions on $g(t)$, such that, in the absence of noise, $S_C(\omega)$ and $S_{CS}(\omega)$ are zero for $\omega = 0$. Suppose we are able to select the baseband pulse $g(t)$ such that

$$\int_{-\infty}^{+\infty} s^2(t;\varepsilon) \sin\left[\frac{2\pi}{T}(t - \varepsilon T)\right] dt = 0 \tag{2-159}$$

irrespective of the data sequence $\{a_k\}$. When (2-159) is true, the following can be derived.

(i) The spectral lines contained in $s^2(t;\varepsilon)$ are orthogonal with respect to $\sin(2\pi(t - \varepsilon T)/T)$. Consequently, the component at the symbol rate is, within a multiplicative constant, equal to $\cos(2\pi(t - \varepsilon T)/T)$.

(ii) Let us denote the in-phase and quadrature components of $s^2(t;\varepsilon)$ with respect to the useful sinusoidal component by $w_C(t;\varepsilon)$ and $w_S(t;\varepsilon)$, respectively:

$$s^2(t;\varepsilon) = \sqrt{2}\,\mathrm{Re}\left[[w_C(t;\varepsilon) + j w_S(t;\varepsilon)]\,\exp\left[j\,\frac{2\pi}{T}(t-\varepsilon T)\right]\right] \tag{2-160}$$

According to Section 1.2, substituting (2-160) into (2-159) yields

$$\begin{aligned} 0 &= \int_{-\infty}^{+\infty} s^2(t;\varepsilon)\,\sin\left[\frac{2\pi}{T}(t-\varepsilon T)\right]\,dt \\ &= -\frac{\sqrt{2}}{2}\int_{-\infty}^{+\infty} w_S(t;\varepsilon)\,dt \end{aligned}$$

Hence, the integral of the quadrature component is zero.

(iii) Because of (ii), both the quadrature spectrum $S_S(\omega)$ and the cross-spectrum $S_{CS}(\omega)$ between the in-phase component $w_C(t;\varepsilon)$ and the quadrature component $w_S(t;\varepsilon)$ are zero at $\omega = 0$. Indeed,

$$\begin{aligned} S_S(0) &= \int_{-\infty}^{+\infty} \langle E[w_S(t;\varepsilon)\,w_S(t+u;\varepsilon)]\rangle_t\,du \\ &= \left\langle E\left[w_S(t;\varepsilon)\int_{-\infty}^{+\infty} w_S(t+u;\varepsilon)\,du\right]\right\rangle_t \\ &= 0 \end{aligned}$$

Similarly,

$$\begin{aligned} S_{CS}(0) &= \int_{-\infty}^{+\infty} \langle E[w_C(t;\varepsilon)\,w_S(t+u;\varepsilon)]\rangle_t\,du \\ &= \left\langle E\left[w_C(t;\varepsilon)\int_{-\infty}^{+\infty} w_S(t+u;\varepsilon)\,du\right]\right\rangle_t \\ &= 0 \end{aligned}$$

Substituting (2-153) into (2-159) yields

$$\int_{-\infty}^{+\infty} s^2(t;\varepsilon)\,\sin\left[\frac{2\pi}{T}(t-\varepsilon T)\right]\,dt = \sum_{m,n} a_m a_{m+n} I_n$$

where

$$I_n = \int_{-\infty}^{+\infty} g(t)\, g(t-nT)\, \sin\left(\frac{2\pi t}{T}\right) dt \tag{2-161}$$

In the following, we look for pulses $g(t)$ which make $I_n = 0$ for all n. For these pulses, (2-159) is fulfilled; consequently, $S_S(0) = S_{SC}(0) = 0$.

Substituting in (2-161) t by $nT/2 + u$, we obtain

$$I_n = (-1)^n \int_{-\infty}^{+\infty} g\left(\frac{nT}{2} + u\right) g\left(-\frac{nT}{2} + u\right) \sin\left(\frac{2\pi u}{T}\right) du \tag{2-162}$$

For even symmetrical or odd symmetrical pulses $g(t)$, it is easily verified that the product $g(nT/2 + u)\, g(-nT/2 + u)$ is an even function of u. Hence, $I_n = 0$ because the integrand in (2-162) is odd. We conclude that (2-159) is fulfilled when the baseband pulse $g(t)$ at the squarer input exhibits even or odd symmetry.

Now we consider the case where (2-157) is fulfilled; hence, the complex envelope $g_L(t)$ of $g(t)$ is real, so that (2-156) holds. Substituting (2-156) into (2-161) yields

$$\begin{aligned} I_n &= 2 \int_{-\infty}^{+\infty} g_L(t)\, g_L(t-nT)\, \cos\left(\frac{\pi t}{T}\right) \cos\left(\frac{\pi}{T}(t-nT)\right) \sin\left(\frac{2\pi t}{T}\right) dt \\ &= (-1)^n \int_{-\infty}^{+\infty} g_L(t)\, g_L(t-nT) \left[\sin\left(\frac{2\pi t}{T}\right) + \frac{1}{2} \sin\left(\frac{4\pi t}{T}\right)\right] dt \end{aligned} \tag{2-163}$$

Noting that (2-157) implies that $G_L(\omega) = 0$ for $|\omega| > \pi/T$, it follows that the Fourier transform of $g_L(t)\, g_L(t-nT)$ is zero for $|\omega| \geq 2\pi/T$. As I_n in (2-163) can be interpreted as a linear combination of Fourier transform values of $g_L(t)\, g_L(t-nT)$, evaluated at $\omega = \pm 2\pi/T$ and $\omega = \pm 4\pi/T$, it follows that $I_n = 0$. We conclude that (2-159) is fulfilled when (2-157) holds, i.e., when $u(\omega)\, G(\omega)$ exhibits complex conjugate symmetry with respect to $\omega = 2\pi/T$, or equivalently, when the complex envelope of the baseband pulse $g(t)$ is real-valued.

Let us summarize our results on self-noise reduction.

(i) In the absence of additive noise, the spectra $S_S(\omega)$ and $S_{CS}(\omega)$ are zero for $|\omega| < 2\pi B$ when the complex envelope of $g(t)$ with respect to $\cos(\pi t/T)$ is real-valued and band-limited to $|\omega| < \pi/T - \pi B$, or equivalently, when $u(\omega)\ G(\omega)$ exhibits complex conjugate symmetry about $\omega = \pi/T$ and $G(\omega) = 0$ for $||\omega| - \pi/T| > \pi/T - \pi B$.

(ii) In the absence of additive noise, the spectra $S_S(\omega)$ and $S_{CS}(\omega)$ are zero for $\omega = 0$, when *either* $g(t)$ exhibits even or odd symmetry, *or* the complex envelope of $g(t)$ with respect to $\cos(\pi t/T)$ is real-valued, or equivalently, $u(\omega)\ G(\omega)$ exhibits complex conjugate symmetry about $\omega = \pi/T$.

Suppose the timing error is caused only by the quadrature component $w_S(t;\varepsilon)$ at the squarer output; this happens for a zero steady-state error at the PLL or for a symmetrical bandpass filter and a zero threshold level. When the condition under (i) is fulfilled, then self-noise does not contribute to the tracking error variance (provided that the closed-loop frequency response of the PLL or the equivalent lowpass frequency response of the bandpass filter are zero for $|\omega| > 2\pi B$); when one of the conditions under (ii) is fulfilled, the self-noise contribution to the timing error variance is proportional to the square (or even a higher power) of the normalized synchronizer bandwidth B_LT. However, when the steady-state error of the PLL is nonzero, or when the bandpass filter is not symmetrical and/or the threshold level is nonzero, also the in-phase component $w_C(t;\varepsilon)$ contributes to the timing error; when in this case one of the conditions under (i) or (ii) is fulfilled, the self-noise contribution to the timing error variance is essentially proportional to B_LT, because the spectrum $S_C(\omega)$ of the in-phase component $w_C(t;\varepsilon)$ is nonzero at $\omega = 0$. This illustrates the importance of keeping small either the steady-state error at the PLL or both the bandpass filter asymmetry and the threshold level.

In most cases, the baseband pulse at the receive filter output is (very close to) an even symmetrical pulse satisfying the first Nyquist criterion, because of the equalization involved (see Section 2.1.2). Hence, because of the even symmetry of the baseband pulse, the receive filter output signal is well conditioned to be applied directly to the squaring synchronizer. However, if one prefers the squaring synchronizer to operate on a PAM signal with a baseband pulse satisfying one of the other conditions under (i) or (ii), the receive filter output should be prefiltered before entering the squaring synchronizer. The need for prefiltering is obvious when we compare in Figure 2-52, in the case of narrowband transmission,

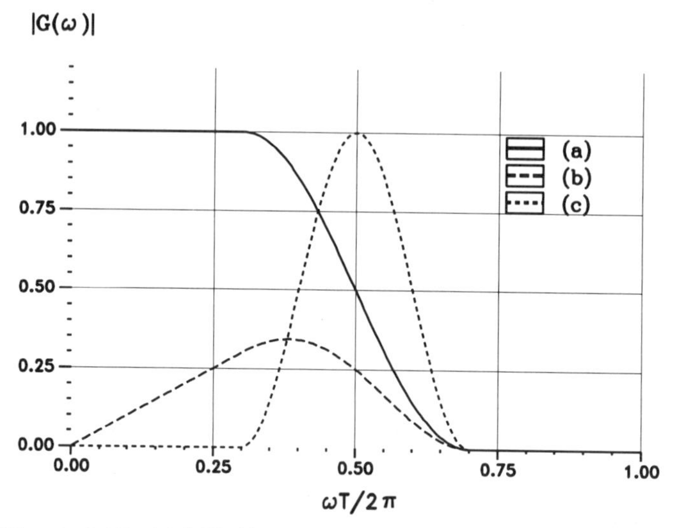

Figure 2-52 (a) $|G(\omega)|$ for Even-Symmetrical Nyquist-I Pulse, (b) $|G(\omega)|$ for Odd Symmetrical Pulse, (c) $|G(\omega)|$ Having Complex Conjugate Symmetry about $\omega = \pi/T$

the magnitudes $|G(\omega)|$ for an even symmetrical Nyquist-I pulse, for an odd symmetrical baseband pulse and for a baseband pulse for which $u(\omega)G(\omega)$ exhibits complex conjugate symmetry with respect to $\omega = \pi/T$ (see Figure 2-52).

Alternative Squaring Synchronizer Implementation

Let us reconsider the case where the useful sinusidal component $\sqrt{2}\ A\ \cos(2\pi(t-\varepsilon T)/T+\psi)$, at the output of the squarer operating on the baseband PAM signal $y(t;\varepsilon)$, is extracted by means of a narrowband bandpass filter, centered at $\omega = 2\pi/T$. The bandpass filter output signal $v_0(t)$ is a sinusoidal signal exhibiting small fluctuations in amplitude and phase, and can be represented as:

$$v_0(t) = \sqrt{2}\, A_0(t)\ \cos\left(\frac{2\pi}{T}(t-\hat{\varepsilon}(t)T)+\psi\right) \tag{2-164}$$

where $\hat{\varepsilon}(t)$ is the estimate of ε, the normalized time delay of $y(t;\varepsilon)$. Denoting the equivalent lowpass frequency response of the bandpass filter by $H_E(\omega)$ and the corresponding impulse response by $h_E(t)$, the bandpass filter output signal is given by

$$v_0(t) = 2\mathrm{Re}\left[z(t)\ \exp\left(j\frac{2\pi t}{T}\right)\right] \tag{2-165}$$

where

$$z(t) = \int_{-\infty}^{+\infty} h_E(t-u)\ v(u;\varepsilon)\ \exp\left(-j\frac{2\pi u}{T}\right)\, du \tag{2-166}$$

$$= \int_{-\infty}^{+\infty} H_E(\omega)\ V\left(w+\frac{2\pi}{T}\ ;\varepsilon\right)\ \exp(j\omega t)\ \frac{d\omega}{2\pi} \tag{2-167}$$

and $V(\omega;\varepsilon)$ denotes the Fourier transform of $v(t;\varepsilon)$. Comparing (2-164) and (2-165), it follows that the timing estimate satisfies the following equation:

$$\hat{\varepsilon}(t) = -\frac{1}{2\pi}\ \arg[z(t)] + \frac{\psi}{2\pi} \tag{2-168}$$

The resulting tracking error variance $E\left[(\hat{\varepsilon}(t)-\varepsilon(t))^2\right]$ is given by (2-144), with $\psi_\eta = 0$ (i.e., for a zero threshold level).

An alternative squaring synchronizer version, inspired by (2-166) and (2-168), is shown in Figure 2-53 (compare also with Section 5.4 where we take a different point-of-view). The synchronizer operates on (quantized) samples of the PAM signal $y(t;\varepsilon)$, taken by a fixed clock operating at rate $1/T_s$, and is well suited for

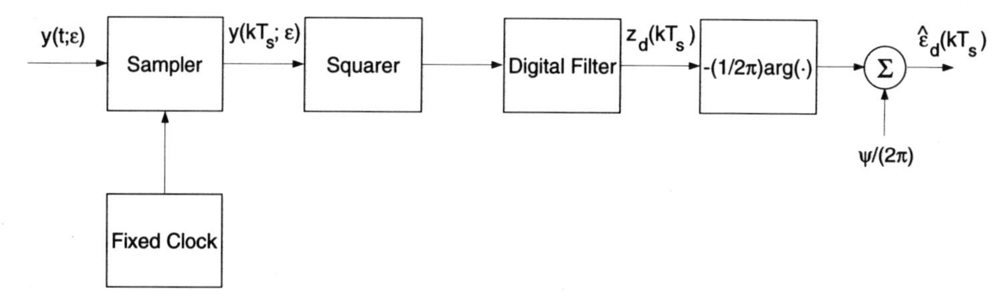

Figure 2-53 Alternative Implementation of Squaring Synchronizer

a fully digital implementation [1]. The timing estimate at instant kT_s is given by

$$\hat{\varepsilon}_d(kT_s) = -\frac{1}{2\pi} \arg\left[z_d(kT_s)\right] + \frac{\psi}{2\pi} \tag{2-169}$$

where

$$\begin{aligned} z_d(kT_s) &= \sum_{m-\infty}^{+\infty} h_d(k-m)\, v(mT_s;\varepsilon)\, \exp\left(-j2\pi m\gamma\right) \\ &= \int_{-\pi/T_s}^{+\pi/T_s} H_d[\exp(\omega T_s)]\, V_d\left[\exp\left(\omega + \frac{2\pi}{T_d};\varepsilon\right)\right]\, \exp\left(j\omega kT_s\right) \frac{d\omega}{2\pi} \end{aligned} \tag{2-170}$$

with $1/T_d = \gamma/T_s$, and

$$V_d\left[\exp\left[j\left(\omega + \frac{2\pi}{T_d};\varepsilon\right)\right]\right] = \sum_{m=-\infty}^{+\infty} V\left(\omega + \frac{2\pi}{T_d} + \frac{2\pi m}{T_s};\varepsilon\right)$$

In (2-170), $\{h_d(m)\}$ is the impulse response of a digital lowpass filter with frequency response $H_d(\omega T_s)$ whose bandwidth B is much smaller than the symbol rate:

$$H_d[\exp(j\omega T_s)] = 0 \quad \text{for} \quad 2\pi B < |\omega| < \pi/T_s$$

In order that (2-170) be the discrete-time version of (2-166), we need $\gamma = T_s/T$, or equivalently $1/T_d = 1/T$. However, as $1/T$ and $1/T_s$ are the frequencies of independently running transmitter and receiver clocks, the value of the ratio T_s/T is not known exactly at the receiver. Hence, in practice we rather have

$$\frac{2\pi}{T_d} = \frac{2\pi}{T} + 2\pi\Delta f_c \qquad \text{with} \qquad \left|\frac{2\pi\Delta f_c}{2\pi}\right| \ll 1 \tag{2-171}$$

where $\Delta f_c T$ represents an inevitable relative frequency offset between the receiver and the transmitter. We call $1/T_d$ the nominal symbol rate, $1/T$ the actual symbol rate, and $1/\gamma$ the nominal number of samples per symbol.

For the sake of implementation, it is useful to select $\gamma = L_1/L_2$, where L_1 and L_2 are (small) integers with no common divisor: in this case, $\exp(j2\pi m\gamma)$ takes on values from a set of size L_2 only.

Let us assume that the sampling at rate $1/T_s$ causes no frequency aliasing of the signal $v(t)\,\exp(-j2\pi t/T_d)$ within the lowpass filter bandwidth, i.e.,

$$H_d(\exp(j\omega T_s))\, V\left[\omega + \frac{2\pi}{T_d} + \frac{2\pi m}{T_s}\right] = 0 \quad \text{for} \quad m \neq 0 \text{ and } |\omega| < \pi/T_s \tag{2-172}$$

Denoting the bandwidth of the squarer output signal $v(t;\varepsilon)$ by W, i.e., $V(\omega;\varepsilon) = 0$ for $|\omega| > 2\pi W$, (2-172) is fulfilled when the sampling rate $1/T_s$ satisfies

$$\frac{1}{T_s} > W + B + \frac{1}{T_d} \tag{2-173}$$

Making use of (2-170), (2-171), and (2-172), (2-169) becomes

$$\hat{\varepsilon}_d(kT_s) = \hat{\varepsilon}(kT_s) + \frac{\Omega k T_s}{2\pi} \tag{2-174}$$

where $\hat{\varepsilon}(kT_s)$ is given by (2-168), with

$$z(t) = \int_{-\pi/T_s}^{+\pi/T_s} H_d \exp[j(\omega T_s - 2\pi \Delta f_c T_s)]\, V\left(\omega + \frac{2\pi}{T}\,;\varepsilon\right) \exp(j\omega t)\, \frac{d\omega}{2\pi}$$

The first term in (2-174) is the estimate $\hat{\varepsilon}$ which would result from a squarer followed by a bandpass filter with an equivalent lowpass frequency response $H_E(\omega)$, given by

$$H_E(\omega) = \begin{cases} H_D(\omega - 2\pi\Delta f_c)\, T_s & |\omega| < \pi/T_s \\ 0 & \text{otherwise} \end{cases} \tag{2-175}$$

The second term in (2-174) increases (for $\Delta f_c > 0$) or decreases (for $\Delta f_c < 0$) linearly with time. The presence of this second term is easily explained when we notice that the synchronizer measures phase with respect to a frequency equal to the nominal symbol rate $1/T_d$ instead of the actual symbol rate $1/T$, because the multiplication of the squarer output samples by $\exp(-j2\pi m\gamma)$ translates to DC the frequency $1/T_d$. Indeed, the useful sinusoidal component at the squarer output can be written as

$$\sqrt{2}\, A \cos\left[\frac{2\pi}{T}\,[t - \varepsilon(t)\, T + \psi]\right] = \sqrt{2}\, A \cos\frac{2\pi}{T_d}\left[\frac{2\pi}{T_d}\,[t - \varepsilon_d(t)\, T_d + \psi]\right]$$

where

$$\varepsilon_d(t) = \varepsilon(t) + \frac{2\pi\Delta f_c t}{2\pi} \tag{2-176}$$

is the normalized time delay of the useful sinusoid represented with respect to the frequency $1/T_d$. Hence, it is the normalized delay $\varepsilon_d(t)$ rather than $\varepsilon(t)$ that is estimated by the synchronizer. As both ε_d in (2-176) and $\hat{\varepsilon}_d$ in (2-174) contain the same term which is linear in time, we obtain $\hat{\varepsilon}_d - \varepsilon_d = \hat{\varepsilon} - \varepsilon$. Hence, the squaring synchronizer from Figure 2-53 yields the same timing error variance as the synchronizer consisting of a squarer, followed by a bandpass filter with an equivalent lowpass frequency response given by (2-175).

In order to reduce the computational complexity of the narrowband digital lowpass filter, we might compute an estimate $\hat{\varepsilon}_d$ only once per M samples instead of once per sample. Also, for even M, we can select

$$h_d(m) = \begin{cases} 1 & -M/2 \le m \le M/2 - 1 \\ 0 & \text{otherwise} \end{cases}$$

yielding

$$\left|H_d(e^{j\omega T_s})\right| = \frac{|\sin(M\omega T_s/2)|}{|\sin(\omega T_s/2)|}$$

The resulting estimates are given by

$$\varepsilon_d(kMT_s) = -\frac{1}{2\pi}\arg[z(kMT_s)] - \frac{\psi}{2\pi}$$

where

$$z(kMT_s) = \sum_{m=0}^{M-1} v_k(mT_s;\varepsilon)\exp(-j2\pi m\gamma) \tag{2-177}$$

and

$$v_k(t;\varepsilon) = v\left(t + \left(k - \frac{1}{2}\right) M\ T_s; \varepsilon\right)$$

Note that $z(kMT_s)$ is the Fourier transform, evaluated at the frequency $\gamma/T_s = 1/T_d$, of a block of M consecutive samples of the sequence $\{v(mT_s;\varepsilon)\}$, centered at the instant kMT_s. The computational complexity of the Fourier transform (2-177) can be considerably reduced by taking $\gamma = 1/4$ (corresponding to a nominal sampling rate of four times the symbol rate), in which case we obtain

$$\text{Re}[z(kMT_s)] = \sum_{m=0}^{(M/2)-1} (-1)^m v_k(2mT_s;\varepsilon)$$

$$\text{Im}[z(kMT_s)] = \sum_{m=0}^{(M/2)-1} (-1)^{m+1} v_k((2m+1)T_s;\varepsilon)$$

so that the Fourier transform computation consists only of additions and subtractions of squarer output samples, which can be performed efficiently by means of pipelining.

Now let us investigate under which conditions $\gamma = 1/4$ is a valid choice. From (2-173) it follows that in order to avoid frequency aliasing, $\gamma = 1/4$ requires $(W + B)\ T_d < 3$, where W and B equal the bandwidth of the squarer output signal $v(t;\varepsilon)$ and of the digital lowpass filter, respectively. In the case of narrowband communication, the bandwidth of the PAM signal $y(t;\varepsilon)$ is between $1/(2T)$ and $1/T$, hence the bandwidth W of the squarer output signal $v(t;\varepsilon)$ satisfies $1/T < W < 2/T$; for the digital lowpass filter, we can conservatively take $B < 1/(2T)$. Using $T \cong T_d$ [see (2-171)], we obtain $(W + B)\ T_d < 5/2 < 3$, so that $\gamma = 1/4$ is valid choice. When the bandwidth of the PAM signal $y(t;\varepsilon)$ exceeds $1/T$ and/or a nonlinearity different from the squarer is used, the bandwidth W of the signal $v(t;\varepsilon)$ at the output of the nonlinearity might be so large that the condition (2-173) is violated for $\gamma = 1/4$. In this case, a smaller value of γ must be selected (which corresponds to a nominal sampling rate higher than four times the symbol rate) in order to avoid frequency aliasing, at the expense of a more complicated computation of the Fourier transform (2-177).

Numerical Example

In order to illustrate the results from the previous sections, we now evaluate the loop noise power spectral density and the timing error variance, in the absence of additive noise, for a squarer followed by a PLL. The squarer operates on a PAM signal with independent binary (± 1) equiprobable data symbols and an odd-symmetrical baseband pulse $g(t)$, given by

$$g(t) = \begin{cases} -\sin\frac{\pi t}{T} & |t| < T \\ 0 & \text{otherwise} \end{cases}$$

and shown in Figure 2-54. This is the case illustrated in Figures 2-12 and 2-13

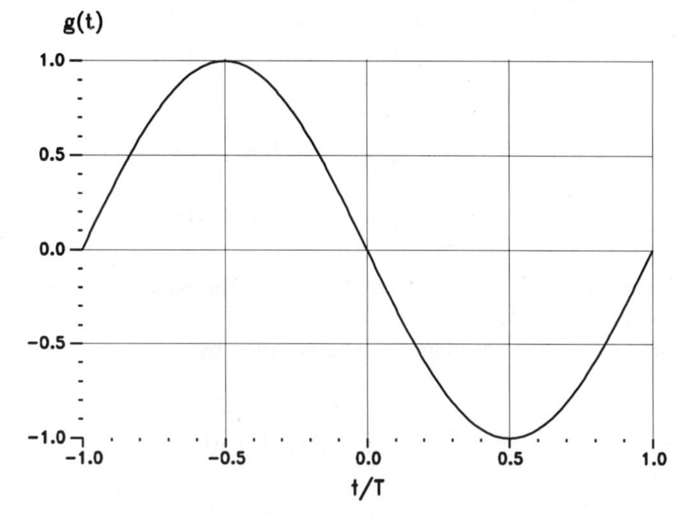

Figure 2-54 Odd-Symmetrical Pulse $g(t)$

[where the PAM signal at the squarer input has been denoted $\dot{y}(t;\varepsilon)$, being the derivative of the PAM signal $y(t;\varepsilon)$ at the receive filter output]. Although $g(t)$ is not a band-limited pulse, the results we will obtain are representative also for narrowband transmission. The advantage of considering a PAM pulse $g(t)$ whose duration is limited to two symbol intervals is that the computations become very simple. For a narrowband pulse which lasts for many symbol intervals, we have to resort either to the much more complicated expressions from the Appendix at the end of this section or to computer simulations; at the end of this section, simulated in-phase and quadrature spectra, resulting from a narrowband pulse $g(t)$ at the squarer input, are presented.

As can be seen from Figure 2-13, the squarer output is nonzero only when a data transition occurs. Hence, the squared PAM signal $v(t;\varepsilon)$ can be written as

$$v(t;\varepsilon) = \sum_m d_m\, q_1\left(t - mT - \frac{T}{2} - \varepsilon T\right)$$

where

$$q_1(t) = \begin{cases} 2\left[1+\cos\frac{2\pi t}{T}\right] & |t| < \frac{T}{2} \\ 0 & \text{otherwise} \end{cases}$$

$$d_m = \begin{cases} 1 & a_{m+1} = -a_m \\ 0 & a_{m+1} = a_m \end{cases}$$

Note that d_m and d_{m+k} are statistically independent for $k \neq 0$. The squared PAM signal is multiplied with the VCO signal of the PLL; this yields the following error signal:

$$\begin{aligned} x(t;\varepsilon,\hat{\varepsilon}) &= \sum_m d_m\, q_1\left(t - mT - \frac{T}{2} - \varepsilon T\right)\ \sin\left[\frac{2\pi}{T}\,(t - \hat{\varepsilon}T)\right] \\ &= -\sum_m d_m\, q_2\left(t - mT - \frac{T}{2} - \varepsilon T; e\right) \end{aligned}$$

where

$$q_2(t;e) = q_{2,C}(t)\ \sin(2\pi e) + q_{2,S}(t)\ \cos(2\pi e)$$

$$q_{2,C}(t) = q_1(t)\ \cos\left(\frac{2\pi t}{T}\right) \qquad q_{2,S}(t) = q_1(t)\ \sin\left(\frac{2\pi t}{T}\right)$$

and $e = \varepsilon - \hat{\varepsilon}$ denotes the normalized timing error. The pulses $q_{2,C}(t)$ and $q_{2,S}(t)$ are shown in Figure 2-55.

The timing error detector characteristic is given by

$$\begin{aligned} \langle E[x(t;\varepsilon,\hat{\varepsilon})]\rangle_t &= -\frac{1}{T}\, E[d_k] \int_{-\infty}^{+\infty} q_2(t;e)dt \\ &= -\frac{1}{2}\ \sin(2\pi e) \end{aligned}$$

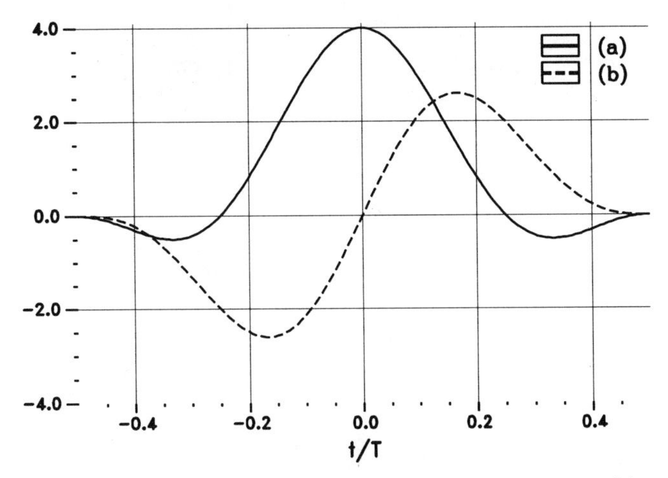

Figure 2-55 (a) The Pulse $q_{2,C}(t)$, (b) The Pulse $q_{2,S}(t)$

from which it follows that the timing error detector slope at $e = 0$ equals $-\pi$. The loop noise is given by

$$N(t; \varepsilon, \hat{\varepsilon}) = \frac{1}{2} \sum_m b_m \, q_2\left(t - mT - \frac{T}{2} - \varepsilon T; e\right)$$

where $b_m = -2\ (d_m - E[d_m])$. Noting that $\{b_m\}$ is a sequence of independent equiprobable binary (± 1) random variables, the loop noise power spectral density $S_N(\omega; e)$ is simply given by

$$S_N(\omega; e) = \frac{1}{4T} |Q_2(\omega; e)|^2$$

where $Q_2(\omega; e)$ is the Fourier transform of $q_2(t; e)$. Taking into account the even and odd symmetry of $q_{2,C}(t)$ and $q_{2,S}(t)$, respectively, $S_N(\omega; e)$ becomes

$$S_N(\omega; e) = S_C(\omega)\ \sin^2(2\pi e) + S_S(\omega)\ \cos^2(2\pi e)$$

where $S_C(\omega)$ and $S_S(\omega)$ are the in-phase and quadrature spectra at the squarer output, which are given by

$$S_C(\omega) = \frac{1}{4T}\ |Q_{2,C}(\omega)|^2 = \frac{1}{16T}\left[Q_1\left(\omega - \frac{2\pi}{T}\right) + Q_1\left(\omega + \frac{2\pi}{T}\right)\right]^2$$

$$S_S(\omega) = \frac{1}{4T}\ |Q_{2,S}(\omega)|^2 = \frac{1}{16T}\left[Q_1\left(\omega - \frac{2\pi}{T}\right) - Q_1\left(\omega + \frac{2\pi}{T}\right)\right]^2$$

In the above, $Q_{2,C}(\omega)$ and $Q_{2,S}(\omega)$ are the Fourier transforms of $q_{2,C}(t)$ and

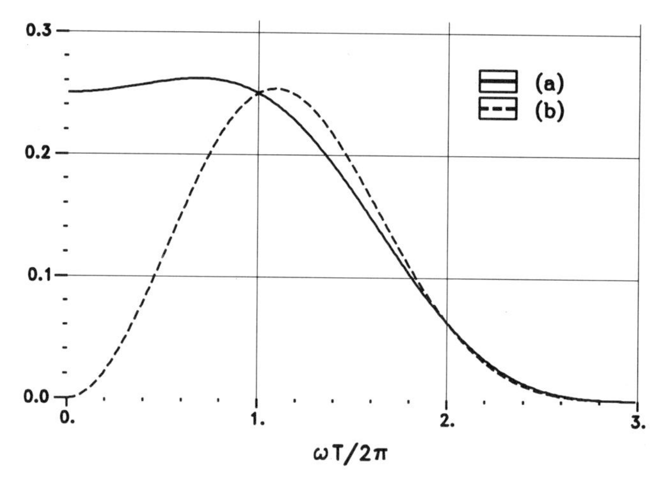

Figure 2-56 (a) In-Phase Spectrum $S_C(\omega)/T$, (b) Quadrature Spectrum $S_S(\omega)/T$

$q_{2,S}(t)$, while $Q_1(\omega)$ is the Fourier transform of $q_1(t)$:

$$Q_1(\omega) = 2Q_0(\omega) + Q_0\left(\omega - \frac{2\pi}{T}\right) + Q_0\left(\omega + \frac{2\pi}{T}\right)$$

with

$$Q_0(\omega) = T\,\frac{\sin(\omega T/2)}{\omega T/2}$$

The in-phase and quadrature spectra $S_C(\omega)$ and $S_S(\omega)$ are shown in Figure 2-56. The in-phase spectrum $S_C(\omega)$ has a lowpass shape [because $q_{2,C}(t)$ is a lowpass pulse], which can be approximated within the loop bandwidth by the constant value $S_C(0)$; consequently, its contribution to the tracking error variance is essentially proportional to the normalized loop bandwidth $B_L T$. The quadrature spectrum $S_S(\omega)$ is zero for $\omega = 0$ [because $q_{2,C}(t)$ exhibits odd symmetry]; consequently, its contribution to the tracking variance is essentially proportional to the square (or even a higher power) of $B_L T$.

For a small steady-state error e_s, the timing error detector slope is close to $-\pi$, and the timing error variance is well approximated by

$$\mathrm{var}[e] = \frac{2S_C(0)\,B_L}{\pi^2}\,\sin^2(2\pi e_s) + \frac{\cos^2(2\pi e_s)}{\pi^2}\int_{-\infty}^{+\infty} |H(\omega)|^2\;S_S(\omega)\,\frac{d\omega}{2\pi}$$

where $H(\omega)$ is the closed-loop frequency response of the PLL. For a perfect second-order loop with large damping and small bandwidth, numerical integration yields

$$\mathrm{var}[e] = \;0.05\,(B_L T)\,\sin^2(2\pi e_s) + 0.045\,(B_L T)^2\cos^2(2\pi e_s)$$

Table 2-2 Timing Error Variance

	var[e]
$B_L T = 10^{-2}$	
$e_s = 0$	4.5×10^{-6}
$e_s = 0.03$	22.0×10^{-6}
$B_L T = 10^{-3}$	
$e_s = 0$	4.5×10^{-8}
$e_s = 0.03$	179.5×10^{-8}

Numerical values for var[e] have been obtained in Table 2-2. Note that a small steady-state error can have a large effect on the timing error variance, especially when the normalized loop bandwidth $B_L T$ is very small.

Finally, we present in Figure 2-57 the simulated in-phase and quadrature spectra $S_C(\omega)$ and $S_S(\omega)$ at the squarer output, in the case where the Fourier transform $G(\omega)$ of the baseband pulse $g(t)$ at the squarer input has a cosine rolloff

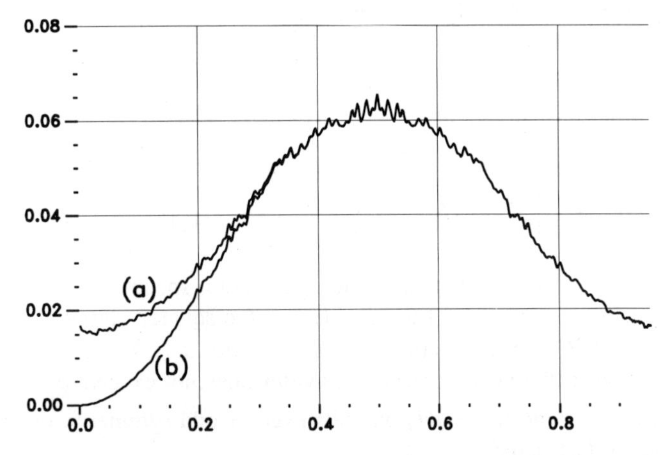

Figure 2-57 (a) In-Phase Spectrum $S_C(\omega)/T$ for Narrowband Transmission (50% Excess Bandwidth), (b) Quadrature Spectrum $S_S(\omega)/T$ for Narrowband Transmission (50% Excess Bandwidth)

amplitude shaping with a 50 percent excess bandwidth and no delay distortion, i.e.,

$$G(\omega) = \begin{cases} T & |\omega T| < \pi(1-\alpha) \\ \frac{T}{2}\left[1 - \sin\left(\frac{|\omega T| - \pi}{2\alpha}\right)\right] & \pi(1-\alpha) < |\omega T| < \pi(1+\alpha) \\ 0 & \text{otherwise} \end{cases}$$

with $\alpha = 0.5$. As in the previous example, we observe that $S_C(\omega)$ has a lowpass shape, whereas $S_S(\omega)$ is zero at $\omega = 0$ [because of the even symmetry of $g(t)$].

Main Points

- The squaring synchronizer is a spectral line generating clock synchronizer. Squaring a noisy PAM signal yields spectral lines at DC and at multiples of the symbol rate. The spectral line at the symbol rate can be extracted by means of a PLL or a narrowband bandpass filter.
- When the bandwidth of the PAM signal to be squared does not exceed $1/(2T)$, the squarer output contains a spectral line at DC only, which evidently carries no timing information. A spectral line at the symbol rate (and at multiples thereof) can be obtained by using a higher-order nonlinearity.
- For narrowband transmission, the bandwidth of the PAM signal is between $1/(2T)$ and $1/T$. In this case the squarer output can contain spectral lines only at DC and at the symbol rate. The spectral line at the symbol rate is caused by the data transitions and disappears in the absence of transitions; therefore, the transmission of long strings of identical symbols should be avoided. The spectral line at the symbol rate becomes stronger when the bandwidth in excess of $1/(2T)$ increases.
- When the Fourier transform $G(\omega)$ of the baseband pulse $g(t)$ satisfies

  $$u\left(\omega + \frac{\pi}{T}\right) G\left(\omega + \frac{\pi}{T}\right) = u\left(-\omega + \frac{\pi}{T}\right) G^*\left(-\omega + \frac{\pi}{T}\right)$$

  and

  $$u\left(\omega + \frac{\pi}{T}\right) G\left(\omega + \frac{\pi}{T}\right) = 0 \quad \text{for} \quad |\omega| > \frac{\pi}{T} - \pi B$$

  then, in the absence of noise, the quadrature spectrum $S_S(\omega)$ and in-phase/quadrature cross-spectrum $S_{CS}(\omega)$ at the squarer output are both zero for $|\omega| < 2\pi B$. Hence, $S_S(\omega)$ and $S_{CS}(\omega)$ do not affect the synchronizer operation when the synchronizer bandwidth does not exceed B.
- When the baseband pulse $g(t)$ exhibits even or odd symmetry, or its Fourier transform $G(\omega)$ satisfies

  $$u\left(\omega + \frac{\pi}{T}\right) G\left(\omega + \frac{\pi}{T}\right) = u\left(-\omega + \frac{\pi}{T}\right) G^*\left(-\omega + \frac{\pi}{T}\right)$$

  then, in the absence of additive noise, $S_S(\omega) = S_{CS}(\omega) = 0$ for $\omega = 0$.

- When the squarer is followed either by a PLL with zero steady-state error or by a symmetrical bandpass filter, and timing is derived from the zero crossings of the bandpass filter output signal, then only the quadrature component at the squarer output contributes to the timing error. In this case, when in the absence of noise $S_S(\omega) = S_{CS}(\omega) = 0$ within the synchronizer bandwidth, self-noise is zero; when in the absence of noise $S_S(\omega) = S_{CS}(\omega) = 0$ for $\omega = 0$, the self-noise contribution to the timing error variance is essentially proportional to the square (or even a higher power) of the normalized loop bandwidth $B_L T$. On the other hand, when either the steady-state error of the PLL is nonzero or the bandpass filter is not symmetrical and/or the threshold level at the bandpass filter output is nonzero, then also the in-phase component at the squarer output contributes to the timing error. In this case, the self-noise contribution to the timing error variance is essentially proportional to $B_L T$.
- An alternative version of the squaring synchronizer consists of a squarer, operating on samples of the noisy PAM signal, followed by Fourier transform computation. This version is well-suited for digital implementation. For narrowband transmission, a nominal sampling rate of four samples per symbol is sufficient, in which case the Fourier transform computation reduces to the addition and subtraction of squarer output samples. This digital version of the squaring synchronizer is also studied in Section 5.4 in a different context.

Bibliographical Notes

These bibliographical notes consider not only spectral line generating synchronizers using a squaring nonlinearity, but also synchronizers with other nonlinearities.

Squaring Nonlinearity

The conditions (2-157)–(2-158) for zero self-noise have first been derived in [1], which investigates the tracking performance resulting from a squarer plus bandpass filter, operating on a prefilterd PAM signal. Also, the paper considers the effect of prefilter bandwidth bandpass filter bandwidth, and bandpass filter detuning on the tracking error variance.

The squarer plus bandpass filter has also been dealt with in [2]. The effects of bandpass filter bandwidth and detuning, along with a nonzero threshold, are studied. It is shown that a baseband PAM pulse with linear delay characteristic (i.e., a pulse with even or odd symmetry with respect to an arbitrary time instant) yields a self-noise power spectral density which is zero for $\omega = 0$ (assuming a bandpass filter which is symmetrical about the symbol rate, and zero threshold level).

Prefiltering the received PAM signal before it enters the squarer has also been examined in [3]. For a given received baseband pulse, the optimum prefilter frequency response has been derived, which eliminates the self-noise and simultaneously minimizes the tracking error variance caused by additive noise.

Expressions for the in-phase and quadrature spectra and the in-phase/quadrature cross-spectrum of the squared baseband PAM signal have been presented in [4], and several interesting properties of these spectra have been derived. Spectra are shown in the case of cosine rolloff baseband PAM pulses; the in-phase spectra have a lowpass shape, whereas the quadrature spectra are zero for $\omega = 0$ (because of the even symmetry of the baseband pulse).

Whereas most papers analyzing the squaring synchronizer consider statistically independent data symbols, the effect of line coding on the tracking performance has been investigated in [5]. It is shown that the considered codes (AMI, 4B3T, MS43, and HDB3) yield a similar tracking error variance, which is upper bounded by the tracking error variance for independent ternary symbols.

The alternative squaring synchronizer implementation from Section 3 has been introduced in [6].

Other Nonlinearities

The tracking error variances for squaring, absolute value, and fourth-power nonlinearities have been compared in [7] by means of computer simulations. It depends on the excess bandwidth and the pulse shape which nonlinearity performs best.

In [8], the tracking performance resulting from the delay-and-multiply nonlinearity is investigated. It is shown that the tracking error variance is only weakly dependent on the value of the delay when the excess bandwidth is small, in which case the performance is nearly the same as for a squaring nonlinearity. Neglecting additive noise, it has been reported in [9] that for several spectral line generating synchronizers the in-phase spectrum at the output of the nonlinearity has a lowpass shape, whereas the quadrature spectrum is (close to) zero at $\omega = 0$. In order to avoid large tracking errors due to self-noise, it is recommended to keep small the steady-state timing error of the PLL following the nonlinearity. For zero steady-state timing error, the tracking error variance is found to be proportional to the square (or even a larger power) of the normalized loop bandwidth $B_L T$.

The in-phase and quadrature spectra and the in-phase/quadrature cross-spectrum, at the output of a fourth-power nonlinearity operating on a noise-free PAM signal, have been evaluated in [10]. These spectra are shown for cosine rolloff pulses, and their shape is according to the general observations made in [9]. Also, it is proved that self-noise is absent in the case of $2N$th order nonlinearity, provided that the Fourier transform $G(\omega)$ of the baseband PAM pulse at the input of the nonlinearity satisfies the following conditions:

$$u\left(\omega + \frac{\pi}{T}\right) G\left(\omega + \frac{\pi}{T}\right) = u\left(-\omega + \frac{\pi}{T}\right) G^*\left(-\omega + \frac{\pi}{T}\right)$$

$$u\left(\omega + \frac{\pi}{T}\right) G\left(\omega + \frac{\pi}{T}\right) = 0 \quad \text{for} \quad |\omega| > \frac{1}{N}\left[\frac{\pi}{T} - \pi B\right]$$

which are a generalization of (2-157) and (2-158).

It has been shown in [11] that for an arbitrary memoryless spectral line generating nonlinearity, the self-noise contribution to the tracking error variance can be made very small when the baseband PAM pulse $g(t)$ is "locally symmetric", i.e.,

$$g(kT+u) = g(kT-u) \qquad |u| < \frac{T}{2}, \quad k = 0, \pm 1, \pm 2, \ldots$$

However, a band-limited finite-energy pulse $g(t)$ cannot be locally symmetric. For a given excess bandwidth, $g(t)$ can be selected to minimize the "local asymmetry". For small excess bandwidths, these optimized pulses approximately satisfy the conditions (2-157)–(2-158), which yield zero self-noise in the case of a squaring nonlinearity.

A quite general analytical method for evaluating the power spectral density of the timing error has been presented in [12], for correlated data symbols (such as resulting from coding), an arbitrary time-limited baseband PAM pulse, and an arbitrary memoryless nonlinearity. However, the computational complexity rapidly increases with increasing PAM pulse duration; hence, the proposed method is less convenient for band-limited transmission.

The "wave difference" symbol synchronizer, resembling a spectral line generating synchronizer, has been introduced in [13]. The timing error detector operates on samples of the received PAM signal, taken at twice the baud rate; its output during the kth transmitted symbol is given by

$$x_k(\varepsilon, \hat{\varepsilon}) = NL[y(kT + \overline{\tau} + \hat{\varepsilon}T + T/4; \varepsilon)] - NL\left[y\left(kT + \overline{\tau} + \hat{\varepsilon}T - \frac{T}{4}; \varepsilon\right)\right] \tag{2-178}$$

where $\overline{\tau}$ is a time shift such that $E[x_k(\varepsilon, \hat{\varepsilon})] = 0$ for $\hat{\varepsilon} = \varepsilon$, and $NL(\cdot)$ is a nonlinear function. A spectral line generating synchronizer, using the nonlinearity $NL(\cdot)$ followed by a PLL, would produce an error signal given by

$$x(t; \varepsilon, \hat{\varepsilon}) = NL[y(t)] \, \sin\left[\frac{2\pi}{T}(t - \overline{\tau} - \hat{\varepsilon}T)\right] \tag{2-179}$$

where $\sin(2\pi(t - \overline{\tau} - \hat{\varepsilon}T)/T)$ denotes the VCO output signal. The timing error detector outputs (2-178) and (2-179) are related by

$$x_k(\varepsilon, \hat{\varepsilon}) = x\left(kT + \overline{\tau} + \hat{\varepsilon}T - \frac{T}{4}; \varepsilon, \hat{\varepsilon}\right) + x\left(kT + \overline{\tau} + \hat{\varepsilon}T + \frac{T}{4}; \varepsilon, \hat{\varepsilon}\right)$$

from which it can be derived that the wave difference symbol synchronizer has essentially the same tracking performance as the spectral line generating synchronizer, provided that the error signal (2-179) of the former is sampled at twice the baud rate (halfway between the zero crossings of the VCO output signal) before entering the loop filter of the PLL. As the wave difference symbol synchronizer operates on samples of the received PAM signal, it is well-suited for digital implementation.

Appendix: Calculation of the Spectra $S_C(\omega)$, $S_S(\omega)$ and $S_{CS}(\omega)$

Denoting by $\sqrt{2}\,A\;\cos\left(2\pi(t-\varepsilon T)/T+\psi\right)$ the useful sinusoid at the squarer output, we introduce a signal $w(t;\beta)$, which is defined as

$$w(t;\beta)=\left[\{v(t;\varepsilon)-E[v(t;\varepsilon)]\}\;\sqrt{2}\;\cos\left[\frac{2\pi}{T}(t-\varepsilon T)+\psi+\beta\right]\right]_{\mathrm{LP}} \tag{2-180}$$

where $v(t;\varepsilon)$ denotes the squarer output, and $[...]_{\mathrm{LP}}$ indicates that the signal between square brackets is passed through an ideal lowpass filter with frequency response $H_{\mathrm{LP}}(\omega)$, given by

$$H_{\mathrm{LP}}(\omega)=\begin{cases}1 & |\omega|\le\pi/T\\ 0 & \text{otherwise}\end{cases}$$

Substituting in (2-180) the values $\beta=0$ and $\beta=\pi/2$, we obtain

$$w(t;0)=[w_C(t;\varepsilon)]_{\mathrm{LP}}$$

$$w(t;\pi/2)=[w_S(t;\varepsilon)]_{\mathrm{LP}}$$

As $w(t;\beta)$ is cyclostationary with period T and band-limited to $|\omega|<\pi/T$, $w(t;\beta)$ is wide-sense stationary (see Section 1.1.2). Consequently, the cross-correlation function $E[w(t;\beta_1)\;w(t+u;\beta_2)]$ does not depend on t and will be denoted as $R_w(u;\beta_1,\beta_2)$. The cross-spectrum $S_w(\omega;\beta_1,\beta_2)$ between $w(t;\beta_1)$ and $w(t;\beta_2)$ is the Fourier transform of $R_w(u;\beta_1,\beta_2)$. The lowpass content of the spectra $S_C(\omega)$, $S_S(\omega)$, and $S_{SC}(\omega)$ is easily obtained from $S_w(\omega;\beta_1,\beta_2)$:

$$|H_{\mathrm{LP}}(\omega)|^2\;S_C(\omega)=S_w(\omega;0,0) \tag{2-181}$$

$$|H_{\mathrm{LP}}(\omega)|^2\;S_S(\omega)=S_w(\omega;\pi/2,\pi/2) \tag{2-182}$$

$$|H_{\mathrm{LP}}(\omega)|^2\;S_{CS}(\omega)=S_w(\omega;0,\pi/2) \tag{2-183}$$

As the one-sided synchronizer bandwidth (in hertz) does not exceed $1/(2T)$, the lowpass content of these spectra is sufficient for determining the timing error variance.

The cross-correlation function $R_w(u;\beta_1,\beta_2)$ and the cross-spectrum $S_w(\omega;,\beta_2)$ can be determined from the autocovariance function $K_v(t,t+u;\varepsilon)$ of the squarer output $v(t;\varepsilon)$, which is defined as

$$K_v(t,t+u;\varepsilon)=E[v(t;\varepsilon)\;v(t+u;\varepsilon)]-E[v(t;\varepsilon)]\;E[v(t+u;\varepsilon)]$$

As $v(t;\varepsilon)$ is cyclostationary with period T, $K_v(t,t+u;\varepsilon)$ is periodic in t with period T, and can be expanded into the following Fourier series:

$$K_v(t,t+u;\varepsilon)=\sum_m k_{v,m}(u)\;\exp\left[j\;\frac{2\pi m}{T}\;(t-\varepsilon T)\right] \tag{2-184}$$

with

$$k_{v,m}(u) = \frac{1}{T} \int_{-T/2}^{+T/2} K_v(t, t+u; \varepsilon) \exp\left[-j\frac{2\pi m}{T}(t - \varepsilon T)\right] dt$$

A similar reasoning as in Section 2.5.2 yields

$$\begin{aligned} R_w(u; \beta_1, \beta_2) = & \left[\left\langle K_v(t, t+u; \varepsilon)\ 2\ \cos\left[\frac{2\pi}{T}(t - \varepsilon T) + \psi + \beta_1\right]\right.\right. \\ & \left.\left. \times \cos\left[\frac{2\pi}{T}(t + u - \varepsilon T) + \psi + \beta_2\right]\right\rangle_t\right]_{\text{LP}} \\ = & \frac{1}{2}\left[k_{v,0}(u)\ \exp\left[-j\left(\frac{2\pi u}{T} + \beta_2 - \beta_1\right)\right]\right. \\ & + k_{v,0}(u)\ \exp\left[j\left(\frac{2\pi u}{T} + \beta_2 - \beta_1\right)\right] \\ & + k_{v,-2}(u)\ \exp\left[j\left(\frac{2\pi u}{T} + 2\psi + \beta_1 + \beta_2\right)\right] \\ & \left. + k_{v,2}(u)\ \exp\left[-j\left(\frac{2\pi u}{T} + 2\psi + \beta_1 + \beta_2\right)\right]\right]_{\text{LP}} \end{aligned}$$

where $\langle ... \rangle_t$ denotes time-averaging with respect to the variable t. Hence, the cross-spectrum $S_w(\omega; \beta_1, \beta_2)$ is given by

$$\begin{aligned} S_w(\omega; \beta_1, \beta_2) = |H_{\text{LP}}(\omega)|^2\ \frac{1}{2}\ & \left[S_{v,0}\left(\omega + \frac{2\pi}{T}\right)\ \exp\left(j\ (\beta_1 - \beta_2)\right)\right. \\ & + S_{v,0}\left(\omega - \frac{2\pi}{T}\right)\ \exp\left(-j\ (\beta_1 - \beta_2)\right) \\ & + S_{v,-2}\left(\omega - \frac{2\pi}{T}\right)\ \exp\left(j\ (2\psi + \beta_1 + \beta_2)\right) \\ & \left. + S_{v,2}\left(\omega + \frac{2\pi}{T}\right)\ \exp\left(-j\ (2\psi + \beta_1 + \beta_2)\right)\right] \end{aligned} \tag{2-185}$$

where $S_{v,m}(\omega)$ is the Fourier transform of $k_{v,m}(u)$.

In the following we compute ψ, $k_{v,m}(u)$, and $S_{v,m}(\omega)$ in the case where the nonlinearity is a squarer, and the input to the squarer is a noisy PAM signal. Using (2-181), (2-182), (2-183), and (2-185), the spectra $S_C(\omega)$, $S_S(\omega)$, and $S_{CS}(\omega)$ for $|\omega| < \pi/T$ can be obtained from ψ, $S_{v,0}(\omega)$ and $S_{v,\pm 2}(\omega)$.

Now we will derive an expression for the $N \times N$, $S \times N$, and $S \times S$ contributions to the coefficients $k_{v,m}(u)$ in the expansion (2-184) and their Fourier transform $S_{v,m}(\omega)$, under the assumption of independent data symbols.

The $N \times N$ Contribution

Making use of the Gaussian nature of the additive noise $n(t)$ at the squarer input, the contribution of the $N \times N$ term to the autocovariance function of the squarer output is

$$\begin{aligned} K_v(t, t+u; \varepsilon) \Big|_{N \times N} &= E\big[n^2(t)\, n^2(t+u)\big] - E\big[n^2(t)\big]\, E\big[n^2(t+u)\big] \\ &= R_n^2(0) + 2\, R_n^2(u) - R_n^2(0) \\ &= 2\, R_n^2(u) \end{aligned}$$

As $n(t)$ is stationary, the $N \times N$ term is also stationary, so its contribution to the autocovariance function $K_v(t, t+u; \varepsilon)$ of the squarer output does not depend on t. Hence,

$$k_{v,0}(u)\,\big|_{N \times N} = 2\, R_n^2(u) : \quad k_{v,m}(u)\,\big|_{N \times N} = 0 \quad \text{for} \quad m \neq 0$$

$$S_{v,0}(\omega)\,\big|_{N \times N} = 2 \int_{-\infty}^{+\infty} S_n(\omega - \nu)\, S_n(\nu)\, \frac{d\nu}{2\pi} \tag{2-186}$$

$$S_{v,m}(\omega)\,\big|_{N \times N} = 0 \quad \text{for} \quad m \neq 0 \tag{2-187}$$

where $S_n(\omega)$, the Fourier transform of $R_n(u)$, is the spectrum of the noise $n(t)$.

The $S \times N$ Contribution

The contribution of the $S \times N$ term to the autocovariance function of the squarer output is given by

$$K_v(t, t+u; \varepsilon)\,\big|_{S \times N} = 4 R_n(u)\, A_2 \sum_m g(t - mT - \varepsilon T)\, g(t + u - mT - \varepsilon T)$$

which is periodic in t with period T. Hence,

$$k_{v,m}(u)\,\big|_{S \times N} = \frac{1}{T}\, 4 R_n(u)\, A_2 \int_{-\infty}^{+\infty} g(t)\, g(t+u)\, \exp\left[-j \frac{2\pi m t}{T}\right] dt$$

$$S_{v,m}(\omega)\,\big|_{S \times N} = \frac{1}{T}\, 4\, A_2 \int_{-\infty}^{+\infty} S_n(\omega - \nu)\, G(\nu)\, G\!\left(\frac{2\pi m}{T} - \nu\right) \frac{d\nu}{2\pi} \tag{2-188}$$

For narrowband transmission, $G(\omega) = 0$ for $|\omega| > 2\pi/T$, yielding $G(\omega) G(\pm 4\pi/T - \omega) = 0$; in this case, $S_{v,\pm 2}(\omega)\,\big|_{S \times N} = 0$, so that only $S_{v,0}(\omega)\,\big|_{S \times N}$ contributes to the low-frequency content of $S_S(\omega)$, $S_C(\omega)$, and $S_{CS}(\omega)$.

The $S \times S$ Contribution

Introducing the function $p_n(t) = g(t)\; g(t - nT)$, the $S \times S$ term can be written as

$$S \times S \text{ term} = \sum_{m,n} a_m \; a_{m+n} \; p_n(t - mT - \varepsilon T)$$

For an arbitrary data symbol distribution, the evaluation of the contribution of the $S \times S$ term to the autocovariance function of the squarer output is straightforward but tedious. Therefore, we make the simplifying assumption that the data symbols are zero-mean (i.e., $A_1 = 0$), which is valid in many cases of practical interest. Under this assumption, we obtain

$$\begin{aligned} K_v(t, t+u; \varepsilon) \,\big|_{S\times S} &= \sum_{m,n,i,j} \left[E[a_m \; a_{m+n} \; a_i \; a_{i+j}] - E[a_m \; a_{m+n}] \; E[\, a_i \; a_{i+j}] \right] \\ &\qquad \times p_n(t - mT - \varepsilon T) \; p_j(t + u - iT - \varepsilon T) \\ &= \left(A_4 - 3A_2^2\right) \sum_m p_0(t - mT - \varepsilon T) \; p_0(t + u - mT - \varepsilon T) \\ &\qquad + 2A_2^2 \sum_{m,n} p_n(t - mT - \varepsilon T) \; p_n(t + u - mT - \varepsilon T) \end{aligned}$$

where we have taken into account that

$$E[a_m \; a_{m+n} \; a_i \; a_{i+j}] = \begin{cases} A_4 & i = m \text{ and } j = n = 0 \\ A_2^2 & (j = n = 0 \text{ and } i \neq m) \text{ or } (i = m \text{ and } j = n \neq 0) \\ & \quad \text{or } \; (i = m + n \text{ and } j = -n \text{ and } i \neq m) \\ 0 & \text{otherwise} \end{cases}$$

$$E[a_m \; a_{m+n}] = \begin{cases} A_2 & n = 0 \\ 0 & \text{otherwise} \end{cases}$$

This yields

$$\begin{aligned} k_{v,m}(u) \,\big|_{S\times S} &= \frac{1}{T} \left(A_4 - 3A_2^2\right) \int_{-\infty}^{+\infty} p_0(t) \; p_0(t+u) \; \exp\left[-j \frac{2\pi mt}{T}\right] dt \\ &\quad + \frac{1}{T} \, 2A_2^2 \sum_n \int_{-\infty}^{+\infty} p_n(t) \; p_n(t+u) \; \exp\left[-j \frac{2\pi mt}{T}\right] dt \end{aligned}$$

$$\begin{aligned} S_{v,m}(\omega) \,\big|_{S\times S} &= \frac{1}{T} \left(A_4 - 3A_2^2\right) P_0(\omega) \; P_0\!\left(\frac{2\pi m}{T} - \omega\right) \\ &\quad + \frac{1}{T} \, A_2^2 \sum_n \; P_n(\omega) \; P_n\!\left(\frac{2\pi m}{T} - \omega\right) \end{aligned} \tag{2-189}$$

where $P_n(\omega)$, given by

$$P_n(\omega) = \int_{-\infty}^{+\infty} G(\omega - \nu)\, G(\nu)\, \exp(-j\nu nT)\, \frac{d\nu}{2\pi} \tag{2-190}$$

is the Fourier transform of $p_n(t)$. The infinite summation over n in the second term of (2-189) can be avoided by substituting (2-190) for $P_n(\omega)$:

$$\sum_n P_n(\omega)\, P_n\left(\frac{2\pi m}{T} - \omega\right) = \iint_{-\infty}^{+\infty} F(\omega, \nu, \mu, m) \left[\sum_n e^{-j(\nu+\mu)nT}\right] \frac{d\nu}{2\pi}\, \frac{d\mu}{2\pi}$$

where

$$F(\omega, \nu, \mu, m) = G(\omega - \nu)\, G(\nu)\, G\left(\frac{2\pi m}{T} - \omega - \mu\right) G(\mu)$$

Making use of the identity

$$\sum_n e^{-j\omega nT} = \frac{2\pi}{T} \sum_k \delta\left(\omega - \frac{2\pi k}{T}\right)$$

where $\delta(\omega)$ denotes a Dirac impulse at $\omega = 0$, we obtain

$$\sum_n P_n(\omega)\, P_n\left(\frac{2\pi m}{T} - \omega\right) = \frac{1}{T} \sum_k \int_{-\infty}^{+\infty} F\left(\omega, \frac{2\pi k}{T} - \mu, \mu, m\right) \frac{d\mu}{2\pi} \tag{2-191}$$

In the case of narrowband transmission, $G(\omega) = 0$ for $|\omega| > 2\pi/T$; hence,

$$F\left(\omega, \frac{2\pi k}{T} - \mu, \mu, m\right) = 0 \quad \text{for} \quad |k| > 1 \text{ or } |k - m| > 1$$

so that the summation over k in (2-191) contains three nonzero terms when $m = 0$ and only one nonzero term when $m = -2$ or $m = 2$.

In summary, the following results have been obtained:

(i) $S_C(\omega)$, $S_S(\omega)$, and $S_{CS}(\omega)$ in terms of $S_w(\omega; \beta_1, \beta_2)$: (2-181), (2-182), and (2-183)

(ii) $S_w(\omega; \beta_1, \beta_2)$ in terms of $S_{v,0}(\omega)$ and $S_{v,\pm 2}(\omega)$: (2-185)

(iii) Contributions to $S_{v,0}(\omega)$ and $S_{v,\pm 2}(\omega)$:

$N \times N$ contribution:	(2-186) and (2-187)
$S \times N$ contribution:	(2-188)
$S \times S$ contribution:	(2-189), (2-190) and (2-191)

2.5.2 Example 2: The Synchronizer with Zero-Crossing Timing Error Detector (ZCTED)

The synchronizer with ZCTED is an error-tracking synchronizer, whose block diagram is shown in Figure 2-14, and whose principle of operation is illustrated in Figure 2-15. The synchronizer adjusts the phase of the square-wave VCO signal such that its negative zero crossings nearly coincide with the zero crossings of the PAM signal $y(t;\varepsilon)$ at the input of the timing error detector. As the nominal zero-crossing instants of the PAM signal in general do not coincide with the decision instants at the receive filter output, the VCO signal must be delayed appropriately before activating the sampler in the decision branch.

The Zero Crossings of the PAM Signal $y(t;\varepsilon)$

The PAM signal $y(t;\varepsilon)$ at the input of the synchronizer is given by

$$y(t;\varepsilon) = \sum_{m} a_m \, g(t - mT - \varepsilon T) + n(t)$$

where $\{a_m\}$ is a stationary sequence of binary antipodal (± 1) data symbols, $g(t)$ is the baseband pulse, and $n(t)$ is stationary noise. We have chosen the origin of time such that $g(t)$ takes on its maximum value at $t = 0$.

A zero crossing of $y(t;\varepsilon)$ occurs in the interval $(kT + \varepsilon T, (k+1)T + \varepsilon T)$ only when the data symbols a_k and a_{k+1} are different, i.e., $a_k a_{k+1} = -1$. This zero-crossing instant, which we denote by t_k, is a random variable, depending on the data symbols $\{a_m\}$ and on the additive noise $n(t)$. We decompose t_k as

$$t_k = kT + \varepsilon T + \tau_k$$

where τ_k is the shift of the zero-crossing instant t_k with respect to the instant $kT + \varepsilon T$, corresponding to the maximum value of the kth transmitted pulse. Expressing that $y(t_k;\varepsilon) = 0$ when $a_k a_{k+1} = -1$, we obtain

$$0 = a_k[g(\tau_k) - g(\tau_k - T)] + \sum_{m \neq 0, m \neq 1} a_{k-m} \, g(\tau_k - mT) + n(kT + \varepsilon T + \tau_k) \tag{2-192}$$

Assuming a large signal-to-noise ratio and rapidly decreasing values of $|g(t)|$ for $|t| > T$, the second and third term in (2-192) are much smaller than $g(\tau_k)$ and $g(\tau_k - T)$. Therefore, τ_k is in the vicinity of the value $\overline{\tau}$, which makes zero the first term of (2-192):

$$g(\overline{\tau}) = g(\overline{\tau} - T) \tag{2-193}$$

The relation (2-193) is illustrated in Figure 2-58. It follows from (2-193) that an

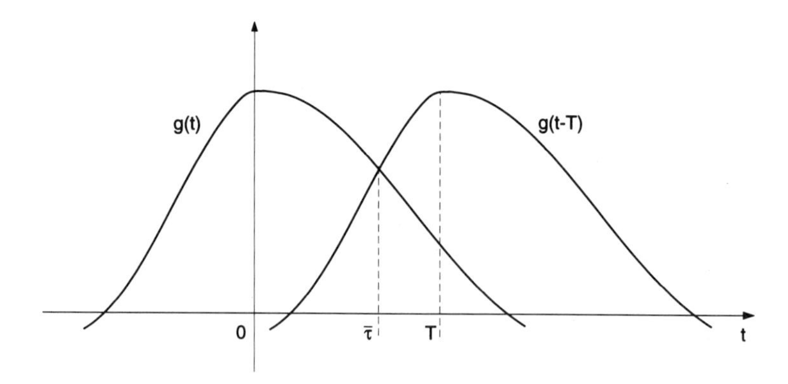

Figure 2-58 Determination of $\overline{\tau}$

even symmetric pulse $g(t)$ yields $\overline{\tau} = T/2$. Decomposing τ_k as

$$\tau_k = \overline{\tau} + w_k T$$

where w_k denotes the random deviation, normalized to the symbol duration T, of τ_k with respect to $\overline{\tau}$, it follows from Figure 2-59 that w_k is well approximated by

$$\begin{aligned} w_k &= \frac{1}{b}\, a_k\; y(kT + \varepsilon T + \overline{\tau}; \varepsilon) \\ &= \frac{1}{b}\left[a_k \sum_{m \neq 0, m \neq 1} a_{k-m}\, g_m + a_k\; n_k \right] \end{aligned} \tag{2-194}$$

which relates the fluctuation of the zero-crossing instants of $y(t;\varepsilon)$ to the data sequence, the baseband pulse, and the noise. In (2-194), g_m and n_k are short-hand

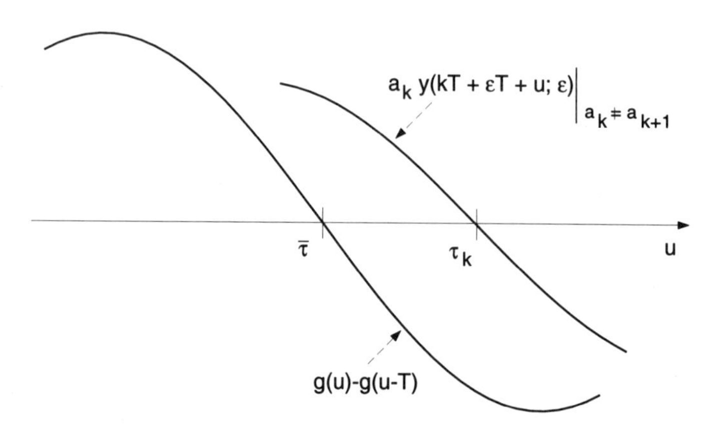

Figure 2-59 Linearization of $\overline{\tau}$

notations for $g(mT+\overline{\tau})$ and $n(kT+\varepsilon T+\overline{\tau})$, respectively, while

$$b = T[g'(\overline{\tau}-T) - g'(\overline{\tau})]$$

is the normalized slope at the instant $kT+\varepsilon T+\overline{\tau}$, and $g'(t)$ is the derivative of $g(t)$. In the case where the data symbols are statistically independent, we obtain

$$\begin{aligned} E[w_k \,|\, a_k a_{k+1} = -1] &= \frac{1}{2}\left(E[w_k \,|a_k = 1, a_{k+1} = -1] + E[w_k \,|\, a_k = -1, a_{k+1} = 1]\right) \\ &= \frac{1}{2}\left(E[w_k \mid a_k = 1] + E[w_k \mid a_k = -1]\right) \\ &= 0 \end{aligned}$$

This indicates that averaging the zero-crossing instant t_k over the noise and the data symbols yields the instant $kT+\varepsilon T+\overline{\tau}$. For correlated data symbols, we might have $E[w_k \mid a_k a_{k+1} = -1] \neq 0$, in which case the average zero-crossing instant is different from $kT+\varepsilon T+\overline{\tau}$.

Tracking Performance

The ZCTED can be built with sequential logic circuits and is mostly used in combination with charge-pump phase-locked loops. A detailed discussion of charge-pump PLLs is presented in Section 2.7 of Volume 1. As the exact analysis of charge-pump PLLs is rather complicated, we resort to the quasi-continuous approximation (see Section 2.7.2 of Volume 1). It should be noted that the quasi-continuous analysis gives no indication about the frequency ripple introduced by the charge-pumps. Fortunately, the results from the quasi-continuous analysis are very close to the exact results, when the normalized loop bandwidth $B_L T$ is in the order of 0.01 or less (which are the $B_L T$ values of practical interest).

The synchronizer is analyzed under open-loop conditions: the loop is opened (say, at the loop filter input), and to the timing error detector we apply the PAM signal $y(t;\varepsilon)$ and a square-wave VCO signal with negative zero crossings at the instants $kT+\hat{\varepsilon}T+\overline{\tau}$, with $|\varepsilon-\hat{\varepsilon}|<1/2$. The statistical properties of the timing error detector output signal are determined under the assumption of statistically independent equiprobable binary (± 1) data symbols; results for nonequiprobable data symbols are presented in Section 2.5.2.

As shown in Figure 2-15, the signal $x(t;\varepsilon,\hat{\varepsilon})$ at the output of the ZCTED consists of pulses with constant magnitude K_1, occurring when the input signal $y(t;\varepsilon)$ crosses the zero level. The pulse width equals the time difference between the actual zero crossing of $y(t;\varepsilon)$ and the negative zero crossing of the VCO signal; the polarity of the pulses is positive (negative) when the zero crossing of $y(t;\varepsilon)$ comes earlier (later) than the negative zero crossing of the VCO signal. Assuming that $a_k a_{k+1} = -1$, a pulse occurs in the interval $(kT+\varepsilon T, kT+T+\varepsilon T)$; its width equals $T(w_k+e)$, where $e=\varepsilon-\hat{\varepsilon}$ and w_k is given by (2-194).

The exact analysis of the synchronizer is very difficult, because it is the width rather than the magnitude of the pulses at the ZCTED output which depends on w_k and e. For loop bandwidths which are much smaller than the symbol rate, the synchronizer operation is essentially determined by the frequency content of the timing error detector output signal near $\omega = 0$. Therefore, we will replace the actual timing error detector output signal by a fictitious signal with the same frequency content near $\omega = 0$, for which the analysis becomes much simpler. This approximation is as follows:

$$x(t; \varepsilon, \hat{\varepsilon}) = \sum_k x_k(\varepsilon, \hat{\varepsilon})\, p(t - kT - \varepsilon T - \overline{\tau})$$

where

$$x_k(\varepsilon, \hat{\varepsilon}) = \int\limits_{kT+\varepsilon T}^{kT+T+\varepsilon T} x(t; \varepsilon, \hat{\varepsilon})\, dt = K_1\, T(w_k + e)\, d_k$$

and $p(t)$ is a unit area lowpass pulse [equivalently, its Fourier transform $P(\omega)$ equals 1 at $\omega = 0$]. As long as $P(\omega)$ is essentially flat within the loop bandwith, the shape of the pulse will turn out not to be important; for example, one could think of $p(t)$ as a rectangular pulse of width T, because in practice $B_L T \ll 1$. The switching variable $d_k = (1 - a_k a_{k+1})/2$ takes the value 1 when there is a data transition (i.e., $a_k \neq a_{k+1}$) and the value 0 otherwise.

The timing error detector characteristic is the average DC component of the timing error detector output signal, for a given value of the normalized timing error. For independent, not necessarily equiprobable symbols, it follows from (2-194) that $E[w_k d_k] = 0$. Indeed,

$$\begin{aligned} E[w_k d_k] &= \frac{1}{2b}\, E\left[(a_k - a_{k+1}) \left[\sum_{m\neq 0, m\neq 1} a_{k-m}\, g_m + n_k\right]\right] \\ &= \frac{1}{2b} \sum_{m\neq 0, m\neq 1} \left(E[a_k]\, E[a_{k-m}] - E[a_{k+1}]\, E[a_{k-m}]\right) g_m \\ &\quad + \frac{1}{2b} \left(E[a_k] - E[a_{k+1}]\right) E[n_k] \\ &= 0 \end{aligned}$$

because of the stationarity of the data symbol sequence. Hence, the timing error detector characteristic is given by

$$\begin{aligned}
\langle E[x(t;\varepsilon,\hat{\varepsilon})]\rangle_t &= K_1 T E[d_k]\, e \left\langle \sum_k p(t - kT - \varepsilon T - \overline{\tau}) \right\rangle_t \\
&= K_1 E[d_k]\, e \int_{-T/2}^{T/2} \sum_k p(t - kT)\, dt \\
&= K_1 E[d_k]\, e \int_{-\infty}^{+\infty} p(t)\, dt \\
&= K_1 E[d_k]\, e
\end{aligned} \tag{2-195}$$

where $\langle ... \rangle_t$ denotes averaging over time, and we have taken into account that the area of $p(t)$ equals 1. We recall that the above expression is valid only for $|e| < 1/2$; for other values of e, the timing error detector characteristic is obtained by periodic extension. Hence, the characteristic has a sawtooth shape with slope $K_1 E[d_k]$. For independent equiprobable data symbols, we obtain

$$\begin{aligned}
E[d_k] &= \frac{1}{2}\, E[1 - a_k a_{k+1}] \\
&= \frac{1}{2}\, (1 - E[a_k]\; E[a_{k+1}]) \\
&= \frac{1}{2}
\end{aligned}$$

because the data symbols have zero mean; consequently, the slope equals $K_1/2$.

The loop noise $N(t;\varepsilon,\hat{\varepsilon})$ represents the statistical fluctuation of the timing error detector output, and is given by

$$\begin{aligned}
N(t;\varepsilon,\hat{\varepsilon}) &= x(t;\varepsilon,\hat{\varepsilon}) - E[x(t;\varepsilon,\hat{\varepsilon})] \\
&= \sum_k N_k(\varepsilon,\hat{\varepsilon})\, p(t - kT - \varepsilon T - \overline{\tau})
\end{aligned} \tag{2-196}$$

where

$$\begin{aligned}
N_k(\varepsilon,\hat{\varepsilon}) &= x_k(\varepsilon,\hat{\varepsilon}) - E[x_k(\varepsilon,\hat{\varepsilon})] \\
&= K_1 T\, [w_k d_k + (d_k - E[d_k])\, e]
\end{aligned} \tag{2-197}$$

It follows from (2-196) and (2-197) that the loop noise consists of two components.

(i) The first component is caused by the random fluctuations $\{w_k\}$ of the zero-crossing instants of the PAM signal $y(t;\varepsilon)$ at the synchronizer input. From (2-194) it follows that this component consists of a self-noise contribution (due to the tails of the pulse $g(t)$ causing ISI) and an additive noise contribution.

(ii) The second component is caused by the random occurence of data transitions, which makes $d_k \neq E[d_k]$ This component is proportional to the normalized timing error e.

In all practical situations, the synchronizer bandwidth is much smaller than the symbol rate. Hence, the timing error is caused only by the lowpass components of the loop noise, which we denote as $N_{\mathrm{LP}}(t;\varepsilon,\hat{\varepsilon})$. The subscript LP denotes lowpass filtering by means of an ideal lowpass filter with a bandwidth of $1/(2T)$. This lowpass filtering yields

$$N_{\mathrm{LP}}(t;\varepsilon,\hat{\varepsilon}) = \sum_k N_k(\varepsilon,\hat{\varepsilon})\, p_{\mathrm{LP}}(t - kT - \varepsilon T - \overline{\tau})$$

where $p_{\mathrm{LP}}(t)$ is the output of an ideal lowpass filter with bandwidth $1/(2T)$, when driven by $p(t)$. As its bandwidth does not exceed $1/(2T)$, $N_{\mathrm{LP}}(t;\varepsilon,\hat{\varepsilon})$ is wide-sense stationary (see Section 1.1.2); consequently, the autocorrelation function $E[N_{\mathrm{LP}}(t;\varepsilon,\hat{\varepsilon})\, N_{\mathrm{LP}}(t+u;\varepsilon,\hat{\varepsilon})]$ does not depend on t, and will be denoted as $R_N(u;e)$. This autocorrelation function is given by

$$R_N(u;e) = \sum_m R_m(e)\, F(t - \varepsilon T - \overline{\tau}, u - mT)$$

where

$$R_m(e) = E[N_k(\varepsilon,\hat{\varepsilon})\, N_{k+m}(\varepsilon,\hat{\varepsilon})]$$
$$F(v,w) = \sum_n p_{\mathrm{LP}}(v - nT)\, p_{\mathrm{LP}}(v + w - nT)$$

The function $F(v,w)$ is periodic in v with period T and can be expanded into a Fourier series. However, as the bandwidth of $p_{\mathrm{LP}}(t)$ does not exceed $1/(2T)$, this Fourier series consists only of its DC term, which depends on w but not on v:

$$\begin{aligned}\text{DC term} &= \frac{1}{T}\int_{-T/2}^{T/2} \sum_n p_{\mathrm{LP}}(v - nT)\, p_{\mathrm{LP}}(v + w - nT)\, dv \\ &= \frac{1}{T}\int_{-\infty}^{+\infty} p_{\mathrm{LP}}(v)\, p_{\mathrm{LP}}(v + w)\, dv\end{aligned}$$

Hence, the autocorrelation function becomes

$$R_N(u;e) = \frac{1}{T}\sum_m R_m(e) \int_{-\infty}^{+\infty} p_{\mathrm{LP}}(v)\, p_{\mathrm{LP}}(v + u - mT)\, dv$$

which does not depend on t. The power spectral density $S_N(\omega;\varepsilon)$ of the lowpass content of the loop noise is the Fourier transform of the autocorrelation function

$R_N(u;e)$:

$$S_N(\omega;e) = \int_{-\infty}^{+\infty} R_N(u;e)\,\exp(-j\omega u)\,du$$

$$= \frac{1}{T}\sum_m R_m(e)\,\exp(-j\omega mT)\int_{-\infty}^{+\infty} p_{LP}(v)\,\exp(j\omega v)\,dv$$

$$\cdot \int_{-\infty}^{+\infty} p_{LP}(u+v-mT)\,\exp(-j\omega(u+v-mT))\,du$$

$$= \frac{1}{T}\,|P_{LP}(\omega)|^2\sum_m R_m(e)\,\exp(-j\omega mT)$$

where

$$P_{LP}(\omega) = \begin{cases} P(\omega) & |\omega| < \pi/T \\ 0 & \text{otherwise} \end{cases}$$

For $\omega = 0$, the loop noise power spectral density is given by

$$S_N(0;e) = \frac{1}{T}\sum_m R_m(e)$$

where we have taken into account that $P_{LP}(0) = P(0) = 1$. When $P(\omega)$ is essentially flat within the loop bandwidth, the specific shape of the pulse $p(t)$ does not affect $S_N(\omega;e_s)$ for $|\omega| < 2\pi B_L$, which is the range of frequencies to which the loop responds.

In order to compute the autocorrelation sequence $\{R_m(e)\}$, we use the following decomposition:

$$N_k(\varepsilon,\hat{\varepsilon}) = \frac{K_1 T}{2}\,[z_1(k) + z_2(k) + z_3(k;e)] \tag{2-198}$$

where, according to (2-197) and (2-194),

$$z_1(k) = \frac{2}{b}\,d_k\,a_k\,n_k = \frac{1}{b}\,(a_k - a_{k+1})\,n_k \tag{2-199}$$

$$z_2(k) = \frac{2}{b}\,d_k\,a_k \sum_{n\neq 0, n\neq -1} a_{k-n}\,g_n = \frac{1}{b}\,(a_k - a_{k+1}) \sum_{n\neq 0, n\neq -1} a_{k-n}\,g_n \tag{2-200}$$

$$z_3(k;e) = 2\,(d_k - E[d_k])\,e \tag{2-201}$$

The components $z_1(k)$ and $z_2(k)$ are caused by the additive noise and the tails of the pulse $g(t)$ at the input of the timing error detector, respectively, whereas the component $z_3(k;e)$ is caused by the random occurrence of the data transitions.

The sequence $\{z_1(k)\}$ is not correlated with $\{z_2(k)\}$ or $\{z_3(k;e)\}$, but $\{z_2(k)\}$ and $\{z_3(k;e)\}$ are correlated. It follows that

$$R_m(e) = \left(\frac{K_1 T}{2}\right)^2 (R_1(m) + R_2(m) + R_3(m;e) + R_{2,3}(m;e) + R_{2,3}(-m;e))$$

where $\{R_1(m)\}$, $\{R_2(m)\}$, and $\{R_3(m;e)\}$ are the autocorrelation sequences of $\{z_1(k)\}$, $\{z_2(k)\}$, and $\{z_3(k;e)\}$, respectively, while $\{R_{2,3}(m;e)\}$ is the cross-correlation sequence of $\{z_2(k)\}$ and $\{z_3(k;e)\}$:

$$R_{2,3}(m;e) = E[z_2(k)\; z_3(k+m;e)]$$

For statistically independent equiprobable data symbols, the following results are obtained:

$$\begin{aligned}
&R_1(0) = \frac{2}{b^2}\, R_n(0); \quad R_1(\pm 1) = -\frac{1}{b^2}\, R_n(T); \quad R_1(m) = 0 \;\text{ for }\; |m| > 1 \\
&R_2(0) = \frac{2}{b^2} \sum_{n \neq 0, n \neq -1} g_n^2\,; \qquad R_2(\pm 1) = -\frac{1}{b^2}\left[g_{-2}\, g_1 + \sum_{|n|>1} g_{n-1}\, g_n\right] \\
&R_2(\pm m) = \frac{1}{b^2}\,(g_m - g_{m-1})\,(g_{-m} - g_{-m-1}) \quad \text{for } m > 1 \\
&R_3(0;e) = e^2; \qquad R_3(m;e) = 0 \;\text{ for }\; m \neq 0 \\
&R_{2,3}(0;e) = 0; \quad R_{2,3}(1;e) = \frac{1}{b}\, g_{-2}\, e; \quad R_{2,3}(-1;e) = -\frac{1}{b}\, g_1\, e; \\
&R_{2,3}(m;e) = 0 \;\text{ for }\; |m| > 1
\end{aligned} \tag{2-202}$$

where $R_n(u)$ is the autocorrelation function of the additive noise at the input of the timing error detector.

The timing error variance is determined by the loop noise power spectral density, evaluated at the stable equilibrium point e_s of the synchronizer. Due to hardware imperfections (such as delay differences of the logic circuits the timing error detector is composed of), a nonzero steady-state error e_s may arise, even in a second-order loop with perfect integrator. Taking into account that for equiprobable symbols the timing error detector slope equals $K_1/2$ irrespective of the value of e_s, the timing error variance is given by

$$\text{var}[e] = \frac{4}{K_1^2} \int_{-\infty}^{+\infty} |H(\omega)|^2\, S_N(\omega; e_s)\, \frac{d\omega}{2\pi} \tag{2-203}$$

where $H(\omega)$ denotes the closed-loop frequency response of the synchronizer. For small synchronizer bandwidths, the behavior of $S_N(\omega; e_s)$ near $\omega = 0$ is important.

When $e_s = 0$, only $\{z_1(k)\}$ and $\{z_2(k)\}$, which are caused by the additive noise and the tails of the baseband pulse $g(t)$ at the timing error detector input, contribute to the loop noise, because $\{z_3(k;0)\}$, which is caused by the random occurrence of data transitions, is identically zero. The corresponding loop noise power spectral density $S_N(\omega;0)$ is given by

$$S_N(\omega;0) = \left(\frac{K_1 T}{2}\right)^2 \frac{1}{T} \left|P_{\mathrm{LP}}(\omega)\right|^2 \left(S_1(\omega) + S_2(\omega)\right)$$

where $S_1(\omega)$ and $S_2(\omega)$ are the power spectral densities of the sequences $\{z_1(k)\}$ and $\{z_2(k)\}$

$$\begin{aligned} S_1(\omega) &= R_1(0) + 2\sum_{m=1}^{\infty} R_1(m)\ \cos(m\omega T) \\ &= \frac{2}{b^2}\left[R_n(0) - R_n(T)\ \cos(\omega T)\right] \\ S_2(\omega) &= R_2(0) + 2\sum_{m=1}^{\infty} R_2(m)\ \cos(m\omega T) \end{aligned}$$

In most cases of practical interest, the autocorrelation of the additive noise is such that $|R_n(T)| \ll R_n(0)$. Hence, $S_1(\omega)$ is essentially flat within the loop bandwidth when $B_L T \ll 1$, and can be approximated by $S_1(0)$, as far as the operation of the synchronizer is concerned; the contribution of $S_1(\omega)$ to the timing error variance is essentially proportional to the normalized loop bandwidth $B_L T$. When the fluctuations of $S_2(\omega)$ within the loop bandwidth are small with respect to $S_2(0)$, also the self-noise contribution [caused by the tails of $g(t)$] to the timing error variance is essentially proportional to the normalized loop bandwidth. In the following section we will show that there exist conditions on $g(t)$ such that $S_2(0) = 0$ (in which case the fluctuation of $S_2(\omega)$ within the loop bandwidth cannot be ignored); for small loop bandwidths, this leads to a substantial reduction of the self-noise contribution to the timing error variance, which then is proportional to the square (or even a higher power) of the normalized loop bandwidth $B_L T$.

When $e_s \neq 0$, the loop noise has an additional component, due to $\{z_3(k;e_s)\}$. Its contribution to the loop noise autocorrelation function consists not only of its autocorrelation sequence $\{R_3(m;e_s)\}$, but also of the cross-correlation sequence $\{R_{2,3}(m;e_s)\}$ between $\{z_2(k)\}$ and $\{z_3(k;e_s)\}$. The power spectral density $S_N(\omega;e_s)$ is given by

$$\begin{aligned} S_N(\omega;e_s) = &\left(\frac{K_1 T}{2}\right)^2 \frac{1}{T} \left|P_{\mathrm{LP}}(\omega)\right|^2 \\ &\times \left(S_1(\omega) + S_2(\omega) + S_3(\omega;e_s) + 2\ \mathrm{Re}[S_{2,3}(\omega;e_s)]\right) \end{aligned} \tag{2-204}$$

where $S_3(\omega;e_s)$ and $S_{2,3}(\omega;e_s)$ are the power spectral density of $\{z_3(k;e_s)\}$ and the cross-power spectral density of $\{z_2(k)\}$ and$\{z_3(k;e_s)\}$:

$$\begin{aligned} S_3(\omega;e) &= R_3(0;e) + 2\sum_{m=1}^{\infty} R_3(m;e)\ \cos(m\omega T) \\ &= e^2 \\ S_{2,3}(\omega;e) &= \sum_{m=-\infty}^{+\infty} R_{2,3}(m;e)\ \exp(-jm\omega T) \\ &= \frac{e}{b}\left[g_{-2}\ \exp(-j\omega T) - g_1\ \exp(j\omega T)\right] \end{aligned}$$

Note that $S_3(\omega; e)$ is flat, while 2 Re$[S_{2,3}(\omega; e)]$, given by

$$2\ \mathrm{Re}[S_{2,3}(\omega; e)] = \frac{2e}{b}\ (g_{-2} - g_1)\ \cos(\omega T)$$

is essentially flat within the loop bandwidth when $B_L T \ll 1$. Hence, their contributions to the timing error variance are proportional to the normalized loop bandwidth $B_L T$.

Finally, we would like to point out that a small steady-state error e_s can have a large effect on the timing error variance, in the case where additive noise is negligible and the tails of $g(t)$ do not contribute to the loop noise power spectral density at $\omega = 0$. Indeed, when $e_s = 0$, the timing error variance is proportional to the square (or a higher power) of $B_L T$, whereas for $e_s \neq 0$, the timing error variance is proportional to $B_L T$. Hence, for small $B_L T$, the timing error variance for $e_s \neq 0$ can be considerably larger than for $e_s = 0$, even when the power $R_2(0)$ of $z_2(k)$ is much larger than $R_3(0) = e_s^2$, the power of $z_3(k; e_s)$. This will be shown in the numerical example discussed below.

Self-Noise Reduction

In this section we show that the baseband pulse $g(t)$ can be selected such that its tails cause no self-noise. Also, we derive a less restrictive condition on $g(t)$, which yields a self-noise spectrum that becomes zero at $\omega = 0$; in this case the contribution of the tails of $g(t)$ to the tracking error variance is proportional to the square (or even a higher power) of the normalized loop bandwidth $B_L T$.

In the absence of additive noise, $\tau_k = \overline{\tau}$ is a solution of (2-192) only when

$$g(\overline{\tau}) = g(\overline{\tau} - T); \quad g(mT + \overline{\tau}) = 0 \quad \text{for} \quad m \neq 0,\ m \neq -1 \tag{2-205}$$

In this case, a zero crossing of $y(t; \varepsilon)$ occurs at $kT + \varepsilon T + \overline{\tau}$, when $a_k a_{k+1} = -1$. As this zero-crossing instant is not influenced by the data sequence, the tails of $g(t)$ do not give rise to any self-noise. Note that the condition (2-205) completely eliminates self-noise, irrespective of the statistics of the binary (± 1) data sequence.

The condition (2-205) is closely related to the *second Nyquist criterion*: a pulse $g_2(t)$ satisfies the second Nyquist criterion when

$$g_2\left(\frac{T}{2}\right) = g_2\left(-\frac{T}{2}\right) = \frac{1}{2}; \quad g_2\left(mT + \frac{T}{2}\right) = 0 \quad \text{for} \quad m \neq 0,\ m \neq -1 \tag{2-206}$$

Comparing (2-205) and (2-206), it follows that $g(t)$ satisfying (2-205) is a time-shifted Nyquist-II pulse, the shift being such that $g(t)$ takes on its maximum value at $t = 0$. Denoting the Fourier transform of $g_2(t)$ by $G_2(\omega)$, the frequency domain equivalent of (2-206) is

$$\sum_m (-1)^m\ G\left[\omega - \frac{2\pi m}{T}\right] = T\ \cos\left[\frac{\omega T}{2}\right]$$

Figure 2-60 shows a band-limited Nyquist-II pulse $g_2(t)$ with 50 percent excess

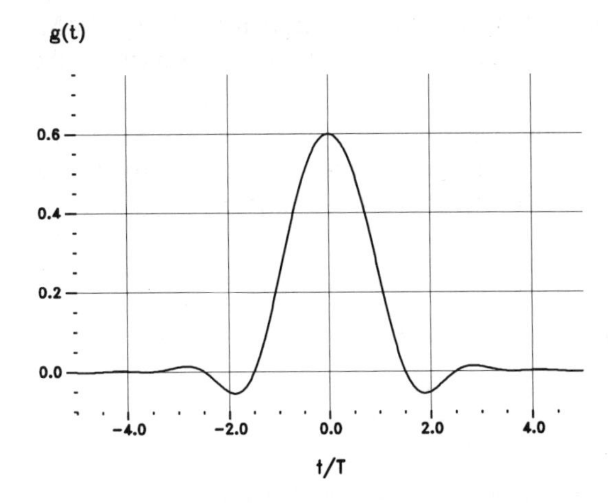

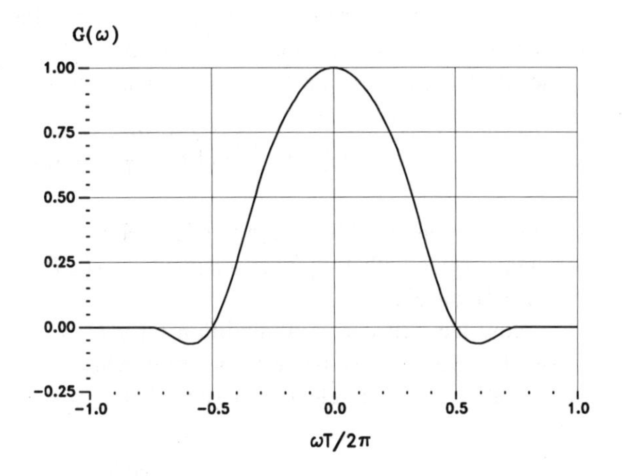

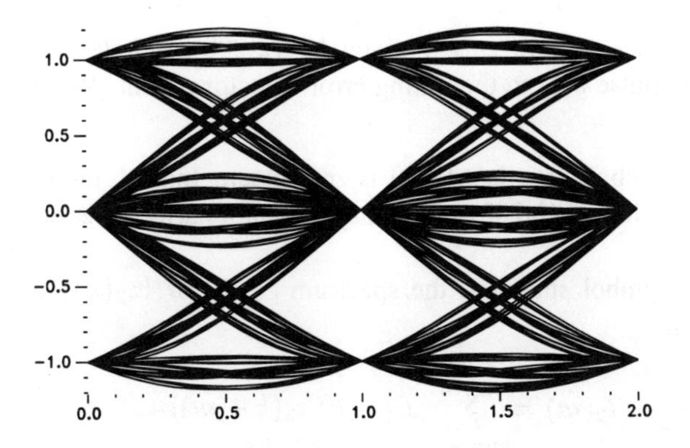

Figure 2-60 (a) Nyquist-II Pulse $g_2(t)$, (b) Fourier Transform $G_2(\omega)$, (c) Eye Diagram for Binary PAM

bandwidth, its Fourier transform $G_2(\omega)$, and the eye diagram for binary PAM. Note from the eye diagram that the zero-crossing instants are independent of the binary data sequence.

In most cases, the baseband pulse at the output of the receive filter (approximately) satisfies the first Nyquist criterion, in order to eliminate (or at least considerably reduce) the ISI at the decision instants. A pulse $g_1(t)$ satisfies the first Nyquist criterion when

$$g_1(0) = 1; \quad g_1(mT) = 0 \quad \text{for} \quad m \neq 0 \tag{2-207}$$

Denoting by $G_1(\omega)$ the Fourier transform of $g_1(t)$, the frequency domain version of (2-207) is

$$\sum_m G\left[\omega - \frac{2\pi m}{T}\right] = T$$

In general, a Nyquist-I pulse does not satisfy the second Nyquist criterion. This implies that only appropriate filtering of the receive filter output before entering the synchronizer can eliminate the self-noise. Assuming that $g_1(t)$ is a Nyquist-I pulse, it is easily verified that $g_2(t)$, given by

$$g_2(t) = \frac{1}{2}\left[g_1\left(t + \frac{T}{2}\right) + g_1\left(t - \frac{T}{2}\right)\right]$$

satisfies the second Nyquist criterion. Hence, passing the receive filter output through a filter with frequency response $\cos(\omega T/2)$ yields a PAM signal with a Nyquist-II baseband pulse.

Now we will derive a condition on $g(t)$, which does not eliminate the self-noise, but yields a self-noise spectrum which becomes zero at $\omega = 0$. More specifically, we will select $g(t)$ such that

$$\sum_{k=-\infty}^{+\infty} z_2(k) = 0 \tag{2-208}$$

irrespective of the data sequence, where $z_2(k)$, given by (2-200), denotes the contribution of the tails of the pulse $g(t)$ to the timing error detector output. When (2-208) holds, then

(i) Irrespective of the data symbol statistics, $z_2(k)$ is zero-mean. Indeed, taking the expectation of both sides of (2-208) and using the stationarity of $z_2(k)$ yields $E[z_2(k)] = 0$.

(ii) Irrespective of the data symbol statistics, the spectrum $S_2(\omega)$ of $\{z_2(k)\}$ is zero at $\omega = 0$. Indeed,

$$\begin{aligned} S_2(0) &= \sum_{m=-\infty}^{+\infty} R_2(m) = \sum_{m=-\infty}^{+\infty} E[z_2(k)\, z_2(k+m)] \\ &= E\left[z_2(k) \sum_{m=-\infty}^{+\infty} z_2(m)\right] = 0 \end{aligned}$$

(iii) Irrespective of the data symbol statistics, the cross-spectrum $S_{2,3}(\omega; e_s)$ of $\{z_2(k)\}$ and $\{z_3(k; e_s)\}$, given by (2-201), is zero at $\omega = 0$. Indeed,

$$S_{2,3}(0; e_s) = \sum_{m=-\infty}^{+\infty} R_{2,3}(m; e_s) = \sum_{m=-\infty}^{+\infty} E[z_2(k-m)\, z_3(k; e_s)]$$

$$= E\left[z_3(k; e_s) \sum_{m=-\infty}^{+\infty} z_2(m) \right] = 0$$

Hence, when (2-208) holds, the tails of $g(t)$ contribute to the loop noise power spectral density but not to the average timing error detector output; this contribution to the loop noise power spectral density is zero at $\omega = 0$. Hence, only the additive noise and a nonzero steady-state error e_s contribute to the loop noise power spectral density at $\omega = 0$.

Using (2-200) in (2-208) yields

$$\sum_{k=-\infty}^{+\infty} z_2(k) = \frac{1}{b} \sum_{m \neq 0, m \neq -1} (\alpha_m - \alpha_{m+1})\, g_m$$

where

$$\alpha_m = \sum_{k=-\infty}^{+\infty} a_k a_{k-m}$$

Noting that $\alpha_m = \alpha_{-m}$, we obtain

$$\sum_{k=-\infty}^{+\infty} z_2(k) = \frac{1}{b} \sum_{m>0} (\alpha_m - \alpha_{m+1})\,(g_m - g_{m-1})$$

Hence, (2-208) holds, irrespective of the data sequence, when

$$g(mT + \overline{\tau}) = g(-mT - T + \overline{\tau}) \qquad m = 0, 1, 2, ... \tag{2-209}$$

Note that (2-209) must be fulfilled also for $m = 0$, because of (2-193) which defines $\overline{\tau}$. The condition (2-209) holds automatically when $g(t)$ is an even symmetric pulse: in this case, $\overline{\tau} = T/2$, so that (2-209) reduces to $g(mT + T/2) = g(-mT - T/2)$ for $m = 0, 1, 2, ...$ As an identically zero self-noise spectrum is a special case of a self-noise spectrum which is zero at $\omega = 0$, the condition (2-205) is a special case of the condition (2-209).

Nonequiprobable Data Symbols

Now we will assume that the binary data symbols are statistically independent, with Prob$[a_k = 1] = p$ and Prob$[a_k = -1] = 1 - p$. The effect of additive noise is neglected.

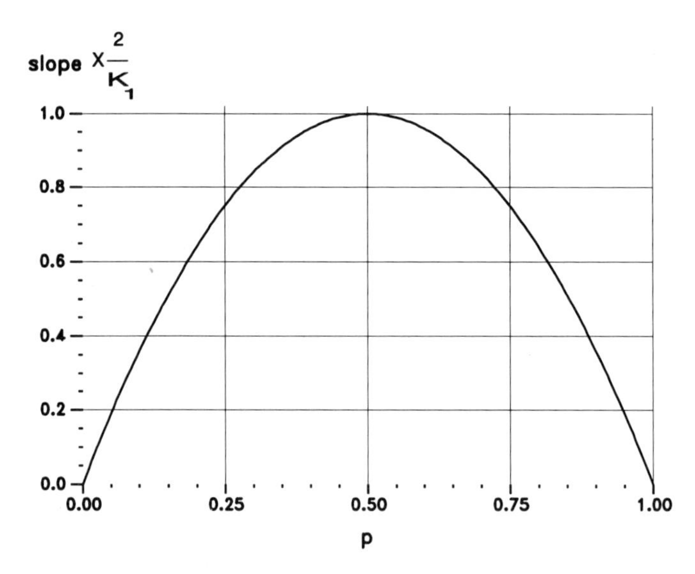

Figure 2-61 Effect of Nonequiprobable Symbols on Timing Error Detector Slope

We have that $E[w_k d_k] = 0$ for independent data symbols. Hence, the timing error detector output is given by (2-195):

$$\begin{aligned}\langle x(t;\varepsilon,\hat{\varepsilon})\rangle_t &= K_1 E[d_k]\; e \\ &= K_1 \frac{1}{2}\;(1 - E[a_k]\; E[a_{k+1}])\; e \\ &= 2\; K_1 p(1-p)\; e\end{aligned}$$

because $E[a_k] = 2p - 1$, irrespective of k. The dependence of the timing error detector slope $2K_1 p(1-p)$ on p is shown in Figure 2-61: the slope is zero for $p = 0$ (i.e., $a_k = -1$ for all k) and for $p = 1$ (i.e., $a_k = 1$ for all k), and takes on its largest value $K_1/2$ for $p = 1/2$.

For an arbitrary pulse $g(t)$, the computation of the loop noise power spectral density $S_N(\omega; e_s)$ is straightforward but quite tedious when $p \neq 1/2$. Instead, we restrict our attention to the case where $g(t)$ satisfies (2-209). According to the previous section, this implies that the contribution of the tails of $g(t)$ to the loop noise power spectral density is zero at $\omega = 0$. Hence, using the decomposition (2-198), only $z_3(k; e_s)$, given by (2-201), contributes to the loop noise power spectral density at $\omega = 0$. In order to evaluate this contribution, we need the autocorrelation sequence $\{R_3(m; e_s)\}$ of $z_3(k; e_s)$. We obtain

$$\begin{aligned}R_3(0; e_s) &= 4\Big[E\left[d_k^2\right] - (E[d_k])^2\Big]\; e_s^2 \\ R_3(\pm 1; e_s) &= 4\Big[E[d_k\; d_{k+1}] - (E[d_k])^2\Big]\; e_s^2 \\ R_3(m; e_s) &= 0 \qquad \text{for} \qquad |m| > 1\end{aligned}$$

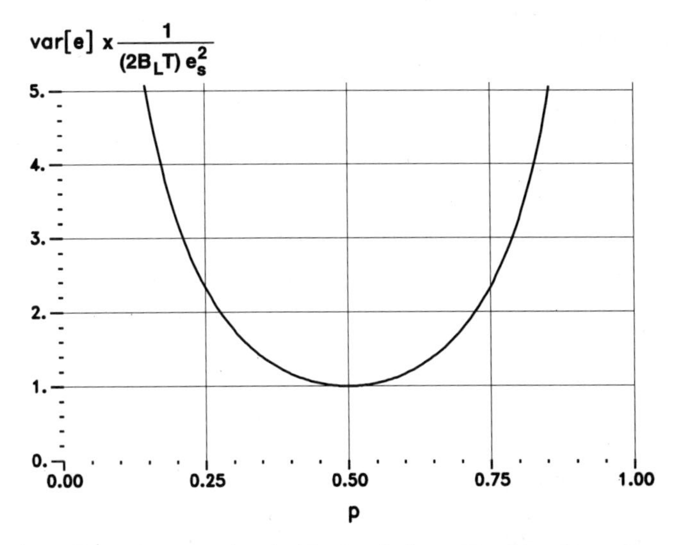

Figure 2-62 Effect of Nonequiprobable Symbols on Tracking Error Detector Slope

The resulting loop noise power spectral density at $\omega = 0$ is given by

$$\begin{aligned} S_N(0; e_s) &= \frac{K_1^2\, T}{4} \left[R_3(0; e_s) + 2R_3(1; e_s)\right] \\ &= \frac{K_1^2\, T}{4} \left[1 - (E[a_k])^2\right] \left[1 + 3(E[a_k])^2\right] e_s^2 \\ &= 4K_1^2\, T\, p\,(1-p)\left(1 - 3p + 3p^2\right)\ e_s^2 \end{aligned}$$

where we have taken into account that

$$E\left[d_k^2\right] = E[d_k] = 2E[d_k\ d_{k+1}] = \frac{1}{2}\left[1 - (E[a_k])^2\right]$$

The resulting timing error variance, caused by the nonzero steady-state error e_s is given by

$$\mathrm{var}[e] = (2B_L T)\ \frac{1 - 3p + 3p^2}{p\,(1-p)}\ e_s^2$$

The dependence on p of the timing error variance is shown in Figure 2-62. This variance reaches its minimum value of $(2B_L T)\ e_s^2$ at $p = 1/2$, and becomes infinitely large for $p = 0$ and $p = 1$. This clearly illustrates that long sequences of identical data symbols are detrimental to symbol synchronization; the occurrence of such sequences should be avoided, for instance, by using a scrambler.

Numerical Example

In order to illustrate the results from previous sections, we now compute the loop noise power spectral density and the timing error variance resulting from a zero-crossing timing error detector, operating on a typical received PAM signal.

System Description

- Pulse shape $g(t)$:
  Cosine rolloff amplitude shaping with linear delay distortion:

$$|G(\omega)| = \begin{cases} T & |\omega T| < \pi(1-\alpha) \\ \frac{T}{2}\left[1 - \sin\left(\frac{|\omega T| - \pi}{2\alpha}\right)\right] & \pi(1-\alpha) < |\omega T| < \pi(1+\alpha) \\ 0 & \text{otherwise} \end{cases}$$

$$\arg\left[G(\omega)\right] = \beta \left[\frac{\omega T}{\pi}\right]^2 \operatorname{sgn}(\omega)$$

  Numerical values:
  $\alpha = 0.5$
  $\beta = 0$ (no delay distortion) and $\beta = 3\pi/4$
- Additive noise $n(t)$: wideband noise

$$R_n(0) = \sigma_n^2 \qquad R_n(mT) = 0 \quad \text{for} \quad m \neq 0$$

  The noise is specified with respect to the signal by the peak-to-noise ratio ρ_{peak}, defined as

$$\rho_{\text{peak}} = \frac{g_{\max}}{\sigma_n}$$

  where $g_{\max}$ is the maximum value of $g(t)$; as we have chosen the origin of time such that $g(t)$ is maximum at $t = 0$, we have $g_{\max} = g(0)$. As a typical value we take $\rho_{\text{peak}} = 26$ dB.
- Independent equiprobable data symbols. This yields $E[d_k] = 1/2$.
- Symbol synchronizer:
  - Zero-crossing timing error detector.
  - Second-order loop with perfect integrator, determined by a damping factor $\zeta = 5$ and a normalized loop bandwidth $B_L T = 0.01$.
  - Steady-state timing error: assuming a symbol rate of 15 Mbit/s and a delay difference of 1 ns in the logic circuits of the timing error detector, we obtain $e_s = 0.015$.

Properties of the PAM Signal

We have numerically obtained the average zero-crossing point $\overline{\tau}$ and the normalized slope $b = (g'(-T + \overline{\tau}) - g'(\overline{\tau}))\ T$ at this zero-crossing point; the results are given in Table 2-3. The dominating pulse samples $g_n = g(nT + \overline{\tau})$, describing the amount of interference at the average zero-crossing instants, are given in Table 2-4. Note that $g(t)$ is an even symmetrical pulse when $\beta = 0$.

Table 2-3 Parameters of Baseband Pulse $g(t)$

	$\beta = 0$	$\beta = 3\pi/4$
$\overline{\tau}/T$	0.5	0.541
b	2.687	2.322
$g_{\max}$	1.	0.92

Loop Noise Power Spectral Density

We first calculate the auto- and cross-correlation functions of the components $z_1(k)$, $z_2(k)$ and $z_3(k;e_s)$ of the loop noise [see the decomposition (2-198)]. The results are given in Table 2-5. Recalling that $z_1(k)$, $z_2(k)$, and $z_3(k;e_s)$ are caused by the additive noise, by the ISI at the average zero-crossing instants and by the nonzero steady-state error, respectively, a comparison of $R_1(0)$, $R_2(0)$, and $R_3(0;e_s)$ reveals that the contribution of ISI to the loop noise *power* is at least one order of magnitude larger than the contributions due to the additive noise and the nonzero steady-state timing error.

Table 2-4 Dominating Baseband Pulse Samples

	$\beta = 0$	$\beta = 3\pi/4$
g_{-4}	5.7×10^{-3}	2.1×10^{-2}
g_{-3}	1.72×10^{-2}	9.6×10^{-2}
g_{-2}	-0.12	-0.343
g_{-1}	0.6	0.592
g_0	0.6	0.592
g_1	-0.12	5.7×10^{-2}
g_2	1.72×10^{-2}	1.6×10^{-2}
g_3	5.7×10^{-3}	9.9×10^{-3}

Table 2-5 Components of Loop Noise Autocorrelation Function

	$\beta = 0$	$\beta = 3\pi/4$
$R_1(0)$	6.96×10^{-4}	8.22×10^{-4}
$R_2(0)$	8.16×10^{-3}	48.50×10^{-3}
$R_2(1)$	-1.45×10^{-3}	9.91×10^{-3}
$R_2(2)$	-2.61×10^{-3}	3.34×10^{-3}
$R_2(3)$	-1.83×10^{-5}	-1.52×10^{-4}
$R_2(4)$	-4.50×10^{-6}	3.5×10^{-5}
$R_3(0; e_s)$	2.25×10^{-4}	2.25×10^{-4}
$R_{2,3}(1; e_s) + R_{2,3}(1; e_s)$	0	-2.58×10^{-3}

Table 2-6 shows the contributions of $z_1(k)$, $z_2(k)$, and $z_3(k; e_s)$ to the loop noise *power spectral density* [see the decomposition (2-204)]. The spectra $S_1(\omega)$ and $S_3(\omega; e_s)$ are flat, while 2 Re$[S_{2,3}(\omega)]$ can be considered as flat within the loop bandwidth, because $\cos(2\pi B_L T) = \cos(\pi/50) = 0.998 \cong 1$. As shown in Figure 2-63, the case $\beta = 0$ yields $S_2(0) = 0$; this is because the even symmetrical pulse $g(t)$ satisfies the condition (2-209) for a self-noise spectrum vanishing at $\omega = 0$. Hence, $S_2(\omega)$ cannot be considered as flat within the loop bandwidth. Figure 2-64 shows $S_2(\omega)$ for $\beta = 3\pi/4$. The pulse $g(t)$ is no longer even symmetrical, so that $S_2(0) \neq 0$. Within the loop bandwidth, $S_2(\omega)$ can be approximated by $S_2(0)$, which equals 74.77×10^{-3}.

Table 2-6 Components of Loop Noise Power Spectral Density

	$\beta = 0$	$\beta = 3\pi/4$
$S_1(\omega)$	6.96×10^{-4}	8.22×10^{-4}
$S_2(\omega)$	Figure 2-62	Figure 2-63
$S_3(\omega; e_s)$	2.25×10^{-4}	2.25×10^{-4}
$2\,\mathrm{Re}[S_{2,3}(\omega; e_s)]$	0	$-5.16 \times 10^{-3} \cos(\omega T)$

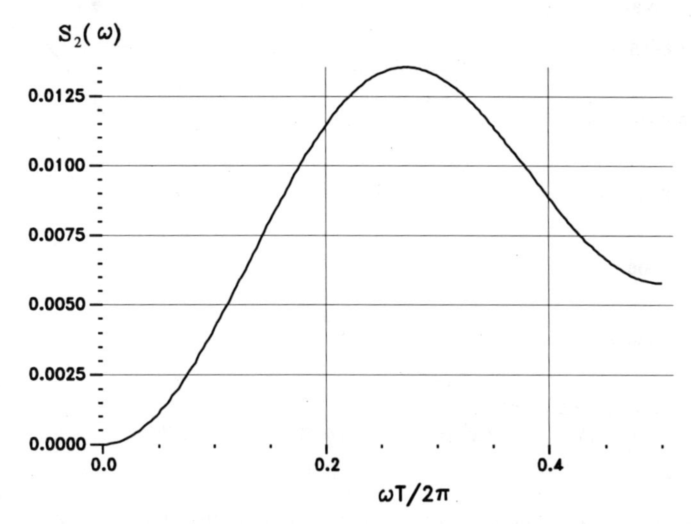

Figure 2-63 Spectrum $S_2(\omega)$ for $\beta = 0$

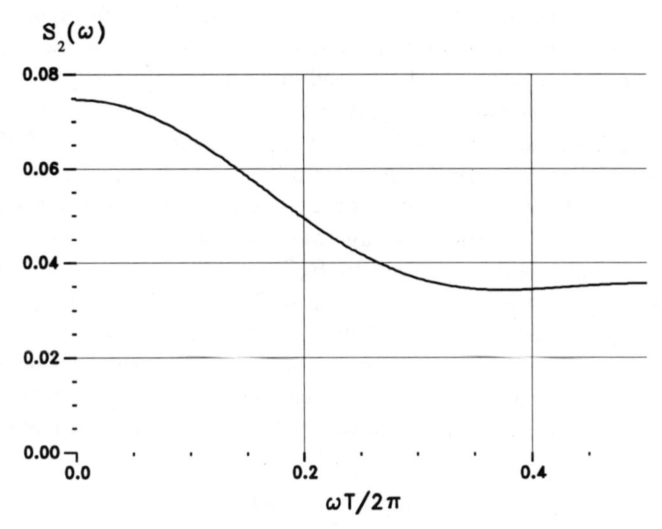

Figure 2-64 Spectrum $S_2(\omega)$ for $\beta = 3\pi/4$

Timing Error Variance

The timing error variance var$[e]$ is given by (2-203). The contributions from $S_1(\omega)$, $S_2(\omega)$, $S_3(\omega; e_s)$ and $2\text{Re}[S_{2,3}(\omega; e_s)]$ are denoted by $\text{var}_1, \text{var}_2, \text{var}_3$, and $\text{var}_{2,3}$, respectively, and are presented in Table 2-7. With the exception of var_2 for $\beta = 0$, all contributions are proportional to the normalized loop bandwidth $B_L T$, because the corresponding spectra are essentially flat within the loop bandwidth.

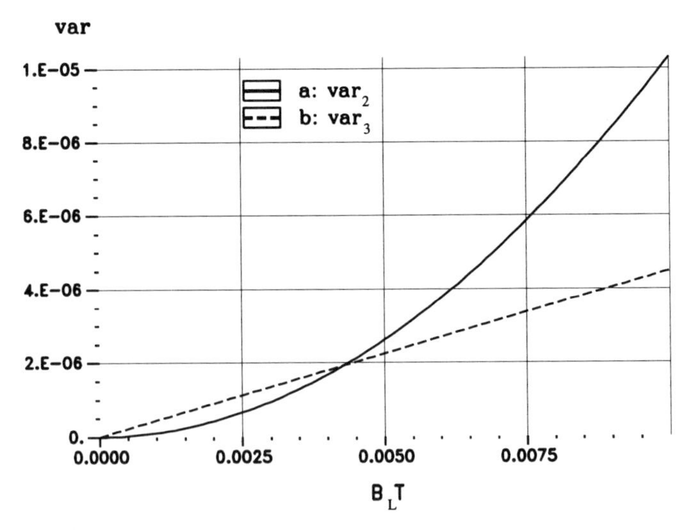

Figure 2-65 Dependence on Loop Bandwidth of var_2, var_3

For $\beta = 0$, the dependence of var_2 on $B_L T$ is shown in Figure 2-65. As the corresponding spectrum $S_2(\omega)$ is zero at $\omega = 0$ (see Figure 2-63), and the closed-loop frequency response of a second-order loop is inversely proportional to ω for large ω, var_2 is essentially proportional to the square of $B_L T$. Also shown in Figure 2-65 is the contribution var_3, which is proportional to $B_L T$. Although the power $R_2(0)$ of the component $z_2(k)$, due to ISI, is about 36 times as large as the power $R_3(0; e_s)$ of the component $z_3(k; e_s)$, due to a nonzero steady-state timing error e_s, var_2 is only 2.3 times as large as var_3 when $B_L T = 0.01$. Moreover, when the loop bandwidth is reduced to $B_L T = 0.001$, var_3 even exceeds var_2 by a factor of about 4.4. The contribution of ISI to the timing error variance is negligible as compared to the contribution due to the nonzero e_s. This example shows that the contribution to the timing error variance, caused by a small nonzero steady-state error e_s, can be much larger than the contribution from ISI, when the loop bandwidth is small and the baseband pulse $g(t)$ exhibits even symmetry.

Table 2-7 Contributions to Timing Error Variance

	$\beta = 0$	$\beta = 3\pi/4$
var_1	13.92×10^{-6}	16.44×10^{-6}
var_2	10.30×10^{-6}	1495.4×10^{-6}
var_3	4.5×10^{-6}	4.5×10^{-6}
$\mathrm{var}_{2,3}$	0	-103.2×10^{-6}
$\mathrm{var}[e]$	28.72×10^{-6}	1413.1×10^{-6}

For $\beta = 3\pi/4$, the loop noise power spectral density near $\omega = 0$ is dominated by the contribution from ISI. Therefore, the effect of the additive noise and a nonzero e_s can be neglected, when compared to the influence of ISI.

Main Points

- The synchronizer with zero-crossing timing error detector tracks the average zero-crossing instants of the PAM signal. The timing error detector generates an error signal only when a data transition occurs. Therefore, the transmission of long strings of identical symbols should be avoided.
- The loop noise consists of three components:
  - A component due to additive noise
  - A self-noise component due to ISI caused by the tails of the baseband pulse $g(t)$
  - A component due to the random occurrence of data transitions; this component is proportional to the steady-state timing error e_s.
- The component due to ISI is eliminated when $g(t)$ is a (time-shifted) Nyquist-II pulse, satisfying

$$g(mT + \overline{\tau}) = 0 \quad \text{for} \quad m \neq 0,\ 1$$

  where $\overline{\tau}$ is defined by

$$g(\overline{\tau}) = g(\overline{\tau} - T)$$

  When this condition is fulfilled, only additive noise and a nonzero steady-state error contribute to the timing error variance; their contributions are proportional to the normalized loop bandwidth $B_L T$.
- A less restrictive condition, which does not eliminate the loop noise component due to ISI but makes zero its contribution to the loop noise power spectral density at $\omega = 0$ is

$$g(mT + \overline{\tau}) = g(-mT - T + \overline{\tau})$$

  which holds for an even symmetrical pulse (in which case $\overline{\tau} = T/2$). When this condition is fulfilled, the contribution to the timing error variance due to ISI is proportional to the square (or even a larger power) of $B_L T$. For $B_L T \ll 1$, this contribution may be considerably smaller than the contribution from a small nonzero steady-state error e_s, the latter being proportional to $B_L T$.

Bibliographical Notes

The tracking performance resulting from the zero-crossing timing error detector has also been considered in [14, 15]. Whereas we obtained the tracking error variance using an "open-loop" approach, the analysis in [14, 15] starts from the closed-loop system equation. Although being linear in the timing error, this system equation is not time-invariant because the timing error detector provides a useful output only when a data transition occurs. The computations in [14] are quite tedious. For small values of the normalized loop bandwidth $B_L T$, the results from [14] reduce to the results given in the discussion of tracking performance above; this validates our open loop approach for realistic values of $B_L T$. In [15], the closed-loop system equation is approximated by a linear time-invariant equation, from which the tracking error variance is easily obtained by applying conventional linear filter theory. The results from [15] coincide with the open-loop results given in the discussion of tracking performance above.

It is shown in [16] that the closed-loop system equation of the synchronizer with zero-crossing timing error detector also applies to a broader class of timing error detectors. A set of conditions is presented under which the tracking error variance reduces to the results obtained from the open-loop approach; basically, these conditions are fulfilled when the normalized loop noise bandwidth $B_L T$ is small.

2.5.3 Example 3: The Mueller and Müller Synchronizer

The Mueller and Müller (M&M) synchronizer is a hybrid discrete-time error-tracking synchronizer using decision feedback, whose block diagram is shown in Figure 2-16, and whose principle of operation is illustrated in Figure 2-17. The PAM signal at the input of the synchronizer is sampled by an adjustable clock operating at the symbol rate. The M&M timing error detector combines the PAM signal samples with the receiver's decisions about the data symbols to form the timing error detector output samples. These error samples are used to bring the sampling clock in synchronism with the PAM signal at the input of the synchronizer.

Tracking Performance

The PAM signal $y(t;\varepsilon)$ at the input of the synchronizer is given by

$$y(t;\varepsilon) = \sum_m a_m \, g(t - mT - \varepsilon T) + n(t) \tag{2-211}$$

where $\{a_m\}$ is a stationary sequence of not necessarily binary data symbols, $g(t)$ is the baseband pulse, and $n(t)$ is stationary noise. We have chosen the origin of time such that $g(t)$ takes on its maximum value at $t = 0$.

The adjustable clock samples the signal $y(t;\varepsilon)$ at the instants t_k, given by

$$t_k = kT + \hat{\varepsilon}T + \overline{\tau} \tag{2-212}$$

where $\overline{\tau}$ is a time shift to be defined soon. The resulting samples are denoted by $y_k(\varepsilon, \hat{\varepsilon})$, with

$$y_k(\varepsilon, \hat{\varepsilon}) = y(kT + \hat{\varepsilon}T + \overline{\tau}; \varepsilon) \tag{2-213}$$

These samples are combined with the receiver's decisions $\{\hat{a}_k\}$, to form the following timing error detector output samples:

$$x_k(\varepsilon,\hat{\varepsilon}) = \hat{a}_{k-1}\; y_k(\varepsilon,\hat{\varepsilon}) - \hat{a}_k\; y_{k-1}(\varepsilon,\hat{\varepsilon}) \tag{2-214}$$

In the following, it will be assumed that the receiver's decisions are correct, i.e., $\hat{a}_k = a_k$ for all k. As far as the synchronizer operation is concerned, this assumption is reasonable when the decision error probability is smaller than about 10^{-2}; the actual error probability on most terrestrial baseband links is smaller than 10^{-2} by several orders of magnitude, because the signal-to-noise ratios are large.

The M&M synchronizer is analyzed under open-loop conditions: the loop is opened (say, at the loop filter input), and the properties of the timing error detector output are determined for sampling instants t_k corresponding to a fixed value of the timing estimate $\hat{\varepsilon}$.

Substituting (2-211) and (2-213) into (2-214), one obtains

$$x_k(\varepsilon,\hat{\varepsilon}) = z_1(k) + z_2(k;e) \tag{2-215}$$

where $e = \varepsilon - \hat{\varepsilon}$ denotes the timing error, and

$$z_1(k) = a_{k-1}\, n_k - a_k\, n_{k-1} \tag{2-216}$$

$$z_2(k;e) = a_{k-1} \sum_m a_{k-m}\, g_m(e) - a_k \sum_m a_{k-1-m}\, g_m(e) \tag{2-217}$$

with

$$n_m = n(mT + \overline{\tau} + \hat{\varepsilon}T)$$

$$g_m(e) = g(mT + \overline{\tau} + eT)$$

The term $z_1(k)$ in (2-215) is zero-mean and caused by additive noise; this term contributes to the loop noise. The term $z_2(k;e)$ in (2-215) contains a useful component $E[z_2(k;e)]$, and a zero-mean self-noise component $z_2(k;e) - E[z_2(k;e)]$, contributing to the loop noise.

For statistically independent data symbols, the timing error detector characteristic is given by

$$\begin{aligned} E[x_k(\varepsilon,\hat{\varepsilon})] &= E[z_2(k;e)] \\ &= \left(A_2 - A_1^2\right)\,[g_1(e) - g_{-1}(e)] \\ &= \mathrm{var}[a_k]\,[g_1(e) - g_{-1}(e)] \end{aligned}$$

where $A_n = E[a_k^n]$ denotes the nth order moment of the data symbols. For a sequence of identical data symbols, $\mathrm{var}[a_k] = 0$. This indicates that the useful timing error detector output becomes small when the data sequence contains long strings of identical symbols; the occurrence of such strings must be avoided, for

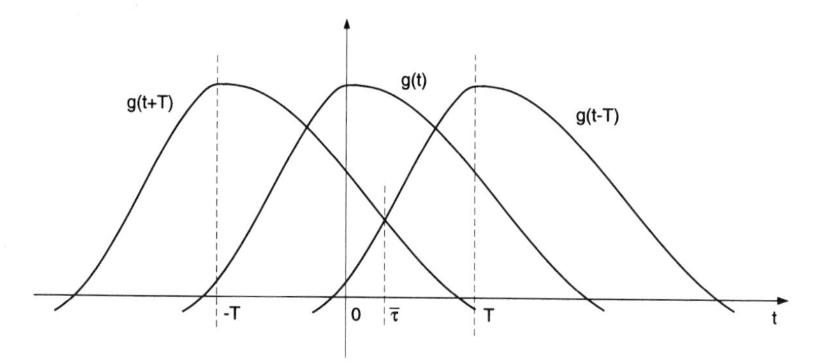

Figure 2-66 Determination of $\overline{\tau}$

the

instance by using a scrambler. Now we define the time shift $\overline{\tau}$ from (2-212) such that $e = 0$ is a stable tracking point; this requires that the timing error detector characteristic becomes zero for $e = 0$, or equivalently,

$$g(T + \overline{\tau}) = g(-T + \overline{\tau})$$

Note that $\overline{\tau} = 0$ when $g(t)$ is an even symmetrical pulse. Figure 2-66 shows how $\overline{\tau}$ can be obtained graphically. The timing error detector slope at $e = 0$ is given by

$$b = \text{var}[a_k] \, [g'(-T + \overline{\tau}) - g'(T + \overline{\tau})] \, T$$

where $g'(t)$ is the derivative of $g(t)$.

The loop noise power spectral density will be evaluated under the assumption of statistically independent data symbols having an even symmetric distribution; as a result, all odd-order moments of the data symbols are zero. The loop noise power spectral density $S_N(\omega; e)$ can be decomposed as

$$S_N(\omega; e) = S_1(\omega) + S_2(\omega; e)$$

where $S_1(\omega)$ and $S_2(\omega; e)$ are due to the additive noise term $z_1(k)$ and self-noise term $z_2(k; e) - E[z_2(k; e)]$ at the timing error detector output. Note that $S_1(\omega)$ does not depend on the timing error e, because the noise $n(t)$ at the synchronizer input is stationary.

The power spectral density $S_1(\omega)$ is given by

$$S_1(\omega) = R_1(0) + 2 \sum_{m=1}^{\infty} R_1(m) \, \cos(m\omega T)$$

where

$$R_1(m) = E[z_1(k) \, z_1(k + m)]$$

From (2-216) we easily obtain

$$\begin{aligned} R_1(0) &= 2\,A_2\,R_n(0) \\ R_1(1) &= -A_2\,R_n(2T) \\ R_1(m) &= 0 \quad \text{for} \quad m > 1 \end{aligned}$$

where $R_n(u) = E[n(t)\,n(t+u)]$ denotes the autocorrelation function of the additive noise $n(t)$ at the synchronizer input. This yields

$$S_1(\omega) = 2A_2\,[R_n(0) - R_n(2T)\,\cos(\omega T)]$$

In most cases of practical interest, $R_n(0) \gg |R_n(2T)|$, so that $S_1(\omega)$ is essentially flat within the loop bandwidth.

The self-noise power spectral density $S_2(\omega; e)$ is given by

$$S_2(\omega; e) = R_2(0; e) + 2\sum_{m=1}^{\infty} R_2(m; e)\,\cos(m\omega T)$$

where

$$R_2(m; e) = E[z_2(k; e)\,z_2(k+m; e)] - \{E[z_2(k; e)]\}^2$$

After some algebra we obtain from (2-217):

$$\begin{aligned} R_2(0; e) &= A_4\left[g_1^2(e) + g_{-1}^2(e)\right] + 2A_2^2 \sum_{|m|>1} g_m^2(e) \\ R_2(1; e) &= -A_4\,g_1(e)\,g_{-1}(e) - A_2^2 \sum_{|m|>1} g_{m-1}(e)\,g_{m+1}(e) \\ R_2(m; e) &= A_2^2[g_{1-m}(e)\,g_{1+m}(e) + g_{-1-m}(e)\,g_{-1+m}(e) - 2g_m(e)\,g_{-m}(e)] \\ &\qquad \text{for} \quad m > 1 \end{aligned}$$

Whether $S_2(\omega; e)$ can be considered as flat within the loop bandwidth depends on the pulse $g(t)$.

For a given steady-state timing error e_s, the timing error variance var[e] is given by

$$\text{var}[e] = \frac{1}{b^2}\,T \int_{-\pi/T}^{\pi/T} |H(\omega)|^2\,S_N(\omega; e_s)\,\frac{d\omega}{2\pi} \tag{2-218}$$

where $H(\omega)$ denotes the closed-loop frequency response of the discrete-time synchronizer. Assuming that both $S_1(\omega)$ and $S_2(\omega; e_s)$ are essentially flat within the loop bandwidth, var[e] is proportional to the normalized loop bandwidth $B_L T$:

$$\text{var}[e] = (2B_L T)\,\frac{S_1(0) + S_2(0; e_s)}{b^2} \tag{2-219}$$

Otherwise, var[e] must be evaluated by means of the general formula (2-218).

Self-Noise Reduction

In this section we show that the baseband pulse $g(t)$ can be selected such that self-noise is completely eliminated when the steady-state error e_s is zero. Also, we derive a less restrictive condition on $g(t)$, which yields a self-noise spectrum that becomes zero at $\omega = 0$; in this case, the tracking error variance due to self-noise is proportional to the square (or even a higher power) of the normalized loop bandwidth $B_L T$.

It is easily seen from eq. (2-217) that $z_2(k; e_s) = 0$ for $e_s = 0$ provided that

$$g(mT + \overline{\tau}) = 0 \quad \text{for} \quad m \neq 0 \tag{2-220}$$

In this case, there is no self-noise at the timing error detector output, irrespective of the statistics of the data sequence: only the additive noise at the input of the synchronizer contributes to the timing error variance. Condition (2-220) requires that $g(t)$ is a Nyquist-I pulse (which is shifted in time such that its maximum value occurs at $t = 0$). In most cases, the baseband pulse at the output of the receive filter (approximately) satisfies the first Nyquist criterion, in order to eliminate or at least substantially reduce the ISI at the decision instants (see Section 1.1.2). In this case, the receive filter output can be applied directly to the M&M synchronizer without additional prefiltering.

Now we will derive a condition on $g(t)$, which does not eliminate the self-noise, but yields a self-noise spectrum which becomes zero at $\omega = 0$. More specifically we will select $g(t)$ such that

$$\sum_{k=-\infty}^{+\infty} z_2(k; e_s) = 0 \quad \text{for} \quad e_s = 0 \tag{2-221}$$

Using a similar reasoning as in the discussion above on self-noise reduction, the following holds when (2-221) is fulfilled:

(i) Irrespective of the data symbol statistics, $z_2(k; e_s)$ is zero-mean for $e_s = 0$.
(ii) Irrespective of the data symbol statistics, the spectrum $S_2(\omega; e_s)$ is zero at $\omega = 0$ for $e_s = 0$.

Hence, when (2-221) holds, $z_2(k; 0)$ contributes to the loop noise power spectral density but not to the average timing error detector output. This contribution to the loop noise power spectral density is zero at $\omega = 0$.

Using (2-217) we obtain

$$\begin{aligned}\sum_{k=-\infty}^{+\infty} z_2(k; 0) &= \sum_{m=-\infty}^{+\infty} (\alpha_{m-1} - \alpha_{m+1})\, g_m(0) \\ &= \sum_{m=-\infty}^{+\infty} \alpha_m \left[g_{m+1}(0) - g_{m-1}(0)\right]\end{aligned}$$

where

$$\alpha_m = \sum_{k=-\infty}^{+\infty} a_k \, a_{k-m}$$

Noting that $\alpha_m = \alpha_{-m}$, we obtain

$$\sum_{k=-\infty}^{+\infty} z_2(k;0) = \alpha_0 \left[g_1(0) - g_{-1}(0)\right] + \sum_{m>0} \alpha_m \left[g_{m+1}(0) - g_{m-1}(0) + g_{-m+1}(0) - g_{-m-1}(0)\right]$$

Hence, (2-221) is fulfilled when

$$g(T+\overline{\tau}) = g(-T+\overline{\tau})$$

$$g(mT+T+\overline{\tau}) + g(-mT+T+\overline{\tau}) = g(mT-T+\overline{\tau}) + g(-mT-T+\overline{\tau}) \qquad m>0$$

The above set of equations is equivalent with

$$g(mT+\overline{\tau}) = g(-mT+\overline{\tau}) \quad \text{for} \quad m = 1, 2, \ldots \tag{2-222}$$

When the condition (2-222) is fulfilled, the self-noise power spectral density $S_2(\omega;0)$ becomes zero at $\omega = 0$, irrespective of the data symbol statistics. The condition (2-222) holds when the baseband pulse $g(t)$ exhibits even symmetry, in which case $\overline{\tau} = 0$.

It is important to note that the condition (2-220) or (2-222) eliminates or substantially reduces the self-noise only when the steady-state timing error e_s is zero. When (2-220) or (2-222) holds but $e_s \neq 0$, then $S_2(0;e_s) \neq 0$ so that the self-noise contribution to the timing error variance is proportional to $B_L T$ instead of being zero [when (2-220) holds and $e_s = 0$] or being proportional to the square (or a higher power) of $B_L T$ [when (2-222) holds and $e_s = 0$]. Hence, the steady-state timing error e_s should be kept small in order not to increase considerably the self-noise contribution to the tracking error variance.

Numerical Example

In order to illustrate the results from the previous sections, we now compute the loop noise power spectral density and the timing error variance resulting from the M&M synchronizer, operating on a properly equalized noisy PAM signal; this equalization results in an even symmetrical Nyquist-I baseband pulse [i.e., $g(t) = g(-t)$ and $g(mT) = 0$ for $m \neq 0$].

System Description

- Pulse shape $g(t)$:

Cosine rolloff amplitude shaping with zero delay distortion:

$$G(\omega) = \begin{cases} T & |\omega T| < \pi(1-\alpha) \\ \frac{T}{2}\left[1 - \sin\left(\frac{|\omega T| - \pi}{2\alpha}\right)\right] & \pi(1-\alpha) < |\omega T| < \pi(1+\alpha) \\ 0 & \text{otherwise} \end{cases}$$

Numerical value: $\alpha = 0.5$.

- Additive noise $n(t)$: wideband noise

$$R_n(0) = \sigma_n^2 \qquad R_n(mT) = 0 \quad \text{for} \quad m \neq 0$$

the noise is specified with respect to the signal by the peak-to-noise ratio ρ_{peak}, defined as

$$\rho_{\text{peak}} = \frac{g_{\max}}{\sigma_n}$$

where $g_{\max}$ is the maximum value of $g(t)$; for the considered $g(t)$, we have $g_{\max} = g(0) = 1$. As a typical value we take $\rho_{\text{peak}} = 30$ dB.

- Independent equiprobable binary (± 1) data symbols. This yields $A_2 = A_4 = 1$.
- Symbol synchronizer:
  Normalized loop bandwidth $B_L T = 0.01$.
  Small steady-state timing error e_s.

Properties of the PAM Signal

The even symmetry of the PAM pulse $g(t)$ yields $\overline{\tau} = 0$. The timing error detector slope, evaluated at $e = 0$, is given by

$$b = 2Tg'(-T) = 1.57$$

The self-noise is determined by the PAM pulse samples $g_m(e_s) = g(mT - e_s T)$ for $m \neq 0$. When the steady-state timing error e_s is zero, we obtain $g_m(e_s) = 0$ for $m \neq 0$, because $g(t)$ is an even symmetrical Nyquist-I pulse; in this case, self-noise is absent. For small nonzero e_s, we use the approximation

$$g_m(e_s) \cong -g'(mT)\, T\, e_s \quad \text{for} \quad m \neq 0 \tag{2-223}$$

where $g'(t)$ is the time-derivative of $g(t)$. The dominating values $g'(mT)T$ are shown in Table 2-8 for $m > 0$; for $m < 0$, use $g'(-mT) = -g'(mT)$.

Loop Noise Power Spectral Density and Tracking Error Variance

Using the results from Section 2.5.3, the loop noise power spectral density $S_1(\omega)$, caused by the additive noise, is given by

$$S_1(\omega) = 2\,\sigma_n^2 = \frac{2}{\rho_{\text{peak}}^2}$$

The spectrum $S_1(\omega)$ is flat, because the noise samples $n(mT)$ are uncorrelated.

Table 2-8 Dominating Pulse Parameters

	$-g'(mT)\,T$	$R_2(m)$
$m = 0$		1.346
$m = 1$	0.758	0.622
$m = 2$	-0.167	0.055
$m = 3$	0.0	0.005
$m = 4$	0.017	
$m = 5$	0.0	
$m = 6$	-0.005	

The self-noise power spectral density $S_2(\omega; e_s)$ is defined as the Fourier transform of the autocorrelation sequence $\{R_2(m; e_s)\}$ of the self-noise contribution to the loop noise. This autocorrelation sequence can be computed using the results from Section 2.5.3. According to the approximation (2-223), $R_2(m; e_s)$ is proportional to e_s^2, and can therefore be written as

$$R_2(m; e_s) = R_2(m)\, e_s^2$$

The dominating values of $R_2(m)$ are shown in Table 2-8 for $m \geq 0$; for $m < 0$, use $R_2(-m) = R_2(m)$. Hence, the corresponding self-noise power spectral density $S_2(\omega; e_s)$ can be written as

$$S_2(\omega; e_s) = S_2(\omega)\, e_s^2$$

where $S_2(\omega)$ is the Fourier transform of the sequence $\{R_2(m)\}$. Figure 2-67

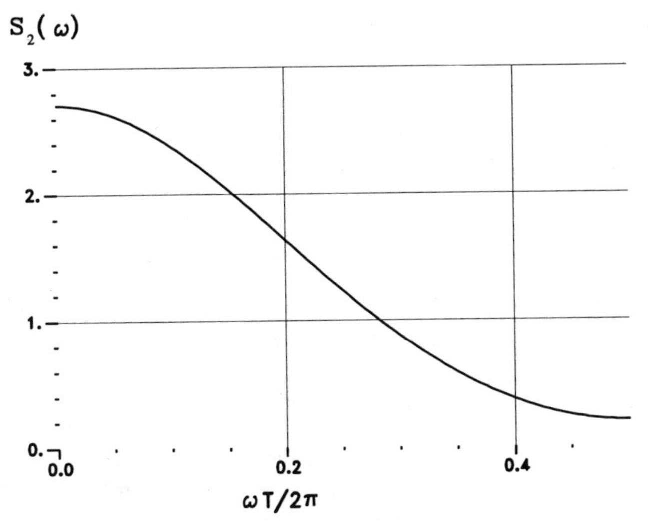

Figure 2-67 Spectrum $S_2(\omega)$

shows how $S_2(\omega)$ depends on ω. Within the loop bandwidth, $S_2(\omega)$ is essentially flat, and can be approximated by $S_2(0)$.

As both $S_1(\omega)$ and $S_2(\omega; e_s)$ are flat within the loop bandwidth, the tracking error variance is well approximated by (2-219). This yields

$$\text{var}[e] = (2B_LT)\left[A\,\rho_{\text{peak}}^{-2} + B\,e_s^2\right]$$

with $A = 0.81$ and $B = 1.10$. When $e_s = 0$, the values $B_LT = 0.01$ and $\rho_{\text{peak}} = 30\,\text{dB}$ yield $\text{var}[e] = 16.2 \times 10^{-6}$; for $e_s = 0.03$, the tracking error variance increases to $\text{var}[e] = 36.0 \times 10^{-6}$. This illustrates that a small steady-state error can have a large effect on the tracking error variance.

Main Points

- The M&M synchronizer uses decision feedback to produce a timing error detector output. In the absence of data transitions, the useful timing error detector output equals zero; therefore, the transmission of long strings of identical symbols must be avoided.

- The loop noise consists of two components:

  - A component due to additive noise
  - A self-noise component due to the tails of the baseband pulse $g(t)$

- The self-noise component is eliminated for a zero steady-state timing error e_s when $g(t)$ is a (time-shifted) Nyquist-I pulse, satisfying

  $$g(mT + \overline{\tau}) = 0 \quad \text{for} \quad m \neq 0$$

  where $\overline{\tau}$ is defined by

  $$g(T + \overline{\tau}) = g(-T + \overline{\tau})$$

  When this condition is fulfilled and $e_s = 0$, only additive noise contributes to the timing error detector. When this condition is fulfilled but $e_s \neq 0$, self-noise is present at the timing error detector output, and its contribution to the timing error variance is proportional to the normalized loop bandwidth B_LT.

- A less restrictive condition, which does not eliminate self-noise but yields for $e_s = 0$, a self-noise spectrum with a null at $\omega = 0$, is

  $$g(mT + \overline{\tau}) = g(-mT + \overline{\tau}) \quad \text{for} \quad m > 0$$

  which holds for an even symmetrical pulse (in which case $\overline{\tau} = 0$). When this condition is fulfilled and $e_s = 0$, the self-noise contribution to the timing error variance is proportional to the square (or even a larger power) of the normalized loop bandwidth $B_L T$. When this condition is fulfilled but $e_s \neq 0$, the self-noise contribution to the timing error variance is proportional to $B_L T$, and, when $B_L T \ll 1$, may be much larger than in the case $e_s = 0$.

Bibliographical Notes

In their seminal paper [17] Mueller and Müller introduced a class of timing error detectors, which operate on samples of a PAM signal, taken at the baud rate. The timing error detector output is given by

$$x_k(\varepsilon,\hat{\varepsilon}) = \sum_{m=0}^{M-1} h_m(\hat{a}_k, \hat{a}_{k-1}, \,\dots\,, \,\hat{a}_{k-M+1})\, y_{k-m}(\varepsilon,\hat{\varepsilon}) \tag{2-224}$$

where $y_n(\varepsilon,\hat{\varepsilon})$ and $\hat{a}_n$ denote the nth baud-rate sample [see (2-213)] and the decision about the nth transmitted symbol, respectively, and $h_m(\ldots)$, $m = 0, \,\dots\,, M-1$, are M arbitrary functions, to be selected by the designer. For example, the timing error detector (2-214) is obtained for $M = 2$, and

$$h_0(\hat{a}_k,\, \hat{a}_{k-1}) = \hat{a}_{k-1} \qquad h_1(\hat{a}_k,\, \hat{a}_{k-1}) = -\hat{a}_k$$

Assuming correct decisions, the timing error detector characteristic $E[x_k(\varepsilon,\hat{\varepsilon})]$, resulting from (2-224), is a linear combination of PAM pulse samples $g(nT + \overline{\tau} + \hat{\varepsilon} - \varepsilon)$, where $\overline{\tau}$ is determined by the condition that the timing error detector characteristic is zero for $\hat{\varepsilon} = \varepsilon$. Once a particular linear combination of T-spaced PAM pulse samples has been specified as the desired timing error detector characteristic, there are several sets of M functions $h_m(\ldots)$ that give rise to this characteristic. Examples with $M = 2$ and $M = 3$ are given, where this freedom is exploited to select a set of M functions yielding a small variance at the timing error detector output. However, from a tracking performance point of view, it is the power spectral density near $\omega = 0$, rather than the variance, of the timing error detector output that should be kept small.

Symbol synchronization in digital subscriber loops, using a baud-rate timing error detector from [17], has been investigated in [18]. The received signal consists of a useful PAM signal from the far-end transmitter and a disturbing echo signal from the near-end transmitter. First, the received signal enters an echo canceler, which subtracts (an estimate of) the echo signal. The echo canceler output signal then enters a decision-feedback equalizer, which reduces the postcursor ISI of the far-end signal. The timing error detector operates on the decision-feedback equalizer output and is designed such that $g(-T + \overline{\tau}) = 0$, which keeps small the precursor ISI of the far-end signal. The advantages of using a baud-rate (instead of higher-rate) timing error detector are as follow:

- Echo cancellation must be performed only at the baud rate; this reduces the implementation complexity of the echo canceler.

- The decision-feedback equalizer (which operates at the baud rate) can precede the timing error detector. Consequently, the ISI at the timing error detector output can be kept small; this enhances the symbol synchronizer performance.

In [17], the timing error detector output (2-224) is evaluated at the baud rate, and the timing estimate is updated accordingly. In the absence of noise, the timing error detector output and the resulting timing estimate exhibit statistical fluctuations due to the random nature of the data sequence. As the magnitude of these fluctuations depends heavily on the transmitted data pattern, it is proposed in [19] to update the timing estimate only at the occurrence of "favorable" data patterns, which yield a very small statistical fluctuation at the timing error detector output, and hence a very small tracking error variance. An example is presented with $M = 3$. However, for such small values of M there are very few favorable patterns, so that the random interval between timing estimate updates can assume large values.

A similar idea as in [19] has been used in [20]. The receiver reads a block of M samples from the PAM signal, taken at the baud rate. Depending on the received data symbol sequence, all pairs of signal samples are selected whose sum or difference $y_m(\varepsilon, \hat{\varepsilon}) \pm y_n(\varepsilon, \hat{\varepsilon})$ can be used as a low variance timing error detector output. The contributions from several pairs within a single block of signal samples are averaged, and used to update the timing estimate; when no such pairs are found, the timing estimate is not updated. Considering $M = 20$, the probability of no estimate updating is negligibly small, so that the updating occurs effectively at a fraction $1/M$ of the baud rate.

Bibliography

[1] L. E. Franks and J. P. Bubrouski, "Statistical Properties of Timing Jitter in a PAM Timing Recovery Scheme," *IEEE Trans. Commun.*, vol. COM-22, pp. 913–920, July 1974.

[2] U. Mengali and G. Pirani, "Jitter Accumulation in PAM Systems," *IEEE Trans. Commun.*, vol. COM-28, pp. 1172–1183, Aug. 1980.

[3] M. Moeneclaey, "Prefilter Optimization for the Filter and Square Synchronizer," *Archiv Elektr. Uebertr.*, vol. 38, pp. 257–261, July 1984.

[4] T. T. Fan, "I and Q Decomposition of Self-Noise in Square-Law Clock Regenerators," *IEEE Trans. Commun.*, vol. COM-36, pp. 1044–1052, Sept. 1988.

[5] A. Del Pistoia, U. Mengali, and R. Molpen, "Effect of Coding on the Jitter Accumulation in Baseband Digital Transmission," *IEEE Trans. Commun.*, vol. COM-30, pp. 1818–1827, Aug. 1982.

[6] M. Oerder and H. Meyr, "Digital Filter and Square Timing Recovery," *IEEE Trans. Commun.*, vol. COM-36, pp. 605–612, May 1988.

[7] A. N. D'Andrea and U. Mengali, "A Simulation Study of Clock Recovery in QPSK and 9 QPRS Systems," *IEEE Trans Commun.*, vol. COM-33, pp. 1139–1142, Oct. 1985.

[8] A. N. D'Andrea and U. Mengali, "Performance Analysis of the Delay-Line Clock Regenerator," *IEEE Trans. Commun.*, vol. COM-34, pp. 321–328, Apr. 1986.

[9] F. M. Gardner, "Self-Noise in Synchronizers," *IEEE Trans. Commun.*, vol. COM-28, pp. 1159–1163, Aug. 1980.

[10] T. T. Fang, "Analysis of Self-Noise in a Fourth-Power Clock Regenerator," *IEEE Trans. Commun.*, vol. COM-39, pp. 133–140, Jan. 1991.

[11] A. N. D'Andrea, U. Mengali, and M. Moro, "Nearly Optimum Prefiltering in Clock Recovery," *IEEE Trans. Commun.*, vol. COM-34, pp. 1081–1088, Nov. 1986.

[12] G. L. Cariolaro and F. Todero, "A General Spectral Analysis of Time Jitter Produced in a Regenerative Repeater," *IEEE Trans. Commun.*, vol. COM-25, pp. 417–426, Apr. 1977.

[13] D. M. O. Agazzi, C.-P. J. Tzeng, D. G. Messerschmitt, and D. A. Hodges, "Timing Recovery in Digital Subscriber Loops," *IEEE Trans. Commun.*, vol. COM-33, pp. 558–569, June 1985.

[14] B. R. Saltzberg, "Timing Recovery for Synchronous Binary Data Transmission," *BSTJ*, vol. 46, pp. 593–662, Mar. 1967.

[15] D. L. Duttweiler, "The Jitter Performance of Phase-Lock Loops Extracting Timing from Baseband Data Waveforms," *BSTJ*, vol. 55, pp. 37–58, Jan. 1976.

[16] E. Roza, "Analysis of Phase-Locked Timing Extraction for Pulse Code Transmission," *IEEE Trans. Commun.*, vol. COM-22, pp. 1236–1249, Sept. 1974.

[17] K. H. Mueller and M. Müller, "Timing Recovery in Digital Synchronous Data Receivers," *IEEE Trans. Commun.*, vol. COM-24, pp. 516–531, May 1976.

[18] C.-P. J. Tzeng, D. A. Hodges, and D. G. Messerschmitt, "Timing Recovery in Digital Subscriber Loops Using Baud-Rate Sampling," *IEEE J. Select. Areas Commun.*, vol. SAC-4, pp. 1302–1311, Nov. 1986.

[19] A. Jennings and B. R. Clarke, "Data-Selective Timing Recovery for PAM Systems," *IEEE Trans. Commun.*, vol. COM-33, pp. 729–731, July 1985.

[20] J. Armstrong, "Symbol Synchronization Using Baud-Rate Sampling and Data-Sequence-Dependent Signal Processing," *IEEE Trans. Commun.*, vol. COM-39, pp. 127–132, Jan. 1991.

PART D
Passband Communication Over Time Invariant Channels

Chapter 3 Passband Transmission

In this chapter we briefly review the fundamentals of passband transmission over a time-invariant channel. In Section 3.1 we describe the transmission methods. In Section 3.2 we introduce channel and transceiver models. In the last section of this chapter we are concerned with the fundamental bounds of the outer receiver (see chapter on introduction and preview), the channel capacity. This bound defines the ultimate transmission rate for error-free transmission. We gain considerable insight by comparing this bound with the performance of any communication system. Furthermore, studying the fundamental bounds on the outer receiver performance (channel capacity) and that of the inner receiver (variance of the parameter estimates) provides a deep understanding of the interdependence between the two parts.

3.1 Transmission Methods

Passband transmission of digital information can be roughly separated into two main classes: noncoherent and coherent. The first class uses so-called *noncoherent* modulation techniques, which do not require an estimate of the carrier frequency and phase.

Noncoherent modulation techniques have significant disadvantages, in particular a power penalty and spectral inefficiency when compared to the second class of techniques employing *coherent* transmission.

The most commonly used member of the class of coherent transmission techniques is pulse-amplitude modulation (PAM). Special cases of passband PAM are phase-shift keying (PSK), amplitude and phase modulation (AM-PM), and quadrature amplitude modulation (QAM). In a pulse-amplitude-modulated passband signal the information is encoded into the complex amplitude values a_n for a single pulse and then modulated onto sinusoidal carriers with the same frequency but a 90° phase difference. PAM is a linear technique which has various advantages to be discussed later.

Another class of coherent modulation techniques discussed is continuous-phase modulation (CPM). Since the information is encoded into the phase, the signal maintains a constant envelope which allows a band-limited signal's amplification without serious spectral spreading of the signal, because the main nonlinear effects of bandpass high-performance amplifiers, namely amplitude-to-amplitude (AM/AM) and amplitude-to-phase (AM/PM) conversion, are avoided.

3.2 Channel and Transceiver Models

3.2.1 Linear Channel Model

The general linear model of a carrier-modulated digital communication system is shown in Figure 3-1. This model adequately describes a number of important channels such as telephone channels and microwave radio transmission.

In Figure 3-1, $u(t)$ represents any type of *linearly* modulated signal in baseband:

$$u(t) = \sum_n a_n g_T(t - nT - \varepsilon_0 T) \qquad \text{(3-1)}$$

The channel symbols $\{a_n\}$ are chosen from an arbitrary signal set over the complex plane and $g_T(t)$ is the impulse response of the pulse-shaping filter. T is referred to as the channel symbol duration or period. The pulse $g_T(t)$ contains a (possibly slowly time varying) time shift ε_0 with respect to the time reference of a hypothetical observer.

In the quaternary PSK modulator, a sign change in both in-phase and quadrature components causes a phase shift of 180°. If only one of the components, either the *I* or *Q* component, changes its sign, a phase shift of $\pm\pi/2$ occurs. This reduces the envelope fluctuations [1 , p. 239ff] which is useful when the signal undergoes nonlinear amplification.

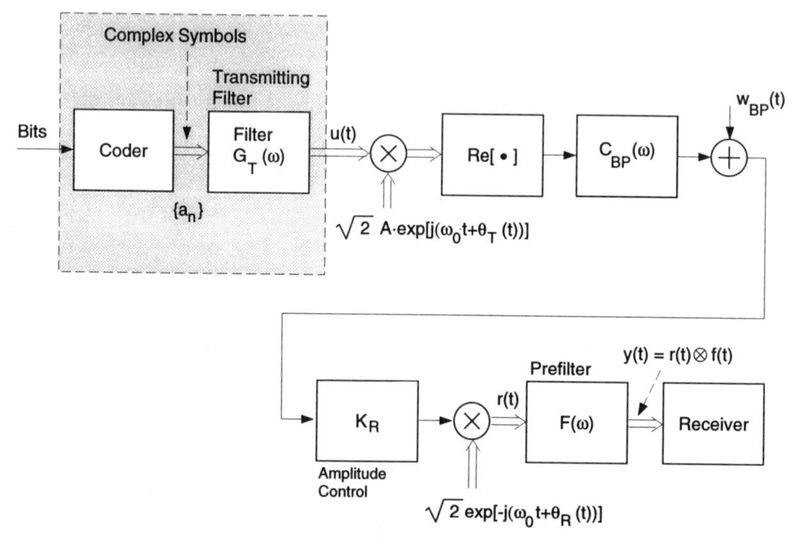

Figure 3-1 Linear Model of Passband Transmission: Linear Modulation

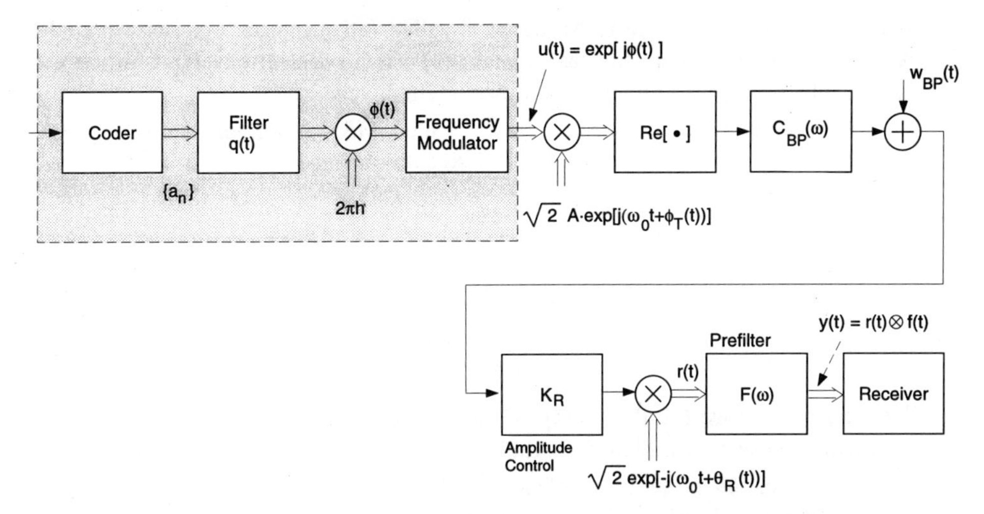

Figure 3-2 Linear Model of Passband Transmission: CPM

When the quadrature component is delayed with respect to the in-phase component by $T/2$ only phase changes of $\pm\pi/2$ occur. This modulation is called offset QPSK (OQPSK) or staggered QPSK. The signal $u(t)$ in (3-1) for OQPSK reads

$$
\begin{aligned}
u(t) = &\left(\sum_n a_n^{(I)} g_T\left(t - 2n\left(\frac{T}{2}\right) - \varepsilon_0 T\right)\right) \\
&+ j\left(\sum_n a_n^{(Q)} g_T\left(t - [2n-1]\left(\frac{T}{2}\right) - \varepsilon_0 T\right)\right)
\end{aligned}
\tag{3-2}
$$

where the channel symbols $a_n^{(I)}$, $a_n^{(Q)}$ take values in the set $\{-1, 1\}$.

For *nonlinear* modulation schemes, the dashed block in Figure 3-1 must be replaced by that of Figure 3-2. For a CPM signal, the signal $u(t)$ is

$$u(t) = \exp\left[j\phi(t)\right] \tag{3-3}$$

The phase is given by

$$\phi(t) = 2\pi h \sum_k a_k q(t - kT - \varepsilon_0 T) + \theta \tag{3-3a}$$

where $\mathbf{a} = \{a_n\}$ is the data sequence of M-ary information symbols selected from the alphabet $\pm 1, \pm 3, ..., \pm(M-1)$, h is the modulation index, and $q(t)$ is some normalized waveform. The waveform $q(t)$ may be represented as the integral of some frequency pulse $v(t)$

$$q(t) = \int_0^t v(\tau)\, d\tau \tag{3-4}$$

If $v(t) = 0$ for $t \geq T$ the CPM signal is called full response, otherwise partial response. The equivalent lowpass signal $u(t)$ has a constant envelope, which is an advantage when a nonlinear amplifier is employed.

Of great practical importance are so-called MSK (minimum shift keying) signals. They are a special case of CPM signals obtained for

$$a_k = \pm 1, \; h = 1/2 \tag{3-5}$$

For MSK the frequency pulse is rectangular

$$v(t) = \begin{cases} 1/2T & 0 \leq t < T \\ 0 & \text{otherwise} \end{cases} \quad \text{(full response)} \tag{3-6}$$

For Gaussian MSK (GMSK) $v(t)$ is the response of the Gaussian filter to a rectangular pulse of duration T (partial response).

An interesting property of MSK signals is that they can be interpreted as linearily modulated passband signals where the quadrature component is delayed with respect to the in-phase component by T [2 , Chapter 6]. The following signal is equivalent to (3-3):

$$\begin{aligned} u(t) = & \left(\sum_n a_{2n} h(t - 2nT - \varepsilon_0 T) \right) \\ & + j \left(\sum_n a_{2n-1} h(t - [2n-1]T - \varepsilon_0 T) \right) \end{aligned} \tag{3-7}$$

with

$$h(t) = \begin{cases} \sin\left(\dfrac{\pi t}{2T}\right) & 0 \leq t \leq 2T \\ 0 & \text{else} \end{cases} \tag{3-8}$$

and a_{2n}, a_{2n-1} take on values ± 1. Notice that the pulse $h(t)$ is time-limited. Hence, $u(t)$ is not band-limited.

The baseband signal is up-converted to passband by an oscillator $\exp[j(\omega_0 t + \theta_T(t))]$. The linear channel has a frequency response $C_{BP}(\omega)$ which models the physical channel frequency response as well as any filtering in the passband region. The signal is subject to wideband additive Gaussian noise $w_{BP}(t)$ with flat spectral density $N_0/2$ (see Section 1.2.3 above and Chapter 3 of Volume 1).

After downconversion by a local oscillator $\exp[-j(\omega_0 t + \theta_R(t))]$ and lowpass filtering $F(\omega)$ in order to remove double-frequency terms we obtain the signal

$$\begin{aligned} y(t) &= K_R[As(t) \otimes f(t) + w(t) \otimes f(t)] \\ &= K_R[As_f(t) + n(t)] \end{aligned} \tag{3-9}$$

where $s_f(t)$ is related to $u(t)$ and $n(t)$ is a normalized complex-valued additive Gaussian noise process (see Table 3-1).

Table 3-1 Signals and Normalizations in the Linear Channel Model

	Signal			
Transmittter	$u(t)$	$=$	$\sum_n a_n\, g_T(t - nT - \varepsilon_0 T)$	Linear modulation
	$u(t)$	$=$	$e^{j\phi(t)}$ $\phi(t)$ $=2\pi h \sum a_k q(t - kT - e_0 T)$	Nonlinear CPM modulation
	$u_{\mathrm{BP}}(t)=$		$\sqrt{2}A\ \mathrm{Re}\{e^{j[\omega_0 t+\theta_T(t)]}\ u(t)\}$	Bandpass signal
Channel	$C(\omega)$	$=$	$u(\omega+\omega_0)C_{\mathrm{BP}}(\omega+\omega_0)$	
Receiver	$r(t)$	$=$	$K_{\mathrm{R}}[As(t)+w(t)]$	
	with			
		$s(t)=$	$\left[u(t)e^{j\theta_T(t)} \otimes c(t)\right]e^{-j\theta_R(t)}$	
		$w_{\mathrm{BP}}(t)=$	$\sqrt{2}\mathrm{Re}\{w(t)e^{j\omega_0 t}\}$	
	$y(t)$	$=$	$r(t)\otimes f(t) = K_R[As_f(t)+n(t)]$	
	with			
		$s_f(t)=$	$s(t)\otimes f(t)$	
		$n(t)=$	$w(t)\otimes f(t)$	
		$r_f(t)=$	$y(t)/(K_R A) = s_f(t) + n(t)/A$	Normalized input signal
Noise		$w_{\mathrm{BP}}(t)$:	Bandpass process with flat spectral density $N_0/2$	

Using the bandpass system model just described, a baseband equivalent model is now derived. The baseband equivalent channel is defined as [see eq. (1–97)]

$$C(\omega) = u(\omega+\omega_0)\, C_{\mathrm{BP}}(\omega+\omega_0) \tag{3-10}$$

Then Figures 3-1 and 3-2 can be simplified to the model shown in Figure 3-3 which still has some traces of the mixer oscillator of the transmitter.

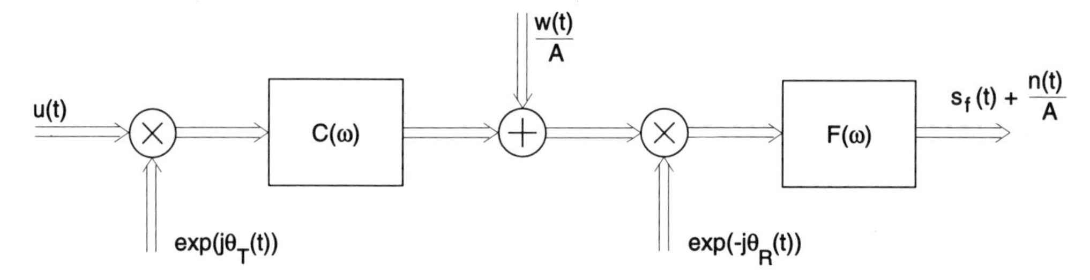

Figure 3-3 Normalized Baseband Model with Arbitrary $\theta_T(t)$ and $\theta_R(t)$

If the phase fluctuations $\theta_T(t)$ have a much smaller bandwidth than the channel frequency response $C(\omega)$, we may interchange the operation of filtering and multiplication. The useful signal $s(t)$ in (3-10) now reads (see Figure 3-4

$$\begin{aligned} s(t) &= \left[u(t)\, e^{j\theta_T(t)} \otimes c(t)\right] e^{-j\theta_R(t)} \\ &\simeq u(t) \otimes c(t)\, e^{j[\theta_T(t)-\theta_R(t)]} \\ &= u(t) \otimes c(t)\, e^{j\theta_0(t)} \end{aligned} \tag{3-11}$$

with $\theta_0(t) = \theta_T(t) - \theta_R(t)$. Summarizing the previous operations, the received signal $r_f(t) = y(t)/(K_R A)$ for linear modulators equals

$$\begin{aligned} r_f(t) &= s_f(t) + \frac{n(t)}{A} \\ &= \sum_n a_n g(t - nT - \varepsilon_0 T)\, e^{j\theta_0(t)} + \frac{n(t)}{A} \end{aligned} \tag{3-12}$$

where $g(t)$ is the pulse form seen by the receiver

$$g(t) = g_T(t) \otimes c(t) \otimes f(t) \tag{3-13}$$

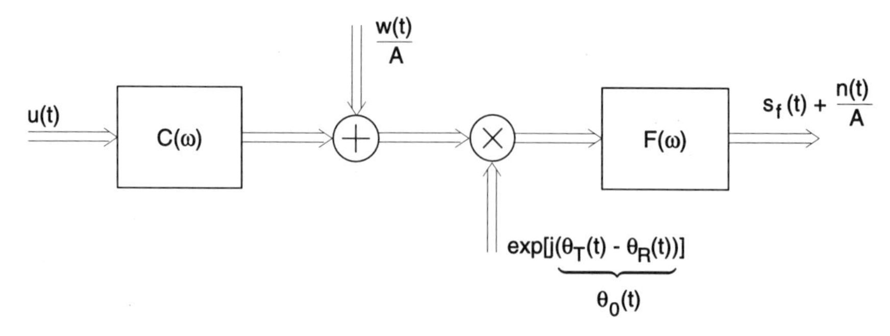

Figure 3-4 Normalized Baseband Model with Slowly Varying Phase $\theta_0(t)$

and $n(t)$ is a noise process with power spectral density

$$S_n(\omega) = |F(\omega)|^2 S_w(\omega) \tag{3-14}$$

The index 0 in $(\varepsilon_0, \theta_0)$ denotes the *actual* (unknown) values of the synchronization parameters which must be estimated by the receiver.

For easy reference, the signals in the channel model are summarized in Table 3-1.

3.2.2 Nonlinear Channel Model

The linear channel model described in the previous section is not applicable to satellite transmission, as shown in Figure 3-5. It consists of two earth stations (Tx and Rx) usually far from each other, connected by a repeater traveling in the sky (satellite) through two radio links. A functional block diagram of the system of Figure 3-5 is shown in Figure 3-6.

The block labeled HPA represents the earth station power amplifier with a nonlinear input-output characteristics of the saturation type. The Tx filter limits the bandwidth of the transmitted signal whereas the input filter in the transponder limits the amount of uplink noise. The block TWT represents the satellite's on-

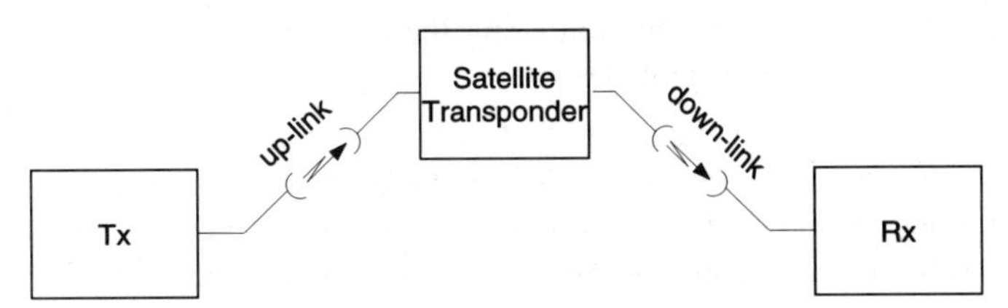

Figure 3-5 Model of a Satellite Link a)

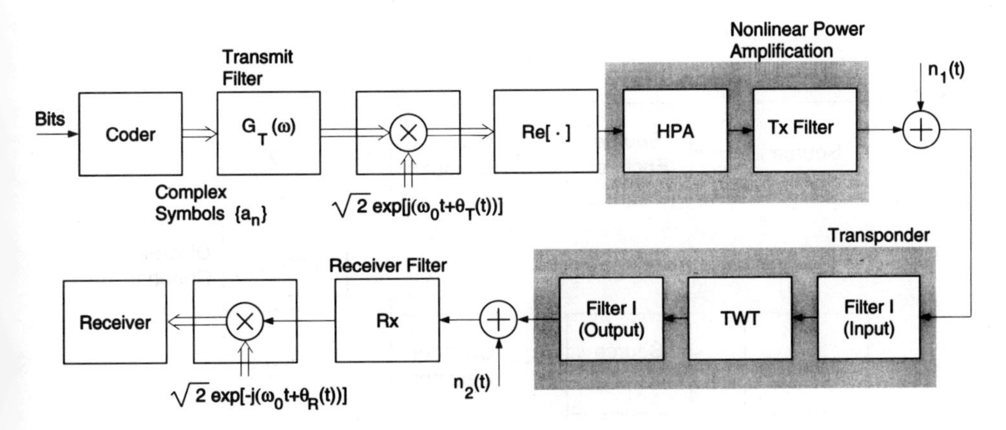

Figure 3-6 Model of a Satellite Link b)

board amplifier. Owing to the satellite's limited power resources, this amplifier is usually operated at or near saturation to obtain maximum power efficiency. The output filter limits the bandwidth of the transmitted signal again whereas the Rx filter limits the downlink noise.

The assessment of the error performance for this (and any other) nonlinear channel is difficult. While refined mathematical tools and general comprehensive theorems are available in the linear case, only very special categories of problems can be analyzed in nonlinear situations. There are mainly two avenues to analyze satellite systems. The first approach makes simplifications of the problems such that an analytical approach is feasible. The other approach uses computer simulation. This later approach is the one taken in this book.

3.3 Channel Capacity of Multilevel/Phase Signals

Information theory answers two fundamental questions:

1. What is the ultimate data compression? (answer: the entropy *H*)
2. What is the ultimate transmission rate of communication? (answer: the channel capacity *C*)

We gain a considerable insight comparing these boundaries with the performance of any communication system design.

The communication model studied in information theory is shown in Figure 3-7. This discrete-time channel is defined to be a system consisting of an input alphabet *X* and an output alphabet *Y* and a probability matrix $p(\mathbf{y} \mid \mathbf{x})$ that expresses the probability of observing the output y given that we send x. If the outcome y_i depends only on the symbol x_i the channel is said to be memoryless.

We consider now the memoryless Gaussian channel when y_i is the sum of input x_i and noise n_i:

$$y_i = x_i + n_i \tag{3-15}$$

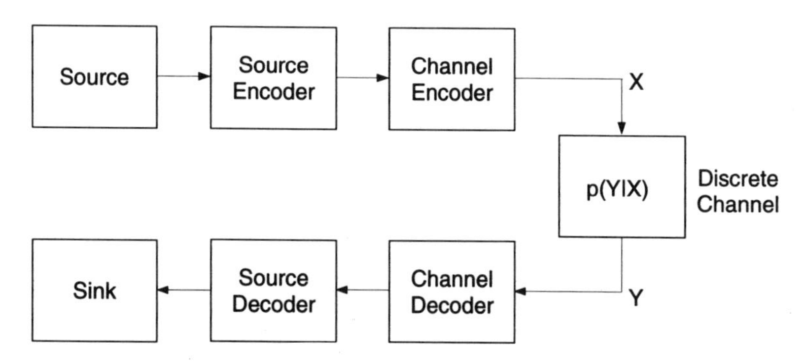

Figure 3-7 Discrete Model of a Communication System

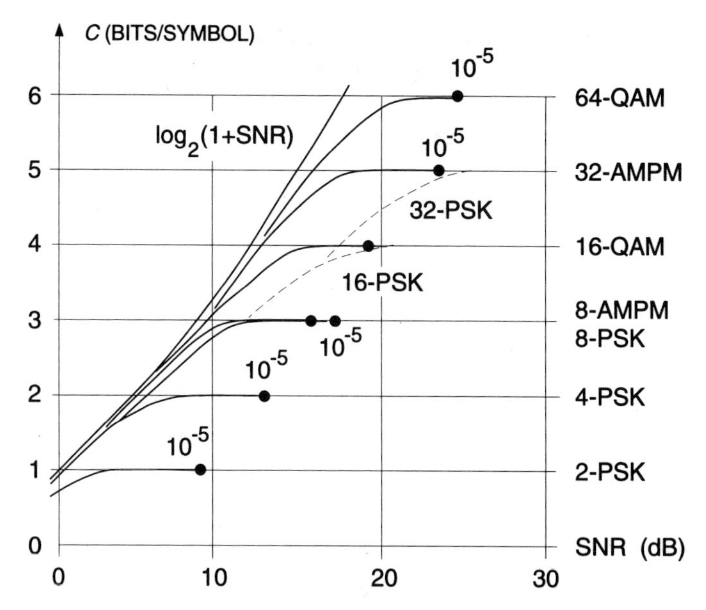

Figure 3-8 Channel Capacity for a Discrete-Valued Input a_i and Continuous-Valued Output y_i

The noise n_i is drawn from an i.i.d. Gaussian distribution with variance σ_n^2. The channel capacity C for this channel can be computed for a discrete input channel signal as well as for a continuous-valued input.

For a discrete input we denote the input by a_i instead of x_i where a_i is a complex-valued PAM symbol. The channel capacity has been computed by Ungerboeck [3] for the case where all symbols are equally probable. The results are displayed in Figure 3-8.

In Figure 3-8 (After Ungerboeck [3], the ultimate bound for a continuous-valued input alphabet x_i is shown:

$$C = \log_2\left(1 + \frac{P}{\sigma_n^2}\right) \tag{3-16}$$

Formula 3-16 is one of the most famous formulas in communication theory. It is important to understand the conditions under which it is valid:

(i) x_i is a complex Gaussian random variable with an average power constraint $E\left[|x_i|^2\right] \leq P$.

(ii) σ_n^2 equals the variance of a complex Gaussian random variable.

The capacity C is measured in bits per channel use, which is the same as bits per symbol transmitted. The capacity is a function of the average signal-to-noise ratio P/σ_n^2 where P is the average power.

Real and imaginary parts of (3-15) each describe a real-valued Gaussian channel. The capacity of a real-valued channel is given by

$$C_D = \frac{1}{2}\log_2\left(1+\frac{P_1}{\sigma_n^2/2}\right) \tag{3-17}$$

where $P_1 = E\{x_i^2\}$, x_i is a real Gaussian random variable and $\sigma_n^2/2$ is the variance of the real noise sample.

The total capacity of both channels is the sum of the capacity of the individual channels. Since the sum is maximized if the available power P is equally split, we obtain

$$\begin{aligned} C = 2C_D &= 2\left(\frac{1}{2}\right)\log_2\left(1+\frac{P/2}{\sigma_n^2/2}\right) \\ &= \log_2\left(1+\frac{P}{\sigma_n^2}\right) \end{aligned} \tag{3-18}$$

which is eq. (3-16)[1].

From Figure 3-8 we also observe that for any SNR we loose very little in capacity by choosing a discrete input alphabet a_i as long as the alphabet is sufficiently large. The higher the SNR the larger the alphabet. This result gives a solid theoretical foundation for the practical use of discrete-valued symbols.

In a real physical channel a time-continuous waveform $s(t, \mathbf{a})$ corresponding to the sequence $\mathbf{a}$ is transmitted. If the signal $s(t, \mathbf{a})$ is band-limited of bandwidth B the sampling theorem tells us that the signal is completely characterized by samples spaced $T = 1/(2B)$ apart. Thus, we can send at most $2B$ symbols per second. The capacity per symbol then becomes the capacity per second

$$\begin{aligned} C_D &= 2B\frac{1}{2}\log_2\left(1+\frac{P}{\sigma_n^2}\right) \qquad \text{(baseband)} \\ &= B\log_2\left(1+\frac{P}{\sigma_n^2}\right) \end{aligned} \tag{3-19}$$

A complex alphabet is always transmitted as a passband signal. Looking at the sampling theorem we notice that $W = 1/T$ symbols per second can be transmitted. Hence,

$$C = W\log_2\left(1+\frac{P}{\sigma_n^2}\right) \qquad \text{(passband)} \tag{3-20}$$

Thus the capacity formula (in bits per second) is applicable to both cases. However, the proper bandwidth definition must be used (see Figure 3-9).

[1]If the symbol a_k is complex, one speaks of two-dimensional modulation. If a_k is real, of one speaks of one-dimensional modulation. The term originates from the viewpoint of interpreting the signals as elements in a vector space.

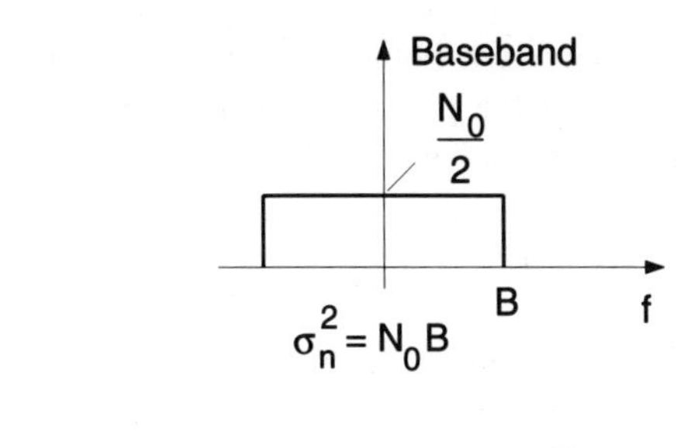

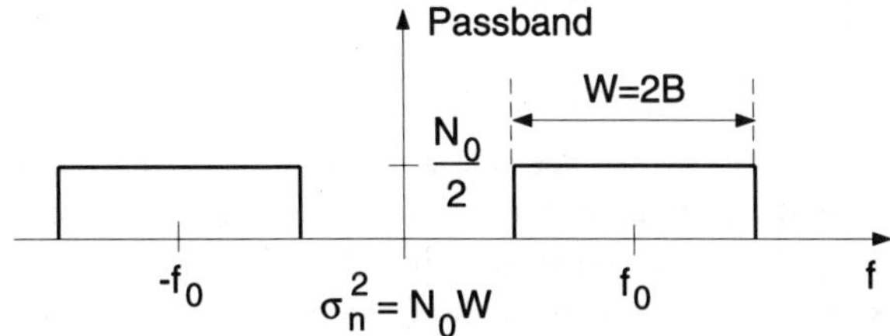

Figure 3-9 Bandwidth Definitions

Introducing the energy E_s per symbol

$$E_s = PT \qquad T: \text{ symbol duration} \tag{3-21}$$

and using

$$\sigma_n^2 = \begin{cases} N_0 W & \text{(passband)} \\ N_0 B & \text{(baseband)} \end{cases} \qquad N_0: \text{power spectral density} \tag{3-22}$$

(see Figure 3-9) as well as (Nyquist sampling conditions) $BT = 1/2$ and $WT = 1$, (3-19) and (3-20) can be written in the form

$$C = \begin{cases} W \log_2 \left(1 + \dfrac{E_s}{N_0}\right) & WT = 1 \\ B \log_2 \left(1 + \dfrac{E_s}{N_0/2}\right) & BT = \dfrac{1}{2} \end{cases} \tag{3-23}$$

In addition to the sequence $\mathbf{a}$ the waveform $s(t, \mathbf{a})$ depends on a set of parameters $\boldsymbol{\theta} = \{\theta_1, \ldots, \theta_L\}$ (representing phase, timing offset, etc.). These parameters are unknown to the receiver. In the simplest case of an AWGN channel these parameters may be assumed to be approximately constant. In order to be able to retrieve the symbol sequence $\mathbf{a}$, these parameters must be estimated. The estimate $\hat{\boldsymbol{\theta}}$ is then used as if it were the true value $\boldsymbol{\theta}$. This is the task of synchronization.

It is by no means evident that the task of synchronization can be accomplished in such a way that the real channel has a capacity which is only slightly smaller than that of the idealized discrete-channel of information theory.

Just as the channel capacity is a bound on the ideal channel, there exists a lower bound on the variance of the parameter estimate $\hat{\theta}$. This bound will be discussed in Chapter 6. Any estimator has a variance larger than this bound. Thus, the channel capacity can at best be approached but never reached. The following examples help to illustrate the issue.

Remark : A truly optimal receiver requires the joint estimation of the data sequence and the parameters $\boldsymbol{\theta}$, see Section 4.3. This optimal receiver is nonrealizable. In any realizable receiver the parameters are estimated separately.

Example 1: 2-PSK Transmission over the Deep-Space Channel

By inspection of Figure 3-8 we see that a 2-PSK modulation comes close to the capacity bound in the region of energy-efficient operation, which is roughly at $0 < E_s/(N_0/2) \leq 1$. It can be shown that the capacity of the AWGN channel using 2-PSK is well approximated by the (unconstrained) capacity (3-19):

$$C = B \log_2 \left(1 + \frac{E_s}{N_0/2}\right) \tag{3-24}$$

The energy per channel use is related to the fundamental quantity energy per bit E_b by

$$P = E_s r = E_b R_b \tag{3-25}$$

where R_b is the rate of information bits per second and r is the rate of coded bits per second. The ratio $R = R_b/r = E_s/E_b$ is the code rate (information bits/coded bits).

Replacing E_s by $E_s = E_b R$ we obtain

$$C = B \log_2 \left(1 + \frac{E_b R}{N_0/2}\right) \tag{3-26}$$

Letting the code rate *R* tend to zero, we can write for (3-26)

$$\begin{aligned} C_\infty &\simeq B \frac{1}{\ln 2} \frac{E_b R}{N_0/2} \\ &= 2BR \frac{1}{\ln 2} \left(\frac{E_b}{N_0}\right) \end{aligned} \tag{3-27}$$

For $R \to 0$ the bandwidth B goes to infinity. Since $2BR = 2(r/2)R = R_b \simeq C_\infty$, we obtain the minimum signal-to-noise ratio:

$$\frac{E_b}{N_0} \simeq \ln 2 \tag{3-28}$$

which is often referred to as the Shannon limit for the AWGN channel. We thus see that by letting the code rate R tend to zero, coded 2-PSK modulation approaches the capacity bound.

But what about synchronization? The information is transmitted in the form

$$u_{\mathrm{BP}}(t) = \sqrt{\frac{2E_s}{T}} \; a_k \cos(\omega_0 t + \theta) \qquad a_k = \pm 1 \tag{3-29}$$

where ω_0 is the carrier frequency and a_k the symbol in the kth T-interval. The signal is coherently demodulated by a PLL. The power at the receiver input equals

$$P_{u_{\mathrm{BP}}} = \frac{E_s}{T} \tag{3-30}$$

The noise power is

$$\sigma_n^2 = N_0 W \tag{3-31}$$

and thus the signal-to-noise ratio (SNR) at the input stage

$$\mathrm{SNR}_i = \frac{E_s}{N_0 W T} \tag{3-32}$$

Since $E_s = E_b R$ and $WT = 1$, we obtain

$$\mathrm{SNR}_i = \frac{E_b R}{N_0} \tag{3-33}$$

Thus, as $R \to 0$, the input signal-to-noise ratio, $\mathrm{SNR}_i \to 0$. Since there is a lower limit which realizable PLLs can cope with, we see that synchronization sets a lower bound on R.

In summary, the limit $R \to 0$ is not meaningful if one takes into account synchronization. It is exactly for this reason that in the Pioneer 9 system a code rate of $R = 1/2$ was selected despite a penalty of (–1.2 dB). For more details the reader is referred to the paper by James Massey [4].

Example 2: Two-Dimensional Transmission at High SNR

For bandwidth-efficient operation more than one bit per symbol is transmitted. Performing the same calculation as in the previous example we obtain

$$\mathrm{SNR}_i = \left(\frac{E_b}{N_0}\right) R \log_2 M_s \tag{3-34}$$

$$M_s : \text{alphabet size}$$

$$R : \text{code rate}$$

Since $R \log_2 M_s > 1$, the synchronization input SNR_i is not the limiting factor. In conclusion, we observe that E_s is the quantity of interest for the *inner* receiver while it is E_b for the *outer* receiver.

Bibliography

[1] S. Benedetto and E. Biglieri and V. Castellani , *Digital Transmission Theory*. Englewood Cliffs, NJ: Prentice-Hall, 1987.

[2] E. A. Lee and D. G. Messerschmitt, *Digital Communication*. Boston: Kluwer Academic, 1994.

[3] G. Ungerboeck, "Channel Coding with Multilevel/Phase Signals," *IEEE Trans. Inf. Theory*, vol. 28, pp. 55–67, Jan. 1982.

[4] J. L. Massey, "Deep Space Communications and Coding: A Marriage Made in Heaven," *Advanced Methods for Satellite and Deep Space Communication (J. Hagenauer, ed.). Springer Verlag*, 1992.

[5] J. L. Massey, "The How and Why of Channel Coding," *Proc. Int. Zürich Seminar on Digital Communication*, pp. 67–73, Mar. 1984.

[6] C. E. Shannon, "A Mathematical Theory of Communications," *BSTJ*, Oct. 1948.

Chapter 4 Receiver Structure for PAM Signals

In this chapter we first discuss the main building blocks of a digital receiver (Section 4.1). The discussion is intended as an introduction to get a qualitative understanding of the design issues involved. In Section 4.2 we are concerned with the question under what conditions the samples of the received signal contain the entire information on the continuous-time signal. The samples obtained under such conditions provide so-called sufficient statistics for the digital receiver to be discussed in Section 4.3. Here we systematically derive the optimum receiver based on the maximum-likelihood criterion. The essential feature of our approach is that the receiver structure is the outcome of a mathematical optimization problem.

4.1 Functional Block Diagram of a Receiver for PAM Signal

This section examines the main building blocks of a data receiver for the purpose of exposing the key functions. It is intended as an introduction for the reader to get a qualitative understanding of the design issues involved which will be discussed in detail in later sections.

As in all receivers the input is an analog signal and the output is a sequence of digital numbers, one per symbol. A typical block diagram of an analog receiver is shown in Figure 4-1. Analog-to-digital conversion (A/D) takes place at the latest stage possible, namely immediately before the detector and decoder, which is always implemented digitally. The other extreme case is a fully digital receiver shown in Figure 4-2 where the signal is A/D converted to baseband immediately after downconversion. The adjective "typical" should be emphasized in both cases, as many variations on the illustrated block diagrams are possible.

For example:

- In a hybrid receiver, some of the building blocks are realized in analog hardware.
- The sequential order of some of the digital signal processing blocks can be interchanged, depending on the realization constraint.

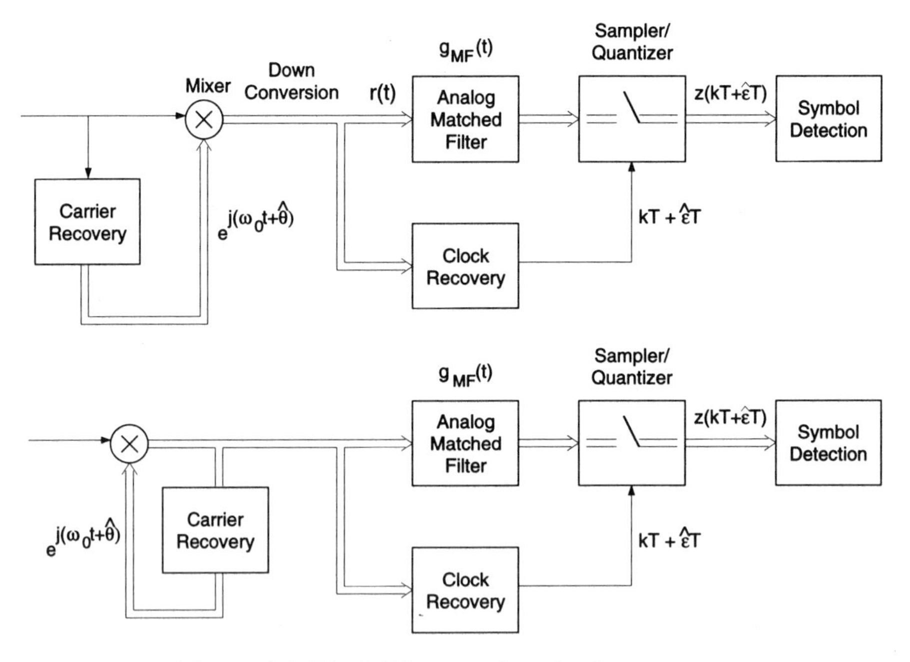

Figure 4-1 Block Diagram of an Analog Receiver

The block diagram is also incomplete, since

- Analog input stages are not shown.
- Important building blocks, e.g., for automatic gain control (AGC), lock detector are omitted for the sake of a concise presentation of fundamental aspects.

In an analog receiver (Figure 4-1) the incoming signal is first coherently demodulated to a complex baseband signal. The phase coherent reference signal $\exp\left[j\left(\omega_0 t + \hat{\theta}(t)\right)\right]$ is generated by a voltage-controlled oscillator (VCO) being controlled by an error signal from a suitable phase error detector. The error signal can either be derived employing the baseband or passband signal, as will be discussed later on. The baseband signal is then further processed in a data filter and subsequently sampled and quantized. The sampling instant is controlled by a timing recovery system which generates exactly one sample per symbol. The order of phase and timing recovery implies that phase recovery must work with an arbitrary timing epoch and data symbol sequence.

A fully digital receiver has an entirely different structure as can be seen from Figure 4-2. In a first stage the signal is downconverted to (approximately) baseband by multiplying it with the complex output of an oscillator. The frequency of the oscillator is possibly controlled by a frequency control loop. As a first major difference to the analog counterpart, the purpose of the downconversion process to baseband is to generate the complex baseband signal (*I*/*Q* components), not to

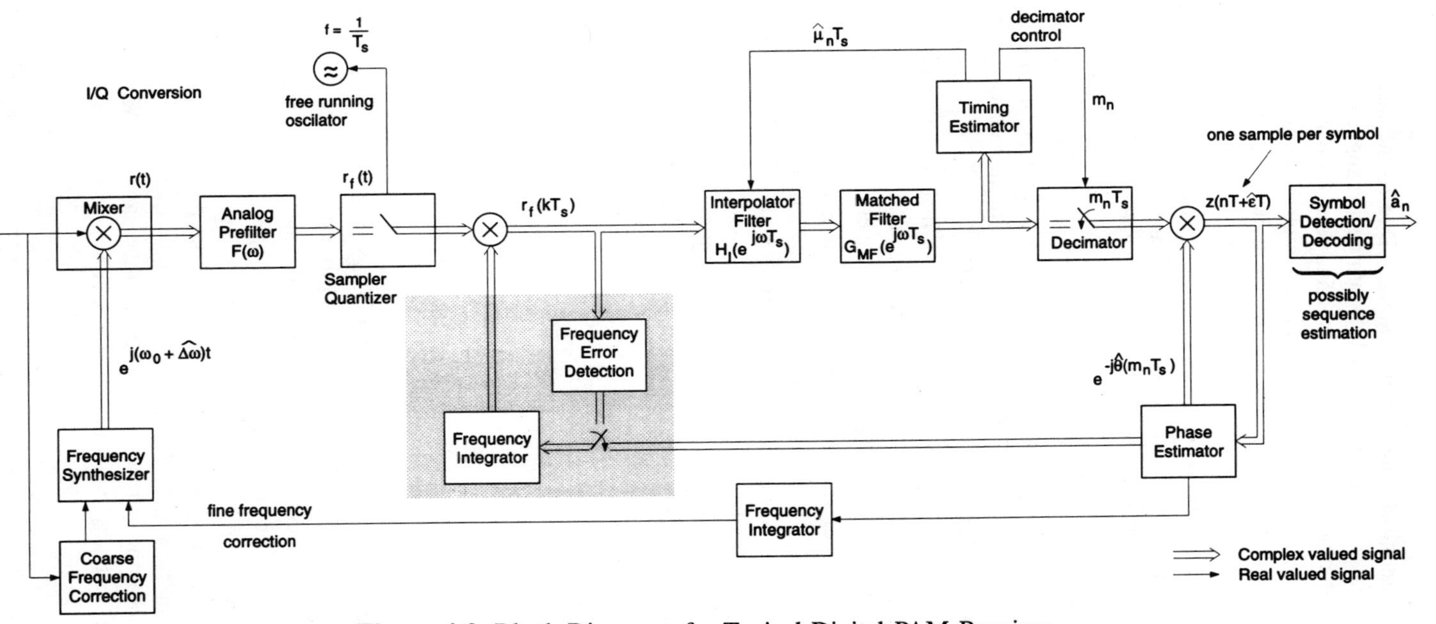

Figure 4-2 Block Diagram of a Typical Digital PAM Receiver

coherently demodulate the bandpass signal. Due to the residual frequency error this complex baseband signal is slowly rotating at an angular frequency equal to the frequency difference between transmitter and receiver oscillators. The signal then enters an analog prefilter $F(\omega)$ before it is sampled and quantized. All subsequent signal processing operations are performed digitally at the fixed processing rate of $1/T_s$ (or fractions of it).

Before we discuss the various signal processing functions in detail we want to draw the attention of the reader to one of the most basic differences between the analog and digital receiver: the digital receiver does *not* need to have a clock frequency equal to the symbol rate $1/T$ as does the analog counterpart. The *only* existing clock rate at the receiver is $1/T_s$ which is unrelated to the symbol rate $1/T$. In other words, the ratio T/T_s is irrational in general; any assumption that T is an exact multiple of T_s oversimplifies the timing recovery problem of a fully digital receiver as we will see in the sequel.

Before we can discuss how to obtain one matched filter output $z(nT + \hat{\varepsilon}T)$ for every symbol transmitted from a system running at rate $1/T_s$, we must examine the analog filtering and sampling/quantizing operation first.

We set the goal that the digital receiver should have no performance loss due to sampling and digital signal processing when compared to the analog counterpart (which we assume to be optimal according to a given criterion). At first sight this seems to be an unattainable goal. Should we not (in the limit) sample infinitely often, $T_s \to 0$, in order to obtain a vanishing discretization error when performing the continuous time filtering operation of the matched filter?

$$\begin{aligned} z(nT) &= \int_{-\infty}^{\infty} r(t)\; g_{\mathrm{MF}}(nT - t)\; dt \\ &= \lim_{\Delta t \to 0} T_s \sum_{i=-\infty}^{\infty} r(iT_s)\; g_{\mathrm{MF}}(nT - iT_s) \qquad \Delta t = \frac{1}{T_s} \end{aligned} \tag{4-1}$$

If we make no further assumption on the signal $r(t)$, this indeed is true and any finite sampling rate would invariably lead to a performance loss.

But since we are mostly concerned with bandwidth efficient transmission, the signals transmitted can be accurately approximated by a band-limited signal. (This will be discussed in Section 4.2.2.) For the moment let us assume the bandwidth of the useful signal $s(t)$ to be B and $s(t)$ passing the filter $F(\omega)$ undistorted. If the signal is sampled at rate $1/T_s \geq 2B$, the sampling theorem tells us that the analog signal can be exactly reconstructed from these samples. We can also show (see Section 4.2) that any analog filtering operation can be exactly represented by a sum

$$\int_{-\infty}^{\infty} s(\tau)\; g(t-\tau)\; d\tau = T_s \sum_{i=-\infty}^{\infty} s(iT_s)\; g(t - iT_s) \tag{4-2}$$

Equation (4-2) is a special case of the Parseval theorem. It is of fundamental importance to digital signal processing since it states the *equivalence of digital and analog* (better: discrete-time and continuous-time) signal processing.

The equivalence argument we have given is correct but incomplete when we are concerned with the optimal detection of a data sequence in the presence of noise. A qualitative argument to explain the problem suffices at this point. The signal $r_f(t)$ at the output of the prefilter $F(\omega)$ is the sum of useful signal $s_f(t)$ plus noise $n(t)$. Since $r_f(t)$ is band-limited the noise $n(t)$ is band-limited, too. The noise samples $\{n(kT_s)\}$, in general, are therefore correlated, i.e., statistically dependent. This implies that they carry information which must be taken into account when further processed in the matched filter. In Section 4.2 we will derive sufficient conditions on the design of the analog prefilter $F(\omega)$ such that the samples $\{r_f(kT_s)\}$ contain all the information contained in the continuous-time signal $r_f(t)$. The reader should have noticed that the symbol rate $1/T$ has not been mentioned. The condition on the prefilter $F(\omega)$ and its corresponding sampling rate $1/T_s$ indeed does not require T/T_s to be rational. In other words: sampling is asynchronous with respect to the transmitter clock.

4.1.1 Timing Recovery

Let us continue on the assumption that the samples $\{r_f(kT_s)\}$ contain all information. Due to a time shift between transmitter and receiver clocks, samples at $t = kT+\hat{\varepsilon}T$ are required. In an analog receiver the solution of this problem is to control the sampling instant of the received signal (see Figure 4-3a). The sampling process is *synchronized* to the symbol timing of the received signal. A modification of this analog solution would be to derive the timing information from the samples of the receiver (instead of the continuous-time signal) and to control the sampling instant. Such a solution is called a *hybrid* timing recovery (Figure 4-3b).

In truly digital timing recovery (Figure 4-3c) there exist only clock ticks at $t = kT_s$ incommensurate with the symbol rate $1/T$. The shifted samples must be obtained from those asynchronous samples $\{r(kT_s)\}$ solely by an algorithm operating on these samples (rather than shifting a physical clock). But this shifting operation is only *one* of two parts of the timing recovery operation. The other part is concerned with the problem of obtaining samples of the matched filter output $z(nT + \hat{\varepsilon}T)$ at symbol rate $1/T$ from the signal samples taken at rate $1/T_s$, as ultimately required for detection and decoding.

It should have become clear why we insist that T and T_s are incommensurate. While it is possible to build extremely accurate clocks, there always exists a small difference between the clock frequency of the two separate and free running clocks at the transmitter and receiver. But even the slightest difference in clock frequencies would (in the long run) cause cycle slips as we will explain next.

The *entire* operation of digital timing recovery is best understood by emphasizing that the *only* time scale available at the receiver is defined by units of T_s and, therefore, the transmitter time scale defined by units of T must be expressed in terms of units of T_s.

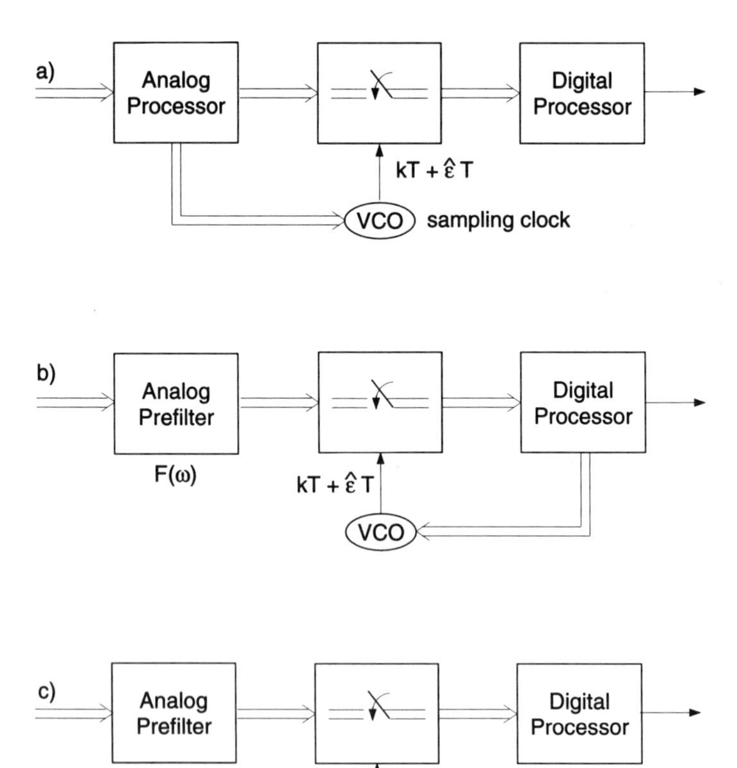

Figure 4-3 Timing Recovery Methods: (a) Analog Timing Recovery: Sampling Is Synchronous with Received Signal at Symbol Rate $1/T$, (b) Hybrid Timing Recovery: Sampling Is Synchronous with Received Signal at Symbol Rate $1/T$, (c) Digital Timing Recovery: Sampling Is Asynchronous with Received Signal; Symbol Rate $1/T$ and Sampling Rate $1/T_s$ Are Incommensurate

Recall that we need samples of the matched filter at $(nT + \hat{\varepsilon}T)$. We write for the argument of these samples

$$nT + \hat{\varepsilon}T = T_s \left[n\,\frac{T}{T_s} + \hat{\varepsilon}\,\frac{T}{T_s} \right] \tag{4-3}$$

The key step is to rewrite the expression in brackets in the form

$$\begin{aligned} \left[n\,\frac{T}{T_s} + \hat{\varepsilon}\,\frac{T}{T_s} \right] &= L_{\text{int}} \left(n\,\frac{T}{T_s} + \hat{\varepsilon}\,\frac{T}{T_s} \right) + \hat{\mu}_n \\ &= m_n + \hat{\mu}_n \end{aligned} \tag{4-4}$$

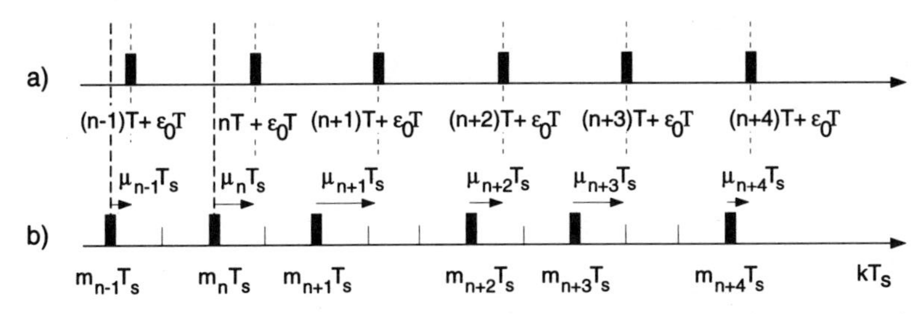

Figure 4-4 (a) Transmitter Time Scale (nT), (b) Receiver Time Scale (kT_s)

where $m_n = L_{\text{int}}(\cdot)$ means the largest integer less than or equal to the real number in the argument and $\hat{\mu}_n$ is a fractional difference.

The situation is illustrated in Figure 4-4 where the transmitter time scale, defined by multiples of T, is shifted by a constant amount of $(\varepsilon_0 T)$, and the time scale at the receiver is defined by multiples of T_s. Two important observations can be made due to the fact that T and T_s are incommensurate. First, we observe that the relative time shift $\mu_n T_s$ is time-variable despite the fact that $(\varepsilon_0 T)$ is constant. Second, we observe that the time instances $m_n T_s$ (when a value of the matched filter is computed) form a completely irregular (though deterministic) pattern on the time axis. This irregular pattern is required in order to obtain an average of exactly T between the output samples of the matched filter, given a time quantization of T_s.

Notice that the timing parameters (μ_n, m_n) are uniquely defined given $\{\varepsilon_0, T, T_s\}$. A "genius" that knew these values could compute the time shift μ_n and decimate the sampling instant. In practice, of course, there is a block labeled timing estimator which estimates $\{\mu_n, m_n\}$ (directly or indirectly via $\hat{\varepsilon}$) based on noisy samples of the received signal. These estimates are then used for further processing as if they were true values. We will discuss methods of interpolation, timing parameter estimation μ_n and decimation control m_n in detail in the following sections.

Summarizing, digital timing recovery comprises two basic functions:

1. ***Estimation***
   The fractional time delay ε_0 has to be estimated. The estimate $\hat{\varepsilon}$ is used as if it were the true value ε_0. The parameters $(m_n, \hat{\mu}_n)$ follow immediately from $\hat{\varepsilon}$ via eq. (4-4).

2. ***Interpolation and Decimation***
   From the samples $\{r_f(kT_s)\}$ a set of samples $\{r_f(kT_s + \hat{\mu}_n T_s)\}$ must be computed. This operation is called *interpolation* and can be performed by a digital, time-variant filter $H_I(e^{j\omega T_s}, \hat{\mu}_n T_s)$. The time shift $\hat{\mu}_n$ is time-variant according to (4-4). The index n corresponds to the nth data symbol.

   Only the subset $\{y(m_n T_s)\} = \{z(nT + \hat{\varepsilon}T)\}$ of values is required for further processing. This operation is called *decimation*.

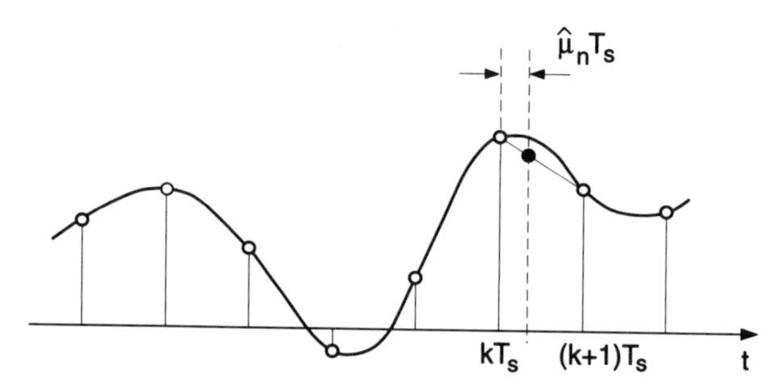

Figure 4-5 Linear Interpolator

Example: Linear Interpolation (Figure 4-5)
The simplest method to obtain a coarse estimate is linear interpolation. Given $\hat{\mu}_n$, we obtain

$$r_f(kT_s + \hat{\mu}_n T_s) \simeq r_f(kT_s) + \hat{\mu}_n[r_f([k+1]T_s) - r_f(kT_s)]$$

$$= \sum_{i=k}^{k+1} c_i \, r_f(iT_s) \qquad \text{with} \quad \begin{cases} c_k = (1-\hat{\mu}_n) \\ c_{k+1} = \hat{\mu}_n \end{cases} \tag{4-5}$$

The coefficients $\{c_i\}$ define a linear time variable transversal filter. As can be seen by inspection of Figure 4-5, there is an interpolation error committed by this simple filter. In Chapter 9 we will discuss more sophisticated types of interpolation filters which asymptotically perform exact interpolation.

4.1.2 Phase Recovery

We pointed out that downconversion still leaves an unknown carrier phase θ_0 relative to a hypothetical observer which results in a rotation of the complex data symbol by $\exp(j\theta_0)$. This phase error can be corrected by multiplying the matched filter output by $\exp(-j\theta_0)$. In reality, the matched filter output is multiplied by an estimate of the phase, $\exp\left(-j\hat{\theta}\right)$. The process of phase recovery thus comprises the basic functions:

1. ***Phase Estimation***

   Phase estimation is performed after the matched filter. Therefore, optimum filtering for phase estimation coincides with optimum filtering for data detection.
   Phase estimation is performed at symbol rate. Since timing recovery is performed before phase recovery, the timing estimation algorithm must either work

   (i) with an arbitrary phase error offset or
   (ii) with a phase estimate (phase-aided timing recovery) or
   (iii) phase and timing are acquired jointly.

2. ***Phase Rotation***

   (i) The samples $z(nT + \hat{\varepsilon}T)$ are multiplied by a complex number $\exp\left(-j\hat{\theta}(nT)\right)$. A small residual frequency error can be compensated by a time-variant phase estimate $\hat{\theta}(nT) = \hat{\theta} + \widehat{\Delta\Omega}(nT)$

   (ii) The samples $z(nT + \hat{\varepsilon}T)\ e^{-j\hat{\theta}}$ are further processed in the detection/decoding unit as if $\left(\hat{\theta}, \hat{\varepsilon}\right)$ were true values under the tacit hypothesis of perfect synchronization.

3. ***Frequency Synchronization***
   Frequency synchronization is an essential task in many applications. In the presence of a (possibly) sizable frequency offset a coarse frequency adjustment in the analog domain is necessary. The reason is that digital timing and phase synchronization algorithms work properly only in the presence of a small residual frequency offset. The remaining frequency offset is then compensated in a second stage as shown in Figure 4-2.

In summary, several characteristics that are essential features of the timing and phase recovery process have been discovered:

1. Phase recovery can be performed after timing recovery. This is the opposite order as found in classical analog receivers.
2. Frequency synchronization is typically done in two steps. A coarse frequency adjustment in the analog domain and a correction of the residual frequency offset in the digital domain.

The receiver structure discussed is a practical one typical for satellite transmission. In satellite communication the filters can be carefully designed to obey the Nyquist criterion. In a well-designed receiver the matched filter output then has negligible intersymbol interference (ISI). For transmission over voice-grade telephony channels the designer no longer has this option since the channel $c(t)$ introduces severe intersymbol interference. The optimal receiver must now use much more complicated mechanisms for detecting the data symbols in the presence of ISI. In Figure 4-6 the structure of a practical, suboptimal receiver using an equalizer is shown.

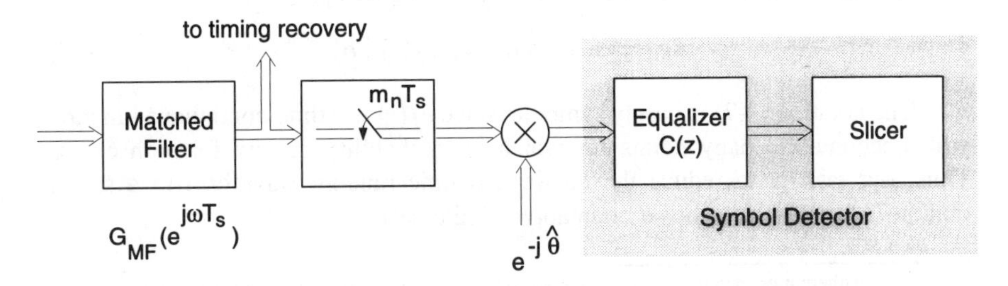

Figure 4-6 Receiver Structure in the Presence of ISI

The task of the equalizer is to minimize the ISI according to a given optimization criterion. An excellent introduction to this subject can be found in the book by Lee and Messerschmitt [1].

Bibliography

[1] E. A. Lee and D. G. Messerschmitt, *Digital Communications*. Boston: Kluwer Academic, 1988.

4.2 Sufficient Statistics for Reception in Gaussian Noise

The goal of this section is to put the qualitative discussion so far on a mathematical basis which will allow us to systematically derive *optimal* receiver structures.

In mathematical abstract terms the problem can be stated as follows: assume that the transmitted signal belongs to a finite set of possible signals. Added to the signal is a disturbance which we will assume to be colored Gaussian noise. The receiver's objective is to estimate the transmitted data from the noisy received signal according to an optimality criterion.

Example

We assume pulse-amplitude modulation transmission (PAM) transmission corrupted by Gaussian noise. The received signal is given by

$$r_f(t) = s_f(t) + \frac{n(t)}{A} = \sum_{n=0}^{N-1} a_n \; g(t - nT - \varepsilon_0 T) e^{j\theta_0} + \frac{n(t)}{A} \tag{4-6}$$

The pulse $g(t)$ is the convolution of $g_T(t)$, the channel response $c(t)$, and the prefilter $f(t)$. For an alphabet size of L the number of possible signals transmitted is N^L.

The criterion of optimality considered in the following is the *maximum-likelihood* (ML) criterion. The ML receiver searches for the symbol sequence $\mathbf{a}$ which has most likely produced the received signal $r_f(t)$.

$$\hat{\mathbf{a}}_{\text{ML}} = \arg \max_{\mathbf{a}} \; p_{r_f}(r_f(t) \mid \mathbf{a}) \tag{4-7}$$

The notation (4-7) is purely symbolic since $r_f(t)$ is a time-continuous function with uncountably many points for which no probability density function exists. Thus, our task is to reduce the continuous time random waveform to a set of random variables (possibly a countably infinite set).[2]

[2] Since the mathematical details can become quite intricate, it is important to intuitively understand the overall structure of this section.

4.2.1 Vector Space Representation of Signals

The notion of a *vector space* will play a key role in our development augmenting our understanding with geometric intuition. The reader not familiar with vector spaces is referred to Chapter 2.6 in [1] for a short and simple introduction.

The waveform $x(t)$ can be understood as an element of a vector space. The set of orthogonal functions $\{\phi_n(t)\}$ serves as *basis* for this vector space while the coefficients $\{x_n\}$ are the *components* of the vector $x(t)$ with respect to this base. The vector $x(t)$ is then represented as a series

$$x(t) = \sum_n x_n \, \phi_n(t) \tag{4-8}$$

where

$$x_n = \int_{T_u}^{T_0} x(t) \, \phi_n(t) \, dt \qquad [T_u, T_0] : \text{ observation interval} \tag{4-9}$$

The equality in (4-8) is understood in the mean-square sense which is defined as the limit

$$\lim_{N \to \infty} E\left[\left| x(t) - \sum_{n=1}^{N} x_n \, \phi_n(t) \right|^2\right] = 0 \qquad T_u \le t < T_0 \tag{4-10}$$

We require that $x(t)$ has finite power (finite length in vector space terminology)

$$E\left[|x(t)|^2\right] < \infty \tag{4-11}$$

If we truncate the series after N terms, we approximate $x(t)$ by $x_N(t)$. Geometrically, we can view $x_N(t)$ as a vector in an N-dimensional subspace which approximates the vector $x(t)$ in the infinite dimensional space. This is illustrated in Figure 4-7. In the figure, the vector space $\mathbf{x}$ is three-dimensional . The subspace

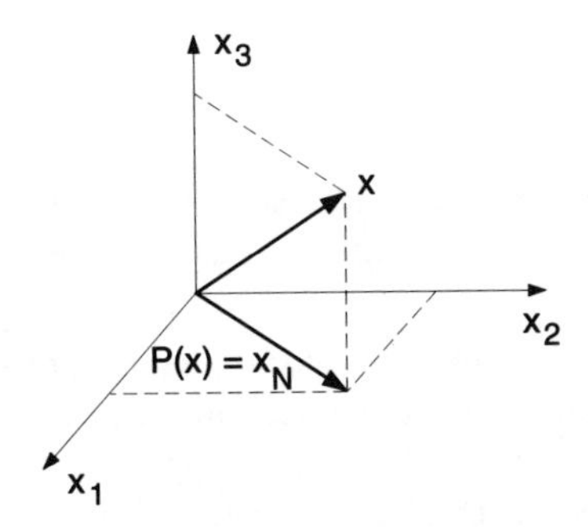

Figure 4-7 Illustration of Approximation of Vector $\mathbf{x}$

is a plane (N=2). The vector closest to $x(t)$ is the projection $P(\mathbf{x})$ onto that plane. We see that the mathematical structure of a vector space allows to meaningfully define optimal approximation of signals.

Since a vector is uniquely defined by its components, given the base, the coefficients (components) $\{x_n\}$ contain *all* the information on $x(t)$. The N-dimensional probability density function (pdf) $p_{x_N}(x_1,\ x_2,\ \ldots,\ x_N)$ therefore is a complete statistical description of the vector $x_N(t)$. In the limit $N \to \infty$, $p_{x_N}(x_1,\ x_2,\ \ldots,\ x_N)$ contains all the information of the random waveform $x(t)$.

We apply this result to the ML receiver. Let us assume we have found a suitable base $\{\phi_n(t)\}$ for the vector space R_f. If the components of the vector $r_f(t)$ are denoted by x_n, then

$$r_{f,N}(t) = \sum_{n=1}^{N} x_n \phi_n(t) \tag{4-12}$$

and the pdf

$$p_{r_{f,N}}(x_1,\ \ldots,\ x_N|\mathbf{a}) \tag{4-13}$$

as well as the limit $N \to \infty$ are mathematically well defined. The pdf (4-13) thus provides the basis for finding the most likely sequence $\mathbf{a}$.

The general solution to finding the pdf $p_{r_{f,N}}(x_1,\ \ldots,\ x_N|\mathbf{a})$ is complicated. The interested reader is referred to van Trees [2, Chapter 3 and Sections 4.2, 4.3]. For our purpose it will be sufficient to consider the special case of *band-limited* signals and an *infinite* observation interval. This restrictive assumption will lead us to a solution perfectly matched to the needs of digital signal processing. The base functions of this vector space are the $\mathrm{si}(x) = \sin(x)/x$ functions,

$$\phi_n(t) = \mathrm{si}\left(2\pi B\left(t - \frac{n}{2B}\right)\right) \qquad B = \frac{1}{2}2T_s \tag{4-14}$$

The components of the vector are obtained by simply sampling the signal $r_f(t)$, resulting in $x_n = r_f(nT_s)$. We do not need to compute an integral (4-9) as in the general case.

4.2.2 Band-Limited Signals

There are a number of reasons why a strictly band-limited signal is an excellent model for signal transmission, despite the fact that the signal is not time-limited (nonphysical).

First, in any bandwidth-efficient transmission the power is concentrated within a bandwidth B while only a negligible amount of power is located outside. This suggests that a band-limited signal is a good approximation to the physical signal. Indeed, the error between a non-band-limited physical signal $x(t)$ approximated by a band-limited signal $x_{\mathrm{BL}}(t)$ obtained by truncating the spectrum B_x equals

$$E\left[|x(t) - x_{\mathrm{BL}}(t)|^2\right] \leq \frac{1}{2\pi} \int\limits_{|\omega|>2\pi B_x} S_x(\omega)\, d\omega \quad \text{for all } t \tag{4-15}$$

and thus can be made arbitrarily small. This can be proven by supposing that the error signal $[x(t)-x_{\mathrm{BL}}(t)]$ is the response of a linear system $[1 - H_{\mathrm{BL}}(\omega)]$ to $x(t)$ where $H_{\mathrm{BL}}(\omega)$ is an ideal lowpass filter:

$$H_{\mathrm{BL}}(\omega) = \begin{cases} 1 & |\omega/2\pi| < B_x \\ 0 & \text{else} \end{cases} \tag{4-16}$$

The mean-square error is given by

$$E\left[x|(t) - x_{\mathrm{BL}}(t)|^2\right] = \frac{1}{2\pi} \int\limits_{-\infty}^{\infty} S_x(\omega)|1 - H_{\mathrm{BL}}(\omega)|^2\, d\omega \tag{4-17}$$

which can obviously be written as

$$E\left[|x(t) - x_{\mathrm{BL}}(t)|^2\right] = \int\limits_{|\omega|>2\pi B_x} S_x(\omega)\, d\omega \tag{4-18}$$

Any band-limited signal can be expanded into a series of the form (4-8) where the base function ϕ_n is given by

$$\phi_n(t) = \frac{\sin[\pi/T_s(t - nT_s)\,]}{\pi/T_s(t - nT_s)} = \mathrm{si}\left[\frac{\pi}{T_s}(t - nT_s)\right] \qquad B_x = \frac{1}{2T_s} \tag{4-19}$$

and the coefficient x_n equals the sample value $x(nT_s)$ of $x(t)$. The function $x(t)$ then has the series representation

$$x(t + \tau) = \sum_{n=-\infty}^{\infty} x(\tau + nT_s)\, \mathrm{si}\left[\frac{\pi}{T_s}(t - nT_s)\right] \tag{4-20}$$

$$\tau: \ \text{arbitrary time shift}$$

If $x(t)$ is a deterministic waveform, then (4-20) is the celebrated sampling theorem.

In the stochastic version of this theorem the equivalence of both sides is to be understood in the mean-square-sense:

$$\lim_{N\to\infty} E\left\{\left|x(t + \tau) - \sum_{n=-N}^{N} x(\tau + nT_s)\, \mathrm{si}\left[\frac{\pi}{T_s}(t - nT_s)\right]\right|^2\right\} = 0 \tag{4-21}$$

For the proof of this theorem the reader is referred to Section 1.3.

In the usual form of the theorem it is assumed that $\tau = 0$. However, the general form (4-20) represents the mathematical basis required for timing recovery. This key observation follows from the following. Since (4-20) is true for all t and τ, it trivially follows by setting $\tau = 0$ and $t = t^{'} + \tau$:

$$x\left(t' + \tau\right) = \sum_{n=-\infty}^{\infty} x(nT_s) \operatorname{si}\left[\frac{\pi}{T_s}\left(t' + \tau - nT_s\right)\right] \tag{4-22}$$

But since $t^{'}$ is just a dummy variable name, the *shift property* of band-limited functions is obtained by equating (4-20) and (4-22):

$$\begin{aligned} x(t+\tau) &= \sum_{n=-\infty}^{\infty} x(\tau + nT_s) \operatorname{si}\left[\frac{\pi}{T_s}(t - nT_s)\right] \\ &= \sum_{n=-\infty}^{\infty} x(nT_s) \operatorname{si}\left[\frac{\pi}{T_s}(t + \tau - nT_s)\right] \end{aligned} \tag{4-23}$$

The shift property states that the signal $x(t+\tau)$ can be either represented by the samples $\{x(\tau + nT_s)\}$ or the samples $\{x(nT_s)\}$.

Now assume that we need the samples $x(kT_s + \tau)$ for further processing. We have two possibilities to obtain these samples. We sample the signal $x(t)$ physically at $t = kT_s + \tau$ [first line of (4-23)]. Alternatively, we sample $x(t)$ at kT_s to obtain the second line of (4-23),

$$x(kT_s + \tau) = \sum_{n=-\infty}^{\infty} x(nT_s) \operatorname{si}\left[\frac{\pi}{T_s}(kT_s + \tau - nT_s)\right] \tag{4-24}$$

which shows how to digitally convert the samples $\{x(kT_s)\}$ to any shifted version $\{x(kT_s + \tau)\}$. We emphasize the word *digitally* since the right-hand side of (4-24) is exactly in the form of a convolutional sum. We can view the sequence on the left side as being generated by a digital interpolator filter with the impulse response

$$h_I(kT_s, \tau) = \operatorname{si}\left[\frac{\pi}{T_s}(kT_s + \tau)\right] \tag{4-25}$$

and input $x(kT_s)$. The two possibilities are illustrated in Figure 4-8.

The frequency response of the interpolating filter is

$$H_I(e^{j\omega T_s}, \tau) = \frac{1}{T_s} \sum_{n=-\infty}^{\infty} H_I\left(\omega - \frac{2\pi}{T_s}n, \tau\right) \tag{4-26}$$

where $H_I(\omega, \tau)$ is the Fourier transform of $\operatorname{si}[\pi/T_s\,(t+\tau)]$:

$$H_I(\omega, \tau) = \begin{cases} T_s \exp(j\omega\tau) & \left|\frac{\omega}{2\pi}\right| < \frac{1}{2T_s} \\ 0 & \text{elsewhere} \end{cases} \tag{4-27}$$

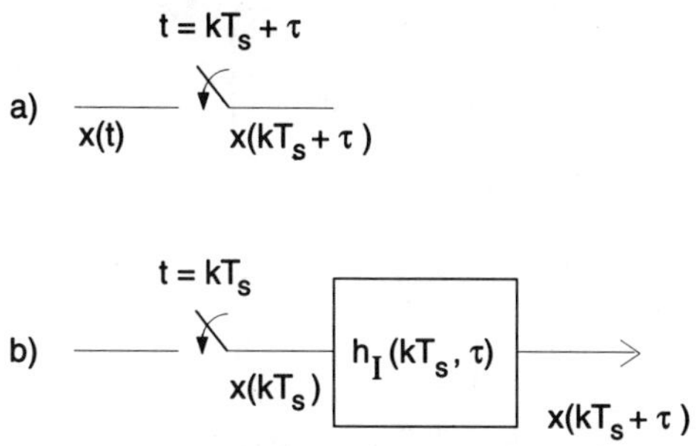

Figure 4-8 Generation of Shifted Samples: (a) Sampling at $t = kT_s + \tau$, (b) by Digital Interpolation

We claimed earlier that there is no penalty associated with digitally processing a continuous-time band-limited signal. Mathematically, the equivalence of analog and digital signal processing can be stated as follows.

4.2.3 Equivalence Theorem

Table 4-1 Equivalence of Digital and Analog Signal Processing I

Linear filtering:

$$\begin{aligned} y(t)|_{t=kT_s} &= \int_{-\infty}^{\infty} h(kT_s - \nu)\, x(\nu)\, d\nu \\ &= T_s \sum_{n=-\infty}^{\infty} h(kT_s - nT_s)\, x(nT_s) \end{aligned} \tag{4-28}$$

Condition:

$H(\omega)$: Band-limited frequency response with $B_h \leq 1/2T_s$

$x(t)$: Band-limited stochastic process with $B_x \leq 1/2T_s$, $E\left[|x(t)|^2\right] < \infty$ or deterministic signal with finite energy

The equivalence theorem is of fundamental importance in digital signal processing. The first line in (4-28) in Table 4-1 represents the form suitable for analog signal processing while the second line represents that for digital signal processing (Figure 4-9).

For the proof of the equivalence theorem we show that the signal $y_d(t)$ reconstructed from the output of the digital filter, $y_d(kT_s)$, equals (in the mean-

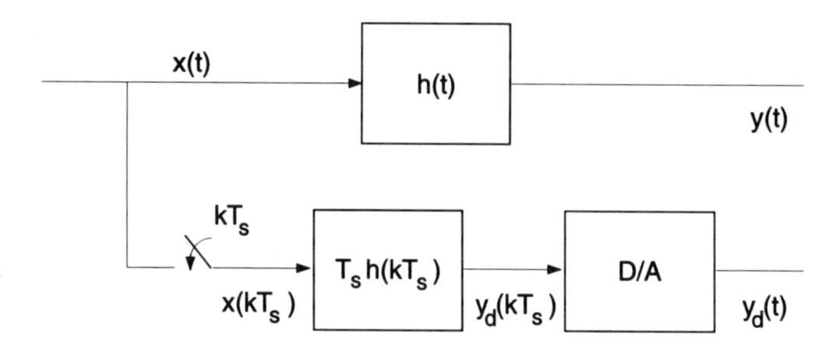

Figure 4-9 Equivalence of Digital and Analog Signal Processing

square sense) the signal $y(t)$ obtained after analog filtering,

$$E\left[|y(t) - y_d(t)|^2\right] = 0 \tag{4-29}$$

The signal $y_d(t)$ is given by

$$\begin{aligned} y_d(t) &= \sum_k y_d(kT_s)\phi_k(t) \\ &= T_s \sum_k \sum_n x(nT_s)h(kT_s - nT_s)\phi_k(t) \end{aligned} \tag{4-30}$$

Inserting (4-30) into (4-29) and taking expected values yields

$$E\left[|y(t) - y_d(t)|^2\right] = R_y(t,t) - R_{yy_d}(t,t) - R_{y^* y_d^*}(t,t) + R_{y_d}(t,t) \tag{4-31}$$

Using the series expansions of correlation functions [equation (1–166)]

$$\begin{aligned} R_x(t_1,t_2) &= \sum_n \sum_m R_x(nT_s, mT_s)\phi_n(t_1)\phi_m(t_2) \\ R_x(t_1, mT_s) &= \sum_n R_x(nT_s, mT_s)\phi_n(t_1) \end{aligned} \tag{4-32}$$

and observing that $\int h(\tau)\phi_n(t-\tau)d\tau = T_s h(t-nT_s)$, the cross-correlations in eq. (4-31) become

$$\begin{aligned} R_y(t,t) &= R_{yy_d}(t,t) = R_{y^* y_d^*}(t,t) = R_{y_d}(t,t) \\ &= T_s^2 \sum_n \sum_m h^*(t-nT_s)h(t-mT_s)R_x(nT_s, mT_s) \end{aligned} \tag{4-33}$$

Hence, the right-hand side of eq. (4-31) equals zero. The two signals $y(t)$ and $y_d(t)$ are thus equal in the mean-square sense.

Despite the fact that a band-limited signal is of infinite duration, it is possible to consider finite estimation intervals. This is done as follows. The truncated series corresponds to a function denoted by

$$\hat{x}_{2K+1}(t) = \sum_{n=-K}^{K} x(nT_s)\ \mathrm{si}\left[\frac{\pi}{T_s}\,(t - nT_s)\right] \tag{4-34}$$

The estimation interval is $T_e = 2\ (K\ T_s)$ but notice that $\hat{x}_{2K+1}(t)$ extends from $(-\infty, \infty)$.

The function has a useful interpretation in the vector space. As any signal, the truncated series represents a vector in the signal space. The components of $\hat{x}_{2K+1}$ with respect to the base vector ϕ_l, $|l| > K$ are zero. The vector $\hat{x}_{2K+1}$ is therefore an approximation in a $(2K+1)$ dimensional subspace to the vector x in the full space. Since the system of base functions $\{\phi_n(t)\}$ is complete, the error vanishes for $K \to \infty$.

4.2.4 Sufficient Statistics

We recall that in our channel model (see Section 3.2.1) we assumed the received signal to be the sum of useful signal plus additive noise. We discussed previously that for bandwidth efficient communication the signal $s(t)$ can be accurately modeled as strictly band-limited to a bandwidth of B. But what about the noise $w(t)$? Going back to Figure 3-4 we recall that the additive noise $w(t)$ has been described as Gaussian with typically much larger bandwidth than the signal and a flat power spectral density within the bandwidth of the useful signal. Any practical receiver comprises a prefilter $F(\omega)$ to remove the spectral portion of the noise lying outside the signal bandwidth B. The signal $r_f(t)$ then equals the sum

$$r_f(t) = s_f(t) + \frac{n(t)}{A} \tag{4-35}$$

where $n(t)$ is colored Gaussian noise with spectral density

$$S_n(\omega) = |F(\omega)|^2\ S_w(\omega) \tag{4-36}$$

Immediately a set of practically important questions arise:

(i) What are the requirements on the prefilter $F(\omega)$ such that no information is lost when signal $r(t)$ is processed by $F(\omega)$?

(ii) In case of a digital receiver we need to know how often the signal must be sampled in order to be able to perform the tasks of data detection, timing and carrier phase recovery solely based on these samples. Furthermore, is it possible to match sampling rate $1/T_s$ and prefilter characteristic $F(\omega)$ such that no performance loss occurs due to time discretization?

To answer these questions we must delve rather deeply into mathematics. But the effort is truly worth it since the abstract mathematical results will provide us with guidelines to synthesize optimal receiver structures not obtainable otherwise. With these preliminary remarks let us start to answer the first question.

It is intuitively clear that the spectral components outside B contain no information and should be removed. To mathematically verify this statement we decompose the signal $r(t)$ into

$$r(t) = \underbrace{s(t) + n_B(t)}_{y_B(t)} + \underbrace{[w(t) - n_B(t)]}_{\tilde{n}(t)} \tag{4-37}$$

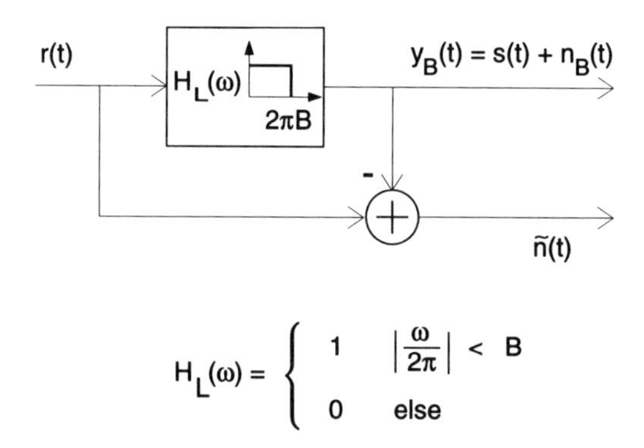

Figure 4-10 Signal Decomposition

where $n_B(t)$ is the part of the total noise which falls into the bandwidth of the useful signal while $\tilde{n}(t)$ equals the spectral components of the noise outside B. The signal decomposition of (4-37) is illustrated in Figure 4-10.

We have to demonstrate two facts. First, it is obvious that the part of the received signal $\tilde{n}(t)$ that we are throwing away is independent of $s(t)$. But we must also demonstrate that $\tilde{n}(t)$ is independent of the noise components which we are keeping. (Otherwise they would be relevant in the sense that they would provide useful information about the noise.) This is readily done as follows. The cross spectral density equals

$$S_{n_B,\tilde{n}}(\omega) = S_w(\omega)\; H_L(\omega)\; [H_L(\omega) - 1] \tag{4-38}$$

Since the product $H_L(\omega)[H_L(\omega) - 1]$ equals zero everywhere, the cross spectrum is zero for all frequencies ω. Hence, the two processes $n_B(t)$ and $\tilde{n}(t)$ are uncorrelated and, by virtue of the Gaussian assumption, statistically independent.

We conclude that $\tilde{n}(t)$ is irrelevant to the decision of which message was transmitted or to the estimation of the synchronization parameters. This is true irrespective of the decision or estimation criterion chosen. Since $y_B(t)$ is strictly band-limited to B, it can be expanded into a series with base functions

$$\phi_n(t) = \mathrm{si}\left[2\pi B\left(t - \frac{n}{2B}\right)\right]$$

Thus, a sufficient statistics can be obtained by prefiltering $r(t)$ with an ideal analog lowpass filter and subsequent sampling at Nyquist rate $1/T_s = 2B$.

This, of course, is a solution of no practical interest since it is impossible to realize (or even closely approximate) such a filter with sufficient accuracy at reasonable complexity. Fortunately, it is possible to generate a sufficient statistic employing a realizable analog filter $F(\omega)$.

We maintain that the samples $\{y(nT_s)\}$ represent sufficient statistics if the following conditions on the analog prefilter $F(\omega)$ and sampling rate $1/T_s$ are satisfied (Figure 4-11).

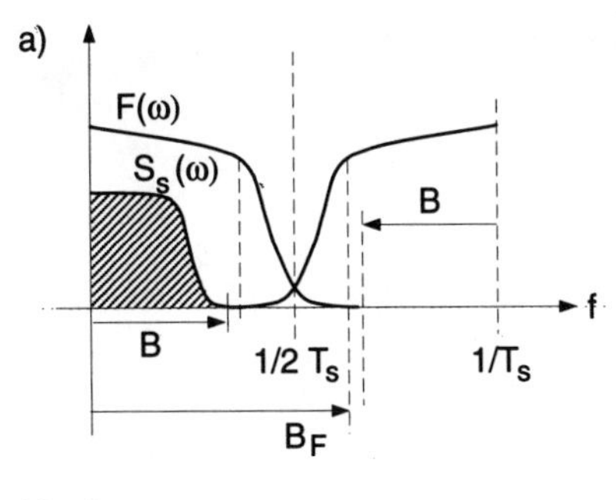

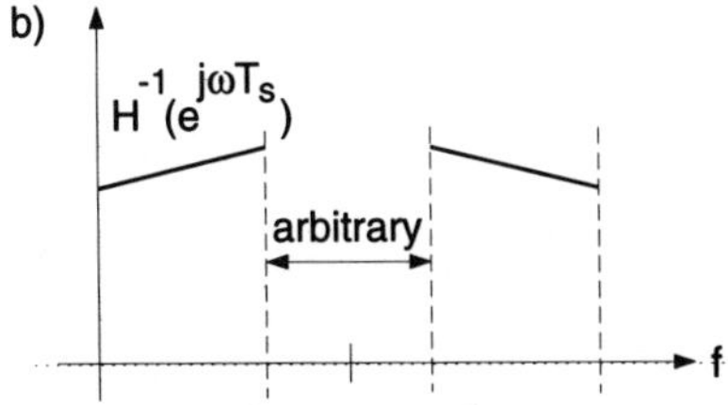

B : bandwidth of the useful signal s(t)

B_F: bandwidth of the prefilter

Figure 4-11 (a) Conditions on the Analog Prefilter, (b) Conditions on the Inverse Filter

Condition for Sufficient Statistics

$$F(\omega) = \begin{cases} \text{arbitrary but unequal to zero} & \left|\frac{\omega}{2\pi}\right| < B \\ \text{arbitrary} & B \le \left|\frac{\omega}{2\pi}\right| < B_F \\ 0 & \left|\frac{\omega}{2\pi}\right| > \frac{1}{T_s} - B \end{cases} \tag{4-39}$$

To prove the assertion we use the concept of reversibility (Figure 4-12). Briefly, the concept of reversibility states that if any preliminary processing employed is reversible it can be made to have no effect on the performance of the system.

We assume the system in the upper branch comprises an ideal lowpass filter of bandwidth B. We recall from our previous discussion that the samples $y(kT_s)$ represent sufficient statistics. Next we show that there exists a digital filter which reverses the effect of the analog prefilter $F(\omega)$, satisfying condition (4-39).

It is tempting to consider the samples $\{r_f(kT_s)\}$ as a representation of the signal $r_f(t)$, but it is faulty reasoning as we learn by inspection of Figure 4-11. The bandwidth B_F of the noise is, in general, larger than half the sampling rate $1/T_s$. Thus, the sampling theorem is *not* applicable.

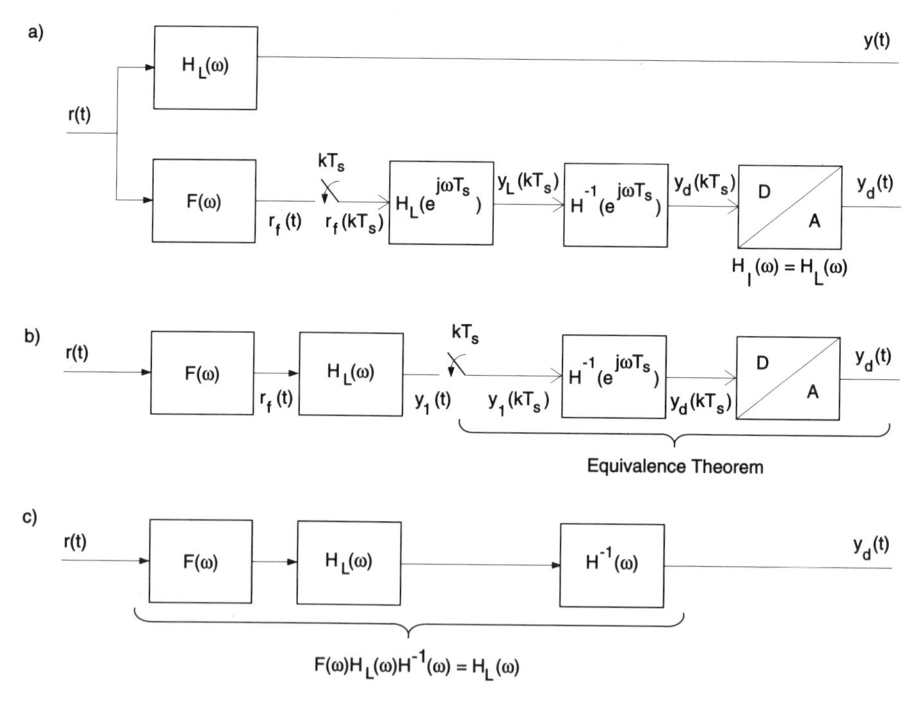

Figure 4-12 Reversibility Steps

Despite that fact we can prove reversibility. We recognize that due to condition (4-39) there is no noise aliasing in the passband of the useful signal. Filtering $r_f(kT_s)$ by the digital lowpass filter $H_L(e^{j\omega T_s})$ is therefore equivalent to filtering $r_f(t)$ in the continuous-time domain by $H_L(\omega)$. In this case we may interchange sampling and filtering. Since the signal $y_1(t)$ is now band-limited to B we can process the samples $y_1(kT_s)$ by a digital filter

$$H^{-1}(e^{j\omega T_s}) = 1/F(\omega) \qquad \left|\frac{\omega}{2\pi}\right| < B \tag{4-40}$$

which reverses the effect of $F(\omega)$ in the passband of the useful signal. Due to the equivalence theorem (4-28) the signal $y_d(t)$ equals $y(t)$ in the mean-square sense. But since $\{y_d(kT_s)\}$ is obtained from $\{r_f(kT_s)\}$, it immediately follows that $\{r_f(kT_s)\}$ is sufficient statistics.

The practical significance of condition (4-39) is that we obtain guidelines on how to trade a simple analog prefilter against the sampling rate $1/T_s$ and possibly elaborate digital signal processing.

4.2.5 Main Points

- *Band-limited signals*
  A band-limited (BL) signal can approximate a physical signal with arbitrary

accuracy. The approximation error is bounded by the power of the BL signal outsid. the bandwidth B of the BL signals [eq. (4-15)].

- *Series representation of* BL *signals:*

$$x(t+\tau) = \sum_n x(nT_s)\text{si}\left[\frac{\pi}{T_s}(t+\tau-nT_s)\right] \tag{4-41}$$

From the series representation it follows that the time-shifted sample $x(kT_s+\tau)$ can be obtained by digital interpolation of the samples $\{x(kT_s)\}$. The frequency response of the digital interpolation filter is

$$\begin{aligned} H_I(e^{j\omega T_s},\tau) &= T_s e^{j\omega\tau} \qquad \text{for } \left|\frac{\omega}{2\pi}\right| < 1/2T_s \\ h_I(kT_s,\tau) &= \text{si}\left[\frac{\pi}{T_s}(kT_s+\tau)\right] \end{aligned} \tag{4-42}$$

Timing recovery can thus be done by digital interpolation. The sample instants are determined by a free-running oscillator. There is no need of controlling a VCO to take samples at $t = kT_s + \tau$.

- *Equivalence of digital and analog signal processing*

$$\begin{aligned} y(t)|_{t=kT_s} &= \int_{-\infty}^{\infty} h(kT_s - v)\, x(v)dv \\ &= T_s \sum_n h(kT_s - nT_s)\, x(nT_s) \end{aligned} \tag{4-43}$$

Digital and analog filters are related as follows:

$$\begin{aligned} h_d(kT_s) &= T_s h(t)|_{t=kT_s} \\ H_d(e^{j\omega T_s}) &= T_s\left[\frac{1}{T_s}\sum_n H\left(\omega - \frac{2\pi n}{T_s}\right)\right] \\ &= H(\omega) \qquad \text{for } \left|\frac{\omega}{2\pi}\right| < 1/2T_s \end{aligned} \tag{4-44}$$

Thus, the two filters have identical frequency responses in the baseband.

- *Sufficient statistics*
All the information of the received signal $r_f(t)$ is contained in the samples $\{r_f(kT_s)\}$. The signal $r_f(t)$ is obtained by filtering $r(t)$ with a generalized anti-aliasing filter $F(\omega)$ [eq. (4-39)]. There is a tradeoff between sampling rate $1/T_s$ and analog prefiltering complexity. Oversampling relaxes the requirements on the rolloff thus leading to simple analog filters.

Bibliography

[1] E. A. Lee and D. G. Messerschmitt, *Digital Communications*. Boston: Kluwer Academic, 1988.

[2] H. L. van Trees, *Detection, Estimation, and Modulation Theory*. New York: Wiley, 1968.

4.3 Optimum ML Receivers

In this section we systematically derive receiver structures based on the ML criterion. This is an example of what we have termed "derived structure approach" in Volume 1. The essential feature of this approach is that the receiver structure is the outcome of the solution of a mathematical optimization problem; no a priori guesses are made.

4.3.1 Receiver Objectives and Synchronized Detection

The ultimate goal of a receiver is to detect the symbol sequence $\mathbf{a}$ in a received signal disturbed by noise with minimal probability of detection error. It is known that this is accomplished when the detector maximizes the a posteriori probability for all sequences $\mathbf{a}$

$$\hat{\mathbf{a}}_{\mathrm{MAP}} = \arg \max_{\mathbf{a}} p\left(\mathbf{a} \mid \mathbf{r}_f\right) \tag{4-45}$$

We can rewrite the a posteriori probability using Bayes's rule:

$$p(\mathbf{a} \mid \mathbf{r}_f) = \frac{p(\mathbf{r}_f \mid \mathbf{a})\, P(\mathbf{a})}{p(\mathbf{r}_f)} \tag{4-46}$$

The MAP (maximum a posteriori) detector therefore needs to know the a priori distribution $P(\mathbf{a})$ and the distribution of $\mathbf{r}_f$ conditioned on the knowledge of the data sequence $\mathbf{a}$. This conditional distribution is completely specified by the noise process. Since $p(\mathbf{r}_f)$ does not depend on the data sequence, maximizing $p(\mathbf{a} \mid \mathbf{r}_f)$ is the same as maximizing $p(\mathbf{r}_f|\mathbf{a})\; P(\mathbf{a})$. For equally probable data sequences then maximizing the a posteriori probability is the same as maximizing the likelihood function $p(\mathbf{r}_f|\mathbf{a})$, and MAP reduces to ML. For the mathematical purist we should mention that conceptually this statement is incorrect. In the ML approach $\mathbf{a}$ is assumed *unknown* but deterministic while in the MAP approach $\mathbf{a}$ is a *random* sequence. However, the result is the same for equally probable sequences. The statement "MAP reduces to ML" should always be understood in this context.

Returning to $p(\mathbf{r}_f|\mathbf{a})$ we notice that the synchronization parameters are absent. As far as detection is concerned they must be considered as unwanted parameters [1, p. 87] which are removed by averaging

$$p(\mathbf{a} \mid \mathbf{r}_f) = P(\mathbf{a}) \int p(\mathbf{r}_f \mid \mathbf{a}, \boldsymbol{\theta})\, p(\boldsymbol{\theta})\; d\boldsymbol{\theta} \tag{4-47}$$

Thus, in an optimal MAP (ML) receiver there exists *no separate synchronization unit.*

The notation in the above equation needs explanation. The symbol $\boldsymbol{\theta}$ denotes a random sequence of synchronization parameters. The function $p(\boldsymbol{\theta})$ is the joint probability density function of this sequence $p(\boldsymbol{\theta}) = p(\boldsymbol{\theta}_0, \boldsymbol{\theta}_1, \boldsymbol{\theta}_2, \ldots)$, where each sample $\boldsymbol{\theta}_k$ may stand for a set of parameters. For example, $\boldsymbol{\theta}_k = [\theta_k, \varepsilon_k]$. The integral (4-47) (weighted average) then is to be understood with respect to the whole sequence and all parameters. The probability density function $p(\boldsymbol{\theta})$ describes the a priori knowledge about the statistical laws governing that sequence. While the truly optimal receiver appears intimidating, suitable approximations will allow us to derive physically implementable receiver structures.

Let us assume the receiver operates at high signal-to-noise ratio. Then the likelihood function weighted by the a priori probabilities becomes concentrated at its maximum:

$$\begin{aligned} &\int p(\mathbf{r}_f|\mathbf{a},\boldsymbol{\theta})\, p(\boldsymbol{\theta})\, d\boldsymbol{\theta} \\ &\propto \max_{\mathbf{a},\boldsymbol{\theta}} \; p(\mathbf{r}_f|\mathbf{a},\boldsymbol{\theta})\, p(\boldsymbol{\theta}) \end{aligned} \tag{4-48}$$

Maximizing the integral leads to the rule

$$\left(\hat{\mathbf{a}}, \hat{\boldsymbol{\theta}}\right) = \arg \max_{\mathbf{a},\boldsymbol{\theta}} \; p(\mathbf{r}_f|\mathbf{a},\boldsymbol{\theta})\, p(\boldsymbol{\theta}) \tag{4-49}$$

As a first important result we observe that the receiver performs a joint detection/estimation. There is no separation between synchronization and detection units. At this point it is necessary to take a closer look at the properties of the synchronization parameters. In particular we have to distinguish between parameters which are essentially static, $\boldsymbol{\theta}_S$, and parameters that may be termed dynamic, $\boldsymbol{\theta}_D$. For static parameters there is no useful probabilistic information except that they are in a given region. Therefore, $\boldsymbol{\theta}_S$ is unknown but nonrandom. In view of eq. (4-49) the joint detection/estimation thus reduces to maximizing the likelihood function $p(\mathbf{r}_f|\mathbf{a},\boldsymbol{\theta}_S)$ with respect to $(\mathbf{a},\boldsymbol{\theta}_S)$:

$$\left(\hat{\mathbf{a}}, \hat{\boldsymbol{\theta}}_S\right) = \arg \max_{\mathbf{a},\boldsymbol{\theta}_S} \; p(\mathbf{r}_f|\mathbf{a},\boldsymbol{\theta}_S) \tag{4-50}$$

For static synchronization parameters, (4-50) defines a joint ML estimation/detection rule. On the other hand, probabilistic information is available for dynamic parameters and should be made use of. Hence joint detection and estimation calls for maximizing

$$\left(\hat{\mathbf{a}}, \hat{\boldsymbol{\theta}}_D\right) = \arg \max_{\mathbf{a},\boldsymbol{\theta}_D} \; p(\mathbf{r}_f|\mathbf{a},\boldsymbol{\theta}_D)\, p(\boldsymbol{\theta}_D) \tag{4-51}$$

Using Bayes's rule we can find a second representation of eq. (4-51),

$$\left(\hat{\mathbf{a}}, \hat{\boldsymbol{\theta}}_D\right) = \arg \max_{\mathbf{a}, \boldsymbol{\theta}_D} \; p(\boldsymbol{\theta}_D | \mathbf{r}_f, \mathbf{a}) \cdot p(\mathbf{r}_f | \mathbf{a}) \tag{4-52}$$

From the second representation it is immediately recognized that $\boldsymbol{\theta}_D$ is a MAP estimate so that one may speak of MAP dynamic parameter estimation and ML data detection. Dynamic parameter estimation will be thoroughly discussed in Part E which deals with fading channels. In Part D we focus on static parameter estimation.

Since there are infinitely many possible values of the synchronization parameters while the number of possible sequences is finite, the most natural joint maximization procedure is to first maximize the joint likelihood function $p(\mathbf{r}_f | \mathbf{a}, \boldsymbol{\theta}_S)$ with respect to $\boldsymbol{\theta}_S$ for each possible $\mathbf{a}$ and then select the sequence $\mathbf{a}$ with the largest likelihood

$$\begin{aligned} \hat{\boldsymbol{\theta}}_S(\mathbf{a}) &= \arg \max_{\boldsymbol{\theta}_S} \; p(\mathbf{r}_f | \mathbf{a}, \boldsymbol{\theta}_S) \\ \Lambda_S(\mathbf{a}) &= p\left(\mathbf{r}_f | \mathbf{a}, \boldsymbol{\theta}_S = \hat{\boldsymbol{\theta}}_S(\mathbf{a})\right) \\ \hat{\mathbf{a}} &= \arg \max_{\mathbf{a}} \; \Lambda_S(\mathbf{a}) \end{aligned} \tag{4-53}$$

The first maximization step yields a conditional synchronization parameter estimate $\hat{\boldsymbol{\theta}}_S(\mathbf{a})$ that is subsequently used in the decision likelihood computation as if it were the true parameter. However, each candidate sequence $\mathbf{a}$ comes with its own synchronization estimate $\hat{\boldsymbol{\theta}}_S(\mathbf{a})$ conditioned on that sequence. Clearly, the optimum joint estimation detection rule is far too complex for a practical application. However, it readily guides us toward practical approximations.

A first approach to separate the joint detection estimation rule into two disjoint tasks is to transmit a preamble $\mathbf{a}_0$ of known symbols prior to sending the useful data. During the preamble the (static) synchronization parameters can be estimated:

$$\hat{\boldsymbol{\theta}}_S(\mathbf{a}_0) = \arg \max_{\boldsymbol{\theta}_S} \; p(\mathbf{r}_f | \mathbf{a} = \mathbf{a}_0, \boldsymbol{\theta}_S) \tag{4-54}$$

The estimation rules which use a known sequence $\mathbf{a}_0$ are called *data-aided* (DA). Subsequent detection of the useful data $\mathbf{a}$ is then done using the estimate $\hat{\boldsymbol{\theta}}_S(\mathbf{a}_0)$ as if it were the true value:

$$\hat{\mathbf{a}} = \arg \max_{\mathbf{a}} \; p\left(\mathbf{r}_f | \mathbf{a}, \boldsymbol{\theta}_S = \hat{\boldsymbol{\theta}}_S(\mathbf{a}_0)\right) \tag{4-55}$$

In order to accommodate ever-present slow variations of the synchronization parameters one may use segments of the decoded sequence to update the synchronization parameter estimate:

$$\hat{\boldsymbol{\theta}}_S(\hat{\mathbf{a}}_L) = \arg \max_{\boldsymbol{\theta}_S} \; p(\mathbf{r}_f \,|\, \mathbf{a} = \hat{\mathbf{a}}_L, \boldsymbol{\theta}_S) \tag{4-56}$$

$$\hat{\mathbf{a}}_L : \quad \text{segment of decoded data}$$

Using segments $\hat{\mathbf{a}}_L$ of the decoded data as if it were the true symbols leads to so-called *decision-directed* (DD) synchronization algorithms. Notice that decision-directed algorithms always assume that an (accurate) initial estimate of $\boldsymbol{\theta}_S$ is available. Only under this condition the synchronization-detection process [eq. (4-56)] works.

Another possibility is to actually perform the averaging operation to remove the data dependency:

$$p(\mathbf{r}_f|\boldsymbol{\theta}_S) = \sum_{\substack{\text{all possible} \\ \text{sequences } \mathbf{a}}} p(\mathbf{r}_f|\mathbf{a}, \boldsymbol{\theta}_S)\; P(\mathbf{a}) \tag{4-57}$$

This leads to the class of *non-data-aided* (NDA) synchronization algorithms,

$$\hat{\boldsymbol{\theta}}_S = \arg \max_{\boldsymbol{\theta}_S} \; p(\mathbf{r}_f|\boldsymbol{\theta}_S) \tag{4-58}$$

Virtually all realizable receiver structures are based on the *principle of synchronized detection* where estimates of the synchronization parameters are used in the detection process as if it were the true values:

$$\hat{\mathbf{a}} = \arg \max_{\mathbf{a}} \; p\Big(\mathbf{r}_f|\mathbf{a}, \boldsymbol{\theta}_S = \hat{\boldsymbol{\theta}}_S\Big) \tag{4-59}$$

The estimates $\hat{\boldsymbol{\theta}}_S$ are obtained by DA, NDA, or DD synchronization algorithms. Systematic derivation of these algorithms will be the topic of Chapter 5.

4.3.2 Optimum ML Receiver for Constant Synchronization Parameters

We consider the detection of a sequence $\mathbf{a}$ of N symbols. The received signal $r_f(t)$ is given by (3-1):

$$r_f(t) = \sum_{n=0}^{N-1} a_n \; g(t - nT - \varepsilon_0 T)\; e^{j\theta_0} + \frac{n(t)}{A} \tag{4-60}$$

where the pulse $g(t)$ is the convolution of the transmitter pulse $g_T(t)$ with the channel impulse response $c(t)$ and the analog prefilter $f(t)$:

$$g(t) = g_T(t) \otimes c(t) \otimes f(t) \tag{4-61}$$

The noise $n(t)$ is characterized by its power spectrum:

$$S_n(\omega) = |F(\omega)|^2 \, S_w(\omega) \tag{4-62}$$

In many cases the noise process can be modeled as a complex random process with flat spectral density N_0 within the passband of the analog prefilter $F(\omega)$. However, this is not essential to the following derivations.

For the time being, the synchronization parameters (θ, ε) are assumed constant. However, we should be aware that the synchronization parameters are always time variant. But the time scale on which changes of these parameters are observable is much larger than the time scale necessary to express the symbol duration. It is therefore reasonable to separate these two time scales. We consider a first time scale which deals with the detection process and which is defined in units of the symbol duration T. The second time scale deals with slowly varying parameter changes of (θ, ε) with time constants much larger than T. Thus, the synchronization parameters may be considered piecewise constant over a number of symbols. This assumption greatly eases the systematic derivation of synchronizers. As we will see, tracking of the slowly varying parameters can be done in a post-processing unit which considers the estimates of the first state as noisy measurements.

After these preliminaries we are now ready to tackle our goal of deriving the ML receiver structure for constant synchronization parameters. We have already done a great deal of preparatory work. It has been shown that the samples $\{r_f(kT_s)\}$ are sufficient statistics. The $(2M+1)$ samples taken symmetrically to the origin

$$\mathbf{r}_{f,2M+1} = \{r_{f,-M}, \dots r_{f,0}, \dots r_{f,M}\} \tag{4-63}$$

are the truncated representation of the sample vector $\mathbf{r}_f$.[3]

Assuming the sequence $\mathbf{a}$ and the synchronization parameters (θ, ε) to be known from (4-60), the likelihood function follows immediately:

$$p(\mathbf{r}_{f,2M+1} \mid \mathbf{a}, \varepsilon, \theta) = p_n(\mathbf{r}_{f,2M+1} - \mathbf{s}_{f,2M+1}) \tag{4-64}$$

which is simply the (multivariant) pdf of the noise sequence $\mathbf{n}$ for the arguments $n_k = r_f(kT_s) - s_f(kT_s)$, $|k| \leq M$. We assume that $n(t)$ is Gaussian with covariance matrix $\mathbf{\Lambda}$ and (provided it exists) inverse matrix $\mathbf{Q}$. The elements of $\mathbf{\Lambda}$ are given by

$$\begin{aligned} \lambda_{k,l} = [\mathbf{\Lambda}]_{k,l} &= E[n(kT_s)\, n^*(lT_s)] \\ &= \frac{1}{A^2}\, R_n([k-l]T_s) \end{aligned} \tag{4-65}$$

[3] However, it would be faulty reasoning to say that $\mathbf{r}_{f,2M+1}$ is a $(2M+1)$-dimensional approximation of the signal $\mathbf{r}_f(t)$. Even for $M \rightarrow \infty$, $\mathbf{r}_f$ does *not* represent the signal $\mathbf{r}_f(t)$ as we explained earlier.

where $R_n(\tau)$ is the correlation function corresponding to the power spectrum $S_n(\omega)$ [eq. (4-62)]. Then

$$\begin{aligned} p(\mathbf{r}_{f,2M+1} \mid \mathbf{a}, \varepsilon, \theta) = &\left[(2\pi)^{2M+1} (\det \mathbf{\Lambda})^{1/2}\right]^{-1} \\ &\times \exp\{-[\mathbf{r}_{f,2M+1} - \mathbf{s}_{f,2M+1}]^H \\ &\times \mathbf{Q}_{2M+1} [\mathbf{r}_{f,2M+1} - \mathbf{s}_{f,2M+1}]\} \end{aligned} \tag{4-66}$$

where $\mathbf{x}^H$ denotes the transposed vector with complex conjugate elements.

Now we can divide $p(\mathbf{r}_{f,2M+1} \mid \mathbf{a}, \varepsilon, \theta)$ by anything that does not depend on the parameters. For $\mathbf{a} = 0$ the received signal equals $r_f(t) = n(t)$. Thus,

$$\begin{aligned} p(\mathbf{r}_{f,2M+1} \mid \mathbf{a}{=}\mathbf{0}) = &\left[(2\pi)^{2M+1} (\det \mathbf{\Lambda})^{1/2}\right]^{-1} \\ &\times \exp\{-\mathbf{r}_{f,2M+1}^H \mathbf{Q}_{2M+1} \mathbf{r}_{f,2M+1}\} \end{aligned} \tag{4-67}$$

We introduce the quotient

$$\frac{p(\mathbf{r}_{f,2M+1} \mid \mathbf{a}, \varepsilon, \theta)}{p(\mathbf{r}_{f,2M+1} \mid \mathbf{a} = \mathbf{0})} \tag{4-68}$$

Substituting (4-66) and (4-67) into this quotient, canceling common terms, letting $M \to \infty$ and taking the logarithm we obtain the log-likelihood function

$$\begin{aligned} L(\mathbf{r}_f | \mathbf{a}, \varepsilon, \theta) &= \mathbf{r}_f^H \mathbf{Q} \mathbf{s}_f + \mathbf{s}_f^H \mathbf{Q} \mathbf{r}_f - \mathbf{s}_f^H \mathbf{Q} \mathbf{s}_f \\ &= 2\,\mathrm{Re}\left[\mathbf{r}_f^H \mathbf{Q} \mathbf{s}_f\right] - \mathbf{s}_f^H \mathbf{Q} \mathbf{s}_f \end{aligned} \tag{4-69}$$

where $\mathbf{r}_f, \mathbf{s}_f$ are the infinite-dimensional sample vectors. For simplicity in notation we have suppressed the parameter set $(\mathbf{a}, \theta, \varepsilon)$ in the signal vector $\mathbf{s}_f$. They will reappear when we now insert the signal samples $s_f(kT_s)$ defined by (4-60) into (4-69).

For the first expression in (4-69) we obtain after some lengthy algebraic manipulation

$$\mathbf{r}_f^H \mathbf{Q}\ \mathbf{s}_f = \sum_{n=0}^{N-1} a_n\, e^{j\theta} \sum_{k=-\infty}^{\infty} r_f^*(kT_s)\, g_{\mathrm{MF}}^*(nT + \varepsilon T - kT_s) \tag{4-70}$$

In Section 4.3.3 we will learn that $g_{\mathrm{MF}}(kT_s)$ is a digital matched filter which for the present discussion we assume to be given. Let us denote the conjugate complex of the second sum in (4-70) by $z(nT + \varepsilon T)$,

$$z(nT + \varepsilon T) = \sum_{k=-\infty}^{\infty} r_f(kT_s)\, g_{\mathrm{MF}}(nT + \varepsilon T - kT_s) \tag{4-71}$$

Formally, the sum has exactly the form of a digital convolution of the input samples $r_f(kT_s)$ with a linear digital filter. The discrete-time impulse response is different

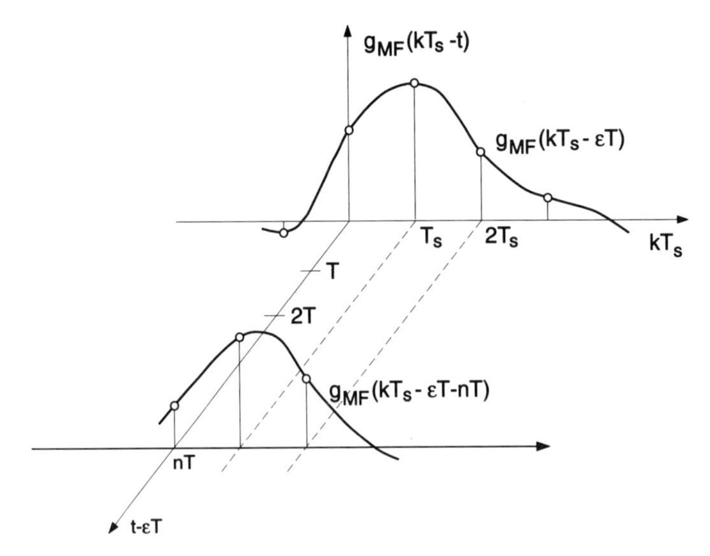

Figure 4-13 Time-Variant, Discrete-Time Impulse Response $g_{\text{MF}}(kT_s - t)$

for every nT, hence, the filter is time-variant. This fact is illustrated by Figure 4-13 where on the horizontal axis the impulse response $g_{\text{MF}}(kT_s - t)$ (all k) is displayed for $t = \varepsilon T$ and $t = nT + \varepsilon T$. For example, to obtain the sample $z(\varepsilon T)$ ($n = 0$), we have to convolve the sampled input signal $r_f(kT_s)$ with an impulse response $g_{\text{MF}}(kT_s - \varepsilon T)$

$$z(\varepsilon T) = \sum_{k=-\infty}^{\infty} r_f(kT_s)\, g_{\text{MF}}(-kT_s + \varepsilon T) \tag{4-72}$$

There are several features in (4-71) which need a thorough discussion. The most intriguing and disturbing fact is the appearance of the symbol duration T in $z(nT + \varepsilon T)$. If the digital convolution of (4-71) is to be interpreted as the output of a (time-variant) digital filter operating at the clock rate $1/T_s$, then *only* outputs at multiples of T_s are available. Under the (oversimplifying) assumption that T is a multiple of T_s, i.e., $T = IT_s$, the output $z(nT + \varepsilon T)$ can be written as

$$\begin{aligned} z(nT + \varepsilon T) &= z([nI]T_s + \varepsilon T) \\ &= \sum_{k=-\infty}^{\infty} r_f(kT_s)\, g_{\text{MF}}(-kT_s + [nI]T_s + \varepsilon T) \end{aligned} \tag{4-73}$$

Then $z(nT + \varepsilon T)$ is the output of the time-invariant digital filter with impulse response $g_{\text{MF}}(kT_s - \varepsilon T)$ taken at $t = (nI)T_s$. We could thus run the digital matched filter at rate $1/T_s$ and take the output $z(kT_s + \varepsilon T)$ at $k = In$ (Figure 4-14). This operation is called *decimation* in digital signal processing.

Clearly the assumption $T = IT_s$ is an unacceptable oversimplification since T is determined by the clock at the transmitter while T_s is determined by an

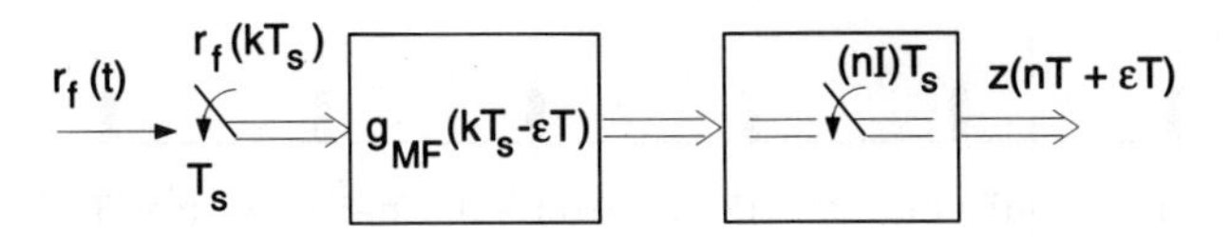

Figure 4-14 Computing $z(nT + \varepsilon T)$ in a Digital Matched Filter Followed by a Decimator; Condition: T is Exactly a Multiple of T_s

independent clock at the receiver. Even with extremely precise clocks we can never assume T and T_s to be commensurate. Indeed, generating an estimate of the transmitter time base at the receiver is what synchronization is all about.

Despite its oversimplification the previous discussion leads into the right direction for a solution to the problem. The key idea is to write the argument of $z(nT + \varepsilon T)$ as the sum

$$nT + \varepsilon T = [m_n + \mu_n]\, T_s \tag{4-74}$$

Here m_n means the largest integer that is less than or equal to the real number on the left side and μ_n is a fractional difference, as illustrated in Figure 4-15. Inserting the right side of (4-74) into (4-71) we obtain

$$z(nT + \varepsilon T) = \sum_{k=-\infty}^{\infty} r_f(kT_s)\, g_{\mathrm{MF}}(-kT_s + m_n T_s + \mu_n T_s) \tag{4-75}$$

The samples of the impulse response shifted by $(\mu_n T_s)$ follow from the shift property of band-limited signals [eq. (4-23)]:

$$\begin{aligned} g_{\mathrm{MF}}(t - \mu_n T_s) &= \sum_{i=-\infty}^{\infty} g_{\mathrm{MF}}(iT_s - \mu_n T_s)\, \mathrm{si}\left[\frac{\pi}{T_s}\,(t - iT_s)\right] \\ &= \sum_{i=-\infty}^{\infty} g_{\mathrm{MF}}(iT_s)\, \mathrm{si}\left[\frac{\pi}{T_s}\,(t - \mu_n T_s - iT_s)\right] \end{aligned} \tag{4-76}$$

For $t = lT_s$ we obtain from the previous equation

$$g_{\mathrm{MF}}(lT_s - \mu_n T_s) = \sum_{i=-\infty}^{\infty} g_{\mathrm{MF}}(iT_s)\, \mathrm{si}\left[\frac{\pi}{T_s}\,(lT_s - \mu_n T_s - iT_s)\right] \tag{4-77}$$

Equation (4-77) is exactly in the form of a convolution sum. We can thus view the sequence $\{g_{\mathrm{MF}}(lT_s - \mu_n T_s)\}$ as being generated by a digital interpolator filter with impulse response

$$h_I(lT_s, \mu_n) = \mathrm{si}\left[\frac{\pi}{T_s}\,(lT_s - \mu_n T_s)\right] \tag{4-78}$$

and input $g_{\mathrm{MF}}(iT_s)$. The interpolator filter is a linear, time-variant filter (due to μ_n) with output denoted $y(kT_s)$. The matched filter is time-invariant.

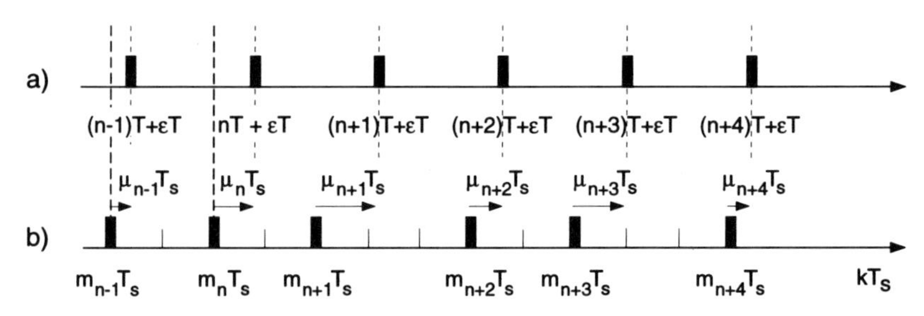

Figure 4-15 Fractional Difference, (a) Transmitter Time Scale (nT), (b) Receiver Time Scale (kT_s)

Unlike conventional digital filters where the input rate equals the output rate, the right-hand side of (4-75) defines a time variable decimation process since only outputs $y(m_n T_s)$ at times $m_n T_s$ are taken:

$$y(m_n T_s) = \sum_{k=-\infty}^{\infty} r_f(kT_s)\, g_{\text{MF}}(-kT_s + m_n\, T_s + \mu_n\, T_s) \tag{4-79}$$

By comparison of (4-79) with the definition of $z(nT + \varepsilon T)$, (4-75), one concludes that

$$z(nT + \varepsilon T) = y(m_n\, T_s) \tag{4-80}$$

Since T/T_s is irrational, the time instances $m_n T_s$ form a completely irregular (though deterministic) pattern on the time axis $\{kT_s\}$ (see Figure 4-15). This irregular pattern is required in order to obtain an average time between an output $z(nT + \varepsilon T)$ of exactly T (given a time quantization of T_s). The samples $y(m_n\, T_s)$ obtained at average rate $1/T$ are then phase rotated to obtain $z(nT + \varepsilon T)\, e^{-j\theta}$.

Figure 4-16 summarizes the digital signal processing required to obtain $z(nT + \varepsilon T)\, e^{-j\theta}$ from the received signal samples $r_f(kT_s)$. The received samples

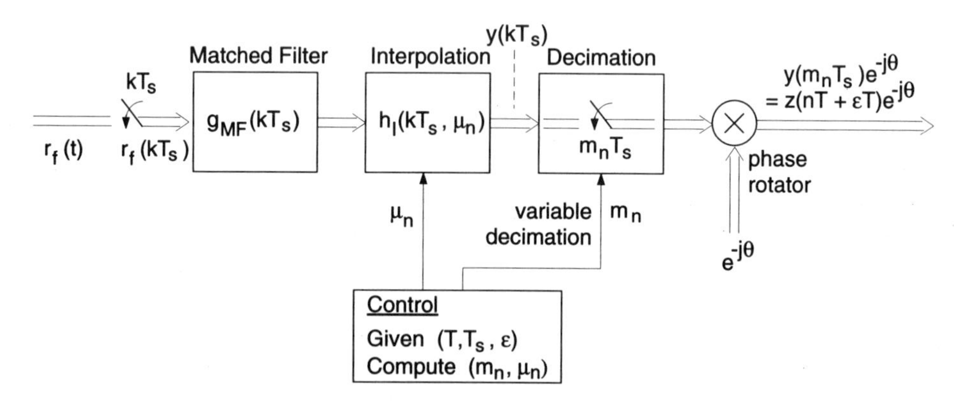

Figure 4-16 Optimum ML Receiver Structure: Computation of the Matched Filter Output

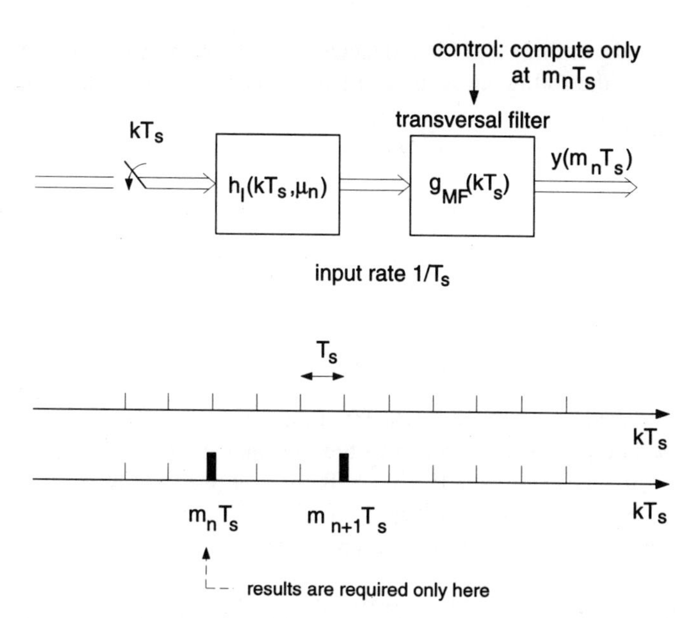

Figure 4-17 Interchanging Position of Digital Interpolator and Matched Filter

$r_f(kT_s)$ are first processed in a digital matched filter followed by the time-variant interpolator, both running at rate $1/T_s$. Since the filters are linear, their position may be interchanged (Figure 4-17). Notice that only the results $y(m_n\, T_s)$ are required for data detection and phase synchronization. The input data rate of the filter is $1/T_s$. It runs at $1/T_s$ but only outputs at instants $m_n T_s$ (average rate $1/T$) are actually computed. In a practical realization $g_{\text{MF}}(kT_s)$ would be realized as an finite impulse response (FIR) filter where the samples are clocked in at rate $1/T_s$. The output is activated at instants $m_n T_s$ only (Figure 4-18). As an interesting detail we remark that in the case where the interpolator precedes the matched filter the output of the interpolator does *not* equal $r_f(kT_s + \mu_n\, T_s)$ since $r_f(t)$ is not band-limited to $B \;\leq\; 1/2\, T_s$.

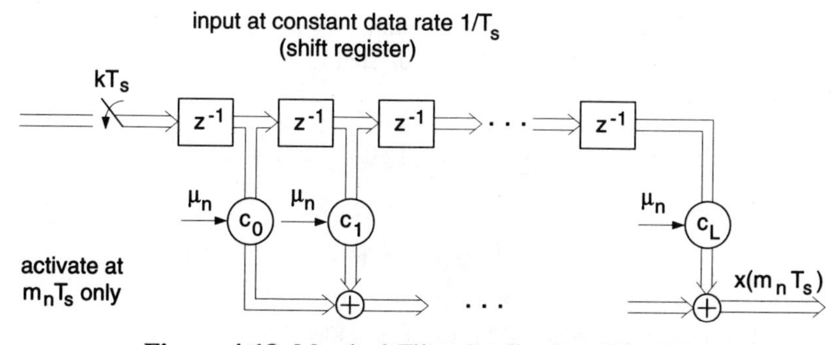

Figure 4-18 Matched Filter Realized as FIR Filter

The second term of (4-69) is independent of the received signal. As will be demonstrated in the following section, it can be written as a quadratic form:

$$\mathbf{s}_f^H \mathbf{Q} \mathbf{s}_f = \mathbf{a}^H \mathbf{H} \mathbf{a} \tag{4-81}$$

with

$$\begin{aligned} h_{m,n} = [\mathbf{H}]_{m,n} &= \sum_{k=-\infty}^{\infty} g(kT_s - mT)\, g_{\mathrm{MF}}(-kT_s + nT) \\ \mathbf{a}^H &= \left(a_0^*, \ldots, a_{N-1}^*\right) \qquad \text{(symbol sequence)} \end{aligned} \tag{4-82}$$

The matrix elements $h_{m,n}$ are independent of the synchronization parameters (θ, ε). They can be interpreted as response to the matched filter $g_{\mathrm{MF}}(kT_s)$ at time nT with $g(kT_s - mT)$ as an input. If $h_{m,n} = 0,\ m \neq n$, we have Nyquist conditions. Defining the vector $\mathbf{z}$ with elements $z_n = z(nT + \varepsilon T)$ and using (4-70) and (4-71), the log-likelihood function can be written in the form

$$L(\mathbf{r}_f \mid \mathbf{a}, \varepsilon, \theta) = 2\,\mathrm{Re}\left[\mathbf{a}^H \mathbf{z}\, e^{-j\theta}\right] - \mathbf{a}^H \mathbf{H} \mathbf{a} \tag{4-83}$$

Alternatively, the log-likelihood function can be written in the form

$$L(\mathbf{r}_f \mid \mathbf{a}, \varepsilon, \theta) = \left[\mathbf{z}\, e^{-j\theta} - \mathbf{H}^{-1}\mathbf{a}\right]^H \mathbf{H} \left[\mathbf{z}\, e^{-j\theta} - \mathbf{H}^{-1}\mathbf{a}\right] \tag{4-84}$$

which shows that the ML estimate is equivalent to a generalized minimum distance rule. For Nyquist pulses the matrix $\mathbf{H}$ is diagonal. Notice that in order to obtain an estimate of the data sequence it is inevitable to jointly estimate the "unwanted" parameter (θ, ε).

The estimator is noncausal and thus an impractical scheme since we have to store the entire received sequence $\{r_f(kT_s)\}$ to perform iteratively the maximum search over the parameter set $(\mathbf{a}, \theta, \varepsilon)$. It requires to compute m_n, μ_n, given T, T_s, for every trial value ε. The samples $\{r_f(kT_s)\}$ are then processed for every (θ, ε) as illustrated by Figure 4-16. With the resulting vector $\mathbf{z}$ the inner product $\mathbf{a}^H \mathbf{z}\ e^{-j\theta}$ is computed.

Before we discuss realizable digital estimators we briefly want to discuss an analog realization. From the equivalence of analog and digital signal processing we obtain

$$\begin{aligned} z(nT + \varepsilon T) &= \sum_{k=-\infty}^{\infty} r_f(kT_s)\, g_{\mathrm{MF}}(-kT_s + nT + \varepsilon T) \\ &= \frac{1}{T_s} \int_{-\infty}^{\infty} r_f(t)\, g_{\mathrm{MF}}(-t + nT + \varepsilon T)\, dt \end{aligned} \tag{4-85}$$

Equation (4-85) can be interpreted in two ways. First, $z(nT + \varepsilon T)$ can be viewed

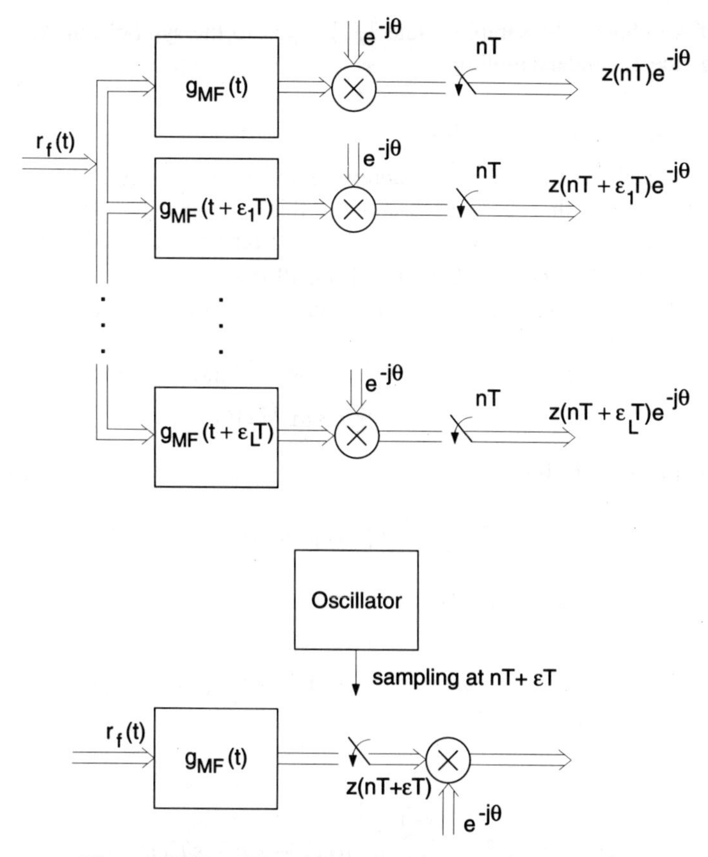

Figure 4-19 Analog Computation of $z(nT + \varepsilon T)$

as the output of an analog matched filter $g_{\mathrm{MF}}(t - \varepsilon T)$ (see Figure 4-19). To find the optimum value, a bank of matched filters with trial values $\varepsilon_i,\ i = 0, ..., L$ must be implemented and sampled at symbol rate $1/T$ to obtain $z(nT + \varepsilon_i T)$. If L is chosen sufficiently large, the discretization of the parameter is negligible. Alternatively, one may use a single filter $g_{\mathrm{MF}}(t)$ and control the sampling instances to occur at $t = nT + \varepsilon T$.

In the literature it is often stated that sampling the analog matched filter $g_{\mathrm{MF}}(t)$ at symbol rate $1/T$ is sufficient statistics. The foregoing discussion has shown that, concerning synchronization, this argument is incorrect since the matched filter output must be computed for *all* possible values $\varepsilon_i, \quad i = 1, ..., L$. It is impossible to estimate ε from the samples $\{z(nT)\}$ as it is from the samples $r_f(kT_s)$. This fact also follows directly from the conditions on the analog prefilter $F(\omega)$ [eq. (4-39)] for sufficient statistics. For bandwidth-efficient transmission the bandwidth B of the signal $s_f(t)$ lies typically between

$$\frac{1}{2T} < B < \frac{1}{T} \qquad (1/T \text{ symbol rate}) \tag{4-86}$$

Now, if we choose the sampling rate $1/T_s$ equal to the symbol rate $1/T$, there is aliasing due to undersampling.

4.3.3 Digital Matched Filter

The matched filter is a key component of any receiver. This section is concerned with a thorough discussion of this component. The derivation is somewhat technical and may be skipped at first reading up to (4-110).

The starting point of our derivation is the $(2M+1)$-dimensional approximation of eq. (4-68). Taking the logarithm and canceling common terms yields

$$\begin{aligned} L(\mathbf{r}_{f,2M+1} \mid \mathbf{a}, \varepsilon, \theta) = & 2 \operatorname{Re}\left[\mathbf{r}^H_{f,\mathbf{2M+1}} \mathbf{Q}_{\,\mathbf{2M+1}}\, \mathbf{s}_{\mathbf{f},\mathbf{2M+1}}\right] \\ & - \mathbf{s}^{\mathbf{H}}_{\mathbf{f},\mathbf{2M+1}}\, \mathbf{Q}_{2M+1}\, \mathbf{s}_{f,2M+1} \end{aligned} \tag{4-87}$$

Defining the vector $\mathbf{d}_{2M+1}$

$$\mathbf{d}_{2M+1} = \mathbf{Q}_{\,2M+1}\, \mathbf{s}_{f,2M+1} \tag{4-88}$$

we obtain for the components d_k [4]

$$d_{k,2M+1} = \sum_{l=-M}^{M} q_{k,l}\, s_l \qquad (2M+1) \text{ approximation} \tag{4-89}$$

Inserting for s_l

$$s_l = [\mathbf{s}_f]_l = \sum_{n=0}^{N-1} a_n\, g(lT_s - nT - \varepsilon T)\, e^{j\theta} \tag{4-90}$$

into (4-89) and letting $M \to \infty$ yields

$$d_k = \sum_{l=-\infty}^{\infty} q_{k,l} \sum_{n=0}^{N-1} a_n\, g(lT_s - nT - \varepsilon T)\, e^{j\theta} \tag{4-91}$$

Changing the order of summation, (4-91) reads

$$d_k = \sum_{n=0}^{N-1} a_n \sum_{l=-\infty}^{\infty} q_{k,l}\, g(lT_s - nT - \varepsilon T)\, e^{j\theta} \tag{4-92}$$

We will prove later on that for $M \to \infty$ the inverse covariance matrix $\mathbf{Q}$ has Toepliz form

$$q_{k,l} = q_{k',l'} \tag{4-93}$$

---

[4] In order to simplify the notation we have dropped the index f in $s_l = [\mathbf{s}_f]_l$.

whenever $k-l=k'-l'$. Using $q_{k,l}=q_{k-l}$ the infinite sum has exactly the form of a digital convolutional sum. We define the matched filter impulse response:

$$g^*_{\mathrm{MF}}(nT+\varepsilon T-kT_s) := \sum_{l=-\infty}^{\infty} q_{k-l}\, g(lT_s-nT-\varepsilon T) \tag{4-94}$$

Using this definition in d_k we then obtain

$$\begin{aligned} \mathbf{r}_f^H \mathbf{Q}\, \mathbf{s}_f &= \mathbf{r}_f^H\, \mathbf{d} \\ &= \sum_{n=0}^{N-1} a_n\, e^{j\theta} \sum_{k=-\infty}^{\infty} r_f^*(kT_s)\, g^*_{\mathrm{MF}}(-kT_s+nT+\varepsilon T) \end{aligned} \tag{4-95}$$

Similarly,

$$\begin{aligned} \mathbf{s}_f^H \mathbf{Q}\, \mathbf{s}_f &= \mathbf{s}_f^H \mathbf{d} \\ &= \sum_{k=-\infty}^{\infty} \underbrace{\left[\sum_{n=0}^{N-1} a_n^*\, e^{-j\theta}\, g^*(kT_s-nT-\varepsilon T)\right]}_{s_k^*} \\ &\qquad \times \underbrace{\left[\sum_{m=0}^{N-1} a_m\, e^{j\theta}\, g^*_{\mathrm{MF}}(-kT_s+mT+\varepsilon T)\right]}_{d_k} \\ &= \sum_{n=0}^{N-1}\left\{a_n^* \sum_{m=0}^{N-1} a_m \left[\sum_{k=-\infty}^{\infty} g^*(kT_s-nT-\varepsilon T)\, g^*_{\mathrm{MF}}(-kT_s+mT+\varepsilon T)\right]\right\} \end{aligned} \tag{4-96}$$

By the equivalence of digital and analog signal processing (4-28) the infinite convolution sum equals

$$\begin{aligned} &\sum_{k=-\infty}^{\infty} g(kT_s-mT-\varepsilon T)\, g_{\mathrm{MF}}(-kT_s+nT+\varepsilon T) \\ &= \frac{1}{T_s}\int_{-\infty}^{\infty} g(t-mT-\varepsilon T)\, g_{\mathrm{MF}}(-t+nT+\varepsilon T)\, dt \\ &= \frac{1}{T_s}\int_{-\infty}^{\infty} g(t-mT)\, g_{\mathrm{MF}}(-t+nT)\, dt \end{aligned} \tag{4-97}$$

From the second line of (4-97) follows that the sum is independent of the timing parameter ε. Introducing the compact notation

$$h_{m,n} = \sum_{k=-\infty}^{\infty} g(kT_s - mT)\, g_{\text{MF}}(-kT_s + nT) \tag{4-98}$$

we can write (4-96) in matrix-vector notation as

$$\mathbf{s}_f^H \mathbf{Q} \mathbf{s}_f = \mathbf{a}^H \mathbf{H}\, \mathbf{a} \tag{4-99}$$

with $h_{m,n} = [\mathbf{H}]_{m,n}$ an element of the $(N \times N)$ matrix $\mathbf{H}$. Notice that this quadratic form is independent of θ and ε. Thus, only the form $\mathbf{r}_f^H \mathbf{Q} \mathbf{s}_f$ plays a role in the optimization of the synchronization parameters.

Our derivation of the log-likelihood function is based on the Toepliz structure of the inverse covariance matrix. This property has allowed us to write d_k as a convolution sum (4-92). We have defined

$$\begin{aligned} &\lambda_{i,j} = [\mathbf{\Lambda}]_{i,j} \quad \text{element of the noise covariance matrix} \\ &q_{l,m} = [\mathbf{Q}]_{l,m} \quad \text{element of the inverse noise covariance matrix} \end{aligned} \tag{4-100}$$

Next we prove the asymptotic Toepliz structure of $\mathbf{Q}$.

Proof: For all M we have

$$\mathbf{\Lambda Q} = \mathbf{I} \tag{4-101}$$

where

$$b_{k,m} = \begin{cases} 1 & k = m \\ 0 & \text{else} \end{cases} \tag{4-102}$$

is an element of the unity matrix $\mathbf{I}$ with

$$b_{k,m} = \sum_{l} \lambda_{k,l}\, q_{l,m} \qquad |\,l\,| \le M \tag{4-103}$$

The covariance matrix $\mathbf{\Lambda}$ is a positive-definite Hermitian matrix. Furthermore, it is a Toepliz matrix. Using the Toepliz form of $\mathbf{\Lambda}$, (4-103) can be written as

$$b_{k,m} = \sum_{l} \lambda_{k-l}\, q_{l,m} \qquad |\,l\,| \le M \tag{4-104}$$

The asymptotic Toepliz form of the inverse matrix $\mathbf{Q}$ may best be proven by applying the z-transform method. The z-transform of $b_{k,m}$ (m fixed) of (4-104) is defined as

$$\begin{gathered} b(z,m) = \sum_k b_{k,m}\, z^{-k} = \sum_k z^{-k} \sum_l \lambda_{k-l}\, q_{l,m} \\ |\,l\,|,\ |\,k\,| \le M \end{gathered} \tag{4-105}$$

Assuming an infinite horizon, $M \to \infty$, we obtain by interchanging the order of summation

$$b(z,m) = \sum_{l=-\infty}^{\infty} q_{l,m}\, z^{-l} \sum_{k=-\infty}^{\infty} \lambda_{k-l}\, z^{-(k-l)} \tag{4-106}$$

Since k runs from $-\infty$ to ∞, the second factor is independent of l. It equals the (two sided) z-transform of the sampled and normalized correlation function $1/A^2\ R_n(kT_s)$:

$$\lambda(z) = \frac{1}{A^2} \sum_{k=-\infty}^{\infty} R_n(kT_s)\, z^{-k} = \frac{1}{A^2}\, S_n(z) \tag{4-107}$$

The sequence $\{b_{k,m}\}$ is nonzero for $k = m$ only. Hence, the z-transform equals

$$b(z,m) = z^{-m} \tag{4-108}$$

The z-transform of $q(z,m)$ is defined as

$$q(z,m) = \sum_{l=-\infty}^{\infty} q_{l,m}\, z^{-l} \tag{4-109}$$

Now, using (4-108) in (4-106) and solving this equation for $q(z,m)$ finally yields

$$q(z,m) = \frac{z^{-m}}{\lambda(z)} = \frac{z^{-m}}{S_n(z)/A^2} \tag{4-110}$$

We recall that $q(z,m)$ is the z-transform of the mth column of the matrix $\mathbf{Q}$. From the right-hand side of (4-110) and the shift theorem of the z-transform it follows that the columns $q_{l,m}$ are identical, apart from a shift m. The (asymptotic) Toepliz form of $\mathbf{Q}$ follows immediately (see Figure 4-20).

After these lengthy mathematical derivations all the results are available to gain some physical insight into the matched filter we formally introduced by (4-94). Previously (in Section 4.3.2) we learned that the output of the matched filter at any time can be generated by a time-invariant matched filter with impulse response $g_{\text{MF}}(kT_s)$ followed by a time-variant interpolator $h_I(kT_s, \mu_n)$. It suffices, therefore, to consider the time-invariant matched filter

$$g^*_{\text{MF}}(-kT_s) = \sum_{l=-\infty}^{\infty} q_{k-l}\, g(lT_s) \tag{4-111}$$

Much understanding can be gained by considering the matched filter in the frequency domain

$$G_{\text{MF}}\left(e^{j\omega T_s}\right) = q\left(e^{j\omega T_s}\right) G^*\left(e^{j\omega T_s}\right) \tag{4-112}$$

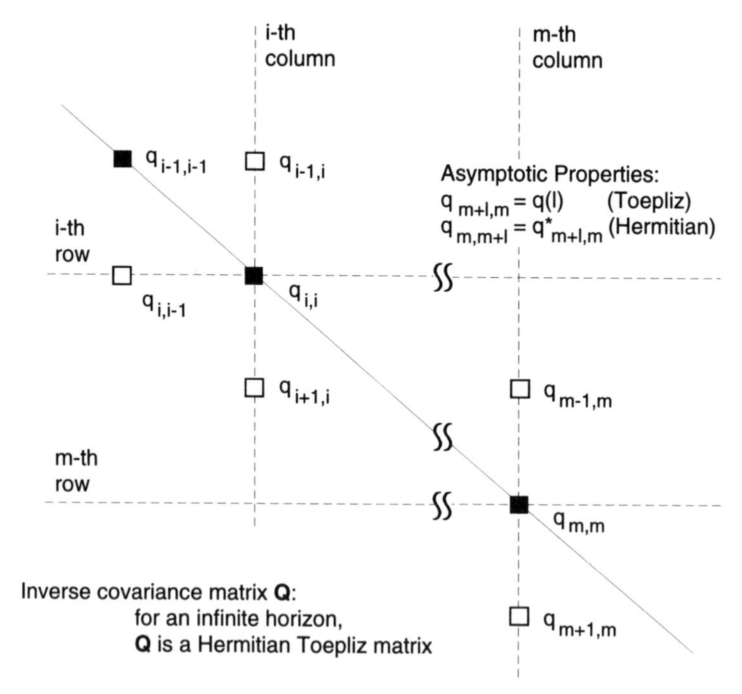

Figure 4-20 Inverse Covariance Matrix **Q**: For an Infinite Horizon, **Q** Is Hermitian and a Toepliz Matrix

where $q\left(e^{j\omega T_s}\right)$ is the Fourier transform of an arbitrary column of the inverse covariance matrix

$$\begin{aligned} q\left(e^{j\omega T_s}\right) &= \sum_{n=-\infty}^{\infty} q_{m+n,m}\; e^{-j\omega T_s n} \\ &= \sum_{n=-\infty}^{\infty} q(n)\; e^{-j\omega T_s n} \end{aligned} \tag{4-113}$$

Using the result (4-110) with $z = e^{j\omega T_s}$ we can replace $q\left(e^{j\omega T_s}\right)$ by

$$q\left(e^{j\omega T_s}\right) = \frac{1}{S_n\left(e^{j\omega T_s}\right)/A^2} \tag{4-114}$$

We notice that the power spectrum $S_n\left(e^{j\omega T_s}\right)$ is real and therefore $q\left(e^{j\omega T_s}\right)$ is real, too. Since $q\left(e^{j\omega T_s}\right)$ is the Fourier transform of $q(n)$, it follows that $q(-n) = q^*(n)$. Finally, from the Toepliz structure of **Q** and the previous observation, we obtain

$$q_{m+n,m} = q(n) \tag{4-115}$$

$$q_{n,m+n} = q^*(-n) \tag{4-116}$$

Hence,

$$q_{m+n,n} = q^*_{n,m+n} \tag{4-117}$$

which shows that **Q** is Hermitian (Figure 4-20).

Employing the Poisson theorem we can relate the power spectrum of the sampled process to that of the continuous-time process $n(t)$:

$$S_n\left(e^{j\omega T_s}\right) = \frac{1}{T_s} \sum_{k=-\infty}^{\infty} S_n\left(\omega - \frac{2\pi}{T_s}\, k\right) \tag{4-118}$$

where [eq. (4-62)]

$$S_n(\omega) = |F(\omega)|^2 \, S_w(\omega) \tag{4-119}$$

is the spectrum of the colored noise at the output of the analog prefilter $F(\omega)$. Replacing q by λ we finally obtain for the frequency response of the digital matched filter (Figure 4-21)

$$G_{\text{MF}}\left(e^{j\omega T_s}\right) = \frac{G^*\left(e^{j\omega T_s}\right)}{1/(T_s A^2) \sum\limits_{k=-\infty}^{\infty} S_n(\omega - 2\pi/T_s k)} \tag{4-120}$$

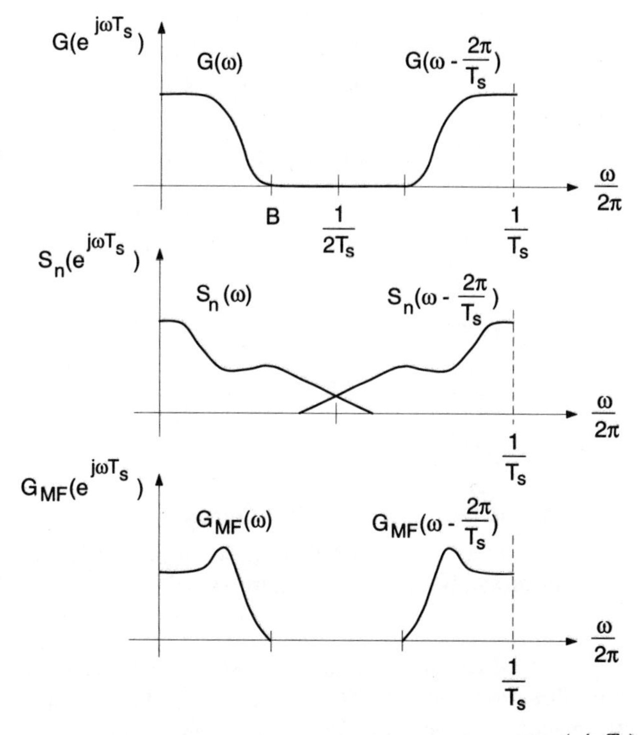

Figure 4-21 Matched Filter Transfer Function $G_{\text{MF}}\left(e^{j\omega T_s}\right)$

Notice that in the baseband

$$G_{\text{MF}}\left(e^{j\omega T_s}\right) = 0 \qquad \text{for} \quad \left|\frac{\omega}{2\pi}\right| > B \tag{4-121}$$

An interesting problem arises when the bandwidth B_f of the noise $n(t)$ is smaller than half the sampling rate: $B_f < (1/2)\ T_s$ (strict oversampling). From the right-hand side of (4-118) we learn that the power spectrum $\lambda\left(e^{j\omega T_s}\right)$ of the sampled noise process has gaps under this condition. As a consequence, $q\left(e^{j\omega T_s}\right)$ [eq. (4-114)] does not exist which in turn implies that the inverse covariance matrix **Q** becomes *singular*. However, for the definition of the matched filter this is irrelevant since only the frequency response of $F(\omega)$ in the interval $|\omega/2\pi| < B$ is of concern. Hence, $F(\omega)$ can be replaced by any filter which coincides with it in the above range and which is arbitrary but nonzero for $B \leq |\omega/2\pi| \leq 1/T_s - B$.

Remark: By construction, the output of the matched filter must be independent of the prefilter $F(\omega)$. If the opposite were true, $\{r_f(kT_s)\}$ would *not* be sufficient statistics. We can easily verify this conjecture. Since

$$\begin{aligned} G(\omega) &= G_T(\omega)C(\omega)F(\omega) \\ S_n(\omega) &= |F(\omega)|^2 S_w(\omega) \end{aligned} \tag{4-122}$$

we obtain (baseband)

$$G^*_{\text{MF}}\left(e^{j\omega T_s}\right) = \frac{[G_T(\omega)C(\omega)F(\omega)]}{1/A^2\ T_s|F(\omega)|^2 S_w(\omega)} \tag{4-123}$$

The useful signal at the MF output has been processed by a filter cascade with frequency response

$$\begin{aligned} H_{\text{tot}}\left(e^{j\omega T_s}\right) &= G_T(\omega)C(\omega)F(\omega)G_{\text{MF}}\left(e^{j\omega T_s}\right) \\ &= |G_T(\omega)|^2|C(\omega)|^2 \frac{1}{1/A^2 T_s S_w(\omega)} \end{aligned} \tag{4-124}$$

The power spectrum of the noise at the output of the matched filter equals (baseband)

$$\begin{aligned} S_{\text{MF}}\left(e^{j\omega T_s}\right) &= \left|G_{\text{MF}}\left(e^{j\omega T_s}\right)\right|^2 S_n\left(e^{j\omega T_s}\right) \\ &= \frac{|G_T(\omega)|^2|C(\omega)|^2}{S_w(\omega)} \frac{1}{1/A^2 T_s} \end{aligned} \tag{4-125}$$

Both outputs are independent of $F(\omega)$.

Signal-to-Noise Ratio at the Matched Filter Output

We compute the signal-to-noise (SNR) ratio when the output of the matched filter equals the sum of symbol plus white noise sample:

$$z_n = a_n + n_n \tag{4-126}$$

Obviously, a detector which has to deal with ISI and colored noise can do no better than under these idealized conditions. The SNR obtained thus serves as a bound on performance. For eq. (4-126) to be true the following conditions are sufficient (Figure 4-22):

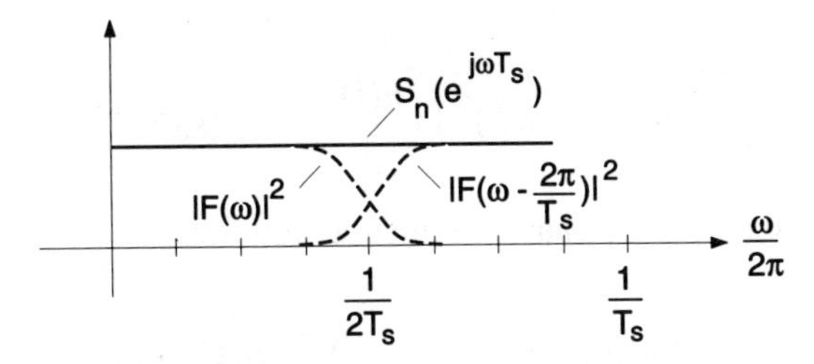

Figure 4-22 Power Spectrum of the Sampled Process $n(kT_s)$ When $|F(\omega)|^2$ Obeys Nyquist Conditions

(i $w(t)$ is a complex process with flat power spectral density N_0.

(ii The prefilter $|F(\omega)|^2$ obeys

$$\sum_k \left| F\left(\omega - \frac{2\pi}{T_s} k\right) \right|^2 = \text{const.} \tag{4-127}$$

(iii Nyquist pulses

$$\begin{aligned} h_{m,n} &= \sum_k g(kT_s - mT) g_{\text{MF}}(nT - kT_s) \\ &= \begin{cases} h_{0,0} & m = n \\ 0 & \text{else} \end{cases} \end{aligned} \tag{4-128}$$

From conditions (i) and (ii) it follows that the power spectrum of the sampled noise $n(kT_s)$ is white:

$$\begin{aligned} S_n\left(e^{j\omega T_s}\right) &= \frac{1}{T_s} \sum_{k=-\infty}^{\infty} \left| F\left(\omega - \frac{2\pi}{T_s} k\right) \right|^2 N_0 \\ &= \frac{N_0}{T_s} \qquad |F(0)| = 1 \end{aligned} \tag{4-129}$$

The variance equals

$$\begin{aligned} \sigma_n^2 &= \frac{T_s}{2\pi} \int_{-\pi/T_s}^{\pi/T_s} S_n\left(e^{j\omega T_s}\right) d\omega \\ &= \frac{N_0}{T_s} \end{aligned} \tag{4-130}$$

From the definition of $G_{\text{MF}}\left(e^{j\omega T_s}\right)$, eq. (4-120), and the previous equation we obtain

$$G_{\text{MF}}\left(e^{j\omega T_s}\right) = \frac{G^*\left(e^{j\omega T_s}\right)}{N_0/(T_s\, A^2)} \tag{4-131}$$

Using

$$r_f(t) = \sum_{n=0}^{N-1} a_n\, g(t-nT) + \frac{n(t)}{A} \qquad (\varepsilon_0 = 0) \tag{4-132}$$

the matched filter output $z(0)$ can be written as follows:

$$\begin{aligned} z(0) &= a_0 \sum_{k=-\infty}^{\infty} g(kT_s)\, g_{\mathrm{MF}}(-kT_s) + \frac{1}{A} \sum_{k=-\infty}^{\infty} n(kT_s)\, g_{\mathrm{MF}}(-kT_s) \\ &= a_0\, h_{0,0} + n_g(0) \end{aligned} \tag{4-133}$$

Due to the Nyquist conditions there is no intersymbol interference, provided we have perfect synchronism, i.e., $\varepsilon_0 = \varepsilon = 0$. [Without any loss of generality we have assumed $n = 0$ in (4-133)].

The variance of the filtered noise

$$n_g(iT) = \frac{1}{A} \sum_{k=-\infty}^{\infty} n(kT_s)\, g_{\mathrm{MF}}(-kT_s + iT) \tag{4-134}$$

is found to be

$$\begin{aligned} \sigma_{n_g}^2 &= \frac{\sigma_n^2}{A^2} \frac{1}{N_0/(A^2 T_s)}\, h_{0,0} \\ &= h_{0,0} \end{aligned} \tag{4-135}$$

The constant $h_{0,0}$ is related to the energy of the pulse $g(t)$ by the Parseval theorem:

$$\begin{aligned} h_{0,0} &= \sum_{k=-\infty}^{\infty} g(kT_s)\, g_{\mathrm{MF}}(-kT_s) \\ &= \left(\frac{A^2\, T_s}{N_0}\right) \sum_{k=-\infty}^{\infty} g(kT_s)\, g^*(kT_s) \\ &= \frac{A^2}{N_0} \int_{-\infty}^{+\infty} g(t)\, g^*(t)\, dt = \frac{A^2 E_g}{N_0} \end{aligned} \tag{4-136}$$

Inserting (4-136) into $\sigma_{n_g}^2$ yields

$$\sigma_{n_g}^2 = \frac{A^2 E_g}{N_0} \tag{4-137}$$

The average energy of the useful signal

$$a_0 \sum_{k=-\infty}^{\infty} g(kT_s)\, g_{\mathrm{MF}}(-kT_s) = a_0\, h_{0,0} \tag{4-138}$$

is

$$E_{\text{av}} = E\left[|a_n|^2\right] |h_{0,0}|^2 \tag{4-139}$$

The ratio of signal to noise is therefore

$$\text{SNR}_{\text{MF}} = \frac{E\left[|a_n|^2\right] A^2 \, E_g}{N_0} \tag{4-140}$$

The SNR is usually written in a different form. The average *energy per symbol* equals

$$E_s = E\left[|a_n|^2\right] A^2 \, E_g \tag{4-141}$$

so that the SNR is given by

$$\text{SNR}_{\text{MF}} = \frac{E_S}{N_0} \tag{4-142}$$

Notice that the result is independent of the sampling rate $1/T_s$. The result is identical to that we would have received for an analog matched filter, again confirming the equivalence of analog and digital signal processing.

4.3.4 Main Points

- *MAP receiver*
  The optimal MAP receiver has no separate synchronization unit. The synchronization parameters are considered as *unwanted* parameters which are removed from the pdf by averaging

$$p(\mathbf{a} \mid \mathbf{r}_f) \propto P(\mathbf{a}) \int p(\mathbf{r}_f \mid \mathbf{a}, \theta, \varepsilon) p(\theta) p(\varepsilon) \, d\theta \, d\varepsilon \tag{4-143}$$

- *ML receiver*
  The ML receiver jointly estimates the synchronization parameters and the data sequence. The receiver comprises a digital matched filter, a time-variant interpolator and decimator, a phase rotation unit and a data sequence estimator. The matched filter operation operates on samples $r_f(kT_s)$. The sample value at the correct sampling instant is obtained by digital interpolation of the matched filter output. The digital interpolator performs the function of a timing recovery circuit. The data sequence estimator and phase recovery circuit operate with one sample per symbol. The samples $z(nT + \varepsilon T)$ ($1/T$ symbol rate) are obtained from the matched filter output by a time-variant decimator.

- *Matched filter*
  Frequency response

$$G_{\mathrm{MF}}\left(e^{j\omega T_s}\right) = \frac{G^*\left(e^{j\omega T_s}\right)}{1/(T_s A^2)\sum\limits_k S_n\left(\omega - \frac{2\pi}{T_s}k\right)} \tag{4-144}$$

with

$$G\left(e^{j\omega T_s}\right) = \frac{1}{T_s}\sum_m G\left(\omega - \frac{2\pi}{T_s}m\right) \tag{4-145}$$

Signal-to-noise ratio at the matched filter output

$$\mathrm{SNR}_{\mathrm{MF}} = \frac{E_s}{N_0} \tag{4-146}$$

with symbol energy

$$E_s = E\left[|a_n|^2\right] A^2 \int\limits_{-\infty}^{\infty} |g(t)|^2 dt \tag{4-147}$$

The output of the matched filter is independent of the sampling rate $1/T_s$ and the frequency response of the generalized anti-aliasing filter $F(\omega)$.

4.3.5 Bibliographical Notes

The classical papers on optimum receivers were published in the 1970s [2]–[5]. They all deal with time-continuous (analog) signal processors. The problem was reconsidered in [6] by restricting the problem to band-limited signals as the basis for (equivalent) optimum receivers employing discrete-time (digital) signal processing. One of the first hardware implementations of a fully digital receiver is reported in [7].

4.3.6 Appendix

Normalized Equations for Nyquist Pulses and Symmetric Prefilters $|F(\omega)|^2$

Inserting

$$r_f(kT_s) = \sum_n a_n\, g(kT_s - nT - \varepsilon_0 T)\, e^{j\theta_0} + \frac{n(kT_s)}{A} \tag{4-148}$$

[eq. (4-6)] into

$$z(nT + \varepsilon T) = \sum_k r_f(kT_s)\, g_{\mathrm{MF}}(nT + \varepsilon T - kT_s) \tag{4-149}$$

[eq. (4-85)] yields

$$\begin{aligned} z(nT+\varepsilon T) &= \sum_k \left[\sum_m a_m\, g(kT_s - mT - \varepsilon_0 T) e^{j\theta_0} + \frac{n(kT_s)}{A} \right] \\ &\quad \times g_{\mathrm{MF}}(nT + \varepsilon T - kT_s) \\ &= \sum_n a_n\, h_{m,n}(\varepsilon - e^{j\theta_0}) + n_g(nT) \end{aligned} \tag{4-150}$$

where $h_{m,n}(\varepsilon - \varepsilon_0)$ is defined by

$$h_{m,n}(\varepsilon - \varepsilon_0) = \sum_k g\,(kT_s - mT - \varepsilon_0 T)\, g_{\mathrm{MF}}(nT + \varepsilon T - kT_s) \tag{4-151}$$

and n_g is the noise process

$$n_g(nT) = \sum_k n(kT_s)\, g_{\mathrm{MF}}(nT + \varepsilon T - kT_s) \tag{4-152}$$

For Nyquist pulses and symmetrical prefilter (conditions 4-127) the noise $n_g(nT)$ is white with variance

$$\sigma^2_{n_g} = \frac{A^2 E_g}{N_0} \tag{4-153}$$

The function $h_{m,n}(\varepsilon - \varepsilon_0)$ for $\varepsilon - \varepsilon_0 = 0$ becomes

$$h_{m,n}(\varepsilon - \varepsilon_0 = 0) = \begin{cases} h_{0,0} & n = m \\ 0 & \text{else} \end{cases} \tag{4-154}$$

Consequently, the matched filter output sample taken at the correct instant $nT + \varepsilon_0 T$ exhibits no ISI.

Normalizing eq. (4-150) by $E\left[|a_n|^2\right] h_{0,0}(0)$ we obtain after some simple algebra (using the quantities defined in Section 4.3.6)

$$\overline{z}(nT + \varepsilon T) = \sum_n \overline{a}_n\, \overline{h}_{m,n}(\varepsilon - \varepsilon_0)\, e^{j\theta_0} + \overline{N}(nT) \tag{4-155}$$

with

$$E\left[|\bar{a}_n|^2\right] = 1 \quad \overline{h}_{0,0}(0) = 1 \quad \sigma^2_{\overline{N}} = \frac{E_s}{N_0} \tag{4-156}$$

The (normalized) likelihood function equals

$$p(\mathbf{r_f}|\mathbf{a}, \theta, \varepsilon|) = \text{const} \times \exp\left\{ -\frac{E_s}{N_0} \sum_n \left[|\overline{a}_n|^2 - 2\,\mathrm{Re}[\overline{a}_n^* \overline{z}(nT + \varepsilon T)] e^{-j\theta} \right] \right\} \tag{4-157}$$

Bibliography

[1] H. L. van Trees, *Detection, Estimation, and Modulation Theory*. New York: Wiley, 1968.

[2] H. Kobayashi, "Simultaneous Adaptive Estimation and Detection Algorithm for Carrier Modulated Data Transmission Systems," *IEEE Trans. Commun.*, vol. COM-19, June 1971.

[3] G. D. Forney, "Maximum-Likelihood Sequence Estimation of Digital Sequences in the Presence of Intersymbol Interference," *IEEE Trans. Inform. Theory*, vol. IT-18, pp. 363–378, May 1972.

[4] G. Ungerboeck, "Adaptive Maximum-Likelihood Receiver for Carrier-modulated Data Transmission Systems," *IEEE Trans. Commun.*, vol. COM-22, May 1974.

[5] D. D. Falconer, "Optimal Reception of Digital Data over the Gaussian Channel with Unknown Delay and Phase Jitter," *IEEE Trans. Inform. Theory*, vol. IT-23, Jan. 1977.

[6] H. Meyr, M. Oerder, and A. Polydoros, "On Sampling Rate, Analog Prefiltering, and Sufficient Statistics for Digital Receivers," *IEEE Trans. Commun.*, vol. 42, pp. 3208–3214, Dec. 1994.

[7] G. Ascheid, M. Oerder, J. Stahl, and H. Meyr, "An All Digital Receiver Architecture for Bandwidth Efficient Transmission at High Data Rates," *IEEE Trans. Commun.*, vol. COM-37, pp. 804–813, Aug. 1989.

Chapter 5 Synthesis of Synchronization Algorithms

In this chapter we derive maximum-likelihood (ML) synchronization algorithms for time and phase. Frequency estimation and synchronization will be treated in Chapter 8.

The algorithms are obtained as the solution to a mathematical optimization problem. The performance criterion we choose is the ML criterion. In analogy to filter design we speak of *synthesis of synchronization algorithms* to emphasize that we use mathematics to find algorithms – as opposed to analyzing their performance.

5.1 Derivation of ML Synchronization Algorithms

Conceptually, the systematic derivation of ML synchronizers is straightforward. The likelihood function must be averaged over the unwanted parameters. For example,

joint estimation of (θ, ε) :

$$p(\mathbf{r}_f | \theta, \varepsilon) = \sum_{\text{all sequences } \mathbf{a}} P(\mathbf{a})\, p(\mathbf{r}_f | \mathbf{a}, \theta, \varepsilon)$$

phase estimation:

$$p(\mathbf{r}_f | \theta) = \int \left[\sum_{\text{all sequences } \mathbf{a}} P(\mathbf{a})\, p(\mathbf{r}_f | \mathbf{a}, \theta, \varepsilon)\, p(\varepsilon) \right] d\varepsilon \tag{5-1}$$

timing estimation:

$$p(\mathbf{r}_f | \varepsilon) = \int \left[\sum_{\text{all sequences } \mathbf{a}} P(\mathbf{a})\, p(\mathbf{r}_f | \mathbf{a}, \theta, \varepsilon)\, p(\theta) \right] d\theta$$

With the exception of a few isolated cases it is not possible to perform these averaging operations in closed form, and one has to resort to approximation techniques. Systematically deriving synchronization algorithms may therefore be understood as the task of finding suitable approximations. The various algorithms are then the result of applying these techniques which can be systematic or ad hoc. The synchronizers can be classified into two main categories:

1. Class DD/DA: Decision-directed (DD) or data-aided (DA)
2. Class NDA: Non-data-aided (NDA)

The classification emerges from the way the data dependency is eliminated. When the data sequence is known, for example a preamble $\mathbf{a}_0$ during acquisition, we speak of *data-aided* synchronization algorithms. Since the sequence $\mathbf{a}_0$ is known, only one term of the sum of eq. (5-1) remains. The joint (θ, ε) estimation rule thus reduces to maximizing the likelihood function $p(\mathbf{r}_f \mid \mathbf{a} = \mathbf{a}_0, \theta, \varepsilon)$:

$$\left(\hat{\theta}, \hat{\varepsilon}\right)_{\text{DA}} = \arg \max_{\theta, \varepsilon} \; p(\mathbf{r}_f \mid \mathbf{a} = \mathbf{a}_0, \theta, \varepsilon) \tag{5-2}$$

When the detected sequence $\hat{\mathbf{a}}$ is used as if it were the true sequence one speaks of *decision-directed* synchronization algorithms. When the probability is high that $\hat{\mathbf{a}}$ is the true sequence $\mathbf{a}_0$, then only one term contributes to the sum of eq. (5-1):

$$\sum_{\text{all sequences } \mathbf{a}} P(\mathbf{a}) p(\mathbf{r}_f | \mathbf{a}, \theta, \varepsilon) \simeq p(\mathbf{r}_f | \mathbf{a} = \hat{\mathbf{a}}, \theta, \varepsilon) \quad (P(\mathbf{a} = \hat{\mathbf{a}}) \simeq 1) \tag{5-3}$$

Thus

$$\left(\hat{\theta}, \hat{\varepsilon}\right)_{\text{DD}} = \arg \max_{\theta, \varepsilon} \; p(\mathbf{r}_f | \mathbf{a} = \hat{\mathbf{a}}, \theta, \varepsilon) \tag{5-4}$$

All DD algorithms require an initial parameter estimate before starting the detection process. To obtain a reliable estimate, one may send a preamble of known symbols.

Class NDA algorithms are obtained if one actually performs (exactly or approximately) the averaging operation.

Example: NDA for BPSK with i.i.d. symbols

$$p(\mathbf{r}_f \mid \theta, \varepsilon) = \prod_{n=0}^{N-1} \left[p(\mathbf{r}_f \mid a_n = 1, \theta, \varepsilon) P(a_n = 1) + p(\mathbf{r}_f \mid a_n = -1, \theta, \varepsilon) P(a_n = -1)\right] \tag{5-5}$$

An analogous classification can be made with respect to the synchronization parameters to be eliminated. For example,

(DD&Dε) : data- and timing directed:

$$p(\mathbf{r}_f | \theta) = p(\mathbf{r}_f | \mathbf{a} = \hat{\mathbf{a}}, \theta, \varepsilon = \hat{\varepsilon})$$

DD, timing independent: (5-6)

$$p(\mathbf{r}_f | \theta) = \int p(\mathbf{r}_f | \mathbf{a} = \hat{\mathbf{a}}, \theta, \varepsilon) \, p(\varepsilon) \, d\varepsilon$$

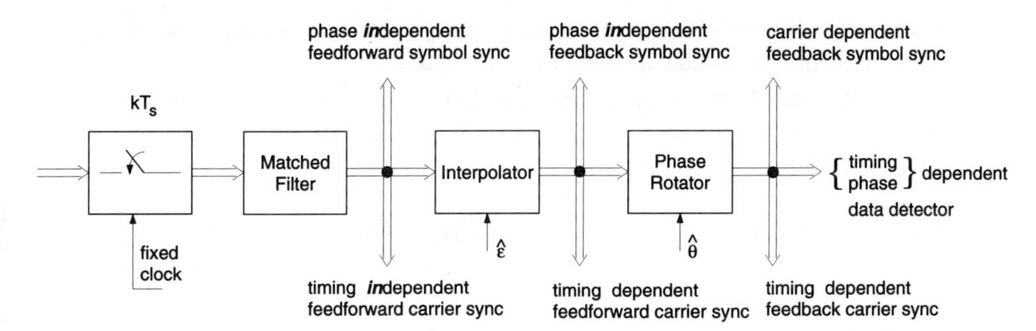

Figure 5-1 Feedforward (FF) and Feedback (FB) Synchronization Algorithms

and

$(\mathrm{DD\&D}\theta)$: data- and phase directed:

$$p(\mathbf{r}_f|\varepsilon) = p\left(\mathbf{r}_f|\mathbf{a} = \hat{\mathbf{a}}, \theta = \hat{\theta}, \varepsilon\right)$$

DD, phase independent: (5-7)

$$p(\mathbf{r}_f|\varepsilon) = \int p(\mathbf{r}_f|\mathbf{a} = \hat{\mathbf{a}}, \theta, \varepsilon)\, p(\theta)\, d\theta$$

The algorithms can be further categorized according to how the timing/phase estimates are extracted from the received signal. We distinguish between algorithms which directly estimate the unknown parameters $(\theta_0, \varepsilon_0)$. Such algorithms are called *feedforward* (FF) because the estimate is derived from the received signal *before* it is corrected in the interpolator (for timing) or the phase rotator (for carrier recovery).

The other category derives an error signal $\hat{e}_\theta = \hat{\theta} - \theta_0$ and $\hat{e}_\varepsilon = \hat{\varepsilon} - \varepsilon_0$, respectively. These algorithms are called *feedback* (FB) because they derive an estimate of the error and feed a corrective signal back to the interpolator or phase rotator, respectively. Feedback structures inherently have the ability to automatically track slowly varying parameter changes. They are therefore also called error-feedback synchronizers.

For illustration, in Figure 5-1 a typical block diagram of a digital receiver is sketched together with the various signals required for FF or FB algorithms. Notice that the position of the various blocks may be interchanged, depending on the application. For example, interpolator and rotator may be interchanged.

When deriving a synchronization algorithm from the ML criterion, one assumes an idealized channel model (to be discussed shortly) and constant parameters, at least for quasi-static channels. In principle, a more realistic channel model and time-variable parameters could be taken into account, but it turns out that this approach is mathematically far too complicated. In view of the often crude approximations made to arrive at a synchronizer algorithm, it makes no sense to consider accurate channel models. Instead, one derives synchronization algorithms under idealized conditions and later on analyzes the performances of these algorithms when used in conjunction with realistic channels.

We assume Nyquist pulses and a prefilter $|F(\omega)|^2$ being symmetric about $1/2T_s$. In this case the likelihood function [see eqs. (4-83), (4-157)] assumes the simplified form

$$p(\mathbf{r}_f \mid \mathbf{a}, \varepsilon, \theta) \propto \exp\left\{ -\frac{1}{\sigma_n^2}\left[2\,\mathrm{Re}\left[\sum_{n=0}^{N-1} |h_{0,0}|^2 |a_n|^2 - 2a_n^* z_n(\varepsilon) e^{-j\theta} \right]\right]\right\} \tag{5-8}$$

with the short-hand notation $z_n(\varepsilon) = z(nT + \varepsilon T)$.

We said earlier that synchronization algorithms can be systematically derived by finding suitable approximations to remove the "unwanted" parameters in the ML function. The result of these approximations is a function $L(\boldsymbol{\theta})$, where $\boldsymbol{\theta}$ denotes the set of parameters to be estimated. The estimate $\hat{\boldsymbol{\theta}}$ is defined as the argument for which $L(\boldsymbol{\theta})$ assumes an extremum. Depending on the definition of $L(\boldsymbol{\theta})$, the extremum can be either a minimum or a maximum:

$$\hat{\boldsymbol{\theta}} = \underset{\boldsymbol{\theta}}{\arg\ \mathrm{extr}}\ L(\boldsymbol{\theta}) \tag{5-9}$$

Strictly speaking, $\hat{\boldsymbol{\theta}}$ is an ML estimate only if the objective function $L(\boldsymbol{\theta})$ is the ML function $p(\mathbf{r}_f \mid \boldsymbol{\theta})$. However, for convenience we frequently speak of ML estimation also in the case that $L(\boldsymbol{\theta})$ is only an approximation to $p(\mathbf{r}_f \mid \boldsymbol{\theta})$.

A first approximation of the likelihood function (5-8) is obtained for large values of N. We have shown in Chapter 4 that the inner product $\mathbf{s}_f^H \mathbf{s}_f = \sum_n |h_{0,0}|^2 |a_n|^2$ is independent of the synchronization parameters. For a sufficiently large N the sum

$$y_N = \sum_{n=0}^{N-1} |a_n|^2 \tag{5-10}$$

is closely approximated by its expected value. This is a consequence of the law of large numbers. We can therefore discard the term $\sum_n |a_n|^2 \rightarrow \sum E\left[|a_n|^2\right] =$ const. from maximization to obtain the objective function

$$L(\mathbf{a}, \theta, \varepsilon) = \exp\left\{ -\frac{2}{\sigma_n^2}\,\mathrm{Re}\left[\sum_{n=0}^{N-1} a_n^* z_n(\varepsilon)\, e^{-j\theta} \right]\right\} \tag{5-11}$$

Several important conclusions can be drawn from this objective function. In most digital receivers timing recovery is done prior to phase recovery. The reason becomes clear by inspection of (5-11). Provided timing is known, *one* sample per symbol of the matched filter output is sufficient for carrier phase estimation and symbol detection. To minimize the computational load in the receiver, carrier phase estimation and correction must be made at the lowest sampling rate possible, which is the symbol rate $1/T$. All digital algorithms for phase estimation to be derived later on will therefore be of the $D\varepsilon$ type running at symbol rate $1/T$. They will be either DD (DA) or NDA.

While the number of different carrier phase estimation algorithms is small, there exists a much larger variety of digital algorithms for timing recovery. This is due to the fact that the number of degrees of freedom in deriving an algorithm is much larger. Most importantly, the sampling rate $1/T_s$ to compute $z_n(\varepsilon)$ can be chosen independently of the symbol rate. At the one extreme the samples $z_n(\varepsilon) = z(nT + \varepsilon T)$ can be obtained by synchronously sampling the output of an analog matched filter $z(t)$ at $t = nT + \varepsilon T$. A digital error feedback algorithm running at rate $1/T$ is used to generate an error signal for the control of an analog voltage-controlled oscillator (VCO) in this hybrid timing recovery system. Using a higher sampling rate $1/T_s > 1/(T(1+\alpha))$ (α: excess bandwidth), the matched filter can be implemented digitally. The samples $z_n(\varepsilon)$ are then obtained at the output of a decimator $z_n(\varepsilon) = z(m_n T_s + \mu_n T_s)$. Timing recovery is performed by a digital error feedback system (FB) or direct estimation (FF) of the timing parameter ε and subsequent digital interpolation. All of them – DD, DA, and NDA – methods are of practical interest.

5.2 Estimator Structures for Slowly Varying Synchronization Parameters

5.2.1 Time Scale Separation

Discussing the optimal receiver in Section 4.3.1 we modeled the synchronization parameters as random processes. The time scale on which changes of these parameters are observable is much coarser than the symbol rate $1/T$. It is therefore reasonable to separate these time scales. We consider a first time scale which operates with units of the symbol duration T to deal with the detection process. The second (slow) time scale deals with the time variation of the parameters $\varepsilon(nT)$ and $\theta(nT)$ with time constants being much larger than T. We can thus consider the synchronization parameters as approximately piecewise constant and estimate these parameters over segments of $M \gg 1$ symbols. The number M of symbols over which the synchronization parameters can be assumed to be approximately constant depends on the stability of the oscillators and the frequency offset.

If the variance of the short-term estimate is much larger than the variance of the synchronization parameter, processing the short-term estimate in a postprocessor which takes advantage of the statistical dependencies between the estimates yields a performance improvement. We have thus arrived at a first practical two-stage structure as shown in Figure 5-2.

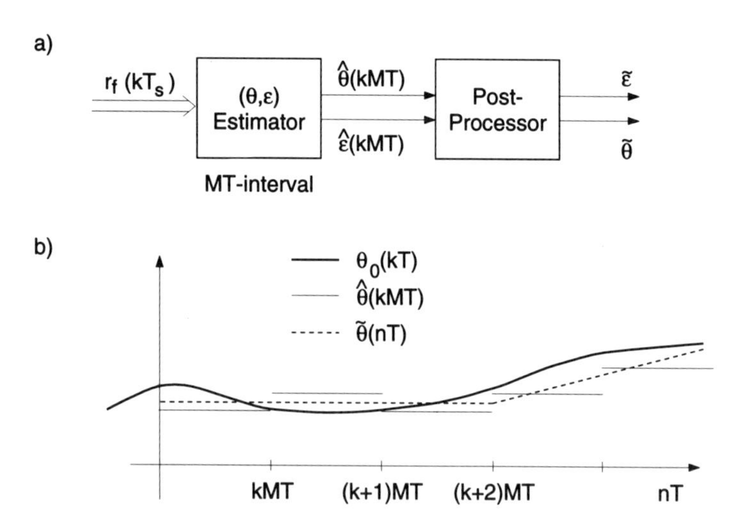

Figure 5-2 (a) Two-Stage Estimator Structure, (b) Time Variation of Parameter

The two-stage estimator has several advantages from a digital signal processing point-of-view. Computation of ML estimates basically is an inner product (feedforward) computation which can be processed in parallel and thus is suitable for VLSI implementation. The more complex processing in the postprocessing stage then runs at a lower rate of $1/(MT)$ compared to $1/T_s > 1/T$ in the first stage.

An alternative to estimating a slowly varying parameter is to generate an estimate of the *error* and use this estimate as an error signal in a feedback system (Figure 5-3).

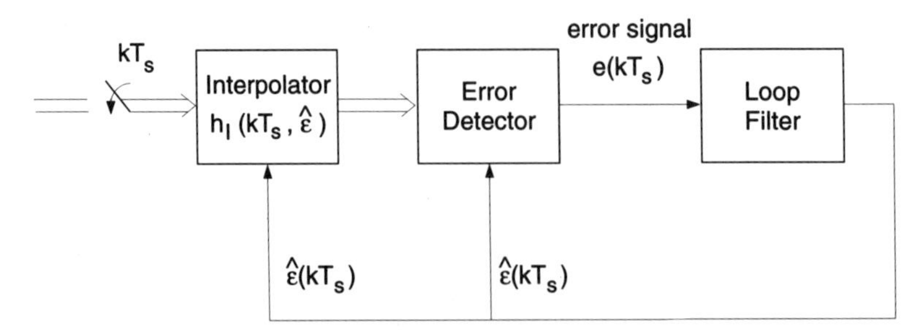

Figure 5-3 Timing Error Feedback System

5.2.2 Truncation and L Symbols Approximation of the Likelihood Estimator[5]

In this section we consider approximations to the optimal estimator with respect to

(i) Estimation of the synchronization parameters using only a subset of $L < N$ symbols out of N transmitted ones

(ii) Estimation of the synchronization parameters using a finite observation interval $\{r_f(kT_s)\},\ a \le k \le b$

In many practical applications we estimate the parameters (θ, ε) over L symbols in a continuous stream of N symbols, $L \ll N$ [case (i)]. The observation interval L is limited by the fact that the parameters are not strictly constant but slowly time variable. L is then chosen such that

$$LT < \tau_\theta, \tau_\varepsilon \tag{5-12}$$

where $\tau_\theta, \tau_\varepsilon$ are the correlation time of the respective processes. Following (4-83) the log-likelihood function for L symbols reads

$$\begin{aligned} L(\mathbf{r}_f \mid \mathbf{a}_L, \theta, \varepsilon) = & 2\ \mathrm{Re}\left[\sum_{n=J}^{J+(L-1)} a_n^*\ z_n(\varepsilon)\ e^{-j\theta}\right] \\ & - \sum_{n=J}^{J+(L-1)} \sum_{l=J}^{J+(L-1)} a_n^*\ a_l\ h_{l,n} \end{aligned} \tag{5-13}$$

Note that the entire received signal $\mathbf{r}_f$ is processed. Only the data vectors $\mathbf{a}$ and $\mathbf{z}$ are replaced by $\mathbf{a}_L$ and $\mathbf{z}_L$, respectively. This remark is important since truncation of the received signal $\mathbf{r}_f$ leads to entirely different degradations [case (ii)] as will be discussed shortly.

The approximation error of choosing a finite interval L among a number of N transmitted symbols is determined next. The received signal equals the sum of useful signal $s(t, \boldsymbol{\theta}_0)$ and noise $n(t)$. Notice that the parameter values $\boldsymbol{\theta}_0 = (\theta_0, \varepsilon_0)$ are the true values to be clearly distinguished from the trial parameters (θ, ε). Inserting $s_f(t) + n(t)$ into (4-71) yields

$$\begin{aligned} z(nT + \varepsilon T) = \sum_{k=-\infty}^{\infty} \Bigg\{ & \underbrace{\left[\sum_{m=0}^{N-1} a_m\ g(kT_s - mT - \varepsilon_0 T)\ e^{j\theta_0} + n(kT_s)\right]}_{r_f(kT_s)} \\ & \times g_{\mathrm{MF}}(-kT_s + nT + \varepsilon T) \Bigg\} \end{aligned} \tag{5-14}$$

[5] Can be omitted at first reading.

Interchanging the summation over k and m yields [again with the short-hand notation $z_n(\varepsilon) = z(nT + \varepsilon T)$]

$$z_n(\varepsilon) = \sum_{m=0}^{N-1} a_m \sum_{k=-\infty}^{\infty} g(kT_s - mT - \varepsilon_0 T)\, g_{\mathrm{MF}}(-kT_s + nT + \varepsilon T)\, e^{j\theta_0} + n_g(nT) \tag{5-15}$$

with the additive noise term

$$n_g(nT) = \sum_{k=-\infty}^{\infty} n(kT_s)\, g_{\mathrm{MF}}(-kT_s + nT + \varepsilon T) \tag{5-16}$$

We note from the last two equations that the entire signal $\mathbf{r}_f$ is processed as expressed by the infinite sum over k. We recognize in the infinite sum the output of the matched filter excited by the pulse $g(kT_s)$. Repeating the steps leading to $h_{m,n}$ of (4-98) we define $h_{m,n}(\varepsilon - \varepsilon_0)$:

$$h_{m,n}(\varepsilon - \varepsilon_0) = \sum_{k=-\infty}^{\infty} g(kT_s - mT - \varepsilon_0 T)\, g_{\mathrm{MF}}(-kT_s + nT + \varepsilon T) \tag{5-17}$$

which depends on the error $(\varepsilon_0 - \varepsilon)$ between the true value and its trial parameter. Notice that for $\varepsilon_0 - \varepsilon = 0$ we recover $h_{m,n}$ of (4-98).

In the L symbol log-likelihood function only the values for $0 \le J \le n < (J+L-1) \le (N-1)$ of $z(nT + \varepsilon T)\, a_n^*$ are summed up. We obtain for the observation-dependent term of (5-13)

$$\sum_{n=J}^{J+(L-1)} a_n^* \sum_{m=0}^{N} a_m h_{m,n}(\varepsilon - \varepsilon_0)\, e^{-j(\theta-\theta_0)} + \sum_{n=J}^{J+(L-1)} a_n^*\, n_g(nT) \tag{5-18}$$

The double sum of the previous equation can be written as a sum over L symbols plus a residue:

$$\begin{aligned} &\sum_{n=J}^{J+(L-1)} \sum_{m=J}^{J+(L-1)} a_n^* a_m h_{m,n}(\varepsilon - \varepsilon_0)\, e^{-j(\theta-\theta_0)} \\ &+ \sum_{n=J}^{J+(L-1)} \sum_{m=0}^{J-1} a_n^* a_m h_{m,n}(\varepsilon - \varepsilon_0)\, e^{-j(\theta-\theta_0)} \\ &+ \sum_{n=J}^{J+(L-1)} \sum_{m=J+L}^{N} a_n^* a_m h_{m,n}(\varepsilon - \varepsilon_0)\, e^{-j(\theta-\theta_0)} \end{aligned} \tag{5-19}$$

We recognize in the first double sum exactly the term that would appear in the log-likelihood function if L symbols were transmitted in total. The second two sums are interfering terms caused by symbols outside the considered L interval. They represent *self-noise* which is present even in the absence of additive noise.

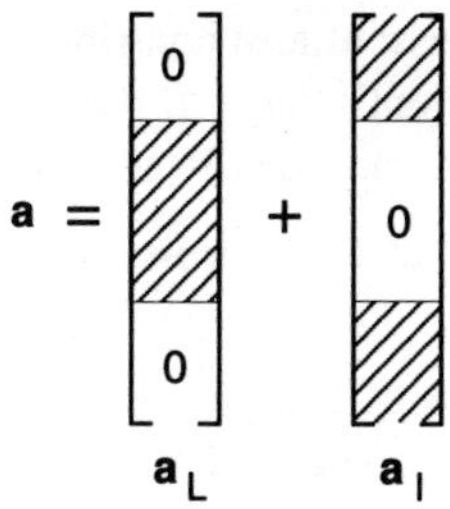

Figure 5-4 Symbol Vector a

Using vector matrix notation the contribution of the self-noise can be seen more clearly. We write the symbol vector $\mathbf{a}$ comprising N symbols as the sum of two disjoint vectors,

$$\mathbf{a} = \mathbf{a}_L + \mathbf{a}_l \tag{5-20}$$

where $\mathbf{a}_L$ contains the symbols for $J \leq n \leq (J+L-1)$ and $\mathbf{a}_l$ the remaining symbols (Figure 5-4).

Leaving aside $e^{-j(\theta-\theta_0)}$, the double sum of (5-19) can then concisely be written in the form

$$\mathbf{a}_L^H \mathbf{H} \mathbf{a}_L + \mathbf{a}_L^H \mathbf{H} \mathbf{a}_l \tag{5-21}$$

The projection of $\mathbf{H}\mathbf{a}_l$ on $\mathbf{a}_L^H$ is marked in Figure 5-5. It is seen that the self-noise contribution depends on the amount of intersymbol interference (ISI) as evidenced by the nonzero off-diagonal terms in the matrix. Let us denote by D

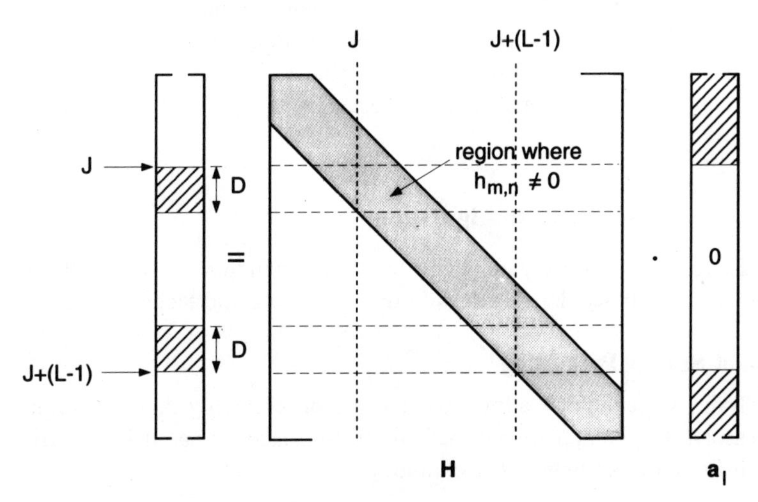

Figure 5-5 Projection of $\mathbf{H}\mathbf{a}_l$ on $\mathbf{a}_L^H$

the number of symbols which essentially contribute to the self-noise. This number is independent of L. For sufficiently large L the self-noise becomes negligible and the performance approaches asymptotically that of the optimal ML estimator.

Further, one recognizes that for Nyquist pulses there is no self-noise contribution for $\varepsilon = \varepsilon_0$.

In case (ii), only a subset of samples $\{r_f(kTs)\}$, $J \le k \le (J+ML-1)$ is processed. The factor M is the integer of the ratio $[T/T_s]_{\text{INT}} = M$. The interval then approximately equals that of received L symbols.

The truncated log-likelihood function is obtained from (4-95) by restricting the summation from J to $(J+LM-1)$:

$$\begin{aligned} &L\left([\mathbf{r}_f]_{\text{trunc}} \mid \mathbf{a}_L, \theta, \varepsilon\right) \\ &=2 \operatorname{Re}\left[\sum_{k=J}^{J+LM-1} r_f^*(kT_s)\, d(kT_s)\right] - \sum_{n=J}^{J+(L-1)} \sum_{l=J}^{J+(L-1)} a_n^*\, a_l\, h_{l,n} \end{aligned} \tag{5-22}$$

with

$$d(kT_s) = \sum_{n=0}^{N-1} a_n\, g_{\text{MF}}^*(-kT_s + nT + \varepsilon T)\, e^{+j\theta} \tag{5-23}$$

In the second term the summation is restricted to L symbols corresponding to the truncated measurement interval. It is the same term as in case (i), eq. (5-13). The first term comprises short-term correlation of segment $[\mathbf{r}_f]_L$ with $[d(kT_s)]_L$. It is important to realize that the L-segment correlation is not equivalent to the matched filter method [compare with (5-13)]. While in both cases a subset of L symbols is considered, in the matched filter approach the entire received signal is processed. Performing the correlation operation defined above, an additional term is introduced by truncating the received signal to *LM* samples. Thus, correlation in general is inferior to matched filtering.

5.2.3 Maximum Search Algorithm

There exist a variety of algorithms for maximum search of the objective function. The choice depends mostly on the bit rate and technology available.

Parallel Search Processing

Today's technology allows the integration of highly complex digital signal processors. The computational load can be managed by parallel processing rather then using an extremely fast technology.

Iterative Search Processing

The maximum search operation can be performed serially. A necessary, but not sufficient condition for the maximum of the objective function is

$$\begin{aligned} \frac{\partial}{\partial \theta} L(\mathbf{r}_f \mid \mathbf{a}, \theta, \varepsilon) \Big|_{\hat{\theta}, \hat{\varepsilon}} &= 0 \\ \frac{\partial}{\partial \varepsilon} L(\mathbf{r}_f \mid \mathbf{a}, \theta, \varepsilon) \Big|_{\hat{\theta}, \hat{\varepsilon}} &= 0 \end{aligned} \tag{5-24}$$

In (5-24) we assume that an estimate of the data sequence is available or that the sequence $\mathbf{a} = \mathbf{a}_0$ is known. For NDA algorithms the argument $\hat{\mathbf{a}}$ is absent.

Since the objective function is a concave function of the parameters (θ, ε) we can apply the gradient technique (or steepest ascent method) to compute the zero of (5-24) if the initial estimates are within the convergence region.

$$\begin{aligned} \hat{\varepsilon}_{k+1} &= \hat{\varepsilon}_k + \alpha_1 \frac{\partial}{\partial \varepsilon} L\left(\mathbf{r}_f \mid \mathbf{a}, \hat{\theta}_k, \hat{\varepsilon}_k\right) \\ \hat{\theta}_{k+1} &= \hat{\theta}_k + \alpha_2 \frac{\partial}{\partial \theta} L\left(\mathbf{r}_f \mid \mathbf{a}, \hat{\theta}_k, \hat{\varepsilon}_k\right) \end{aligned} \tag{5-25}$$

$$\alpha_i : \quad \text{convergence parameter}$$

with the short-hand notation $\partial/\partial x \;\; L(\mathbf{r}_f|\hat{x}) = [\partial/\partial x \;\; L(\mathbf{r}_f|x)]_{x=\hat{x}}$.

Notice that the received data over a segment of L symbols is processed repeatedly which requires that the data are stored in memory. This places no problem with today's digital technology. The iterative search is of particular interest to achieve acquisition with known symbols during a training period.

5.2.4 Error Feedback Systems

Error feedback systems adjust the synchronization parameters using an error signal. The error signal is obtained by differentiating an objective function and computing the value of the derivative for the most recent estimates $\hat{\theta}_n, \hat{\varepsilon}_n$,

$$\begin{aligned} &\frac{\partial L}{\partial \varepsilon}\left(\hat{\mathbf{a}}, \theta = \hat{\theta}_n, \varepsilon = \hat{\varepsilon}_n\right) \\ &\frac{\partial L}{\partial \theta}\left(\hat{\mathbf{a}}, \theta = \hat{\theta}_n, \varepsilon = \hat{\varepsilon}_n\right) \end{aligned} \tag{5-26}$$

For causality reasons, the error signal may depend only on previously reviewed symbols $\mathbf{a}_n$ (which are assumed to be known). The error signal is used to predict the new estimate:

$$\begin{aligned} \hat{\varepsilon}_{n+1} &= \hat{\varepsilon}_n + \alpha_\varepsilon \frac{\partial}{\partial \varepsilon} L\left(\mathbf{a}_n, \hat{\theta}_n, \hat{\varepsilon}_n\right) \\ \hat{\theta}_{n+1} &= \hat{\theta}_n + \alpha_\theta \frac{\partial}{\partial \theta} L\left(\mathbf{a}_n, \hat{\theta}_n, \hat{\varepsilon}_n\right) \end{aligned} \tag{5-27}$$

We readily recognize in eq. (5-27) the estimate of a first-order discrete-time error feedback system where $(\alpha_\varepsilon, \alpha_\theta)$ determines the loop bandwidth. Higher-order tracking systems can be implemented employing appropriate loop filters.

The error signal can always be decomposed into a useful signal plus noise. For ε and similarly for θ we obtain

$$\begin{aligned} \frac{\partial L}{\partial \varepsilon}\left(\mathbf{a}_n, \hat{\theta}_n, \hat{\varepsilon}_n\right) =& E\left[\frac{\partial}{\partial \varepsilon} L\left(\hat{\mathbf{a}}_n, \hat{\theta}_n, \hat{\varepsilon}_n\right)\right] \\ & \underbrace{+\frac{\partial L}{\partial \varepsilon}\left(\hat{\mathbf{a}}_n, \hat{\theta}_n, \hat{\varepsilon}_n\right) - E\left[\frac{\partial}{\partial \varepsilon} L\left(\hat{\mathbf{a}}_n, \hat{\theta}_n, \hat{\varepsilon}_n\right)\right]}_{\text{zero-mean noise process}} \end{aligned} \tag{5-28}$$

The useful signal depends nonlinearly on the error $\left(\hat{\theta}_n - \theta_0\right)$ and $(\hat{\varepsilon}_n - \varepsilon_0)$. We speak of a *tracking mode operation* of an error feedback system when the error is sufficiently small. The useful signal in (5-28) must be zero for zero error in order to provide an unbiased estimate. The process of bringing the system from its initial state to the tracking mode is called *acquisition*. Acquisition is a nonlinear phenomenon (see Volume 1, Chapter 4).

We observe some similarities between the maximum search and error feedback systems. In both cases we use the derivative of the objective function to derive an error signal. However, we should be aware of the fundamental differences. The maximum search algorithm processes the entire signal to iteratively converge to the final estimate. The feedback control systems, on the other hand, operate in real time using only that segment of the signal which was received in past times.

5.2.5 Main Points

- *Classification of algorithms*
  We distinguish between algorithms which assume the symbol sequence to be known and the obverse: The first class is called decision-directed (DD) or data-aided (DA), the obverse NDA. We further distinguish between feedforward (FF) and feedback (FB) structures.
- Synchronization parameters are slowly varying. They are estimated using one of the two structures:
  - *Two-stage tracker*
    The synchronization parameters are approximately piecewise constant. They are estimated over an interval of L symbols in a first stage.

Postprocessing of the short-term estimates is done in a second stage. This stage exploits the statistical dependencies between the short-term estimates of the first stage.

- *Error feedback system*
  An automatic control system is used to track a slowly varying parameter.

- We process the *entire* received signal $r_f(kT_s)$ but use only a subset of $L \ll N$ symbols out of N transmitted ones for estimation. This produces self-noise. If we process only a segment of samples $r_f(kT_s)$ we generate additional noise terms.

Bibliography

[1] M. Oerder, *Algorithmen zur Digitalen Taktsynchronisation bei Datenübertragung.* Düsseldorf: VDI Verlag, 1989.

5.3 NDA Timing Parameter Estimation

The objective function for the synchronization parameters (θ, ε) is given by eq. (5-11):

$$\begin{aligned} L(\mathbf{a}, \theta, \varepsilon) &= \exp\left\{ -\frac{2}{\sigma_n^2} \operatorname{Re}\left[\sum_{n=0}^{N-1} a_n^* \, z_n(\varepsilon)\, e^{-j\theta} \right] \right\} \\ &= \prod_{n=0}^{N-1} \exp\left\{ -\frac{2}{\sigma_n^2} \operatorname{Re}\left[a_n^* \, z_n(\varepsilon)\, e^{-j\theta} \right] \right\} \end{aligned} \tag{5-29}$$

In a first step we derive data and phase-independent timing estimators. The estimate of ε is obtained by removing the unwanted parameters $\mathbf{a}$ and θ in eq. (5-29).

To remove the data dependency we have to multiply (5-29) by $P({}^i a)$, where $({}^i a)$ is the ith of M symbols, and sum over the M possibilities. Assuming independent and equiprobable symbols the likelihood function reads

$$L(\theta, \varepsilon) = \prod_{n=0}^{N-1} \sum_{i=1}^{M} \exp\left\{ -\frac{2}{\sigma_n^2} \operatorname{Re}\left[{}^i a_n^* \, z_n(\varepsilon)\, e^{-j\theta} \right] P({}^i a) \right\} \tag{5-30}$$

There are various avenues to obtain approximations to (5-30). Assuming M-PSK modulation with $M > 2$, the probabilities

$$P({}^i a) = \frac{1}{M} \qquad \text{for } {}^i a = e^{j 2\pi i/M} \qquad i = 1, \ldots, M \tag{5-31}$$

can be approximated by a continuous-valued probability density function (pdf) of

$e^{j\alpha}$ where α has a uniform distribution over $(-\pi, \pi)$:

$$\begin{aligned} L(\theta,\varepsilon) &\simeq \prod_{n=0}^{N-1} \int_{-\pi}^{\pi} \exp\left\{-\frac{2}{\sigma_n^2} \operatorname{Re}\left[z_n(\varepsilon)\, e^{j(-\alpha-\theta)}\right]\right\} d\alpha \\ &= \prod_{n=0}^{N-1} \int_{-\pi}^{\pi} \exp\left\{-2\left|\frac{z_n(\varepsilon)}{\sigma_n^2}\right| \operatorname{Re}\left[e^{j(-\alpha-\theta+\arg z_n(\varepsilon))}\right]\right\} d\alpha \\ &= \prod_{n=0}^{N-1} \int_{-\pi}^{\pi} \exp\left\{-\frac{2}{\sigma_n^2} |z_n(\varepsilon)| \cos\left(-\alpha-\theta+\arg z_n(\varepsilon)\right)\right\} d\alpha \end{aligned} \tag{5-32}$$

Since $\cos(\cdot)$ is integrated over a full period of 2π, it is independent of θ and $\arg z_n(\varepsilon)$:

$$\begin{aligned} L_1(\varepsilon) &= \prod_{n=0}^{N-1} \int_{-\pi+\theta-\arg z_n(\varepsilon)}^{\pi+\theta-\arg z_n(\varepsilon)} \exp\left\{-\frac{2}{\sigma_n^2} |z_n(\varepsilon)| \cos(x)\right\} dx \\ &= \prod_{n=0}^{N-1} I_0\left(\frac{|z_n(\varepsilon)|}{\sigma_n^2/2}\right) \end{aligned} \tag{5-33}$$

where $I_0(\cdot)$ is the Bessel function of the first kind of order zero. It is quite interesting that in this approximation the distribution of the phase θ is of no concern. Thus, by averaging over the symbols we have achieved phase independency. But also notice that maximization requires knowledge of the signal-to-noise ratio σ_n^2.

Let us in a second approach *first* average over the phase to obtain a data-dependent algorithm:

$$\begin{aligned} L_2(\mathbf{a},\varepsilon) &\simeq \prod_{n=0}^{N-1} \int_{-\pi}^{\pi} \exp\left\{-\frac{2}{\sigma_n^2} |z_n(\varepsilon)a_n^*| \operatorname{Re}\left[e^{j(-\arg a_n-\theta+\arg z_n(\varepsilon))}\right]\right\} d\theta \\ &= \prod_{n=0}^{N-1} I_0\left(\frac{|z_n(\varepsilon)\, a_n^*|}{\sigma_n^2/2}\right) \end{aligned} \tag{5-34}$$

Notice that the result is the same for all phase modulations (M-PSK) (since $|a_n| = \text{const.}$), but not for M-QAM. In order to obtain an NDA synchronization algorithm for M-QAM, we would have to average over the symbols which is not possible in closed form.

But the objective functions of (5-33) and (5-34) can be further simplified by a series expansion of the modified Bessel function. Taking the logarithm, expanding into a series

$$I_0(x) \simeq 1 + \frac{x^2}{2} \qquad |x| \ll 1 \tag{5-35}$$

and discarding any constant irrelevant for the estimation yields

$$\begin{aligned}
\text{NDA:} \quad \hat{\varepsilon} &= \arg\max_{\varepsilon} \; L_1(\varepsilon) \\
&\simeq \arg\max_{\varepsilon} \; \sum_{n=0}^{N-1} |z_n(\varepsilon)|^2 \\
\text{DA:} \quad \hat{\varepsilon} &= \arg\max_{\varepsilon} \; L_2(\mathbf{a}, \varepsilon) \\
&\simeq \arg\max_{\varepsilon} \; \sum_{n=0}^{N-1} |z_n(\varepsilon)|^2 |a_n|^2
\end{aligned} \tag{5-36}$$

For M-PSK ($|a_n|^2 = \text{const.}$) both algorithms are identical.

Let us now explore a totally different route. We want to eliminate the data dependency in (5-29). This requires averaging over the symbols, if possible at all. Furthermore, it requires knowledge of σ_n^2 (operating point) which is not available in general. The algorithm would possibly be sensitive to this operating point. Both problems can be circumvented by considering the limit of the likelihood function (5-29) for low signal-to-noise ratio (SNR), $\sigma_n^2 \gg 1$. For this purpose we expand the exponential (5-29) function into a Taylor series:

$$\begin{aligned}
&\exp\left\{ -\frac{2}{\sigma_n^2} \operatorname{Re}\left[\sum_{n=0}^{N-1} a_n^* \, z_n(\varepsilon) \, e^{-j\theta} \right] \right\} \\
&\quad \simeq 1 - \frac{2}{\sigma_n^2} \operatorname{Re}\left[\sum_{n=0}^{N-1} a_n^* \, z_n(\varepsilon) \, e^{-j\theta} \right] + \left(\frac{1}{\sigma_n^2} \right)^2 \left[\sum_{n=0}^{N-1} \operatorname{Re}\left[a_n^* \, z_n(\varepsilon) \, e^{-j\theta} \right] \right]^2
\end{aligned} \tag{5-37}$$

We next average every term in the series with respect to the data sequence. For an i.i.d. data sequence we obtain for the first term

$$\begin{aligned}
&\sum_{n=0}^{N-1} E_a\left[\operatorname{Re}\left(a_n^* \, z_n(\varepsilon) \, e^{-j\theta} \right) \right] \\
&= \sum_{n=0}^{N-1} \operatorname{Re}\left[E_a[a_n^*] \, z_n(\varepsilon) \, e^{-j\theta} \right] \\
&= 0
\end{aligned} \tag{5-38}$$

since $E[a_n^*] = 0$. For the second term we obtain

$$\left[\sum_{n=0}^{N-1} \mathrm{Re}\left[a_n^*\, z_n(\varepsilon)\, e^{-j\theta}\right]\right]^2 = \frac{1}{4}\left[\sum_{n=0}^{N-1} a_n^*\, z_n(\varepsilon)\, e^{-j\theta} + a_n\, z_n^*(\varepsilon)\, e^{j\theta}\right]^2$$

$$= \frac{1}{4}\sum_{n\neq m}^{N-1}\sum_{m}^{N-1}\left[a_n^* a_m^* z_n(\varepsilon) z_m(\varepsilon) e^{-j2\theta} + a_n a_m z_n^*(\varepsilon) z_m^*(\varepsilon) e^{+j2\theta} + 2a_n^* a_m z_n(\varepsilon) z_m^*(\varepsilon)\right]$$

$$+ \frac{1}{4}\sum_{n=0}^{N-1}\left[(a_n^*)^2 (z_n(\varepsilon))^2 e^{-j2\theta} + (a_n)^2 (z_n^*(\varepsilon))^2 e^{+j2\theta} + 2|a_n|^2 |z_n(\varepsilon)|^2\right]$$

(5-39)

Now, taking expected values of (5-39) with respect to the data sequence assuming i.i.d. symbols yields ($E[a_n] = 0$):

$$E_a\left[\mathrm{Re}\sum_{n=0}^{N-1}\left[a_n^*\, z_n(\varepsilon)\, e^{-j\theta}\right]\right]^2 = \frac{1}{2}\sum_{n=0}^{N-1} E\left[|a_n|^2\right] |z_n(\varepsilon)|^2 + \frac{1}{2}\sum_{n=0}^{N-1} \mathrm{Re}\left[E\left[a_n^2\right]\, (z_n^*(\varepsilon))^2\, e^{j2\theta}\right] \tag{5-40}$$

Using (5-40) leads to the following objective function:

$$\begin{aligned} L(\theta, \varepsilon) = &\sum_{n=0}^{N-1} E\left[|a_n|^2\right] |z_n(\varepsilon)|^2 \\ &+ \mathrm{Re}\left[\sum_{n=0}^{N-1} E\left[a_n^2\right] (z_n^*(\varepsilon))^2\, e^{-j2\theta}\right] \end{aligned} \tag{5-41}$$

Averaging over an uniformly distributed phase yields the noncoherent (NC) timing estimator

$$L_1(\varepsilon) = \sum_{n=0}^{N-1} |z_n(\varepsilon)|^2 \qquad \text{(NC)} \tag{5-42}$$

which is the same as equation (5-36) (NDA).

Equation (5-41) serves as a basis for the *joint* non-data-aided estimation of phase and timing. The phase estimate

$$\hat{\theta} = \frac{1}{2}\arg\left\{\sum_{n=0}^{N-1} E\left[a_n^2\right]\, (z_n^*(\varepsilon))^2\right\} \tag{5-43}$$

maximizes the second sum of (5-41) for *any* ε, since the sum

$$\sum_{n=0}^{N-1} E\left[a_n^2\right] \left(z_n^*(\varepsilon)\right)^2 e^{-j2\hat{\theta}} \tag{5-44}$$

becomes a real number. Consequently, the estimate $\hat{\varepsilon}$ is obtained by maximization of the absolute value

$$\hat{\varepsilon} = \arg\max_{\varepsilon} \sum_{n=0}^{N-1} E\left[|a_n|^2\right] |z_n(\varepsilon)|^2 + \left| E\left[a_n^2\right] \sum_{n=0}^{N-1} z_n^2(\varepsilon) \right| \tag{5-45}$$

This is an interesting result since it shows that the two-dimensional search for (θ, ε) can be reduced to a one-dimensional search for ε by maximizing the objective function

$$\begin{aligned} L(\varepsilon) &= \sum_{n=0}^{N-1} E\left[|a_n|^2\right] |z_n(\varepsilon)|^2 + \left| E\left[a_n^2\right] \sum_{n=0}^{N-1} z_n^2(\varepsilon) \right| \\ &= E\left[|a_n|^2\right] \sum_{n=0}^{N-1} |z_n(\varepsilon)|^2 + \left| E\left[a_n^2\right] \sum_{n=0}^{N-1} z_n^2(\varepsilon) \right| \end{aligned} \tag{5-46}$$

Incidentally, we have found another form of a non-data-aided/phase-independent algorithm. More precisely, the algorithm we found is an *implicit* phase-directed algorithm. This is because (5-46) is independent of the trial parameter θ. See Figure 5-6.

Comparing (5-46) and (5-42) we observe that the noncoherent (NC) timing estimator does not depend on the signal constellation while the implicitly coherent (IC) estimator does so. In the following we restrict our attention to two classes of signal constellations of practical importance. The first contains the one-dimensional (1D) constellation comprising real-valued data for which $E\left[a_n^2\right] = 1$ (normalized). The second class consist of two-dimensional (2D) constellations which exhibit a $\pi/2$ rotational symmetry, this yields $E\left[a_n^2\right] = 0$. Hence, for $\pi/2$ rotational symmetric 2D constellations, the IC and NC synchronizers are identical, but for 1D constellations they are different.

Main Points

Conceptually, the approach to obtain estimators for the synchronization parameters is straightforward. The likelihood (*not* the log-likelihood function!) must be averaged over the unwanted parameters. This is possible only in a few isolated cases in closed form. One must resort to approximations.

In this chapter various approximation techniques are introduced to derive NDA

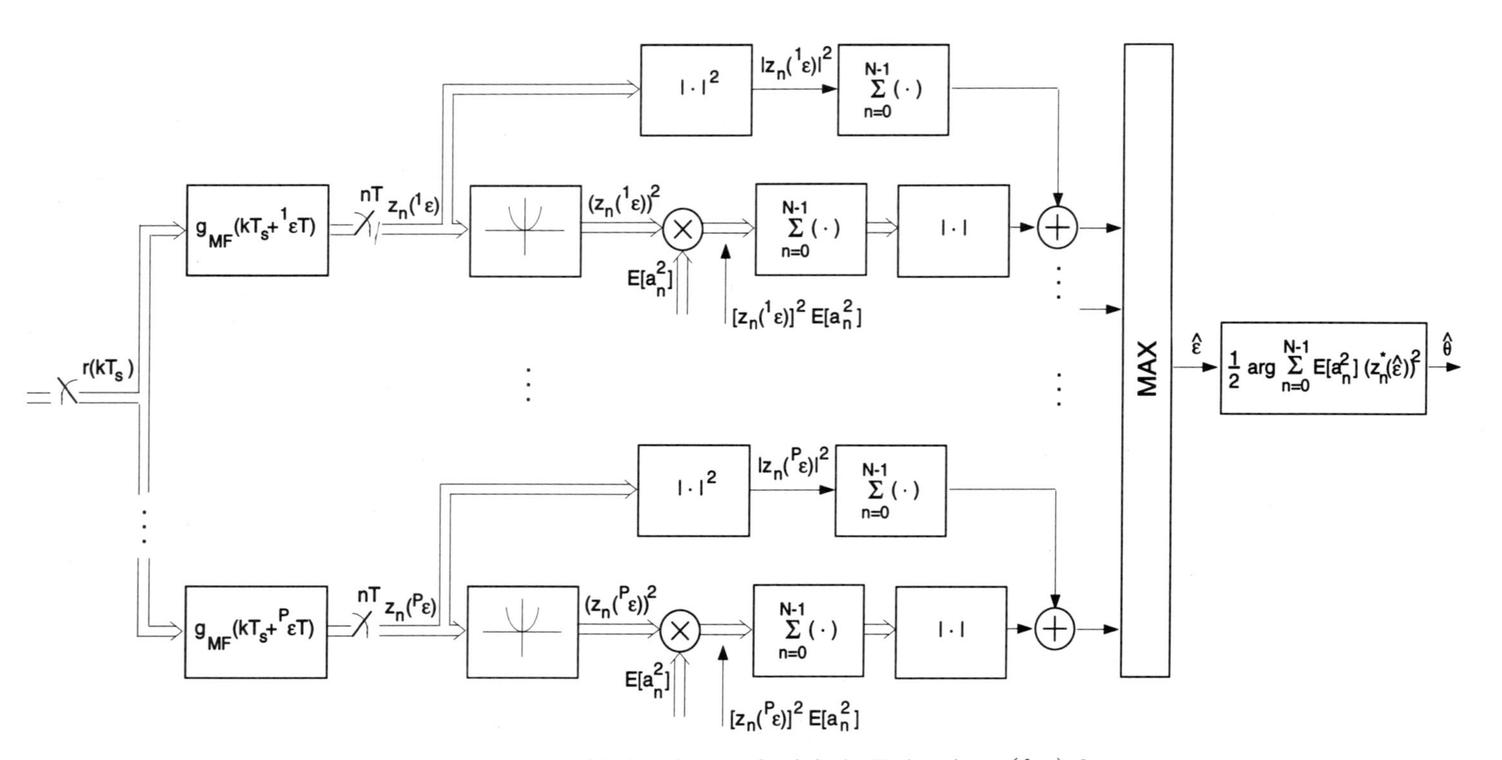

Figure 5-6 Non-Data-Aided Estimator for jointly Estimationg (θ, ε) for Arbitrary PAM Signals: (a) First Part: Phase-Directed Timing Estimator, (b) Upper Branch only: Non-Phase-Directed

timing estimators. The practically most important result is a phase-independent algorithm

$$\hat{\varepsilon} = \arg \max_{\varepsilon} \sum_{n} |z_n(\varepsilon)|^2 \tag{5-47}$$

The algorithm works for M-QAM and M-PSK signaling methods.

5.4 NDA Timing Parameter Estimation by Spectral Estimation

In the previous section the unknown timing parameter was obtained by a maximum search (see Figure 5-6). It is interesting and surprising that one can circumvent this maximum search.

We consider the objective function (5-42),

$$L(\varepsilon) = \sum_{l=-L}^{L} |z(lT + \varepsilon T)|^2 \tag{5-48}$$

We assume a symmetric observation interval $[-L, L]$. This choice is solely to simplify some of the following computations. For a sufficiently large number N of transmitted symbols ($N \gg L$) the process $|z(lT + \varepsilon T)|^2$ is (almost) cyclostationary in the observation interval. A cyclostationary process has a Fourier series representation

$$\begin{aligned} |z(lT + \varepsilon T)|^2 &= \sum_{n=-\infty}^{\infty} c_n^{(l)} \, e^{j \frac{2\pi}{T} n T \varepsilon} \\ &= \sum_{n=-\infty}^{\infty} c_n^{(l)} \, e^{j 2\pi n \varepsilon} \end{aligned} \tag{5-49}$$

where the coefficients $c_n^{(l)}$ are random variables defined by

$$\begin{aligned} c_n^{(l)} &= \frac{1}{T} \int_0^T |z(lT + \varepsilon T)|^2 \, e^{-j(2\pi/T)nT\varepsilon} \, d(\varepsilon T) \\ &= \int_0^1 |z(lT + \varepsilon T)|^2 \, e^{-j 2\pi n \varepsilon} \, d\varepsilon \end{aligned} \tag{5-50}$$

The objective function (5-48) equals the *time average* of the cyclostationary process $|z(lT + \varepsilon T)|^2$ over the interval $[-L, L]$. This time average can be obtained by computing the Fourier coefficients $c_n^{(l)}$ in every T interval and averaging over $2L+1$ values:

$$\begin{aligned}\sum_{l=-L}^{L} |z(lT+\varepsilon T)|^2 &= \sum_{l=-L}^{L} \left[\sum_{n=-\infty}^{\infty} c_n^{(l)} \, e^{j2\pi n\varepsilon} \right] \\ &= \sum_{n=-\infty}^{\infty} c_n \, e^{j2\pi n\varepsilon}\end{aligned} \tag{5-51}$$

with

$$c_n = \sum_{l=-L}^{L} c_n^{(l)} \tag{5-52}$$

We will show later on that only three coefficients, namely $\{c_{-1}, c_0, c_1\}$ have nonzero mean. Thus

$$\sum_{l=-L}^{L} |z(lT+\varepsilon T)|^2 = c_0 + 2\,\mathrm{Re}\left[c_1 \, e^{j2\pi\varepsilon}\right] + \underbrace{\sum_{|n|\geq 2} 2\,\mathrm{Re}\left[c_n \, e^{j2\pi n\varepsilon}\right]}_{\substack{\text{zero-mean disturbance} \\ \text{for all values of } \varepsilon}} \tag{5-53}$$

By definition, the ML estimate $\hat{\varepsilon}$ is that value of ε for which (5-48) assumes its maximum:

$$\sum_{l=-L}^{L} |z(lT+\hat{\varepsilon} T)|^2 > \sum_{l=-L}^{L} |z(lT+\varepsilon T)|^2 \qquad \forall \varepsilon \tag{5-54}$$

On the other hand, by the Fourier series representation of (5-53) we see that $\hat{\varepsilon}$ is also defined by

$$\hat{\varepsilon} = \arg\max_{\varepsilon} \left(c_0 + 2\,\mathrm{Re}\left[c_1 \, e^{j2\pi\varepsilon}\right]\right) \tag{5-55}$$

Since c_0 and the absolute value $|c_1|$ are independent of ε (to be shown later) the maximum of (5-55) is assumed for

$$\hat{\varepsilon} = -\frac{1}{2\pi} \arg c_1 \tag{5-56}$$

It is quite remarkable that no maximum search is needed to find $\hat{\varepsilon}$ since it is *explicitly* given by the argument of the Fourier coefficient c_1. The coefficient c_1 is defined by a summation of $(2L+1)$ integrals. The question is whether the integration can be equivalently replaced by a summation, since only digital algorithms are of interest here. This is indeed so, as can be seen from the following plausibility arguments (an exact proof will be given later).

The output of the matched filter is band-limited to $B_z = (1/2T)(1+\alpha)$. Since squaring the signal doubles the bandwidth, the signal $|z(t)|^2$ is limited to twice this value. Provided the sampling rate $1/T_s$ is such that the sampling theorem is fulfilled for $|z(t)|^2$ [and not only for $z(t)$],

$$B_{|z|^2} = \frac{1}{T}(1+\alpha) < \frac{1}{2T_s} \tag{5-57}$$

the coefficients c_1, c_0 can be computed by a discrete Fourier transform (DFT). Let us denote by the integer M_s the (nominal) ratio between sampling and symbol rate, $M_s = T/T_s$. For the samples taken at kT_s we obtain

$$\begin{aligned} c_1 &= \sum_{l=-L}^{L} \int_0^1 |z(lT+\varepsilon T)|^2 e^{-j2\pi\varepsilon} d\varepsilon \\ &= \sum_{l=-L}^{L} \left[\frac{1}{M_s} \sum_{k=0}^{M_s-1} |z([lM_s+k]T_s)|^2 e^{-j(2\pi/M_s)k} \right] \end{aligned} \tag{5-58}$$

A particularly simple implementation is found for $M_s = 4$ (Figure 5-7). For this value we obtain a multiplication-free realization of the estimator:

$$c_1 = \sum_{l=-L}^{L} \left[\frac{1}{M_s} \sum_{k=0}^{3} |z([4l+k]T_s)|^2 (-j)^k \right] \qquad e^{-j\frac{2\pi}{4}k} = (-j)^k \tag{5-59}$$

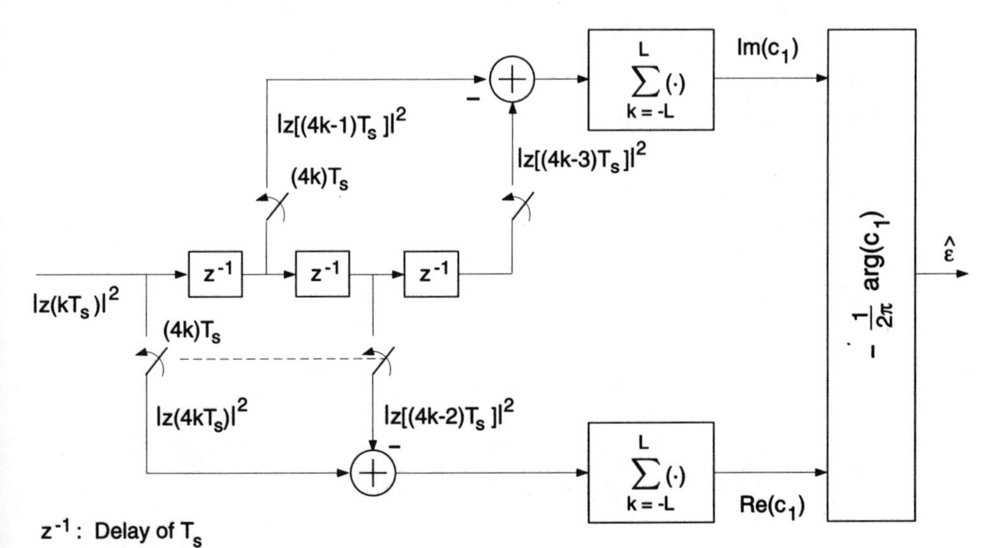

Figure 5-7 Timing Estimation by Spectral Analysis; $M_s = T/T_s = 4$

Splitting up in real and imaginary part yields

$$\begin{aligned} \text{Re}\,\{c_1\} &= \frac{1}{M_s} \sum_{l=-L}^{L} |z(4lT_s)|^2 - |z[(4l+2)T_s]|^2 \\ \text{Im}\,\{c_1\} &= \frac{1}{M_s} \sum_{l=-L}^{L} -|z[(4l+1)T_s]|^2 + |z[(4l+3)T_s]|^2 \end{aligned} \tag{5-60}$$

We next prove that $\hat{\varepsilon} = -1/2\pi \arg c_1$ is an unbiased estimate. The proof is rather technical and may be omitted at first reading.

Proof: We assume

(i) Sampling rate $1/T_s > (2/T)(1+\alpha)$ (twice the rate required for the data path).
(ii) The ratio T/T_s is an integer $M_s = T/T_s$.
(iii) i.i.d. data.
(iv) $g(-t) = g(t)$ (real and symmetric pulse).

The matched filter output $z(lT_s)$ equals

$$z(lT_s) = \sum_{k=-\infty}^{\infty} r_f(kT_s)\, g^*(kT_s - lT_s) \tag{5-61}$$

Replacing in the previous equation the samples $r_f(kT_s)$ by

$$r_f(kT_s) = \sum_{n=-N}^{N} a_n\, g(kT_s - nT - \varepsilon_0 T) + n(kT_s) \tag{5-62}$$

(ε_0 true unknown value), we obtain

$$\begin{aligned} z(lT_s) &= \sum_{n=-N}^{N} \left[a_n \sum_{k=-\infty}^{\infty} g(kT_s - nT - \varepsilon_0 T)\, g^*(kT_s - lT_s) \right] + m(kT_s) \\ &= \sum_{n=-N}^{N} [a_n\ h(lT_s - nT - \varepsilon_0 T)] + m(kT_s) \end{aligned} \tag{5-63}$$

where $m(kT_s)$ is filtered noise, and the function $h(\cdot)$ is defined by the inner sum over k

$$\begin{aligned} h(lT_s - nT - \varepsilon_0 T) &= \sum_{k=-\infty}^{\infty} g(kT_s - nT - \varepsilon_0 T)\, g^*(kT_s - lT_s) \\ &= \frac{1}{T_s} \int_{-\infty}^{\infty} g(x - nT - \varepsilon_0 T)\, g^*(x - lT_s)\, dx \end{aligned} \tag{5-64}$$

The last line of the previous equation is a consequence of the equivalence theorem of analog and digital signal processing [assumption (i)].

Squaring $z(lT_s)$ and subsequently taking expected values with respect to i.i.d. data and noise yields

$$E\left[|z(lT_s)|^2\right] = E\left[|a_n|^2\right] \sum_{n=-N}^{N} |h(lT_s - nT - \varepsilon_0 T)|^2 + P_n \tag{5-65}$$

with $P_n \geq 0$ the additive noise contribution being independent of ε_0.

If the number of symbols is sufficiently large, we commit a negligible error when the summation in the previous equation runs from $-\infty$ to $+\infty$ (instead of $[-N, N]$). The infinite sum

$$h_p(t) = \sum_{n=-\infty}^{\infty} |h(t - nT - \varepsilon_0 T)|^2 \tag{5-66}$$

represents a periodic function $h_p(t)$ which can be expanded into a Fourier series where $h_p(lT_s)$ are the samples of $h_p(t)$ taken at $t = lT_s$,

$$h_p(t) = \sum_{i=-\infty}^{\infty} d_i \; e^{j2\pi it/T} \tag{5-67}$$

The coefficients d_i are related to the spectrum of $|h(t - \varepsilon_0 T)|^2$ by the Poisson theorem

$$\begin{aligned} d_i &= \frac{1}{T} \int_{-\infty}^{+\infty} |h(t - \varepsilon_0 T)|^2 \; e^{-j2\pi it/T} \; dt \\ &= \frac{1}{T} \; e^{-j2\pi i(\varepsilon_0 T)/T} \int_{-\infty}^{+\infty} |h(t)|^2 \; e^{-j2\pi it/T} \; dt \end{aligned} \tag{5-68}$$

Since the pulse $g(t)$ is band-limited to

$$B_g = \frac{1}{2T}(1 + \alpha) \tag{5-69}$$

so is $h(t)$. Since squaring the signal $h(t)$ doubles the bandwidth, the spectrum of $|h(t)|^2$ is limited to twice this bandwidth:

$$B_{|h|^2} = \frac{1}{T}(1 + \alpha) \tag{5-70}$$

The integral in (5-68) equals the value of the spectrum of $|h(t)|^2$ at $f = i(1/T)$.

Thus, only three coefficients d_{-1}, d_0, d_1 are nonzero ($\alpha < 1$) in the Fourier series (Figure 5-8):

$$\begin{aligned} E\left[|z(t)|^2\right] &= E\left[|a_n|^2\right] \sum_{n=-\infty}^{\infty} |h(t - nT - \varepsilon_0 T)|^2 \\ &= \sum_{i=-1}^{1} E\left[|a_n|^2\right] d_i \; e^{j(2\pi/T)\,it} \end{aligned} \tag{5-71}$$

with d_0, d_1, d_{-1} given by (5-68). Due to the band limitation of $E\left[|z(t)|^2\right]$ the Fourier coefficients d_i and $E\left[|z(kT_s)|^2\right]$ are related via the discrete Fourier transform. The coefficients d_0 and d_1, respectively, are given by

$$\begin{aligned} d_0 &= \frac{1}{M_s} \sum_{k=0}^{M_s-1} E\left[|z(kT_s)|^2\right] \\ d_1 &= \frac{1}{M_s} \sum_{k=0}^{M_s-1} E\left[|z(kT_s)|^2\right] e^{-j(2\pi/M_s)k} \end{aligned} \tag{5-72}$$

Comparing with the coefficient c_n of (5-58) we see

$$d_1 = \text{const.}E[c_1] \qquad d_0 = \text{const.}E[c_0] \qquad \begin{aligned} d_n &= E[c_n] = 0 \\ n &\neq \{0, +1, -1\} \end{aligned} \tag{5-73}$$

as conjectured.

From (5-68) it follows that $E[c_0]$ is independent of ε:

$$E[c_0] = \text{const.} \int_{-\infty}^{+\infty} |h(t)|^2 dt \tag{5-74}$$

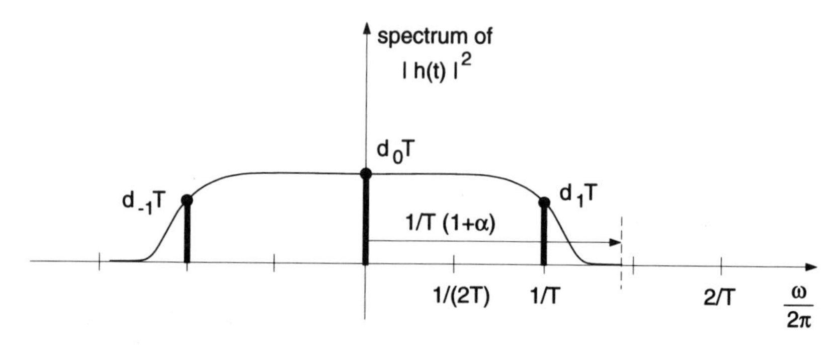

Figure 5-8 Spectrum of $|h(t)|^2$ and Coefficients d_i

For c_1 we find

$$E[c_1] = \text{const.}\ e^{-j2\pi\varepsilon_0} \int_{-\infty}^{+\infty} |h(t)|^2 e^{-j(2\pi/T)t} dt \tag{5-75}$$

If the pulse $g(t)$ is real and symmetric, then $|h(t)|^2 = |h(-t)|^2$ and

$$\arg\ E[c_1] = -2\pi\varepsilon_0 \tag{5-76}$$

Thus

$$\hat{\varepsilon} = -\frac{1}{2\pi} \arg c_1 \tag{5-77}$$

is an unbiased estimate.

Main Points

One can circumvent the maximum search for $\hat{\varepsilon}$ by viewing $|z_n(\varepsilon)|^2$ as a *periodic* $(1/T)$ timing wave. The argument of the first coefficient c_1 of the Fourier series representation of $|z_n(\varepsilon)|^2$ yields an unbiased estimate

$$\hat{\varepsilon} = -\frac{1}{2\pi} \arg c_1 \tag{5-78}$$

with

$$c_1 = \sum_{l=-L}^{L} \left[\frac{1}{M_s} \sum_{k=0}^{M_s-1} |z[(lM_s + k)T_s]|^2 \, e^{-j \frac{2\pi}{M_s} k} \right] \tag{5-79}$$

$$(M_s = T/T_s \text{ integer})$$

In the inner sum the Fourier coefficient $c_1^{(l)}$ for the lth time interval of length T is computed. The final estimate for c_1 is obtained by averaging the result over $(2L+1)$ intervals.

A particularly simple realization is obtained for four samples per symbol, $M_s = 4$. Because in this case we have $\exp(-j2\pi k/M_s) = (-j)^k$, a multiplication-free realization is possible.

Since the algorithm provides a *unique* estimate of ε, it is hang-up free. For this reason and because of its simplicity it is often used in practice.

Bibliography

[1] M. Oerder and H. Meyr, "Digital Filter and Square Timing Recovery," *IEEE Trans. Commun.*, vol. COM-36, pp. 605–612, May 1988.

5.5 DA (DD) Timing Parameter Estimators

Replacing in eq. (5-29) the trial parameter a_n and the phase θ by their estimates yields a DD phase-independent algorithm:

$$L\left(\hat{\mathbf{a}}, \varepsilon, \hat{\theta}\right) = \exp\left\{ -\frac{2}{\sigma_n^2} \operatorname{Re}\left[\sum_{n=0}^{N-1} \hat{a}_n^* \; z_n(\varepsilon) \; e^{-j\hat{\theta}} \right] \right\} \tag{5-80}$$

This algorithm finds application when phase synchronization is done prior to timing recovery. The computation of the objective function can again be computed in parallel. Instead of the nonlinearity to remove the data dependency we now have the multiplication by the symbols (see Figure 5-9).

We next consider the joint estimation of (θ, ε),

$$L(\hat{\mathbf{a}}, \varepsilon, \theta) = \exp\left\{ -\frac{2}{\sigma_n^2} \operatorname{Re}\left[\sum_{n=0}^{N-1} \hat{a}_n^* \; z_n(\varepsilon) \; e^{-j\theta} \right] \right\} \tag{5-81}$$

The two-dimensional search over (θ, ε) can be reduced to a one-dimensional one by defining

$$\mu(\varepsilon) = \sum_{n=0}^{N-1} \hat{a}_n^* \; z_n(\varepsilon) \tag{5-82}$$

We then have

$$\max_{\varepsilon, \theta} \; \operatorname{Re}\left[\mu(\varepsilon) \; e^{-j\theta}\right] \; = \max_{\varepsilon, \theta} \; |\mu(\varepsilon)| \; \operatorname{Re}\left[e^{-j(\theta - \arg \mu(\varepsilon))}\right] \tag{5-83}$$

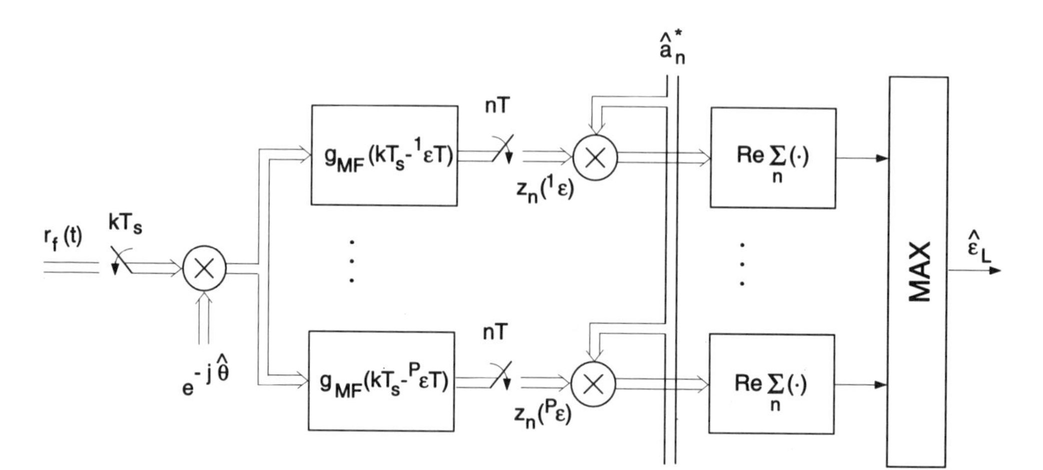

Figure 5-9 Decision-Directed Timing Estimation Using the Phase-Corrected Signal

The joint maximum is found by first maximizing the absolute value of $\mu(\varepsilon)$ (which is independent of θ). The second factor

$$\operatorname{Re}\left[e^{-j(\theta-\arg\mu(\varepsilon))}\right] \tag{5-84}$$

is maximized to a value of one by $\theta = \arg\ \mu(\varepsilon)$. Hence, for timing estimation we only need to maximize

$$\hat{\varepsilon} = \arg\ \max_{\varepsilon}\ |\mu(\varepsilon)| \tag{5-85}$$

The carrier phase estimate $\hat{\theta}$ can then be computed directly as

$$\hat{\theta} = \arg\ \mu(\hat{\varepsilon}) \tag{5-86}$$

In a practical realization the summation is confined to $L \ll N$ symbols.

Main Points

The two-dimensional search for jointly estimating (θ, ε) can always be reduced to a one-dimensional search (Figure 5-10):

$$\begin{aligned} \hat{\varepsilon} &= \max_{\varepsilon} \left| \sum_{n=0}^{N-1} \hat{a}_n^* \, z_n(\varepsilon) \right| \\ \hat{\theta} &= \arg \sum_{n=0}^{N-1} \hat{a}_n^* \, z_n(\hat{\varepsilon}) \end{aligned} \tag{5-87}$$

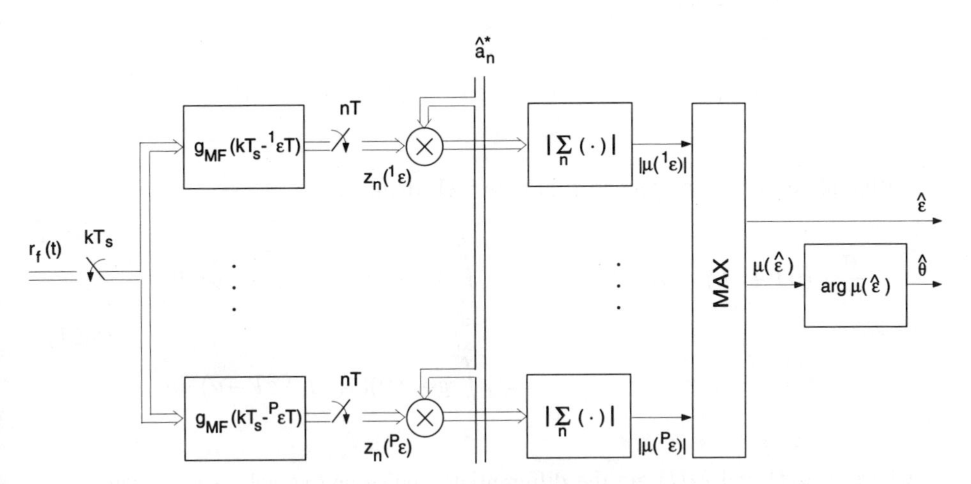

Figure 5-10 DA (DD) Joint (θ, ε) Estimator

Bibliography

[1] G. Ascheid and H. Meyr, "Maximum Likelihood Detection and Synchronization by Parallel Digital Signal Processing," *Proceedings of the International Conference GLOBECOM*, pp. 804–810, 1993.

5.6 Timing Error Feedback Systems at Higher Than Symbol Rate

5.6.1 DD and Phase-Directed Timing Recovery

We differentiate the log-likelihood function (5-29) with respect to the parameter ε

$$\frac{\partial}{\partial \varepsilon} L\left(\hat{\mathbf{a}}, \hat{\theta}, \varepsilon\right) \propto -\frac{2}{\sigma_n^2} \operatorname{Re}\left[\sum_{n=0}^{N-1} \hat{a}_n^* \frac{\partial}{\partial \varepsilon} z(nT + \varepsilon T)\, e^{-j\hat{\theta}}\right] \tag{5-88}$$

Since the summation is performed in the loop filter we drop the summation to obtain the error signal at time nT (see Section 5.2.4):

$$x(nT) = \operatorname{Re}\left[\hat{a}_n^* \frac{\partial}{\partial \varepsilon} z(nT + \varepsilon T)\Big|_{\varepsilon=\hat{\varepsilon}} e^{-j\hat{\theta}}\right] \tag{5-89}$$

Our definition of the error signal is formal since we have not yet explained how to compute the differential $\partial\left[z_n(\varepsilon)\right]/\partial\varepsilon$. To find the answer let us for a moment forget that we intend to process the information digitally. Because of the equivalence of digital and analog signal processing, we know that $z(nT + \varepsilon T)$ can be obtained by processing the incoming signal $r_f(t)$ with an analog matched filter $g_{\mathrm{MF}}(t)$ which is sampled at $t = nT + \varepsilon T$:

$$z(nT + \varepsilon T) = \int_{-\infty}^{\infty} g_{\mathrm{MF}}(nT + \varepsilon T - \nu)\, r_f(\nu)\, d\nu \tag{5-90}$$

Differentiation with respect to ε is now well defined:

$$\begin{aligned} \frac{\partial}{\partial \varepsilon} z\,(nT{+}\varepsilon T) = \dot{z}(t)\,\Big|_{t=nT+\varepsilon T} &= \int_{-\infty}^{\infty} \dot{g}_{\mathrm{MF}}(nT{+}\varepsilon T{-}\nu) r_f(\nu)\, d\nu \\ &= \int_{-\infty}^{\infty} g_{\mathrm{MF}}(\nu) \dot{r}_f(nT{+}\varepsilon T{-}\nu)\, d\nu \end{aligned} \tag{5-91}$$

where $\dot{g}_{\mathrm{MF}}(t)$ and $\dot{r}_f(t)$ are the differentiated pulse and signal, respectively.

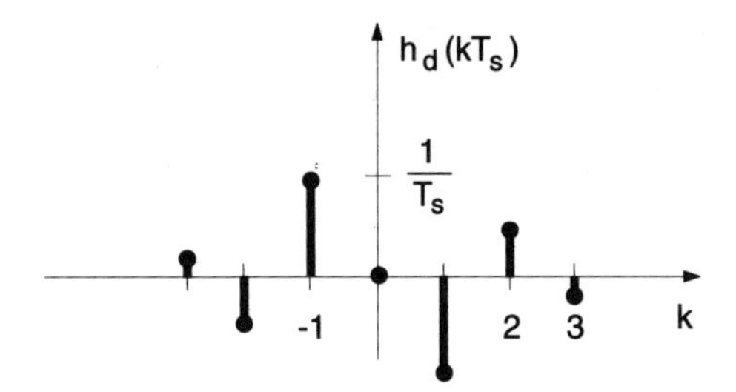

Figure 5-11 Impulse Response of the Digital Differentiator

In books of numerical mathematics, differentiation of functions is considered as a problematic operation. Fortunately, in our case differentiation is well behaved since it is always done in conjunction with matched filtering, which is a lowpass process.

Due to the equivalence of digital and analog signal processing, differentiation can be done in the digital domain. The impulse response of the digital differentiator is defined in the frequency domain by

$$H_d\left(e^{j\omega T_s}\right) = j\omega T_s \qquad \left|\frac{\omega}{2\pi}\right| \leq \frac{\pi}{T_s} \tag{5-92}$$

The impulse response $h_d(kT_s)$ is the inverse Fourier transform of $H_d\left(e^{j\omega T_s}\right)$:

$$h_d(kT_s) = \frac{T_s}{2\pi} \int\limits_{-\pi/T_s}^{\pi/T_s} H_d\left(e^{j\omega T_s}\right) e^{j\omega k T_s}\, d\omega \tag{5-93}$$

After some elementary algebraic manipulations we find

$$h_d(kT_s) = \begin{cases} 0 & k = 0 \\ \dfrac{1}{kT_s}\,(-1)^k & \text{else} \end{cases} \tag{5-94}$$

(see Figure 5-11). The digital differentiator produces exactly the derivative $\dot{z}(nT + \varepsilon T)$; there is no approximation involved as usual in numerical differentiation. One error signal per symbol is produced.

In Figure 5-12 we show a first structure of a digital timing error detector. The signal is sampled at a rate $1/T_s$, processed in a digital matched filter followed by a digital interpolator. These are the familiar building blocks we have encountered earlier. The signal path then branches in two directions. The upper path is the familiar one, leading to the data detector (after being decimated). In the lower path the signal is processed in the digital differentiator $h_d(kT_s)$ and subsequently decimated. Interpolation and decimation are controlled and determined by the

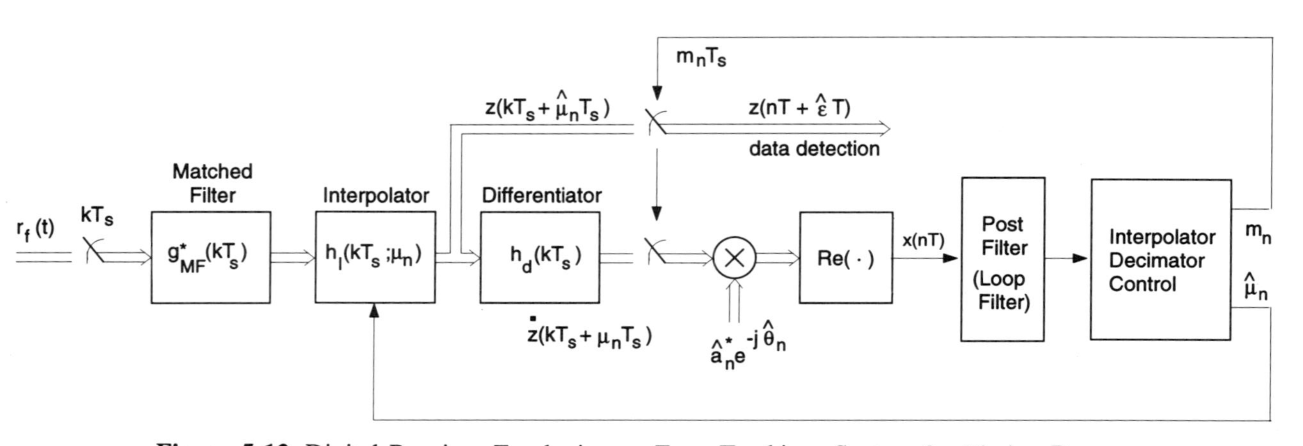

Figure 5-12 Digital Receiver Employing an Error-Tracking System for Timing Recovery. The timing recovery is decision-directed (DD) and assumes a known phase θ

fractional time delay $\hat{\mu}_n$ and basepoint m_n. We recall from Section 4.1 that these quantities are related to the desired time instant $(nT + \hat{\varepsilon}T)$ by

$$\begin{aligned} &nT + \hat{\varepsilon}T = m_n T_s + \hat{\mu}_n T_s \\ &m_n : \text{ largest integer less than or equal } nT + \hat{\varepsilon}T \end{aligned} \tag{5-95}$$

A detailed discussion of the (m_n, μ_n) computation is postponed to Section 9.2.

The structure shown in Figure 5-12 is not necessarily the one most favorable for implementation. Since matched filter, interpolator, and differentiator are linear filters, their position can be interchanged and possibly merged, thus reducing the complexity of the receiver. While a digital differentiator (or an ideal interpolator) alone would be very difficult to approximate at reasonable complexity, considering the linear filtering operation comprising matched filter, interpolator, and digital differentiator as an entity may lead to a far simpler solution at virtually negligible implementation loss. An example of such an integrated design approach is discussed in Chapter 10.

Example: Two-Point Differentiator for $T/T_s = 2$
The simplest approximation to the differentiator for $T/T_s = 2$ is

$$h_{d_2}(kT_s) = \begin{cases} 1 & k = -1 \\ -1 & k = 1 \\ 0 & \text{else} \end{cases} \tag{5-96}$$

Thus, the differential quotient is approximated by taking the differences. The signal $x(nT)$ (for $T/T_s = 2$) then equals

$$x(nT) = \text{Re}\left\{\hat{a}_n^* \left[z\left(nT + \frac{T}{2} + \hat{\varepsilon}\, T\right) - z\left(nT - \frac{T}{2} + \hat{\varepsilon}\, T\right)\right] e^{-j\hat{\theta}}\right\} \tag{5-97}$$

The stable tracking point for the error signal (5-97) is found by taking expected values over the noise and data. For an i.i.d. data sequence we find (verify)

$$E[x(nT)] = \text{Re}\left\{E\left[|a_n|^2\right]\left[h\left(\frac{T}{2} + [\hat{\varepsilon} - \varepsilon_0]T\right) - h\left(-\frac{T}{2} + [\hat{\varepsilon} - \varepsilon_0]T\right)\right] e^{j(\theta_0 - \hat{\theta})}\right\} \tag{5-98}$$

Remark: We introduced the notation

$$\begin{aligned} &h([n-m]T + [\hat{\varepsilon} - \varepsilon_0]T) \\ &= \sum_{k=-\infty}^{\infty} g(kT_s - mT - \varepsilon_0 T)\, g_{\text{MF}}(nT + \hat{\varepsilon}T - kT_s) \end{aligned} \tag{5-99}$$

which is more descriptive in describing the mean error function of tracking systems.

The notation $h_{n,m}([\hat{\varepsilon}-\varepsilon_0]T)$ for the right-hand side of (5-99) is appropriate if $h_{n,m}([\hat{\varepsilon}-\varepsilon_0]T)$ is an element of a matrix $\mathbf{H}$.

For symmetric pulses the mean error signal equals zero for $e = (\hat{\varepsilon}-\varepsilon_0) = 0$. The signal is sampled at the maximum of the peak point. If we replace the differential quotient by

$$z(nT + \hat{\varepsilon}\, T) - z([n-1]T + \hat{\varepsilon}T) \tag{5-100}$$

we obtain

$$x(nT) = \text{Re}\left\{ \hat{a}_n^* \, [z(nT + \hat{\varepsilon}\, T) - z([n-1]T + \hat{\varepsilon}T)] \, e^{-j\hat{\theta}} \right\} \tag{5-101}$$

Taking the expected value yields

$$E[x(nT)] = \text{Re}\left\{ E\left[|a_n|^2\right] [h(eT) - h(-T+eT)] \, e^{j[\theta_0 - \hat{\theta}]} \right\} \tag{5-102}$$

The error signal $E[x(nT)]$ is zero for $e = 1/2$ which corresponds to a value $\hat{\varepsilon} = \varepsilon_0 + 1/2$. Thus, the sample of the matched filter required for detection must be shifted by $T/2$:

$$z\left(nT + \left[\hat{\varepsilon} - \frac{1}{2}\right] T \right) \quad \text{sample required for detection} \tag{5-103}$$

The algorithm (5-101) would also work for T-spaced samples if the clock is physically adjusted to generate samples at $t = nT + \hat{\varepsilon}T$. The two tracking algorithms are equivalent for symmetric Nyquist pulses but perform substantially different for nonideally filtered signals.

There is a subtle point hidden in the algorithm of eq. (5-97). It works with two samples per symbol. We therefore have to map the $\{kT/2\}$ time axis onto the (receiver) $\{kT_s\}$ axis to obtain the matched filter output:

$$z\left(nT + \frac{T}{2} + \hat{\varepsilon}T \right) = z\left([2n+1]\frac{T}{2} + 2\hat{\varepsilon}\frac{T}{2} \right) \tag{5-104}$$

Basepoint m_k and fractional delay μ_k are therefore defined for $T/2$ intervals:

$$k\frac{T}{2} + (2\hat{\varepsilon})\frac{T}{2} = m_k T_s + \mu_k T_s \tag{5-105}$$

A second decimation takes place to produce one error signal per symbol. This process is completely slaved to the first one: we take precisely (for $T/2$) every second basepoint, $m_{n=2n}$, to produce an error detector output. A detailed discussion of the entire control operation is postponed to Section 9.2.

5.6.2 NDA Timing Recovery

From the non-data-aided objective function (5-36) we can derive a simple timing error algorithm. Differentiating with respect to ε yields

$$\begin{aligned}\frac{\partial L(\varepsilon)}{\partial \varepsilon} &= \frac{\partial}{\partial \varepsilon} \sum_{n=0}^{N-1} |z(nT + \varepsilon T)|^2 \\ &= \sum_{n=0}^{N-1} 2 \operatorname{Re}\{z(nT + \varepsilon T)\, \dot{z}^*(nT + \varepsilon T)\}\end{aligned} \tag{5-106}$$

Since summation is performed in the loop filter, we obtain for the error signal (setting $\varepsilon = \hat{\varepsilon}$ in the above equation)

$$x(nT) = \operatorname{Re}\{z(nT + \hat{\varepsilon}\, T)\, \dot{z}^*(nT + \hat{\varepsilon}\, T)\} \tag{5-107}$$

Differentiation can be approximated by taking symmetric differences. Equation (5-96) is a good approximation for $T_s = T/4$ and a reasonable one for $T_s = T/2$. For $T/2$ we obtain the error signal

$$x(nT) = \operatorname{Re}\left\{ z(nT{+}\hat{\varepsilon}T)\left[z^*\left(nT{+}\frac{T}{2}{+}\hat{\varepsilon}T\right) - z^*\left(nT{-}\frac{T}{2}{+}\hat{\varepsilon}T\right)\right]\right\} \tag{5-108}$$

The mean value of the error signal taken with respect to the noise and symbol sequence is found to be

$$\begin{aligned}E[x(nT)] = \operatorname{Re}\Bigg\{ \sum_{m=0}^{N-1} h([n{-}m]T{-}eT) \Bigg[& h^*\left([n{-}m]T{-}eT{-}\frac{T}{2}\right) \\ & - h^*\left([n{-}m]T{-}eT{+}\frac{T}{2}\right)\Bigg]\Bigg\}\end{aligned} \tag{5-109}$$

with timing error $e = \varepsilon_0 - \hat{\varepsilon}$. For symmetric pulses the mean error signal vanishes for $e = 0$. Thus, stable tracking point and sampling instant of the matched filter output used for detection coincide.

Remark: The same comments with respect to interpolation/decimation control as in Section 5.6.1 apply here.

5.6.3 Main Points

A timing error signal is obtained by differentiating the objective function. Both DD and NDA algorithms are derived. The differentiation is approximated by taking symmetric differences.

Bibliography

[1] F. Gardner, "A BPSK/QPSK Timing-Error Detector for Sampled Receivers," *IEEE Trans. Commun.*, vol. COM-34, pp. 423–429, May 1986.

[2] M. Oerder and H. Meyr, "Derivation of Gardner's Timing Error Detector from the Maximum Likelihood Principle," *IEEE Trans. Commun.*, vol. COM-35, pp. 684–685, June 1987.

5.7 Timing Error Feedback Systems at Symbol Rate

The preceding material has demonstrated that for $\alpha < 1$ excess bandwidth two samples per symbol are sufficient for timing recovery. From that reasoning it is tempting to conclude that two samples are necessary as well as sufficient.

This conclusion is erroneous. There are *error* detector algorithms (both DA and NDA) whereby timing can be recovered from the T-spaced samples of the *analog* receiver filter which need not necessarily be a matched filter. The sampling instant is controlled by a numerically controlled oscillator (NCO) (see Figure 5-13) which performs *synchronous* sampling. If the matched filter is realized digitally, there is no way around using the higher sampling rate $1/T_s$ to satisfy the sufficient statistic conditions. Sampling at symbol rate $1/T$ does not provide the information required for timing recovery.

In practice, quite frequently the analog filter is not matched to the signal. Timing recovery must be achieved from symbol-rate samples of a possibly severely distorted signal. Typically, the dominating impairment is phase distortion resulting in heavily skewed pulses $g(t)$. Our approach in this section will be as follows. We will first derive symbol-rate timing recovery algorithms from analog matched filter outputs. In a second step we will modify these algorithms to cope with nonideally filtered signals.

The matched filter output $z(mT + \varepsilon T)$ equals

$$z_m(\varepsilon) = \sum_{k=-\infty}^{\infty} r_f(kT_s)\, g_{\mathrm{MF}}(mT + \varepsilon T - kT_s) \tag{5-110}$$

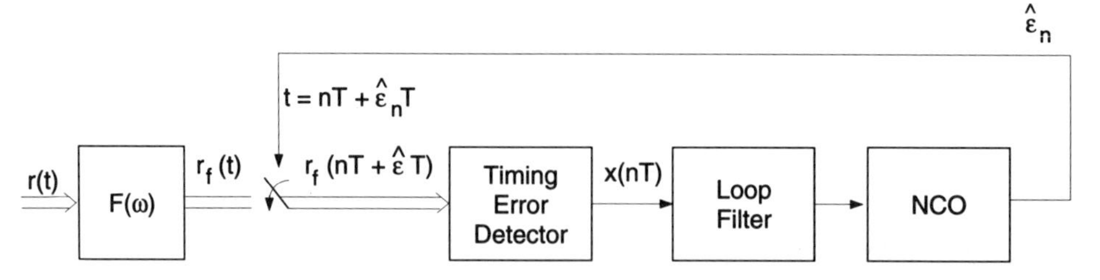

Figure 5-13 Synchronous Sampling in an Error Feedback System. Structure of timing error feedback system operating at symbol rate 1/T

with the (phase-corrected) received signal

$$r_f(kT_s) = \sum_{n=0}^{N-1} a_n \, g(kT_s - nT - \varepsilon_0 T) \, e^{j(\theta_0 - \hat{\theta})} + n(kT_s) \tag{5-111}$$

Notice that we distinguish between the *true* (unknown) parameters $(\varepsilon_0, \theta_0)$ and the trial parameters (ε, θ) in the last two equations. In the sequel we assume that an accurate phase estimate $\hat{\theta}$ is available, thus $\exp\left[j\left(\theta_0 - \hat{\theta}\right)\right] \simeq 1$. Inserting the right-hand side of (5-111) into (5-110) yields

$$\begin{aligned} z_m(\varepsilon) &= \sum_{n=0}^{N-1} a_n \sum_{k=-\infty}^{\infty} [g(kT_s - nT - \varepsilon_0 \, T) + n(kT_s)] \, g_{\mathrm{MF}}(mT + \varepsilon T - kT_s) \\ &= \sum_{n=0}^{N-1} a_n \; h_{n,m}(\varepsilon - \varepsilon_0) + n_g(mT) \end{aligned} \tag{5-112}$$

where $h_{n,m}(\varepsilon - \varepsilon_0)$ is defined as

$$h_{n,m}(\varepsilon - \varepsilon_0) = \sum_{k=-\infty}^{\infty} g(kT_s - nT - \varepsilon_0 T) g_{\mathrm{MF}}(mT + \varepsilon T - kT_s) \tag{5-113}$$

For a symmetric prefilter $F(\omega)$ it was shown that the noise process

$$n_g(mT) = \sum_{k=-\infty}^{\infty} n(kT_s) \, g_{\mathrm{MF}}(mT + \varepsilon T - kT_s) \tag{5-114}$$

is white. We can now *formally* write a log-likelihood function for the samples $\{z(nT)\}$ over the L-sample segment:

$$L\left(\mathbf{z}_L \mid \mathbf{a}, \varepsilon, \theta\right) = \text{const.} + \sum_{m=J}^{J+L-1} \left| z_m(\varepsilon) - \sum_{n=0}^{N-1} a_n \; h_{n,m}(\varepsilon - \varepsilon_0) \right|^2 \tag{5-115}$$

For the present it should not bother us that the log-likelihood function is purely formal since it contains the *true* value ε_0. Since we are interested in error feedback algorithms, we consider only values of ε close to the true value ε_0. This will allow us to approximate $h_{n,m}(\varepsilon - \varepsilon_0)$ in the vicinity of ε_0. For data-aided timing estimation the log-likelihood function reads

$$L\left(\mathbf{z}_L \mid \hat{\mathbf{a}}, \varepsilon, \theta\right) = \text{const.} + \sum_{m=J}^{J+L-1} \left| z_m(\varepsilon) - \sum_{n=0}^{N-1} \hat{a}_n \; h_{n,m}(\varepsilon - \varepsilon_0) \right|^2 \tag{5-116}$$

Eliminating all terms irrelevant to optimization from (5-116), we obtain the objective function

$$\begin{aligned} L_1\left(\mathbf{z}_L \mid \hat{\mathbf{a}}, \varepsilon, \theta\right) = & \left(\sum_{m=J}^{J+L-1} \left|z_m(\varepsilon)\right|^2\right) \\ & + 2\text{Re}\left\{\sum_{m=J}^{J+L-1} z_m(\varepsilon) \sum_{n=0}^{N-1} \hat{a}_n^*\, h_{n,m}^*(\varepsilon-\varepsilon_0)\right\} \end{aligned} \tag{5-117}$$

The error signal is obtained by differentiating (5-117) with respect to ε and evaluating the function at $\varepsilon = \hat{\varepsilon}$:

$$\begin{aligned} \frac{\partial}{\partial \varepsilon} L_1\left(\mathbf{z}_L \mid \hat{\mathbf{a}}, \varepsilon, \theta\right) = & \, 2\text{Re}\left\{\sum_{m=J}^{J+L-1} \dot{z}_m(\varepsilon)\, z_m(\varepsilon)\right\} \\ & + 2\text{Re}\left\{\sum_{m=J}^{J+L-1} \dot{z}_m(\varepsilon) \sum_{n=0}^{N-1} \hat{a}_n^*\, h_{n,m}^*(\varepsilon-\varepsilon_0)\right. \\ & \left. + \sum_{m=J}^{J+L-1} z_m(\varepsilon) \sum_{n=0}^{N-1} \hat{a}_n^* \dot{h}_{n,m}^*\,(\varepsilon-\varepsilon_0)\right\} \end{aligned} \tag{5-118}$$

A class of particularly important algorithms can be derived from the third line of (5-118) which works with T-spaced samples of the matched filter, while the first two lines require suitable approximation of the derivative $\dot{z}(mT + \varepsilon T)$. We thus neglect the first two lines of (5-118). Doing so we later will have to verify that this approximation indeed leads to a working timing error signal.

Since we are concerned with an error feedback system, only estimates $\hat{\varepsilon}$ close to the true values are of interest. We thus approximate

$$\dot{h}_{n,m}(\varepsilon - \varepsilon_0) \simeq \dot{h}_{n,m}(0) \tag{5-119}$$

For simplicity we use the simpler notation $\dot{h}_{n,m} = \dot{h}_{n,m}(0)$ from now on. Hence, the error signal at time l for the L samples from $l - (L-1)$ to l becomes

$$x(lT) = \text{Re}\left\{\sum_{m=l-(L-1)}^{l} z_m(\hat{\varepsilon}) \sum_{n=0}^{N-1} \hat{a}_n^*\, \dot{h}_{n,m}^*\right\} \tag{5-120}$$

As the above result stands it gives little insight in what really happens. It helps to introduce a vector-matrix notation.

The sum over m can be seen as the mth component of the signal vector $\mathbf{s}_h$:

$$[\mathbf{s}_h]_m = \left[\hat{\mathbf{a}}^T \dot{\mathbf{H}}\right]_m^* = \sum_{n=0}^{N-1} \dot{h}_{n,m}^*\, \hat{a}_n^* \tag{5-121}$$

The entire error signal then is an inner product of the truncated vector

$$\mathbf{z}_L^T = \underbrace{\left(0 \,\ldots\, 0 \;\; z_{l-(L-1)} \;\cdots\; z_l \;\; 0 \,\ldots\, 0\right)}_{N-\text{dimensional}} \tag{5-122}$$

and the signal vector $\mathbf{s}_h$:

$$x(lT) = \mathrm{Re}\left\{ \underbrace{\hat{\mathbf{a}}^H \dot{\mathbf{H}}^*}_{\mathbf{s}_h^T} \mathbf{z}_L \right\} \tag{5-123}$$

For a causal system only symbols a_i, $i \leq l$ are available. For symmetry reasons to become clear soon we also truncate $\hat{\mathbf{a}}$ at $i < l-(L-1)$. In short, we use the symbol set $\hat{\mathbf{a}}_L^T = \left(\hat{a}_{l-(L-1)} \;\cdots\; \hat{a}_l\right)$ corresponding to the sample set $\mathbf{z}_L^T = \left(z_{l-(L-1)} \;\cdots\; z_l\right)$ to form the error signal. The $(N \times N)$ matrix $\dot{\mathbf{H}}$ can then also be truncated to $(L \times L)$ elements (Figure 5-14). We obtain a causal error signal

$$x(lT) = \mathrm{Re}\left\{ \hat{\mathbf{a}}_L^H \dot{\mathbf{H}}_L^* \mathbf{z}_L \right\} \tag{5-124}$$

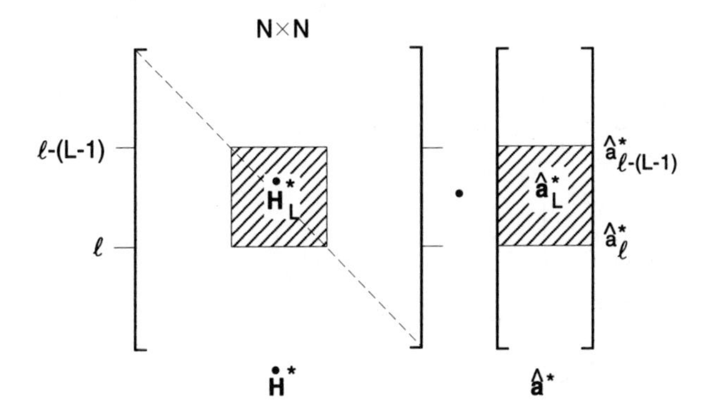

$$\underbrace{\begin{bmatrix} c_0(\ell) \\ c_1(\ell) \\ \vdots \\ c_{L-1}(\ell) \end{bmatrix}}_{\mathbf{c}(\ell)} = \underbrace{\begin{bmatrix} 0 & \dot{h}_{0,1} & \dot{h}_{0,2} & \cdots & \\ \dot{h}_{1,0} & 0 & \dot{h}_{1,2} & & \\ \vdots & & 0 & & \\ & & & 0 & \\ & & & & 0 \end{bmatrix}}_{\dot{\mathbf{H}}_L^*} \cdot \underbrace{\begin{bmatrix} \hat{a}_{\ell-(L-1)} \\ \vdots \\ \hat{a}_\ell \end{bmatrix}}_{\hat{\mathbf{a}}_L^*}$$

Figure 5-14 Truncation of Matrix $\dot{\mathbf{H}}^*$ and Vector $\hat{\mathbf{a}}^*$ to Generate a Causal Error Signal

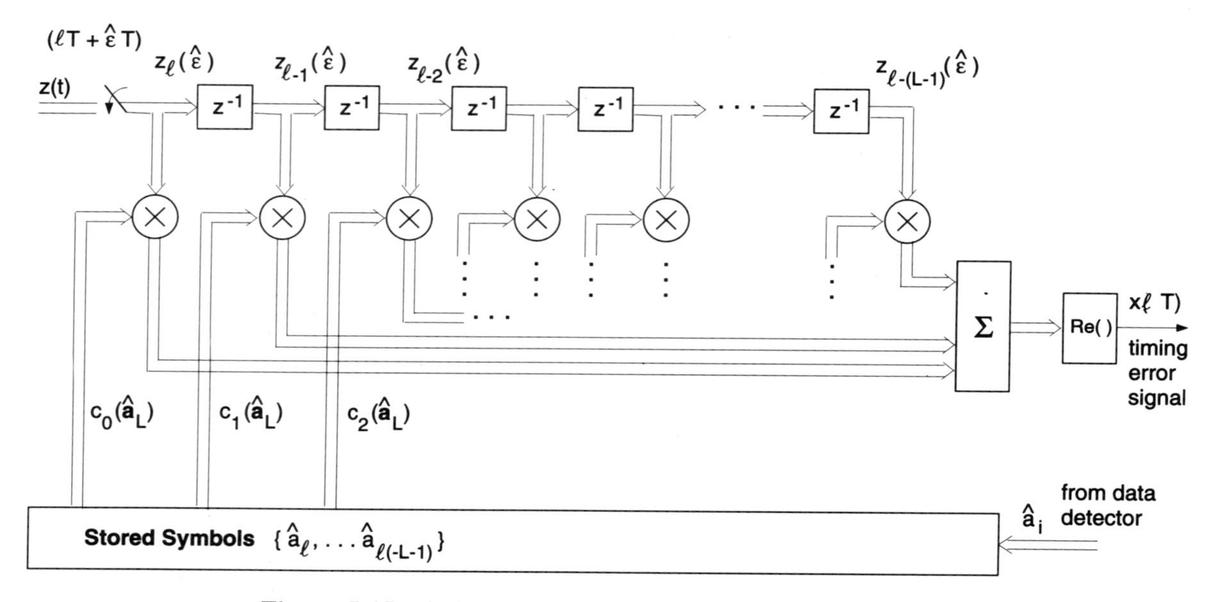

Figure 5-15 Timing Error Signal as Output of a Transversial Filter with Time Variable Coefficients

The matrix $\dot{\mathbf{H}}$ possesses skew symmetry

$$\begin{aligned} \mathrm{Re}\left\{\dot{h}_{i,j}\right\} &= -\mathrm{Re}\left\{\dot{h}_{j,i}\right\} \\ \mathrm{Im}\left\{\dot{h}_{i,j}\right\} &= \mathrm{Im}\left\{\dot{h}_{j,i}\right\} \end{aligned} \tag{5-125}$$

For real pulses $g(t)$ the matrix becomes real.

The timing error detector has a useful interpretation as the output of a time variable transversal filter shown in Figure 5-15 where the coefficients of the taps are given by

$$c_i = \left[\hat{\mathbf{a}}_L^H \dot{\mathbf{H}}_L^*\right]_i \qquad 0 \le i \le L-1 \tag{5-126}$$

Example: For $L = 2$ we obtain (see Figure 5-16)

$$\begin{aligned} x(lT) &= \mathrm{Re}\left\{ (\hat{a}_{l-1}^* \quad \hat{a}_l^*) \begin{pmatrix} 0 & \dot{h}_{0,1}^* \\ \dot{h}_{1,0}^* & 0 \end{pmatrix} \begin{pmatrix} z_{l-1}(\hat{\varepsilon}) \\ z_l(\hat{\varepsilon}) \end{pmatrix} \right\} \\ &= \mathrm{Re}\left\{ z_{l-1}(\hat{\varepsilon})\, \hat{a}_l^*\, \dot{h}_{0,1}^* + z_l(\hat{\varepsilon})\, \hat{a}_{l-1}^*\, \dot{h}_{1,0}^* \right\} \end{aligned} \tag{5-127}$$

For a real pulse $g(t)$ the matrix $\dot{\mathbf{H}}$ is real. With $-\dot{h}_{1,0} = \dot{h}_{0,1}$ we obtain

$$x(lT) = \mathrm{const.Re}\left\{ z_{l-1}(\hat{\varepsilon})\, \hat{a}_l^* - z_l(\hat{\varepsilon})\, \hat{a}_{l-1}^* \right\} \tag{5-128}$$

This algorithm was first proposed by Mueller and Müller in their seminal paper [1] and has been widely used in telephony modem applications.

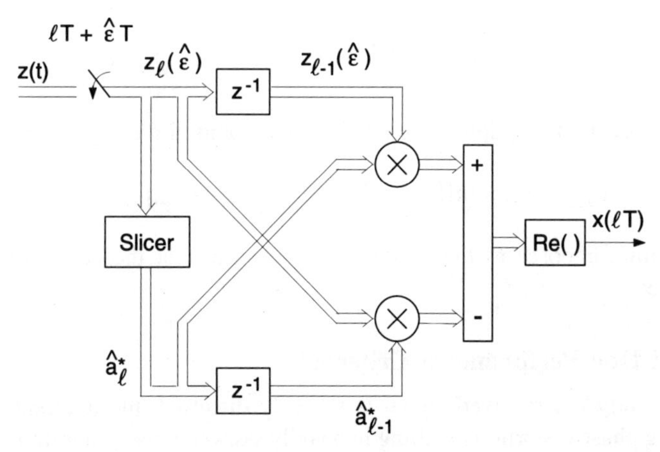

Figure 5-16 Mueller and Müller Algorithm

It is instructive to consider the error signal for zero timing error. Ideally, we would expect $e = 0$ for $(\varepsilon_0 - \hat{\varepsilon}) = 0$ and no additive noise. This is indeed what we obtain for Nyquist pulses. The (noise-free) matched filter output $z_m(\hat{\varepsilon})$ equals

$$z_m(\hat{\varepsilon}) = \sum_{n=0}^{N-1} a_n h_{n,m}(\hat{\varepsilon} - \varepsilon_0) \tag{5-129}$$

For Nyquist pulses and zero timing error $z_m(\varepsilon_0) = a_m h_{0,0}$. Furthermore, assuming error-free detection $\hat{\mathbf{a}}_L = \mathbf{a}_L$ we obtain for the error signal

$$x(lT) = 2\,\mathrm{Re}\Big\{(\mathbf{a}_L)^H\ \dot{\mathbf{H}}^*\ (\mathbf{a}_L)\Big\} \tag{5-130}$$

Due to the skew symmetry, $x(lT) = 0$ for any sequence $\mathbf{a}$.

Let us next consider the (noise-free) error signal $x(lT)$ as a function of the timing error:

$$\begin{aligned} z_{l-1}\ \hat{a}_l^* &= \sum_n \hat{a}_l^*\ a_n\ h_{n,l-1}(\hat{\varepsilon} - \varepsilon_0) \\ &= |a_l|^2\ h_{l,l-1}(\hat{\varepsilon} - \varepsilon_0) + \sum_{n \neq l} \hat{a}_l^*\ a_n\ h_{n,l-1}(\hat{\varepsilon} - \varepsilon_0) \\ &= |a_l|^2\ h(-T + [\hat{\varepsilon} - \varepsilon_0]) + \sum_{n \neq l} \hat{a}_l^*\ a_n\ h_{n,l-1}(\hat{\varepsilon} - \varepsilon_0) \end{aligned} \tag{5-131}$$

and

$$\begin{aligned} z_l\ \hat{a}_{l-1}^* &= \sum_n \hat{a}_{l-1}^*\ a_n\ h_{n,l}(\hat{\varepsilon} - \varepsilon_0) \\ &= |a_{l-1}|^2\ h_{l-1,l}(\hat{\varepsilon} - \varepsilon_0) + \sum_{n \neq l-1} (\cdot) \\ &= |a_{l-1}|^2\ h(T + [\hat{\varepsilon} - \varepsilon_0]) + \sum_{n \neq l-1} (\cdot) \end{aligned} \tag{5-132}$$

For independent and equiprobable symbols we obtain the mean error signal

$$E[x(lT)] = \text{const.}[h([\hat{\varepsilon} - \varepsilon_0]T - T) - h([\hat{\varepsilon} - \varepsilon_0]T + T)] \tag{5-133}$$

The algorithm chooses as the optimum sampling instant the center of the pulse, as expected.

5.7.1 Ad Hoc Performance Criteria

Frequently the received signal is severely distorted, the dominating impairment being phase distortion resulting in heavily skewed pulses, and timing recovery must be achieved with these pulses. Equation (5-124) then serves as basis for the design of an appropriate timing error detector by choosing the coefficients vector $\mathbf{c}_L$:

$$c_i = [\mathbf{c}_L]_i = F_i\left(\hat{a}_{l-(L-1)}, \ldots, \hat{a}_l\right) \tag{5-134}$$

according to a suitable ad hoc criterion.

We first consider the less severe case of amplitude distortion. When the pulses $g(t)$ are symmetric and the timing recovery algorithm devised for the Nyquist pulses still works due to the symmetry, the sampling instant is chosen at the maximum, as for Nyquist pulses. The difference lies solely in the fact that the ISI causes self-noise distortion at the correct timing instant $\hat{\varepsilon} = \varepsilon_0$.

For heavily skewed pulses, the symmetric-pulse-type algorithms must be modified since they will perform very poorly. We refer to the paper by Müller and Mueller [1] for a detailed discussion.

5.7.2 Main Points

A class of decision-directed (DD) timing error detectors for use in a hybrid timing recovery circuit is derived. The error detector works with samples from the output of an analog filter taken at symbol rate $1/T$. The synchronous sampling process is controlled by an NCO.

The error detector has the form of a finite impulse response (FIR) filter with time-variant coefficients which depend on the last L received symbols.

Bibliography

[1] K. H. Mueller and M. Müller, "Timing Recovery in Digital Synchronous Data Receivers," *IEEE Trans. Commun.*, vol. COM-24, pp. 516–531, May 1976.

5.8 (DD&Dε) Carrier Phasor Estimation and Phase Error Feedback

Replacing ε and $\mathbf{a}$ by its estimates, the objective function (5-11) reads

$$L(\hat{\mathbf{a}}, \hat{\varepsilon}, \theta) = \text{Re}\left\{ \sum_{n=0}^{N-1} \hat{a}_n^* z_n(\hat{\varepsilon}) e^{-j\theta} \right\} \tag{5-135}$$

The phasor estimator needs one synchronized matched filter output sample $z_n(\hat{\varepsilon})$ and a detected symbol $\hat{a}_n$. The objective function is maximized by the phasor

$$e^{j\hat{\theta}} = e^{j \arg \sum\limits_n \hat{a}_n^* \cdot z_n(\hat{\varepsilon})} \tag{5-136}$$

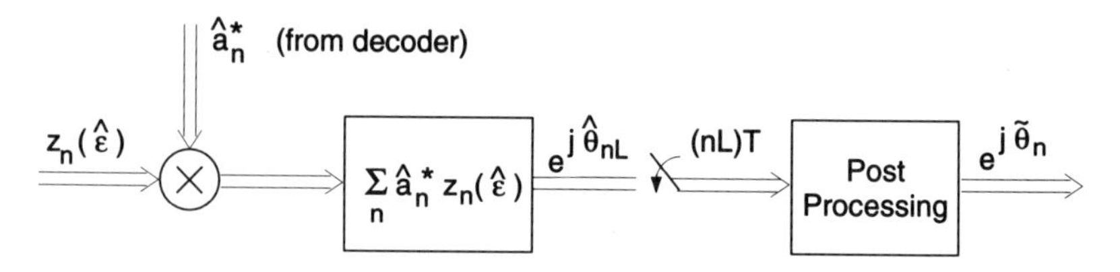

Figure 5-17 Carrier Phasor Estimator (Feedforward)

Equation (5-135) thus defines the ML estimation of a phasor and *not* of the scalar quantity θ. The estimate $\exp\left(j\hat{\theta}\right)$ is unique: for any phasor $\sum_n \hat{a}_n^* z_n(\hat{\varepsilon})$ there exists exactly one phasor $\exp\left(j\hat{\theta}\right)$ which maximizes the objective function. The estimation of a phasor is frequently referred to as *planar filtering* (Figure 5-17).

The property of uniqueness is lost in a phase error feedback system as will be demonstrated next. A phase error signal is readily obtained by differentiating (5-135) with respect to θ. Since the summation is performed in the loop filter, we obtain for the error signal $x(nT) = x_n$

$$x_n = \operatorname{Im}\left[\hat{a}_n^* z_n(\hat{\varepsilon}) e^{-j\hat{\theta}}\right] \tag{5-137}$$

(one signal per T seconds is generated). Since we are using the imaginary part of the phasor *only*, there is no unique error signal:

$$\operatorname{Im}\left[e^{j\hat{\theta}}\right] = \operatorname{Im}\left[e^{j\left(\pi-\hat{\theta}\right)}\right] \tag{5-138}$$

The error detector signal is further processed in a loop filter. An update of the phase estimate is performed in the digital integrator,

$$\hat{\theta}_{n+1} = \hat{\theta}_n + K_1 e_n \tag{5-139}$$

For high SNR the estimate $\hat{a}_n$ is generated in the feedback loop applying a slicer to the signal $z_n(\hat{\varepsilon}) \exp\left(-j\hat{\theta}_n\right)$. For a sufficiently small phase error we obtain

$$z_n(\hat{\varepsilon}) e^{-j\hat{\theta}_n} = a_n h_{0,0}\, e^{j\left(\theta_0 - \hat{\theta}_n\right)} + \text{noise} \approx a_n + \text{noise} \tag{5-140}$$

(Figure 5-18). For multi-level signals the slice operation requires prior amplitude control.

Operation of the digital phase tracker (DPLL) is quite similar to that of an analog PLL. Since the DPLL is nonlinear, it is plagued by the same phenomena as its analog counterpart (see Volume 1).

Example: First-order digital PLL
We assume noise-free operation, perfect timing, and known symbols (DA). Under

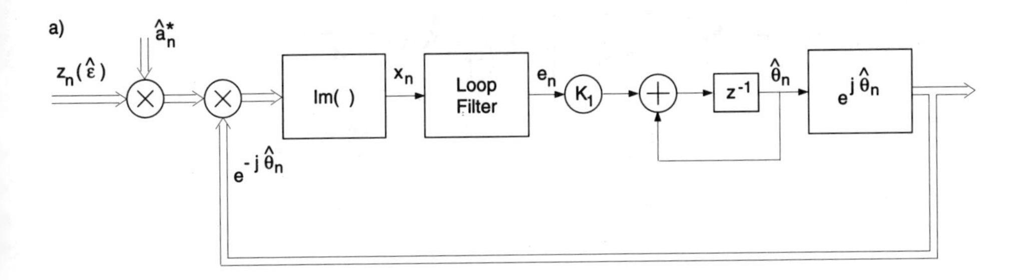

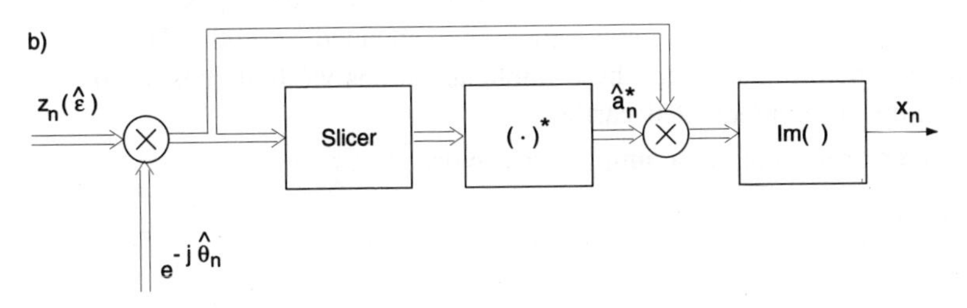

Figure 5-18 (a) Carrier Phase Error Feedback System (Digital PLL), (b) DD Phase Error Detector

these conditions the matched filter output equals

$$z_n = a_n e^{j\theta_0} \tag{5-141}$$

Using (5-137) the error signal x_n reads

$$\begin{aligned} x_n &= \text{Im}\left[|a_n|^2 e^{j(\theta_0 - \hat{\theta}_n)}\right] \\ &= |a_n|^2 \sin\left(\theta_0 - \hat{\theta}_n\right) \end{aligned} \tag{5-142}$$

where $\phi_n = \theta_0 - \hat{\theta}_n$ is the phase error. The nonlinear equation of operation of the digital PLL is obtained by inserting (5-141) into (5-142):

$$\hat{\theta}_{n+1} = \hat{\theta}_n + K_1 \sin\phi_n \qquad K_1 \text{ loop constant} \tag{5-143}$$

5.9 Phasor-Locked Loop (PHLL)

Looking at the digital phase-locked loop depicted in Figure 5-18 we see that the incoming signal is multiplied by the phasor $\exp\left(j\hat{\theta}\right)$. The idea is therefore quite obvious to track the incoming phasor rather than the phase. The structure of

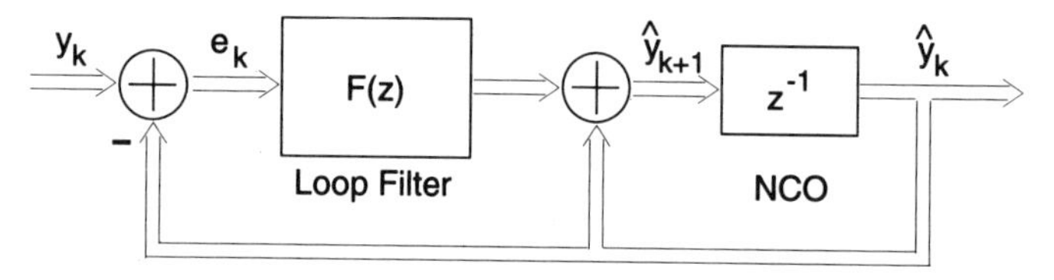

Figure 5-19 Phasor-Locked Loop

such an error feedback system is shown in Figure 5-19 and is given the acronym *PHLL* for **ph**asor **l**ocked **l**oop. The control system is *linear* as opposed to an analog or digital PLL. A phasor estimator may be implemented for both NDA and DD algorithms, but before discussing these implementations we first consider the problem of tracking an unmodulated carrier.

The phasor error detector is a simple subtractor,

$$x_k = y_k - \hat{y}_k \tag{5-144}$$

The error signal is subsequently processed in a (complex) loop filter $F(z)$ with impulse response f_k. The output of the loop filter serves to update the phasor according to

$$\hat{y}_{k+1} = \hat{y}_k + f_k \otimes x_k \tag{5-145}$$

Since the PHLL is a linear circuit, the transfer function from y to $\hat{y}$ is readily established as

$$G(z) = \frac{F(z)}{z - 1 + F(z)} \tag{5-146}$$

For a constant input phasor the steady-state error for a first-order loop, $F(z) = \alpha$ (scalar $0 < \alpha < 1$), is zero as is readily verified. For a rotating phasor $\exp(j\Delta\omega T)$ there does not exist a causal loop filter $F(z)$ for which the steady-state error vanishes. This can be demonstrated as follows. The incoming phasor y_k coincides with the estimate $\hat{y}_k$ (up to a constant) only if the closed-loop transfer function $G(z{=}\exp(j\Delta\omega T))$ is real, which implies that the impulse response is symmetric: $g_k = g_{-k}$. By inspection of $G(z)$ [eq. (5-146)] we see that there exists no loop filter $F(z)$ to meet this condition.

The steady-state phase error for a rotating phasor $\exp(j\Delta\omega T)$ equals the argument of $G(z)$ taken at $\exp(j\Delta\omega T)$

$$\arg G\left(e^{j\Delta\omega T}\right) \simeq -\frac{\Delta\omega T}{\alpha} \qquad \Delta\omega T \ll 1 \tag{5-147}$$

A possible frequency offset must therefore be corrected to a small residual amount in a frequency estimator unit preceding the PHLL.

First-Order PHLL

With $F(z) = \alpha$ (scalar) the loop equation (5-145) for a first-order PHLL reads

$$\begin{aligned}\hat{y}_{k+1} &= \hat{y}_k + \alpha(y_k - \hat{y}_k)\\ &= \hat{y}_k(1-\alpha) + \alpha y_k\end{aligned} \tag{5-148}$$

The step response for a constant reference phasor

$$y_k = y_0 \tag{5-149}$$

is readily obtained in closed form:

$$\hat{y}_k = \hat{y}_0(1-\alpha)^k + y_0\alpha\sum_{i=1}^{k-1}(1-\alpha)^i \tag{5-150}$$

Since the PHLL is a *linear* circuit, convergence to the reference value occurs for any initial estimate $\hat{y}_0$

$$\hat{y}_\infty = y_0 \tag{5-151}$$

Consider the situation where the phasors $\hat{y}_0$ and y_0 are real with opposite sign. The phase error signal of the PLL is zero ($\sin\pi = 0$) and the PLL remains in the false lockpoint. For the PHLL, on the other hand, there exists an error signal which forces the estimate $\hat{y}_k$ to eventually coincide with the reference phasor y_0. This is a fundamental difference to the classical PLL where convergence is dependent on the initial state (see Volume 1, Chapter 4). This difference can be explained by realizing that the PHLL uses both the real and the imaginary parts of the phasor while the PLL uses only the imaginary part.

For a modulated carrier both NDA- and DD-PHLLs exist. Figure 5-20 shows the block diagram of DD-PHLL.

The modulation is removed by multiplying the received signal with the conjugate complex symbol decisions. The resulting phasor is subsequently treated

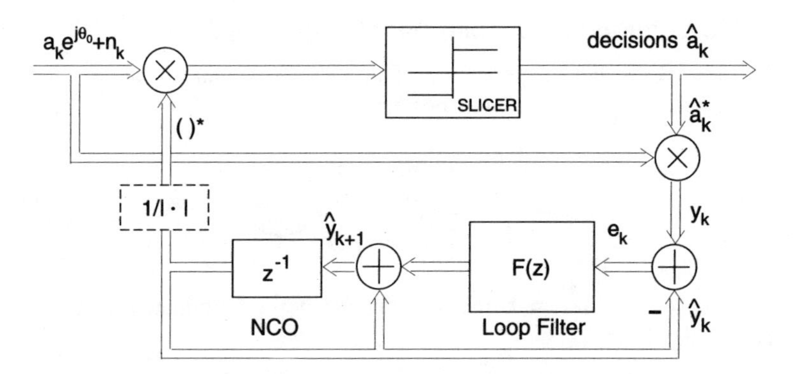

Figure 5-20 Decision-Directed Phasor-Locked Loop

as the observation y_k and filtered by $G(z)$. Amplitude normalization, if necessary, is done in the dashed block. Due to the multiplication with the symbol estimate $\hat{a}_k$ the loop is nonlinear. The concept based upon this idea has been developed by Gardner [1].

1. Although hang-up-free in the linear case, hang-ups may occur due to the slicer nonlinearity: in case the phase error moves toward the bounds of the decision device, the output of the filter can become zero, i.e., a false zero can appear under certain circumstances.
2. For QAM the estimated phasor requires amplitude normalization (as indicated in Figure 5-20).

Bibliography

[1] F. M. Gardner, "Hangup in Phase-Lock Loops," *IEEE Trans. Commun.*, vol. COM-25, pp. 1210–1214, Oct. 1977.

5.10 NDA Carrier Phasor Estimation and Phase Error Feedback Systems for M-PSK

In general, we observe some performance degradation of the synchronizer when ignoring a reliable estimate of the data values. For multiamplitude signals (QAM), which always operate at high SNR, DD phase recovery is exclusively employed. For M-PSK signals the situation is different. There exist NDA circuits which can be employed when no reliable data estimate exists, for example, at low SNR. For analog implementations NDA methods are almost exclusively used in order to minimize complexity, however, for digital implementation this argument is no longer valid.

5.10.1 NDA Phasor Estimation for M-PSK Signals

To remove the data dependency of an M-PSK signal at the matched filter output, we take $z_n(\hat{\varepsilon})$ to the Mth power:

$$\begin{aligned} z_n^M(\hat{\varepsilon}) &= \left[a_n e^{j\theta_0} + m_n\right]^M \\ &= a_n^M e^{jM\theta_0} + \underbrace{m_n'}_{\text{noise}} \end{aligned} \tag{5-152}$$

Since $a^M = \left(e^{j2\pi l/M}\right)^M = 1$, the data dependency is eliminated, and we obtain a signal of the form

$$z_n^M(\hat{\varepsilon}) = e^{jM\theta_0} + m_n' \tag{5-153}$$

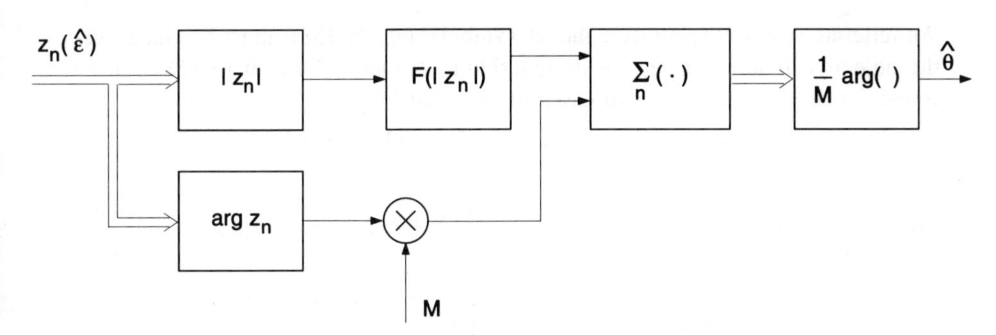

Figure 5-21 Viterbi & Viterbi Algorithm for NDA Phase Recovery

where $m_n^{'}$ is a noise term resulting from the exponentiation. Replacing $\hat{a}_n^* z_n(\hat{\varepsilon})$ by $(z_n(\hat{\varepsilon}))^M$ the objective function (5-135) becomes

$$L(\theta) = \text{Re}\left\{\sum_n (z_n(\hat{\varepsilon}))^M e^{-j\theta M}\right\} \tag{5-154}$$

It is maximized for one phasor

$$\exp\left(j\hat{\theta}M\right) = \exp\left[j \arg \sum_n (z_n(\hat{\varepsilon}))^M\right] \tag{5-155}$$

Notice that the exponentiation of $z_n(\hat{\varepsilon})$ causes an M-fold ambiguity. If $\hat{\theta}$ is a solution maximizing (5-154), so is $\left(\hat{\theta} + 2\pi l/M\right)$ for $l = 0, \ldots, M-1$ since

$$e^{j\left(\hat{\theta}+2\pi l/M\right)M} = e^{j\hat{\theta}M} \tag{5-156}$$

This ambiguity must be resolved by means of differential precoding.

The algorithm can be generalized to

$$F(|z_n(\hat{\varepsilon})|)e^{j \arg z_n(\hat{\varepsilon})M} \tag{5-157}$$

where $F(|\cdot|)$ is an arbitrary function. The algorithm has been derived and analyzed by A. J. Viterbi and A. M. Viterbi [1] (Figure 5-21).

5.10.2 NDA Phase Error Feedback Systems

In this section we treat the elimination of the data dependency by averaging. For known timing and independent data symbols the objective function (5-11) reads

$$L(\theta, \hat{\varepsilon}) = \prod_{n=0}^{N-1} \left[\frac{1}{M}\sum_{i=0}^{M-1} \exp\left\{-\frac{2}{\sigma_n^2}\text{Re}\left[a_n^{*(i)} z_n(\hat{\varepsilon})e^{-j\theta}\right]\right\}\right] \tag{5-158}$$

$a_n^{(i)}$: ith symbol value at time nT

We reiterate that averaging over the M symbols, [eq. (5-158)] must be taken over the objective function $L(\cdot)$, *not* its logarithm. A closed form of (5-158) can be found for M=2 and M=4, two cases of practical interest.

Inserting $a_n^{(0)} = 1, a_n^{(1)} = -1$ into (5-158) yields for 2-PSK:

$$\begin{aligned} L(\theta,\hat{\varepsilon}) &= \frac{1}{2}\prod_{n=0}^{N-1}\left[\exp\left\{-\frac{2}{\sigma_n^2}\text{Re}\big[z_n(\hat{\varepsilon})e^{-j\theta}\big]\right\} + \exp\left\{+\frac{2}{\sigma_n^2}\text{Re}\big[z_n(\hat{\varepsilon})e^{-j\theta}\big]\right\}\right] \\ &= \frac{1}{2}\prod_{n=0}^{N-1}\cosh\left(\frac{2}{\sigma_n^2}\,\text{Re}\big[z_n(\hat{\varepsilon})e^{-j\theta}\big]\right) \end{aligned} \tag{5-159}$$

where $\cosh(\cdot)$ is the hyperbolical cosine. Taking logarithm *after* having taken expected values, we obtain the objective function

$$L_1(\theta,\hat{\varepsilon}) = \sum_{n=0}^{N-1}\ln\,\cosh\left(\frac{2}{\sigma_n^2}\,\text{Re}\big[z_n(\hat{\varepsilon})e^{-j\theta}\big]\right) \tag{5-160}$$

The effect of the averaging operation is best understood by considering the noiseless case

$$\text{Re}\big[z_n(\hat{\varepsilon})e^{-j\theta}\big] = a_n\cos(\theta-\theta_0) \qquad a_n = \pm 1 \tag{5-161}$$

Since the data symbol a_n assumes the two values $a_n = \pm 1$ and since $\ln\cosh(x)$ is a symmetric and increasing function for $|x|$, the value $\hat{\theta} = \theta_0$ maximizes the likelihood function, irrespective of the actual value of the symbol. Notice that if the argument of the maximum is $\hat{\theta}$, then $\hat{\theta}+\pi$ is also a maximum. (We have a π ambiguity.)

A phase error signal is obtained by differentiating (5-160) with respect to θ:

$$\frac{\partial}{\partial\theta}L_1(\theta,\hat{\varepsilon})\bigg|_{\theta=\hat{\theta}} = \frac{2}{\sigma_n^2}\text{Im}\Big[z_n(\hat{\varepsilon})e^{-j\hat{\theta}}\Big]\,\tanh\left(\frac{2}{\sigma_n^2}\text{Re}\Big[z_n(\hat{\varepsilon})e^{-j\hat{\theta}}\Big]\right) \tag{5-162}$$

For optimum operation σ_n^2 must be known since it appears in the argument of the $\tanh(\cdot)$.

It is instructive to consider the two extreme cases of low and high SNR. For high SNR $\left(\sigma_n^2 \to 0\right)$ we may approximate

$$\tanh(x) \simeq \text{sign}(x)$$

to obtain the error signal

$$x_n^{(1)} = \text{Im}\Big[z_n(\hat{\varepsilon})e^{-j\hat{\theta}}\Big]\text{sign}\left(\frac{2}{\sigma_n^2}\text{Re}\Big[z_n(\hat{\varepsilon})e^{-j\hat{\theta}}\Big]\right) \tag{5-163}$$

But since [see (5-161)]

$$\operatorname{sign}\left(\frac{2}{\sigma_n^2}\operatorname{Re}\left[z_n(\hat{\varepsilon})e^{-j\hat{\theta}}\right]\right) = \hat{a}_n \tag{5-164}$$

(for small phase error) we obtain exactly the same expression as (5-137) for the digital PLL.

For low SNR ($\sigma_n^2 \gg 1$), the $\tanh(x)$ is approximated by $\tanh(x) \simeq x$. The error signal (5-162) now becomes

$$x_n^{(2)} = \operatorname{Im}\left[z_n(\hat{\varepsilon})e^{-j\hat{\theta}}\right]\operatorname{Re}\left[z_n(\hat{\varepsilon})e^{-j\hat{\theta}}\right] \tag{5-165}$$

which is equivalent to

$$x_n^{(2)} = \underbrace{\left[\frac{z_n(\hat{\varepsilon})e^{-j\hat{\theta}} - z_n^*(\hat{\varepsilon})e^{j\hat{\theta}}}{2j}\right]}_{\operatorname{Im}(\cdot)}\underbrace{\left[\frac{z_n(\hat{\varepsilon})e^{-j\hat{\theta}} + z_n(\hat{\varepsilon})e^{j\hat{\theta}}}{2}\right]}_{\operatorname{Re}(\cdot)} \tag{5-166}$$

$$= \left[\frac{z_n^2(\hat{\varepsilon})e^{-j2\hat{\theta}} - \left(z_n^2(\hat{\varepsilon})e^{-j2\hat{\theta}}\right)^*}{2j}\right]$$

The error signal is zero for

$$\hat{\theta} = \frac{1}{2}\ \arg\ z_n(\hat{\varepsilon}) + k\pi \tag{5-167}$$

The estimate $\hat{\theta}$ has a twofold ambiguity, compare with eq. (5-156).

The results obtained in this example are of general interest. For high SNR a slicer (hard limiter) produces a reliable estimate $\hat{a}_n$. Since no tracker can perform better than working with known symbols, it is evident that decision-directed methods approach that optimum for high SNR. For low SNR no reliable data estimate is produced by a hard limiter. It is therefore advantageous to use a device which makes better use of the information contained in the signal samples. The expression

$$\operatorname{Re}\left[z_n(\hat{\varepsilon})e^{-j\hat{\theta}}\right] = a_n \cos\left(\theta_0 - \hat{\theta}\right) \quad + \text{noise} \tag{5-168}$$

may be regarded as a "soft-decision" output. The example illustrates that removing unwanted parameters by averaging is always optimal.

The derivation for 4-PSK is quite similar. For $a_n = e^{j2\pi l/4}$, $l = 0, 1, 2, 3,$ we obtain for the averaged log-likelihood function

$$L_1(\theta, \hat{\varepsilon}) = \sum_{n=0}^{N-1}\left[\ln\ \cosh\left(\frac{2}{\sigma_n^2}\operatorname{Re}\left[z_n(\hat{\varepsilon})e^{-j\hat{\theta}}\right]\right) + \ln\ \cosh\left(\frac{2}{\sigma_n^2}\operatorname{Im}\left[z_n(\hat{\varepsilon})e^{-j\hat{\theta}}\right]\right)\right] \tag{5-169}$$

and for the error signal

$$
\begin{aligned}
x_n = \quad & \mathrm{Im}\left[z_n(\hat{\varepsilon})e^{-j\hat{\theta}}\right] \tanh\left(\frac{2}{\sigma_n^2}\mathrm{Re}\left[z_n(\hat{\varepsilon})e^{-j\hat{\theta}}\right]\right) \\
& - \mathrm{Re}\left[z_n(\hat{\varepsilon})e^{-j\hat{\theta}}\right] \tanh\left(\frac{2}{\sigma_n^2}\mathrm{Im}\left[z_n(\hat{\varepsilon})e^{-j\hat{\theta}}\right]\right)
\end{aligned}
\tag{5-170}
$$

For high SNR we obtain the difference between the output of two 2-PSK phase error detectors:

$$
\begin{aligned}
x_n = \quad & \mathrm{Im}\left[z_n(\hat{\varepsilon})e^{-j\hat{\theta}}\right] \mathrm{sign}\left(\frac{1}{\sigma_n^2}\mathrm{Re}\left[z_n(\hat{\varepsilon})e^{-j\hat{\theta}}\right]\right) \\
& - \mathrm{Re}\left[z_n(\hat{\varepsilon})e^{-j\hat{\theta}}\right] \mathrm{sign}\left(\frac{1}{\sigma_n^2}\mathrm{Im}\left[z_n(\hat{\varepsilon})e^{-j\hat{\theta}}\right]\right)
\end{aligned}
\tag{5-171}
$$

Notice that (5-171) is identical to the DD 4-PSK tracker algorithm of (5-137).

For 2-PSK we approximated $\tanh(x) \simeq x$ for small SNR. If the same approximation were made for 4-PSK in (5-170), the error signals would be identically zero, $x_n \equiv 0$, a useless result. A remedy is found if the second term in the Taylor series expansion for $\tanh(x)$ is taken

$$
\tanh x \simeq x - \frac{x^3}{3}
\tag{5-172}
$$

A fourth-order nonlinerarity is introduced which leads to a useful error signal

$$
\begin{aligned}
x_n = \quad & \mathrm{Im}\left[z_n(\hat{\varepsilon})e^{-j\hat{\theta}}\right]\mathrm{Re}\left[z_n(\hat{\varepsilon})e^{-j\hat{\theta}}\right] \\
& \times \left(\mathrm{Re}\left[z_n(\hat{\varepsilon})e^{-j\hat{\theta}}\right] - \mathrm{Im}\left[z_n(\hat{\varepsilon})e^{-j\hat{\theta}}\right]\right)
\end{aligned}
\tag{5-173}
$$

The right-hand side of the previous equation is shown to be the same as

$$
x_n = \left[\frac{z_n^4(\hat{\varepsilon})e^{-j4\hat{\theta}} + \left(z_n^4(\hat{\varepsilon})e^{-j4\hat{\theta}}\right)^*}{2j}\right]
\tag{5-174}
$$

Therefore

$$
\hat{\theta} = \frac{1}{4}\ \arg\ z_n^4(\hat{\varepsilon}) + \frac{2\pi}{4}k \qquad k = 0,\ldots,3
\tag{5-175}
$$

The estimate has a 4-fold ambiguity.

For higher-order M-PSK it appears impossible to obtain a simple closed form for the error signal (5-158). Notice, however, that the Viterbi & Viterbi (V&V) method is applicable to any M. In view of the fact that larger M-PSK ($M \geq 8$) constellations require a higher SNR, thus making DD algorithms applicable, it seems not fruitful to pursue the averaging approach further.

5.10.3 Main Points

Since multilevel signals (QAM) operate at high SNR, decision-directed algorithms are exclusively used. For M-PSK data independent (NDA) algorithms have been derived. The V&V algorithm is a feedforward algorithm which works for all M-PSK signals.

For M=2 and 4–PSK the averaging operation over the data can be expressed in closed form. The two examples nicely illustrate the fact that averaging is *always* optimum. The optimum phase error feedback systems involve rather complicated nonlinear operations which depend on knowledge of the SNR. For the two extreme cases of low and high SNR the operation can be greatly simplified. The resulting phase error detectors are DD for high SNR and "soft-deciding" for low SNR.

All NDA phase error feedback systems suffer from an M-fold phase ambiguity which must be taken care of by proper precoding of the data and a "phase-unwrapping" unit.

Bibliography

[1] A. J. Viterbi and A. M. Viterbi, "Nonlinear Estimation of PSK Modulated Carrier Phase with Application to Burst Digital Transmission," *IEEE Trans. Info. Theory*, vol. IT-32, pp. 543–551, July 1983.

5.11 Phase and Timing Recovery for Nonlinear Modulation Schemes

Nonlinear modulation methods with the notable exception of MSK (minimum shift keying) and G-MSK (Gaussian MSK) [1, 2] have been used very seldom in practice since they seem to offer no performance advantage over linear modulation.

This is even true for transmission over nonlinear channels where a constant envelope is advantageous. As shown by Rapp [3] the bit error rate of M-PSK is surprisingly insensitive to the nonlinear distortion and heavy filtering.

Receivers for coherent CPM (continuous phase modulation) are far more complex to implement than receivers for linear modulation. Synchronization is difficult to achieve. For a thorough discussion the reader is referred to the work by D'Andrea and co-workers [4] and the books by Huber [5] and Anderson and Sundberg [6] where additional references can be found.

Joint phase and timing recovery for CPM signals have been first discussed by de Buda in [7]. This algorithm was further analyzed in [8] and [9]. Based on this technique, further developments were made by D'Andrea and co-workers in [4, 10]. In [11] an algorithm and a single-chip implementation of an MSK receiver is described.

Timing synchronization is discussed in [12] where a digital feedback algorithm based on a suitable nonlinearity is devised. The same nonlinearity is used in [13] for joint timing and frequency synchronization. This algorithm will be discussed in Section 8.6. Performance analysis for offset QPSK (O-QPSK) is documented in [14]. Timing synchronization is also described in [15] where the effects of a flat fading channel are taken in consideration. The maximum-likelihood approach to phase synchronization of CPFSK signals is investigated by Kam [16]. The Cramér-Rao bound for CPM signals has been derived by Moeneclaey [17].

The appropriate description of a phase-modulated signal as linear PAM is presented in the paper by Laurant [18]. The results are interesting as they allow to apply the results for linear modulation to nonlinear modulation. The dissertation by Mehlan [19] features the modem design for vehicle-to-vehicle communication systems employing MSK modulation.

Bibliography

[1] CEPT/CCH/GSM, ed., *GSM Recommendation*, 1988. Draft Standard 5.01–5.10.

[2] E. T. S. Institute, *Digital European Cordless Telecommunications Common Interface*. Tech. Rep., ETSI, F-06561 Valbonne Cedex, France, 1991.

[3] C. Rapp, "Coded and Uncoded M-PSK and CPM Signals on Nonlinear Bandpass Channels: An Approach to a Fair Comparison," *IEEE Proc. of the Global Telecommunications Conference*, pp. 720–724, 1990.

[4] A. D'Andrea, U. Mengali, and R. Reggiannini, "Carrier Phase and Clock Recovery for Continouos Phase Modulated Signals," *IEEE Trans. Commun.*, vol. 35, pp. 1095–1101, Oct. 1987.

[5] J. Huber, *Trelliscodierung. Grundlagen und Anwendungen in der digitalen Übertragungstechnik.* Springer, 1992.

[6] J. B. Anderson, T. Aulin and C. E. Sundberg, *Digital Phase Modulation.* Plenum Press, 1986.

[7] R. de Buda, "Coherent Demodulation of Frequency-Shift Keying with Low Deviation Ratio," *IEEE Trans. Commun.*, vol. 20, pp. 429–435, June 1972.

[8] R. Matyas, "Effect of Noisy Phase References on Coherent Detection of FFSK Signals," *IEEE Trans. Commun.*, vol. 26, pp. 807–815.

[9] J. R. Cruz and R. S. Simpson, "Minimum-Shift Keying Signal Detection with Noisy Reference Signals," *IEEE Trans. Commun.*, vol. 26, pp. 896–902.

[10] A. N. D'Andrea, "Symbol Timing Estimation with CPM Modulation," *IEEE Trans. Communication*, vol. 44, pp. 1362–1372, Oct. 1996.

[11] H. Suzuki, Y. Yamao and H. Kikuchi, "A Single Chip MSK Coherent Demodulator for Mobile Radio Transmission," *IEEE Trans. on Vehicular Techn.*, vol. 34, pp. 157–168, Nov. 1985.

[12] A. N. D'Andrea, U. Mengali, and R. Reggiannini, "A Digital Approach to Clock Recovery in Generalized Minimum Shift Keying," *IEEE Trans. on Vehicular Techn.*, vol. 39, pp. 227–234, Aug. 1990.

[13] R. Mehlan, Y. Chen, and H. Meyr, "A Fully Digital Feedforward MSK Demodulator with Joint Frequency Offset and Symbol Timing Estimation for Burst Mobile Radio," *IEEE Trans. Vehicular Techn.*, Nov. 1993.

[14] T. Jesupret, M. Moeneclaey, and G. Ascheid, *Digital Demodulator Synchronization.* Tech. Rep. Final Report to ESTEC contract no. 8437/89/NL/RE, 1991.

[15] F. Patenaude, John H. Lodge, and P. A. Galko, "Symbol Timing Tracking for Continuous Phase Modulation over Fast Flat-Fading Channels," *IEEE Trans. on Vehicular Techn.*, pp. 615–626, Aug. 1991.

[16] P. Y. Kam, "Maximum Likelihood Carrier Phase Recovery for Coherently Orthogonal CPFSK Signals," *IEEE Trans. Commun.*, vol. 38, pp. 397–398, Apr. 1990.

[17] M. Moeneclaey, "The Joint Carrier and Symbol Synchronizability of Continuous Phase Modulated Waveforms," *Proceedings of the International Communications Conference*, 1986.

[18] P. A. Laurant, "Exact and Approximate Construction of Digital Phase Modulations by Superposition of Amplitude Modulated Pulses (AMP)," *IEEE Trans. Commun.*, vol. COM 34, pp. 150–160.

[19] R. Mehlan, *Ein Transceiverkonzept für die Interfahrzeugkommunikation bei 64 GHz.* PhD thesis, RWTH Aachen, 1995.

Chapter 6 Performance Analysis of Synchronizers

During data transmission the synchronizer provides an estimate which most of the time exhibits small fluctuations about the true value. The synchronizer is operating in the tracking mode. The performance measure of this mode is the variance of the estimate.

In Section 6.1 and the appendix (Section 6.2) we derive a lower bound on the variance of these estimates. This bound will allow us to compare the variance of practical estimators to that of the theoretical optimum and thus assess the implementation loss.

In Section 6.3 we compute the variance of carrier and symbol synchronizers of practical interest. The tracking performance is first computed under the assumption that the parameters are constant over the memory of the synchronizer. Later on we relax this prescription to investigate the effect of small random fluctuations (oscillator phase noise) and of a small frequency offset.

Occasionally, noise or other disturbances push the estimate away from the stable tracking point into the domain of attraction of a neighboring stable tracking point. This event is called cycle slip (Section 6.4). Cycle slips have a disastrous effect since they affect many symbols. Their probability of occurrence must be at least a few orders of magnitude less frequent than the bit error rate. Cycle slipping is a highly nonlinear phenomenon which defies exact mathematical formalism in many cases. One must resort to computer simulation.

At the start of signal reception the synchronizer has no knowledge about the value of the parameters. During a start-up phase the synchronizer reduces the initial uncertainty to a small steady-state error. This process is called acquisition. To efficiently use the channel, the acquisition time should be short. In Section 6.5 we discuss various methods to optimize the acquisition process.

6.1 Bounds on Synchronization Parameter Errors

In this section we compute bounds on the variance of the estimation errors of synchronization parameters. These bounds will allow us to compare the variance of practical synchronizers to that of the theoretical optimum and thus assess the implementation loss.

We consider the task of joint estimation of a frequency offset Ω, a phase θ and a timing delay ε. All parameters are unknown but nonrandom. The signal samples $s_f(kT_s)$ are given by

$$s_f(kT_s) = \sum_n a_n g(kT_s - nT - \varepsilon T)\, e^{j\theta} e^{j\Omega k T_s} \tag{6-1}$$

The received signal is disturbed by additive noise

$$r_f(kT_s) = s_f(kT_s) \; + \frac{n(kT_s)}{A} \tag{6-2}$$

Equations (6-1) and (6-2) are those of the transmission model introduced in Section 4.3 with the additional property that the frequency offset Ω is introduced into the model.

Since the parameters are unknown but deterministic, the lower bound on the variance is determined by the elements of the inverse Fisher information matrix $\mathbf{J}^{-1}$ which we introduced in Section 1.4. The elements of $\mathbf{J}$ equal

$$J_{il} = -E\left[\frac{\partial^2 \ln p(\mathbf{r}_f|\boldsymbol{\theta})}{\partial\theta_i\, \partial\theta_l}\right] \tag{6-3}$$

We recall that $E[\,\cdot\,]$ indicates averaging over the noise, and $\boldsymbol{\theta}$ is a set of parameters $\boldsymbol{\theta} = \{\theta_1, \ldots, \theta_K\}$.

For the log-likelihood function [eq. (4–69)] we obtain, up to a constant,

$$\ln p(\mathbf{r}_f|\boldsymbol{\theta}) = 2\,\mathrm{Re}\{\mathbf{r}_f^H \mathbf{Q} \mathbf{s}_f\} - \mathbf{s}_f^H \mathbf{Q} \mathbf{s}_f \tag{6-4}$$

Taking the derivative of (6-4) with respect to the trial parameter θ_i we obtain

$$\begin{aligned}\frac{\partial \ln p(\mathbf{r}_f|\boldsymbol{\theta})}{\partial\theta_i} &= 2\,\mathrm{Re}\left\{\mathbf{r}_f^H\, \mathbf{Q}\, \frac{\partial \mathbf{s}_f}{\partial\theta_i}\right\} - 2\,\mathrm{Re}\left\{\mathbf{s}_f^H\, \mathbf{Q}\, \frac{\partial \mathbf{s}_f}{\partial\theta_i}\right\} \\ &= 2\,\mathrm{Re}\left\{\left[\mathbf{r}_f^H - \mathbf{s}_f^H\right] \mathbf{Q}\, \frac{\partial \mathbf{s}_f}{\partial\theta_i}\right\}\end{aligned} \tag{6-5}$$

Taking the second derivative with respect to θ_l leads to

$$\frac{\partial^2 \ln p(\mathbf{r}_f|\boldsymbol{\theta})}{\partial\theta_i\, \partial\theta_l} = 2\,\mathrm{Re}\left\{\left[\mathbf{r}_f^H - \mathbf{s}_f^H\right] \mathbf{Q}\, \frac{\partial \mathbf{s}_f}{\partial\theta_i\, \partial\theta_l} - \frac{\partial \mathbf{s}_f^H}{\partial\theta_l}\, \mathbf{Q}\, \frac{\partial \mathbf{s}_f}{\partial\theta_i}\right\} \tag{6-6}$$

Finally we observe that

$$E[\mathbf{r}_f - \mathbf{s}_f] = E\left[\mathbf{n}_f\right] = 0 \tag{6-7}$$

since on the average the trial signal $\mathbf{s}_f$ must equal the received signal $\mathbf{r}_f$. Therefore

$$E\left[\frac{\partial^2 \ln p(\mathbf{r}_f|\boldsymbol{\theta})}{\partial\theta_i\, \partial\theta_l}\right] = E\left[2\,\mathrm{Re}\left\{-\frac{\partial \mathbf{s}_f^H}{\partial\theta_l}\, \mathbf{Q}\, \frac{\partial \mathbf{s}_f}{\partial\theta_i}\right\}\right] \tag{6-8}$$

But since the right-hand side of (6-7) contains nonrandom quantities, the expectation operation is the quantity itself:

$$E\left[\frac{\partial^2 \ln p(\mathbf{r}_f|\boldsymbol{\theta})}{\partial\theta_i\,\partial\theta_l}\right] = 2\,\mathrm{Re}\left\{-\frac{\partial \mathbf{s}_f^H}{\partial\theta_l}\,\mathbf{Q}\,\frac{\partial \mathbf{s}_f}{\partial\theta_i}\right\} = -J_{il} \tag{6-9}$$

The variance of the estimation error $\left[\theta_i - \hat{\theta}_i(\mathbf{r}_f)\right]$ obeys the inequality

$$\mathrm{var}\left[\theta_i - \hat{\theta}_i(\mathbf{r}_f)\right] \geq J^{ii} \tag{6-10}$$

where J^{ii} is an element of the inverse matrix $\mathbf{J}^{-1}$.

We recall that $\hat{\theta}_i(\mathbf{r}_f)$ is any unbiased estimator. The parameters $\boldsymbol{\theta} = \{\theta, \varepsilon, \Omega\}$ appear in $\mathbf{s}_f$ in a nonlinear manner and thus the estimator error cannot be expressed as the weighted sum of the partial derivatives of $\ln p(\mathbf{r}_f|\boldsymbol{\theta})$ with respect to the various parameters. But since this condition [see eq. (1–211)] is necessary and sufficient for an efficient estimate, we conclude that there exists no such estimator for the synchronization parameters.

It is, however, intuitively plausible that for high SNR or a large number of data symbols the performance of the synchronization parameter estimator should come close to the theoretical optimum. We next want to prove this assertion and quantitatively investigate the asymptotic behavior of the joint synchronization parameter estimator.

Consider again the general parameter vector $\boldsymbol{\theta} = (\theta_1, \ldots, \theta_K)$. For a high SNR or a large number of symbols the ML estimate $\hat{\boldsymbol{\theta}}$ will be clustered tightly around the point $\boldsymbol{\theta}_0 = (\theta_1, \ldots, \theta_K)_0$ of the actual parameter values. We therefore develop the log-likelihood function into a Taylor series retaining terms up to second order:

$$\begin{aligned}\ln\, p(\mathbf{r}_f|\boldsymbol{\theta}) \simeq{}& \ln\, p(\mathbf{r}_f|\boldsymbol{\theta})|_{\boldsymbol{\theta}_0} + \sum_{k=1}^{K} \frac{\partial \ln\, p(\mathbf{r}_f|\boldsymbol{\theta})}{\partial\theta_k}\bigg|_{\boldsymbol{\theta}_0} (\theta_k - \theta_{k,0}) \\ &+ \frac{1}{2}\sum_{k,l=1}^{K} \frac{\partial^2 \ln\, p(\mathbf{r}_f|\boldsymbol{\theta})}{\partial\,\theta_k\,\partial\,\theta_l}\bigg|_{\boldsymbol{\theta}_0} (\theta_k - \theta_{k,0})(\theta_l - \theta_{l,0}) \\ &+ \text{terms of higher order}\end{aligned} \tag{6-11}$$

The ML estimate $\hat{\theta}_i$ then satisfies

$$\frac{\partial \ln p(\mathbf{r}_f|\boldsymbol{\theta})}{\partial\,\theta_i}\bigg|_{\boldsymbol{\theta}_0} + \sum_{l=1}^{K} \frac{\partial^2 \ln\, p(\mathbf{r}_f|\boldsymbol{\theta})}{\partial\theta_i\,\partial\theta_l}\bigg|_{\boldsymbol{\theta}_0} \left(\hat{\theta}_l - \theta_{l,0}\right) = 0 \tag{6-12}$$

because

$$\max_{\boldsymbol{\theta}}\ \ln p(\mathbf{r}_f|\boldsymbol{\theta}) = \ln p(\mathbf{r}_f|\boldsymbol{\theta})|_{\hat{\boldsymbol{\theta}}} \tag{6-13}$$

Using

$$\ln\, p(\mathbf{r}_f|\boldsymbol{\theta}) = 2\,\mathrm{Re}\{\mathbf{r}_f^H \mathbf{Q}\mathbf{s}_f\} - \mathbf{s}_f^H \mathbf{Q}\mathbf{s}_f \tag{6-14}$$

yields

$$\begin{aligned}\frac{\partial \ln p(\mathbf{r}_f|\boldsymbol{\theta})}{\partial \theta_k}\bigg|_{\theta_0} &= 2\,\mathrm{Re}\left\{\mathbf{r}_f^H\,\mathbf{Q}\frac{\partial \mathbf{s}_f}{\partial \theta_k}\bigg|_{\theta_0}\right\} - 2\,\mathrm{Re}\left\{\mathbf{s}_f^H\,\mathbf{Q}\frac{\partial \mathbf{s}_f}{\partial \theta_k}\bigg|_{\theta_0}\right\}\\ &= 2\,\mathrm{Re}\left\{\mathbf{n}_f^H\,\mathbf{Q}\frac{\partial \mathbf{s}_f}{\partial \theta_k}\bigg|_{\theta_0}\right\}\end{aligned} \tag{6-15}$$

Because $\mathbf{r}_f = \mathbf{s}_f|_{\theta_0} + \mathbf{n}_f$, we have

$$\begin{aligned}\frac{\partial^2 \ln p(\mathbf{r}_f|\boldsymbol{\theta})}{\partial \theta_k \partial \theta_l}\bigg|_{\theta_0} &= 2\,\mathrm{Re}\left\{\mathbf{r}_f^H\,\mathbf{Q}\,\frac{\partial^2 \mathbf{s}_f}{\partial \theta_k \partial \theta_l}\bigg|_{\theta_0}\right\} - 2\,\mathrm{Re}\left\{\mathbf{s}_f^H\,\mathbf{Q}\,\frac{\partial^2 \mathbf{s}_f}{\partial \theta_k \partial \theta_l}\bigg|_{\theta_0}\right\}\\ &\quad - 2\,\mathrm{Re}\left\{\frac{\partial \mathbf{s}_f^H}{\partial \theta_l}\,\mathbf{Q}\,\frac{\partial \mathbf{s}_f}{\partial \theta_k}\bigg|_{\theta_0}\right\}\end{aligned} \tag{6-16}$$

$$\frac{\partial^2 \ln p(\mathbf{r}_f|\boldsymbol{\theta})}{\partial \theta_k \partial \theta_l}\bigg|_{\theta_0} = -2\,\mathrm{Re}\left\{\frac{\partial \mathbf{s}_f^H}{\partial \theta_l}\,\mathbf{Q}\,\frac{\partial \mathbf{s}_f}{\partial \theta_k}\bigg|_{\theta_0}\right\} + 2\,\mathrm{Re}\left\{\mathbf{n}_f^H\,\mathbf{Q}\,\frac{\partial^2 \mathbf{s}_f}{\partial \theta_k \partial \theta_l}\bigg|_{\theta_0}\right\} \tag{6-17}$$

Because of weak noise, the second term can be neglected. Inserting (6-17) into (6-12) we obtain

$$\frac{\partial \ln p(\mathbf{r}_f|\boldsymbol{\theta})}{\partial \theta_k}\bigg|_{\theta_0} = \sum_{l=1}^{K} \mathrm{Re}\left\{\frac{\partial \mathbf{s}_f^H}{\partial \theta_l}\,\mathbf{Q}\,\frac{\partial \mathbf{s}_f}{\partial \theta_k}\bigg|_{\theta_0}\right\}(\theta_l - \theta_{l,0}) \tag{6-18}$$

We immediately recognize that the coefficients in the last term are the elements of the Fisher matrix $\mathbf{J}$ (eq. 6-9) so that

$$\frac{\partial \ln p(\mathbf{r}_f|\boldsymbol{\theta})}{\partial \theta_k}\bigg|_{\theta_0} = \left[\mathbf{J}\left(\hat{\boldsymbol{\theta}} - \boldsymbol{\theta}_0\right)\right]_k \tag{6-19}$$

Solving the last equation for the ML estimate we obtain

$$\left(\hat{\boldsymbol{\theta}} - \boldsymbol{\theta}_0\right) = \mathbf{J}^{-1}\begin{pmatrix} \dfrac{\partial \ln p(\mathbf{r}_f|\boldsymbol{\theta})}{\partial \theta_1}\bigg|_{\theta_0} \\ \vdots \\ \dfrac{\partial \ln p(\mathbf{r}_f|\boldsymbol{\theta})}{\partial \theta_K}\bigg|_{\theta_0} \end{pmatrix} \tag{6-20}$$

which has exactly the form of the necessary and sufficient condition for the ML estimate to be *efficient* [eq. (1–211)]. Furthermore, the estimates are asymptotically Gaussian distributed. This follows directly from (6-15) and (6-19). The right-hand side of (6-15) represents a weighted sum of Gaussian random variables for every k. Hence, $\hat{\theta}_k$ is a Gaussian random variable with mean $\theta_{k,0}$ (and thus unbiased).

We now want to apply these general results to the problem of jointly estimating the synchronization parameter set $\boldsymbol{\theta} = \{\theta, \varepsilon, \Omega\}$. The computation of the elements of the Fisher information matrix $\mathbf{J}$ (6-9) is conceptually simple but technically somewhat tricky and for this reason delegated to the appendix (Section 6.2).

In the following two examples we determine the Fisher information matrix for two different estimation problems. While in the first example we consider the joint estimation of the synchronization parameter set $\boldsymbol{\theta} = \{\Omega, \theta, \varepsilon\}$, we assume in the second example that the frequency offset Ω is known when jointly estimating the parameter set $\boldsymbol{\theta} = \{\theta, \varepsilon\}$. The dimension of the Fisher information matrix is determined by the number of unknown parameters. Therefore, the Fisher information matrix is of dimension 3×3 in the first example and is reduced to a 2×2 matrix in the second example. For the special case that only one parameter requires estimation, the Fisher information matrix degenerates to a scalar, and the bound is the Cramér-Rao bound known from the one-dimensional problem.

Example 1

We consider the Fisher information matrix for the joint estimation $\boldsymbol{\theta} = \{\Omega, \theta, \varepsilon\}$. We assume

(i) Independent noise samples $n(kT_s)$. The inverse covariance matrix $\mathbf{Q}$ is thus a diagonal matrix with elements $1/\sigma^2 = \left(N_0/A^2T_s\right)^{-1}$.

(ii) The signal pulse $g(t)$ is real.

(iii) Random data.

Condition (i) is merely for convenience; all results also apply for colored noise. Due to conditions (ii) and (iii) the cross terms $J_{\Omega\varepsilon}$ and $J_{\varepsilon\theta}$ vanish (see Section 6.2). Notice that $g(t)$ stands for the overall pulse shape of the transmission path. The requirement for $g(t)$ to be real therefore imposes demands on the transmitter as well as receiver filter and on the other units such as mixers and oscillator accuracy. Under these conditions, the Fisher matrix equals

$$\mathbf{J} = \begin{pmatrix} J_{\Omega\Omega} & J_{\Omega\theta} & 0 \\ J_{\theta\Omega} & J_{\theta\theta} & 0 \\ 0 & 0 & J_{\varepsilon\varepsilon} \end{pmatrix} \tag{6-21}$$

It can be verified that the lower bounds are given by the following expressions:

$$\begin{aligned} \operatorname{var}\left\{\Omega_0 - \hat{\Omega}\right\} \geq J^{\Omega\Omega} &= \frac{1}{J_{\Omega\Omega} - J_{\Omega\theta}^2/J_{\theta\theta}} \\ &= \left[\frac{2\,E_s}{N_0}\right]^{-1} \left[\frac{(N-1)^2 N}{12}\,T^2 \right. \\ &\quad \left. + \frac{N}{E_g}\int\limits_{-\infty}^{+\infty} t^2\,|g(t)|^2\,dt - N\left(\frac{1}{E_g}\int\limits_{-\infty}^{+\infty} t\,|g(t)|^2\right)^2 dt\right]^{-1} \\ &= \left[\frac{2\,E_s}{N_0}\right]^{-1} \frac{12}{(N-1)^2\,N\,T^2}\,\frac{1}{1+O(N^{-2})} \end{aligned} \tag{6-22}$$

and

$$\operatorname{var}\left\{\theta_0 - \hat{\theta}\right\} \ge J^{\theta\theta} = \frac{1}{J_{\theta\theta} - J_{\Omega\theta}^2 / J_{\Omega\Omega}} = \left[\frac{2\,E_s}{N_0}\right]^{-1} \frac{1}{N}$$

$$\times \left[1 - \frac{\left[\varepsilon_0 T + \frac{1}{E_g} \int\limits_{-\infty}^{+\infty} t\,|g(t)|^2\,dt\right]^2}{\frac{(N-1)^2}{12} T^2 + T^2 \varepsilon_o^2 + \frac{1}{E_g} \int\limits_{-\infty}^{+\infty} t^2 |g(t)|^2 dt + \frac{2T\varepsilon_0}{E_g} \int\limits_{-\infty}^{+\infty} t\,|g(t)|^2 dt}\right]^{-1}$$

$$= \left[\frac{2\,E_s}{N_0}\right]^{-1} \frac{1}{N} \frac{1}{1 - O(N^{-2})} \tag{6-23}$$

and

$$\begin{aligned} \operatorname{var}\{\varepsilon_0 - \hat{\varepsilon}\} &\ge J^{\varepsilon\varepsilon} \\ &= \frac{1}{J_{\varepsilon\varepsilon}} \\ &= \left[\frac{2\,E_s}{N_0}\right]^{-1} \frac{1}{N} \frac{\int\limits_{-\infty}^{+\infty} |G(\omega)|^2\,d\omega}{T^2 \int\limits_{-\infty}^{+\infty} \omega^2\,|G(\omega)|^2\,d\omega} \end{aligned} \tag{6-24}$$

where $O(N^{-2})$ is an expression in the order of N^{-2} and E_s/N_0 is the signal-to-noise ratio at the matched filter output, see (4–142).

For large N we can approximate $J^{\Omega\Omega}$ and $J^{\theta\theta}$ by

$$\operatorname{var}\left\{\Omega_0 - \hat{\Omega}\right\} \ge J^{\Omega\Omega} = \left[\frac{2\,E_s}{N_0}\right]^{-1} \frac{12}{(N-1)^2 N} \tag{6-25}$$

and

$$\operatorname{var}\left\{\theta_0 - \hat{\theta}\right\} \ge J^{\theta\theta} = \left[\frac{2\,E_s}{N_0}\right]^{-1} \frac{1}{N} \tag{6-26}$$

Since $J_{\Omega\theta}^2 / J_{\Omega\Omega} \ge 0$ and $J_{\Omega\theta}^2 / J_{\theta\theta} \ge 0$, the expressions for $J^{\Omega\Omega}$ (6-22) and $J^{\theta\theta}$ (6-23) exemplify that the estimation error variances for θ and Ω are always larger than the variance obtained when only a single parameter is estimated, because in this case the variance of the estimation error is the inverse of the corresponding entry in the main diagonal of $\mathbf{J}$. Given that the Fisher information matrix is diagonal, the estimation error variances of a joint estimation are equal to those obtained when only a single parameter requires estimation. It is evident from the above consideration that we should try to force as many cross-terms of $\mathbf{J}$ to zero as possible. This can be accomplished, for example, by a proper selection of the pulse shape.

The lower bound on the timing error variance [eq. (6-24)] is not influenced by the fact that all synchronization parameters have to be jointly estimated, since the cross coupling terms $J_{\Omega\varepsilon}$ and $J_{\theta\varepsilon}$ vanish under the above conditions. Therefore, the timing error variance is the same as the variance we would expect to find if we had to estimate the timing when phase and frequency were known. $J_{\varepsilon\varepsilon}$ is strongly dependent on the pulse shape, as we would expect for timing recovery. The variance of the timing error increases with decreasing excess bandwidth.

A last comment concerns the dependency of $\text{var}\left\{\theta_0 - \hat{\theta}\right\}$ on ε. This dependency expresses the fact that there exists an optimal timing offset ε for the estimation of this parameter. However, the influence is weak since it decreases with $O\left(N^{-2}\right)$.

Example 2

Now we assume that the frequency offset Ω_0 is known. Then the Fisher information matrix is reduced to a 2×2 matrix. Applying the same conditions as in the previous example and using the results of the appendix (Section 6.2), we get the following diagonal matrix:

$$\mathbf{J} = \begin{bmatrix} J_{\theta\theta} & 0 \\ 0 & J_{\varepsilon\varepsilon} \end{bmatrix} \tag{6-27}$$

and for the lower bounds it follows

$$\text{var}\left\{\theta_0 - \hat{\theta}\right\} \geq J^{\theta\theta} = \frac{1}{J_{\theta\theta}} \quad = \left[\frac{2\,E_s}{N_0}\right]^{-1} \frac{1}{N} \tag{6-28}$$

and

$$\text{var}\{\varepsilon_0 - \hat{\varepsilon}\} \geq J^{\varepsilon\varepsilon} = \frac{1}{J_{\varepsilon\varepsilon}} \quad = \left[\frac{2\,E_s}{N_0}\right]^{-1} \frac{1}{N} \; \frac{\int\limits_{-\infty}^{+\infty} |G(\omega)|^2 \, d\omega}{T^2 \int\limits_{-\infty}^{+\infty} \omega^2 \, |G(\omega)|^2 \, d\omega} \tag{6-29}$$

Following the above argumentation we can conclude that it is possible for a joint estimation of θ and ε to have the same jitter variances as those we would expect to obtain by estimating ε or θ individually, given that all other parameters are known. Due to the diagonal structure of the Fisher information matrix (6-27) we can in addition conclude that the dynamics of the phase and timing error are uncoupled, at least in the steady state, if the synchronizer structure in the joint estimation case meets the lower bounds determined by (6-28) and (6-29).

6.2 Appendix: Fisher Information Matrix

The Fisher information matrix is given by

$$\mathbf{J} = \begin{bmatrix} J_{\Omega\Omega} & J_{\Omega\theta} & J_{\Omega\varepsilon} \\ J_{\theta\Omega} & J_{\theta\theta} & J_{\theta\varepsilon} \\ J_{\varepsilon\Omega} & J_{\varepsilon\theta} & J_{\varepsilon\varepsilon} \end{bmatrix} \tag{6-30}$$

with entries $J_{\theta_i \theta_l}$ [eq. (6-9)]:

$$J_{\theta_i \theta_l} = -E\left[\frac{\partial^2}{\partial\theta_i \partial\theta_l} \ln p(\mathbf{r}_f \mid \boldsymbol{\theta})\right] = 2 \,\mathrm{Re}\left\{\frac{\partial \mathbf{s}_f^H}{\partial \theta_l} \mathbf{Q} \frac{\partial \mathbf{s}_f}{\partial \theta_i}\right\} \tag{6-31}$$

where θ_i, θ_l are elements of the synchronization parameter set $\boldsymbol{\theta} = \{\Omega, \theta, \varepsilon\}$. From (6-31) it follows that $J_{\theta_i \theta_l} = J_{\theta_l \theta_i}$. Throughout this appendix we assume that the second condition given by (4–127) [symmetric prefilter $|F(\omega)|^2$] is satisfied so that $\mathbf{Q}$ is a diagonal matrix with entry $A^2 T_s / N_0$:

$$\mathbf{Q} = \frac{1}{N_0/(T_s A^2)} \mathbf{I} \tag{6-32}$$

($\mathbf{I}$ is the identity matrix). Note that this still leaves the pulse shape $g(t)$ unspecified. With

$$s_f(k\,T_s) = \sum_{n=-(N-1)/2}^{n=(N-1)/2} a_n\, g(k\,T_s - nT - \varepsilon T)\, e^{j(\Omega k T_s + \theta)} \tag{6-33}$$

from (4–96) it follows that

$$\begin{aligned} &\frac{\partial \mathbf{s}_f^H}{\partial \theta_l} \mathbf{Q} \frac{\partial \mathbf{s}_f}{\partial \theta_i} \\ &= \frac{N_0}{A^2\, T_s} \sum_{k=-\infty}^{\infty} \frac{\partial}{\partial \theta_l}\left(\sum_{m=-(N-1)/2}^{(N-1)/2} a_m^*\, g^*(kT_s - mT - \varepsilon T)\, e^{-j(\Omega k T_s + \theta)}\right) \\ &\times \frac{\partial}{\partial \theta_i}\left(\sum_{n=-(N-1)/2}^{(N-1)/2} a_n\, g(kT_s - nT - \varepsilon T)\, e^{+j(\Omega k T_s + \theta)}\right) \end{aligned} \tag{6-34}$$

holds. For reasons to become clear later on we assume a symmetrical interval $n \in [-(N-1)/2,\ (N-1)/2]$ for the transmission of the $(N+1)$ symbols. We discuss the case of statistically independent known symbols.

Performing the operations in (6-34) for the various combinations of parameters yields the elements $J_{\theta_i \theta_l} = J_{\theta_l \theta_i}$. We begin with $J_{\Omega\Omega}$.

1. ***Calculation of*** $J_{\Omega\Omega}$:

$$J_{\Omega\Omega} = 2 \,\mathrm{Re}\left\{\frac{\partial \mathbf{s}_f^H}{\partial \Omega} \frac{\partial \mathbf{s}_f}{\partial \Omega} \frac{1}{N_0/(A^2 T_s)}\right\} \tag{6-35}$$

With

$$\frac{\partial \mathbf{s}_f^H}{\partial \Omega} \frac{\partial \mathbf{s}_f}{\partial \Omega} = \sum_{k=-\infty}^{\infty} \left(\sum_{m=-(N-1)/2}^{(N-1)/2} a_m^* \, g^*(kT_s - mT - \varepsilon T) \, e^{-j(\theta + \Omega k T_s)} \, (-jkT_s) \right) \times \left(\sum_{n=-(N-1)/2}^{(N-1)/2} a_n \, g(kT_s - nT - \varepsilon T) \, e^{+j(\theta + \Omega k T_s)} \, (jkT_s) \right) \tag{6-36}$$

and applying the Parseval theorem (4–97) to the outer sum leads to

$$\frac{\partial \mathbf{s}_f^H}{\partial \Omega} \frac{\partial \mathbf{s}_f}{\partial \Omega} = \sum_{m=-(N-1)/2}^{(N-1)/2} a_m^* \sum_{m=-(N-1)/2}^{(N-1)/2} a_n \, \frac{1}{T_s} \int_{-\infty}^{\infty} t^2 \, g^*(t - mT - \varepsilon T) \, g(t - nT - \varepsilon T) \, dt \tag{6-37}$$

Using the convolution and differentiation theorem of the Fourier transform, the right-hand side of the above equation can be expressed in the frequency domain as

$$\int_{-\infty}^{\infty} t^2 \, g(t - nT - \varepsilon T) \, g^*(t - mT - \varepsilon T) \, dt = \frac{1}{2\pi} \int \left[-\frac{\partial^2}{\partial^2 \omega} \left(G(\omega) \, e^{-j\omega(nT + \varepsilon T)} \right) \right] \left[G^*(\omega) \, e^{j\omega(mT + \varepsilon T)} \right] d\omega \tag{6-38}$$

Partial integration yields

$$\begin{aligned} & \frac{1}{2\pi} \int \left[\frac{\partial}{\partial \omega} \left(G(\omega) \, e^{-j\omega(nT + \varepsilon T)} \right) \right] \left[\frac{\partial}{\partial \omega} \left(G^*(\omega) \, e^{j\omega(mT + \varepsilon T)} \right) \right] d\omega \\ & = \frac{1}{2\pi} \int \left\{ \left| \frac{\partial}{\partial \omega} G(\omega) \right|^2 e^{-j\omega(n-m)T} \right. \\ & \quad - j(nT + \varepsilon T) \, G(\omega) \left(\frac{\partial}{\partial \omega} G^*(\omega) \right) e^{-j\omega(n-m)T} \\ & \quad + j(mT + \varepsilon T) \left(\frac{\partial}{\partial \omega} G(\omega) \right) G^*(\omega) \, e^{-j\omega(n-m)T} \\ & \quad \left. + (nT + \varepsilon T)(mT + \varepsilon T) \, |G(\omega)|^2 \, e^{-j\omega(n-m)T} \right\} d\omega \end{aligned} \tag{6-39}$$

Since $G(\omega)$ and $\partial/\partial\omega\, G(\omega)$ vanish at the boundaries,[1] inserting (6-39) into (6-35) yields for statistically independent symbols

$$\begin{aligned} J_{\Omega\Omega} =& \frac{E\left[|a_n|^2\right]}{N_0/A^2 T_s} \frac{2}{T_s\, 2\pi} \\ &\times \left\{ N \int \left| \frac{\partial}{\partial\omega} G(\omega) \right|^2 d\omega \right. \\ &+ 2\, \mathrm{Im}\left[\left(N\,\varepsilon + \sum_{n=1}^{N} n \right) T \int \left(\frac{\partial}{\partial\omega} G^*(\omega) \right) G(\omega)\, d\omega \right] \\ &+ \sum_{n=1}^{N} n^2\, T^2 \int |G(\omega)|^2\, d\omega \\ &+ N\,\varepsilon^2\, T^2 \int |G(\omega)|^2\, d\omega \\ &\left. + 2\,\varepsilon \sum_{n=1}^{N} n\, T^2 \int |G(\omega)|^2\, d\omega \right\} \end{aligned} \tag{6-40}$$

In obtaining (6-40) we made use of the fact that for large N the double summation over $\sum_{n=1}^{N} \sum_{m=1}^{N} a_n^*\, a_m$ tends to its mean value $\sum_{n=1}^{N} E[a_n^*\, a_n]$; this is a consequence of the strong law of large numbers. The inverse of $J_{\Omega\Omega}$ [eq. (6-40)] can be interpreted as the Cramér-Rao bound for frequency estimation, provided that all other synchronization parameters (θ, ε) are *known*. Now it becomes clear why in (6-33) we took a symmetrical summation interval, because for this interval the frequency bound described by $J_{\Omega\Omega}^{-1}$ attains its maximum value [1].

$J_{\Omega\Omega}$ can be written as

$$\begin{aligned} J_{\Omega\Omega} =& \frac{E\left[|a_n|^2\right]}{N_0/(A^2 T_s)} \frac{2}{T_s\, 2\pi} \\ &\times \left\{ N \int \left| \frac{\partial}{\partial\omega}\, G(\omega) \right|^2 d\omega \right. \\ &+ 2\, \mathrm{Im}\left[N\,\varepsilon\, T \int \left(\frac{\partial}{\partial\omega}\, G^*(\omega) \right) G(\omega)\, d\omega \right] \\ &+ \frac{(N^2-1)\, N}{12}\, T^2 \int |G(\omega)|^2\, d\omega \\ &\left. + N\,\varepsilon^2\, T^2 \int |G(\omega)|^2\, d\omega \right\} \end{aligned} \tag{6-41}$$

[1] Notice that by this condition the ideally band-limited signal with rectangular frequency response $G(\omega)$ is excluded.

With E_s denoting the symbol energy and with the abbreviation E_g for the energy of the pulse $g(t)$:

$$E_s = E[a_n^* \, a_n] \, A^2 E_g = \frac{E[a_n^* a_n]}{2\pi} \, A^2 \int |G(\omega)|^2 \, d\omega \tag{6-42}$$

(6-41) becomes

$$\begin{aligned} J_{\Omega\Omega} = \frac{E_s}{N_0} \, 2 \Bigg\{ & \frac{(N^2-1)N}{12} \, T^2 + T^2 \, N \, \varepsilon^2 \\ & + \frac{N}{2\pi E_g} \int \left| \frac{\partial}{\partial\omega} G(\omega) \right|^2 d\omega \\ & + \frac{2N\varepsilon T}{2\pi E_g} \, \mathrm{Im}\left[\int \left(\frac{\partial}{\partial \omega} \, G^*(\omega) \right) G(\omega) \, d\omega \right] \Bigg\} \end{aligned} \tag{6-43}$$

Occasionally, the equivalent time-domain formulation of $J_{\Omega\Omega}$

$$\begin{aligned} J_{\Omega\Omega} = \frac{E_s}{N_0} \, 2 \Bigg\{ & \frac{(N^2-1) \, N}{12} \, T^2 + T^2 \, N \, \varepsilon^2 \\ & + \frac{N}{E_g} \int t^2 \, |g(t)|^2 \, dt + \frac{2N\varepsilon T}{E_g} \int t \, |g(t)|^2 \, dt \Bigg\} \end{aligned} \tag{6-44}$$

is more convenient for discussion and interpretation of $J_{\Omega\Omega}$.

We observe

a. The assumption of statistically independent data symbols is critical for the validity of the result (6-43). In particular, the bound is invalid for certain periodic patterns which may feign a frequency offset which is not discernible from a true offset Ω_0.

b. The dependency of $J_{\Omega\Omega}$ on ε expresses the fact that there exists an optimal time offset for the estimation of Ω. $J_{\Omega\Omega}$ depends on both the magnitude and on the sign of ε. If we demand that $g(t)$ is real and even, then the sign dependence vanishes as expected in a symmetrical problem where a preferential direction cannot exist. The ε dependence is of order N and therefore becomes rapidly negligible compared to the term of order N^3.

c. The dominant term of $O(N^3)$ is independent of the pulse shape. Only for small N the pulse shape comes into play with order $O(N)$. The term belonging to $\partial G(\omega)/\partial\omega$ is proportional to the rolloff factor $1/\alpha$ (for Nyquist pulses) as can readily be verified.

2. ***Calculation of*** $J_{\varepsilon\varepsilon}$: $J_{\varepsilon\varepsilon}$ is given by

$$J_{\varepsilon\varepsilon} = 2 \ \mathrm{Re} \left\{ \frac{\partial \mathbf{s}_f^H}{\partial \varepsilon} \frac{\partial \mathbf{s}_f}{\partial \varepsilon} \frac{1}{N_0/(A^2 T_s)} \right\} \tag{6-45}$$

Applying the Parseval theorem we obtain the following expression after some straightforward manipulation:

$$\begin{aligned}
\frac{\partial \mathbf{s}_f^H}{\partial \varepsilon} \frac{\partial \mathbf{s}_f}{\partial \varepsilon} &= \sum_{n=-N-1/2}^{N-1/2} E[a_n^* \ a_n] \\
&\times \left(\sum_{k=-\infty}^{k=\infty} \dot{g}^*(kT_s - nT - \varepsilon T) \ T \ e^{-j(\theta + kT_s\Omega)} \ \dot{g}(kT_s - nT - \varepsilon T) \ T \ e^{j(\theta + kT_s\Omega)} \right) \\
&= \sum_{n=-(N-1)/2}^{(N-1)/2} E[a_n^* \ a_n] \ \frac{T^2}{T_s} \int_{-\infty}^{\infty} \dot{g}^*(t - nT - \varepsilon T) \ \dot{g}(t - nT - \varepsilon T) \ dt \\
&= \sum_{n=-(N-1)/2}^{(N-1)/2} E[a_n^* \ a_n] \ \frac{T^2}{2\pi \ T_s} \int_{-\infty}^{\infty} (-j\omega) \ G^*(\omega) \ e^{j\omega nT} \ (j\omega) \ G(\omega) \ e^{-j\omega nT} \ d\omega
\end{aligned} \tag{6-46}$$

where $\dot{g}(t)$ denotes derivation with respect to t. Inserting (6-46) into (6-45) yields

$$J_{\varepsilon\varepsilon} = \frac{E_s}{N_0} 2N \frac{T^2 \int_{-\infty}^{\infty} \omega^2 |G(\omega)| \ d\omega}{\int_{-\infty}^{\infty} |G(\omega)| \ d\omega} \tag{6-47}$$

Notice:

a. In obtaining (6-47) we assumed statistically independent data. If this assumption is violated, the above bound does not hold. Assuming the extreme case of identical symbols and $N \to \infty$, the signal $s_f(kT_s)$ (6-33) becomes a periodic function with period $1/T$. But since the signal is band-limited to $B < 1/T$, $s_f(kT_s)$ becomes a constant from which no timing information can be obtained. As a consequence $J_{\varepsilon\varepsilon}$ equals zero, and the variance becomes infinite.[2]
b. $J_{\varepsilon\varepsilon}$ is independent of Ω and θ.
c. In the case of Nyquist pulses, $J_{\varepsilon\varepsilon}$ is minimum for zero rolloff ($\alpha = 0$). This is as expected as the pulses decrease more rapidly with increasing excess bandwidth.

---

[2] In most cases cross terms can be neglected.

3. ***Calculation of*** $J_{\theta\theta}$: $J_{\theta\theta}$ is given by

$$J_{\theta\theta} = 2 \ \mathrm{Re} \left\{ \frac{\partial \mathbf{s}_f^H}{\partial \theta} \frac{\partial \mathbf{s}_f}{\partial \theta} \frac{1}{N_0/(A^2 T_s)} \right\} \tag{6-48}$$

Inserting (6-34) yields

$$\begin{aligned} &\frac{\partial \mathbf{s}_f^H}{\partial \theta} \frac{\partial \mathbf{s}_f}{\partial \theta} \frac{1}{N_0/(A^2 T_s)} \\ &= \frac{1}{N_0/(A^2 T_s)} \sum_m a_m^* \sum_n a_n \sum_{k=-\infty}^{\infty} g^*(kT_s - mT - \varepsilon T) \, g(kT_s - nT - \varepsilon T) \, dt \end{aligned} \tag{6-49}$$

Employing the Parseval theorem yields

$$J_{\theta\theta} = 2 \ \mathrm{Re} \left\{ \sum_m a_m^* \sum_n a_n \ h_{m,n} \right\} \tag{6-50}$$

with (4–97)

$$h_{m,n} = \frac{1}{T_s} \int_{-\infty}^{\infty} \frac{1}{N_0/(A^2 T_s)} g^*(t - mT) \, g(t - nT) \, dt \tag{6-51}$$

4. ***Remarks***

a. While for $J_{\Omega\Omega}$ and $J_{\varepsilon\varepsilon}$ random data were assumed, $J_{\theta\theta}$ exists for any sequence $\{a_n\}$. This is a consequence of the fact that the quadratic form (6-50) is positive definite.

b. This is most clearly seen if $h_{m,n}$ has Nyquist shape. Then

$$\begin{aligned} J_{\theta\theta} &= 2 \ \mathrm{Re} \left\{ \sum_n |a_n|^2 \ h_{0,0} \right\} \\ &= 2 \sum_n |a_n|^2 \frac{1}{N_0/A^2} E_g \quad = 2 \ N \frac{E_s}{N_0} \end{aligned} \tag{6-52}$$

where E_g and E_s are the pulse and symbol energies, respectively.

c. For random data, by the law of large numbers we obtain

$$J_{\theta\theta} = 2 \ \mathrm{Re} \left\{ N \ E \left[|a_n|^2 \right] \ h_{0,0} \right\} \tag{6-53}$$

which is the true irrespective whether $h_{m,n}$ is of Nyquist type or not. However, note the different reasoning which lead to the same result:

$$J_{\theta\theta} = \frac{E_s}{N_0} \ 2N \tag{6-54}$$

5. ***Calculation of the cross terms*** $J_{\varepsilon\theta}$, $J_{\varepsilon\Omega}$ ***and*** $J_{\theta\Omega}$: The details of the calculations are omitted, since the underlying computational principles are identical to those of $J_{\Omega\Omega}$, $J_{\varepsilon\varepsilon}$, and $J_{\theta\theta}$.

a. ***Calculation of the cross term*** $J_{\varepsilon\theta}$:

$$J_{\varepsilon\theta} = 2 \,\mathrm{Re} \left\{ \frac{\partial \mathbf{s}_f^H}{\partial \theta} \frac{\partial \mathbf{s}_f}{\partial \varepsilon} \frac{1}{N_0/(A^2 T_s)} \right\} \tag{6-55}$$

We assume a random sequence and large N such that the cross term $\{a_m^* \, a_n\}$, $m \neq n$ can be neglected. With

$$\begin{aligned}
&\frac{\partial \mathbf{s}_f^H}{\partial \theta} \frac{\partial \mathbf{s}_f}{\partial \varepsilon} \\
&= \sum_{n=-(N-1)/2}^{n=(N-1)/2} E[a_n a_n^*] \frac{1}{T_s} \int_{-\infty}^{\infty} (-j)\,(-T)\, g^*(t-nT-\varepsilon T)\, \dot{g}(t-nT-\varepsilon T)\, dt \\
&= \sum_{n=-(N-1)/2}^{n=(N-1)/2} E[a_n a_n^*] \frac{jT}{T_s} \frac{1}{2\pi} \int (j\omega)\, |G(\omega)|^2 \, d\omega \\
&= \sum_{n=-(N-1)/2}^{n=(N-1)/2} E[a_n a_n^*] \frac{-T}{T_s} \frac{1}{2\pi} \int \omega \, |G(\omega)|^2 \, d\omega
\end{aligned} \tag{6-56}$$

we obtain

$$J_{\varepsilon\theta} = \; -2\, N \, \frac{E[a_n a_n^*]}{N_0/A^2} \frac{T}{2\pi} \int \omega \, |G(\omega)|^2 \, d\omega \tag{6-57}$$

with the Fourier transform pair

$$h(t) \circ\!\!-\!\!\bullet \frac{|G(w)|^2}{N_0/A^2} \tag{6-58}$$

$J_{\varepsilon\theta}$ reads in the time domain

$$\begin{aligned}
J_{\varepsilon\theta} &= 2NE\left[|a_n|^2\right] \frac{j\,T}{2\pi} \int_{-\infty}^{\infty} (j\omega)\, |G(\omega)|^2 \, d\omega \\
&= 2NE\left[|a_n|^2\right] T \,\mathrm{Re}\left\{ j\, \dot{h}(0) \right\} \\
&= -2NE\left[|a_n|^2\right] T \,\mathrm{Im}\left\{ \dot{h}(0) \right\}
\end{aligned} \tag{6-59}$$

We observe

(i) $J_{\varepsilon\theta}$ vanishes for $|G(-\omega)|^2 = |G(\omega)|^2$. This implies $G^*(-\omega) = G(\omega)$ which is satisfied if $g(t)$ is real.

(ii) Random data were assumed.

b. ***Calculation of the cross term*** $J_{\varepsilon\Omega}$:

$$J_{\varepsilon\Omega} = 2 \text{ Re} \left\{ \frac{\partial \mathbf{s}_f^H}{\partial \Omega} \frac{\partial \mathbf{s}_f}{\partial \varepsilon} \frac{1}{N_0/(A^2 T_s)} \right\} \tag{6-60}$$

We assume a random sequence. With the following expressions:

$$\begin{aligned} &\frac{\partial \mathbf{s}_f^H}{\partial \Omega} \frac{\partial \mathbf{s}_f}{\partial \varepsilon} \\ &= \sum_{n=-(N-1)/2}^{n=(N-1)/2} E[a_n a_n^*] \frac{1}{T_s} \int_{-\infty}^{\infty} (-jt)\,(-T)\, g^*(t-nT-\varepsilon T)\, \dot{g}(t-nT-\varepsilon T)\, dt \\ &= \sum_{n=-(N-1)/2}^{n=(N-1)/2} E[a_n a_n^*]\, jT \frac{1}{T_s} \int_{-\infty}^{\infty} (t+nT+\varepsilon)\, g^*(t)\, \dot{g}(t)\, dt \\ &= \sum_{n=-(N-1)/2}^{n=(N-1)/2} E[a_n a_n^*]\, jT \frac{1}{T_s} \int_{-\infty}^{\infty} (t+\varepsilon)\, g^*(t)\, \dot{g}(t)\, dt \end{aligned} \tag{6-61}$$

we can write $J_{\varepsilon\Omega}$ as

$$J_{\varepsilon\Omega} = 2\, N\, \frac{E[a_n a_n^*]}{N_0 A^2} (-T) \text{ Im} \left\{ \int_{-\infty}^{\infty} (t+\varepsilon)\, g^*(t)\, \dot{g}(t)\, dt \right\} \tag{6-62}$$

Again this entry vanishes for real pulses and random data for any ε.

c. ***Calculation of the cross term*** $J_{\theta\Omega}$: $J_{\theta\Omega}$ is given by

$$J_{\theta\Omega} = 2 \text{ Re} \left\{ \frac{\partial \mathbf{s}_f}{\partial \Omega} \frac{\partial \mathbf{s}_f}{\partial \theta} \frac{1}{N_0/(A^2 T_s)} \right\} \tag{6-63}$$

With the following expressions and assuming random data

$$\begin{aligned} &\frac{\partial \mathbf{s}_f^H}{\partial \Omega} \frac{\partial \mathbf{s}_f}{\partial \theta} \\ &= \sum_{n=-(N-1)/2}^{n=(N-1)/2} E[a_n\, a_n^*] \frac{1}{T_s} \int_{-\infty}^{\infty} (-jt)\,(j)\, g^*(t-nT-\varepsilon T)\, g(t-nT-\varepsilon T)\, dt \\ &= \sum_{n=-(N-1)/2}^{n=(N-1)/2} E[a_n\, a_n^*] \frac{1}{T_s} \int_{-\infty}^{\infty} (t+nT+\varepsilon)\, g^*(t)\, g(t)\, dt \\ &= \sum_{n=-(N-1)/2}^{n=(N-1)/2} E[a_n\, a_n^*] \frac{1}{T_s} \int_{-\infty}^{\infty} (t+\varepsilon)\, g^*(t)\, g(t)\, dt \\ &= \sum_{n=-(N-1)/2}^{n=(N-1)/2} E[a_n\, a_n^*] \frac{1}{T_s} \left\{ \varepsilon E_g + \frac{-j}{2\pi} \int_{-\infty}^{\infty} \left(\frac{\partial}{\partial \omega} G^*(\omega) \right) G(\omega)\, d\omega \right\} \end{aligned} \tag{6-64}$$

we can write $J_{\theta\Omega}$ as

$$J_{\theta\Omega} = 2N \, \frac{E_s}{N_0/A^2} \left\{ \varepsilon T + \frac{1}{2\pi \; E_g} \, \mathrm{Im} \left[\int\limits_{-\infty}^{\infty} \left(\frac{\partial}{\partial \omega} G^*(\omega) \right) G(\omega) \; d\omega \right] \right\} \tag{6-65}$$

or in the time domain as follows:

$$J_{\theta\Omega} = 2 \; N \, \frac{E_s}{N_0} \left\{ \varepsilon T \; + \; \frac{1}{E_g} \int\limits_{-\infty}^{\infty} t \; |g(t)|^2 \; dt \right\} \tag{6-66}$$

If $|g(t)| = |g(-t)|$ or equivalently $g^*(-t) = g(t)$, then $J_{\theta\Omega}$ vanishes for $\varepsilon = 0$.

Bibliography

[1] D. C. Rife and R. R. Boorstyn, "Multiple-Tone Parameter Estimation from Discrete-Time Observations," *Bell Syst. Tech. J.*, vol. 55, pp. 1121–1129, Nov. 1976.

[2] M. H. Meyers and L. E. Franks, "Joint Carrier Phase and Symbol Timing Recovery for PAM Systems," *IEEE Trans. Commun.*, vol. COM-28, pp. 1121–1129, Aug. 1980.

[3] D. C. Rife and R. R. Boorstyn, "Single-Tone Parameter Estimation from Discrete-Time Observations," *IEEE Trans. Inf. Theory*, vol. IT-20, pp. 591–599, Sept. 1974.

[4] M. Moeneclaey, "A Simple Lower Bound on the Linearized Performance of Practical Symbol Synchronizers," *IEEE Trans. Commun.*, vol. COM-31, pp. 1029–1032, Sept. 1983.

[5] C. N. Georghiades and M. Moeneclaey, "Sequence Estimation and Synchronization from Nonsynchronized Samples," *IEEE Trans. Inf. Theory*, vol. IT-37, pp. 1649–57, Nov. 1991.

[6] M. Moeneclaey, "A Fundamental Lower Bound on the Performance of Joint Carrier and Bit Synchronizers," *IEEE Trans. Commun.*, vol. COM-32, pp. 1007–1012, Sept. 1984.

6.3 Tracking Performance of Carrier and Symbol Synchronizers

6.3.1 Introduction

When operating in a steady state, carrier and symbol synchronizers provide estimates which most of the time exhibit small fluctuations with respect to the true synchronization parameters; the synchronizers are operating in the tracking mode.

Assuming unbiased estimates, a convenient measure of the tracking performance is the synchronization error variance, which should be small in order to keep the associated degradation of other receiver functions such as data detection within reasonable limits. As synchronization errors are small during tracking, the system equations describing the synchronizers can be linearized with respect to the stable operating point. This simplifies the evaluation of the synchronization error variance, which is then referred to as the linearized tracking performance. We will evaluate the linearized tracking performance of various carrier and symbol synchronizers operating on the complex envelope $r_f(t)$ of a carrier-modulated PAM signal given by

$$r_f(t) = \sum_m a_m \, g(t - mT - \varepsilon_0 T) \, e^{j\theta_0} \quad + \frac{n(t)}{A} \tag{6-67}$$

The operation to the synchronizer is discussed in terms of the matched filter samples

$$z(nT + \hat{\varepsilon}T) = \sum_m a_m h\left[(n-m)T + (\hat{\varepsilon} - \varepsilon_0)T\right] e^{j\theta_0} + N(nT) \tag{6-68}$$

We use the normalized form of Section 4.3.6 (appendix), eq. (4–155), which holds for a symmetric prefilter and Nyquist pulses. In this case the symbol unit variance $E\left[|a_n|^2\right] = 1$, $h_{0,0}(\hat{\varepsilon} - \varepsilon_0 = 0) = 1$, and the complex noise process $N(nT)$ is white with independent real and imaginary parts, each having a variance of $N_0/2E_s$. Since

$$h_{m,n}(0) = \begin{cases} 1 & m = n \\ 0 & \text{else} \end{cases} \tag{6-69}$$

the matched filter output (ideal timing) exhibits no ISI:

$$z(nT + \varepsilon_0 T) = a_n \, e^{j\theta} + N(nT) \tag{6-70}$$

The likelihood function [see eq. (4–157)] for the normalized quantities reads

$$p(\mathbf{r}_f | \mathbf{a}, \theta, \varepsilon) = c \, \exp\left\{ -\frac{E_s}{N_0} \left[\sum_{n=0}^{N-1} \left(|a_n|^2 - 2\,\mathrm{Re}\left[a_n^* z_n(\varepsilon) \, e^{-j\theta}\right] \right) \right] \right\} \tag{6-71}$$

The linearized tracking performance will be computed under the assumption that the synchronization parameters are constant over the memory of the synchronizers. This restriction is relaxed in Section 6.3.8, where the effects of a small zero-mean random fluctuation (oscillator phase noise) and of a small frequency offset are investigated.

6.3.2 Tracking Performance Analysis Methods

Here we outline the methods for analyzing the linear tracking performance of feedback and feedforward synchronizers.

Feedback Synchronizers

Error feedback synchronizers make use of a (phase or timing) error detector whose output gives a noisy indication about the instantaneous synchronization error. The (filtered) error detector output is used to update the estimate of the synchronizer such that the magnitude of the error detector output is reduced.

The analysis of feedback synchronizers for baseband PAM transmission has already been considered in Section 2.3.1. This method of analysis also pertains to feedback carrier and symbol synchronizers for carrier-modulated signals.

Let us consider a feedback carrier synchronizer. We denote the phase error detector output – when receiving the kth data symbol – by $x_\theta\left(k;\hat{\theta}\right)$, where $\hat{\theta}$ is the estimate of the true carrier phase θ_0. The phase error detector characteristic $g_\theta(\phi)$ is defined as the average of the phase error detector output:

$$g_\theta(\phi) = E\left[x_\theta\left(k;\hat{\theta}\right)\right] \tag{6-72}$$

where $\phi = \theta_0 - \hat{\theta}$ denotes the carrier phase error. It is assumed that $g_\theta(0) = 0$, so that, with appropriate feedback, $\phi = 0$ is a stable equilibrium point. The zero-mean loop noise $N_\theta\left(k;\hat{\theta}\right)$ when receiving the kth data symbol is defined as

$$N_\theta\left(k;\hat{\theta}\right) = x_\theta\left(k;\hat{\theta}\right) - g_\theta\left(\theta_0 - \hat{\theta}\right) \tag{6-73}$$

Its autocorrelation function $R_\theta(k;\phi)$ and power spectral density $S_\theta(\exp(j\omega T);\phi)$ depend only on the synchronization error ϕ:

$$R_\theta(k;\phi) = E\left[N_\theta\left(m;\hat{\theta}\right) N_\theta\left(m+k;\hat{\theta}\right)\right] \tag{6-74}$$

$$S_\theta(\exp(j\omega T);\phi) = \sum_{k=-\infty}^{\infty} R_\theta(k;\phi)\exp(-j\omega kT) \tag{6-75}$$

Linearizing the feedback synchronizer about the stable equilibrium point $\phi = 0$, the carrier phase error $\phi(k)$ when receiving the kth data symbol is given by

$$\phi(k) = -\frac{1}{K_\theta}\sum_m h_\theta(k-m) N_\theta(m;\theta_0) \tag{6-76}$$

As $E[N_\theta(m;\theta_0)] = 0$, the carrier phase estimate is unbiased. The resulting linearized tracking performance is given by

$$\mathrm{var}[\phi] = \frac{1}{K_\theta^2}\,T\int_{-\pi/T}^{\pi/T} \left|H_\theta(\exp(j\omega T))\right|^2 S_\theta(\exp(j\omega T);0)\,\frac{d\omega}{2\pi} \tag{6-77}$$

In (6-76) and (6-77), $K_\theta = (d/d\phi)\,g_\theta(\phi)|_{\phi=0}$ denotes the slope of the phase error detector characteristic at the stable equilibrium point $\phi = 0$, $H_\theta(\exp(j\omega T))$

is the closed-loop transfer function of the carrier synchronizer, and $\{h_\theta(m)\}$ is the corresponding impulse response, i.e., the inverse Fourier transform of $H_\theta(\exp(j\omega T))$.

The above analysis method for carrier synchronizers also applies to symbol synchronizers. One only has to make in (6-72) to (6-77) the following obvious substitutions:

$$
\begin{array}{rclrcl}
\theta_0 & \to & \varepsilon_0 & \hat{\theta} & \to & \hat{\varepsilon} \\
\phi & \to & e & H_\theta(\exp(j\omega T)) & \to & H_\varepsilon(\exp(j\omega T)) \\
x_0\left(k;\hat{\theta}\right) & \to & x_\varepsilon(k;\hat{\varepsilon}) & N_\theta\left(k;\hat{\theta}\right) & \to & N_\varepsilon(k;\hat{\varepsilon}) \\
g_\theta(\phi) & \to & g_\varepsilon(e) & S_\theta(\exp(j\omega T);\phi) & \to & S_\varepsilon(\exp(j\omega T);e) \\
K_\theta & \to & K_\varepsilon & h_\theta(m) & \to & h_\varepsilon(m)
\end{array}
$$

where "→" means "is replaced by." The quantities ε_0, $\hat{\varepsilon}$ and $e = \varepsilon - \hat{\varepsilon}$ are the actual time delay, the time-delay estimate, and the timing error (all normalized with respect to the symbol interval T). The timing error detector output $x_\varepsilon(k;\hat{\varepsilon})$ can be written as the sum of the timing error detector characteristic $g_\varepsilon(e)$ and the zero-mean loop noise $N_\varepsilon(k;\hat{\varepsilon})$. The loop noise has an autocorrelation function $R_\varepsilon(k;e)$ and a power spectral density $S_\varepsilon(\exp(j\omega T);e)$. The timing error detector slope at the stable equilibrium point $e = 0$ equals K_ε. The closed-loop transfer function of the symbol synchronizer and the associated impulse response are denoted by $H_\varepsilon(\exp(j\omega T))$ and $\{h_\varepsilon(m)\}$, respectively.

Feedforward Synchronizers

A feedforward carrier synchronizer produces an estimate $\hat{\theta}$ which maximizes some objective function $L(\theta)$ over the trial value θ of the carrier phase estimate:

$$L\left(\hat{\theta}\right) = \max_{\theta} L(\theta) \tag{6-78}$$

The objective function depends not only on the trial value θ of the carrier phase estimate, but also on the received signal $r_f(t)$; in most cases, $L(\theta)$ is derived from the likelihood function or a suitable approximation thereof. Because of the dependence on $r_f(t)$, $L(\theta)$ is a random function; hence, different realizations of $r_f(t)$ give rise to different carrier phase estimates.

Under normal operating conditions, the estimate $\hat{\theta}$ is close to the true carrier phase value θ_0, so that the following truncated Taylor series expansion is sufficiently accurate for all values of θ within a small interval containing both θ_0 and $\hat{\theta}$:

$$L(\theta) \simeq L(\theta_0) + (\theta - \theta_0)L'(\theta_0) + \frac{1}{2}(\theta - \theta_0)^2 L''(\theta_0) \tag{6-79}$$

where $L'(\theta_0)$ and $L''(\theta_0)$ are the first and second derivative of $L(\theta)$ with respect to θ, evaluated at $\theta = \theta_0$. Expressing that $L(\theta)$ from (6-79) becomes maximum for $\theta = \hat{\theta}$ yields

$$\hat{\theta} - \theta_0 = -\frac{L'(\theta_0)}{L''(\theta_0)} \tag{6-80}$$

In all cases of practical interest, the statistical fluctuation of $L''(\theta_0)$ with respect to its mean value is much smaller than this mean value. Hence, $L''(\theta_0)$ can be safely replaced by $E[L''(\theta_0)]$, where $E[\cdot]$ denotes averaging with respect to the data symbols and the additive noise contained within the received signal $r_f(t)$. This yields

$$\hat{\theta} - \theta_0 = -\frac{L'(\theta_0)}{E[L''(\theta_0)]} \tag{6-81}$$

Usually, the objective function is selected such that $E[L(\theta)]$ becomes maximum for $\theta = \hat{\theta}$; hence, $E[L'(\theta_0)] = 0$ so that the carrier phase estimate is unbiased. From (6-81) we obtain

$$\text{var}[\phi] = \frac{E\left[(L'(\theta_0))^2\right]}{(E[L''(\theta_0)])^2} \tag{6-82}$$

where $\phi = \theta_0 - \hat{\theta}$.

The above method of analysis for feedforward carrier synchronizers also applies to feedforward symbol synchronizers. Obvious modifications of (6-81) and (6-82) yield

$$\hat{\varepsilon} - \varepsilon_0 = -\frac{L'(\varepsilon_0)}{E[L''(\varepsilon_0)]} \tag{6-83}$$

$$\text{var}[e] = \frac{E\left[(L'(\varepsilon_0))^2\right]}{(E[L''(\varepsilon_0)])^2} \tag{6-84}$$

where $e = \varepsilon - \hat{\varepsilon}$ is the timing error and $L(\varepsilon)$ denotes the objective function to be maximized by the timing estimate.

Correspondence Between Feedforward and Feedback Synchronizers

For many feedforward synchronizers, the objective function to be maximized equals the sum of individual contributions which have identical statistical properties. Restricting our attention to carrier synchronizers (a similar derivation is valid for symbol synchronizers), the objective function is given by

$$L(\theta) = \sum_{k=1}^{K} L(k;\theta) \tag{6-85}$$

where $L(k;\theta)$ denotes the contribution corresponding to the instant of reception of the kth data symbol and K is the duration of the observation window measured in symbol intervals; the statistical properties of $L(k;\theta)$ do not depend on the value of k. The resulting phase error ϕ is given by

$$\phi = \frac{1}{E[L''(k;\theta_0)]} \frac{1}{K} \sum_{k=1}^{K} L'(k;\theta_0) \tag{6-86}$$

which follows from (6-81).

Alternatively, one could try to maximize $L(\theta)$ from (6-85) by means of a feedback carrier synchronizer, which uses the following phase error detector output $x_\theta\left(k;\hat{\theta}\right)$:

$$x_\theta\left(k;\hat{\theta}\right) = L'\left(k;\hat{\theta}\right) \tag{6-87}$$

The idea is that, in the steady state, the feedback synchronizer tries to make zero the derivative of $L(k;\theta)$. At the stable equilibrium point $\phi = 0$, the slope K_θ of the phase error detector characteristic and the loop noise $N_\theta(k;\theta_0)$ are given by

$$K_\theta = E[L''(k;\theta_0)] \tag{6-88}$$

$$N_\theta(k;\theta_0) = L'(k;\theta_0) \tag{6-89}$$

Hence, it follows from (6-76) that the corresponding phase error is given by

$$\phi(k) = -\frac{1}{E[L''(k;\theta_0)]} \sum_{m} h_\theta(k-m)\, L'(m,\theta_0) \tag{6-90}$$

where $h_\theta(m)$ denotes the closed-loop impulse response of the carrier synchronizer.

Comparing (6-86) and (6-90), it follows that (within a minus sign) the feedforward phase error (6-86) can be interpreted as a feedback phase error, provided that the closed-loop impulse response $\{h_\theta(m)\}$ in (6-90) is selected as

$$h_\theta(m) = \begin{cases} 1/K & m = 0,1,\ldots,K-1 \\ 0 & \text{otherwise} \end{cases} \tag{6-91}$$

Hence, the tracking performance of the feedforward synchronizer maximizing the objective function (6-85) is the same as for the corresponding feedback synchronizer using the phase error detector (6-87) and having a closed-loop impulse response given by (6-91).

Interaction Between Carrier and Symbol Synchronizers

Several carrier synchronizers make use of a timing estimate, and several symbol synchronizers make use of a carrier phase estimate. Hence, carrier and symbol synchronizers in general do not operate independently of each other.

In the case of a feedback carrier synchronizer which makes use of a timing estimate (the same reasoning holds for a feedback symbol synchronizer which makes use of a carrier phase estimate), the phase error detector characteristic depends not only on the phase error ϕ but also on the timing error e, and will therefore be denoted as $g_\theta(\phi;e)$. Linearizing $g_\theta(\phi;e)$ about the stable equilibrium point $\phi = e = 0$ yields

$$g_\theta(\phi;e) = C_{\theta\phi}\phi + C_{\theta e}e \tag{6-92}$$

where

$$C_{\theta\alpha} = \frac{\partial}{\partial\alpha} g_\theta(\phi;e)\mid_{\phi=e=0} \qquad \alpha \in \{\phi, e\} \tag{6-93}$$

Hence, unless $C_{\theta e} = 0$, it follows that the linearized tracking performance of the feedback carrier synchronizer is affected by the timing error. When $C_{\theta e} = 0$, the linearized tracking performance can be evaluated under the assumption of perfect timing, i.e., $e = 0$. A sufficient condition yielding $C_{\theta e} = 0$ is that $g_\theta(\phi;e)$ is zero for $\phi = 0$, irrespective of the timing error e.

In the case of a feedforward carrier synchronizer which makes use of a timing estimate (the same reasoning holds for a feedforward symbol synchronizer which makes use of a carrier phase estimate) the objective function to be maximized with respect to the trial value θ of the carrier phase estimate depends also on the timing estimate. Assuming that the objective function is of the form

$$L(\theta;\mathbf{e}) = \sum_{k=1}^{K} L(k;\theta;e(k)) \tag{6-94}$$

where $\mathbf{e}$ is the vector of timing errors corresponding to the symbols within the observation window, the tracking performance of the feedforward synchronizer is the same as for a feedback synchronizer with the closed-loop impulse response (6-91) and using the phase error detector output $x_\theta\left(k;\hat{\theta};e\right)$, which is given by

$$x_\theta\left(k;\hat{\theta};e\right) = \frac{\partial}{\partial\theta} L(k;\theta;e)\mid_{\theta=\hat{\theta}} \tag{6-95}$$

The corresponding phase error detector characteristic is

$$g_\theta(\phi;e) = \frac{\partial}{\partial\theta} E[L(k;\theta;e)]\mid_{\theta=\hat{\theta}=\theta_0-\phi} \tag{6-96}$$

Linearizing $g_\theta(\phi;e)$ about the stable equilibrium point $\phi = e = 0$, it follows that the linearized tracking performance is not affected by the timing error, provided that

$$\frac{\partial^2}{\partial\theta\partial e} E[L(k;\theta;e)]\mid_{\substack{\theta=\theta_0 \\ e=0}} = 0 \tag{6-97}$$

When (6-97) holds, the feedforward carrier synchronizer can be analyzed as if perfect timing (i.e., $\hat{\varepsilon} = \varepsilon_0$) were available. A sufficient condition for (6-97) to be satisfied is that $E[L(k;\theta;e)]$ is maximum for $\theta = \theta_0$, irrespective of the timing error e.

For all feedback and feedforward carrier and symbol synchronizers to be considered in Sections 6.3.3 to 6.3.6, we have verified that in the tracking mode there is no interaction between the carrier and symbol synchronizers; basically, this is due to the fact that the baseband pulse at the matched filter output is real and even. Consequently, carrier synchronizers will be analyzed under the assumption of perfect timing, and symbol synchronizers will be analyzed under the assumption of perfect carrier recovery.

6.3.3 Decision-Directed (DD&Dε) Carrier Synchronization

The feedforward version (DD&Dε) of the decision-directed maximum-likelihood phase synchronizer maximizes over θ the function $L(\theta)$, given by Section 5.8:

$$L(\theta) = \sum_{k=1}^{K} \mathrm{Re}[\hat{a}_k^* z_k \exp(-j\theta)] \tag{6-98}$$

where $\hat{a}_k$ is the receiver's decision about the kth data symbol, and z_k is a short-hand notation for $z(kT + \hat{\varepsilon}T)$, the sample of the matched filter output signal taken at the estimate $kT + \hat{\varepsilon}T$ of the kth decision instant. The carrier phase estimate $\hat{\theta}$ which maximizes $L(\theta)$ from (6-98) is given by

$$\hat{\theta} = \arg\left(\sum_{k=1}^{K} \hat{a}_k^* \, z_k\right) \tag{6-99}$$

where $\arg(\cdot)$ denotes the argument function. Hence, instead of performing a search over θ to maximize $L(\theta)$, one can directly compute the feedforward estimate according to (6-99). The feedforward (DD&Dε) carrier synchronizer needs one synchronized matched filter output sample per symbol.

The feedback version of the (DD&Dε) carrier synchronizer makes use of a phase error detector whose output $x_\theta\left(k;\hat{\theta}\right)$ is given by

$$x_\theta\left(k;\hat{\theta}\right) = \mathrm{Im}\left[\hat{a}_k^* \, z_k \, \exp\left(-j\hat{\theta}\right)\right] \tag{6-100}$$

which is the derivative with respect to the kth term in (6-98). The feedback (DD&Dε) carrier synchronizer needs one synchronized matched filter output sample per symbol.

In the following, we will investigate the linearized tracking performance of the feedback version of the carrier synchronizer. As indicated in Section 6.3.2 in the discussion of feedforward and feedback synchronizers, the performance of the feedforward synchronizer can be derived from the result for the feedback synchronizer. According to the discussion of the interaction between carrier and symbol synchronizers, we will assume that $\hat{\varepsilon} = \varepsilon_0$.

When dealing with decision-directed synchronizers, we will make the simplifying assumption that the receiver's decision $\hat{a}_k$ is not affected by additive noise. Hence, for sufficiently small synchronization errors, this assumption yields $\hat{a}_k = a_k$. For example, in the case of M-PSK and in the absence of additive noise, one obtains

$$\hat{a}_k = a_k \exp(j2\pi m/M) \qquad (2m-1)\pi/M < \phi < (2m+1)\pi/M \tag{6-101}$$

which indicates that $\hat{a}_k = a_k$ for $|\phi| < \pi/M$.

Taking (6-70) into account, (6-100) with $\hat{\varepsilon} = \varepsilon_0$ yields

$$x_\theta\left(k;\hat{\theta}\right) = \mathrm{Im}\left[\left|a_k^2\right| \exp(j\phi) + a_k^* N_k\right] \tag{6-102}$$

where

$$N_k = N(kT + \varepsilon_0 T)\, \exp\left(-j\hat{\theta}\right) \tag{6-103}$$

Note that $\{N_k\}$ is a sequence of complex-valued statistically independent Gaussian random variables with statistically independent real and imaginary parts, each having a variance of $N_0/(2E_s)$. The resulting phase error detector characteristic $g_\theta(\phi)$ is given by

$$g_\theta(\phi) = E\left[x_\theta\left(k;\hat{\theta}\right)\right] = \sin(\phi) \tag{6-104}$$

where we have taken into account that $E\left[\left|a_k^2\right|\right] = 1$. Note that $g_\theta(0) = 0$, so that $\phi = 0$ is a stable equilibrium point. The phase error detector slope K_θ and loop noise $N_\theta(k;\theta_0)$ at the stable equilibrium point $\phi = 0$ are

$$K_\theta = 1 \tag{6-105}$$

$$N_\theta(k;\theta_0) = \mathrm{Im}[a_k^* N_k] \tag{6-106}$$

Because of the statistical properties of $\{N_k\}$, the loop noise at the stable equilibrium point $\phi = 0$ is white, and its power spectral density satisfies

$$S_\theta(\exp(j\omega T); 0) = \frac{N_0}{2E_s} \tag{6-107}$$

Consequently, it follows from (6-77) that

$$\mathrm{var}[\phi] = (2B_L T)\frac{N_0}{2E_s} \tag{6-108}$$

where B_L is the one-sided loop bandwidth of the carrier synchronizer:

$$\begin{aligned} 2B_L T &= T \int_{-\pi/T}^{\pi/T} \left|H_\theta(\exp(j\omega T))\right|^2 \frac{d\omega}{2\pi} \\ &= \sum_m h_\theta^2(m) \end{aligned} \tag{6-109}$$

In view of (6-91), the linearized tracking performance of the feedforward version of the (DD&Dε) carrier synchronizer is obtained by replacing in (6-108) the quantity $2B_L T$ by $1/K$.

It is important to note that the linearized tracking error variance (6-108) equals the Cramér-Rao bound, which means that the synchronizer achieves the theoretically optimum performance. However, the result (6-108) has been obtained under the simplifying assumption that additive noise does not affect the receiver's decisions about the data symbols. The effect of decision errors on the tracking performance will be considered in Section 6.3.7.

6.3.4 Non-Decision-Aided Carrier Synchronization

The considered non-decision-aided (NDA) carrier synchronizer maximizes the low-SNR limit of the likelihood function [eq. (5–154)] after taking the expected value $E_{\mathbf{a}}[\cdot]$ with respect to the data sequence. The samples of the matched filter are taken at $nT + \hat{\varepsilon}T$. In the case of a signal constellation with a symmetry angle of $2\pi/M$, the resulting NDA-ML carrier synchronizer operates on the matched filter output samples z_k, raised to the Mth power; we recall that the samples z_k are taken at the estimated decision instants $kT + \hat{\varepsilon}T$.

The feedforward version of the NDA carrier synchronizer maximizes over θ the function $L(\theta)$, given by

$$L(\theta) = \sum_{k=1}^{K} \mathrm{Re}\left[E\left[(a_k^*)^M\right] z_k^M \exp\left(-jM\theta\right)\right] \tag{6-110}$$

The carrier phase estimate $\hat{\theta}$ which maximizes $L(\theta)$ is given by

$$\hat{\theta} = \frac{1}{M} \arg\left(E\left[(a_k^*)^M\right] \sum_{k=1}^{K} z_k^M\right) \tag{6-111}$$

Hence, the feedforward estimate can be computed directly, without actually performing a search over θ. The feedforward NDA carrier synchronizer needs one synchronized matched filter output sample per symbol.

The feedback version of the NDA–ML carrier synchronizer uses a phase error detector, whose output $x_\theta\left(k;\hat{\theta}\right)$ is given by

$$x_\theta\left(k;\hat{\theta}\right) = \mathrm{Im}\left[E\left[(a_k^*)^M\right] z_k^M \exp\left(-jM\hat{\theta}\right)\right] \tag{6-112}$$

which is proportional to the derivative with respect to θ of the kth term in (6-110). The feedback NDA carrier synchronizer needs one synchronized matched filter output sample per symbol.

In the following, the linearized tracking performance of the feedback version of the NDA carrier synchronizer will be investigated. According to Section 6.3.2, it will be assumed that perfect timing is available, i.e., $\hat{\varepsilon} = \varepsilon_0$. The performance of the feedforward synchronizer can be derived from the result for the feedback synchronizer.

Taking (6-70) into account, (6-112) with $\hat{\varepsilon} = \varepsilon_0$ yields

$$x_\theta\left(k;\hat{\theta}\right) = \mathrm{Im}\left[E\left[(a_k^*)^M\right](a_k \exp(j\phi) + N_k)^M\right] \tag{6-113}$$

where N_k is given by (6-103). Taking into account that

$$E[(N_k)^m] = 0 \quad \text{for} \quad m \geq 1 \tag{6-114}$$

it follows from (6-113) that

$$g_\theta(\phi) = E\left[x_\theta\left(k;\hat{\theta}\right)\right] = \left|E\left[a_k^M\right]\right|^2 \sin(M\phi) \tag{6-115}$$

which indicates that the phase error detector characteristic is sinusoidal with period $2\pi/M$. Note that $g_\theta(0) = 0$, so that $\phi = 0$ is a stable equilibrium point. The phase error detector slope K_θ at $\phi = 0$ equals

$$K_\theta = M\left|E\left[a_k^M\right]\right|^2 \tag{6-116}$$

Using a binomial series expansion in (6-113), the loop noise at $\phi = 0$ can be written as

$$N_\theta(k;\theta_0) = \mathrm{Im}\left[E\left[(a_k^*)^M\right]\sum_{m=0}^{M} C_{M,m}\, N_k^m\, a_k^{M-m}\right] \tag{6-117}$$

where

$$C_{M,m} = \frac{M!}{m!(M-m)!} \tag{6-118}$$

This loop noise is white, and it can be verified from

$$E[(N_k)^m (N_k^*)^n] = 0 \qquad \text{for } m \neq n \tag{6-119}$$

that the $(M+1)$ terms of (6-117) are uncorrelated. Hence, the loop noise power spectral density is given by

$$\begin{aligned} S_\theta(\exp(j\omega T);0) =& E\left[\left(\mathrm{Im}\left[a_k^M E\left[(a_k^*)^M\right]\right]\right)^2\right] \\ &+ \frac{1}{2}\left|E\left[a_k^M\right]\right|^2 \sum_{m=1}^{M} (C_{M,m})^2 m! \left[\frac{N_0}{E_s}\right]^m E\left[|a_k|^{2M-2m}\right] \end{aligned} \tag{6-120}$$

where we have made use of

$$E\left[|N_k|^{2m}\right] = m!\left[\frac{N_0}{E_s}\right]^m \tag{6-121}$$

The first term in (6-120) is a self-noise term, which does not depend on E_s/N_0. The remaining terms, which represent the contribution from additive noise, consist of powers of N_0/E_s. For moderate and high values of E_s/N_0, the terms with $m \geq 2$ in (6-117) can be ignored. This yields

$$\text{var}[\phi] \approx (2B_L T)\left[A\frac{N_0}{2E_s} + B\right] \tag{6-122}$$

where

$$A = \frac{E\left[|a_k|^{2M-2}\right]}{\left|E\left[a_k^M\right]\right|^2} \tag{6-123}$$

$$B = \frac{E\left[\left(\text{Im}\left[a_k^M E\left[(a_k^*)^M\right]\right]\right)^2\right]}{M^2\left|E\left[a_k^M\right]\right|^4} \tag{6-124}$$

The tracking performance of the feedforward version of the NDA carrier synchronizer is obtained by replacing $2B_L T$ by $1/K$.

Taking into account that $E\left[|a_k|^2\right] = 1$, application of the Schwarz inequality to (6-123) yields $A \geq 1$. Hence, comparing (6-122) with (6-108), it follows that for moderate and large E_s/N_0 the tracking performance of the NDA carrier synchronizer is not better than that of the DD carrier synchronizer. Numerical performance results will be presented in Section 6.3.7.

Another important NDA carrier synchronizer is the "Viterbi and Viterbi" (V&V) carrier synchronizer (Section 5.10.1), which can be viewed as a generalization of the NDA carrier synchronizer. The V&V synchronizer will be discussed in Section 6.3.10.

6.3.5 Decision-Directed (DD) Symbol Synchronization

In this section we consider two types of decision-directed (DD) symbol synchronizers, i.e., the decision-directed maximum-likelihood (DD-ML) symbol synchronizer (introduced in Section 5.5) and the Mueller and Müller (M&M) symbol synchronizer. Both synchronizers make use of the carrier phase estimate and of the receiver's decisions. According to Section 6.3.2, it will be assumed that a perfect carrier phase estimate is available. Both synchronizers will be analyzed under the simplifying assumption that the receiver's decisions are not affected by additive noise; this implies that, for small timing errors, the receiver's decisions will be assumed to be correct. The effect of decision errors on the tracking performance will be considered in Section 6.3.7.

Another DD symbol synchronizer, i.e., the data-transition tracking loop (DTTL), will be briefly discussed in Section 6.3.10.

Decision-Directed ML Symbol Synchronizer

The feedforward version of the DD-ML symbol synchronizer maximizes over ε the function $L(\varepsilon)$, given by (see Section 5.5)

$$L(\varepsilon) = \sum_{k=1}^{K} \mathrm{Re}\left[\hat{a}_k^* \, z(kT + \varepsilon T) \, \exp\left(-j\hat{\theta}\right)\right] \tag{6-125}$$

This maximization over ε must be performed by means of a search.

The feedback version of the DD-ML symbol synchronizer uses a timing error detector, whose output $x_\varepsilon(k; \hat{\varepsilon})$ is given by (Section 5.6.1)

$$x_\varepsilon(k; \hat{\varepsilon}) = T \, \mathrm{Re}\left[\hat{a}_k^* \, z'(kT + \hat{\varepsilon} T) \, \exp\left(-j\hat{\theta}\right)\right] \tag{6-126}$$

where $z'(t)$ is the derivative (with respect to time) of $z(t)$.

In the following, we will investigate the tracking performance of the feedback synchronizer; from the obtained result, the performance of the feedforward synchronizer can be derived.

Taking (6-68) into account, (6-126) with $\hat{a}_k = a_k$ and $\hat{\theta} = \theta_0$ yields

$$x_\varepsilon(k; \hat{\varepsilon}) = T \, \mathrm{Re}\left[a_k^* \sum_m a_{k-m} h'(mT - eT) + a_k^* N_k'\right] \tag{6-127}$$

where $e = \varepsilon_0 - \hat{\varepsilon}$ denotes the timing error, $h'(t)$ is the time derivative of the matched filter output pulse $h(t)$, and

$$N_k' = N'(kT + \hat{\varepsilon} T) \exp\left(-j\theta\right) \tag{6-128}$$

where $N'(t)$ is the time derivative of the noise $N(t)$ at the matched filter output. It follows from (6-128) and the statistical properties of $N(t)$, determined in Section 6.3.1, that

$$E\left[|N_k'|^2\right] = -h''(0)(N_0/E_s) \tag{6-129}$$

where $h''(0)$ is the second derivative of $h(t)$, taken at $t = 0$. The timing error detector characteristic $g_\varepsilon(e)$ is given by

$$g_\varepsilon(e) = E[x_\varepsilon(k;\hat{\varepsilon})] = Th'(-eT) \tag{6-130}$$

As $h(t)$ is maximum for $t = 0$, it follows that $g_\varepsilon(0) = 0$, so that $e = 0$ is a stable equilibrium point. The resulting timing error detector slope K_ε at $e = 0$ is

$$K_\varepsilon = \ -h''(0)T^2 \tag{6-131}$$

The loop noise $N_\varepsilon(k;\varepsilon_0)$ at $e = 0$ consists of two uncorrelated terms:

$$N_\varepsilon(k;\varepsilon_0) = N_{\varepsilon,n}(k;\varepsilon_0) + N_{\varepsilon,s}(k;\varepsilon_0) \tag{6-132}$$

where

$$N_{\varepsilon,n}(k;\varepsilon_0) = T\ \text{Re}[a_k^* N_k'] \tag{6-133}$$

$$N_{\varepsilon,s}(k;\varepsilon_0) = T\ \text{Re}\left[a_k^* \sum_{m\neq 0} a_{k-m} h'(mT)\right] \tag{6-134}$$

The first term in (6-132) is caused by additive noise; the second term represents self-noise.

As the data symbols are statistically independent, the power spectral density $S_{\varepsilon,n}(\exp(j\omega T);0)$ of the additive noise contribution $N_{\varepsilon,n}(k;\varepsilon_0)$ to the loop noise is flat. Using (6-129), we obtain

$$S_{\varepsilon,n}(\exp(j\omega T;0)) = \left(-h''(0)T^2\right)\frac{N_0}{2E_s} \tag{6-135}$$

Denoting by e_n the contribution of $N_{\varepsilon,n}(k;\varepsilon_0)$ to the timing error, it follows from (6-133) and (6-131) that

$$\text{var}[e_n] = (2B_LT)\frac{1}{\left(-h''(0)\right)^2 T^2}\frac{N_0}{2E_s} \tag{6-136}$$

where B_L denotes the loop bandwidth. In the case of the feedforward synchronizer which maximizes $L(\varepsilon)$ from (6-125), $2B_LT$ must be replaced by $1/K$. Note that (6-136) equals the Cramér-Rao bound; this indicates that the DD-ML symbol synchronizer achieves optimum tracking performance as far as the contribution from additive noise to the timing error is concerned. However, recall that our analysis assumes correct decisions, i.e., $\hat{a}_k = a_k$; for low E_s/N_0, decision errors cause a reduction of the timing error detector slope, which yields an increase of the tracking error variance as compared to the Cramér-Rao bound.

Now we concentrate on the power spectral density of the self-noise contribution $N_{\varepsilon,s}(k;\varepsilon_0)$ to the loop noise $N_\varepsilon(k;\varepsilon_0)$. This power spectral density will be evaluated via the self-noise autocorrelation function $R_{\varepsilon,s}(n) =$

$E[N_{\varepsilon,s}(k)N_{\varepsilon,s}(k+n)]$. For complex-valued data symbols with $E\left[a_k^2\right] = 0$ (this holds for M-PSK with $M > 2$ and for QAM) and using the fact that $h'(t)$ is an odd function of time, we obtain

$$R_{\varepsilon,s}(0) = \frac{1}{2} \sum_{m \neq 0} \left(h'(mT)T\right)^2 \tag{6-137}$$

$$R_{\varepsilon,s}(n) = -\frac{1}{2} \left(h'(nT)T\right)^2 \qquad n \neq 0 \tag{6-138}$$

For real-valued data symbols (as with BPSK or M-PAM constellations) the self-noise autocorrelation function is twice as large as for complex-valued data symbols with $E\left[a_k^2\right] = 0$. Use of (6-137), (6-138), and (6-75) yields

$$S_{\varepsilon,s}(\exp(j\omega T); 0) = \sum_{m>0} [1-\cos(m\omega T)](h'(mT)T)^2 \tag{6-139}$$

It is easily verified from (6-139) that the self-noise power spectral density becomes zero at $\omega = 0$, and, hence, cannot be considered as approximately flat within the loop bandwidth. Consequently, the tracking error variance caused by self-noise depends not only on the loop bandwidth, but also on the shape of the closed-loop transfer function. Denoting by e_s the self-noise contribution to the timing error, we show in Section 6.3.11 that, for small synchronizer bandwidths,

$$\text{var}[e_s] = K_F(2B_LT)^2 \frac{1}{K_\varepsilon^2} \sum_{m>0} m(h'(mT)\,T)^2 \tag{6-140}$$

where K_ε is given by (6-131), and the value of K_F depends on the type of closed-loop transfer function [the closed-loop impulse response (6-91) is equivalent to feedforward synchronization]:

$$K_F = \begin{cases} 1 & \text{feedforward synchonization} \\ 2 & \text{first-order feedback synchronizer} \\ 2\left[4\zeta^2/(1+4\zeta^2)\right]^2 & \text{second-order feedback synchronizer} \end{cases} \tag{6-141}$$

Note that the tracking error variance caused by self-noise is proportional to the square of the loop bandwidth (replace $2B_LT$ by $1/K$ for feedforward synchronization).

The total tracking error variance equals the sum of the additive noise contribution (6-136) and the self-noise contribution (6-140). Numerical performance results will be presented in Section 6.3.7.

Mueller and Müller (M&M) Symbol Synchronizer

The Mueller and Müller (M&M) symbol synchronizer (Section 5.7) is a feedback synchronizer, whose timing error detector output $x_\varepsilon(k;\hat{\varepsilon})$ is given by

$$x_\varepsilon(k;\hat{\varepsilon}) = \text{Re}\left[\left(\hat{a}_{k-1}^* z(kT+\hat{\varepsilon}T) \; - \; \hat{a}_k^* z((k-1)T+\hat{\varepsilon}T)\right)\exp\left(-j\hat{\theta}\right)\right] \tag{6-142}$$

The M&M synchronizer needs only one synchronized sample per symbol, taken at the estimated decision instant.

Let us investigate the tracking performance of the M&M synchronizer. Assuming that $\hat{a}_k = a_k$ for all k and $\hat{\theta} = \theta_0$, it follows from (6-68) that

$$x_\varepsilon(k;\hat{\varepsilon}) = \mathrm{Re}\left[\sum_m \left(a^*_{k-1} a_{k-m} - a^*_k a_{k-1-m}\right) h(mT - eT) + \left(a^*_{k-1} N_k - a^*_k N_{k-1}\right)\right] \tag{6-143}$$

where N_k is defined by (6-103). This yields the following timing error detector characteristic:

$$g_\varepsilon(e) = E[x_\varepsilon(k;\hat{\varepsilon})] = h(T - eT) - h(-T - eT) \tag{6-144}$$

As $h(t)$ is an even function of time, it follows that $g(0) = 0$, so that $e = 0$ is a stable equilibrium point. The resulting timing error detector slope at $e = 0$ is

$$K_\varepsilon = 2h'(-T)T \tag{6-145}$$

The loop noise $N_\varepsilon(k;\varepsilon_0)$ at $e = 0$ is given by

$$N_\varepsilon(k;\varepsilon_0) = \mathrm{Re}\left[a^*_{k-1} N_k - a^*_k N_{k-1}\right] \tag{6-146}$$

Note that $N_\varepsilon(k;\varepsilon_0)$ contains no self-noise; this is due to the fact that the matched filter output pulse $h(t)$ is a Nyquist-I pulse, i.e., $h(mT) = \delta_m$. As both $\{a_k\}$ and $\{N_k\}$ are sequences of statistically independent random variables, it follows that the loop noise $N_\varepsilon(k;\varepsilon_0)$ is white. The loop noise power spectral density $S_\varepsilon(\exp(j\omega T);0)$ equals

$$S_\varepsilon(\exp(j\omega T);0) = \frac{N_0}{E_s} \tag{6-147}$$

The corresponding tracking error performance is given by

$$\mathrm{var}[e] = (2B_L T)\frac{1}{2\left(h'(-T)\,T\right)^2}\,\frac{N_0}{2E_s} \tag{6-148}$$

Numerical performance results will be presented in Section 6.3.7.

6.3.6 Non-Decision-Aided Symbol Synchronizer

The feedforward version of the non-decision-aided maximum-likelihood (NDA–ML) symbol synchronizer maximizes over ε the function $L(\varepsilon)$, given by (Section 5.6.2)

$$L(\varepsilon) = \sum_{k=1}^{K} \left|z(kT + \varepsilon T)\right|^2 \tag{6-149}$$

The feedback version of the NDA symbol synchronizer uses a timing error detector, whose output $x_\varepsilon(k;\hat{\varepsilon})$ is given by [eq. (6-106)]

$$x_\varepsilon(k;\hat{\varepsilon}) = T \,\mathrm{Re}[z^*(kT+\hat{\varepsilon}T)z'(kT+\hat{\varepsilon}T)] \tag{6-150}$$

where $z'(t)$ is the derivative (with respect to time) of $z(t)$. The timing error detector output $x_\varepsilon(k;\hat{\varepsilon})$ is proportional to the derivative, with respect to ε, of the kth term in (6-149). The feedback NDA symbol synchronizer needs two synchronized samples per symbol, taken at the estimated decision instants: one at the matched filter output and one at the derivative matched filter output.

In the following, we will investigate the tracking performance of the feedback synchronizer; from the obtained result, the performance of the feedforward synchronizer can be derived.

Taking (6-68) into account, (6-150) yields

$$\begin{aligned} &x_\varepsilon(k;\hat{\varepsilon}) \\ &= T\,\mathrm{Re}\left[\left(\sum_m a_{k-m}^* \; h(mT-eT) + N_k^*\right)\left(\sum_m a_{k-m} \; h'(mT-eT) + N_k'\right)\right] \end{aligned} \tag{6-151}$$

where

$$N_k = N(kT+\hat{\varepsilon}T)\exp(-j\theta_0) \tag{6-152}$$

$$N_k' = N'(kT+\hat{\varepsilon}T)\exp(-j\theta_0) \tag{6-153}$$

and $h'(t)$ is the derivative of $h(t)$ with respect to time. For later use, we derive from (6-152) and (6-153) the following correlations:

$$E[N_k^* \; N_{k+m}] = \frac{N_0}{E_s}\delta_m \tag{6-154}$$

$$E\left[N_k'^* \; N_{k+m}'\right] = \frac{N_0}{E_s}(-h''(mT)) \tag{6-155}$$

$$E\left[N_k^* \; N_{k+m}'\right] = \frac{N_0}{E_s}h'(mT) \tag{6-156}$$

where $h''(t)$ is the second derivative of $h(t)$ with respect to time. The timing error detector characteristic $g_\varepsilon(e)$ is given by

$$g_\varepsilon(e) = E[x_\varepsilon(k;\hat{\varepsilon})] = \sum_m h(mT-eT)\,h'(mT-eT)\,T \tag{6-157}$$

which can be shown to become sinusoidal in e with period 1 when $H(\omega)=0$ for $|\omega| > 2\pi/T$, i.e., $\alpha < 1$ excess bandwidth. As $h(t)$ and $h'(t)$ are even and odd functions of time, respectively, it follows that $g_\varepsilon(0)=0$, so that $e=0$ is a stable equilibrium point. The resulting timing error detector slope at $e = 0$ is

$$K_\varepsilon = \left(-h''(0)\,T^2\right) - \sum_m \left(h'(mT)\,T\right)^2 \tag{6-158}$$

The loop noise $N_\varepsilon(k;\varepsilon_0)$ at $e = 0$ consists of three uncorrelated terms:

$$N_\varepsilon(k;\varepsilon_0) = N_{\varepsilon,N\times N}(k;\varepsilon_0) + N_{\varepsilon,S\times N}(k;\varepsilon_0) + N_{\varepsilon,S\times S}(k;\varepsilon_0) \tag{6-159}$$

where

$$N_{\varepsilon,N\times N}(k;\varepsilon_0) = T\ \mathrm{Re}[N_k^*\ \ N_k'] \tag{6-160}$$

$$N_{\varepsilon,S\times N}(k;\varepsilon_0) = T\ \mathrm{Re}\left[a_k^*\ N_k' + \sum_m a_{k-m}\ h'(mT) N_k^*\right] \tag{6-161}$$

$$N_{\varepsilon,S\times S}(k;\varepsilon_0) = T\ \mathrm{Re}\left[a_k^* \sum_m a_{k-m}\ h'(mT)\right] \tag{6-162}$$

and the subscripts $N \times N$, $S \times N$, and $S \times S$ refer to noise × noise term, signal × noise term, and signal × signal term, respectively.

The autocorrelation function $R_{\varepsilon,N\times N}(k;\varepsilon_0)$ of the $N \times N$ contribution to the loop noise is given by

$$R_{\varepsilon,N\times N}(0) = \frac{1}{2}\left(-h''(0)\,T\right)^2 \left[\frac{N_0}{E_s}\right]^2 \tag{6-163}$$

$$R_{\varepsilon,N\times N}(nT) = -\frac{1}{2}\left(h'(nT)\,T\right)^2 \left[\frac{N_0}{E_s}\right]^2 \tag{6-164}$$

The corresponding loop noise power spectral density $S_{\varepsilon,N\times N}(\exp(j\omega T);0)$ is

$$S_{\varepsilon,N\times N}(\exp(j\omega T);0) = \frac{1}{2}\left[\frac{N_0}{E_s}\right]^2 \left[\left(-h''(0)\,T^2\right) - \sum_m \left(h'(mT)\,T\right)^2 \cos(m\omega T)\right] \tag{6-165}$$

Note that $S_{\varepsilon,N\times N}(\exp(j\omega T);0)$ is not flat. However, as $S_{\varepsilon,N\times N}(\exp(j\omega T);0)$ is nonzero at $\omega = 0$, it can be approximated by its value at $\omega = 0$, provided that the loop bandwidth is so small that the variation of $S_{\varepsilon,N\times N}(\exp(j\omega T);0)$ within the loop bandwidth can be neglected. This yields

$$\mathrm{var}[e_{N\times N}] = (2B_L T)\frac{2}{\left(-h''(0)\,T^2\right) - \sum_m \left(h'(mT)\,T\right)^2}\left[\frac{N_0}{2E_s}\right]^2 \tag{6-166}$$

where $e_{N\times N}$ denotes the timing error caused by the $N \times N$ contribution to the loop noise. The performance of the feedforward version is obtained by replacing $2B_LT$ by $1/K$.

The autocorrelation function $R_{\varepsilon,S\times N}(k;\varepsilon_0)$ of the $S \times N$ contribution to the loop noise is given by

$$R_{\varepsilon,S\times N}(0) = \frac{1}{2}\left[\left(-h''(0)\,T^2\right) + \sum_m \left(h'(mT)\,T\right)^2\right]\frac{N_0}{E_s} \tag{6-167}$$

$$R_{\varepsilon,S\times N}(nT) = -\left(h'(nT)\,T\right)^2\frac{N_0}{E_s} \tag{6-168}$$

The corresponding loop noise power spectral density $S_{\varepsilon,S\times N}(\exp(j\omega T);0)$ is

$$\begin{aligned} &S_{\varepsilon,S\times N}(\exp(j\omega T);0) \\ &=\frac{1}{2}\frac{N_0}{E_s}\left(-h''(0)\,T^2\right) + \frac{1}{2}\frac{N_0}{E_s}\left[\sum_m \left(h'(mT)\,T\right)^2\left(1-2\cos(m\omega T)\right)\right] \end{aligned} \tag{6-169}$$

Approximating $S_{\varepsilon,S\times N}(\exp(j\omega T);0)$ by its value at $\omega = 0$, we obtain

$$\mathrm{var}[e_{S\times N}] = (2B_LT)\frac{1}{\left(-h''(0)\,T^2\right) - \sum_m \left(h'(mT)\,T\right)^2}\frac{N_0}{2E_s} \tag{6-170}$$

where $e_{S\times N}$ denotes the timing error caused by the $S \times N$ contribution to the loop noise. The performance of the feedforward version is obtained by replacing $2B_LT$ by $1/K$.

The self-noise term $N_{\varepsilon,S\times S}(k;\varepsilon_0)$ from (6-162) is the same as the self-noise term $N_{\varepsilon,s}(k;\varepsilon_0)$ given by (6-134). Hence, denoting by $e_{S\times S}$ the timing error caused by the $S \times S$ contribution to the loop noise, $\mathrm{var}[e_{S\times S}]$ is given by the right-hand side of (6-140), with K_ε given by (6-158). Notice that (6-140) has been derived for complex-valued data symbols with $E\left[a_k^2\right] = 0$; for real-valued data symbols the result from (6-140) should be multiplied by 2.

The total timing error variance equals the sum of the $N\times N$, $S\times N$ and $S\times S$ contributions, given by (6-166), (6-170) and (6-140) with K_ε given by (6-158). Numerical performance results will be presented in Section 6.3.7.

Other NDA symbol synchronizers, such as the Gardner synchronizer and the digital filter and square synchronizer, are briefly discussed in Section 6.3.10.

6.3.7 Tracking Performance Comparison

Carrier Synchronizers

For moderate and large E_s/N_0, the linearized tracking error variances (6-108) and (6-122) resulting from the DD and the NDA carrier synchronizers, respectively, are well approximated by

$$\mathrm{var}[\phi] \approx (2B_LT)\left[A\frac{N_0}{2E_s} + B\right] \tag{6-171}$$

The approximation involves neglecting the effect of decision errors (for the DD synchronizer) and of higher-order noise terms (for the NDA synchronizer). For the DD synchronizer, $A = 1$ and $B = 0$, in which case (6-171) equals the Cramér-Rao bound. For the NDA synchronizer, A and B are given by (6-123) and (6-124), respectively.

Figure 6-1 shows the linearized tracking performance of the DD carrier synchronizer for QPSK, taking decision errors into account, and compares this result with the Cramér-Rao bound (CRB).

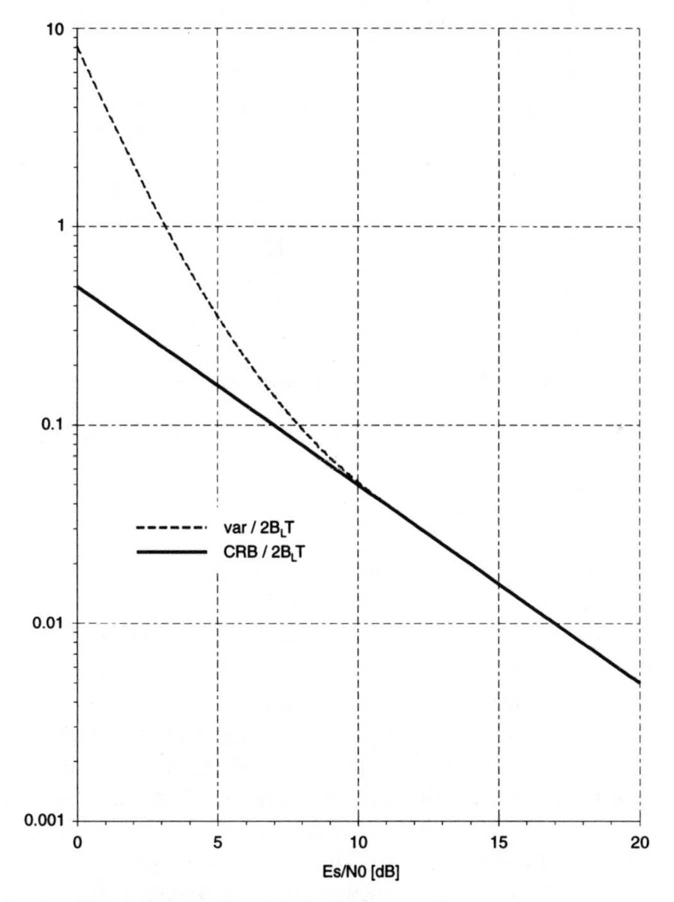

Figure 6-1 Linearized Tracking Performance of DD-ML Carrier Synchronizer for QPSK

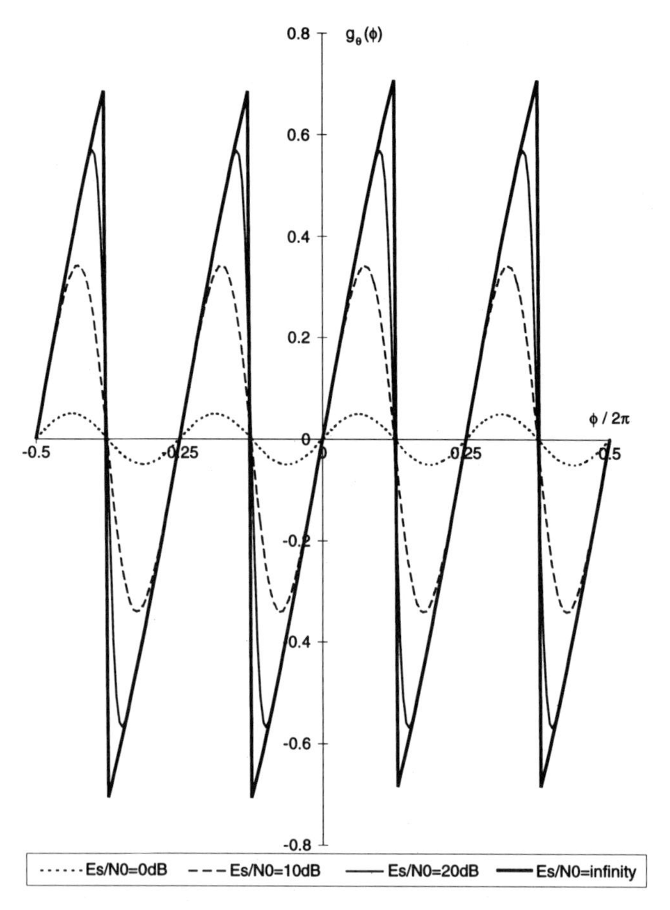

Figure 6-2 DD Phase Error Detector Characteristic Taking Decision Errors into Account

The difference between both curves is caused only by the additive noise affecting the receiver's decisions. Figure 6-2 shows the phase error detector characteristic $g_\theta(\phi)$ for QPSK, with the effect of decision errors included.

For small E_s/N_0, $g_\theta(\phi)$ becomes essentially sinusoidal; this agrees with the observations made in Section 3.2.2 of Volume 1. Also, we notice that $g_\theta(\phi)$ for QPSK is periodic in ϕ with period $\pi/2$ rather than 2π. Indeed, when the signal constellation has a symmetry angle of $2\pi/M$ (for QPSK, we have $M = 4$), the statistics of the received signal do not change when the data symbols a_k are replaced by $a_k \exp(j2\pi/M)$ and the carrier phase θ_0 is replaced by $\theta_0 - 2\pi/M$. Hence, the statistics of any phase error detector output are periodic in ϕ with period $2\pi/M$, when for each valid data sequence $\{a_k\}$ the sequence $\{a_k \exp(j2\pi/M)\}$

also represents a valid data sequence. We observe from Figure 6-2 that the decision errors reduce the phase error detector slope at $\phi = 0$, which in turn increases the tracking error variance. Denoting this slope corresponding to a given E_s/N_0 by $K_\theta(E_s/N_0)$, Figure 6-3 shows the ratio $K_\theta(E_s/N_0)/K_\theta(\infty)$.

Let us consider the NDA carrier synchronizer for two signal constellations of practical interest, i.e., the M-PSK constellation and the square N^2–QAM constellation. These constellations have symmetry angles of $2\pi/M$ and $\pi/2$, respectively, so the corresponding synchronizers use the Mth power (for M-PSK) and 4-th power (for QAM) nonlinearities, respectively. For M-PSK, (6-123) and (6-124) yield $A = 1$ and $B = 0$, so that (6-171) reduces to the CRB. This indicates that, for large E_s/N_0, the NDA synchronizer yields optimum tracking performance

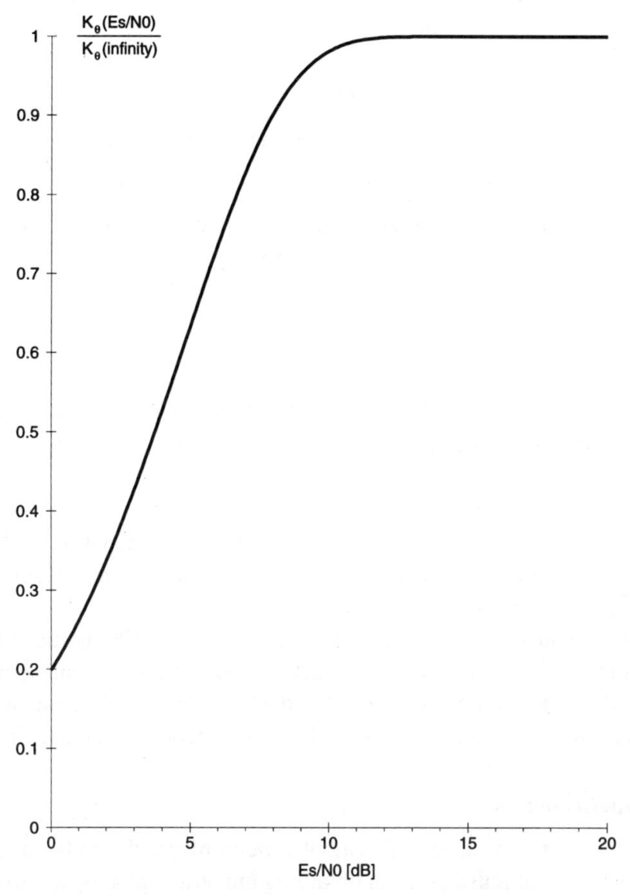

Figure 6-3 Phase Error Detector Slope

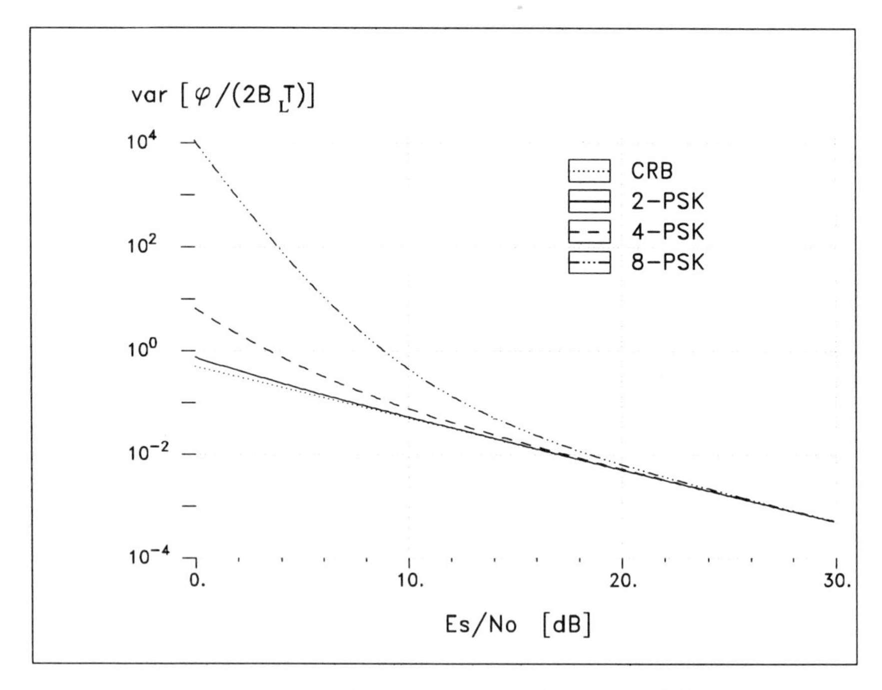

Figure 6-4 Linearized Tracking Performance of NDA Carrier Synchronizer for M-PSK

in the case of M-PSK constellations. Figure 6-4 shows the tracking performance for M-PSK, taking into account all terms from (6-120), and compares it with the CRB.

The degradation with respect to the CRB is caused by the terms with $m \geq 2$ in (6-120). It increases with increasing size of the constellation and decreasing E_s/N_0. The tracking performance for N^2-QAM, taking into account all terms from (6-120), is shown in Figure 6-5.

The performance for 4–QAM is the same as for 4–PSK, because the constellations are identical. For $N^2 > 4$, the tracking performance is considerably worse than the CRB [basically because $A > 1$ and $B > 0$ in (6-171)], and worsens with increasing constellation size. Note that a limiting performance exists for $N^2 \to \infty$.

Symbol Synchronizers

The tracking performance of symbol synchronizers depends on the shape of the received baseband pulse $g(t)$. In obtaining numerical results, it will be assumed that $g(t)$ is such that the pulse $h(t)$ at the output of the matched filter is a raised

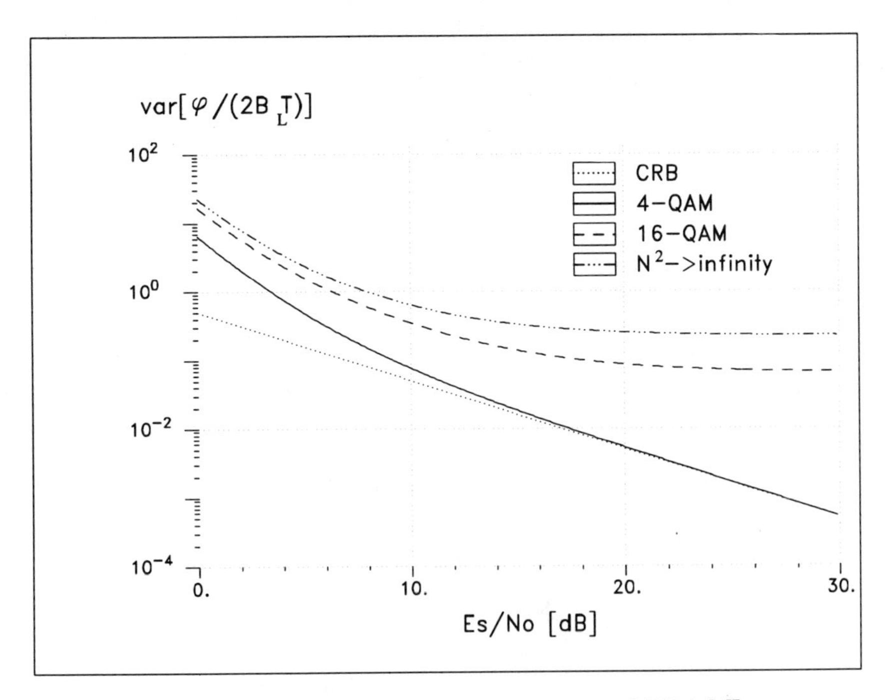

Figure 6-5 Linearized Tracking Performance of NDA-ML Carrier Synchronizer for N^2–QAM

cosine pulse, i.e.,

$$h(t) = \frac{\sin(\pi t/T)}{\pi t/T} \frac{\cos(\alpha \pi t/T)}{1 - 4\alpha^2 t^2/T^2} \tag{6-172}$$

where $\alpha \in (0, 1)$ represents the rolloff factor.

For moderate and large E_s/N_0, the timing error variance is well approximated by

$$\text{var}[e] \simeq (2B_L T) A(\alpha) \frac{N_0}{2E_s} + K_F (2B_L T)^2 B(\alpha) \tag{6-173}$$

The first term in (6-173) is proportional to the synchronizer bandwidth and represents an approximation of the contribution from additive noise; the approximation involves neglecting the effect of decision errors (DD-ML and M&M synchronizers) or of the noise × noise contribution (NDA synchronizer). The second term in (6-173) is proportional to the square of the synchronizer bandwidth and represents the self-noise contribution; the quantity K_F depends on the closed-loop transfer

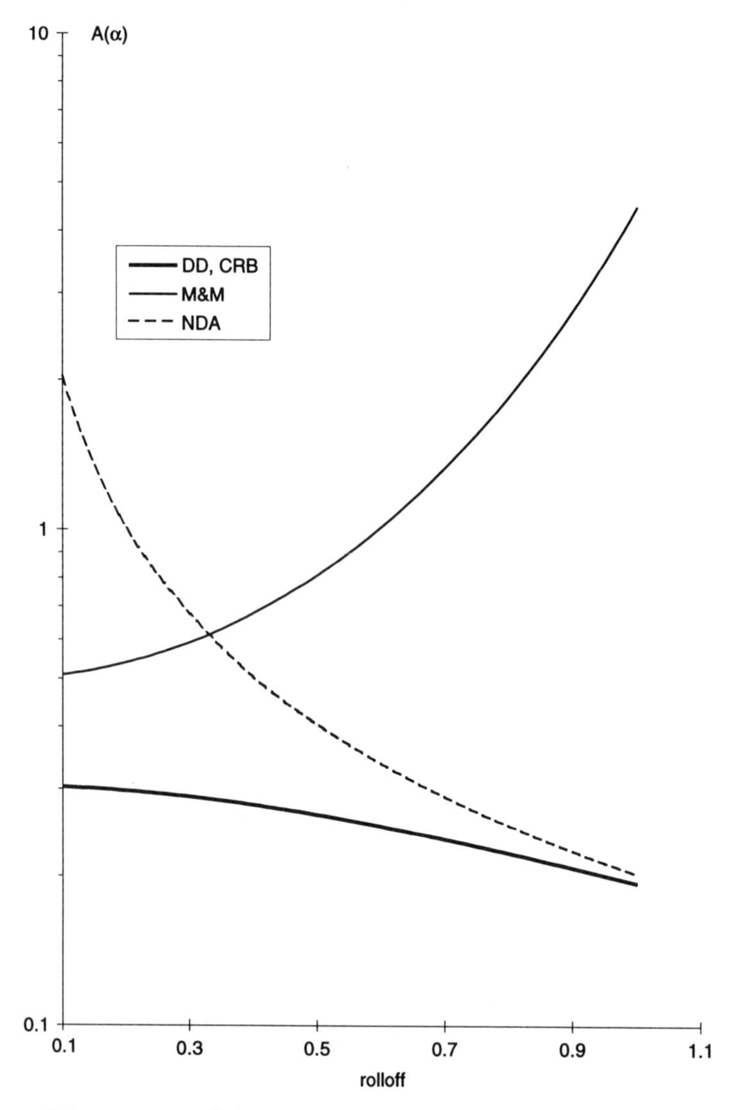

Figure 6-6 Quantity $A(\alpha)$ for the DD-ML, M&M, and NDA Synchronizers

function. The factors $A(\alpha)$ and $B(\alpha)$ incorporate the effect of the shape of the matched filter output pulse $g(t)$.

Figure 6-6 shows the quantity $A(\alpha)$ for the DD-ML, the M&M and the NDA symbol synchronizers.

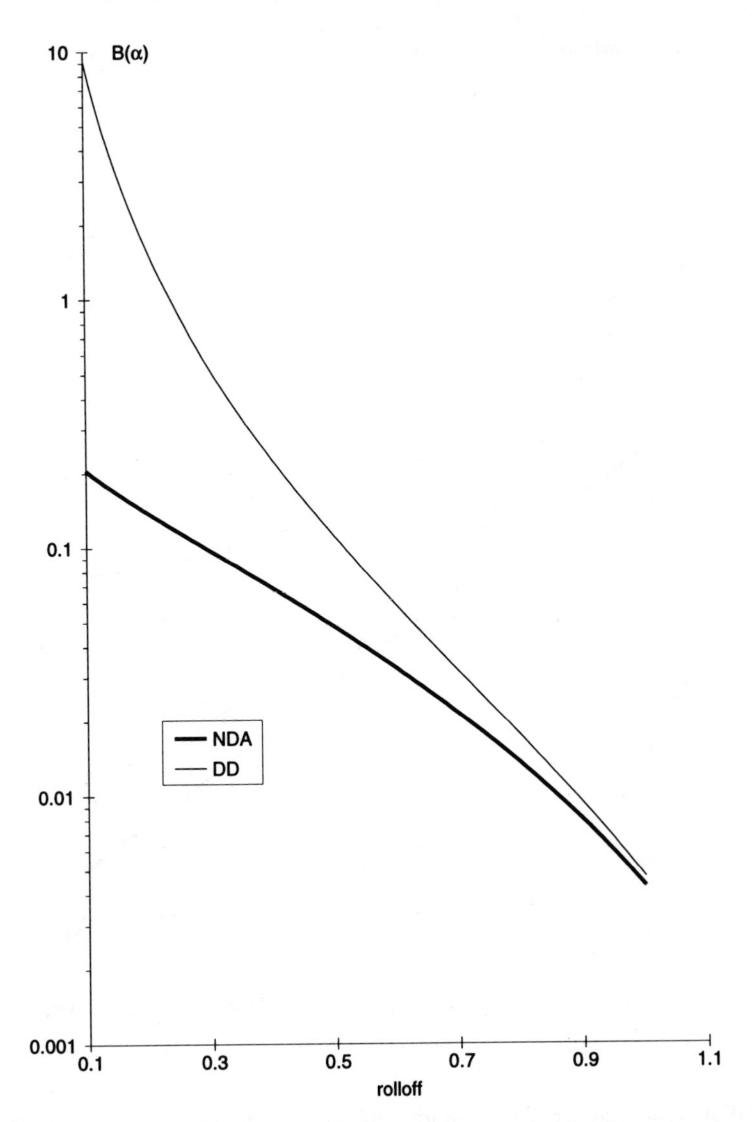

Figure 6-7 Quantity $B(\alpha)$ for the DD-ML and NDA-ML Synchronizers

The DD-ML synchronizer yields the smallest $A(\alpha)$, which corresponds to the CBR. The M&M and NDA synchronizers behave poorly for large and small values of α, respectively. The quantity $B(\alpha)$ is shown in Figure 6-7 for the DD-ML and the NDA symbol synchronizers; for the M&M synchronizer, self-noise is absent, i.e., $B(\alpha) = 0$.

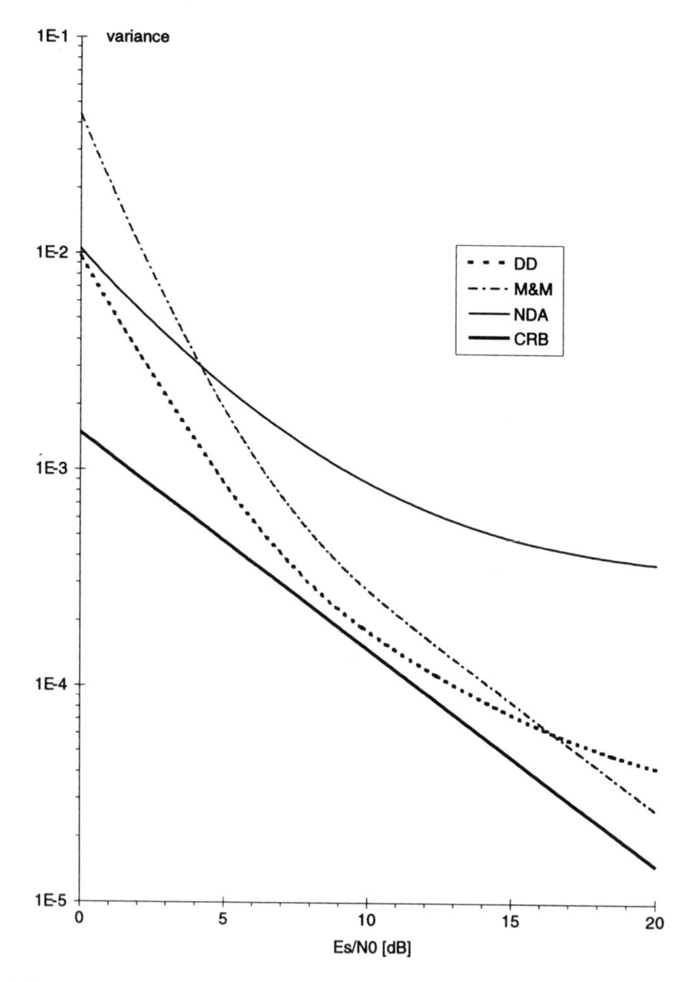

Figure 6-8 Tracking Error Variance for DD-ML, M&M, and NDA Synchronizers

The NDA synchronizer yields more self-noise than the DD-ML synchronizer, especially for small values of α; for both synchronizers the self-noise decreases with increasing rolloff, due to the faster decay of the baseband pulse $h(t)$ at the matched filter output.

The approximation (6-173) of the timing error variance ignores the effect of decision errors and of the noise × noise contribution, so that (6-173) underestimates the tracking error variance at low E_s/N_0. Figure 6-8 shows the tracking error variance when these effects are taken into account, along with the CBR; it is assumed that $\alpha = 0.2$, $2B_LT = 10^{-2}$, and $K_F = 2$.

We observe that the timing error variance of the DD-ML synchronizer exceeds

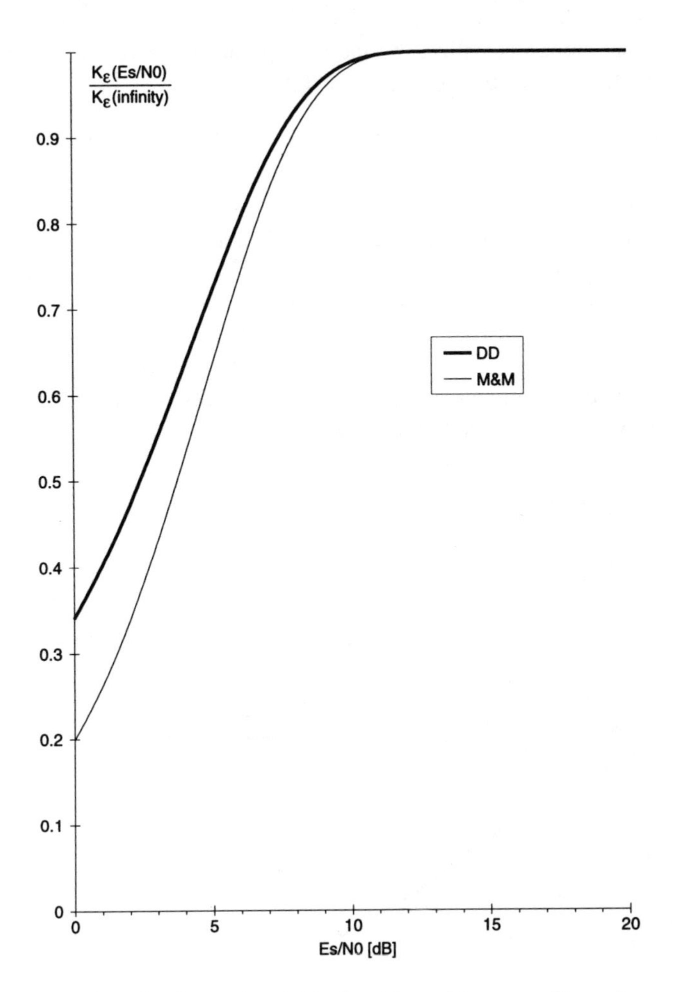

Figure 6-9 Normalized Timing Error Detector Slope for Decision-Directed Synchronizers

the CBR [which equals the first term in (6-173) in the case of the DD-ML synchronizer], especially at low E_s/N_0 (where the effect of decision errors becomes important) and high E_s/N_0 (where self-noise becomes important). At moderate E_s/N_0, the synchronizer yielding the larger value of $A(\alpha)$ yields the larger tracking error variance; this is confirmed by Figure 6-6 which indicates that $A(\alpha)\,\big|_{\text{DD-ML}} < A(\alpha)\,\big|_{\text{M\&M}} < A(\alpha)\,\big|_{\text{NDA}}$ for $\alpha < 0.33$. At high E_s/N_0, it is the quantity $B(\alpha)$ that determines the tracking performance; note from Figure 6-7 that $0 = B(\alpha)\,\big|_{\text{M\&M}} < B(\alpha)\,\big|_{\text{DD-ML}} < B(\alpha)\,\big|_{\text{NDA}}$, which is reflected by the tracking performance at high E_s/N_0 shown in Figure 6-8. In the case of the DD synchronizers, operation at low E_s/N_0 yields decision errors which reduce the timing error detector slope and increase the tracking error variance; Figure 6-9 shows the actual slope normalized by the asymptotic (large E_s/N_0) slope for the DD-ML and M&M synchronizers at $\alpha = 0.2$.

In the case of the NDA symbol synchronizer, the noise × noise contribution to the timing error variance is inversely proportional to the square of E_s/N_0 and hence cannot be neglected at low E_s/N_0.

The additive noise contribution and the self-noise contribution to the timing error variance are proportional to $2B_LT$ and $(2B_LT)^2$, respectively. Consequently, these contributions are affected differently when the synchronizer bandwidth is changed. Figure 6-10 illustrates that modifying the synchronizer bandwidth has a larger impact at high E_s/N_0 (where self-noise dominates) than at low E_s/N_0 (where additive noise dominates); these numerical results are obtained for the NDA synchronizer with $\alpha = 0.2$ and $K_F = 2$.

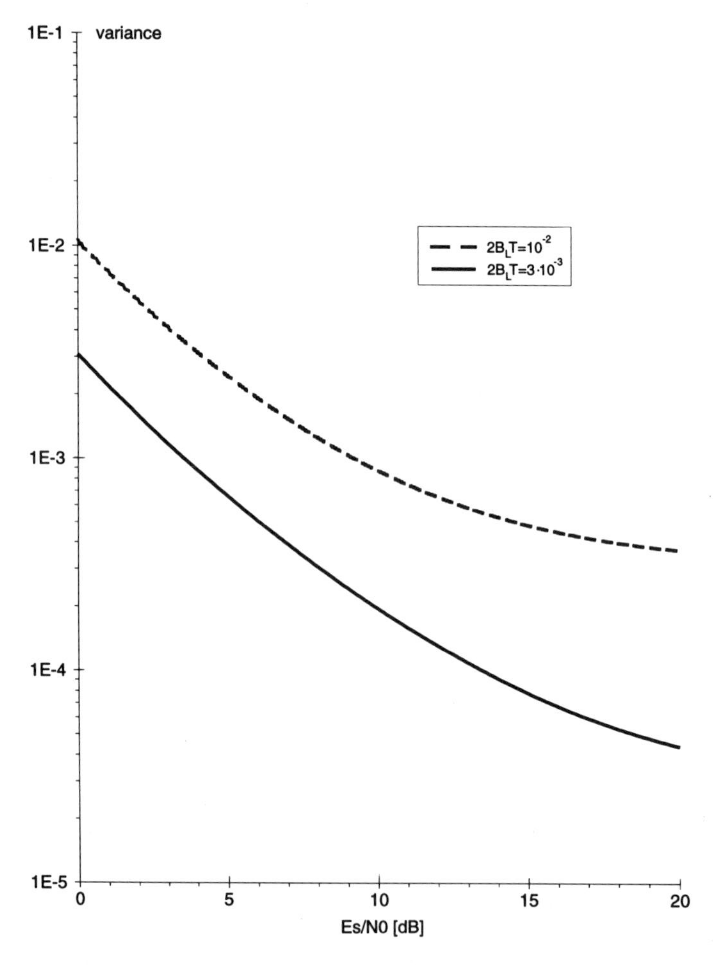

Figure 6-10 Tracking Error Variance for NDA Synchronizers

6.3.8 Effect of Time-Varying Synchronization Parameters

Until now, we have assumed that the estimation parameters carrier phase and time delay remain constant. In practice, however, the actual synchronization parameters vary with time, because of oscillator phase noise and frequency offsets. In the following, we consider the tracking of a time-varying carrier phase; the results are qualitatively the same in the case of tracking a time-varying time delay.

Carrier Phase Noise

When carrier phase noise is present, the actual carrier phase $\theta_0(k)$ corresponding to the kth data symbol can be modeled as

$$\theta_0(k) = \theta_c + \Delta(k) \tag{6-174}$$

where θ_c is a constant phase and $\{\Delta(k)\}$ is a sequence of stationary zero-mean random variables representing the phase noise.

In the case of a feedback synchronizer, we use a similar reasoning as in Section 3.6 of Volume 1 to obtain the following expression for the carrier phase estimate:

$$\hat{\theta}(k) = \sum_m h_\theta(k-m)\,\theta_0(m) \quad = \theta_c + \sum_m h_\theta(k-m)\,\Delta(m) \tag{6-175}$$

where $\{h_\theta(m)\}$ denotes the closed-loop impulse response. Hence, the carrier phase estimate is a lowpass-filtered version of the actual carrier phase; this indicates that the high-frequency components of the phase noise cannot be tracked.

Now we consider the case of a feedforward carrier synchronizer. We restrict our attention to the NDA synchronizer; the same result also holds for the DD synchronizer. Neglecting additive noise and self-noise, the carrier phase estimate is given by

$$\begin{aligned}\hat{\theta} &= \frac{1}{M}\arg\left(\frac{1}{K}\sum_{k=1}^{K}\exp\left(jM\theta(k)\right)\right) \\ &= \theta_c + \frac{1}{M}\arg\left(\frac{1}{K}\sum_{k=1}^{K}\exp\left(jM\Delta(k)\right)\right) \qquad \text{for } |\theta_c| \le \frac{\pi}{M}\end{aligned} \tag{6-176}$$

For $M\Delta(k) << 1$, linearization of the function $\arg(\cdot)$ yields

$$\hat{\theta} \simeq \theta_c + \frac{1}{K}\sum_{k=1}^{K}\Delta(k) = \frac{1}{K}\sum_{k=1}^{K}\theta_0(k) \tag{6-177}$$

This indicates that the carrier phase estimate is the arithmetical average of the K carrier phase values within the observation window. Note that the feedforward estimate (6-177) can be written in the form (6-175) of the feedback estimate, with the closed-loop impulse response given by (6-91). Therefore we restrict our attention to the feedback synchronizer.

Denoting the carrier phase error by $\phi(k) = \theta_0(k) - \hat{\theta}(k)$, it follows from (6-175) that

$$\text{var}[\phi(k)] = \frac{T}{2\pi} \int_{-\pi/T}^{\pi/T} |1 - H_\theta(\exp(j\omega T))|^2 S_\Delta(\exp(j\omega T))\, d\omega \tag{6-178}$$

where $H_\theta(\exp(j\omega T))$ is the closed-loop transfer function and $S_\Delta(\exp(j\omega T))$ is the phase noise power spectral density. The result (6-178) is also valid for feedforward synchronization, provided that the Fourier transform of the impulse response (6-91) is inserted for $H_\theta(\exp(j\omega T))$. As the phase noise components outside the synchronizer bandwidth cannot be tracked, decreasing the synchronizer bandwidth increases the tracking error variance caused by phase noise.

We conclude that feedback and feedforward synchronizers behave similarly in the presence of phase noise. The synchronization error variance caused by phase noise increases with decreasing synchronizer bandwidth. On the other hand, the error variance caused by additive noise and self-noise decreases with decreasing synchronizer bandwidth. Hence, the synchronizer bandwidth value is a compromise between tracking the phase noise and suppressing additive noise and self-noise.

Carrier Frequency Offset

When a carrier frequency offset is present, the received complex envelope $r_f(t)$ is given by

$$r_f(t) = \sum_m a_m\, g(t - mT - \varepsilon_0 T) \exp(j\Omega_0 t + j\theta_c') + \frac{n(t)}{A} \tag{6-179}$$

where Ω denotes the frequency offset in radians per second, and θ_c' is a constant phase. Applying $r_f(t)$ to the matched filter with impulse response $g(-t)$ and sampling at the instants $kT + \varepsilon_0 T$ yields samples $z(kT + \varepsilon_0 T)$, given by

$$z(kT + \varepsilon_0 T) = \sum_m a_{k-m}\, h(mT; \Omega_0) \exp(j\Omega_0 kT + j\theta_c) + N(kT + \varepsilon_0 T) \tag{6-180}$$

where

$$h(mT; \Omega_0) = \int_{-\infty}^{+\infty} g(mT + u)\, g(u)\, \exp(j\Omega_0 u)\, du \tag{6-181}$$

$$\theta_c = \theta_c' + \Omega \varepsilon_0 T \tag{6-182}$$

For $\Omega_0 = 0$, (6-181) reduces to (6-70) because $h(mT; 0) = \delta_m$. For $\Omega_0 \neq 0$, comparison of (6-180) and (6-70) indicates that a frequency offset Ω_0 gives rise to the following effects:

(i) For $\Omega_0 \neq 0$, the received signal is ill-centered with respect to the matched filter. This gives rise to a distorted matched filter output pulse (as compared to the case $\Omega_0 = 0$), which introduces intersymbol interference [i.e., $h(mT; \Omega_0) \neq \delta_m$ for $\Omega_0 \neq 0$].

(ii) The carrier phase $\theta_0(k) = \Omega_0 kT + \theta_c$ to be estimated is a linear function of the symbol index k. Unlike the case of carrier phase noise, $\{\theta_0(k)\}$ cannot be represented as the sum of a constant phase and a zero-mean stationary random sequence.

When the frequency offset is so small that the variation of the carrier phase over an interval equal to the effective duration of the baseband pulse $g(t)$ – which equals a few symbol intervals – is negligible, $h(mT; \Omega_0)$ is well approximated by $h(mT; 0)$, i.e., $h(mT; \Omega_0) \approx \delta_m$. This yields

$$z(kT + \varepsilon_0 T) \simeq a_k \exp\left(j\Omega_0 kT + j\theta_c\right) + N(kT + \varepsilon_0 T) \tag{6-183}$$

When the frequency offset is too large, so that (6-183) is no longer valid, the intersymbol interference at the output of the matched filter is likely to cause an unacceptable degradation of the receiver performance. For such large frequency offsets a frequency correction must be made before the signal enters the matched filter. Frequency estimation will be discussed in Chapter 8. Denoting now by Ω_0 the residual frequency offset after nonideal frequency correction, Ω_0 is usually small enough for (6-183) to be valid.

A similar reasoning as in Section 2.3.2 of Volume 1 indicates that a frequency offset gives rise to a steady-state phase offset in the case of a feedback synchronizer with a first-order loop; this phase offset becomes larger with increasing frequency offset and decreasing loop bandwidth. The occurrence of a phase offset can be avoided by using a second- (or higher-) order loop with perfectly integrating loop filter; in this case, the tracking performance of the feedback synchronizer is independent of the value Ω_0 of the frequency offset.

Now we investigate the effect of a small frequency offset on the tracking performance of a feedforward synchronizer. We consider the NDA-ML carrier synchronizer for M-PSK, which produces a carrier phase estimate given by

$$\hat{\theta} = \frac{1}{M} \arg\left(\frac{1}{K}\sum_{k=-L}^{L} z_k^M\right) \tag{6-184}$$

where z_k is a short-hand notation for $z(kT + \varepsilon_0 T)$, given by (6-183), and $K = 2L + 1$ is the length of the observation window. Note that θ_c from (6-183) denotes the carrier phase corresponding to the data symbol a_0 at the center of the observation window, i.e., at $k = 0$. Using a binomial series expansion, we obtain

$$\frac{1}{K}\sum_{k=-L}^{L} z_k^M = \exp\left(jM\theta_c\right)\left[F(K; M\Omega_0 T) + \frac{1}{K}\sum_{k=-L}^{L} D_k\right] \tag{6-185}$$

where

$$F(K;x) = \frac{1}{K}\sum_{k=-L}^{L} \exp(jkx) \; = \frac{1}{K}\frac{\sin(Kx/2)}{\sin(x/2)} \leq 1 \qquad \text{(6-186)}$$

$$D_k = \sum_{m=0}^{M-1} C_{M,m}\; a_k^m \exp(jkm\Omega_0 T) N_k^{M-m} \qquad \text{(6-187)}$$

N_k is a short-hand notation for $N(kT + \varepsilon_0 T)\exp(-j\theta_c)$ and $C_{M,m}$ is defined as in (6-118). Linearization of the function $\arg(\cdot)$ yields

$$\hat{\theta} = \theta_c + \frac{1}{F(K;M\Omega_0 T)}\frac{1}{K}\sum_{k=-L}^{L}\frac{1}{M}\,\mathrm{Im}[D_k] \qquad \text{(6-188)}$$

As $E[D_k] = 0$, it follows from (6-188) that the carrier phase estimate resulting from the feedforward synchronizer is unbiased with respect to the carrier phase at the center of the observation window. If $\hat{\theta}$ is used as an estimate for $\theta_0(k)$ with $k \neq 0$, a bias equal to $k\Omega_0 T$ occurs; this bias assumes its largest values of $\pm(K-1)\Omega_0 T/2$ at the edges of the observation window. As $\{D_k\}$ is a sequence of uncorrelated random variables, we obtain from (6-188)

$$E\left[\left(\hat{\theta} - \theta_c\right)^2\right] = \frac{1}{KF^2(K;M\Omega_0 T)}\,\frac{E\left[|D_k|^2\right]}{2M^2} \qquad \text{(6-189)}$$

where

$$E\left[|D_k|^2\right] = \sum_{m=1}^{M}\left(C_{M,m}\right)^2 m! \left[\frac{N_0}{E_s}\right]^m \qquad \text{(6-190)}$$

It is important to note that $E\left[|D_k|^2\right]$ does not depend on Ω_0. Hence, a frequency offset Ω_0 increases the tracking error variance by a factor $1/F^2(K;M\Omega_0 T)$, which equals one for $\Omega_0 = 0$, increases with K, $\Omega_0 T$, and M when $\Omega_0 \neq 0$, and becomes infinite when $KM\Omega_0 = 2\pi$.

It follows from (6-189) that the dependence of the tracking error variance on the observation window length K is contained in the factor $K^{-1}F^{-2}(K;M\Omega_0 T)$. This factor is shown in Figure 6-11 as a function of K for various values of $M\Omega_0 T$.

Clearly, an optimum value of K exists for $\Omega_0 \neq 0$: the useful component and the variance after averaging the Mth power nonlinearity output samples over the observation window are both decreasing with increasing K. The optimum value of K which minimizes the tracking error variance for a given value of $M\Omega_0 T$ is denoted K_0, and satisfies

$$\frac{1}{K_0}\sin^2\left(\frac{K_0 M\Omega_0 T}{2}\right) = \max_K\;\frac{1}{K}\sin^2\left(\frac{KM\Omega_0 T}{2}\right) \qquad \text{(6-191)}$$

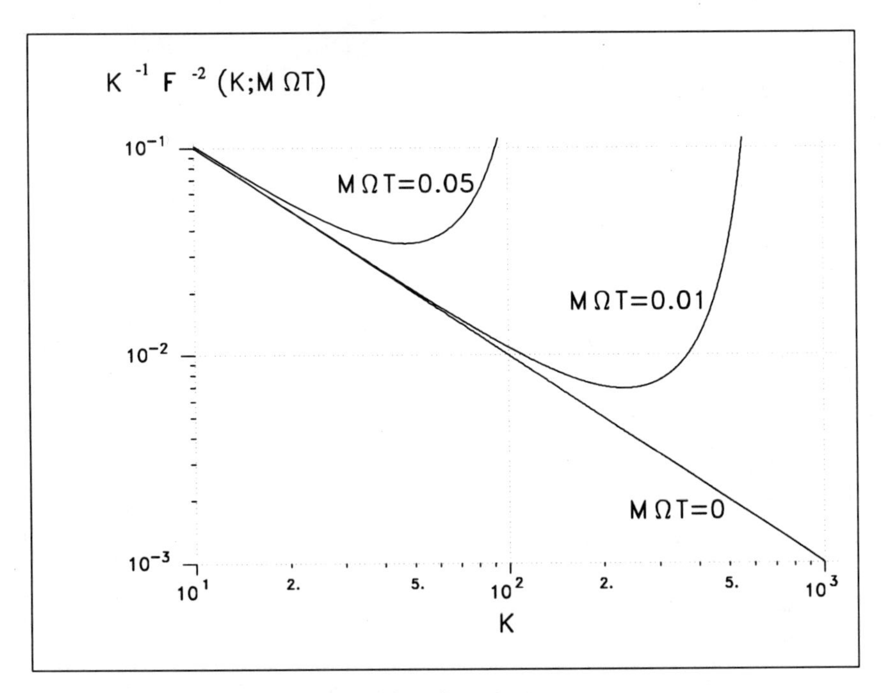

Figure 6-11 Factor $K^{-1}F^{-2}(K;M\Omega_0 T)$ as Function of K

Equivalently, K_0 is given by $K_0 = 2x_0/(M\Omega_0 T)$, with

$$\frac{1}{x_0}\sin^2(x_0) = \max_x \ \frac{1}{x}\sin^2(x) \tag{6-192}$$

Figure 6-12 shows a plot of the function.

We obtain:

$$x_0 = 1.1655 \qquad\qquad \sin^2(x_0)/x_0 = 0.7246 \tag{6-193}$$

Hence, the optimum value K_0 for a given $M\Omega_0 T$ is

$$K_0 = \frac{2.331}{M\Omega_0 T} \tag{6-194}$$

When the frequency offset $f_0 = \Omega_0/(2\pi)$ equals 0.1 percent of the symbol rate $1/T$, the optimum values K_0 are 185, 93, and 47 for M = 2, 4, and 8, respectively.

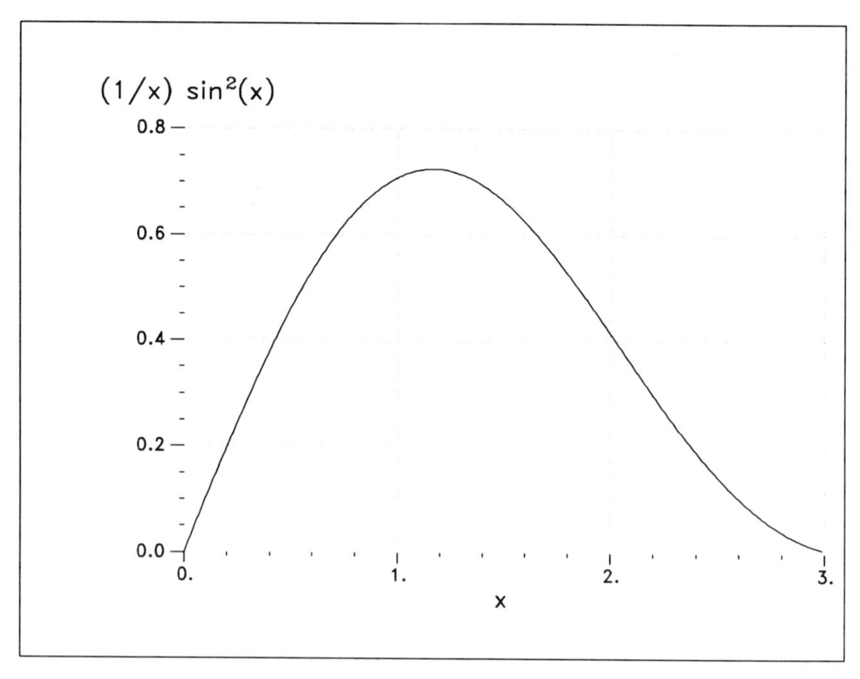

Figure 6-12 Function $\sin^2(x)/x$

The minimum variance of the estimate is obtained by substituting (6-194) into (6-189). Taking (6-193) into account, this yields

$$
\begin{aligned}
E\left[\left(\hat{\theta}-\theta_c\right)^2\right] &= \frac{\sin^2(M\Omega_0 T/2)}{M\Omega_0 T}\,\frac{1}{0.7246}\,\frac{E\left[|D_k|^2\right]}{2M^2} \qquad (K=K_0) \\
&\simeq \frac{M\Omega_0 T}{1.45}\,\frac{E\left[|D_k|^2\right]}{2M^2} \\
&= \frac{1.61}{K_0}\,\frac{E\left[|D_k|^2\right]}{2M^2}
\end{aligned}
\tag{6-195}
$$

where we have made use of $\sin(x) \approx x$ for $|x| \ll 1$. Hence, the error variance for M-PSK in the presence of a small frequency offset Ω_0 is proportional to $\Omega_0 T$ and to the size M of the constellation; this variance is a factor of 1.61 (or about 2 dB) worse than the variance corresponding to the case where the actual frequency offset is zero, but an observation window with $K = K_0$ has been selected, with K_0 related to an assumed nonzero frequency offset Ω_0 .

The presence of a frequency offset Ω_0 sets a lower limit (6-195) on the achievable phase etimate variance of the NDA synchronizer, and, hence, also on the bit error rate (BER) degradation caused by random carrier phase errors. In Chapter 7 it is shown that the BER degradation (in decibels) for M-PSK, caused by random carrier phase errors, is well approximated by

$$\begin{aligned} \frac{\text{BER degradation}}{\text{dB}} &= \frac{10}{\ln(10)}\left[1+\frac{2E_s}{N_0}\cos^2\left(\frac{\pi}{M}\right)\right]E\left[\left(\hat{\theta}-\theta_c\right)^2\right] \\ &\simeq \frac{10}{\ln(10)}\frac{2E_s}{N_0}\cos^2\left(\frac{\pi}{M}\right)E\left[\left(\hat{\theta}-\theta_c\right)^2\right] \end{aligned} \qquad (6\text{-}196)$$

where the approximation is valid for $M > 2$ and moderate to large E_s/N_0. For moderate and large E_s/N_0, we can ignore the terms with $m > 1$ in (6-190), in which case the minimum tracking error variance reduces to

$$E\left[\left(\hat{\theta}-\theta_c\right)^2\right] \simeq \frac{M\Omega_0 T}{1.45}\frac{N_0}{2E_s} \qquad (K = K_0,\ \text{large } E_s/N_0) \qquad (6\text{-}197)$$

Insertion of the minimum tracking error variance (6-197) into (6-196) yields the minimum BER degradation in decibels:

$$\frac{\text{BER degradation}}{\text{dB}} = \frac{10}{\ln(10)}\frac{1}{1.45}M\Omega_0 T\cos^2\left(\frac{\pi}{M}\right) \qquad (6\text{-}198)$$

For a minimum BER degradation of 0.1 dB, the allowable frequency offsets $f_0 T = \Omega_0 T/(2\pi)$ are only 0.27 percent for $M = 4$ and 0.08 percent for $M = 8$; the corresponding optimum observation window lengths are $K_0 = 35$ for $M = 4$ and $K_0 = 59$ for $M = 8$.

A similar investigation can be made for the feedforward DD carrier synchronizer. The result is that for $\Omega_0 \neq 0$, the feedforward estimate is unbiased only with respect to the carrier phase of the data symbol at the center of the observation window and has a bias of $k\Omega_0 T$ with respect to the carrier phase of a data symbol, which is k positions away from the center; this situation is identical to the case of the Mth power feedforward synchronizer. As compared to the case $\Omega_0 = 0$, a nonzero frequency offset Ω_0 yields an increase of the variance by a factor of $1/F^2(K;\Omega_0 T)$, where $F(K;x)$ is given by (6-186); this situation is different from the case of the NDA feedforward synchronizer, where the factor equals $1/F^2(K;M\Omega_0 T)$. As a result, the optimum length K_0 of the observation window equals

$$K_0 = \frac{2.331}{\Omega_0 T} \qquad (6\text{-}199)$$

which is a factor of M larger than for the NDA feedforward carrier synchronizer. For example, when the frequency offset $f_0 = \Omega_0/(2\pi)$ equals 0.1 percent of the symbol rate $1/T$, the optimum value K_0 equals 371, irrespective of the constellation. The resulting minimum variance for moderate and large E_s/N_0 is

$$E\left[\left(\hat{\theta}-\theta_c\right)^2\right] = \frac{\Omega_0 T}{1.45}\frac{N_0}{2E_s} \qquad (K = K_0,\ \text{large } E_s/N_0)$$
$$= \frac{1.61}{K_0}\frac{N_0}{2E_s} \tag{6-200}$$

which is smaller than for the feedforward NDA-ML synchronizer by a factor of M. Using (6-200) in the formula (6-196) for the BER degradation yields

$$\frac{\text{BER degradation}}{\text{dB}} = \frac{10}{\ln(10)}\frac{1}{1.45}\Omega_0 T \cos^2\left(\frac{\pi}{M}\right) \tag{6-201}$$

For a minimum BER degradation of 0.1 dB, the allowable frequency offsets $f_0T = \Omega_0 T/(2\pi)$ are 1.06 percent for $M = 4$ and 0.62 percent for $M = 8$; the corresponding optimum observation window lengths are the same as for the Mth power synchronizer: $K_0 = 35$ for $M = 4$ and $K_0 = 59$ for $M = 8$.

We have shown that feedback and feedforward synchronizers behave quite differently in the presence of a nonzero frequency offset Ω_0. When the feedback synchronizer has a loop filter with a perfect integrator, the tracking performance is not affected by the frequency offset. In the case of a feedforward synchronizer, the frequency offset Ω_0 yields an increase of the tracking error variance as compared to $\Omega_0 = 0$. The observation window size K can be optimized to yield minimum variance and minimum associated BER degradation. In order to obtain reasonable values of the BER degradation, it is required that the frequency offset is a small fraction of the symbol rate, i.e., in the order of 0.1 percent for the Mth power synchronizer and 1 percent for the decision-directed maximum-likelihood synchronizer. When the frequency offset exceeds these small values, the variance of the feedforward carrier-phase estimate is too large. A carrier-phase estimate with lower variance can be obtained by combining the feedforward estimates that result from successive observation windows. Alternatively, a sufficiently accurate frequency correction must be applied before the averaging over the observation window is performed, so that the residual frequency offset at the input of the averaging filter does not exceed the required small fraction of the symbol rate.

6.3.9 Main Points

We have investigated the tracking performance of various carrier and symbol synchronizers under the assumption of constant unknown synchronization parameters, by linearizing the system equations about the stable operating point and computing the resulting synchronization error variance. As far as the tracking performance is concerned, we have shown that a feedforward synchronizer is equivalent with a feedback synchronizer having a specific closed-loop transfer function. When operating in the tracking mode, the considered carrier and symbol synchronizers do not interact after linearization of the system equations, when the baseband pulse at the matched filter output is real and even.

We have analyzed the DD and NDA carrier synchronizers. For both synchronizers, the phase error variance is proportional to the synchronizer bandwidth. The DD carrier synchronizer is free of self-noise and operates close to the Cramér-Rao bound. Deviations from the Cramér-Rao bound are caused by erroneous decisions on the received data symbols. The phase error variance resulting from the NDA carrier synchronizer consists of an additive noise contribution and a self-noise contribution. In the case of M-PSK, the NDA carrier synchronizer is free of self-noise and operates close to the Cramér-Rao bound; the degradation with respect to the Cramér-Rao bound is caused by higher-order noise terms at the output of the Mth power nonlinearity. In the case of N^2–QAM with $N > 2$, the NDA carrier synchronizer yields self-noise, and the additive noise contribution is considerably larger than the Cramér-Rao bound; both contributions increase with increasing constellation size.

- We have investigated the DD, M&M, and NDA symbol synchronizers, assuming a cosine rolloff pulse at the matched filter output. The additive noise contribution to the timing error variance is proportional to the synchronizer bandwidth, whereas the self-noise contribution is proportional to the square of the synchronizer bandwidth.
- The additive noise contribution for the DD symbol synchronizer is close to the Cramér-Rao bound, and slightly decreases with increasing rolloff; the degradation with respect to the Cramér-Rao bound is caused by the erroneous decisions on the transmitted data symbols. The self-noise contribution for the DD symbol synchronizer decreases with increasing rolloff.
- The additive noise contribution for the NDA synchronizer is considerably larger than the Cramér-Rao bound for small rolloff, decreases with increasing rolloff, and reaches a value close to the Cramér-Rao bound for 100% rolloff. The NDA synchronizer yields a self-noise contribution which is much larger than for the DD synchronizer when the rolloff is small; the self-noise contribution decreases with increasing rolloff, and reaches a value close to the DD self-noise contribution for 100% rolloff.
- The M&M symbol synchronizer is free of self-noise; its additive noise contribution at small rolloff is between the DD-ML and NDA noise contributions, and further increases with increasing rolloff.

When the unknown synchronization parameters are no longer constant but exhibit a small zero-mean random fluctuation (such as caused by oscillator phase noise), feedback and feedforward synchronizers still behave in a similar way. The synchronizers cannot track the frequency components of random fluctuations that fall outside the synchronizer bandwidth. Therefore, the synchronizer bandwidth value is a compromise between the ability to track random fluctuations of the synchronization parameters (this requires a large synchronizer bandwidth) and the reduction of the effect of additive noise and self-noise (this requires a small synchronizer bandwidth).

In the case of a small frequency offset between the oscillators at the receiver and the transmitter, feedback and feedforward synchronizers behave quite differently. When the loop filter of the feedback synchronizer contains a perfect integrator, a frequency offset has no effect on the tracking performance. In the presence of a small frequency offset, a feedforward synchronizer yields an estimate which is unbiased only with respect to the synchronization parameter at the center of the observation window; the resulting estimation error variance is larger than it would be if no frequency offset were present. The size of the observation window which minimizes the error variance is inversely proportional to the frequency offset. The allowable frequency offset which is a very small fraction of the symbol rate keeps the BER degradation within reasonable limits.

6.3.10 Bibliographical Notes

The linearized tracking performance of various carrier and symbol synchronizers that are motivated by the maximum-likelihood criterion has received considerable attention in the literature [1]–[11]. All these synchronizers operate on samples of the filtered received complex envelope, taken at the symbol rate or a small multiple thereof, which makes them suitable for a digital implementation.

The methods is discussed in Sections 6.3.2 for analyzing feedback and feedforward synchronizers have also been used in [5] and [9]. In the same papers, the conditions discussed in Section 6.3.2 yield carrier and symbol synchronizers, that do not interact in the tracking mode.

The NDA carrier synchronizer from Section 6.3.4, which maximizes the low E_s/N_0 limit of the likelihood function averaged of the data symbol sequence, uses an Mth power nonlinearity, where $2\pi/M$ is the angle of rotational symmetry of the constellation. This result has been shown for BPSK ($M = 2$) and QPSK ($M = 4$) in [12], for M-PSK in [11], and for general rotationally symmetric constellations (such as QAM for which $M = 4$) in [13].

Some simplifications of the feedback DD carrier synchronizer from Section 6.3.3 have been proposed in [14] for PSK and QAM constellations, and the resulting tracking performance has been investigated in [15].

The Viterbi and Viterbi (V&V) feedforward carrier synchronizer for M-PSK, introduced in [16], is an important generalization of the feedforward NDA carrier synchronizer from Section 6.3.4. The V&V carrier phase estimate is given by

$$\hat{\theta} = \frac{1}{M} \arg \left(\sum_{k=1}^{K} F(|z_k|) \exp\left(jM \arg\left(z_k\right)\right) \right) \tag{6-202}$$

where $F(x)$ is a nonlinearity which can be optimized to yield minimum phase error variance. For $F(x) = x^M$, the V&V synchronizer reduces to the NDA-ML synchronizer from Section 6.3.4. It is shown in [16] that the phase error variance for large E_s/N_0 converges to the Cramér-Rao bound, for any function $F(x)$ with $F(1) \neq 0$. Hence, the selection of the function $F(x)$ determines the contribution of higher-order noise terms to the tracking error variance, and has most effect on the tracking performance at low and moderate E_s/N_0. The optimum nonlinearity $F(x)$ which minimizes the phase error variance has been derived in [18], and turns out to be dependent on E_s/N_0. The following limits for the optimum $F(x)$ are obtained:

$$F(x) = \begin{cases} x^M & E_s/N_0 \rightarrow 0 \\ x & E_s/N_0 \rightarrow \infty, \text{ no frequency offset} \\ 1 & E_s/N_0 \rightarrow \infty, \text{ frequency offset} \end{cases} \tag{6-203}$$

Hence, the NDA-ML carrier synchronizer from Section 6.3.4 is optimum for very small E_s/N_0. From an implementation point of view, a function $F(x)$ which does not depend on E_s/N_0 is to be preferred; it is shown in [17] that for QPSK with zero frequency offset, the function $F(X) = x^2$ yields a tracking performance which is very close to the performance corresponding to the optimum function $F(x)$, for $E_s/N_0 > -3$ dB.

The feedback DD-ML symbol synchronizer from Section 6.3.5 requires the computation of two samples per symbol, both taken at the estimated decision instant: these samples are $z(kT + \hat{\varepsilon}T)$ at the matched filter output (needed for making the decision $\hat{a}_k$) and $z'(kT + \hat{\varepsilon}T)$ at the output of the derivative matched filter. From the implementation point of view, it is interesting to consider feedback symbol synchronizers that do not need the derivative matched filter. One such synchronizer is the Mueller and Müller (M&M) synchronizer from Section 6.3.5, which uses only the matched filter output samples, taken at the estimated decision instants. The M&M synchronizer and other decision-directed symbol synchronizers operating at the symbol rate have been introduced in [18]; these synchronizers and some modifications thereof are also considered in Section 2.5.3 in the context of baseband PAM transmission. Symbol synchronizers operating at the symbol rate are very attractive in those applications where the filtering in front of the synchronizer is computationally demanding, so that the number of samples per symbol to be produced by this filtering should be kept to a minimum; an example is data transmission over the digital subscriber line, where the filtering consists of echo cancellation and equalization which are both adaptive. A disadvantage of the M&M synchronizer with respect to the DD symbol synchronizer is its larger additive noise contribution to the tracking error variance, especially for large rolloff.

The tracking performance can be improved by using two instead of one matched filter output samples per symbol, i.e., the samples $z(kT + \hat{\varepsilon}T)$ at the estimated decision instants and $z(kT + \hat{\varepsilon}T + \tau/2)$ halfway between estimated decision instants. An example of a decision-directed symbol synchronizer operating a two matched filter output samples per symbol is the data-transition tracking loop

(DTTL), whose timing error detector output is given by

$$x_\varepsilon(k;\hat{\varepsilon}) = \text{Re}\left[\left(\hat{a}_k^* - \hat{a}_{k+1}^*\right) z(kT + \hat{\varepsilon}T + \tau/2) \exp\left(-j\hat{\theta}\right)\right] \tag{6-204}$$

which is the extension to narrowband communication of the DTTL with $\xi_0 = 1/2$ from [3], originally intended for communication with rectangular baseband pulses. The additive noise contribution to the timing error variance resulting from the DTTL synchronizer (6-204) is only slightly larger than for the DD symbol synchronizer, whereas the self-noise contribution is smaller [19].

The feedback NDA symbol synchronizer from Section 6.3.6 also needs the samples $z(kT + \hat{\varepsilon}T)$ and $z'(kT + \hat{\varepsilon}T)$. The implementation of the derivative matched filter can be circumvented by using the Gardner symbol synchronizer introduced in [20]. The Gardner symbol synchronizer needs two matched filter output samples per symbol (i.e., at and halfway between estimated decision instants), and uses the following timing error detector output

$$x_\varepsilon(k;\hat{\varepsilon}) = \text{Re}[(z^*(kT + \hat{\varepsilon}T) - z^*(kT + T + \hat{\varepsilon}T)) z(kT + \hat{\varepsilon}T + T/2)] \tag{6-205}$$

The additive noise contribution to the timing error variance resulting from the Gardner symbol synchronizer (6-205) is only slightly larger than for the NDA symbol synchronizer, whereas the self-noise contribution is smaller [19].

The NDA carrier synchronizer from Section 6.3.4 makes use of the timing estimate. Timing-independent NDA carrier synchronizers have been considered in [3] in the case of communication with rectangular baseband pulses, but can also be used with narrowband baseband pulses. The resulting synchronizers are the so-called Mth power synchronizers or Costas loops, and operate in continuous time. The feedforward version of the timing-independent NDA carrier synchronizer, maximizes with respect to θ the objective function $L(\theta)$, given by

$$L(\theta) = \text{Re}\left[\int_0^{KT} \left(r_f(t)e^{-j\theta}\right)^M dt\right] \tag{6-206}$$

This yields the estimate

$$\hat{\theta} = \frac{1}{M} \arg\left(\int_0^{KT} (r_f(t))^M dt\right) \tag{6-207}$$

The feedback version of the timing-independent NDA carrier synchronizer uses the phase error detector output $x_\theta\left(t;\hat{\theta}\right)$, given by

$$x_\theta\left(t;\hat{\theta}\right) = \frac{1}{M} \text{Im}\left[\left(r_f(t)e^{-j\hat{\theta}}\right)^M\right] \tag{6-208}$$

The tracking performance of the timing-independent NDA carrier synchronizer for narrowband M-PSK has been investigated in [11] and was found to be worse than for the NDA synchronizer from Section 6.3.4, because its additive noise contribution is larger than the Cramér-Rao bound and self-noise is introduced. In a digital implementation, $r_f(t)$ is replaced by a sequence $\{r_f(kT_s)\}$ of nonsynchronized samples taken at a rate $1/T_s$ which is sufficiently large to avoid aliasing in the synchronizer bandwidth after the Mth power nonlinearity.

The DD symbol synchronizers from Section 6.3.5 all make use of the carrier phase estimate. A carrier-independent feedforward maximum-likelihood DD symbol synchronizer is obtained by maximizing over ε the function $L(\varepsilon)$, given by (see eq. 5–85)

$$L(\varepsilon) = \left| \sum_{k=1}^{K} \hat{a}_k^* z(kT + \varepsilon T) \right| \tag{6-209}$$

where $z(t)$ is the matched filter output signal; this maximization is to be performed by means of a search. This algorithm is well suited for receivers using differential data detection, which does not involve a carrier phase estimate at all. It has been shown in [19] that for moderate and large E_s/N_0, the tracking performance is essentially the same as for the DD synchronizer from Section 6.3.5, which makes use of the carrier phase estimate.

6.3.11 Appendix: Self-Noise Contribution to Timing Error Variance

It follows from (6-77) and (6-139) that the self-noise contribution to the timing error variance is given by

$$\text{var}[e_s] = \frac{1}{K_\varepsilon^2} \sum_{m>0} A(m)(h'(mT)\,T)^2 \tag{6-210}$$

where

$$A(m) = T \int_{-\pi/T}^{\pi/T} |H_\varepsilon(\exp{(j\omega T)})|^2 (1 - \cos{(m\omega T)})\, \frac{d\omega}{2\pi} \tag{6-211}$$

Denoting the inverse Fourier transform of $|H_\varepsilon(\exp{(j\omega T)})|^2$ by $h_{\varepsilon,2}(m)$, and taking into account that $h_{\varepsilon,2}(m) = h_{\varepsilon,2}(-m)$, (6-211) yields

$$A(m) = h_{\varepsilon,2}(0) - h_{\varepsilon,2}(m) \tag{6-212}$$

Further, $h_{\varepsilon,2}(m)$ can be expressed as

$$h_{\varepsilon,2}(m) = \sum_{n} h_\varepsilon(n)\, h_\varepsilon(m+n) \tag{6-213}$$

where $\{h_\varepsilon(m)\}$ is the closed-loop impulse response.

A feedforward synchronizer is equivalent with a feedback synchronizer having a closed-loop impulse response given by (6-91). This yields

$$h_{\varepsilon,2}(m) = \begin{cases} (K - |m|)/K^2 & |m| \leq K \\ 0 & \text{otherwise} \end{cases} \tag{6-214}$$

Hence, for $m > 0$,

$$A(m) = \begin{cases} m/K^2 & m \leq K \\ 1/K & \text{otherwise} \end{cases} \tag{6-215}$$

In most cases of practical interest, K is so large that $h'(mT)T \approx 0$ for $m > K$. When this holds, the summation in (6-210) is not affected when $A(m)$ is approximated for all $m > 0$ by

$$A(m) = \frac{m}{K^2} = K_F(2B_LT)^2 m \tag{6-216}$$

where $2B_LT = h_{\varepsilon,2}(0) = 1/K$ and $K_F = 1$.

In the case of a feedback synchronizer, the closed-loop impulse response $\{h_\varepsilon(m)\}$ can be viewed as a sequence of samples, taken at a rate $1/T$, of a causal continuous-time function $p(t)$, i.e., $h_\varepsilon(m) = p(mT)$. Combining (6-213) and (6-212) then yields

$$\begin{aligned} A(m) &= \sum_{n \geq 0} p(nT)[p(nT) - p(nT + mT)] \\ &\simeq \frac{1}{T} \int_0^\infty p(t)[p(t) - p(t + mT)]\ dt \end{aligned} \tag{6-217}$$

Approximating the above summation by an integral is valid when the loop bandwidth is much smaller than the symbol rate. Because of the small loop bandwidth, $h_\varepsilon(m)$ decreases very slowly with m as compared to $h'(mT)T$. Hence, for those integers m yielding values of $h'(mT)\,T$ that contribute significantly to (6-210), the following approximation holds:

$$p(t) - p(t + mT) \simeq -p'(t)mT \qquad t \geq 0 \tag{6-218}$$

Inserting (6-218) into (6-217) yields

$$\begin{aligned} A(m) &\simeq -m \int_0^\infty p(t)p'(t)\,dt \\ &= \frac{1}{2}\,m\,p^2(0) \\ &= \frac{1}{2}\,m\,h_\varepsilon^2(0) \end{aligned} \tag{6-219}$$

As $h_\varepsilon(0)$ is essentially proportional to $2B_LT$, with the constant of proportionality depending on the specific shape of the closed-loop transfer function, $A(m)$ can be expressed as

$$A(m) = K_F(2B_LT)^2 m \tag{6-220}$$

where K_F depends on the shape of the closed-loop transfer function.

In the case of a narrowband first-order loop, the closed-loop transfer function is well approximated by

$$H_\varepsilon(\exp(j\omega T)) \simeq \frac{1}{1+j\omega\tau} \qquad |\omega T| < \pi \tag{6-221}$$

This yields

$$h_\varepsilon(0) \simeq \frac{T}{\tau} \qquad 2B_LT \simeq \frac{T}{2\tau} \tag{6-222}$$

Equating (6-219) and (6-220) yields $K_F = 2$.

In the case of a narrowband second-order loop with perfectly integrating loop filter, the closed-loop transfer function is well approximated by

$$H_\varepsilon(\exp(j\omega T)) \simeq \frac{\omega_n^2 + 2j\zeta\omega_n\omega}{\omega_n^2 + 2j\zeta\omega_n\omega - \omega^2} \qquad |\omega T| < \pi \tag{6-223}$$

This yields

$$H_\varepsilon(0) \simeq 2\zeta\omega_n T \qquad 2B_LT \simeq \frac{1+4\zeta^2}{4\zeta}\omega_n T \tag{6-224}$$

Equating (6-219) and (6-220) yields

$$K_F = 2\left[\frac{4\zeta^2}{1+4\zeta^2}\right]^2 \tag{6-225}$$

Bibliography

[1] R. D. Gitlin and J. Salz, "Timing Recovery in PAM Systems," *BSTJ*, vol. 50, pp. 1645–1669, May 1971.

[2] H. Kobayashi, "Simultaneous Adaptive Estimation and Decision Algorithm for Carrier Modulated Data Transmission Systems," *IEEE Trans. Commun.*, vol. COM-19, pp. 268–280, June 1971.

[3] W. C. Lindsey and M. K. Simon, *Telecommunication Systems Engineering*. Englewood Cliffs, NJ: Prentice-Hall, 1973.

[4] L. E. Franks, "Acquisition of carrier and timing data–i" in new directions in signal processing in communication and control," no. (J. K. Skwirzinski). Leiden, The Netherlands: Noordhoff, 1979.

[5] U. Mengali, "Joint Phase and Timing Acquisition in Data-Transmission," *IEEE Trans. Commun.*, vol. COM-25, pp. 1174–1185, Oct. 1977.

[6] L. E. Franks, *Timing Recovery Problems in Data Communication.*

[7] M. K. Simon, "On the Optimality of the MAP Estimation Loop for Carrier Phase Tracking BPSK and QPSK Signals," *IEEE Trans. Commun.*, vol. COM-28, pp. 158–165, Jan. 1979.

[8] L. E. Franks, "Carrier and Bit Synchronization in Data Communications–A Tutorial Review," *IEEE Trans. Commun.*, vol. COM-28, pp. 1107–1121, Aug. 1980.

[9] M. H. Meyers and L. E. Franks, "Joint Carrier Phase and Symbol Timing Recovery for PAM Systems," *IEEE Trans. Commun.*, vol. COM-28, pp. 1121–1129, Aug. 1980.

[10] M. Moeneclaey, "Synchronization Problems in PAM Systems," *IEEE Trans. Commun.*, vol. COM-28, pp. 1130–1136, Aug. 1980.

[11] A. N. D'Andrea, U. Mengali, and R. Reggiannini, "Carrier Phase Recovery for Narrow-Band Polyphase Shift Keyed Signals," *Alta Frequenza*, vol. LVII, pp. 575–581, Dec. 1988.

[12] M. K. Simon, "Optimum Receiver Structures for Phase-Multiplexed Modulations," *IEEE Trans. Commun.*, vol. COM-26, pp. 865–872, June 1978.

[13] M. Moeneclaey and G. De Jonghe, "ML-Oriented NDA Carrier Synchronization for General Rotationally Symmetric Signal Constellations," *IEEE Trans. Commun.*

[14] A. Leclert and P. Vandamme, "Universal Carrier Recovery Loop for QASK and PSK Signal Sets," *IEEE Trans. Commun.*, vol. COM-31, pp. 130–136, Jan. 1983.

[15] S. Morodi and H. Sari, "Analysis of Four Decision-Feedback Carrier Recovery Loops in the Presence of Intersymbol Interference," *IEEE Trans. Commun.*, vol. COM-33, pp. 543–550, June 1985.

[16] A. J. Viterbi and A. M. Viterbi, "Nonlinear Estimation of PSK Modulated Carrier Phase with Application to Burst Digital Transmission," *IEEE Trans. Inform. Theory*, vol. IT-32, pp. 543–551, July 1983.

[17] B. E. Paden, "A Matched Nonlinearity for Phase Estimation of a PSK-Modulated Carrier," *IEEE Trans. Inform. Theory*, vol. IT-32, pp. 419–422, May 1986.

[18] K. H. Mueller and M. Müller, "Timing Recovery in Digital Synchronous Data Receivers," *IEEE Trans. Commun.*, vol. COM-24, pp. 516–531, May 1976.

[19] T. Jesupret, M. Moeneclaey, and G. Ascheid, "Digital Demodulator Synchronization, Performance Analysis," *Final Report to ESTEC Contract No. 8437/89/NL/RE*, June 1991.

[20] F. Gardner, "A BPSK/QPSK Timing-Error Detector for Sampled Receivers," *IEEE Trans. Commun.*, vol. COM-34, pp. 423–429, May 1986.

6.4 Cycle Slipping

6.4.1 Introduction

When the signal constellation is invariant under a rotation of angle p, and random data is transmitted, the carrier synchronizer cannot distinguish between an angle θ and an angle $\theta + kp$, with $k = \pm 1, \pm 2, \ldots$. As a result, the carrier synchronizer has infinitely many stable operating points, which are spaced by p. For M-PAM, M^2-QAM, and M-PSK constellations, we have $p = \pi$, $p = \pi/2$ and $p = 2\pi/M$, respectively.

Similarly, because the received signal is cyclostationary with period T, the symbol synchronizer cannot distinguish between the normalized delays ε and $\varepsilon + kp$, with $k = \pm 1, \pm 2, \ldots$ and $p = 1$. Hence, the symbol synchronizer has infinitely many stable operating points, which are spaced by p. Most of the time, the (carrier phase or timing) estimate exhibits small random fluctuations about a stable operating point: the synchronizer is in the tracking mode. Occasionally, noise or other disturbances push the estimate away from the current stable operating point, into the domain of attraction of a neighboring stable operating point. This phenomenon is called a cycle slip. After this, the estimate remains for a long time in the close vicinity of the new operating point, until the next slip occurs. This is illustrated in Figure 6-13.

Cycle slipping is a highly nonlinear phenomenon. An exact theoretical analysis is not possible for many cases, so that one must resort to approximations. Computer simulations provide an alternative to theoretical analysis; however, under normal operating conditions cycle slips should have a probability of occurrence which is at least a few orders of magnitude smaller than the decision error probability for perfect synchronization, which implies that computer simulations of cycle slips are extremely time consuming.

6.4.2 Effect of Cycle Slips on Symbol Detection

During a carrier phase slip, the carrier phase error takes on large values, so that a burst of symbol errors occurs. After the slip, the receiver's carrier phase reference differs by $\pm p$ from the carrier phase reference before the slip. During a slip of the symbol synchronizer, the timing error is very large, so that a burst of symbol errors occurs. After the slip, the receiver's symbol count is wrong by ± 1 symbol (repetition or deletion of a symbol). Obviously, the receiver's carrier phase reference or symbol count has to be corrected.

The occurrence and the direction of a slip can be detected by monitoring a known synchronization word, which is, at regular intervals, inserted into the symbol stream to be transmitted. After a carrier phase slip, the known symbols of the synchronization word are found to have the wrong phase, as compared to the receiver's carrier phase reference. After a timing slip, the synchronization word is found at the wrong position, as compared to the receiver's symbol count. Once

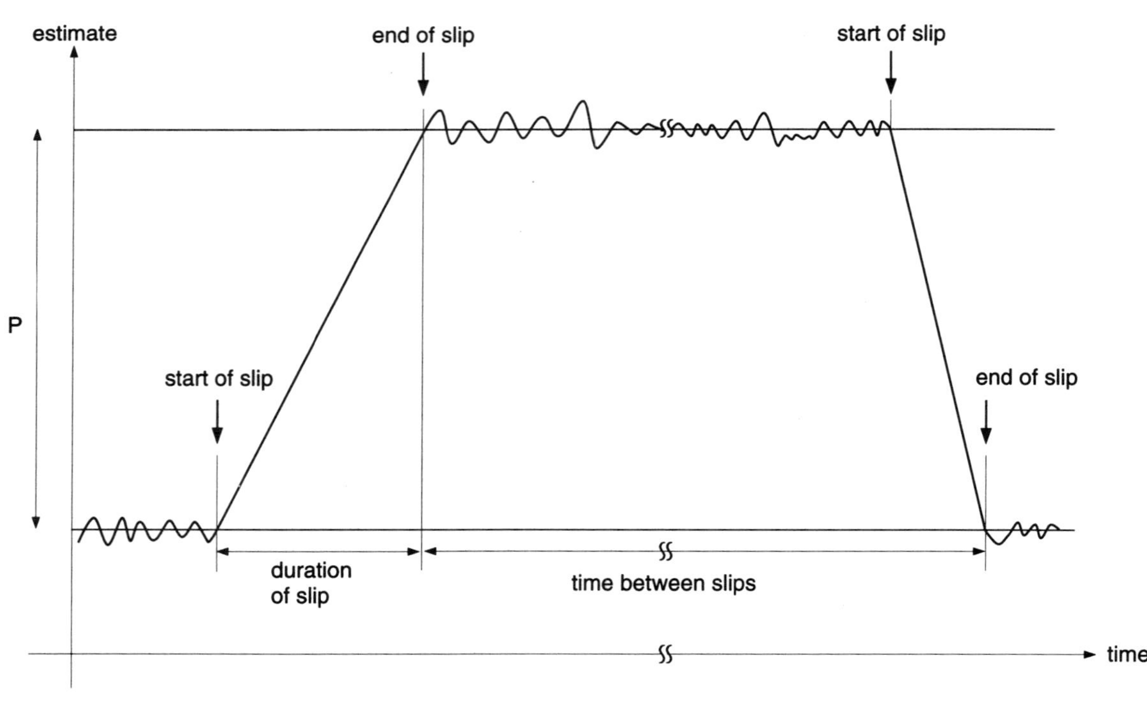

Figure 6-13 Illustration of Cycle Slipping

the slip has been detected, the receiver's carrier phase reference or symbol count is corrected before the detection of the symbols following the synchronization word. Hence, the effect of a slip extends until the synchronization word which allows to detect the slip. On the average, it takes about half the number of symbols between synchronization words before a slip is detected and the carrier phase reference or symbol count corrected. Hence, each slip gives rise to a large number of decision errors. This implies that the cycle slip probability should be at least a few orders of magnitude smaller than the decision error probability for perfect synchronization, in order to avoid that the error performance is dominated by cycle slipping.

The effect of a carrier phase slip on the symbol detection can be limited to the duration of the actual slip by using differential encoding/decoding of the information symbols; in this case the phase of the nth information symbol is the phase difference between the $(n + 1)$st and the nth transmitted symbol. At the receiver, the transmitted symbols are detected coherently, and the phase of the information symbols is obtained as the phase difference between consecutively detected symbols; hence, the incorrect carrier phase reference after a carrier phase slip has no effect on the reliability of the recovered information symbols.

6.4.3 Cycle Slips in Feedback Synchronizers

Let us denote ψ as the synchronization error where $\psi = \phi$ for carrier synchronization, $\psi = e$ for symbol synchronization. In the case of feedback synchronization, the (phase or timing) error detector characteristic and the loop noise power spectral density are both periodic in ψ with period p. In the case of symbol synchronization we have $p = 1$, whereas for carrier synchronization p equals the symmetry angle of the signal constellation ($p = \pi$ for M-PAM, $p = \pi/2$ for M^2-QAM, and $p = 2\pi/M$ for M-PSK).

Cycle slips in feedback synchronizers occur when the loop noise occasionally pushes the estimate away from its stable equilibrium, into the domain of attraction of a neighboring stable equilibrium point; the estimate remains in the vicinity of this new equilibrium point, until the next slip occurs.

Fokker-Planck theory is a powerful tool for investigating cycle slips in feedback synchronizers. Strictly speaking, Fokker-Planck theory applies only to continuous-time systems, whereas nowadays interest has moved from analog continuous-time synchronizers to discrete-time synchronization algorithms, which can be implemented fully digitally. However, when the loop bandwidth of the discrete-time synchronizer is much smaller than the rate at which the carrier phase or timing estimates are updated, a continuous-time synchronizer model can be derived, such that the estimates resulting from the discrete-time synchronizer model are samples of the estimate resulting from the continuous-time model, taken at the update rate $1/T_u$; in many cases of practical interest, the update rate equals the symbol rate. In the appendix (Section 6.4.7) it is shown that the relation between the discrete-time model and the equivalent continuous-time model is as shown in

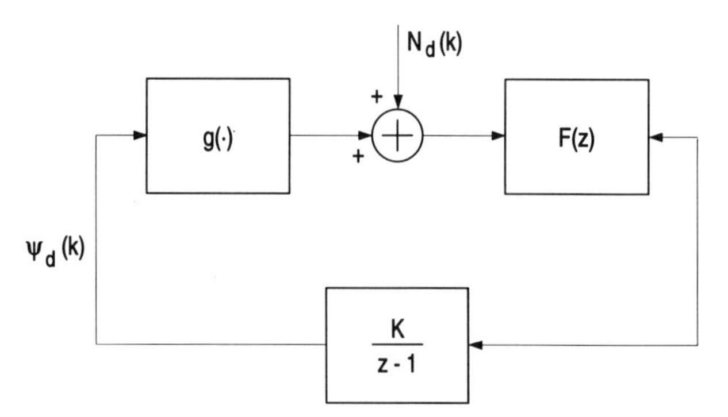

Figure 6-14 Discrete-Time Synchronizer Model

Figures 6-14 and 6-15, where the subscripts *d* and *c* refer to the discrete-time and the continuous-time model, respectively. The loop noise power spectral density in the continuous-time model equals T_u times the loop noise power spectral density in the discrete-time model; note that in general this power spectral density is a periodic function of the synchronization error. The cycle slip performance of the discrete-time synchronizer can, at least in principle, be obtained by applying Fokker-Planck theory to the equivalent continuous-time model from Figure 6-15.

Converting a specific synchronization algorithm into the discrete-time and continuous-time models from Figures 6-14 and 6-15 requires the evaluation of the (carrier phase or timing) error detector characteristic and the loop noise power spectral density as a function of the synchronization error. Deriving an analytical expression for these quantities is often a difficult task, especially in the case of decision-directed algorithms, where a synchronization error introduces quadrature interference or ISI which affect the decision error probability. Alternatively, these quantities can be obtained by means of computer simulations, as outlined in Section 2.3.7.

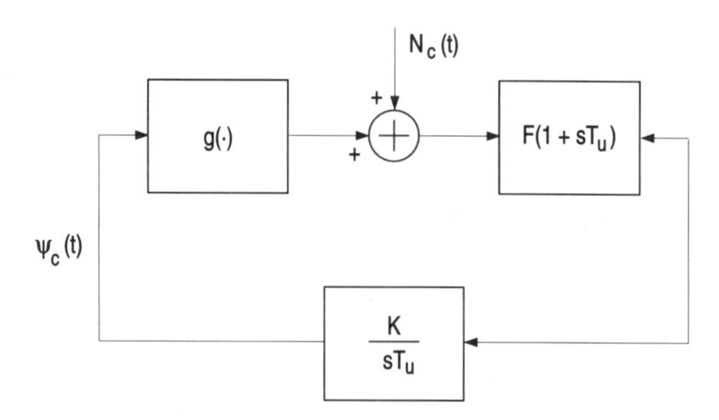

Figure 6-15 Equivalent Continuous-Time Synchronizer Model

At moderate and high E_s/N_0, we expect the cycle slip probability to be many orders of magnitude below the decision error probability, so that cycle slips can be considered as insignificant. At low E_s/N_0, cycle slips are much more frequent, and might have a serious impact on the performance of the receiver; for this mode of operation, an accurate estimate of the normalized cycle slip rate is required. Fortunately, at low E_s/N_0, simple approximations exist for both the error detector characteristic and the loop noise power spectral density as a function of the synchronization error, irrespective of the specific synchronizer algorithm:

- At low E_s/N_0, the error detector characteristic becomes sinusoidal in the synchronization error with period p. The argumentation for this is the same as in Volume 1, Sections 3.2.2 and 7.1.

- At low E_s/N_0, the loop noise power spectral density becomes independent of the synchronization error. This is because the loop noise in the absence of a useful signal at the input of the receiver depends only on the additive Gaussian noise, and the statistics of a phase-rotated sample of the complex envelope of this noise depend neither on the angle of rotation nor on the sampling instant. In addition, the loop noise power spectral density is essentially flat over a frequency interval in the order of the (small) loop bandwidth.

Hence, as far as the cycle slip performance at low E_s/N_0 is concerned, the carrier or symbol synchronizer can be approximated by a PLL with white loop noise and a sinusoidal error detector characteristic with period p; the synchronization error and the loop SNR (which is defined as the inverse of the linearized tracking error variance) are denoted by ψ and ρ, respectively. Defining $\psi' = 2\pi\psi/p$ and $\rho' = p^2\rho/(4\pi^2)$, ψ' can be viewed as the synchronization error resulting from a PLL with sinusoidal error detector characteristic having period 2π, operating at a loop SNR ρ'. Consequently, we can borrow from the vast literature on conventional PLLs to obtain cycle slip results at low E_s/N_0 for any synchronizer.

Fokker-Planck theory yields closed-form analytical results for the cycle slip rate only in the case of first-order loops (see Volume 1, Section 11.1.6). Sophisticated, series-expansion-based numerical methods are needed to determine cycle slip rates for higher-order loops (see Volume 1, Chapter 12). In many applications, a second-order loop is used: it has a more favorable response to frequency offsets than a first-order loop, while avoiding the potential stability problems of third-order and higher-order loops. For a second-order loop with perfect integrator, the steady-state performance is independent of the frequency offset. As the integrator in the loop filter provides a noisy frequency offset estimate, the second-order loop yields a larger cycle slip rate than a first-order loop operating at zero frequency offset. When the damping factor ζ of the second-order loop increases, the frequency offset estimate becomes less noisy, and the cycle slip rate decreases toward the cycle slip rate of a first-order loop with zero frequency offset. A quantitative comparison of cycle slip rates in first-order and second-order PLLs is provided in Volume 1, Section 6.3.3.

In the case of zero frequency offset, the normalized mean time between slips of a first-order PLL with sinusoidal error detector characteristic (with a period of 2π) and white loop noise is given by

$$2B_L E[T_{\text{slip}}] = \pi^2 \rho' I_0^2(\rho') \qquad \text{(period } 2\pi\text{)} \tag{6-226}$$

where ρ' is the loop SNR of the PLL, B_L is the one-sided loop noise bandwidth, $E[T_{\text{slip}}]$ is the mean time between slips, $1/E[T_{\text{slip}}]$ is the cycle slip rate, and $I_0(x)$ is the modified Bessel function of zero order:

$$I_0(x) = \frac{1}{2\pi} \int_{-\pi}^{+\pi} \exp(x \cos\theta)\, d\theta \tag{6-227}$$

For large ρ', the following asymptotic approximation holds:

$$2B_L E[T_{\text{slip}}] \simeq \frac{\pi}{2} \exp(2\rho') \qquad \text{(period } 2\pi,\ \rho' \gg 1\text{)} \tag{6-228}$$

It has been verified in Volume 1, Section 11.1.10, that the approximation (6-228) is accurate for $2B_L E[T_{\text{slip}}] > 10$, which is the range of interest for most applications.

As mentioned before, for a low E_S/N_0 the cycle slip performance of any carrier or clock synchronizer is well approximated by the cycle slip performance of a PLL with sinusoidal phase error detector characteristic (with the appropriate period p) and white loop noise. In the case of a first-order loop with zero frequency error, the corresponding normalized mean time between slips is given by (6-226) or approximated by (6-227) , with ρ' replaced by $p^2\rho/(4\pi^2)$:

$$\begin{aligned} 2B_L E[T_{slip}] &= \frac{p^2 \rho}{4} I_0^2\left(\frac{p^2\rho}{4\pi^2}\right) && \text{(period } p\text{)} \\ &\simeq \frac{\pi}{2} \exp\left(\frac{p^2\rho}{2\pi^2}\right) && \text{(period } p,\ \rho \gg 1\text{)} \end{aligned} \tag{6-229}$$

This indicates that the cycle slip performance is degraded by 20 log($2\pi/p$) dB in loop SNR, as compared to the PLL with the period 2π sinusoidal characteristic. This is intuitively clear: when the distance p between stable equilibrium points decreases, cycle slips become more likely.

Figure 6-16 shows the normalized mean time between slips as a function of the loop SNR ρ, for various values of p. In the case of symbol synchronization, the period p of the timing error detector characteristic equals 1, irrespective of the signal constellation. In the case of carrier synchronization, the phase error detector characteristic has a period p equal to the symmetry angle of the signal constellation: $p = \pi$, $\pi/2$, $\pi/4$, and $\pi/8$ for M-PAM, M^2-QAM, 8-PSK, and 16-PSK, respectively. We observe that a reduction of p by a factor of 2 yields a deterioration of the cycle slip performance by 6 dB in loop SNR.

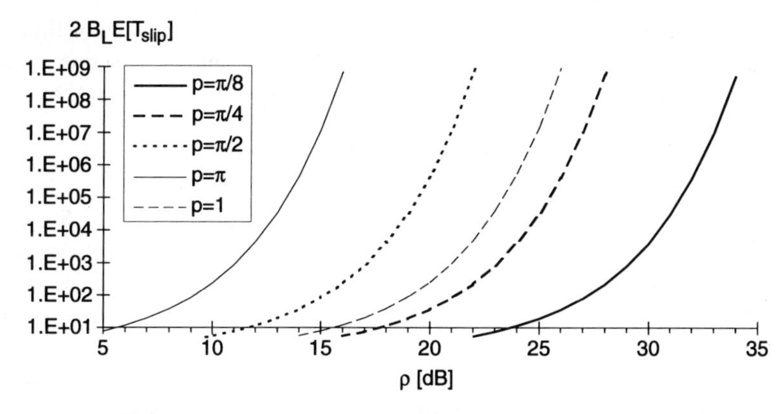

Figure 6-16 Normalized Mean Time Between Slips

As the mean time between slips increases exponentially with the loop SNR ρ, and ρ depends on the normalized loop bandwidth $B_L T$, the mean time between slips is very sensitive to the normalized loop bandwidth. This is illustrated in Table 6-1, which shows the mean time between slips for $p = \pi/2$ as a function of $B_L T$: a decrease of $B_L T$ by a factor of 10 increases the mean time between slips by many orders of magnitude. These results have been obtained under the assumption that the loop SNR is inversely proportional to the normalized loop bandwidth; this assumption is valid when the tracking error variance is dominated by the additive noise contribution. When the tracking error variance is dominated by the self-noise contribution, the loop SNR might be inversely proportional with the second or even a higher power of the normalized loop bandwidth, in which case the dependence of the mean time between slips on the normalized loop bandwidth is even much stronger than suggested by Table 6-1.

6.4.4 Cycle Slips in Feedforward Synchronizers

In the case of feedforward synchronization, the received signal, or a filtered version thereof, is segmented into (possibly overlapping) blocks; for each block,

Table 6-1 Dependence of Mean Time Between Slips on Loop Bandwidth for $p = \pi/2$

$B_L T$	ρ [dB]	$E[T_{\text{slip}}]/T$	$E[T_{\text{slip}}]@T = 1\mu s$
1×10^{-3}	25	1.2×10^{-20}	3.8 million years
3×10^{-3}	20.2	1.3×10^{-8}	1.1 minutes
1×10^{-2}	15	4.1×10^{-3}	4.1 ms
3×10^{-2}	10.2	9.7×10^{-1}	97 μs
1×10^{-1}	5	9.8×10^{0}	9.8 μs

an estimate of the synchronization parameter (carrier phase or symbol timing) is determined as the trial value which maximizes some suitable function, such as the likelihood function or an approximation thereof. Estimates corresponding to different blocks are made independently of each other. The values of the resulting estimates are restricted to the basic interval $(-p/2,\ p/2)$, where $p = 1$ for symbol synchronization and p equals the symmetry angle of the signal constellation in the case of carrier synchronization. Because of this restriction, cycle slips cannot occur, but on the other hand the feedforward estimates are not always able to "follow" the dynamics of the synchronization parameter. For example, in the case of 4-PSK with a small carrier frequency offset, Figure 6-17 shows the increasing carrier phase θ to be estimated, along with a typical trajectory of the feedforward carrier phase estimate $\hat{\theta}$ which is restricted to the interval $(-\pi/4,\ \pi/4)$: obviously the estimation algorithm cannot handle a carrier frequency offset.

The feedforward estimates resulting from the successive blocks are to be post-processed in order to obtain estimates that follow the dynamics of the synchronization parameter to be estimated. The task of the post-processing is to "unwrap" the feedforward estimates and, optionally, reduce the variance of the unwrapped estimates. The structure of the post-processing can range from very simple to quite sophisticated, involving Kalman filtering to exploit a priori knowledge about the statistics of the synchronization parameters to be estimated. Figure 6-17 shows that the estimate $\tilde{\theta}$, resulting from unwrapping the feedforward estimate $\hat{\theta}$, closely follows the true phase trajectory θ.

The post-processing used for unwrapping is shown in Figure 6-18. The output $\tilde{\theta}(k)$ equals $\hat{\theta}(k) + mp$, where the integer m is selected such that $|\ \tilde{\theta}(k) - \tilde{\theta}(k-1)\ | \leq p/2$. Note that the post-processing introduces feedback.

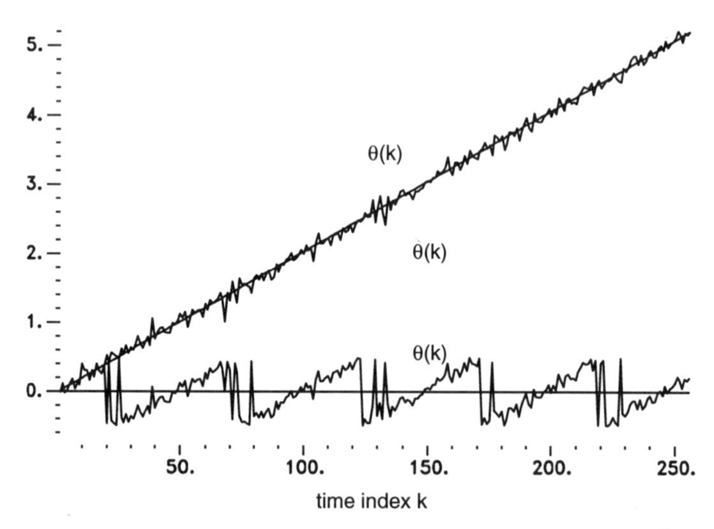

Figure 6-17 Trajectories of Feedforward Carrier Phase Estimate $\hat{\theta}$, of Unwrapped Carrier Phase Estimate $\tilde{\theta}$, and of True Carrier Phase θ (4–PSK)

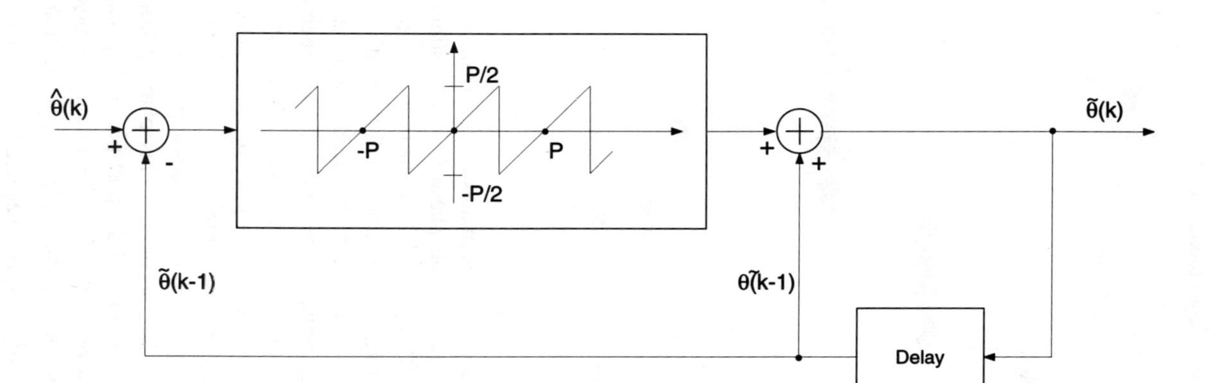

Figure 6-18 Post-Processing Structure for Phase Unwrapping

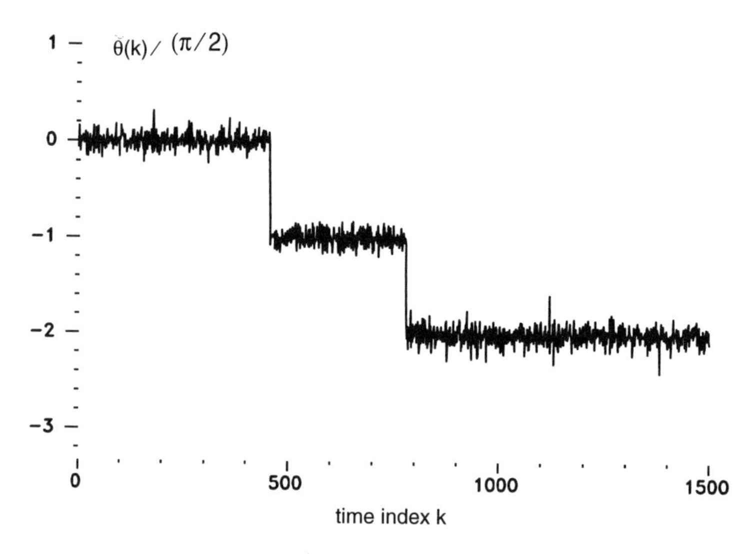

Figure 6-19 Cycle Slipping at Post-Processor Output (4–PSK)

Post-processing the feedforward estimates $\hat{\theta}$ creates the possibility for cycle slips to occur. Let us assume that the phase θ to be estimated is a constant, belonging to the interval $(-p/2,\ p/2)$. Most of the time, the feedforward estimates $\hat{\theta}$ and the post-processing estimates $\tilde{\theta}$ exhibit small fluctuations about the correct phase θ. Occasionally, the noise affects the feedforward estimates in such a way that the post-processing structure interprets them as being estimates of an increasing (decreasing) carrier phase θ. As a result, the post-processing estimates $\tilde{\theta}$ increase (decrease) and leave the interval $(-p/2,\ p/2)$. After this, the feedforward estimates $\hat{\theta}$ again fluctuate in the close vicinity of the true phase θ, but the carrier phase estimates $\tilde{\theta}$ at the output of the post-processing are now close to $\theta + p$ (or close to $\theta - p$): a cycle slip has occurred. This phenomenon is obvious from Figure 6-19, which shows a trajectory of the post-processing output, in the case of 4-PSK with a constant carrier phase $\theta = 0$ and assuming the post-processing from Figure 6-18. The mechanism causing the cycle slips is illustrated in Figure 6-20.

The decrease from $\hat{\theta}(2) = \theta$ to $\hat{\theta}(3)$ gives rise to an increase from $\tilde{\theta}(2) = \hat{\theta}(2)$ to $\tilde{\theta}(3) = \hat{\theta}(3) + p$, because $|\ \hat{\theta}(3) + p - \hat{\theta}(2)\ | < |\ \hat{\theta}(3) - \hat{\theta}(2)\ |$. The increases from $\hat{\theta}(3)$ to $\hat{\theta}(4)$ and $\hat{\theta}(4)$ to $\hat{\theta}(5) = \theta$ give rise to increases from $\tilde{\theta}(3) = \hat{\theta}(3) + p$ to $\tilde{\theta}(4) = \hat{\theta}(4) + p$ and $\tilde{\theta}(4) = \hat{\theta}(4) + p$ to $\tilde{\theta}(5) = \hat{\theta}(5) + p = \theta + p$. After this, $\hat{\theta}$ remains at θ, whereas $\tilde{\theta}$ remains at $\theta + p$.

Quantitative results on cycle slipping in feedforward synchronizers are hard to obtain analytically, because of the highly nonlinear nature of the phenomenon. As the post-processing output $\tilde{\theta}$ cannot be approximated by the output of a white-noise-driven system with only a few (one or two) state variables, Fokker-Planck

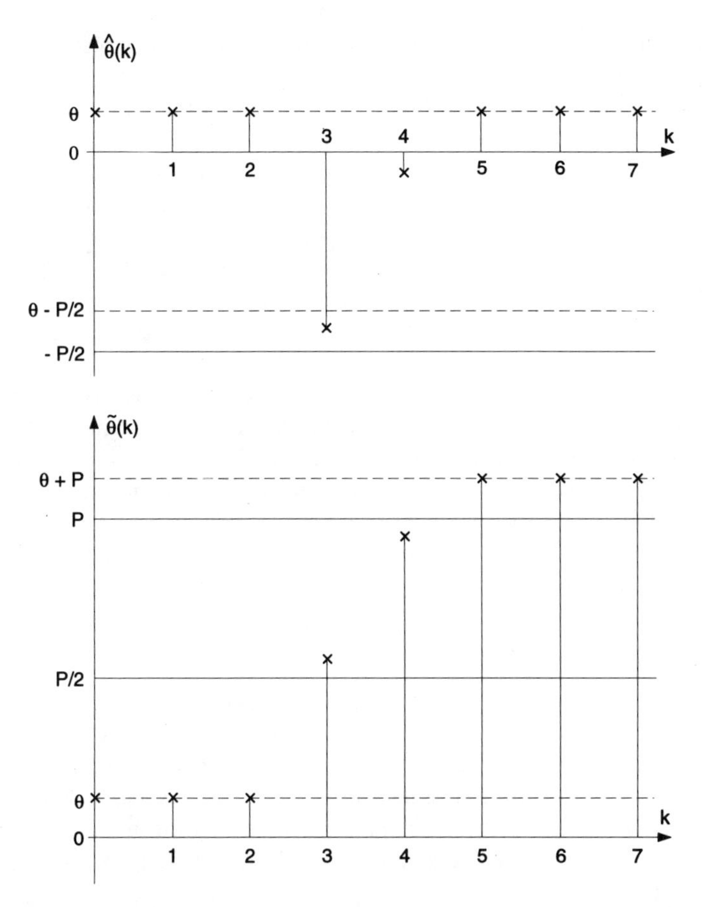

Figure 6-20 Illustration of Mechanism That Causes Cycle Slipping

results from the first-order or second-order PLL cannot be adopted. A few research results are commented upon in the bibliographical notes (Section 6.4.6).

6.4.5 Main Points

- Occasionally, noise causes the estimate resulting from a synchronizer to move from a stable operating point to a neighboring stable operating point; this phenomenon is called a cycle slip.
- Cycle slips cause the receiver's carrier phase reference or symbol count to be erroneous after the occurrence of a slip. As a result, a burst of symbol decision errors occurs, until the carrier phase reference or symbol count have been corrected. The occurrence of a slip and its direction can be derived from

monitoring a known synchronization word which has been inserted at regular intervals into the symbol stream to be transmitted.

- Cycle slips in feedback synchronizers occur when the loop noise pushes the estimate of the synchronization parameter away from a stable equilibrium point, into the domain of attraction of a neighboring stable equilibrium point. Cycle slipping can be investigated by means of Fokker-Planck theory. The mean time between slips increases exponentially with the loop SNR, and, as such, is very sensitive to the value of the loop bandwidth. The larger the period of the synchronizer's error detector characteristic, the larger the mean time between slips, because the stable equilibrium points are further away from each other.
- Cycle slips in feedforward synchronizers occur because of the feedback introduced in the post-processing of the feedforward estimates; such post-processing with feedback is necessary to be able to follow the dynamics of the synchronization parameter to be estimated. Analytical investigation of cycle slipping is very complicated, because Fokker-Planck theory is not the appropriate tool. Only few results are available; some of them are discussed in the bibliographical notes (Section 6.4.6).

6.4.6 Bibliographical Notes

The mean time between slips has been evaluated in [1] for various feedback carrier and symbol synchronization algorithms operating on narrowband BPSK and (O)QPSK, assuming a first-order loop, and, whenever possible, taking into account the correct expressions for the error detector characteristic and the loop noise power spectral density. In [2], the cycle slipping has been investigated in a similar way for feedback carrier synchronizers operating on various (O)QPSK modulations.

Cycle slips in Mth power feedforward carrier synchronizers have been investigated in [3], in the case of M-PSK transmission and assuming the post-processing from Figure 6-18. The mean time between slips was found to depend exponentially on the signal-to-noise ratio at the output of the averaging filter of the feedforward synchronizer, which implies that the bandwidth of the averaging filter has a very large effect on the cycle slip performance. Also, the mean time between slips decreases with the number M of constellation points. The feedforward Mth power synchronizer exhibits more cycle slips than its feedback counterpart; in this comparison, the noise bandwidth of the averaging filter of the feedforward synchronizer equals the loop noise bandwidth of the feedback synchronizer.

Several structures for post-processing the feedforward estimates can be envisaged, but only one of them has been considered in Section 6.4.4. Other structures can be found in Section 6.5.4; these structures perform not only the unwrapping of the feedforward estimates, but also reduce the variance of the unwrapped estimates. The minimum variance is obtained by applying Kalman filtering, in which case the post-processing becomes time-varying. The cycle slipping for first-order post-processing has been investigated in [4].

6.4.7 Appendix: Approximating a Discrete-Time System by a Continuous-Time System

Let us consider a discrete-time system with an $N \times 1$ state vector $\mathbf{x}_d(k)$, satisfying the following state equation:

$$\mathbf{x}_d(k+1) = \mathbf{x}_d(k) + \mathbf{G}(\mathbf{x}_d(k)) + \mathbf{B}\mathbf{n}_d(k) \tag{6-230}$$

where $\mathbf{G}(\cdot)$ is an $N \times 1$ vector function, $\mathbf{n}_d(k)$ is an $M \times 1$ noise vector, and $\mathbf{B}$ is an $N \times M$ matrix; a block diagram of this system is shown in Figure 6-21. The system bandwidth is assumed to be very small with respect to the update rate $1/T_u$, so that the state vector changes very little from $\mathbf{x}_d(k)$ to $\mathbf{x}_d(k+1)$. The bandwidth of $\mathbf{n}_d(k)$ is assumed to be much larger than the bandwidth of $\mathbf{x}_d(k)$; consequently, $\mathbf{n}_d(k)$ can be considered as a discrete-time process whose power spectral density matrix is essentially constant over frequency. This power spectral density matrix is given by

$$\sum_m \mathbf{R}_d(m) \exp(-j\omega T_u) \simeq \sum_m \mathbf{R}_d(m) \tag{6-231}$$

where $\mathbf{R}_d(m)$ is the $M \times M$ autocorrelation matrix of $\mathbf{x}_d(k)$. Now we show that $\mathbf{x}_d(k)$ can be approximated by the sample, at instant kT_u, of the state vector $\mathbf{x}_d(k)$ of a continuous-time system, which satisfies

$$\frac{d}{dt}\mathbf{x}_c(t) = \frac{1}{T_u}[\mathbf{G}(\mathbf{x}_c(t)) + \mathbf{B}\mathbf{n}_c(t)] \tag{6-232}$$

A block diagram of this system is shown in Figure 6-22. The noise $\mathbf{n}_c(t)$ has a bandwidth which is much larger than the system bandwidth; consequently, its power spectral density matrix can be considered as constant over frequency. This power spectral density matrix is given by

$$\int_{-\infty}^{+\infty} \mathbf{R}_c(u) \exp(-j\omega u)\, du \simeq \int_{-\infty}^{+\infty} \mathbf{R}_c(u) du \tag{6-233}$$

where $\mathbf{R}_c(u)$ is the $M \times M$ autocorrelation matrix of $\mathbf{n}_c(t)$. The power spectral density matrix of $\mathbf{n}_c(t)$ is selected to be T_u times the power spectral density matrix of $\mathbf{n}_d(k)$. From (6-232) it follows that

$$\mathbf{x}_c(kT_u+T_u) = \mathbf{x}_c(kT_u) + \frac{1}{T_u}\int_{kT_u}^{kT_u+T_u} \mathbf{G}(\mathbf{x}_c(t))\, dt + \frac{1}{T_u}\int_{kT_u}^{kT_u+T_u} \mathbf{B}\mathbf{n}_c(t)\, dt \tag{6-234}$$

As the bandwidth of the discrete-time system from Figure 6-21 is very small with respect to the update rate $1/T_u$, the same holds for the continuous-time system from Figure 6-22. Hence, the following approximation is valid:

$$\frac{1}{T_u}\int_{kT_u}^{kT_u+T_u} \mathbf{G}(\mathbf{x}_c(t))\, dt \simeq \mathbf{G}(\mathbf{x}_c(kT_u)) \tag{6-235}$$

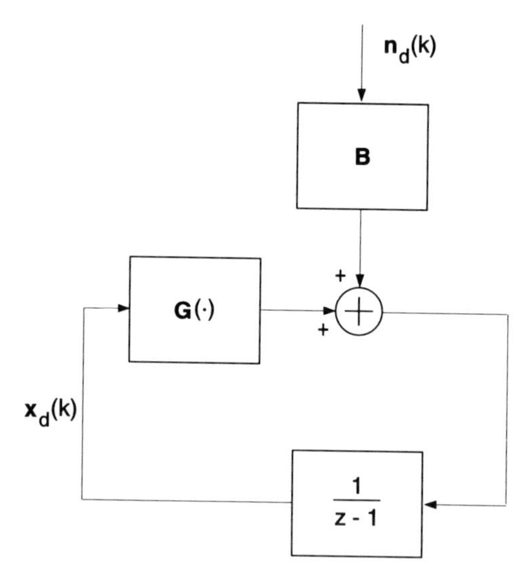

Figure 6-21 Discrete-Time Feedback System

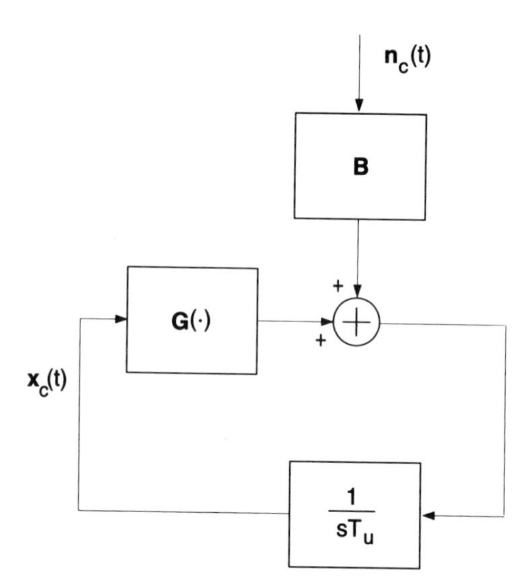

Figure 6-22 Equivalent Continuous-Time Feedback System

Substitution of (6-235) into (6-234) indicates that $\mathbf{x}_d(k)$ and $\mathbf{x}_c(kT_u)$ exhibit the same dynamics. Now let us investigate the $M \times 1$ noise vector $\mathbf{N}(k)$, defined by

$$\mathbf{N}(k) = \frac{1}{T_u} \int_{kT_u}^{kT_u+T_u} \mathbf{n}_c(t)\, dt \tag{6-236}$$

As $\mathbf{n}_c(t)$ can be viewed as a nearly white process with a power spectral density matrix given by (6-233), it follows that $\mathbf{N}(k)$ and $\mathbf{N}(k+m)$ are essentially uncorrelated for $m \neq 0$. Hence, the power spectral density matrix of $\mathbf{N}(k)$ can be considered as constant over frequency, and is given by

$$E\left[\mathbf{N}(k)\mathbf{N}^H(k)\right] = \frac{1}{T_u} \int_{-\infty}^{+\infty} \mathbf{R}_c(u)\, du \tag{6-237}$$

which, by construction, equals the power spectral density matrix (6-231) of $\mathbf{n}_d(t)$. The conclusion is that the narrowband discrete-time system from Figure 6-21 can be replaced by the narrowband continuous-time system from Figure 6-22, by simply performing the following operations:

(i) Replace the discrete-time integrator, with transfer function $1/(z-1)$ in the z-domain, by a continuous-time integrator, with transfer function $1/(sT_u)$ in the Laplace domain. Formally, this corresponds to substituting the operator z by the operator $1 + sT_u$.

(ii) Replace the discrete-time noise by continuous-time noise, with the latter having a power spectral density matrix equal to T_u times the power spectral density matrix of the former.

Bibliography

[1] T. Jesupret, M. Moeneclaey, and G. Ascheid, "Digital Demodulator Synchronization, Performance Analysis," *Final Report to ESTEC Contract No. 8437/89/NL/RE*, June 1991.

[2] A. N. D'Andrea and M. L. Luise, "Cycle Slipping in Serial Receivers for Offset Binary Modulations," *IEEE Trans Commun.*, vol. COM-38, pp. 127–131, Jan. 1990.

[3] G. De Jonghe and M. Moeneclaey, "The Effect of the Averaging Filter on the Cycle Slipping of NDA Feedforward Carrier Synchronizers for MPSK," *Proc. ICC '92, Chicago*, pp. 365–369, June 1992.

[4] G. De Jonghe and M. Moeneclaey, "Cycle Slip Analysis of the NDA FF Carrier Synchronizer Based on the Viterbi & Viterbi Algorithm," *Proc. ICC '94*, pp. 880–884, May 1994.

6.5 Acquisition of Carrier Phase and Symbol Timing

6.5.1 Introduction

At the start of signal reception, the synchronizer has no knowledge about the synchronization parameter values. After some processing of the received signal, the synchronizer is able to deliver accurate estimates of the synchronization parameters that are needed for reliable data detection. This transition from a large initial uncertainty about the synchronization parameters to a small steady-state estimation error variance is called *acquisition.*

We will assume that the frequency offset between transmitter and receiver is small, so that only the slowly time-varying carrier phase and symbol timing have to be acquired. In the case of a large frequency offset, this implies that a separate frequency estimation algorithm has acquired the carrier frequency, and that frequency correction has been applied to the signal entering the carrier phase and symbol synchronizers.

Two distinct modes of operation will be considered, i.e., the continuous mode and the burst mode. In the continuous mode, data are sent continuously from the transmitter to the receiver. The burst mode is typical of time-division multiple access (TDMA), where several users share the capacity of the communication channel by transmitting bursts of data in nonoverlapping time intervals. A further distinction will be made between short burst operation and long burst operation. In the case of short burst communication, the number of data symbols per burst is so small, that the carrier phase and symbol timing can be considered as constant over the entire burst; in the case of long burst communication, the fluctuation of the synchronization parameters over the burst cannot be ignored. For a synchronizer operating in short burst mode, it is sufficient to produce a single carrier phase and timing estimate per burst, and use these parameters for detecting all data symbols within the burst. A synchronizer operating in long burst mode has to make multiple carrier phase and symbol timing estimates per burst, in order to track the variations of the synchronization parameters; in this respect, operation in long burst mode is similar to operation in continuous mode.

During acquisition, the estimates produced by the synchronizer are not accurate enough to perform reliable data detection, so that information symbols transmitted during acquisition are lost. Therefore, the transmitter sends a preamble of training symbols which precedes the actual information symbols; at the end of the preamble, the acquisition should be completed so that reliable data detection can start. The content of the preamble affects the acquisition performance; for the acquisition of timing, it is advisable that the preamble contains many data symbol transitions, so that a lot of timing information is available in the received signal during the preamble. Preamble symbols consume bandwidth and power that are not used to convey digital information. In order to keep a high efficiency, the ratio of the number of preamble symbols to the total number of transmitted symbols should be small.

In burst mode communication, the synchronization parameters (especially the carrier phase) can change considerably between bursts from the same user, so that these parameters must be acquired again for each burst. In the case of long burst communication, the power and bandwidth efficiency are only marginally affected when each transmitted burst contains a preamble whose duration is sufficiently long (say, in the order of a hundred symbols) for most synchronizers to acquire the synchronization parameters. In the case of short burst communication, the information sequence per burst might be only about a hundred symbols long, so that for a high efficiency there should be only a very short preamble or preferably no preamble at all. Therefore, the acquisition time requirements on synchronizers for short burst operation are very stringent.

In some TDMA applications via satellite, symbol timing is derived from a master clock which is distributed to all users; the users adjust their burst transmission instants such that they receive their own bursts in synchronism with the received master clock. For these applications, the correct symbol timing is available at the beginning of each burst, so that only the carrier phase must be acquired.

6.5.2 Feedback Synchronizers

Feedback synchronizers make use of (phase and timing) error detectors, which provide a noisy indication about the sign and magnitude of the instantaneous error between the actual synchronization parameter values and their estimates. From the error detector output signals, a (small) correction term is derived, and is applied to the synchronization parameter estimates so that the magnitude of the estimation error is reduced. At the start of signal reception, the estimation error magnitude might be large; during acquisition this magnitude gradually decreases because of the feedback. Acquisition is completed when the estimation error is close to a stable equilibrium point.

Coupling between Feedback Carrier and Symbol Synchronizers

Several feedback carrier synchronizers make use of a timing estimate, and several feedback symbol synchronizers make use of a carrier phase estimate; this is certainly the case for decision-directed synchronizers [because both the carrier phase and the symbol timing are needed for (coherent) detection], but also for some non-data-aided synchronizers. Consequently, in general the dynamics of carrier and symbol synchronizers are coupled. Although for most synchronizers this coupling is negligible in the tracking mode because linearization of the system equations about the stable equilibrium point yields uncoupled dynamics, the coupling can no longer be ignored during acquisition, where estimation errors might be so large that linearization does no longer apply. Usually, the effect of a timing error on the average phase error detector output and of a phase error on the average timing error detector output is to reduce the magnitude of the error detector output. This means that the coupling reduces the restoring force which drives the estimates toward a stable equilibrium point, so that acquisition is slowed down. Acquisition performance is hard to predict, because the analysis of coupled nonlinear dynamical systems that are affected by noise is very difficult; acquisition performance can be studied by means of computer simulations, but many simulations would be needed to study the effect of the initial condition, as the coupling increases the dimension of the state space.

The coupling between the dynamics of the carrier and symbol synchronizer can be avoided by using a carrier synchronizer which operates independently of the timing estimate [a candidate algorithm is the discrete-time version of (6-209)] and a symbol synchronizer which does not use the carrier phase estimate [candidate algorithms are the NDA algorithm (6-150) and the Gardner algorithm (6-205)]. In this case, both synchronizers simultaneously operate on the received signal, and the carrier phase and symbol timing are acquired independently. The time needed to acquire both the carrier phase and the symbol timing is the maximum of the acquisition times for the individual synchronization parameters.

The situation is still favorable when only one of the synchronizers operates completely independently of the other. Let us consider the case where the symbol synchronizer does not use the carrier phase estimate, but the carrier synchronizer needs a timing estimate (this is the case for most carrier synchronizers). The acquisition of the symbol timing is not affected by the carrier synchronizer. During the symbol timing acquisition, the carrier phase acquisition process is hardly predictable, because it is influenced by the instantaneous value of the symbol timing estimate. However, when the symbol timing acquisition is almost completed, the carrier synchronizer uses a timing estimate which is close to the correct timing; from then on, the carrier phase acquisition is nearly the same as for correct timing, and is essentially independent of the symbol synchronizer operation. The time needed to acquire both the carrier phase and the symbol timing is approximated by the sum of the acquisition time for the symbol timing (independent of the carrier phase) and the acquisition time for the carrier phase (assuming correct timing).

When the receiver knows in advance the training symbols contained in the preamble, this knowledge can be exploited during acquisition by means of a data-aided (DA) instead of a decision-directed (DD) synchronization algorithm. The DA algorithm uses the correct training symbols instead of the receiver's decisions about the training symbols (which might be unreliable during acquisition, especially for large initial synchronization errors); the magnitude of the average error detector output is larger for the DA algorithm than for the DD algorithm, so that acquisition is enhanced. On the other hand, a coupling is introduced when both the carrier and the symbol synchronizer use DA algorithms, because most feedback DA carrier synchronizers (symbol synchronizers) need a timing estimate (a carrier phase estimate).

From the above, we conclude that when at least one synchronizer operates independently of the other, the approximate time to acquire both the carrier phase and the symbol timing can be expressed in terms of the acquisition times of the individual synchronizers, assuming no coupling.

Acquisition Performance of Feedback Synchronizers

In the following, we consider the acquisition of the carrier synchronizer, assuming perfect timing; a similar reasoning applies to a symbol synchronizer, assuming perfect carrier phase estimation.

The acquisition performance of a conventional PLL with sinusoidal phase error detector characteristic has been investigated in Chapters 4 and 5 of Volume 1. Similar conclusions can be drawn for synchronizers with arbitrary (phase or timing) error detector characteristics, when there is no coupling between carrier phase estimation and symbol timing estimation. We restrict our attention to feedback carrier synchronizers having a sinusoidal phase error detector characteristic with period $2\pi/M$. This includes the DD synchronizer operating at low E_S/N_0 and the NDA synchronizer, with $2\pi/M$ denoting the symmetry angle of the constellation; when the DD synchronizer uses known preamble symbols instead of the receiver's decisions (i.e., DA instead of DD operation), the phase error detector characteristic is sinusoidal with period 2π, irrespective of the symmetry angle of the constellation and the value of E_S/N_0. The loop SNR is denoted by ρ, and defined as $\rho = 1/\text{var}[\phi]$, where var$[\phi]$ is the linearized steady-state phase error variance. The acquisition performance for a sinusoidal characteristic with period $2\pi/M$ (in the absence of noise) is briefly summarized below.

A first-order loop can acquire the carrier phase without slipping cycles when the magnitude of the carrier frequency offset f_0 does not exceed the pull-in frequency $\Delta\omega_p$; when $\mid \Delta\omega \mid > |\Delta\omega_p|$, the loop slips cycles continuously without ever achieving lock. The pull-in frequency F_p is related to the parameters of the loop by

$$\Delta\omega_p T = \frac{4B_L T}{M} \tag{6-238}$$

where $B_L T$ denotes the loop bandwidth normalized to the symbol rate $1/T$. The pull-in frequency is proportional to the loop bandwidth, and inversely proportional to M. When $\mid \Delta\omega \mid < |\Delta\omega_p|$, a nonzero frequency offset gives rise to a steady-state phase error ϕ_S, given by $\phi_S = (\arcsin{(F/F_p)})/M$. If this steady-state error is too large, a second-order loop with a perfectly integrating loop filter could be used; this yields zero steady-state error, at the expense of a somewhat longer acquisition time for the same loop bandwidth.

The acquisition time in a first-order loop depends critically on the initial phase error. When the initial phase error is close to an unstable equilibrium point, the average restoring force is very small, and the phase error remains in the vicinity of the unstable equilibrium point for an extended period of time; this phenomenon is called hangup. Assuming a uniformly distributed initial phase error, the average acquisition time T_{acq} in the absence of noise is well approximated by

$$\frac{T_{\text{acq}}}{T} = \frac{1}{2B_L T} \tag{6-239}$$

However, when hangup occurs, the acquisition time is considerably larger than this average time. The probability that acquisition has not been achieved within a time equal to five times (ten times) the average acquisition time T_{acq} is about 10^{-4} (about 10^{-8}). The presence of noise slightly increases the average acquisition time. Hangup can be avoided and the average acquisition time reduced by means of the acquisition aid described in Section 5.1 of Volume 1.

When the carrier synchronizer must handle a carrier frequency offset which is larger than $\Delta\omega_p$ from (6-238), a higher-order loop must be used; in the following we restrict our attention to the second-order loop with perfectly integrating loop filter, for which the pull-in frequency is infinitely large. When the frequency offset at the input of the second-order loop is much larger than the loop bandwidth, unaided frequency acquisition (pull-in) is a slow process which is susceptible to noise; this yields excessively long acquisition times. In the absence of noise, the acquisition time T_{acq} for $\mid F \mid >> B_L$ is well approximated by

$$\frac{T_{\text{acq}}}{T} = \frac{\pi^2 \left(4\zeta^2+1\right)^3}{256\zeta^4} \, \frac{1}{B_L T} \left[\frac{\Delta\omega T}{2\pi \, B_L T}\right]^2 M^2 \tag{6-240}$$

where ζ denotes the damping factor of the loop; for $\Delta\omega/2\pi = 10B_L$, $\zeta = 1$, and $M = 4$, (6-240) yields an acquisition time which is 15421 times as large as T_{acq} from (6-239). Acquisition performance can be improved by using acquisition aids, such as sweeping or a separate frequency error detector (see Section 5.2 of Volume 1). In the case of sweeping, the loop SNR determines the maximum sweep rate. When $\rho > 10M^2$, the maximum rate of change of the locally generated carrier frequency corresponds to

$$T^2 \frac{d^2\hat{\theta}}{dt^2} = 0.4 \, \frac{64\zeta^2}{\left(1+4\zeta^2\right)^2} \left(B_L T\right)^2 \frac{1}{M} \tag{6-241}$$

The resulting acquisition time T_{acq} is approximated by

$$\frac{T_{\text{acq}}}{T} = 5\pi \, \frac{\left(1+4\zeta^2\right)^2}{64\zeta^2} \, \frac{1}{B_L T} \, \frac{|\Delta\omega/2\pi| T}{B_L T} \, M \tag{6-242}$$

For $\Delta\omega = 10B_L$, $\zeta = 1$, and $M = 4$, (6-242) yields an acquisition time which is about 31 times smaller than T_{acq} for unaided acquisition [see (6-240)], but still about 490 times larger than T_{acq} for a first-order loop with $\mid \Delta\omega \mid < \Delta\omega_p$ [see (6-239)].

In the case of symbol synchronization, the clock frequency is usually known quite accurately, so that a first-order loop can be used. Let us assume a sinusoidal timing error detector characteristic; this includes all quadratic symbol synchronizer algorithms (such as the NDA algorithm and Gardner's algorithm) operating on a useful signal whose lowpass bandwidth does not exceed $1/T$. Then the pull-in frequency $\Delta\omega_p$ is given by (6-238) with $M = 1$. As for carrier synchronization, the average acquisition time is approximated by (6-239); hangup occurs when the initial timing estimate is close to an unstable equilibrium point. The principle of the acquisition aid from Section 5.1 of Volume 1 can also be applied to symbol synchronizers.

From the above considerations and the results from Chapters 4 and 5 of Volume 1, it is concluded that the acquisition time is in the order of the inverse of the loop bandwidth (say, in the order of a hundred symbols) when the frequency offset is small, but might be considerably longer when the frequency offset is much larger than the loop bandwidth. Because of the long training sequence required, feedback synchronizers are suitable for acquiring the carrier phase and the symbol timing only when operating in continuous mode and long burst mode, but not in short burst mode.

Lock Detectors

During tracking, feedback synchronizers might use different system parameters (or even a different algorithm) than during acquisition. Therefore, a reliable indication is needed about whether or not the synchronizer is in lock: when the synchronizer gets locked, it enters the tracking mode, whereas it reenters the acquisition mode when lock is lost. A few examples of different parameters or algorithms during acquisition and tracking are given below:

- During acquisition, a larger synchronizer bandwidth is used in order to obtain fast acquisition. During tracking, the bandwidth is reduced in order to suppress the loop noise.
- Some acquisition aids (such as sweeping) must be switched off during tracking, in order to avoid a steady-state estimation error (see Section 5.2.1 of Volume 1).
- During acquisition, non-decision-aided algorithms are selected for carrier and symbol synchronization, in order to avoid strong coupling between both synchronizers. During tracking, decision-directed algorithms yielding smaller tracking error variance are used.

An unambiguous lock indication cannot be derived just from the phase or timing error detector output. Indeed, although an in-lock condition corresponds to a small average detector output, during pull-in or hangup the average detector output is small as well. Therefore, an additional circuit, called lock detector, is needed for an unambiguous indication. In the following, we give a few examples of lock detectors for carrier and symbol synchronizers. We denote by $y_\theta(k;\hat{\theta})$ and $y_\varepsilon(k;\hat{\varepsilon})$ the lock detector outputs for the carrier and symbol synchronizers, respectively, with k indicating the symbol count. A reliable lock indication is obtained by lowpass filtering the lock detector outputs. The useful component at the lowpass filter output equals $E[y_\theta(k;\hat{\theta})]$ for carrier synchronization or $E[y_\varepsilon(k;\hat{\varepsilon})]$ for symbol synchronization, where $E[\cdot]$ denotes statistical expectation with respect to noise and data symbols. The variance at the output of the lowpass filter is essentially proportional to the filter bandwidth; however, for a small filter bandwidth it takes a long transient before the filter output is in the steady state, which in turn increases the acquisition time.

Carrier synchronization:

1.

$$y_\theta\left(k;\hat{\theta}\right) = \mathrm{Re}\left[\hat{a}_k^* z_k \exp\left(-j\hat{\theta}\right)\right] \tag{6-243}$$

$$E\left[y_\theta\left(k;\hat{\theta}\right)\right] \propto \cos\left(\phi\right) \tag{6-244}$$

2.

$$y_\theta\left(k;\hat{\theta}\right) = \mathrm{Re}\left[z_k^M \exp\left(-jM\hat{\theta}\right)\right] \tag{6-245}$$

$$E\left[y_\theta\left(k;\hat{\theta}\right)\right] \propto \cos\left(M\phi\right) \tag{6-246}$$

Symbol synchronization:

1.

$$y_\varepsilon(k;\hat{\varepsilon}) = \mathrm{Re}[z^*(kT{+}T/4{+}\hat{\varepsilon}T)z'(kT{+}T/4{+}\hat{\varepsilon}T)] \tag{6-247}$$

$$E[y_\varepsilon(k;\hat{\varepsilon})] \propto \cos\left(2\pi e\right) \tag{6-248}$$

2.

$$\begin{aligned} y_\varepsilon(k;\hat{\varepsilon}) = \mathrm{Re}\big[&\{z^*(kT{+}T/4{+}\hat{\varepsilon}T) - z^*(kT{+}5T/4{+}\hat{\varepsilon}T)\} \\ &\times \{z(kT{+}3T/4{+}\hat{\varepsilon}T)\}\big] \end{aligned} \tag{6-249}$$

$$E[y_\varepsilon(k;\hat{\varepsilon})] \propto \cos\left(2\pi e\right) \tag{6-250}$$

3.

$$y_\varepsilon(k;\hat{\varepsilon}) = |z(kT{+}\hat{\varepsilon}T)|^2 \tag{6-251}$$

$$E[y_\varepsilon(k;\hat{\varepsilon})] = A + B\cos\left(2\pi e\right) \qquad A > 0,\ \ B > 0 \tag{6-252}$$

In the above, ϕ and e denote the phase error and the timing error, $z(t)$ is the matched filter output signal, and z_k is a short-hand notation for $z(kT + \hat{\varepsilon}T)$. The results (6-244) and (6-246) for carrier synchronization assume perfect timing; in addition, (6-244) assumes that either known preamble signals are used (DA operation) or the receiver's decisions about the data symbols are correct (DD operation). The results (6-248), (6-250), and (6-252) for symbol synchronization assume narrowband communication, i.e., the equivalent lowpass bandwidth of the transmitted signal does not exceed $1/T$.

In the above examples, the average of the lock detector output signals is maximum for zero synchronization error, i.e., when the synchronizer is in lock. A binary in-lock/out-of-lock indication is obtained by comparing the lowpass filter output signal with an appropriate threshold.

6.5.3 Feedforward Synchronizers for Short Burst Operation

In this section we consider the case of time division multiple access (TDMA) with short bursts of data. Because of the short bursts, it is sufficient to make for each burst a single estimate of the carrier phase and the symbol timing. The time between bursts from a given user is much longer than the duration of a burst; this means that the receiver has a lot of time available for processing a burst, before the next burst from the same user arrives. It will be assumed that the short TDMA bursts do not contain training symbols, in order not to waste bandwidth and power. Because of their long acquisition time, feedback algorithms are not suited for short burst operation. Hence, we restrict our attention to feedforward algorithms.

When operating in short burst mode, feedforward synchronizers yield for each individual burst a single estimate of the synchronization parameters, by processing each burst independently of the bursts already received. As the synchronizer has no memory from burst to burst, the mean-square synchronization error is the same for all bursts. Consequently, there is no transient, so that acquisition is said to be "immediate."

The feedforward carrier phase and symbol timing estimates are obtained by maximizing a suitable objective function over the trial values of the carrier phase and time delay. When the maximization over at least one synchronization parameter can be carried out independently of the other synchronization parameter, the joint maximization reduces to two one-dimensional maximizations, instead of a computationally more demanding two-dimensional search. The smallest computational load is obtained when these one-dimensional maximizations can be carried out by means of direct computation instead of a search. A few examples of combining various feedforward carrier and symbol synchronizers are given below.

Using NDA feedforward algorithms for both carrier and symbol synchronization, the receiver could operate as follows:

- The received burst is processed a first time to derive the symbol timing by means of an NDA algorithm which does not use a carrier phase estimate. Candidate algorithms are the NDA algorithm maximizing the objective function (5-48) by means of a search, and the digital filter and square algorithm yielding the symbol timing estimate (5-56) by means of direct computation.
- The received burst is processed a second time to derive a carrier phase estimate by means of an NDA algorithm which makes use of the timing estimate obtained in the previous step. Candidate algorithms are the NDA and V&V algorithms, which yield a carrier phase estimate by directly computing (5-155) or (5-157), respectively.

- The received burst is processed a third time to make the symbol decisions (coherent detection), using the carrier phase and symbol timing estimates obtained in the previous steps.

The last two steps in the above algorithm can be replaced by a single step, using the feedforward DD carrier synchronizer (which for most E_s/N_0 values of practical interest yields a smaller tracking error variance than the NDA and V&V carrier synchronizers). The carrier phase estimate and the data symbol decisions are obtained by maximizing the function $L(\theta)$, given by

$$L(\theta) = \sum_{k=1}^{K} \mathrm{Re}\left[\hat{a}_k^*(\theta) z_k e^{-j\theta}\right] \tag{6-253}$$

where z_k is the matched filter output sample taken at the estimated decision instant $kT + \hat{\varepsilon}T$ resulting from the first step of the algorithm, and $\hat{a}_k(\theta)$ is the receiver's decision of the kth data symbol a_k, corresponding to the trial value θ of the carrier phase estimate. The maximization of $L(\theta)$ is to be performed by means of a search. The final data symbol decisions are obtained as a by-product of the maximization of $L(\theta)$: they are given by $\left\{\hat{a}_k\left(\hat{\theta}\right)\right\}$, where $\hat{\theta}$ is the carrier phase estimate maximizing $L(\theta)$.

For most E_s/N_0 values of practical interest, the timing error variance resulting from the above feedforward NDA symbol synchronization algorithms can be reduced by using feedforward DD symbol synchronization. For feedforward DD joint carrier and symbol synchronization, the objective function to be maximized over the trial values θ and ε is

$$L(\theta, \varepsilon) = \sum_{k=1}^{K} \mathrm{Re}\left[\hat{a}_k^*(\theta, \varepsilon) z(kT+\varepsilon T) e^{-j\theta}\right] \tag{6-254}$$

where $z(t)$ is the matched filter output signal, and $\hat{a}_k(\theta, \varepsilon)$ is the receiver's decision of the kth data symbol a_k, corresponding to the trial values θ and ε. The maximization of $L(\theta, \varepsilon)$ is to be performed by means of a two-dimensional search. The final data symbol decisions are obtained as a by-product of the maximization of $L(\theta, \varepsilon)$; they are given by $\left\{\hat{a}_k\left(\hat{\theta}, \hat{\varepsilon}\right)\right\}$, where $\hat{\theta}$ and $\hat{\varepsilon}$ are the synchronization parameter estimates maximizing the objective function $L(\theta, \varepsilon)$.

It is possible to avoid a two-dimensional search and still use DD algorithms for both carrier synchronization and symbol synchronization by introducing intermediate decisions obtained from differential detection. The receiver could operate as follows:

- The received burst is processed a first time to derive the symbol timing by means of a DD algorithm which does not use a carrier phase estimate. A candidate algorithm is the one which maximizes $|\mu(\varepsilon)|$ of Section 5.5. This objective function is given by:

$$L(\varepsilon) = \left| \sum_{k=1}^{K} \hat{a}_k^*(\varepsilon) z(kT+\varepsilon T) \right| \tag{6-255}$$

where $\hat{a}_k(\varepsilon)$ denotes the receiver's decision about the symbol a_k, obtained from differential detection (which does not involve a carrier phase estimate) and corresponding to the trial value ε of the timing. The maximization of $L(\varepsilon)$ is to be performed by means of a search.

- The carrier phase is obtained using

$$\hat{\theta} = \arg \mu(\hat{\varepsilon}) = \arg \sum_{k=0}^{K-1} \hat{a}_k^* z_k(\hat{\varepsilon}) \tag{6-256}$$

which uses the symbol timing and the receiver's decisions obtained in the previous step.

- The received burst is processed a third time in order to make more reliable symbol decisions. These decisions are obtained by performing coherent detection, using the carrier phase and symbol timing estimates from the previous steps.

The feedforward estimates involving a search can be obtained by discrediting the synchronization parameter(s) over which the search is to be performed into a finite number of values, evaluating the objective function for all discrete parameter values, and selecting that value of the carrier phase and/or the symbol timing estimate yielding the largest value of the objective function; the discretization step imposes a lower limit on the achievable synchronization error variance. Alternatively, the maximization of the objective function could be carried out by applying standard numerical methods, which iteratively look for the synchronization parameter(s) which maximize the objective function. As an example of such an iterative method, we consider the maximization of the objective function $L(\theta)$, given by (6-253). Starting from an initial carrier phase estimate $\hat{\theta}(0)$, the estimate $\hat{\theta}(i+1)$ after the ith iteration is given by

$$\hat{\theta}(i+1) = \hat{\theta}(i) + \alpha_2 \arg\left[\sum_{k=1}^{K} \hat{a}_k^*\left(\hat{\theta}(i)\right) z_k \exp\left(-j\hat{\theta}(i)\right) \right] \tag{6-257}$$

α_2 : convergence parameter

It should be noted that an unfortunate choice of the initial estimate $\hat{\theta}(0)$ may give rise to unreliable decisions $\hat{a}_k^*\left(\hat{\theta}(0)\right)$. In this case the second term of (6-257) has a small deterministic component but a large noise component, so that many iterations are needed before satisfactory convergence is obtained: a phenomenon similar to hangup in feedback synchronizers occurs.

6.5.4 Feedforward Synchronizers for Long Burst Operation and Continuous Operation

When the TDMA bursts are so long that the synchronization parameters cannot be considered as essentially constant over the burst duration, the receiver needs several estimates per burst so that the variation of the synchronization parameters over the burst can be tracked. The situation is similar in the case of continuous operation: here, too, successive estimates are needed to track the fluctuation of the synchronization parameters with time.

Feedforward synchronization for long burst operation and continuous operation involves dividing the received signal into (possibly overlapping) blocks, that are short enough to consider the synchronization parameters as constant over a block. Then a single carrier and symbol timing estimate is made for each block, using the same feedforward algorithms as for short TDMA bursts. This means that the estimates corresponding to a given block do not take into account the values of the estimates corresponding to previous blocks. However, as the resulting feedforward estimates are restricted to a basic interval ($|\hat{\theta}| < \pi/M$ for constellations with a symmetry angle of $2\pi/M$, and $|\hat{\varepsilon}| < 1/2$), they cannot follow all variations of the actual synchronization parameters; for example, a linearly increasing carrier phase, corresponding to a small carrier frequency offset, cannot be tracked. In fact, the feedforward estimates can be considered as estimates of the synchronization parameters (reduced modulo the basic interval). This problem is solved by unwrapping the feedforward estimates, yielding final estimates that can assume values outside the basic interval.

The unwrapping can be accomplished in the following way. Suppose we are presented a sequence $\left\{\hat{\theta}(i)\right\}$ of feedforward carrier phase estimates which are restricted to the basic interval $[-\pi/M, \pi/M)$; the argument i refers to the block count. The unwrapped estimates $\left\{\tilde{\theta}(i)\right\}$ to be derived from $\left\{\hat{\theta}(i)\right\}$ satisfy the following equation:

$$\tilde{\theta}(i) = \hat{\theta}(i) + N(i)\frac{2\pi}{M} \tag{6-258}$$

where $N(i)$ is an integer, to be determined from the following restriction that we impose on the unwrapped estimates $\left\{\tilde{\theta}(i)\right\}$:

$$\left|\tilde{\theta}(i+1) - \tilde{\theta}(i)\right| < \pi/M \tag{6-259}$$

This restriction is motivated by the fact that we expect the differences between carrier phases in adjacent blocks to be small. This yields the following solution:

$$\tilde{\theta}(i+1) = \tilde{\theta}(i) + \text{SAW}\left(\hat{\theta}(i+1) - \tilde{\theta}(i)\right) \tag{6-260}$$

where $\text{SAW}(x)$ is the sawtooth function with period $2\pi/M$, displayed in Figure 6-23. The unwrapping is performed by means of the post-processing structure shown in Figure 6-24, which operates on the feedforward carrier phase estimates $\left\{\hat{\theta}(i)\right\}$.

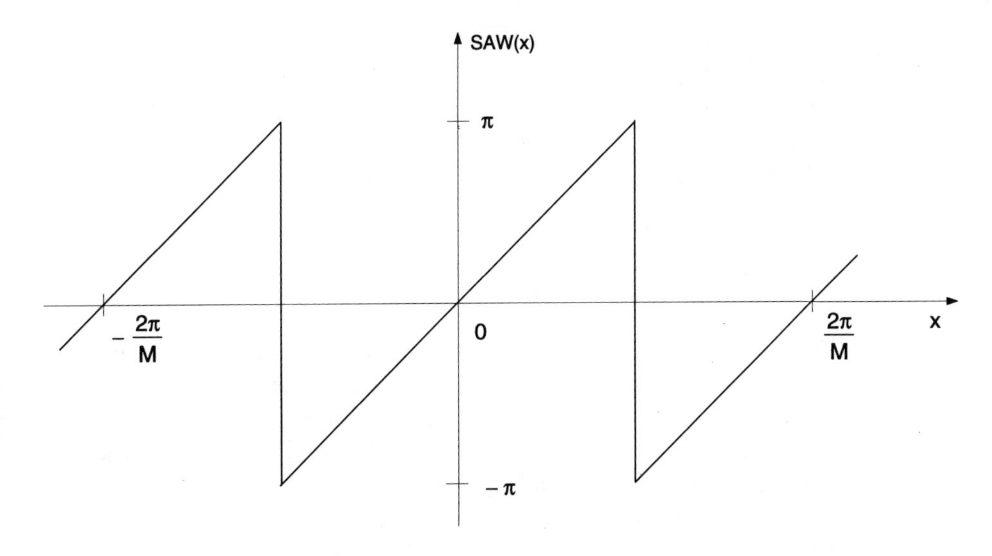

Figure 6-23 Sawtooth Function SAW(x)

In the case of symbol synchronization, a similar reasoning applies; we only have to substitute $2\pi/M$ and π/M by *1* and *1/2*, respectively.

It is obvious from Figure 6-24 that the post-processing introduces feedback. Because of the nonlinearity SAW(·), a countably infinite number of stable and unstable equilibrium points exists. Therefore, we have to investigate the acquisition behavior and the possibility of hangup. Let us consider the case where $\tilde{\theta}(i) = \hat{\theta}(i) = \theta_a$ for $i < 0$, and $\hat{\theta}(i) = \theta_b$ for $i \geq 0$, with both θ_a and θ_b in the interval $[-\pi/M, \pi/M)$. Then it follows from (6-260) that $\hat{\theta}(i) = \theta_c$ for $i \geq 0$, with θ_c given by

$$\theta_c = \begin{cases} \theta_b & -\pi/M \leq \theta_b - \theta_a < \pi/M \\ \theta_b + 2\pi/M & \theta_b - \theta_a < -\pi/M \\ \theta_b - 2\pi/M & \theta_b - \theta_a \geq \pi/M \end{cases} \tag{6-261}$$

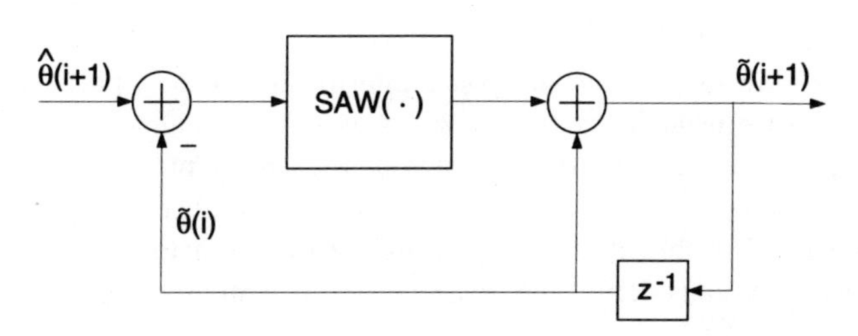

Figure 6-24 Unwrapping Post-Processing Structure

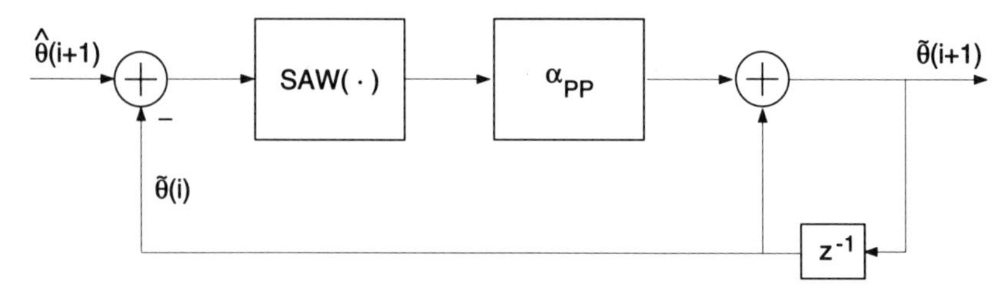

Figure 6-25 First-Order Filtering in Post Processing

This means that at $i = 0$ the post-processor output $\tilde{\theta}(i)$ instantaneously jumps from its initial value θ_a to its final value θ_c; there is no acquisition transient, and hangup does not occur.

The post-processing structure from Figure 6-24 unwraps the feedforward carrier phase estimates by adding a suitable multiple of $2\pi/M$, but does not further reduce the variance of the feedforward estimates. This variance can be reduced by providing additional filtering in the post-processing. Let us consider the case of first-order filtering, shown in Figure 6-25; note that $\alpha_{\text{pp}} = 1$ yields the structure from Figure 6-24.

Linearizing the sawtooth function yields a loop noise bandwidth B_L which is determined by $2B_LT_0 = \alpha_{\text{pp}}/(2 - \alpha_{\text{pp}})$, where T_0 is the time interval between successive feedforward estimates; in the case of nonoverlapping blocks, T_0 equals the duration of a block. Hence, the smaller α_{pp}, the smaller the variance of the estimates $\left\{\tilde{\theta}(i)\right\}$ at the output of the post-processing structure. When applying a phase step and assuming small α_{pp}, it takes a long acquisition time before the estimate $\tilde{\theta}(i)$ is close to its steady-state value; when the phase step is about π/M in magnitude, hangup occurs. Figure 6-26 compares the phase error trajectories at the output of the post-processing, resulting from $\alpha_{\text{pp}} = 1$ (unwrapping only) and $\alpha_{\text{pp}} = 0.1$ (unwrapping and filtering), assuming 4-PSK modulation.

For $\alpha_{\text{pp}} = 1$, there is no acquisition transient, and steady-state operation is achieved immediately; the random fluctuations of the phase error are caused by random noise. For $\alpha_{\text{pp}} = 0.1$, an acquisition transient occurs; in the steady-state, the random phase error fluctuations are smaller than for $\alpha_{\text{pp}} = 1$, because of the smaller bandwidth of the post-processing structure. Figure 6-27 shows the occurrence of a hangup for $\alpha_{\text{pp}} = 0.1$, assuming 4-PSK modulation; the initial phase error is in the vicinity of $\pi/4$, which corresponds to an unstable equilibrium point of the post-processing structure.

It is possible to provide additional filtering in the post-processing and yet avoid hangups by using a technique which is called "planar filtering". Instead of operating on the feedforward estimates of the carrier phase or symbol timing, the planar filtering structure operates on complex numbers (i.e., phasors in the complex plane) that are related to the feedforward estimates. Let us explain the planar filtering by considering the NDA carrier synchronizer; a similar reasoning

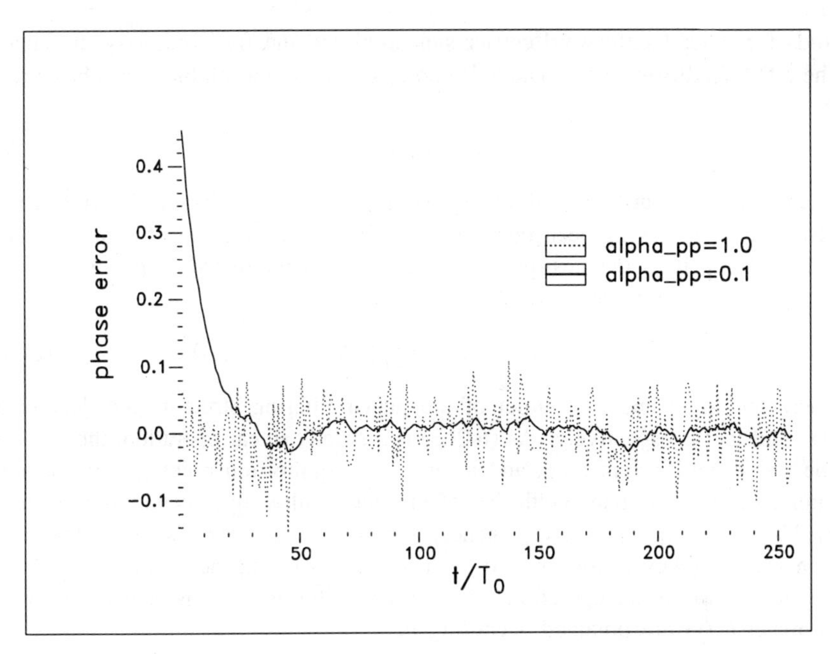

Figure 6-26 Phase Error Trajectories (4–PSK)

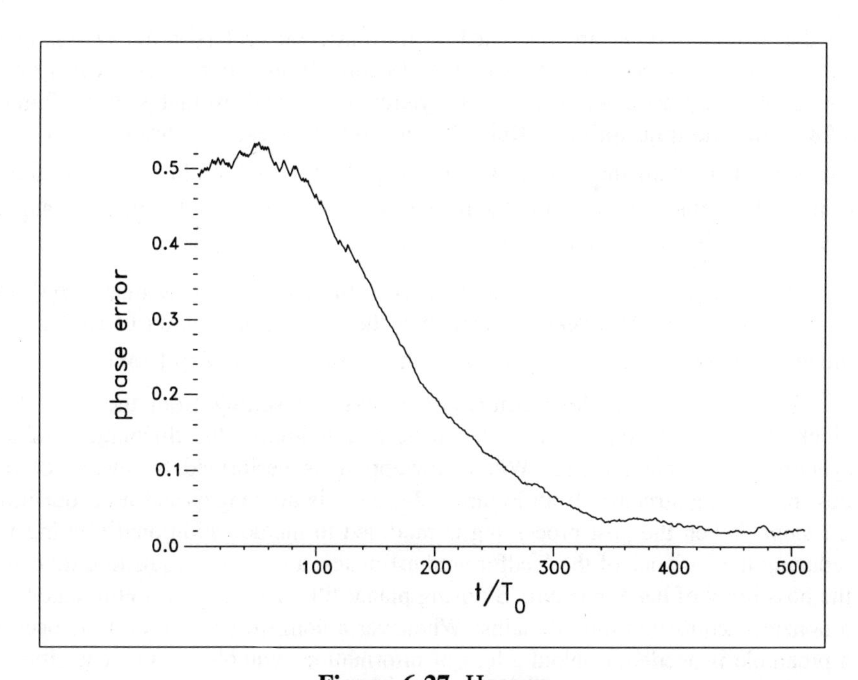

Figure 6-27 Hangup

holds for other feedforward carrier synchronizers and for symbol synchronizers. The NDA feedforward estimate $\hat{\theta}(i)$ corresponding to the ith block can be denoted by

$$\hat{\theta}(i) = \frac{1}{M} \arg(v(i)) \tag{6-262}$$

where $v(i)$ is the arithmetical average over the considered block of matched filter output samples having been raised to the Mth power. Instead of computing $\hat{\theta}(i)$ and applying it to the post-processing circuit, planar filtering operates on $v(i)$. First-order planar filtering yields

$$w(i+1) = w(i) + \alpha_{\text{PP}}(v(i+1) - w(i)) \tag{6-263}$$

where the input and the output of the planar filtering structure are denoted by $v(i)$ and $w(i)$, respectively. As (6-263) is a strictly linear operation, there is only one stable equilibrium point and no unstable equilibrium points, so that hangup cannot occur. The bandwidth B_L of the planar filtering structure is determined by $2B_LT_0 = \alpha_{\text{PP}}/(2-\alpha_{\text{PP}})$; hence, for small α_{PP}, $w(i)$ is much less noisy than $v(i)$. However, for small α_{PP}, it still takes a long acquisition time before reaching steady-state operation. After planar filtering, provisional carrier phase estimates $\tilde{\theta}'(i)$ are obtained according to

$$\tilde{\theta}'(i) = \frac{1}{M} \arg(w(i)) \tag{6-264}$$

As the estimates $\tilde{\theta}'(i)$ are restricted to the basic interval $[-\pi/M, \pi/M)$, they have to be unwrapped by means of the structure from Figure 6-24 (which does not involve any acquisition transient), yielding the final estimates $\tilde{\theta}(i)$. Figure 6-28 shows the trajectories of Re$[w(i)]$ and Im$[w(i)]$ resulting from (6-263) and compares them with the trajectories of $\cos\left(M\tilde{\theta}(i)\right)$ and $\sin\left(M\tilde{\theta}(i)\right)$ resulting from (6-260), that correspond to the hangup shown in Figure 6-27; clearly, hangup does not occur when using planar filtering.

Planar filtering can also be applied when the feedforward estimates $\hat{\theta}(i)$ are obtained by means of a search rather than the direct computation of (6-262); in this case the planar filtering operates on the phasors $v(i) = \exp\left(jM\hat{\theta}(i)\right)$.

We conclude that the feedforward estimates resulting from the individual blocks must be unwrapped, in order to be able to follow the fluctuations of the synchronization parameters. When unwrapping is performed by means of the post-processing structure from Figure 6-24, there is no hangup and no acquisition transient. When the post-processing is modified to include additional filtering for reducing the variance of the feedforward estimates, a nonzero acquisition time and the possibility of hangup result. By using planar filtering, hangup is eliminated but a nonzero acquisition time remains. Whenever a nonzero acquisition time occurs, a preamble is needed to avoid a loss of information symbols during acquisition

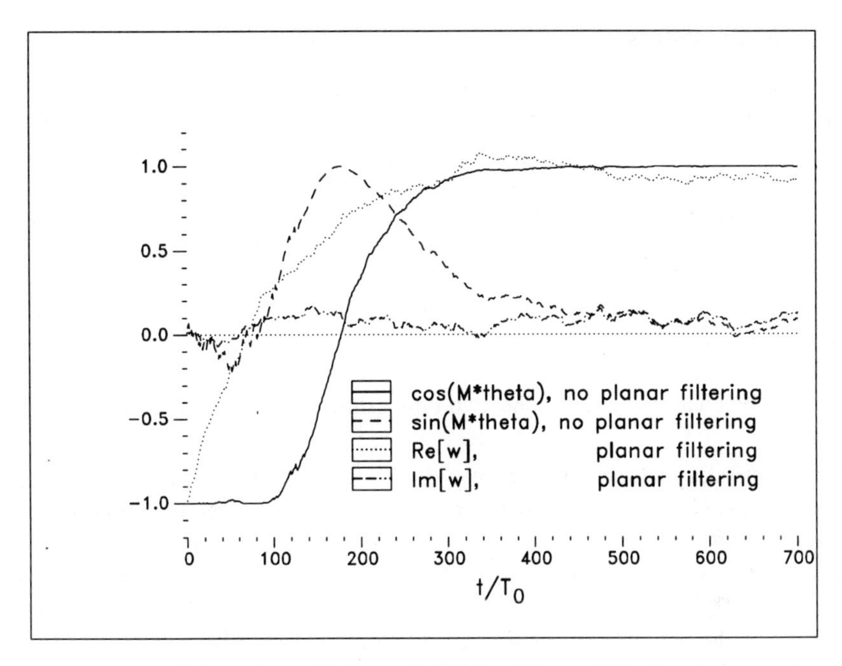

Figure 6-28 Trajectories Using Planar Filtering

Until now we have assumed that the algorithms for estimating the synchronization parameters for a given block are the same as for short burst TDMA operation. However, we have pointed out in Section 6.5.3 that a large amount of computations per block is required when the estimates are to be determined by means of a search rather than direct computation. This amount can be considerably reduced when the estimates are obtained iteratively, but instead of processing each individual block repeatedly until convergence of the estimate, a new block is processed with each new iteration; hence, the processing per block is reduced to the computation of a single iteration. For example, in the case of carrier synchronization, the estimate $\hat{\theta}(i)$ resulting from processing the signal $\mathbf{r}(i)$ of the ith block can be represented as

$$\hat{\theta}(i) = \theta_{\text{pr}}(i) + (f)(\mathbf{r}(i), \theta_{\text{pr}}(i)) \tag{6-265}$$

where $\theta_{\text{pr}}(i)$ is a prediction of the carrier phase for the ith block, based upon previous estimates $\hat{\theta}(m)$ with $m < i$, and $(f)(.,.)$ is an increment computed during the ith block. The drawback of this approach is, as for feedback synchronizers, the occurrence of an acquisition transient and the possibility of hangup. Hence, in order not to lose information symbols during acquisition, a preamble is needed. During this preamble, DA algorithms can be used.

6.5.5 Main Points

For the acquisition of the carrier phase and the symbol timing, it is advantageous that there is no strong coupling between the carrier and the symbol synchronizers. In the case of feedback synchronization, this coupling slows down the acquisition. In the case of feedforward synchronization, the coupling increases the computational requirements because a two-dimensional maximization over the carrier phase and the symbol timing must be performed. Therefore, it is recommended that at least one synchronizer operates completely independently of the other. As a result, the time to acquire both the carrier phase and the symbol timing by means of feedback synchronizers can be expressed in terms of the acquisition times of the individual synchronizers, while for feedforward synchronization two one-dimensional (instead of one two-dimensional) maximizations must be performed.

Feedback synchronizers need very long acquisition times when the frequency offset is much larger than the loop bandwidth (pull-in) or when the initial synchronization error is close to an unstable equilibrium point (hangup); these acquisition times can be reduced by using acquisition aids. The acquisition time is in the order of the inverse loop bandwidth (say, about a hundred symbols) when the frequency offset is small. Because of the long preamble needed, feedback synchronizers are suited only for operation in continuous mode or for TDMA with long bursts, but not for TDMA with short bursts.

Feedforward synchronizers operating on short bursts have to provide only one carrier phase estimate and timing estimate per burst. The bursts need no preamble, and each single short burst is processed individually until its synchronization parameter estimates are obtained. As the synchronization error variance is the same for all bursts, there is no transient, so that steady-state operation is achieved already with the estimates for the first received burst; therefore, acquisition is said to be "immediate". However, the amount of processing per burst depends strongly on the types of synchronization algorithms. The least processing is required when at least one synchronizer operates independently of the other, and for both synchronizers the estimates are obtained by direct computation. The most processing is required when the coupling between algorithms is such that a two-dimensional search over the carrier phase and the symbol timing is to be performed; this is the case when both synchronization algorithms make use of coherently detected data symbols. When an iterative method is used to determine the estimate, an unfortunate choice of the initial value of the estimate may yield a very slow convergence towards the final estimate: a hangup-like phenomenon occurs.

When feedforward synchronizers operate in continuous mode or long burst mode, the received signal is subdivided into short blocks; during each block, the synchronization parameters can be considered as being constant. The feedforward synchronizers then provide one carrier phase estimate and one symbol timing estimate per block, in much the same way as for short burst operation. In order to be able to follow the dynamics of the actual synchronization parameters, the estimates resulting from the successive blocks must be unwrapped. This unwrapping can be accomplished by means of a post-processing structure which necessarily introduces feedback. When the post-processing does not reduce the variance of the individual feedforward estimates, acquisition is immediate and hangup does not occur, in spite of the introduced feedback. When the post-processing also performs additional filtering in order to reduce the variance of the individual feedforward estimates, the feedback gives rise to an acquisition transient, and hangup might occur; because of the nonzero acquisition time, a preamble is needed. The hangup can be eliminated by using planar filtering; however, planar filtering still yields a nonzero acquisition time. A nonzero acquisition time with the possibility of hangup also occurs when the estimates are computed iteratively as for short burst operation, with the only difference that for each new iteration a new block is processed, instead of processing the same block repeatedly until convergence occurs.

6.5.6 Bibliographical Notes

Here we briefly mention some results from the literature that are related to the acquisition of the carrier phase and the symbol timing.

An interesting discussion on hangups in a PLL can be found in [1], [2], [3]. For a continuous phase error detector characteristic, hangup is caused by the restoring force near the unstable equilibrium point being very small; loop noise does not cause prolonged stays around the unstable equilibrium point. On the other hand, for a phase error detector characteristic with a discontinuity at the unstable equilibrium point (such as a sawtooth characteristic), the restoring force near the unstable equilibrium point is large, but hangup still occurs when noise is present. This can be explained by observing that, in the presence of noise, the average phase error detector output is no longer discontinuous at the unstable equilibrium point (see also Section 3.2.2 from Volume 1); the larger the noise, the smaller the slope at the unstable equilibrium point, and the smaller the restoring force. PLLs with a discontinuous phase error detector characteristic determine the instantaneous phase error by measuring the time between the zero crossings of the input signal and the voltage-controlled oscillator (VCO) signal. An alternative explanation of hangups in PLLs with a discontinuous phase error detector characteristic is the following: because of noise, the zero crossings of the input signal fall randomly to either side of the discontinuity when the initial phase error corresponds to the unstable equilibrium point, so that the average restoring force is small.

Lock detector performance for BPSK has been investigated in [4]. The considered lock detectors take the difference of either the square or the absolute value of the in-phase and quadrature components after phase rotation. The probability of lock detection has been determined, taking into account the random fluctuations of the carrier phase estimate when the synchronizer is in lock. These random fluctuations affect the performance of the lock detector: for a loop SNR of 15 dB, a degradation of 1.5 dB in E_s/N_0 occurs, as compared to the performance corresponding to infinite loop SNR.

Post-processing of the feedforward carrier phase and symbol timing estimates for reducing their variance has been discussed in [5] and [6]. When the parameters to be estimated can be modeled as a Gauss-Markov process, the use of a Kalman filter is proposed; in the case of a first-order Gauss-Markov process, this yields the post-processing structure from Figure 6-25, with α_{pp} not constant but varying with time. In a practical implementation, α_{pp} could be restricted to only a few values, e.g., a larger value during acquisition and a smaller one during tracking. Planar filtering for avoiding hangup has been proposed in [5]; this involves filtering phasors instead of the synchronization parameter estimates, and taking the argument of the phasor after planar filtering for obtaining the synchronization parameter estimate. In [7], it is shown that a decision-directed feedforward estimate of the phasor $(\cos\theta, \sin\theta)$ is sufficient for detecting M-PSK symbols; the resulting receiver makes no explicit carrier phase estimate.

Bibliography

[1] F. M. Gardner, "Hangup in Phase-Lock Loops," *IEEE Trans. Commun.*, vol. COM-25, pp. 1210–1214, Oct. 1977.

[2] H. Meyr and L. Popken, "Phase Acquisition Statistics for Phase-Locked Loops," *IEEE Trans. Commun.* , vol. COM-28, pp. 1365–1372, Aug. 1980.

[3] F. M. Gardner, "Equivocation as a Cause of PLL Hangup," *IEEE Trans. Commun.*, vol. COM-30, pp. 2242–2243, Oct. 1982.

[4] A. Mileant and S. Hinedi, "Lock Detection in Costas Loops," *IEEE Trans. Commun*, vol. COM-40, pp. 480–483, Mar. 1992.

[5] M. Oerder and H. Meyr, "Digital Filter and Square Timing Recovery," *IEEE Trans. Commun.*, vol. COM-36, pp. 605–612, May 1988.

[6] G. Ascheid, M. Oerder, J. Stahl, and H. Meyr, "An All Digital Receiver Architecture for Bandwidth Efficient Transmission at High Data Rates," *IEEE Trans. Commun.*, vol. COM-37, pp. 804–813, Aug. 1989.

[7] P. Y. Kam, "Maximum Likelihood Carrier Phase Recovery for Linear Suppressed-Carrier Digital Data Modulations," *IEEE Trans. Commun.*, vol. COM-34, pp. 522–527, June 1986.

Chapter 7 Bit Error Rate Degradation Caused by Random Tracking Errors

7.1 Introduction

For coherent detection of digitally modulated signals, the receiver must be provided with accurate carrier phase and symbol timing estimates; these estimates are derived from the received signal itself by means of a synchronizer. The bit error rate (BER) performance under the assumption of perfect synchronization is well documented for various modulation formats [1–5]. However, in practice the carrier phase and timing estimates exhibit small random fluctuations (jitter) about their optimum values; these fluctuations give rise to a BER degradation as compared to perfect synchronization. It is important to know this BER degradation in terms of the accuracy of the estimates provided by the synchronizer, so that the synchronizer can be designed to yield a target BER degradation (which should not exceed about 0.2 dB for most applications).

For various linear modulation formats (M-PSK, M-PAM, and M^2-QAM) we evaluate the BER degradation caused by random carrier phase and timing errors. In Section 7.7 we show that the results also apply for the practically important case of coded transmission. For nonlinear modulation and coded transmission we refer to the bibliographical notes in Section 7.9.

7.2 ML Detection of Data Symbols

Figure 7-1 conceptually shows how a maximum-likelihood (ML) decision about the symbol sequence $\{a_k\}$ is obtained. The matched filter output is sampled at the instant $kT + \hat{\varepsilon}T$ where $\hat{\varepsilon}$ denotes the estimate of the normalized time delay ε_0. The matched filter output samples are rotated counterclockwise over an angle $\hat{\theta}$ which is an estimate of the unknown carrier phase θ. The receiver's decision about the transmitted sequence is the data sequence which maximizes the ML function [eq. 4–84] when the trial parameters ε, θ are replaced by their estimates

$$L\left(\mathbf{r}_f \mid \mathbf{a}, \hat{\varepsilon}, \hat{\theta}\right) = \left[\mathbf{z}e^{-j\hat{\theta}} - \mathbf{H}^{-1}\mathbf{a}\right]^H \mathbf{H}\left[\mathbf{z}e^{-j\hat{\theta}} - \mathbf{H}^{-1}\mathbf{a}\right] \tag{7-1}$$

The last equation shows that in general the ML symbol $\hat{a}_k$ cannot be obtained by a symbol-by-symbol decision but that the entire sequence must be considered due

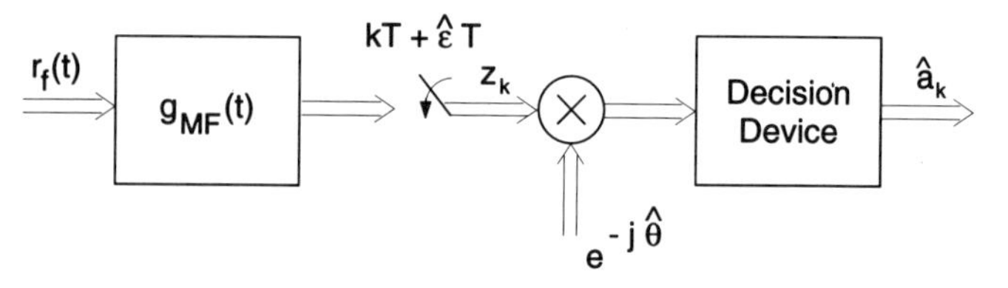

Figure 7-1 Conceptual Receiver Performing Maximum-Likelihood Decision

to intersymbol interference (nondiagonal **H**). The decision rule also applies when in addition the data sequence is convolutionally encoded. If the real-valued pulse $g(t)$ is selected such that $g(t)$ and $g(t-mT)$ are orthogonal (Nyquist condition), then the matrix **H** becomes diagonal and a symbol-by-symbol decision is possible. Using the normalization of Section 4.3.6 the matrix **H** is the identity matrix. The receiver's decision $\hat{a}_k$ about the transmitted symbol a_k then is the data symbol which has the smallest Euclidean distance to the sample $z_k e^{-j\hat{\theta}}$ at the input of the decision device

$$\left|\hat{a}_k - z_k e^{-j\hat{\theta}}\right|^2 = \min \left|a_k - z_k e^{-j\hat{\theta}}\right|^2 \tag{7-2}$$

Assuming perfect synchronization, the BER performance of the ML receiver is well documented in the literature for various modulation formats [1–5]. However, in the presence of synchronization errors, the BER performance deteriorates. In the following we determine the BER degradation caused by random synchronization errors.

7.3 Derivation of an Approximate Expression for BER Degradation

In this section we derive an expression for the BER degradation caused by synchronization errors, which is valid irrespective of the specific modulation format. We first restrict our attention to either carrier phase errors in the presence of perfect timing recovery, or timing errors in the presence of perfect carrier recovery. At the end of this section, we consider the BER degradation when both carrier phase errors and timing errors are present.

Let us introduce the notation ψ for the synchronization error: $\psi = \phi = \theta_0 - \hat{\theta}$ in the case of a carrier phase error, whereas $\psi = e = \varepsilon_0 - \hat{\varepsilon}$ in the case of a timing error. The BER degradation D (detection loss), measured in decibels, is defined as the increase of E_s/N_0, required to maintain the same BER as the receiver without synchronization errors. Hence, the BER degradation D at a bit error rate value BER_0 is given by

$$D = 10 \log\left(\sigma_0^2/\sigma^2\right) \quad [\mathrm{dB}] \tag{7-3}$$

where σ^2 and σ_0^2 are determined by

$$\text{BER}_0 = P(0;\sigma_0) = E_\psi[P(\psi;\sigma)] \tag{7-4}$$

In (7-4), $P(\psi;\sigma)$ represents the conditional BER, corresponding to a synchronization error ψ, at $E_s/N_0 = 1/(2\sigma^2)$, and $E_\psi[\cdot]$ denotes averaging over the synchronization error ψ. Brute force numerical evaluation of the BER degradation can be quite time consuming, because of the averaging over the synchronization error ψ, the computation of $P(\psi;\sigma)$ for many values of ψ, and the iterations required to obtain σ at a given value of σ_0. However, in most applications the BER degradation caused by synchronization errors should not exceed about 0.2 dB; in the following we will derive an approximate expression for the BER degradation, which is very accurate for small degradations.

For small synchronization errors, $P(\psi;\sigma)$ can be approximated by a truncated Taylor series expansion around $\psi = 0$. This yields

$$\begin{aligned} E_\psi[P(\psi;\sigma)] &\simeq E_\psi\left[P(0;\sigma) + \psi P^{(\psi)}(0;\sigma) + \frac{1}{2}\,\psi^2 P^{(\psi\psi)}(0;\sigma)\right] \\ &= [P(0;\sigma)] + \frac{1}{2}\,\text{var}[\psi]\,P^{(\psi\psi)}(0;\sigma) \end{aligned} \tag{7-5}$$

where $P^{(\psi)}(.;.)$ and $P^{(\psi\psi)}(.;.)$ denote single and double differentiation with respect to ψ, and $\text{var}[\psi]$ is the variance of the synchronization error ψ; the second line of (7-5) assumes that ψ is a zero-mean random variable, i.e. $E_\psi[\psi] = 0$. For small degradations, the second line of (7-5) can be approximated by a truncated Taylor series expansion about $\sigma = \sigma_0$. Keeping only linear terms in $(\sigma - \sigma_0)$ and $\text{var}[\psi]$, we obtain

$$E_\psi[P(\psi;\sigma_0)] \simeq P(0;\sigma_0) + (\sigma-\sigma_0)\,P^{(\sigma)}(0;\sigma_0) + \frac{1}{2}\,\text{var}[\psi]\,P^{(\psi\psi)}(0;\sigma_0) \tag{7-6}$$

where $P^{(\sigma)}(.;.)$ denotes differentiation with respect to σ. Taking (7-6) into account, (7-4) yields

$$\frac{\sigma}{\sigma_0} = 1 - \frac{1}{2\sigma_0}\,\frac{P^{(\psi\psi)}(0;\sigma_0)}{P^{(\sigma)}(0;\sigma_0)}\,\text{var}[\psi] \tag{7-7}$$

Hence, for small $\text{var}[\psi]$, the BER degradation is well approximated by

$$\begin{aligned} D &= -20\,\log(\sigma/\sigma_0) \\ &\simeq \frac{10}{\ln(10)}\,\frac{P^{(\psi\psi)}(0;\sigma_0)}{\sigma_0 P^{(\sigma)}(0;\sigma_0)}\,\text{var}[\psi] \qquad [\text{dB}] \end{aligned} \tag{7-8}$$

which indicates that the (small) BER degradation, measured in decibels, is essentially proportional to the tracking error variance $\text{var}[\psi]$, and independent of the specific shape of the probability density of the synchronization error ψ.

When both carrier phase errors and timing errors are present, a similar reasoning can be followed to obtain an approximate expression for the BER degradation. Assuming that the carrier phase error and the timing error are uncorrelated, it turns out that the BER degradation, measured in decibels, equals the sum of the BER degradations caused by carrier phase errors and timing errors individually.

7.4 M-PSK Signal Constellation

In the case of M-PSK, the data symbols a_k take values from the alphabet $A = \{\exp(j2\pi m/M) | m = 0, \ldots, M-1\}$. The ML receiver from Figure 7-1 decides $\hat{a}_k = \exp(j2\pi m/M)$ when the argument of the sample $z_k e^{-j\hat{\theta}}$ at the input of the decision device is in the interval $((2m-1)\,\pi/M,\ (2m+1)\pi/M)$.

The BER for M-PSK depends on how the transmitter maps blocks of $\log_2 M$ bits onto constellation points. The minimum BER is achieved when blocks of $\log_2 M$ bits that are mapped onto constellation points being nearest neighbors in the Euclidean sense, differ by one bit only; this is called Gray mapping. Figure 7-2 shows an example of Gray mapping for 8-PSK.

In the case of Gray mapping, the conditional bit error rate $P(\psi;\sigma)$ is well approximated by

$$P(\psi;\sigma) \simeq \frac{1}{\log_2 M} \text{Prob}(\hat{a}_k \neq a_k \mid \psi) \tag{7-9}$$

the approximation being that each symbol error gives rise to one bit error only; this approximation is accurate at moderate and large E_s/N_0, where the major part

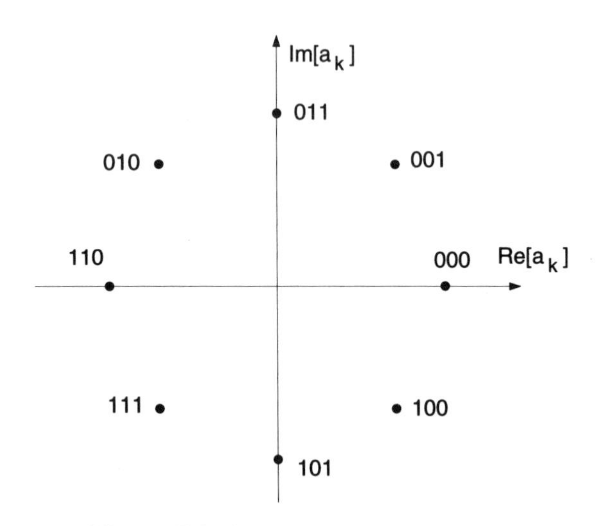

Figure 7-2 Gray Mapping for 8-PSK

of the erroneous decisions of the receiver corresponds to a detected symbol being a nearest neighbor of the transmitted symbol. For M-PSK, the conditional symbol error rate is given by

$$\begin{aligned}\text{Prob}(\hat{a}_k \neq a_k \mid \psi) &= \text{Prob}(\hat{a}_k \neq 1 \mid a_k = 1, \psi) \\ &= \text{Prob}\left(\arg\left(z_k e^{-j\hat{\theta}}\right) \notin \left(-\frac{\pi}{M}, \frac{\pi}{M}\right) | a_k = 1, \psi\right)\end{aligned} \tag{7-10}$$

because of the rotational symmetry of the constellation and the constant magnitude of the data symbols. Let us decompose the sample $z_k e^{-j\hat{\theta}}$ at the input of the decision device as

$$z_k e^{-j\hat{\theta}} = s(\psi) + w_k \tag{7-11}$$

where w_k is the Gaussian additive noise component, with zero-mean independent real and imaginary parts, each having a variance $\sigma^2 = N_0/(2E_s)$, and $s(\psi)$ is the signal component, corresponding to a synchronization error ψ and a transmitted symbol $a_k = 1$; note that $s(0) = 1$ because the matched filter output is normalized to $h(0) = 1$. As a result, the conditional symbol error rate is approximately given by

$$\text{Prob}(\hat{a}_k \neq a_k \mid \psi) = \begin{cases} E_{\mathbf{a}}[Q(d_1(\psi)/\sigma)] & M = 2 \\ E_{\mathbf{a}}[Q(d_1(\psi)/\sigma) + Q(d_2(\psi)/\sigma)] & M > 2 \end{cases} \tag{7-12}$$

where $E_{\mathbf{a}}[\cdot]$ denotes averaging over all data symbols a_n (with $n \neq k$) that contribute to $s(\psi)$, and where

$$Q(x) = \int_x^{\infty} \frac{1}{\sqrt{2\pi}} \exp\left(\frac{-u^2}{2}\right) du \tag{7-13}$$

is the area under the tail of the Gaussian probability density function, while

$$d_1(\psi) = \text{Im}[s(\psi) \exp(j\pi/M)] \tag{7-14}$$

and

$$d_2(\psi) = -\text{Im}[s(\psi) \exp(-j\pi/M)] \tag{7-15}$$

equal the distances of the signal point $s(\psi)$ to the decision boundaries at angles $-\pi/M$ and π/M, respectively, as indicated in Figure 7-3.

The result (7-12) is exact for $M = 2$, and a close upper bound for $M > 2$ at moderate and large E_s/N_0. Assuming that the probability density of the synchronization error ψ is an even function, it can be verified that $d_1(\psi)$ and $d_2(\psi)$ have identical statistical properties. In this case the average bit error rate $E_\psi[P(\psi;\sigma)]$ is not affected when $d_2(\psi)$ is replaced by $d_1(\psi)$ in (7-12). Hence,

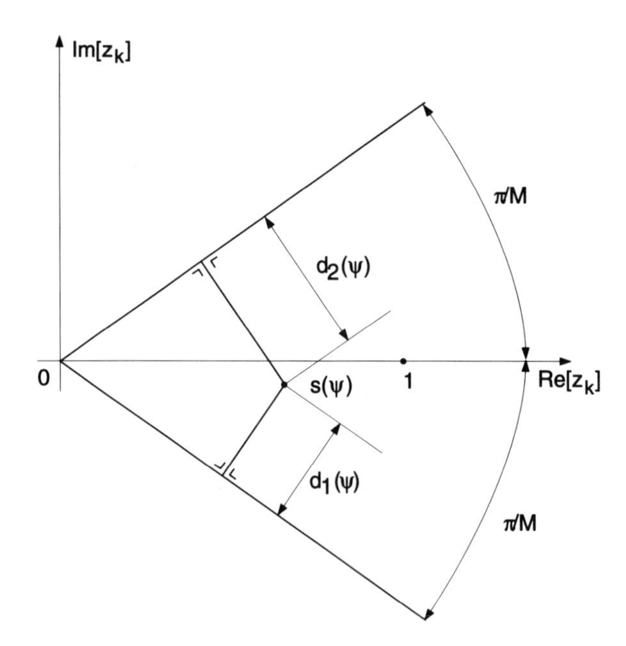

Figure 7-3 Illustration of $d_1(\psi)$ and $d_2(\psi)$

as far as the evaluation of $E_\psi[P(\psi;\sigma)]$ is concerned, the following expression can be used for $P(\psi;\sigma)$:

$$P(\psi;\sigma) = \frac{1}{\log_2 M}\, f(M)\, E_{\mathbf{a}}\left[Q\left(\frac{d(\psi)}{\sigma}\right)\right] \tag{7-16}$$

with

$$d(\psi) = \mathrm{Im}[s(\psi)\exp(j\pi/M)] \tag{7-17}$$

and

$$f(M) = \begin{cases} 1 & M = 2 \\ 2 & M > 2 \end{cases} \tag{7-18}$$

As $s(0) = 1$, the bit error rate in the absence of synchronization errors is given by

$$P(0;\sigma) = \frac{1}{\log_2 M}\, f(M)\, Q\left(\frac{\sin(\pi/M)}{\sigma}\right) \tag{7-19}$$

The expressions (7-16) and (7-19) are correct for $M = 2$ and $M = 4$, and a good approximation at moderate and large E_s/N_0 for other values of M.

Substituting (7-16) in the general expression (7-8) for the BER degradation, and making use of

$$\frac{d^2}{dx^2}\, Q(x) = -x\frac{d}{dx}\, Q(x) \tag{7-20}$$

one obtains

$$D \simeq \frac{10}{\ln(10)} \left(A + \frac{B}{\sigma_0^2} \right) \mathrm{var}[\psi] \qquad [\mathrm{dB}] \tag{7-21}$$

where

$$A = \frac{1}{d(0)}\, E_{\mathbf{a}}\left[-d^{(\psi\psi)}(0) \right] \qquad B = E_{\mathbf{a}}\left[\left(d^{(\psi)}(0) \right)^2 \right] \tag{7-22}$$

and the superscripts ψ and $\psi\psi$ denote single and double differentiation with respect to the synchronization error ψ.

Although the BER degradation formula (7-21) for M-PSK has been obtained in a rather formal way, an interesting interpretation is given below.

Considering complex numbers as phasors, the average BER for M-PSK equals $f(M)/\log_2 M$ times the probability that the projection of the sample $z_k e^{-j\hat{\theta}}$ on $j\exp(-j\pi/M)$ is negative. The projection of the noise component w_k of $z_k e^{-j\hat{\theta}}$ is zero-mean and Gaussian with a variance σ^2; the projection of the signal component $s(\psi)$ of $z_k e^{-j\hat{\theta}}$ equals $d(\psi)$, which for small ψ can be expanded in a truncated Taylor series expansion:

$$d(\psi) = d(0) + \psi\, d^{(\psi)}(0) + \frac{1}{2}\, \psi^2\, d^{(\psi\psi)}(0) \tag{7-23}$$

The first term in (7-23) is the projection of the signal component in the absence of synchronization errors. The second term is a zero-mean disturbance, with a variance equal to B var$[\psi]$ [see (7-22)]; this second term adds to the projection of the noise component w_k of $z_k e^{-j\hat{\theta}}$, yielding a total disturbance with variance equal to $\sigma^2 + B$ var$[\psi]$. The third term has a mean equal to $-Ad(0)$ var$[\psi]/2$ (see [7-22)] which reduces the effect of the first term of (7-23); for small ψ, the fluctuation of this third term can be neglected as compared to the second term of (7-23). When $\sigma^2 \gg B$ var$[\psi]$, the total disturbance is approximately Gaussian, in which case the average BER for M-PSK is given by

$$E_{\psi}[P(\psi;\sigma)] = \frac{1}{\log_2 M}\, f(M)\, Q\left(\frac{d(0)(1 - (A/2)\,\mathrm{var}[\psi])}{\sqrt{\sigma^2 + B\,\mathrm{var}[\psi]}} \right) \tag{7-24}$$

where the square of the argument of the function $Q(\cdot)$ in (7-24) equals the signal-to-noise ratio of the projection of $z_k e^{-j\hat{\theta}}$ on $j\exp(-j\pi/M)$. Taking into account that

$$P(0;\sigma_0) = \frac{1}{\log_2 M}\, f(M)\, Q\left(\frac{d(0)}{\sigma_0} \right) \tag{7-25}$$

it follows that the BER degradation D, defined by (7-3) and (7-4), is obtained by simply equating the arguments of the function $Q(\cdot)$ in (7-24) and (7-25). For small values of var$[\psi]$ this yields

$$\frac{\sigma^2}{\sigma_0^2} \simeq 1 - \left(A + \frac{B}{\sigma_0^2} \right) \mathrm{var}[\psi] \tag{7-26}$$

so that the BER degradation in decibels is approximated by

$$\begin{aligned} D = & -10 \log \left(\frac{\sigma^2}{\sigma_0^2} \right) \\ & \simeq \frac{10}{\ln(10)} \left(A + \frac{B}{\sigma_0^2} \right) \mathrm{var}[\psi] \qquad [\mathrm{dB}] \end{aligned} \tag{7-27}$$

which is the same as (7-21). Hence, we conclude that the BER degradation caused by synchronization errors is due to the combined effect of a reduction of the useful signal (this is accounted for by the quantity A) and an increase of the variance (this is accounted for by the quantity B) at the input of the decision device. The latter effect becomes more important with increasing E_s/N_0, i.e., at smaller values of the ideal BER.

In the following, we evaluate the quantities A and B from (7-22), in the cases of carrier phase errors ($\psi = \phi$) and timing errors ($\psi = e$), respectively.

7.4.1 Carrier Phase Errors

When the synchronization error ψ equals the carrier phase error ϕ, and timing is perfect, we obtain

$$\begin{aligned} s(\phi) &= \exp(j\phi) \\ &= \cos\phi + j \sin\phi \end{aligned} \tag{7-28}$$

$$\begin{aligned} d(\phi) &= \sin(\phi + (\pi/M)) \\ &= \cos\phi \ \sin(\pi/M) + \sin\phi \ \cos(\pi/M) \end{aligned} \tag{7-29}$$

Using the above in (7-22) yields

$$A = 1 \qquad B = \cos^2(\pi/M) \tag{7-30}$$

Note that $B = 0$ for $M = 2$.

It follows from (7-28) that a carrier phase error affects the signal component $s(\phi)$ at the input of the decision device by a reduction of the signal component ($\cos\phi \leq 1$) and the introduction of a zero-mean disturbance $j \sin\phi$. The useful component and the disturbance are along the real axis and imaginary axis, respectively. The BER degradation is determined by $d(\phi)$ from (7-29), which is the projection of $s(\phi)$ on $j \exp(-j\pi/M)$.

7.4.2 Timing Errors

When the synchronization error ψ equals the timing error e, and carrier synchronization is perfect, we obtain

$$d(e) = \mathrm{Im}\left[\left(h(eT) + \sum_{m \neq 0} a_{k-m} \, h(mT - eT) \right) \exp\left(\frac{j\pi}{M} \right) \right] \tag{7-31}$$

Using the above in (7-22) yields

$$A = -h''(0)\,T^2 \qquad B = \begin{cases} \sum\limits_m \left(h'(mT)T\right)^2 & M = 2 \\ \frac{1}{2}\sum\limits_m \left(h'(mT)T\right)^2 & M > 2 \end{cases} \tag{7-32}$$

where $h'(x)$ and $h''(x)$ denote the first and second derivative of $h(x)$, respectively. For $M = 2$, the value of B is twice as large as for $M > 2$, because $E\left[a_k^2\right] = 1$ for $M = 2$, but $E\left[a_k^2\right] = 0$ for $M > 2$.

Equation (7-31) shows that a timing error yields a reduction of the useful component $(h(eT) \leq 1)$ at the input of the decision device, and introduces intersymbol interference (ISI), which acts as an additional disturbance.

7.5 M-PAM and M^2-QAM Signal Constellations

In the case of M-PAM, the data symbols a_k are real-valued and are denoted as $a_k = a_{R,k} + j0$. The symbols $a_{R,k}$ take values from the alphabet $A = \{\pm\Delta, \pm 3\Delta, \ldots, \pm(M-1)\Delta\}$, where the value of Δ is selected such that $E\left[a_{R,k}^2\right] = 1$. Taking into account that

$$\frac{2}{M}\sum_{m=1}^{M/2} (2m-1)^2 = \frac{1}{3}\left(M^2 - 1\right) \tag{7-33}$$

it follows that

$$\Delta = \sqrt{\frac{3}{M^2 - 1}} \tag{7-34}$$

Based upon the sample $z_k e^{-j\hat{\theta}}$ at the input of the decision device, a decision $\hat{a}_{R,k}$ about the symbol $a_{R,k}$ is made, according to the following decision rule:

$$\hat{a}_{R,k} = \begin{cases} (2m-1)\Delta & \text{for} \quad (2m-2)\Delta < \mathrm{Re}\left[z_k e^{-j\theta}\right] < 2m\Delta; \\ & \qquad -(M/2-1) < m < M/2 \\ (M-1)\Delta & \text{for} \quad (M-2)\Delta < \mathrm{Re}\left[z_k e^{-j\theta}\right] \\ (M-1)\Delta & \text{for} \quad \mathrm{Re}\left[z_k e^{-j\theta}\right] < -(M-2)\Delta \end{cases} \tag{7-35}$$

In the following we will assume that blocks of $\log_2 M$ bits, which are mapped onto constellation points being nearest neighbors, differ by one bit only (Gray mapping); Figure 7-4 illustrates the Gray mapping for 4-PAM.

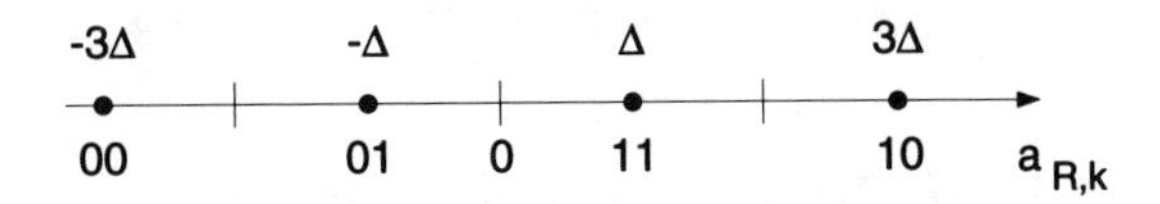

Figure 7-4 Gray Mapping for 4-PAM

In the case of M^2-QAM, the data symbols are complex-valued and are denoted as $a_k = a_{R,k} + j a_{I,k}$; $a_{R,k}$ and $a_{I,k}$ are the in-phase symbols and quadrature symbols, respectively. The symbols $a_{R,k}$ and $a_{I,k}$ are statistically independent, and both take values from the M-PAM alphabet; this yields a square M^2-QAM constellation for the complex symbols a_k. The requirement that $E\left[|a_k|^2\right] = 1$ gives rise to

$$\Delta = \sqrt{\frac{3}{2(M^2-1)}} \tag{7-36}$$

yielding $E\left[a_{R,k}^2\right] = E\left[a_{I,k}^2\right] = 1/2$. The decision rule for the in-phase symbols $a_{R,k}$ is the same as for M-PAM; the decision rule for the quadrature symbols $a_{I,k}$ is the same as for the in-phase symbols a_k, but with $\text{Re}\left[z_k e^{-j\hat{\theta}}\right]$ replaced by $\text{Im}\left[z_k e^{-j\hat{\theta}}\right]$. In the following we will assume that blocks of $2 \log_2 M$ bits that are mapped onto constellation points being nearest neighbors, differ by one bit only (Gray mapping); Figure 7-5 illustrates the Gray mapping for 16-QAM. Under this assumption, the in-phase decisions $\hat{a}_{R,k}$ and quadrature decisions $\hat{a}_{I,k}$ yield the

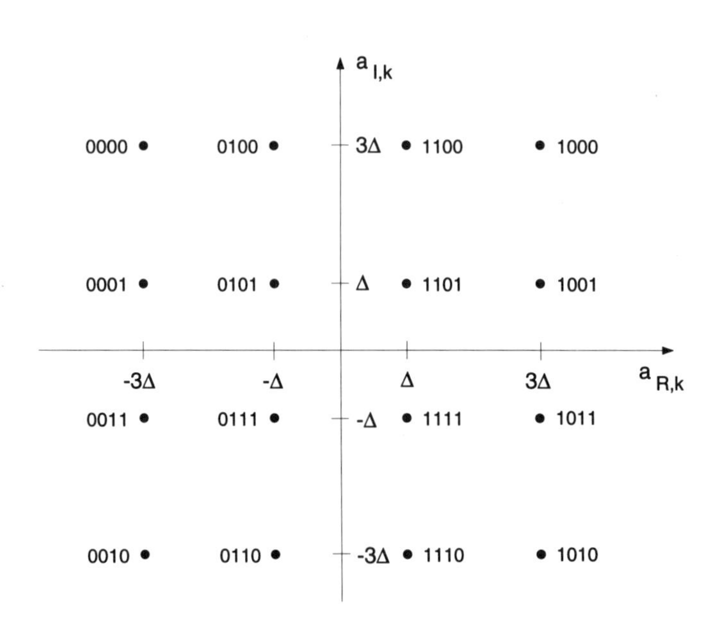

Figure 7-5 Gray Mapping for 16-QAM

same bit error rate, which equals the bit error rate of the M^2-QAM transmission system. Hence, we have to consider only the bit error rate resulting from the decision $\hat{a}_{R,k}$.

Taking the Gray mapping into account, the conditional bit error rate for both M-PAM and M^2-QAM is approximately given by

$$\begin{aligned} P(\psi;\sigma) &= \frac{1}{\log_2 M}\ \text{Prob}[\hat{a}_{R,k}\neq a_{R,k}\mid\psi] \\ &= \frac{1}{\log_2 M}\,\frac{2}{M}\sum_{m=1}^{M/2} P_s(\psi;\sigma;2m-1) \end{aligned} \tag{7-37}$$

the approximation being that each symbol error gives rise to one bit error only. In (7-37),

$$P_s(\psi;\sigma;2m-1) = \text{Prob}[\hat{a}_{R,k}\neq a_{R,k}\mid a_{R,k}=(2m-1)\Delta;\ \psi] \tag{7-38}$$

denotes the symbol error rate, conditioned on the transmitted in-phase symbol and on the synchronization error. Equation (7-37) takes into account that $P_s(\psi;\sigma;2m-1)=P_s(\psi;\sigma;-2m+1)$, so that we can restrict our attention to positive in-phase symbols.

Let us denote by $s(\psi;2m-1)$ the signal component of the sample $z_k e^{-j\hat{\theta}}$ at the input of the decision device, corresponding to a synchronization error ψ and a transmitted in-phase symbol $a_{R,k} = (2m-1)\Delta$; the additive noise component w_k of $z_k e^{-j\hat{\theta}}$ is Gaussian, with zero-mean independent real and imaging parts, each having a variance $\sigma^2 = N_0/(2E_s)$. Hence,

$$P_s(\psi;\sigma;2m-1) = E_{\mathbf{a}}\left[Q\left(\frac{d(\psi;2m-1)}{\sigma}\right)+Q\left(\frac{2\Delta-d(\psi;2m-1)}{\sigma}\right)\right] \tag{7-39}$$

for $m = 1,2,\ldots,M/2-1$, and

$$P_s(\psi;\sigma;M-1) = E_{\mathbf{a}}\left[Q\left(\frac{d(\psi;M-1)}{\sigma}\right)\right] \tag{7-40}$$

where, as shown in Figure 7-6,

$$d(\psi;2m-1) = \text{Re}[s(\psi;2m-1)]\ -\ (2m-2)\Delta \tag{7-41}$$

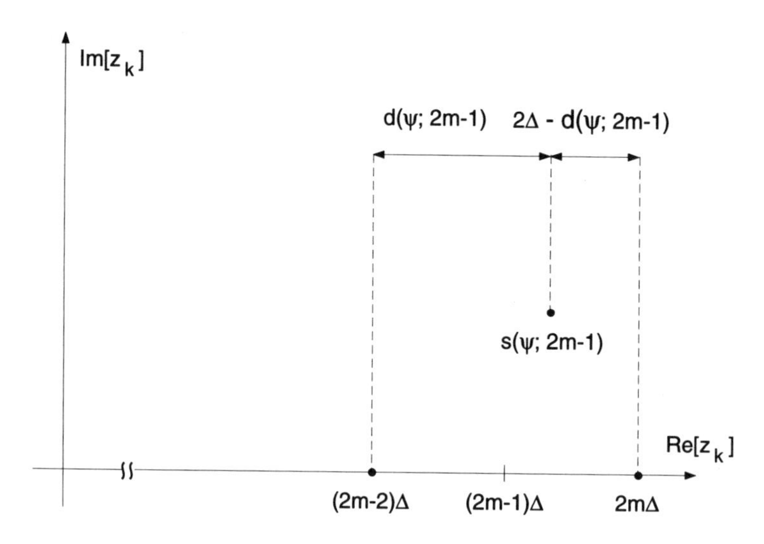

Figure 7-6 Illustration of $d(\psi; 2m-1)$

In (7-39) and (7-40), $E_{\mathbf{a}}[\cdot]$ denotes averaging over all in-phase and quadrature symbols that contribute to $s(\psi; 2m-1)$, with the exception of the in-phase symbol $a_{R,k} = (2m-1)\Delta$. Note that $d(0; 2m-1) = \Delta$, because $s(0; 2m-1) = (2m-1)\Delta$. The bit error rate in the case of perfect synchronization (i.e., $\psi = 0$) is obtained from (7-37), (7-39), and (7-40) as

$$P(0; \sigma) = \frac{1}{\log_2 M} \, \frac{2(M-1)}{M} \, Q\left(\frac{\Delta}{\sigma}\right) \tag{7-42}$$

Evaluating the BER degradation in decibels, given by (7-8), one obtains

$$D \simeq \frac{10}{\ln(10)} \left(A + \frac{B}{\sigma_0^2}\right) \text{var}[\psi] \qquad [\text{dB}] \tag{7-43}$$

where

$$A = \frac{1}{(M-1)\Delta} \, E_{\mathbf{a}}\left[-d^{(\psi\psi)}(0; M-1)\right] \tag{7-44}$$

$$B = \frac{1}{M-1} \left(E_{\mathbf{a}}\left[\left(d^{(\psi)}(0; M-1)\right)^2\right] + 2 \sum_{m=1}^{M/2-1} E_{\mathbf{a}}\left[\left(d^{(\psi)}(0; 2m-1)\right)^2\right] \right) \tag{7-45}$$

and the superscripts ψ and $\psi\psi$ denote single and double differentiation with respect to the synchronization error ψ. Using a similar reasoning as for M-PSK, it can be verified that the quantities A and B reflect the reduction of the useful component of $\text{Re}\left[z_k e^{-j\hat{\theta}}\right]$ and the increase of the variance of $\text{Re}\left[z_k e^{-j\hat{\theta}}\right]$, respectively.

In the following, we evaluate the quantities A and B from (7-44) and (7-45), in the cases of carrier phase errors ($\psi = \phi$) and timing errors ($\psi = e$), respectively.

7.5.1 Carrier Phase Errors

When the synchronization error ψ equals the carrier phase error ϕ, and timing is perfect, we obtain

$$d(\phi, 2m-1) = \begin{cases} (2m-1)\Delta \cos\phi - (2m-2)\Delta & \text{M} - \text{PAM} \\ (2m-1)\Delta \cos\phi - (2m-2)\Delta - a_{I,k}\sin\phi & \text{M}^2 - \text{PAM} \end{cases} \tag{7-46}$$

Using (7-46) in (7-44) and (7-45) yields

$$A = 1 \qquad B = \begin{cases} 0 & M - \text{PAM} \\ \frac{1}{2} & M^2 - \text{PAM} \end{cases} \tag{7-47}$$

In the case of M-PAM, a carrier phase error yields only a reduction of the useful component at the input of the decision device ($\cos\phi \leq 1$), whereas for M^2-QAM also a zero-mean quadrature interference term $-a_{I,k}\sin\phi$ occurs, which acts as an additional disturbance.

7.5.2 Timing Errors

When the synchronization error ψ equals the timing error e, and carrier synchronization is perfect, we obtain

$$d(e; 2m-1) = (2m-1)\Delta h(eT) + \sum_{m \neq 0} a_{R,k-m} h(mT - eT) - (2m-2)\Delta \tag{7-48}$$

Using (7-48) in (7-44) and (7-45) yields

$$A = -h''(0)T^2 \qquad B = \begin{cases} \sum_m (h'(mT)T)^2 & \text{M} - \text{PAM} \\ \frac{1}{2}\sum_m (h'(mT)T)^2 & \text{M}^2 - \text{PAM} \end{cases} \tag{7-49}$$

The difference between M-PAM and M^2-QAM with respect to the value of B comes from the fact that $E\left[a_{R,k}^2\right] = 1$ for M-PAM and $E\left[a_{R,k}^2\right] = \frac{1}{2}$ for M^2-QAM.

It is clear from (7-48) that a timing error reduces the useful component at the input of the decision device ($h(eT) \leq 1$) and additionally introduces ISI.

7.6 Examples

Let us first check the accuracy of the approximate expression (7-8) for the BER degradation, by considering a specific case. Restricting our attention to carrier phase errors, we compute from (7-3) and (7-4) the true BER degradation for 4-PSK, and compare it with the result obtained from (7-21) and (7-30). The average bit error rate $E_\phi[P(\phi;\sigma)]$ for 4-PSK is given by

$$E_\phi[P(\phi;\sigma)]=E_\phi\left[Q\left[\frac{\cos\phi-\sin\phi}{\sqrt{2}\,\sigma}\right]\right] \tag{7-50}$$

In order to carry out the statistical expectation indicated in (7-50), an assumption must be made about the probability density function $p(\phi)$ of the carrier phase error. We consider two probability density functions, i.e., the Tikhonov probability density function and the uniform probability density function. The Tikhonov probability density is the density of the modulo $2\pi/M$ reduced phase error of a PLL with sinusoidal phase error detector characteristic (with period $2\pi/M$) and white loop noise (see Volume 1, Section 11.1.6), and is given by

$$p(\phi)=\begin{cases} M/(2\pi I_0(\rho/M^2))\,\exp((\rho/M^2)\cos(M\phi)) & |\phi|<\frac{\pi}{M} \\ 0 & \text{otherwise}\end{cases} \tag{7-51}$$

where $1/\rho$ is the linearized phase error variance (ρ is the loop signal-to-noise ratio of the PLL) and $I_0(x)$ is the zeroth-order modified Bessel function:

$$I_0(x)=\frac{1}{2\pi}\int_{-\pi}^{+\pi}\exp\left(x\cos\phi\right)\,d\phi \tag{7-52}$$

For large ρ, the actual variance var$[\phi]$ is very close to $1/\rho$. The uniform probability density function (which is most unlikely to result from any practical synchronizer!) is given by

$$p(\phi)=\begin{cases}1/(2\phi) & |\phi|<\phi_0 \\ 0 & \text{otherwise}\end{cases} \tag{7-53}$$

where ϕ_0 is selected such that the Tikhonov probability density function (7-51) and the uniform probability density function (7-53) yield the same phase error variance var$[\phi]$; hence, ϕ_0 and ρ are related by

$$\int_{-\pi/M}^{+\pi/M}\phi^2\frac{M}{2\pi I_0(\rho/M^2)}\exp\left(\frac{\rho}{M^2}\cos\left(M\phi\right)\right)d\phi=\frac{1}{3}\phi_0^2 \tag{7-54}$$

Figure 7-7 shows the actual BER degradation at $\text{BER}_0=10^{-2}$ and $\text{BER}_0=10^{-6}$ for 4-PSK, corresponding to the Tikhonov and the uniform phase error

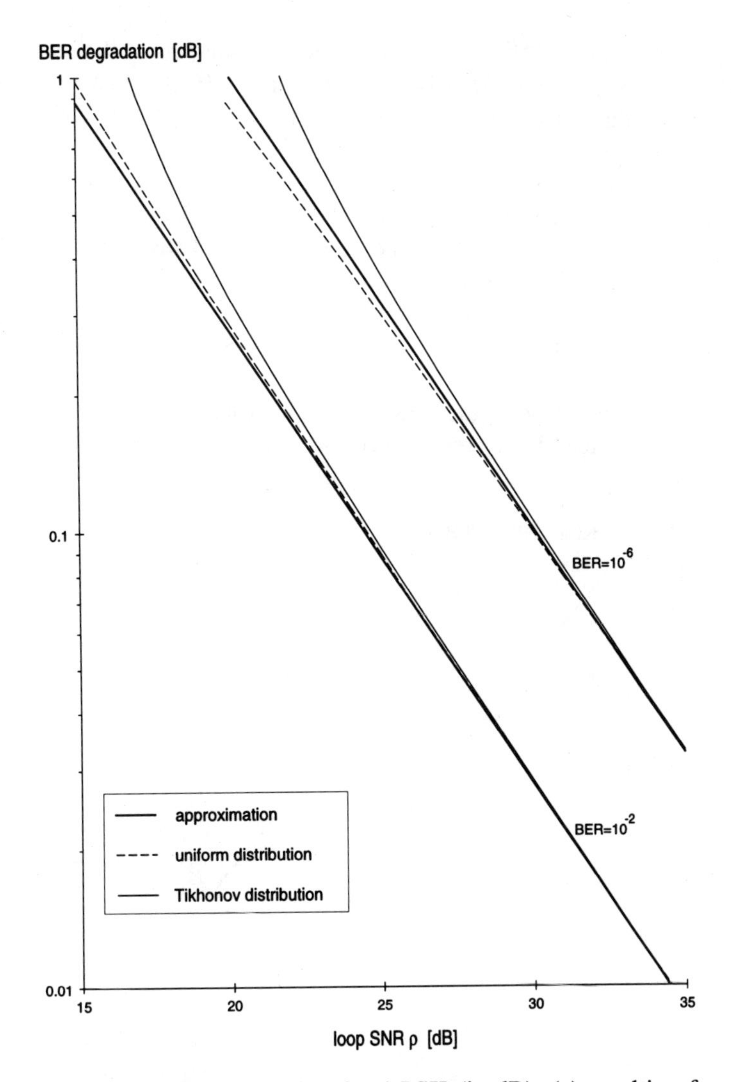

Figure 7-7 BER Degradation for 4-PSK (in dB): (a) resulting from uniform phase error distribution, (b) resulting from Tikhonov phase error distribution, (c) approximate BER degradation

probability densities; also shown is the approximate BER degradation resulting from (7-21) and (7-30), with $\text{var}[\phi]$ replaced by $1/\rho$. We observe that for large ρ the actual BER degradation becomes independent of the specific shape of the phase error probability density function and converges to the approximate BER degradation, which is inversely proportional to ρ.

In the following, we use the approximate formula (7-8) to evaluate the BER degradation in decibels, caused by carrier phase errors and timing errors, at $\mathrm{BER}_0 = 10^{-2}$ and 10^{-6}, in the case of the following constellations: M-PSK (with $M = 2, 4, 8, 16$), M-PAM (with $M = 2, 4$), and M^2-QAM ($M = 2, 4$). The corresponding values of $E_s/N_0\,[= 1/(2\sigma_0^2)]$ and E_b/N_0, where E_s and E_b denote the received energy per symbol and per bit, respectively, are shown in Table 7-1. The 2-PSK (or BPSK) constellation is identical to the 2-PAM constellation, and the 4-PSK (or QPSK) constellation is identical to the 4-QAM constellation. The M^2-QAM constellation needs the same E_b/N_0 as (or 3 dB more E_s/N_0) than the M-PAM constellation in order to achieve the same BER_0. The shape of the baseband pulse $h(t)$ at the output of the matched filter affects the BER degradation in the case of timing errors; in the sequel it will be assumed that $h(t)$ is a cosine rolloff pulse.

Figures 7-8 and 7-9 show the BER degradation caused by carrier phase errors, at $\mathrm{BER}_0 = 10^{-2}$ and 10^{-6}, respectively, as a function of $\rho = 1/\mathrm{var}[\phi]$; the

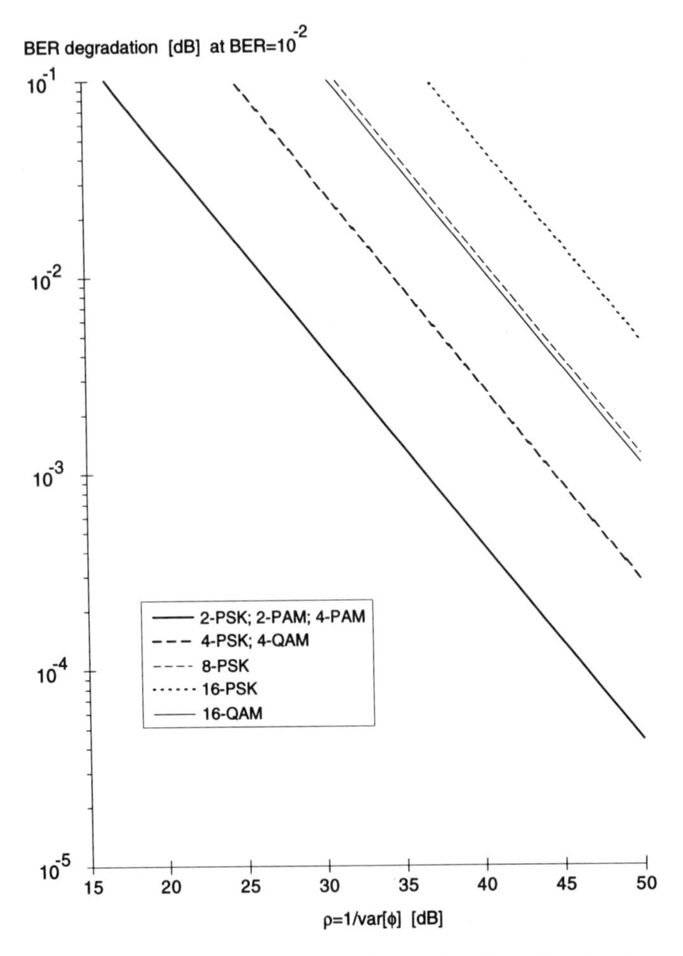

Figure 7-8 BER Degradation (in dB) due to Random Carrier Phase Errors ($\mathrm{BER}_0 = 10^{-2}$)

Table 7-1 Values of E_b/N_0 and E_s/N_0 Yielding $\text{BER}_0 = 10^{-2}$ and 10^{-6}

	$\text{BER}_0 = 10^{-2}$		$\text{BER}_0 = 10^{-6}$	
	E_b/N_0 $[dB]$	E_s/N_0 $[dB]$	E_b/N_0 $[dB]$	E_s/N_0 $[dB]$
2-PSK, 2-PAM	4.32	4.32	10.53	10.53
4-PSK, 4-QAM	4.32	7.33	10.53	13.54
8-PSK	7.29	12.06	13.95	18.72
16-PSK	11.42	17.43	18.44	24.46
4-PAM	7.88	10.89	14.40	17.41
16-QAM	7.88	13.90	14.40	20.42

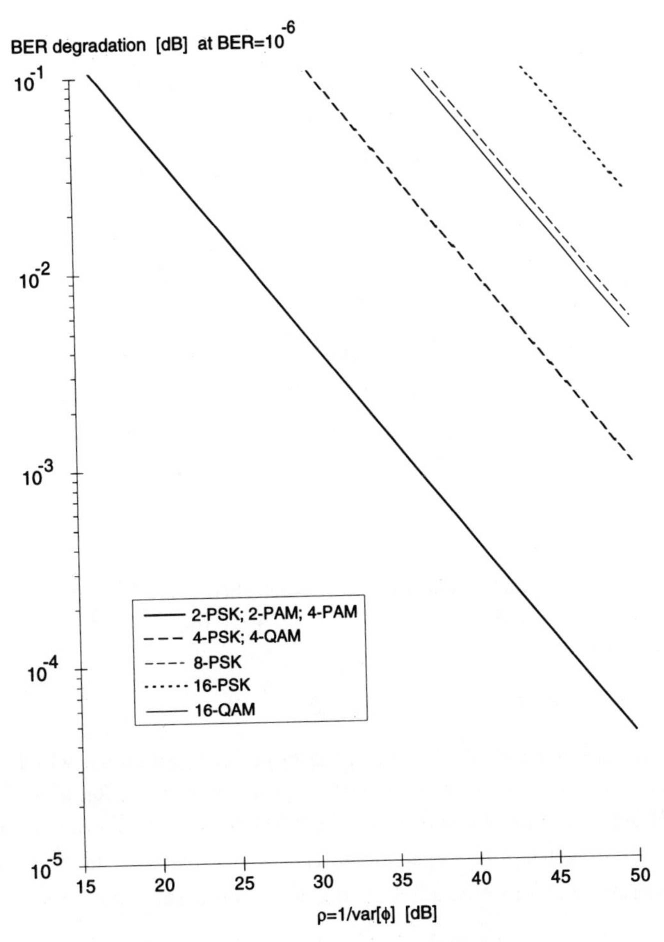

Figure 7-9 BER Degradation (in dB) due to Random Carrier Phase Errors ($\text{BER}_0 = 10^{-6}$)

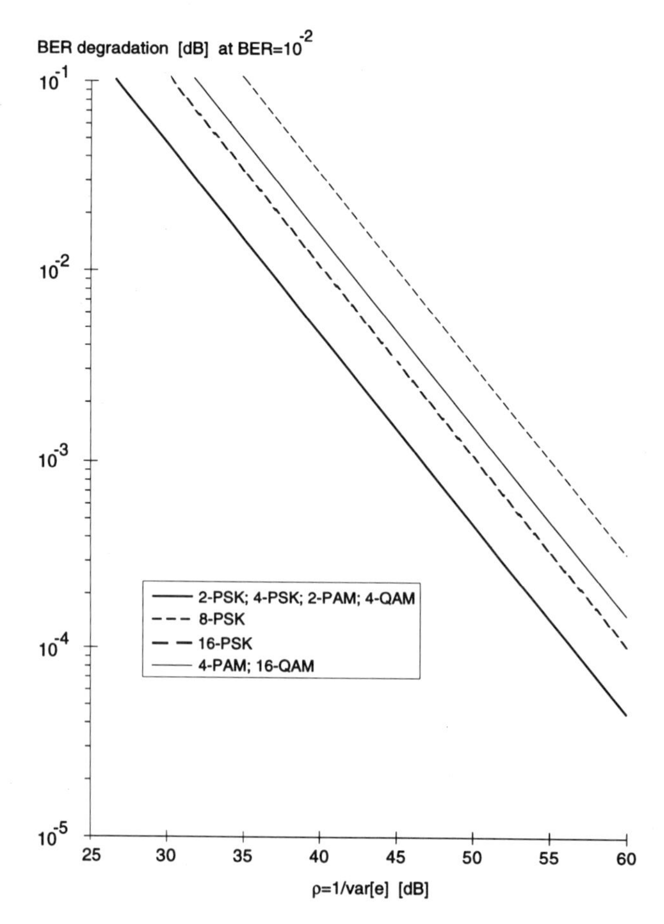

Figure 7-10 BER Degradation (in dB) due to Random Symbol Timing Errors ($\text{BER}_0 = 10^{-2}$, rolloff = 50%)

BER degradation caused by timing errors is shown in Figures 7-10 and 7-11, for $\text{BER}_0 = 10^{-2}$ and 10^{-6}, respectively, as a function of $\rho = 1/\text{var}[e]$ and assuming a 50% rolloff.

The following observations are made:

For given values of BER_0 and ρ, the larger constellations, which need larger values of E_b/N_0 to achieve a given bit error rate of BER_0, give rise to larger BER degradations. Indeed, when the number of constellation points increases under the restriction $E\left[|a_k|^2\right] = 1$, the Euclidean distance between them decreases, and the constellation becomes more sensitive to synchronization errors.

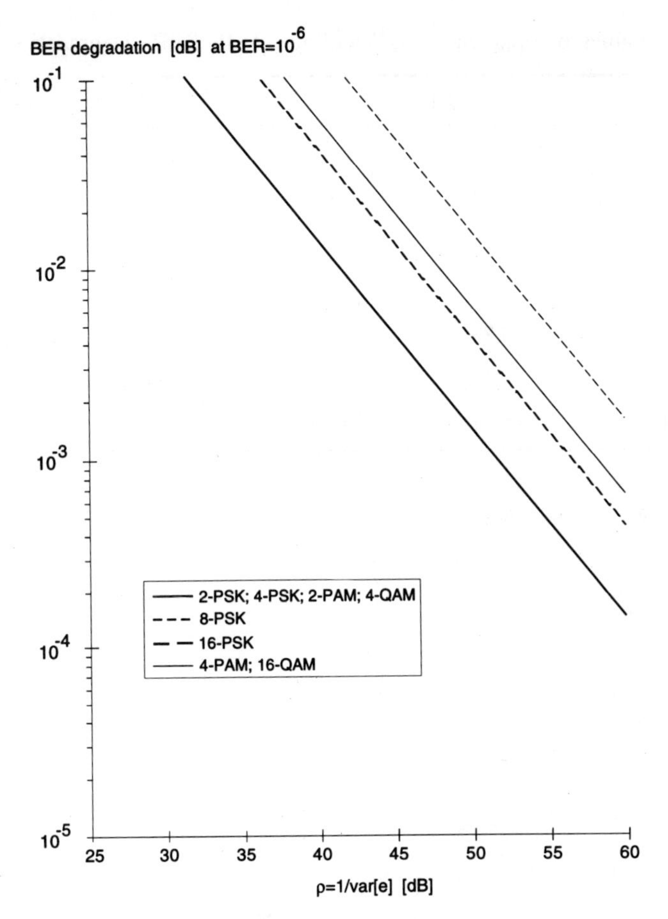

Figure 7-11 BER Degradation (in dB) due to Random Symbol Timing Errors ($\text{BER}_0 = 10^{-6}$, rolloff = 50%)

- In the case of carrier phase errors, M-PAM yields a smaller BER degradation than M^2-QAM, although both constellations need the same E_b/N_0 to achieve the same BER_0. This is explained by noting that a carrier phase error gives rise to quadrature interference in the case of M^2-QAM, but not for M-PAM.
- In the case of timing errors, M-PAM and M^2–QAM yield the same BER degradation. Indeed, when carrier synchronization is perfect, the real and imaginary parts of the sample at the input of the decision device for M^2-QAM have the same statistical properties as the real part of the sample at the input of the decision device for M-PAM, assuming that both modulations have the same E_b and, hence, the same value of BER_0.
- The BER degradation increases with decreasing BER_0 (or decreasing σ_0^2),

Table 7-2 Values of ϕ_{rms} and e_{rms} Yielding 0.1 dB BER_0 Degradation

	$\text{BER}_0 = 10^{-2}$		$\text{BER}_0 = 10^{-6}$	
	ϕ_{rms} [deg.]	e_{rms} [%]	ϕ_{rms} [deg.]	e_{rms} [%]
2-PSK, 2-PAM	8.64	4.60	8.64	2.63
4-PSK, 4-QAM	3.41	4.60	1.78	2.63
8-PSK	1.62	3.05	0.76	1.51
16-PSK	0.83	1.74	0.37	0.79
4-PAM	8.64	2.53	8.64	1.25
16-QAM	1.71	2.53	0.82	1.25

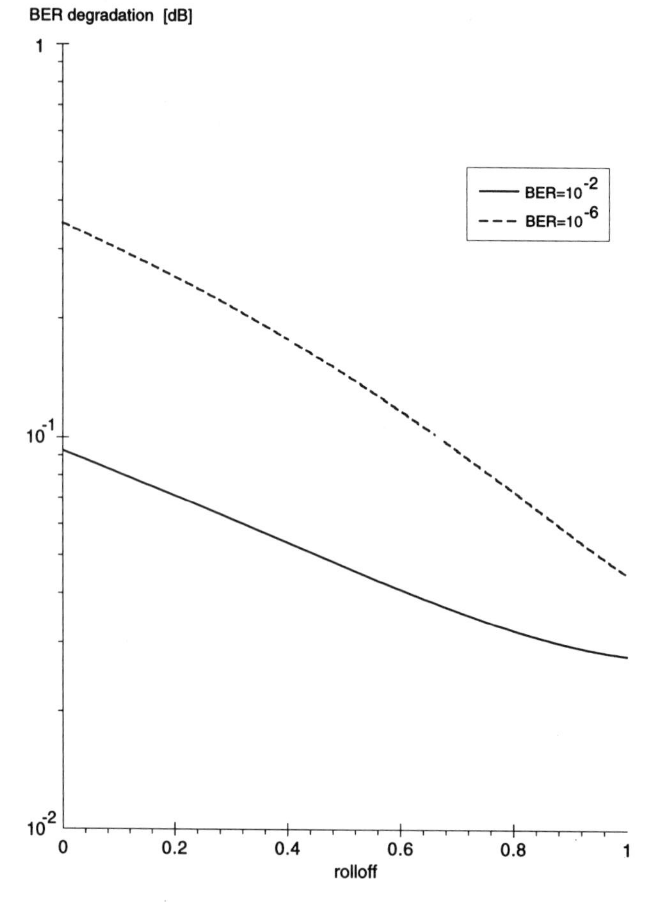

Figure 7-12 BER Degradation (in dB) due to Random Symbol Timing Errors (4-PSK, ρ=30 dB)

because the synchronization errors yield an increase of the variance at the input of the decision device, which becomes relatively more important when the variance of the additive noise decreases. However, for M-PAM, the only effect of a carrier phase error ϕ is a multiplication by $\cos\phi$ of the useful component at the input of the decision device, so that for M-PAM the BER degradation caused by carrier phase errors depends neither on BER_0 nor on M.

For the various constellations, Table 7-2 shows the root-mean-square (rms) phase and timing errors, each yielding a 0.1–dB BER degradation; the results for the rms timing error assume a 50 percent cosine rolloff pulse at the output of the matched filter.

Finally, we investigate the effect of the rolloff on the BER degradation caused by timing errors. Figure 7-12 shows the BER degradation in the case of 4-PSK, for operating BER values of 10^{-2} and 10^{-6}, and assuming that $\rho = 30$ dB. When the rolloff increases, the following two effects occur: the useful component $h(eT)$ decreases (which tends to increase the BER) but also the ISI decreases (which tends to decrease the BER). From Figure 7-12 it follows that the latter effect dominates.

7.7 Coded Transmission

Today most practical communication systems employ some form of coding. This raises the question of how random carrier phase and timing errors affect coded systems.

As for uncoded transmission, the effect of random synchronization errors on coded transmission is twofold:

1. The useful component of $z_k e^{-j\hat{\theta}}$ is reduced by a factor of $1 - A/2\ \text{var}(\psi)$. This reduction is exactly the same as for uncoded transmission.
2. The samples of $z_k e^{-j\hat{\theta}}$ are affected by an additional interference, consisting of quadrature interference (caused by phase errors) or ISI (caused by timing errors). This additional interference is the same function of the data symbols as for uncoded transmission. However, as the statistics of the data symbols are determined by the specific code used, the statistical properties of this additional interference are not exactly the same as for uncoded transmission.

From the above considerations it follows that the BER degradation for coded transmission depends on the specific code used. However, a first estimate of this BER degradation is obtained by simply assuming that the statistics of the additional interference are the *same* as for uncoded transmission. Under this assumption, the BER degradation for coded transmission is independent of the code, and equals the BER degradation of an uncoded system with the same constellation and the same E_s/N_0 (with E_s denoting the energy per coded symbol).

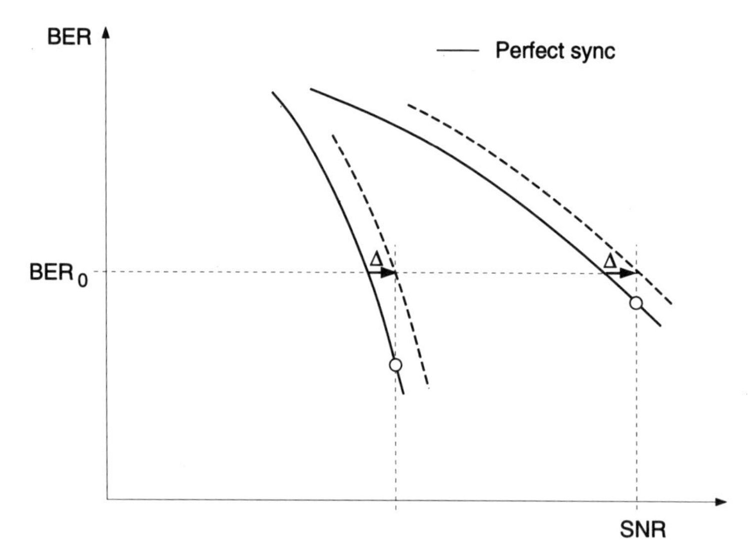

Figure 7-13 BER Degradation for Coded Transmission

Notice in Figure 7-13 that for a *given* E_s/N_0 the same degradation D causes a much larger BER increase in the case of coded transmission.

Some references dealing with the effect of synchronization errors on coded transmission are mentioned in Section 7.9

7.8 Main Points

- In the presence of random synchronization errors, a BER degradation occurs as compared to perfect synchronization. This degradation is caused by a reduction of the useful component and an increase of the variance at the input of the decision device.
- For a small estimate variance, the BER degradation (in decibels) caused by random synchronization errors is essentially proportional to the estimate variance, and independent of the specific synchronization error probability density.
- In general, the BER degradation increases with the size of the constellation and decreases with the operating BER.
- The BER caused by timing errors depends on the shape of the baseband pulse at the matched filter output. Assuming a cosine rolloff pulse, the BER degradation decreases with increasing rolloff, because of the reduced ISI.

7.9 Bibliographical Notes

In the following, we give an overview of a representative literature selection on the BER degradation caused by synchronization errors.

Perfect Synchronization

Exact and approximate expressions of upper and lower bounds on the BER for various modulation formats in the case of perfect synchronization can be found in many textbooks and articles, e.g., in [1–5].

Carrier Phase Errors

In the case of carrier phase jitter, several authors have evaluated the average BER by taking the expectation of the conditional BER (or its approximation) with respect to the carrier phase error [1, 6–10]; in most cases, the carrier phase error is assumed to have a Tikhonov or a (truncated) Gaussian probability density function. Using a somewhat similar reasoning as in Section 7.3, an approximate expression for the BER degradation for BPSK and QPSK has been derived in [6].

For M-PSK, M-PAM, and M^2-QAM, the BER degradation caused by carrier phase errors does not depend on the shape of the baseband pulse. This is no longer true when using offset QPSK (OQPSK), where the quadrature symbols are delayed by half a symbol interval with respect to the in-phase symbols. The BER degradation for OQPSK has been investigated in [6] and [9, 10] in the case of rectangular (NRZ) and square root cosine rolloff transmit pulses, respectively.

Timing Errors

Various authors [11–14] have considered the BER in the presence of a fixed, not necessarily small timing error. When the timing error is not small, the truncated Taylor series expansion method from Section 7.3 is no longer accurate, so that the conditional BER has to be evaluated in a different way. As timing errors give rise to ISI, we are faced with the more general problem of computing the BER in the presence of additive noise and ISI. This problem has received considerable attention in the literature, and various bounds on and approximations of the BER have been proposed [11–15], which avoid the time-consuming brute force averaging in (7-12), (7-39), or (7-40) over all symbols that contribute to the ISI.

The BER in the presence of random timing errors has been obtained in [1, 16] by computing the expectation of (a bound on) the conditional BER with respect to the timing error, assuming a Tikhonov probability density function.

Nonlinear Modulation and Coded Linear Modulation

Until now, we have restricted our attention to uncoded linear modulation. Several authors have investigated the effect of carrier phase errors on the BER in the case of continuous-phase modulation (CPM), which is a nonlinear modulation [17], and of trellis-coded M-PSK.

The BER for CPM in the presence of a fixed carrier phase offset has been considered in [18].

The BER for trellis-coded M-PSK has been investigated in [19] for a fixed carrier phase error, in [20] for a random carrier phase error and without interleaving of the data symbols, and in [21] for a random carrier phase error and interleaving of the data symbols.

Bibliography

[1] W. C. Lindsey and M. K. Simon, *Telecommunication Systems Engineering*. Englewood Cliffs, NJ: Prentice-Hall, 1973.

[2] J. G. Proakis, *Digital Communications*. New York: McGraw-Hill, 2nd ed., 1989.

[3] C. M. Chie, "Bounds and Approximation for Rapid Evaluation of Coherent MPSK Error Probabilities," *IEEE Trans. Commun.*, vol. COM-33, pp. 271–273, Mar. 1985.

[4] P. J. Lee, "Computation of the Bit Error Rate of Coherent M-ary PSK with Gray Code Bit Mapping," *IEEE Trans. Commun.*, vol. COM-34, pp. 488–491, May 1986.

[5] M. I. Irshid and I. S. Salous, "Bit Error Probability for Coherent M-ary PSK Systems," *IEEE Trans. Commun.*, vol. COM-39, pp. 349–352, Mar. 1991.

[6] S. A. Rhodes, "Effect of Noisy Phase Reference on Coherent Detection of Offset-QPSK Signals," *IEEE Trans. Commun.*, vol. COM-22, pp. 1046–1054, Aug. 1974.

[7] V. K. Prabhu, "Imperfect Carrier Recovery Effect on Filtered PSK Signals," *IEEE Trans. Aerospace Electronic Syst.*, vol. AES-14, pp. 608–615, July 1978.

[8] N. Liskov and R. Curtis, "Performance of Coherent Phase and Amplitude Digital Modulations with Carrier Recovery Noise," *IEEE Trans. Commun.*, vol. COM-35, pp. 972–976, Sept. 1987.

[9] F. Fan and L. M. Li, "Effect of Noisy Phase Reference on Coherent Detection of Band-Limited Offset-QPSK Signals," *IEEE Trans. Commun.*, vol. COM-38, pp. 156–159, Feb. 1990.

[10] T. Jesupret, M. Moeneclaey, and G. Ascheid, "Digital Demodulator Synchronization, Performance Analysis," *Final Report to ESTEC Contract No. 8437/89/NL/RE*, June 1991.

[11] B. R. Saltzberg, "Intersymbol Interference Error Bounds with Application to Ideal Bandlimited Signaling," *IEEE Trans. Inform. Theory*, vol. IT-14, pp. 563–568, July 1968.

[12] E. Ho and Y. Yeh, "A New Approach for Evaluating the Error Probability in the Presence of Intersymbol Interference and Additive Gaussian Noise," *Bell Syst. Tech. J.*, vol. 49, pp. 2249–2265, Nov. 1970.

[13] F. E. Glave, "An Upper Bound on the Probability of Error Due to Intersymbol Interference for Correlated Digital Signals," *IEEE Trans. Inform. Theory*, vol. IT-18, pp. 356–363, May 1972.

[14] K. Metzger, "On the Probability Density of Intersymbol Interference," *IEEE Trans. Commun.*, vol. COM-35, pp. 396–402, Apr. 1987.

[15] V. K. Prabhu, "Modified Chernoff Bounds for PAM Systems with Noise and Interference," *IEEE Trans. Inform. Theory*, vol. IT-28, pp. 95–100, Jan. 1982.

[16] M. Moeneclaey, "The Influence of Four Types of Symbol Synchronizers on the Error Probability of a PAM Receiver," *IEEE Trans. Commun.*, vol. COM-32, pp. 1186–1190, Nov. 1984.

[17] J. B. Anderson, T. Aulin, and C.-E. Sundberg, *Digital Phase Modulation*. New York: Plenum, 1986.

[18] R. A. Di Fazio, "An Approximation for the Phase Offset Sensitivity of Viterbi Detected Continous Phase Modulation Signals," *IEEE Trans. Commun.*, vol. COM-36, pp. 1324–1326, Dec. 1988.

[19] J. Hagenauer and C.-E. Sundberg, "On the Performance Evaluation of Trellis-Coded 8–PSK Systems with Carrier Phase Offset," vol. Archiv Elektr. Übertr. AEU-42, pp. 274–284, May 1988.

[20] H. Leib and S. Pasupathy, "Trellis-Coded MPSK with Reference Phase Errors," *IEEE Trans. Commun.*, vol. COM-35, pp. 888–900, Sept. 1987.

[21] E. Zehavi and G. Kaplan, "Phase Noise Effects on M-ary PSK Trellis Codes," *IEEE Trans. Commun.*, vol. COM-39, pp. 373–379, Mar. 1991.

Chapter 8 Frequency Estimation

8.1 Introduction / Classification of Frequency Control Systems

In this chapter we are concerned with frequency estimation. We could have studied this problem earlier in Chapter 5 by including an additional parameter Ω in the set $\boldsymbol{\theta} = \{\theta, \varepsilon, \mathbf{a}\}$. The main reason why we choose to include a separate chapter is that for a sizeable frequency offset Ω we must *first* compensate this frequency offset before the other parameters $\{\theta, \varepsilon, \mathbf{a}\}$ can be estimated. This implies that frequency offset estimation algorithms must work independently of the values of the other parameters. The operation of the algorithms is nondata aided and nonclock aided. The only exception occurs for small frequency offset $(\Omega T) \ll 1$. In this case, timing-directed algorithms are possible.

This chapter is organized as follows. We first discuss in Section 8.1.1 the channel model modifications necessary to include the estimation of Ω. In Section 8.2 we derive estimators which work independently of the other parameters $\{\theta, \varepsilon, \mathbf{a}\}$. In a familiar way we obtain feedback algorithms by differentiating the likelihood function with respect to the parameter Ω (Section 8.3). The algorithms of the first two sections operate on samples $\{r_f(kT_s)\}$ which are sufficient statistics. If the frequency offset is restricted to small values, roughly $|\Omega T| < 0.15$, timing can be recovered *prior* to frequency compensation. Given the timing, frequency estimators are developed which work at symbol rate $1/T$. These algorithms have superior tracking performance compared to the algorithms operating with samples $\{r_f(kT_s)\}$. Direct frequency estimators are discussed in Section 8.4. The corresponding error-feedback algorithms are studied in Section 8.5. In Section 8.6 frequency estimation for MSK signals is studied. In summary, the rate-$1/T_s$ algorithms can be regarded as coarse acquisition algorithms reducing the frequency offset to small fractions of the symbol rate. If necessary, timing-directed algorithms with improved accuracy can be employed in a second stage running at symbol rate $1/T$.

8.1.1 Channel Model and Likelihood Function

We refer to the linear channel model of Figure 3-1 and Table 3-1. The equivalent baseband model is shown in Figure 3–3. The input signal to the channel is given by $u(t)e^{j\theta_T(t)}$. In the presence of a frequency offset Ω we model the phase process $\theta_T(t)$ as the sum

$$\theta_T(t) = \Omega t + \theta \tag{8-1}$$

(θ constant phase offset). The input signal to the channel is then given by

$$u(t)e^{j(\Omega t+\theta)} \tag{8-2}$$

We require that the frequency response of the channel $C(f)$ and that of the prefilter $F(f)$ are flat within the frequency range

$$|\omega| \leq 2\pi B + |\Omega_{\max}| \tag{8-3}$$

when B is the (one-sided) bandwidth of the signal $u(t)$, and $\pm\Omega_{\max}$ is the maximum frequency uncertainty (see Figure 8-1). Only under this condition the signal $s_f(t,\Omega)$ can be written as

$$\begin{aligned} s_f(t,\Omega) &= u(t)\; e^{j\Omega t} \otimes [c(t) \otimes f(t)] \\ &= u(t) \otimes c(t) \otimes f(t)\; e^{j\Omega t} \end{aligned} \tag{8-4}$$

Thus, $s_f(t,\Omega)$ is the frequency-translated signal $s_f(t)\; e^{j\Omega t}$:

$$\begin{aligned} s_f(t,\Omega) &= \sum_{n=0}^{N-1} a_n\; g(t - nT - \varepsilon T) e^{j(\Omega t+\theta)} \\ \text{with} \quad g(t) &= g_T(t) \otimes c(t) \otimes f(t) \end{aligned} \tag{8-5}$$

We recognize the signal previously employed for estimating ε and θ is multiplied by $e^{j\Omega t}$. We should be aware that the model is valid *only* under the condition of a flat wideband channel $C(\omega)$ and prefilter $F(\omega)$. If this condition is violated, the frequency estimator algorithms presented here produce a biased estimate. This issue will be discussed when we analyze the performance of the frequency estimators.

Assuming a symmetrical prefilter $|F(\omega)|^2$ about $1/2T_s$ and Nyquist pulses, the normalized likelihood function is given by eq. (4–157). Using the signal definition

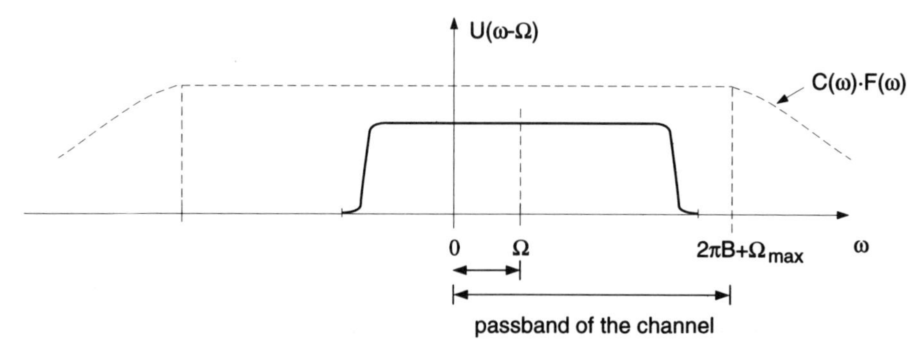

Figure 8-1 Frequency Translation and Passband of the Channel $G(\omega)$ and Prefilter $F(\omega)$

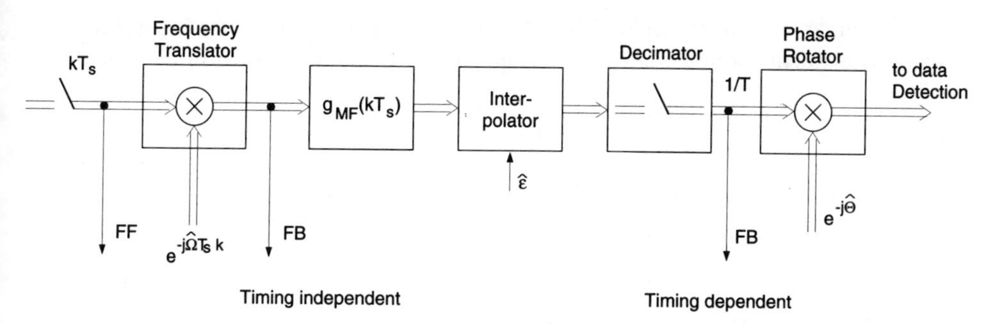

Figure 8-2 Classification of ML Frequency Estimation Algorithms: Typical Signal Flow

(8-5) the matched filter output in the presence of a frequency uncertainty is given by

$$z_n(\varepsilon, \Omega) = \sum_{k=-\infty}^{\infty} r_f(kT_s)\, e^{-j\Omega T_s k}\; g_{\mathrm{MF}}(nT + \varepsilon T - kT_s) \tag{8-6}$$

To obtain $z_n(\varepsilon, \Omega)$ the signal is first multiplied by $e^{-j\Omega T_s k}$. This operation must be performed for each realization Ω_i of the trial parameter Ω. Subsequently each signal is filtered by the matched filter. The sampling rate (see Figure 8-1) must obey the inequality

$$\frac{1}{2T_s} > B + \frac{|\Omega_{\max}|}{2\pi} \tag{8-7}$$

8.1.2 Classification of Algorithms

The typical chain of signal processing operations in the presence of a frequency offset Ω is shown in Figure 8-2. The received signal is *first* multiplied by $\exp(-j\hat{\Omega}kT_s)$ which results in a frequency translation of $\hat{\Omega}$ of the signal. Subsequently the signal flow is the same as discussed in the absence of a frequency offset.

The only exception to the signal flow diagram shown in Figure 8-2 occurs if the frequency offset is small, $|\Omega T| \ll 1$. In this case it may be advantageous to *first* recover timing and perform frequency estimation in a timing-directed mode (compare Figure 8-3).

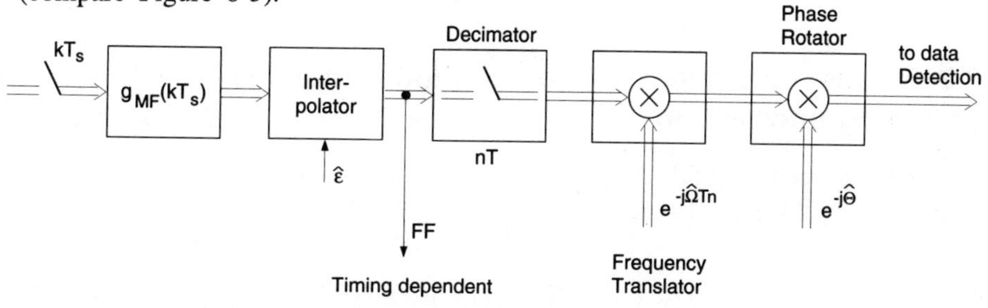

Figure 8-3 Classification of ML Frequency Estimation Algorithms: Alternative Signal Flow for $|\Omega T| \ll 1$

Then, as we will see in Section 8.4, it is convenient to introduce a mathematically equivalent formulation of (8-6). Equation (8-6) can be written in the form

$$z_n(\varepsilon, \Omega) = e^{-jn\Omega T} \sum_{k=-\infty}^{\infty} r_f(kT_s)\, g_{\text{MF}}(nT + \varepsilon T - kT_s) e^{-j\Omega(kT_s - nT)} \tag{8-8}$$

The sum describes a convolution of the received signal $r_f(kT_s)$ with a filter with impulse response

$$g_\Omega(kT_s) = g_{\text{MF}}(kT_s) e^{j\Omega T_s k} \tag{8-9}$$

and frequency response

$$G_\Omega(e^{j\omega T_s}) = G_{\text{MF}}\left(e^{j(\omega-\Omega)T_s}\right) \tag{8-10}$$

The approximation of $g_\Omega(kT_s)$ by $g_{\text{MF}}(kT_s)$ as it is suggested in Figure 8-3 is discussed later on in Section 8.4.

Besides the above distinction between *timing-directed* (Dε) and *non-timing-directed* (NDε), the algorithms employed in the frequency synchronization unit can be further classified as discussed earlier (NDA, DA, DD).

Remark: Before we are going into further detail of the frequency synchronization schemes, it should be emphasized that in practice (8-7) and the relationship $|\Omega_{\text{max}}|/2\pi < B_{f1} - B$ determine the maximal resolvable frequency offset that any structure (Figures 8-2 and 8-3) can cope with. The value B_{f1} stands for the (one-sided) frequency range of the analog prefilter, where the signal passes undistorted. If the key parameters (prefilter shape and sampling rate) are fixed, the pull-in range – determined by the maximal manageable frequency offset – can only be increased by an additional control loop for the analog oscillator in front of the analog prefilter.

8.1.3 Main Points

- The typical signal processing chain is to correct the frequency offset in front of the matched filter.
- The frequency estimation algorithms can be classified as Dε, NDε, DA, and NDA.
- There are open-loop (FF) and closed-loop (FB) structures.

8.2 Frequency Estimator Operating Independently of Timing Information

Starting point of our discussion is the low signal-to-noise ratio (SNR) approximation of the likelihood function [eq. (5-48)]

$$L(\varepsilon, \Omega) = \sum_{l=-L}^{L} |z(lT + \varepsilon T, \Omega)|^2 \tag{8-11}$$

for the joint estimation of (ε, Ω). We will show in this section that there exists an unbiased estimator of Ω which requires no knowledge of ε. In other words, we can first estimate the frequency offset Ω, compensate for Ω, and then estimate the remaining synchronization parameters.

To qualitatively understand why this separation is possible we recall the main result of the section on timing parameters estimation via spectral estimation, (see Section 5.4). In that section, we expanded the timing wave $|z(lT+\varepsilon T, \Omega)|^2$ into a Fourier series. Due to the band-limitation of $|z(lT+\varepsilon T, \Omega)|^2$ only three coefficients $\{c_{-1}, c_0, c_1\}$ of the Fourier series have a nonzero mean. Therefore, *only* these three coefficients need to be estimated while the remaining terms of the Fourier series contribute to the random disturbance only

$$\sum_{l=-L}^{L} |z(lT+\varepsilon T, \Omega)|^2 = c_0 + 2\,\mathrm{Re}\left[c_1 e^{j2\pi\varepsilon}\right] + \underbrace{\sum_{|n|\geq 2} c_n\, e^{j2\pi n\varepsilon}}_{\text{random disturbance}} \tag{8-12}$$

As will be shown, the expected value of c_0 depends on Ω but is independent of ε. Furthermore, $E[c_0]$ is shown to be maximum if Ω assumes the true value Ω_0. Hence, the value $\hat{\Omega}$ which maximizes the coefficient $c_0(\Omega)$ is an unbiased estimate:

$$\hat{\Omega} = \arg\ \max_{\Omega}\ c_0(\Omega) \tag{8-13}$$

Remark: Despite the fact that $E[|c_1|]$ is also a function of Ω, we may seek the maximum of the likelihood function (8-11) by maximizing $c_0(\Omega)$. This, of course, is possible only because (8-13) *alone* provides an unbiased estimate of Ω.

We maintain that under the following conditions:

(i) The sampling rate fulfills $1/T_s > 2(1+\alpha)/T$ (twice the rate required for the data path).
(ii) The ratio $T/T_s = M_s$ is an integer.
(iii) i.i.d. data $\{a_n\}$,

the following sum

$$\hat{\Omega} = \arg\ \max_{\Omega} \sum_{l=-LM_s}^{LM_s-1} |z(lT_s, \Omega)|^2 \tag{8-14}$$

defines an unbiased estimate. It is remarkable that *no* conditions on the pulse $g(t)$ are required in order to get an unbiased estimate. But be aware that our derivation requires that the transmission pulse $g(t)$ be known at the receiver.

Much of the discussion that follows is similar to that on timing parameters estimation via spectral estimation (Section 5-4). The reader is therefore urged to reconsult Chapter 5 in case the following discussion is found to be too concise.

The matched filter output $z(lT_s, \Omega)$ in the presence of a frequency offset Ω is given by

$$z(lT_s, \Omega) = \sum_{k=-\infty}^{\infty} r_f(kT_s)\, e^{-j\Omega T_s k}\, g_{\text{MF}}(lT_s - kT_s) \tag{8-15}$$

Replacing in the previous equation the received signal samples $r_f(kT_s)$ by

$$r_f(kT_s) = \sum_{n=-N}^{N} a_n g(kT_s - nT - \varepsilon_0 T)\, e^{j\Omega_0 T_s k} + n(kT_s) \tag{8-16}$$

we obtain

$$\begin{aligned} z(lT_s, \Omega) = \sum_{n=-N}^{N} a_n \Bigg[\sum_{k=-\infty}^{\infty} g(kT_s - nT - \varepsilon_0 T)\, e^{j(\Omega_0 - \Omega) T_s k} \\ \times g_{\text{MF}}(lT_s - kT_s) \Bigg] + m(lT_s) \end{aligned} \tag{8-17}$$

where $m(lT_s)$ is a filtered noise process. Notice that we follow our usual convention to label the true parameters by index zero (i.e., ε_0, Ω_0), while the trial parameters are denoted by ε and Ω. Using

$$\begin{aligned} h(lT_s - \varepsilon_0 T - nT, \Delta\Omega) \\ &= \sum_{k=-\infty}^{\infty} g(kT_s - nT - \varepsilon_0 T)\, e^{j(\Omega_0 - \Omega) T_s k}\, g_{\text{MF}}(lT_s - kT_s) \\ &= \frac{1}{T_s} \int_{-\infty}^{\infty} g(x - nT - \varepsilon_0 T)\, e^{j(\Omega_0 - \Omega) x}\, g_{\text{MF}}(lT_s - x)\, dx \end{aligned} \tag{8-18}$$

we can write for $z(lT_s, \Omega)$

$$z(lT_s; \Omega) = \sum_{n=-N}^{N} a_n h(lT_s - nT - \varepsilon_0 T, \Delta\Omega) + m(lT_s) \tag{8-19}$$

with $\Delta\Omega = \Omega_0 - \Omega$.

Squaring $z(lT_s, \Omega)$ and subsequently taking expected values with respect to i.i.d. data and noise, we obtain

$$E\left[|z(lT_s, \Omega)|^2 \right] = E\left[|a_n|^2 \right] \sum_{n=-N}^{N} |h(lT_s - nT - \varepsilon_0 T, \Delta\Omega)|^2 + P_n \tag{8-20}$$

with $P_n \geq 0$ additive noise contribution. If the number of symbols $(2N + 1)$ is sufficiently large, then the error committed by running the sum over an infinite interval is negligible. The expected value then is a periodic function

$$h_p(t, \Delta\Omega) = \sum_{n=-\infty}^{\infty} |h(t - nT - \varepsilon_0 T, \Delta\Omega)|^2 \tag{8-21}$$

which can be represented by a Fourier series

$$h_p(t, \Delta\Omega) = \sum_{i=-\infty}^{\infty} d_i \; e^{j2\pi ti/T} \tag{8-22}$$

The coefficients d_i are related to the spectrum of $|h(t - \varepsilon_0 T, \Delta\Omega)|^2$ by the Poisson theorem

$$\begin{aligned} d_i &= \frac{1}{T} \int_{-\infty}^{+\infty} |h(t - \varepsilon_0 T, \Delta\Omega)|^2 \; e^{-j2\pi it/T} \, dt \\ &= \frac{1}{T} \, e^{-j\frac{2\pi}{T}i\varepsilon_0 T} \int_{-\infty}^{+\infty} |h(t, \Delta\Omega)|^2 \; e^{-j2\pi it/T} \, dt \end{aligned} \tag{8-23}$$

From the definition of $h(t, \Delta\Omega)$ the spectrum of $h(t, \Delta\Omega)$ is found to be

$$\begin{aligned} &H(\omega, \Delta\Omega) = G(\omega) \; G^*(w - \Delta\Omega) \quad G(\omega) \text{ is band} - \text{limited to } B \quad [\text{Hz}] \\ &B = \frac{1}{2T}(1 + \alpha) \end{aligned} \tag{8-24}$$

(α excess bandwidth). Since squaring the signal $h(t, \Delta\Omega)$ doubles the bandwidth, the spectrum of $|h(t, \Delta\Omega)|^2$ is limited to twice this value

$$B_{|h|^2} = \frac{1}{T}(1 + \alpha) \tag{8-25}$$

From this follows that only the coefficients d_{-1}, d_0, d_1 are nonzero in the Fourier series. Hence, the expected value of $E\left[|z(lT_s, \Delta\Omega)|^2\right]$ can be written in the form

$$E\left[|z(lT_s, \Omega)|^2\right] = d_0 + 2 \, \mathrm{Re}\left[d_1 e^{j2\pi l/M_s}\right] \tag{8-26}$$

where the coefficients d_0, d_1 are defined by (8-23), and $M_s = T/T_s$ is an integer.

We have now everything ready to prove that the estimate of $\hat{\Omega}$ of (8-12) is indeed unbiased. First, let us take expected value of the sum (8-14)

$$\sum_{l=-LM_s}^{LM_s - 1} E\left[|z(lT_s, \Omega)|^2\right] \tag{8-27}$$

Next, replace $E\left[|z(lT_s, \Delta\Omega)|^2\right]$ in the previous equation by the right-hand side of (8-26)

$$\sum_{l=-LM_s}^{LM_s-1} E\left[|z(lT_s,\Omega)|^2\right] = \sum_{l=-LM_s}^{LM_s-1} \left\{d_0 + 2\,\mathrm{Re}\left[d_1\, e^{j2\pi l/M_s}\right]\right\} = (2LM_s)\, d_0 \tag{8-28}$$

since

$$\sum_{l=-LM_s}^{LM_s-1} e^{j2\pi l/M_s} = 0 \tag{8-29}$$

The coefficient d_0 is a function of Ω but *independent* of ε

$$d_0 = \frac{1}{T}\int_{-\infty}^{+\infty} |h(t,\Delta\Omega)|^2 dt \tag{8-30}$$

Using the Parseval formula, the integral can be expressed as

$$\int_{-\infty}^{+\infty} |h(t,\Delta\Omega)|^2 dt = \frac{1}{2\pi}\int_{-\infty}^{+\infty} |H(\omega,\Delta\Omega)|^2 d\omega = \frac{1}{2\pi}\int_{-\infty}^{+\infty} |G(\omega-\Delta\Omega)|^2\, |G(\omega)|^2 d\omega \tag{8-31}$$

The last equality results from replacing $H(\omega,\Delta\Omega)$ by the right-hand side of (8-24). From the Schwarz inequality it follows that d_0 (8-31) is maximum for $\Delta\Omega = \Omega_0 - \Omega = 0$. Hence

$$\hat{\Omega} = \arg\max_{\Omega} \sum_{l=-LM_s}^{LM_s-1} |z(lT_s,\Omega)|^2 \tag{8-32}$$

is an unbiased estimate (Figure 8-4).

Regarding the manageable frequency uncertainty Ω_0 we refer to Section 8.1.1 and the relation (8-7) and obtain

$$\frac{|\Omega_{\max}|}{2\pi} < \frac{1}{2T_s} - B \tag{8-33}$$

Note that, employing this structure, the magnitude of the frequency offset we can cope with is only limited by the analog prefilter shape and the sampling rate.

We noticed [see eq. (8-30)] that the joint estimation of (ε,Ω) can be *decoupled*. In a first stage Ω is obtained by maximizing the sum (8-14) with respect to

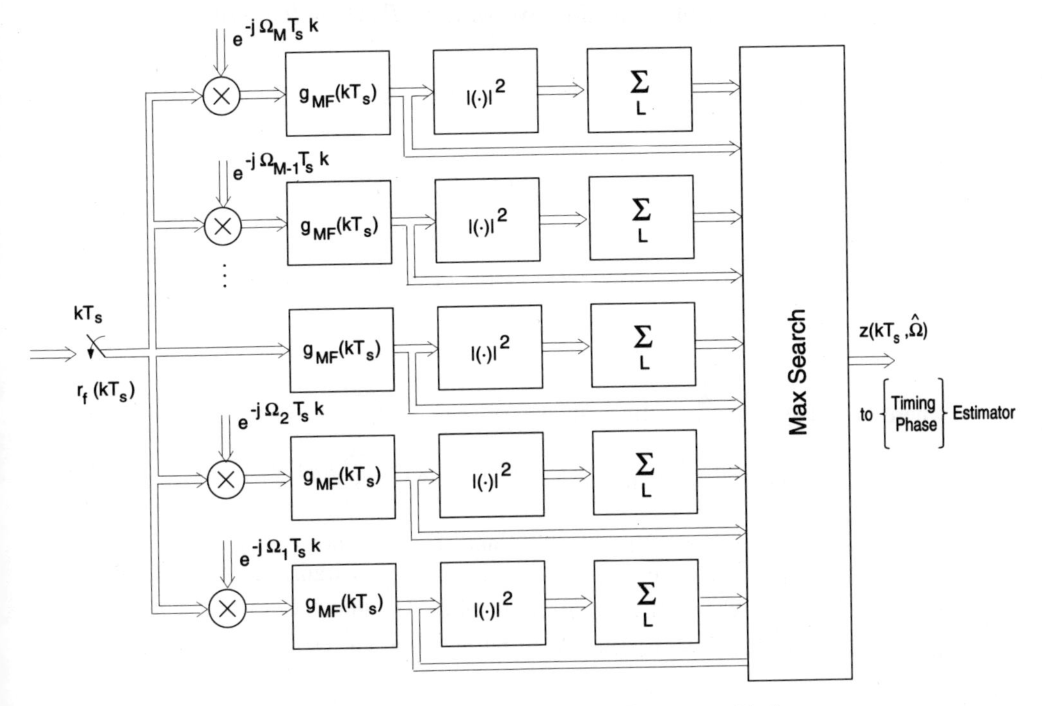

Figure 8-4 Maximum-Searching NDA Frequency Estimator

Ω. The timing parameter ε can subsequently be found as

$$\hat{\varepsilon} = \arg \sum_{l=-LM_s}^{LM_s-1} \left| z\left(lT_s, \hat{\Omega}\right) \right|^2 e^{-j 2\pi l / M_s} \qquad \text{(8-34)}$$

$$\left(\hat{\Omega} - \text{directed estimator}\right)$$

Remark: In order for $\hat{\varepsilon}$ to be unbiased, $g(t)$ must be real and symmetric [compare with eq. (5–77)]. No conditions on the pulse shape are imposed for an unbiased $\hat{\Omega}$. But keep in mind that our derivation requires that the transmission pulse $g(t)$ be known at the receiver.

8.2.1 Frequency Estimation via Spectrum Analysis

Consider the estimation rule of (8-14):

$$\hat{\Omega} = \arg \max_{\Omega} \sum_{l=-L\,M_s}^{l=L\,M_s} |z(lT_s, \Omega)|^2 \qquad \text{(8-35)}$$

introduced in the previous section. A mathematically equivalent form leads to a different realization of the estimator. We write $z(lT_s, \Omega)$ in the form

$$\begin{aligned} z(lT_s, \Omega) &= \sum_{k=-\infty}^{\infty} r_f(kT_s)\, e^{-j\Omega kT_s} g_{\mathrm{MF}}(lT_s - kT_s) \\ &= e^{-j\Omega lT_s} \sum_{k=-\infty}^{\infty} r_f(kT_s) g_{\mathrm{MF}}(lT_s - kT_s)\; e^{-j\Omega(k-l)T_s} \\ &= e^{-j\Omega lT_s} \sum_{k=-\infty}^{\infty} r_f(kT_s) g_{\Omega}(lT_s - kT_s) \\ &= e^{-j\Omega lT_s} z_{\Omega}(lT_s) \end{aligned} \tag{8-36}$$

where $g_\Omega(kT_s)$ is the impulse response of

$$g_\Omega(kT_s) = g_{\mathrm{MF}}(kT_s)\, e^{j\Omega kT_s} \;\circ\!\!-\!\!\bullet\; G^*\left(e^{j(\omega-\Omega)T_s}\right) \tag{8-37}$$

A block diagram corresponding to (8-35) and (8-36) is shown in Figure 8-5. The samples $z_{\Omega_j}(kT_s)$ are obtained as output of a frequency translated matched filter, $G^*\left(e^{j(\omega-\Omega_j)kT_s}\right)$, where Ω_j is the jth trial value. The number of parallel branches in the matched filter bank is determined by the required resolution $\Delta\Omega = \frac{|\Omega_{\max}|}{N}$.

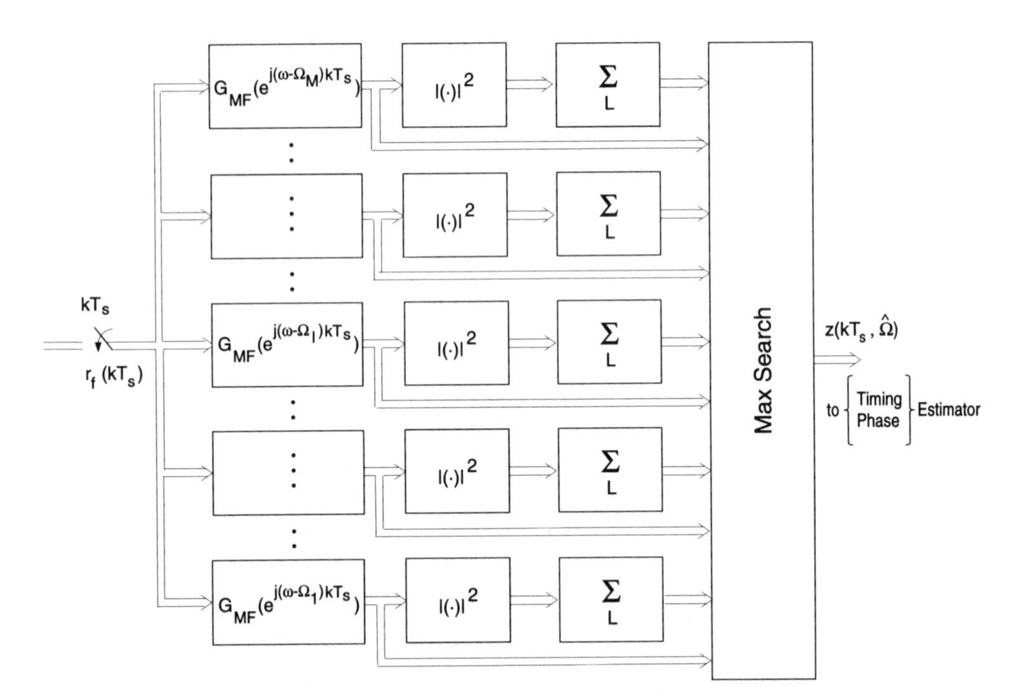

Figure 8-5 Block Diagram of the Estimator

Since

$$
\begin{aligned}
|z(lT_s, \Omega)| &= \left|e^{j\Omega l T_s}\right| |z_\Omega(lT_s)| \\
&= |z_\Omega(lT_s)|
\end{aligned}
\tag{8-38}
$$

the multiplication of the output $z_{\Omega_j}(lT_s)$ by a complex exponent $e^{-j\Omega l T_s}$ can be avoided. The squared absolute value $\left|z_{\Omega_j}(kT_s)\right|^2$ is averaged and the maximum of all branches is determined.

The above estimator structure can be interpreted as a device that analyzes the power spectral density of a segment $t \in \{-T_E/2, T_E/2\}$ of the received signal $r_f(kT_s)$. We assume that the estimation interval is so large that

$$
\sum_l \left|z_{\Omega_j}(lT_s)\right|^2 \simeq \frac{1}{T_s} \int_{-T_E/2}^{T_E/2} \left|z_{\Omega_j}(t)\right|^2 dt
\tag{8-39}
$$

Then the Parseval theorem (approximately) applies

$$
\begin{aligned}
X_j &= \frac{1}{T_s} \int_{-T_E/2}^{T_E/2} \left|z_{\Omega_j}(t)\right|^2 dt \\
&= \frac{1}{2\pi} \int_{-\infty}^{\infty} |R_{T_E}(\omega)|^2 \; |G(\omega - \Omega_j)|^2 \, d\omega
\end{aligned}
\tag{8-40}
$$

where $R_{T_E}(\omega)$ is the spectrum of the signal segment

$$
R_{T_E}(\omega) = \int_{-T_E/2}^{T_E/2} r_f(t) \, e^{-j\omega t} \, dt
\tag{8-41}
$$

and $G(\omega - \Omega_j)$ is the baseband part of the frequency response of the jth filter. The expression for X_j equals the energy of the signal segment $r_{T_E}(t)$ passed through the frequency translated matched filter $G(\omega - \Omega_j)$. The energy is maximum (on the average) for that value of Ω_j for which the signal spectrum best fits the passband of $G(\omega - \Omega_j)$.

Since Figure 8-5 is merely a different implementation of the algorithm described by Figure 8-4 the maximum manageable frequency uncertainty Ω_0 is again given by

$$
\frac{|\Omega_{\text{max}}|}{2\pi} < \frac{1}{2T_s} - B
\tag{8-42}
$$

compare with (8-33).

In a digital realization X_j is computed using the FFT (Fast Fourier Transform) algorithm. The number N_{FFT} of samples $r(kT_s)$ is determined by the specific

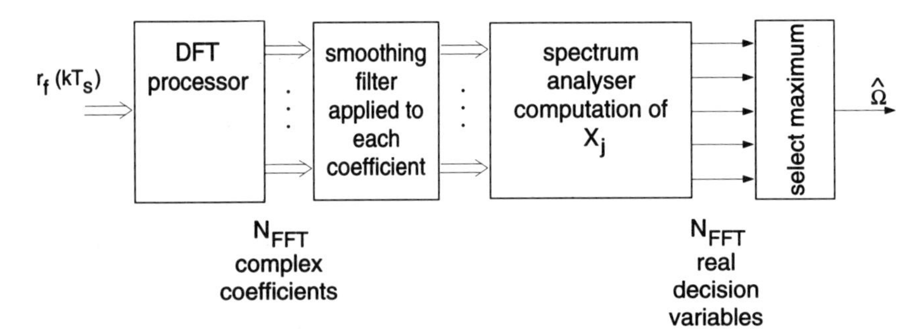

Figure 8-6 Block Diagram of the DFT Analyzer, X_j [compare (8-40) and (8-44)]

frequency resolution

$$\Delta f = \frac{1}{T_s \, N_{\text{FFT}}} \tag{8-43}$$

To obtain a reliable estimator the result of several DFT (Discrete Fourier Transform) spectra must be averaged. The resulting structure is shown in Figure 8-6.

The operation required to generate a frequency estimate is illustrated in Figure 8-6.

(i) N_{av} spectra are generated by the DFT using N_{av} nonoverlapping sequences $\{r_f(kT_s)\}$ of length N_{FFT}.
(ii) These N_{av} spectra are accumulated in the next operation unit.
(iii) In the spectrum analyzer an estimate of Ω_0 is generated via

$$\hat{\Omega}_j = \arg \max_j \sum_{m=-(N_{\text{FFT}}/2-1)}^{m=N_{\text{FFT}}/2} \left| \hat{R}_{T_E}(m \, \Delta f) \right|^2 |G(m \, \Delta f - (j \, \Delta f))|^2 \tag{8-44}$$

It can be easily verified that the implementation of Figure 8-6 is much simpler than that of Figure 8-5.

Provided an unlimited resolution of the spectrum analyzer the above structure provides an unbiased estimate; this is proven using the same reasoning as for the structure of Figure 8-4. Nevertheless, the performance of the above structure should not be measured in terms of an estimate variance or estimation bias alone. A performance measure for a maximum seeking algorithm is the probability that an estimation error exceeds a given threshold.

Table 8-1 shows some simulation results. Parameters are N_{FFT}, N_{av}, the ratio T/T_s, and the SNR defined as E_s/N_0. The numbers in the right-hand columns denote how often the estimation error $\Delta\Omega$ exceeds a given threshold as a function of the SNR. The results obtained for SNR=200 dB indicate that the above algorithm suffers from self-noise phenomena, too. The self-noise influence can be lowered by increasing the number of the FFT spectra to be averaged. This, in turn, means that the estimation length and the acquisition length, respectively, are prolonged.

Table 8-1 Number of Trials where $\left|\frac{\hat{\Omega}T-\Omega_0 T}{2\pi}\right| > 12.5\%$. The total number of trials is 10,000 for each entry.

			SNR			
N_{av}	N_{FFT}	T/T_s	4 dB	7 dB	10 dB	200 dB
8	16	4	1349	457	168	22
16	16	4	305	47	9	1
32	16	4	27	0	0	0
64	16	4	0	0	0	0

8.2.2 Frequency Estimation via Phase Increment Estimation

Up to now we were concerned with maximum-seeking frequency estimators. The basic difference of the present estimator compared to the maximum-seeking algorithms is that we *directly* estimate the parameter Ω as the argument of a rotating phasor $\exp(j\Omega t)$. We discuss a non-data-aided, non-timing-directed feedforward (NDA-NDε-FF) approach.

We start from the basic formulation of the likelihood function introduced in Chapter 4:

$$L(\Omega,\theta,\varepsilon,\mathbf{a}) = \sum_{k=-\infty}^{\infty} 2\,\mathrm{Re}\{r_f(kT_s)\, s_f^*(kT_s)\} \tag{8-45}$$

where $s_f(kT_s)$ is given by

$$s_f(k\,T_s)= e^{j(\Omega k T_s+\theta)} \sum_n a_n\; g(k\,T_s-nT-\varepsilon T) \tag{8-46}$$

and $r_f(kT_s)$ is given by

$$r_f(k\,T_s) = s_f(kT_s) + n(kT_s) \tag{8-47}$$

Neglect for a moment the influence of the modulation in (8-46). We recognize that the phase, $\theta(kT_s) = \theta + \Omega k T_s$, is increased by an amount of $\Delta\theta(kT_s) = \Omega T_s$ between two sampling instants. Planar filtering of a phasor $\exp(j\,\Delta\,\theta(kT_s))$ yields an unbiased estimate of Ω, as will be proved below.

Now let us conjecture the following approximation:

$$s_f(kT_s) \approx e^{j\Omega T_s}\; s_f((k-1)T_s) \tag{8-48}$$

Obviously, the larger the ratio T/T_s is, the better this approximation becomes.

Also, for high SNR we have

$$s_f(kT_s) \approx r_f(kT_s) \tag{8-49}$$

Inserting (8-48) and (8-49) into (8-45) we get

$$\begin{aligned} L_1(\Omega, \theta, \varepsilon, a) &= \sum_{k=-\infty}^{\infty} \text{Re}\{e^{-j\Omega T_s}\ r_f(kT_s)\ r_f^*((k-1)T_s)\} \\ &= \text{Re}\left\{e^{-j\Omega T_s}\ |Y|\ e^{+j\ \arg\{Y\}}\right\} \end{aligned} \tag{8-50}$$

with

$$Y = \sum_{k=-\infty}^{\infty} r_f(kT_s)\ r_f^*((k-1)T_s) \tag{8-51}$$

Since the expression Y is independent of the trial parameter Ω, the maximum of (8-50) is obtained via

$$\hat{\Omega}T_s = \arg\left\{\sum_{k=-\infty}^{\infty} r_f(kT_s)\ r_f^*((k-1)T_s)\right\} \tag{8-52}$$

Equation (8-52) defines a very simple estimation rule, where neither a search operation nor any determination of the other elements of the parameter set is required.

In a practical realization the summation is truncated to L_F symbols

$$\hat{\Omega}T = \frac{T}{T_s} \arg\left\{\sum_{\substack{L_F(T/T_s) \\ \text{samples}}} r_f(kT_s)\ r_f^*((k-1)T_s)\right\} \tag{8-53}$$

A block diagram of the estimator is shown in Figure 8-7.

In the sequel we analyze the performance of the algorithm (8-53). We start by proving that the algorithm has (for all practical purposes) a negligible bias. Rather than averaging $\hat{\Omega}T$ we take the expected value of the complex phasor $r_f(kT_s)\ r_f^*((k-1)T_s)$. If the argument of $E\left[r_f(kT_s)\ r_f^*((k-1)T_s)\right]$ can be

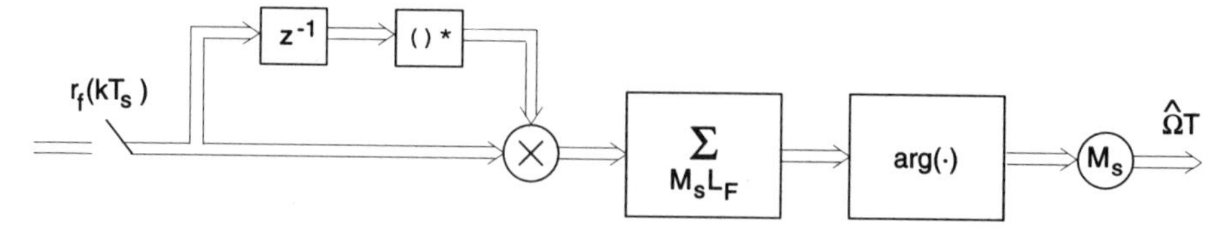

Figure 8-7 Block Diagram of the Direct Estimator Structure

shown to be equal $\Omega_0 T$, then it follows that the algorithm is unbiased. Writing $r_f(kT_s)$ as the sum of useful signal plus noise we obtain after some straightforward algebraic steps

$$\begin{aligned}
&\sum_{k=-M_s L_F}^{M_s L_F} E\left[r_f(kT_s)\, r_f^*((k-1)T_s)\right] \\
&= \sum_{k=-M_s L_F}^{M_s L_F} e^{j\Omega T_s} \left\{ \sum_{n=-N}^{N} E\left[|a_n|^2\right] g(kT_s - nT) g^*((k-1)T_s - nT) + R_n(T_s) \right\}
\end{aligned} \tag{8-54}$$

where $R_n(T_s)$ is the correlation function of the complex noise process $n(kT_s)$, $(2L_F + 1)$ is the number of symbols in the observation interval, and $2N + 1$ the number of symbols transmitted.

For a properly designed prefilter $F(\omega)$ the noise samples are uncorrelated, so $R_n(T_s) = 0$. It remains to evaluate the first term. Interchanging the order of summation we obtain

$$\sum_{n=-N}^{N} E\left[|a_n|^2\right] \sum_{k=-M_s L_F}^{M_s L_F} g(kT_s - nT) g^*((k-1)T_s - nT) \tag{8-55}$$

In all reasonable applications the number of transmitted symbols is larger than the estimation interval, i.e., $N \gg L_F$. Therefore, the correlation-type algorithm suffers from self-noise caused by truncation of the estimation interval. This effect was previously discussed in Section 5.2.2. If the number D of symbols which essentially contribute to the self-noise is much smaller than L_F (i.e., $D \ll L_F$), these contributions to the self-noise effects are small. The summation over kT_s then equals approximately

$$\begin{aligned}
\sum_k g(kT_s - nT)\, g^*((k-1)T_s - nT) &\approx \frac{1}{T_s} \int_{-\infty}^{\infty} g(t)\, g^*(t - T_s)\, dt \\
&= h_g(T_s)
\end{aligned} \tag{8-56}$$

This expression is clearly independent of the other estimation parameters $\{\theta, \varepsilon, \mathbf{a}\}$. Provided $h_g(t)$ is real, which is the case for a real pulse $g(t)$, the algorithm (8-53) yields a (nearly) unbiased estimate for a sufficiently large estimation interval. Simulation results confirmed the assertion. If $g(t)$ is known, the condition on $g(t)$ can be relaxed [compare with the Remark following eq. (8-34)] because the bias caused by a particular filter shape $g(t)$ is known a priori and therefore can be compensated.

Our next step is to compute the variance of the estimate:

$$\operatorname{var}\left[\hat{\Omega} T\right] = \left(\frac{T}{T_s}\right)^2 \frac{\operatorname{var}\left[\operatorname{Im}\left\{ \sum\limits_{L_F(T/T_s)} r_f(kT_s)\, r_f^*((k-1)T_s) \right\}\right]}{\left(E\left[\operatorname{Re}\left\{ \sum\limits_{L_F(T/T_s)} r_f(kT_s)\, r_f^*((k-1)T_s) \right\}\right]\right)^2} \tag{8-57}$$

It proves advantageous to decompose the variance into

$$\text{var}\left[\hat{\Omega}T\right] = \sigma^2_{S\times S} + \sigma^2_{S\times N} + \sigma^2_{N\times N} \tag{8-58}$$

which reflects the fact that the frequency jitter is caused by (signal × signal), (signal × noise), and (noise × noise) interactions. Using similar approximations as outlined in Section 8.3.2, the expectations of (8-57) can be calculated. For BPSK and QPSK modulation we get the expressions listed below.

Example

1. (signal × signal) nonzero only for QPSK, $\sigma_{S\times S} = 0$ for BPSK:

$$\sigma_{S\times S} = \frac{(T/T_s)^2}{2(2\pi)^2\, T_s^2\, h_g^2(T_s)}\,\frac{1}{L_F}\,\left(\left(1-\frac{\alpha}{2}\right) - (I_1 + 2\,I_2)\right) \tag{8-59}$$

2. (signal × noise) for QPSK and BPSK:

$$\sigma_{S\times N} = \frac{(T/T_s)^2}{(2\pi)^2\, T_s^2\, h_g^2(T_s)}\,\frac{1}{L_F}\,(1 - T_s\, h_g(2T_s))\,\frac{1}{E_s/N_0} \tag{8-60}$$

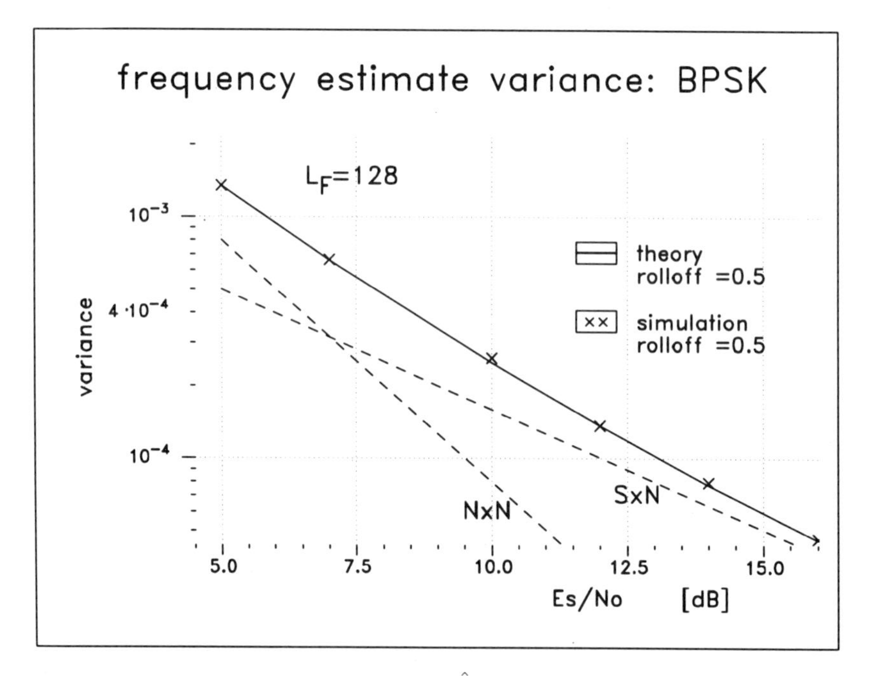

Figure 8-8 Estimation Variance of $\hat{\Omega}T/2\pi$ for BPSK Transmission, $T/T_s = 4$, Estimation Length L_F;
dotted lines: S×N, N×N contribution for $L_F = 128$ and $\alpha = 0.5$

3. (noise × noise) for QPSK and BPSK:

$$\sigma_{N\times N} = \frac{(T/T_s)^3}{2\,(2\pi)^2\, T_s^2\, h_g^2(T_s)}\,\frac{1}{L_F}\,\left(\frac{1}{E_s/N_0}\right)^2 \tag{8-61}$$

where I_1 is given by

$$I_1 = \frac{T_s^2}{\pi}\int_0^{(1+\alpha)/2T} H_g(\omega)\; H_g(\omega)\cos\left(2\omega T_s\right)\; d\omega \tag{8-62}$$

and I_2 is given by

$$I_2 = \frac{T_s^2}{2\pi}\int_0^{(1+\alpha)/2T} H_g(\omega)\; H_g(\omega + 2\pi/T)\sin\left(2\omega T_s\right)\; d\omega \tag{8-63}$$

where $H_g(\omega)$ is the Fourier transform of $h_g(t)$. Figures 8-8 and 8-9 show some simulated variances for various estimation lengths and various rolloff factors. The transmission pulse was a root-raised cosine filter whose energy was normalized to

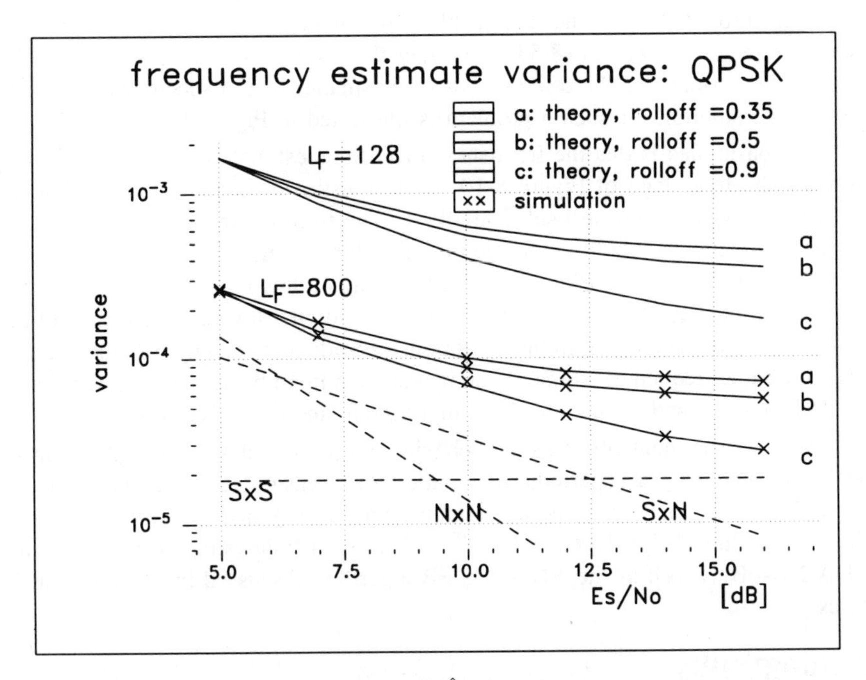

Figure 8-9 Estimation Variance of $\hat{\Omega}T/2\pi$ for QPSK Transmission, $T/T_s = 4$, Estimation Length L_F;
dotted lines: S×S, S×N, N×N contribution for $L_F = 800$ and $\alpha = 0.9$

unity. The good agreement between analytically obtained curves and the simulation proves the approximations to be valid.

Further simulations not reported here confirm that the estimate properties are independent of the parameter set $\{\theta_0, \varepsilon_0\}$, as was predicted from our theoretical considerations, and that the estimator can be applied to higher-order modulation schemes such as 8-PSK and M-QAM.

The maximal frequency offset $|\Omega_{\max}|$ the estimator can cope with is upper bounded by

$$\frac{|\Omega_{\max}|}{2\pi} < \frac{1}{2\,T_s} \tag{8-64}$$

Strictly speaking, the maximum frequency offset $|\Omega_{\max}|$ is the minimum of $|\Omega_{\max}|$ given by (8-64) and $|\Omega_{\max}|$ fulfilling (8-7). For example, if the analog prefilter has a frequency range B_{f_1} where the signal passes undistorted, the maximal permissible frequency offset is limited to

$$\frac{|\Omega_{\max}|}{2\pi} < B_{f_1} - \frac{1}{2T}(1+\alpha) \tag{8-65}$$

in the case of a root-raised cosine transmission pulse with rolloff α. If a larger offset must be handled, the analog prefilter bandwidth as well as the sampling rate must be increased. This has the effect that the variance is increased for two reasons: First, because the variance of the noise contribution in the sampled signal $r_f(kT_s)$ is directly proportional to the bandwidth of the analog prefilter; second, because of the prefactor (T/T_s) in (8-57). Possibly the estimation interval becomes too large if the variance of the estimate has to be smaller than a specified value. An appropriate solution for such a problem is illustrated in Figure 8-10.

The basic idea is that the frequency offset to be estimated is lowered step by step. In the first stage of Figure 8-10 a coarse estimate is generated. Although this coarse estimate may still substantially deviate from the true value, the quality of this first estimate should guarantee that no information is lost if the frequency-adjusted samples $r_f(kT_s)\,e^{-j\hat{\Omega}kT_s}$ are fed to a digital lowpass filter. Note that the decimation rate T_{s_1}/T_{s_2} and the bandwidth of the digital lowpass filter have to be properly selected according to the conditions on sufficient statistics. In the second stage the same frequency estimator can operate in a more benign environment than in the first stage with respect to the sampling rate and the noise power.

The last comment concerns the behavior of the estimator if, instead of statistically independent data, periodic data pattern are transmitted. Simulation results show that the estimation properties remain intact in the case of an unmodulated signal or a dotted signal $[a_m = (-1)^m]$. This is an interesting result since the NDA-ND-FB as well as the NDεA-Dε-FB algorithm discussed in [1] fail in such cases.

Bibliography

[1] F. M. Gardner, "Frequency Detectors for Digital Demodulators Via Maximum Likelihood Derivation," *ESA-ESTEC Final Report: Part 2, ESTEC Contract No 8022-88-NL-DG*, March 1990.

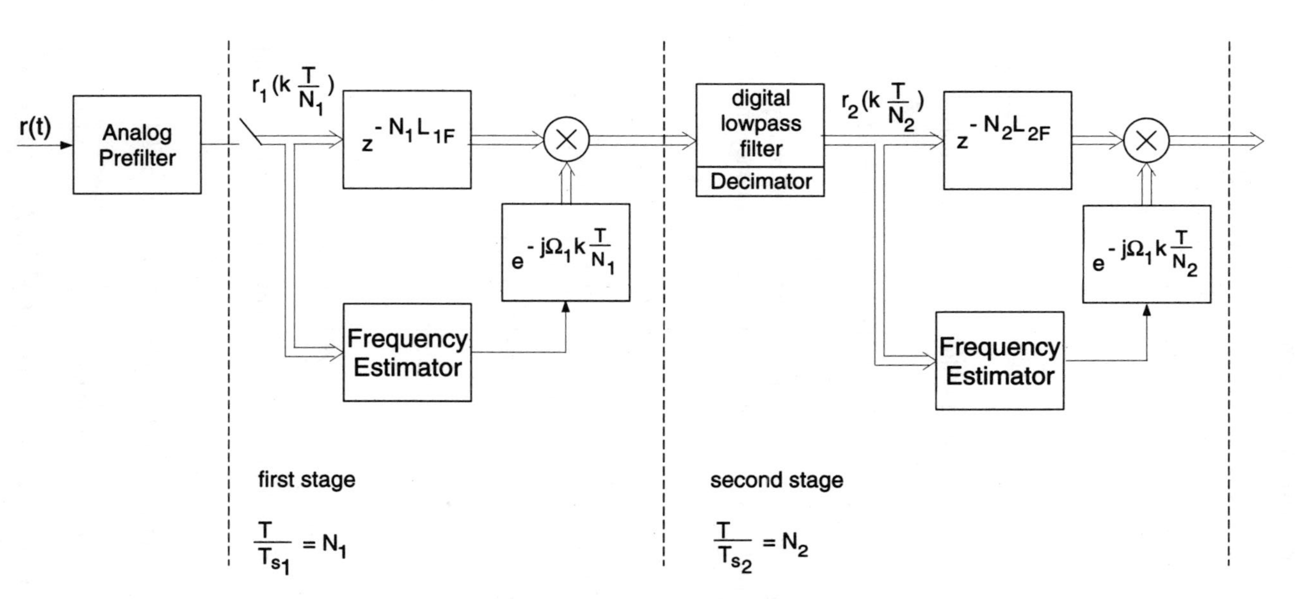

Figure 8-10 Block Diagram of a Two-Stage Frequency Estimator

8.3 Frequency Error Feedback Systems Operating Independently of Timing Information

To obtain a frequency error signal we follow the familiar pattern. We differentiate the log-likelihood function of Section 8.2 [eq. 8-13] with respect to Ω to obtain (neglecting an irrelevant factor of 2):

$$x(kT_s) = \mathrm{Re}\left[z(kT_s,\Omega)\,\frac{\partial}{\partial\Omega} z^*(kT_s,\Omega)\right]\Big|_{\Omega=\hat{\Omega}_{k|k-1}} \tag{8-66}$$

Recall that we have shown that (8-4) is independent of ε. Therefore, $x(kT_s)$ is independent of ε also.

The derivative of the matched filter (MF) output equals

$$\frac{\partial}{\partial\Omega} z(kT_s,\Omega) = \sum_{l=-\infty}^{\infty} (-jlT_s) r_f(lT_s) e^{-j\Omega l T_s}\, g_{\mathrm{MF}}(kT_s - lT_s) \tag{8-67}$$

The time-weighting of the received signal can be avoided by applying the weighting to the matched filter. We define a time-invariant frequency matched filter (FMF) with impulse response:

$$g_{\mathrm{FMF}}(kT_s) = g_{\mathrm{MF}}(kT_s) jkT_s \tag{8-68}$$

The output of this filter with $r_f(kT_s)$ as input is

$$\begin{aligned} z_{\mathrm{FMF}}(kT_s,\Omega) &= \sum_{l=-\infty}^{\infty} r_f(lT_s)\, e^{-j\Omega T_s l}\, g_{\mathrm{FMF}}(kT_s - lT_s) \\ &= \sum_{l=-\infty}^{\infty} r_f(lT_s)\, e^{-j\Omega T_s l}\, g_{\mathrm{MF}}(kT_s - lT_s)(-j)(lT_s - kT_s) \\ &= \frac{\partial}{\partial\Omega} z(kT_s,\Omega) + jkT_s\, z(kT_s,\Omega) \end{aligned} \tag{8-69}$$

The last equality follows from the definition of $\partial/\partial\Omega(z(kT_s,\Omega))$ and of $z(kT_s,\Omega)$.

Replacing

$$\frac{\partial}{\partial\Omega} z(kT_s,\Omega) = z_{\mathrm{FMF}}(kT_s,\Omega) - jkT_s\, z(kT_s,\Omega) \tag{8-70}$$

in (8-66), we obtain the error signal

$$\begin{aligned} x(kT_s) &= \mathrm{Re}\{z(kT_s,\Omega)[z^*_{\mathrm{FMF}}(kT_s,\Omega) + jkT_s z^*(kT_s,\Omega)]\} \\ &= \mathrm{Re}\Big\{z(kT_s,\Omega)\, z^*_{\mathrm{FMF}}(kT_s,\Omega) + jkT_s|z(kT_s,\Omega)|^2\Big\} \\ &= \mathrm{Re}\{z(kT_s,\Omega)\, z^*_{\mathrm{FMF}}(kT_s,\Omega)\} \end{aligned} \tag{8-71}$$

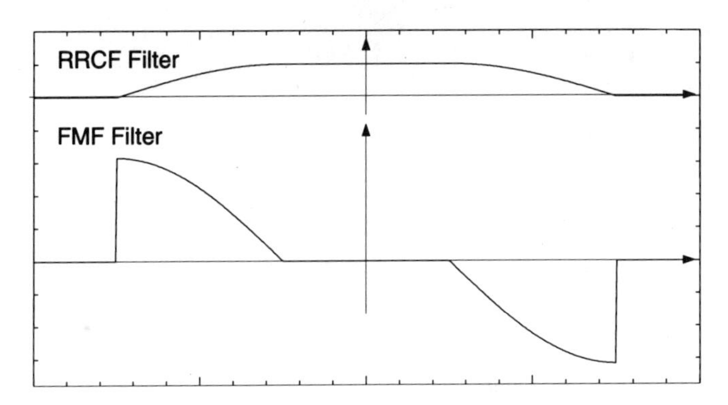

Figure 8-11 Frequency Response of the Digital Frequency Matched Filter $G_{\text{FMF}}\left(e^{j\omega T_s}\right)$ for a Root-Raised Cosine Pulse $G(\omega)$, $\alpha = 0.5$

The descriptive term "frequency matched filter" was introduced by [1] and is readily apprcciated when we consider the frequency response of the filter:

$$\begin{aligned} G_{\text{FMF}}\left(e^{j\omega T_s}\right) &= \sum_{k=-\infty}^{\infty} (jkT_s)\, g_{\text{MF}}(kT_s) e^{-j\omega T_s k} \\ &= \frac{\partial}{\partial \omega} G_{\text{MF}}\left(e^{j\omega T_s}\right) \end{aligned} \tag{8-72}$$

Thus, the frequency matched filter $g_{\text{FMF}}(kT_s)$ is the derivative of the frequency response of the signal matched filter. A dual situation was found for the timing matched filter which is the time derivative of the signal matched filter in the time domain and weighted by $(j\omega)$ in the frequency domain.

Example

The frequency response of the frequency matched filter $G_{\text{FMF}}\left(e^{j\omega T_s}\right)$ is shown in Figure 8-11 for a root-raised cosine signal pulse.

The operation of the frequency matched filter is best explained if we write the error signal in the mathematically equivalent form of a power difference.

$$\begin{aligned} &4\,\text{Re}\{z(kT_s,\Omega)\; z^*_{\text{FMF}}(kT_s,\Omega)\} \\ &\quad = \left|z(kT_s,\Omega) + z_{\text{FMF}}(kT_s,\Omega)\right|^2 - \left|z(kT_s,\Omega) - z_{\text{FMF}}(kT_s,\Omega)\right|^2 \\ &\quad = \quad \left|r_f(kT_s)e^{-j\Omega T_s k} \otimes \left[g_{\text{MF}}(kT_s) + g_{\text{FMF}}(kT_s)\right]\right|^2 \\ &\quad\quad - \left|r_f(kT_s)e^{-j\Omega T_s k} \otimes \left[g_{\text{MF}}(kT_s) - g_{\text{FMF}}(kT_s)\right]\right|^2 \end{aligned} \tag{8-73}$$

The first term in the difference equals the signal power at the output of the filter with impulse response $g_{\text{MF}}(kT_s) + g_{\text{FMF}}(kT_s)$ and input $r_f(kT_s)\; e^{-j\Omega T_s k}$. The second term equals the output power of a filter $g_{\text{MF}}(kT_s) - g_{\text{FMF}}(kT_s)$ when the same signal is applied.

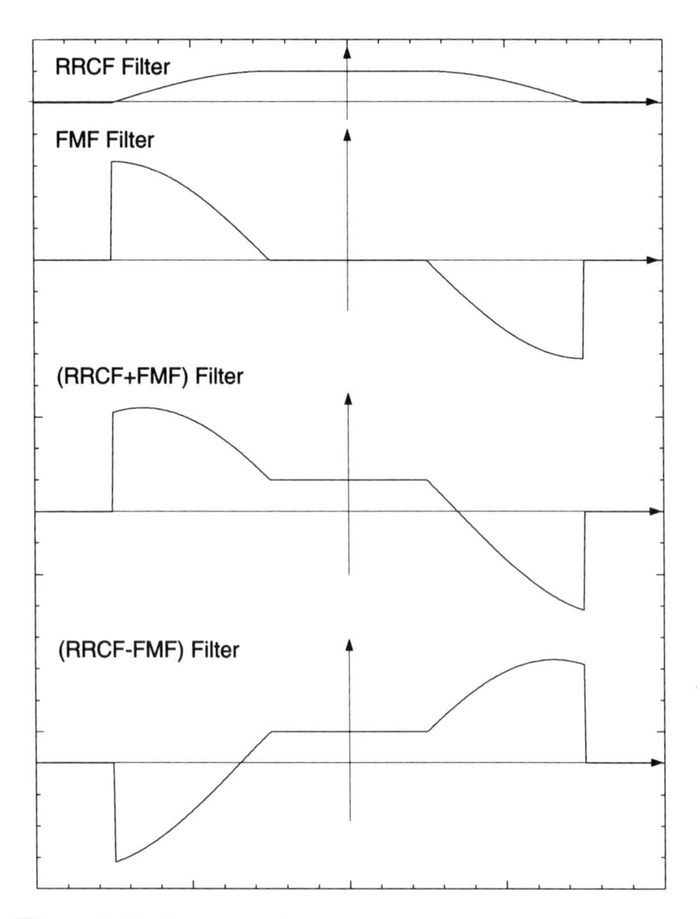

Figure 8-12 Frequency Response of the Filters Performing the Power Difference of (8-73), $\alpha = 0.5$
(RRCF stands for root-raised cosine filter, and FMF for the corresponding frequency matched filter)

For the sake of a concise explanation we consider a root-raised cosine pulse. The two filters in (8-73) differ in the rolloff frequency range only, see Figures 8-11 and 8-12. The power difference remains unchanged if we replace the two filters by filters with a frequency response which is nonzero only in these regions (compare Figure 8-13). Furthermore, the two filters may be simplified appropriately. The simplified filter $H_P\left(e^{j\omega T_s}\right)$ passes those frequencies which lie in the positive rolloff region, while the filter $H_N\left(e^{j\omega T_s}\right)$ passes the negative frequencies.

Let us see now what happens when a signal $r_f(t)$ shifted by Ω_0 is applied to these filters. Due to the symmetry of the filter frequency response and the signal spectrum, the output of the two filters is identical for $\Omega_0 = 0$.

For all other values of $|\Omega_0| < 2B$ an error signal is generated. The maximum frequency uncertainty $|\Omega_{\max}|$ the algorithm can cope with is thus

$$|\Omega_{\max}| = \frac{2\pi}{T}(1+\alpha) \tag{8-74}$$

with rolloff factor α. Strictly speaking, the maximum frequency uncertainty is the minimum of $|\Omega_{\max}|$ given by (8-74) and

$$\frac{|\Omega_{max}|}{2\pi} < \frac{1}{2T_s} - B \tag{8-75}$$

[compare (8-7)].

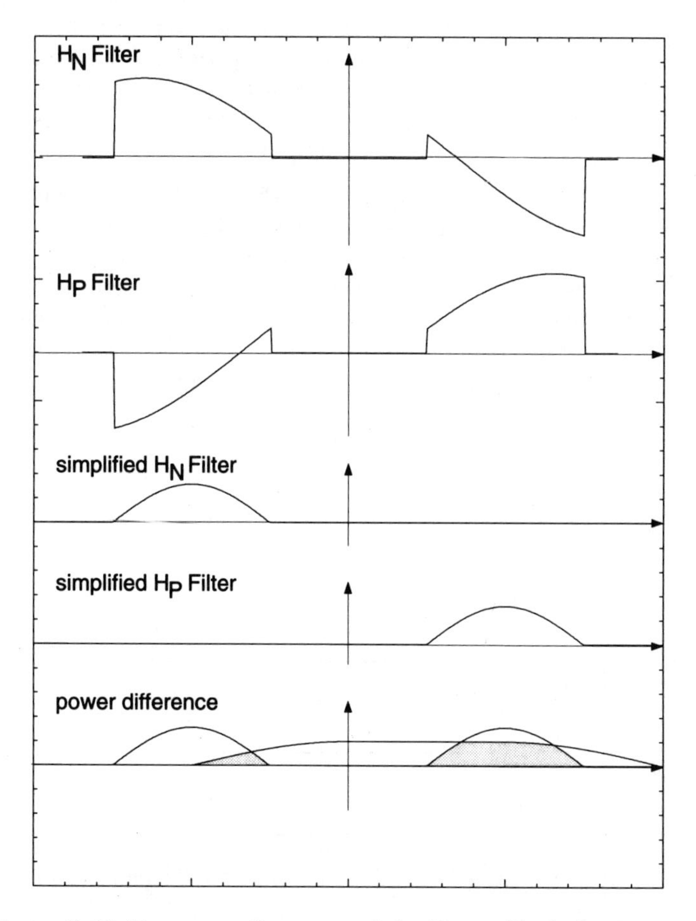

Figure 8-13 Frequency Response of the Power Equivalent Filters $H_N\left(e^{j\omega T_s}\right)$ and $H_P\left(e^{j\omega T_s}\right)$.
The frequency error signal equals the difference of the shaded areas.

An S-curve expression is obtained by taking the expected value of the error signal with respect to the random data and the noise,

$$E[x(kT_s)] = E[\mathrm{Re}\{z(kT_s, \Omega)\, z^*_{\mathrm{FMF}}(kT_s, \Omega)\}] \tag{8-76}$$

For the case that all filters are tailored to a root-raised cosine transmission pulse it can be shown from (8-76) and (8-73) that the error signal has a piecewise defined structure. The location of the piece boundaries, and even the number of pieces depends upon the excess bandwidth factor α which is obvious from the above interpretation of the frequency error detector in the frequency domain. The formulas in Table 8-2 are valid for $\alpha \leq 0.5$.

The expressions in Table 8-2 and the simulation results of Figure 8-14 present the error signal for the case that the transmission pulse energy is normalized to unity and all normalization of the MF and FMF was performed in such a way that the fully synchronized output of the MF holds $z(n) = a_n + N(n)$, with a_n the transmitted symbol and $N(n)$ the filtered noise with variance $\mathrm{var}[N(n)] = N_0/E_s$.

From the foregoing discussion it is clear that the filters $H_N(e^{j\omega T_s})$ and $H_P(e^{j\omega T_s})$ defined by the signal and frequency matched filter can be approximated by a set of simple filters which perform a differential power measurement. The resulting algorithms are known as *dual filter* and *mirror image filter* in the literature [1]–[3]. They were developed ad hoc. Again, it is interesting that these algorithms can be derived systematically from the maximum-likelihood principle by making suitable approximations. Conditions on the filter $H_N(e^{j\omega T_s})$ and $H_P(e^{j\omega T_s})$ to produce an unbiased estimate, and, preferably, a pattern-jitter-free error signal will be discussed later on.

Table 8-2 S Curve of the NDε FED; S Curve Is Odd Symmetric about Zero, $\Delta\Omega = \Omega_0 - \hat{\Omega} = 2\pi\Delta f$

	$E[x(kT_s)]$
$0 \leq \Delta f T < \alpha$	$\frac{1}{2}\left\{\sin^2\left(\frac{\pi\Delta f T}{2\alpha}\right) + \frac{\pi}{4}\left(1 - \frac{\Delta f T}{\alpha}\right)\sin\left(\frac{\pi\Delta f T}{\alpha}\right)\right\}$
$\alpha \leq \Delta f T < 1 - \alpha$	$\frac{1}{2}$
$1 - \alpha \leq \Delta f T < 1$	$\frac{1}{4}\left\{1 + \sin^2\left(\frac{\pi}{2\alpha}[\Delta f T - 1]\right) - \frac{\pi}{4}\left(1 + \frac{\Delta f T - 1}{\alpha}\right)\sin\left(\frac{\pi}{\alpha}[\Delta f T - 1]\right)\right\}$
$1 \leq \Delta f T < 1 + \alpha$	$\frac{1}{4}\left\{\cos^2\left(\frac{\pi}{2\alpha}[\Delta f T - 1]\right) - \frac{\pi}{4}\left(1 - \frac{\Delta f T - 1}{\alpha}\right)\sin\left(\frac{\pi}{\alpha}[\Delta f T - 1]\right)\right\}$

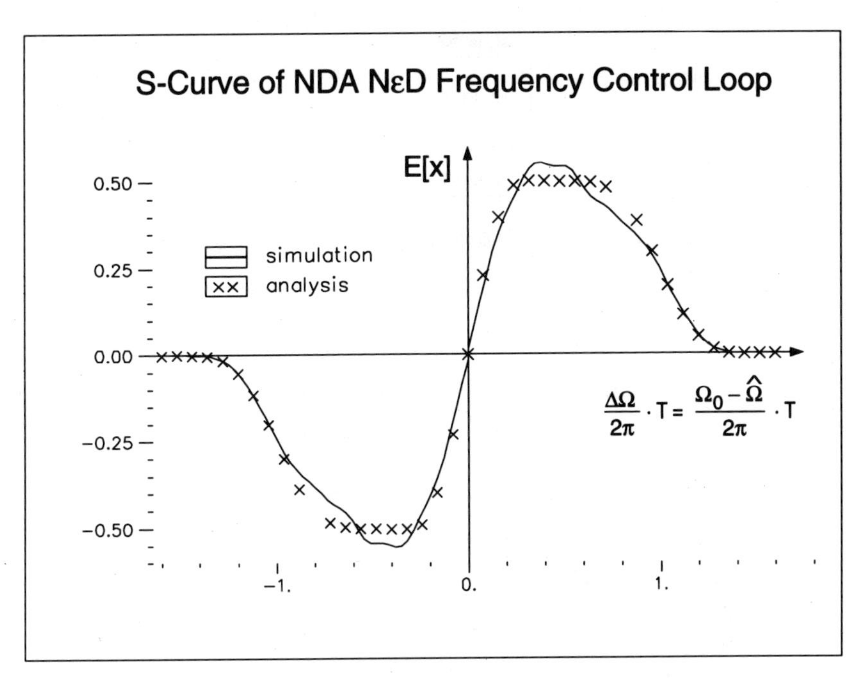

Figure 8-14 S Curve of the NDA NDε Frequency Control Loop for $\alpha = 0.5$ and $E\left[|a|^2\right] = 1$; Data of Table 8-2

8.3.1 Tracking Performance Analysis

In this section we analyze the tracking performance of the algorithm. In a first step, we discuss the self-noise phenomena. Self-noise phenomena are the reasons why estimates are disturbed although thermal noise is absent. As a result, this leads to an irreducible degradation of the estimate properties. Such an impact can only be mitigated if the algorithm itself is modified.

The algorithm of Figure 8-15 serves as an example of how to analyze self-noise effects of a tracking loop. The first step was carried out in the previous section by the determination of the S curve. We next determine the variance of the estimate. We follow the methodology outlined in Section 6.3 where we have shown that the variance can be approximated by

$$\mathrm{var}\left(\hat{\Omega}T\right) = \frac{2B_L T}{K_{\Omega}^2} \; S_x\left(e^{j2\pi fT}\right)\Big|_{f=0} \tag{8-77}$$

where K_{Ω} is the slope of the S curve at the origin, $2B_L T$ is the equivalent two-sided loop bandwidth, and $S_x\left(e^{j2\pi fT}\right)$ stands for the power spectral density of the frequency error output. The approximation in (8-77) is valid for a small loop bandwidth and a nearly flat power spectral density about $f = 0$.

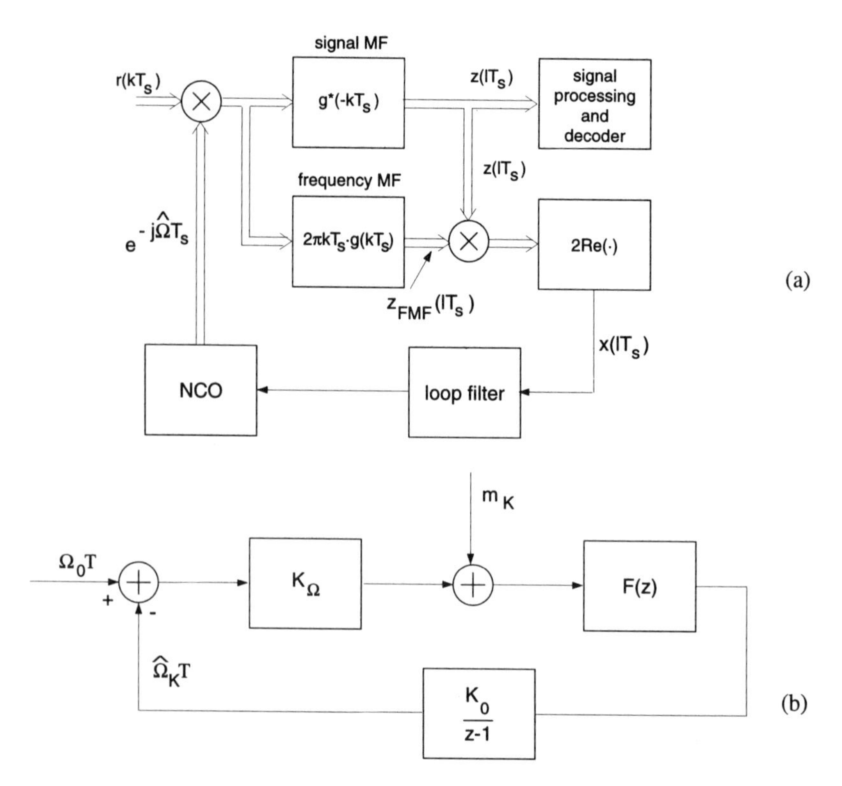

Figure 8-15 (a) Block Diagram of a NDA-NDε Frequency Control Loop, (b) Equivalent Linearized Model

In Figure 8-16 the variance of self-noise $\sigma^2_{S\times S}$ is plotted versus the normalized bandwidth $2B_LT$. We observe that the variance increases approximately linearly with $2B_LT$. This implies that the spectrum of the self-noise and thus the entire spectrum $S_x\left(e^{j2\pi fT}\right)$ is nearly flat within the loop bandwidth. This observation was used to analytically assess the self-noise contribution.

In Figure 8-17 the total variance $\sigma^2_{\hat{\Omega}}$ is plotted versus E_s/N_0. Also indicated are the analytical results of the self-noise variance $\sigma^2_{S\times S}$. (The details of its calculation are delegated to Section 8.3.2.) We observe that the algorithm suffers from strong self-noise disturbances in the moderate and high SNR region. It should therefore only be used to acquire initial frequency lock in the absence of a timing information. In the tracking mode, timing-directed algorithms as discussed in Sections 8.4 and 8.5 are preferable since they can be designed to be free of self-noise (pattern jitter).

Finally we concisely touch the question of acquisition time. As for any feedback loop, the time to acquire lock is a statistical parameter depending on the frequency uncertainty region and the SNR.

To get a rough guess about the time to acquire, the acquisition length L_{aq} can be assessed from a linearized model of the tracking loop where the frequency

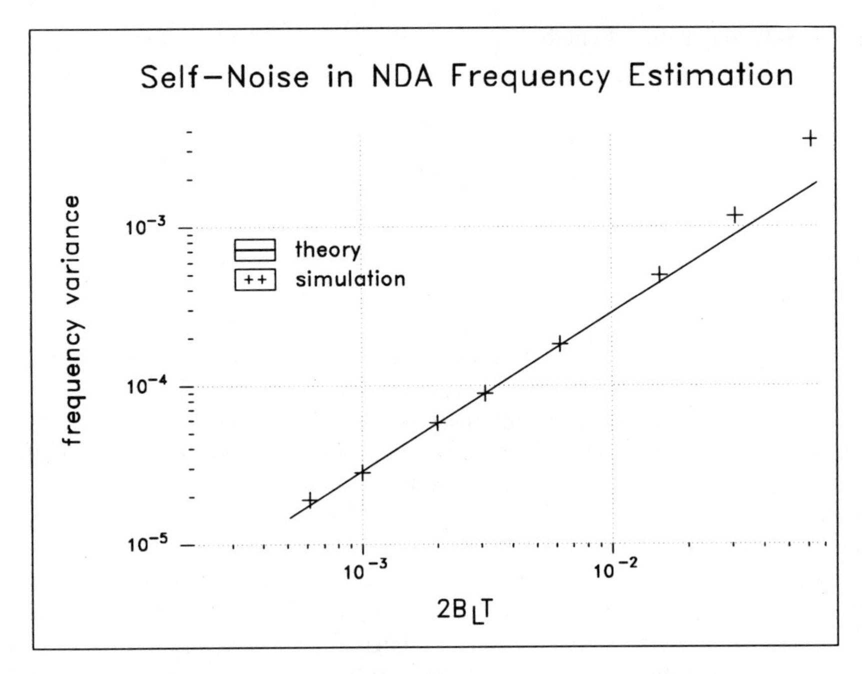

Figure 8-16 Self-Noise of the NDA NDε FB Algorithm, QPSK Transmission

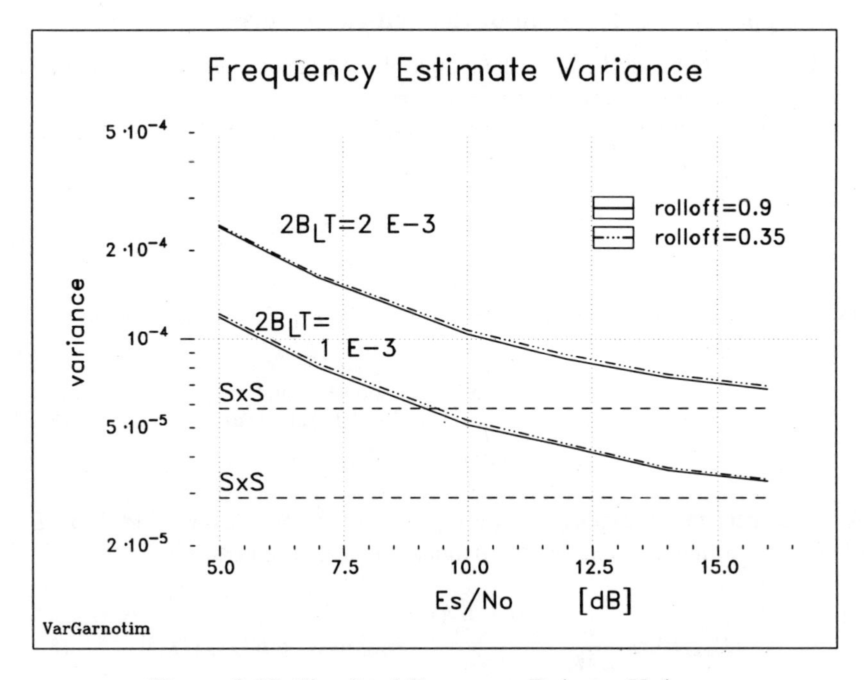

Figure 8-17 Simulated Frequency Estimate Variances

Table 8-3 Acquisition Length L_{aq}

$2B_L T$	3.13×10^{-2}	1.5×10^{-2}	6×10^{-3}	3.13×10^{-3}
L_{aq} (symbols)	150	300	750	1500

detector characteristic in Figure 8-14 is stepwise linearized. In Table 8-3 we have summarized the acquisition lengths, measured in required symbol intervals, for different loop bandwidths. The calculations and the simulation, respectively, were performed for an acquisition process in the noiseless case where the initial frequency offset was set to $\Omega_0 T/2\pi = 1$, and we defined the acquisition process to be successfully terminated if the difference between the estimated frequency offset and the true frequency offset $\Delta fT = (1/2\pi)|\Omega_0 T - \hat{\Omega}T|$ was less than 0.01. These results should give only a rough guess about the order of magnitude of the length of the acquisition process. The exact analysis of the loop acquisition time requires more sophisticated mathematical tools as mentioned in Chapter 4 of Volume 1.

Only if $|\Omega_{\text{max}}|T \ll 1$ and the time to acquire is uncritical should feedback structures be employed. For the case of a burst transmission, open-loop structures are to be used. Here for a given estimation interval the probability of exceeding a threshold can be computed. The time to acquire lock is then constant and identical to the estimation interval.

8.3.2 Appendix: Calculation of the Self-Noise Term

It can easily be verified that the output of the frequency error detector $x(lT_s)$ is a cyclostationary process, because its statistics are invariant to a shift of the time origin by multiples of the symbol interval T. Therefore, the power spectrum density at the origin is

$$S_x\left(e^{j2\pi fT}\right)\Big|_{f=0} = \sum_{k=-\infty}^{k=\infty} R_{xx}(k) \tag{8-78}$$

where the autocorrelation function $R_{xx}(k)$ is given by the time average of the correlation function

$$R_{xx}(k) = \frac{1}{T} \int_{-T/2}^{T/2} E[x(t+kT_s)x(t)]\ dt \tag{8-79}$$

To get a manageable expression, we replace in the above expression T by LT, where L is an integer, and we approximate the above integral by

$$R_{xx}(k) \simeq \frac{1}{LM_s} \sum_{l=-[(L-1)/2]M_s}^{[(L-1)/2]M_s} E[x(lT_s+kT_s)\,x(lT_s)] \tag{8-80}$$

where we assumed that L is sufficiently large and $M_s = T/T_s$. Thus we have to find

$$S_x\left(e^{j2\pi fT}\right)\Big|_{f=0} = S_x = \frac{1}{L\,M_s}\sum_{k=-\infty}^{\infty}\;\sum_{l=-[(L-1)/2]M_s}^{[(L-1)/2]M_s} E[x(lT_s + kT_s)\,x(lT_s)] \tag{8-81}$$

Changing the variables, $k' = l + k$, we get

$$\begin{aligned} S_x &= \frac{1}{L\,M_s}\sum_{k'=-\infty}^{\infty}\;\sum_{l=-[(L-1)/2]M_s}^{[(L-1)/2]M_s} E[x(k'T_s)\,x(lT_s)] \\ &= \frac{1}{L\,M_s}\sum_{k=-\infty}^{\infty}\;\sum_{l=-[(L-1)/2]M_s}^{[(L-1)/2]M_s} E[x(kT_s)\,x(lT_s)] \end{aligned} \tag{8-82}$$

where we have suppressed in the second line the apostrophe ($k' \to k$) for the sake of simplifying the notation.

For the frequency error detector signal output signal we get

$$\begin{aligned} x(lT_s) &= \mathrm{Re}\left\{\left(\sum_{m,N} a_m\;h(lT_s - mT)\right)\left(\sum_{m,N} a_m^*\;h_{\mathrm{FMF}}^*(lT_s - mT)\right)\right\} \\ &= \underbrace{\left(\sum_{m,N} a_{Q,m}\;h(lT_s - mT)\right)\left(\sum_{m,N} a_{I,m}\;jh_{\mathrm{FMF}}^*(lT_s - mT)\right)}_{A(lT_s)} \\ &\quad - \underbrace{\left(\sum_{m,N} a_{I,m}\;h(lT_s - mT)\right)\left(\sum_{m,N} a_{Q,m}\;jh_{\mathrm{FMF}}^*(lT_s - mT)\right)}_{B(lT_s)} \end{aligned} \tag{8-83}$$

where the operator $\sum_{m,N}$ is the short-hand notation for $\sum_{m=-(N-1)/2}^{m=(N-1)/2}$ and $a_m = a_{I,m} + j\;a_{Q,m}$.

Inserting (8-83) in (8-82) we have to take the expectations of four different product terms P_1, P_2, P_3, and P_4.

Considering the expression

$$P_1 = \frac{1}{LM_s}\sum_{k,\infty}\;\sum_{l,(LM_s)} E[A(lT_s)\,A(kT_s)] \tag{8-84}$$

we demonstrate exemplarily how the expressions can be numerically calculated. We start with

$$\begin{aligned}P_1 = & \frac{-1}{LM_s} \sum_{l,(LM_s)} \sum_{k,\infty} E\Bigg[\sum_{m_1,N} \sum_{m_2,N} \sum_{m_3,N} \sum_{m_4,N} \\ & \times a_{Q,m_1}\, h(lT_s - m_1 T)\, a_{I,m_2}\, h^*_{\text{FMF}}(lT_s - m_2 T) \\ & \times a_{Q,m_3}\, h(kT_s - m_3 T)\, a_{I,m_4}\, h^*_{\text{FMF}}(kT_s - m_4 T)\Bigg]\end{aligned} \tag{8-85}$$

where we introduced the variable set $\{m_1, m_2, m_3, m_4\}$ to indicate the link between a particular sum sign and its sum term. Next, we exploit the common property of uncorrelated symbols with $E[a_{I,m}\, a_{Q,n}] \equiv 0$ for all m, n and $E[a_{I,m}\, a_{I,n}] \neq 0$ only for $m = n$, and $E[a_{Q,m}\, a_{Q,n}] \neq 0$ analogously. After interchanging the order of summation we arrive at

$$\begin{aligned}P_1 = & \frac{-1}{LM_s} \sum_{m_1,N} \sum_{m_2,N} E\left[a^2_{Q,m_1}\right] E\left[a^2_{I,m_2}\right] \\ & \times \sum_{l,(LM_s)} h(lT_s - m_1 T) h^*_{\text{FMF}}(lT_s - m_2 T) \\ & \times \sum_{k,\infty} h(kT_s - m_1 T) h^*_{\text{FMF}}(kT_s - m_2 T)\end{aligned} \tag{8-86}$$

Defining the function

$$h_{\text{MF, FMF}}(nT) = T_s \sum_{l,LM_s} h(lT_s - m_1 T)\, h^*_{\text{FMF}}(lT_s - m_2 T) \tag{8-87}$$

we get

$$P_1 = \frac{-1}{LM_s} \sum_{m_1,N} \sum_{m_2,N} E\left[a^2_{Q,m_1}\right] E\left[a^2_{I,m_2}\right] \frac{1}{T_s^2}\, h^2_{\text{MF, FMF}}((m_1 - m_2)T) \tag{8-88}$$

The next step toward a closed-loop expression is to employ the *Poisson sum formula* given by

$$\sum_{n=-\infty}^{\infty} y(t + nT) = \frac{1}{T} \sum_{n=-\infty}^{\infty} Y\left(n \frac{2\pi}{T}\right) e^{jn(2\pi/T)t} \tag{8-89}$$

In doing so we have to assume that N in (8-88) is sufficiently large. Then we get

$$\begin{aligned}P_1 = & \frac{-1}{LM_s} \sum_{m_2,N} E\left[a_Q^2\right] E\left[a_I^2\right] \frac{1}{T_s^2} \frac{1}{T} \\ & \times \left\{ \mathfrak{F}\left\{h^2_{\text{MF, FMF}}(t)\right\}_{|f=0} + \mathfrak{F}\left\{h^2_{\text{MF, FMF}}(t)\right\}_{|f=1/T} e^{(2\pi/T)m_2 T} \right. \\ & \left. + \ \mathfrak{F}\left\{h^2_{\text{MF, FMF}}(t)\right\}_{|f=-1/T} e^{-(2\pi/T)m_2 T} \right\}\end{aligned} \tag{8-90}$$

where we make use of the fact that the spectrum of $h^2_{\text{MF, FMF}}(t)$ vanishes for $f \geq \frac{2}{T}$, and finally we obtain

$$
\begin{aligned}
P_1 = & \frac{-1}{L\, M_s}\; N\; E\big[a_Q^2\big] E\big[a_I^2\big]\; \frac{1}{T_s^2}\, \frac{1}{T} \\
& \times \left\{ \mathfrak{F}\{h^2_{\text{MF, FMF}}(t)\}_{|f=0} + 2\; \mathfrak{F}\{h^2_{\text{MF, FMF}}(t)\}_{|f=1/T} \right\}
\end{aligned}
\tag{8-91}
$$

$\mathfrak{F}(\cdot)$ gives the Fourier coefficient of $(\cdot)$. At a first glance the numerical value of (8-91) depends on the length of the transmitted symbol sequence. But we conjecture that $N/L = 1$ which coincides with the interpretation that an observation interval of length $(M_s L)T_s$ contains approximately $L = N$ relevant symbols if we neglect all side effects at the boundary of the observation interval.

If all other terms of (8-85) are calculated analogously, then a closed-form expression for the estimate variance can be obtained. The analytical $S{\times}S$ contributions in Figures 8-16 and 8-17 were obtained via the above calculation. If, additionally, additive white Gaussian noise should be taken into account, exactly the same procedure can be applied. Besides the $S{\times}S$ the variance comprises $S{\times}N$ and $N{\times}N$ contributions. But it is left to the reader to analyze them. In Section 8.5 we demonstrate for a different frequency control loop how such terms can be analytically assessed. Therefore, here we have restricted ourselves to the analysis of the $S{\times}S$ contribution.

8.3.3 Discussion of NDA and NDε Algorithms

Non-data-aided and non-timing-directed frequency estimators operate in a poor noise environment. This is because the noise variance is proportional to the (one-sided) bandwidth B_f of the prefilter which, typically, is considerably wider than the signal bandwidth. Roughly, the SNR at the prefilter output is $2B_f T$ times smaller than at the matched filter output. As a consequence, these algorithms demand long averaging intervals. Also, the resolution is limited and the algorithms presented here suffer from self-noise. These drawbacks can only be mitigated by a considerably increased implementation complexity and long averaging intervals. In [4] Andrea and Mengali discuss strategies to minimize self-noise. They propose a design criterion on frequency error detectors which allows the minimization of the part of the power spectrum of the self-noise contribution which lies within the loop bandwidth. If the $S \times N$ and the $N \times N$ contributions are minimized, too, the advantage of such a structure is that only one algorithm is required for both the acquisition and the tracking task.

A different remedy to the problem is to employ a two-stage approach as depicted in Figure 8-10. The advantage of the two-stage structure is that for each stage an algorithm can be tailored to the specific task performed in this particular stage. The first stage has the task to rapidly acquire a coarse frequency estimate. During acquisition an algorithm with a large acquisition range and a short acquisition time is required whereas the tracking behavior is of no concern. The second stage is then optimized for tracking performance, since a large acquisition range is no longer required. Frequency algorithms suited for the second stage of a two-stage acquisition scheme are introduced in Section 8.4. Common to these algorithm is that they need correct timing. Timing can be recovered even in the presence of a residual frequency offset [5].

8.3.4 Main Points

- Frequency estimation algorithm which operate *independently* of correct timing are discussed.
- NDε algorithms can be derived from the ML theory.
- The maximum manageable frequency offset is determined by the analog prefilter shape and the sampling rate.
- There are FF and FB structures. Their tracking performance is comparable.
- Generally, NDA algorithms require statistically independent symbols. The phase increment estimator (Section 8.2.2) also works with an unmodulated carrier and a dotted signal.
- A real and symmetric pulse is required for the algorithm in Section 8.2.2 to obtain an unbiased estimate $\hat{\Omega}$. The other algorithms impose *no* such conditions
- The low-SNR approximation [eq. (8-35)]

$$L(\varepsilon, \Omega) = \sum_{l=-L}^{L} |z(lT + \varepsilon T),\ \Omega|^2 \tag{8-92}$$

  is the basis for all algorithms discussed.
- The joint (ε, Ω) estimation can be decoupled as follows:

$$\begin{aligned} \hat{\Omega} &= \arg\ \max_{\Omega} \sum_{l=-LM_s}^{LM_s-1} |z(lT_s, \Omega)|^2 \\ \hat{\varepsilon} &= \arg\ \left\{ \sum_{l=-LM_s}^{LM_s-1} \left|z\left(lT_s, \hat{\Omega}\right)\right|^2 e^{-j2l/M_s} \right\} \end{aligned} \tag{8-93}$$

  ($\hat{\Omega}$-directed timing recovery). The estimates $(\hat{\Omega}, \hat{\varepsilon})$ are unbiased. Conditions: random data, real, and symmetric pulse $g(t)$ [not necessary for $\hat{\Omega}$ alone to be unbiased, but $g(t)$ has then to be known].
- In Section 8.2.1 the estimate $\hat{\Omega}$ [eq. (8-32)] is obtained by filtering the received signal with a frequency-translated matched filter $G_{\text{MF}}\left(e^{j(\omega - \Omega_l)T_s}\right)$ where Ω_l is the trial parameter. The filter with maximum energy at the output corresponds to $\hat{\Omega}$.
- In Section 8.2.2 $\hat{\Omega}$ is obtained by estimating the phase increment (ΩT_s):

$$\hat{\Omega}T_s = \arg\left\{ \sum_{L_s} r_f(k\,T_s)\, r_f^*((k-1)T_s) \right\} \tag{8-94}$$

The estimate is unbiased for a real and symmetric pulse $g(t)$. The estimator variance contains a term due to self-noise. The algorithm works for random data and unmodulated or dotted signals.

- A frequency error-tracking system is obtained in the usual way by differentiating the log-likelihood function (Section 8.3). Error signal:

$$x(kT_s) = \text{Re}\left\{z(kT_s, \Omega)\; z^*_{\text{FMF}}(kT_s, \Omega)\right\} \tag{8-95}$$

The frequency matched filter response $G_{\text{FMF}}\left(e^{j\omega T_s}\right)$ equals

$$G_{\text{FMF}}\left(e^{j\omega T_s}\right) = \frac{\partial}{\partial \omega}\, G_{\text{MF}}\left(e^{j\omega T_s}\right) \tag{8-96}$$

The error signal can be interpreted as differential power measurement output. Simple approximations of the frequency matched filter can be used to produce the error signal. The error signal $x(kT_s)$ is unbiased. It contains self-noise. Conditions: random data, real, and symmetric pulse $g(t)$.

- All NDA NDε algorithms discussed suffer from self-noise. Therefore, they should be employed to rapidly acquire a coarse estimate. These estimates can serve as input to a second stage which is optimized for tracking performance: see the next Section 8.4.
- If the transmission pulse shape is unknown it is worthwhile to note that the condition for obtaining an unbiased estimate via one of the following methods – the direct estimation scheme or any type of a power difference measurement system – are the same. This is because the requirement that $h_g(\cdot)$ has to be real (condition for the direct estimator via eq. 8-52) corresponds in the frequency domain to the requirement that the power spectrum of the transmitted signal has to be symmetric about the origin (condition for the estimators via the power difference measurement).

Bibliography

[1] F. M. Gardner, "Frequency Detectors for Digital Demodulators Via Maximum Likelihood Derivation," *ESA-ESTEC Final Report: Part 2, ESTEC Contract No 8022-88-NL-DG*, March 1990.

[2] F. D. Natali, "AFC Tracking Performance," *IEEE COM*, vol. COM-32, pp. 935–944, Aug. 1984.

[3] T. Alberty and V. Hespelt, "A New Pattern Jitter Free Frequency Error Detector," *IEEE COM*, Feb. 1989.

[4] A. N. Andrea, and U. Mengali, "Design of Quadricorrelators for Automatic Frequency Control Systems," *IEEE COM*, vol. COM-41, pp. 988–997, June 1993.

[5] H. Sari and S. Moridi, "New Phase and Frequency Detectors for Carrier Recovery in PSK and QAM Systems," *IEEE COM*, vol. COM-36, pp. 1035–1043, Sept. 1988.

8.4 Frequency Estimators Operating with Timing Information

In this section we consider frequency estimators which operate on samples taken at symbol rate $1/T$. We assume that timing has been established *prior* to frequency synchronization. This implies that the frequency offset must not be arbitrarily large because the allowed frequency offset range is now no longer determined solely by the capture range of the frequency estimation algorithm itself but also by the capability of the timing synchronization algorithm to recover timing in the presence of a frequency offset. Therefore, the structures derived in this section are useful when the frequency offset is limited to about $|\Omega T/2\pi| \leq 0.15$. This is not a severe restriction because the normalized frequency offset is commonly less than 0.1. If this cannot be guaranteed a coarse frequency estimate of a first stage operating independently of timing information (see Sections 8.2 and 8.3) is required.

In the next section we first derive different estimators classified according to the common classes of data-aided, non-data-aided, and decision-directed algorithms. Then the problem of the joint estimation of the frame position and the frequency offset is shortly addressed. In Section 8.4.2 we discuss the performance of the algorithms in presence of additive white Gaussian noise and finish up with the presentation of the entire feedforward carrier synchronization structure comprising a phase and a frequency estimator.

8.4.1 Derivation of the Frequency Estimation Algorithm

We use for our derivation the second formulation (8-8) of the matched filter output function. Since we assume timing to be known, we resort to a simplified notation $z_n(\Omega)$ instead of $z_n(\hat{\varepsilon}, \Omega)$:

$$z_n(\Omega) = e^{-jn\Omega T} \sum_{k=-\infty}^{\infty} r_f(kT_s)\, g_{\text{MF}}(nT - kT_s)\, e^{-j\Omega(kT_s - nT)} \tag{8-97}$$

The expressions in the sum of (8-97) equals the output of the frequency-translated matched filter

$$\tilde{z}_n(\Omega) = \sum_{k=-\infty}^{\infty} r_f(kT_s)\, g_{\Omega}(nT - kT_s) \tag{8-98}$$

with

$$g_{\Omega}(kT_s) = g_{\text{MF}}(kT_s)\, e^{j\Omega kT_s} \tag{8-99}$$

The tilde has been used to highlight the different meaning of $\tilde{z}_n(\Omega)$ and $z_n(\Omega)$:

$$\tilde{z}_n(\Omega) = z_n(\Omega) \quad e^{jn\Omega T} \tag{8-100}$$

Using this formulation we get for the log-likelihood function

$$\text{Re}\left\{\mathbf{r}_f^T\ \mathbf{s}_f^*\right\} = \text{Re}\left\{ \sum_{n=-(N-1)/2}^{n=(N-1)/2} a_n^*\, e^{-j\theta}\, e^{-jnT\Omega} \tilde{z}_n \right\} \tag{8-101}$$

For small frequency offsets we approximate $g_\Omega(kT_s)$ by the matched filter $g_{\Omega=0}(kT_s) = g_{\text{MF}}(kT_s)$. We will discuss the validity of this simplification later on.

First, we consider the *data-aided* case. Data-aided operation means we have known symbols, for example, from a preamble. They are denoted by $\{a_{0,n}\}$. The ML function requires the joint estimation of $\{\Omega, \theta\}$,

$$\left\{\hat{\Omega}, \hat{\theta}\right\} = \arg\ \max_{\Omega,\theta} \sum_{n=-(N-1)/2}^{n=(N-1)/2} a_{0,n}^*\, e^{-j\theta}\, e^{-jnT\Omega}\, z_n \tag{8-102}$$

However, the two-dimensional search over $\{\Omega, \theta\}$ can be reduced to a one-dimensional search:

$$\begin{aligned}&\arg\ \max_{\Omega,\theta}\ \text{Re}\left\{\sum_{n=-(N-1)/2}^{n=(N-1)/2} a_{0,n}^*\, e^{-j\theta}\, e^{-jnT\Omega}\, z_n\right\}\\ =&\arg\ \max_{\Omega,\theta}\ |Y(\Omega)|\text{Re}\left\{e^{-j(\theta-\arg\{Y(\Omega)\})}\right\}\end{aligned} \tag{8-103}$$

with the obvious notation

$$Y(\Omega) = \sum_{n=-(N-1)/2}^{n=(N-1)/2} a_{0,n}^*\, e^{-jnT\Omega}\, z_n \tag{8-104}$$

The joint maximum is found by first maximizing the absolute value of $Y(\Omega)$ which is *independent* of θ. The second factor $\text{Re}\left\{e^{-j(\theta-\arg(Y(\Omega)))}\right\}$ is maximized by choosing

$$\hat{\theta} = \arg\left(Y\left(\hat{\Omega}\right)\right) \tag{8-105}$$

(An analogous result was found in the case of joint (θ, ε) estimation discussed in Section 5.5.) Thus for frequency estimation we only need to maximize

$$\hat{\Omega} = \arg\ \max_{\Omega}\ |Y(\Omega)| \tag{8-106}$$

Maximization of $|Y(\Omega)|$ is equivalent to the maximization of $(Y(\Omega)\, Y^*(\Omega))$. A sufficient condition for the maximum is that the derivative of $|Y(\Omega)|$ with respect to Ω equals zero:

$$\begin{aligned}&\frac{\partial}{\partial\Omega}(Y(\Omega)\, Y^*(\Omega))\\ =&\frac{\partial}{\partial\Omega}\left\{\left(\sum_{n=-(N-1)/2}^{n=(N-1)/2} a_{0,n}^*\, e^{-jnT\Omega}\, z_n\right)\left(\sum_{n=-(N-1)/2}^{n=(N-1)/2} a_{0,n}^*\, e^{-jnT\Omega}\, z_n\right)^*\right\}\\ =&2\ \text{Re}\left\{\left(\sum_{n=-(N-1)/2}^{n=(N-1)/2} a_{0,n}^*\, (-jnT)\, e^{-jnT\Omega}\, z_n\right)\left(\sum_{n=-(N-1)/2}^{n=(N-1)/2} a_{0,n}^*\, e^{-jnT\Omega}\, z_n\right)^*\right\}\\ =&2\ \text{Re}\left\{\sum_{n=-(N-1)/2}^{n=(N-1)/2} a_{0,n}^*\, (-jnT)\, e^{-jnT\Omega}\, Y^*(\Omega)\right\}\end{aligned} \tag{8-107}$$

After some obvious algebraic manipulations of (8-107) we get

$$\begin{aligned} 0 &= \frac{\partial}{\partial \Omega}\left(Y(\Omega)Y^*(\Omega)\right)\Big|_{\Omega=\hat{\Omega}} \\ &= 2\left|Y\left(\hat{\Omega}\right)\right| \mathrm{Re}\left\{\sum_{n=-(N-1)/2}^{n=(N-1)/2} a_{0,n}^* \; (-jnT)\, z_n\, e^{-jnT\hat{\Omega}} \underbrace{e^{j\,\arg\left(Y^*(\hat{\Omega})\right)}}_{e^{-j\hat{\theta}}}\right\} \end{aligned} \tag{8-108}$$

Since $|Y(\Omega)| \neq 0$, we need only to consider the $\mathrm{Re}\{\cdot\}$ expression of (8-108). By reordering the terms of the sum in (8-108) and introducing the short-hand notation

$$b_n = \frac{1}{2}\left[\frac{N^2-1}{4} - n(n+1)\right] \tag{8-109}$$

we get

$$\begin{aligned} 0 &= \mathrm{Re}\left\{\sum_{n=-(N-1)/2}^{n=(N-1)/2} e^{-j\hat{\theta}} jb_n \left(a_{0,n+1}^*\, z_{n+1}\, e^{-j(n+1)T\hat{\Omega}} - a_{0,n}^*\, z_n\, e^{-jnT\hat{\Omega}}\right)\right\} \\ &= \mathrm{Re}\left\{\sum_{n=-(N-1)/2}^{n=(N-1)/2} jb_n a_{0,n}^* \frac{z_n}{|z_n|}\, e^{-jnT\hat{\Omega}}\, e^{-j\hat{\theta}} \left[\frac{a_{0,n+1}^*}{a_{0,n}^*}\, z_{n+1}\, z_n^*\, e^{-j\hat{\Omega}T} - 1\right]\right\} \end{aligned} \tag{8-110}$$

The expression before the bracket can be simplified. For high SNR the matched filter output equals approximately

$$z_n \simeq a_{0,n}\, e^{j\theta_0}\, e^{j\Omega_0 T n} \tag{8-111}$$

Using this result we obtain for

$$\begin{aligned} & a_{0,n}^* \frac{z_n}{|z_n|}\, e^{-jnT\hat{\Omega}}\, e^{-j\hat{\theta}} \\ & = |a_{0,n}|^2 \frac{z_n}{|z_n|}\, e^{j(\Omega_0-\hat{\Omega})nT}\, e^{j(\theta_0-\hat{\theta})} \approx 1 \end{aligned} \tag{8-112}$$

Thus

$$\begin{aligned} 0 &= \mathrm{Re}\left\{\sum_{n=-(N-1)/2}^{n=(N-1)/2} jb_n \left(\frac{a_{0,n+1}^*}{a_{0,n}^*}\, z_{n+1}\, z_n^*\, e^{-jT\hat{\Omega}} - 1\right)\right\} \\ &= \mathrm{Im}\left\{\sum_{n=-(N-1)/2}^{n=(N-1)/2} b_n\, e^{-jT\hat{\Omega}} \left(\frac{a_{0,n+1}^*}{a_{0,n}^*}\, z_{n+1}\, z_n^*\right)\right\} \end{aligned} \tag{8-113}$$

From this equation a frequency estimate is directly obtained by

$$\hat{\Omega}T = \arg\left\{\sum_{n=-(N-1)/2}^{n=(N-1)/2} b_n \frac{a_{0,n+1}^*}{a_{0,n}^*} \left[z_{n+1}\, z_n^*\right]\right\} \tag{8-114}$$

The algorithm performs a *weighted* average over the phase increments $\arg(z_{n+1}\, z_n^*)$ of two successive samples. The weighting function b_n is maximum in the center of the estimation interval and decreases quadratically toward the boundaries. The maximum phase increment that can be uniquely attributed to a frequency offset is π. A larger positive phase increment cannot be distinguished from a negative one of value $2\pi - \hat{\Omega}T$. Thus, the capture range of the algorithm equals

$$\frac{|\Omega T|}{2\pi} < \frac{1}{2} \tag{8-115}$$

This algorithm can be viewed as belonging to either the class of the rotational frequency detector described by Messerschmitt [1] or to that of the cross-product detector described by Natali [2]. It is similar to the algorithms proposed by Tretter [3], Classen [4], and Kay [5] for estimating a single frequency in a noisy environment.

The *decision-directed* variant of the estimator coincides with (8-114) if the known symbols are replaced by (tentative) decisions.

Taking the complex number $(a_{0,n+1}^*/a_{0,n}^*)(z_{n+1}\, z_n^*)$ to the Mth power yields the NDA version of the algorithms for M-PSK:

$$\hat{\Omega}\, T = \frac{1}{M} \arg \left\{ \sum_{n=-(N-1)/2}^{n=(N-1)/2} b_n \left(z_{n+1}\, z_n^*\right)^M \right\} \tag{8-116}$$

We will later see that the following generalization, which resembles the generalization proposed by Viterbi and Viterbi [6] for a carrier phase estimation rule, is useful:

$$\begin{aligned}\hat{\Omega}\, T = \frac{1}{M} \arg \Bigg\{ & \sum_{m=-(L-1)/2}^{m=(L-1)/2} d_m\, F(|z_{n+m+1}|)\, F(|z_{n+m}|) \\ & \times e^{jM\,(\arg\{z_{n+m+1}\} - \arg\{z_{n+m}\})} \Bigg\}\end{aligned} \tag{8-117}$$

where d_m is an arbitrary filter function, $F(|z_n|)$ an arbitrary nonlinearity, and L is the estimation length with $L < N$. The frequency offset which the estimation algorithm can cope with is upper bounded by $|\Omega T/2\pi| \leq \frac{1}{2M}$. A block diagram of the algorithm is shown in Figure 8-18.

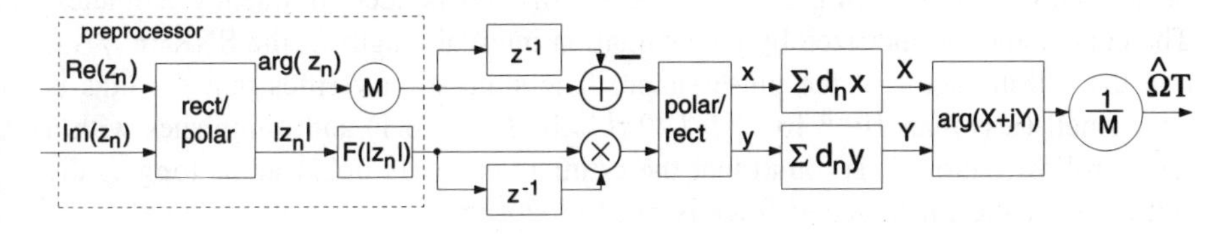

Figure 8-18 Block Diagram of the Feedforward Frequency Estimation Structure

8.4.2 Performance in the Presence of Noise

In this section, mean and variance of the frequency estimate are analytically investigated and the results are verified by simulation. The variances are compared with the Cramér-Rao bound. First estimators employing the optimal filter with coefficients b_n are dealt with and then other filter functions which are easier to implement are addressed.

It is conceptually useful to split z_n into a useful part and a random part. The useful part consists only of the unknown parameter set $\{\theta_0, \Omega_0\}$ and the symbol a_n corresponding to the sample interval nT

$$z_n = e^{j(\Omega_0\, nT + \theta_0)} \left(a_n\, h_\Omega(0T) + \underbrace{\sum_{l \neq n} a_l\;\, e^{j\Omega_0 (l-n)T} h_\Omega((l-n)T)}_{\text{ISI random}} \right) + \underbrace{n(nT)}_{\text{noise}} \tag{8-118}$$

where $h_\Omega(nT)$ is here defined as $h_\Omega(nT) = \sum\limits_{k=-\infty}^{\infty} g(kT_s)\, e^{j\Omega_0 kT_s}\, g_{\mathrm{MF}}(nT - kT_s)$. Recall that $z_n(\Omega)$ was approximated by $z_n(\Omega = 0)$ for small frequency offset, see remarks following eq. (8-101). The product of two consecutive samples is

$$\begin{aligned} z_n\; z^*_{n-1} = e^{j\,\Omega_0 T} \big(&a_n\; h_\Omega(0)\, a^*_{n-1} h^*_\Omega(0) \\ &+ F\{a_n h_\Omega(0T),\; a_{n-1} h_\Omega(0T), \mathrm{ISI},\; \mathrm{noise}\}\big) \end{aligned} \tag{8-119}$$

where $F\{\cdot\}$ is the abbreviated notation for the remaining contributions.

The mean of the estimate is obtained by taking the argument of the expected value of the phasor $\left(z_n\; z^*_{n-1}\right)^M$

$$E[\Delta\Omega T] = \frac{1}{M}\, \arg\left\{ \sum b_n\; E\left[\left(z_n\; z^*_{n-1} \right)^M \right] \right\} \tag{8-120}$$

It can be shown from (8-119) and (8-120) that the noise does not introduce bias. A bias can arise only from the ISI term caused by the mismatched filter. This bias can be analytically assessed if we average over the random data a_n which we assume to be statistically independent. However, it seems to be more efficient to resort to simulation. Figure 8-19 shows the difference between the estimated frequency offset and the true value versus the frequency offset Δf. Each point has been obtained by averaging over 10^4 statistically independent frequency estimates. The curves are parametrized by the estimation interval length L, the SNR=E_s/N_0, and the rolloff factor of the pulse shape. The simulation verifies that the bias is very small (less than 10^{-3} for $|\Omega_0 T/2\pi| = |\Delta f T| < 0.1$) and diminishes with a larger rolloff factor. It is found that the estimate is nearly unbiased as long as the influence of the mismatched filter is not too strong.

We evaluate and discuss the variance for $\Delta f = \Omega_0/2\pi = 0$, and verify that

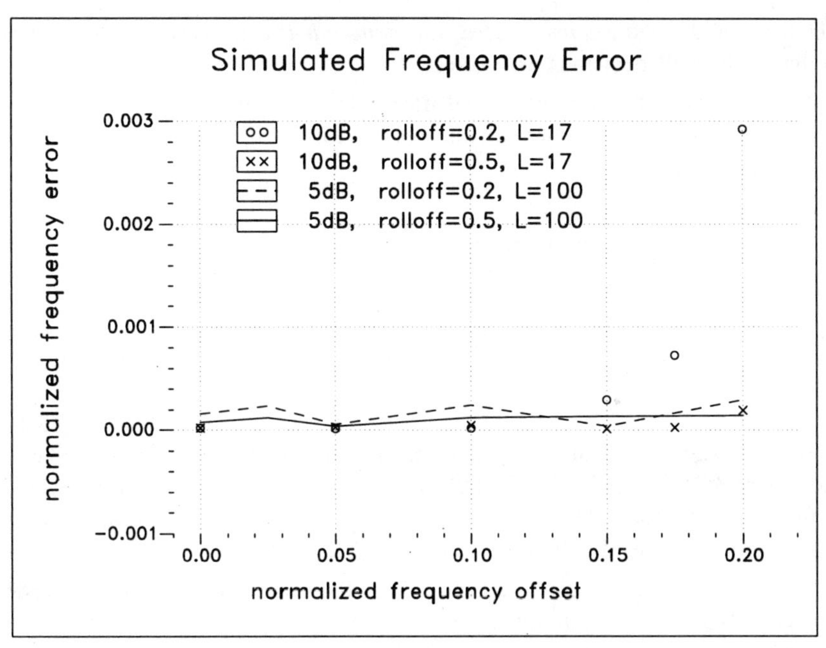

Figure 8-19 Simulated Frequency Error $\left|\frac{\Omega_0-\hat{\Omega}}{2\pi}T\right|$ (BPSK Transmission)

the properties for $\Delta f \neq 0$ remain unchanged by means of simulations. We get

$$\operatorname{var}\left[M\hat{\Omega}\,T\right] \approx \frac{E\left[\left(\operatorname{Im}\left\{\sum\limits_{L} b_n\ \left(z_n\ z_{n-1}^*\right)^M\right\}\right)^2\right]}{\left(E\left[\operatorname{Re}\left\{\sum\limits_{L} b_n\ \left(z_n\ z_{n-1}^*\right)^M\right\}\right]\right)^2} \tag{8-121}$$

This latter approximation is valid since the imaginary part of $\sum\limits_{L} b_n\ \left(z_n\ z_{n-1}^*\right)^M$ has zero mean (a condition which coincides with the demand to be unbiased) and because the variances of both imaginary and real part are small compared to the squared real mean.

Applying this approximation we obtain the following expression after a straightforward calculation:[1]

$$\operatorname{var}\left[\hat{\Omega}T\right] = \frac{12}{L(L^2-1)}\,\frac{1}{2E_s/N_0} + \frac{12}{5}\,\frac{1}{L}\,\frac{L^2+1}{L^2-1}\,\frac{1}{\left(2E_s/N_0\right)^2} \tag{8-122}$$

[1] The MF was tailored to a root-raised cosine transmission pulse in such a way that for the synchronized MF output holds $z(nT) = a_n + N(n)$ with a_n as transmitted symbol with $E\left[|a_n|^2\right] = 1$ and $N(n)$ is the filtered noise with variance $\operatorname{var}[N(n)] = N_0/E_s$.

which is valid for the *unmodulated*, the *data-aided*, and *decision-directed* cases (provided that all decisions are correct).

Analytical results for modulated transmission can be obtained only at great effort. Exemplarily, we mention the results for BPSK for which we find the following expression

$$\begin{aligned}\operatorname{var}\left[\hat{\Omega}T\right] =& \frac{12}{L(L^2-1)}\left(\frac{1}{2E_s/N_0}+\frac{1}{(2E_s/N_0)^2}\right)\\ &+\frac{12}{5}\,\frac{1}{L}\,\frac{L^2+1}{L^2-1}\left(\frac{4}{(2E_s/N_0)^2}+\frac{8}{(2E_s/N_0)^3}+\frac{4}{(2E_s/N_0)^4}\right)\end{aligned} \tag{8-123}$$

Figure 8-20 shows both analytical and simulation results for the unmodulated cases, and Figure 8-21 for the BPSK and QPSK cases. The results show that the behavior of the frequency estimate variance is well approximated by the above expression also in the presence of a frequency offset $\Delta f\ T = 0.1$.

The solid line in Figures 8-20 and 8-21 is the leading term of the Cramér-Rao bound (CRB) (6–22) for frequency estimation.

We observe:

1. The CRB is only met at high SNR. The algorithm is thus asymtotically efficient.
2. The region of practical interest, however, are at low to moderate SNRs. Here we see a very large deviation from the CRB. For the data-aided case (8-122)

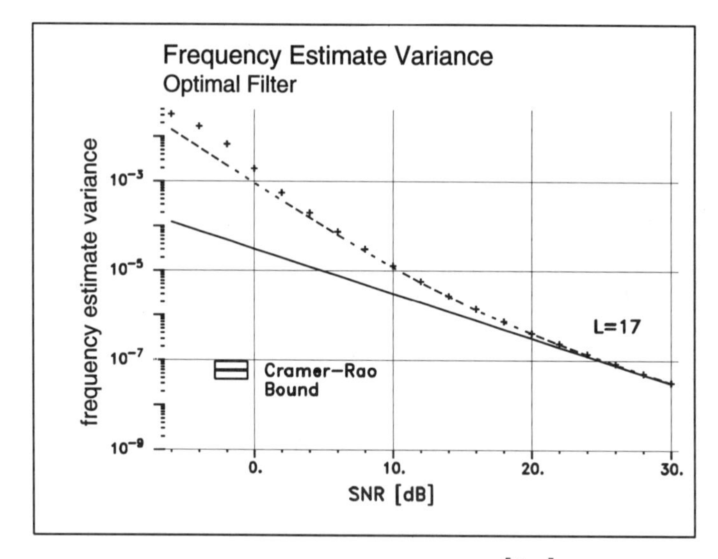

Figure 8-20 Frequency Estimate Variance $\operatorname{var}\left[\frac{\hat{\Omega}T}{2\pi}\right]$; unmodulated carrier, $\Omega_0 T/2\pi = \Delta f T = 0.1$, $\alpha = 0.2$, SNR=E_s/N_0

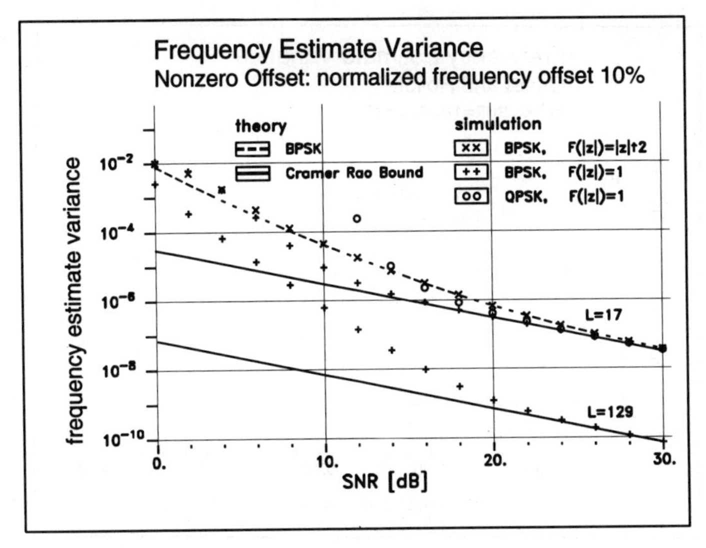

Figure 8-21 Frequency Estimate Variance var$\left[\frac{\hat{\Omega}T}{2\pi}\right]$; optimal filter, SNR=$E_s/N_0$

the dominant term in the variance decreases quadratically with $(1/\text{SNR})^2$. The situation is far worse for MPSK modulation, due to the nonlinear operation performed on the noise. For BPSK the dominant term decreases with $(1/\text{SNR})^4$ while for QPSK the algorithm is no longer applicable.

3. Unfortunately, at low SNR the dominant term decreases only linearly with L thus requiring large estimation intervals.
4. A remedy for this behavior is to consider a multiple of the phase increment, $(D\Omega T)$, as discussed in Section 8.4.3 (D-spaced estimator).

The performance of the estimator can be improved by choosing an appropriate nonlinearity $F(|z_n|)$. If we select the nonlinearity $F(|\ z_n|) = 1$, the simulations (Figure 8-21) reveal that the variances (shown for BPSK and QPSK) now cling more tightly to the Cramér-Rao bound.

Applying (8-121) when an integrate and dump filter is employed (instead of the filter function b_n), for the unmodulated case we find

$$\text{var}\left[\hat{\Omega}T\right] = \frac{2}{L^2}\,\frac{1}{2E_s/N_0} + \frac{2}{L}\,\frac{1}{\left(2E_s/N_0\right)^2} \tag{8-124}$$

and for BPSK

$$\begin{aligned}\text{var}\left[\hat{\Omega}T\right] &= \frac{2}{L^2}\left(\frac{1}{2E_s/N_0} + \frac{1}{\left(2E_s/N_0\right)^2}\right) \\ &+ \frac{8}{L}\left(\frac{1}{\left(2E_s/N_0\right)^2} + \frac{2}{\left(2E_s/N_0\right)^3} + \frac{1}{\left(2\ E_s/N_0\right)^4}\right)\end{aligned} \tag{8-125}$$

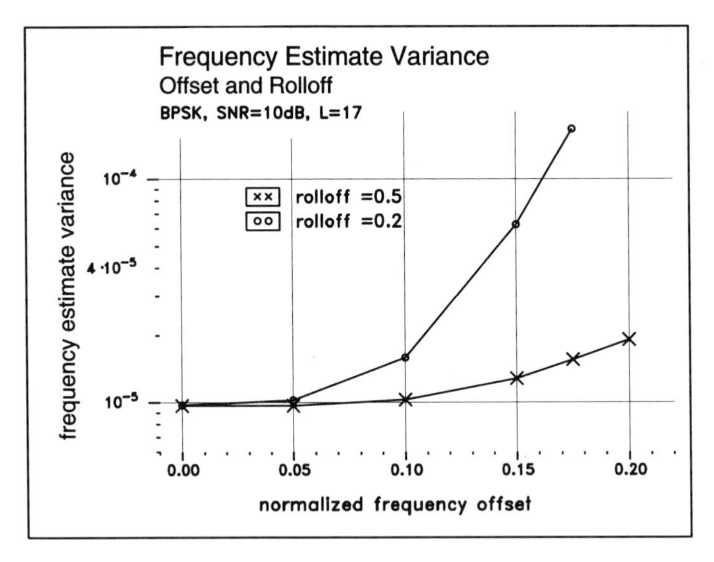

Figure 8-22 Influence of the Rolloff and of $|\Omega_0 T|$ on the Estimate Variance $\mathrm{var}\left[\frac{\hat{\Omega}T}{2\pi}\right]$, SNR=$E_s/N_0$

A comparison between (8-122) and (8-124), (8-123) and (8-125), respectively, reveals that all expressions decrease with $1/L$ for low SNR and large L. Only at high SNR the optimal filter prevails against the integrate and dump filter. Since the implementation of the integrate and dump filter is significantly less complicated compared to that of the optimal filter with the coefficients b_n, and since the performance improvement is small, an integrate and dump will be used in practice.

A last comment concerns the influence of the mismatched filter in case of a frequency offset. Strictly speaking, all analytically obtained variance expressions are only valid for the case $\Omega_0 = 0$. The simulation results and especially the simulations reported in Figure 8-22 indicate that the applied approximations do indeed hold. The variance is influenced mainly by the degradation of the expected value of the real part of $\sum_L b_n \left(z_n\ z_{n-1}^*\right)^M$ in (8-121). This value becomes smaller in the case of $\Omega_0 T \neq 0$ because $h_\Omega(0T)$ decreases with increasing frequency offset. But for frequency offsets which are small in comparison to the symbol rate the degradation remains very small. An intuitive explanation of why the influence of the ISI terms [compare (8-118)] is so small is that these contributions are averaged out if the estimation length is properly chosen; see Figure 8-19.

8.4.3 Miscellaneous Modifications of DA Estimators

Joint Frame and Frequency Synchronization

Common to all NDA frequency estimators is their poor performance for low and medium SNR values. The reason is that due to the nonlinear operation required to remove the modulation the noise is strongly increased.

For TDMA applications with short data packets, the acquisition time length of common frequency estimation structures may be too long (for a large modulation index $M \geq 4$ and moderate E_s/N_0 <7 dB) if the algorithm has to operate solely on random data. To overcome this problem associated with non-data-aided (NDA) frequency estimation a typical packet format used for burst mode transmission employs a preamble at the start of the packet. Besides a unique word (UW) used for frame synchronization such a preamble typically incorporates a clock recovery acquisition sequence and an unmodulated carrier sequence for assisting the carrier frequency estimation. This preamble is followed by the random data part. However, in order to increase the transmission efficiency a separate acquisition preamble for each of the synchronizers is inefficient.

A remedy to this problem is to employ a data-aided (DA) frequency estimation

$$\hat{\Omega T} = \arg\left\{ \sum_{L_F} \left[z_{n+1}\, z_n^* \right] \left(a_{0,n+1}^*\, a_{0,n} \right) \right\} \tag{8-126}$$

where the data from the UW are used for both frame synchronization and frequency estimation as well. But DA operation is only possible after frame synchronization has been established, because the relative position of the sequence $\{a_{0,n}\}$ to the sequence $\{z_n\}$ must be known exactly if we want to make use of the known symbols $a_{0,n}$. For this reason, frame sync must be performed *prior* to the DA operation or at least simultaneously.

This requirement can be relaxed if the symbols of the preamble exhibit some kind of periodicity. For example, if an unmodulated sequence or a $(+1, -1, +1, \ldots)$ modulated BPSK preamble is employed, then the exact position of the preamble within the received sequence is not required, since the product of $\left(a_{0,n+1}^*\, a_{0,n}\right)$ is constant and known a priori. Not being required to know the exact position leads to a reduction in the channel transmission efficiency since now the preamble must contain a specific carrier acquisition sequence.

The following approach overcomes this deficiency. The symbols of the unique word are differentially encoded in such a way that the resulting sequence $\{d_n\} = \left\{a_{n+1}^*\, a_n\right\}$ possesses the property of commonly used sequences for frame detection. Now frequency estimation can be performed simultaneously with frame detection as follows:

1. The estimate of the starting position μ is obtained via

$$\hat{\mu} = \arg \max_{\mu} \left| \sum_{l=1}^{L_{UW}} \left[z_{\mu-l}\, z_{\mu-l-1}^* \right] d_l \right| \tag{8-127}$$

2. The frequency estimate is then determined by

$$\hat{\Omega T} = \arg \left\{ \sum_{l=1}^{L_{UW}} \left[z_{\mu-l}\, z_{\mu-l-1}^* \right] d_l \right\} |\mu = \hat{\mu} \tag{8-128}$$

The performance of this simple correlation rule for obtaining the frame start position is not optimum from the point of view of separate frame synchronization.

But the approach is a robust method to limit the required training symbol sequence length.

D-Spaced Estimator

The performance of the frequency estimate can be significantly improved if the product $\left(\left[z_{\mu+l}\; z^*_{\mu+l-1}\right] d_l\right)$ is replaced by $\left(\left[z_{\mu+l}\; z^*_{\mu+l-D}\right] d_l\right)$:

$$\hat{\Omega T} = \frac{1}{D} \arg \left\{ \sum_{l=1}^{L_{UW}} \left[z_{\mu-l}\; z^*_{\mu-l-D}\right] d_l \right\} |\mu = \hat{\mu} \tag{8-129}$$

The underlying hypothesis is that the variance of the estimate does not depend on the magnitude of the frequency value to be estimated. Therefore, the idea is to generate an estimate of a virtual frequency offset $D\Omega_0$ which is a multiple of the frequency offset Ω_0. The value D must be a positive integer and $(D\Omega_0 T < \pi)$ has to be fulfilled. The estimate variance of $\hat{\Omega T}$ is given by

$$\text{var}\left[\hat{\Omega T}\right] = \frac{1}{D^2}\left(\frac{D}{L^2_{\text{UW}}} \frac{2}{(2E_s/N_0)} + \frac{1}{L_{\text{UW}}} \frac{2}{(2E_s/N_0)^2}\right) \tag{8-130}$$

(Note the sequence $\{a_n\}$ belonging to the UW now has to be chosen in such a way that $\left\{a^*_{0,n+1}\; a_{0,n+1-D}\right\} = \{d_n\}$.)

Without having increased the implementation effort we have thus reduced the estimate variance by at least a factor of D [compare (8-130) and Figure 8-23]. The performance of the frame detection remains unchanged.

The same principle can be applied to NDA estimators provided that the frequency offset obeys $|\Omega_0 T| < 1/(2MD)$.

8.4.4 Example of an All-Feedforward Carrier Synchronization Structure

Figure 8-24 shows the block diagram of a complete synchronizer chain for small frequency offset $|\Omega T/2\pi| < 0.15$. Timing is recovered in the presence of a frequency offset employing the square and filter algorithm of Section 5.4. Next, the frequency offset is estimated and compensated for by means of a timing-directed algorithm (see Section 8.4.3) operating at symbol rate. The carrier phase is recovered by a generalized form of the ML FF algorithm (see Section 5.10.1):

$$\hat{\theta}_n = \frac{1}{M} \arg \left(\sum_{l=-L_V}^{l=L_V} F|z_{n+l}|\, e^{jM\left(\arg(z_{n+l}) - (n+l)\hat{\Omega T}\right)} \right) + i\frac{2\pi}{M} \tag{8-131}$$

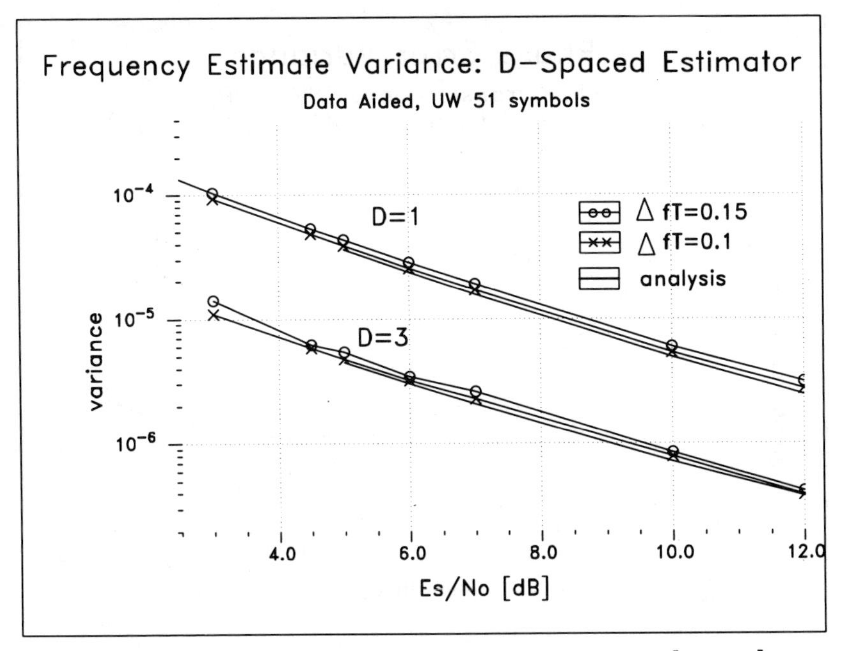

Figure 8-23 Variance of the Estimated Frequency var$\left[\hat{\Omega}T/2\pi\right]$

Note that the phase estimation algorithm has to cope with a residual frequency offset. Therefore, the estimation length has to be optimized with respect to the expected phase fluctuation resulting from the residual frequency offset $\Delta\Omega = \left(\Omega_0 - \hat{\Omega}\right)$ or from oscillator imperfections. Under high SNR approximation the optimal length L_P for a given Δf can be shown to be bounded by

$$2L_P + 1 < \frac{1}{2M \,\Delta\, fT} \tag{8-132}$$

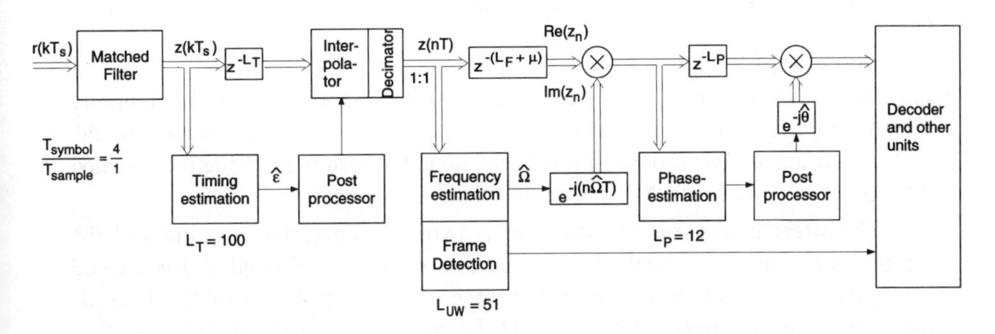

Figure 8-24 Demodulator Block Diagram; $L_{(.)}$ denotes estimation interval

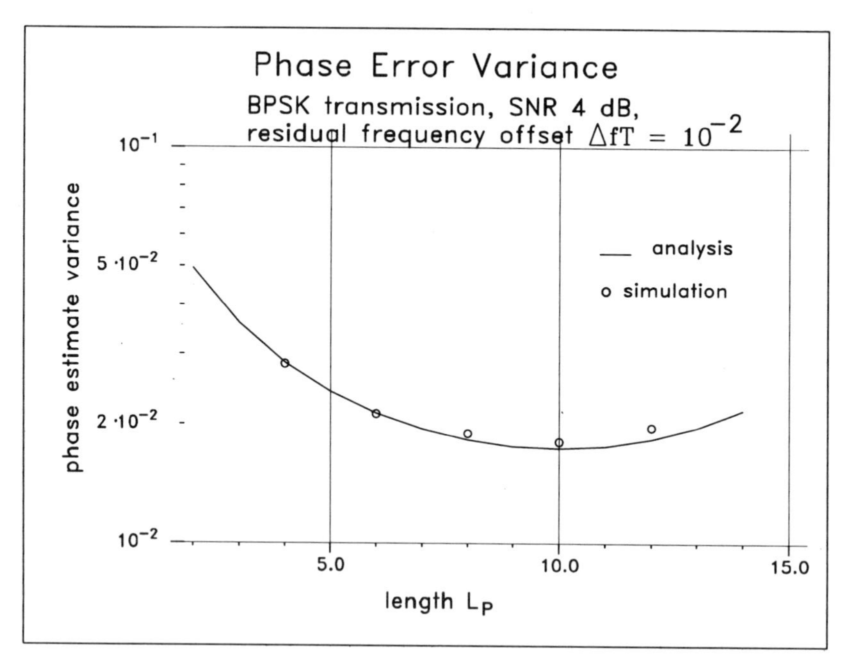

Figure 8-25 Phase Estimate Variance, $\Delta f\ T = 10^{-2}$

where M is the modulation index. (For low SNR the result [6, eq. A13] should be used.) Figure 8-25 shows the dependence of var $\left[\hat{\theta}\right]$ on the parameter L_P for a given SNR and a fixed frequency estimate error Δf. The minimum of the curve indicates the "best" L_P. The theoretical results were obtained by using the results of [6, eq. A13].

The post-processing structure shown in Figure 8-24 is necessary to perform the phase unwrapping, which is necessary as a result of the restricted interval $[-\pi/M,\ \pi/M]$ of the phase estimator; see Section 5.10.1. A detailed analysis of the above synchronization structure is found in [7].

Simulation Results

For an AWGN channel the bit error rate (BER) performance of differentially encoded QPSK transmission is shown in Figure 8-26. The degradation of the BER is not due to synchronization imperfections but results from the mismatched filter (difference between MF bound and the —×— curve).

The effect of the mismatched filter is twofold. First, the magnitude of the samples taken at the optimal sampling instant is reduced. Second, ISI is created [compare (8-118)] which is the main source of the degradation of the BER. If the degradation is intolerable (i.e., if $|\Omega_0 T/2\pi| > 0.1$ and the rolloff $\alpha < 0.5$) a separate decoder path is required where the frequency offset of the samples is compensated before they enter the matched filter. The BER curve —o— in Figure

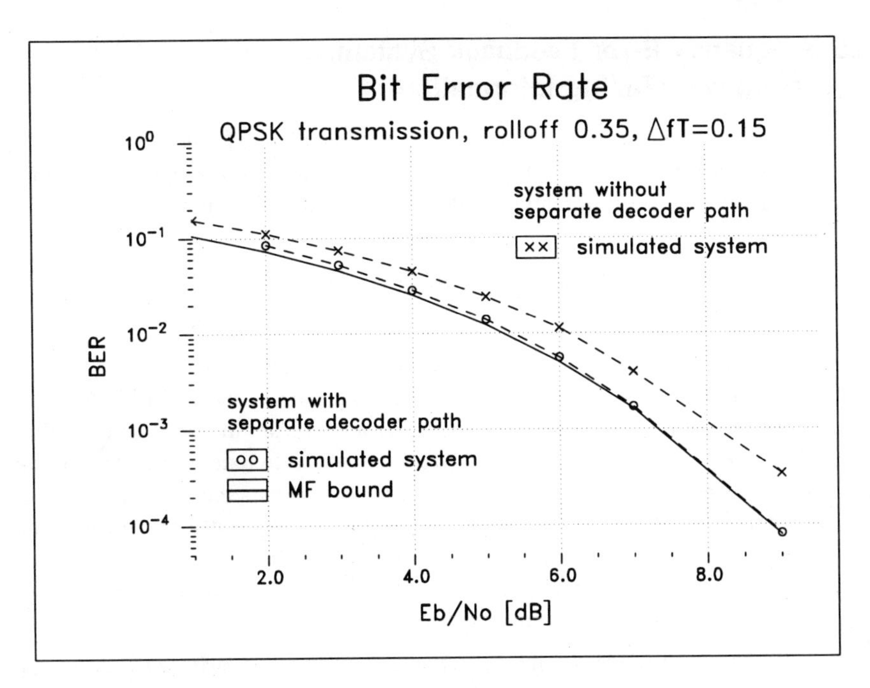

Figure 8-26 Bit Error Rate of the Demodulator of Figure 8-24

8-26 demonstrates that by an appropriate choice of the synchronization parameters the performance of the demodulator is near the theoretical optimum for $E_b/N_0 \geq 3$ dB.

Bibliography

[1] D. G. Messerschmitt, "Frequency Detectors for PLL Acquisition in Timing and Carrier Recovery," *IEEE COM*, vol. COM-27, pp. 1288–1295, Sept. 1979.

[2] F. D. Natali, "AFC Tracking Performance," *IEEE COM*, vol. COM-32, pp. 935–944, Aug. 1984.

[3] S. A. Tretter, "Estimating the Frequency of a Noisy Sinusoid by Linear Regession," *IEEE Trans. Info. Theory*, pp. 823–835, Nov. 1985.

[4] F. Classen, H. Meyr and P. Sehier, "Maximum Likelihood Open Loop Carrier Synchronizer for Digital Radio," *Proceedings ICC' 93*, pp. 493–497, 1993.

[5] S. Kay, "A Fast and Accurate Single Frequency Estimator," *IEEE Trans. Acoust., Speech, Signal Processing*, vol. ASSP-37, pp. 1987–1990, Dec. 1989.

[6] A. Viterbi and A. Viterbi, "Nonlinear Estimation of PSK-Modulated Carrier Phase with Application to Burst Digital Transmission," *IEEE Trans. Info. Theory*, pp. 543–551, July 1983.

[7] F. Classen, H. Meyr and P. Sehier, "An All Feedforward Synchronization Unit for Digital Radio," *Proceedings VTC' 93*, pp. 738–741, 1993.

8.5 Frequency Error Feedback Systems Operating with Timing Information

In this section we are concerned with timing directed frequency feedback tracking systems running at symbol rate $1/T$. A frequency error signal is obtained in the familiar way by differentiating the log-likelihood function

$$\frac{\partial}{\partial \Omega} \sum_{N} |z(nT + \varepsilon T, \Omega)|^2 \tag{8-133}$$

The derivation is analogous to that given for non-timing-directed frequency error feedback system. The resulting frequency tracking loop is shown in Figure 8-27. The signal and frequency matched filters of Figure 8-27 are the same as in Section 8.3; see eq. (8-68). There are, however, fundamental differences compared to the non-directed frequency control loop of Section 8.3. First, the impulse response of the filters $g_{\text{MF}}(kT_s, \varepsilon)$ and $g_{\text{FMF}}(kT_s, \varepsilon)$ and consequently the error signal $x(nT; \varepsilon)$ are a function of the timing parameter ε. Second, an error signal is produced only at symbol rate $1/T$. Both properties have a profound influence on the loop behavior as will be shown below.

In the tracking mode we may assume that timing is nearly perfect, $\hat{\varepsilon} \simeq \varepsilon_0$. This is not the case in the acquisition mode where no knowledge about the correct timing parameter is available. Therefore, understanding of the loop behavior as a function of an arbitrary value of the timing error $\Delta\varepsilon$ is necessary. In Figure 8-28 the S curve is plotted as a function of $\Delta\varepsilon$. All filter functions correspond to a transmission pulse having a root-raised cosine spectrum shape with $\alpha \leq 0.5$. The pulse energy of $g(t)$ is normalized to 1.

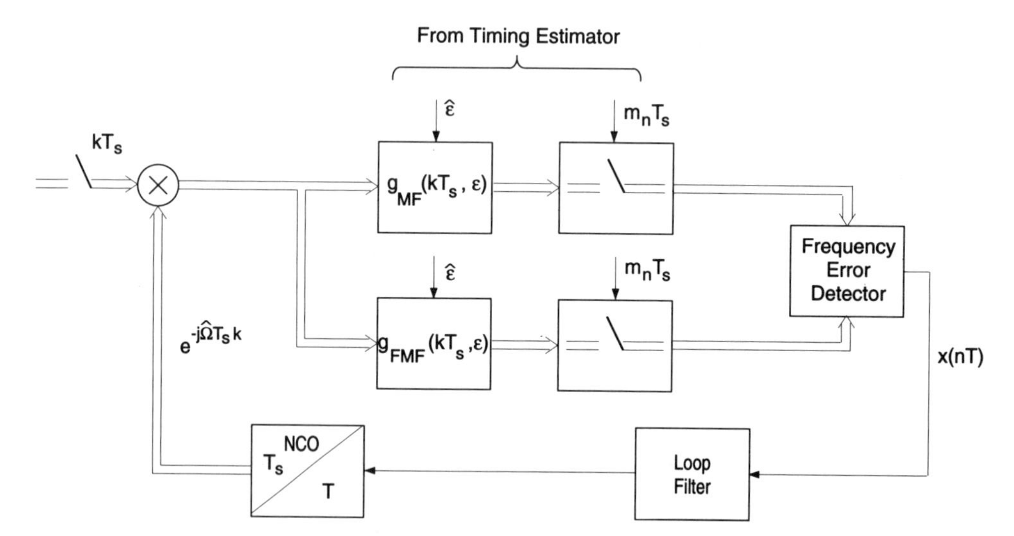

Figure 8-27 Block Diagram of a NDA Dε Frequency Control Loop

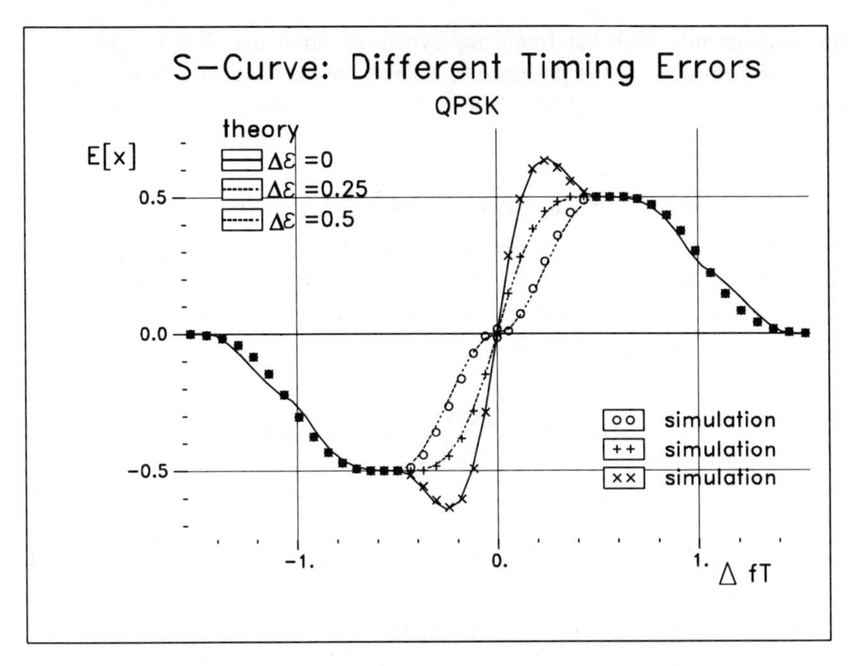

Figure 8-28 S curve for Different Timing Errors ε, $E_g = 1$

For all values of $\Delta\varepsilon$ displayed in Figure 8-28 the sign of the error signal is correct in the mean and its magnitude is larger than zero. There exist no false lock points and the loop will always pull into the only equilibrium point. The S curve for $\Delta\varepsilon = 0.5$ requires special attention. Using Table 8-4 it is readily verified that the slope of the S curve at the origin becomes zero. Thus, the loop will not pull in for this value of $\Delta\varepsilon$. However, the timing recovery scheme which

Table 8-4 S Curve of the Dε FED; S curve Is Odd Symmetric about Zero

	$E[x(nT)]$
$0 \le \Delta fT < \alpha$	$\frac{1}{2}\left\{\sin^2\left(\frac{\pi\Delta fT}{2\alpha}\right) + \frac{\pi}{4}\left(1 - \frac{\Delta fT}{\alpha}\right)\sin\left(\frac{\pi\Delta fT}{\alpha}\right)\left(1 + \cos\left(\frac{2\pi\varepsilon}{T}\right)\right)\right\}$
$\alpha \le \Delta fT < 1-\alpha$	$\frac{1}{2}$
$1-\alpha \le \Delta fT < 1$	$\frac{1}{4}\left\{1+\sin^2\left(\frac{\pi}{2\alpha}[\Delta fT - 1]\right) - \frac{\pi}{4}\left(1 + \frac{\Delta fT-1}{\alpha}\right)\sin\left(\frac{\pi}{\alpha}[\Delta fT - 1]\right)\right\}$
$1 \le \Delta fT < 1+\alpha$	$\frac{1}{4}\left\{\cos^2\left(\frac{\pi}{2\alpha}[\Delta fT - 1]\right) - \frac{\pi}{4}\left(1 - \frac{\Delta fT-1}{\alpha}\right)\sin\left(\frac{\pi}{\alpha}[\Delta fT-1]\right)\right\}$

runs concurrently with the frequency synchronization unit will be able to settle to equilibrium even in the presence of a small frequency error. Subsequently, the frequency control loop will be able to pull in and complete the acquisition process.

8.5.1 Performance in the Presence of Additive Noise

Before we analyze the performance in the presence of noise we derive conditions on the pulse shape $g(t)$ for the algorithm to be free of self-noise. We assume perfect timing, noise-free conditions, and $\Omega_0 = 0$. Under these conditions the matched filter output equals

$$z(nT) = \sum_{m=-(N-1)/2}^{(N-1)/2} a_m h((m-n)T) \tag{8-134}$$

with

$$h((m-n)T) = \sum_{k=-\infty}^{\infty} g(kT_s - nT)\; g_{\text{MF}}(mT - kT_s) \tag{8-135}$$

The frequency matched filter output equals

$$z_{\text{FMF}}(nT) = \sum_{l=-(N-1)/2}^{(N-1)/2} a_l h_{\text{FMF}}((l-n)T) \tag{8-136}$$

with

$$h_{\text{FMF}}((l-n)T) = \sum_{k=-\infty}^{\infty} g(kT_s - nT) g_{\text{FMF}}(lT - kT_s) \tag{8-137}$$

The error signal

$$x(nT) = \text{Re}\{z(nT)\; z^*_{\text{FMF}}(nT)\} \tag{8-138}$$

becomes identically zero if at least one of the terms of the above product vanishes. Since the matched filter output is unequal to zero, the only hope of avoiding self-noise is that the frequency matched filter output is identically zero at all times nT. This is the case if we can show that

$$h_{\text{FMF}}(lT) = 0 \qquad \forall\, l \tag{8-139}$$

[besides the trivial solution $h_{\text{FMF}}(t) \equiv 0$].

Now

$$h_{FMF}(lTs) = \frac{T}{2\pi} \int_{-\pi/T}^{\pi/T} H_{\text{FMF}}(e^{j\omega T})\; e^{j\omega T l}\; d\omega \tag{8-140}$$

The only solution is the trivial one $H_{\mathrm{FMF}}\left(e^{j\omega T}\right) \equiv 0$ for all ω. The spectrum of the sampled filter response $h_{\mathrm{FMF}}(lT_s)$ can be expressed as

$$H_{\mathrm{FMF}}\left(e^{j\omega T}\right) = \frac{1}{T} \sum_{m=-\infty}^{\infty} H_{\mathrm{FMF}}\left(\omega - \frac{2\pi}{T}\, m\right) \tag{8-141}$$

If the *time-continuous* frequency response $H_{\mathrm{FMF}}(\omega)$ possesses odd symmetry about π/T,

$$H_{\mathrm{FMF}}\left(\omega - \frac{\pi}{T}\right) = -H_{\mathrm{FMF}}\left(\omega + \frac{\pi}{T}\right) \tag{8-142}$$

and vanishes outside the baseband ($|\omega/2\pi| < 1/T$), then $h(lT) \equiv 0$ for all l. The frequency response $H_{\mathrm{FMF}}(\omega)$ is the product of

$$H_{\mathrm{FMF}}(\omega) = G(\omega)\ \ G_{\mathrm{FMF}}(\omega) \tag{8-143}$$

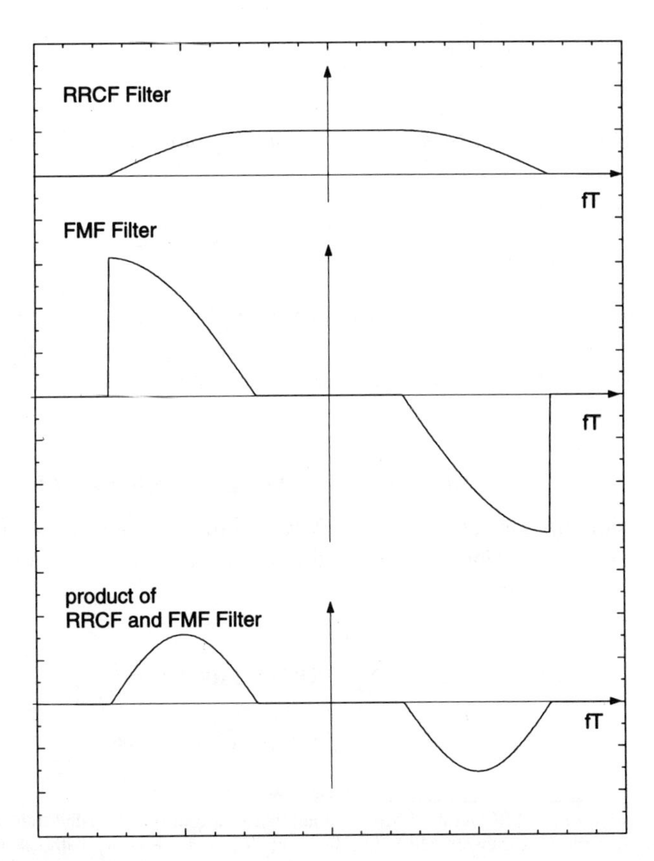

Figure 8-29 Transfer Function of G_{RRC}, G_{FMF}, and H_{FMF} with $\alpha = 0.5$

It can easily be verified that (8-142) is fulfilled for a root-raised cosine filter characteristic (see Figure 8-29).

The performance of the loop in the presence of noise is evaluated next. We assume perfect timing and a root-raised cosine pulse.

The error signal in the presence of noise equals

$$x(nT) = \mathrm{Re}\left\{ \left(a_n + \sum_{k=-\infty}^{\infty} n(kT_s)\, g_{\mathrm{MF}}(-kT_s + nT) \right) \times \left(\sum_{k=-\infty}^{\infty} n^*(kT_s)\, g^*_{\mathrm{FMF}}(nT - kT_s) \right) \right\} \tag{8-144}$$

The expected value with respect to noise equals zero, as required for an unbiased estimate. Since the frequency error is free of self-noise, the variance comprises the two terms $\sigma^2_{S \times N}$ and $\sigma^2_{N \times N}$. The variance is given by

$$\mathrm{var}\left[\frac{\hat{\Omega}\, T}{2\pi}\right] = \frac{2 B_L T}{K_{\Omega}^2}\, S_x\left(e^{j\omega T}\right)\Big|_{\omega=0} \tag{8-145}$$

After some lengthy and tedious algebra we find exemplarily for QPSK

$$\mathrm{var}\left[\frac{\hat{\Omega} T}{2\pi}\right] = \sigma^2 \approx \frac{2 B_L T}{K_{\Omega}^2}\,\frac{\pi^2}{8\alpha}\left(\frac{1}{\mathrm{SNR}} + \frac{1}{\mathrm{SNR}^2}\right) \tag{8-146}$$

where K_{Ω}^2 is the slope of the S curve in the origin, SNR=E_s/N_0 and α is the rolloff factor. Simulation results in Figure 8-30 closely agree with the analytical result. Similar results can be obtained for higher order modulations.

8.5.2 Appendix: Computation of var$\{\Omega\}$ as a Function of E_s/N_0

We consider the case of M-PSK and M-QAM transmission and assume perfect timing and $\Omega_0 = 0$. Under these conditions the matched filter output in the presence of noise[2] equals

$$z(nT) = a_n + \sum_{k=-\infty}^{\infty} n(kT_s)\; g_{\mathrm{MF}}(nT - kT_s) \tag{8-147}$$

[2] The normalization of the MF and the FMF were tailored to a root-raised cosine transmission pulse in such a way that the synchronized MF output holds $z(nT) = a_n + N(n)$ with a_n as transmitted symbol with $E\left[|a_n|^2\right] = 1$ and $N(n)$ is the filtered noise with variance var$[N(n)] = N_0/E_s$.

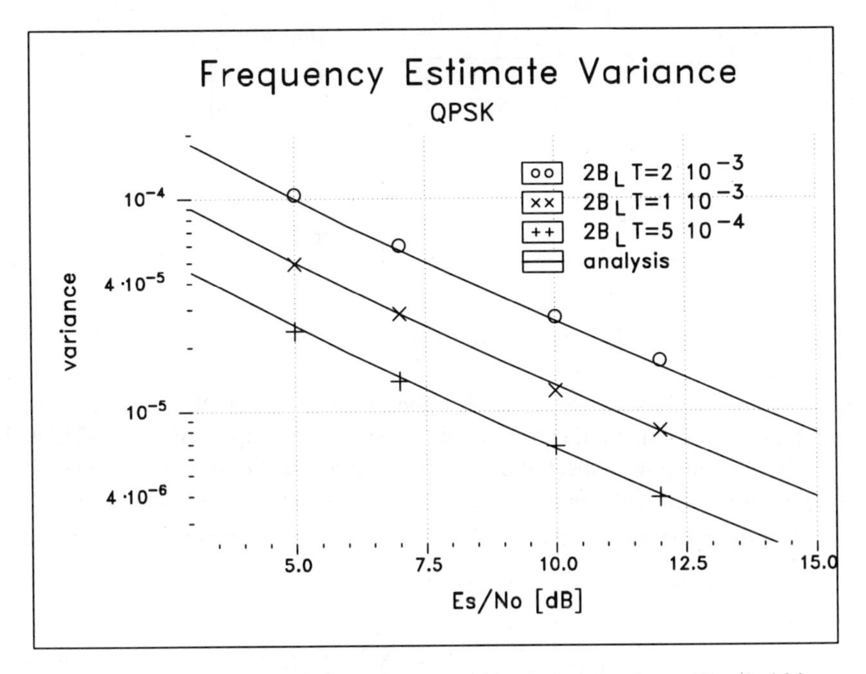

Figure 8-30 Simulated Variance Parametrized by the Loop Bandwidth; $T/T_s = 4$ and MF and FMF with 41 Taps Each (QPSK transmission)

and the frequency matched filter output

$$z_{\text{FMF}}(nT) = 0 + \sum_{k=-\infty}^{\infty} n(kT_s)\; g_{\text{FMF}}(nT - kT_s) \tag{8-148}$$

The output of the frequency error detector is given by

$$\begin{aligned} x(nT) = \text{Re}\Bigg\{ &\left(a_n + \sum_{k=-\infty}^{\infty} n(kT_s)\, g_{\text{MF}}(nT - kT_s)\right) \\ &\times \left(\sum_{k=-\infty}^{\infty} n^*(kT_s)\, g^*_{\text{FMF}}(nT - kT_s)\right)\Bigg\} \end{aligned} \tag{8-149}$$

The variance is determined using a linearized model of the frequency control loop and is given by

$$\sigma^2 = \frac{2B_L T}{K_\Omega^2}\; S_x\!\left(e^{j2\pi fT}\right)\Big|_{f=0} \tag{8-150}$$

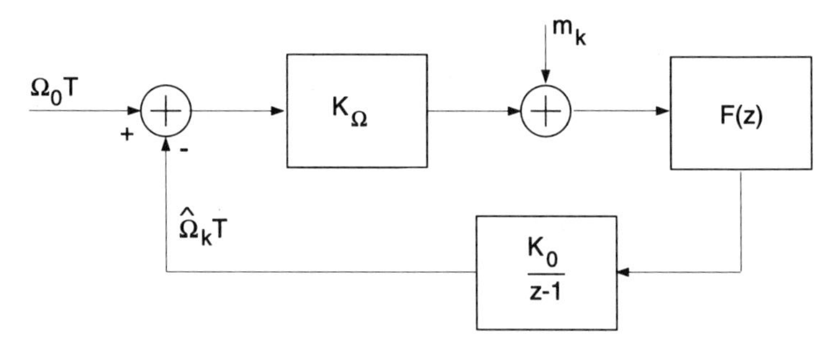

Figure 8-31 Linearized Model of the Frequency Control Loop

where K_Ω is the slope of the S curve in the origin, $2B_L T$ is the equivalent two-sided loop bandwidth of the transfer function of the loop in Figure 8-31, and $S_x\left(e^{j2\pi fT}\right)$ is the power spectral density of the frequency error output. Generally, (8-77) is only valid if the power spectral density is flat about the origin.

The power spectrum density $S_x\left(e^{j2\pi fT}\right)\big|_{f=0}$ can be calculated from the sum of the autocorrelation functions $R_{xx}(iT)$ via

$$S_x\left(e^{j2\pi fT}\right)\big|_{f=0} = \sum_{i=-\infty}^{\infty} R_{xx}(iT) \tag{8-151}$$

where the autocorrelation function of the frequency error detector output $x(lT)$ is given by

$$R_{xx}(i) = E[x(lT + iT)\, x(lT)] \tag{8-152}$$

The tedious determination of many values of the autocorrelation function must be performed if, as in our case, we cannot assume in advance that the power spectrum is flat, i.e., $R_{xx}(iT) \equiv 0 \quad \forall\ i \neq 0$. Below, we will see that it is the Nyquist properties of the transmitted pulse and the property of the frequency matched filter which lead to the self-noise-free characteristic results in a comfortable flat spectrum and thus confirm that our approach in (8-77) is valid.

Using (8-149) we start from the following expression of the frequency error detector output:

$$\begin{aligned}
2x(nT) &= z(n) z^*_{\text{FMF}}(n) + z^*(n) z_{\text{FMF}}(n) \\
&= a_n \sum_{k=-\infty}^{\infty} n^*(kT_s)\, g^*_{\text{FMF}}(nT - kT_s) && (1) \\
&\quad + \sum_{k=-\infty}^{\infty} n(kT_s)\, g_{\text{MF}}(nT - kT_s) \sum_{k=-\infty}^{\infty} n^*(kT_s)\, g^*_{\text{FMF}}(nT - kT_s) && (2) \\
&\quad + a^*_n \sum_{k=-\infty}^{\infty} n(kT_s)\, g_{\text{FMF}}(nT - kT_s) && (3) \\
&\quad + \sum_{k=-\infty}^{\infty} n^*(kT_s)\, g^*_{\text{MF}}(nT - kT_s) \sum_{k=-\infty}^{\infty} n(kT_s)\, g_{\text{FMF}}(nT - kT_s) && (4)
\end{aligned} \tag{8-153}$$

Inserting (8-153) into (8-152) we get 16 terms. Most of them vanish due to the assumed independency between the noise and the transmitted symbols.

As an example we will now consider the expressions which determine the $S\times N$ part. We have to take the expectation of the following expressions with respect to the data and the noise:

$$E_{S\times N} = E[x(nT)\, x((n+i)T)]_{|S\times N} \tag{8-154}$$

whereby only the expressions (1) and (3) in (8-153) give a contribution to $E_{S\times N}$. For all $i \neq 0$, the above equation equals zero since we assumed the transmitted symbols to be uncorrelated. For $i = 0$ we obtain

$$E_{S\times N} = \frac{1}{4}\left\{\sum |g_{\text{FMF}}|^2(nT - kT_s)\left(2\text{Re}\{a^{*2}n^2\} + 2|a|^2|n|^2\right)\right\} \tag{8-155}$$

The numerical value of (8-155) can easily be calculated in the frequency domain using the Poisson sum formula.

Applying the above approach to all of the other terms we find that all contributions to $R_{xx}(iT)$ are equal to zero for $i \neq 0$. Therefore we can conclude that the power spectral density $S_x\left(e^{j2\pi fT}\right)$ is flat and the expression for $\sigma^2 = \text{var}\left[\hat{\Omega}\text{T}/2\pi\right]$ in (8-77) is valid.

8.5.3 Main Points

- Frequency estimation algorithms which require *correct* timing are discussed. Both DA and NDA types exist.
- There are two basic types of frequency detectors. The first is based on the examination of the power spectrum (Section 8.5) of the received signal, and the second estimates, in essence, the phase increment between two subsequent samples (Section 8.4).
- For small frequency offsets all synchronization operations can be performed behind the matched filter.
- Removing the data dependence by a power operation is not recommended for high-order modulation schemes.
- Performance: The algorithms are asymptotically efficient. But at SNR regions of practical interest there exist large deviations from the CRB. The estimate is (nearly) unbiased for small $|\Omega T|/2\pi < 0.15$.
- There exist sufficient joint frame and frequency algorithms.

8.6 Frequency Estimators for MSK Signals

The MSK signal [eq. (3-3)] contains a term which is proportional to the frequency offset Ω

$$s(t) = \exp\{j[\Omega t + \phi(t - \varepsilon T)]\} \tag{8-156}$$

with

$$\phi(t) = 2\pi h \sum a_n \phi(t - nt) \tag{8-157}$$

The MSK signal is not band-limited. The analog prefilter must therefore be wide enough in order to pass most of the energy of the signal $s(t)$ in the presence of a frequency uncertainty Ω.

The received signal is sampled at rate $1/T_s$ to obtain $r_f(kT_s)$. By $M_s = T/T_s$ we denote the integer relating symbol and sampling rate. We introduce the double notation

$$r_{k,i} = r(t)\,\Big|_{t=(kM_s+i)T_s} \tag{8-158}$$

with $0 \le i \le M_s - 1$. The phase increment over one symbol T is obtained by multiplying $r_{k,i}$ by $r^*_{k-1,i}$

$$c_{k,i} = r_{k,i} r^*_{k-1,i} \tag{8-159}$$

$c_{k,i}$ is a function of the random data sequence $\{\mathbf{a}\}$, the timing parameter ε, and the frequency offset Ω. To remove the data dependency, $c_{k,i}$ must be processed in a suitable nonlinearity. Here squaring of $c_{k,i}$ is sufficient as will be seen shortly. The sampled signal $r_{k,i}$ is

$$\begin{aligned} r_{k,i} &= e^{j[\phi(kT+(i/M_s)T-\varepsilon_0 T)+\Omega_0(k+(i/M_s))T+\theta_0]} + n_{k,i} \\ (\Omega_0, \varepsilon_0) &: \text{ true, unknown parameter values} \end{aligned} \tag{8-160}$$

To verify whether $c_{k,i}$ produces an unbiased non-data-aided (NDA) estimate we have to take the expected value of $(c_{k,i})^2$ with respect to data and noise. The noise samples at symbol rate $1/T$ are assumed to be white Gaussian noise with variance $\sigma_n^2 = E\left[|n_k|^2\right] = E_s/N_0$. The data are assumed to be independent and identically distributed (i.i.d.).

Inserting the sampled signal $r_{k,i}$ into $(c_{k,i})^2$ yields

$$c^2_{k,i} = e^{j2\Delta\phi_{k,i}}\, e^{j2\Omega_0 T} + \text{ terms with noise} \tag{8-161}$$

with the phase increment $\Delta\phi_{k,i} = \phi(kT - \Delta\varepsilon_i T) - \phi((k-1)T - \Delta\varepsilon_i T)$ and the timing error

$$\begin{aligned} \Delta\varepsilon_i &= \varepsilon_0 - \frac{i}{M_s} \\ &= \varepsilon_0 - \varepsilon_i \end{aligned} \tag{8-162}$$

Since the frequency pulse of the MSK signal is of duration T, only three symbols are present in the phase increment $\Delta\phi_{k,i}$:

$$\begin{aligned} \Delta\phi_{k,i} &= 2\pi h\{(1/2)a_{k-2} - a_{k-2}q((1-\Delta\varepsilon_i)T) + a_{k-1}q((1-\Delta\varepsilon_i)T)\},\ \Delta\varepsilon_i > 0 \\ &= 2\pi h\{(1/2)a_{k-1} - a_{k-1}q(-\Delta\varepsilon_i T) + a_k q(-\Delta\varepsilon_i T)\},\ \Delta\varepsilon_i < 0 \end{aligned} \tag{8-163}$$

Taking the expected value with respect to noise and data a straightforward calculation shows that all terms of (8-161) involving noise vanish. With i.i.d. data symbols we obtain the following result:

$$\begin{aligned} E\{c_{k,i}^2\} &= (1/4)\left[e^{j2\pi h} + e^{-j2\pi h}\right] e^{j2\Omega_0 T} \\ &+ (1/4)\left[e^{j4\pi h[2q((1-\Delta\varepsilon_i)T)-1/2]} + e^{-j4\pi h[2q((1-\Delta\varepsilon_i)T)-1/2]}\right] e^{j2\Omega_0 T} \end{aligned} \tag{8-164}$$

Using in the previous equation

$$\begin{aligned} q(t) &= \frac{1}{2T} t \qquad\qquad 0 \le t \le T \\ h &= 1/2 \end{aligned} \tag{8-165}$$

we finally obtain

$$E\{c_{k,i}^2\} = -(1/2)[1 + \cos(2\pi\Delta\varepsilon_i)] e^{j2\Omega_0 T} \tag{8-166}$$

We observe:

1. The absolute value of the expected value is independent of Ω_0. Its maximum is obtained for $|\Delta\varepsilon_i| \to \min$

$$\max_i \left\{ E\left\{|c_{k,0}|^2\right\}, E\left\{|c_{k,1}|^2\right\}, \ldots, E\left\{|c_{k,M_s-1}|^2\right\} \right\} \to |\Delta\varepsilon_i|_{\min} \tag{8-167}$$

2. The argument of $E\left\{c_{k,i}^2\right\}$ is a function of the frequency offset Ω_0 only.

From these observations we deduce that the (ε, Ω) estimation can be decoupled. The estimator has the following structure. The preprocessed signal $c_{k,i}^2$ at rate $T_s = T/M_s$, $0 \le i \le M_s - 1$ is demultiplexed in M_s parallel data streams $c_{k,0}^2, \ldots, c_{k,M_s-1}^2$. The expected value operation is approximated by a time average in M_s parallel filters

$$\nu_{k,i} = \frac{1}{L} \sum_{n=0}^{L-1} c_{n,i}^2 \qquad\qquad 0 \le i \le M_s - 1 \tag{8-168}$$

From (8-167) follows that the sample position i which produces the maximum of $|\nu_{k,i}|$ is the estimate of ε_0 with the minimum residual timing error $\Delta\varepsilon_i$. The index i thus plays the role of a quantized trial parameter $\Delta\varepsilon_i$. Due to the quantization the estimate $\hat{\varepsilon}$ is in general biased with a maximum bias of

$$E[\Delta\varepsilon_i] \le \frac{1}{2M_s} \tag{8-169}$$

The argument of the filter output $\nu_{k,i}$ where i is the position with minimum residual timing error is an unbiased estimate for the frequency offset Ω:

$$\begin{aligned} &\hat{\Omega} T = 1/2 \arg\{\nu_{k,i}\} \\ &i: \text{ location of } |\Delta\varepsilon_i|_{\min} \end{aligned} \tag{8-170}$$

Figure 8-32 shows the structure of the (ε, Ω) digital feedforward demodulator

8.6.1 Performance Analysis

Using standard techniques the variance of the frequency estimate can be analytically obtained under the following idealized conditions:

$$\begin{aligned} &1) \quad \Omega_0 = 0 \\ &2) \quad \Delta\varepsilon_i = 0 \\ &3) \quad \text{matched filter reception} \end{aligned} \tag{8-171}$$

The estimate $\hat{\Omega}$ is given by

$$\begin{aligned} \hat{\Omega}T &= \frac{1}{2} \arg \left\{ \frac{1}{L} \sum_{k=-(L-1)/2}^{(L-1)/2} c_{k,i}^2 \right\} \\ &= \frac{1}{2} \arg \left\{ \frac{1}{L} \sum_k \operatorname{Re}\left(c_{k,i}^2\right) + j \operatorname{Im}\left(c_{k,i}^2\right) \right\} \end{aligned} \tag{8-172}$$

If the timing error is zero, the mean of the real part of $\nu_{k,i}$ is much larger than the mean and the variance of the imaginary part. Then, the same following approximation as introduced in Section 8.4 [eq. (8-121)] holds:

$$\operatorname{var}\left(\hat{\Omega}T\right) \approx \frac{1}{4} \frac{E\left\{(\operatorname{Im}(\nu_{k,i}))^2\right\}}{E\{\operatorname{Re}(\nu_{k,i})\}^2} \tag{8-173}$$

After some tedious algebraic operations one finds

$$\begin{aligned} \operatorname{var}\left(\frac{\hat{\Omega}T}{2\pi}\right) = \left(\frac{1}{2\pi}\right)^2 \Bigg\{ &\frac{2}{L^2} \left[\frac{1}{2E_s/N_0} + \frac{1}{(2E_s/N_0)^2}\right] \\ &+ \frac{8}{L} \left[\frac{1}{(2E_s/N_0)^2} + \frac{2}{(2E_s/N_0)^3} + \frac{1}{(2E_s/N_0)^4}\right] \Bigg\} \end{aligned} \tag{8-174}$$

It is interesting to note that the expression (8-174) coincides with the results obtained for the BPSK NDA estimator in Section 8.4, eq. (8-125). But be aware that in the above calculation we did not take into account the ISI effects which arise if the analog prefilter cuts away the spectrum of the MSK signal beyond $\omega > 2\pi B$, and we neglect that the noise taken at symbol rate is correlated in case of a matched filter reception of an MSK signal.

Additional Remarks

The bit error performance of the demodulated structure of Figure 8-32 for differentially coherent reception is discussed in [1] taking into account the distortion

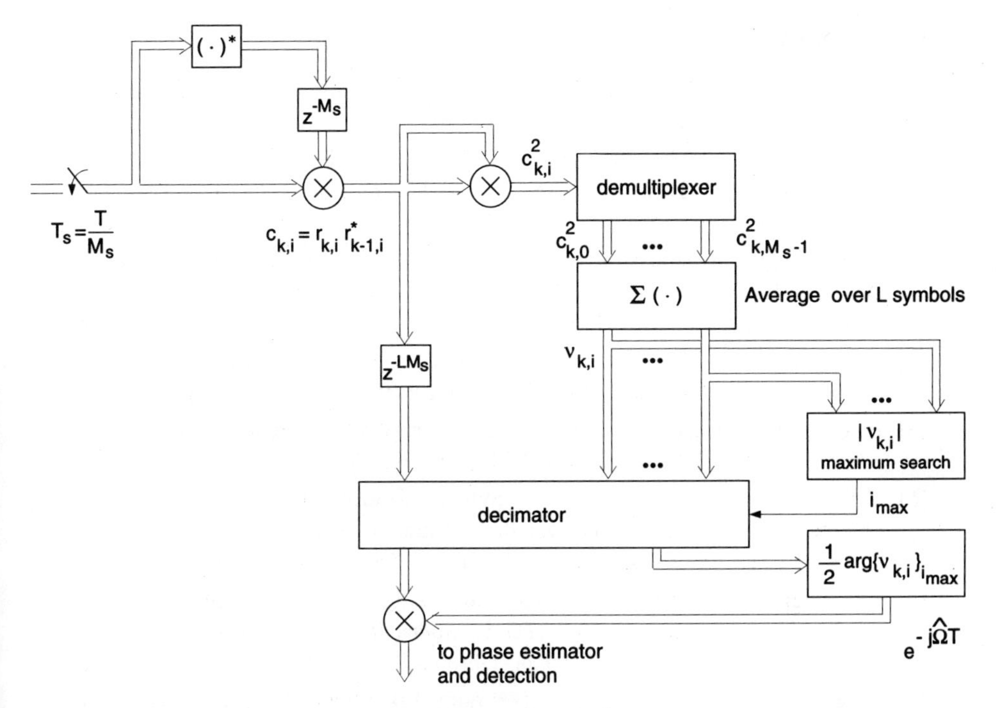

Figure 8-32 Feedforward Estimator for Timing and Frequency Offset

caused by the analog prefilter for small BT products as well as the imperfect timing estimate. The paper addresses the AWGN and Fading channel.

Main Points

- Exemplarily it was shown that the same methods – which were derived for linear modulated signals – can be applied to MSK signals.
- For MSK signals the (ε, Ω) estimation can be decoupled.
- Whereas the frequency offset is unbiased, the timing estimation suffers from a bias.

8.7 Bibliographical Notes

Multi carrier systems are of great practical interest for transmission over frequency-selective-fading channels [2, 3]. A good introduction into the subject can be found in the paper by Bingham [4]. Additional references are listed below.

Exact frequency estimation is of particular importance. The Ph.D. dissertation of Ferdinand Claßen [5] thoroughly discusses the algorithm design and performance analysis of a digital OFDM receiver. It includes a bibliography containing major references.

Bibliography

[1] R. Mehlan, *Ein Transceiverkonzept für die Interfahrzeugkommunikation bei 64 GHz*. PhD thesis, RWTH Aachen, 1995.

[2] ETSI, "Digital Audio Broadcasting (DAB) to Mobile, Portable and Fixed Receivers: Final Draft pr ETS 300 401," *Technical Report, European Telecommunications Standards Institute*, Nov. 1994.

[3] ETSI, "Digital Broadcasting Systems for Television, Sound and Data Services: Draft pr ETS 300 744," *Technical Report, European Telecommunications Standards Institute*, May 1996.

[4] J. A. Bingham, "Multicarrier Modulation for Data Transmission: An Idea Whose Time Has Come," *IEEE Comm. Magazine*, pp. 5–14, May 1990.

[5] F. Claßen, *Systemkomponenten für eine terrestrische digitale mobile Breitbandübertragung*. PhD thesis, RWTH Aachen, 1996.

[6] F. Claßen and H. Meyr, "Frequency Synchronization for OFDM Systems Suitable for the Communication over the Frequency Selective Fading Channel," *Proc. of VTC' 94, Stockholm, Sweden*, pp. 1655–1659, June 1994.

[7] F. Claßen and H. Meyr, "Synchronization Algorithms for an OFDM System for Mobile Communications," *ITG-Fachbericht 130 Codierung für Quelle, Kanal und Übertragung, München, Germany*, Oct. 1994.

[8] F. Daffara and A. Chouly, "Maximum Likelihood Frequency Detectors for Orthogonal Multicarrier Systems," *Proc. ICC'93, Geneva, Switzerland*, pp. 761–765, May 1993.

[9] F. Daffara and A. Chouly, "A New Frequency Detector for Orthogonal Multicarrier Transmission Techniques," *Proc. VTC95, Chicago, USA*, pp. 804–809, July 1995.

[10] P. Hoeher, "TCM on Frequency Selective Land Mobile Fading Channels," *Proc. Tirrenia Int'l Workshop on Digital Communications*, Sept. 1991.

[11] P. Moose, "A Technique for Orthogonal Frequency Division Multiplexing Frequency Offset Correction," *IEEE Trans. Commun.*, vol. 43, pp. 2908–2914, Oct. 1994.

[12] H. Sari, G. Karam, and Isabelle Jeanclaude, "Transmission Techniques for Digital Terrestrial TV Broadcasting," *IEEE Communications Magazine*, pp. 100–109, Feb. 1995.

[13] T. Pollet, M. Van Bladel, and M. Moeneclaey, "BER Sensitivity of OFDM Systems to Carrier Frequency Offset and Wiener Phase Noise," *IEEE Trans. Commun.*, vol. 43, pp. 191–193, Mar. 1995.

[14] T. M. Schmidl and D. C. Cox, "Low-Overhead, Low-Complexity [Burst] Synchronization for OFDM," *Proc. of ICC'96, Dallas*, June 1996.

Chapter 9 Timing Adjustment by Interpolation

In this chapter we focus on digital interpolation and interpolator control. In Section 9.1 we discuss approximations to the ideal interpolator. We first consider FIR filters which approximate the ideal interpolator in the mean square sense. A particularly appealing solution for high rate applications will be obtained if the dependency of each filter tap coefficient on the fractional delay is approximated by a polynomial in the fractional delay. It is shown that with low-order polynomials excellent approximations are possible.

In Section 9.2 we focus on how to determine the basepoint m_n and fractional delay μ_n, considering either a timing error feedback system or a timing estimator.

9.1 Digital Interpolation

The task of the interpolator is to compute intermediate values between signal samples $x(kT_s)$. The ideal linear interpolator has a frequency response (Section 4.2.2):

$$H_I\left(e^{j\omega T_s}, \mu T_s\right) = \frac{1}{T_s} \sum_{n=-\infty}^{\infty} H_I\left(\omega - \frac{2\pi}{T_s} n, \mu T_s\right) \tag{9-1}$$

with

$$H_I(\omega, \mu T_s) = \begin{cases} T_s \exp(j\omega\mu T_s) & |\omega/2\pi| < 1/2T_s \\ 0 & \text{elsewhere} \end{cases} \tag{9-2}$$

and is shown in Figure 9-1.

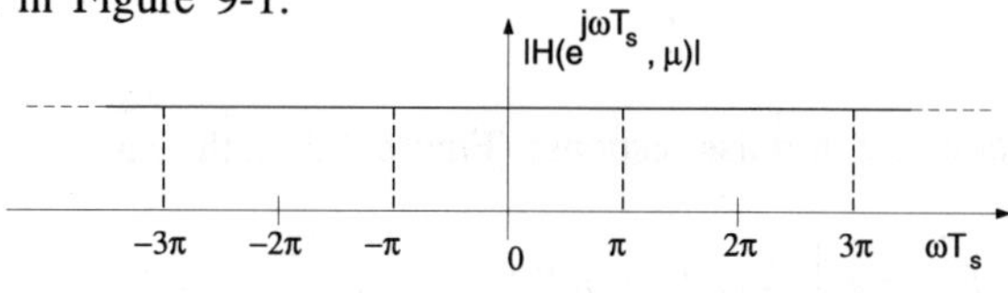

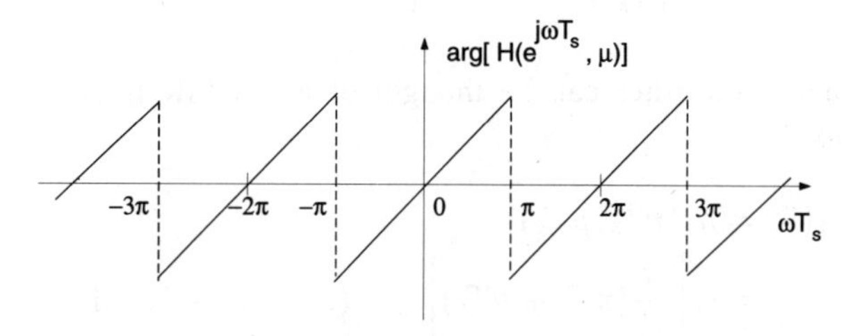

Figure 9-1 Frequency Response of the Ideal Interpolator

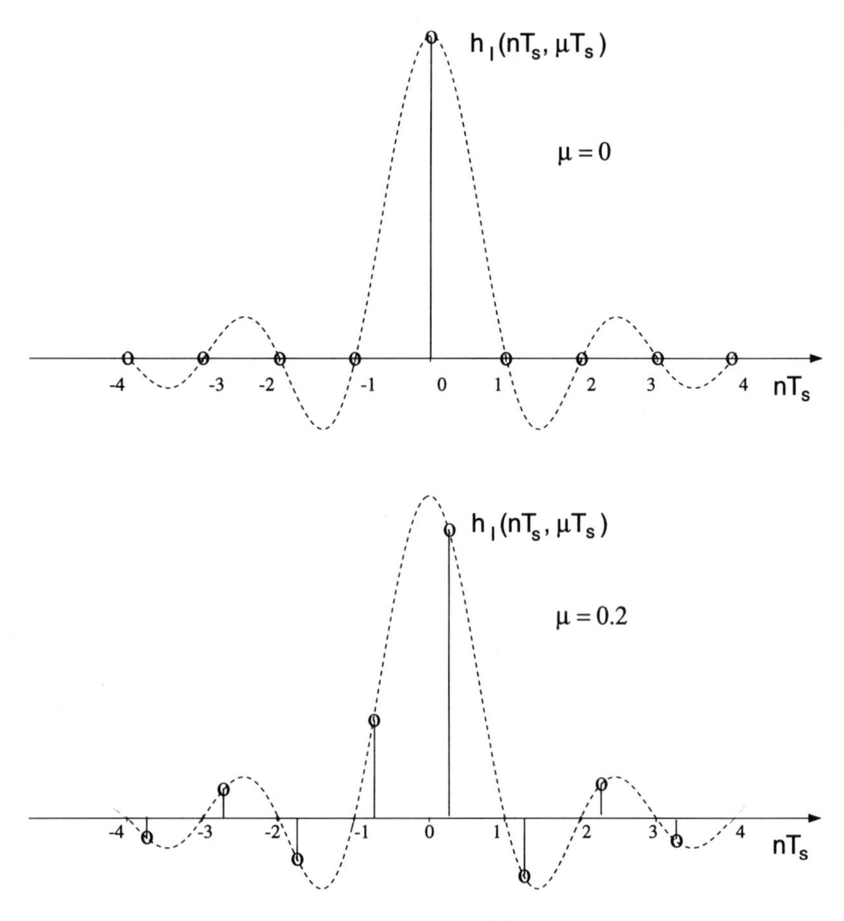

Figure 9-2 Impulse Response of the Ideal Interpolator

The corresponding impulse response (Figure 9-2) is the sampled $\mathrm{si}(x)$ function

$$h_I(nT_s, \mu T_s) = \mathrm{si}\left[\frac{\pi}{T_s}(nT_s + \mu T_s)\right] \qquad (n = \ldots, -1, 0, 1, \ldots) \tag{9-3}$$

Conceptually, the filter can be thought of as an FIR filter with an infinite number of taps

$$\begin{aligned} h_n(\mu) &= h_I(nT_s, \mu T_s) \\ &= \mathrm{si}\left[\frac{\pi}{T_s}(nT_s + \mu T_s)\right] \qquad (n = \ldots, -1, 0, 1, \ldots) \end{aligned} \tag{9-4}$$

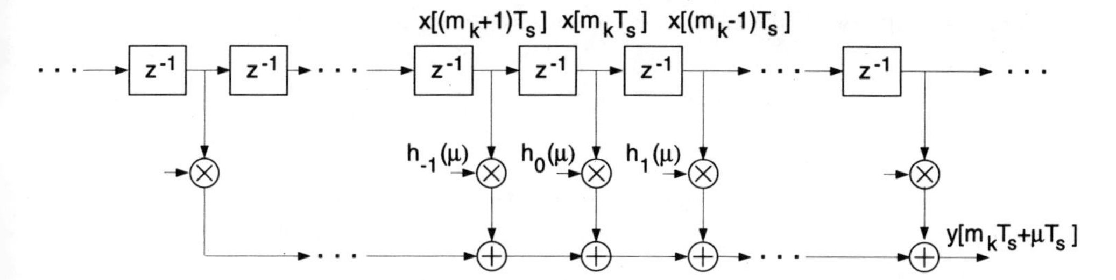

Figure 9-3 FIR Filter Structure of the Ideal Interpolator

The taps are a function of μ. For a practical receiver the interpolator must be approximated by a finite-order FIR filter

$$H\left(e^{j\omega T_s}, \mu\right) = \sum_{n=-I_1}^{I_2} h_n(\mu) e^{-j\omega T_s n} \tag{9-5}$$

In Figures 9-3 and 9-4, the FIR filter structures of the ideal and the fourth-order interpolating filter, respectively, are shown.

The filter performs a linear combination of the $(I_1 + I_2 + 1)$ signal samples $x(nT_s)$ taken around the basepoint m_k:

$$y(m_k T_s + \mu T_s) = \sum_{n=-I_1}^{I_2} x[(m_k - n)T_s]\, h_n(\mu) \tag{9-6}$$

9.1.1 MMSE FIR Interpolator

The filter coefficients $h_n(\mu)$ must be chosen according to a criterion of optimality. A suitable choice is to minimize the quadratic error between the

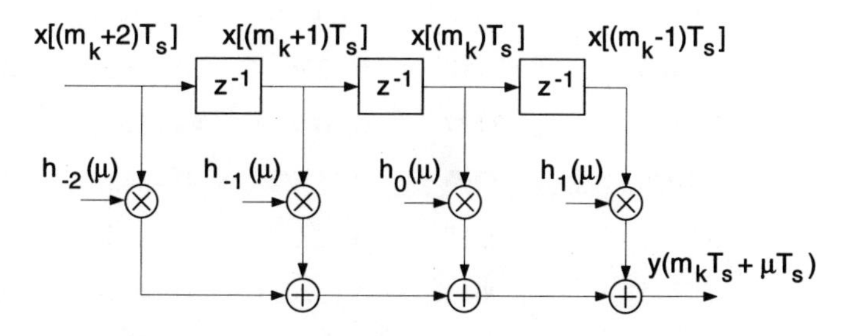

Figure 9-4 Fourth-Order Interpolating Filter

impulse response of the ideal interpolator and its approximation.

$$e^2(\mu) = \int_{-2\pi B}^{2\pi B} \left| e^{j\omega T_s \mu} - \sum_{n=-I_1}^{I_2} h_n(\mu) e^{-j\omega T_s n} \right|^2 d\omega \quad \rightarrow \min. \tag{9-7}$$

where B is the one-sided signal bandwidth. The optimization is performed within the passband of the signal $x(t)$. No attempt is made to constrain the frequency response outside B.

The result of the optimization for $\mu = 0$ and $\mu = 1$ are immediately evident. For $\mu = 0$ the integrand becomes zero in the entire interval for

$$h_n(0) = \begin{cases} 1 & n = 0 \\ 0 & \text{else} \end{cases} \tag{9-8}$$

For $\mu = 1$ we obtain

$$h_n(1) = \begin{cases} 1 & n = -1 \\ 0 & \text{else} \end{cases} \tag{9-9}$$

From (9-6) we learn that for these two values the interpolated function equals the sampled values of the input signal as was to be expected:

$$\begin{aligned} y(m_k T_s) &= x(m_k T_s) \\ y[(m_k + 1)T_s] &= x[(m_k + 1)T_s] \end{aligned} \tag{9-10}$$

How should the intervals I_1 and I_2 be chosen? It seems plausible that maximum accuracy is obtained when interpolation is performed in the center interval. From this it follows that the number of samples should be even with $I_1 = N$ and $I_2 = N-1$. The reader interested in mathematical details is referred to the paper by Oetken [1].

Example: MMSE FIR Interpolator with 8 Taps

Table 9-1 Coefficients of the MMSE interpolator with $B = 1/4T_s$ and $N = 4$

n \ μ	0.1	0.2	0.3	0.4	0.5
-4	-0.00196	-0.00376	-0.00526	-0.00630	-0.00678
-3	0.01091	0.02118	0.02990	0.03622	0.03946
-2	-0.03599	-0.07127	-0.10281	-0.12758	-0.14266
-1	0.10792	0.22795	0.35522	0.48438	0.60984
0	0.96819	0.90965	0.82755	0.72600	0.60984
1	-0.06157	-0.10614	-0.13381	-0.14545	-0.14266
2	0.01481	0.02662	0.03480	0.03908	0.03946
3	-0.00240	-0.00439	-0.00582	-0.00663	-0.00678

The values for $\mu > 0.5$ are obtained from those for $\mu < 0.5$ from

$$h_l(1-\mu) = h_{-l-1}(\mu) \tag{9-11}$$

A plot of the tap values for $\mu = 0.1,\ 0.25$, and 0.5 is shown in Figure 9-5. The plot can be combined to Figure 9-6.

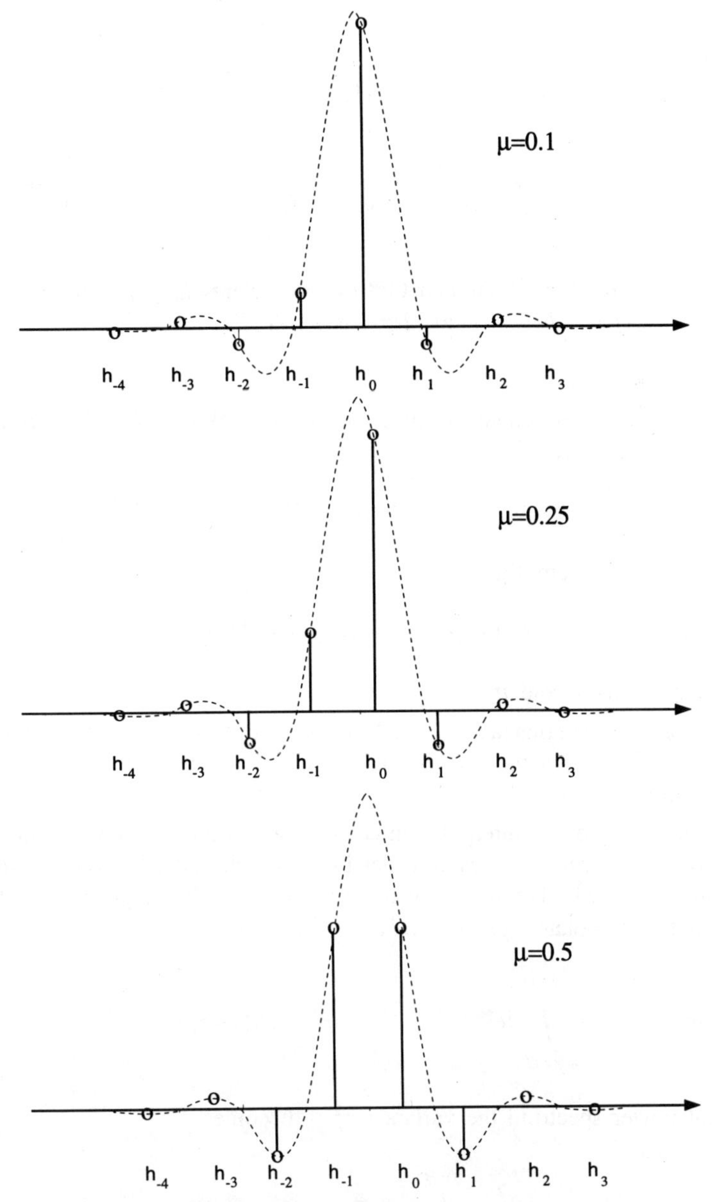

Figure 9-5 Plot of Tap Coefficients $h_n(\mu)$ for Various Values of μ. Parameter $BT_s = 0.25$

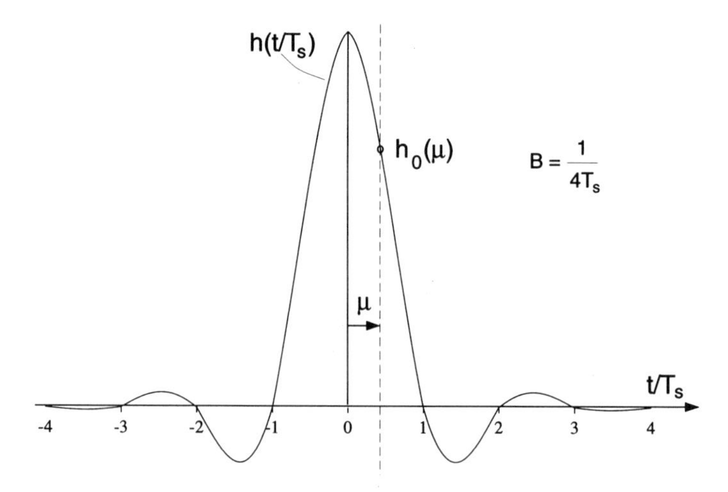

Figure 9-6 Combined Plot of Coefficients $h_n(\mu)$, Where $h_n(\mu) = h(t/T_s)$ at $t = (n+\mu)T_s$

The values $h_n(\mu)$ equal the function $h(t/T_s)$ taken at $t = (n+\mu)T_s$. The normalized delay error

$$e_D(\omega,\mu) = \mu\omega T_s - \arg H_{\text{opt}}\left(e^{j\omega T_s},\mu\right) \tag{9-12}$$

and the normalized amplitude error

$$e_A(\omega,\mu) = 1 - \left|H_{\text{opt}}\left(e^{j\omega T_s},\mu\right)\right| \tag{9-13}$$

are plotted versus ω and μ in Figure 9-7.

We observe level plateaus for the frequency range of interest. Since no attempt was made to constrain the out-of-band response, a steep error increase can be observed for $\omega > 2\pi B$ $(B = 1/4T_s)$.

The accuracy of the interpolation is a function of the interpolator filter as well as the signal spectrum. Let us consider the stochastic signal $x(nT_s)$ with power spectrum $S_x\left(e^{j\omega T_s}\right)$. The mean square error between the output of the ideal and approximate interpolator, given $x(nT_s)$ as input, is

$$\sigma^2_{e(\mu)} = \frac{T_s}{2\pi}\int_{-2\pi B}^{2\pi B} \left|e^{j\omega T_s \mu} - H_{\text{opt}}\left(e^{j\omega T_s},\mu\right)\right|^2 S_x\left(e^{j\omega T_s}\right)d\omega \tag{9-14}$$

For a flat power spectrum the variance $\sigma^2_{e(\mu)}$ becomes

$$\sigma^2_{e(\mu)} = \frac{\sigma_x^2}{4\pi B}\int_{-2\pi B}^{2\pi B} \left|e^{j\omega T_s \mu} - H_{\text{opt}}\left(e^{j\omega T_s},\mu\right)\right|^2 d\omega \tag{9-15}$$

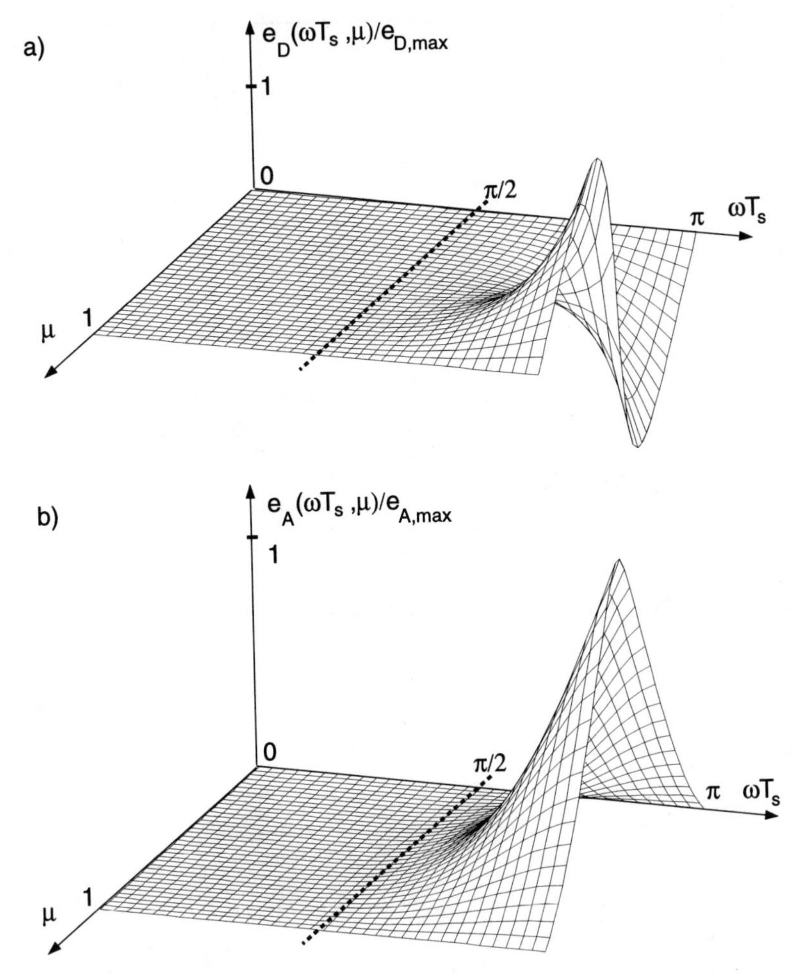

Figure 9-7 (a) Normalized Delay Error; (b) Normalized Amplitude Error

where σ_x^2 is the variance of the input signal. The function has been numerically evaluated to obtain the fractional delay $\mu_{\max}$ corresponding to the worst-case $\sigma_{e,\max}^2$. In Figure 9-8 we have plotted the number of filter taps $2N$ for various values of $\sigma_{e,\max}^2$ as a function of the useful bandwidth B.

From Figure 9-8 it is possible to select the order of the FIR interpolating filter to meet a specified error requirement. This seems an appropriate performance measure for communication applications. Since the values of μ are equiprobable the variance σ_e^2 averaged over all possible values of μ is also of interest:

$$\overline{\sigma}_e^2 = \frac{\sigma_x^2}{4\pi B} \int\limits_0^1 \int\limits_{-2\pi B}^{2\pi B} \left| e^{j\omega T_s \mu} - H_{\text{opt}}\left(e^{j\omega T_s}, \mu\right) \right|^2 d\omega \; d\mu \tag{9-16}$$

A plot of this quantity is shown in Figure 9-9.

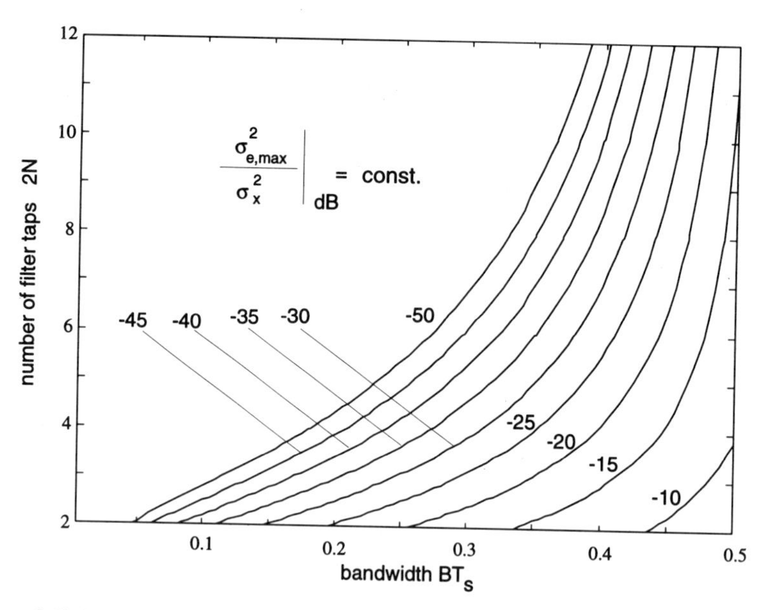

Figure 9-8 Number of Taps versus Normalized Bandwidth for Constant $\sigma^2_{e,\max}$

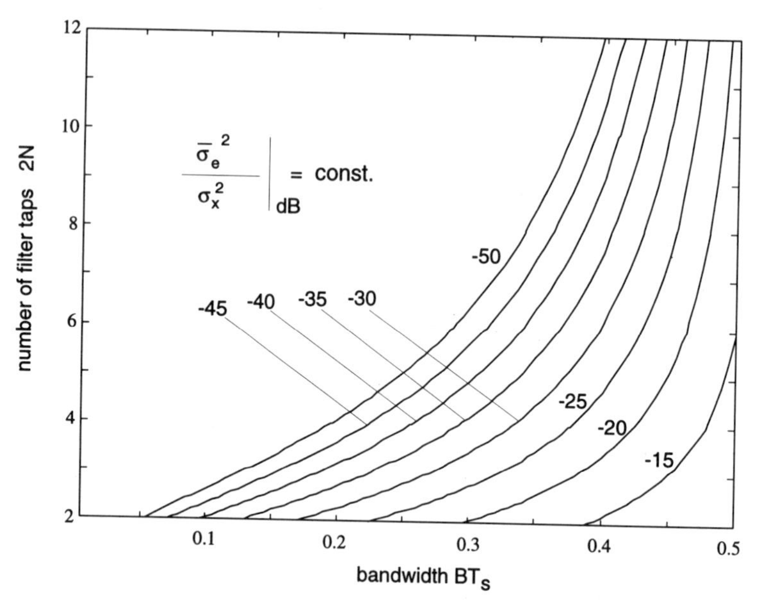

Figure 9-9 Number of Taps versus Normalized Bandwidth for Constant $\overline{\sigma}^2_e$

The $2N$ coefficients of the FIR filters must be precomputed and stored in memory for a number of L possible values $\mu_l = l/L$. As a consequence, the recovered clock suffers from a maximum discretization error of $L/2$.

Assume the coefficients $h_n(\mu_l)$ are represented by a word length of W bits. For each interpolation these $2N$ words must be transferred to the filter structure. The complexity of the transfer thus can easily become the limiting factor in a VLSI realization. An alternative in high-speed applications is to use a bank of parallel filters where each filter corresponds to a different quantized value of μ.

Implementation complexity depends upon the structure parameters, viz. order of filter $2N$, time discretization L, and word length W.

9.1.2 Polynomial FIR Interpolator

We approximate each coefficient $h_n(\mu)$ by a (possibly different) polynomial in μ of degree $M(n)$:

$$h_n(\mu) = \sum_{m=0}^{M(n)} c_m(n)\, \mu^m \tag{9-17}$$

For a $2N$th order FIR filter the

$$2N + \sum_{n=-N}^{N-1} M(n) \tag{9-18}$$

coefficients are obtained by minimizing the quadratic error:

$$\int_0^1 \int_{-2\pi B}^{2\pi B} \left| e^{j\omega T_s \mu} - \sum_{n=-N}^{N-1} \left[\sum_{m=0}^{M(n)} c_m(n) \mu^m \right] e^{-jn\omega T_s} \right|^2 d\omega\, d\mu \quad \rightarrow \min. \tag{9-19}$$

Notice that we optimize the quadratic error averaged over all fractional delays μ. Compare this with (9-7) where a set of optimal coefficient was sought for every μ. We impose the following constraints on (9-19),

$$h_n(0) = \begin{cases} 1 & \text{for } n = 0 \\ 0 & \text{else} \end{cases} \qquad h_n(1) = \begin{cases} 1 & \text{for } n = -1 \\ 0 & \text{else} \end{cases} \tag{9-20}$$

Since we restrict the function $h_n(\mu)$ to be of polynomial type, the quadratic error of the polynomial interpolator will be larger than for the MMSE interpolator discussed previously, although it can be made arbitrarily small by increasing the degree of the polynomial. Since the polynomial interpolator performs worse, why then consider it at all? The main reason is that the polynomial interpolator can be implemented very efficiently in hardware as will be seen shortly.

For simplicity (though mathematically not necessary) we assume that all polynomials have the degree of $M(n) = M$. Inserting for $h_n(\mu)$ the polynomial expression of (9-17) the FIR transfer function reads

$$\begin{aligned} H(z,\mu) &= \sum_{n=-N}^{N-1} h_n(\mu)\, z^{-n} \\ &= \sum_{n=-N}^{N-1} \left[\sum_{m=0}^{M} c_m(n)\, \mu^m \right] z^{-n} \end{aligned} \tag{9-21}$$

Interchanging summation we obtain

$$H(z,\mu) = \sum_{m=0}^{M} \mu^m \left[\sum_{n=-N}^{N-1} c_m(n)\, z^{-n} \right] \tag{9-22}$$

For every degree m of the polynomial the expression in the squared brackets describes a time-invariant FIR filter which is independent of μ:

$$H_m(z) = \sum_{n=-N}^{N-1} c_m(n) z^{-n} \tag{9-23}$$

The polynomial interpolator can thus be realized as a bank of M parallel FIR filters where the output of the mth branch is first multiplied by μ^m and then summed up. This structure was devised by Farrow [2].

The Farrow structure is attractive for a high-speed realization. The $(M+1)$ FIR filters with constant coefficients can be implemented very efficiently as VLSI circuits. Only one value for μ must be distributed to the M multipliers which can be pipelined (see Figure 9-10). The basic difference of the polynomial interpolator with respect to the MMSE interpolator is that the coefficients are computed in real time rather than taken from a table.

Example: Linear Interpolator

The simplest polynomial interpolator is obtained for $M = 1$ and $N = 1$. The four coefficients are readily obtained from the constraints

$$h_n(0) = \begin{cases} 1 & n = 0 \\ 0 & n = -1 \end{cases} \qquad h_n(1) = \begin{cases} 0 & n = 0 \\ 1 & n = -1 \end{cases} \tag{9-24}$$

The interpolator (Figure 9-11) performs a linear interpolation between two samples,

$$y(kT_s) = x(kT_s) + \mu[x((k+1)T_s) - x(kT_s)] \tag{9-25}$$

The coefficients $c_m(n)$ for a set of parameter values of practical interest are tabulated Section 9.1.4.

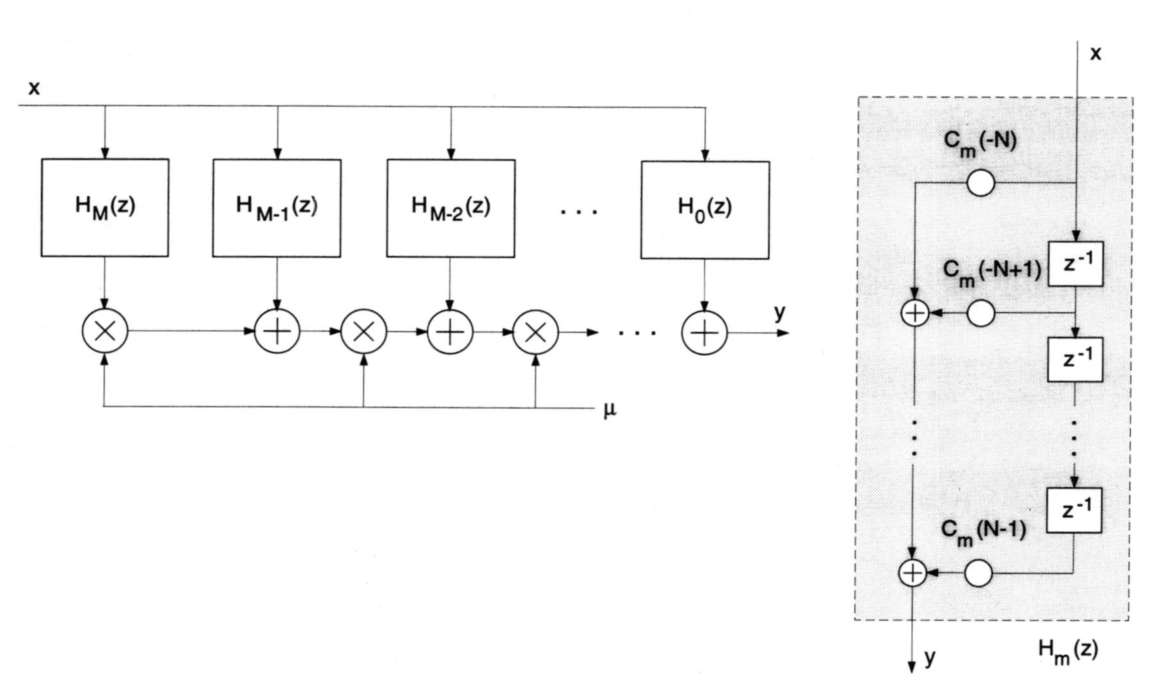

Figure 9-10 Functional Diagram of the Farrow Structure of the Polynomial Interpolator

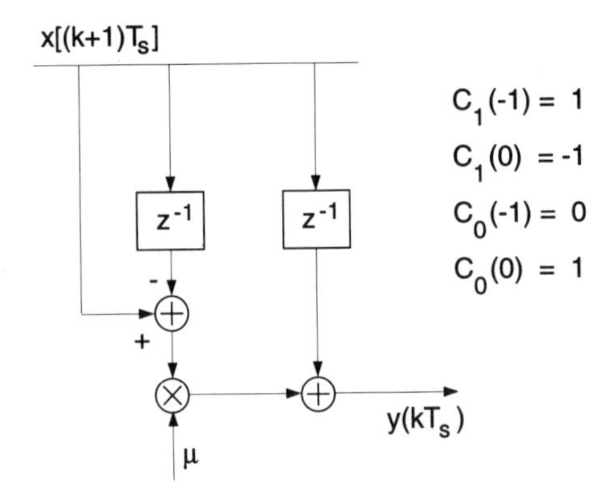

Figure 9-11 Linear Interpolator

The number of taps of the polynomial interpolator versus normalized bandwidth is plotted in Figure 9-12 for various values of σ_e^2.

A number of interesting conclusions can be drawn from this figure. For a given signal bandwidth BT_s there is a trade-off between signal degradation and signal processing complexity, which is roughly estimated here by the number of taps $2N$ and polynomial degree M. It can be seen that for a signal bandwidth $BT_s = 0.25$ already $2N = 2$ taps (independent of the polynomial order) will suffice to produce less than -20 dB signal degradation. For $BT_s = 0.45$ and -20 dB signal degradation the minimal number of taps is $2N = 6$ with a polynomial order

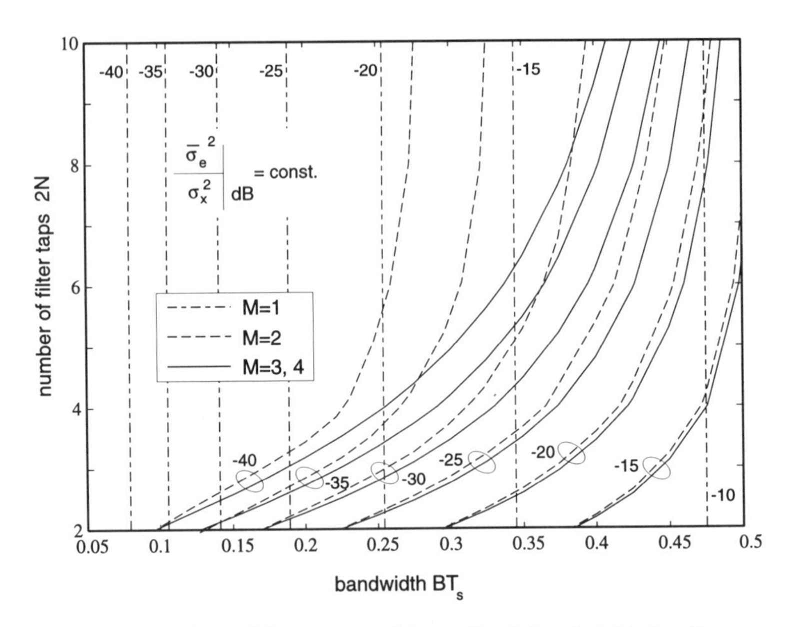

Figure 9-12 Number of Taps versus Normalized Bandwidth for Constant $\overline{\sigma}_e^2$

$M > 2$. In the plotted domain $(2N \leq 10)$ almost no difference between results for the third- and fourth-order polynomial has been observed. The independence of the linear interpolator of the number of taps $2N$ is a consequence of the constraints (9-24).

In Figure 9-13 the variance $\overline{\sigma}_e^2/\sigma_x^2$ is plotted versus the excess bandwidth α for a random signal with raised cosine spectrum. For most cases a linear interpolation between two samples will be sufficient. Doubling the number of taps is more effective than increasing the order of the polynomial to reduce the variance.

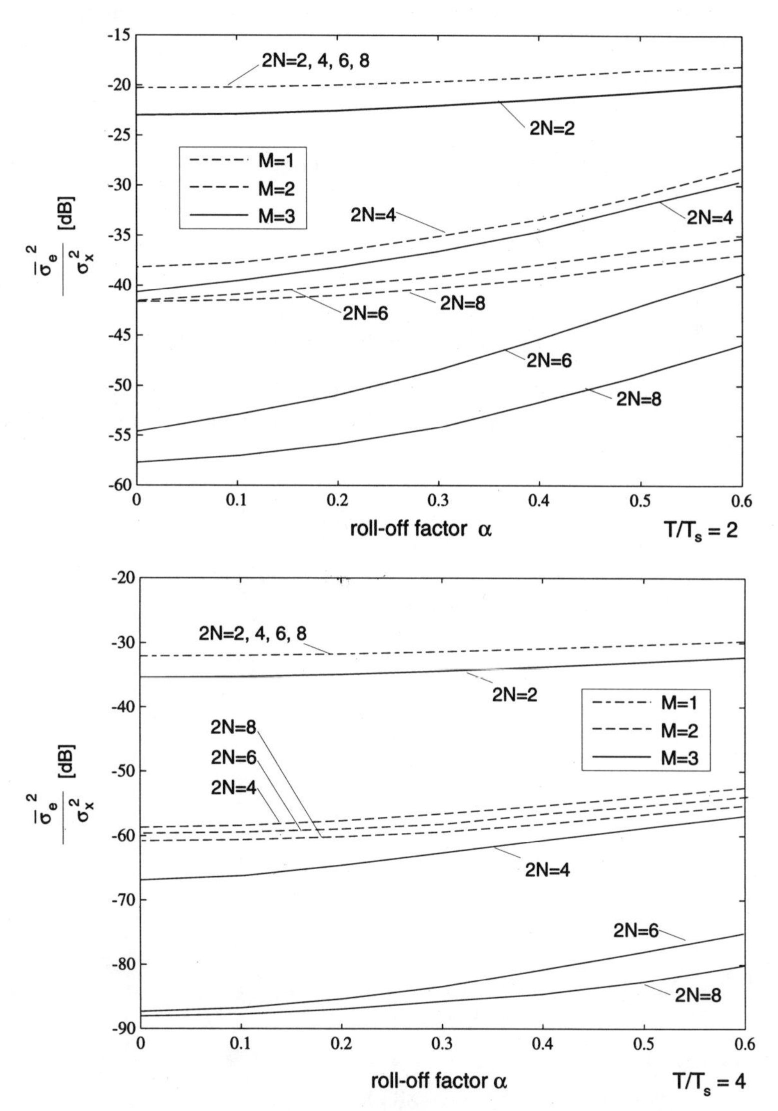

Figure 9-13 SNR Degradation versus Rolloff Factor α for (a) $T/T_s = 2$ and (b) $T/T_s = 4$

9.1.3 Classical Lagrange Interpolation

In numerical mathematics the task of interpolation for our purposes can be stated as follows. Given a function $x(t)$ defined for $t_{-(N-1)}, \ldots, t_0, \ldots, t_N$, find a polynomial $P(t)$ of degree $(2N-1)$ which assumes the given values $x(t_n)$,

$$P(t_n) = x(t_n) \qquad (n = -(N-1), \ldots, -1, 0, 1, \ldots, N) \tag{9-26}$$

(see Figure 9-14). In general, the points t_n do not need to be equidistant, nor does the number of points have to be an even number $2N$. As shown in any textbook on numerical mathematics the solution to this problem is unique. There exists a Lagrange polynomial of degree $2N-1$:

$$P(t) = \sum_{n=-(N-1)}^{N} \lambda_n \left[\left(t - t_{-(N-1)}\right) \cdots (t - t_{n-1})(t - t_{n+1}) \cdots (t - t_N) \right] x\,(t_n) \tag{9-27}$$

with

$$\lambda_n = \frac{1}{\left(t_n - t_{-(N-1)}\right) \cdots (t_n - t_{n-1})(t_n - t_{n+1}) \cdots (t_n - t_N)} \tag{9-28}$$

Using the definition of (9-27) it is easily verified that for every t_n, we have $P(t_n) = x(t_n)$ as required. The polynomial $P(t)$ is a linear combination of the values $x(t_n)$,

$$P(t) = \sum_{n=-(N-1)}^{N} q_n(t) x(t_n) \tag{9-29}$$

with the so-called Lagrange coefficients

$$q_n(t) = \lambda_n \left[\left(t - t_{-(N-1)}\right) \cdots (t - t_{n-1})(t - t_{n+1}) \cdots (t - t_N) \right] \tag{9-30}$$

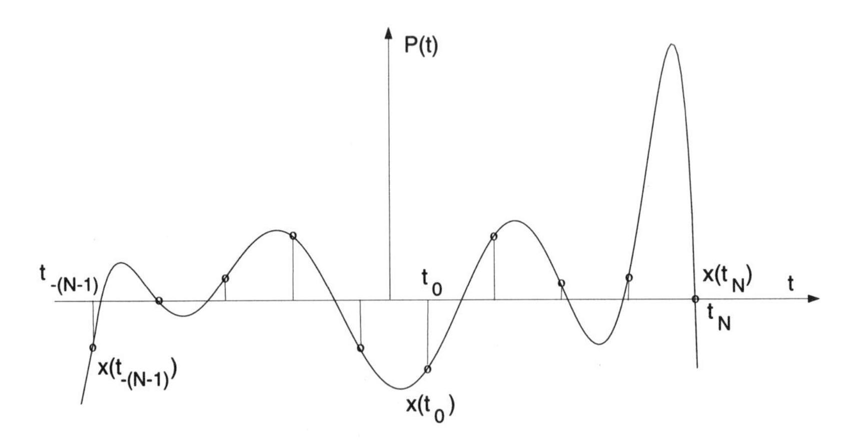

Figure 9-14 Lagrange Interpolation

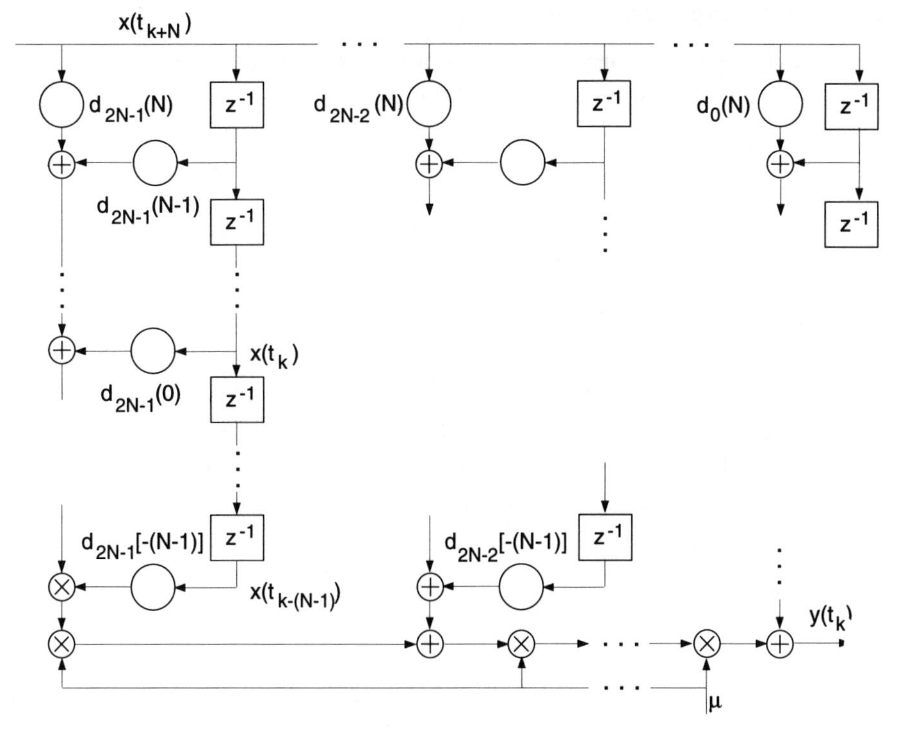

Figure 9-15 Farrow Structure of Lagrange Interpolation

Since the FIR filter also computes a linear combination of sample values $x(kT_s)$, it is necessary to point out the differences between the two approaches.

As we are only interested in the interpolated values in the central interval $0 \leq t \leq 1$, we set $t = \mu$ in the definition of the Lagrange coefficients $q_n(t)$. Every $q_n(\mu)$ is a polynomial in μ which can be written as

$$q_n(\mu) = \sum_{m=0}^{2N-1} d_m(n)\, \mu^m \tag{9-31}$$

Inserting into (9-29) we obtain

$$\begin{aligned} P(\mu) &= \sum_{n=-(N-1)}^{N} \left[\sum_{m=0}^{2N-1} d_m(n)\, \mu^m \right] x(t_n) \\ &= \sum_{m=0}^{2N-1} \mu^m \left[\sum_{n=-(N-1)}^{N} d_m(n)\, x(t_n) \right] \end{aligned} \tag{9-32}$$

Thus, from (9-32) we learn that $P(\mu)$ can be computed as the output of a Farrow structure (see Figure 9-15). However, there are some basic differences to the polynomial interpolator:

- The degree of the Lagrange interpolator always equals the number of samples minus one. The array $d_m(n)$ is thus quadratic.
- For the polynomial interpolator the degree of the polynomial M can be chosen independently of the number of samples. The array $c_m(n)$ is, in general, not quadratic.
- There is a single polynomial valid for the entire range $t_{-(N-1)} \leq t < t_N$. For the polynomial interpolator, to each tap $h_n(\mu)$ there is associated a different polynomial valid just for this tap.
- The coefficients of the Lagrange interpolator are completely specified given $x(t_n)$, while the polynomial interpolator coefficients are the result of an optimization. Thus, even for quadratic polynomial interpolator structures they have nothing in common with the Lagrange array coefficients $d_m(n)$.

For more details on Lagrange interpolation see the work of Schafer and Rabiner [3].

Example: Cubic Lagrange Interpolator

Using (9-31), we get

$$\begin{aligned} q_{-1}(\mu) &= -\frac{1}{6}\mu^3 + \frac{1}{2}\mu^2 - \frac{1}{3}\mu \\ q_0 \;\; (\mu) &= \frac{1}{2}\mu^3 - \mu^2 - \frac{\mu}{2} + 1 \\ q_1 \;\; (\mu) &= -\frac{1}{2}\mu^3 + \frac{\mu^2}{2} + \mu \\ q_2 \;\; (\mu) &= \frac{1}{6}\mu^3 - \frac{1}{6}\mu \end{aligned} \tag{9-33}$$

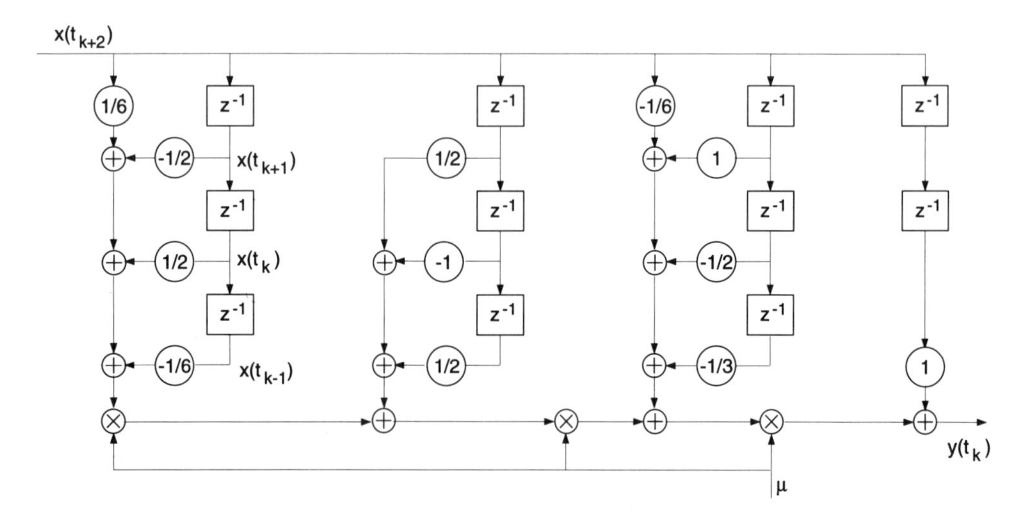

Figure 9-16 Farrow Structure of a Cubic Lagrange Interpolator

9.1.4 Appendix

In Tables 9–2 to 9–9 Farrow coefficients for various N, M, and $BT_s = 1/4$ are listed.

Table 9-2 *N*=1 *M*=2

$n\backslash m$	0	1	2
-1	1	-0.80043	-0.19957
0	0	1.19957	-0.19957

Table 9-3 *N*=2 *M*=2

n \ m	0	1	2
-2	0	-0.36881	-0.36881
-1	1	-0.65696	-0.34303
0	0	1.34303	-0.34303
1	0	-0.36881	0.36881

Table 9-4 *N*=2 *M*=3

n \ m	0	1	2	3
-2	0	-0.48124	0.70609	-0.22485
-1	1	-0.33413	-1.31155	0.64567
0	0	1.02020	0.62547	-0.64567
1	0	-0.25639	0.03154	0.22485

Table 9-5 *N*=3 *M*=2

n \ m	0	1	2
-3	0	0.09317	-0.09317
-2	0	-0.49286	0.49286
-1	1	-0.59663	-0.40337
0	0	1.40337	-0.40337
1	0	-0.49286	0.49286
2	0	0.09317	-0.09317

Table 9-6 N=3 M=3

n \ m	0	1	2	3
-3	0	0.10951	-0.14218	0.03267
-2	0	-0.64994	0.96411	-0.31416
-1	1	-0.20510	-1.57796	0.78306
0	0	1.01184	0.77122	-0.78306
1	0	-0.33578	0.02162	0.31416
2	0	0.07683	-0.04416	-0.03267

Table 9-7 N=4 M=2

n \ m	0	1	2
-4	0	-0.02646	0.02646
-3	0	0.15446	-0.15446
-2	0	-0.56419	0.56419
-1	1	-0.56437	-0.43563
0	0	1.43563	-0.43563
1	0	-0.56419	0.56419
2	0	0.15446	-0.15446
3	0	-0.02646	0.02646

Table 9-8 N=4 M=3

n \ m	0	1	2	3
-4	0	-0.02971	0.03621	-0.00650
-3	0	0.18292	-0.23982	0.05691
-2	0	-0.74618	1.11018	-0.36399
-1	1	-0.13816	-1.71425	0.85241
0	0	1.00942	0.84299	-0.85241
1	0	-0.38219	0.01819	0.36399
2	0	0.12601	-0.06911	-0.05690
3	0	-0.02321	0.01670	0.00650

Table 9-9 *N*=4 *M*=4

n \ m	0	1	2	3	4
-4	0	-0.02584	0.01427	0.02964	-0.01807
-3	0	0.16286	-0.12615	-0.13032	0.09361
-2	0	-0.70773	0.89228	-0.00510	-0.17945
-1	1	-0.16034	-1.58857	0.64541	0.10350
0	0	0.98724	0.96867	-1.05941	0.10350
1	0	-0.34374	-0.19971	0.72289	-0.17945
2	0	0.10595	0.04457	-0.24413	0.09361
3	0	-0.01933	-0.00524	0.04265	-0.01807

9.2 Interpolator Control

This section focuses on the control of the interpolator. The task of the interpolator control is to determine the basepoint m_n and the corresponding fractional delay μ_n based on the output of a timing estimator or of a timing *error* estimator. The control algorithm will be different for the two cases.

9.2.1 Interpolator Control for Timing Error Feedback Systems

Error detectors produce an error signal at symbol rate $1/T$ using fractionally spaced samples $kT_I = kT/M_I$ (M_I integer). Most error detectors work with two or four samples per symbol. Since the sampling rate $1/T_s$ is not an exact multiple of the symbol rate, the samples $\{kT_I\}$ have to be mapped onto the time scale $\{kT_s\}$ of the receiver. Mathematically, this is done by expressing the time instant $kT_I + \varepsilon_I T_I$ by multiples of T_s plus fractional rest,

$$\begin{aligned} kT_I + \varepsilon_I T_I &= L_{\text{INT}}\left[kT_I + \varepsilon_I T_I\right] T_s + \mu_k T_s \\ &= m_k T_s + \mu_k T_s \end{aligned} \tag{9-34}$$

$$L_{\text{INT}}(x): \quad \text{largest integer} \ \leq x$$

In eq. (9-34) ε_I is defined with respect to T_I. The relation between (T, ε) and (T_I, ε_I), the basepoint m_k and the fractional delay μ_k is illustrated in Figure 9-17.

From eq. (9-34) it follows that for every sample one has to compute the corresponding basepoint m_k and the fractional delay μ_k in order to obtain the interpolated matched filter output $z(kT_I + \varepsilon_I T_I) = z(m_k T_s + \mu_k T_s)$. Interpolator and decimator are shown as two separate building blocks in the block diagram of Figure 9-18. In a practical implementation interpolator and decimator are realized

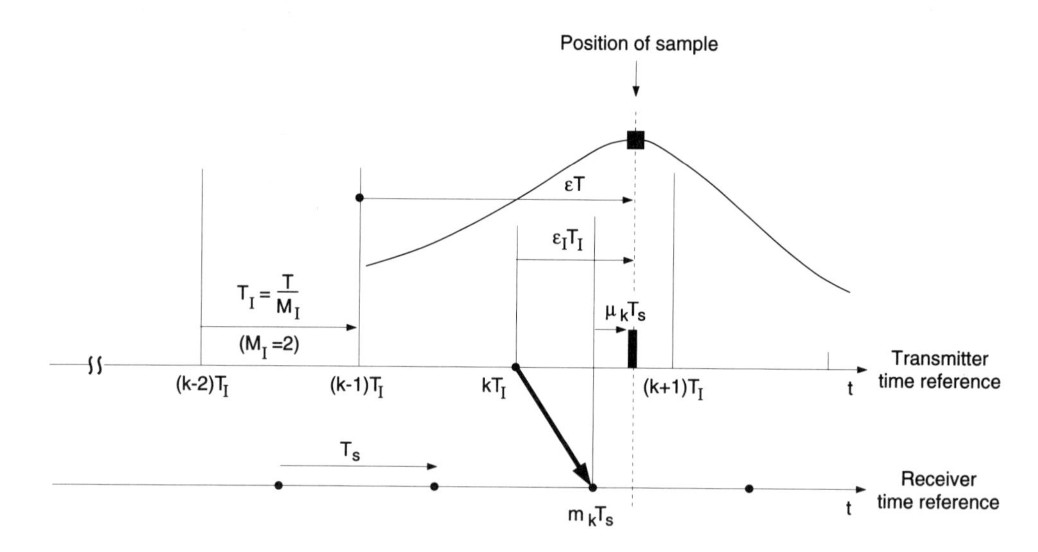

Figure 9-17 Definition of (ε_I, T_I) and Their Relation to the Receiver Time Axis $\{kT_s\}$.
Mapping of kT_I onto $m_k T_s$ is exemplarily shown for one sample and M_I=2

as an FIR filter with time-variant coefficients. A new sample is read into the filter at constant rate $1/T_s$, while a new output is computed only at the basepoints $m_k T_s$ (Figure 9-19).

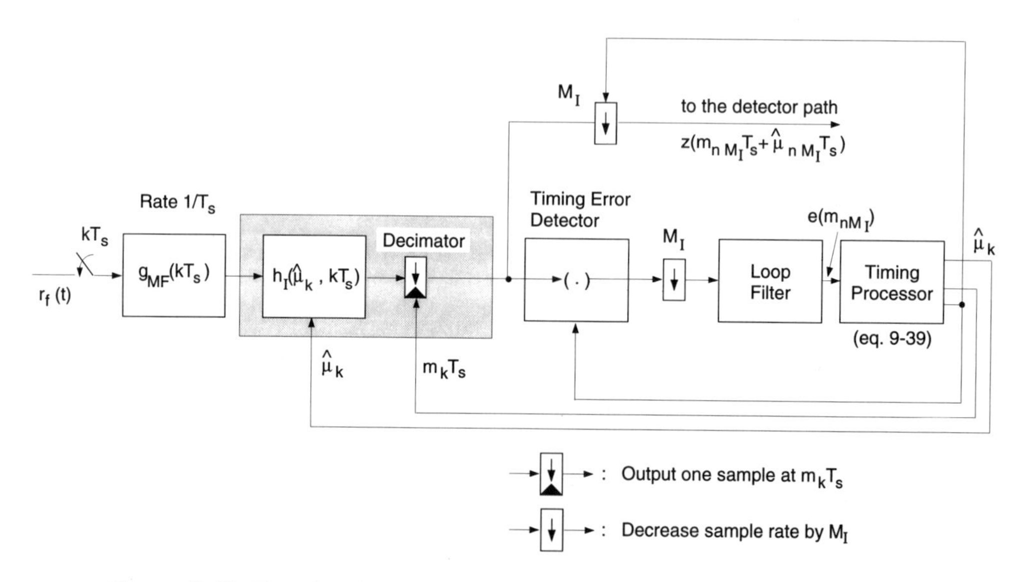

Figure 9-18 Functional Block Diagram of a Timing Error Feedback System

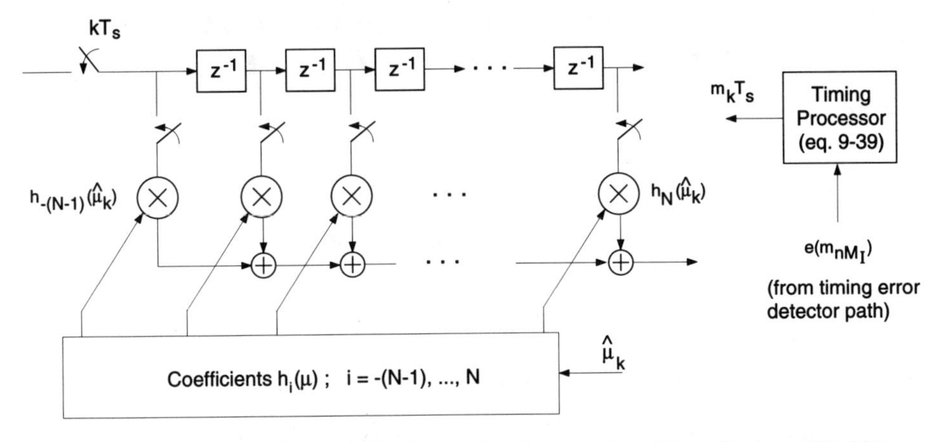

Figure 9-19 Interpolator and Decimator Implemented as Time-Variant FIR Filter

Based on the fractionally spaced samples the error detector produces an error signal at symbol rate. Therefore, a second decimation is performed in the error detector. This second decimation process is slaved to the first one. It selects *exactly* every M_Ith basepoint m_k ($k = nM_I$), to output *one* error signal.

The error detector signal is further processed in a loop filter. The output of the loop filter $e(m_{nM_I})$ is used to adjust the control word of the timing processor

$$\begin{aligned} w(m_{nM_I}) &= w\big(m_{(n-1)M_I}\big) + K_o\, e\big(m_{(n-1)M_I}\big) \\ K_o :&\quad \text{constant} \end{aligned} \tag{9-35}$$

In the absence of any disturbance the control word assumes its correct value:

$$w = \frac{T_I}{T_s} = \frac{1}{M_I}\left(\frac{T}{T_s}\right) \tag{9-36}$$

Basepoint and fractional delay are recursively computed in the timing processor as follows. We express $kT_I + \varepsilon_I T_I$ as function of (m_k, μ_k), see eq. (9-34). The next sample at $(k+1)T_I + \varepsilon_I T_I$ is then given by

$$\begin{aligned} (k+1)\, T_I + \varepsilon_I T_I &= m_k T_s + \mu_k T_s + T_I \\ &= m_k T_s + \left(\mu_k + \frac{T_I}{T_s}\right) T_s \end{aligned} \tag{9-37}$$

Replacing in the previous equation the unknown ratio T_I/T_s by its estimate $w(m_k)$ we obtain

$$(k+1)\, T_I + \varepsilon_I T_I = m_k T_s + L\text{INT}\, [\hat{\mu}_k + w(m_k)]\, T_s + [\hat{\mu}_k + w(m_k)]_{\text{mod}1}\, T_s \tag{9-38}$$

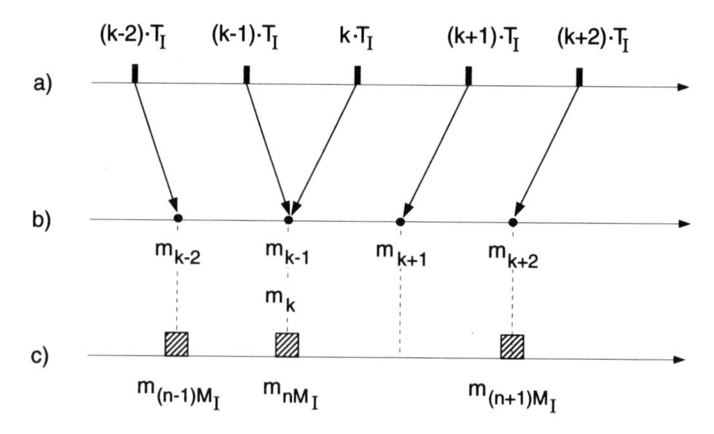

Figure 9-20 Basepoint Computation and Symbol Rate Decimation for $T_I/T_s \simeq 1$ and $M_I = 2$. (a) Transmitter time scale, (b) basepoints m_k, and (c) symbol-rate decimated basepoints m_{nM_I}.

From this readily follows the recursion for the estimates:

$$
\begin{aligned}
m_{k+1} &= m_k + L_{\text{INT}}\left[\hat{\mu}_k + w(m_k)\right] \\
\hat{\mu}_{k+1} &= \left[\hat{\mu}_k + w(m_k)\right]_{\text{mod}1}
\end{aligned}
\tag{9-39}
$$

Figure 9-20 illustrates the basepoint computation and symbol rate decimation. If the ratio T_I/T_s is substantially larger than 1, then $L_{\text{INT}}(\hat{\mu}_k + w(m_k)) \geq 1$ with high probability. Thus, every sample kT_I is mapped onto a different basepoint, i.e., $m_{k+1} \neq m_k$. The situation is different for $T_I/T_s \simeq 1$ which occurs frequently in practice. For example, $M_I = 2$ and twofold oversampling yields $T_I \simeq T_s$. The event that $L_{\text{INT}}(x) = 0$ occurs frequently. In this case two consecutive samples kT_I and $(k+1)T_I$ are mapped onto the same basepoint, $m_{k+1} = m_k$. This means that for the same set of input samples in the interpolator (see Figure 9-19) two different output values must be computed: one for the set of coefficients corresponding to $\hat{\mu}_k$ and the other one for $\hat{\mu}_{k+1}$.

We recall that the symbol rate decimation process selects every M_Ith basepoint. Since in Figure 9-20 $m_{k-1} = m_k$, we find different increments between basepoints

$$
\begin{aligned}
m_k - m_{k-2} &= 1 \\
m_{k+2} - m_k &= 2
\end{aligned}
\tag{9-40}
$$

The control word $w(m_k)$ in eq. (9-39) is changed only every symbol

$$
w(m_k) = w(m_{nM_I}) \quad \text{for } nM_I \leq k < (n+1)M_I \tag{9-41}
$$

The symbol rate basepoints m_{nM_I} are obtained by incrementing a modulo M_I counter for every basepoint computation performed,

$$c_{k+1} = [c_k + 1]_{\text{mod} M_I} \tag{9-42}$$

and by generating a select pulse for every full cycle.

Remark: In a hardware realization we would select a ratio of $T_I/T_s = 1+\Delta > 1$ (where Δ is a small number) to avoid $L_{\text{INT}}(x) = 0$ with high probability. In the rare case where $L_{\text{INT}}(x) = 0$ is due to noise, one selects $m_{k+1} = m_k + 1$ and sets $\hat{\mu}_k = 0$.

Example

We compute the increments of $m_{k+1} - m_k$ and μ_k for two different ratios of T_I/T_s (see Table 9–10):

$$\begin{aligned} &\text{a)} \quad T_I/T_s = 1.575 \\ &\text{b)} \quad T_I/T_s = 1 - 10^{-5} \end{aligned} \tag{9-43}$$

If T_I/T_s is nominally an integer, the fractional delay changes very slowly. In a hardware implementation it can be kept constant over a large number of symbols which simplifies the hardware.

Table 9-10 Basepoint and Fractional Delay Computation for Two Different T_I/T_s Values

	$T_I/T_s = 1.575, \quad \mu_0 = 0$		$T_I/T_s = 1 - 10^{-5}, \quad \mu_0 = 4.1 \cdot 10^{-5}$	
n	$m_{k+1} - m_k$	μ_k	$m_{k+1} - m_k$	μ_{k+1}
0	1	0.575	1	0.000031
1	2	0.150	1	0.000021
2	1	0.725	1	0.000011
3	2	0.300	1	0.000001
4	1	0.875	0	0.999991
5	2	0.450	1	0.999981
6	2	0.025	1	0.999971
7	1	0.600	1	0.999961

For low data rate application we can derive the correct sampling instant from the (much higher) clock of the digital timing processor. Assume the signal is sampled at rate $(1/T)M_I$. The timing processor selects one of M_c possible timing instants in the T_I interval $T/M_I = T_I = M_c T_c$. If M_c is chosen sufficiently large so that the maximum time discretization error $T/(M_c M_I)$ is negligible, we can entirely dispense with interpolation. The processor simply computes the basepoints m_k as multiples of T_c:

$$m_{k+1} = m_k + w(m_k) \tag{9-44}$$

The control word $w(m_k)$ is an integer with nominal value

$$w = T_I/T_c = M_c \tag{9-45}$$

The basepoint computation can be done by a numerically controlled oscillator (NCO) which basically is a down-counter presettable to an integer $w(m_k)$. The control word is updated at symbol rate. Since the clock in the receiver and transmitter are very accurate and stable, the loop bandwidth of the feedback system can be chosen small. Therefore, the increment $K_o e(m_{nM_I})$ per cycle is at most one. It therefore suffices to compare it with a threshold and possibly increment/decrement the control word by one. For $k = nM_I$

$$w(m_{nM_I}) = w\left(m_{(n-1)M_I}\right) + \Delta\left[K_o e\left(m_{(n-1)M_I}\right)\right] \tag{9-46}$$

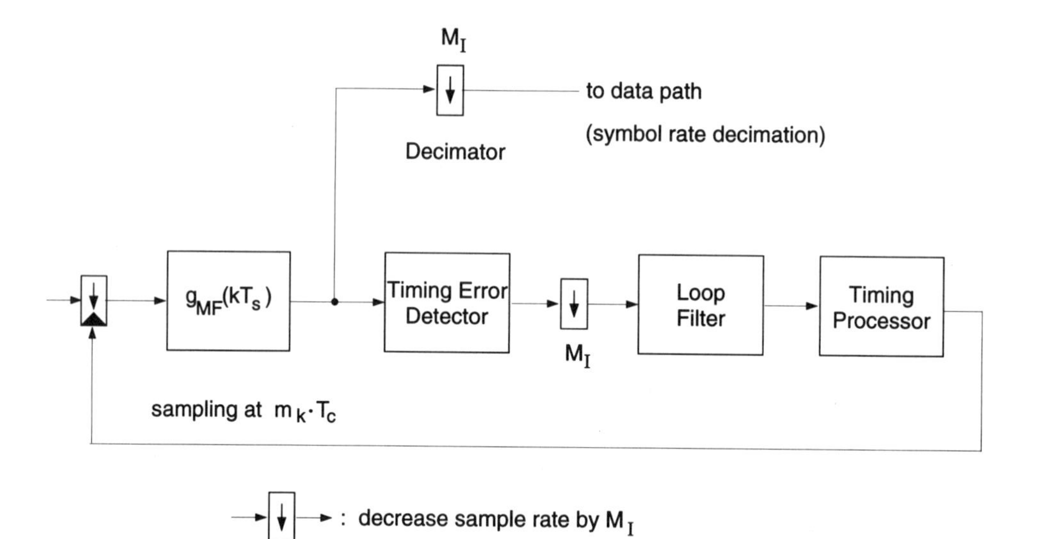

Figure 9-21 Digital Timing Recovery when the Sampling Instant Is Determined by a Numerically Controlled Oscillator (NCO)

where

$$\Delta(x) = \begin{cases} 1 & x \geq 1 \\ 0 & |x| < 1 \\ -1 & x \leq -1 \end{cases} \tag{9-47}$$

A block diagram of the timing error feedback system is shown in Figure 9-21. Because of its simplicity this solution finds widespread application.

9.2.2 Interpolator Control for Timing Estimators

The timing estimator directly provides us with an estimate of the relative time shift ε normalized to the symbol rate $1/T$. Consider

$$nT + \varepsilon T = m_n T_s + \mu_n T_s \tag{9-48}$$

and replace the trial parameter by its estimate $\hat{\varepsilon}_{n-1}$ computed in the previous T interval:

$$nT + \hat{\varepsilon}_{n-1} T = m_n T_s + \hat{\mu}_n T_s \tag{9-49}$$

Now consider the next symbol

$$(n+1)\, T + \hat{\varepsilon}_n\, T \tag{9-50}$$

Adding $\hat{\varepsilon}_{n-1}$ and immediately subtracting this quantity, we can write for (9-50)

$$(n+1)\, T + \hat{\varepsilon}_n\, T = nT + \hat{\varepsilon}_{n-1} T + T + (\hat{\varepsilon}_n - \hat{\varepsilon}_{n-1})\, T \tag{9-51}$$

Replacing $nT + \hat{\varepsilon}_{n-1} T$ by the right-hand side of (9-49) yields

$$\begin{aligned} (n+1)\, T + \hat{\varepsilon}_n\, T &= m_n T_s + \hat{\mu}_n T_s + T + (\hat{\varepsilon}_n - \hat{\varepsilon}_{n-1})\, T \\ &= \left[m_n + \hat{\mu}_n + \frac{T}{T_s} \left[1 + (\hat{\varepsilon}_n - \hat{\varepsilon}_{n-1}) \right] \right] T_s \end{aligned} \tag{9-52}$$

Since ε is restricted to values $0 \leq \varepsilon < 1$ special care must be taken when $\hat{\varepsilon}_n$ crosses the boundaries at 0 or 1. Assume that $\hat{\varepsilon}_{n-1}$ is close to 1 and that the next estimate $\hat{\varepsilon}_n$ from the timing estimator is a very small number. The difference $\Delta\hat{\varepsilon}_n = (\hat{\varepsilon}_n - \hat{\varepsilon}_{n-1})$ then would be close to 1. But since ε_n is very slowly changing, the increment $\Delta\hat{\varepsilon}_n = (\hat{\varepsilon}_n - \hat{\varepsilon}_{n-1})$ is almost always a number of very small magnitude, $|\Delta\hat{\varepsilon}_n| \ll 1$. From this we conclude that the large negative difference is an artifact caused by crossing the boundary of $\hat{\varepsilon}_n$ at 1. To properly accommodate this effect we have to compute $\mathrm{SAW}(\hat{\varepsilon}_n - \hat{\varepsilon}_{n-1})$ in (9-52) where

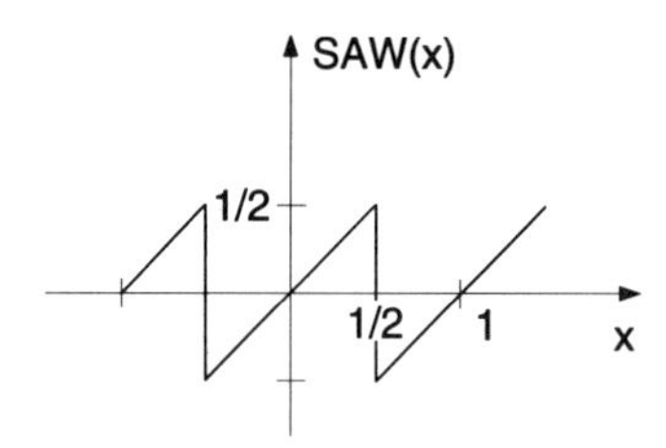

Figure 9-22 Sawtooth Function

$\mathrm{SAW}(x)$ is the sawtooth function with period 1 displayed in Figure 9-22. Then we readily obtain the recursions

$$\begin{aligned} m_{n+1} &= m_n + L_{\mathrm{INT}}\left[\hat{\mu}_n + \frac{T}{T_s}\left(1+\mathrm{SAW}(\hat{\varepsilon}_n - \hat{\varepsilon}_{n-1})\right)\right] \\ \hat{\mu}_{n+1} &= \left[\hat{\mu}_n + \frac{T}{T_s}(1+\mathrm{SAW}(\hat{\varepsilon}_n - \hat{\varepsilon}_{n-1}))\right]_{\mathrm{mod}\ 1} \end{aligned} \tag{9-53}$$

Remark: for T/T_s the nominal value is used.

9.2.3 Tracking Performance and Decimation Revisited

When the tracking performance was calculated in Chapter 6, we explicitly assumed that an error signal is available at symbol rate. The random disturbance of the error detector output was modeled as stationary process with power spectrum $S_N\left(e^{-j\omega T}\right)$. But now consider the actual implementation of an error feedback system. The output samples of the error detector are irregularly spaced on the $\{kT_s\}$ axis. What possible meaning can be ascribed to a power spectrum of such a sequence and what is the relation to the power spectrum $S_N\left(e^{-j\omega T}\right)$? Do we have to take the time-variant decimation process into account when analyzing the tracking performance?

Fortunately, this is not the case. By definition, every output of the decimator represents a sample $x(nT + \hat{\varepsilon}_n T)$. It is irrelevant at what exact physical time $m_n T_s$ the sample value $x(m_n T_s) = x(nT + \hat{\varepsilon}_n T)$ is computed. Thus, the sequence $\{x(m_n T_s)\}$, which is regularly spaced on the physical time scale $\{kT_s\}$, corresponds to a *regularly* spaced sequence $x(nT + \hat{\varepsilon}_n T)$.

9.2.4 Main Points

- Timing recovery using the samples $r(kT_s)$ of a free running oscillator can be done by digital interpolation. The fractional time delay μ_k and the basepoint m_k are recursively computed.
- Digital interpolators for high data rate applications are realized as a bank of FIR filters (Farrow structure). The number of filters equals the degree of the interpolating polynomial plus 1. The number of taps of the FIR filter is always even.

- For virtually all practical applications 4-tap FIR filters and a quadratic polynomial will be sufficient.
- For a sufficiently large number of samples per symbol ($T/T_s > 2$) even linear interpolation between two samples will be sufficient.
- If the sampling instant can be derived from the clock running at a much higher frequency, no interpolation is required. The timing processor selects one among $M_c \gg 1$ possible values in a time interval $T_I = T/M_I$.

9.2.5 Bibliographical Notes

Interpolation and decimation are discussed in the book by Crochiere and Rabiner [4]. The rate conversion is restricted to the ratio of two integers. Timing recovery and interpolation require an noninteger decimation rate. The first papers to address this issue appear to be [5]–[7]. The ESA report by Gardner [8] made us aware of the fact that timing recovery comprises the two tasks of interpolation (in the strict mathematical sense) and decimator control; see also [9, 10]. Digital receivers capable of operating at variable data rates are of interest in digital broadcasting systems [11]. They require an interactive design of anti-aliasing filter, matched filter, and interpolation/decimation control [12, 13]. A recent publication [14] discusses the various areas of application of fractional time delays. The paper provides a very readable and interesting overview of seemingly unrelated topics. It contains a comprehensive list of references.

Bibliography

[1] G. Oetken, "A New Approach for the Design of Digital Interpolating Filters," *IEEE Trans. Acoust., Speech, Signal Processing*, vol. ASSP-27, pp. 637–643, Dec. 1979.

[2] C. W. Farrow, "A Continuously Variable Digital Delay Element," *Proceedings IEEE Intern. Symp. Circuits Syst., Espoo, Finland*, pp. 2641–2645, June 1988.

[3] R. W. Schafer and L. R. Rabiner, "A Digital Signal Processing Approach to Interpolation," *Proceedings IEEE*, vol. 61, pp. 692–702, June 1973.

[4] R. E. Crochiere and L. R. Rabiner, *Multirate Signal Processing*. Englewood Cliffs; New York: Prentice Hall, 1983.

[5] F. Takahata et. al., "A PSK Group Modem for Satellite Communication," *IEEE J. Selected Areas Commun.*, vol. SAC-5, pp. 648–661, May 1987.

[6] M. Oerder, G. Ascheid, R. Haeb, and H. Meyr, "An All Digital Implementation of a Receiver for Bandwidth Efficient Communication," *Proceedings of the European Signal Processing Conference III, I.T. Young et al. (eds), Amsterdam, Elsevier*, pp. 1091–1094, 1986.

[7] G. Ascheid, M. Oerder, J. Stahl, and H. Meyr, "An All Digital Receiver Architecture for Bandwidth Efficient Transmission at High Data Rates," *IEEE Trans. Commun.*, vol. COM-37, pp. 804–813, Aug. 1989.

[8] F. M. Gardner, *Timing Adjustment via Interpolation in Digital Demodulators*. European Space Agency. ESTEC Contract No. 8022/88/NL/DG. Final Report, June 1991.

[9] F. M. Gardner, "Interpolation in Digital Modems – Part I: Fundamentals," *IEEE Trans. Commun.*, vol. 41, pp. 501–507, Mar. 1993.

[10] F. M. Gardner, "Interpolation in Digital Modems – Part II: Implementation and Performance," *IEEE Trans. Commun.*, vol. COM-41, pp. 998–1008, June 1993.

[11] European Telecommunications Institute, "Digital Broadcasting System for Television, Sound and Data Services; Framing Structure, Channel Coding and Modulation for 11/12 GHz Satellite Services," Aug. 1994. Draft DE/JTC-DVB-6, ETSI Secretariat.

[12] U. Lambrette, K. Langhammer, and H. Meyr, "Variable Sample Rate Digital Feedback NDA Timing Synchronization," *Proceedings IEEE GLOBECOM London*, vol. 2, pp. 1348–1352, Nov. 1996.

[13] G. Karam, K. Maalej, V. Paxal, and H. Sari, "Variable Symbol-Rate Modem Design for Cable and Satellite Broadcasting," *Signal Processing in Telecommunications, Ezio Biglieri and Marco Luise (eds.), Milan, Italy*, pp. 244–255, 1996.

[14] T. Laakso, V. Välimäki, Matti Karjalainen, and U. K. Laine, "Splitting the Unit Delay," *IEEE Signal Processing Magazine*, vol. 13, pp. 30–60, Jan. 1996.

[15] H. Meyr, M. Oerder, and A. Polydoros, "On Sampling Rate, Analog Prefiltering and Sufficient Statistics for Digital Receivers," *IEEE Trans. Commun.*, pp. 3208–3214, Dec. 1994.

[16] V. Zivojnovic and H. Meyr, "Design of Optimum Interpolation Filters for Digital Demodulators," *IEEE Int. Symposium Circuits Syst., Chicago, USA*, pp. 140–143, 1993.

Chapter 10 DSP System Implementation

This chapter is concerned with the implementation of digital signal processing systems. It serves the purpose to make the algorithm designer aware of the strong interaction between algorithm and architecture design.

Digital signal processing systems are an assembly of heterogeneous hardware components. The functionality is implemented in both hardware and software subsystems. A brief overview of DSP hardware technology is given in Section 10.1. Design time and cost become increasingly more important than chip cost. A look at hardware-software co-design is done in Section 10.2. Section 10.3 is devoted to quantization issues. In Sections 10.4 to 10.8 an ASIC (application-specific integrated circuit) design of a fully digital receiver is discussed. We describe the design flow of the project, the receiver structure, and the decision making for its building blocks. The last two sections are bibliographical notes on Viterbi and Reed-Solomon decoders.

10.1 Digital Signal Processing Hardware

Digital signal processing systems are an assembly of heterogeneous subsystems. The functionality is implemented in both hardware and software subsystems. Commonly found hardware blocks are shown in Figure 10-1.

There are two basic types of processors available to the designer: a programmable general-purpose digital signal processor (DSP) or a microprocessor. A general-purpose DSP is a software-programmable integrated circuit used for

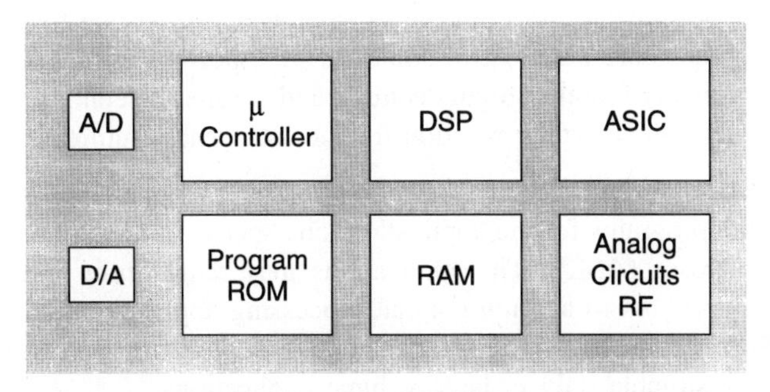

A/D: Analog-to-Digital Converter, D/A: Digital-to-Analog Converter

Figure 10-1 Hardware Components of a Digital Signal Processing System

speech coding, modulation, channel coding, detection, equalization, and associated modem tasks such as frequency, symbol timing and phase synchronization as well as amplitude control. Moreover, a DSP is preferably used with regard to flexibility in applications and ability to add new features with minimum re-design and re-engineering. Microprocessors are usually used to implement protocol stacks, system software, and interface software. Microprocessors are better suited to perform the non-repetitive, control-oriented input/output operations as well as all housekeeping chores.

ASICs are used for various purposes. They are utilized for high-throughput tasks in the area of digital filtering, synchronization, equalization, and channel decoding. An ASIC is often likely to provide also the glue logic to interface components. In some systems, the complete digital receiver is implemented as an ASIC, coupled with a microcontroller. ASICs have historically been used because of their lower power dissipation per function. In certain applications like spread-spectrum communications, digital receiver designs require at least partial ASIC solutions in order to execute the wideband processing functions such as despreading and code synchronization. This is primarily because the chip-rate processing steps cannot be supported by current general-purpose DSPs.

Over the last few years, as manufacturers have brought to market first- and second-generation digital cellular and cordless solutions, programmable general-purpose digital signal processors are slowly being transformed into "accelerator-assisted DSP-microcontroller" hybrids. This transformation is a result of the severe pressure being put on reducing the power consumption. As firmware solutions become finalized, cycle-hungry portions of algorithms (e.g., equalizers) are being "poured into silicon", using various VLSI architectural ideas. This has given rise to, for example, new DSPs with hardware accelerators for Viterbi decoding, vectorized processing, and specialized domain functions. The combination of programmable processor cores with custom data-path accelerators within a single chip offers numerous advantages: performance improvements due to time-critical computations implemented in accelerators, reduced power consumption, faster internal communication between hardware and software, field programmability due to the programmable cores and lower total system cost due to a single DSP chip solution. Such core-based ASIC solutions are especially attractive for portable applications typically found in digital cellular and cordless telephony, and they are likely to become the emerging solution for the foreseeable future.

If a processor is designed by jointly optimizing the architecture, the instruction set and the programs for the application, one speaks of an application-specific integrated processor (ASIP). The applications may range from a small number of different algorithms to an entire signal processing application. A major drawback of ASIPs is that they require an elaborate support infrastructure which is economically justifiable only in large-volume applications.

The decision to implement an algorithm in software or as a custom data-path (accelerator) depends on many issues. Seen from a purely computational power

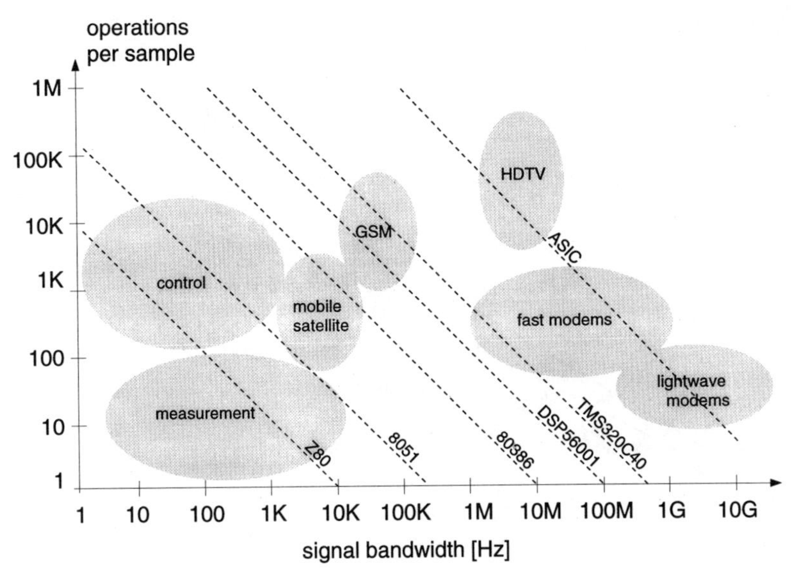

Figure 10-2 Complexity versus Signal Bandwidth Plot

point-of-view, algorithms can be categorized according to the two parameters signal bandwidth and number of operations per sample. The first parameter defines a measure of the real-time processing requirement of the algorithm. The second provides a measure of complexity of the algorithm. In Figure 10-2 we have plotted complexity versus bandwidth on a double logarithmic scale. A straight line in the graph corresponds to a processing device that performs a given number of instructions per second. Applications in the upper-right corner require massive parallel processing and pipelining and are the exclusive domain of ASICs. In contrast, in the lower-left corner the signal bandwidth is much smaller than the clock rate of a VLSI chip. Hence, hardware resources can be shared and the programmable processor is almost always the preferred choice. For the region between these two extremes resource sharing is possible either using a processor or an ASIC. There are no purely computational power arguments in favor of either one of the two solutions. The choice depends on other issues such as time-to-market and capability profile of the design team, to mention two examples.

The rapid advance of microelectronic is illustrated by Figure 10-3. The complexity of VLSI circuits (measured in number of gates) increases tenfold every 6 years. This pattern has been observed for memory components and general-purpose processors over the last 20 years and appears to be true also for the DSP. The performance measured in MOPS (millions of operations per second) is related to the chip clock frequency which follows a similar pattern. The complexity of software implemented in consumer products increases tenfold every 4 years [1].

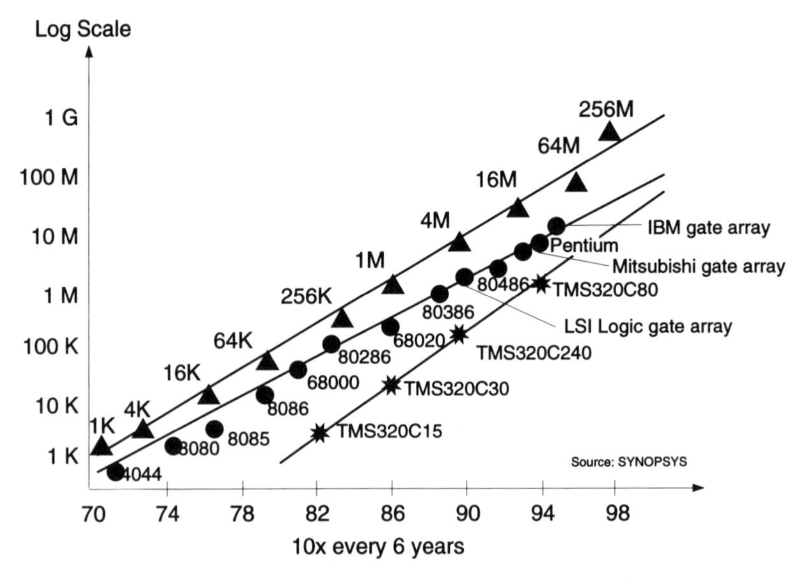

Figure 10-3 Complexity of VLSI Circuits

10.2 DSP System Hardware-Software Co-Design

The functionality in a DSP system is implemented in both hardware and software subsystems. But even within the software portions there is diversity. Control-oriented processes (protocols) have different characteristics than data-flow-oriented processes (e.g., filtering). A DSP system design therefore not only mixes hardware design with software design but also mixes design styles within each of these categories.

One can distinguish between two opposing philosophies for system level design [2]. One is the unified approach which seeks a consistent semantics for the specification of the complete system. The other is a heterogeneous approach which seeks to combine semantically disjoint subsystems. For the foreseeable future the latter appears to be the feasible approach. In the heterogeneous approach, for example, the design automation tool for modeling and analysis of algorithms is tightly coupled with tools for hardware and software implementation. This makes it possible to explore algorithm/architecture trade-offs in a joint optimization process, as will be discussed in the design case study of Section 10.4.

The partitioning of the functionality into hardware and software subsystems is guided by a multitude of (often conflicting) goals. For example, a software implementation is more flexible than a hardware implementation because changes in the specification are possible in any design phase. On the negative side we mentioned the higher power consumption compared to an ASIC solution which is a key issue in battery-operated terminals. Also, for higher volumes an ASIC is more cost effective.

Design cost and time become increasingly more important than chip processing costs. On many markets product life cycles will be very short. To compete successfully, companies will need to be able to turn system concepts into silicon quickly. This puts high priority on computerized design methodology and tools in order to increase the productivity of engineering design teams.

10.3 Quantization and Number Representation

In this section we discuss the effect of finite word-lengths in digital signal processing. There are two main issues to be addressed. First, when sampling was considered so far, we assumed that the samples were known with infinite precision, which of course is impossible. Each sample must therefore be approximated by a binary word. The process where a real number is converted to a finite binary word is called *quantization.*

Second, when in digital processing the result of an operation contains more bits than can be handled by the process downstream, the word length must be reduced. This can be done either by rounding, truncation, or clipping.

For further reading, we assume that the reader is familiar with the basics of binary arithmetics. As a refresher we suggest Chapter 9.0 to 9.2 of the book by Oppenheim and Schafer [3].

A quantizer is a zero-memory nonlinear device whose output x_{out} is related to the input x_{in} according to

$$x_{\text{out}} = q_i \qquad \text{if} \quad x_i \le x_{\text{in}} < x_{i+1} \tag{10-1}$$

where q_i is an output number that identifies the input interval $[x_i,\ x_{i+1})$. Uniform quantization is the most widely used law in data signal processing and the only one discussed here.

All uniform quantizer characteristics have the staircase form shown in Figure 10-4. They differ in the number of levels, the limits of operation, and the location

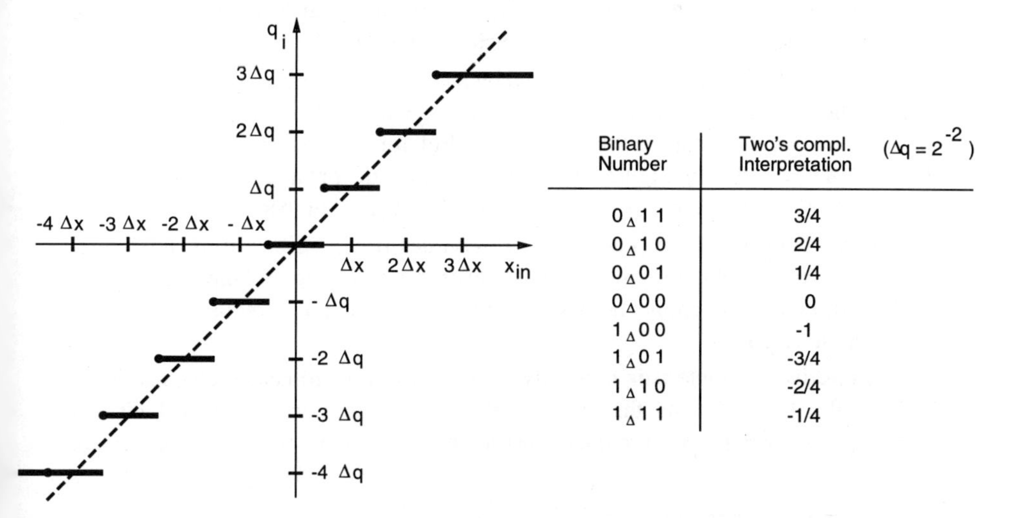

Binary Number	Two's compl. Interpretation ($\Delta q = 2^{-2}$)
$0_\Delta 1 1$	3/4
$0_\Delta 1 0$	2/4
$0_\Delta 0 1$	1/4
$0_\Delta 0 0$	0
$1_\Delta 0 0$	-1
$1_\Delta 0 1$	-3/4
$1_\Delta 1 0$	-2/4
$1_\Delta 1 1$	-1/4

Figure 10-4 Uniform Quantizer Characteristic with b= 3 Bit. Rounding to the nearest level is employed. The binary number is interpreted as 2's complement.

of the origin. Every quantizer has a finite range that extends between the input limits $x_{\text{min}}, x_{\text{max}}$. Any input value exceeding the limits is clipped:

$$\begin{aligned} x_{\text{out}} &= q_{\text{min}} \qquad \text{if} \quad x_{\text{in}} < x_{min} \\ x_{\text{out}} &= q_{\text{max}} \qquad \text{if} \quad x_{\text{in}} \geq x_{\text{max}} \end{aligned} \tag{10-2}$$

The number of levels is chosen as $L = 2^b$ or $L = 2^b - 1$. Any output word is then described by a binary word of b bits. The operation is called *b-bit quantization.*

The meaning of Δq is arbitrary. For example, one can equally well interpret Δq as an integer or as a binary fraction 2^{-b}.

A quantizer exhibits small-scale nonlinearity within its individual steps and large-scale nonlinearity if operated in the saturation range. The amplitude of the input signal has to be controlled to avoid severe distortion of the signal from either nonlinearity. The joint operation of the analog-to-digital (A/D) converter and the AGC is of crucial importance to the proper operation of any receiver. The input amplitude control of the A/D converter is known as loading adjustment.

Quantizer characteristics can be categorized as possessing midstep or midriser staircases, according to their properties in the vicinity of zero input. Each has its own advantages and drawbacks and is encountered extensively in practice.

In Fig. 10-4 a mid-step characteristic with an even number of levels $L = 2^3$ is shown. The characteristic is obtained by *rounding* the input value to the nearest quantization level. A 2's complement representation of the binary numbers is used in this example. The characteristic exhibits a dead zone at the origin. When an error detector possesses such a dead zone, the feedback loop tends to instability. This characteristic is thus to be avoided in such applications. The characteristic is asymmetric since the number -1 is represented but not the number $+1$. If the quantizer therefore operates in both saturation modes, then it will produce a nonzero mean output despite the fact that the input signal has zero mean. The characteristic can easily be made symmetric by omitting the most negative value.

A different characteristic is obtained by *truncation.* Truncation is the operation which chooses the largest integer less than or equal $(x_{\text{in}}/\Delta x)$. For example, $x_{\text{in}}/\Delta x = 0.8$ we obtain $\text{INT}(0.8) = 0$. But for $x_{\text{in}}/\Delta x = -3.4$ we obtain $\text{INT}(-3.4) = -4$. In Figure 10-5 the characteristic obtained by truncation is shown. A 2's complement representation of the binary numbers is used.

This characteristic is known as *offset* quantizer[3]. Since it is no longer symmetric, it will bias the output signals even for small signal amplitudes. This leads to difficulties in applications where a zero mean output is required for a zero mean input signal.

A midriser characteristic is readily obtained from the truncated characteristics in Fig. 10-5 by increasing the word length of the quantizer output by 1 bit and choosing the LSB (least significant bit) identical 1 for all words (Figure 10-6).

[3] In practical A/D converters the origin can be shifted by adjusting an input offset voltage.

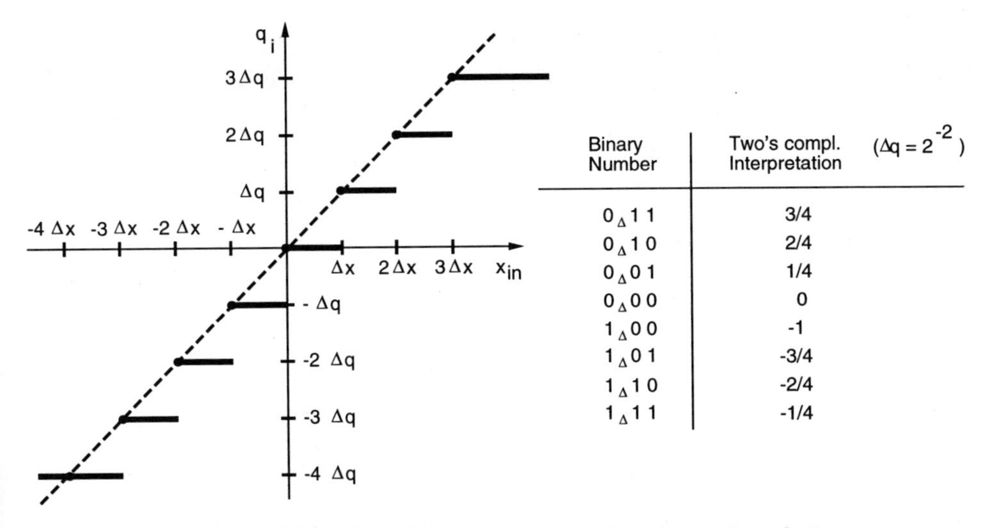

Binary Number	Two's compl. Interpretation ($\Delta q = 2^{-2}$)
$0_\Delta 1\,1$	3/4
$0_\Delta 1\,0$	2/4
$0_\Delta 0\,1$	1/4
$0_\Delta 0\,0$	0
$1_\Delta 0\,0$	-1
$1_\Delta 0\,1$	-3/4
$1_\Delta 1\,0$	-2/4
$1_\Delta 1\,1$	-1/4

Figure 10-5 Offset Quantizer Employing Truncation. b=3; binary number interpreted as 2's complement.

Notice that the extra bits need not to be produced in the physical A/D converter but can be added in the digital processor after A/D conversion.

The midriser characteristic is symmetric and has no dead zone around zero input. A feedback loop with a discontinuous step at zero in its error detector will dither about the step which is preferable to limit cycles induced by a dead zone.

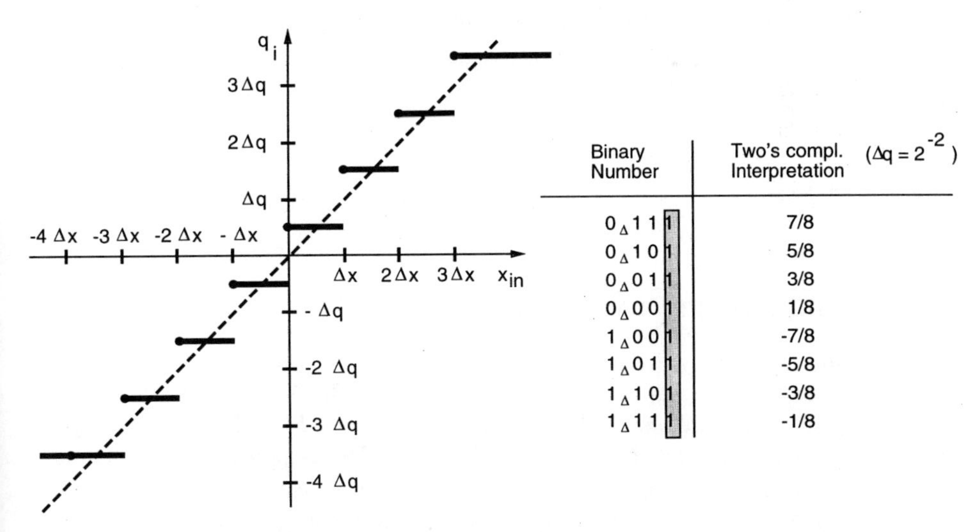

Binary Number	Two's compl. Interpretation ($\Delta q = 2^{-2}$)
$0_\Delta 1\,1\,1$	7/8
$0_\Delta 1\,0\,1$	5/8
$0_\Delta 0\,1\,1$	3/8
$0_\Delta 0\,0\,1$	1/8
$1_\Delta 0\,0\,1$	-7/8
$1_\Delta 0\,1\,1$	-5/8
$1_\Delta 1\,0\,1$	-3/8
$1_\Delta 1\,1\,1$	-1/8

Figure 10-6 Midriser Characteristic Obtained by Adding an Additional Bit to the Quantizer Output

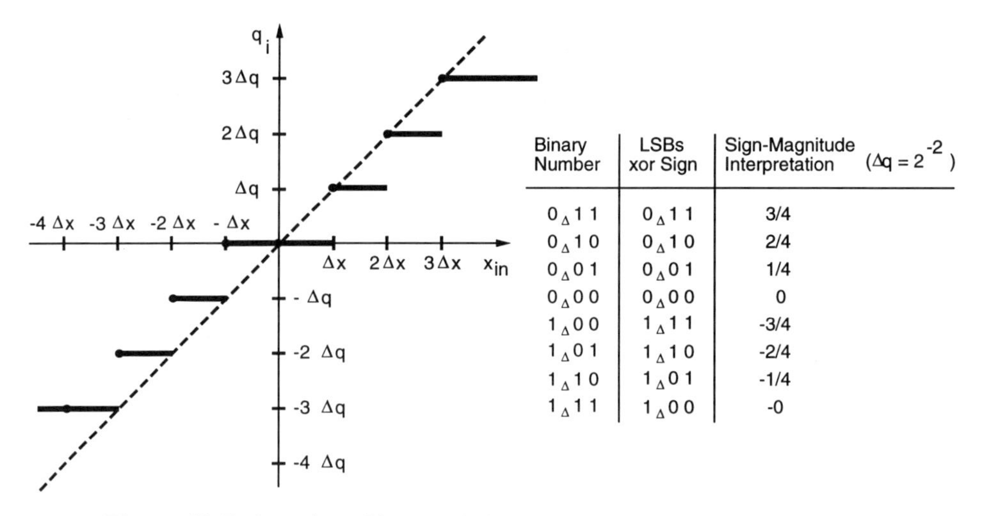

Binary Number	LSBs xor Sign	Sign-Magnitude Interpretation ($\Delta q = 2^{-2}$)
$0_{\Delta}11$	$0_{\Delta}11$	3/4
$0_{\Delta}10$	$0_{\Delta}10$	2/4
$0_{\Delta}01$	$0_{\Delta}01$	1/4
$0_{\Delta}00$	$0_{\Delta}00$	0
$1_{\Delta}00$	$1_{\Delta}11$	-3/4
$1_{\Delta}01$	$1_{\Delta}10$	-2/4
$1_{\Delta}10$	$1_{\Delta}01$	-1/4
$1_{\Delta}11$	$1_{\Delta}00$	-0

Figure 10-7 Quantizer Characteristic Obtained by Magnitude Truncation

The effect of the number representation on the quantizer characteristic is illustrated in Figure 10-7. In some applications it is advantageous to employ a sign-magnitude representation.

Figure 10-7 shows the characteristic by truncation of the magnitude of an input signal. This is no longer a uniform quantizer, the center interval has double width.

When the result of an operation contains more bits than can be handled downstream, the result must be shortened. The effect of rounding, truncation, or clipping is different for the various number representations.

The resulting quantization characteristic is analogous to that obtained earlier with the exception that now both input and output are discretized. However, quantizing discrete values is more susceptible to causing biases than quantizing continuous values. Thus quantizing discrete values should be performed even more carefully.

10.4 ASIC Design Case Study

In this case study we describe the design of a complete receiver chip for digital video broadcasting over satellite (DVB-S)[4]. The data rate of DVB is in the order of 40 Msymbols/s. The chip was realized in 0.5 μ CMOS technology with a (maximum) clock frequency of 88 MHz. The complexity of operation and the symbol rate locates it in the right upper corner of the complexity versus bandwidth plot of Figure 10-2. We outline the design flow of the project, the receiver structure, and the rationale of the decision making for its building blocks.

10.4.1 Implementation Loss

In an ASIC realization the chip area is in a first approximation proportional to the word length. Since the cost of a chip increases rapidly with the area, choosing quantization parameters is a major task.

The finite word length representation of numbers in a properly designed digital receiver ideally has the same effect as an additional white noise term. The resulting decrease of the signal-to-noise ratio is called the *implementation loss*. A second degradation with respect to perfect synchronization is caused by the variance of the synchronization parameter estimates and was called *detection loss* (see Chapter 7). The sum of these two losses, D_{total}, is the decrease of the signal-to-noise ratio with respect to a receiver with perfect synchronization and perfect implementation. It is exemplarily shown in Figure 10-8 for an 8-PSK trellis coded modulation [5].

The left curve shows the BER for a system with perfect synchronization and infinite precision arithmetics while the dotted line shows the experimental results. It is seen that the experimental curve is indeed approximately obtained by shifting the perfect system performance curve by D_{total} to the right.

Quantization is a nonlinear operation. It exhibits small-scale nonlinearity in its individual steps and large-scale nonlinearity in the saturation range. Its effect depends on the specific algorithm, it cannot be treated in general terms.

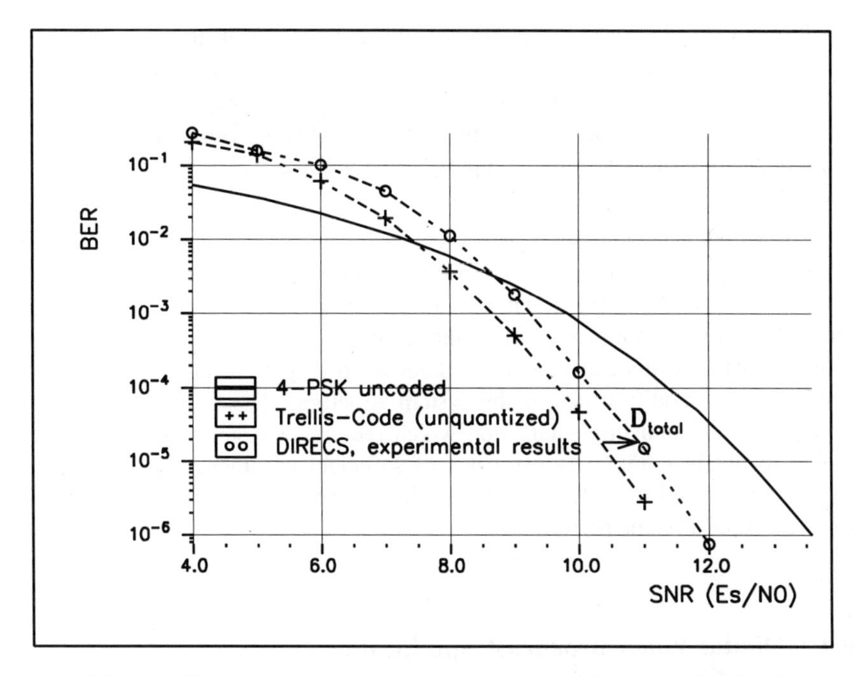

Figure 10-8 Loss D_{total} of the Experimental Receiver DIRECS

In a digital receiver the performance measure of interest is the bit error rate. We are allowed to nonlinearly distort the signal as long as the processed signal represents a sufficient statistics for detection of acceptable accuracy. For this reason, quantization effects in digital receivers are distinctly different than in other areas of digital signal processing (such as audio signal processing), which require a virtually quantization-error free representation of the analog signal.

10.4.2 Design Methodology

At this level of complexity, system simulation is indispensable to evaluate the performance characteristics of the system with respect to given design alternatives. The design framework should provide the designer with a flexible and efficient environment to explore the alternatives and trade-offs on different levels of abstraction. This comprises investigations on the

- structural level, e.g., joint or separate carrier and timing synchronization
- algorithmic level, e.g., various estimation algorithms
- implementation level, e.g., architectures and word lengths

There is a strong interaction between these levels of abstraction. The principal task of a system engineer is to find a compromise between implementation complexity and system performance. Unfortunately, the complexity of the problem prevents formalization of this optimization problem. Thus, practical system design is partly based on rough complexity estimates and experience, particularly at the structural level.

Typically a design engineer works hierarchically to cope with the problems of a complex system design. In a first step structural alternatives are investigated. The next step is to transform the design into a model that can be used for system simulation. Based on this simulation, algorithmic alternatives and their performance are evaluated. At first this can be done without word length considerations and may already lead to modifications of the system structure. The third step comprises developing the actual implementation which requires a largely fixed structure to be able to obtain complexity estimates of sufficient accuracy. At this step bit-true modeling of all imperfections due to limited word lengths is indispensable to assess the final system performance.

10.4.3 Digital Video Broadcast Specification

The main points of the DVB standard are summarized in Table 10-1:

Table 10-1 Outline of the DVB Standard

Modulation QPSK with Root-Raised Cosine Pulses (excess bandwidth $\alpha = 0.35$) and Gray-Encoding	
Convolutional Channel Coding	
Convolutional Code (CC)	$R = 1/2,\ 2/3,\ 3/4,\ 5/6,\ 7/8$
Reed-Solomon Code	$n = 204,\ k = 188$ over GF(8)
Operating Point	
E_b^{CC}/N_0	4.2, ..., 6.15 dB (depending on the code rate R)
BER behind CC	2×10^{-4}
Example for Symbol Rates	20, ..., 44 Msymbols/s

The data rate is not specified as a single value but suggested to be within a range of [18 to 68] Mb/s. The standard defines a concatenated coding scheme consisting of an inner convolutional code and an outer Reed-Solomon (RS) block code.

Figure 10-9 displays the bit error rate versus E_b/N_0 after the convolutional

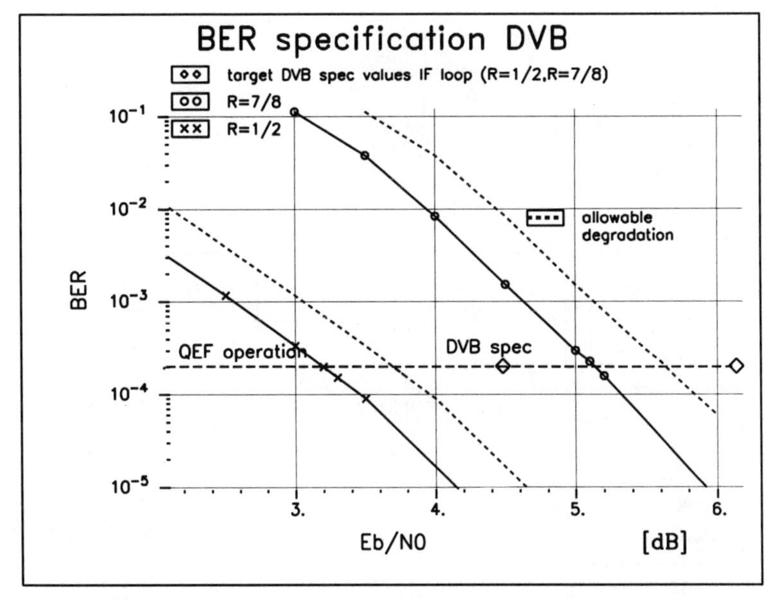

Figure 10-9 Bit Error Specification of the DVB Standard

decoder for the two code rates of $R = 1/2$ and $R = 7/8$ under the assumption of perfect synchronization and perfect convolutional decoder implementation. The output of the outer RS code is supposed to be quasi-error-free (one error per hour). The standard specifies a BER of 2×10^{-4} at $E_b/N_0 = 4.2$ dB for $R = 1/2$ and at $E_b/N_0 = 6.15$ dB for code rate $R = 7/8$. This leaves a margin of 1 dB (see Figure 10-9) for the implementation loss of the complete receiver. This loss must also take into account the degradation due to the analog front end (AGC, filter, oscillator for down conversion). In the actual implementation the total loss was equally split into 0.5 dB for the analog front end and 0.5 dB for the digital part.

10.4.4 Receiver Structure

Figure 10-10 gives a structural overview of the receiver. A/D conversion is done at the earliest point of the processing chain. The costly analog parts are thus reduced to the minimum radio frequency components. Down conversion and sampling is accomplished by free-running oscillators.

In contrast to analog receivers where down conversion and phase recovery is performed simultaneously by a PLL, the two tasks are separated in a digital receiver. The received signal is first down converted to baseband with a free-running oscillator at approximately the carrier frequency f_c. This leaves a small residual normalized frequency offset Ω.

The digital part consists of the timing and phase synchronization units, the Viterbi and RS-decoder, frame synchronizer, convolutional deinterleaver, descram-

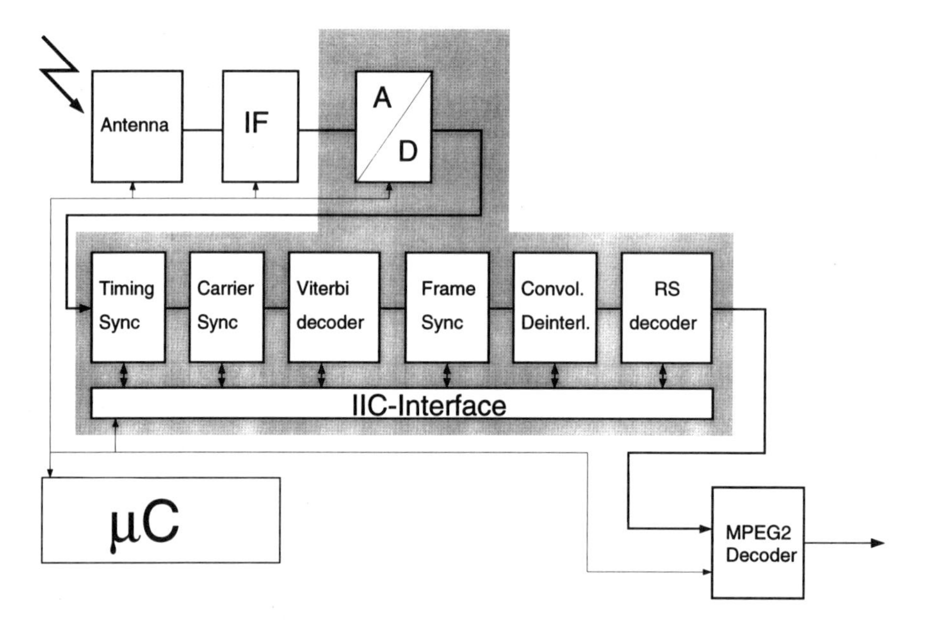

Figure 10-10 Block Diagram of the DVB Receiver

bler, and the MPEG decoder for the video data. A micro controller interacts via I^2C bus with the functional building blocks. It controls the acquisition process of the synchronizers and is used to configure the chip.

10.4.5 Input Quantization

The input signal to the A/D converter comes from a noncoherent AGC (see Volume 1, p. 278); the signal-to-noise ratio is unknown to the A/D converter. We must consider both large-scale and small-scale quantization effects.

An A/D converter can be viewed as a series connection of a soft limiter and a quantizer with an infinite number of levels. We denote the normalized overload level of the limiter by

$$V(\rho_i) = \frac{C_c(\rho_i)}{\sqrt{P_s + P_n}} \tag{10-3}$$

with

$C_c(\rho_i)$: threshold of the soft limiter

P_s: signal power, P_n: noise power

$\rho_i = P_s/P_n$, signal-to-noise ratio of the input signal

Threshold level C_c, interval width Δx, and word length b are related by (Figure 10-11)

$$C_c + \Delta x = 2^{b-1}\Delta x \tag{10-4}$$

Two problems arise:

1. What is the optimum overload level $V(\rho_i)$?
2. Since the signal-to-noise ratio is unknown, the sensitivity of the receiver performance with respect to a mismatch of the overload level $V(\rho_i)$ must be determined.

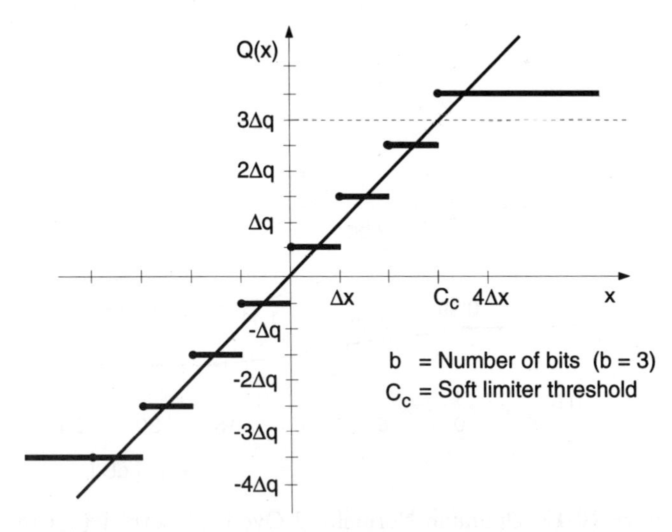

Figure 10-11 A/D Conversion Viewed as Series Connection of Soft Limiter and Infinitely Extended Quantizer

It is instructive to first consider the simple example of a sinusoidal signal plus Gaussian noise. We determine $V(\rho_i)$ subject to the optimization criterion (the selection of this criterion will be justified later on)

$$E[Q(x|C_c, b) - x]^2 \rightarrow \min \tag{10-5}$$

with $Q(x|C_c, b)$ the uniform midriser quantizer characteristic with parameters (C_c, b). In eq. (10-5) we are thus looking for the uniform quantizer characteristic which minimizes the quadratic error between input signal and quantizer output.

The result of the optimization task is shown in Figure 10-12. In this figure the overload level is plotted versus ρ_i with word length b as parameter. With $P_s = A^2/2$, A: amplitude of the sinusoidal signal, we obtain for high SNR

$$V(\rho_i) \simeq \frac{C_c(\rho_i)}{\sqrt{P_s}} = \left(\frac{C_c(\rho_i)}{A}\right)\sqrt{2} \qquad \rho_i \gg 1 \tag{10-6}$$

For large word length b we expect that the useful signal passes the limiter undistorted, i.e., $C_c(\rho_i)/A \simeq 1$ and $V(\rho_i) \simeq \sqrt{2}$. For a 4-bit-quantization we obtain $V(\rho_i) \simeq 1.26$ which is close to $\sqrt{2}$. The value of $V(\rho_i)$ decreases with decreasing word length. The overload level increases at low ρ_i. For a sufficiently fine quantization $V(\rho_i)$ becomes larger than the amplitude of the useful signal in order to pass larger amplitude values due to noise.

We return to the optimality criterion of eq. (10-5) which was selected in order not to discard information prematurely. In the present case this implies to pass the input signal amplitude undistorted to the ML decoder. It is well known that the ML decoder requires soft decision inputs for optimum performance. The bit error rate increases rapidly for hard-quantized inputs. We thus expect minimizing the

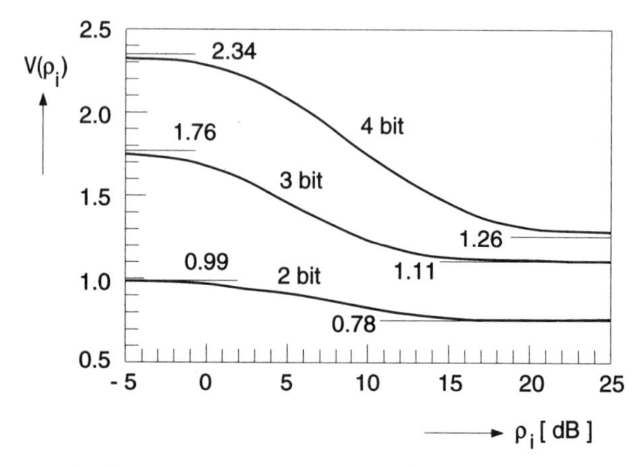

Figure 10-12 Optimum Normalized Overload Level $V(\rho_i)$ for a Sinusoidal Signal plus Gaussian Noise.
Parameter is word length b.

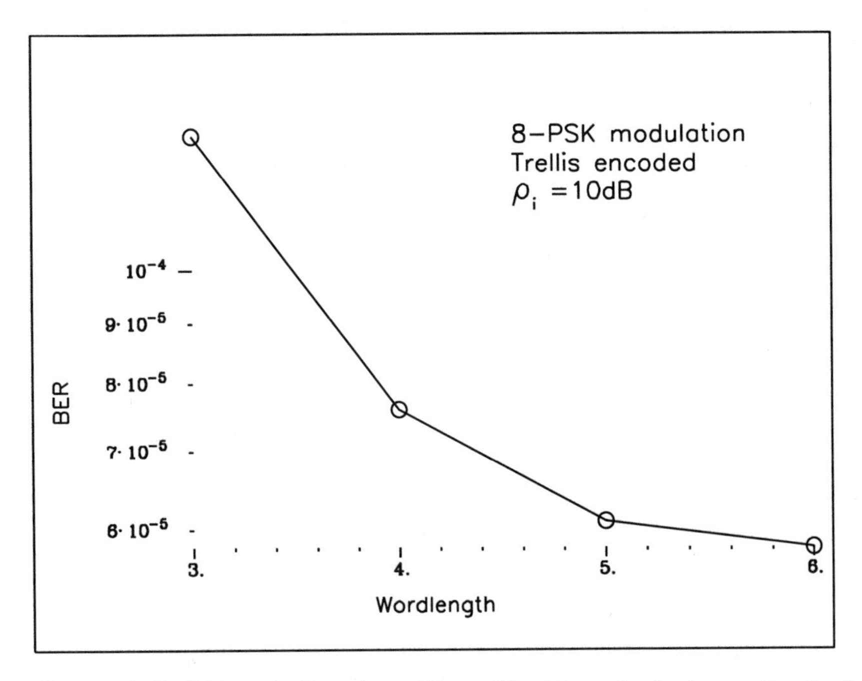

Figure 10-13 BER as a Function of Input Word Length; Optimum Overload Factor $V(\rho_i)$; Perfect Synchronization and Decoder Implementation

mean square error of eq. (10-5) to be consistent with the goal of minimizing the bit error rate. (This is indeed true as will be demonstrated shortly.)

If one optimizes the output SNR of the A/D converter instead, it can be shown that this is *not* the case. The overload factor which maximizes the output SNR is far from the optimum factor for minimizing the BER. The reason is that for maximum output SNR the input signal undergoes strong clipping which, in turn, results in near hard-quantized symbol decisions for the ML detector.

The small scale effects of input quantization are shown exemplarily in Figure 10-13 for an 8-PSK modulation over an additive Gaussian noise channel [5]. From this figure we conclude that a 4-bit quantization is sufficient and a 5-bit quantization is practically indistinguishable from an infinite precision representation.

The bit error performance of Figure 10-13 assumes an optimum overload factor $V(\rho_i)$. To determine the sensitivity of the bit error rate to a deviation from the optimum value a computer experiment was carried out. The result is shown in Figure 10-14. The clipping level is normalized to the square root of the signal power, $\sqrt{P_s}$. The BER in Figure 10-14 is plotted for the two smallest values of E_b/N_0. The input signal-to-noise ratio, E_s/N_0, is related to E_b/N_0 via eq. (3–33):

$$\frac{E_s}{N_0} = \left(\frac{E_b}{N_0}\right) R \log_2 M \tag{10-7}$$

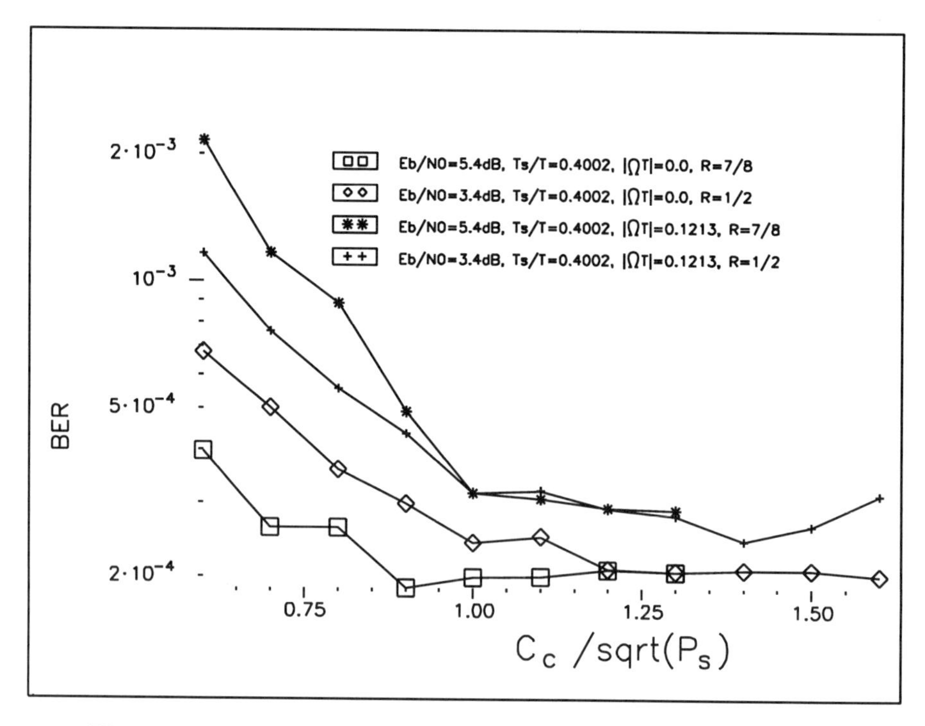

Figure 10-14 BER as a Function of the Normalized Clipping Level $C_c/\sqrt{P_s}$. Sampling ratio $T_s/T = 0.4002$. Residual frequency offsets $|\Omega T| = 0.1213$ and $|\Omega T| = 0$. Code rates $R = 7/8$ and $R = 1/2$.

For both input values of E_s/N_0 the results are plotted for zero and maximum residual frequency offset, $|(\Omega T)|_{\text{max}} = 0.1213$. The sampling rate is $T_s/T = 0.4002$.

The BER is minimal for a value larger than 1. A design point of

$$\frac{C_c}{\sqrt{P_s}} \simeq 1.25 \tag{10-8}$$

was selected. The results show a strong asymmetry with respect to the sign of the deviation from the optimum value. Clipping $\left(C_c/\sqrt{P_s}\right) < \left(C_c/\sqrt{P_s}|_{\text{opt}}\right)$ strongly increases the BER since the ML receiver is fed with hard-quantized input signals which degrades its performance. The opposite case, $\left(C_c/\sqrt{P_s}\right) > \left(C_c/\sqrt{P_s}|_{\text{opt}}\right)$, is far less critical since it only increases the quantization noise by increasing the resolution Δx to

$$\Delta x = \Delta x|_{\text{opt}} \left(\frac{C_c}{C_{c,\text{opt}}}\right) \tag{10-9}$$

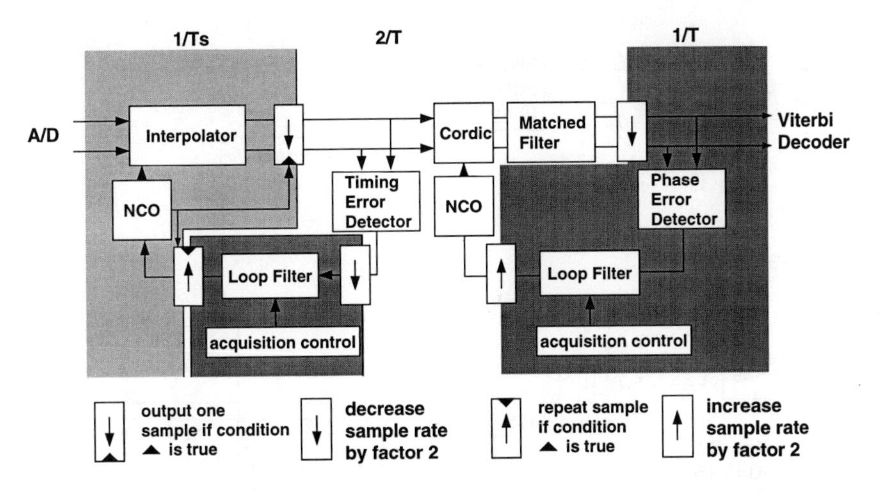

Figure 10-15 Synchronizer Structure

10.4.6 Timing and Phase Synchronizer Structure

The first step in the design process is the selection of a suitable synchronizer structure. Timing and phase synchronizer are separated (see Figure 10-15), which avoids interaction between these units. From a design point of view this separation is also advantageous, since it eases performance analysis and thus reduces design time and test complexity. An error feedback structure for both units was chosen for the following reasons: video broadcasting data is transmitted as a continuous stream. Only an initial acquisition process which is not time-critical has to be performed. For tracking purposes error feedback structures are well suited and realizable with reasonable complexity. Among the candidate algorithms which were initially considered for timing recovery was the square and filter algorithm (Section 5.4). The algorithm works independently of the phase. It delivers an unambiguous estimate, requires no acquisition unit, and is simple to implement. This ease of implementation, however, exists only for a known nominal ratio of $T/T_s = 4$. Since the ratio T/T_s is only known to be in the interval $[2; 2.5]$, the square and filter algorithm is ruled out for this application.

10.4.7 Digital Phase-Locked Loop (DPLL) for Phase Synchronization

The detailed block diagram of the DPLL is shown in Figure 10-16. In this figure the input word length of the individual blocks and the output truncation operations are shown. The word length of the DPLL are found by bit-true computer simulation. The notation used is summarized in Figure 10-17 below.

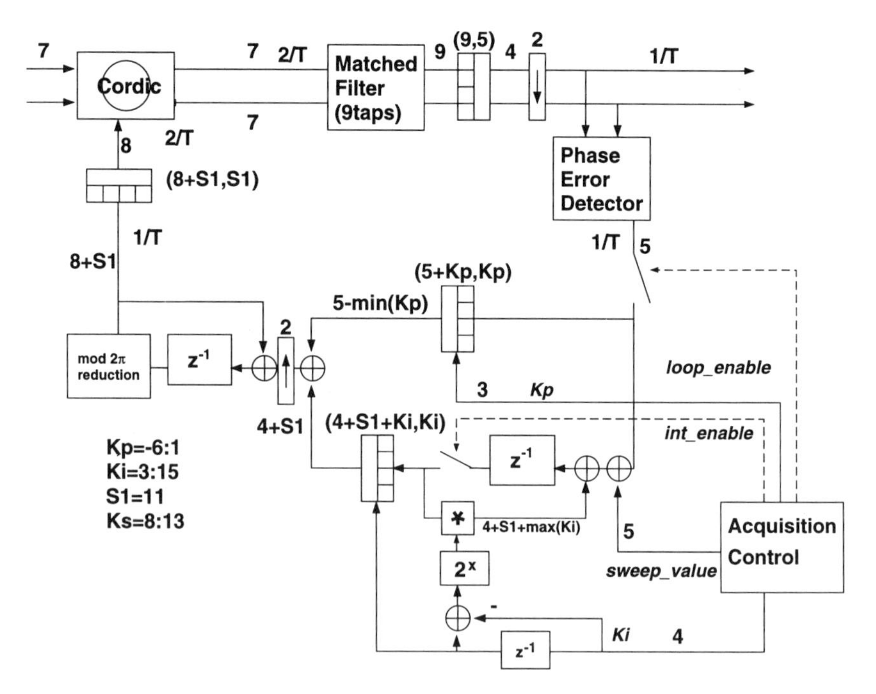

Figure 10-16 Block Diagram of the DPLL for Carrier Phase Synchronization

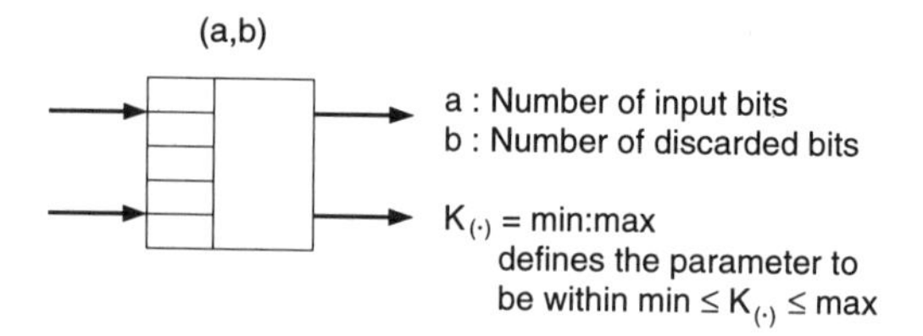

Figure 10-17 Notation Used in Figure 10-16

We next discuss the functional building block in some detail. The incoming signal is multiplied by a rotating phasor $\exp\left[j\left(\Omega kT/2+\hat{\theta}_k\right)\right]$ by means of a CORDIC algorithm [6, 7], subsequently filtered in the matched filter and decimated to symbol rate. One notices that the matched filter is placed inside the closed loop. This is required to achieve a sufficiently large SNR at the phase error detector input. The loop filter has a proportional plus integral path. The output rate of the loop filter is doubled to $2/T$ by repeating each value which is subsequently accumulated in the NCO. The accumulator is the digital equivalent to the integrator of the VCO of an analog PLL. The modulo 2π reduction (shown as a separate block) of the accumulator is automatically performed by the adder if one uses a 2's complement number representation. The DPLL is brought into lock by applying a sweep value to the accumulator in the loop filter. The sweep value and the closing of the loop after detecting lock is controlled by the block *acquisition control.*

Matched Filter

The complex-valued matched filter is implemented as two equivalent real FIR filters with identical coefficients. To determine the number of taps and the word length of the filter coefficients, Figure 10-18 is helpful. The lower part shows the number of coefficients which can be represented for a given numerical value of the center tap. As an example, assume a center value of $h_0 = 15$ which can be represented by a 5 bit word in 2's complement representation. From Figure 10-18 it follows that the number of nonzero coefficients is nine. Increasing the word length of h_0 to 6, the maximum number of coefficients is nine for $h_0 \leq 21$ and 14 for $21 < h_0 \leq 31$. The quadratic approximation error is shown in the upper part of Figure 10-18, again as a function of the center tap value. The error decays slowly for values $h_0 > 15$ while the number of filter taps and the word length increase rapidly. As a compromise between accuracy and complexity the design point of Figure 10-18 and Table 10-2 was chosen.

In the following we will explain the detailed considerations that lead to the hardware implementation of the matched filter. This example serves the purpose to illustrate the close interaction between the algorithm and architecture design found in DSP systems.

A suitable architecture of an FIR filter is the transposed direct form (Figure 10-19). Since the coefficients are fixed, the area-intensive multipliers can be replaced by sequences of shift-and-add operations. As we see in Figure 10-20, each "1" of a coefficient requires an adder. Thus, we encounter an additional constraint on the system design: choosing the coefficient in a way that results in the minimum number of "1".

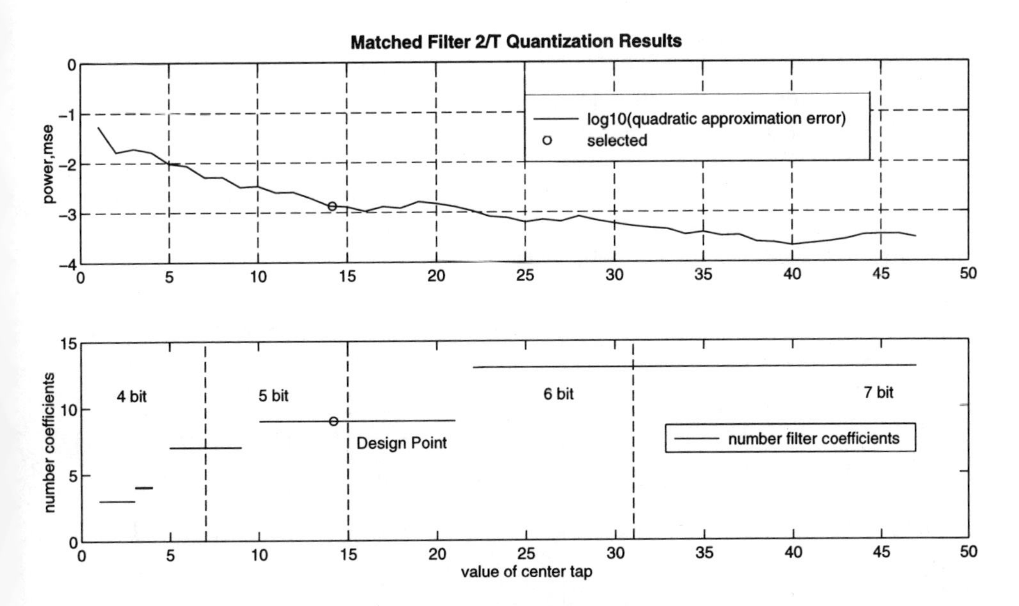

Figure 10-18 Matched Filter Approximation

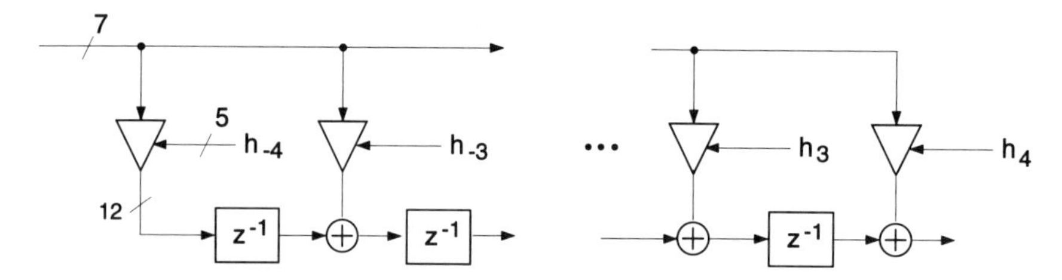

Figure 10-19 FIR Filter in Transposed Direct Form

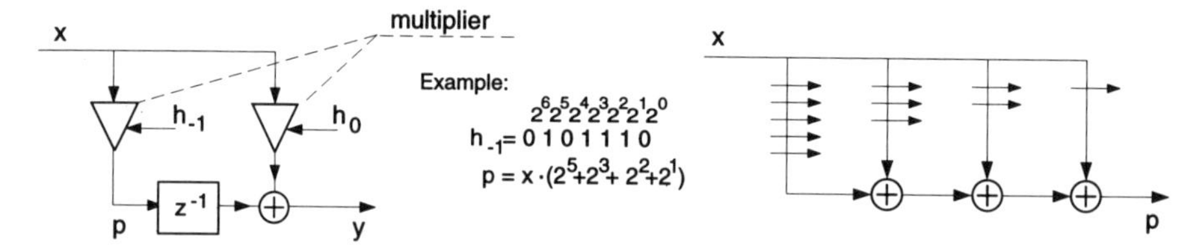

Figure 10-20 Replacement of Multipliers by Shift-and-Add Operations

The number of 1's can be further reduced by changing the number representation of the coefficients. In filters with variable coefficients (e.g., in the interpolator filter) this can be performed by Booth encoding [8]. For fixed coefficient filters each sequence of 1's (without the most significant bit) in a 2's complement number can be described in a canonical signed digit (CSD) representation with at most two 1's

$$\sum_{i=N}^{M} 1 \times 2^i = 1 \times 2^{M+1} - 1 \times 2^N \tag{10-10}$$

This leads to the matched filter coefficients in CSD format shown in Table 10-2.

As a result, we need nine adders to implement a branch of the matched filter. For these adders different implementation alternatives exist. The simplest is the carry ripple adder (Figure 10-21). It consists of full adder cells. This adder requires the smallest area but is the slowest one since the critical path consists

Table 10-2 Matched Filter Coefficients

Coefficient	Numerical Value	2's complement	Canonical Signed Digit Representation
h_0	14	0 1 1 1 0	1 0 0 -1 0
$h_{-1} = h_1$	8	0 1 0 0 0	0 1 0 0 0
$h_{-2} = h_2$	-1	1 1 1 1 1	0 0 0 0 -1
$h_{-3} = h_3$	-2	1 1 1 1 0	0 0 0 -1 0
$h_{-4} = h_4$	1	0 0 0 0 1	0 0 0 0 1

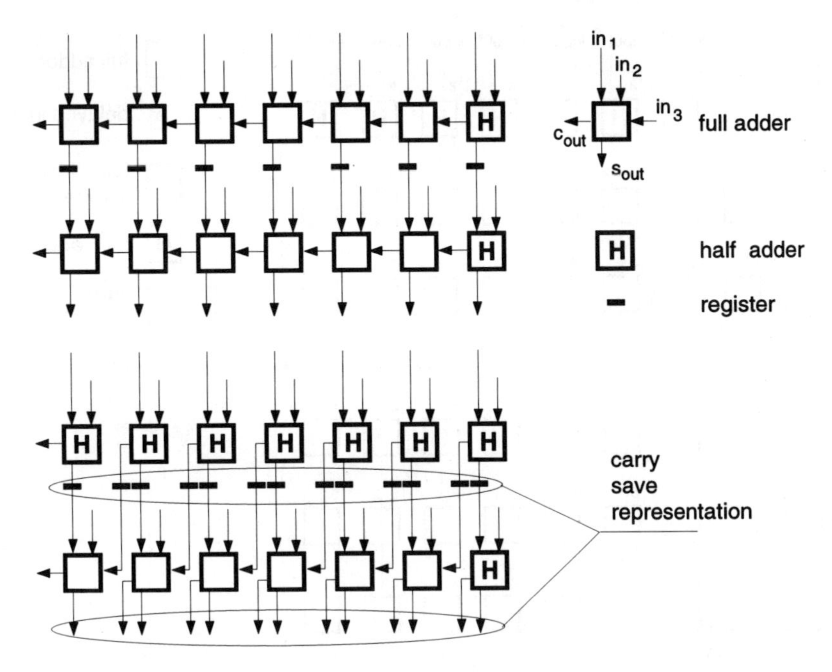

Figure 10-21 Carry Ripple versus Carry Save Adder

of the entire carry path. By choosing an alternative number representation of the intermediate results, we obtain a speedup at a slightly increased area: the carry output of each adder is fed into the inputs of the following filter tap. This carry save format is a redundant number representation since each numerical value can be expressed by different binary representations. Now, the critical path consists of one full adder cell only.

The word length of the intermediate results increases from left to right in Figure 10-19. This implies a growing size of the adders. By reordering the partial products or the "bitplanes" [9, 10, 11] in such a way that the smallest intermediate results are added first, the increase of word length is the smallest possible. Thus, the silicon area is minimized. Taking into account that the requirements on the processing speed and the silicon technology allow to place three bitplanes between two register slices, we get the structural block diagram of the matched filter depicted in Figure 10-22.

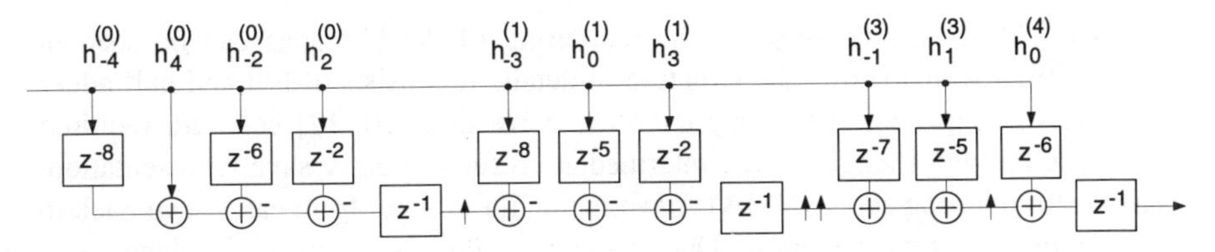

Figure 10-22 Block Diagram of One Branch of the Matched Filter

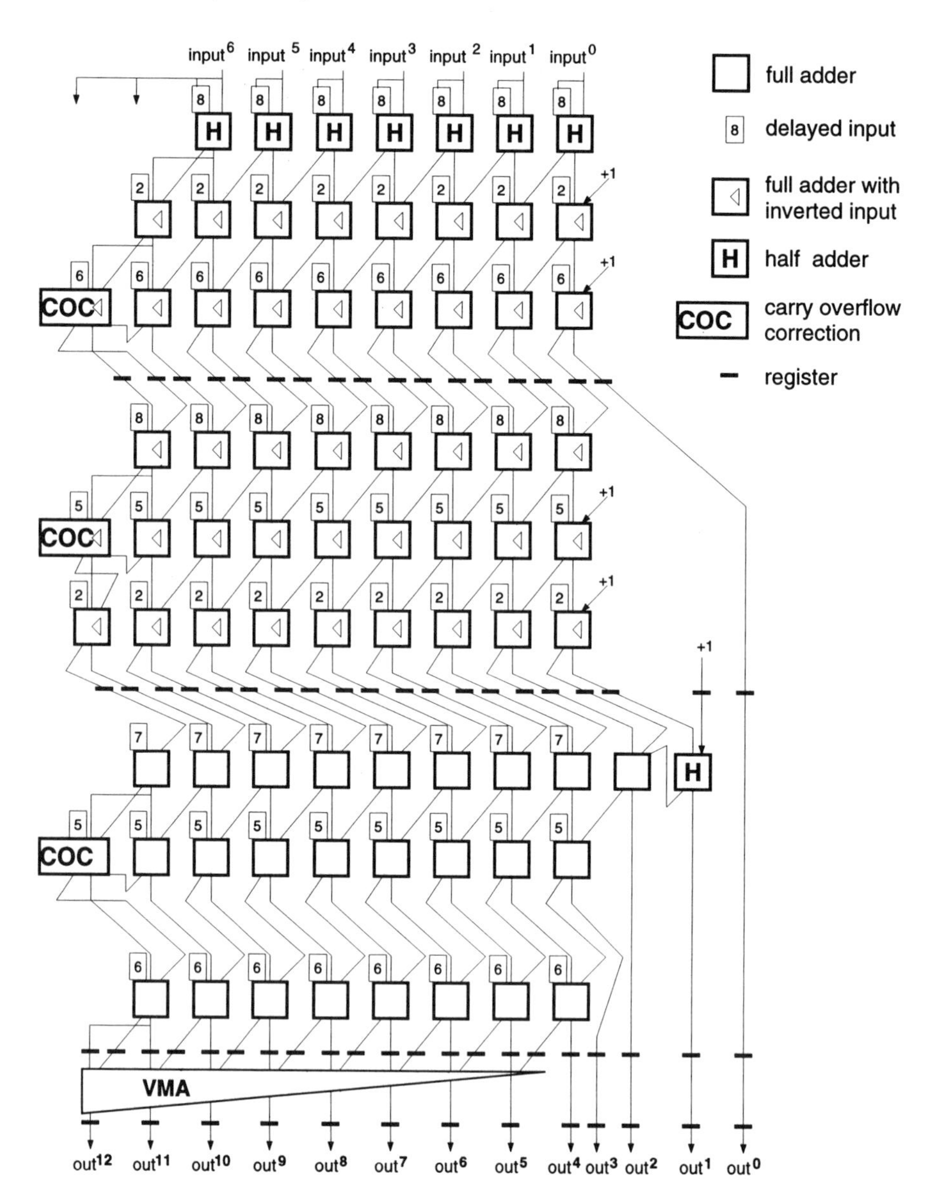

Figure 10-23 Detailed Block Diagram of Matched Filter

An additional advantage of the reordering of the bitplanes can be seen in Figure 10-23 which shows the structure in detail. It consists of full and half adder cells and of registers. The carry overflow correction [10, 12] cells are required to reduce the word length of the intermediate results in carry save representation. The vector merging adder (VMA) converts the filter output from carry save back to 2's complement representation. Due to the early processing of the bitplanes with the smallest numerical coefficient values, the least significant bits are computed first and may be truncated without side effects. The word length of the VMA is decreased as well.

Phase Error Detector

A decision-directed detector with hard quantized decisions is used (see Section 5.8):

$$g\left(\theta_0 - \hat{\theta}\right) = \text{Im}\left[\hat{a}_n^* z_n(\hat{\varepsilon}) e^{-j\hat{\theta}}\right] \tag{10-11}$$

with

$$\hat{a}_n = \text{sign}\left\{\text{Re}\left[z_n(\hat{\varepsilon}) e^{-j\hat{\theta}}\right]\right\} + j\ \text{sign}\left\{\text{Im}\left[z_n(\hat{\varepsilon}) e^{-j\hat{\theta}}\right]\right\} \tag{10-12}$$

Inserting the hard quantized symbols $\hat{a}_n$ into the previous equation we obtain

$$g\left(\theta_0 - \hat{\theta}\right) = \text{Re}(\hat{a}_n)\text{Im}\left[z_n(\hat{\varepsilon}) e^{-j\hat{\theta}}\right] - \text{Im}(\hat{a}_n)\text{Re}\left[z_n(\hat{\varepsilon}) e^{-j\hat{\theta}}\right] \tag{10-13}$$

The characteristic $g\left(\theta_0 - \hat{\theta}\right)$ is plotted in Figure 10-24. It varies with the signal-to-noise ratio. The slope at the origin becomes smaller with decreasing E_s/N_0 (see also Section 6.3). Since the detector takes the difference between two noisy samples, it is sensitive to quantization. For this reason an 8-bit internal word length was found to be necessary for an acceptable performance (see below).

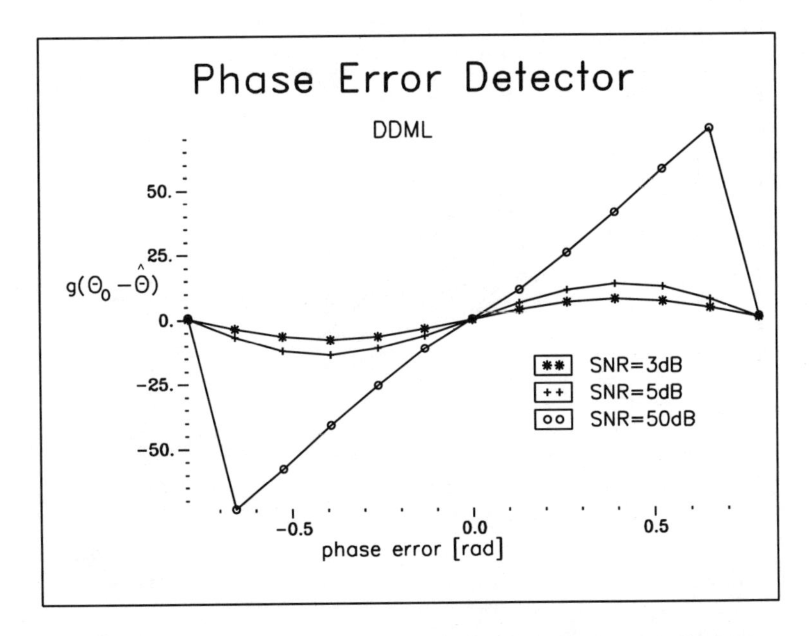

Figure 10-24 Phase Error Detector with Hard Quantized Decisions

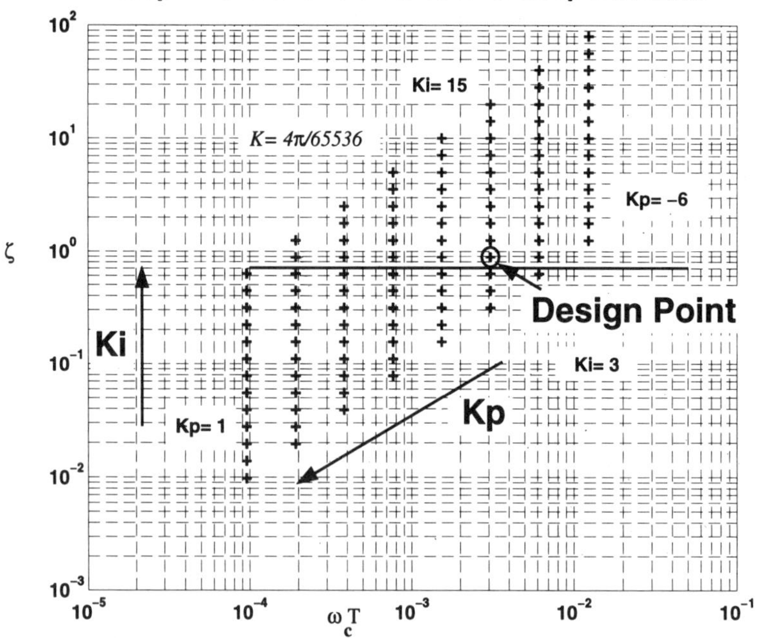

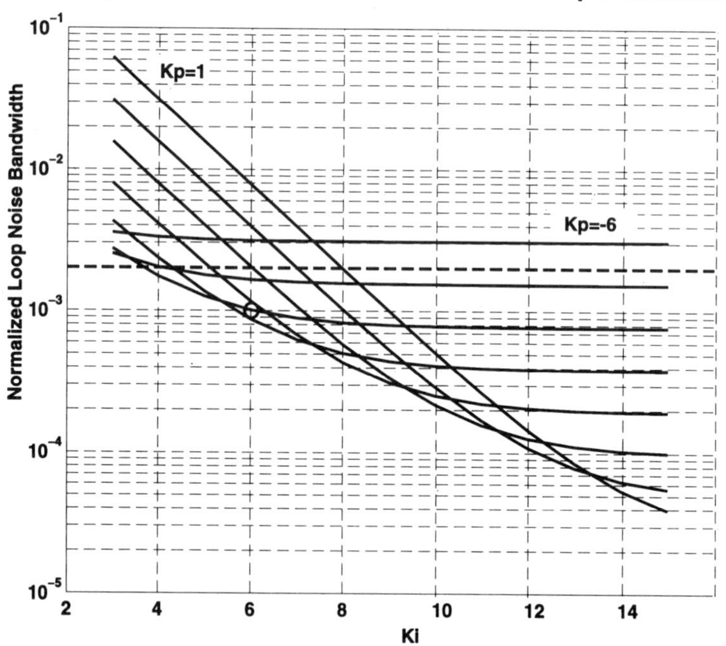

Figure 10-25 Loop Parameters as a Function of the Quantized Loop Constants K_I and K_p

Design of the DPLL

Since the loop bandwidth B_L is much smaller than the symbol rate, $B_L T = 10^{-3}$, a quasi-time-continuous design (see Volume 1) approach is taken. In Figure 10-25 the PLL parameters $B_L T$, $\omega_c T$ are displayed as a function of the quantized coefficients:

$$\begin{aligned} K_p &\in [-6, 1] \qquad (3 \text{ bit}) \\ K_I &\in [3, 15] \qquad (4 \text{ bit}) \end{aligned} \tag{10-14}$$

For example, for $K_p = -4$, $K_I = 6$ we obtain $B_L T \simeq 10^{-3}$, $\xi \simeq 0.8$.

Notice: K_p and K_I in Figure 10-16 are the negative 2's logarithms of the quantity used in the continuous realization (see Figure 10-16).

Loop Filter and NCO Accumulator

In order to avoid overflow effects, the accumulator word length of the filter and NCO must be large. The units are implemented with a word length of 25 bits and of 19 bits, respectively.

Tracking Performance

The normalized $\operatorname{var}\left(\hat{\theta}\right)/2B_L T$ is plotted in Figure 10-26. At the design point $E_s/N_0 = 3$ dB the variance is approximately 8.5 dB above the CR-bound. Of this difference, 7 dB are due to incorrect decisions of the DD phase error detector

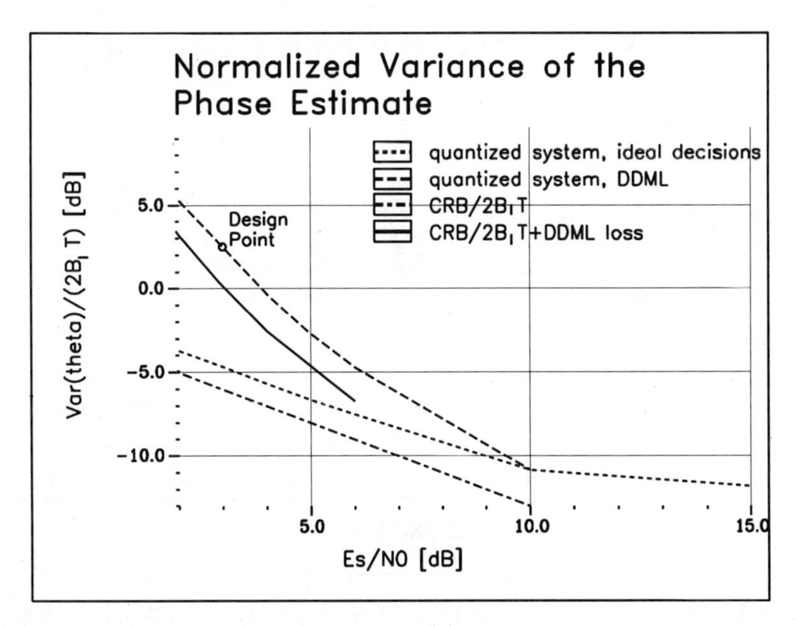

Figure 10-26 Normalized Variance $\operatorname{var}\left(\hat{\theta}\right)/2B_L T$ as a Function of E_s/N_0

(compare also with Figure 6-1), and 1.5 dB is attributed to numerical effects. To validate the design a computer experiment was run assuming known symbols in the DD algorithm. The resulting variance is about 1.5 dB above the CR-bound, as predicted. By inspection of Figure 6-4 we find that an NDA algorithm for QPSK offers no performance improvement over the DD algorithm.

Due to erroneous decisions the slope of the phase error characteristic decreases (Figures 6-3 and 10-24) with decreasing E_s/N_0. For this reason the loop bandwidth also varies with E_s/N_0, as shown in Figure 10-27. In the same figure we have plotted the variance $\mathrm{var}\left(\hat{\theta}\right)$ (*not* normalized to $2B_LT$) taking the E_s/N_0 dependence of the loop bandwidth into account. Notice that it is the variance $\mathrm{var}\left(\hat{\theta}\right)$ which affects the demodulation process.

Remark: At the operating point of $E_s/N_0 = 3$ dB the difference in performance between the algorithm using correct decisions and the DD algorithm is 4.5 dB, compared to 8.5 dB for the corresponding normalized quantities in the previous figure. The seeming discrepancy is readily resolved by noting that the slope of the phase error detector characteristic is different in the two cases.

The variance is lower-bounded by the quantization of the phase in the CORDIC at $(2\pi/256)^2/12$. For any practical application this value is negligibly small.

Using formula (6-231) for the cycle slip rate, one computes for $B_LT = 10^{-2}$ and $E_s/N_0 = 3$ dB approximately one cycle slip per 1.6 days.

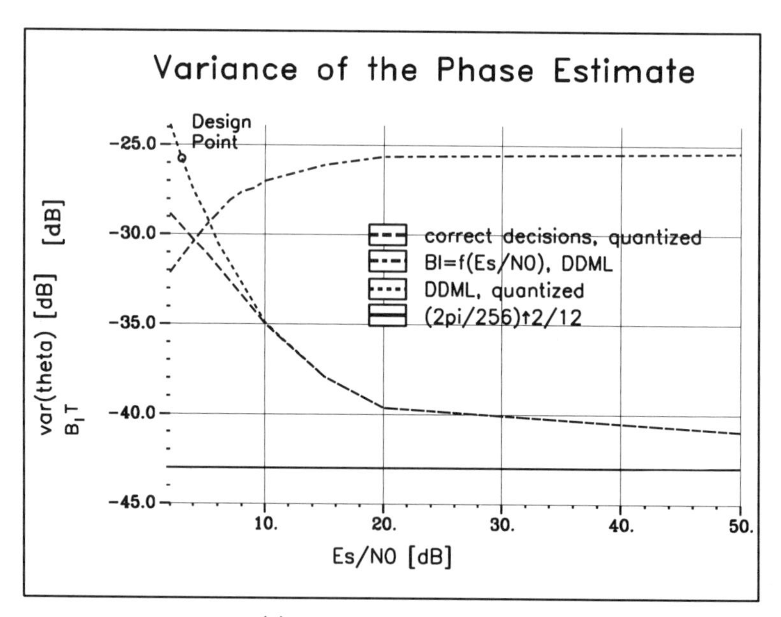

Figure 10-27 Variance $\mathrm{var}\left(\hat{\theta}\right)$ and Loop Bandwidth B_LT as a Function of E_s/N_0

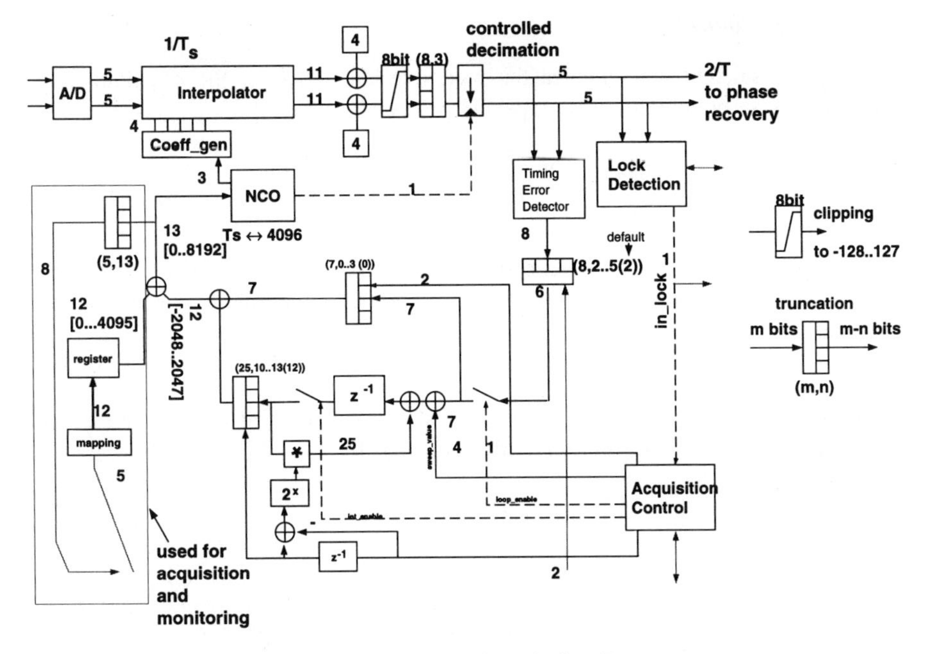

Figure 10-28 DPLL for Timing Recovery

We recall that to accommodate for phase noise effects (Volume 1, Chapter 3.6) the bandwidth must be larger than a minimum value, which depends on the oscillator characteristic and the symbol rate.

10.4.8 Timing Recovery

Figure 10-28 shows the block diagram of the DPLL used for timing recovery. In the block diagram the input word lengths of every block and the clipping and truncation operation at the output are shown.

In the sequel we discuss the building blocks of the DPLL. The interpolator is implemented as a 4-tap FIR filter with time-variant coefficients. The word length of the filter coefficients, $h_I(\mu)$, is 4 bits. Their values for the 8 different fractional delays μ were determined by minimizing the quadratic approximation error. To compensate for the bias at the interpolator output (due to the A/D characteristic) the value 4 is added.

The NDA timing error detector works with two samples per symbol (see Section 5.6.2). The loop filter has a proportional plus integral type path in order to realize a second-order closed-loop filter transfer function. Similarly as for the carrier recovery, DPLL-aided acquisition is used to bring the loop into lock.

Interpolator Control

For the sake of convenience the recursive equations for the basepoint and fractional delay computation are repeated below [Eq. (9-39)]:

$$\begin{aligned} m_{k+1} &= m_k + L_{\text{INT}}\left[\hat{\mu}_k + w(m_k)\right] \\ \hat{\mu}_{k+1} &= \left[\hat{\mu}_k + w(m_k)\right]_{\text{mod}1} \end{aligned} \tag{10-15}$$

The control word $w(m_n T_s)$ is an estimate of the ratio T_I/T_s $(T_I = T/2)$. The $L_{\text{INT}}(\cdot)$ and fractional rest computation are performed in the following way. Define the function

$$\eta(m_k, i) = \hat{\mu}_k + w(m_k) - i; \quad i = 0, 1, 2, ... \tag{10-16}$$

At the basepoint m_k the value $\eta(m_k, 0)$ is stored in an N-bit register. At every T_s cycle the value of the register is decremented by 1,

$$\eta(m_k, i+1) = \eta(m_k, i) - 1 \tag{10-17}$$

As long as $\eta(m_k, i) > 1$, there obviously exists an integer $L_{\text{INT}}(\cdot) > m_k + i$. Hence, the criterion for obtaining the next basepoint m_{k+1} is $\eta(m_k, i_{\text{min}}) < 1$, where i_{min} is the smallest integer for which the condition is fulfilled. Thus, the decrementation operation is continued until the condition $\eta(m_k, i_{\text{min}}) < 1$ is detected. By definition, the register content $\eta(m_k, i_{\text{min}})$ equals $\hat{\mu}_{k+1}$.

The operations are continued for m_{k+1} with the initial value

$$\eta(m_{k+1}, 0) = \eta(m_k, i_{\text{min}}) + w(m_{k+1}) \tag{10-18}$$

The nominal ratio T_I/T_s is always larger than one to guarantee that no two basepoints coincide, i.e., $m_{k+1} \neq m_k$ in the noiseless case. Noise may rarely cause the condition $\eta(m_k, 0) < 1$. In such a case the result is considered false and is replaced by $\eta(m_k, 0) = 1$. Hence, $\hat{\mu}_{k+1} = 0$ and $m_{k+1} = m_k + 1$.

The following numerical values apply to the DVB chip. The sampling interval is quantized to $N = 12$ bits. A (quantized) fractional delay μ is thus represented by one of the $2^{12} = 4096$ integer numbers. The intermediate time interval $T_I = T/2$ ranges from $[4096, 5120]$ depending on the ratio $1.0 < T_I/T_s \leq 1.2$. The $N = 12$ bit word for μ is truncated to 3 bits to obtain one of the 8 possible values used in the interpolator.

Constraints on the T/T_s Ratio

After interpolation a controlled decimation is carried out. Therefore it may occur that signal and noise spectra, which do not interfere before the decimation process, do aliase after decimation. In order to satisfy the conditions on sufficient statistics (Section 4.2.4) the following relations must hold [13]:

$$\text{B}_\text{a} \leq \left(\frac{T}{T_s}\right)\left[\frac{1+\alpha}{2T} + |\Omega|\right] \tag{10-19}$$

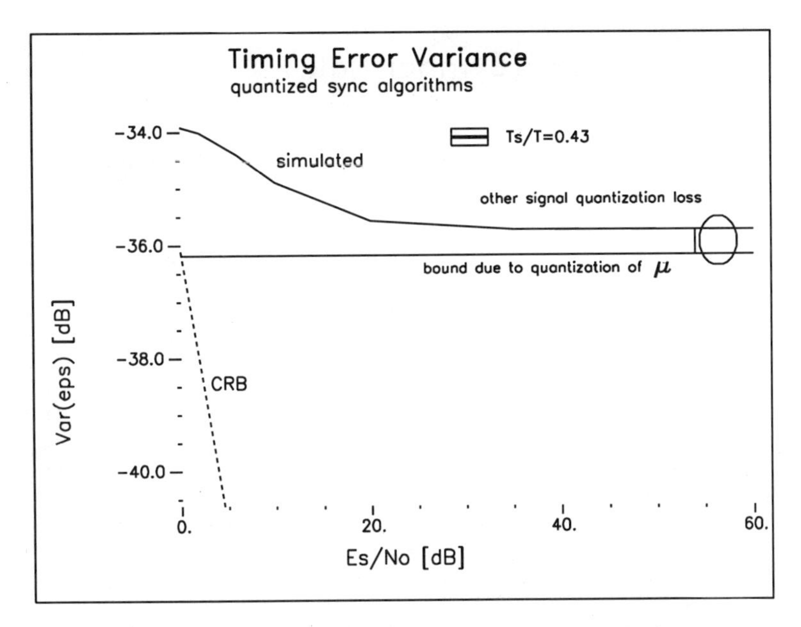

Figure 10-29 Timing Error Variance versus E_s/N_0

with BaW the bandwidth of the analog prefilter, α the excess bandwidth, and $|\Omega T|$ the maximum frequency offset normalized to the symbol rate.

Tracking Performance

In Figure 10-29 the variance $\text{var}(\hat{\varepsilon})$ is plotted versus E_s/N_0. At the design point of $E_s/N_0 = 3.4$ dB, the CRB equals -40.5 dB. One part of the difference of 6.5 dB is due to the fact that the NDA algorithm does not reach the CRB while the other part is due to numerical effects. From Figure 6-6 we obtain for the NDA algorithm a variance of $\text{var}(\hat{\varepsilon}) = -36.8$ dB for perfect implementation. This leaves a 2.8-dB loss which must be attributed to quantization. Using Figure 10-29 we can further analyze the structure of the quantization loss. The variance is lower bounded by the quantization of μ into 8 levels,

$$\text{var}(\hat{\varepsilon}) \geq \left(\frac{T_s}{8T}\right)^2 \frac{1}{12} \tag{10-20}$$

At high SNR the variance is mostly due to this quantization. We see that the asymptotic value of the variance is slightly above the lower bound. We therefore can attribute most of the loss to the quantization of μ and a small part of 0.45 dB to other word length effects. The decomposition at large E_s/N_0 merely serves as a qualitative mark. For small values of E_s/N_0 we observe an increase of the quantization loss which is caused by complicated nonlinear effects in the estimator.

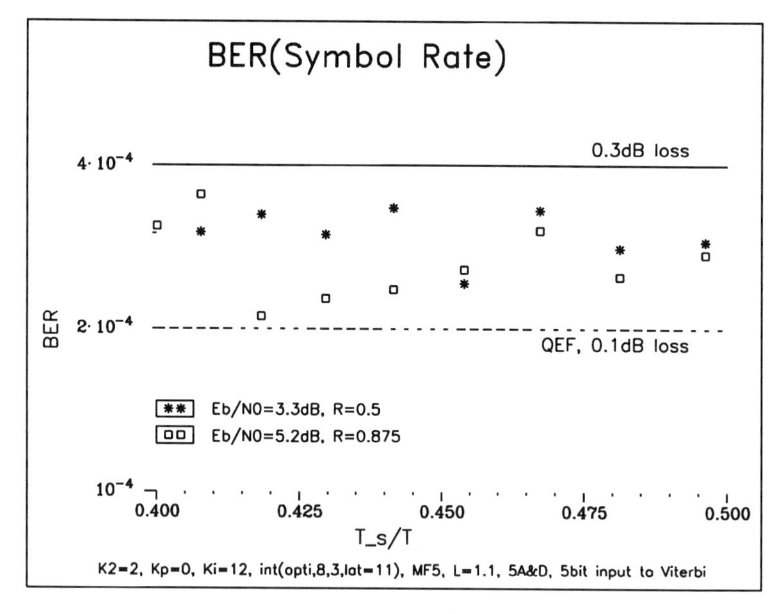

Figure 10-30 Bit Error Rate versus T_s/T

In Figure 10-30 the BER is plotted for the code rates $R = 1/2$ and $R = 7/8$, respectively, as a function of the ratio T_s/T. The operating points are

$$E_s/N_0 = \begin{cases} E_b/N_0 & R = 1/2 \\ (E_b/N_0)R\log_2 M & R = 7/8,\ M = 4 \end{cases} \tag{10-21}$$

(perfect carrier synchronization). We observe that the bit error rate is practically independent of the ratio T_s/T for both code rates. It varies between 4×10^{-4} and 2×10^{-4}. The loss D_ε is the difference of E_s/N_0 to achieve a bit error rate of 4×10^{-4} and 2×10^{-4}, respectively.

10.5 Bit Error Performance of the DVB Chip

The bit error performance for the code rates of $R = 1/2$ and $R = 7/8$ of the receiver is shown in Figure 10-31. The solid lines belong to the hypothetical receiver with perfect synchronization and implementation. The dotted curves belong to the inner receiver comprising carrier synchronization, timing recovery, and convolutional decoder (see Figure 10-10). The performance is obtained by simulating the bit-true models of the functional building blocks. A perfect analog front end is assumed. The output of the convolutional decoder is input to the Reed-Solomon decoder, after frame synchronization and deinterleaving is carried out. For a BER $\leq 2 \times 10^{-4}$ of the convolutional decoder the Reed-Solomon decoder output is quasi error-free.

The total loss D_{total} is below the specified 0.5 dB for both code rates. This leaves approximately 0.5 dB for the imperfection of the analog front end.

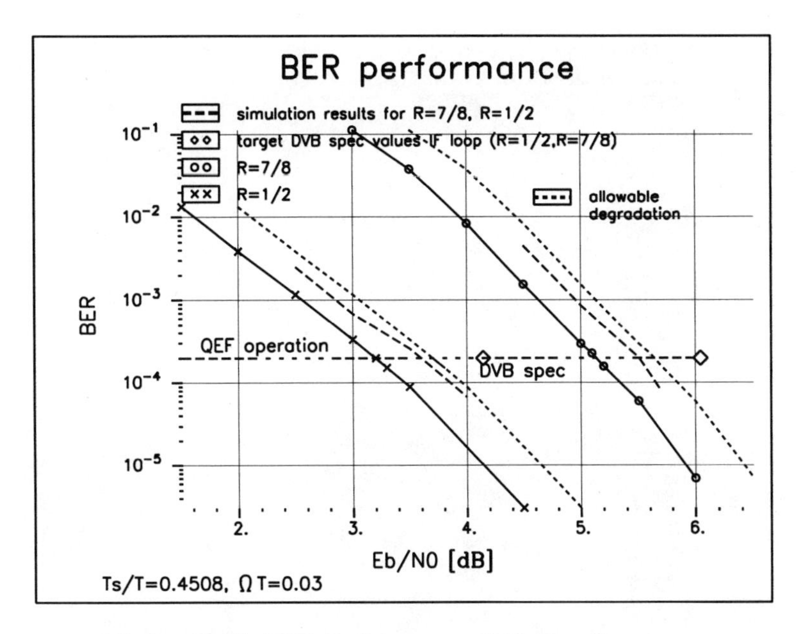

Figure 10-31 BER Performance of Bit-True Receiver

10.6 Implementation

The circuit is manufactured in a 3.3 Volt advanced $0.5\mu m$ CMOS technology with a power consumption below 1.5 W and fits into a P-QFP-64 package. The maximum allowable clock rate is 88 MHz. Table 10-3 gives a complexity breakdown of the data-path components. A photo of the device is shown in Figure 10-32.

Table 10-3 Complexity Breakdown

Block	Cell Area	RAM	# Lines of VHDL
Timing/Carrier Synchronizer	32.5%	—	7000
Viterbi Decoder (incl. node synchronizer)	40.0%	83.8%	4000
Frame Synchronizer	1.7%	—	700
Deinterleaver	2.2%	8.4%	640
Reed-Solomon Decoder	22.9%	7.8%	5600
Descrambler	0.7%	—	360

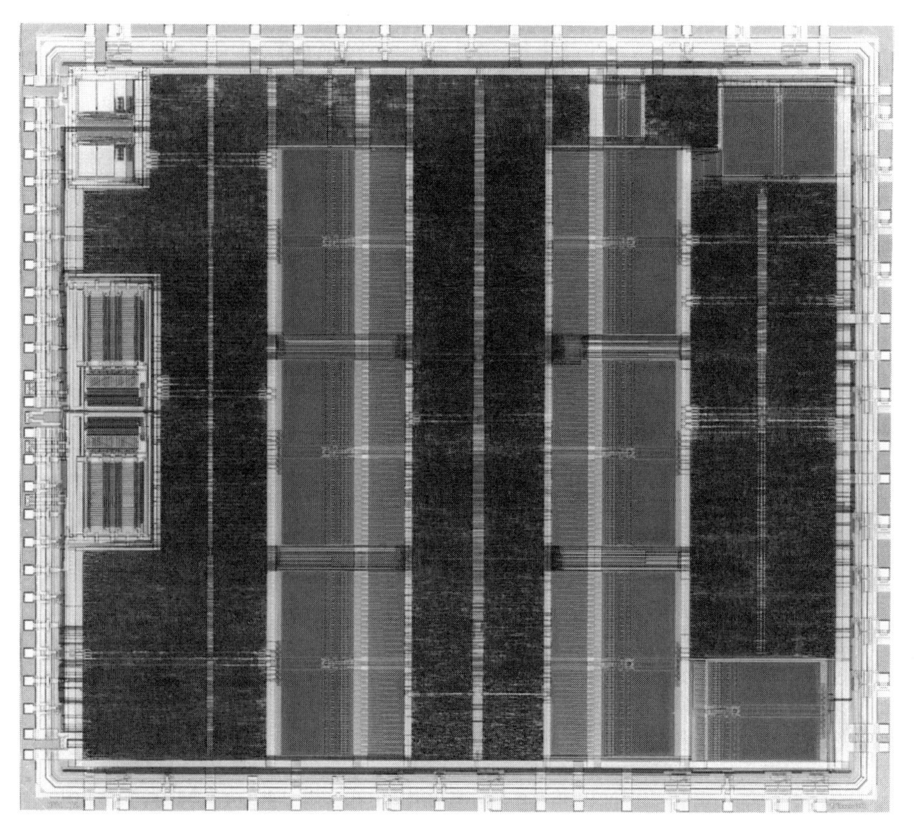

Figure 10-32 Chip Photograph of the DVB Receiver

10.7 CAD Tools and Design Methodology

The design flow is illustrated by Figure 10-33.

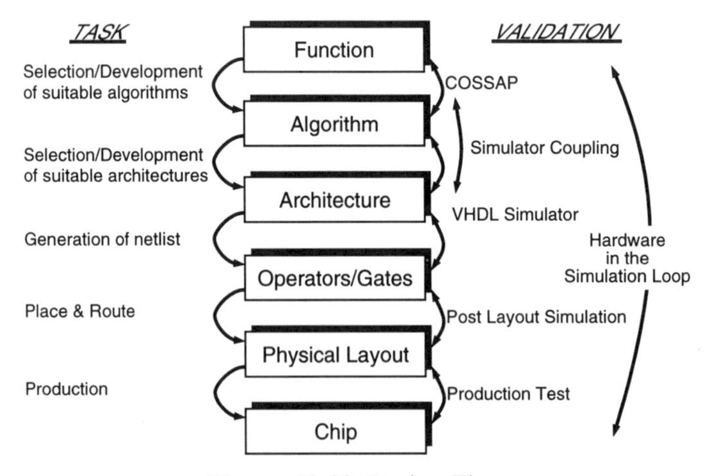

Figure 10-33 Design Flow

10.7.1 Algorithm Design

In a first step a model of the channel and a floating-point model of the functional units under considerations are simulated to obtain the performance of a perfect implementation. This performance serves as a benchmark against which the finite word length implementation is compared.

During the design process it may become necessary to modify or replace algorithms. The interactive algorithm design process is completed when the system performance meets the specification. The process of comparing the specifications against the performance is called validation.

Computer simulation is the experimental side of the algorithm design process. The experimental results must always be compared to the theoretical results, such as variance of estimates or the Cramér Rao bound. Only in such a way one can eliminate inevitable modeling or simulation errors or identify deficiencies of approximations to optimal algorithms. As a typical example, consider the performance evaluation of the carrier phase synchronizer in Section 10.4.7.

In the next step the floating models are successively replaced by finite word length bit-true models of the units to be implemented. The performance of the bit-true models is exactly that of the actual VLSI implementation. In this design cycle no modifications of the algorithms should be necessary. Since the level of abstraction is far more detailed, the simulation time greatly increases. Therefore, only finite word length effects are evaluated. Various concepts exist for modeling and simulating the bit-true behavior [15]–[19]. The replacement of a floating-point model by a finite word length model is a difficult, error-prone and time-consuming task which requires much experience. Only very recently computer-aided design (CAD) tool support become available [20]–[22].

The CAD tool used in the DVB receiver design was COSSAP™.[4] COSSAP™ has an extensive telecommunication library which allows floating-point as well as bit-true simulation of an algorithm. The block diagram editor of COSSAP™ and the stream-driven simulation paradigm allow efficient modeling and simulation of dynamic data flow.

Remark: Among other tools, SPW™ of Alta Group of Cadence offers similar capabilities.

10.7.2 Architecture Design

For each of the building blocks in Figure 10-10 an architecture was developed. The architectural description was done in the VHDL [23] language. A rough indicator of the modeling complexity is the number of lines of VHDL code, given in Table 10-3. The functional verification of the VHDL model against the bit-true and cycle-true COSSAP reference was done by coupling two simulation engines. Via an interface [24] COSSAP block simulation data is fed into the VHDL simulator.

[4] COSSAP™: Registered Trademark of SYNOPSYS Inc., Mountain View, California.

The result to the input is transferred back into the COSSAP environment and compared with the reference output.

For logic synthesis and estimation in terms of silicon area and minimum clock frequency the SYNOPSYS tool suite (Rev. 3.0) was used.

The area of design methodology and CAD tools is rapidly advancing. The design flow described is state of the art of 1996. The reader is urged to update this information.

Remark: Recently (1996) a new generation of commercial CAD tools emerged which allows the synthesis of an architecture from a behavioral description. The advantage of this approach is that only one model must be developed. An example of such a tool is the Behavioral Compiler™.[5]

10.8 Topics Not Covered

Simulation of communication system is a demanding discipline. It is the subject of the books by Jeruchim, Balaban, and Shanmugan, [25] and the books by Gardner and Baker [26] which are highly recommended reading.

Hardware-software co-design is extremely important for meeting design goals. The area is rapidly evolving due to the expansion of the CAD tools support to co-design problems. An in-depth treatment of the various aspects can be found in the proceedings of the first Hardware/Software Co-Design Workshop in Tremezzo, 1995 [27, 28].

The design of decoders (the outer receiver) is regrettably outside the scope of this book. For the reader interested in this fascinating area we provide bibliographical notes at the end of this chapter.

Bibliography

[1] B. Wilkie et. al., "Silicon or software," *IEEE VTC News*, pp. 16–17, Aug. 1995.

[2] W.-T. Chang, A. Kalavade, and E. A. Lee, "Effective heterogeneous design and co-simulation," *Proceedings of the NATO ASI on Hardware/Software Co-Design, Tremezzo, Italy*, vol. 310, pp. 187–212, June 1995.

[3] A. V. Oppenheim and R. W. Shafer, *Digital Signal Processing*. Englewood Cliffs, NJ: Prentice-Hall, 1975.

[4] ETSI, "Digital Broadcasting System for Television, Sound and Data Services; Framing Structure, Channel Coding, and Modulation for 11/12 GHz Satellite Services," *Draft DE/JTC-DVB-6, ETSI, F-06921 Sophia Antipolis, France*, Aug. 1994.

[5] Behavioral Compiler™ is a trademark of SYNOPSYS, Inc., Mountain View, California.

[5] O. J. Joeressen, M. Oerder, R. Serra, and H. Meyr, "DIRECS: System Design of a 100 Mbit/s Digital Receiver," *IEE Proceedings-G*, vol. 139, pp. 220–230, Apr. 1992.

[6] J. E. Volder, "The CORDIC Trigonometric Computing Technique," *IRE Trans. El. Comp.*, vol. EC-8, pp. 330–334, Sept. 1959.

[7] H. Dawid and H. Meyr, "The Differential CORDIC Algorithm: Constant Scale Factor Redundant Implementation without correcting Iterations," *IEEE Trans. Computers*, vol. 45, Mar. 1996.

[8] A. D. Booth, "A Signed Binary Multiplication Technique," *Quart. J. Mech. Appl. Math.*, vol. 4, pp. 236–240, 1951.

[9] P. B. Denyer and D. Myers, "Carry-Save Arrays for VLSI Signal Processing," *Proceedings of first Int. Conf. VLSI (Edinburgh)*, pp. 151–160, Aug. 1981.

[10] T. G. Noll, "Semi-Systolic Maximum Rate Transversal Filters with Programmable Coefficients," *Systolic Arrays, W. Müller et. al., (eds.)*, pp. 103–112, 1987.

[11] M. Vaupel and H. Meyr, "High Speed FIR-Filter Architectures with Scalable Sample Rates," *Proceedings ISCAS, London, IEEE*, pp. 4127–4130, May 1994.

[12] T. Noll, "Carry-Save Architectures for High-Speed Digital Signal Processing," *J. VLSI Signal Processing*, vol. 3, pp. 121–140, June 1991.

[13] U. Lambrette, K. Langhammer, and H. Meyr, "Variable Sample Rate Digital Feedback NDA Timing Synchronization," *Proceedings IEEE GLOBECOM London*, vol. 2, pp. 1348–1352, Nov. 1996.

[14] J. S. Walter, "A Unified Algorithm for Elementary Functions," in *in AFIPS Spring Joint Computer Conference*, pp. 379–385, 1971.

[15] Electronic Research Laboratory, UC Berkeley, *Almagest: Ptolemy User's Manual*, 1991.

[16] Cadence Design Systems, 919 E. Hillsdale Blvd., Foster City, CA 94404, USA, *SPW User's Manual.*

[17] Synopsys Inc., 700 East Middlefield Road, Mountain View, CA 94043–4033, USA, *Cossap User Guide.*

[18] Angeles Systems, *Vanda-Design Environment for DSP Systems*, 1994.

[19] Mentor Graphics, 1001 Ridder Park Drive, San Jose, CA 95131, USA, *DSP Station User's Manual.*

[20] M. Willems, V. Bürsgens, and H. Meyr, "FRIDGE: Tool Supported Fixed-Point Code Generation from Floating-Point Programs Based on Interpolative Approach," *Proceedings of the DAC*, 1997. accepted for publication.

[21] M. Willems, V. Bürsgens, H. Keding, and H. Meyr, "An Interactive Fixed-Point Code Generation Environment for HW/SW-CoDesign," *Proceedings of the ICASSP, München, Germany*, pp. 687–690, Apr. 1997.

[22] W. Sung and K. Kum, "Word-Length Determination and Scaling Software for a Signal Flow Block Diagram," *Proceedings of ICASSP '94*, pp. II 457–460, Apr. 1994.

[23] IEEE, "Standard VHDL Language Reference Manual," *IEEE Std.* , vol. 1076, 1987.

[24] P. Zepter and K. ten Hagen, "Using VHDL with Stream Driven Simulators for Digital Signal Processing Application," *Proceedings of EURO-VHDL '91, Stockholm*, 1991.

[25] M. C. Jeruchim, P. Balaban, and K. S. Shanmugan, *Simulation of Communication Systems.* Plenum, 1994.

[26] F. M. Gardner and J. D. Baker, *Simulation Techniques: Models of Communication Signals and Processes.* New York: Wiley, 1997.

[27] *Proceedings of the NATO ASI on Hardware/Software Co-Design, Tremezzo, Italy*, vol. 310, June 1995.

[28] H. Meyr, T. Grötker, and O. Mauss, *Concurrent HW/SW Design for Telecommunication Systems: A Product Development Perspective.* Proceedings of NATO Advanced Study Institute on Hardware/Software Co-Design. Boston: Kluwer Academic, 1995.

10.9 Bibliographical Notes on Convolutional Coding and Decoding

In 1967 Viterbi [1] devised a decoding algorithm for convolutional codes which is named after him. A comprehensive discussion about the Viterbi algorithm is given by Forney [2]. The theoretical aspects of the algorithm are well covered in the standard textbooks on communication theory, whereas the literature on Viterbi decoder (VD) implementation is widely scattered in journals. In the following a guide to access the literature on implementation is provided. The Viterbi Decoder consists of three main building blocks: a branch metric computation unit, a unit for updating the path metric (add-compare select unit, ACSU), and a unit for the survivor sequence generation (survivor memory unit, SMU). Since the branch metric computation unit usually is very small[6] the ACSU and SMU unit are the candidates for optimizing the implementation of the VD.

10.9.1 High-Rate Viterbi Decoders

The bottleneck for any high-rate VD is the recursive add-compare-select computation. Fettweis and Meyr [3, 4] showed that this bottleneck can be broken by introducing parallelism at three different levels. At the *algorithmic level* the algorithm may be modified to decode k data blocks in parallel [3, 5]–[8]. This results in a k-fold increase of the throughput of the Viterbi decoder at about k-fold silicon area.

[6] For rate 1/2 decoders and architectures not parallelized at the block-level.

Parhi showed that parallelism can also be introduced at the *block level* [9]. Here, each step in the recursive ACS update processes k trellis steps which leads to k decoded bits per clock cycle. Therefore, a k-fold improvement of data throughput can be achieved. The introduction of parallelism at this level may also be derived from linear matrix algebra on a (maximum,+)-semi-ring [3, 10, 11] which is basically equivalent to the transitive closure calculation of graph theory. Some of the lock-ahead architectures that these methods are based upon canbe traced back to Kogge and Stone. [12, 13] For convolutional codes with a low number of states $N \leq 4$ this method can be implemented at an acceptable increase in hardware cost. For the practically relevant convolutional codes with $N = 64$ states it results in an exponential increase (factor 2^k) in silicon area. Therefore, the method is not economical for $k > 2$. For the practically relevant and area-efficient $k = 2$ case the architecture has been rediscovered in [14] (radix-4 ACS architecture). Black and Meng [14] report a highly optimized full-custom implementation of this architecture for $N = 32$ states.

At the *bit level* parallelism may be introduced by employing a redundant number system, for example, by using carry-save ACS processing elements [15]. Basically, this leads to a critical path that is independent of the word-length w and thus has a potential of $O(w)$ throughput improvement. But since the word-length of the ACS unit usually requires less than 10 bits (for the currently used codes) the increase in throughput is only about 30 to 50 percent. This speedup comes at the expense of an increased ACS unit silicon area (by a factor 2 to 3). A cascadable 600 Mb/s 4-state Viterbi decoder chip that simultaneously exploits all three levels of parallelism was developed by Dawid et al. [10, 16].

10.9.2 Area-Efficient Implementations for Low to Medium Speed

The traditional high-speed architecture, the node-parallel (NP) architecture, uses one processing element (PE) per state of the trellis. Due to its simplicity it is often used even if the computational power is not fully utilized in the decoder. This mismatch of implemented and required processing power results in a waste of silicon area. Scalable architectures where the number of processing elements can be selected to match the data rate are reported in [17]–[26]. Furthermore, these architectures make use of pipeline-interleaving that allows the use of highly efficient deeply pipelined processor elements.

Gulak and Shwedyk [17, 18] showed that an architecture analogous to the well-known *cascade architecture* for the FFT can be used for the ACS computations of Viterbi decoders. The comparatively low processor utilization of about 50 percent of the original design was later improved to more than 95 percent by Feygin et. al. [21] (generalized folded cascade architectures). Since each processing stage contains just one butterfly ACS processor (2 node equations are processed in parallel at each stage) and the implementation consists of up to $\log_2(N)/2$ stages, the maximum data rate is given by $f_{\mathrm{ACS}} \log_2(N)/N$ (f_{ACS}: throughput of a pipelined ACS processing element). Thus the application of this architectures is limited to the lower speed region. Since the processor utilization is slightly less than 100 percent, a complicated rate conversion unit (e.g., FIFOs) is needed for most applications.

Another area-efficient architecture for the medium-speed region has been proposed by Shung et al. [22, 23] and Bitterlich et al. [25, 27]. The architecture proposed in [22, 23] can be used for a very wide range of trellis types; a heuristic algorithm is suggested to construct schedules that lead to high processor utilization. [25, 27] is restricted to the practically important shuffle-exchange type trellises that are used by most of todays telecommuncation systems (e.g. DVB [28] and DAB [29]). Here a mathematical formalism is introduced that allows the direct calculation of the schedules. In contrast to the cascade VD, the trellis-pipeline interleaving (TPI) [25, 27] architecture processes several ACS node equations of a single trellis step in parallel. Data rates of $f_{\text{ACS}}2^k/N$ can be achieved ($k = 1, ..., \log_2(N)$). The main deficiency of these architectures is that the parameter k which determines the amount of resource sharing and thus the data rate is tied to the selected pipelining degree of the ACS processing elements. Furthermore, like cascade VDs, rate conversion circuits are often needed. A class of architectures that overcomes these problems is presented in [30]. Moreover [30] provides a mathematical framework to understand node-parallel, cascade and TPI architectures as special cases of an underlying unified concept which allows the explicit derivation of the required control circuits (''schedules''). Layout results using these concepts are reported in [26].

Another direction of research is the reduction of power dissipation. This is in particular important for mobile receivers. Kubuta et. al. [31]introduced the scarce-state-transition Viterbi decoder architecture which leads to significant reduction of power dissipation by reducing the number of state transitions in the ACSU for operation in the high SNR region [32, 33].

10.9.3 SMU Architectures

The optimization of the survivor-memory unit is another important topic. Two main SMU architecture classes are known: register-exchange and RAM traceback architecture. The register-exchange architectures implement the traceback algorithm by using N registers of L bits (N: number of states, L: survivor depth). For each decoded bit $N \times L$ flip-flops are concurrently updated, which leads to a high-power consumption for medium-to-large constraint length codes. The other alternative, the RAM traceback SMU [34], overcomes this problem and is often the superior solution for medium to high constraint length codes because it also consumes less silicon area. Cypher and Shung [35] introduced and analyzed a unified concept of RAM traceback architectures which allows multiple, different ''speed'' read and write traceback pointers that can be applied to a wide range of applications. Black and Meng recently intorduced as an alternative to the classical trace-back methods the trace-forward method. [36] This method leads to a reduction of silicon area income applications

10.9.4 Decoding of Concatenated Codes

An area of recent research has focused on the decoding of *concatenated* codes. The idea behind concatenated coding is that one complex code may be substitutable by two concatenated (simpler) codes with the same total error correction capability but hopefully with reduced total decoding costs [37]. At the sender site two (or more) cascaded convolutional coders are used.

Decoding of two concatenated codes can be achieved by cascaded decoders at the receiver site [38]. Since the second decoder usually requires soft input for an acceptable decoding performance, this naturally leads to the development of *soft-output* decoders. Two major classes are currently under discussion for concatenated convolutional codes: maximum-a-posteriori (MAP) decoders and implementations of the soft-output Viterbi algorithm (SOVA).

The SOVA was originally proposed by Hagenauer and Höher [39]. The implementation of the SOVA is more complex than a hard-output VD: in addition to all the data processing required by the hard-output VA, the SOVA requires the generation of reliability information for the decoded output which leads to significantly more internal processing. Joeressen and Meyr [40] showed that an implementation of the SOVA for practically relevant cases (16 states, rate 1/2) requires twice the area of a hard-output Viterbi decoder . An implementation of a concatenated convolutional coding system consisting of two 16-state convolutional codes is reported to be about as complex as a 64-state hard-output Viterbi decoder [40]–[42] which has about the same error correcting potential. It has been reported by Dawid [43] and Meyr that the MAP algorithm can be implemented at about the same hardware complexity as well.

10.9.5 Punctured Codes

Current telecommunication standards, e.g., digital video broadcasting (DVB [28]) or digital audio broadcasting (DAB [29]) usually support a variety of different code rates (run-time selectable). The decoder hardware is often directly influenced by the codes used because the specific code determines the interconnection structure of the ACS and SMU unit for high-speed VDs. Therefore selecting ''unrelated'' codes for each rate would lead to the implementation of a separate decoder hardware for each rate which would result in a waste of silicon area. The same decoder hardware can be *shared* if the different codes are derived from a single code by puncturing, i.e. by systematically removing some bits from the coded output [44, 45]. In comparison with the base code this leads to codes of higher rate. Higher-rate codes are preferable when the full error correction potential of the base code is not needed. Usually a rate 1/2 code is punctured to obtain rate 2/3, 3/4, 5/6, and 7/8 codes [28, 29, 46].

Bibliography

[1] A. Viterbi, "Error Bounds for Convolutional Coding and an Asymptotically Optimum Decoding Algorithm," *IEEE Trans. Inform. Theory*, vol. IT 13, pp. 260–269, Apr. 1967.

[2] G. Forney, "On the Viterbi Algorithm," *Proceedings IEEE*, vol. 61, pp. 268–278, Mar. 1973.

[3] G. Fettweis and H. Meyr, "High-speed Viterbi Processor: A Systolic Array Solution," *IEEE J. Sel. Areas in Commun*, vol. SAC-8, pp. 1520–1534, Oct. 1990.

[4] G. Fettweis and H. Meyr, "High-speed Parallel Viterbi Decoding," *IEEE Communications Magazine*, pp. 46–55, May 1991.

[5] T. Thapar, "Application of Block Processing to the Design of High-speed Viterbi Detectors," *Tech.. Rep., IBM, San Jose, CA*, Dec. 1986.

[6] H. Thapar and J. Cioffi, "A Block Processing Method for Designing High-speed Viterbi Detector," *IEEE Int. Conf. Commun.*, vol. 2, pp. 1096–1100, June 1985.

[7] H.-D. Lin and D. Messerschmitt, "Improving the Iteration Bound of Finite State Machines," *Proceedings Int. Symp. Circuits Systems*, pp. 1328–1331, 1989.

[8] H.-D. Lin and D. Messerschmitt, "Algorithms and Architectures for Concurrent Viterbi Decoding," *IEEE Int. Conf. Commun. (ICC)*, vol. 2, pp. 836–840, June 1989.

[9] U. U. Parhi, "Algorithm Information Techniques for Concurrent Processors," *Proceedings IEEE*, vol. 77, Dec. 1989.

[10] G. Fettweis, H. Dawid, and H. Meyr, "Minimized Method Viterbi Decoding: 600 Mbit/s per Chip," *Proceedings IEEE Global Commun. Conf. (GLOBECOM)*, vol. 3, pp. 1712–1716, Dec. 1990.

[11] G. Fettweis and H. Meyr, "Feedforward Architectures for Parallel Viterbi Decoding," *Kluwer J. VLSI Signal Processing*, vol. 3, no. 1 & 2, 1991.

[12] P. M. Kogge and H. S. Stone, "A Parallel Algorithm for the Efficient Solution of a General Class of Recurrence Equations," *IEEE Tans. Com.*, vol. C-22, pp. 786–792, Aug. 1973.

[13] P. M. Kogge, "Parallel Solution of Recurrence Problems," *IBM J. Res, Develop.*, vol. 18, pp. 138–148, Mar. 1974.

[14] Black and T. Meng, "'A 140 Mb/s, 32-State, Radix-4 Viterbi Decoder," *IEEE J. Solid State Circuits*, vol. 27, no. 12, pp. 1877–1885, 1992.

[15] G. Fettweis and H. Meyr, "A 100 Mbit/s Viterbi-Decoder Chip: Novel Architecture and its Realization," *Proceedings IEEE Int. Conf. Commun. (ICC)*, vol. 2, pp. 463–467, Apr. 1990.

[16] H. Dawid, G. Fettweis, and H. Meyr, "A CMOS IC for *G*bit/s Viterbi Decoding," *IEEE Trans. VLSI Syst.*, June 1995.

[17] P. Gulak and E. Shwedyk, "VLSI Structures for Viterbi Receivers: Part I – General Theory and Applications," *IEEE J. Sel. Areas Commun.*, vol. SAC-4, pp. 142–154, Jan. 1986.

[18] P. Gulak and E. Shwedyk, "VLSI Structures for Viterbi Receivers: Part II – Encoded MSK Modulation," *IEEE J. Sel. Areas Commun.*, vol. SAC-4, pp. 155–159, Jan. 1986.

[19] P. Gulak and T. Kailath, "Locally Connected VLSI Architectures for the Viterbi Algorithm," *IEEE J. Sel. Areas Commun.*, vol. SAC-6, pp. 527–537, Apr. 1988.

[20] C. B. Shung, H.-D. Lin, P. H. Siegel, and H. K. Thapar, "Area-Efficient Architectures for the Viterbi Algorithm," *IEEE Global Telecommun. Conf. (GLOBECOM)*, vol. 3, pp. 1787–1793, Dec. 1991.

[21] G. Feygin, G. Gulak, and P. Chow, "Generalized Cascade Viterbi Decoder – A Locally Connected Multiprocessor with Linear Speed-up," *ICASSP, Phoenix, Arizona*, pp. 1097–1100, 1991.

[22] C. B. Shung, H.-D. Lin, R. Cypher, P. H. Siegel, and H. K. Thapar, "Area-Efficient Architectures for the Viterbi Algorithm–Part I: Theory," *IEEE Trans. Comm.*, vol. 41, pp. 636–644, Apr. 1993.

[23] C. B. Shung, H.-D. Lin, R. Cypher, P. H. Siegel, and H. K. Thapar, "Area-Efficient Architectures for the Viterbi Algorithm–Part II: Applications," *IEEE Trans. Comm.*, vol. 41, pp. 802–807, May 1993.

[24] J. Sparso, H. N. Jorgensen, E. Paaske, S. Pedersen, and T. Rübner-Petersen, "Area-Efficient Topology for VLSI Implementation of Viterbi Decoders and Other Shuffle-Exchange Type Structures," *IEEE Journal of Solid-State Circuits*, vol. 26, pp. 90–97, Feb. 1991.

[25] S. Bitterlich, H. Dawid, and H. Meyr, "Boosting the Implementation Efficiency of Viterbi Decoders by Novel Scheduling Schemes," *IEEE Global Telecommun. Conf. (GLOBECOM), Orlando, Florida, IEEE*, pp. 1260–1264, Dec. 1992.

[26] S. J. Bitterlich, B. Pape, and H. Meyr, "Area Efficient Viterbi Decoder Macros," *Proceedings of the 20th European Solid State Circuits Conference, Ulm, Germany, Editon Frontières, Gif-sur-Yvette, Cedex, France*, Sept. 1994.

[27] H. Dawid, S. Bitterlich, and H. Meyr, "Trellis Pipeline-Interleaving: A Novel Method for Efficient Viterbi-Decoder Implementation," *Proceedings IEEE ISCAS*, pp. 1875–1878, 1992.

[28] European Telecommunication Standard, "Digital Broadcasting System for Television, Sound and Data Services; Framing Structure, Channel Coding and Modulation for 11/12 GHz Satellite Services," Aug. 1995.

[29] European Telecommunication Standard ETS 300 401, "Radio Broadcasting Systems; Digital Audio Broadcasting (DAB) to Mobile, Portable and Fixed Receivers," *Sophia Antipolis Cedex, France: ETSI*, 1995.

[30] S. Bitterlich and H. Meyr, "Efficient Scalable Architectures for Viterbi Decoders," *International Conf. Application Specific Array Processors (ASAP), Venice, Italy, IEEE Computer Society Press*, Oct. 1993.

[31] S. Kubota, K. Ohtani, and S. Kato, "High-Speed and High-Coding-Gain Viterbi Decoder with Low Power Consumption Employing SST (Scarce State Transition) Scheme," *Electronic Lett.*, vol. 22, pp. 491–493, Apr. 1986.

[32] T. Ishitani and et. al., "A Scarce-State-Transition Viterbi Decoder VLSI for Bit Error Correction," *IEEE J. Solid-State Circuits*, vol. SC-22, Aug. 1987.

[33] S. Kubota and S. Kato, "Viterbi Decoder VLSI Implementation and its Applications," *Abstracts & Papers of IEEE Int. Workshop Microelectronics and Photonics in Communications, Cape God, MA*, pp. II.1.1–8, June 1989.

[34] C. M. Rader, "Memory Management in a Viterbi Decoder," *IEEE Trans. Commun.*, vol. COM-29, pp. 1399–1401, Sept. 1981.

[35] R. Cypher and B. Shung, "Generalized Trace Back Techniques for Survivor Memory Management in the Viterbi Algorithm," *IEEE Global Telecommun. Conf. (GLOBECOM), San Diego, IEEE*, vol. 2, pp. 1318–1322, Dec. 1990.

[36] P. J. Black and Y. H.-Y. Meng, "Hybrid Survivor Path Architectures for Viterbi Decoders," *Proceedings CASSP*, pp. I–433–I–436, 1993.

[37] J. Hagenauer, "Soft-In/Soft-Out: The Benefits of Using Soft Values in All Stages of Digital Receivers," *3rd International Workshop on DSP Techniques applied to Space Communications*, pp. 7.1–7.15, ESTEC, Sept. 1992.

[38] J. Hagenauer and P. Höher, "Concatenated Viterbi-Decoding," *Fourth Swedish-Soviet International Workshop on Information Theory, Gotland, Sweden*, pp. 29–33, Aug. 1989.

[39] J. Hagenauer and P. Höher, "A Viterbi Algorithm with Soft Outputs and It's Application," *Proceedings of the IEEE Global Telecommunications Conference GLOBECOM, Texas, Dallas, IEEE*, pp. 47.1.1–47.1.7, Nov. 1989.

[40] O. J. Joeressen and H. Meyr, "A 40 Mbit/s Soft Output Viterbi Decoder," *Proceedings of the 20th European Solid State Circuits Conference, Ulm, Germany, Editon Frontières, Gif-sur-Yvette, Cedex, France*, pp. 216–219, Sept. 1994.

[41] O. J. Joeressen, M. Vaupel, and H. Meyr, "High-Speed VLSI Architectures for Soft-Output Viterbi Decoding," *Journal of VLSI Signal Processing*, vol. 8, pp. 169–181, Oct. 1994.

[42] O. J. Joeressen and H. Meyr, "A 40 Mbit/s Soft Output Viterbi Decoder," *IEEE Journal of Solid-State Circuits*, vol. 30, pp. 812–818, July 1995.

[43] H. Dawid, R. Hakenes, and H. Meyr, "Scalable Architectures for High Speed Channel Decoding,," *VLSI Signal Processing VII, J. Rabaey, P. Chau, and J. Eldon, (eds.), IEEE*, 1994.

[44] Y. Yasuda, K. Kashiki, and Y. Hirata, "High-Rate Punctured Convolutional Codes for Soft Decision," *IEEE Trans. Commun.*, vol. COM-32, pp. 315–319, Mar. 1984.

[45] J. Hagenauer, "Rate-Compatible Punctured Convolutional Codes (RCPC Codes) and Their Applications," *IEEE Trans. Commun.*, vol. 36, pp. 389–400, Apr. 1988.

[46] A. J. Viterbi, J. K. Wolf, E. Zehavi, and R. Padovani, "A Pragmatic Approach to Trellis-Coded Modulation," *IEEE Communications Magazine*, pp. 11–19, July 1989.

10.10 Bibliographical Notes on Reed-Solomon Decoders

Binary cyclic codes making use of finite-field algebra were introduced first by Bose and Ray-Chaudhuri [1, 2] and Hocquenghem [3] and named BCH codes in honor of their inventors. Reed and Solomon [4] extended these BCH codes allowing symbols to be elements of $\mathrm{GF}(q^m)$. Due to the generally nonbinary nature of their symbols, Reed-Solomon (RS) codes are more applicable to burst error correction.

Kasami, Lin, Peterson [5, 6] and later independently Wolf [7] discovered the so-called extended Reed-Solomon codes. Two information symbols can be added to an RS code of length $n = q^m - 1$ without reducing its minimum distance. The extended RS code has length $n + 2$ and the same number of redundancy symbols as the original code.

10.10.1 Conventional Decoding Methods

Since BCH codes and RS codes are closely related, decoding algorithms for both codes are very similar. Nevertheless, RS decoding is computationally much more expensive, since all operations need to be performed in the Galois field $\mathrm{GF}(q^m)$. Therefore, early decoding algorithms were primarily developed for BCH codes and later extended and applied to RS codes.

Generally, BCH/RS decoding is performed in three or four steps:

- Re-encoding in order to compute the syndromes
- Solving the key equation in order to find an error-locator polynomial
- Determining the error locations by finding the roots of the error-locator polynomial
- Computing the corresponding error values (necessary for RS codes only)

Solving the key equation is the most sophisticated problem in this decoding procedure and has been the subject of numerous papers concerning BCH/RS decoding.

The first decoding algorithm for binary BCH codes was developed by Peterson [8]. Several refinements and generalizations [9]–[11] led to the well-known

Berlekamp-Massey algorithm [12, 13], where the solution of the key equation is regarded as a problem equivalent to shift-register synthesis. A computationally efficient version of the Berlekamp-Massey algorithm is presented by Lin and Costello [14], where some circuitry for Galois field arithmetic and RS re-encoding is proposed as well.

Completely different approaches to RS decoding have been pursued by Sugiyama et. al., Blahut, and Welch & Berlekamp. Sugiyama et. al. [15] use the Euclidean algorithm to find the GCD (greatest common divisor) of two integers to solve the key equation. Blahut [16] employs finite-field transforms in order to solve the key-equation in the frequency domain. Finally, the "remainder decoding algorithm" by Welch and Berlekamp [17] introduces a different key-equation which allows the preceding expensive computation of the syndromes to be eliminated. The Welch-Berlekamp algorithm has been further refined by Liu [18] and more recently by Morii and Kasahara [19].

10.10.2 Soft-Decision RS Decoding

If some reliability information about the symbols at the RS decoder input is available, exploiting this additional information increases the error correction performance. Decoding algorithms that take account of such reliability information are known as "soft-input" or "soft-decision" decoding algorithms. The most simple kind of reliability information is a binary flag which states if a received symbol is "fully reliable" or "unreliable". Unreliable symbols are denoted "erasures". For decoding, the value of a symbol marked as an erasure is discarded, the symbol is treated as an error with known location.

These basic ideas are found in Forney's generalized minimum distance (GMD) decoding algorithm [20]. The symbols of an input code word are sorted according to their reliabilities. The least reliable symbols are successively declared as erasures, the remaining symbols are subject to conventional error correction. A Reed-Solomon code with $k = 2t + e$ redundancy symbols can correct up to e erasures and t errors simultaneously.

Due to the enormous progress in VLSI technology soft-decision algorithms hithereto rendered too complex have become practically feasible. Recent approaches of Welch and Berlekamp [21] as well as Doi et. al. [22] are adaptations of Forney's GMD algorithm to the particular problem of soft-decision RS decoding. More combined error-erasure correcting RS decoding algorithms are presented by Araki et. al. [23], Kötter [24], and Sorger [25]. A refined algorithm proposed by Vardy and Be'ery [26] allows an exploitation of bit-level reliability information rather than symbol reliabilities. Recently, Berlekamp [27] has presented a new soft-decision RS decoding algorithm which is capable of correcting up to $k + 1$ symbol erasures, where k is the number of redundancy symbols. This is one additional correctable symbol erasure than allowed by previous algorithms.

10.10.3 Concatenated Codes

Forney [28] examined the employment of concatenated codes in order to increase error-correction performance. Many variations in concatenating codes are possible. An effective and popular approach is to use a convolutional code (applying Viterbi decoding) as inner code and a Reed-Solomon code as outer code, as proposed by Odenwalder [29]. Zeoli [30] modified Odenwalder's basic concatenation scheme such that additional reliability information is passed from the inner soft-output Viterbi decoder to an outer soft-decision RS decoder. A similar soft-decision concatenated coding scheme has been presented by Lee [31].

Bibliography

[1] R. C. Bose and D. K. Ray-Chaudhuri, "On A Class of Error Correcting Binary Group Codes," *Information Control*, vol. 3, pp. 68–79, Mar. 1960.

[2] R. C. Bose and D. K. Ray-Chaudhuri, "Further Results on Error Correcting Binary Group Codes," *Information Control*, vol. 3, pp. 279–290, Sept. 1960.

[3] A. Hocquenghem, "Codes Correcteurs d'Erreurs," *Chiffres*, vol. 2, pp. 147–156, 1959.

[4] I. S. Reed and G. Solomon, "Polynominal Codes over Certain Finite Fields," *J. Soc. Industrial and Appl. Math.*, vol. 8, pp. 300–304, June 1960.

[5] T. Kasami, S. Lin, and W. W. Peterson, "Some Results on Weight Distributions of BCH Codes," *IEEE Trans. Infor. Theory*, vol. 12, p. 274, Apr. 1966.

[6] T. Kasami, S. Lin, and W. W. Peterson, "Some Results on Cyclic Codes Which Are Invariant under the Affine Group." Scientific Report AFCRL-66-622, Air Force Cambridge Research Labs., Bedford, Mass., 1966.

[7] J. K. Wolf, "Adding Two Information Symbols to Certain Nonbinary BCH Codes and Some Applications," *Bell Syst. Tech. J.*, vol. 48, pp. 2405–2424, 1969.

[8] W. W. Peterson, "Encoding and Error-Correction Procedures for Bose-Chaudhuri Codes," *IRE Trans. Inform. Theory*, vol. 6, pp. 459–470, Sept. 1960.

[9] R. T. Chien, "Cyclic Decoding Procedures for BCH Codes," *IEEE Trans. Infor. Theory*, vol. 10, pp. 357–363, Oct. 1964.

[10] G. D. Forney, Jr., "On Decoding BCH Codes," *IEEE Trans. Infor. Theory*, vol. 11, pp. 549–557, Oct. 1965.

[11] J. L. Massey, "Step-by-Step Decoding of the Bose-Chaudhuri-Hocquenghem Codes," *IEEE Trans. Infor. Theory*, vol. 11, pp. 580–585, Oct. 1965.

[12] E. R. Berlekamp, *Algebraic Coding Theory*. New York: McGraw-Hill, 1968.

[13] J. L. Massey, "Shift Register Synthesis and BCH Decoding," *IEEE Trans. Infor. Theory*, vol. 15, pp. 122–127, Jan. 1969.

[14] S. Lin and D. J. Costello, *Error Control Coding - Fundamentals and Applications*. Englewood Cliffs, NJ: Prentice-Hall, 1995.

[15] Y. Sugiyama, M. Kasahara, S. Hirasawa, and T. Namekawa, "A Method for

Solving Key Equation for Decoding Goppa Codes," *Information Control*, vol. 27, pp. 87–99, 1975.

[16] R. E. Blahut, "Transform Techniques for Error Control Codes," *IBM J. Res. Dev.*, vol. 23, May 1979.

[17] L. R. Welch and E. R. Berlekamp, "Error Correction for Algebraic Block Codes." U. S. Patent No. 4,633,470, 1983.

[18] T. H. Liu, *A New Decoding Algorithm for Reed-Solomon Codes*. PhD thesis, Univ. of Southern California, Los Angeles, CA, 1984.

[19] M. Morii and M. Kasahara, "Generalized Key-Equation of Remainder Decoding Algorithm for Reed-Solomon Codes," *IEEE Trans. Infor. Theory*, vol. 38, pp. 1801–1807, Nov. 1992.

[20] G. D. Forney, Jr., "Generalized Minimum Distance Decoding," *IEEE Trans. Infor. Theory*, vol. 12, pp. 125–131, Apr. 1966.

[21] L. R. Welch and E. R. Berlekamp, "Soft Decision Reed-Solomon Decoder." U. S. Patent No. 4,821,268, Apr. 1987. (assigned to Kodak).

[22] N. Doi, H. Imai, M. Izumita, and S. Mita, "Soft Decision Decoding for Reed-Solomon Codes," in *Proc. IEEE GLOBECOM*, (Tokyo), pp. 2090–2094, IEEE, 1987.

[23] K. Araki, M. Takada, and M. Morii, "The Generalized Syndrome Polynominal and its Application to the Efficient Decoding of Reed-Solomon Codes Based on GMD Criterion," in *Proc. Int. Symp. on Information Theory*, (San Antonio, TX), p. 34, IEEE, Jan. 1993.

[24] R. Koetter, "A New Efficient Error-Erasure Location Scheme in GMD Decoding," in *Proc. Int. Symp. on Information Theory*, (San Antonio, TX), p. 33, IEEE, Jan. 1993.

[25] U. K. Sorger, "A New Reed-Solomon Code Decoding Algorithm Based on Newton's Interpolation," *IEEE Trans. Infor. Theory*, vol. 39, pp. 358–365, Mar. 1993.

[26] A. Vardy and Y. Be'ery, "Bit-Level Soft-Decision Decoding of Reed-Solomon Codes," *IEEE Trans. Commun.*, vol. 39, pp. 440–444, Mar. 1991.

[27] E. R. Berlekamp, "Bounded Distance + 1 Soft-Decision Reed-Solomon Decoding," vol. 42, pp. 704–720, May 1996.

[28] G. D. Forney, Jr., *Concatenated Codes*. PhD thesis, Dept. of Elec. Engrg., M.I.T., Cambridge, Mass., June 1966.

[29] J. P. Odenwalder, *Optimal Decoding of Convolutional Codes*. PhD thesis, University of California, Los Angeles, 1970.

[30] G. W. Zeoli, "Coupled Decoding of Block-Convolutional Concatenated Codes," *IEEE Trans. Commun.*, vol. 21, pp. 219–226, Mar. 1973.

[31] L. N. Lee, "Concatenated Coding Systems Employing a Unit-Memory Convolutional Code and a Byte-Oriented Decoding Algorithm," *IEEE Trans. Commun.*, vol. 25, pp. 1064–1074, Oct. 1977.

PART E
Communication over Fading Channels

Chapter 11 Characterization, Modeling, and Simulation of Linear Fading Channels

11.1 Introduction

In order to meet the ever-increasing need for both increased mobility and higher quality of a larger selection of services, wireless radio transmission of digital information such as digitized speech, still or moving images, written messages, and other data plays an increasingly important role in the design and implementation of mobile and personal communication systems [1, 2].

Nearly all radio channels of interest are more or less *time-variant* and *dispersive* in nature. However, many electromagnetic environments, e.g., satellite or line-of-sight (LOS) microwave channels, may often be regarded as effectively time-invariant. In such cases, receiver structures, including synchronizers that have been derived for static channels (see the material of the preceding chapters), may be applied.

On the other hand, when environments such as the land-mobile (LM), satellite-mobile (SM), or ionospheric shortwave (high-frequency, HF) channels exhibit significant signal variations on a short-term time scale, this *signal fading* affects nearly every stage of the communication system. Throughout this part of the book, we shall focus on *linear* modulation formats. Large variations of received signal levels caused by fading put additional strain on linear digital receiver components; the resolution of A/D converters and the precision of digital signal processing must be higher than in the case of static channels. More importantly, deep signal fades that may occur quite frequently must be bridged by applying *diversity techniques*, most often explicit or implicit *time* diversity (provided, e.g., by retransmission protocols or the use of appropriate channel coding with interleaving), *antenna*, *frequency*, *spatial*, and/or *polarization* diversity [3]. Moreover, if the channel dispersion results in intersymbol interference (ISI), this must be counteracted by means of an (adaptive) *equalizer*. Finally, transmission over fading channels necessitates specifically designed *synchronizer structures* and algorithms that are, in general, substantially different from those for static channels.

Following the ideas outlined in previous chapters, we are primarily interested in synchronizers that are mathematically *derived* in a systematic manner, based upon a suitable model of all signals and systems involved [4]. In particular, adequate modeling of the fading channel is of highest concern. Since the channel variations as observed by the receiver appear to be random, the channel model will most often be a *statistical* one. Furthermore, as synchronizers primarily have to cope with short-term variations of quantities such as amplitude(s) and phase(s) of received signals, it often suffices to assume *stationary* statistical channel properties, at least over a reasonably short time frame.

11.2 Digital Transmission over Continuous-Time and Discrete-Equivalent Fading Channels

11.2.1 Transmission over Continuous-Time Fading Channels

In digital communications over linear channels, the baseband-equivalent transmitted signal $s(t)$ is a train of transmitter shaping filter impulse responses $g_T(t-kT)$, delayed by integer multiples k of symbol duration T and weighted by complex-valued M-PSK or M-QAM data symbols a_k:

$$s(t) = \sum_k a_k \, g_T(t-kT) \tag{11-1}$$

Throughout this part of the book, we are concerned with strictly band-limited radio frequency (RF) communications. Therefore, all signals and systems are understood to be complex-valued lowpass-equivalent envelope representations of their passband counterparts; the label L denoting lowpass equivalence in earlier chapters is dropped here. All lowpass envelope signals and systems are taken to refer to the *transmitter* carrier $\sqrt{2}\cos(\omega_0 t)$ so that the bandpass transmitted signal $\sqrt{2}\,\mathrm{Re}\big[s(t)\cdot e^{j\omega_0 t}\big]$ is centered about the transmitter carrier frequency ω_0. The physical fading channel – just as any lowpass-equivalent linear system – can thus be characterized by the complex-valued time-variant fading channel impulse response (CIR) $c(\tau;t)$ or equivalently its Fourier transform with respect to the delay variable τ, the instantaneous channel transfer function $C(\omega;t)$ valid at time instant t. Most radio channels are characterized by *multipath propagation* where a number of reflected or scattered radio rays arrive at the receiving end. Such a typical scattering scenario is illustrated in Figure 11-1 for the example of a mobile radio environment. Unless obstructed, the LOS ray (dashed line) arrives first at the receiver, while the other rays (solid lines) are reflected from various objects in the environs. Each of the rays is characterized by a distinct attenuation (amplitude "gain"), a phase shift, and a propagation delay. The former two are jointly expressed by a complex-valued gain factor $c_n(t)$ where $\alpha_n(t) = |\, c_n(t)\,|$ is the time-variant amplitude gain and $\varphi_n(t) = \arg\{c_n(t)\}$ the random phase shift. Here, the delays $\tau_n(t)$ are taken to be relative to the propagation delay τ_p of the first arriving ray (usually the LOS ray, if present). The propagation delay is related to the propagation distance d_p between transmitter and receiver by

$$\tau_p = \frac{d_p}{c} = 3.33 \frac{d_p}{[km]}\ \mu s \tag{11-2}$$

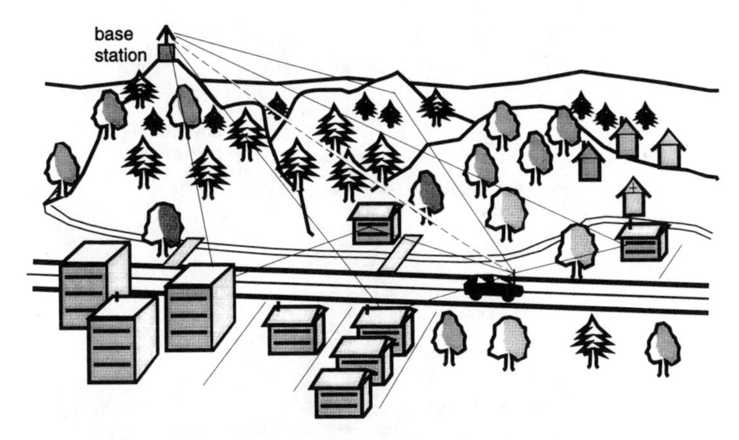

Figure 11-1 Typical Scattering Scenario in Mobile Radio Communications

where c is the speed of light. Usually, $\tau_n(t)$ and τ_p change only slowly with time; the instantaneous differential delays $\tau_n = \tau_n(t)$ can thus be assumed to remain stationary within a reasonably short time frame so that they may be indexed in natural order, i.e., $0 = \tau_0 \leq \tau_1 \leq \ldots \leq \tau_{N-1} = \tau_{\max}$. The physical channel impulse response, including the propagation delay τ_p, is then expressed as the superposition of a number N (which may be virtually infinite) of weighted and delayed Dirac pulses:

$$c_p(\tau;t) = \sum_{n=0}^{N-1} c_n(t)\, \delta(\tau - [\tau_p + \tau_n]) \tag{11-3}$$

In digital communications where transmitter and receiver clock phases may be different, a fractional receiver clock delay (or clock advance, if negative) $\tau_c = \varepsilon_c T$ has to be added to the physical path delays $(\tau_p + \tau_n)$. Assuming equal transmitter and receiver clock rates $1/T$ (the rationale for this is discussed below), the relative timing offset ε_c is stationary and in the range $-0.5 < \varepsilon_c \leq 0.5$. The propagation delay τ_p may now be expressed in terms of integer and fractional multiples of symbol duration T:

$$\tau_p = [L_p + \varepsilon_c + \varepsilon]\, T \tag{11-4}$$

with L_p an integer such that the fractional extra delay (or advance) ε is also in the range $-0.5 < \varepsilon \leq 0.5$. From the illustration in Figure 11-2 it is seen that ε is the fractional delay of the first arriving multipath ray with respect to the nearest receiver symbol clock tick.

For the purpose of receiver design, it is convenient to introduce the channel impulse response in terms of the receiver timing reference:

$$\begin{aligned} c_\varepsilon(\tau;t) &= c_p(\tau + [L_p + \varepsilon_c]T;t) \\ &= \sum_{n=0}^{N-1} c_n(t)\, \delta(\tau - [\varepsilon T + \tau_n]) \end{aligned} \tag{11-5}$$

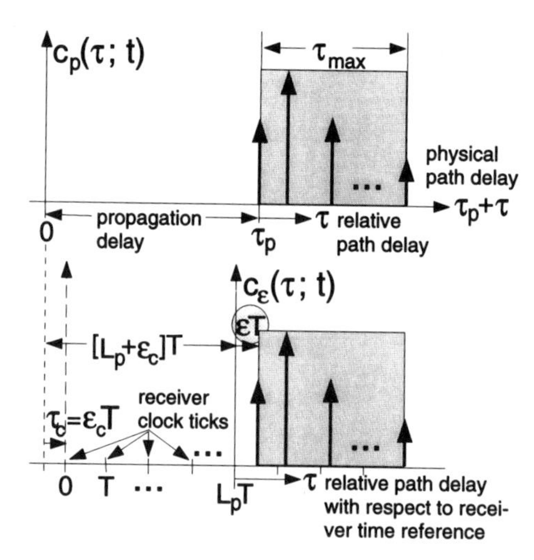

Figure 11-2 Transmitter and Receiver Time Scales

(see Figure 11-2). Since the propagation delay τ_p is in general a noninteger multiple of symbol duration T, the timing delay ε may assume any value in the range $-0.5 < \varepsilon \leq 0.5$ even in the case of perfect match between transmitter and receiver clocks ($\varepsilon_c = 0$). Hence, the "start" of the channel impulse response (first arriving ray) may be offset by up to half a symbol interval with respect to the receiver timing reference.

From the channel model of (11-5), the various receiver synchronization tasks are readily identified. Being concerned about coherent or differentially coherent reception only, the existence of randomly varying complex-valued path weights $c_n(t)$ necessitates some kind of *carrier recovery*, i.e., phase synchronization, and, in addition, *amplitude (gain) control* when amplitude-sensitive modulation formats are employed. The differential multipath and timing delays τ_n and ε, respectively, call for some sort of *timing synchronization*. If the channel is nonselective ($\tau_n \ll T$, see below), this can be accomplished by means of estimation and compensation of the timing delay ε (Chapters 4 and 5). In the case of selective channels, however, a filtered and sampled version of the channel impulse response $c_\varepsilon(\tau; t)$ must be estimated and compensated for by means of equalization. The latter case will receive much attention in the remainder of the book.

Apart from the random phase shift introduced by the channel itself, imperfect transmitter and receiver oscillators may give rise to a sizeable – often nonrandom but unknown – frequency shift. It is assumed here that if very large offsets in the order or in excess of the symbol rate $1/T$ occur, these are taken care of by a coarse frequency synchronization stage in the receiver front end ([5], Chapter 8). Following the guidelines established in Chapter 8, we shall henceforth assume small and moderate frequency shifts in the range $(\Omega T)/(2\pi) \leq 0.1 - 0.15$, i.e., the received signal spectrum may be shifted by up to 10–15 percent of the symbol rate. Taking ΩT into account in the signal model and incorporating the constant carrier phase shift θ of Chapter 8 into the complex-valued path weights $c_n(t)$, the information-bearing signal $s(t)$ [eq. (11-1)] being transmitted through the channel yields the received signal shifted in frequency through the rotating phasor $e^{j\Omega t}$:

$$\begin{aligned} r(t) &= e^{j\Omega t}\left[\sum_{n=0}^{N-1} c_n(t)\, s(t-\varepsilon T-\tau_n)\right] + n(t) \\ &= e^{j\Omega t}\left[\sum_{n=0}^{N-1} c_n(t)\left(\sum_k a_k\, g_T(t-\varepsilon T-\tau_n-kT)\right)\right] + n(t) \\ &= e^{j\Omega t}\left[\sum_k a_k \underbrace{\left(\sum_{n=0}^{N-1} c_n(t)\, g_T([t-kT]-[\varepsilon T+\tau_n])\right)}_{h_\varepsilon(\tau=t-kT\,;\,t)}\right] + n(t) \\ &= e^{j\Omega t}\left[\sum_k a_k\, h_\varepsilon(\tau=t-kT\,;\,t)\right] + n(t) \end{aligned} \tag{11-6}$$

with $h_\varepsilon(\tau;t)$ the time-variant *effective* channel impulse response, including transmitter filtering and a fractional timing delay. The effective CIR $h_\varepsilon(\tau;t)$ and its transfer function $H_\varepsilon(\omega;t)$ may be expanded as follows:

$$\begin{aligned} h_\varepsilon(\tau;t) &= \sum_{n=0}^{N-1} c_n(t)\, g_T(\tau-[\varepsilon T+\tau_n]) \\ &= c(\tau;t) * g_T(\tau) * \delta(\tau-\varepsilon T) && c(\tau;t) = \sum_{n=0}^{N-1} c_n(t)\,\delta(\tau-\tau_n) \\ &= c_\varepsilon(\tau;t) * g_T(\tau) && c_\varepsilon(\tau;t) = c(\tau;t) * \delta(\tau-\varepsilon T) \\ &= h(\tau;t) * \delta(\tau-\varepsilon T) && h(\tau;t) = c(\tau;t) * g_T(\tau) \\ H_\varepsilon(\omega;t) &= \sum_{n=0}^{N-1} c_n(t)\, G_T(\omega)\, e^{-j\omega[\varepsilon T+\tau_n]} \\ &= C(\omega;t)\, G_T(\omega)\, e^{-j\omega\varepsilon T} && C(\omega;t) = \sum_{n=0}^{N-1} c_n(t)\, e^{-j\omega\tau_n} \\ &= C_\varepsilon(\omega;t)\, G_T(\omega) && C_\varepsilon(\omega;t) = C(\omega;t)\, e^{-j\omega\varepsilon T} \\ &= H(\omega;t)\, e^{-j\omega\varepsilon T} && H(\omega;t) = C(\omega;t)\, G_T(\omega) \end{aligned} \tag{11-7}$$

where $*$ denotes the convolution operator, and $c(\tau;t)$, $h(\tau;t)$ the physical and effective CIRs, respectively, taking into account differential delays only, thus disregarding propagation and timing delays. This expansion is useful for the purpose of channel modeling and simulation (Section 11.3) since the effects of physical channel (fading, dispersion), transmitter filtering, and timing offset (propagation, receiver clock) can be attributed to the constituents $c(\tau;t)$, $g_T(\tau)$, and $\delta(\tau-\varepsilon T)$, respectively.

We remark that the definitions of channel impulse response used here do not include receive filtering and thus are different from that of the earlier parts of the book where $h(\tau;t)\circ\!\!-\!\!\bullet H(\omega;t)$ was meant to denote the cascade of transmit filter, physical channel, *and* receive filter. Here, the additive noise $n(t)$ is taken to be white Gaussian (AWGN) with power spectral density N_0, although in reality $n(t)$ may be dominated by co-channel interference (CCI) in interference-limited environments. Moreover, $n(t)$ is correlated via filtering by the anti-aliasing filter $F(\omega)$. However, the flatness condition ($F(\omega)=1$; $|\omega|\leq B_r$, see below) imposed on $F(\omega)$ leaves the noise undistorted within the bandwidth of interest, so that it is immaterial whether the effect of $F(\omega)$ is considered or not.

Since spectrum is a most valuable resource especially in multiple-access environments, narrowband signaling using tightly band-limited transmitter pulse shaping filters is a necessity. This, by the way, also applies to CDMA communications where the term "narrowband" is taken to refer to the chip rate instead of the symbol rate. Tight pulse shaping also helps in suppressing adjacent channel interference (ACI). Hence, we assume that the filter $g_T(\tau)\circ\!\!-\!\!\bullet G_T(\omega)$ can be approximated with sufficient accuracy as being strictly band-limited to (two-sided) RF bandwidth B, so that the effective channel $H_\varepsilon(\omega;t) = H(\omega;t)\,\mathrm{e}^{-j\omega\varepsilon T}$ [eq. (11-7)] is also strictly band-limited to B. A common choice is a transmitter filter with root-raised-cosine transfer function [3]

$$G_T(\omega) = T\begin{cases} 1 & |\omega| < (1-\alpha)\dfrac{\pi}{T} \\ \cos\left[\dfrac{T}{4\alpha}\left(|\omega|-(1-\alpha)\dfrac{\pi}{T}\right)\right] & (1-\alpha)\dfrac{\pi}{T} \leq |\omega| < (1+\alpha)\dfrac{\pi}{T} \\ 0 & |\omega| \geq (1+\alpha)\dfrac{\pi}{T} \end{cases}$$

$$g_T(\tau) = \frac{1}{1-\left(4\alpha\frac{\tau}{T}\right)^2}\left[(1-\alpha)\,\mathrm{si}\left((1-\alpha)\pi\frac{\tau}{T}\right) + \frac{4\alpha}{\pi}\cos\left((1+\alpha)\pi\frac{\tau}{T}\right)\right] \tag{11-8}$$

where the filter energy

$$E_{g_T} = \int_{-\infty}^{+\infty} g_T^2(\tau)\,d\tau = \frac{1}{2\pi}\int_{-\infty}^{+\infty} |G_T(\omega)|^2 d\omega = T \tag{11-9}$$

is equal to the symbol interval T. The filter and thus the channel is strictly band-limited to the (two-sided) bandwidth $B = (1+\alpha)\,(1/T)$ with bandwidth expansion (or rolloff) factor $0 < \alpha < 1$. As discussed in Section 4.3, the (energy-normalizing) received matched filter for the special case of nonselective channels (AWGN, no frequency shift) is given by $G_{\mathrm{MF}}(\omega) = (1/T)\,G_T^*(\omega)$. Hence, the cascade $G(\omega) = G_T(\omega)\,G_{\mathrm{MF}}(\omega)$ of pulse shaping and pulse matched filters equals

$G(\omega) = (1/T)\,|G_T(\omega)|^2$, i.e.,

$$G(\omega) = T\begin{cases} 1 & |\omega| < (1-\alpha)\frac{\pi}{T} \\ \frac{1}{2}\left(1+\cos\left[\frac{T}{2\alpha}\left(|\omega|-(1-\alpha)\frac{\pi}{T}\right)\right]\right) & (1-\alpha)\frac{\pi}{T} \leq |\omega| < (1+\alpha)\frac{\pi}{T} \\ 0 & |\omega| \geq (1+\alpha)\frac{\pi}{T} \end{cases}$$

$$g(\tau) = \mathrm{si}\left(\pi\frac{\tau}{T}\right)\frac{\cos\left[\alpha\pi(\tau/T)\right]}{1-4\left[\alpha(\tau/T)\right]^2} \tag{11-10}$$

is a raised-cosine pulse satisfying the Nyquist condition on ISI-free transmission [eq. (2–17)]. Notice, however, that in baseband communications (Chapter 2) B has been defined as the one-sided bandwidth while here at passband B is taken to denote the *two-sided* RF bandwidth.

As explained above, the frequency content of the received signal $r(t)$ is allowed to be offset in frequency due to oscillator imperfections by up to a certain maximum value $\Omega_{\max}$ so that, after downconversion, the receiver anti-aliasing filter $F(\omega)$ in front of the D/A converter must leave the received input signal undistorted within the frequency range $|\omega| \leq 2\pi B_r/2 = (2\pi B + \Omega_{\max})/2$. Only if Ω is very small or has been effectively compensated for by a preceding frequency controlling stage operating on the time-continuous signal in front of $F(\omega)$, the widening of the receiver input frequency range may be neglected in the design of $F(\omega)$.

From eq. (11-7) it is observed that the effective channel $H_\varepsilon(\omega;t) = H(\omega;t)\,e^{-j\omega\varepsilon T}$ exhibits a more or less frequency-dependent transmission characteristic. The degree of selectivity is dependent upon the physical channel $C_\varepsilon(\omega;t) = C(\omega;t)\,e^{-j\omega\varepsilon T}$ *and* the transmission bandwidth B. In particular, the channel transfer function is effectively *frequency-nonselective* (*nonselective* or *flat*) within B if $e^{-j\omega_{\max}\tau_{\max}} \approx 1$, where $\omega_{\max} = B/2$ will be in the order of the symbol rate $1/T$ in bandwidth-limited environments. Hence, the channel is nonselective when the dispersion (span of ray transmission delays) satisfies $\tau_{\max} \ll T$ (in receiver design, nonselectivity can often be safely assumed if $\tau_{\max} < 0.1\,T$), while the channel is *frequency-selective* when the dispersion is comparable to or in excess of the symbol duration T.

In case the channel is nonselective, the effective channel transfer function and impulse response collapse to

$$\begin{aligned} H_\varepsilon(\omega;t) &= G_T(\omega)\left(\sum_{n=0}^{N-1} c_n(t)\,e^{-j\omega(\varepsilon T)}\,\underbrace{e^{-j\omega\tau_n}}_{\approx 1 \text{ in } B}\right) \\ &\approx G_T(\omega)\left(\sum_{n=0}^{N-1} c_n(t)\right)e^{-j\omega(\varepsilon T)} \qquad \text{(flat fading)} \\ &= G_T(\omega)\,c(t)\,e^{-j\omega(\varepsilon T)} \\ h_\varepsilon(\tau;t) &= c(t)\,g_T(\tau-\varepsilon T) \end{aligned} \tag{11-11}$$

Thus all (nonresolvable) path weights $c_n(t)$ merge into a single weight $c(t)$ termed *multiplicative distortion* (MD) [6], and all path delays up to $\tau_{\max} \ll T$ can be taken to be effectively zero so that one is left with the timing delay εT. The received signal can then be written as

$$
\begin{aligned}
r(t) &= e^{j\Omega t}\left[\sum_k a_k\, c(t)\, g_T(t-\varepsilon T-kT)\right] + n(t) \qquad \text{(flat fading)} \\
&= \underbrace{e^{j\Omega t}\, c(t)}_{c_\Omega(t)} \left[\sum_k a_k\, g_T(t-\varepsilon T-kT)\right] + n(t)
\end{aligned}
\tag{11-12}
$$

Very small frequency offsets $|\Omega T| \ll 1$ are sometimes included in the dynamical channel MD process model to yield the combined *frequency-channel MD* process $c_\Omega(t) = e^{j\Omega t}\, c(t)$. If all sync parameters $[\Omega, \varepsilon, c(t)]$ were known in advance, one would be able to process the received signal using the (ideal) energy-normalizing *frequency channel matched filter*:

$$
\begin{aligned}
H_{\mathrm{MF},\Omega}(\omega;t) &= e^{-j\Omega t}\, H_{MF}(\omega;t) \qquad \text{(selective and flat fading)} \\
&= e^{-j\Omega t}\, \frac{1}{T} H_\varepsilon^*(\omega;t) \\
&= e^{-j\Omega t}\, C^*(\omega;t)\, \frac{1}{T} G_T^*(\omega)\, e^{+j\omega(\varepsilon T)} \\
&= \left[e^{-j\Omega t}\, c^*(t)\right] \left[\frac{1}{T} G_T^*(\omega)\, e^{+j\omega(\varepsilon T)}\right] \qquad \text{(flat fading)} \\
h_{\mathrm{MF},\Omega}(\tau;t) &= e^{-j\Omega t}\, h_{\mathrm{MF}}(\tau;t) \qquad \text{(selective and flat fading)} \\
&= e^{-j\Omega t}\, \frac{1}{T} h_\varepsilon^*(-\tau;t) \\
&= e^{-j\Omega t} \left[c^*(-\tau;t) * \frac{1}{T} g_T^*(-\tau) * \delta(\tau+\varepsilon T)\right] \\
&= \left[e^{-j\Omega t}\, c^*(t)\right] \left[\frac{1}{T} g_T^*(-\tau) * \delta(\tau+\varepsilon T)\right] \qquad \text{(flat fading)}
\end{aligned}
\tag{11-13}
$$

[see also eq. (11-7)], with frequency compensation (back-rotation of the complex phasor $e^{j\Omega t}$) via the term $e^{-j\Omega t}$ and channel matched filtering by $h_{\mathrm{MF}}(\tau;t)$, which, in the case of flat fading, comprises phase correction (randomly varying channel phase $\varphi(t) = \arg[c(t)]$) through $c^*(t)$, pulse matched filtering by $g_{\mathrm{MF}}(\tau) = (1/T)\, g_T^*(-\tau)$, and timing delay compensation via $\delta(\tau + \varepsilon T)$.

Notice that, for *perfect* frequency channel matched filtering, the order of operations cannot be interchanged, i.e., frequency and phase correction are to be performed *prior to* pulse matched filtering. Obviously, large frequency offsets and fast channel variations call for the received signal to be shifted in frequency such that its spectrum matches that of the pulse MF. However, considering only (residual) frequency offsets and channel fading bandwidths being small relative to the bandwidth B, the transmission model and receiver design for flat fading channels can be substantially simplified by attempting to compensate for frequency and phase *following* the (known and fixed) pulse MF $g_{\mathrm{MF}}(\tau) \circ\!\!-\!\!\bullet G_{\mathrm{MF}}(\omega)$, thus avoiding the (ideal but a priori unknown) frequency channel matched filter. The pulse MF can then be implemented either as part of the analog front end [e.g., by combining it with the analog prefilter: $G_{\mathrm{MF}}(\omega) = F(\omega)$] or as a digital filter following $F(\omega)$ and A/D conversion. The output of the pulse MF is then written as

$$
\begin{aligned}
z(t) &= g_{\mathrm{MF}}(t) * r(t) \\
&= \int_{-\infty}^{\infty} \frac{1}{T} g_T^*(-u)\, e^{j\Omega(t-u)}\, c(t-u) \left[\sum_k a_k\, g_T([t-u]-\varepsilon T-kT)\right] du \;+ m(t) \\
&\approx e^{j\Omega t}\, c(t) \left[\sum_k a_k \int_{-\infty}^{\infty} e^{-j\Omega u} \frac{1}{T} g_T^*(-u)\, g_T([t-u]-\varepsilon T-kT)\, du\right] \;+ m(t) \\
&= \quad e^{j\Omega t}\, c(t) \left[\sum_k a_k\, g(t-\varepsilon T-kT)\right] \qquad + m(t) \\
&\quad \underbrace{-e^{j\Omega t}\, c(t) \left[\sum_k a_k \int_{-\infty}^{\infty} \underbrace{\left(1-e^{-j\Omega u}\right)}_{\text{small if } \Omega u \ll 1} \frac{1}{T} g_T^*(-u)\, g_T([t-u]-\varepsilon T-kT)\, du\right]}_{\text{small if } \Omega T \ll 1}
\end{aligned}
\tag{11-14}
$$

where $m(t) = g_{\mathrm{MF}}(t) * n(t)$ is the filtered noise with power spectral density $S_m(\omega) = (1/T)\,|G_T(\omega)|^2\, N_0 = N_0\, G(\omega)$ and autocorrelation $R_m(t) = N_0\, g(t)$. Since the vast majority of systems operating over fading channels are designed such that the fading rates remain well below the symbol rate $1/T$, the approximation $c(t-u) \approx c(t)$ can be taken to be valid within the duration of the pulse $g_T(t)$ whose main lobe spans the region $-T < t < T$. The third term of eq. (11-14) is identified as the distortion resulting from mismatched filtering by using $g_{\mathrm{MF}}(\tau)$ – instead of $e^{-j\Omega t}\, g_{\mathrm{MF}}(\tau)$ – prior to frequency correction. As discussed in Section 8.4 in the context of AWGN channels, this term is small if the relative frequency offset is well below 1. Hence, the pulse matched filter output may be well approximated by

$$
z(t) \underset{(\Omega T \text{ small})}{\approx} \underbrace{e^{j\Omega t}\, c(t)}_{c_\Omega(t)} \left[\sum_k a_k\, g(t-\varepsilon T-kT)\right] + m(t) \qquad \text{(flat fading)}
\tag{11-15}
$$

Figure 11-3 summarizes the discussion above and illustrates the channel transmission models for both frequency-selective and nonselective fading channels. As already mentioned, interchanging frequency correction and pulse matched filtering (as shown in the figure) is allowable only for small relative offsets of up to

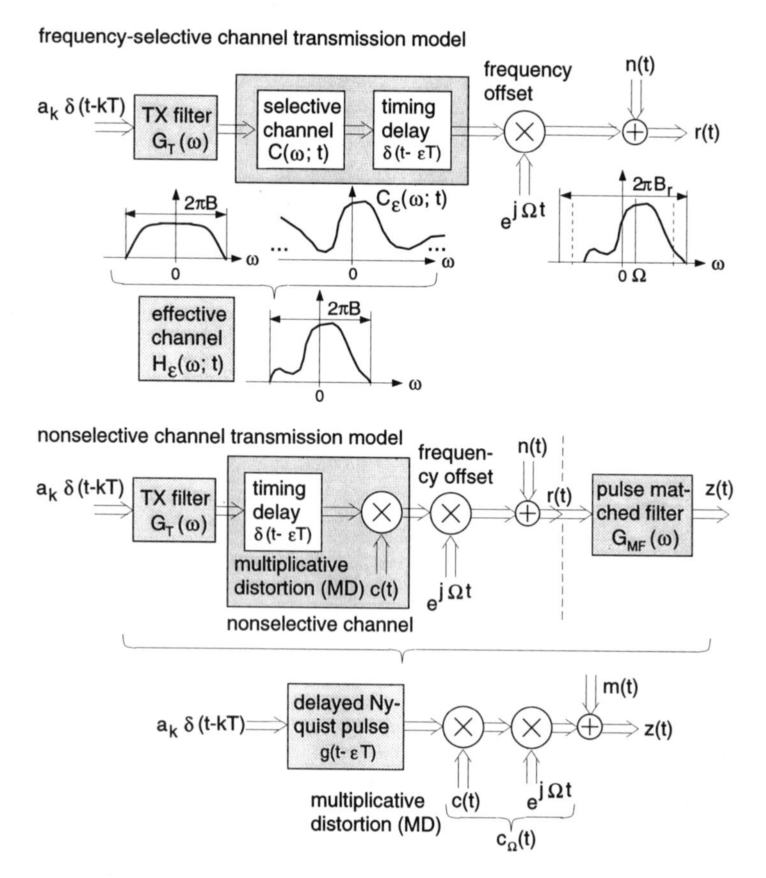

Figure 11-3 Linear Fading Channel Transmission Model

10–15 percent. If this cannot be guaranteed, a separate frequency synchronizer must be employed in front of $G_{\mathrm{MF}}(\omega)$. Frequency-selective channel matched filtering is more sensitive against frequency offsets so that, if the (time-variant, a priori unknown) channel matched filter $H_{\mathrm{MF}}(\omega;t) = G_{\mathrm{MF}}(\omega)\, C^*(\omega;t)$ is used for near-optimal reception (Chapter 13), frequency synchronization prior to matched filtering is generally advisable unless the frequency shift is in the order of, or smaller than, the channel fading rate.

11.2.2 Discrete-Equivalent Transmission Behavior of Fading Channels

In all-digital receiver implementations, the received signal $r(t)$ [eq. (11-6)] should be sampled as early as possible in the receiver processing chain. In order to fully preserve the information content, a minimum sampling rate of $(1/T_s)_{\min} = B_r = (1+\alpha)\,(1/T) + (\Omega_{\max}/2\pi)$ is required (see Figure 11-3). This, however, would necessitate an ideal lowpass anti-aliasing filter $F(\omega)$ with (one-sided) bandwidth $B_r/2$. Also, $(1/T_s)_{\text{min}}$ would, in general, be incommensurate with the symbol rate $1/T$. However, considering small frequency shifts Ω_{max} and typical pulse shaping rolloff factors α ranging between about 0.2 and 0.7, a nominal sampling frequency of $1/T_s = 2/T$ may be chosen. This also allows for a smooth transition between pass- and stopband and thereby easier implementation of the anti-aliasing filter $F(\omega)$.

While in practice the sampling frequency of a free-running receiver clock will never be *exactly* equal to $2/T$ (see Chapter 4), the variation in timing instants resulting from slightly incommensurate rates can nevertheless be assumed to remain small over a reasonably short time interval. This is especially true for fading channels where information is most often transferred in a block- or packet-like fashion. Over the duration of such blocks, the relative timing delay εT can therefore be assumed to be stationary.

Of course, there are many variations on the theme of sampling. For instance, the received signal may be sampled at rates higher than $2/T$, say $8/T$, in order to make the anti-aliasing filter simpler (higher cutoff frequency, smoother rolloff). In that case, however, the sampled signal may contain unwanted noise and adjacent channel interference. Digital lowpass filtering and subsequent decimation then yields a signal of rate $2/T$. Alternatively, one may downconvert the received signal to some intermediate (or "audio") band, sample the (filtered) mixer output at a high rate using a *single* A/D converter, perform digital downconversion to baseband, and finally decimate to rate $2/T$.

Assuming *double-spaced* sampling at rate $2/T$, the sampled received signal $r(t)$ [eq. (11-6), including frequency offset] can be expressed as

$$
\begin{aligned}
r_k^{(i)} &= r(t=[2k+i]T_s) \quad = r\left(t=\left[k+\frac{i}{2}\right]T\right) \qquad (i=0,1) \\
&= e^{j\Omega T(k+i/2)}\left[\sum_n a_n\, h_\varepsilon\left(\tau=\left[k+\frac{i}{2}\right]T-nT;\; t=\left[k+\frac{i}{2}\right]T\right)\right] \\
&\quad + n\left(t=\left[k+\frac{i}{2}\right]T\right) \\
&= e^{j\Omega T(k+i/2)}\left[\sum_n a_n\, h^{(i)}_{\varepsilon,k-n;\,k}\right] + n_k^{(i)}
\end{aligned}
\tag{11-16}
$$

with indices $i = 0, 1$ denoting samples taken at timing instants kT (integer multiples of T) and $kT + T/2$ (half-integer multiples of T), respectively. From eq. (11-16), the discrete-equivalent dispersive channel impulse response (including receiver timing offset) is identified as

$$
h^{(i)}_{\varepsilon,n;\,k} = h_\varepsilon\left(\tau=\left[n+\frac{i}{2}\right]T;\; t=\left[k+\frac{i}{2}\right]T\right) \tag{11-17}
$$

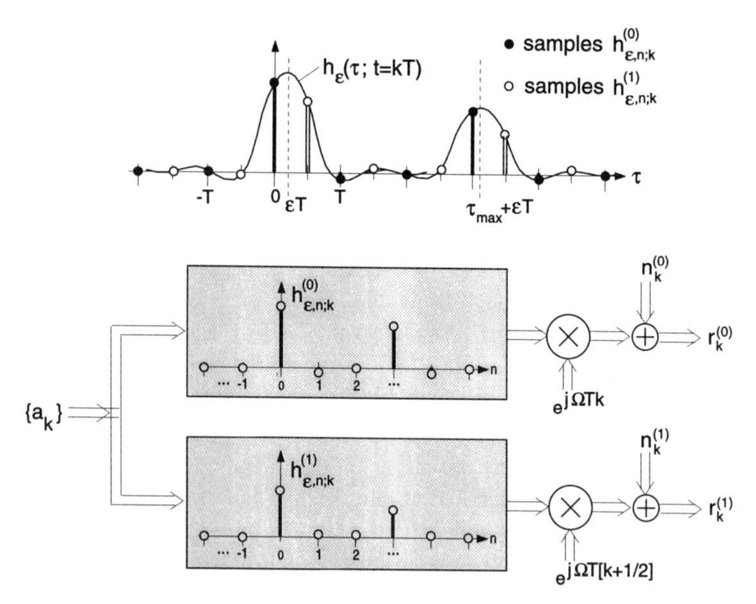

Figure 11-4 Discrete-Equivalent Channel Transmission Model for Frequency-Selective Fading Channels

The channel therefore manifests itself as if it were sampled in the delay and time domains, both at rate $2/T$. Furthermore, the peculiar indexing in eqs. (11-16) and (11-17) suggests demultiplexing the received signal into the two partial signals $r_k^{(0)}$ and $r_k^{(1)}$, respectively. Each of these partial signals $r_k^{(i)}$ is dependent on its own partial channel $h^{(i)}_{\varepsilon,n;\,k}$ while being independent from the other partial channel. Therefore, the transmission system can be modeled as two separate systems (the partial channels $h^{(i)}_{\varepsilon,n;\,k}$), both being fed by the same input signal (the symbol stream $\{a_k\}$) and producing the two partial received signals. The sampled noise processes $n_k^{(i)}$ in eq. (11-16) can be viewed as individually uncorrelated (see note on noise properties above), but the processes $n_k^{(0)}$ and $n_k^{(1)}$ are, in general, mutually correlated through the action of the anti-aliasing filter $F(\omega)$. The discrete-equivalent partial channel transmission model thus obtained is illustrated in Figure 11-4 for the example of a two-ray channel. This model is quite convenient since all discrete partial signals and systems are the result of sampling in the delay and time domains at the same rate, viz. the *symbol* rate $1/T$ (instead of $2/T$ as before). If necessary, this partitioning technique can be easily extended to sampling at higher multiples of the symbol rate.

In the case of nonselective fading channels, the transmission model can be simplified considerably. Observing eq. (11-11), the sampled channel impulse response [eq. (11-17)] is written as

$$\begin{aligned} h^{(i)}_{\varepsilon,n;k} &= h_\varepsilon\left(\tau=\left[n+\frac{i}{2}\right]T;\ t=\left[k+\frac{i}{2}\right]T\right) \\ &= \underbrace{c\left(t=\left[k+\frac{i}{2}\right]T\right)}_{c_k^{(i)}}\ \underbrace{g_T\left(\tau=\left[n+\frac{i}{2}\right]T-\varepsilon T\right)}_{g^{(i)}_{T\varepsilon,n}} \qquad \text{(flat fading)} \\ &= c_k^{(i)}\ g^{(i)}_{T\varepsilon,n} \end{aligned} \tag{11-18}$$

For slow and moderate fading rates, the approximation $c(t=[k+0.5]T) \approx c(t=kT)$ and therefore $c_k^{(1)} \approx c_k^{(0)}$ for the MD process holds. The digital-equivalent time-invariant filter $g^{(i)}_{T\varepsilon,n} = g_T(\tau=[n+i/2]T-\varepsilon T)$ is the sampled transmitter pulse response shifted by the fractional timing delay εT. Sampling the received signal $r(t)$ [eq. (11-12)] at rate $2/T$ then yields

$$\begin{aligned} r_k^{(i)} &= r\left(t=\left[k+\frac{i}{2}\right]T\right) \qquad (i=0,1) \\ &= e^{j\Omega T(k+i/2)}\left[\sum_n a_n\, h^{(i)}_{\varepsilon,k-n;\,k}\right] + n_k^{(i)} \\ &= \underbrace{e^{j\Omega T(k+i/2)}\, c_k^{(i)}}_{c^{(i)}_{\Omega,k}}\left[\sum_n a_n\, g^{(i)}_{T\varepsilon,k-n}\right] + n_k^{(i)} \qquad \text{(flat fading)} \end{aligned} \tag{11-19}$$

As discussed above, pulse matched filtering further simplifies the flat-fading transmission model. One may implement $g_{\mathrm{MF}}(\tau)$ either as an analog filter and sample its output $z(t)$ [eq. (11-15)], or, equivalently, apply the (partial) digital pulse MF $g^{(i)}_{\mathrm{MF},n} = (1/T)\, g_T^*(\tau = -[n+i/2]T)$ to filter the sampled (partial) signal $r_k^{(i)}$ [approximation of eq. (11-19)]. The sampled pulse MF output thus becomes

$$\begin{aligned} z_k^{(i)} &= z\left(t=\left[k+\frac{i}{2}\right]T\right) \\ &\underset{\left(\substack{\Omega T \\ \text{small}}\right)}{\approx} \underbrace{e^{j\Omega T[k+i/2]}\, c_k^{(i)}}_{c^{(i)}_{\Omega,k}}\left[\sum_n a_n\, g\left(\tau=\left[k+\frac{i}{2}\right]T-\varepsilon T-nT\right)\right] \\ &\qquad + m\left(t=\left[k+\frac{i}{2}\right]T\right) \\ &= \underbrace{e^{j\Omega T[k+i/2]}\, c_k^{(i)}}_{c^{(i)}_{\Omega,k}}\left[\sum_n a_n\, g^{(i)}_{\varepsilon,k-n}\right] + m_k^{(i)} \qquad \text{(flat fading)} \end{aligned} \tag{11-20}$$

where $g^{(i)}_{\varepsilon,n} = g(\tau{=}[n{+}i/2]T{-}\varepsilon T)$ is the sampled Nyquist pulse delayed by εT. The autocorrelation of the partial noise process $m^{(i)}_k$ and the cross correlation between the two partial processes $m^{(0)}_k$ and $m^{(1)}_k$ are given by

$$\begin{aligned} R_{m^{(i)}}(n) &= E\left[m^{(i)}_k m^{(i)*}_{k+n}\right] && = R_{m^{(i)}}(t{=}nT) \\ &= N_0\, g(\tau{=}nT) && = N_0\, \delta_n \\ R_{m^{(0)},m^{(1)}}(n) &= E\left[m^{(0)}_k m^{(1)*}_{k+n}\right] && = E[m(t{=}kT)\, m^*(t{=}kT{+}[n{+}1/2]T)] \\ &= R_m(t{=}[n{+}1/2]T) && = N_0\, g(\tau{=}[n{+}1/2]T) \end{aligned} \tag{11-21}$$

respectively.[1] Therefore, $m^{(0)}_k$ and $m^{(1)}_k$ are individually white noise processes with variance N_0, however, mutually coupled through pulse matched filtering.

Often the timing parameter ε is known, either via initial timing acquisition or from continuous tracking during steady-state operation. In fact, on flat fading channels, tracking of the timing phase may be accomplished using the same algorithms as for AWGN channels (see preceding chapters), since the degradation in performance (compared with tracking on static channels) remains small [7]. Then ε may be compensated for by digital interpolation (Chapters 4 and 9) or by physically adjusting the sampling clock such that $\varepsilon = 0$. With quasi-perfect timing recovery, the MF output can be decimated down to symbol rate $1/T$ without loss of information, so that, of the two partial signals $z^{(i)}_k$ and partial MD processes $c^{(i)}_k$, only $z^{(0)}_k = z_k$ and $c^{(0)}_k = c_k$ remain, respectively. In addition, we have $g^{(0)}_{(\varepsilon=0),n} = \delta_n$ due to Nyquist (ISI-free) pulse shaping and energy-normalizing matched filtering. Then the transmission model for the decimated pulse MF output boils down to

$$z_k \underset{\begin{pmatrix}\Omega T \text{ small;} \\ \varepsilon=0\end{pmatrix}}{=} \underbrace{e^{j\Omega T k}\, c_k}_{c_{\Omega,k}}\, a_k + m_k \qquad \begin{pmatrix}\text{flat fading,} \\ \text{perfect timing}\end{pmatrix} \tag{11-22}$$

where $m_k = m^{(0)}_k$ is white additive noise with variance N_0. Hence, the equivalent flat fading channel model for small frequency offsets and perfect timing consists of just a memory-free but time-variant multiplicative distortion $c_{\Omega,k}$ and a discrete AWGN process with variance N_0. The discrete-equivalent flat fading channel transmission models for unknown and known/compensated timing delay, respectively, are illustrated in Figure 11-5.

[1] Notice that, in this part of the book, the cross correlation between two random sequences x_k and y_k is defined as $R_{x,y}(n) = E[x_k\, y^*_{k+n}]$ (= complex conjugate of the cross-correlation definition in the previous chapters). By virtue of this redefinition, cross correlation matrices of sequences of random vectors can be expressed more elegantly in terms of a Hermitian transpose: $\mathbf{R}_{x,y}(n) = E[\mathbf{x}_k\, \mathbf{y}^H_{k+n}]$.

flat fading transmission model for unknown timing parameter ε

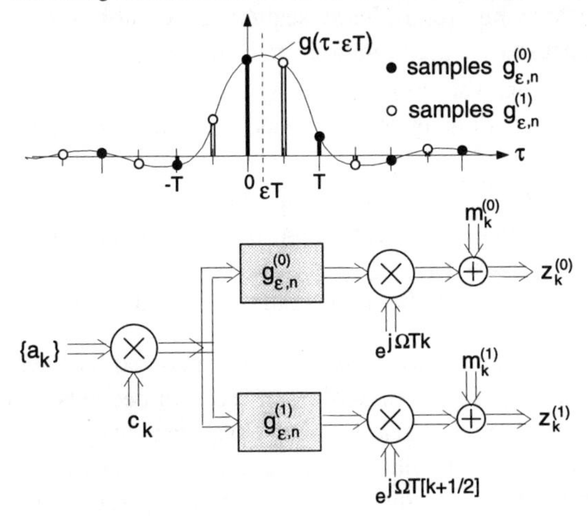

flat fading transmission model for known/compensated timing parameter ε

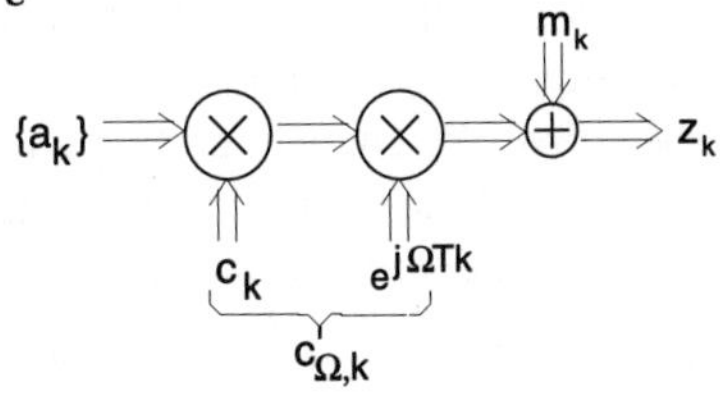

Figure 11-5 Discrete-Equivalent Channel Transmission Models for Flat Fading Channels

11.2.3 Statistical Characterization of Fading Channels

Up to now, we have been concerned with the transmission model regarding the channel delay profile or, equivalently, the degree of frequency selectivity, i.e., the characteristics of $c(\tau;t) \circ\!\!-\!\!\bullet C(\omega;t)$ in the τ and ω domains, respectively. We now turn our attention to the *time variations* of fading channels, i.e., the variations of $c(\tau;t) \circ\!\!-\!\!\bullet C(\omega;t)$ in the t domain. These are caused by variations of inhomogeneous media (ionosphere, atmospheric refraction), by moving obstacles along the propagation path, or by movements of the radio terminals (see Figure 11-1). The physical mechanisms that make up a fading process may have very different rates of change. Three distinct time scales of fading can be identified, so that one can distinguish between the following three broad categories of signal fading:

- **Long-term** (*large-area* or *global*) signal fading: slow variations of *average* signal strength, caused by varying distances between terminals leading to changes in free-space attenuation (mobile or personal radio), by the variability of the ionization and curvature of reflecting ionospheric layers (shortwave radio), slowly varying tropospheric scattering conditions (VHF and UHF range), losses caused by precipitation, and the like.

- **Medium-term** signal variations, typically caused by occasional obstruction of the LOS path (shadowing by buildings, hills, etc.) in mobile or satellite mobile radio.
- **Short-term** (*small-area* or *local*) signal fading: relatively fast variations of amplitude and phase of information-bearing signals picked up by the receiver, typically caused by rapid succession of instants with constructive and destructive interference between scattered or reflected rays.

Long- or medium-term signal variations are often modeled as *lognormal* fading, i.e., the short-term average signal strength, expressed in decibels, is taken to be a Gaussian random variable with a certain mean (long-term average signal strength) and variance (measure of fluctuation about the long-term average) [8]. Long- or medium-term fading determines the channel availability (or outage probability) and thus strongly affects the choice of transmission protocols and, to some lesser extent, the error control coding scheme. However, it is the "fastest" of the above three fading mechanisms that has a most profound impact on the design of transmission systems and digital receivers. From the viewpoint of receiver design – encompassing error-corrective channel coding and decoding, modulation, equalization, diversity reception, and synchronization – it is therefore necessary (and often sufficient) to focus on the *short-term* signal fading.

Unfortunately, attempting to achieve a *deterministic* mapping of the time-varying electromagnetic scenario onto the instantaneous channel impulse response $c(\tau;t)$ would be a very ambitious endeavor since it necessitates *fine-grain modeling* of the entire scattering scenario, including relevant parameters such as terrain (geological structure, buildings, vegetation, ground absorption and reflection), atmosphere (temperature, pressure, humidity, precipitation, ionization), constellation of obstacles along the propagation path(s), transmitting and receiving antennas (near and far field), etc. This, however, is most often impossible since some, if not all, relevant scattering parameters are usually unknown. Notice also that tiny variations in the scattering scenario may have a tremendous impact on the instantaneous channel transmission behavior. For instance, path-length variations as small as a fraction of the wavelength, caused, e.g., by rustling leaves, may give rise to large phase shifts of scattered rays. On the other hand, deterministic *ray tracing* modeling of the CIR $c(\tau;t)$ may be feasible for some indoor environments and very high carrier frequencies (e.g., 60 GHz) where the propagation characteristics can be obtained from the geometrical and material properties, using the rules of quasi-optical ray transmission and reflection. The results thus obtained are expected to be more accurate than using the WSSUS statistical model (discussed below) whose validity is restricted to a small area (in indoor environments a few square centimeters). In the cellular mobile arena where fine-grain modeling is not feasible, ray tracing is used for determining long-term *averages* of channel conditions for the purpose of cellular planning. At any rate, using ray tracing methods requires a lot of expertise and computational power, and when it comes to exploring the characteristics of hithereto unknown channels, the predictions made by ray tracing are often cross-checked against empirical results from measurement campaigns.

From the viewpoint of digital communications, it is seldom feasible nor necessary to trace every detail of the scattering scenario. Rather, one resorts to *statistical* modeling of the short-term channel variations [9]. The construction of statistical models and finding their parameters is accomplished based either on measurements alone (empirical model), on a simplified model of the physical scenario (coarse-grain or analytical model), or a combination of both. Usually, it is assumed that the random fading processes are *wide-sense stationary* (WSS), i.e., these processes are sufficiently characterized by their means and covariances. Furthermore, the elementary rays [weights $c_n(t)$] that constitute the channel are assumed to undergo mutually *uncorrelated scattering* (US), which is plausible since individual rays can often be attributed to distinct physical scatterers. *Wide-sense stationary, uncorrelated scattering* (WSSUS) fading process models [9, 10] thus have long been a widely accepted standard.

Fundamental Short-Term Statistical Parameters of Fading Channels

The short-term statistics of a fading channel are completely characterized by a single basic statistical function, viz. the *scattering function.* All other parameters describing the statistical properties of $c(\tau;t) \circ\!\!-\!\!\bullet C(\omega;t)$ can be derived from this basic function. The scattering function is one of four statistically equivalent correlation functions in the time and frequency domains:

- Spaced-time spaced-frequency correlation function

$$R_C(\Delta\omega;\Delta t) = E[C(\omega;t)\,C^*(\omega+\Delta\omega;\ t+\Delta t)] \tag{11-23}$$

- Spaced-time delay correlation function

$$R_c(\tau;\Delta t) = E[c(\tau;t)\,c^*(\tau,\ t+\Delta t)] \tag{11-24}$$

- Spaced-frequency Doppler power spectrum (psd)

$$S_C(\Delta\omega;\psi) = \int_{-\infty}^{\infty} R_C(\Delta\omega;\Delta t)\, e^{-j\psi(\Delta t)}\, d(\Delta t) \tag{11-25}$$

- Delay Doppler power spectrum = scattering function

$$S_c(\tau;\psi) = \int_{-\infty}^{\infty} R_c(\tau;\Delta t)\, e^{-j\psi(\Delta t)}\, d(\Delta t) \tag{11-26}$$

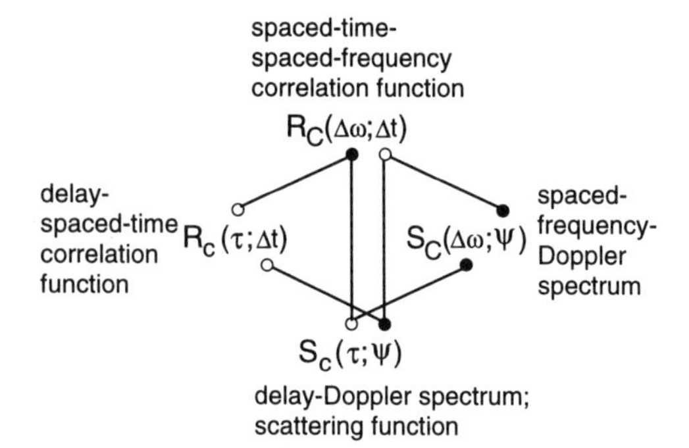

Figure 11-6 Fourier Transform Relations between Equivalent Statistical Functions

where $\psi = 2\pi\lambda$ is the angular frequency variable (λ = frequency variable) of the Doppler spectra. The Fourier transform relations between the above four statistically equivalent functions are depicted in Figure 11-6.

The elementary ray weight processes

$$c_n(t) = \xi_n \, e^{j(\psi_{D_n} t + \theta_n)} \quad = \xi_n \, e^{j(2\pi \Lambda_{D_n} t + \theta_n)} \tag{11-27}$$

that constitute the physical channel [eq. (11-7), disregarding propagation and clock timing delays] are characterized by gain factors ξ_n, Doppler shifts ψ_{D_n}, and phase shifts θ_n that may be assumed fixed during very short time intervals. Most often the number N of these processes is virtually infinite so that the $c_n(t)$ have infinitesimal gains. Invoking the WSSUS assumption, the spaced-time delay correlation and scattering functions [eqs. (11-24) and (11-26)] become

$$\begin{aligned} R_c(\tau; \Delta t) &= \sum_{n=0}^{N-1} \left(\xi_n^2\right) e^{j\psi_{D_n} \Delta t} \, \delta(\tau - \tau_n) \\ S_c(\tau; \psi) &= \sum_{n=0}^{N-1} \left(\xi_n^2\right) \delta(\psi - \psi_{D_n}) \, \delta(\tau - \tau_n) \end{aligned} \tag{11-28}$$

respectively. Each elementary ray manifests itself as a point (τ_n, ψ_{D_n}) in the delay–Doppler plane (τ, ψ), and the multitude of such rays make up a quasi-continuous two–dimensional function $S_c(\tau; \psi)$.

From measurements one often observes that the elementary rays form distinct *clusters* in some areas of the delay-Doppler plane. Let us distinguish between clusters by denoting $S_{c_m}(\tau; \psi)$ the partial scattering function of the mth cluster, so that the total spaced-time delay correlation and scattering functions

$$\begin{aligned} R_c(\tau; \Delta t) &= \sum_{m=0}^{M-1} R_{c_m}(\tau; \Delta t) \\ S_c(\tau; \psi) &= \sum_{m=0}^{M-1} S_{c_m}(\tau; \psi) \end{aligned} \tag{11-29}$$

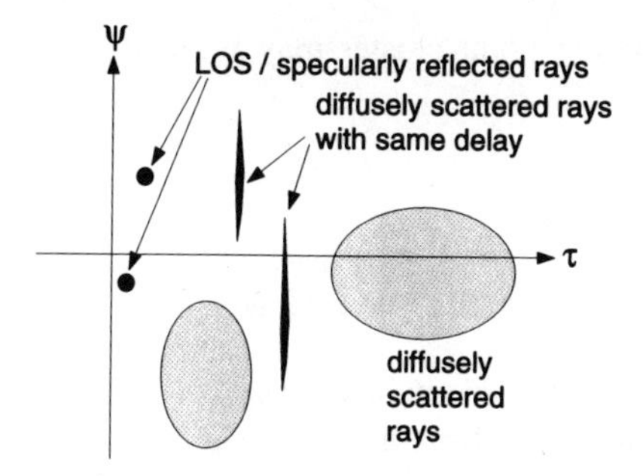

Figure 11-7 Types of Clusters in the Delay-Doppler Plane

can also be expressed as the superposition of M cluster spaced-time delay correlation and scattering functions $R_{c_m}(\tau;\Delta t)$ and $S_{c_m}(\tau;\psi)$, respectively. Further denote by $\mathcal{R}_m$ the region in the delay-Doppler plane (τ,ψ) for which $S_{c_m}(\tau;\psi)$ is nonzero, and by $\mathcal{N}_m$ the set of indices n for which the elementary rays $c_n(t)$ belong to the mth cluster. Typically, three types of ray clustering can be distinguished:

- Strong clustering about a single point $\mathcal{R}_m = (\tau_m, \psi_{D_m})$,
- Clustering in an oblong region $\mathcal{R}_m$ with nearly equal propagation delays $\tau_m \approx \tau_n$ for all $n \in \mathcal{N}_m$, and
- Weak clustering in an extended region $\mathcal{R}_m$.

This clustering scenario is illustrated by Figure 11-7.

The type of clustering is determined by the underlying physical scattering scenario, in particular, the spatial distribution and material properties of scatterers (hence, the strengths and angles of incidence of scattered rays) and the velocity of terminals (or scatterers). If some knowledge on this scattering scenario is available, scattering functions for certain typical environments (urban, suburban, etc.) may be derived from this physical model.

Strong clustering about a single point $\mathcal{R}_m = (\tau_m, \psi_{D_m})$ in the (τ,ψ) plane is attributed to either the LOS ray or specular (quasi-optical) reflection from a nearly *singular* scattering point on smooth surfaces such as buildings, asphalt, water surfaces, or tall mountains. Although the scattering scenario is changing continuously due to the motion of radio terminals, obstacles along the propagation path or the scatterer itself, this scattering scenario may often be regarded as "frozen," at least over a short duration of time, so that the scattering point remains stationary for some time. Specular reflection from point scatterers results in wavefronts that are approximately coherent, i.e., the elementary rays add

[2] For simplicity's sake, the complex-valued path weights $c_m(t)$, Doppler shifts ψ_{D_m}, and phases θ_m of the mth cluster are given the same variable names as their elementary counterparts. In order to avoid confusion, indices m are taken to refer to clusters whereas indices n refer to elementary rays.

constructively. Thus, the point cluster may be modeled essentially like a single coherent path:

$$c_m(t) = \sum_{n=0}^{N_m-1} c_n(t) \underset{\begin{pmatrix}\text{LOS or specularly}\\ \text{reflected path}\end{pmatrix}}{=} \alpha_m\, e^{j(\psi_{D_m} t + \theta_m)} \tag{11-30}$$

where N_m is the number of elementary rays constituting the mth point cluster.[2] Depending on the motion of terminals or scatterers and the direction of arrival, the path may bear a sizeable Doppler shift $\psi_{D_m} = 2\pi\Lambda_{D_m} \approx \psi_{D_n}$. Assuming phase coherence ($\theta_m \approx \theta_n$), the path gain factor – which is almost time-invariant – becomes $\alpha_m \approx \sum_{n=1}^{N_m-1} \xi_n$. Hence, the LOS / specular path model consists of a rotating phasor with fixed amplitude α_m and coherent phase $\varphi_m(t) = \psi_{D_m} t + \theta_m$.

Frequently, scattering takes place over a large area of rough (relative to the wavelength) or irregular surfaces such as vegetation or objects close to the antennas. Such scattering is not specular but *diffuse*, and a cluster of diffusely scattered rays is composed of a multitude of individual rays which exhibit less directivity and no phase coherence. In some cases of interest, all incoherent rays of a cluster have (approximately) equal propagation delays $\tau_m \approx \tau_n$ (Figure 11-7), so that the respective cluster weight

$$\begin{aligned} c_m(t) = \sum_{n=0}^{N_m-1} c_n(t) &\underset{\begin{pmatrix}\text{diffuse scattering,}\\ \text{delays almost equal}\end{pmatrix}}{=} \sum_{n=0}^{N_m-1} \xi_n\, e^{j(\psi_{D_n} t + \theta_n)} \\ &= \alpha_m(t)\, e^{j\varphi_m(t)} \end{aligned} \tag{11-31}$$

representing an oblong region $\mathcal{R}_m$ in the (τ, ψ) plane, effectively becomes a complex-Gaussian random process. Its individual Doppler spectrum and average power can be determined from the quasi-continuous scattering function as

$$\begin{aligned} S_{c_m}(\psi) &= \int_{\mathfrak{R}_m} S_c(\tau;\psi)\, d\tau \\ \rho_{c_m} &= R_{c_m}(0) = E\left[|c_m(t)|^2\right] \\ &= \frac{1}{2\pi}\int_{\mathfrak{R}_m} S_c(\tau;\psi)\, d\tau\, d\psi \;=\; \frac{1}{2\pi}\int_{\mathfrak{R}_m} S_{c_m}(\psi)\, d\psi \end{aligned} \tag{11-32}$$

If the channel comprises M such clusters with distinct delays τ_m, the spaced-time delay correlation and scattering functions [eqs. (11-24) and (11-26)] simplify to

$$\begin{aligned} R_c(\tau;\Delta t) &= \sum_{m=0}^{M-1} R_{c_m}(\Delta t)\, \delta(\tau - \tau_m) \\ S_c(\tau;\psi) &= \sum_{m=0}^{M-1} S_{c_m}(\psi)\, \delta(\tau - \tau_m) \end{aligned} \tag{11-33}$$

where

$$R_{c_m}(\Delta t) = E[c_m(t)\, c_m^*(t+\Delta t)]$$
$$S_{c_m}(\psi) = \int_{-\infty}^{\infty} R_{c_m}(\Delta t)\, e^{j\psi(\Delta t)}\, d(\Delta t) \tag{11-34}$$

are the spaced-time correlation function and Doppler power spectrum of the mth cluster process $c_m(t)$, respectively.

Derived Short-Term Statistical Parameters of Fading Channels

The probability density function as well as other functions and parameters that characterize the dynamic behavior of fading channels can be derived from the scattering function.

If a cluster represents a multitude of incoherent elementary rays, the *density function* of its weight process $c(t) = \alpha(t)\, e^{j\varphi(t)}$ (for clarity, the index m is dropped in the pdf's) is complex Gaussian with uniformly distributed phase $\varphi(t)$, Rayleigh-distributed amplitude $\alpha(t)$ and exponentially distributed energy (power) $E(t) = \alpha^2(t)$:

$$\begin{aligned} p(\alpha) &= 2\frac{\alpha}{\overline{E}}\, e^{-\alpha^2/\overline{E}} \\ p(E) &= \frac{1}{\overline{E}}\, e^{-E/\overline{E}} \end{aligned} \tag{11-35}$$

respectively, where $\overline{E} = E\{\alpha^2\}$ is the average energy (power) of the cluster process. Notice that the cluster weight can be attributed an energy since it is (part of) a system. When viewed as a random process, $c(t)$ can be attributed a power.

Particularly in the case of flat fading channels, a cluster of rays having the same delay may comprise both a specular/LOS and a diffuse component, so that the composite weight process $c(t)$ [eq. (11-11)]

$$\begin{aligned} c(t) &= c_s(t) + c_d(t) \\ &= \alpha_s\, e^{j\varphi_s(t)} + \alpha_d(t)\, e^{j\varphi_d(t)} \\ &= \alpha_s\, e^{j(\psi_{D_s} t + \theta_s)} + \alpha_d(t)\, e^{j\varphi_d(t)} \\ &= \alpha(t)\, e^{j\varphi(t)} \end{aligned} \tag{11-36}$$

obeys a Rician density

$$\begin{aligned} p(\alpha) &= 2\frac{\alpha}{\overline{E}_d} \exp\left\{-\left(\frac{\alpha^2 + E_s}{\overline{E}_d}\right)\right\} I_0\left(2\sqrt{E_s}\frac{\alpha}{\overline{E}_d}\right) \\ &= 2\frac{\alpha}{\overline{E}_d} \exp\left\{-\left(\frac{\alpha^2}{\overline{E}_d} + K\right)\right\} I_0\left(2\sqrt{K}\frac{\alpha}{\sqrt{\overline{E}_d}}\right) \\ p(E) &= \frac{1}{\overline{E}_d} \exp\left\{-\frac{E + E_s}{\overline{E}_d}\right\} I_0\left(2\frac{\sqrt{E_s}}{\overline{E}_d} \cdot \sqrt{E}\right) \end{aligned} \tag{11-37}$$

with $E_s = \alpha_s^2$ and $\overline{E}_d = E\{\alpha_d^2\}$ the (average) energies of the specular and diffuse components, respectively, $E = E(t) = |c(t)|^2$ the total energy of $c(t)$, and $K = E_s/\overline{E}_d$ the ratio between the energies of the specular and diffuse components (K factor).

The dynamic behavior of fading channels is often expressed in terms of one-dimensional (thus coarser) characteristics:

- Spaced-time correlation function

$$R_C(\Delta t) = E[C(\omega;t)\, C^*(\omega;\ t+\Delta t)] \quad = R_C(\Delta\omega=0;\Delta t) \tag{11-38}$$

- Spaced-frequency correlation function

$$R_C(\Delta\omega) = E[C(\omega;t)\, C^*(\omega+\Delta\omega;\ t)] \quad = R_C(\Delta\omega;\Delta t=0) \tag{11-39}$$

- Power delay profile

$$\begin{aligned} R_c(\tau) &= E[c(\tau;t)\, c^*(\tau,\ t)] \qquad = R_c(\tau;\Delta t=0) \\ &= \frac{1}{2\pi}\int_{-\infty}^{\infty} S_c(\tau;\psi)\, d\psi \end{aligned} \tag{11-40}$$

- Doppler spectrum

$$\begin{aligned} S_c(\psi) &= \int_{-\infty}^{\infty} R_C(\Delta t)\, e^{-j\psi(\Delta t)}\, d(\Delta t) \quad = S_C(\Delta\omega=0;\psi) \\ &= \int_{-\infty}^{\infty} S_c(\tau;\psi)\, d\tau \end{aligned} \tag{11-41}$$

The *power delay profile* $R_c(\tau)$ contains the necessary information on the channel dispersion and thus the amount of ISI to be expected for a given symbol duration T. The *spaced-time correlation function* $R_C(\Delta t) \circ\!\!-\!\!\bullet S_c(\psi)$ and Doppler spectrum shed light on the channel dynamics, i.e., fastness of fading.

From the above one-dimensional functions, a number of characteristic parameters can be extracted:

- Channel coherence time

$$T_{\text{coh}} = \text{rms}[R_C(\Delta t)] \tag{11-42}$$

- Channel coherence bandwidth

$$B_{\text{coh}} = \frac{1}{2\pi}\,\text{rms}[R_C(\Delta\omega)] \tag{11-43}$$

- rms channel delay spread

$$\tau_D = \text{rms}[R_c(\tau)] \tag{11-44}$$

- Maximum channel delay spread

$$\tau_{\max} = \tau_{N-1} - \tau_0 \tag{11-45}$$

- Channel Doppler shift

$$\Lambda_D = \frac{1}{2\pi} E[S_c(\psi)] \tag{11-46}$$

- Channel Doppler spread

$$\sigma_D = \frac{1}{2\pi} \mathrm{rms}[S_c(\psi)] \tag{11-47}$$

- Channel spread factor

$$S_c = \tau_D \, \sigma_D \tag{11-48}$$

where

$$\overline{x} = E[x] = \frac{\int_{-\infty}^{\infty} x \, f(x)\, dx}{\int_{-\infty}^{\infty} f(x)\, dx} \qquad \mathrm{rms}[x] = \sqrt{\frac{\int_{-\infty}^{\infty} (x-\overline{x})^2 \, f(x)\, dx}{\int_{-\infty}^{\infty} f(x)\, dx}} \tag{11-49}$$

denote the *mean* and *root mean square* values of a function$f(x)$, respectively. The *coherence time* T_{coh} and *coherence bandwidth* B_{coh}, respectively, indicate the time and frequency shifts over which a channel is essentially correlated. The *delay spread* τ_D is linked with the coherence bandwidth via $\tau_D \approx 1/B_{\mathrm{coh}}$. However, in receiver design, the *maximum* delay spread $\tau_{\max} = \tau_{N-1} - \tau_0$ is of higher relevance than the (one-sided) rms value τ_D of the power delay profile. The *Doppler shift* Λ_D (in Hertz) is the global frequency offset introduced by the physical channel itself (not by an oscillator offset), and the *Doppler spread* σ_D (in Hertz) is the (one-sided) rms bandwidth of the Doppler spectrum. It is linked with the coherence time via the approximate relation $\sigma_D \approx 1/T_{\mathrm{coh}}$. In a set of clusters with M distinct delays τ_m, the scattering function of eq. (11-33) comprises M cluster Doppler spectra $S_{c_m}(\psi)$ [eq. (11-34)] from which *cluster* Doppler shifts and spreads may be extracted:

$$\begin{aligned} \Lambda_{D_m} &= \frac{1}{2\pi} E[S_{c_m}(\psi)] \\ \sigma_{D_m} &= \frac{1}{2\pi} \mathrm{rms}[S_{c_m}(\psi)] \end{aligned} \tag{11-50}$$

Occasionally, the Doppler shift and Doppler spread are lumped together to yield an "efficient" (larger) Doppler spread that can be used as a global measure of the degree of channel fading. If the Doppler spectrum $S_c(\psi)$ is strictly band-limited (e.g., in the case of mobile radio channels), a more appropriate measure is the channel cutoff frequency (*Doppler frequency*) λ_D (in Hertz). Finally, the channel *spread factor* $S_c = \tau_D \, \sigma_D$ is a measure of overall fading channel quality;

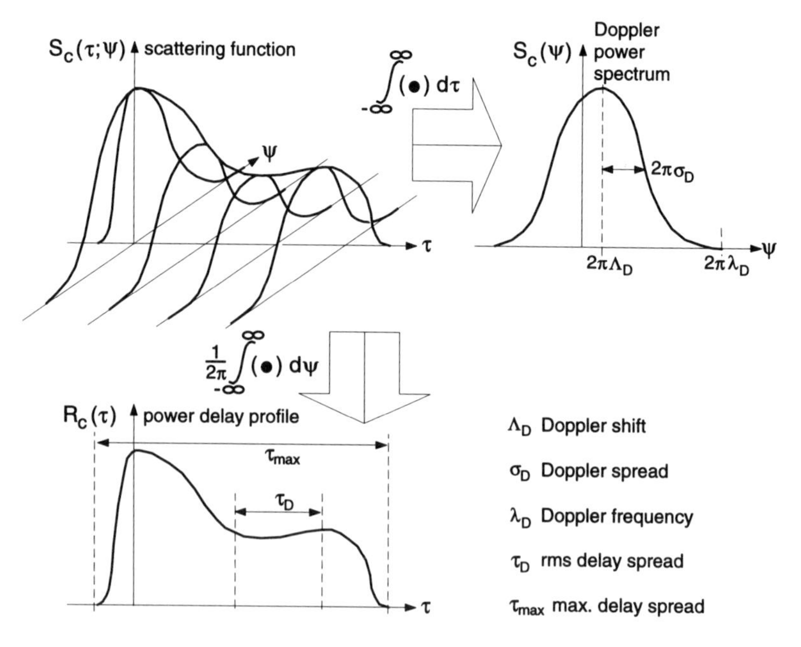

Figure 11-8 Example of Scattering Function, Power Delay Profile and Doppler Spectrum

if the channel is *underspread* ($S_c \ll 1$), the fading is slow with respect to the dispersion so that coherent transmission is possible if suitable antifading techniques are applied. On the other hand, if the channel is *overspread* (S_c in the order or in excess of 1), the channel changes significantly over the duration of its impulse response, so that, in general, only noncoherent transmission is possible.

An example of a typical scattering function for a channel with a virtually infinite number of paths, together with the power delay profile, Doppler spectrum and some important parameters, are visualized in Figure 11-8.

When designing countermeasures against fading (e.g., providing for a fade margin or selecting a suitable channel coding scheme) or assessing the outage probability of a system operating over fading channels, the rate of occurrence and the duration of "deep" fades are of interest. Consider again a fading process $c(t)$ (cluster index m has been dropped). The *level crossing rate* $n(C)$ is the average number of crossings (both upward and downward) of the amplitude process $\alpha(t) = |c(t)|$ with a certain fading threshold C. If C is small, this corresponds with the average fading rate. The average time elapsing from a downward to the next upward crossing of the fading amplitude $\alpha(t)$ with level C is the *average fade duration* $d(C)$. The parameters $n(C)$ and $d(C)$ are given by

- Level crossing rate

$$n(C) = \int_0^{\infty} \dot{\alpha}\, p(\alpha = C, \dot{\alpha})\, d\dot{\alpha} \tag{11-51}$$

- Average fade duration

$$d(C) = \frac{1}{n(C)} \int_0^{C} p(\alpha)\, d\alpha \tag{11-52}$$

The two-dimensional pdf $p(\alpha, \dot{\alpha})$ may be evaluated by assuming that the amplitude process $\alpha(t)$ and the process $\dot{\alpha}(t) = d\alpha(t)/dt$ of its derivative are uncorrelated, i.e., $p(\alpha, \dot{\alpha}) = p(\alpha)\, p(\dot{\alpha})$. The pdf $p(\dot{\alpha})$ can then be computed with the help of $p(\alpha)$ and the spaced-time correlation function $R_C(\Delta t)$.

Statistical Parameters of Some Important Fading Channels

In this section, the most important properties of some technically relevant fading channels are summarized.

The *geostationary satellite channel* exhibits some scintillation caused by movements of the ionospheric penetration point of microwave rays [11]. Under mild scintillation conditions, the maximum dispersion $\tau_{\max}$ remains well below 10 ns and the Doppler spread σ_D below 1 Hz. When severe scintillation conditions are present (caused, e.g., by ionospheric turbulences, low elevation, or the use of low VHF frequencies), the Doppler spread may rise up to 10 Hz. In the case that the satellite is not geostationary, one may also have to cope with very large Doppler shifts Λ_D. Usually, the LOS ray is dominant (unless obstructed), and the fading is Rician with large K factor. Because of the strong LOS component, very mild selectivity and slow fading, the satellite channel can often be well approximated by an AWGN channel model, possibly including a frequency shift [12]. Provision of a small fade margin is usually sufficient for reliable communications.

The *microwave line-of-sight* channel [13, 14] may exhibit fading caused by objects near the LOS ray, by ground reflections, or by atmospheric layering and anomalies of the atmospheric refractive index. Superrefraction (very large refractive index) may give rise to multipath propagation and long-distance co-channel interference. On the other hand, subrefraction (very small refractive index) is likely to cause deep fades of the LOS ray. Under such conditions, dispersions of up to 10–100 ns, Doppler spreads of several Hertz and signal fades of 20 dB and more may occur, so that large fade margins and the use of space diversity and possibly adaptive equalization should be considered. Under normal propagation conditions, the dispersion and Doppler remain below 2 ns and 1 Hz, respectively.

The *tropospheric scattering* channel used for high-power narrowband (up to several hundred kHz) transmission in microwave bands up to 10 GHz [15] features a continuum of randomly moving scattering points in the troposphere and therefore Rayleigh fading. For medium ranges up to 200 km, the dispersion and Doppler are below 1 μs and between 0.1 and 5 Hz (50 Hz in the case of aircraft echos), respectively, while for large distances the dispersion may be as large as a few milliseconds.

The *ionospheric shortwave* channel (high frequency [HF] channel) in the 2–to–30–MHz frequency range is used for power-efficient narrowband communications over very large distances [16]. Despite its small bandwith of typically 3 kHz, the channel may be strongly frequency-selective. Solar ionization of approximately horizontally stratified layers in the upper atmosphere cause radio rays to be either absorbed or refracted. Absorption is due to collisions of electrons, especially in the D and F1 layers, while refraction is due to reradiation from electrons that are set into motion by the electric field of the incoming radio wavefront. Refracted rays may thus be reflected back to the ground; single- and multihop reflection from one or more ionospheric layers and the ground can give rise to considerable multipath propagation. Up to five or more clusters with distinct (relative) path delays τ_m may be present. During daytime, reflections commonly occur from the D layer (70–90 km above the ground), E layer (90–130 km), F1 layer (130–210 km), and F2 layer (250-400 km), all characterized by distinct "critical" frequencies – rays with vertical incidence are reflected up to the critical frequency – of f_0D=0.1–0.3 MHz, $f_0E\approx$3 MHz, $f_0F1\approx$4 MHz, and f_0F2=6–12 MHz. During nighttime, only the F layer (300 km above) is sufficiently ionized with critical frequency $f_0F\approx$2 MHz. The maximal usable frequency (MUF) is given by $f_0X/\sin(\alpha_X)$ where α_X is the elevation angle of the radio ray reflected from a layer X. During daytime, the low end of usable frequencies for a desired link is determined by the D absorption, while at nighttime transmission is possible over all HF bands up to the MUF.

The ionospheric HF channel is subject to annual, seasonal, and diurnal *long-term* variations. In addition, the critical frequencies are dependent on the solar cycle (11-year cycle of sunspot activity). The *short-term* fading is quasi-stationary over up to 10 minutes or more. Occasionally, ionospheric disturbances may increase the absorption, e.g., sudden ionospheric disturbance (SID) due to solar X-ray flare, or polar cap absorption (PCA) resulting from solar protons being guided toward polar regions. Also, ionospheric storms due to solar particles being deflected toward auroral zones ("aurora," spread-F) can give rise to flutter fading. Apart from the time and frequency selectivity resulting from skywave propagation, the HF channel is characterized by additive noise originating from a number of diverse sources [16]. Atmospheric noise, for the most part due to lightning discharges, is the predominant natural noise for frequencies up to 10 MHz. Above 10 MHz, extraterrestrial noise is dominant. Man-made noise from motors, power lines, ignition systems, etc. may predominate in urban and suburban areas. Moreover, co- and adjacent-channel interference (voice, smallband tones, teletype, broadcast stations, etc.) by legitimate users or jammers may be the strongest source of interference (up to 30 dB above the ambient noise), especially during nighttime.

Each cluster or ionospheric *mode*, being a superposition of a multitude of elementary physical rays, exhibits a relatively small dispersion (20–200 μs). Under multipath propagation conditions, however, the total dispersion is dependent upon the range and may be as large as 5 ms. Occasionally, weak rays may be delayed by up to 8 or 10 ms, e.g., under auroral conditions. The cluster Doppler spectra $S_{c_m}(\psi)$ are usually Gaussian-shaped, sometimes with two lobes (magneto-ionic components) [17, 18]. Doppler shift and spreads range between 0.01 and 0.5 Hz for monomode propagation, 0.1–1 Hz for multipath, and 5–10 Hz or more for aurora conditions.

The *land-mobile* (LM) channel [19] for small- and wideband transmission at frequencies below 10 GHz exhibits more or less frequency-dependent multipath transmission (see Figure 11-1), depending on the bandwidth, range (up to 20–50 km), and the terrain. The short-term stationarity assumption is valid within a mobile moving distance of about 10–50 wavelengths. Scattered rays with small excess delays τ_n (with respect to the LOS ray which may or may not be present) can often be attributed to diffuse scatterers such as the ground or objects near the antennas (e.g., parts of the vehicle whereupon the antenna is mounted). Such diffuse Rayleigh-distributed rays tend to arrive from all directions and often cluster near the origin of the (τ, ψ) plane. On the other hand, rays having large extra delays are often due to specular reflection from large objects such as mountains or tall buildings and therefore tend to form more pronounced or even strongly peaked clusters. Depending on the maximum relative speed v between the mobile terminals, the Doppler spectrum is strictly band-limited to the Doppler frequency $\lambda_D = f_0\,(v/c)$ (f_0 carrier frequency, c speed of light). Under the assumption of isotropic scattering from all directions, the diffuse cluster Doppler spectra become U-shaped (Jakes spectrum [20]):

$$S_{c_m}(\psi) \underset{\begin{pmatrix}\text{diffuse scattering,}\\ \text{delays almost equal}\end{pmatrix}}{=} \rho_m \frac{1}{\pi\lambda_D} \frac{1}{\sqrt{1-\left(\frac{\psi}{2\pi\lambda_D}\right)^2}} \qquad (|\psi| \le 2\pi\lambda_D) \tag{11-53}$$

where $\rho_m = R_{c_m}(\Delta t = 0)$ denotes the average energy (power) of the mth cluster. On the other hand, *specular* reflection results in sharply peaked Doppler spectra [strong isolated peaks of $S_c(\tau;\psi)$], i.e., the cluster Doppler spectrum virtually reduces to a single Dirac peak with delay τ_m, weight ρ_m, and Doppler shift $\psi_{D_m} = 2\pi\Lambda_{D_m}$ [see eq. (11-30)] where Λ_{D_m} is bounded by the Doppler frequency λ_D.

In strongly irregular terrain (urban, dense vegetation), rays suffer from high absorption so that the dispersion is small (typically 1–3 μs). In suburban environments and in hilly terrain, the power-delay profile $R_c(\tau)$ drops off roughly exponentially, and significant dispersion usually spans less than 15–20 μs, occasionally up to 30 μs or more. In mountaineous areas, the dispersion of specularly reflected rays may be as large as 150 μs ; such extreme values, however, can be avoided and the dispersion limited to about 20 μs by suitable cell planning [21]. In flat rural areas, on the other hand, the dispersion is usually very small (below 1 μs), and the fading tends to be more Rician due to the often-present strong LOS ray.

The *land-mobile satellite* channel for smallband transmission (typically 5 kHz) at frequencies around 1 GHz is characterized by frequent *shadowing* of the LOS ray by obstacles such as buildings, tunnels, bridges, etc. The absorption and duration of shadowing events can be quite large. The effect of shadowing is often described by a Gilbert-Elliot model with the two states "good channel" (no shadowing, strong LOS ray, nonselective Rice fading) and "bad channel" (deep fade because of LOS ray obstruction, only scattered rays, Rayleigh fading). The *aeronautical satellite* channel used for smallband signals (bandwidth in kilohertz range) is also nonselective. Ground reflection (usually a single reflected ray) leads to Rician fading. Due to the large aircraft velocities, Doppler shifts in the order of 1 kHz and Doppler spreads of up to 200 Hz may be present.

The *indoor radio* channel becomes increasingly important for micro- and pico-cellular communications inside of offices, plants, private homes, etc. The channel is stongly nonstationary; long periods of stable propagation conditions may be interrupted by short-term fading disturbances caused by movements of terminals, antennas, people, or objects along the radio rays. Such events result in absorption between 10 and 30 dB. The dispersion usually remains below 100 ns, in large buildings occasionally up to 1 μs. The Doppler spread ranges from less than 10 Hz (fixed terminals) to values up to 100 Hz (moving terminals).

The most relevant statistical parameters of important fading channels are summarized in Table 11-1 for typical values of carrier frequency f_0 and symbol rate $1/T$. Given the symbol rate, the dispersion and fading parameters can be normalized to the symbol duration and rate, respectively, i.e., $\tau' = \tau/T$ (dispersion variable) and $\sigma' = \sigma T$, $\lambda' = \lambda T$, etc. (fading variable). In addition, Table 11-1 lists critical issues that must be given special attention when designing digital communication systems for those channels.

11.2.4 Main Points

- Physical multipath fading channels can be modeled as continuous-time systems with channel impulse response (CIR) and channel transfer function (CTF) $c(\tau;t) \circ\!\!-\!\!\bullet C(\omega;t)$, respectively. Including the effects of band-limited TX pulse shaping filtering $g_T(\tau) \circ\!\!-\!\!\bullet G_T(\omega)$ and a delay ε (propagation and timing) into the channel model yields the effective CIR and band-limited CTF $h_\varepsilon(\tau;t) \circ\!\!-\!\!\bullet H_\varepsilon(\omega;t) = C(\omega;t)\, G_T(\omega)\, e^{-j\omega\varepsilon T}$ [eq. (11-7)]. A fading channel may either be *frequency-nonselective (flat)* or *frequency-selective* within the bandwidth of interest. In the flat fading case the relevant physical CIR and CTF reduce to the *multiplicative distortion* $c(t)$.

- Sampling of the received signal at a rate larger than the symbol rate, typically at rate $2/T$, yields received samples which may be partitioned into two T—spaced *partial* signals $r_k^{(i)}$ ($i = 0, 1$). Associated with each of these

Table 11-1 Parameters of Some Important Fading Channels

	Typical Applications		Dispersion	Fading	
Channel	f_0	$1/T$	$\tau'_{\max}$	Λ'_D, σ'_D	critical issues
Microwave LOS	5 GHz	5 MBd	1	$<10^{-6}$	attenuation, dispersion
Ionospheric HF	10 MHz	2 kBd	1 to 10	10^{-3}	dispersion, fading, addit. noise
Troposcatter	5 GHz	100 kBd	$\ll 1$	$<10^{-3}$	fading
Land-mobile radio	1 GHz	200 kBd	1 to 5	10^{-3}	dispersion, fading, interference
Land-mobile satellite	1 GHz	2 kBd	$\ll 1$	0.1	shadowing, Doppler
Aero.-mobile satellite	1 GHz	2 kBd	$\ll 1$	up to 1	Doppler
Indoor radio	1 GHz	200 kBd	<1	10^{-4}	attenuation

signals is a T-spaced partial *discrete-equvialent channel* $h^{(i)}_{\varepsilon,n;k}$. Depending on whether the channel is selective or flat, and on whether pulse matched filtering without or with perfect timing sync has been employed (flat fading), the following *transmission models* apply:

$$
\begin{aligned}
&\text{Selective fading, rate} - 2/T \text{ sampling :}\\
&r_k^{(i)} = e^{j\Omega T\left(k+\frac{i}{2}\right)} \left[\sum_n a_n\, h^{(i)}_{\varepsilon,k-n;\,k}\right] \qquad + n_k^{(i)} \qquad (i=0,1)\\
&\text{Flat fading, rate} - 2/T \text{ sampling :}\\
&r_k^{(i)} = e^{j\Omega T\left(k+\frac{i}{2}\right)}\, c_k^{(i)} \left[\sum_n a_n\, g^{(i)}_{T\varepsilon,k-n}\right] \qquad + n_k^{(i)}\\
&\text{Flat fading, pulse} - \text{MF output, imperfect timing, rate} - 2/T :\\
&z_k^{(i)} = e^{j\Omega T\left(k+\frac{i}{2}\right)}\, c_k^{(i)} \left[\sum_n a_n\, g^{(i)}_{\varepsilon,k-n}\right] \qquad + m_k^{(i)}\\
&\text{Flat fading, pulse} - \text{MF output, perfect timing, rate} - 1/T :\\
&z_k = e^{j\Omega T k}\, c_k\, a_k \qquad + m_k
\end{aligned}
\tag{11-54}
$$

with $g^{(i)}_{T\varepsilon,n}$ the sampled TX pulse shaping filter and $g^{(i)}_{\varepsilon,n}$ the sampled cascade of TX and RX pulse shaping filters.

- Long- and short-term fading processes are usually characterized in terms of certain *statistical* parameters. *Short-term fading*, being most important for receiver design, is governed by the *scattering function*, i.e., the delay Doppler power spectrum $S_c(\tau;\psi)$. Often, *clustering* of multipath rays in the $(\tau;\psi)$ plane is observed. The kind of clustering also determines the fading amplitude density $p(\alpha)$: diffuse scattering (NLOS) yields *Rayleigh* fading, while the presence of a LOS path or specular reflection leads to *Rice* fading. From the scattering function a number of secondary statistics can be derived, including the power delay profile $R_c(\tau)$ and the Doppler spectrum $S_c(\psi)$.
- The basic properties of some important fading channels, including satellite, microwave LOS, ionospheric shortwave, land-mobile, and indoor, have been reviewed.

11.3 Modeling and Simulation of Discrete-Equivalent Fading Channels

When designing communication systems for fading channels, it is important to be able to assess and verify system performance during the entire design phase, long before actually implementing the system in hardware and performing *in situ* field tests. For a large selection of important channels, theoretical studies and extensive measurements have been conducted over years or decades. From these investigations, the properties of many channels are well understood and their statistical parameters known with sufficient accuracy. Therefore, statistical channel models can be constructed and implemented in hard- or software. Software models are particularly useful since a realization of a random channel process may be reproduced arbitrarily often, whereby a comparison between different receivers is possible even if simulation time is limited, and it is possible to emulate a wide range of well-defined channel conditions, in particular worst-case conditions that occur very rarely in nature.

11.3.1 Direct Filtering Approach to Selective Fading Channel Simulation

There are several possibilities of constructing a simulation model of a fading channel being characterized by the scattering function. Depending on the type of multipath ray clustering (Figure 11-7), one or the other approach is more suitable. The *direct* filtering approach is suited for selective fading channels which exhibit no weak clustering in extended regions $\mathcal{R}_m$ but a finite number M of clusters in oblong or point regions $\mathcal{R}_m$ having *distinct delays* τ_m (Figure 11-7). Each such cluster process $c_m(t)$ is best modeled individually, and all contributions are then superponed to form a realization of the random process of the time-variant *effective* partial channel impulse response $h^{(i)}_{\varepsilon,n;k}$, sampled at rate $1/T$ in both the τ and time domains. In the context of channel simulation, it is often more appropriate to first generate a realization of the sequence of $T_s = (T/2)$-spaced channel impulse responses at sample rate $1/T_s = 2/T$ and then split the CIRs so obtained into partial CIRs to be used for T-spaced filtering of the symbol data stream:

$$\begin{aligned} h_{\varepsilon,\nu;l} &= h_\varepsilon(\tau=\nu T_s;\ t=lT_s) \\ &= h_\varepsilon(\tau=[\nu/2]T;\ t=[l/2]T) \\ &= \begin{cases} h^{(0)}_{\varepsilon,n=[\nu/2];\ k=[l/2]} & (\nu \text{ even}) \\ h^{(1)}_{\varepsilon,n=[(\nu-1)/2];\ k=[(l-1)/2]} & (\nu \text{ odd}) \end{cases} \\ h^{(i)}_{\varepsilon,n;k} &= h_\varepsilon(\tau=[n+i/2]T;\ t=[k+i/2]T) \\ &= h_{\varepsilon,\nu=[2n+i];\ l=[2k+i]} \end{aligned} \tag{11-55}$$

with tap indices $\nu = 2n+i$ and time indices $l = 2k+i$.[3] From eqs. (11-7) and (11-17) we then have

$$\begin{aligned} h_{\varepsilon,\nu;l} &= \sum_{m=0}^{M-1} \underbrace{c_m\left(t=\frac{l}{2}T\right)}_{c_{m;l}} \cdot \underbrace{g_T\left(\tau=\frac{\nu}{2}T-\varepsilon T-\tau_m\right)}_{g_{T\varepsilon,\nu}(\tau_m)} \\ &= \sum_{m=0}^{M-1} c_{m;l}\ g_{T\varepsilon,\nu}(\tau_m) \\ h^{(i)}_{\varepsilon,n;k} &= \sum_{m=0}^{M-1} \underbrace{c_m(t=[k+i/2]T)}_{c^{(i)}_{m;k}}\ \underbrace{g_T(\tau=[n+i/2]T-\varepsilon T-\tau_m)}_{g^{(i)}_{T\varepsilon,n}(\tau_m)} \\ &= \sum_{m=0}^{M-1} c^{(i)}_{m;k}\ g^{(i)}_{T\varepsilon,n}(\tau_m) \end{aligned} \tag{11-56}$$

The samples $h_{\varepsilon,\nu;l}$ of $T_s = (T/2)$-spaced channel taps may be collected in the CIR vector:

$$\begin{aligned} \mathbf{h}_{\varepsilon;l} &= (\ldots\ h_{\varepsilon,-1;l}\ \ h_{\varepsilon,0;l}\ \ h_{\varepsilon,1;l}\ \ldots)^T \\ &= \sum_{m=0}^{M-1} c_{m;l}\ \underbrace{(\ldots\ g_{T\varepsilon,-1}(\tau_m)\ \ g_{T\varepsilon,0}(\tau_m)\ \ g_{T\varepsilon,1}(\tau_m)\ \ldots)^T}_{\mathbf{g}_{T\varepsilon}(\tau_m)=\mathbf{g}_{T\varepsilon,m}} \\ &= \sum_{m=0}^{M-1} c_{m;l}\ \mathbf{g}_{T\varepsilon,m} \end{aligned} \tag{11-57}$$

---

[3] For simplicity, indices ν and l are taken to refer to the $T_s = (T/2)$-spaced channel, while n and k refer to the T-spaced channel model.

where the vector $\mathbf{g}_{T\varepsilon,m}$ represents the delayed (by $\varepsilon T + \tau_m$) and sampled $(T/2)$-spaced partial transmitter filter impulse response.

For practical reasons, it is advisable to separate the effects of differential delays τ_m and timing delays εT in the channel simulator, i.e., the digital equivalent of $h_\varepsilon(\tau;t) = h(\tau;t)\ \delta(\tau - \varepsilon T)$ [eq. (11-7)] should be implemented as the cascade of two filters, the first (time-variant) filter $h_{\nu;l}$ representing the effective channel $h(\tau;t)$, taking into account differential ray delays τ_m only, and the second (fixed) filter $h_{I\varepsilon,\nu}$ (digital interpolator, see Chapters 4 and 9) representing a delay element:

$$
\begin{aligned}
h_{\varepsilon,\nu;l} &= \sum_{m=0}^{M-1} c_{m;l}\ \underbrace{\left[g_{T,\nu}(\tau_m) * \mathrm{si}(\pi[\nu - 2\varepsilon])\right]}_{g_{T\varepsilon,\nu}(\tau_m)} \\
&= \underbrace{\left(\sum_{m=0}^{M-1} c_{m;l}\ g_{T,\nu}(\tau_m)\right)}_{h_{\nu;l}} * \underbrace{\mathrm{si}(\pi[\nu - 2\varepsilon])}_{h_{I\varepsilon,\nu}} \\
&= h_{\nu;l} * h_{I\varepsilon,\nu}
\end{aligned}
\tag{11-58}
$$

The channel tap processes to be modeled by the channel simulator kernel are thus collected in the CIR vector:

$$
\begin{aligned}
\mathbf{h}_l &= (\dots\ h_{-1;l}\ \ h_{0;l}\ \ h_{1;l}\ \dots)^T \\
&= \sum_{m=0}^{M-1} c_{m;l}\ \underbrace{(\dots\ g_{T,-1}(\tau_m)\ \ g_{T,0}(\tau_m)\ \ g_{T,1}(\tau_m)\ \dots)^T}_{\mathbf{g}_T(\tau_m) = \mathbf{g}_{T,m}} \\
&= \sum_{m=0}^{M-1} c_{m;l}\ \mathbf{g}_{T,m}
\end{aligned}
\tag{11-59}
$$

where $\mathbf{g}_{T,m}$ represents the delayed (by τ_m) and sampled $(T/2)$-spaced partial transmitter filter impulse response.

We remark that the channel simulator of eq. (11-59) – as well as the other types of simulators (operating at sample rate $1/T_s$) discussed below – is also appropriate for filtering *nonlinearly* modulated signals $s_l = s(t = lT_s)$. Then the pulse shaping filter $G_T(\omega)$ — whose delayed replicas are represented by $\mathbf{g}_{T,m}$ — must be chosen such that the filter is transparent (flat frequency response) within the bandwidth B_s of the signal power spectrum $S_s(\omega)$, i.e., $(1-\alpha)/T \geq B_s$ [assuming $1/T_s = 2/T$, see eq. (11-8)].

In any realizable simulator, $\mathbf{h}_l$ must be truncated to some finite-dimensional CIR vector. Care should be taken with such truncation since all relevant taps to the left and right of the impulse response (see Figure 11-4) are to be included in the CIR vector; this is especially important in the case of small bandwidth expansion factors α where the root-raised cosine impulse response $g_{T,\nu}(\tau_m) = g_T(\tau = [\nu/2]T - \tau_m)$ [eq. (11-8)] has strong pre- and postcursors about its center of gravity (delay τ_m).

While the vectors $\mathbf{g}_{T,m}$ are fixed once α and the set of M cluster delays $\{\tau_m\}$ have been set, realizations of the time-variant cluster processes $c_{m;l}$ (rate $2/T$) must correctly represent the fading statistics, i.e., the density function (pdf) and power spectrum (psd) should match the cluster pdf and the discrete-time cluster Doppler spectrum $S_{c_m}(\psi'')$, respectively, where $\psi'' = \psi\, T_s = \psi\, (T/2)$ denotes the Doppler frequency variable normalized to the *sample* rate $1/T_s = 2/T$.[4]

In the case of *diffuse* scattering, $c_{m;l}$ is a *Rayleigh* fading complex-Gaussian random process. The *direct filtering* simulator, also termed *quadrature amplitude modulation fading simulator* [20], then consists of a number of branches where each process $c_{m;l}$ is simulated by applying complex white Gaussian noise (WGN) $w_{m;l}$ of unity power to a digital filter with z-transform $T_m(z)$ (the filter output is still complex Gaussian, hence the fading is Rayleigh-distributed) whose squared spectrum $T_m(z = e^{j\psi''})$ matches or at least approximates the (normalized) cluster Doppler spectrum $S_{c_m}(\psi'')$ (Wiener-Lee theorem [22]):

$$T_m\left(z = e^{j\psi''}\right) \overset{!}{\simeq} \sqrt{S_{c_m}(\psi'')} \tag{11-60}$$

Whenever one or more of the diffuse cluster parameters $\rho_m = R_{c_m}(0)$, Λ_{D_m}, or σ_{D_m} change, the filter $T_m(z)$ of eq. (11-60) must be redesigned. This can be circumvented – and thus the simulator made more versatile – by using a *fixed* unity-energy filter matched to the *shape* of the cluster Doppler spectrum. In order to be able to adjust to variable Doppler shifts and spreads, this filter is designed for zero Doppler shift and some Doppler spread $\sigma_0 \gg \sigma_D$ being much larger than any Doppler that occurs in practice. As a side effect, this also improves on the numerical stability thanks to the larger filter bandwidth. The actual cluster process $c_{m;l}$ is then simulated by first driving $T_m(z)$ by complex WGN $w_{m;l}$ of power ρ_m, then scaling down the tap Doppler spread by linearly interpolating the filter output (interpolation rate factor $I_m = \sigma_0/\sigma_{D_m}$) and finally multiplying by the tap Doppler shift phasor $e^{j2\pi\Lambda''_{D_m} l}$.

The fading processes of the LOS ray or a point cluster of *specularly* reflected rays can be modeled directly by *phase modulation fading simulator* branches [20] where the ray weights $c_{m;l}$ are represented by rotating phasors $c_{m;l} = \alpha_m e^{j(\psi''_{D_m} l + \theta_m)}$ with nonfading amplitudes α_m, relative Doppler shifts $\psi''_{D_m} = 2\pi\Lambda''_{D_m}$ [eq. (11-30)], and uniformly distributed phases θ_m.

The structure of the direct filtering simulator for a total of M LOS/specular and diffuse clusters with distinct delays τ_m is visualized in Figure 11-9. The T_s-spaced cluster or ray processes $c_{m;l}$ – which have been generated by either a quadrature amplitude or a phase modulation fading simulator branch – are used for weighting the respective delayed and sampled pulse shaping filter vectors $\mathbf{g}_{T,m}$. All M such weighted vectors are finally superponed according to eq. (11-59) to

[4] Here, all double-primed quantities are defined to be normalized to sample rate $1/T_s = 2/T$.

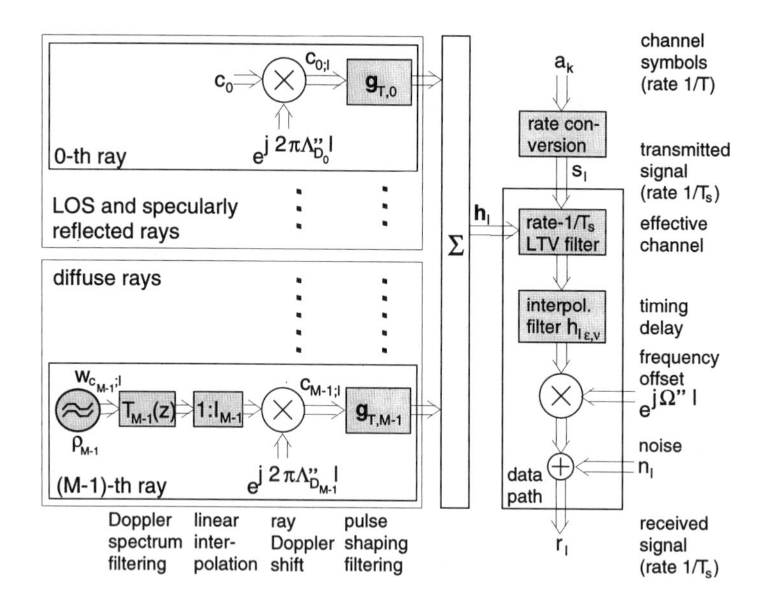

form the effective CIR vector $\mathbf{h}_l$. In the data path of the simulator, the channel symbols a_k are first converted to the T_s-spaced transmitted signal

$$s_l = \begin{cases} a_{k=l/2} & l \text{ even} \\ 0 & l \text{ odd} \end{cases} \tag{11-61}$$

[digital equivalent of Dirac pulse train $a_k\ \delta(t-kT)$; linear modulation] and then convolved with $\mathbf{h}_l$ in a linear time-variant (LTV) filter. The timing delay εT is taken care of by the digital interpolation filter $h_{I\varepsilon,\nu}$ [eq. (11-58)]. If present, a global oscillator frequency shift Ω is simulated by multiplying the output signal by the phasor $e^{j\Omega'' l}$. Finally, the noise process n_l is added to yield the received signal:

$$\begin{aligned} r_l &= e^{j\Omega'' l} \left(\sum_\nu s_{l-\nu} \underbrace{\left[h_{\nu;l} * h_{I\varepsilon,\nu}\right]}_{h_{\varepsilon,\nu;l}} \right) + n_l \\ &\simeq e^{j\Omega'' l} \left(s_l * h_{\nu;l} * h_{I\varepsilon,\nu} \right) + n_l \end{aligned} \tag{11-62}$$

which may be split into partial signals $r_k^{(i)}$ [eq. (11-16)], if necessary.

In the case of *flat fading* with imperfect timing, pulse matched filtering and small frequency offset $\Omega' \ll 1$, the channel model of eq. (11-20) applies, and it therefore suffices to simulate the $T/2$-spaced MD process $c_{\Omega,l} = e^{j\Omega'' l}\ c_l$. If the timing error has been perfectly compensated for, the simple model of eq. (11-22) remains so that only the T-spaced MD process $c_k = c_k^{(0)} = c_{l=2k}$ has to be generated by the simulator. In the case that the MD process is Rician, the MD may be modeled as the superposition of a specular/LOS and a diffuse component according to eq. (11-11).

Of course, there are many variations on direct filtering channel simulation. For instance, a continuous cluster Doppler spectrum $S_{c_m}(\psi'')$ – which can be viewed as the *average* distribution of Doppler shifts – may reduce to a peak with an *instantaneous* Doppler shift Λ''_{D_m} that is essentially constant over the duration of very short time frames of interest, typically TDMA bursts in mobile radio having a spatial duration of a few wavelengths. Then diffuse Rayleigh-distributed fading clusters may be represented by phase modulation fading simulator branches as if they were specularly reflected point clusters. The sequence of Doppler shifts $\{\psi''_{D_m}\}$ taken to be valid for the transmission bursts can then be generated by Monte Carlo simulation as a realization of a random sequence with pdf $p_m(\psi'') \propto S_{c_m}(\psi'')$.

In practice, the intended application may necessitate the channel simulator to be modified. For example, when the received signal is undersampled (rate $1/T$ instead of $2/T$; entails a loss of information due to aliasing but may be appropriate in low-noise environments), only a T-spaced (partial) channel simulator needs to be implemented. In many receivers, the analog anti-aliasing filter $F(\omega)$ is lumped together with the pulse matched filter $g_{\mathrm{MF}}(\tau) \circ\!\!-\!\!\bullet G_{\mathrm{MF}}(\omega)$ (at IF band or baseband) for improved noise and interference rejection [$G_{\mathrm{MF}}(\omega)$ is band-limited with bandwidth B while, in general, $F(\omega)$ may be more wideband]. In that case, the pulse MF must be included into the filter vectors, i.e., the $\mathbf{g}_{T,m}$ (Figure 11-9) are to be replaced by $\mathbf{g}_m$ representing the cascade of transmitter and pulse MF responses with elements $g_{m,\nu} = g_{T,\nu} * (1/T) g^*_{T,-\nu}$.

When focusing on the *outer* transmission system (e.g., designing a coding scheme or assessing equalizer and decoder performance), another interesting channel simulator may be devised by including the optimum (channel-dependent) receiver filter into the channel model. Depending on the kind of equalizer, this optimum filter is either the channel matched filter (Ungerboeck's receiver [23]) or the channel whitening matched filter (Forney's canonical receiver [24] being discussed in Section 13.3). Then the tap fading processes of the resulting T-spaced *equivalent* channel [25, 26], as seen by the equalizer or decoder (Section 13.3.3), can also be simulated by filtering. Since this simulator is based on perfect channel estimation and ideal whitening (matched) filtering, it is not applicable to the development and performance evaluation of synchronizers (channel estimators) and nonideal channel-dependent receiver filters.

11.3.2 Transformed Filtering Approach to Selective Fading Channel Simulation

The *direct* filtering channel simulator concept is well suited for channels that exhibit but a few multipath clusters with distinct delays τ_m. For some channels, however, the number M of such clusters is large, or a virtually infinite number

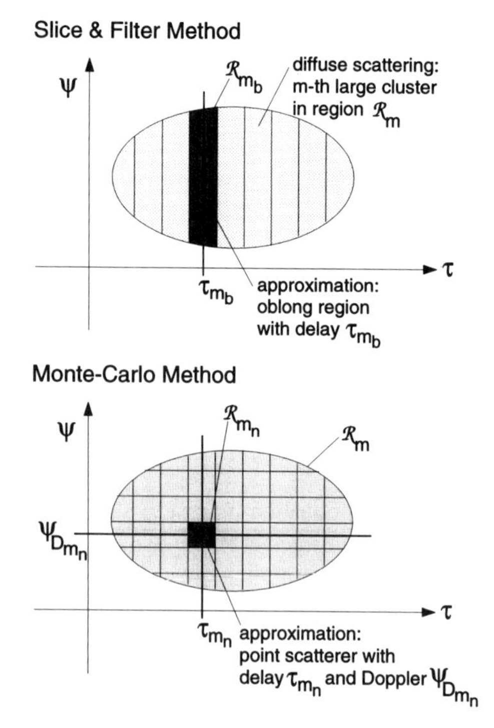

Figure 11-10 Methods of Simulating Large Scattering Clusters

of scattered rays form a large cluster in the (τ, ψ)-plane (Figure 11-7) so that the cluster scattering function $S_{c_m}(\tau; \psi)$ is nonzero in an extended region $\mathcal{R}_m$. The corresponding cluster spaced-time delay correlation $R_{c_m}(\tau; \Delta t)$ and thus the cluster delay profile $R_{c_m}(\tau)$ become quasi-continuous in the delay variable τ. Obviously, the direct filtering simulator is not readily applicable; a different representation of such clusters is needed.

Methods of Simulating Large Clusters of Scattered Rays

One possible solution of simulating large clusters consists in approximating $S_{c_m}(\tau; \psi)$ by artificially "discretizing" $S_{c_m}(\tau; \psi)$ in τ direction, i.e., chopping $S_{c_m}(\tau; \psi)$ into a number B of slices $S_{c_{m_b}}(\psi) = \int_{\mathcal{R}_{m_b}} S_{c_m}(\tau; \psi) d\tau$ where $\mathcal{R}_{m_b}$ is the bth bin with center delay τ_{m_b}. This is illustrated in Figure 11-10 (upper part). The so formed artificial oblong "clusters" with Doppler spectra $S_{c_{m_b}}(\psi)$ can now be simulated by generating WGN processes $w_{c_{m_b},l}$ with powers $\rho_{m_b} = \int S_{c_{m_b}}(\psi) d\psi$, followed by filtering [filters $T_{m_b}(z)$ approximating $\sqrt{S_{c_{m_b}}(\psi'')}$] in fading simulator branches just as in the direct filtering simulator of Figure 11-9. Obviously, there exists a trade-off between simulator complexity and accuracy: fine-grain cluster partitioning necessitates quite a large number of equivalent "clusters" to be implemented, while coarse-grain partitioning may lead to the fading statistics no longer being represented correctly.

Another representation of fading channels with extended clusters can be devised by interpreting the scattering function as a measure of the *probability density* of scattered rays in the (τ, ψ)-plane, i.e., $p(\tau; \psi) \propto S_c(\tau; \psi)$. This is readily understood by recalling that each elementary ray manifests itself as a point in the delay Doppler plane with infinitesimaly small power ξ_n^2, Doppler shift ψ_{D_n}, delay τ_n, and random phase θ_n [see eq. (11-28)]. In the so-called *Monte-Carlo model* – an example is illustrated in Figure 11-10 (lower part) – the multitude of N_m elementary ray processes that make up the quasi-continuous two-dimensional cluster scattering function $S_{c_m}(\tau; \psi)$ is replaced by a *finite* number $\overline{N}_m$ of discrete processes $c_{m_n,l} = \alpha_m \; e^{j(\psi''_{D_{m_n}} l + \theta_{m_n})} = e^{j\psi''_{D_{m_n}} l} \; (\alpha_{m_n} \cdot e^{j\theta_{m_n}})$ with delays τ_{m_n}. Each such path is modeled as if it were a specularly reflected ray; it is taken to represent a small region $\mathcal{R}_{m_n}$ about the point $(\tau_{m_n}, \psi_{D_{m_n}})$.

In the simulator, the set of path parameters $(\tau_{m_n}, \psi_{D_{m_n}}, \alpha_{m_n}, \theta_{m_n})$ ($n = 0, \ldots, \overline{N}_m - 1$ paths per cluster) is chosen *offline* before the simulation is started. There are many possible ways of parameter selection. For instance, the set of points $(\tau_{m_n}, \psi_{D_{m_n}})$ may be chosen randomly such that the resulting point scatter diagram reflects the clustering of rays in the (τ, ψ)-plane. Alternatively, one may select $(\tau_{m_n}, \psi_{D_{m_n}})$ in some deterministic way, e.g., by forming equidistant grids in the cluster regions $\mathcal{R}_m$ (Figure 11-10). Likewise, there are several strategies of choosing the cluster weights $(\alpha_{m_n} \; e^{j\theta_{m_n}})$ pertaining to $(\tau_{m_n}, \psi_{D_{m_n}})$. For example, the gains α_{m_n} may be matched to the scattering function in a deterministic way, e.g., by integrating $S_{c_m}(\tau; \psi)$ over the region $\mathcal{R}_{m_n}$. Alternatively, the gains may be chosen randomly as a realization of a complex Gaussian random variable whose average power $E[\alpha_{m_n}^2]$ matches the profile $S_{c_m}(\tau; \psi)$ in region $\mathcal{R}_{m_n}$. The uniformly distributed phases $\theta_{m_n} \in [0, 2\pi)$ are chosen randomly and independently from one another. Notice that, once the parameter set has been fixed, the simulated channel trajectory during runtime is in fact *deterministic*. It has been shown that a couple of hundered so-formed discrete processes yield good approximations to Rayleigh fading clusters [27, 28]. This simulator concept makes it very easy to switch between totally different channels (e.g., frequency hopping) just by selecting a new set of parameters $(\tau_{m_n}, \psi_{D_{m_n}}, \alpha_{m_n}, \theta_{m_n})$.

The Transformed Filtering Channel Simulator

In digital communications, only a *finite* number of $T_s = (T/2)$-spaced taps $h_{\nu;l} = h(\tau = \nu T_s; t = l T_s)$ of the discrete-equivalent CIR needs to be simulated. This motivates a different approach to selective fading channel simulaton, here termed *transformed filtering* approach, where it is desired to establish a mapping from a cluster scattering function $S_{c_m}(\tau; \psi)$ of the physical channel onto the statistics of the respective collection of sampled tap processes $h_{\nu;l}^{(m)}$. The resulting transformed filtering simulator thus circumvents the need for generating a large number of diffuse rays, yet does not compromise the correct representation of channel statistics, in particular, the rate and duration of deep fades. Depending on the ray clustering profile, the transformed model may lead to a substantial reduction in the number of fading simulator branches. This advantage, however, is paid for by a less flexible simulator whose coefficients must be calculated anew whenever the clustering profile changes. Hence, the transformed filtering approach is suited for simulating certain well-defined and widely used channel scenarios such as the hilly terrain (HT) or typical urban (TU) mobile channels as specified by CEPT [29, 30].

The derivation starts with expressing the temporal and intertap correlations of the processes $h^{(m)}_{\nu;l}$ pertaining to the mth extended cluster in terms of the cross-correlation function:[5]

$$\begin{aligned} R_{h_\nu h_{\nu+\mu}}(\Delta t=\Delta l\, T_s) &= E\left[h_{\nu;\,l}\, h^*_{\nu+\mu;\,l+\Delta l}\right] \\ &\underset{\left(\text{US property}\right)}{=} \sum_{n\in\mathcal{N}_m} E\left[c_{n;l}\, c^*_{n;l+\Delta l}\right] \left[g_T\left(\nu-\tau''_n\right) g_T\left(\mu+\left[\nu-\tau''_n\right]\right)\right] \\ &= \sum_{n\in\mathcal{N}_m} \left(\xi^2_n\right) e^{-j\psi_{D,n}(\Delta t=\Delta l T_s)} \left[g_T\left(\nu-\tau''_n\right) g_T\left(\mu+\left[\nu-\tau''_n\right]\right)\right] \end{aligned} \tag{11-63}$$

with $g_T(\tau'') = g(\tau/T_s)$. The link between the cross-correlation function $R_{h_\nu h_{\nu+\mu}}(\Delta t)$ and $S_c(\tau'';\psi)$ can be established via reformulating the cross-power spectrum:

$$\begin{aligned} S_{h_\nu h_{\nu+\mu}}(\psi) &= \int_{-\infty}^{\infty} R_{h_\nu h_{\nu+\mu}}(\Delta t)\, e^{-j\psi(\Delta t)}\, d(\Delta t) \\ &= \sum_{n\in\mathcal{N}_m} \left(\xi^2_n\right) \delta(\psi-\psi_{D_n}) \left[g_T\left(\nu-\tau''_n\right) g_T\left(\mu+\left[\nu-\tau''_n\right]\right)\right] \\ &= \int_{-\infty}^{\infty} \underbrace{\sum_{n\in\mathcal{N}_m} \left(\xi^2_n\right) \delta(\psi-\psi_{D_n})\, \delta\left(\tau''-\tau''_n\right)}_{S_c(\tau'';\psi)} \left[g_T(\nu-\tau'')\, g_T(\mu+[\nu-\tau''])\right] d\tau'' \\ &= \int_{-\infty}^{\infty} \left[g_T(\nu-\tau'')\, g_T(\mu+[\nu-\tau''])\right] S_c(\tau'';\psi)\, d\tau'' \end{aligned} \tag{11-64}$$

Since the fading is assumed to be slow with respect to the symbol rate so that $S_c(\tau'';\psi)$ is nonzero only for small $\psi \ll \pi(1/T_s)$, the spectrum $S_{h_\nu h_{\nu+\mu}}(e^{j\psi''})$ of the *sampled* autocorrelation $R_{h_\nu h_{\nu+\mu}}(\Delta l) = R_{h_\nu h_{\nu+\mu}}(\Delta t = \Delta l T_s)$ satisfies $S_{h_\nu h_{\nu+\mu}}(e^{j\psi''}) = S_{h_\nu h_{\nu+\mu}}(\psi'' = \psi T_s)$ with high accuracy. Transforming

[5] In order to avoid notational overloading, the cluster index m is dropped in this derivation.

$S_{h_\nu h_{\nu+\mu}}(e^{j\psi''})$ back to the time domain yields

$$
\begin{aligned}
R_{h_\nu h_{\nu+\mu}}(\Delta l) &= E\left[h_{\nu;\,l}\, h^*_{\nu+\mu;\,l+\Delta l}\right] \\
&= \frac{1}{2\pi} \int\limits_{-\pi}^{\pi} S_{h_\nu h_{\nu+\mu}}\left(e^{j\psi''}\right) e^{j\psi''(\Delta l)}\, d\psi'' \\
&= \frac{1}{2\pi} \int\limits_{-\pi}^{\pi} \int\limits_{-\infty}^{\infty} [g_T(\nu-\tau'')\, g_T(\mu + [\nu-\tau''])]\, S_c\left(\tau''; e^{j\psi''}\right) e^{j\psi''(\Delta l)}\; d\tau''\, d\psi''
\end{aligned}
\tag{11-65}
$$

The third line in eq. (11-65) involves the (cluster) scattering function $S_c(\tau''; e^{j\psi''})$ of the T_s-spaced channel which also satisfies $S_c(\tau''; e^{j\psi''}) \approx S_c(\tau''; \psi'' = \psi T_s)$ with high accuracy.

The sampled cross-correlation function of eq. (11-65) as well as the cross-power spectrum may be split into two terms:

$$
\begin{aligned}
R_{h_\nu h_{\nu+\mu}}(\Delta l) &= \rho_{\nu,\nu+\mu}\; \alpha_{\nu,\nu+\mu}(\Delta l) \\
S_{h_\nu h_{\nu+\mu}}\left(e^{j\psi''}\right) &= \rho_{\nu,\nu+\mu}\; \alpha_{\nu,\nu+\mu}\left(e^{j\psi''}\right)
\end{aligned}
\tag{11-66}
$$

respectively, where $\rho_{\nu,\nu+\mu} = R_{\nu,\nu+\mu}(\Delta l = 0)$ is the cross-correlation between taps and $\alpha_{\nu,\nu+\mu}(\Delta l)$ the temporal cross-correlation profile normalized to unity, i.e., $\alpha_{\nu,\nu+\mu}(\Delta l = 0) = (1/[2\pi]) \int_{-\pi}^{\pi} \alpha_{\nu,\nu+\mu}(e^{j\psi''})\; d(\psi'') = 1$.

In principle, $R_{h_\nu h_{\nu+\mu}}(\Delta l) \circ\!\!-\!\!\bullet S_{h_\nu h_{\nu+\mu}}(e^{j\psi''})$ can be evaluated numerically for any kind of (cluster) scattering function $S_c(\tau''; e^{j\psi''})$ via eqs. (11-65) and (11-64). In general, however, all intertap cross-power spectra $S_{h_\nu h_{\nu+\mu}}(e^{j\psi''})$ may have distinct profiles $\alpha_{\nu,\nu+\mu}(e^{j\psi''})$, thus rendering their modeling quite an involved task.

For this reason, let us consider the important *special case* of the cluster scattering function being *separable* in the delay and Doppler domains, i.e.,

$$
S_c\left(\tau''; e^{j\psi''}\right) \underset{\left(\substack{\text{special case:} \\ S_c \text{ separable}}\right)}{=} R_c(\tau'')\, \alpha_c\left(e^{j\psi''}\right)
\tag{11-67}
$$

where $\alpha_c(\Delta l) \circ\!\!-\!\!\bullet \alpha_c(e^{j\psi''})$ is again normalized to unity energy. The tap cross-correlation functions and cross-power spectra of eq. (11-66) then simplify to

$$
\begin{aligned}
R_{h_\nu h_{\nu+\mu}}(\Delta l) &= \rho_{\nu,\nu+\mu}\, \alpha_c(\Delta l) \\
S_{h_\nu h_{\nu+\mu}}\left(e^{j\psi''}\right) &= \rho_{\nu,\nu+\mu}\, \alpha_c\left(e^{j\psi''}\right)
\end{aligned}
\tag{11-68}
$$

The tap cross-correlation coefficients $\rho_{\nu,\nu+\mu}$ can be determined via eqs. (11-64) and (11-67):

$$
\begin{aligned}
S_{h_\nu h_{\nu+\mu}}\left(e^{j\psi''}\right) &= \int_{-\infty}^{\infty} \left[g_T(\nu-\tau'')\, g_T(\mu+[\nu-\tau''])\right] S_c(\tau'';\psi'')\ d\tau'' \\
&= \left[\int_{-\infty}^{\infty} \left[g_T(\nu-\tau'')\, g_T(\mu+[\nu-\tau''])\right] R_c(\tau'')\ d\tau''\right] \alpha_c\left(e^{j\psi''}\right)
\end{aligned}
\tag{11-69}
$$

Comparison with eq. (11-68) immediately yields

$$
\rho_{\nu,\nu+\mu} = \int_{-\infty}^{\infty} \left[g_T(\nu-\tau'')\, g_T(\mu+[\nu-\tau''])\right] R_c(\tau'')\ d\tau'' \tag{11-70}
$$

As expected, $\rho_{\nu,\nu+\mu}$ is a function of the pulse shaping filter response $g_T(\tau'')$ and the delay power profile $R_c(\tau'')$.

Let us now arrange the tap cross-correlation functions and cross-power spectra (pertaining to the mth cluster) in infinite-dimensional matrices:

$$
\mathbf{R}_h(\Delta l) = \begin{pmatrix} \ddots & & \vdots & & \\ & R_{h_{-1}h_{-1}}(\Delta l) & R_{h_{-1}h_0}(\Delta l) & R_{h_{-1}h_1}(\Delta l) & \\ \cdots & R_{h_0 h_{-1}}(\Delta l) & R_{h_0 h_0}(\Delta l) & R_{h_0 h_1}(\Delta l) & \cdots \\ & R_{h_1 h_{-1}}(\Delta l) & R_{h_1 h_0}(\Delta l) & R_{h_1 h_1}(\Delta l) & \\ & & \vdots & & \ddots \end{pmatrix}
$$

$$
\mathbf{S}_h\left(e^{j\psi''}\right) = \begin{pmatrix} \ddots & & \vdots & & \\ & S_{h_{-1}h_{-1}}\left(e^{j\psi''}\right) & S_{h_{-1}h_0}\left(e^{j\psi''}\right) & S_{h_{-1}h_1}\left(e^{j\psi''}\right) & \\ \cdots & S_{h_0 h_{-1}}\left(e^{j\psi''}\right) & S_{h_0 h_0}\left(e^{j\psi''}\right) & S_{h_0 h_1}\left(e^{j\psi''}\right) & \cdots \\ & S_{h_1 h_{-1}}\left(e^{j\psi''}\right) & S_{h_1 h_0}\left(e^{j\psi''}\right) & S_{h_1 h_1}\left(e^{j\psi''}\right) & \\ & & \vdots & & \ddots \end{pmatrix}
\tag{11-71}
$$

Thanks to the separation property [eqs. (11-67) and (11-68)], these matrices simplify to

$$
\begin{aligned}
\mathbf{R}_h(\Delta l) &= \mathbf{R}_h \cdot \alpha_c(\Delta l) \\
\mathbf{S}_h\left(e^{j\psi''}\right) &= \mathbf{R}_h \cdot \alpha_c\left(e^{j\psi''}\right)
\end{aligned}
\tag{11-72}
$$

where

$$\mathbf{R}_h = \begin{pmatrix} \ddots & & \vdots & & \\ & \rho_{-1,-1} & \rho_{-1,0} & \rho_{-1,1} & \\ \cdots & \rho_{0,-1} & \rho_{0,0} & \rho_{0,1} & \cdots \\ & \rho_{1,-1} & \rho_{1,0} & \rho_{1,1} & \\ & & \vdots & & \ddots \end{pmatrix} \tag{11-73}$$

is the (infinite-dimensional) symmetric covariance matrix of the (likewise infinite-dimensional) discrete-equivalent channel tap vector

$$\begin{aligned} \mathbf{R}_h = \mathrm{cov}[\mathbf{h}_l] &= E\left[\mathbf{h}_l \cdot \mathbf{h}_l^H\right] \\ &= E\left[\begin{pmatrix} \vdots \\ h_{-1;l} \\ h_{0;l} \\ h_{1;l} \\ \vdots \end{pmatrix} \begin{pmatrix} \cdots & h^*_{-1;l} & h^*_{0;l} & h^*_{1;l} & \cdots \end{pmatrix}\right] \end{aligned} \tag{11-74}$$

A cluster scattering function representing a number N of elementary rays leads to a rank-N covariance matrix $\mathbf{R}_h$. In extended clusters of diffusely scattered rays we have $N \longrightarrow \infty$ so that $\mathbf{R}_h$ is of infinite rank. The average powers of tap processes $h_{\nu;l}$, viz. $\rho_\nu = \rho_{\nu,\nu} = E[| h_{\nu;l} |^2]$ are arranged along the main diagonal of $\mathbf{R}_h$. The off-diagonal elements $\rho_{\nu,\nu+\mu}$ of $\mathbf{R}_h$ reflect the mutual coupling of the sampled tap processes introduced by pulse shaping.

In order to synthesize channel tap processes $h_{\nu;l}$ with these statistical properties in an efficient way, consider the unitary spectral decomposition of $\mathbf{R}_h$:

$$\mathbf{R}_h = \mathbf{U} \cdot \mathbf{\Lambda} \cdot \mathbf{U}^T \quad = \sum_p \lambda_p \cdot \mathbf{u}_p \mathbf{u}_p^T \tag{11-75}$$

with diagonal eigenvalue matrix $\mathbf{\Lambda} = \mathrm{diag}(\lambda_0, \lambda_1, \ldots)$ and unitary eigenvector matrix $\mathbf{U} = (\mathbf{u}_0\ \mathbf{u}_1\ \ldots)$. From this decomposition it follows that the set of *correlated* complex Gaussian random processes $\{h_{\nu;l}\}$ (collected in vector $\mathbf{h}_l$) can be generated by filtering a set of *uncorrelated* complex Gaussian processes $\{p_{p;l}\}$ [collected in vector $\mathbf{p}_l = (p_{0;l}\ p_{1;l}\ \ldots)^T$] by means of a unitary filtering network $\mathbf{U}$, i.e.,

$$\mathbf{h}_l = \mathbf{U} \cdot \mathbf{p}_l \tag{11-76}$$

The processes $p_{p;l}$ have average powers λ_p so that $\mathrm{cov}[\mathbf{p}_l] = \mathbf{\Lambda}$ and therefore

$$\begin{aligned} \mathrm{cov}[\mathbf{h}_l] &= E\left[\mathbf{h}_l \cdot \mathbf{h}_l^H\right] \\ &= E\left[\mathbf{U} \cdot \mathbf{p}_l \cdot \mathbf{p}_l^H \cdot \mathbf{U}^T\right] \\ &= \mathbf{U} \cdot \mathbf{\Lambda} \cdot \mathbf{U}^T \\ &= \mathbf{R}_h \end{aligned} \tag{11-77}$$

as desired.

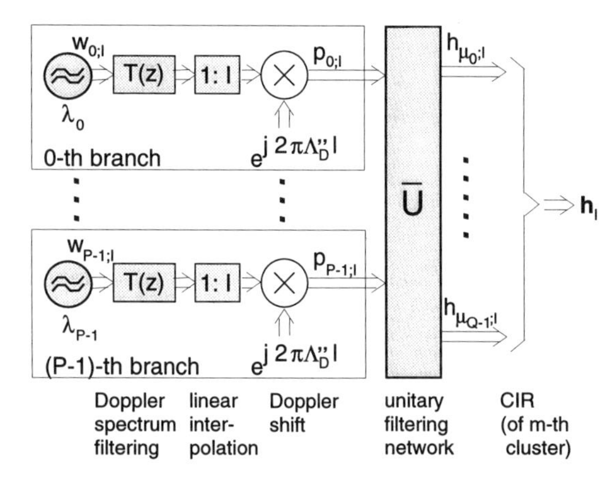

Figure 11-11 Transformed Filtering Selective Fading Channel Simulator (mth Cluster)

In a practical simulator, vectors $\mathbf{p}_l$ and $\mathbf{h}_l$ must be truncated to some finite dimension, i.e.,

$$\begin{aligned}\overline{\mathbf{p}}_l &= \left(p_{0;l}\ \ p_{1;l}\ \ \ldots\ \ p_{P-1;l}\right)^T \\ \overline{\mathbf{h}}_l &= \left(h_{\nu_0;l}\ \ h_{\nu_1;l}\ \ \ldots\ \ h_{\nu_{Q-1};l}\right)^T\end{aligned} \tag{11-78}$$

with dimensions P and Q, respectively. The number P of processes $p_{p;l}$ should be chosen such that their average powers, contained in the respective truncated eigenvalue matrix $\overline{\mathbf{\Lambda}} = \text{diag}(\lambda_0, \ldots, \lambda_{P-1})$, are much larger than the discarded eigenvalues. Likewise, the number Q of tap processes $h_{\nu_q;l}$ to be synthesized should be chosen such that their tap powers $\rho_{\nu_q} = \rho_{\nu_q,\nu_q}$, contained along the main diagonal of the respective truncated correlation matrix $\overline{\mathbf{R}}_h$, are much larger than the powers of "weak" taps disregarded by the simulator.

The resulting transformed filtering simulator for a channel cluster whose cluster scattering function can be separated according to eq. (11-67) is shown in Figure 11-11.

The data path (not included in the figure) is the same as in Figure 11-9. The P processes $p_{p;l}$ with powers λ_p are generated by filtering mutually independent WGN processes $w_{p;l}$ of powers λ_p by unity-energy Doppler spectrum shaping filters $T(z)$ approximating the widened (interpolation factor I) cluster Doppler spectrum $S_c(e^{j(\psi'' I)})$ that may have been shifted in frequency by its negative Doppler shift $-\psi''_{D_m}$ such that it is roughly centered about the origin. Backshifting the cluster Doppler spectrum by $\psi''_{D_m} = 2\pi\Lambda''_{D_m}$ may then be accomplished following linear interpolation (rate factor I). Realizations of processes $p_{p;l}$ thus obtained are delivered to the $Q \times M$ filtering matrix $\overline{\mathbf{U}}$ whose output processes are statistically correct realizations of relevant channel tap processes $h_{\nu_q;l}$. If necessary, the Q-tap vector $\mathbf{h}_l$ so obtained may again be split into two vectors $\mathbf{h}_k^{(i)}$ containing the taps of the two partial channels. When more than one cluster is present, all M cluster CIRs $\mathbf{h}_l^{(m)}$ generated by filtering simulator branches are finally superponed to form the realization $\mathbf{h}_l$ of the total CIR process.

Example: Transformed Filtering Simulation of Land-Mobile Channels

As an example of the transformed filtering simulator, consider the land-mobile channel model whose cluster scattering functions can often be separated according to eq. (11-67). In mobile radio, clusters with small delays τ'' tend to exhibit a U-shaped Jakes Doppler spectrum $\alpha_c(e^{j\psi''})$ [eq. (11-53)], while clusters with large delays are likely to have more pronounced spectra, e.g., Gaussian-shaped spikes.

In the following examples, the CIR is sampled in τ direction at rate $1/T_s$=270 kHz. In contrast to the GSM recommendations [partial response Gaussian minimum shift keying (GMSK) signal: rate $1/T = 1/T_s$=270 kHz, bandwidth $\approx$200 kHz], we assume *linear* (PSK or QAM) modulation with root-raised cosine Nyquist filtering ($\alpha = 0.5$) and a symbol rate of $1/T$=135 kBd (2 bits per symbol, signal bandwidth $B = (1+\alpha)(1/T)\approx$200 kHz) [31]. As pointed out earlier, this simulator can also be used for nonlinearly modulated signals occupying a bandwidth not exceeding that of the flat region of the pulse matched filter $g_{\mathrm{MF}}(\tau)\circ\!\!-\!\!\bullet G_{\mathrm{MF}}(\omega)$, here $(1-\alpha)(1/T)\approx$77.5 kHz. By doubling the sampling rate and tightening the bandwidth factor to $\alpha = 0.25$, the channel simulator can be modified so as to accommodate the full GSM signal bandwidth of 200 Hz.

Let us consider the two types of delay-power profiles $R_c(\tau'')$ specified by CEPT for GSM channel modeling at 900 MHz [29, 30]. The first model features 12 (or 6) discrete densely spaced oblong Jakes clusters, and the second model is specified in terms of one or two continuous exponentials.

The average powers ρ_m and delays τ''_m of $M = 12$ oblong clusters of the first delay-power profile [30] are listed in Table 11-2 for the hilly terrain (HT) and typical urban (TU) GSM channels, along with the parameters (average tap powers ρ_{ν_q}, eigenvalues λ_p) of the respective transformed filtering models. Of course, the 12 clusters may well be modeled by direct filtering, but notice from Table 11-2 that, by virtue of transformed filtering, the number of filtering branches (Figure 11-11) is reduced from $M = 12$ to $P = 4$ and $P = 3$ for the HT and TU channels, respectively. Here, the matrices have been truncated such that spurious channel taps $h_{\nu_q;l}$ with powers ρ_{ν_q} below a threshold of 0.001 × the strongest tap power, as well as spurious eigenvalues $\lambda_p \leq 0.001 \times \lambda_0$ were eliminated. Notice that in high-noise environments the simulator can be further simplified by neglecting taps and eigenvalues whose contributions are buried in noise. For instance, by raising the threshold to 0.01, only (P, Q)=(4,9) and (2,6) filtering branches and taps remain to be implemented for the HT and TU channel simulators, respectively.

Continuous delay-power profiles $R_c(\tau'')$ as specified by COST [29] comprise M clusters of weighted exponentials:

$$R_c(\tau'') = \sum_{m=0}^{M-1} \underbrace{\rho_m \frac{1}{\overline{\tau}''_m} e^{-\left(\tau''-\tau''_m\right)/\overline{\tau}''_m} \epsilon\left(\tau'' - \tau''_m\right)}_{R_{c_m}(\tau'')} \tag{11-79}$$

Table 11-2 Parameters of Direct and Transformed Filtering GSM-HT and TU Channel Simulation Models

GSM-HT, $M=12$, Jakes Doppler		Transformed, $Q=11$, $P=4$			GSM-TU, $M=12$, Jakes Doppler		Transformed, $Q=9$, $P=3$		
ρ_m	τ''_m	ν_q	ρ_{ν_q}	λ_p	ρ_m	τ''_m	ν_q	ρ_{ν_q}	λ_p
0.0264	0.	-3	0.0006	0.8395	0.0904	0.	-4	0.0006	0.9474
0.0418	0.0542	-2	0.0094	0.1168	0.1138	0.0542	-3	0.0011	0.0532
0.0662	0.108	-1	0.0499	0.0323	0.2271	0.108	-2	0.0094	0.0027
0.1050	0.163	0	0.4739	0.0113	0.1433	0.163	-1	0.0853	
0.2640	0.217	1	0.3213		0.1138	0.217	0	0.5702	
0.2640	0.542	2	0.0095		0.0718	0.325	1	0.3111	
0.1050	0.615	3	0.0187		0.0454	0.379	2	0.0178	
0.0418	4.065	4	0.0703		0.0718	0,488	3	0.0067	
0.0332	4.119	5	0.0400		0.0571	0.650	4	0.0009	
0.0264	4.282	6	0.0054		0.0286	0.813			
0.0166	4.662	7	0.0009		0.0180	0.867			
0.0105	5.420				0.0227	1.355			

with average cluster power ρ_m, delay τ''_m [where $\epsilon(\tau'')$ denotes the unit step function], and time constant $\overline{\tau}''_m$ of the mth exponentially decaying cluster profile. In order to keep this example simple, the distinct cluster Doppler spectra (including Gaussian-shaped spikes) specified by COST are replaced by Jakes spectra. The cluster parameters are listed in Table 11-3 for the HT, TU, and BU (bad urban) channels according to COST [29], along with the average tap powers ρ_{ν_q} and eigenvalues λ_p of the respective transformed filtering simulators. Again, only $P = 4$ filtering branches and between $Q = 4$ and 12 taps need to be implemented to adequately represent these channels. In high-noise environments (threshold 0.01), these figures boil down to (P, Q)=(2,6), (2,4), and (3,7) for the COST-HT, TU, and BU channel simulators, respectively. This example illustrates the benefits, both in terms of modeling accuracy and simulator efficiency, that can be gained by employing the transformed filtering technique for modeling the effect of "critical" extended clusters in the scattering function.

11.3.3 Main Points

- System performance evaluation via *simulation* necessitates software channel simulators which accurately represent the statistics of a multipath fading channel. This may be accomplished by a number of techniques, including the direct and transformed filtering approaches which have been discussed in more detail. In the *direct filtering* model, each cluster fading process $c_m(t)$ with delay τ_m is modeled and generated individually. This technique is best suited to channels featuring a small number of clusters with distinct delays.
- Channels which exhibit a quasi-continuous delay profile due to diffuse scattering can be represented by a *transformed filtering* model, given that the

Table 11-3 Parameters of Transformed Filtering COST-HT and BU Channel Simulation Models

COST-HT, $M=2$, Jakes Doppler	Transformed, $Q=12$, $P=4$			COST-BU, $M=2$, Jakes Doppler	Transformed, $Q=10$, $P=4$		
	ν_q	ρ_{ν_q}	λ_p		ν_q	ρ_{ν_q}	λ_p
$R_{c_0}(\tau'')$: ρ_0=0.88, τ_0''=0, $\bar{\tau}_0''$=0.077 $R_{c_1}(\tau'')$: ρ_1=0.12, τ_1''=4.05, $\bar{\tau}_1''$=0.27	-4	0.0007	0.8726	$R_{c_0}(\tau'')$: ρ_0=0.67, τ_0''=0, $\bar{\tau}_0''$=0.27 $R_{c_1}(\tau'')$: ρ_1=0.33, τ_1''=1.35, $\bar{\tau}_1''$=0.27	-4	0.0004	0.7114
	-3	0.0016	0.1132		-3	0.0007	0.2666
	-2	0.0069	0.0065		-2	0.0065	0.0184
	-1	0.1162	0.0043		-1	0.0577	0.0019
	0	0.5592			0	0.3803	
	1	0.1849			1	0.3455	
	2	0.0043			2	0.1788	
	3	0.0113			3	0.0235	
	4	0.0659			4	0.0045	
	5	0.0423			5	0.0004	
	6	0.0024					
	7	0.0010					

COST-TU, $M=2$, Jakes Doppler	Transformed, $Q=4$, $P=4$		
	ν_q	ρ_{ν_q}	λ_p
$R_c(\tau'')$: ρ=1 τ''=0 $\bar{\tau}''$=0.27	-1	0.0082	0.7878
	0	0.7358	0.1120
	1	0.1543	0.0057
	2	0.0080	0.0012

(cluster) scattering function is separable in the delay and Doppler domains. The infinitely many elementary ray processes of such a cluster contribute to the discrete-equivalent channel tap vector process $\mathbf{h}_l$, and these contributions are synthesized through a *spectral decomposition* of the covariance matrix $\mathbf{R}_h$ of $\mathbf{h}_l$. This technique yields high modeling accuracy and improved simulator efficiency for some important channels of interest.

11.4 Bibliographical Notes

Meanwhile, a large body of literature exists on channel characterization and modeling. Material on flat fading channels being characterized by multiplicative distortion is found, e.g., in [32, 33, 6, 34]. Statistical modeling of short-term channel variations is detailed in [9, 35, 36, 19, 10], and deriving the scattering function from a physical channel model is the topic of [35, 37, 8].

For further information on microwave LOS channels, the reader is referred to [13, 14]. Land-mobile/personal radio channels, land-mobile satellite channels, and aeronautical satellite channels are covered in [35, 38, 25, 19, 39, 10], [40, 34, 41], and [42, 43], respectively. Material on ionospheric shortwave channels and the various sources of interference is found, e.g., in [44, 45, 16, 46, 17, 18] and [47, 48, 49, 50, 51]. Important characteristics of indoor radio channels are found in [52, 53, 54, 55]. When designing receivers for new emerging applications such as wireless ATM or terrestrial digital video broadcasting (DVB-T) [56], it is important to consult latest results on channel characterization and modeling.

Hard- or software implementations of statistical channel models are detailed in [57, 13, 25, 41]. Material on the direct filtering approach and on simulating extended clusters is found in [25, 26, 20] and [31, 26, 27, 28, 58], respectively.

Bibliography

[1] R. Steele, "An Update on Personal Communications," *IEEE Communications Magazine*, vol. 30, no. 12, pp. 30–31, Dec. 1992.

[2] B. Frankel, "Wireless Data Communications – Taking Users into the Future." ICC/Supercomm Joint Plenary, Chicago, IL, June 18, 1992.

[3] J. G. Proakis, *Digital Communications*. New York: McGraw-Hill, 1989.

[4] H. Meyr, "Signal Processing in Digital Receivers: Model, Algorithm, Architecture," in *Proc. ITG Fachbericht 107*, (Nürnberg, Germany, VDE Verlag), pp. 325–337, Apr. 1989.

[5] H. Meyr and G. Ascheid, *Synchronization in Digital Communications – Volume 1*. New York: Wiley, 1990.

[6] R. Häb and H. Meyr, "A Systematic Approach to Carrier Recovery and Detection of Digitally Phase-Modulated Signals on Fading Channels," *IEEE Trans. Commun.*, vol. 37, no. 7, pp. 748–754, July 1989.

[7] G. S. Liu and C. H. Wei, "Timing Recovery Techniques for Digital Cellular Radio with pi/4-DQPSK Modulation," in *Proc. ICC*, (Chicago, IL), pp. 319–323, June 1992.

[8] A. Mehrotra, *Cellular Radio Performance Engineering*. Boston: Artech House, 1994.

[9] P. A. Bello, "Characterization of Randomly Time-Variant Linear Channels," *IEEE Trans. Commun. Syst.*, vol. 11, p. 360, 1963.

[10] J. D. Parsons and A. S. Bajwa, "Wideband Characterization of Fading Mobile Radio Channels," *IEE Proc.*, vol. 129, Pt. F, no. 2, pp. 95–101, Apr. 1982.

[11] J. J. Spilker, *Digital Communications by Satellite*. Englewood Cliffs, NJ: Prentice-Hall, 1977.

[12] G. Ascheid, M. Oerder, J. Stahl, H. Meyr

[13] L. J. Greenstein, "A Multipath Fading Channel Model for Terrestrial Digital Radio Systems," *IEEE Trans. Commun.*, vol. 26, pp. 1247–1250, Aug. 1978.

[14] C. A. Siller, "Multipath Propagation," *IEEE Commun. Magazine*, vol. 22, no. 2, pp. 6–15, Feb. 1984.

[15] S. Stein, "Fading Channel Issues in System Engineering," *IEEE J. Sel. Areas Commun.*, vol. 5, pp. 68–89, Feb. 1987.

[16] N. M. Maslin, *HF Communications: a Systems Approach*. New York: Plenum, 1987.

[17] R. A. Shepherd and J. B. Lomax, "Frequency Spread in Ionospheric Radio Propagation," *IEEE Trans. Commun. Technol.*, vol. 15, no. 2, pp. 268–275, Apr. 1967.

[18] C. C. Watterson, W. D. Juroshek, and W. D. Bensema, "Experimental Confirmation of an HF Channel Model," *IEEE Trans. Commun. Technol.*, vol. 18, pp. 792–803, Dec. 1970.

[19] W. C. Y. Lee, *Mobile Communications Engineering*. New York: McGraw-Hill, 1982.

[20] W. C. Jakes, *Microwave Mobile Communications*. New York: Wiley, 1974.

[21] A. R. Potter, "Implementation of PCNs Using DCS1800," *IEEE Communications Magazine*, vol. 30, no. 12, pp. 32–36, Dec. 1992.

[22] H. D. Lüke, *Signalübertragung*. Springer-Verlag, 1985.

[23] G. Ungerboeck, G., "Adaptive Maximum-Likelihood Receiver for Carrier-Modulated Data Transmission Systems," *IEEE Commun. Magazine*, vol. 22, no. 5, pp. 624–636, May 1974.

[24] G. D. Forney, "Maximum-Likelihood Sequence Estimation of Digital Sequences in the Presence of Intersymbol Interference," *IEEE Trans. Inform. Theory*, vol. 18, no. 3, pp. 363–378, May 1972.

[25] P. Höher, *Kohärenter Empfang trelliscodierter PSK-Signale auf frequenzselektiven Mobilfunkkanälen – Entzerrung, Decodierung und Kanalparameterschätzung (in German)*. PhD thesis, Kaiserslautern University of Technology, 1990.

[26] P. Höher, "A Statistical Discrete-Time Model for the WSSUS Multipath Channel," *IEEE Trans. Vehicular Technol.*, vol. 41, no. 4, pp. 461–468, Nov. 1992.

[27] A. Mueller, "Monte-Carlo Modelling and Simulation of Multipath Mobile Radio Channels," in *Proc. Aachener Kolloquium für Signaltheorie*, (Aachen, Germany), pp. 341–346, Mar. 1994.

[28] M. Pätzhold, U. Killat, Y. Shi, and F. Laue, "A Discrete Simulation Model for the WSSUS Multipath Channel Derived from a Specified Scattering Function," in *Proc. Aachener Kolloquium für Signaltheorie, VDE Verlag*, (Aachen, Germany), pp. 347–351, Mar. 1994.

[29] CEPT/COST 207 WG1, "Proposal on Channel Transfer Functions to be used in GSM Tests late 1986." COST 207 TD (86)51 Rev. 3, Sept. 1986.

[30] CEPT/ETSI, "GSM recommendations 5.x," 1988.

[31] S. A. Fechtel, "A Novel Approach to Modeling and Efficient Simulation of Frequency-Selective Fading Channels," *IEEE Trans. Sel. Areas Commun.*, vol. 11, no. 3, pp. 422–431, Apr. 1993.

[32] A. Aghamohammadi, *Adaptive Phase Synchronization and Automatic Gain Control of Linearly-Modulated Signals on Frequency-Flat Fading Channels.* PhD thesis, Aachen University of Technology, 1989.

[33] A. Aghamohammadi, H. Meyr, and G. Ascheid, "Adaptive Synchronization and Channel Parameter Estimation Using an Extended Kalman Filter," *IEEE Trans. Commun.*, vol. 37, no. 11, pp. 1212–1219, Nov. 1989.

[34] J. Hagenauer and E. Lutz, "Forward Error Correction Coding for Fading Compensation in Mobile Satellite Channels," *IEEE J. Sel. Areas Commun.*, vol. 5, no. 2, pp. 215–225, Feb. 1987.

[35] W. R. Braun and U. Dersch, "A Physical Mobile Radio Channel Model," *IEEE Trans. Vehicular Technol.*, vol. 40, no. 2, pp. 472–482, May 1991.

[36] R. S. Kennedy, *Fading Dispersive Communication Channels.* New York: Wiley, 1969.

[37] U. Dersch and R. J. Rüegg, "Simulations of the Time and Frequency Selective Outdoor Mobile Radio Channel," *IEEE Trans. Vehicular Technol.*, vol. 42, no. 3, pp. 338–344, Aug. 1993.

[38] S. C. Gupta *et al.*, "Land Mobile Radio Systems: a Tutorial Exposition," *IEEE Commun. Magazine*, vol. 23, no. 6, pp. 34–45, June 1985.

[39] R. W. Lorenz and G. Kadel, "Propagation Measurements Using a Digital Channel Sounder Matched to the GSM System Bandwidth," in *Proc. ICC*, (Denver, CO), pp. 548–552, June 1991.

[40] D. Divsalar and M. K. Simon, "Trellis Coded Modulation for 4800-9600 bit/s Transmission over a Fading Mobile Satellite Channel," *IEEE J. Sel. Areas Commun.*, vol. 5, no. 2, pp. 162–175, Feb. 1987.

[41] E. Lutz, D. Cygan, M. Dippold, F. Dolainsky, and W. Pape, "The Land Mobile Satellite Communication Channel: Recording, Statistics, and Channel Model," *IEEE Trans. Vehicular Technol.*, vol. 40, no. 2, pp. 375–386, May 1991.

[42] F. Davarian, "Digital Modulation for an Aeronautical Mobile Satellite System," in *Proc. Globecom*, (Tokyo, Japan), pp. 1642–1649, Nov. 1987.

[43] A. Neul, J. Hagenauer, W. Papke, F. Dolainsky, and F. Edbauer, "Aeronautical Channel Characterization based on Measurement Flights," in *Proc. Globecom*, (Tokyo, Japan), pp. 1654–1659, Nov. 1987.

[44] B. Goldberg, "300khz-30mhz mf/hf," *IEEE Trans. Commun. Technol.*, vol. 14, pp. 767–784, Dec. 1966.

[45] M. Haines, "A Statistical Channel Model for Adaptive HF Communications via a Severely Disturbed Ionosphere." RADC in-house report, Dec. 1983.

[46] H. N. Shaver, B. C. Tupper, and J. B. Lomax, "Evaluation of a Gaussian HF Channel Model," *IEEE Trans. Commun. Technol.*, vol. 15, no. 1, pp. 79–88, Feb. 1967.

[47] A. A. Giordano and F. Haber, "Modeling of Atmospheric Noise," *Radio Science*, vol. 7, no. 11, pp. 1011–1023, Nov. 1972.

[48] D. Middleton, "Statistical-Physical Models of Urban Radio-Noise Environments, Part I: Foundations," *IEEE Trans. Electromag. Compat.*, vol. 14, no. 2, pp. 38–56, May 1972.

[49] D. Middleton, "Canonical and Quasi-Canonical Probability Models of Class A Interference," *IEEE Trans. Electromag. Compat.*, vol. 25, no. 2, pp. 76–106, May 1983.

[50] R. A. Shepherd, "Measurements of Amplitude Probability Distributions and Power of Automobile Ignition Noise at HF," *IEEE Trans. Vehic. Technol.*, vol. 23, no. 3, pp. 72–83, Aug. 1974.

[51] M. P. Shinde and S. N. Gupta, "A Model of HF Impulsive Atmospheric Noise," *IEEE Trans. Electromag. Compat.*, vol. 16, no. 2, pp. 71–75, May 1974.

[52] R. J. C. Bultitude, "Measurement, Characterization and Modeling of Indoor 800/900 MHz Radio Channels for Digital Communications," *IEEE Commun. Magazine*, vol. 25, no. 6, pp. 5–12, June 1987.

[53] R. Ganesh and K. Pahlavan, "Statistics of Short-Time Variations of Indoor Radio Propagation," in *Proc. ICC*, (Denver, CO), pp. 1–5, June 1991.

[54] S. J. Howard and K. Pahlavan, "Statistical Autoregressive Models for the Indoor Radio Channel," in *Proc. ICC*, (Singapore), pp. 1000–1006, April 1990.

[55] E. Zollinger, "A Statistical Model for Wideband Inhouse Radio Channels," in *Proc. Melecon*, (Lisboa, Portugal), Apr. 1989.

[56] DVB Project Office, "Digital Video Broadcasting: Framing Structure, Channel Coding and Modulation for Digital Terrestrial Television." DVB Document A012, June 1996.

[57] L. Ehrman, L. B. Bates, J. F. Eschle, and Kates, J. M., "Real-time Software Simulation of the HF Radio Channel," *IEEE Trans. Commun.*, vol. 30, pp. 1809–1817, Aug. 1982.

[58] H. Schulze, "Stochastic Models and Digital Simulation of Mobile Radio Channels (in German)," in *Proc. Kleinheubacher Berichte*, vol. 32, pp. 473–483, 1989.

Chapter 12 Detection and Parameter Synchronization on Fading Channels

In this chapter, we are concerned with optimal receiver structures and near-optimal algorithms for data detection and sync parameter estimation on both flat and frequency-selective fading channels. Emphasis is placed on the concept of *synchronized detection* where the sync parameters are explicitly estimated and then used for detection. Based on this mathematical framework, optimal estimator-detector receivers for joint detection and synchronization are derived. This chapter focuses on the methodology and fundamental insights rather than on details of implementation; these are addressed in the following chapters on realizable receiver structures and fading channel estimation.

12.1 Fading Channel Transmission Models and Synchronization Parameters

In this section, the transmission models and synchronization parameters of interest are briefly reviewed and put in a mathematical framework suitable for detection and synchronization. Based on these models, optimal joint detection and sync parameter estimation strategies and algorithms are systematically derived. Even though the optimal algorithms as well as most of their simplified versions admittedly suffer from an extremely high complexity, the ideas, strategies, and valuable insights worked out here provide a universal framework that can be applied to the systematic development of both the data-aided (DA) and non-data-aided (NDA) synchronization algorithms covered in Chapters 14 and 15.

As before, we are concerned with *linearly* modulated, possibly encoded QAM or PSK symbol sequences. Although of practical importance, nonlinear modulations such as MSK (Section 3.2) are not discussed here, but the basic principles of deriving optimal detectors and synchronizers remain the same.

As detailed in Chapter 11 [eqs. (11-16), (11-19), (11-20), and (11-22), respectively], the following transmission models are of interest:

Selective fading:

$$r_k^{(i)} = e^{j\Omega'(k+i/2)} \left[\sum_n a_n \, h_{k-n;\,k}^{(i)} \right] + n_k^{(i)}$$

Flat fading:

$$r_k^{(i)} = \underbrace{e^{j\Omega'(k+i/2)} \, c_k^{(i)}}_{c_{\Omega,k}^{(i)}} \left[\sum_n a_n \, g_{T,k-n}^{(i)}(\varepsilon) \right] + n_k^{(i)}$$

Flat fading, pulse MF output, imperfect timing:

$$z_k^{(i)} \underset{(\Omega' \text{ small})}{\approx} \underbrace{e^{j\Omega'(k+\frac{i}{2})} \, c_k^{(i)}}_{c_{\Omega,k}^{(i)}} \left[\sum_n a_n \, g_{k-n}^{(i)}(\varepsilon) \right] + m_k^{(i)} \tag{12-1}$$

Flat fading, pulse MF output, perfect timing:

$$z_k \underset{(\varepsilon \to 0)}{\approx} \underbrace{e^{j\Omega' k} \, c_k}_{c_{\Omega,k}} \, a_k + m_k$$

with noise autocorrelation functions [eq. (11-21)] $R_n^{(i)}(n) = N_0 \delta_n$ (AWGN) and $R_m^{(i)}(n) = N_0 \delta_n$, $R_{m^{(0)},m^{(1)}}(n) = N_0 g(\tau = [n+0.5]T)$.

We consider the transmission of an isolated block of N symbols

$$\mathbf{a} = (a_0 \; a_1 \; \ldots \; a_{N-1})^T \tag{12-2}$$

starting at reference time index $\mu = 0$, where μ may be unknown at the receiver. The sequence length N may range from a small number (short data packet) up to near infinity (quasi-continuous transmission). Some of the symbols a_k may be known (training); the unknown (random data) symbols may be uncoded or coded and possibly interleaved. Depending on the channel transmission model and the availability of prior knowledge, the sync parameters $\boldsymbol{\theta}$ of interest may encompass the start-of-frame index μ, the (relative) frequency offset Ω', the set $\mathbf{h}$ of time-variant selective channel impulse response vectors or the set $\mathbf{c}$ of flat fading channel weights, the timing offset ε, and carrier phase φ. Since the channel is band-limited, the (partial) channel impulse response (CIR) vector $\mathbf{h}_k^{(i)}$ as defined in eq. (11-59) comprises, in theory, infinitely many samples $h_{n,k}^{(i)}$. The same applies to the sampled pulse shaping filter response $g_{T,n}^{(i)}(\varepsilon)$ and the concatenation $g_n^{(i)}(\varepsilon)$ of pulse shaping and matched filters.

While it is important to represent weak taps (channel pre- and postcursors) in the channel simulator (Section 11.3), the derivation of manageable detection and synchronization algorithms most often requires that the number of sync parameters to be considered is kept at a minimum. In this vein, the CIR model used for the derivation of receiver algorithms must be *truncated* to some finite length L (symbol intervals) so that the (partial) CIR vector of interest becomes

$$\mathbf{h}_k^{(i)} = \left(h_{0;k}^{(i)} \; h_{1;k}^{(i)} \; \dots \; h_{L;k}^{(i)} \right)^T \tag{12-3}$$

The $T/2$-spaced CIR vector $\mathbf{h}_k$ may be expressed in a short-hand notation as the "sum" ($\oplus$) of the two partial CIR vectors:

$$\begin{aligned} \mathbf{h}_k &= \mathbf{h}_k^{(0)} \oplus \mathbf{h}_k^{(1)} \\ &= \left(h_{0;k} \; h_{1;k} \; h_{2;k} \; h_{3;k} \; \dots \; h_{2L;k} \; h_{2L+1;k} \right)^T \\ &= \left(h_{0;k}^{(0)} \; h_{0;k}^{(1)} \; h_{1;k}^{(0)} \; h_{1;k}^{(1)} \; \dots \; h_{L;k}^{(0)} \; h_{L;k}^{(1)} \right)^T \end{aligned} \tag{12-4}$$

The set of all CIR vectors $\mathbf{h}_k$ ($k = 0, 1, \dots, N-1, \dots$) may then be collected in a vector $\mathbf{h}$ encompassing all channel tap processes that are relevant for the detection of sequence $\mathbf{a}$. Likewise, the (single) weight process trajectory of a flat fading channel can be collected in a vector:

$$\mathbf{c} = \left(c_0 \; c_1 \; \dots \; c_{N-1} \right)^T \tag{12-5}$$

where it has been assumed that the fading process does not change significantly during a symbol interval, i.e., $c_k^{(1)} \approx c_k^{(0)} = c_k$.

Using this notation, the sync parameters of interest for flat and selective channels can be summarized in Table 12-1.

Table 12-1 Synchronization Parameters for Flat and Selective Channels

Channel model	Sync parameters $\boldsymbol{\theta}$
Selective fading	$\mu,\ \Omega',\ \mathbf{h} = (\mathbf{h}_0^T \; \mathbf{h}_1^T \; \dots \; \mathbf{h}_{N-1}^T \; \dots)^T$
(selective fixed)	$\mu,\ \Omega',\ \mathbf{h} = \mathbf{h}^{(0)} \oplus \mathbf{h}^{(1)}$ $(\mathbf{h}_k = \mathbf{h})$
Flat fading	$\mu,\ \Omega',\ \varepsilon,\ \mathbf{c} = \{c_0 \; c_1 \; \dots \; c_{N-1}\}$
(flat fixed, i.e., AWGN)	$\mu,\ \Omega',\ \varepsilon,\ \varphi = \arg[c]$ $(c_k = c)$

The parameters $\mu, \Omega', \varepsilon, \varphi$ are assumed to be invariant over the block length N. Notice that the dimensionality of the sync parameter vector $\boldsymbol{\theta}$ may be very large; for instance, in the case of a selective fading channel, at least $(N-1)2(L+1)$ channel taps [$2(L+1)$ channel tap processes of duration $\geq N-1$ symbol intervals] are to be estimated.

A *sufficient statistic* for detection and synchronization is given by the collection of all received samples that are dependent on the data $\mathbf{a}$. Since the channel memory is assumed to be finite, the true sufficient statistic, i.e., the infinite-length received sequence $\{r_{-\infty}^{(i)}, \ldots, r_{+\infty}^{(i)}\}$, can also be truncated to a finite-length vector:

$$
\begin{aligned}
&\text{Receiver input signal: (flat fading: } L=0\text{)}\\
&\mathbf{r} = \mathbf{r}^{(0)} \oplus \mathbf{r}^{(1)} \qquad \text{with} \quad \mathbf{r}^{(i)} = \left(r_0^{(i)} \; r_1^{(i)} \; \ldots \; r_{N+L-1}^{(i)} \right)^T \\
&\text{Pulse MF output signal, imperfect timing:}\\
&\mathbf{z} = \mathbf{z}^{(0)} \oplus \mathbf{z}^{(1)} \qquad \text{with} \quad \mathbf{z}^{(i)} = \left(z_0^{(i)} \; z_1^{(i)} \; \ldots \; z_{N-1}^{(i)} \right)^T \\
&\text{Pulse MF output signal, perfect timing:}\\
&\mathbf{z} = \left(z_0 \; z_1 \; \ldots \; z_{N-1} \right)^T
\end{aligned}
\tag{12-6}
$$

Notice that, for the special case of perfect timing, the pulse MF output is free of intersymbol interference (ISI) so that $\mathbf{z}$ becomes a true (not approximate) sufficient statistic in this case. Note that the exact dimension of the observation vector depends on the particular application and the sync parameters of interest. For instance, if the start-of-frame instant μ [which has implicitely been assumed known in eq. (12-6)] is unknown but bounded within a finite interval $-\mu_0 \leq \mu \leq \mu_0$ about zero, additional samples of the received signal have to be included into the observation vector which then ranges from $r_{-\mu_0}^{(i)}$ to $r_{\mu_0+(N+L)-1}^{(i)}$ (receiver input signal). At any rate, the observation may be expressed concisely as a function of the data sequence $\mathbf{a}$ and the sync parameters $\boldsymbol{\theta}$, plus additive random noise $\mathbf{n}$ or $\mathbf{m}$:

$$
\begin{aligned}
\mathbf{r} &= f_r(\mathbf{a}, \boldsymbol{\theta}) + \mathbf{n} \qquad &\text{(receiver input)}\\
\mathbf{z} &= f_z(\mathbf{a}, \boldsymbol{\theta}) + \mathbf{m} \qquad &\text{(pulse MF output)}
\end{aligned}
\tag{12-7}
$$

The observation is linearly dependent on the data $\mathbf{a}$ but may be nonlinearly dependent on some sync parameters $\boldsymbol{\theta}$.

To further illustrate the construction of transmission models, we will now take a closer look at some important special cases of vector-matrix formulations of the observation.

12.1.1 Flat Fading Channel Transmission

As a first example, consider the (relatively simple) case of *flat fading* channel transmission with pulse matched filtering and perfect timing error compensation, with known start-of-frame instant $\mu=0$ but (possibly) unknown frequency Ω' and unknown flat fading process vector $\mathbf{c}$. Via eqs. (12-1) and (12-6), the (T-spaced) observation $\mathbf{z}$ can be expressed in terms of the N-dim. data vector $\mathbf{a}$, an $N \times N$-dim. diagonal channel matrix $\mathbf{C}$ and an $N \times N$-dim. diagonal frequency offset matrix $\mathbf{W}(\Omega')$:

$$
\begin{aligned}
\mathbf{z} &= \mathbf{M}(\boldsymbol{\theta})\ \mathbf{a}\ + \mathbf{m} \\
\underbrace{\begin{pmatrix} z_0 \\ z_1 \\ \vdots \\ z_{N-1} \end{pmatrix}}_{\mathbf{z}} &= \underbrace{\begin{pmatrix} e^{j\Omega'0} & & & \\ & e^{j\Omega'1} & & \\ & & \ddots & \\ & & & e^{j\Omega'(N-1)} \end{pmatrix}}_{\mathbf{W}(\Omega')} \\
& \quad \underbrace{\begin{pmatrix} c_0 & & & \\ & c_1 & & \\ & & \ddots & \\ & & & c_{N-1} \end{pmatrix}}_{\mathbf{C}} \underbrace{\begin{pmatrix} a_0 \\ a_1 \\ \vdots \\ a_{N-1} \end{pmatrix}}_{\mathbf{a}} + \mathbf{m} \\
&= \mathbf{W}(\Omega')\ \mathbf{C}\ \mathbf{a}\ + \mathbf{m}
\end{aligned}
\tag{12-8}
$$

The transmission matrix $\mathbf{M}(\boldsymbol{\theta})$ of eq. (12-8) with sync parameter vector $\boldsymbol{\theta} = \left(\Omega' \ \mathbf{c}^T\right)^T$ is linearly dependent on the channel $\mathbf{c}$. Hence, the observation $\mathbf{z}$ can likewise be expressed in terms of the N-dim. channel process vector $\mathbf{c}$, an $N \times N$-dim. diagonal data matrix $\mathbf{A}$ and (as above) the $N \times N$-dim. frequency offset matrix $\mathbf{W}(\Omega')$:

$$
\begin{aligned}
\underbrace{\begin{pmatrix} z_0 \\ z_1 \\ \vdots \\ z_{N-1} \end{pmatrix}}_{\mathbf{z}} &= \underbrace{\begin{pmatrix} e^{j\Omega'0} & & & \\ & e^{j\Omega'1} & & \\ & & \ddots & \\ & & & e^{j\Omega'(N-1)} \end{pmatrix}}_{\mathbf{W}(\Omega')} \\
& \quad \cdot \underbrace{\begin{pmatrix} a_0 & & & \\ & a_1 & & \\ & & \ddots & \\ & & & a_{N-1} \end{pmatrix}}_{\mathbf{A}} \cdot \underbrace{\begin{pmatrix} c_0 \\ c_1 \\ \vdots \\ c_{N-1} \end{pmatrix}}_{\mathbf{c}} + \mathbf{m} \\
&= \mathbf{W}(\Omega')\ \mathbf{A}\ \mathbf{c}\ + \mathbf{m}
\end{aligned}
\tag{12-9}
$$

The observation being linearly dependent on some of the synchronization parameters $\boldsymbol{\theta}$ is an important special case since then the powerful concepts of optimal linear parameter estimation theory [1, 2] can be applied.

12.1.2 Selective Fading Channel Transmission

Consider now *selective fading* channel transmission, again with known start-of-frame instant $\mu = 0$, (possibly) unknown frequency Ω', and unknown channel impulse response process vector $\mathbf{h}$ with (maximum) delay spread L. Using eqs. (12-1) and (12-6), the observation $\mathbf{r}$ can be expressed in terms of the N-dim. data vector $\mathbf{a}$, a $(2[N+L] \times N)$-dim. banded channel transmission matrix $\mathbf{H}$ and a $(2[N+L] \times 2[N+L])$-dim. diagonal frequency offset matrix $\mathbf{W}(\Omega')$:

$$
\begin{aligned}
\mathbf{r} &= \mathbf{r}^{(0)} \oplus \mathbf{r}^{(1)} \\
&= \mathbf{M}(\boldsymbol{\theta}) \cdot \mathbf{a} + \mathbf{n} \\
&= \underbrace{\left(\mathbf{W}^{(0)}(\Omega') \oplus \mathbf{W}^{(1)}(\Omega')\right)}_{\mathbf{W}(\Omega')} \cdot \underbrace{\left(\mathbf{H}^{(0)} \oplus \mathbf{H}^{(1)}\right)}_{\mathbf{H}} \cdot \mathbf{a} + \mathbf{n}
\end{aligned}
$$

$$
\mathbf{r}^{(i)} = \begin{pmatrix} r_0 \\ r_1 \\ \vdots \\ r_{N-1} \\ \vdots \\ r_{N+L-1} \end{pmatrix}^{(i)} = \underbrace{e^{j\Omega'(i/2)} \begin{pmatrix} e^{j\Omega' 0} & & & \\ & e^{j\Omega' 1} & & \\ & & \ddots & \\ & & & e^{j\Omega'(N+L-1)} \end{pmatrix}}_{\mathbf{W}^{(i)}(\Omega')}
$$

$$
\cdot \underbrace{\begin{pmatrix} h_{0,0} & & & & & \\ h_{1,1} & h_{0,1} & & & & \\ \vdots & \vdots & \ddots & & & \\ h_{L,L} & h_{L-1,L} & \ddots & \ddots & & \\ & h_{L,L+1} & \ddots & \ddots & \ddots & \\ & & \ddots & \ddots & \ddots & h_{0,N-2} & \\ & & & \ddots & \ddots & h_{1,N-1} & h_{0,N-1} \\ & & & & \ddots & \vdots & \vdots \\ & & & & & h_{L,N+L-2} & h_{L-1,N+L-2} \\ & & & & & & h_{L,N+L-1} \end{pmatrix}^{(i)}}_{\mathbf{H}^{(i)}} \cdot \underbrace{\begin{pmatrix} a_0 \\ a_1 \\ \vdots \\ a_{N-1} \end{pmatrix}}_{\mathbf{a}} + \mathbf{n}^{(i)}
$$

$$
= \mathbf{W}^{(i)}(\Omega')\, \mathbf{H}^{(i)}\, \mathbf{a} + \mathbf{n}^{(i)} \tag{12-10}
$$

As above, the short-hand operator $\oplus$ denotes the "sum" $\mathbf{X} = \mathbf{X}^{(0)} \oplus \mathbf{X}^{(1)}$ of two matrices (or vectors) $\mathbf{X}^{(0)}$, $\mathbf{X}^{(1)}$ in the sense the elements of $\mathbf{X}^{(0)}$, $\mathbf{X}^{(1)}$ are copied to appropriate positions of the new matrix $\mathbf{X}$ with appropriate dimensions. For instance, $\mathbf{H}$ is constructed from $\mathbf{H}^{(0)}$, $\mathbf{H}^{(1)}$ by copying the rows 0, 1, 2,... of $\mathbf{H}^{(0)}$ ($\mathbf{H}^{(1)}$) to rows 0, 2, 4,... (1, 3, 5,...) of $\mathbf{H}$. Similarly, $\mathbf{W}$ is constructed from $\mathbf{W}^{(0)}$, $\mathbf{W}^{(1)}$ by copying the main diagonal elements of $\mathbf{W}^{(0)}$ ($\mathbf{W}^{(1)}$) to positions 0, 2, 4,... (1, 3, 5,...) of the main diagonal of $\mathbf{W}$.

In eq. (12-10), the transmission matrix $\mathbf{M}(\boldsymbol{\theta})$ with sync parameter vector $\boldsymbol{\theta} = \left(\Omega' \ \mathbf{h}^T\right)^T$ is, of course, nonlinearly dependent on the frequency shift Ω' but linearly dependent on the channel $\mathbf{h}$ (here arranged in matrix $\mathbf{H}$). This linear dependence motivates a second, equivalent transmission model where the observation $\mathbf{r}$ is expressed in terms of the $2[L+1]N$-dim. CIR process vector $\mathbf{h}$, an

$(2[N+L] \times 2[L+1]N)$-dim. data matrix $\mathbf{A}$ and (as above) the $(2[N+L] \times 2[N+L])$-dim. frequency offset matrix $\mathbf{W}(\Omega')$:

$$
\begin{aligned}
\mathbf{r} &= \mathbf{r}^{(0)} \oplus \mathbf{r}^{(1)} \\
&= \mathbf{W}(\Omega') \cdot \underbrace{\left(\mathbf{A}^{(0)} \oplus \mathbf{A}^{(1)}\right)}_{\mathbf{A}} \cdot \underbrace{\left(\mathbf{h}^{(0)} \oplus \mathbf{h}^{(1)}\right)}_{\mathbf{h}} + \mathbf{n} \\
\mathbf{r}^{(i)} &= \begin{pmatrix} r_0 \\ r_1 \\ \vdots \\ r_{N-1} \\ \vdots \\ r_{N+L-1} \end{pmatrix}^{(i)} = \underbrace{e^{j\Omega'(i/2)} \begin{pmatrix} e^{j\Omega' 0} & & & \\ & e^{j\Omega' 1} & & \\ & & \ddots & \\ & & & e^{j\Omega'(N+L-1)} \end{pmatrix}}_{\mathbf{W}^{(i)}(\Omega')} \\
&\cdot \underbrace{\begin{pmatrix} a_0 & & & & & \\ & \ddots & a_1 & & & \\ & a_0 & \ddots & \ddots & & \\ & & a_1 & \ddots & a_{N-1} & \\ & & & \ddots & \ddots & \\ & & & & & a_{N-1} \end{pmatrix}^{(i)}}_{\mathbf{A}^{(i)}} \cdot \underbrace{\begin{pmatrix} h_{0,0} \\ \vdots \\ h_{L,L} \\ \vdots \\ h_{0,N-1} \\ \vdots \\ h_{L,N+L-1} \end{pmatrix}^{(i)}}_{\mathbf{h}^{(i)}} + \mathbf{n}^{(i)} \\
&= \mathbf{W}^{(i)}(\Omega')\, \mathbf{A}^{(i)}\, \mathbf{h}^{(i)} + \mathbf{n}^{(i)}
\end{aligned}
\tag{12-11}
$$

where the $(2[N+L] \times 2[L+1]N)$-dim. matrix $\mathbf{A}$ is constructed from the $([N+L] \times [L+1]N)$-dim. matrix $\mathbf{A}^{(i)}$ by doubling the size of the $([L+1] \times [L+1])$-dim. diagonal submatrices $a_k \cdot \mathbf{I}_{L+1}$ $(k = 0, \ldots, N-1)$ contained in $\mathbf{A}^{(i)}$ to dimension $(2[L+1] \times 2[L+1])$ and abutting them just like in $\mathbf{A}^{(i)}$, but with an offset of two rows between abutted submatrices.

Of course, eqs. (12-10) and (12-11) are just two of many possible ways of representing the observation resulting from transmission through a selective fading channel; vector and matrix arrangements should be selected such that they best suit the application and detection / synchronization problem at hand.

12.1.3 Main Points

- Vector/matrix models for fading channel transmission have been established which yield vector output processes $\mathbf{z}$ (flat fading channel, including pulse matched filtering and perfect timing sync) and $\mathbf{r}$ (selective channel). By inspection of these models, the sync parameter vector $\boldsymbol{\theta}$ of interest is easily identified.

- Of particular interest are the flat and selective fading channel trajectories $\mathbf{c}$ and $\mathbf{h}$, respectively. In this case, the observation ($\mathbf{z}$ or $\mathbf{r}$) is of the form

$$\begin{aligned} \mathbf{z} &= \mathbf{W}(\Omega')\, \mathbf{A}\, \mathbf{c} \ + \mathbf{m} \\ \mathbf{r} &= \mathbf{W}(\Omega')\, \mathbf{A}\, \mathbf{h} \ + \mathbf{n} \end{aligned} \tag{12-12}$$

and thus linearly dependent on the sync parameters $\mathbf{c}$ and $\mathbf{h}$, respectively.

12.2 Optimal Joint Detection and Synchronization

Up to the present, only relatively few attempts have been made to apply the concepts of joint data detection and parameter synchronization to (static or fading) dispersive channels. The basic idea of joint detection and channel estimation is a – conceptually simple but computationally very complex – exhaustive search for the "overall best fit" between the "model output" (hypothetical data sequence transmitted over its associated hypothetical channel) and the observation (received signal), most often aided by some side information on the channel dynamics.

Recently, several approaches to joint detection and synchronization, in particular selective channel estimation, have been investigated. Most of these assume that the fading is very slow so that the CIR $\mathbf{h}$ can be taken as *time-invariant* within a sufficiently large time interval. Neglecting oscillator frequency offsets, one may attempt to maximize the joint likelihood function of $(\mathbf{a}, \mathbf{h})$:

$$\left(\hat{\mathbf{a}}, \hat{\mathbf{h}}\right)_{\text{ML}} = \arg \max_{\mathbf{a},\mathbf{h}} \ p(\mathbf{r}|\mathbf{a}, \mathbf{h}) \tag{12-13}$$

(maximum likelihood, ML) which, in the case of white noise, is equivalent to minimizing the least squares (LS) error between model output $\hat{\mathbf{r}} = \hat{\mathbf{A}}\ \hat{\mathbf{h}}$ and observation $\mathbf{r}$:

$$\left(\hat{\mathbf{a}}, \hat{\mathbf{h}}\right)_{\text{LS}} = \arg \min_{\mathbf{a},\mathbf{h}} \ ||\mathbf{r} - \mathbf{A}\mathbf{h}||^2 \tag{12-14}$$

The minimization may be accomplished by first computing the channel estimate $\hat{\mathbf{h}}(\mathbf{a})$ for each possible symbol hypothesis $\mathbf{a}$ (within the finite interval where $\mathbf{h}$ is stationary) and then select the one which best fits the observation (minimum LS error) [3]. In order to restrict the per-symbol computational effort to some upper limit, suboptimal algorithms usually search for the optimum in a recursive manner. For instance, the *generalized Viterbi algorithm* [4] retains a fixed number of "best" data sequences (*survivors*), together with their associated channel estimates, after progressing from time instant $(k–1)$ to k. Generating a distinct channel estimate for every single survivor is sometimes referred to as *per-survivor processing* (PSP) [5, 6, 7]. The number of channel estimates may be reduced further by computing only those $\hat{\mathbf{h}}(\mathbf{a})$ that are based on the assumption of binary data $\mathbf{a}$ (*reduced constellation approach* [8]).

An alternative scheme termed *quantized channel approach* [9] is based on a finite set of candidate channels $\mathbf{h}$ that is selected before the joint detection/estimation algorithm is started. In principle, ML or LS detection and estimation is performed by detecting the optimal data sequence $\hat{\mathbf{a}}(\mathbf{h})$ associated with each candidate $\mathbf{h}$ and then selecting the one with the smallest LS error. During recursive Viterbi data detection, the candidate channels themselves should be adapted (by driving the residual error to a minimum) so that they converge against the (unquantized) true value.

Simulation results [10, 8] indicate that convergence of such joint detection/estimation algorithms can be remarkably fast, e.g., within 50 to several hundred iterations. Hence, such algorithms have the potential to tolerate some degree of fading and therefore remain to be a *hot topic* for further research.

12.2.1 The Bayesian Approach to Joint Detection and Synchronization

Let us now focus on joint detection and estimation in the context of truly *fading* channels. For the following discussion of Bayesian detection and estimation techniques, let $\mathbf{r}$ denote the observation in general (either $\mathbf{r}$ or the pulse MF output $\mathbf{z}$). Then optimal maximum a posteriori (MAP) detection of the data $\mathbf{a}$ contained in $\mathbf{A}$ calls for maximizing the probability of $\mathbf{a}$, conditioned on the observation $\mathbf{r}$,

$$\hat{\mathbf{a}}_{\mathrm{MAP}} = \arg \max_{\mathbf{a}} \; P(\mathbf{a}|\mathbf{r}) \tag{12-15}$$

As the probability $P(\mathbf{a} \mid \mathbf{r})$ and the likelihood function $p(\mathbf{r} \mid \mathbf{a})$ are linked via Bayes's rule $P(\mathbf{a}|\mathbf{r}) = [P(\mathbf{a})/p(\mathbf{r})] \cdot p(\mathbf{r}|\mathbf{a})$, we have

$$\hat{\mathbf{a}}_{\mathrm{ML}} = \arg \max_{\mathbf{a}} \; p(\mathbf{r}|\mathbf{a}) \underset{\left(\substack{P(\mathbf{a}) \text{ equal} \\ \text{for all } \mathbf{a}}\right)}{=} \arg \max_{\mathbf{a}} \; P(\mathbf{a}|\mathbf{r}) = \hat{\mathbf{a}}_{\mathrm{MAP}} \tag{12-16}$$

so that MAP detection is equivalent to ML detection if all possible sequences $\mathbf{a}$ are equally likely, an assumption which is generally made in practice.

As has already been pointed out in Section 4.3, ML detection can be performed directly through maximizing the likelihood function $p(\mathbf{r}|\mathbf{a})$ without the need for any kind of synchronization. However, we are most interested in receiver structures which follow the concept of synchronized detection introduced in Chapter 4. Thus, a sync parameter estimate $\hat{\boldsymbol{\theta}}$ must be formed and subsequently used for detection as if it were the true parameter. All joint estimation-detection structures of this chapter and, more importantly, virtually all realizable receiver structures are based on synchronized detection.

The sync parameters $\boldsymbol{\theta}$ can be brought into play by reformulating the likelihood $p(\mathbf{r}|\mathbf{a})$:

$$\begin{aligned} p(\mathbf{r}|\mathbf{a}) &= \int p(\mathbf{r}, \boldsymbol{\theta}|\mathbf{a})\, d\boldsymbol{\theta} \\ &= \int p(\mathbf{r}|\mathbf{a}, \boldsymbol{\theta}) \cdot p(\boldsymbol{\theta}|\mathbf{a})\, d\boldsymbol{\theta} \end{aligned} \tag{12-17}$$

Following the same high-SNR argument as in Section 4.3, maximizing the integrand

$$\begin{aligned} p(\mathbf{r}, \boldsymbol{\theta}|\mathbf{a}) &= p(\mathbf{r}|\mathbf{a}, \boldsymbol{\theta}) \cdot p(\boldsymbol{\theta}|\mathbf{a}) \\ &= p(\boldsymbol{\theta}|\mathbf{r}, \mathbf{a}) \cdot p(\mathbf{r}|\mathbf{a}) \qquad \left(\begin{matrix}\text{first}\\ \text{representation}\end{matrix}\right) \\ &\propto p(\mathbf{r}|\mathbf{a}, \boldsymbol{\theta}) \cdot p(\boldsymbol{\theta}) \qquad \left(\begin{matrix}\text{second}\\ \text{representation}\end{matrix}\right) \end{aligned} \tag{12-18}$$

yields the joint estimation-detection rule

$$\left(\hat{\mathbf{a}}_{\mathrm{ML}}, \hat{\boldsymbol{\theta}}\right) = \arg \max_{\mathbf{a}, \boldsymbol{\theta}} \; p(\mathbf{r}, \boldsymbol{\theta}|\mathbf{a}) \tag{12-19}$$

In the important case that all random quantities are Gaussian, this high-SNR approximation is actually the *optimal* estimation-detection rule, regardless of the SNR (see Sections 12.2.2 and 12.2.6).

At this point, it is necessary to take a closer look at the properties of the sync parameters $\boldsymbol{\theta}$ to be estimated. In particular, we distinguish between parameters $\boldsymbol{\theta}_S$ that are essentially *static*, such as $\mu, \varepsilon, \varphi$ or time-invariant channels $\mathbf{h}$, and parameters $\boldsymbol{\theta}_D$ that may be termed *dynamic* in the sense that they are taken from time-variant processes, such as flat or selective fading channel process vectors $\mathbf{c}$ or $\mathbf{h}$, respectively. Usually, there is little or no probabilistic information on static parameters other than that they are in a given region. Therefore, $p(\boldsymbol{\theta}_S)$ does not exist or can be assumed constant within a finite region. In view of the second representation of eq. (12-18), joint detection and static parameter estimation thus reduces to maximizing the joint likelihood function $p(\mathbf{r}|\mathbf{a},\boldsymbol{\theta}_S)$ with respect to $(\mathbf{a}, \boldsymbol{\theta}_S)$ (see introduction above). On the other hand, probabilistic side information on dynamic parameters is usually available and should be made use of. Hence, joint detection and dynamic parameter estimation calls for maximizing either the first or second representation of the density function $p(\mathbf{r}, \boldsymbol{\theta}_D|\mathbf{a})$ [eq. (12-18)]:

Static sync parameters:

$$\left(\hat{\mathbf{a}}, \hat{\boldsymbol{\theta}}_S\right) = \arg \max_{\mathbf{a}, \boldsymbol{\theta}_S} \; p(\mathbf{r}|\mathbf{a}, \boldsymbol{\theta}_S)$$

Dynamic sync parameters:

$$\begin{aligned}\left(\hat{\mathbf{a}}, \hat{\boldsymbol{\theta}}_D\right) &= \arg \max_{\mathbf{a}, \boldsymbol{\theta}_D} \; p(\boldsymbol{\theta}_D|\mathbf{r}, \mathbf{a}) \cdot p(\mathbf{r}|\mathbf{a}) \qquad \left(\begin{matrix}\text{first}\\ \text{representation}\end{matrix}\right) \\ &= \arg \max_{\mathbf{a}, \boldsymbol{\theta}_D} \; p(\mathbf{r}|\mathbf{a}, \boldsymbol{\theta}_D) \cdot p(\boldsymbol{\theta}_D) \qquad \left(\begin{matrix}\text{second}\\ \text{representation}\end{matrix}\right)\end{aligned} \tag{12-20}$$

It is immediately recognized that $\hat{\boldsymbol{\theta}}_S$ is an ML estimate and $\hat{\boldsymbol{\theta}}_D$ a MAP estimate, so that one may speak of ML detection with MAP (dynamic) or ML (static) parameter synchronization.

Since there are infinitely many possible realizations of sync parameters $\boldsymbol{\theta}$ whereas the number of possible sequences $\mathbf{a}$ is finite, the most natural joint maximization procedure consists in first maximizing the joint likelihood $p(\mathbf{r}|\mathbf{a},\boldsymbol{\theta}_S)$ with respect to $\boldsymbol{\theta}_S$ (ML) or the conditional pdf $p(\boldsymbol{\theta}_D \mid \mathbf{r}, \mathbf{a})$ with respect to $\boldsymbol{\theta}_D$ (MAP) for each of the possible $\mathbf{a}$, and then selecting the sequence $\mathbf{a}$ with the largest likelihood:

ML estimation of static sync parameters;
joint likelihood for decision:

$$\begin{aligned}\hat{\boldsymbol{\theta}}_S(\mathbf{a}) &= \arg \max_{\boldsymbol{\theta}_S} \; p(\mathbf{r} \mid \mathbf{a}, \boldsymbol{\theta}_S) \\ \Lambda_S(\mathbf{a}) &= p\Big(\mathbf{r} \mid \mathbf{a},\, \boldsymbol{\theta}_S = \hat{\boldsymbol{\theta}}_S(\mathbf{a})\Big)\end{aligned} \tag{12-21}$$

$$\hat{\mathbf{a}} = \arg \max_{\mathbf{a}} \; \Lambda_S(\mathbf{a})$$

MAP estimation of dynamic sync parameters;
conditional pdf for decision:

$$\begin{aligned}\hat{\boldsymbol{\theta}}_D(\mathbf{a}) &= \arg \max_{\boldsymbol{\theta}_D} \; p(\boldsymbol{\theta}_D \mid \mathbf{r}, \mathbf{a}) \\ \Lambda_D(\mathbf{a}) &= p\Big(\mathbf{r} \mid \mathbf{a},\, \boldsymbol{\theta}_D = \hat{\boldsymbol{\theta}}_D(\mathbf{a})\Big) \cdot p\Big(\boldsymbol{\theta}_D = \hat{\boldsymbol{\theta}}_D(\mathbf{a})\Big)\end{aligned} \tag{12-22}$$

$$\hat{\mathbf{a}} = \arg \max_{\mathbf{a}} \; \Lambda_D(\mathbf{a})$$

The first maximization step yields a conditional sync parameter estimate $\hat{\theta}(\mathbf{a})$ that is subsequently used in the decision likelihood computation as if it were the true parameter. Hence, this procedure resembles the concept of synchronized detection. Here, however, each candidate sequence $\mathbf{a}$ carries its own sync estimate $\hat{\theta}(\mathbf{a})$ conditioned on that sequence.

12.2.2 Optimal Linear Estimation of Static and Gaussian Dynamic Synchronization Parameters

In the previous parts of this book, the estimation of fixed sync parameters $\boldsymbol{\theta}_S$ has been discussed in great detail. Naturally, we shall focus on fading channels with dynamic sync parameters $\boldsymbol{\theta}_D$. On a number of occasions, however, the quasi-stationarity assumption on the sync parameters holds (e.g., in the context of blockwise channel acquisition), so that we will also deal with estimating static parameters $\boldsymbol{\theta}_S$.

In this section, it is assumed that the observation is linearly dependent on $\boldsymbol{\theta}$ (this applies to both static and dynamic parameters), i.e., the observation is of the form

$$\mathbf{r} = \mathbf{A} \cdot \boldsymbol{\theta} + \mathbf{n} \tag{12-23}$$

with an appropriately defined data matrix $\mathbf{A}$ and zero-mean complex Gaussian noise $\mathbf{n}$ with covariance matrix $\mathbf{R}_n$. Notice that the transmission models of eqs. (12-9) and (12-11) comply with this linearity assumption when there is no frequency offset. Concerning dynamic sync parameters $\boldsymbol{\theta}_D$, we shall further assume that these follow a multivariate complex *Gaussian* distribution with known mean vector $\boldsymbol{\mu}_D = E[\boldsymbol{\theta}_D]$ and Hermitian covariance matrix $\mathbf{R}_D = E\left[(\boldsymbol{\theta}_D - \boldsymbol{\mu}_D) \cdot (\boldsymbol{\theta}_D - \boldsymbol{\mu}_D)^H\right]$. Under these assumptions the likelihood / density functions for ML static / MAP dynamic parameter estimation — as well as the likelihood function for detection — are also Gaussian.

ML Static Parameter Estimation

In the case of static parameters, only the noise is random, and the likelihood function becomes

$$\begin{aligned} p(\mathbf{r}|\mathbf{a}, \boldsymbol{\theta}_S) &= N(\mathbf{A}\boldsymbol{\theta}_S, \mathbf{R}_n) \\ &\propto \exp\left\{-[\mathbf{r} - \mathbf{A}\boldsymbol{\theta}_S]^H \cdot \mathbf{R}_n^{-1} \cdot [\mathbf{r} - \mathbf{A}\boldsymbol{\theta}_S]\right\} \end{aligned} \tag{12-24}$$

where $N(\boldsymbol{\mu}, \boldsymbol{\Sigma})$ denotes the (complex multivariate) Gaussian density function of a random vector with mean $\boldsymbol{\mu}$ and covariance matrix $\boldsymbol{\Sigma}$. Maximizing $p(\mathbf{r}|\mathbf{a},\boldsymbol{\theta}_S)$ calls for minimizing the quadratic form (metric)

$$\begin{aligned} m(\mathbf{a}, \boldsymbol{\theta}_S) &= [\mathbf{r} - \mathbf{A}\boldsymbol{\theta}_S]^H \cdot \mathbf{R}_n^{-1} \cdot [\mathbf{r} - \mathbf{A}\boldsymbol{\theta}_S] \\ &= \boldsymbol{\theta}_S^H \cdot \left[\mathbf{A}^H \mathbf{R}_n^{-1} \mathbf{A}\right] \cdot \boldsymbol{\theta}_S - \left[\mathbf{r}^H \mathbf{R}_n^{-1} \mathbf{A}\right] \cdot \boldsymbol{\theta}_S \\ &\quad - \boldsymbol{\theta}_S^H \cdot \left[\mathbf{A}^H \mathbf{R}_n^{-1} \mathbf{r}\right] + \left[\mathbf{r}^H \mathbf{R}_n^{-1} \mathbf{r}\right] \end{aligned} \tag{12-25}$$

By setting the derivative of $m(\mathbf{a},\boldsymbol{\theta}_S)$

$$\frac{\partial}{\partial \boldsymbol{\theta}_S} m(\mathbf{a},\boldsymbol{\theta}_S) = \boldsymbol{\theta}_S^H \cdot \left[\mathbf{A}^H \mathbf{R}_n^{-1} \mathbf{A}\right] - \left[\mathbf{r}^H \mathbf{R}_n^{-1} \mathbf{A}\right] = \left(\left[\mathbf{A}^H \mathbf{R}_n^{-1} \mathbf{A}\right] \cdot \boldsymbol{\theta}_S - \left[\mathbf{A}^H \mathbf{R}_n^{-1} \mathbf{r}\right]\right)^H \tag{12-26}$$

with respect to $\boldsymbol{\theta}_S$ to zero, one immediately obtains the static sync parameter ML estimate and its associated error covariance:

$$\begin{aligned} \hat{\boldsymbol{\theta}}_S(\mathbf{a}) &= \underbrace{\left[\mathbf{A}^H \mathbf{R}_n^{-1} \mathbf{A}\right]^{-1}}_{\boldsymbol{\Sigma}_S(\mathbf{a})} \cdot \left[\mathbf{A}^H \mathbf{R}_n^{-1} \mathbf{r}\right] \\ \boldsymbol{\Sigma}_S(\mathbf{a}) &= \left[\mathbf{A}^H \mathbf{R}_n^{-1} \mathbf{A}\right]^{-1} \end{aligned} \tag{12-27}$$

respectively.

MAP Dynamic Parameter Estimation

In the case of dynamic parameters, both the parameters $\boldsymbol{\theta}_D$ and the additive noise $\mathbf{n}$ are Gaussian random variables. MAP parameter estimation calls for maximizing the pdf $p(\boldsymbol{\theta}_D \mid \mathbf{r}, \mathbf{a})$ [eq. (12-21)], conditioned on the observation $\mathbf{r}$, with respect to $\boldsymbol{\theta}_D$.

When two random vectors $\mathbf{x}$ and $\mathbf{y}$ with means $\boldsymbol{\mu}_x$, $\boldsymbol{\mu}_y$, covariances $\boldsymbol{\Sigma}_x$, $\boldsymbol{\Sigma}_y$ and cross covariance $\boldsymbol{\Sigma}_{xy}$ are jointly Gaussian, the conditional pdf $p(\mathbf{x} \mid \mathbf{y})$ is also Gaussian, i.e., $p(\mathbf{x} \mid \mathbf{y}) = N\left(\boldsymbol{\mu}_{x|y}, \boldsymbol{\Sigma}_{x|y}\right)$, with mean and covariance [11]

$$\begin{aligned} \boldsymbol{\mu}_{x|y} &= \boldsymbol{\mu}_x + \boldsymbol{\Sigma}_{xy} \boldsymbol{\Sigma}_y^{-1} (\mathbf{y} - \boldsymbol{\mu}_y) \\ \boldsymbol{\Sigma}_{x|y} &= \boldsymbol{\Sigma}_x - \boldsymbol{\Sigma}_{xy} \boldsymbol{\Sigma}_y^{-1} \boldsymbol{\Sigma}_{xy}^H \end{aligned} \tag{12-28}$$

respectively. With the correspondences $\mathbf{x} = \boldsymbol{\theta}_D$, $\mathbf{y} = \mathbf{r}$, the auto- and cross covariances evaluate as

$$\begin{aligned} \boldsymbol{\Sigma}_x &= \boldsymbol{\Sigma}_\theta = E\left[(\boldsymbol{\theta}_D - \boldsymbol{\mu}_D) \cdot (\boldsymbol{\theta}_D - \boldsymbol{\mu}_D)^H\right] = \mathbf{R}_D \\ \boldsymbol{\Sigma}_{xy} &= \boldsymbol{\Sigma}_{\theta r} = E\left[(\boldsymbol{\theta}_D - \boldsymbol{\mu}_D) \cdot (\mathbf{r} - \mathbf{A}\boldsymbol{\mu}_D)^H\right] = \mathbf{R}_D \cdot \mathbf{A}^H \\ \boldsymbol{\Sigma}_y &= \boldsymbol{\Sigma}_r = E\left[(\mathbf{r} - \mathbf{A}\boldsymbol{\mu}_D) \cdot (\mathbf{r} - \mathbf{A}\boldsymbol{\mu}_D)^H\right] = \mathbf{A} \cdot \mathbf{R}_D \cdot \mathbf{A}^H + \mathbf{R}_n \end{aligned} \tag{12-29}$$

Recalling the well-known fact that the optimal MAP estimate obtained by maximizing $p(\mathbf{x} \mid \mathbf{y})$ is equivalent to the conditional mean

$$\hat{\mathbf{x}}_{\mathrm{MAP}} = E[\mathbf{x} \mid \mathbf{y}] = \boldsymbol{\mu}_{x|y} \tag{12-30}$$

and that its error covariance equals the conditional covariance matrix $\boldsymbol{\Sigma}_{x|y}$ [11], the MAP estimate of a dynamic parameter $\boldsymbol{\theta}_D$ (which is also conditioned on a here) and its error covariance matrix become

$$\begin{aligned}
\hat{\boldsymbol{\theta}}_D(\mathbf{a}) &= \boldsymbol{\mu}_x + \boldsymbol{\Sigma}_{xy}\boldsymbol{\Sigma}_y^{-1}(\mathbf{y}-\boldsymbol{\mu}_y) \\
&= \boldsymbol{\mu}_D + \mathbf{R}_D\mathbf{A}^H\left(\mathbf{A}\mathbf{R}_D\mathbf{A}^H + \mathbf{R}_n\right)^{-1}(\mathbf{r}-\mathbf{A}\boldsymbol{\mu}_D) \\
&= \quad \left[\mathbf{R}_D\mathbf{A}^H\left(\mathbf{A}\mathbf{R}_D\mathbf{A}^H + \mathbf{R}_n\right)^{-1}\right]\cdot\mathbf{r} \\
&\quad + \left[\mathbf{I} - \mathbf{R}_D\mathbf{A}^H\left(\mathbf{A}\mathbf{R}_D\mathbf{A}^H + \mathbf{R}_n\right)^{-1}\mathbf{A}\right]\cdot\boldsymbol{\mu}_D \\
\boldsymbol{\Sigma}_D(\mathbf{a}) &= \mathbf{R}_D - \mathbf{R}_D\mathbf{A}^H\left(\mathbf{A}\mathbf{R}_D\mathbf{A}^H + \mathbf{R}_n\right)^{-1}\mathbf{A}\mathbf{R}_D
\end{aligned} \tag{12-31}$$

respectively.

We remark that MAP estimation *formally* collapses to ML estimation by setting $\mathbf{R}_D^{-1}$ to zero. This claim – which is not apparent from eq. (12-31) – is proven by using the equivalent expression $\hat{\boldsymbol{\theta}}_D(\mathbf{a}) = \boldsymbol{\Sigma}_D(\mathbf{a}) \cdot \left[\mathbf{A}^H\mathbf{R}_n^{-1}\mathbf{r} + \mathbf{R}_D^{-1}\boldsymbol{\mu}_D\right]$ established below [eq. (12-41)], inserting $\boldsymbol{\Sigma}_D(\mathbf{a}) = \left[\mathbf{A}^H\mathbf{R}_n^{-1}\mathbf{A} + \mathbf{R}_D^{-1}\right]^{-1}$ [eq. (12-40)] and then setting $\mathbf{R}_D^{-1} = \mathbf{0}$ or $\mathbf{R}_D = \infty\cdot\mathbf{I}$. Hence, static parameters – for which statistical side information is not available – can formally be cast into the framework of dynamic parameter estimation by interpreting these parameters as if they were mutually uncorrelated and of infinite power.

Decision Metric for Synchronized Detection

Having obtained a sync parameter estimate $\hat{\boldsymbol{\theta}}(\mathbf{a})$ and its error covariance matrix $\boldsymbol{\Sigma}(\mathbf{a})$ associated with a (hypothetical) data sequence $\mathbf{a}$, the estimate $\hat{\boldsymbol{\theta}}(\mathbf{a})$ can now be inserted into the decision metric $\Lambda(\mathbf{a})$ of eq. (12-21) in order to perform *synchronized ML detection* using the ML or MAP sync parameter estimate.

Since all random quantities are Gaussian, we have

$$\begin{aligned}
p(\mathbf{r}|\mathbf{a},\boldsymbol{\theta}_S) &= N(\mathbf{A}\boldsymbol{\theta}_S,\mathbf{R}_n) \\
&\propto \exp\left\{-[\mathbf{r}-\mathbf{A}\boldsymbol{\theta}_S]^H\cdot\mathbf{R}_n^{-1}\cdot[\mathbf{r}-\mathbf{A}\boldsymbol{\theta}_S]\right\} \\
p(\mathbf{r}|\mathbf{a},\boldsymbol{\theta}_D) &= N(\mathbf{A}\boldsymbol{\theta}_D,\mathbf{R}_n) \\
&\propto \exp\left\{-[\mathbf{r}-\mathbf{A}\boldsymbol{\theta}_D]^H\cdot\mathbf{R}_n^{-1}\cdot[\mathbf{r}-\mathbf{A}\boldsymbol{\theta}_D]\right\} \\
p(\boldsymbol{\theta}_D) &= N(\boldsymbol{\mu}_D,\mathbf{R}_D) \\
&\propto \exp\left\{-[\boldsymbol{\theta}_D-\boldsymbol{\mu}_D]^H\cdot\mathbf{R}_D^{-1}\cdot[\boldsymbol{\theta}_D-\boldsymbol{\mu}_D]\right\}
\end{aligned} \tag{12-32}$$

where the inverse matrices $\mathbf{R}_n^{-1}$, $\mathbf{R}_D^{-1}$ are understood to be pseudoinverses in the case that $\mathbf{R}_n$ and/or $\mathbf{R}_D$ are singular. This is not just an academic subtlety but arises frequently in practice whenever the underlying processes (especially the fading channel processes) are band-limited.

Via eqs. (12-21) and (12-32), the metrics $m(\mathbf{a},\boldsymbol{\theta}_S) \propto -\ln[p(\mathbf{r}\mid\mathbf{a},\boldsymbol{\theta}_S)]$ and $m(\mathbf{a},\boldsymbol{\theta}_D) \propto -\ln[p(\mathbf{r}\mid\mathbf{a},\boldsymbol{\theta}_D)\cdot p(\boldsymbol{\theta}_D)]$ for ML detection with ML/MAP synchronization, respectively, can be expressed as

$$\begin{aligned}
&\text{Static sync parameters:}\\
&m(\mathbf{a},\boldsymbol{\theta}_S) = [\mathbf{r} - \mathbf{A}\boldsymbol{\theta}_S]^H \mathbf{R}_n^{-1} [\mathbf{r} - \mathbf{A}\boldsymbol{\theta}_S]\\
&\text{Dynamic sync parameters:}\\
&m(\mathbf{a},\boldsymbol{\theta}_D) = [\mathbf{r} - \mathbf{A}\boldsymbol{\theta}_D]^H \mathbf{R}_n^{-1} [\mathbf{r} - \mathbf{A}\boldsymbol{\theta}_D] + [\boldsymbol{\theta}_D - \boldsymbol{\mu}_D]^H \mathbf{R}_D^{-1} [\boldsymbol{\theta}_D - \boldsymbol{\mu}_D]
\end{aligned} \tag{12-33}$$

Inserting the ML/MAP parameter estimates of eqs. (12-27) and (12-31) into eq. (12-33) immediately yields the decision metrics $m(\mathbf{a}) = m\left(\mathbf{a},\, \boldsymbol{\theta} = \hat{\boldsymbol{\theta}}\right)$ (to be minimized):

$$\begin{aligned}
&\text{Static sync parameters:}\\
&m_S(\mathbf{a}) = \left[\mathbf{r} - \mathbf{A}\hat{\boldsymbol{\theta}}_S(\mathbf{a})\right]^H \mathbf{R}_n^{-1} \left[\mathbf{r} - \mathbf{A}\hat{\boldsymbol{\theta}}_S(\mathbf{a})\right]\\
&\text{Dynamic sync parameters:}\\
&m_D(\mathbf{a}) = \left[\mathbf{r} - \mathbf{A}\hat{\boldsymbol{\theta}}_D(\mathbf{a})\right]^H \mathbf{R}_n^{-1} \left[\mathbf{r} - \mathbf{A}\hat{\boldsymbol{\theta}}_D(\mathbf{a})\right]\\
&\qquad\qquad + \left[\hat{\boldsymbol{\theta}}_D(\mathbf{a}) - \boldsymbol{\mu}_D\right]^H \mathbf{R}_D^{-1} \left[\hat{\boldsymbol{\theta}}_D(\mathbf{a}) - \boldsymbol{\mu}_D\right]
\end{aligned} \tag{12-34}$$

These metrics can be reformulated by expanding the estimate $\hat{\boldsymbol{\theta}}(\mathbf{a})$. For static sync parameters, expanding eq. (12-34) (ML) yields

$$\begin{aligned}
m_S(\mathbf{a}) = {} & \underbrace{\left[\mathbf{r}^H \mathbf{R}_n^{-1} \mathbf{r}\right]}_{\text{indep. of } \mathbf{a}} + \left[\hat{\boldsymbol{\theta}}_S^H(\mathbf{a}) \cdot \underbrace{\mathbf{A}^H \mathbf{R}_n^{-1} \mathbf{A}}_{\boldsymbol{\Sigma}_S^{-1}(\mathbf{a})} \cdot \hat{\boldsymbol{\theta}}_S(\mathbf{a})\right]\\
& - \left[\hat{\boldsymbol{\theta}}_S^H(\mathbf{a}) \cdot \mathbf{A}^H \mathbf{R}_n^{-1} \mathbf{r}\right] - \left[\mathbf{r}^H \mathbf{R}_n^{-1} \mathbf{A} \cdot \hat{\boldsymbol{\theta}}_S(\mathbf{a})\right]
\end{aligned} \tag{12-35}$$

By reformulating the term

$$\begin{aligned}
\mathbf{r}^H \mathbf{R}_n^{-1} \mathbf{A} \cdot \hat{\boldsymbol{\theta}}_S(\mathbf{a}) &= \mathbf{r}^H \mathbf{R}_n^{-1} \mathbf{A} \cdot \left[\mathbf{A}^H \mathbf{R}_n^{-1} \mathbf{A}\right]^{-1} \cdot \mathbf{A}^H \mathbf{R}_n^{-1} \mathbf{r}\\
&= \underbrace{\mathbf{r}^H \mathbf{R}_n^{-1} \mathbf{A} \cdot \left[\mathbf{A}^H \mathbf{R}_n^{-1} \mathbf{A}\right]^{-1}}_{\hat{\boldsymbol{\theta}}_S^H(\mathbf{a})} \cdot \underbrace{\left[\mathbf{A}^H \mathbf{R}_n^{-1} \mathbf{A}\right]}_{\boldsymbol{\Sigma}_S^{-1}(\mathbf{a})}\\
&\quad \cdot \underbrace{\left[\mathbf{A}^H \mathbf{R}_n^{-1} \mathbf{A}\right]^{-1} \cdot \mathbf{A}^H \mathbf{R}_n^{-1} \mathbf{r}}_{\hat{\boldsymbol{\theta}}_S(\mathbf{a})}\\
&= \mathbf{r}^H \mathbf{R}_n^{-1} \mathbf{A} \cdot \boldsymbol{\Sigma}_S(\mathbf{a}) \cdot \mathbf{A}^H \mathbf{R}_n^{-1} \mathbf{r}\\
&= \hat{\boldsymbol{\theta}}_S^H(\mathbf{a}) \cdot \boldsymbol{\Sigma}_S^{-1}(\mathbf{a}) \cdot \hat{\boldsymbol{\theta}}_S(\mathbf{a})
\end{aligned} \tag{12-36}$$

one observes that the last three terms of eq. (12-35) are equal. Furthermore, the first term can be dropped since it is independent of **a**. Hence, the decision metric based on ML sync can be equivalently expressed as

$$
\begin{aligned}
m_S(\mathbf{a}) &= \mathbf{r}^H \mathbf{R}_n^{-1} \mathbf{A} \cdot \underbrace{\left[\mathbf{A}^H \mathbf{R}_n^{-1} \mathbf{A}\right]^{-1}}_{\boldsymbol{\Sigma}_S(\mathbf{a})} \cdot \mathbf{A}^H \mathbf{R}_n^{-1} \mathbf{r} \\
&= \hat{\boldsymbol{\theta}}_S^H(\mathbf{a}) \cdot \underbrace{\left[\mathbf{A}^H \mathbf{R}_n^{-1} \mathbf{A}\right]}_{\boldsymbol{\Sigma}_S^{-1}(\mathbf{a})} \cdot \hat{\boldsymbol{\theta}}_S(\mathbf{a}) \qquad \to \max_{\mathbf{a}}
\end{aligned}
\tag{12-37}
$$

The same procedure can likewise be applied to dynamic sync parameters. Expanding eq. (12-34) (MAP) yields

$$
\begin{aligned}
m_D(\mathbf{a}) = \; & \underbrace{\left[\mathbf{r}^H \mathbf{R}_n^{-1} \mathbf{r} + \boldsymbol{\mu}_D^H \mathbf{R}_D^{-1} \boldsymbol{\mu}_D\right]}_{\text{indep. of } \mathbf{a}} + \hat{\boldsymbol{\theta}}_D^H(\mathbf{a}) \cdot \left[\mathbf{A}^H \mathbf{R}_n^{-1} \mathbf{A} + \mathbf{R}_D^{-1}\right] \cdot \hat{\boldsymbol{\theta}}_D(\mathbf{a}) \\
& - \hat{\boldsymbol{\theta}}_D^H(\mathbf{a}) \cdot \left[\mathbf{A}^H \mathbf{R}_n^{-1} \mathbf{r} + \mathbf{R}_D^{-1} \boldsymbol{\mu}_D\right] - \left[\mathbf{A}^H \mathbf{R}_n^{-1} \mathbf{r} + \mathbf{R}_D^{-1} \boldsymbol{\mu}_D\right]^H \cdot \hat{\boldsymbol{\theta}}_D(\mathbf{a})
\end{aligned}
\tag{12-38}
$$

In order to obtain a metric expression similar to that of eq. (12-37), let us invoke the matrix inversion lemma [12, 2]:

$$
\begin{aligned}
\mathbf{E} &= \mathbf{A}^H \mathbf{B}^{-1} \mathbf{A} + \mathbf{C}^{-1} \\
\mathbf{E}^{-1} &= \mathbf{C} - \mathbf{C}\mathbf{A}^H \left[\mathbf{B} + \mathbf{A}\mathbf{C}\mathbf{A}^H\right]^{-1} \mathbf{A}\mathbf{C} \\
&= \mathbf{D} - \mathbf{D}[\mathbf{C} + \mathbf{D}]^{-1}\mathbf{D} \qquad \text{with} \quad \mathbf{D} = \left[\mathbf{A}^H \mathbf{B}^{-1} \mathbf{A}\right]^{-1}
\end{aligned}
\tag{12-39}
$$

whereby the following identities are established:

$$
\begin{aligned}
\boldsymbol{\Sigma}_D(\mathbf{a}) &= \left[\mathbf{A}^H \mathbf{R}_n^{-1} \mathbf{A} + \mathbf{R}_D^{-1}\right]^{-1} \qquad = \left[\boldsymbol{\Sigma}_S^{-1} + \mathbf{R}_D^{-1}\right]^{-1} \\
&= \mathbf{R}_D - \mathbf{R}_D \mathbf{A}^H \left[\mathbf{A}\mathbf{R}_D\mathbf{A}^H + \mathbf{R}_n\right]^{-1} \mathbf{A}\mathbf{R}_D \qquad \begin{pmatrix}\text{first} \\ \text{identity}\end{pmatrix} \\
&= \boldsymbol{\Sigma}_S - \boldsymbol{\Sigma}_S[\boldsymbol{\Sigma}_S + \mathbf{R}_D]^{-1}\boldsymbol{\Sigma}_S \qquad \begin{pmatrix}\text{second} \\ \text{identity}\end{pmatrix}
\end{aligned}
\tag{12-40}
$$

where the second identity is valid provided that $\boldsymbol{\Sigma}_S = \boldsymbol{\Sigma}_S(\mathbf{a}) = \left[\mathbf{A}^H \mathbf{R}_n^{-1} \mathbf{A}\right]^{-1}$ exists. From eq. (12-40), the term $\mathbf{A}^H \mathbf{R}_n^{-1} \mathbf{A} + \mathbf{R}_D^{-1}$ is identified as the inverse error covariance matrix $\boldsymbol{\Sigma}_D^{-1}$. Furthermore, with the help of eq. (12-31), the expression

$$
\begin{aligned}
&\boldsymbol{\Sigma}_D(\mathbf{a}) \cdot \left[\mathbf{A}^H \mathbf{R}_n^{-1}\mathbf{r} + \mathbf{R}_D^{-1}\boldsymbol{\mu}_D\right] \\
&= \left[\mathbf{R}_D - \mathbf{R}_D\mathbf{A}^H\left(\mathbf{A}\mathbf{R}_D\mathbf{A}^H + \mathbf{R}_n\right)^{-1}\mathbf{A}\mathbf{R}_D\right] \cdot \left[\mathbf{A}^H \mathbf{R}_n^{-1}\mathbf{r} + \mathbf{R}_D^{-1}\boldsymbol{\mu}_D\right] \\
&= \mathbf{R}_D\mathbf{A}^H\mathbf{R}_n^{-1}\mathbf{r} + \boldsymbol{\mu}_D - \mathbf{R}_D\mathbf{A}^H\left(\mathbf{A}\mathbf{R}_D\mathbf{A}^H + \mathbf{R}_n\right)^{-1}\mathbf{A}\mathbf{R}_D\mathbf{A}^H\mathbf{R}_n^{-1}\mathbf{r} \\
&\quad - \mathbf{R}_D\mathbf{A}^H\left(\mathbf{A}\mathbf{R}_D\mathbf{A}^H + \mathbf{R}_n\right)^{-1}\mathbf{A}\boldsymbol{\mu}_D \\
&= \boldsymbol{\mu}_D + \mathbf{R}_D\mathbf{A}^H\Bigg[\underbrace{\left(\mathbf{R}_n^{-1} - \left(\mathbf{A}\mathbf{R}_D\mathbf{A}^H + \mathbf{R}_n\right)^{-1}\mathbf{A}\mathbf{R}_D\mathbf{A}^H\mathbf{R}_n^{-1}\right)}_{\left(\mathbf{A}\mathbf{R}_D\mathbf{A}^H + \mathbf{R}_n\right)^{-1}\underbrace{\left[\left(\mathbf{A}\mathbf{R}_D\mathbf{A}^H + \mathbf{R}_n\right)\mathbf{R}_n^{-1} - \mathbf{A}\mathbf{R}_D\mathbf{A}^H\mathbf{R}_n^{-1}\right]}_{\mathbf{I}}} \cdot \mathbf{r} \\
&\qquad\qquad - \left(\mathbf{A}\mathbf{R}_D\mathbf{A}^H + \mathbf{R}_n\right)^{-1}\mathbf{A}\boldsymbol{\mu}_D\Bigg] \\
&= \boldsymbol{\mu}_D + \mathbf{R}_D\mathbf{A}^H\left(\mathbf{A}\mathbf{R}_D\mathbf{A}^H + \mathbf{R}_n\right)^{-1}\left(\mathbf{r} - \mathbf{A}\boldsymbol{\mu}_D\right) \\
&= \hat{\boldsymbol{\theta}}_D(\mathbf{a})
\end{aligned}
\tag{12-41}
$$

is identified as the MAP estimate $\hat{\boldsymbol{\theta}}_D(\mathbf{a})$. With these identities, the last term of eq. (12-38) can be expressed as

$$
\begin{aligned}
&\left[\mathbf{A}^H \mathbf{R}_n^{-1}\mathbf{r} + \mathbf{R}_D^{-1}\boldsymbol{\mu}_D\right]^H \cdot \hat{\boldsymbol{\theta}}_D(\mathbf{a}) \\
&\quad = \underbrace{\left[\mathbf{A}^H \mathbf{R}_n^{-1}\mathbf{r} + \mathbf{R}_D^{-1}\boldsymbol{\mu}_D\right]^H \cdot \boldsymbol{\Sigma}_D(\mathbf{a})}_{\hat{\boldsymbol{\theta}}_D^H(\mathbf{a})} \cdot \boldsymbol{\Sigma}_D^{-1}(\mathbf{a}) \cdot \hat{\boldsymbol{\theta}}_D(\mathbf{a}) \\
&\quad = \hat{\boldsymbol{\theta}}_D^H(\mathbf{a}) \cdot \boldsymbol{\Sigma}_D^{-1}(\mathbf{a}) \cdot \hat{\boldsymbol{\theta}}_D(\mathbf{a})
\end{aligned}
\tag{12-42}
$$

so that, once again, the last three terms of eq. (12-38) are equal. This leads to the equivalent metric based on MAP sync:

$$
\begin{aligned}
m_D(\mathbf{a}) &= \left[\mathbf{A}^H \mathbf{R}_n^{-1}\mathbf{r} + \mathbf{R}_D^{-1}\boldsymbol{\mu}_D\right]^H \cdot \underbrace{\left[\mathbf{A}^H \mathbf{R}_n^{-1}\mathbf{A} + \mathbf{R}_D^{-1}\right]^{-1}}_{\boldsymbol{\Sigma}_D(\mathbf{a})} \cdot \left[\mathbf{A}^H \mathbf{R}_n^{-1}\mathbf{r} + \mathbf{R}_D^{-1}\boldsymbol{\mu}_D\right] \\
&= \hat{\boldsymbol{\theta}}_D^H(\mathbf{a}) \cdot \underbrace{\left[\mathbf{A}^H \mathbf{R}_n^{-1}\mathbf{A} + \mathbf{R}_D^{-1}\right]}_{\boldsymbol{\Sigma}_D^{-1}(\mathbf{a})} \cdot \hat{\boldsymbol{\theta}}_D(\mathbf{a}) \qquad \to \max_{\mathbf{a}}
\end{aligned}
\tag{12-43}
$$

In summary, the decision metrics can be expressed in terms of an inner product of the ML or MAP sync parameter estimate $\hat{\boldsymbol{\theta}}(\mathbf{a})$, whichever applies.

So far, the results on synchronized detection – especially the detection part – have been derived using the high-SNR approximation of eq. (12-21). In the linear Gaussian case, however, the ML decision rule of eq. (12-43) [which is based

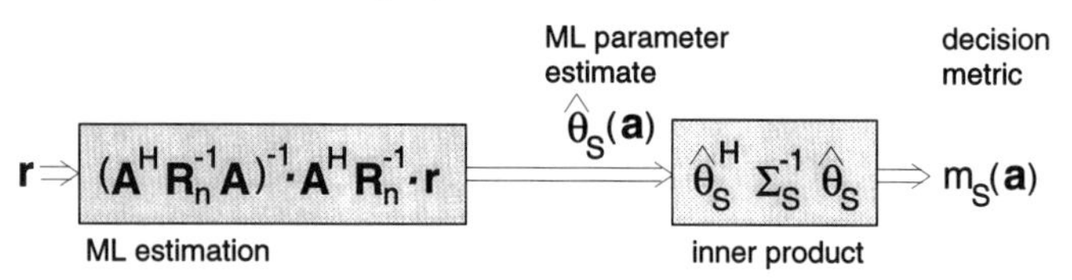

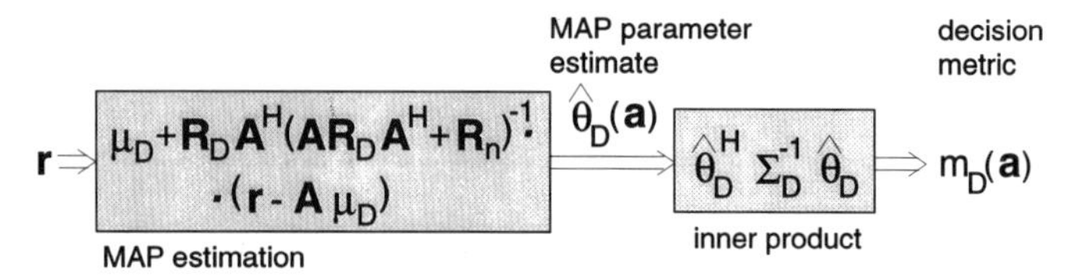

Figure 12-1 Generation of Decision Metrics via ML/MAP Sync Parameter Estimation

on MAP estimation according to eq. (12-31)] and, as a consequence, all results on synchronized detection established in this chapter, are not approximative but indeed optimal at all SNR. The proof for this claim is given in Section 12.2.6.

To summarize the discussion of joint detection and estimation so far, Figure 12-1 displays the generation of the ML decision metric based on ML/MAP sync parameter estimation. For joint detection and ML estimation, the optimal decision metric $m_S(\mathbf{a})$ is obtained by first generating the ML estimate $\hat{\theta}_S(\mathbf{a})$ [eq. (12-27)] and then forming an inner product according to eq. (12-37). Likewise, for joint detection and MAP estimation, the optimal decision metric $m_D(\mathbf{a})$ is obtained by first generating the MAP estimate $\hat{\theta}_D(\mathbf{a})$ [eq. (12-31)] and then forming an inner product according to eq. (12-43).

Due to its tremendous computational complexity, the full optimal joint detection and estimation procedure involving ML detection based on ML [eqs. (12-27) and (12-37)] or MAP [eqs. (12-31) and (12-43)] sync parameter estimation will most often not be feasible to implement in practice, especially when dynamic (fading) parameters are involved. Nevertheless, it is instructive to notice that there exists a closed-form optimal solution to the joint estimation-detection problem in a fading environment just as for nonfading transmission and reception. The optimal receiver for fading channels is derived by applying the same standard elements of estimation and detection theory that have been used for nonfading transmission. Furthermore, the expressions of eqs. (12-27) and (12-37) [ML] and (12-31) and (12-43) [MAP] can serve as the basis for deriving simplified receivers that can actually be implemented.

MAP Dynamic Sync Parameter Estimation via ML Estimation

In the case that the ML estimate of a dynamic parameter θ_D exists (the MAP estimate always exists), it is interesting to establish a link between the ML and MAP estimates as well as their error covariances [eqs. (12-27) and (12-31)].

From eq. (12-40) one notices that the error covariance matrix $\mathbf{\Sigma}_S$ of ML estimation also appears in the context of MAP dynamic parameter estimation. This observation gives rise to a relationship that can be established between MAP and ML linear Gaussian sync parameter estimation, provided that $\mathbf{\Sigma}_S$ and thus the ML estimate exists. As we know, the MAP estimate $\hat{\boldsymbol{\theta}}_D(\mathbf{a})$ [eq. (12-31)] is the optimal dynamic parameter estimate which exploits prior knowledge of some statistical channel parameters. Suppose now that we choose not to consider this prior knowledge for the moment, the dynamic channel parameters are estimated as if they were static, i.e., one ends up with the ML estimate $\hat{\boldsymbol{\theta}}_S(\mathbf{a})$ of dynamic parameters $\boldsymbol{\theta}_D$. The expressions for $\hat{\boldsymbol{\theta}}_S(\mathbf{a})$ and $\mathbf{\Sigma}_S(\mathbf{a})$ of eq. (12-27) are then valid not only for (optimal) ML estimation of static parameters but likewise for (suboptimal) ML estimation of dynamic sync parameters.

The existence of an ML estimate of dynamic parameters is by no means guaranteed. In the flat fading case ($\boldsymbol{\theta}_D = \mathbf{c}$), matrices $\mathbf{A}$, $\mathbf{R}_n^{-1}$ [eq. (12-9)] are $N \times N$ square so that $\mathbf{A}^H \mathbf{R}_n^{-1} \mathbf{A}$ is also $N \times N$ square and, assuming nonzero data symbols a_k, full rank and thus invertible. Hence, the ML estimate $\hat{\boldsymbol{\theta}}_S = \hat{\mathbf{c}}_S$ of the *flat* fading channel $\boldsymbol{\theta}_D = \mathbf{c}$ exists. On the other hand, in the selective fading case ($\boldsymbol{\theta}_D = \mathbf{h}$), matrix $\mathbf{A}$ is $2(N{+}L)\times 2(L{+}1)N$ and $\mathbf{R}_n^{-1}$ $2(N{+}L)\times 2(N{+}L)$ [eq. (12-11)] so that $\mathbf{A}^H \mathbf{R}_n^{-1} \mathbf{A}$ is $2(L{+}1)N \times 2(L{+}1)N$ square but not full rank. Hence, an ML estimate $\hat{\boldsymbol{\theta}}_S = \hat{\mathbf{h}}_S$ of the *selective* fading channel $\boldsymbol{\theta}_D = \mathbf{h}$ does not exist. This is intuitively clear since $2(L{+}1)N$ sync parameters $\mathbf{h}$ cannot be estimated from only $2(N{+}L)$ (with $L > 1$) observed samples $\mathbf{r}$ without side information.

Case 1: ML Estimate $\hat{\theta}_S$ of θ_D Exists (Flat Fading)

In this case, dynamic sync parameters can also be viewed as if they were static, i.e., $\boldsymbol{\theta}_S = \boldsymbol{\theta}_D = \boldsymbol{\theta}$. Now the second identity of eq. (12-40) reveals that the MAP error covariance matrix $\mathbf{\Sigma}_D(\mathbf{a})$ can be obtained from the ML error covariance matrix $\mathbf{\Sigma}_S(\mathbf{a})$ by a linear matrix operation:

$$\mathbf{\Sigma}_D(\mathbf{a}) = \underbrace{\left[\mathbf{I} - \mathbf{\Sigma}_S(\mathbf{a}) \cdot [\mathbf{R}_D + \mathbf{\Sigma}_S(\mathbf{a})]^{-1}\right]}_{\mathbf{N}(\mathbf{a})} \cdot \mathbf{\Sigma}_S(\mathbf{a}) \tag{12-44}$$

A similar relationship can be established between the MAP and ML estimates via eqs. (12-27) and (12-41):

$$\begin{aligned}
\hat{\boldsymbol{\theta}}_D(\mathbf{a}) &= \underbrace{\left[\mathbf{A}^H \mathbf{R}_n^{-1} \mathbf{A} + \mathbf{R}_D^{-1}\right]^{-1}}_{\mathbf{\Sigma}_D(\mathbf{a})} \cdot \left[\mathbf{A}^H \mathbf{R}_n^{-1} \mathbf{r} + \mathbf{R}_D^{-1} \boldsymbol{\mu}_D\right] \\
&= \mathbf{\Sigma}_D(\mathbf{a}) \cdot \left[\mathbf{A}^H \mathbf{R}_n^{-1} \mathbf{r}\right] + \mathbf{\Sigma}_D(\mathbf{a}) \cdot \left[\mathbf{R}_D^{-1} \boldsymbol{\mu}_D\right] \\
&= \mathbf{N}(\mathbf{a}) \cdot \underbrace{\mathbf{\Sigma}_S(\mathbf{a}) \cdot \left[\mathbf{A}^H \mathbf{R}_n^{-1} \mathbf{r}\right]}_{\hat{\boldsymbol{\theta}}_S(\mathbf{a})} + \underbrace{\left[\mathbf{\Sigma}_D(\mathbf{a}) \cdot \mathbf{R}_D^{-1}\right]}_{\mathbf{M}(\mathbf{a})} \cdot \boldsymbol{\mu}_D \\
&= \mathbf{N}(\mathbf{a}) \cdot \hat{\boldsymbol{\theta}}_S(\mathbf{a}) \; + \; \mathbf{M}(\mathbf{a}) \cdot \boldsymbol{\mu}_D
\end{aligned} \tag{12-45}$$

Therefore, the MAP parameter estimate $\hat{\boldsymbol{\theta}}_D(\mathbf{a})$ can be obtained from the ML estimate $\hat{\boldsymbol{\theta}}_S(\mathbf{a})$ of the *same* parameters by premultiplication with a matrix $\mathbf{N}(\mathbf{a})$, followed by the addition of the weighted mean. From eq. (12-44) it is observed that the MAP error covariance matrix $\mathbf{\Sigma}_D(\mathbf{a})$ can be obtained from the ML error covariance matrix $\mathbf{\Sigma}_S(\mathbf{a})$ by premultiplication with the *same* matrix $\mathbf{N}(\mathbf{a})$. This relationship may be termed the *separation property* of MAP dynamic parameter estimation in the sense that MAP estimation can be performed by the two-step procedure of (i) computing the ML estimate from the observation and (ii) computing the MAP estimate from the ML estimate:

$$\begin{array}{ccccc} \text{observation} & & \text{ML estimate} & & \text{MAP estimate} \\ \mathbf{r} & \longrightarrow & \hat{\boldsymbol{\theta}}_S(\mathbf{a}) & \longrightarrow & \hat{\boldsymbol{\theta}}_D(\mathbf{a}) \\ & \begin{array}{c}\text{premultiply by} \\ \mathbf{\Sigma}_S(\mathbf{a})\cdot\mathbf{A}^H\mathbf{R}_n^{-1}\end{array} & & \begin{array}{l}\text{premultiply by } \mathbf{N}(\mathbf{a}), \\ \text{add weighted mean } \mathbf{M}(\mathbf{a})\cdot\boldsymbol{\mu}_D\end{array} & \end{array} \tag{12-46}$$

The transition matrix $\mathbf{N}(\mathbf{a})$ that links both the ML/MAP estimates and the ML/MAP error covariances, as well as the weighting matrix $\mathbf{M}(\mathbf{a})$ are given by

$$\begin{aligned} \mathbf{N}(\mathbf{a}) &= \mathbf{I} - \mathbf{\Sigma}_S(\mathbf{a})\cdot[\mathbf{R}_D + \mathbf{\Sigma}_S(\mathbf{a})]^{-1} \\ &= \mathbf{I} - \left[\mathbf{A}^H\mathbf{R}_n^{-1}\mathbf{A}\right]^{-1}\cdot\left(\mathbf{R}_D + \left[\mathbf{A}^H\mathbf{R}_n^{-1}\mathbf{A}\right]^{-1}\right)^{-1} \\ \mathbf{M}(\mathbf{a}) &= \mathbf{\Sigma}_D(\mathbf{a})\cdot\mathbf{R}_D^{-1} \\ &= \mathbf{N}(\mathbf{a})\cdot\mathbf{\Sigma}_S(\mathbf{a})\cdot\mathbf{R}_D^{-1} \\ &= \mathbf{I} - \mathbf{R}_D\mathbf{A}^H\left[\mathbf{A}\mathbf{R}_D\mathbf{A}^H + \mathbf{R}_n\right]^{-1}\mathbf{A} \end{aligned} \tag{12-47}$$

As indicated above, ML estimation does not require any knowledge of channel parameters other than the covariance matrix $\mathbf{R}_n$ of the additive noise. In the case of AWGN, not even the noise power needs to be known since then the matrix by which the observation must be multiplied collapses to $\mathbf{\Sigma}_S(\mathbf{a})\cdot\mathbf{A}^H\mathbf{R}_n^{-1} = \left[\mathbf{A}^H\mathbf{R}_n^{-1}\mathbf{A}\right]^{-1}\cdot\mathbf{A}^H\mathbf{R}_n^{-1} = \left[\mathbf{A}^H\mathbf{A}\right]^{-1}\cdot\mathbf{A}^H$. On the other hand, matrices $\mathbf{N}(\mathbf{a})$ and $\mathbf{M}(\mathbf{a})$ needed for MAP estimation incorporate the prior knowledge about the channel, as quantified by the mean $\boldsymbol{\mu}_D$ and covariance $\mathbf{R}_D$. The reduction in the MAP error covariance with respect to the ML error covariance (elements along the main diagonal of $\mathbf{\Sigma}_D$ and $\mathbf{\Sigma}_S$, respectively) is also determined by the transition matrix $\mathbf{N}(\mathbf{a})$ and thus the channel statistics.

Case 2: ML Estimate $\hat{\boldsymbol{\theta}}_S$ of $\boldsymbol{\theta}_D$ Does Not Exist (Selective Fading)

In this case, the entire vector of dynamic sync parameters cannot be viewed as if it were static. However, one may try and find a smaller *subset* $\boldsymbol{\theta}_S \subset \boldsymbol{\theta}_D = \boldsymbol{\theta}$ of $\boldsymbol{\theta}_D$ for which an ML estimate $\hat{\boldsymbol{\theta}}_S$ exists. For example, from the vector $\mathbf{h}$ of N selective channel impulse responses $\mathbf{h}_k$ ($k = 0, \ldots, N$–1), one may select a number $\overline{N}$ of channel "probes" $\mathbf{h}_{k_K}$ ($K = 0, \ldots, \overline{N}-1$) such that the number $\overline{N}\cdot 2(L+1)$ of unknowns does not exceed the number $2(N+L)$ of observed samples. Then MAP estimation can again be performed using the ML estimate $\hat{\boldsymbol{\theta}}_S$. However, in order to make sure that the MAP estimate $\hat{\boldsymbol{\theta}}_D$ so obtained is actually optimal, the subset $\boldsymbol{\theta}_S$ must be chosen tacitly such that ML estimation does not entail a loss of information with respect to the observation $\mathbf{r}$. In other words, the ML estimate

$\hat{\theta}_S$ must be a sufficient statistic for MAP parameter estimation. These and other aspects are discussed in greater detail in Section 12.2.4.

12.2.3 Joint Detection and Estimation for Flat Fading Channels

Let us now further explore joint detection and channel estimation for the – seemingly simple – case of flat fading channels. Here, zero frequency offset and perfect pulse matched filtering (implying perfect timing compensation) is assumed. We remark that close-to-perfect timing estimation may be accomplished by the following procedure: (i) store the entire received signal (or a sufficiently long section thereof); (ii) in a first processing pass, acquire a timing estimate by applying a non-data-aided (NDA) timing estimation algorithm just as for nonfading channels (Chapter 5); and (iii) in a second pass, perform timing-compensating matched filtering on the entire stored received signal. Thus no data is lost during the timing acquisition phase. Then the model of eq. (12-9) [with $\mathbf{W}(\Omega') = \mathbf{I}$] applies:

$$\underbrace{\begin{pmatrix} z_0 \\ z_1 \\ \vdots \\ z_{N-1} \end{pmatrix}}_{\mathbf{z}} = \underbrace{\begin{pmatrix} a_0 & & & \\ & a_1 & & \\ & & \ddots & \\ & & & a_{N-1} \end{pmatrix}}_{\mathbf{A}} \cdot \underbrace{\begin{pmatrix} c_0 \\ c_1 \\ \vdots \\ c_{N-1} \end{pmatrix}}_{\mathbf{c}} + \mathbf{m} \tag{12-48}$$

with dynamic sync parameter vector $\boldsymbol{\theta}_D = \mathbf{c}$ representing the flat fading channel process over the duration of the message $\mathbf{a}$, and AWGN $\mathbf{m}$ with covariance matrix $\mathbf{R}_m = N_0 \cdot \mathbf{I}$. Further assume that the dynamic channel process $\{c_k\}$ is the superposition of a real-valued line-of-sight (LOS) component $c_{s,k} = \alpha_s$ (zero Doppler shift) with fixed power $\rho_s = |\alpha_s|^2$ and a complex Gaussian process $c_{d,k}$ with average power ρ_d made up of a cluster of elementary scattered rays [see eq. (11-36)]. The channel process therefore is Rician with K-factor $K = \rho_s/\rho_d$. The total average channel process power $\rho = \rho_s + \rho_d$ is assumed to be unity, as well as the average symbol energy $\overline{E}_a = E\{|a_k|^2\}$.

On mobile channels, the scattered component $c_{d,k}$ is often modeled to exhibit the U-shaped Jakes Doppler spectrum $S_{c_d}(\psi')$ of eq. (11-53). The Jakes spectrum, which is based on the isotropic scattering assumption, accentuates instantaneous Doppler shifts near the cutoff frequency λ'_D. As we shall see later in this section, the actual shape of the Doppler spectrum has no noticeable effect on the estimator performance so that, for the purpose of receiver design, the Jakes Doppler spectrum may as well be replaced by an *ideal lowpass* spectrum with the same cutoff frequency λ'_D:

Jakes Doppler spectrum and respective autocorrelation:

$$S_{c_d}\left(e^{j\psi'}\right) = \rho_d \cdot \begin{cases} \dfrac{1}{\pi\lambda'_D} \cdot \dfrac{1}{\sqrt{1-(\psi'/(2\pi\lambda'_D))^2}} & \text{if } |\psi'| \le 2\pi\lambda'_D \\ 0 & \text{otherwise} \end{cases}$$

$$R_{c_d}(m) = \rho_d \cdot \alpha_d(m) \qquad \text{with} \quad \alpha_d(m) = \mathrm{J}_0(2\pi\lambda'_D \cdot m)$$

Ideal lowpass approximation:

$$S_{c_d}\left(e^{j\psi'}\right) = \rho_d \cdot \begin{cases} \dfrac{1}{2\lambda'_D} & \text{if } |\psi'| \leq 2\pi\lambda'_D \\ 0 & \text{otherwise} \end{cases}$$
$$R_{c_d}(m) = \rho_d \cdot \alpha_d(m) \qquad \text{with} \quad \alpha_d(m) = \mathrm{si}(2\pi\lambda'_D \cdot m) \tag{12-49}$$

We remark that the lowpass approximation is further motivated in the case that the receiver performs some kind of frequency offset estimation and/or Doppler tracking in front of the detection and channel estimation stage. The Doppler tracking stage attempts to counteract instantaneous Doppler shifts so that the channel estimator is left with *residual* channel variations whose Doppler spectrum tends to be more pronounced about zero.

At any rate, the dynamic sync parameter vector $\boldsymbol{\theta}_D = \mathbf{c}$ is characterized by its mean $\boldsymbol{\mu}_D = \mathbf{c}_s = \alpha_s \cdot \mathbf{1} = \alpha_s \cdot (1\;1\;\dots\;1)^T$ (LOS component) and covariance matrix

$$\begin{aligned} \mathbf{R}_D &= E\left[(\boldsymbol{\theta}_D - \boldsymbol{\mu}_D)\cdot(\boldsymbol{\theta}_D - \boldsymbol{\mu}_D)^H\right] \qquad = E\left[\mathbf{c}_d \cdot \mathbf{c}_d^H\right] \\ &= \rho_d \cdot \begin{pmatrix} \alpha_d(0) & \dots & \alpha_d(N-1) \\ \vdots & \ddots & \vdots \\ \alpha_d(N-1) & \dots & \alpha_d(0) \end{pmatrix} \end{aligned} \tag{12-50}$$

As a word of caution we remark that, due to the band-limitation of the underlying random process, the matrix $\mathbf{R}_D$ is very badly conditioned so that the numerical inversion of $\mathbf{R}_D$ should be avoided.

Making use of prior knowledge of the channel statistics $\boldsymbol{\mu}_D$ (thus the K-factor), $\mathbf{R}_D$ and the noise power N_0, the optimal MAP channel estimator, conditioned on data sequence $\mathbf{a}$, and its error covariance matrix become [eq. (12-31)]

$$\begin{aligned} \hat{\boldsymbol{\theta}}_D(\mathbf{a}) &= \underbrace{\left[\mathbf{A}^H\mathbf{R}_m^{-1}\mathbf{A} + \mathbf{R}_D^{-1}\right]^{-1}}_{\boldsymbol{\Sigma}_D(\mathbf{a})} \cdot \left[\mathbf{A}^H\mathbf{R}_m^{-1}\mathbf{z} + \mathbf{R}_D^{-1}\boldsymbol{\mu}_D\right] \\ \boldsymbol{\Sigma}_D(\mathbf{a}) &= \left[\mathbf{A}^H\mathbf{R}_m^{-1}\mathbf{A} + \mathbf{R}_D^{-1}\right]^{-1} \end{aligned} \tag{12-51}$$

Noting that $\mathbf{A}^H\mathbf{R}_m^{-1}\mathbf{A} = (1/N_0)\cdot\mathbf{P}(\mathbf{a})$ where $\mathbf{P}(\mathbf{a}) = \mathbf{A}^H\mathbf{A} = \mathrm{diag}(p_0, p_1, \dots, p_{N-1})$ is a diagonal matrix containing the symbol powers $p_k = |\, a_k\,|^2$ along its main diagonal, the transition matrices $\mathbf{N}(\mathbf{a})$ and $\mathbf{M}(\mathbf{a})$ [eq. (12-47)] can be reformulated using the second identity of eq. (12-40):

$$\begin{aligned} \mathbf{N}(\mathbf{a}) &= \mathbf{I} - \boldsymbol{\Sigma}_S(\mathbf{a})\cdot[\mathbf{R}_D + \boldsymbol{\Sigma}_S(\mathbf{a})]^{-1} \\ &= \mathbf{I} - N_0\mathbf{P}^{-1}(\mathbf{a})\cdot\left[\mathbf{R}_D + N_0\mathbf{P}^{-1}(\mathbf{a})\right]^{-1} \\ &= \frac{1}{N_0}\cdot\left[\mathbf{I} - \mathbf{R}_D\cdot\left[\mathbf{R}_D + N_0\mathbf{P}^{-1}(\mathbf{a})\right]^{-1}\right]\cdot\mathbf{R}_D\cdot\mathbf{P}(\mathbf{a}) \\ \mathbf{M}(\mathbf{a}) &= \mathbf{N}(\mathbf{a})\cdot\boldsymbol{\Sigma}_S(\mathbf{a})\cdot\mathbf{R}_D^{-1} \\ &= \mathbf{I} - \mathbf{R}_D\cdot\left(\mathbf{R}_D + N_0\mathbf{P}^{-1}(\mathbf{a})\right)^{-1} \end{aligned} \tag{12-52}$$

Matrices $\mathbf{N}(\mathbf{a})$ and $\mathbf{M}(\mathbf{a})$ are seen to be dependent on the symbol powers $p_k =$

$| a_k |^2$ but not the symbol phases, and, of course, dependent on the channel parameters K, λ'_D, N_0. Therefore, the ML and MAP flat fading Rician channel estimates and covariances can be cast into the form:

$$
\begin{aligned}
&\text{ML channel estimate and error covariance:} \\
&\hat{\boldsymbol{\theta}}_S(\mathbf{a}) = \left[\boldsymbol{\Sigma}_S(\mathbf{a}) \cdot \mathbf{A}^H \mathbf{R}_m^{-1}\right] \cdot \mathbf{z} \qquad = \mathbf{P}^{-1}(\mathbf{a}) \cdot \mathbf{A}^H \cdot \mathbf{z} \\
&\boldsymbol{\Sigma}_S(\mathbf{a}) = \left[\mathbf{A}^H \mathbf{R}_m^{-1} \mathbf{A}\right]^{-1} \qquad = N_0 \cdot \mathbf{P}^{-1}(\mathbf{a}) \\
&\text{MAP channel estimate and error covariance:} \\
&\hat{\boldsymbol{\theta}}_D(\mathbf{a}) = \mathbf{N}(\mathbf{a}, K, \lambda_D', N_0) \cdot \hat{\boldsymbol{\theta}}_S(\mathbf{a}) + \boldsymbol{\mu}(\mathbf{a}, K, \lambda_D', N_0) \\
&\boldsymbol{\Sigma}_D(\mathbf{a}) = \mathbf{N}(\mathbf{a}, K, \lambda_D', N_0) \cdot \boldsymbol{\Sigma}_S(\mathbf{a})
\end{aligned}
\tag{12-53}
$$

with matrix $\mathbf{N}(\bullet)$, weighted mean vector $\boldsymbol{\mu}(\bullet) = \mathbf{M}(\bullet) \cdot \boldsymbol{\mu}_D = \alpha_s \cdot \mathbf{M}(\bullet) \cdot \mathbf{1}$ ($\rho_s = \alpha_s^2 = K/(1+K)$), and channel covariance matrix $\mathbf{R}_D(K, \lambda'_D, N_0)$ [eq. (12-50) with $\rho_d = 1/(1+K)$] given by

$$
\begin{aligned}
\mathbf{N}(\mathbf{a}, K, \lambda'_D, N_0) &= \mathbf{I} - N_0 \mathbf{P}^{-1}(\mathbf{a}) \cdot \left[\mathbf{R}_D(K, \lambda'_D) + N_0 \mathbf{P}^{-1}(\mathbf{a})\right]^{-1} \\
\boldsymbol{\mu}(\mathbf{a}, K, \lambda'_D, N_0) &= \sqrt{\frac{K}{K+1}} \cdot \left[\mathbf{I} - \mathbf{R}_D(K, \lambda'_D) \cdot \left[\mathbf{R}_D(K, \lambda'_D) + N_0 \mathbf{P}^{-1}(\mathbf{a})\right]^{-1}\right] \cdot \mathbf{1} \\
\mathbf{R}_D(K, \lambda'_D) &= \frac{1}{K+1} \cdot \begin{pmatrix} \alpha_d(0) & \dots & \alpha_d(N-1) \\ \vdots & \ddots & \vdots \\ \alpha_d(N-1) & \dots & \alpha_d(0) \end{pmatrix}
\end{aligned}
\tag{12-54}
$$

According to eq. (12-53), the (conditional) ML channel estimate $\hat{\boldsymbol{\theta}}_S(\mathbf{a})$ is computed by attempting to remove the effect of modulation from the observation $\mathbf{z}$; premultiplication by $\mathbf{A}^H$ removes the phase modulation, and premultiplication by $\mathbf{P}^{-1}(\mathbf{a})$ compensates for the amplitude variations. The error covariance in the ML estimate is the noise power N_0 weighted by the inverse (hypothetical) symbol powers. The MAP channel estimate $\hat{\boldsymbol{\theta}}_D(\mathbf{a})$ is obtained from $\hat{\boldsymbol{\theta}}_S(\mathbf{a})$ by premultiplication with matrix $\mathbf{N}(\bullet)$, being effectively a smoothing operation based on the knowledge of the channel dynamics and the noise level, followed by the addition of vector $\boldsymbol{\mu}(\bullet)$ based on the knowledge of the LOS path strength.

The decision metric $m(\mathbf{a})$ [eq. (12-37)] associated with ML flat fading estimation,

$$
\begin{aligned}
m_S(\mathbf{a}) &= \hat{\boldsymbol{\theta}}_S^H(\mathbf{a}) \cdot \boldsymbol{\Sigma}_S^{-1}(\mathbf{a}) \cdot \hat{\boldsymbol{\theta}}_S(\mathbf{a}) \qquad \underset{\mathbf{a}}{\rightarrow \max} \\
&= \left(\mathbf{z}^H \mathbf{A} \mathbf{P}^{-1}\right) \cdot \left(N_0 \mathbf{P}^{-1}\right)^{-1} \cdot \left(\mathbf{P}^{-1} \mathbf{A}^H \mathbf{z}\right) \\
&\propto \mathbf{z}^H \cdot \underbrace{\mathbf{A} \mathbf{P}^{-1} \mathbf{A}^H}_{\mathbf{I}} \cdot \mathbf{z} \qquad = \mathbf{z}^H \cdot \mathbf{z}
\end{aligned}
\tag{12-55}
$$

is seen to be useless since it does not depend on the hypothetical symbols $\mathbf{a}$. This is because no knowledge about the channel statistics is used; rather, the ML dynamic channel estimator assumes that all symbols $\mathbf{a}$ as well as all channel trajectories $\mathbf{c}$ are equally likely. For instance, the noise-free observation sequence

$\mathbf{z} = \{1, -1, 1, -1\}$ (very short message length $N = 4$) may have been produced by symbols $\mathbf{a} = \{1, -1, 1, -1\}$ transmitted through channel $\mathbf{c} = \{1, 1, 1, 1\}$, or by symbols $\mathbf{a} = \{1, 1, 1, 1\}$ transmitted through $\mathbf{c} = \{1, -1, 1, -1\}$, or by any other symbol sequence. The intuitive notion of a slowly fading (or piecewise constant) channel process such as $\mathbf{c} = \{1, 1, 1, 1\}$ being more likely than an erratic trajectory $\mathbf{c} = \{1, -1, 1, -1\}$ is part of the prior knowledge which is not made use of by the ML algorithm. Hence, joint ML detection and ML estimation of dynamic channel parameters is not possible. To circumvent this, one may resort to data-aided (DA) dynamic channel estimation where *known* pilot symbols are transmitted (hence, the channel $\mathbf{c}$ remains the only random process to be estimated, see Section 14.2). Otherwise, ML channel estimation leads to a meaningful decision metric only when the channel is static or can be taken to be static for the duration of data blocks.

The decision metric based on MAP channel estimation, on the other hand, is dependent on $\mathbf{a}$, provided that the channel dynamics (most often lowpass) is different from the data symbol statistics (most often white). It can be evaluated either via eq. (12-34) or eq. (12-43). For example, from eq. (12-34) one obtains the metric

$$\begin{aligned} m_D(\mathbf{a}) &= \left[\mathbf{z} - \mathbf{A}\hat{\boldsymbol{\theta}}_D\right]^H \mathbf{R}_n^{-1} \left[\mathbf{z} - \mathbf{A}\hat{\boldsymbol{\theta}}_D\right] + \left[\hat{\boldsymbol{\theta}}_D - \boldsymbol{\mu}_D\right]^H \mathbf{R}_D^{-1} \left[\hat{\boldsymbol{\theta}}_D - \boldsymbol{\mu}_D\right] \\ &\propto \left\|\mathbf{z} - \mathbf{A}\hat{\boldsymbol{\theta}}_D\right\|^2 + N_0 \left[\mathbf{X}\hat{\boldsymbol{\theta}}_S - \mathbf{y}\right]^H \mathbf{R}_D \left[\mathbf{X}\hat{\boldsymbol{\theta}}_S - \mathbf{y}\right] \quad \underset{\mathbf{a}}{\rightarrow \min} \\ \text{with} \quad \mathbf{X} &= \frac{1}{N_0}\left[\mathbf{I} - \left(\mathbf{R}_D + N_0 \mathbf{P}^{-1}\right)^{-1} \mathbf{R}_D\right] \mathbf{P} \\ \mathbf{y} &= \left(\mathbf{R}_D + N_0 \mathbf{P}^{-1}\right)^{-1} \cdot \boldsymbol{\mu}_D \end{aligned} \tag{12-56}$$

in terms of the squared norm of the error vector $\mathbf{z} - \mathbf{A}\hat{\boldsymbol{\theta}}_D(\mathbf{a})$ [same as with ML estimation, just $\hat{\boldsymbol{\theta}}_S(\mathbf{a})$ replaced by $\hat{\boldsymbol{\theta}}_D(\mathbf{a})$] plus an additional term. Alternatively, from eq. (12-43) and with the help of eqs. (12-53), (12-54), and (12-55) one obtains the equivalent metric:

$$\begin{aligned} m_D(\mathbf{a}) &= \hat{\boldsymbol{\theta}}_D^H \cdot \boldsymbol{\Sigma}_D^{-1} \cdot \hat{\boldsymbol{\theta}}_D \quad \underset{\mathbf{a}}{\rightarrow \max} \\ &= \left[\mathbf{N}\hat{\boldsymbol{\theta}}_S + \boldsymbol{\mu}\right]^H \cdot (\mathbf{N}\boldsymbol{\Sigma}_S)^{-1} \cdot \left[\mathbf{N}\hat{\boldsymbol{\theta}}_S + \boldsymbol{\mu}\right] \\ &= -\hat{\boldsymbol{\theta}}_S^H \left[\mathbf{R}_D + N_0 \mathbf{P}^{-1}\right]^{-1} \hat{\boldsymbol{\theta}}_S + \left(\hat{\boldsymbol{\theta}}_S + \boldsymbol{\mu}\right)^H \left(\frac{1}{N_0}\mathbf{P}\right) \left(\hat{\boldsymbol{\theta}}_S + \boldsymbol{\mu}\right) + \underbrace{\boldsymbol{\mu}^H \mathbf{R}_D^{-1} \boldsymbol{\mu}}_{\left(\substack{\text{indep.} \\ \text{of } \mathbf{a}}\right)} \\ &\propto \hat{\boldsymbol{\theta}}_S^H \left[\mathbf{R}_D + N_0 \mathbf{P}^{-1}\right]^{-1} \hat{\boldsymbol{\theta}}_S - \left(\hat{\boldsymbol{\theta}}_S + \boldsymbol{\mu}\right)^H \left(\frac{1}{N_0}\mathbf{P}\right) \left(\hat{\boldsymbol{\theta}}_S + \boldsymbol{\mu}\right) \quad \underset{\mathbf{a}}{\rightarrow \min} \\ &= \hat{\boldsymbol{\theta}}_S^H \underbrace{\left[\mathbf{R}_D + N_0 \mathbf{P}^{-1}\right]^{-1}}_{\frac{1}{N_0}\mathbf{P}(\mathbf{I} - \mathbf{N})} \hat{\boldsymbol{\theta}}_S - \frac{1}{N_0}\left[2\left|\boldsymbol{\mu}^H \left(\mathbf{A}^H \mathbf{z}\right)\right| + \boldsymbol{\mu}^H \mathbf{P} \boldsymbol{\mu}\right] \\ &\propto \hat{\boldsymbol{\theta}}_S^H \cdot (\mathbf{P}\mathbf{N}) \cdot \hat{\boldsymbol{\theta}}_S + \left[2\left|\boldsymbol{\mu}^H \left(\mathbf{A}^H \mathbf{z}\right)\right| + \boldsymbol{\mu}^H \mathbf{P} \boldsymbol{\mu}\right] \quad \underset{\mathbf{a}}{\rightarrow \max} \end{aligned} \tag{12-57}$$

whose second term is seen to vanish in the important special case that the channel process is zero-mean.

Let us now investigate some instructive examples of ML/MAP flat fading channel estimation. Consider M-PSK modulated symbols $\mathbf{a}$ so that $\mathbf{P} = \mathbf{I}$ and $\mathbf{N}(\bullet)$, $\boldsymbol{\mu}(\bullet)$ become independent of $\mathbf{a}$. Given the message length N and channel parameters K, λ'_D, N_0, matrices $\mathbf{R}_D, \mathbf{N}$ and vector $\boldsymbol{\mu}$ are computed according to eqs. (12-53) and (12-54). As a consequence of the separation property, the ratio between the error covariances in the MAP estimate $\hat{\theta}_D(\mathbf{a})$ and the ML estimate $\hat{\theta}_S(\mathbf{a})$ is given directly by the main diagonal entries of matrix $\mathbf{N}$ [eqs. (12-53) and (12-54)]:

$$\mathbf{r}(\mathbf{N}) = \begin{pmatrix} r_0(\mathbf{N}) \\ r_1(\mathbf{N}) \\ \vdots \\ r_{N-1}(\mathbf{N}) \end{pmatrix} = \begin{pmatrix} (\mathbf{N})_{00} \\ (\mathbf{N})_{11} \\ \vdots \\ (\mathbf{N})_{N-1,N-1} \end{pmatrix} \tag{12-58}$$

In general, the error covariance ratios $r_k(\mathbf{N})$ are distinct for each intramessage tap index $k = 0, 1, \ldots, N-1$.

Figure 12-2 displays the reduction in tap error covariance that can be gained by performing MAP channel estimation for tap indices $k = 0, \ldots, 50$ (half of the message length of $N = 101$). Several interesting conclusions can be drawn from these results. First, notice that Rayleigh and Rician ($K = 10$) channels yield almost the same error covariances, except for tap indices near the margins (start or end) of the message, very fast fading ($\lambda'_D = 0.1$) combined with strong noise ($N_0 = 0.1$, i.e., $E_s/N_0 = 10$ dB), and very large K-factors far beyond 10 (not shown in the figure). Thus, exploiting the knowledge of the K-factor does not lead to very much

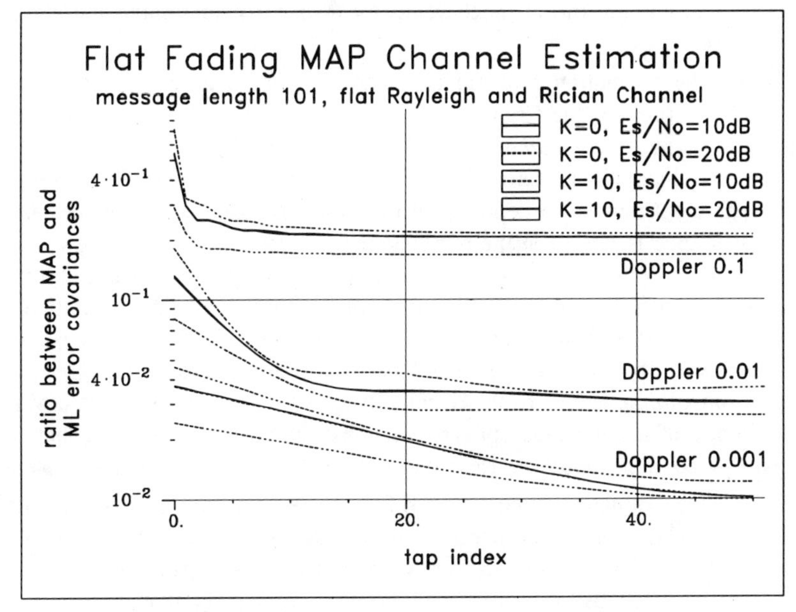

Figure 12-2 MAP Flat Fading Channel Estimation: Error Covariance vs. Intramessage Position

of an improvement; this is also supported by the fact that the weighted mean vector $\boldsymbol{\mu}$ is found to be much smaller than the mean $\boldsymbol{\mu}_D$ of the channel process itself, except for the aforementioned extreme cases. Hence, for quasi-optimal estimation it usually suffices to assume Rayleigh fading unless the LOS component is very strong. This also simplifies the estimator and the computation of the decision metric [the second term of eq. (12-57) is disposed of].

From Figure 12-2, the MAP estimation gain is observed to increase with decreasing Doppler – obviously, the out-of-band noise can be suppressed more effectively when the bandwidth of the desired signal (here: channel trajectory) is small. In the limiting case of zero Doppler when the channel is quasi-static within the message length, the MAP estimator performs a simple averaging operation thus reducing the error covariance by a factor of N. Hence, with N=101, the error floor is $r_{k,\min} \approx 0.01$; in Figure 12-2, the factor r_k is seen to approach this minimum level near the center of the message and for low Doppler λ'_D=0.001.

While the ML error covariance is dependent on the noise level only, the additional improvement gained by MAP estimation is seen to be very much dependent on the fading rate λ'_D, further on the sequence length N, intrapacket tap position k, and, to a much smaller extent, on the SNR.

As an interesting side remark, note that the estimation error covariance is independent of the particular choice of the transmitted M-PSK sequence. In contrast to selective channel estimation (Section 12.2.4), there is no constraint on the statistical properties (e.g., whiteness) of the data $\mathbf{a}$ simply because the channel is memoryless. The effect of the modulation can easily be "undone" – provided that the hypothesis $\mathbf{a}$ is correct – in the ML estimation part of the receiver [premultiplication by matrix $\mathbf{A}^H$, see eq. (12-53)] so that the MAP estimation part reduces to estimating the channel trajectory $\boldsymbol{\theta} = \mathbf{c}$ by smoothing its own noisy samples $\hat{\boldsymbol{\theta}}_S = \hat{\mathbf{c}}_S$.

Since the estimation problem is linear and Gaussian, the MAP estimate is *efficient*, i.e., it attains the Cramér-Rao bound [1]. Figure 12-2 therefore gives the error performance of *optimal* dynamic system identification that is achievable when the system (channel) is excited by a finite-length unity-power (PSK) sequence. Clearly, this optimum also depends on the intramessage position of the tap. The error covariance is largest at the margins of the message (channel taps c_0, c_{N-1}). In the limit as $N \longrightarrow \infty$, this corresponds to optimal *filtering* where the infinite future (past) up to the present of both the observation $\mathbf{z}$ and the excitation $\mathbf{a}$ are known. The error in channel taps c_1, c_{N-2} (smaller than the error in c_0, c_{N-1}) then corresponds to optimal *one-step smoothing*, the error in taps c_2, c_{N-3} to optimal *two-step smoothing*, and so forth. Eventually as $N \longrightarrow \infty$, the tap error covariance in the center of the message converges against the absolute optimum that can be attained by ∞-*step smoothing* for a particular Doppler frequency and noise level.

Since the evaluation of eq. (12-53) for $N \longrightarrow \infty$ is not feasible, the minimal error covariance is best evaluated in the frequency domain. The optimal filter (smoother) for estimating the stationary process $\{c_k\}$ with psd $S_c(e^{j\psi'})$ [eq. (12-49)] from a noisy observation $\hat{c}_k = c_k + m_k$ – remember that the effect of the modulation is attempted to be undone in the ML estimation part – with noise psd $S_m(e^{j\psi'}) = N_0$ is the well-known *Wiener filter* [13]:

$$W\left(e^{j\psi'}\right) = \frac{S_c\left(e^{j\psi'}\right)}{S_c(e^{j\psi'}) + S_m(e^{j\psi'})} = \frac{S_c\left(e^{j\psi'}\right)}{S_c(e^{j\psi'}) + N_0} \tag{12-59}$$

with mean-square error:

Any filter $W\left(e^{j\psi'}\right)$:

$$\zeta = \frac{1}{2\pi}\int_{-\pi}^{\pi}\left[\left(1-W\left(e^{j\psi'}\right)\right)^2 S_c\left(e^{j\psi'}\right) + W^2\left(e^{j\psi'}\right)S_m\left(e^{j\psi'}\right)\right]d\psi'$$

Optimal Wiener filter $W\left(e^{j\psi'}\right)$:

$$\zeta = \frac{1}{2\pi}\int_{-\pi}^{\pi}\frac{S_c\left(e^{j\psi'}\right)S_m\left(e^{j\psi'}\right)}{S_c(e^{j\psi'})+S_m(e^{j\psi'})}d\psi' = \frac{1}{2\pi}\int_{-\pi}^{\pi}W\left(e^{j\psi'}\right)S_m\left(e^{j\psi'}\right)d\psi' \tag{12-60}$$

Recognizing that the error covariance in the ML estimate is the inverse average SNR per symbol $N_0 = 1/\overline{\gamma}_s$ ($\overline{\gamma}_s = \overline{E}_s/N_0 = 1/N_0$), evaluating the minimal error covariance (= error covariance of MAP estimation for $N \longrightarrow \infty$) according to eqs. (12-59) and (12-60) yields the following ratios $r(\lambda'_D, \overline{\gamma}_s)$ between optimal MAP and ML error covariances for the Jakes and rectangular fading spectra of eq. (12-49), respectively:

Jakes fading psd, respective Wiener filter:

$$r(\lambda'_D, \overline{\gamma}_s) = 2\lambda'_D \underbrace{\int_0^1 \frac{1}{1 + (\pi\lambda'_D/\overline{\gamma}_s)\sqrt{1-x^2}}dx}_{I(\lambda'_D,\overline{\gamma}_s)}$$

Rectangular fading psd, rectangular Wiener filter;

Jakes fading psd, suboptimal rectangular filter:

$$r(\lambda'_D, \overline{\gamma}_s) = 2\lambda'_D \underbrace{\frac{1}{1 + 2\lambda'_D/\overline{\gamma}_s}}_{F(\lambda'_D,\overline{\gamma}_s)} \tag{12-61}$$

The first result is valid for the case that the fading spectrum is in fact Jakes and the Wiener filter is optimally matched to this Jakes fading spectrum. The second equation applies to the two cases that (i) the fading spectrum is in fact rectangular and the Wiener filter is optimally matched to this rectangular fading spectrum and (ii) the fading spectrum is in fact Jakes but the Wiener filter is matched to

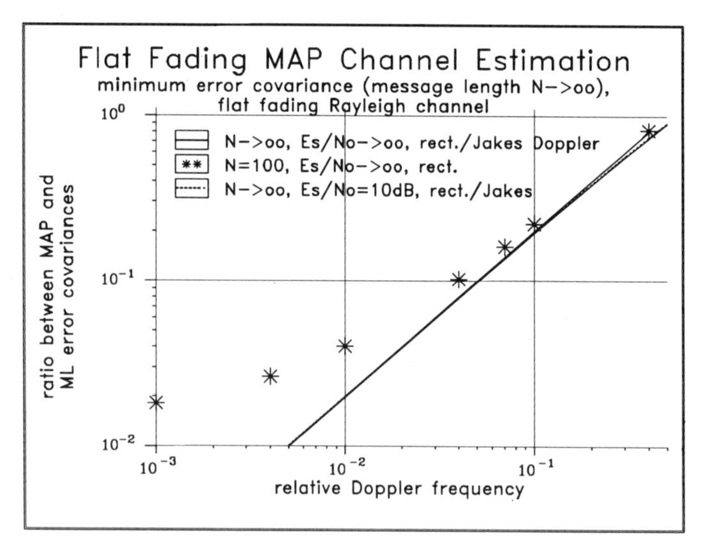

Figure 12-3 MAP Flat Fading Channel Estimation: Minimal Error Covariance vs. Doppler Frequency

the rectangular fading spectrum (therefore suboptimal). Evaluating the integral $I(\lambda'_D, \overline{\gamma}_s)$ and the factor $F(\lambda'_D, \overline{\gamma}_s)$, one finds that:

- $I(\lambda'_D, \overline{\gamma}_s)$ is only slightly smaller than $F(\lambda'_D, \overline{\gamma}_s)$
- $I(\lambda'_D, \overline{\gamma}_s)$, $F(\lambda'_D, \overline{\gamma}_s)$ are close to unity except for very fast fading and strong noise.

This is also illustrated by Figure 12-3 where the ratio $r(\lambda'_D, \overline{\gamma}_s)$ between MAP and ML error covariances is plotted against the (relative) Doppler frequency λ'_D. The ratio is nearly independent of the SNR $\overline{\gamma}_s$, and the curves for Jakes and rectangular fading spectra are indistinguishable. In summary, optimal MAP fading estimation yields error covariance reduction factors

$$r(\lambda'_D, \overline{\gamma}_s) \approx r(\lambda'_D) \approx 2\,\lambda'_D \tag{12-62}$$

(= bandwidth of the T-spaced fading channel process) at all relevant SNR, regardless of whether the Jakes model or its rectangular approximation [eq. (12-49)] is adopted in the estimator design.

It is interesting to note that the matrix $\mathbf{N}$ for $N \longrightarrow \infty$ can be obtained by transforming the Wiener filter [eq. (12-59)] into the time domain. For instance:

Rectangular Wiener filter:

$$\begin{aligned} W\left(e^{j\psi'}\right) &= \begin{cases} \frac{1}{1+2\lambda'_D/\overline{\gamma}_s} & \text{if } |\psi'| \le 2\pi\lambda'_D \\ 0 & \text{else} \end{cases} \\ w_k &= \frac{1}{1+2\lambda'_D/\overline{\gamma}_s}\, 2\lambda'_D \;\text{si}(2\pi\lambda'_D\, k) \end{aligned} \tag{12-63}$$

leads to the ∞-dim. Toeplitz matrix

$$\mathbf{N} = \begin{pmatrix} \ddots & \ddots & \ddots & & \\ \ddots & w_0 & w_1 & w_2 & \\ \ddots & w_1 & w_0 & w_1 & \ddots \\ & w_2 & w_1 & w_0 & \ddots \\ & & \ddots & \ddots & \ddots \end{pmatrix} \tag{12-64}$$

Since the cascade of two ideal lowpasses remains to be an ideal lowpass, we further have $\mathbf{N}^H \cdot \mathbf{N} = [1/(1+2\lambda'_D/\overline{\gamma}_s)] \cdot \mathbf{N}$, so that the expression for the decision metric [eq. (12-57) with $\mathbf{P} = \mathbf{I}$ and $\boldsymbol{\mu} = 0$] can be simplified to

$$\begin{aligned} m_D(\mathbf{a}) &= \hat{\boldsymbol{\theta}}_S^H(\mathbf{a}) \cdot \mathbf{N} \cdot \hat{\boldsymbol{\theta}}_S(\mathbf{a}) \\ &\underset{\left(\substack{\text{matrix } \mathbf{N} \text{ ideal} \\ \text{lowpass; } N \to \infty}\right)}{\propto} \hat{\boldsymbol{\theta}}_S^H(\mathbf{a}) \cdot \mathbf{N}^H \cdot \mathbf{N} \cdot \hat{\boldsymbol{\theta}}_S(\mathbf{a}) \\ &= \left\| \hat{\boldsymbol{\theta}}_D(\mathbf{a}) \right\|^2 \end{aligned} \tag{12-65}$$

The generation of decision metrics is illustrated in Figure 12-4. This decision rule calls for selecting that sequence $\mathbf{a}$ for which the MAP channel estimate $\hat{\boldsymbol{\theta}}_D(\mathbf{a})$

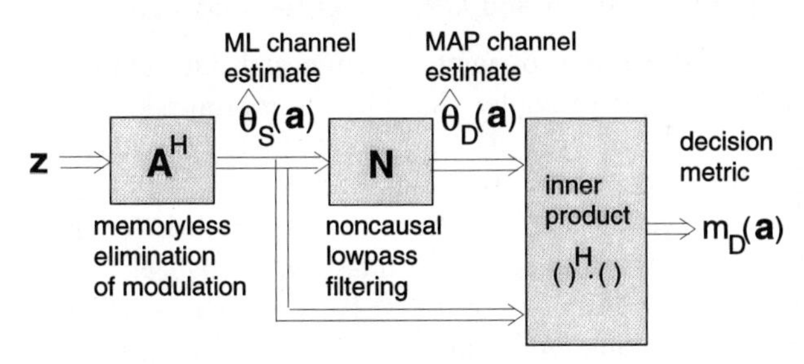

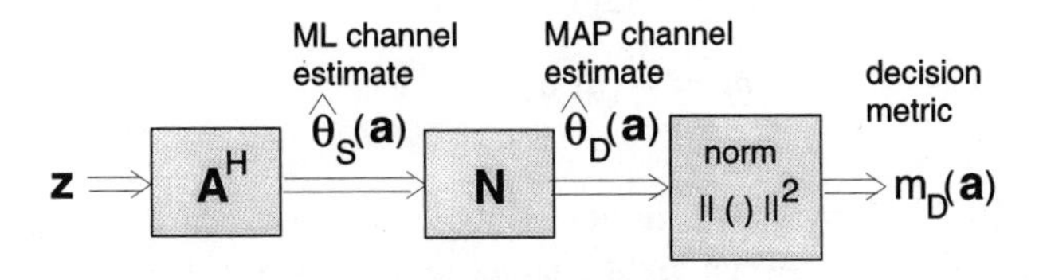

Figure 12-4 Generation of PSK Symbol Decision Metrics for Flat Fading Channels

has maximum energy. This is intuitively appealing since $\hat{\theta}_D(\mathbf{a})$ is generated by filtering the ML estimate

$$\begin{aligned} \hat{\theta}_S(\mathbf{a}) = \mathbf{A}^H \cdot \mathbf{z} &= \mathbf{A}^H \cdot (\mathbf{A}_0 \cdot \mathbf{c}_0 + \mathbf{m}) \\ &= \left(\mathbf{A}^H \mathbf{A}_0\right) \cdot \mathbf{c}_0 + \mathbf{A}^H \mathbf{m} \end{aligned} \tag{12-66}$$

($\mathbf{A}_0$ and $\mathbf{c}_0$ denote the "true" symbol matrix and channel, respectively) whose first term $(\mathbf{A}^H \cdot \mathbf{A}_0) \cdot \mathbf{c}_0$ passes the Wiener lowpass filter undistorted only if the symbol hypothesis is correct, i.e., $(\mathbf{A}^H \cdot \mathbf{A}_0) = \mathbf{I}$, so that, leaving aside the noise effect, the energy of the MAP channel estimate $\hat{\theta}_D(\mathbf{a} = \mathbf{a}_0)$ based on the correct symbol hypothesis is largest. In all other cases, the trajectory $(\mathbf{A}^H \cdot \mathbf{A}_0) \cdot \mathbf{c}_0$ contains high-frequency components that are cut off by the lowpass filter thus leading to a reduction in the energy of $\hat{\theta}_D(\mathbf{a})$.

As a side remark, truly *optimal* channel estimation and metric generation (Figure 12-4) calls for *noncausal* lowpass filtering, in the limit ∞-dim. smoothing. This implies that all *recursive-type* joint detection and estimation strategies based on the received sequence $\mathbf{z}_k$ up to the "present" (time index k) are clearly *suboptimal* in that they neglect the information contained in the "future" received samples. However, for nearly optimal decision and estimation, the smoother lag can be fixed to some finite value D (depending on the fading rate, see Figure 12-2) so that recursive schemes can make a *quasi-optimal* decision on the symbol sequence $\mathbf{a}_k$ up to the "present" based on the received sequence $\mathbf{z}_{k+D}$ up to future time index $k + D$.

Example of Joint Detection and Estimation for Flat Fading Channels

Wrapping up the discussion of joint detection and flat fading channel estimation, let us study a simple example. Consider the transmission of a very short binary data message $\mathbf{a} = \{a_0,\ a_1,\ a_2,\ a_3\}$ ($a_k \in \{-1, 1\}$) with length $N = 4$ over a flat Rayleigh channel ($K = 0$). Suppose that $\mathbf{z} = \{+1, +1, +1, +1\}$ has been received, an observation that may have been generated, for instance, by a message $\mathbf{a} = \{1, 1, 1, 1\}$ transmitted over a quasi-static channel $\mathbf{c} = \{1, 1, 1, 1\}$, or $\mathbf{a} = \{-1, -1, -1, -1\}$ transmitted over $\mathbf{c} = \{-1, -1, -1, -1\}$. The phase ambiguity [message/channel pairs $(\mathbf{a}, \mathbf{c})$ and $(-\mathbf{a}, -\mathbf{c})$ yield the same observation] may be resolved by differential precoding and decoding:

$$\begin{aligned} a_k &= a_{k-1}\,(2b_k - 1) \\ b_k &= \frac{1}{2}\,(a_k\, a_{k-1} + 1) \end{aligned} \tag{12-67}$$

(input bits $b_k \in \{0, 1\}$, $k = 1, 2, 3$, initial symbol $a_0 = +1$), so that the receiver effectively has to decide between eight distinct sequences $\mathbf{a} = \{+1,\ a_1,\ a_2,\ a_3\}$ corresponding to $N–1 = 3$ source bits $\mathbf{b} = \{b_1,\ b_2,\ b_3\}$. As discussed above, from the eight ML channel estimates $\hat{\theta}_S(\mathbf{a}) = \hat{\mathbf{c}}_S(\mathbf{a}) = \mathbf{A}^H \mathbf{z} = \left[a_0^* z_0\ \ldots\ a_3^* z_3\right]^T$, the respective MAP channel estimates can be obtained via $\hat{\theta}_D(\mathbf{a}) = \hat{\mathbf{c}}_D(\mathbf{a}) =$

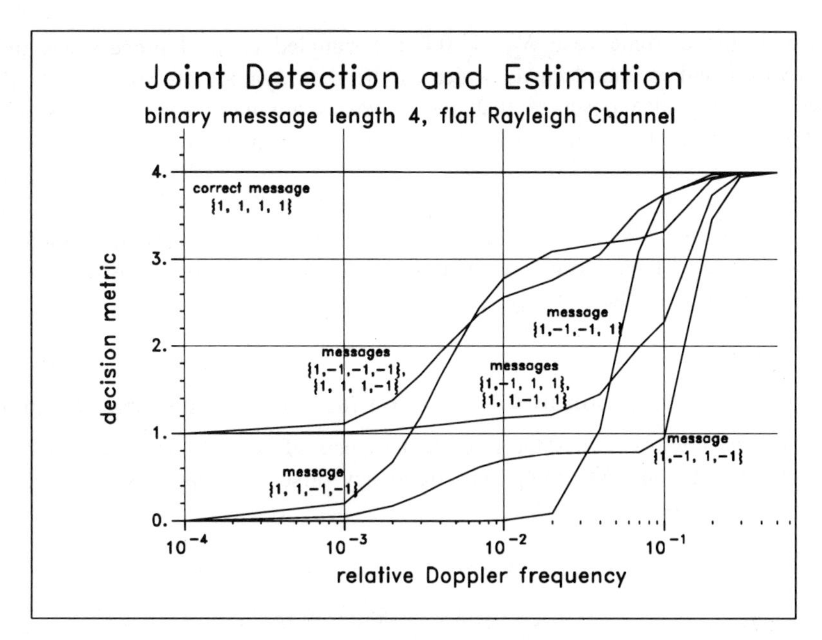

Figure 12-5 Joint Detection and MAP Flat Fading Channel Estimation: Metrics vs. Doppler Frequency

$\mathbf{N}\cdot\hat{\boldsymbol{\theta}}_S(\mathbf{a})$. Here we assume a low noise level of $N_0 = 0.001$ ($\overline{\gamma}_s = 30$ dB) in the computation of $\mathbf{N}$. The eight decision metrics are best evaluated via eq. (12-57) with $\mathbf{P} = \mathbf{I}$ and $\boldsymbol{\mu} = 0$ resulting in $m_D(\mathbf{a}) = \hat{\boldsymbol{\theta}}_S^H(\mathbf{a}) \cdot \mathbf{N}\cdot\hat{\boldsymbol{\theta}}_S(\mathbf{a}) = \hat{\boldsymbol{\theta}}_S^H(\mathbf{a}) \cdot \hat{\boldsymbol{\theta}}_D(\mathbf{a})$ (Figure 12-4).

Figure 12-5 displays the eight metrics $m_D(\mathbf{a})$ versus the relative Doppler frequency λ'_D for this example. For quasi-static channels (Doppler below $\lambda'_D = 10^{-3}$), the matrix $\mathbf{N}$ is found to be very close to (1/4)×$\mathbf{1}$ (where $\mathbf{1}$ is the 4×4 all-1 matrix) so that the metrics $m_D(\mathbf{a})$

$$\begin{aligned} m_D(\mathbf{a})\big|_{\lambda'_D \to 0} &\approx \sum_{k=0}^{3} \underbrace{\hat{c}^*_{S;k}(a_k)}_{a_k z_k^*} \cdot \left(\frac{1}{4} \sum_{j=0}^{3} \underbrace{\hat{c}_{S;j}(a_j)}_{a_j^* z_j} \right) \\ &\underset{\left(\substack{\text{observation } \mathbf{z}=(1\,1\,1\,1)^T \\ \text{binary symbols } \mathbf{a}}\right)}{=} \frac{1}{4} \cdot \left(\sum_{k=0}^{3} a_k \right)^2 \end{aligned} \tag{12-68}$$

assume values 4 (correct symbol hypothesis), 1, or 0 (incorrect hypotheses) for very small Doppler. The corresponding decision noise margin of 3 is seen to shrink with rising Doppler until it becomes zero for $\lambda'_D = 0.5$ where all metrics converge to 4 and thus become indistinguishable. This is to be expected since in the extreme case $\lambda'_D = 0.5$ the sampled channel process has unity bandwidth and is therefore as white as the data symbol process. From Figure 12-5, one also observes that some metrics are more sensitive to channel variations than others; in this example, these correspond to symbol sequences $\mathbf{a} = \{1,\ -1,\ -1,\ -1\}, \{1,\ 1,\ -1,\ -1\}, \{1,\ 1,\ 1,\ -1\}$ with only one transition $+1 \rightarrow -1$ or $-1 \rightarrow +1$ between successive symbols. The respective ML channel estimates $\hat{\boldsymbol{\theta}}_S(\mathbf{a})$, which likewise exhibit only one such sign transition, are therefore "more lowpass" than "erratic" estimates with two or three sign transitions, hence a larger portion of their energy passes the lowpass filter $\mathbf{N}$.

12.2.4 Joint Detection and Estimation for Selective Fading Channels

Let us now explore some important aspects of joint detection and *selective* channel estimation. Assuming zero frequency offset, the transmission model of eq. (12-11) reduces to

$$\begin{aligned} \mathbf{r} &= \mathbf{r}^{(0)} \oplus \mathbf{r}^{(1)} \\ &= \underbrace{\left(\mathbf{A}^{(0)} \oplus \mathbf{A}^{(1)}\right)}_{\mathbf{A}} \cdot \underbrace{\left(\mathbf{h}^{(0)} \oplus \mathbf{h}^{(1)}\right)}_{\mathbf{h}} + \mathbf{n} \\ \mathbf{r}^{(i)} &= \mathbf{A}^{(i)} \cdot \mathbf{h}^{(i)} + \mathbf{n}^{(i)} \end{aligned} \tag{12-69}$$

with the $2[N+L]$-dim. observation $\mathbf{r}$, $2[L+1]N$-dim. CIR process vector $\mathbf{h}$, and $(2[N+L] \times 2[L+1]N)$-dim. data matrix $\mathbf{A}$. One immediately recognizes that the ML estimate [eq. (12-27)]

$$\hat{\boldsymbol{\theta}}_S(\mathbf{a}) = \left[\mathbf{A}^H \mathbf{R}_n^{-1} \mathbf{A}\right]^{-1} \cdot \left[\mathbf{A}^H \mathbf{R}_n^{-1} \mathbf{r}\right] \tag{12-70}$$

of the dynamic process $\hat{\boldsymbol{\theta}}_D = \mathbf{h}$ does *not exist* simply because the number $2[L+1]N$ of sync parameters to be estimated exceeds the number $2N$ of observed samples (for $L \geq 1$ the matrix $\mathbf{A}^H \mathbf{R}_n^{-1} \mathbf{A}$ is not invertible).

There are two ways of resolving this dilemma. First, one may attempt to *directly* compute the MAP estimate from the observation via eq. (12-31):

$$\begin{aligned} \hat{\boldsymbol{\theta}}_D(\mathbf{a}) &= \underbrace{\left[\mathbf{R}_D - \mathbf{R}_D \mathbf{A}^H \left(\mathbf{A}\mathbf{R}_D \mathbf{A}^H + \mathbf{R}_n\right)^{-1} \mathbf{A}\mathbf{R}_D\right]}_{\boldsymbol{\Sigma}_D(\mathbf{a})} \cdot \left[\mathbf{A}^H \mathbf{R}_n^{-1} \mathbf{r}\right] \\ &= \mathbf{R}_D \mathbf{A}^H \left(\mathbf{A}\mathbf{R}_D \mathbf{A}^H + \mathbf{R}_n\right)^{-1} \cdot \mathbf{r} \end{aligned} \tag{12-71}$$

with the $2[L+1]N$-dim. dynamic parameter covariance matrix $\mathbf{R}_D = E[\mathbf{h} \cdot \mathbf{h}^H]$. As motivated above in the context of flat fading channel estimation, $\boldsymbol{\mu}_D = 0$ is assumed. Due to the large vector/matrix dimensions, the evaluation of eq. (12-71) quickly becomes unfeasible already for moderate message and channel lengths N and L, respectively.

A second, more elegant approach can be devised by computing the MAP estimate *indirectly* via the ML estimate of some intermediate parameters. In order to keep the analysis manageable, the vector estimate of the entire channel trajectory

$$\mathbf{h} = \begin{pmatrix} \mathbf{h}_0 \\ \mathbf{h}_1 \\ \vdots \\ \mathbf{h}_{N-1} \end{pmatrix} \quad \text{with} \quad \mathbf{h}_k = \mathbf{h}_k^{(0)} \oplus \mathbf{h}_k^{(1)} \quad \text{and} \quad \mathbf{h}_k^{(i)} = \begin{pmatrix} h_{0,k}^{(i)} \\ \vdots \\ h_{L,k+L}^{(i)} \end{pmatrix} \tag{12-72}$$

should be computed by first forming the individual $2[L+1]$-dim. CIR estimates $\hat{\mathbf{h}}_k(\mathbf{a})$, valid at particular time instants k, and then arranging the set $\{\hat{\mathbf{h}}_k(\mathbf{a})\}$ in a vector so as to form an estimate of the entire trajectory $\hat{\mathbf{h}}(\mathbf{a})$.

As above [eq. (12-28)], the optimal estimate $\hat{\mathbf{h}}_k(\mathbf{a})$ is the expected value of $\mathbf{h}_k$, i.e., the conditional mean of $\mathbf{h}_k$, given the observation $\mathbf{r}$ and the data $\mathbf{a}$:

$$\begin{aligned} \hat{\theta}_{D,k}(\mathbf{a}) &= \hat{\mathbf{h}}_k(\mathbf{a}) \\ &= \boldsymbol{\mu}_{h|r} = \underbrace{\boldsymbol{\mu}_h}_{0} + \boldsymbol{\Sigma}_{hr} \cdot \boldsymbol{\Sigma}_r^{-1} \cdot \left(\mathbf{r} - \underbrace{\boldsymbol{\mu}_r}_{0} \right) \\ &= \boldsymbol{\Sigma}_{hr} \cdot \boldsymbol{\Sigma}_r^{-1} \cdot \mathbf{r} \\ \boldsymbol{\Sigma}_{D,k} &= \boldsymbol{\Sigma}_{h|r} = \boldsymbol{\Sigma}_h - \boldsymbol{\Sigma}_{hr} \cdot \boldsymbol{\Sigma}_r^{-1} \cdot \boldsymbol{\Sigma}_{hr}^H \end{aligned} \tag{12-73}$$

We start evaluating the matrices $\boldsymbol{\Sigma}_h$, $\boldsymbol{\Sigma}_{hr}$, and $\boldsymbol{\Sigma}_r$ analogous to eq. (12-29). The channel autocorrelation matrix $\boldsymbol{\Sigma}_h$, as defined by eqs. (11-73) and (11-74), is given by

$$\begin{aligned} \boldsymbol{\Sigma}_h &= \text{cov}[\mathbf{h}_k] = E\left[\mathbf{h}_k \cdot \mathbf{h}_k^H\right] \\ &= \mathbf{R}_h(0) = \mathbf{R}_h = \begin{pmatrix} \rho_0 & \rho_{0,1} & \cdots & \rho_{0,2L+1} \\ \rho_{1,0} & \rho_1 & & \rho_{1,2L+1} \\ \vdots & & \ddots & \vdots \\ \rho_{2L+1,0} & \rho_{2L+1,1} & \cdots & \rho_{2L+1} \end{pmatrix} \end{aligned} \tag{12-74}$$

with tap powers $\rho_\nu = E[|\, h_{\nu;k}\, |^2]$ $(\nu = 0, \ldots, 2L+1)$ and intertap covariances $\rho_{\nu,\nu+\mu} = E[h_{\nu;k}\, h_{\nu+\mu;k}^*]$. The latter are often set to zero $(\mu \neq 0)$ for the purpose of channel estimation since they are small (except for neighboring taps) and difficult to estimate in practice – the tap powers ρ_ν, on the other hand, can be estimated easily. Then the channel autocorrelation becomes diagonal: $\mathbf{R}_h = \text{diag}\{\rho_0\ \rho_1\ \cdots\ \rho_{2L+1}\}$.

As mentioned above, we wish to perform MAP channel estimation in an elegant way via ML estimation of some intermediate parameters. This can be accomplished by *partitioning* the received signal $\mathbf{r}$ (somewhat artificially) into a number $\overline{N}$ of blocks $\mathbf{r}_K$ $(K = 0, \ldots, \overline{N}-1)$. Figure 12-6 displays an example of such a partition, together with the definition of some indices. The K-th block $\mathbf{r}_K$ of length R_K (in symbol intervals) starts at time index $k = N_K$ and ends at $k = N_K + R_K - 1$. Each block $\mathbf{r}_K$ depends on the data symbol vector $\mathbf{a}_K = (\, a_{N_K-L}\ \cdots\ a_{N_K+R_K-1}\,)^T$. Due to the channel memory spanning L symbol intervals, data blocks $\mathbf{a}_K$ pertaining to adjacent received blocks $\mathbf{r}_K$ overlap by L symbols (see Figure 12-6).

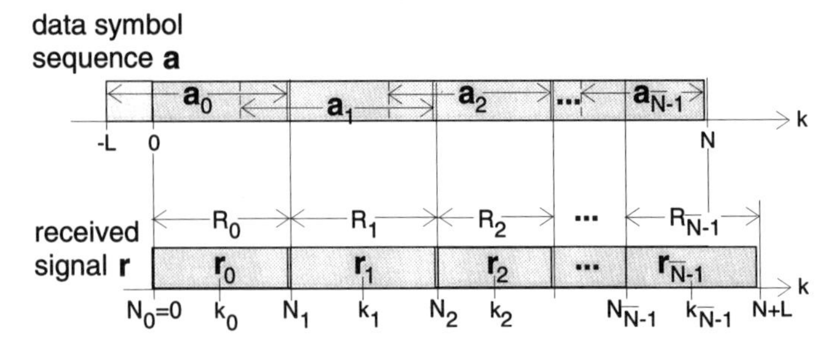

Figure 12-6 Partitioning of the Received Signal into Blocks

The reason for introducing block partitioning is that the channel can now be safely assumed to be piecewise *stationary* within such a block $\mathbf{r}_K$, at least for the purpose of deriving a MAP channel estimator based on intermediate ML estimates. This will turn out to be justified in practice (Section 15.2), provided that the blocks are sufficiently short and the channel is well underspread, i.e., the fading is *slow* (relative Doppler λ_D' well below $1/R_K$). Then the CIRs within a block are approximately equal to the CIR in the center of the block, i.e., $\mathbf{h}_k \approx \mathbf{h}_{k_K}$ for all $k = N_K, \ldots, N_K + R_K - 1$, where the *reference index* k_K (see Figure 12-6) is the nearest integer to $N_K + (R_K - 1)/2$. The (partial) channel transmission model [eq. (12-1) with $\Omega = 0$] with the channel memory confined to L symbol intervals

$$r_k^{(i)} = \sum_{n=k-L}^{k} a_n \, h_{k-n;\,k}^{(i)} + n_k^{(i)} = \sum_{n=0}^{L} h_{n;\,k}^{(i)} \, a_{k-n} + n_k^{(i)} \tag{12-75}$$

then translates into the simplified block transmission model

$$\underbrace{\begin{pmatrix} r_{N_K} \\ r_{N_K+1} \\ \vdots \\ r_{N_K+R_K-1} \end{pmatrix}^{(i)}}_{\mathbf{r}_K^{(i)}} = \underbrace{\begin{pmatrix} a_{N_K} & \cdots & a_{N_K-L} \\ a_{N_K+1} & \cdots & a_{N_K+1-L} \\ \vdots & & \vdots \\ a_{N_K+R_K-1} & \cdots & a_{N_K+R_K-1-L} \end{pmatrix}}_{\mathbf{A}_K^{(i)}} \cdot \underbrace{\begin{pmatrix} h_{0;k_K} \\ h_{1;k_K} \\ \vdots \\ h_{L;k_K} \end{pmatrix}^{(i)}}_{\mathbf{h}_{k_K}^{(i)}} + \mathbf{n}_K^{(i)} \tag{12-76}$$

with the R_K-dim. (partial) received vector $\mathbf{r}_K^{(i)}$, the $R_K \times [L+1]$-dim. block data matrix $\mathbf{A}_K^{(i)}$, and the $[L+1]$-dim. (partial) CIR vector $\mathbf{h}_{k_K}^{(i)}$. As before, the corresponding $(T/2)$-spaced vectors and matrices $\mathbf{r}_K$, $\mathbf{A}_K$, $\mathbf{h}_{k_K}$, and $\mathbf{n}_K$ with doubled dimensions can be constructed by abutting their partial counterparts (short-hand operator $\oplus$).

With these definitions, we can now proceed in evaluating the matrices $\mathbf{\Sigma}_{hr}$ and $\mathbf{\Sigma}_r$ needed for optimal estimation [eq. (12-73)]. Via substituting $\mathbf{r}_K = \mathbf{A}_K \cdot \mathbf{h}_{k_K} + \mathbf{n}_K$, the cross-correlation matrix $\mathbf{\Sigma}_{hr}$ between channel $\mathbf{h}_k$ and observation $\mathbf{r}$ can be well approximated by

$$\begin{aligned}\mathbf{\Sigma}_{hr} = E\left[\mathbf{h}_k \cdot \mathbf{r}^H\right] &= E\left[\mathbf{h}_k \cdot \left(\mathbf{r}_0^H \vdots \ldots \vdots \mathbf{r}_{\overline{N}-1}^H\right)\right] \\ &\approx \left(\mathbf{R}_h(k-k_0)\cdot\mathbf{A}_0^H \vdots \ldots \vdots \mathbf{R}_h\left(k-k_{\overline{N}-1}\right)\cdot\mathbf{A}_{\overline{N}-1}^H\right)\end{aligned} \tag{12-77}$$

Likewise, expanding the observation autocorrelation matrix $\mathbf{\Sigma}_r$ yields

$$\begin{aligned}\mathbf{\Sigma}_r = E\left[\mathbf{r} \cdot \mathbf{r}^H\right] &= E\left[\begin{pmatrix}\mathbf{r}_0 \\ \vdots \\ \mathbf{r}_{\overline{N}-1}\end{pmatrix} \cdot \left(\mathbf{r}_0^H \vdots \ldots \vdots \mathbf{r}_{\overline{N}-1}^H\right)\right] \\ &= \begin{pmatrix}\mathbf{R}_{0,0} & \ldots & \mathbf{R}_{0,\overline{N}-1} \\ \vdots & \ddots & \vdots \\ \mathbf{R}_{\overline{N}-1,0} & \ldots & \mathbf{R}_{\overline{N}-1,\overline{N}-1}\end{pmatrix}\end{aligned} \tag{12-78}$$

where the block autocorrelation matrices are well approximated by

$$\begin{aligned}\mathbf{R}_{K_1,K_2} &= E\left[\mathbf{r}_{K_1} \cdot \mathbf{r}_{K_2}^H\right] \\ &= \mathbf{A}_{K_1} \cdot E\left[\mathbf{h}_{k_{K_1}} \cdot \mathbf{h}_{k_{K_2}}^H\right] \cdot \mathbf{A}_{K_2}^H + E\left[\mathbf{n}_{K_1} \cdot \mathbf{n}_{K_2}^H\right] \\ &\approx \mathbf{A}_{K_1} \cdot \mathbf{R}_h(k_{K_1} - k_{K_2}) \cdot \mathbf{A}_{K_2}^H + \delta_{K_1,K_2} \cdot \mathbf{R}_{n,K_1}\end{aligned} \tag{12-79}$$

with $\mathbf{R}_{n,K}$ the noise covariance matrix of the Kth block.

These expressions for $\mathbf{\Sigma}_h$, $\mathbf{\Sigma}_{hr}$, and $\mathbf{\Sigma}_r$ can now be inserted into eq. (12-73), which yields

$$\begin{aligned}&\hat{\theta}_{D,k}(\mathbf{a}) = \hat{\mathbf{h}}_k(\mathbf{a}) \\ &= \left(\mathbf{R}_h(k-k_0)\cdot\mathbf{A}_0^H \vdots \mathbf{R}_h(k-k_1)\cdot\mathbf{A}_1^H \vdots \ldots \vdots \mathbf{R}_h\left(k-k_{\overline{N}-1}\right)\cdot\mathbf{A}_{\overline{N}-1}^H\right) \\ &\cdot\left[\begin{pmatrix}\mathbf{A}_0\mathbf{R}_h(k_0-k_0)\mathbf{A}_0^H & \ldots & \mathbf{A}_0\mathbf{R}_h\left(k_0-k_{\overline{N}-1}\right)\mathbf{A}_{\overline{N}-1}^H \\ \vdots & \ddots & \vdots \\ \mathbf{A}_{\overline{N}-1}\mathbf{R}_h\left(k_{\overline{N}-1}-k_0\right)\mathbf{A}_0^H & \ldots & \mathbf{A}_{\overline{N}-1}\mathbf{R}_h\left(k_{\overline{N}-1}-k_{\overline{N}-1}\right)\mathbf{A}_{\overline{N}-1}^H\end{pmatrix} + \mathbf{R}_n\right]^{-1} \cdot \begin{pmatrix}\mathbf{r}_0 \\ \vdots \\ \mathbf{r}_{\overline{N}-1}\end{pmatrix}\end{aligned} \tag{12-80}$$

Unfortunately, quantities which may be interpreted as intermediate ML estimates cannot be identified from this expression. In Section 12.2.6 (appendix B), eq. (12-80) is shown to be equivalent to the following form of the MAP estimate:

$$
\begin{aligned}
\hat{\theta}_{D,k}(\mathbf{a}) &= \hat{\mathbf{h}}_k(\mathbf{a}) \\
&= \left(\mathbf{R}_h(k-k_0) \vdots \ldots \vdots \mathbf{R}_h\left(k-k_{\overline{N}-1}\right)\right) \\
&\cdot \left[\begin{pmatrix} \mathbf{R}_h(k_0-k_0) & \ldots & \mathbf{R}_h\left(k_0-k_{\overline{N}-1}\right) \\ \vdots & \ddots & \vdots \\ \mathbf{R}_h\left(k_{\overline{N}-1}-k_0\right) & \ldots & \mathbf{R}_h\left(k_{\overline{N}-1}-k_{\overline{N}-1}\right) \end{pmatrix} + \begin{pmatrix} \left[\mathbf{A}_0^H \mathbf{R}_{n,0}^{-1} \mathbf{A}_0\right]^{-1} & & \\ & \ddots & \\ & & \left[\mathbf{A}_{\overline{N}-1}^H \mathbf{R}_{n,\overline{N}-1}^{-1} \mathbf{A}_{\overline{N}-1}\right]^{-1} \end{pmatrix}\right]^{-1} \\
&\cdot \begin{pmatrix} \left[\mathbf{A}_0^H \mathbf{R}_{n,0}^{-1} \mathbf{A}_0\right]^{-1} \cdot \left[\mathbf{A}_0^H \mathbf{R}_{n,0}^{-1}\, \mathbf{r}_0\right] \\ \vdots \\ \left[\mathbf{A}_{\overline{N}-1}^H \mathbf{R}_{n,\overline{N}-1}^{-1} \mathbf{A}_{\overline{N}-1}\right]^{-1} \cdot \left[\mathbf{A}_{\overline{N}-1}^H \mathbf{R}_{n,\overline{N}-1}^{-1}\, \mathbf{r}_{\overline{N}-1}\right] \end{pmatrix}
\end{aligned}
\tag{12-81}
$$

under the mild assumption that the noise correlation across adjacent blocks is neglected, i.e., the noise covariance matrix $\mathbf{R}_n$ reduces to the direct sum of block noise covariance matrices $\mathbf{R}_{n,K}$. In Section 12.2.6 (appendix B), it is also shown that the error covariance matrix $\mathbf{\Sigma}_{D,k} = \mathbf{\Sigma}_h - \mathbf{\Sigma}_{hr} \cdot \mathbf{\Sigma}_r^{-1} \cdot \mathbf{\Sigma}_{hr}^H$ [eq. (12-73)] associated with the MAP estimate becomes

$$
\begin{aligned}
\mathbf{\Sigma}_{D,k}(\mathbf{a}) &= \mathbf{R}_h - \left(\mathbf{R}_h(k-k_0) \vdots \ldots \vdots \mathbf{R}_h\left(k-k_{\overline{N}-1}\right)\right) \\
&\cdot \left[\begin{pmatrix} \mathbf{R}_h(k_0-k_0) & \ldots & \mathbf{R}_h\left(k_0-k_{\overline{N}-1}\right) \\ \vdots & \ddots & \vdots \\ \mathbf{R}_h\left(k_{\overline{N}-1}-k_0\right) & \ldots & \mathbf{R}_h\left(k_{\overline{N}-1}-k_{\overline{N}-1}\right) \end{pmatrix} + \begin{pmatrix} \left[\mathbf{A}_0^H \mathbf{R}_{n,0}^{-1} \mathbf{A}_0\right]^{-1} & & \\ & \ddots & \\ & & \left[\mathbf{A}_{\overline{N}-1}^H \mathbf{R}_{n,\overline{N}-1}^{-1} \mathbf{A}_{\overline{N}-1}\right]^{-1} \end{pmatrix}\right]^{-1} \\
&\cdot \begin{pmatrix} \mathbf{R}_h^H(k-k_0) \\ \vdots \\ \mathbf{R}_h^H\left(k-k_{\overline{N}-1}\right) \end{pmatrix}
\end{aligned}
\tag{12-82}
$$

Comparing eq. (12-81) with (12-70), one observes that the entries of the vector on the right-hand side of eq. (12-81) resemble the expression for the (nonexisting) ML selective channel estimate of eq. (12-70), with the only difference that the "global" matrices $\mathbf{A}$, $\mathbf{R}_n$, and vector $\mathbf{r}$ have been replaced by their "block" counterparts $\mathbf{A}_K$, $\mathbf{R}_{n,K}$, and $\mathbf{r}_K$, respectively. Therefore, the vector entries of eq. (12-81) are identified as the desired intermediate ML estimates, viz. ML block channel estimates $\hat{\mathbf{h}}_K$ which, together with their error covariances, are given by

$$
\begin{aligned}
\hat{\theta}_{S,K}(\mathbf{a}_K) = \hat{\mathbf{h}}_K(\mathbf{a}_K) &= E\left[\mathbf{h}_{k_K} \mid \begin{matrix}\text{no channel} \\ \text{statistics used}\end{matrix}\right] \\
&= \underbrace{\left[\mathbf{A}_K^H \mathbf{R}_{n,K}^{-1} \mathbf{A}_K\right]^{-1}}_{\mathbf{\Sigma}_{S,K}(\mathbf{a}_K)} \cdot \left[\mathbf{A}_K^H \mathbf{R}_{n,K}^{-1}\, \mathbf{r}_K\right] \\
\mathbf{\Sigma}_{S,K}(\mathbf{a}_K) &= \left[\mathbf{A}_K^H \mathbf{R}_{n,K}^{-1} \mathbf{A}_K\right]^{-1}
\end{aligned}
\tag{12-83}
$$

The entire intermediate ML estimate $\hat{\boldsymbol{\theta}}_S(\mathbf{a})$ and its error covariance $\boldsymbol{\Sigma}_S(\mathbf{a})$ are then given by the direct sum of all acquired estimates $\hat{\boldsymbol{\theta}}_{S,K}(\mathbf{a}_K) = \hat{\mathbf{h}}_K(\mathbf{a}_K)$ and of their error covariance matrices $\boldsymbol{\Sigma}_{S,K}(\mathbf{a}_K)$, respectively:

$$
\begin{aligned}
\hat{\boldsymbol{\theta}}_S(\mathbf{a}) &= \begin{pmatrix} \hat{\boldsymbol{\theta}}_{S,0}(\mathbf{a}_0) \\ \vdots \\ \hat{\boldsymbol{\theta}}_{S,\overline{N}-1}(\mathbf{a}_{\overline{N}-1}) \end{pmatrix} = \begin{pmatrix} \hat{\mathbf{h}}_0(\mathbf{a}_0) \\ \vdots \\ \hat{\mathbf{h}}_{\overline{N}-1}(\mathbf{a}_{\overline{N}-1}) \end{pmatrix} \\
&= \underbrace{\begin{pmatrix} \left[\mathbf{A}_0^H \mathbf{R}_{n,0}^{-1} \mathbf{A}_0\right]^{-1} & & \\ & \ddots & \\ & & \left[\mathbf{A}_{\overline{N}-1}^H \mathbf{R}_{n,\overline{N}-1}^{-1} \mathbf{A}_{\overline{N}-1}\right]^{-1} \end{pmatrix}}_{\boldsymbol{\Sigma}_S(\mathbf{a})} \cdot \begin{pmatrix} \mathbf{A}_0^H \mathbf{R}_{n,0}^{-1} \mathbf{r}_0 \\ \vdots \\ \mathbf{A}_{\overline{N}-1}^H \mathbf{R}_{n,\overline{N}-1}^{-1} \mathbf{r}_{\overline{N}-1} \end{pmatrix} \\
\boldsymbol{\Sigma}_S(\mathbf{a}) &= \begin{pmatrix} \boldsymbol{\Sigma}_{S,0}(\mathbf{a}_0) & & \\ & \ddots & \\ & & \boldsymbol{\Sigma}_{S,\overline{N}-1}\left(\mathbf{a}_{\overline{N}-1}\right) \end{pmatrix} \\
&= \begin{pmatrix} \left[\mathbf{A}_0^H \mathbf{R}_{n,0}^{-1} \mathbf{A}_0\right]^{-1} & & \\ & \ddots & \\ & & \left[\mathbf{A}_{\overline{N}-1}^H \mathbf{R}_{n,\overline{N}-1}^{-1} \mathbf{A}_{\overline{N}-1}\right]^{-1} \end{pmatrix}
\end{aligned}
\tag{12-84}
$$

Each estimate $\hat{\mathbf{h}}_K(\mathbf{a}_K)$ is a quasi-static approximation to the (slowly varying) channel state over the duration of the Kth block. For further processing, $\hat{\mathbf{h}}_K(\mathbf{a}_K)$ may be regarded valid at the reference index k_K corresponding to the center of the block. Thus, $\hat{\mathbf{h}}_K(\mathbf{a}_K)$ may also be termed a (noisy but otherwise unbiased) channel *snapshot* taken from the dynamic channel trajectory $\{\mathbf{h}_0, \ldots, \mathbf{h}_{N-1}\}$. Each snapshot $\hat{\mathbf{h}}_K(\mathbf{a}_K)$ is dependent on the block observation $\mathbf{r}_K$, its own (hypothetical) block data sequence $\mathbf{a}_K$ having an impact on $\mathbf{r}_K$ (see Figure 12-6), and the noise covariance. Notice, however, that $\hat{\mathbf{h}}_K(\mathbf{a}_K)$ is *not* dependent of the fading channel parameters; the only prerequisite of unbiased snapshot acquisition is that the estimation window must be wide enough to accommodate all $L+1$ relevant (partial) channel taps.

Of course, all snapshots must exist, i.e., the matrices $\mathbf{A}_K^H \mathbf{R}_{n,K}^{-1} \mathbf{A}_K$ must all be invertible for a particular data hypothesis $\mathbf{a}$. Even if this is the case, the error covariances $\boldsymbol{\Sigma}_{S,K}(\mathbf{a}_K)$ of snapshots (and thus the error covariance of the final estimate) are strongly dependent on the properties of the data $\mathbf{a}$. Clearly, this is a major drawback of joint detection and *selective* channel estimation when compared to joint detection and *flat* channel estimation where $\boldsymbol{\Sigma}_S(\mathbf{a})$ is *independent* of the statistical properties of $\mathbf{a}$ (except for the symbol magnitudes). Here, on the other hand, it is, in principle, necessary to select an appropriate block partitioning

(Figure 12-6) and compute $\mathbf{\Sigma}_{S,K}(\mathbf{a}_K)$ for each and every hypothesis $\mathbf{a}$. For some (however pathological) data sequences, such a partitioning does not exist at all; think, for example, of the all-1 sequence for which $\mathbf{A}_K^H \mathbf{R}_{n,K}^{-1} \mathbf{A}_K$ is a rank-1 matrix (assuming white noise), no matter which kind of partitioning is selected. However, the rationale for this discussion will become apparent in the context of the data-aided channel estimation techniques discussed in Section 15.2.

Let us now proceed to the second step of MAP selective channel estimation. Via eqs. (12-81) and (12-84), the desired MAP estimate $\hat{\boldsymbol{\theta}}_{D,k}(\mathbf{a}) = \hat{\mathbf{h}}_k(\mathbf{a})$ can be computed from the acquired ML estimate $\hat{\boldsymbol{\theta}}_S(\mathbf{a})$ by a vector/matrix operation:

$$
\begin{aligned}
\hat{\boldsymbol{\theta}}_{D,k}(\mathbf{a}) &= \hat{\mathbf{h}}_k(\mathbf{a}) \\
&= \left(\mathbf{R}_h(k-k_0) \vdots \ldots \vdots \mathbf{R}_h\left(k-k_{\overline{N}-1}\right) \right) \\
&\cdot \left[\begin{pmatrix} \mathbf{R}_h(k_0-k_0) & \ldots & \mathbf{R}_h\left(k_0-k_{\overline{N}-1}\right) \\ \vdots & \ddots & \vdots \\ \mathbf{R}_h\left(k_{\overline{N}-1}-k_0\right) & \ldots & \mathbf{R}_h\left(k_{\overline{N}-1}-k_{\overline{N}-1}\right) \end{pmatrix} + \begin{pmatrix} \left[\mathbf{A}_0^H \mathbf{R}_{n,0}^{-1} \mathbf{A}_0\right]^{-1} & & \\ & \ddots & \\ & & \left[\mathbf{A}_{\overline{N}-1}^H \mathbf{R}_{n,\overline{N}-1}^{-1} \mathbf{A}_{\overline{N}-1}\right]^{-1} \end{pmatrix} \right]^{-1} \\
&\cdot \underbrace{\begin{pmatrix} \hat{\mathbf{h}}_0(\mathbf{a}_0) \\ \vdots \\ \hat{\mathbf{h}}_{\overline{N}-1}\left(\mathbf{a}_{\overline{N}-1}\right) \end{pmatrix}}_{\hat{\boldsymbol{\theta}}_S(\mathbf{a})}
\end{aligned}
\tag{12-85}
$$

This again reflects the *separation property* of MAP dynamic parameter estimation, as illustrated by eq. (12-46). Analogous to the notation of eq. (12-45), the matrix by which the ML estimate is to be premultiplied may be abbreviated by $\mathbf{N}_k(\mathbf{a})$ (here, however, dependent on the intrapacket index k for which the MAP channel estimate is to be generated). Then the mapping ML $\longrightarrow$ MAP can be written in its concisest form as:

$$
\hat{\boldsymbol{\theta}}_{D,k}(\mathbf{a}) = \mathbf{N}_k(\mathbf{a}) \cdot \hat{\boldsymbol{\theta}}_S(\mathbf{a}) \tag{12-86}
$$

with the $2(L+1) \times \overline{N} \cdot 2(L+1)$-dim. transition matrix

$$
\begin{aligned}
\mathbf{N}_k &= \left(\mathbf{R}_h(k-k_0) \vdots \ldots \vdots \mathbf{R}_h\left(k-k_{\overline{N}-1}\right) \right) \\
&\cdot \left[\begin{pmatrix} \mathbf{R}_h(k_0-k_0) & \ldots & \mathbf{R}_h\left(k_0-k_{\overline{N}-1}\right) \\ \vdots & \ddots & \vdots \\ \mathbf{R}_h\left(k_{\overline{N}-1}-k_0\right) & \ldots & \mathbf{R}_h\left(k_{\overline{N}-1}-k_{\overline{N}-1}\right) \end{pmatrix} + \underbrace{\begin{pmatrix} \left[\mathbf{A}_0^H \mathbf{R}_{n,0}^{-1} \mathbf{A}_0\right]^{-1} & & \\ & \ddots & \\ & & \left[\mathbf{A}_{\overline{N}-1}^H \mathbf{R}_{n,\overline{N}-1}^{-1} \mathbf{A}_{\overline{N}-1}\right]^{-1} \end{pmatrix}}_{\mathbf{\Sigma}_S(\mathbf{a})} \right]^{-1}
\end{aligned}
\tag{12-87}
$$

The expression for the MAP error covariance, given by eqs. (12-111) [Section 12.2.6 (appendix B)] and (12-82), likewise contains the matrix $\mathbf{N}_k(\mathbf{a})$:

$$\mathbf{\Sigma}_{D,k}(\mathbf{a}) = \mathbf{R}_h - \mathbf{N}_k(\mathbf{a}) \cdot \begin{pmatrix} \mathbf{R}_h^H(k-k_0) \\ \vdots \\ \mathbf{R}_h^H\left(k-k_{\overline{N-1}}\right) \end{pmatrix} \tag{12-88}$$

Here, however, $\mathbf{\Sigma}_{D,k}(\mathbf{a})$ is no more given by the relation $\mathbf{\Sigma}_{D,k}(\mathbf{a}) = \mathbf{N}_k(\mathbf{a}) \cdot \mathbf{\Sigma}_S(\mathbf{a})$ [eq. (12-44)] simply because $\mathbf{\Sigma}_S(\mathbf{a})$ and $\mathbf{\Sigma}_{D,k}(\mathbf{a})$ have different dimensions so that $\mathbf{N}_k(\mathbf{a})$ is not square.

To summarize this discussion on the concept of joint detection and selective channel estimation, it has been shown in this section that the MAP estimate $\hat{\boldsymbol{\theta}}_D(\mathbf{a}) = \hat{\mathbf{h}}(\mathbf{a})$ of the entire selective channel trajectory can in principle be computed by the following two-step procedure:

1. **ML channel snapshot acquisition**:
   mapping $(\mathbf{r}, \mathbf{a}) \longrightarrow \hat{\boldsymbol{\theta}}_S(\mathbf{a})$
   via mappings of blocks $(\mathbf{r}_K, \mathbf{a}_K) \longrightarrow \hat{\boldsymbol{\theta}}_{S,K}(\mathbf{a}_K) = \hat{\mathbf{h}}_K(\mathbf{a}_K)$ [eq. (12-83)], without using any knowledge on channel statistics
2. **MAP channel estimation** from the ML estimate:
   mapping $\hat{\boldsymbol{\theta}}_S(\mathbf{a}) \longrightarrow \hat{\boldsymbol{\theta}}_D(\mathbf{a})$
   via mappings $\hat{\boldsymbol{\theta}}_S(\mathbf{a}) \longrightarrow \hat{\boldsymbol{\theta}}_{D,k}(\mathbf{a})$ for all $k = 0, \ldots, N-1$ [eq. (12-86)], using the knowledge on channel statistics.

This two-step procedure is illustrated in Figure 12-7. We remark that the sequence of snapshots $\left\{\hat{\boldsymbol{\theta}}_{S,K}(\mathbf{a}_K)\right\} = \left\{\hat{\mathbf{h}}_K(\mathbf{a}_K)\right\}$ shown in Figure 12-7 in fact follows an

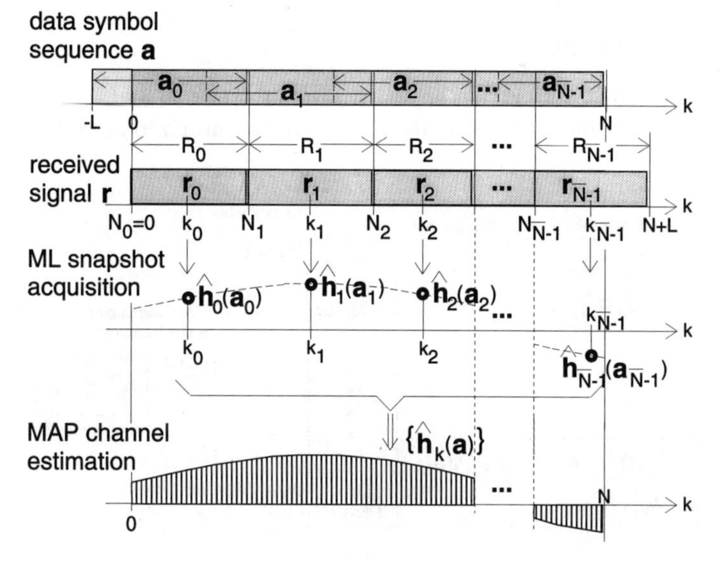

Figure 12-7 Two-Step MAP Estimation of Selective Fading Channels

erratic trajectory for wrong data hypotheses $\mathbf{a}$. Only in the special case of correct (apart from a constant phase shift) data hypotheses $\mathbf{a}$ the snapshots are unbiased estimates which follow the smooth trajectory of the (lowpass) fading process.

Having obtained conditional MAP estimates and associated error covariances of the entire channel trajectory for all data hypotheses $\mathbf{a}$, the set of decision metrics may be evaluated according to eq. (12-43):

$$m_D(\mathbf{a}) = \hat{\theta}_D^H(\mathbf{a}) \cdot \mathbf{\Sigma}_D^{-1}(\mathbf{a}) \cdot \hat{\theta}_D(\mathbf{a}) \qquad \underset{\mathbf{a}}{\rightarrow \max} \tag{12-89}$$

To summarize this section, the generation of decision metrics based on MAP selective channel estimation via ML snapshot acquisition is visualized in Figure 12-8. The detection part is identical to that of Figure 12-1 displaying the more general case of detection based on MAP sync parameter estimation, while the channel estimation part is now split into the two subtasks of blockwise ML snapshot acquisition and MAP channel estimation from these snapshots. Analogous to the flat fading case [eq. (12-65) and Figure 12-4], the operations performed on the sequence of conditional snapshots $\left\{\hat{\theta}_{S,K}(\mathbf{a}_K)\right\} = \left\{\hat{\mathbf{h}}_K(\mathbf{a}_K)\right\}$ [here: filtering by matrices $\mathbf{N}_k(\mathbf{a})$ and computation of inner products $\hat{\theta}_D^H(\mathbf{a}) \cdot \mathbf{\Sigma}_D^{-1}(\mathbf{a}) \cdot \hat{\theta}_D(\mathbf{a})$], can be interpreted as lowpass filtering operations. The decision metrics $m_D(\mathbf{a})$ may thus be viewed as measures of how well the conditional snapshot sequence $\left\{\hat{\theta}_{S,K}(\mathbf{a}_K)\right\} = \left\{\hat{\mathbf{h}}_K(\mathbf{a}_K)\right\}$ follows a smooth lowpass trajectory. Even though conceptually simple, the detection part entails an even higher computational burden than selective fading channel estimation. Therefore, we shall not pursue this kind of detection further but turn to realizable detection and channel estimation techniques in Chapters 13 to 15.

12.2.5 Main Points

- By virtue of the Bayesian approach to joint detection and estimation, basic receiver structures following the concept of synchronized detection on fading

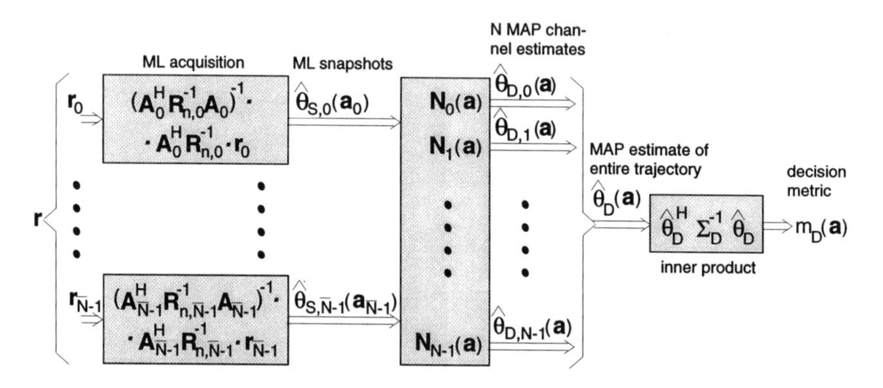

Figure 12-8 Generation of Decision Metrics via MAP Selective Channel Estimation

channels have been derived. In the linear Gaussian case, the optimal decision metric

$$\begin{aligned} m_S(\mathbf{a}) &= \hat{\boldsymbol{\theta}}_S^H(\mathbf{a}) \cdot \boldsymbol{\Sigma}_S^{-1}(\mathbf{a}) \cdot \hat{\boldsymbol{\theta}}_S(\mathbf{a}) \\ m_D(\mathbf{a}) &= \hat{\boldsymbol{\theta}}_D^H(\mathbf{a}) \cdot \boldsymbol{\Sigma}_D^{-1}(\mathbf{a}) \cdot \hat{\boldsymbol{\theta}}_D(\mathbf{a}) \end{aligned} \tag{12-90}$$

[eqs. (12-37) and (12-43)] is expressible in terms of the optimal ML static or MAP dynamic parameter estimate:

$$\begin{aligned} \hat{\boldsymbol{\theta}}_S(\mathbf{a}) &= \left(\mathbf{A}^H \mathbf{R}_n^{-1} \mathbf{A}\right)^{-1} \cdot \mathbf{A}^H \mathbf{R}_n^{-1} \mathbf{r} \\ \hat{\boldsymbol{\theta}}_D(\mathbf{a}) &= \boldsymbol{\mu}_D + \mathbf{R}_D \mathbf{A}^H \left(\mathbf{A} \mathbf{R}_D \mathbf{A}^H + \mathbf{R}_n\right)^{-1} \cdot (\mathbf{r} - \mathbf{A} \boldsymbol{\mu}_D) \end{aligned} \tag{12-91}$$

[eqs. (12-27) and (12-31)].

- In the case that the ML estimate of dynamic parameters exists, the estimator part can be partitioned into two distinct steps: (i) ML estimation from the observation $\mathbf{r}$ [eq. (12-27)], disregarding any knowledge on channel statistics, followed by (ii) MAP estimation from the ML estimate $\hat{\boldsymbol{\theta}}_S(a) \rightarrow \hat{\boldsymbol{\theta}}_D(a)$, using the knowledge on channel statistics:

$$\hat{\boldsymbol{\theta}}_D(\mathbf{a}) = \mathbf{N}(\mathbf{a}) \cdot \hat{\boldsymbol{\theta}}_S(\mathbf{a}) \; + \; \mathbf{M}(\mathbf{a}) \cdot \boldsymbol{\mu}_D \tag{12-92}$$

[eq. (12-45)].

- Concerning the estimator part of joint detection and estimation for *flat* Rician channels, MAP channel estimation via ML estimation is always possible. The ML→MAP gain ratio r_k is almost independent of the noise and, in the case of virtually infinite message length, very close to the fading bandwidth $2\lambda'_D$. Simplified – yet quasi-optimal – channel estimation is achieved by matching the estimator to Rayleigh fading with an ideal – or near-ideal – lowpass Doppler spectrum. In the case of M-PSK signaling, both channel estimation and detection can be further simplified.
- Concerning the estimator part of joint detection and estimation for *selective* slowly fading channels, MAP channel estimation via ML estimation is accomplished by partitioning the received signal into a set of $\overline{N}$ blocks $\mathbf{r}_K = \mathbf{A}_K \cdot \mathbf{h}_{k_K} + \mathbf{n}_K$ and performing the two-step procedure of (i) ML acquisition of the set of $\overline{N}$ channel snapshots $\hat{\mathbf{h}}_K(\mathbf{a}_K)$

$$\begin{aligned} \hat{\boldsymbol{\theta}}_{S,K}(\mathbf{a}_K) = \hat{\mathbf{h}}_K(\mathbf{a}_K) &= \left(\mathbf{A}_K^H \mathbf{R}_{n,K}^{-1} \mathbf{A}_K\right)^{-1} \cdot \mathbf{A}_K^H \mathbf{R}_{n,K}^{-1} \cdot \mathbf{r}_K \\ \hat{\boldsymbol{\theta}}_S(\mathbf{a}) &= \left(\hat{\boldsymbol{\theta}}_{S,0}^T(\mathbf{a}_0) \; \ldots \; \hat{\boldsymbol{\theta}}_{S,\overline{N}-1}^T\left(\mathbf{a}_{\overline{N}-1}\right)\right)^T \end{aligned} \tag{12-93}$$

[eq. (12-83)], followed by (ii) MAP estimation of the entire channel trajectory $\hat{\mathbf{h}}(\mathbf{a})$ (set of N CIR vectors) from the set of snapshots

$$\begin{aligned} \hat{\boldsymbol{\theta}}_{D,k}(\mathbf{a}) = \hat{\mathbf{h}}_k(\mathbf{a}) &= \mathbf{N}_k(\mathbf{a}) \cdot \hat{\boldsymbol{\theta}}_S(\mathbf{a}) \\ \hat{\boldsymbol{\theta}}_D(\mathbf{a}) = \hat{\mathbf{h}}(\mathbf{a}) &= \left(\hat{\boldsymbol{\theta}}_{D,0}^T(\mathbf{a}) \; \ldots \; \hat{\boldsymbol{\theta}}_{D,N-1}^T(\mathbf{a})\right)^T \end{aligned} \tag{12-94}$$

[eqs. (12-81) and (12-86)].

12.2.6 Appendices

Appendix A: Proof of Optimality of the Joint Detection and Estimation Metric in the Gaussian Case

The results on joint detection and estimation established in this chapter have been derived using the high-SNR approximation of eq. (12-21). We now wish to prove that, in the Gaussian case, these results are indeed not an approximation but optimal at all SNR. This is shown by relating the decision rule for synchronized detection [i.e., the ML decision rule of eq. (12-43) based on MAP estimation according to eq. (12-31)] to the optimal decision rule. The latter is derived from the likelihood function for the Gaussian case [14],

$$\begin{aligned} p(\mathbf{r}|\mathbf{a}) &= N(\bar{\mathbf{r}}, \mathbf{R}_r) \\ &\propto \exp\left\{-[\mathbf{r}-\bar{\mathbf{r}}]^H \cdot \mathbf{R}_r^{-1} \cdot [\mathbf{r}-\bar{\mathbf{r}}]\right\} \\ &= \exp\left\{-\left[\mathbf{r} - \underbrace{\mathbf{A}\boldsymbol{\mu}_D}_{\bar{\mathbf{r}}=E[\mathbf{r}]}\right]^H \cdot \underbrace{\left(\mathbf{A}\mathbf{R}_D\mathbf{A}^H + \mathbf{R}_n\right)^{-1}}_{\mathbf{R}_r^{-1}} \cdot [\mathbf{r}-\mathbf{A}\boldsymbol{\mu}_D]\right\} \end{aligned} \tag{12-95}$$

which can, in theory, be evaluated without the need for explicit parameter estimation. Reformulating $\mathbf{R}_r^{-1}$ via the matrix inversion lemma [eq. (12-39)] and taking the logarithm yelds the optimal log-likelihood metric (to be maximized),

$$\begin{aligned} m_{D,\text{opt}}(\mathbf{a}) =& -[\mathbf{r}-\mathbf{A}\boldsymbol{\mu}_D]^H \cdot \left(\mathbf{A}\mathbf{R}_D\mathbf{A}^H + \mathbf{R}_n\right)^{-1} \cdot [\mathbf{r}-\mathbf{A}\boldsymbol{\mu}_D] \\ =& -[\mathbf{r}-\mathbf{A}\boldsymbol{\mu}_D]^H \cdot \left(\mathbf{R}_n^{-1} - \mathbf{R}_n^{-1}\mathbf{A}\underbrace{\left[\mathbf{R}_D^{-1} + \mathbf{A}^H\mathbf{R}_n^{-1}\mathbf{A}\right]^{-1}}_{\boldsymbol{\Sigma}_D(\mathbf{a})}\mathbf{A}^H\mathbf{R}_n^{-1}\right) \cdot [\mathbf{r}-\mathbf{A}\boldsymbol{\mu}_D] \end{aligned} \tag{12-96}$$

On the other hand, expanding the metric of eq. (12-43) yields

$$\begin{aligned} m_D(\mathbf{a}) =& \underbrace{\left[\mathbf{A}^H\mathbf{R}_n^{-1}\mathbf{r} + \mathbf{R}_D^{-1}\boldsymbol{\mu}_D\right]^H}_{\hat{\boldsymbol{\theta}}_D^H(\mathbf{a})} \cdot \boldsymbol{\Sigma}_D \cdot \underbrace{\left[\mathbf{A}^H\mathbf{R}_n^{-1}\mathbf{r} + \mathbf{R}_D^{-1}\boldsymbol{\mu}_D\right]}_{\hat{\boldsymbol{\theta}}_D(\mathbf{a})} \\ =& \left([\mathbf{r}-\mathbf{A}\boldsymbol{\mu}_D]^H\mathbf{R}_n^{-1}\mathbf{A} + \boldsymbol{\mu}_D^H\underbrace{\left(\mathbf{A}^H\mathbf{R}_n^{-1}\mathbf{A} + \mathbf{R}_D^{-1}\right)}_{\boldsymbol{\Sigma}_D^{-1}}\right) \cdot \boldsymbol{\Sigma}_D \\ & \cdot \left(\mathbf{A}^H\mathbf{R}_n^{-1}[\mathbf{r}-\mathbf{A}\boldsymbol{\mu}_D] + \boldsymbol{\Sigma}_D^{-1}\boldsymbol{\mu}_D\right) \\ =& [\mathbf{r}-\mathbf{A}\boldsymbol{\mu}_D]^H\mathbf{R}_n^{-1}\mathbf{A}\boldsymbol{\Sigma}_D\mathbf{A}^H\mathbf{R}_n^{-1}[\mathbf{r}-\mathbf{A}\boldsymbol{\mu}_D] + \boldsymbol{\mu}_D^H\boldsymbol{\Sigma}_D^{-1}\boldsymbol{\mu}_D \\ & + [\mathbf{r}-\mathbf{A}\boldsymbol{\mu}_D]^H\mathbf{R}_n^{-1}\mathbf{A}\boldsymbol{\mu}_D + \boldsymbol{\mu}_D^H\mathbf{A}^H\mathbf{R}_n^{-1}[\mathbf{r}-\mathbf{A}\boldsymbol{\mu}_D] \end{aligned} \tag{12-97}$$

The last two terms can be reformulated as

$$\begin{aligned} & [\mathbf{r}-\mathbf{A}\boldsymbol{\mu}_D]^H\mathbf{R}_n^{-1}\mathbf{A}\boldsymbol{\mu}_D + \boldsymbol{\mu}_D^H\mathbf{A}^H\mathbf{R}_n^{-1}[\mathbf{r}-\mathbf{A}\boldsymbol{\mu}_D] \\ &= \mathbf{r}^H\mathbf{R}_n^{-1}\mathbf{A}\boldsymbol{\mu}_D - 2\cdot\boldsymbol{\mu}_D^H\mathbf{A}^H\mathbf{R}_n^{-1}\mathbf{A}\boldsymbol{\mu}_D + \boldsymbol{\mu}_D^H\mathbf{A}^H\mathbf{R}_n^{-1}\mathbf{r} \\ &= -[\mathbf{r}-\mathbf{A}\boldsymbol{\mu}_D]^H\mathbf{R}_n^{-1}[\mathbf{r}-\mathbf{A}\boldsymbol{\mu}_D] - \boldsymbol{\mu}_D^H\mathbf{A}^H\mathbf{R}_n^{-1}\mathbf{A}\boldsymbol{\mu}_D + \mathbf{r}^H\mathbf{R}_n^{-1}\mathbf{r} \end{aligned} \tag{12-98}$$

so that $m_D(\mathbf{a})$ can be expressed such that one of its terms is identified as the optimal metric $m_{D,\text{opt}}(\mathbf{a})$ [eq. (12-96)]:

$$\begin{aligned}
m_D(\mathbf{a}) &= [\mathbf{r}-\mathbf{A}\boldsymbol{\mu}_D]^H \mathbf{R}_n^{-1}\mathbf{A}\boldsymbol{\Sigma}_D\mathbf{A}^H\mathbf{R}_n^{-1}[\mathbf{r}-\mathbf{A}\boldsymbol{\mu}_D] + \boldsymbol{\mu}_D^H\boldsymbol{\Sigma}_D^{-1}\boldsymbol{\mu}_D \\
&\quad - [\mathbf{r}-\mathbf{A}\boldsymbol{\mu}_D]^H\mathbf{R}_n^{-1}[\mathbf{r}-\mathbf{A}\boldsymbol{\mu}_D] - \boldsymbol{\mu}_D^H\mathbf{A}^H\mathbf{R}_n^{-1}\mathbf{A}\boldsymbol{\mu}_D + \mathbf{r}^H\mathbf{R}_n^{-1}\mathbf{r} \\
&= \underbrace{[\mathbf{r}-\mathbf{A}\boldsymbol{\mu}_D]^H\left(\mathbf{R}_n^{-1}\mathbf{A}\boldsymbol{\Sigma}_D\mathbf{A}^H\mathbf{R}_n^{-1} - \mathbf{R}_n^{-1}\right)[\mathbf{r}-\mathbf{A}\boldsymbol{\mu}_D]}_{m_{D,\text{opt}}(\mathbf{a})} \\
&\quad + \boldsymbol{\mu}_D^H\left(\boldsymbol{\Sigma}_D^{-1} - \mathbf{A}^H\mathbf{R}_n^{-1}\mathbf{A}\right)\boldsymbol{\mu}_D + \mathbf{r}^H\mathbf{R}_n^{-1}\mathbf{r} \\
&= m_{D,\text{opt}}(\mathbf{a}) + \boldsymbol{\mu}_D^H\left(\underbrace{\mathbf{R}_D^{-1} + \mathbf{A}^H\mathbf{R}_n^{-1}\mathbf{A}}_{\boldsymbol{\Sigma}_D^{-1}} - \mathbf{A}^H\mathbf{R}_n^{-1}\mathbf{A}\right)\boldsymbol{\mu}_D + \mathbf{r}^H\mathbf{R}_n^{-1}\mathbf{r} \\
&= m_{D,\text{opt}}(\mathbf{a}) + \underbrace{\boldsymbol{\mu}_D^H\mathbf{R}_D^{-1}\boldsymbol{\mu}_D + \mathbf{r}^H\mathbf{R}_n^{-1}\mathbf{r}}_{\text{independent of } \mathbf{a}} \\
&\propto m_{D,\text{opt}}(\mathbf{a})
\end{aligned} \tag{12-99}$$

Hence, in the Gaussian case, the decision metric $m_D(\mathbf{a})$ of synchronized detection is equivalent to the optimal metric $m_{D,\text{opt}}(\mathbf{a})$.

Appendix B: Derivation of the MAP Channel Estimate from an Intermediate ML Estimate

In this appendix, we wish to reformulate the MAP estimate such that an ML estimate of intermediate parameters can be identified. With the abbreviations (used only in this appendix) $\mathbf{R}_{K;k} = \mathbf{R}_h(k-k_K)$ and $\mathbf{R}_{K_1,K_2} = \mathbf{R}_h(k_{K_1}-k_{K_2})$, the MAP estimate according to eq. (12-80) is of the form

$$\begin{aligned}
\hat{\boldsymbol{\theta}}_{D,k}(\mathbf{a}) &= \hat{\mathbf{h}}_k(\mathbf{a}) \\
&\approx \left(\mathbf{R}_{0;k}\cdot\mathbf{A}_0^H \,\vdots\, \mathbf{R}_{1;k}\cdot\mathbf{A}_1^H \,\vdots\, \ldots \,\vdots\, \mathbf{R}_{\overline{N}-1;k}\cdot\mathbf{A}_{\overline{N}-1}^H\right) \\
&\quad\cdot\left[\begin{pmatrix} \mathbf{A}_0\mathbf{R}_{0,0}\mathbf{A}_0^H & \ldots & \mathbf{A}_0\mathbf{R}_{0,\overline{N}-1}\mathbf{A}_{\overline{N}-1}^H \\ \vdots & \ddots & \vdots \\ \mathbf{A}_{\overline{N}-1}\mathbf{R}_{\overline{N}-1,0}\mathbf{A}_0^H & \ldots & \mathbf{A}_{\overline{N}-1}\mathbf{R}_{\overline{N}-1,\overline{N}-1}\mathbf{A}_{\overline{N}-1}^H \end{pmatrix} + \mathbf{R}_n\right]^{-1}\cdot\begin{pmatrix}\mathbf{r}_0 \\ \vdots \\ \mathbf{r}_{\overline{N}-1}\end{pmatrix} \\
&= \left(\mathbf{R}_{0;k} \,\vdots\, \ldots \,\vdots\, \mathbf{R}_{\overline{N}-1;k}\right)\cdot\begin{pmatrix}\mathbf{A}_0 & & \\ & \ddots & \\ & & \mathbf{A}_{\overline{N}-1}\end{pmatrix}^H \\
&\quad\cdot\left[\underbrace{\begin{pmatrix}\mathbf{A}_0 & & \\ & \ddots & \\ & & \mathbf{A}_{\overline{N}-1}\end{pmatrix}}_{\mathbf{A}}\underbrace{\begin{pmatrix}\mathbf{R}_{0,0} & \cdots & \mathbf{R}_{0,\overline{N}-1} \\ \vdots & \ddots & \vdots \\ \mathbf{R}_{\overline{N}-1,0} & \cdots & \mathbf{R}_{\overline{N}-1,\overline{N}-1}\end{pmatrix}}_{\mathbf{R}}\underbrace{\begin{pmatrix}\mathbf{A}_0 & & \\ & \ddots & \\ & & \mathbf{A}_{\overline{N}-1}\end{pmatrix}^H}_{\mathbf{A}^H} + \mathbf{R}_n\right]^{-1}\cdot\begin{pmatrix}\mathbf{r}_0 \\ \vdots \\ \mathbf{r}_{\overline{N}-1}\end{pmatrix} \\
&= \left(\mathbf{R}_{0;k} \,\vdots\, \ldots \,\vdots\, \mathbf{R}_{\overline{N}-1;k}\right)\cdot\mathbf{A}^H\cdot\underbrace{\left[\mathbf{A}\cdot\mathbf{R}\cdot\mathbf{A}^H + \mathbf{R}_n\right]^{-1}}_{\mathbf{E}^{-1}}\cdot\begin{pmatrix}\mathbf{r}_0 \\ \vdots \\ \mathbf{r}_{\overline{N}-1}\end{pmatrix}
\end{aligned} \tag{12-100}$$

By invoking the matrix inversion lemma [eq. (12-39)],

$$\begin{aligned} \mathbf{E} &= \mathbf{A}^H\mathbf{B}^{-1}\mathbf{A} + \mathbf{C}^{-1} \\ \mathbf{E}^{-1} &= \mathbf{C} - \mathbf{C}\mathbf{A}^H\left[\mathbf{B} + \mathbf{A}\mathbf{C}\mathbf{A}^H\right]^{-1}\mathbf{A}\mathbf{C} \end{aligned} \tag{12-101}$$

one obtains for the term $\mathbf{E}^{-1}$

$$\begin{aligned} \mathbf{E}^{-1} &= \left[\mathbf{A}\cdot\mathbf{R}\cdot\mathbf{A}^H + \mathbf{R}_n\right]^{-1} \\ &= \mathbf{R}_n^{-1} - \mathbf{R}_n^{-1}\mathbf{A}\cdot\underbrace{\left[\mathbf{R}^{-1} + \mathbf{A}^H\mathbf{R}_n^{-1}\mathbf{A}\right]^{-1}}_{\mathbf{F}^{-1}}\cdot\mathbf{A}^H\mathbf{R}_n^{-1} \end{aligned} \tag{12-102}$$

Reformulating the term $\mathbf{F}^{-1}$ in the same manner yields

$$\begin{aligned} \mathbf{F}^{-1} &= \left[\mathbf{R}^{-1} + \mathbf{A}^H\mathbf{R}_n^{-1}\mathbf{A}\right]^{-1} \\ &= \left(\mathbf{A}^H\mathbf{R}_n^{-1}\mathbf{A}\right)^{-1} - \left(\mathbf{A}^H\mathbf{R}_n^{-1}\mathbf{A}\right)^{-1} \\ &\quad \cdot\left[\mathbf{R} + \left(\mathbf{A}^H\mathbf{R}_n^{-1}\mathbf{A}\right)^{-1}\right]^{-1}\cdot\left(\mathbf{A}^H\mathbf{R}_n^{-1}\mathbf{A}\right)^{-1} \\ &= \left(\mathbf{A}^H\mathbf{R}_n^{-1}\mathbf{A}\right)^{-1}\cdot\left(\mathbf{I} - \left[\mathbf{R}+\left(\mathbf{A}^H\mathbf{R}_n^{-1}\mathbf{A}\right)^{-1}\right]^{-1}\cdot\left(\mathbf{A}^H\mathbf{R}_n^{-1}\mathbf{A}\right)^{-1}\right) \end{aligned} \tag{12-103}$$

so that the expression $\mathbf{A}^H\cdot\mathbf{E}^{-1}$ is evaluated as

$$\begin{aligned} \mathbf{A}^H\mathbf{E}^{-1} &= \mathbf{A}^H\mathbf{R}_n^{-1} - \mathbf{A}^H\mathbf{R}_n^{-1}\mathbf{A}\cdot\underbrace{\left[\mathbf{R}^{-1} + \mathbf{A}^H\mathbf{R}_n^{-1}\mathbf{A}\right]^{-1}}_{\mathbf{F}^{-1}}\cdot\mathbf{A}^H\mathbf{R}_n^{-1} \\ &= \mathbf{A}^H\mathbf{R}_n^{-1} - \left(\mathbf{I} - \left[\mathbf{R}+\left(\mathbf{A}^H\mathbf{R}_n^{-1}\mathbf{A}\right)^{-1}\right]^{-1}\cdot\left(\mathbf{A}^H\mathbf{R}_n^{-1}\mathbf{A}\right)^{-1}\right)\cdot\mathbf{A}^H\mathbf{R}_n^{-1} \\ &= \left[\mathbf{R}+\left(\mathbf{A}^H\mathbf{R}_n^{-1}\mathbf{A}\right)^{-1}\right]^{-1}\cdot\left(\mathbf{A}^H\mathbf{R}_n^{-1}\mathbf{A}\right)^{-1}\cdot\mathbf{A}^H\mathbf{R}_n^{-1} \end{aligned} \tag{12-104}$$

Ignoring the noise correlation across adjacent blocks, the noise covariance matrix $\mathbf{R}_n$ and its inverse can be approximated by the direct sum of block noise covariance matrices $\mathbf{R}_{n,K}$ and their inverses, respectively. Hence, the matrix $\mathbf{A}^H\cdot\mathbf{R}_n^{-1}$ and the inverse matrix $\left(\mathbf{A}^H\cdot\mathbf{R}_n^{-1}\cdot\mathbf{A}\right)^{-1}$ can be simplified to

$$\begin{aligned} \mathbf{A}^H\mathbf{R}_n^{-1} &\approx \underbrace{\begin{pmatrix} \mathbf{A}_0 & & \\ & \ddots & \\ & & \mathbf{A}_{\overline{N}-1} \end{pmatrix}^H}_{\mathbf{A}^H} \underbrace{\begin{pmatrix} \mathbf{R}_{n,0}^{-1} & & \\ & \ddots & \\ & & \mathbf{R}_{n,\overline{N}-1}^{-1} \end{pmatrix}}_{\approx\mathbf{R}_n^{-1}} \\ &= \begin{pmatrix} \mathbf{A}_0^H\mathbf{R}_{n,0}^{-1} & & \\ & \ddots & \\ & & \mathbf{A}_{\overline{N}-1}^H\mathbf{R}_{n,\overline{N}-1}^{-1} \end{pmatrix} \end{aligned} \tag{12-105}$$

$$
(\mathbf{A}^H\mathbf{R}_n^{-1}\mathbf{A})^{-1} \approx \left(\underbrace{\begin{pmatrix} \mathbf{A}_0 & & \\ & \ddots & \\ & & \mathbf{A}_{\overline{N}-1} \end{pmatrix}^H}_{\mathbf{A}^H} \underbrace{\begin{pmatrix} \mathbf{R}_{n,0}^{-1} & & \\ & \ddots & \\ & & \mathbf{R}_{n,\overline{N}-1}^{-1} \end{pmatrix}}_{\approx \mathbf{R}_n^{-1}} \underbrace{\begin{pmatrix} \mathbf{A}_0 & & \\ & \ddots & \\ & & \mathbf{A}_{\overline{N}-1} \end{pmatrix}}_{\mathbf{A}} \right)^{-1}
$$
$$
= \begin{pmatrix} \left[\mathbf{A}_0^H\mathbf{R}_{n,0}^{-1}\mathbf{A}_0\right]^{-1} & & \\ & \ddots & \\ & & \left[\mathbf{A}_{\overline{N}-1}^H\mathbf{R}_{n,\overline{N}-1}^{-1}\mathbf{A}_{\overline{N}-1}\right]^{-1} \end{pmatrix} \tag{12-106}
$$

respectively. Then the MAP estimate of eqs. (12-80) and (12-100) can be equivalently expressed as

$$
\hat{\theta}_{D,k}(\mathbf{a}) = \hat{\mathbf{h}}_k(\mathbf{a})
$$
$$
\approx \left(\mathbf{R}_{0;k} \vdots \ldots \vdots \mathbf{R}_{\overline{N}-1;k}\right) \cdot \underbrace{\mathbf{A}^H\left[\mathbf{A}\cdot\mathbf{R}\cdot\mathbf{A}^H + \mathbf{R}_n\right]^{-1}}_{\mathbf{A}^H\mathbf{E}^{-1}} \cdot \begin{pmatrix} \mathbf{r}_0 \\ \vdots \\ \mathbf{r}_{\overline{N}-1} \end{pmatrix}
$$
$$
= \left(\mathbf{R}_{0;k} \vdots \ldots \vdots \mathbf{R}_{\overline{N}-1;k}\right) \cdot \left[\mathbf{R} + (\mathbf{A}^H\mathbf{R}_n^{-1}\mathbf{A})^{-1}\right]^{-1} \cdot (\mathbf{A}^H\mathbf{R}_n\mathbf{A})^{-1} \cdot \mathbf{A}^H\mathbf{R}_n^{-1} \cdot \begin{pmatrix} \mathbf{r}_0 \\ \vdots \\ \mathbf{r}_{\overline{N}-1} \end{pmatrix} \tag{12-107}
$$

$$
\hat{\theta}_{D,k}(\mathbf{a}) \approx \left(\mathbf{R}_{0;k} \vdots \ldots \vdots \mathbf{R}_{\overline{N}-1;k}\right)
$$
$$
\cdot \left[\begin{pmatrix} \mathbf{R}_{0,0} & \ldots & \mathbf{R}_{0,\overline{N}-1} \\ \vdots & \ddots & \vdots \\ \mathbf{R}_{\overline{N}-1,0} & \ldots & \mathbf{R}_{\overline{N}-1;\overline{N}-1} \end{pmatrix} + \begin{pmatrix} \left[\mathbf{A}_0^H\mathbf{R}_{n,0}^{-1}\mathbf{A}_0\right]^{-1} & & \\ & \ddots & \\ & & \left[\mathbf{A}_{\overline{N}-1}^H\mathbf{R}_{n,\overline{N}-1}^{-1}\mathbf{A}_{\overline{N}-1}\right]^{-1} \end{pmatrix} \right]^{-1}
$$
$$
\cdot \begin{pmatrix} \left[\mathbf{A}_0^H\mathbf{R}_{n,0}^{-1}\mathbf{A}_0\right]^{-1} & & \\ & \ddots & \\ & & \left[\mathbf{A}_{\overline{N}-1}^H\mathbf{R}_{n,\overline{N}-1}^{-1}\mathbf{A}_{\overline{N}-1}\right]^{-1} \end{pmatrix} \cdot \begin{pmatrix} \mathbf{A}_0^H\mathbf{R}_{n,0}^{-1}\cdot\mathbf{r}_0 \\ \vdots \\ \mathbf{A}_{\overline{N}-1}^H\mathbf{R}_{n,\overline{N}-1}^{-1}\cdot\mathbf{r}_{\overline{N}-1} \end{pmatrix} \tag{12-108}
$$

from which eq. (12-81) follows immediately.

The error covariance pertaining to the MAP estimate is easily evaluated by observing that the MAP estimate $\ddot{\theta}_{D,k}(\mathbf{a}) = \mathbf{\Sigma}_{hr} \cdot \mathbf{\Sigma}_r^{-1} \cdot \mathbf{r}$ and the MAP error covariance matrix $\mathbf{\Sigma}_{D,k} = \mathbf{\Sigma}_h - \mathbf{\Sigma}_{hr} \cdot \mathbf{\Sigma}_r^{-1} \cdot \mathbf{\Sigma}_{hr}^H$ [both given by eq. (12-73)] contain the same matrix expression $\mathbf{\Sigma}_{hr} \cdot \mathbf{\Sigma}_r^{-1}$ Furthermore, we have $\mathbf{\Sigma}_h = \mathbf{R}_h = \mathbf{R}_h(0)$ and

$$\mathbf{\Sigma}_{hr}^H = \begin{pmatrix} \mathbf{A}_0 \cdot \mathbf{R}_{0;k}^H \\ \vdots \\ \mathbf{A}_{\overline{N}-1} \cdot \mathbf{R}_{\overline{N}-1;k}^H \end{pmatrix} \tag{12-109}$$

[eq. (12-74)]. Hence, the MAP error covariance matrix $\mathbf{\Sigma}_{D,k}$ is given by the expression for the MAP channel estimate, eq. (12-108), with block observations $\mathbf{r}_K$ replaced by the entries $\mathbf{A}_K \cdot \mathbf{R}_K^H$ of $\mathbf{\Sigma}_{hr}^H$. As a result, the (approximation to the) MAP error covariance matrix becomes

$$\begin{aligned}\mathbf{\Sigma}_{D,k}(\mathbf{a}) \approx \;& \mathbf{R}_h - \begin{pmatrix} \mathbf{R}_{0;k} \vdots \dots \vdots \mathbf{R}_{\overline{N}-1;k} \end{pmatrix} \\ & \cdot \left[\begin{pmatrix} \mathbf{R}_{0,0} & \dots & \mathbf{R}_{0,\overline{N}-1} \\ \vdots & \ddots & \vdots \\ \mathbf{R}_{\overline{N}-1,0} & \dots & \mathbf{R}_{\overline{N}-1,\overline{N}-1} \end{pmatrix} + \begin{pmatrix} \left[\mathbf{A}_0^H \mathbf{R}_{n,0}^{-1} \mathbf{A}_0\right]^{-1} & & \\ & \ddots & \\ & & \left[\mathbf{A}_{\overline{N}-1}^H \mathbf{R}_{n,\overline{N}-1}^{-1} \mathbf{A}_{\overline{N}-1}\right]^{-1} \end{pmatrix} \right]^{-1} \\ & \cdot \begin{pmatrix} \left[\mathbf{A}_0^H \mathbf{R}_{n,0}^{-1} \mathbf{A}_0\right]^{-1} & & \\ & \ddots & \\ & & \left[\mathbf{A}_{\overline{N}-1}^H \mathbf{R}_{n,\overline{N}-1}^{-1} \mathbf{A}_{\overline{N}-1}\right]^{-1} \end{pmatrix} \cdot \begin{pmatrix} \mathbf{A}_0^H \mathbf{R}_{n,0}^{-1} \mathbf{A}_0 \cdot \mathbf{R}_{0;k} \\ \vdots \\ \mathbf{A}_{\overline{N}-1}^H \mathbf{R}_{n,\overline{N}-1}^{-1} \mathbf{A}_{\overline{N}-1} \cdot \mathbf{R}_{\overline{N}-1;k} \end{pmatrix}\end{aligned} \tag{12-110}$$

$$\begin{aligned}\mathbf{\Sigma}_{D,k}(\mathbf{a}) = \;& \mathbf{R}_h - \begin{pmatrix} \mathbf{R}_{0;k} \vdots \dots \vdots \mathbf{R}_{\overline{N}-1;k} \end{pmatrix} \\ & \cdot \left[\begin{pmatrix} \mathbf{R}_{0,0} & \dots & \mathbf{R}_{0,\overline{N}-1} \\ \vdots & \ddots & \vdots \\ \mathbf{R}_{\overline{N}-1,0} & \dots & \mathbf{R}_{\overline{N}-1,\overline{N}-1} \end{pmatrix} + \begin{pmatrix} \left[\mathbf{A}_0^H \mathbf{R}_{n,0}^{-1} \mathbf{A}_0\right]^{-1} & & \\ & \ddots & \\ & & \left[\mathbf{A}_{\overline{N}-1}^H \mathbf{R}_{n,\overline{N}-1}^{-1} \mathbf{A}_{\overline{N}-1}\right]^{-1} \end{pmatrix} \right]^{-1} \\ & \cdot \begin{pmatrix} \mathbf{R}_{0;k} \\ \vdots \\ \mathbf{R}_{\overline{N}-1;k} \end{pmatrix}\end{aligned} \tag{12-111}$$

12.3 Bibliographical Notes

For fundamentals of estimation and detection theory, the reader is referred to the textbooks [11, 12, 1, 2, 13]. Recent approaches to joint detection and estimation are found in [10, 5, 3, 8, 9]. An early paper by Kailath [14] gives a good introduction into optimal receiver structures for fading channels.

Bibliography

[1] J. L. Melsa and D. L. Cohn, *Decision and Estimation Theory*. New York: McGraw-Hill, 1978.

[2] L. L. Scharf, *Statistical Signal Processing*. New York: Addison-Wesley, 1991.

[3] N. Seshadri, "Joint Data and Channel Estimation Using Fast Blind Trellis Search Techniques," in *Proc. Globecom*, (San Diego, CA), pp. 1659–1663, Dec. 1990.

[4] N. Seshadri and C. E. W. Sundberg, "Generalized Viterbi Algorithms for Error Detection with Convolutional Codes," in *Proc. Globecom*, (Dallas, TX), pp. 1534–1537, Nov. 1989.

[5] J. Lin, F. Ling, and J. G. Proakis, "Joint Data and Channel Estimation for TDMA Mobile Channels," in *Proc. PIMRC*, (Boston, MA), pp. 235–239, Oct. 1992.

[6] R. Raheli, A. Polydoros, and C. K. Tzou, "Per-Survivor Processing," *Digital Signal Processing*, vol. 3, pp. 175–187, July 1993.

[7] R. Raheli, A. Polydoros, and C. K. Tzou, "Per-Survivor Processing: A General Approach to MLSE in Uncertain Environments," *IEEE Trans. Commun.*, vol. 43, pp. 354–364, Feb./Mar./Apr. 1995.

[8] N. Seshadri, "Joint Data and Channel Estimation Using Blind Trellis Search Techniques," *IEEE Trans. Commun.*, vol. 42, no. 2/3/4, pp. 1000–1011, Mar./Apr./May 1994.

[9] E. Zervas, J. G. Proakis, and V. Eyuboglu, "A 'Quantized' Channel Approach to Blind Equalization," in *Proc. ICC*, (Chicago, IL), pp. 1539–1543, June 1992.

[10] R. A. Iltis, J. J. Shynk, and K. Giridhar, "Bayesian Algorithms for Blind Equalization Using Parallel Adaptive Filtering," *IEEE Trans. Commun.*, vol. 42, no. 2/3/4, Feb./Mar./Apr. 1994.

[11] B. D. O. Anderson and J. B. Moore, *Optimal Filtering*. Englewood Cliffs, NJ: Prentice-Hall, 1979.

[12] S. Haykin, *Adaptive Filter Theory*. Englewood Cliffs, NJ: Prentice-Hall, 1986.

[13] H. L. van Trees, *Detection, Estimation, and Modulation Theory – Part I*. New York: Wiley, 1968.

[14] T. Kailath, "Optimum Receivers for Randomly Varying Channels," in *Proc. 4-th London Symposium on Information Theory, Butterworth Scientific Press, London*, pp. 109–122, 1961.

Chapter 13 Receiver Structures for Fading Channels

In this chapter, realizable receiver structures for synchronized detection on flat and selective fading channels are derived and discussed. Keeping the interaction between synchronization and detection paths at a minimum results in receivers of low complexity. As motivated below, we shall concentrate on the so-called inner receiver, its components, and the necessary synchronization tasks.

13.1 Outer and Inner Receiver for Fading Channels

In compliance with Shannon's third lesson learned from information theory: "make the channel look like a Gaussian channel" [1], we have distinguished between an outer and inner receiver [2] in the introduction of this book. The sole – yet by no means trivial – task of the inner receiver has been identified as attempting to provide the decoder with a symbol sequence that is essentially corrupted by white Gaussian noise only. If this can be accomplished, the inner transmission system serves as a "good" channel for the outer transmission system, i.e., the source and (possibly) channel decoding system.

If the channel is *nonfading* and Gaussian, the inner receiver aims at delivering an optimally preprocessed and synchronized signal, typically the matched filter output sequence $\hat{a}_k = a_k + m_k$ serving as an estimate of the symbol sequence and sometimes termed "soft decision". If the inner receiver were perfect (i.e., perfect frequency, timing and phase synchronization, matched filtering, and decimation to symbol rate), no loss of information would entail. In practice, the imperfections (sync errors, word-length effects, etc.) translate into a slight increase of the noise power as seen by the outer receiver, i.e., a small SNR loss.

In the case of *fading* channels, the situation is more intriguing, since synchronization and prefiltering alone do not suffice to establish a nonfading Gaussian inner transmission system. Even if the timing and the fading trajectory[6] $\{c_k\}$ (multiplicative distortion, MD) were perfectly known, the compensated pulse matched

[6] For this more general discussion, it is immaterial whether flat or selective fading is assumed, except for the discussion on multipath diversity resulting from frequency selectivity (see below).

filter (MF) output sequence [uncompensated signal eq. (11-22) divided by the MD]

$$\begin{aligned}&\text{Uncompensated pulse MF output:}\\&z_k = c_k\, a_k + m_k\\&\text{Compensated pulse MF output (soft decision):}\\&\hat{a}_k = \frac{z_k}{c_k} \quad = a_k + \left(\frac{m_k}{c_k}\right)\end{aligned} \tag{13-1}$$

remains to be dependent on the fading via the time-varying instantaneous noise power $(N_0/|\, c_k\, |^2)$. Hence, the noise is AWGN as desired, but the SNR per symbol, $\gamma_{s;k} = |\, c_k\, |^2\ \overline{\gamma}_s$ with $\overline{\gamma}_s = 1/N_0$, is *time-varying* and may deviate strongly from the average SNR per symbol $\overline{\gamma}_s$. During very deep fades, the decoder faces a signal $\hat{a}_k$ that is buried in noise and therefore useless.

Conceptually, there are two basic ways to resolve this dilemma. The first, very popular solution is to leave the structure of the inner receiver unchanged [i.e., it delivers the soft decisions $\hat{a}_k$ of eq. (13-1)] and thus accept that the inner transmission system does not reproduce a stationary AWGN channel, but have the inner receiver generate an additional signal, termed *channel state information* (CSI), that is used to aid the outer decoding system. Of course, the CSI should be matched to the particular decoding system. When trellis-coded modulation is employed, the optimal CSI is given by the sequence of instantaneous MD powers $|\, c_k\, |^2$ which are used to weigh the branch metrics in the Viterbi decoder [3]. In the case of block decoding, it often suffices to form a coarse channel quality measure, e.g., a binary erasure sequence derived from $|\, c_k\, |^2$ by means of a simple threshold decision [4].

The second principal solution consists in augmenting the structure of the inner transmission system in such a way that it approaches compliance with Shannon's lesson above. Since in most cases of interest the transmitter does not have knowledge of the instantaneous fading power and thus cannot adaptively match the transmit powers and rates to the short-term channel conditions, the inner transmission system must provide for a means of levelling out the signal variations introduced by fading without knowing in advance which signal segments are affected by deep fades. This can be accomplished very effectively by providing for explicit or implicit forms of signal *diversity* [5]. By virtue of its *averaging* effect, diversity aids in "bridging" deep signal fades so that the outer decoding system faces a signal whose disturbance resembles stationary AWGN to a much higher degree than without diversity. Explicit forms of diversity include *D-antenna* (or spatial) diversity (message is transmitted over D distinct independently fading physical paths $c_{d,k}$), *frequency* diversity (message is transmitted simultaneously over several channels spaced in frequency), and *time* diversity (message is repeatedly transmitted). The latter two methods suffer from a very low bandwidth efficiency and are therefore not pursued further. Implicit forms of diversity include *multipath* diversity (message is transmitted over several independently fading resolvable paths $c_{m,k}$ of a frequency-selective channel), and *time* diversity provided by channel coding (each uncoded message bit b_i is mapped onto a number of channel symbols a_k which may then be further spread in time by means of interleaving so that they undergo mutually independent fading). These implicit forms of diversity are particularly interesting since they have the potential of achieving large diversity gains without compromising the power and bandwidth efficiency.

To illustrate the beneficial effect of diversity, consider the – grossly simplified – scenario of combined Dth-order antenna diversity reception (D channels $c_{d;k}$), ideal equalization of M resolvable multipaths of equal average gain (M "channels" $c_{m;k}$), and ideal Cth-order coding diversity (C "channels" $c_{c;k}$), provided, e.g., by trellis-coded modulation where each (uncoded) message bit b_i affects, via the encoding law, C consecutive symbols a_i. By virtue of interleaving, the symbols a_i are then mapped onto channel symbols a_k which undergo independent fading. The parameter C, sometimes termed *effective code length* (ECL), can therefore be viewed as the order of diversity resulting from this kind of channel coding. As a result, a total of $L = D\ M\ C$ independent "paths" $c_{l;k}$ $(l = 1, \ldots, L)$ contribute to the received signal. A discrete-equivalent transmission model thus consists of L parallel branches with *flat* fading path gains $c_{l;k}$ plus additive noise processes $m_{l;k}$ making up L received signals $z_{l;k}$. Under the assumptions of perfect timing, equal average path gain powers $E\{|\ c_{l;k}\ |^2\} = 1/L$, and branch noise powers $E\{|\ m_{l;k}\ |^2\} = N_0$, the *optimal combiner* [5] – modeling the antenna diversity combiner (or any other kind of diversity combiner), ideal equalizer, and channel decoder – forms a weighted sum of the received signals $z_{l;k}$. As shown below [eqs. (13-15) and (13-17)], the combining operation and optimal weights $q_{l;k}$ are given by

$$\begin{aligned} \hat{a}_k &= \sum_{l=1}^{L} q_{l;k} \cdot z_{l;k} \\ q_{l;k} &= \frac{c^*_{l;k}}{\left(\sum\limits_{l=1}^{L} |c_{l;k}|^2\right)} = \frac{c^*_{l;k}}{|c_k|^2} \end{aligned} \tag{13-2}$$

respectively. The entire Lth-order diversity transmission system can then be modeled as an equivalent system consisting of a single path weight of 1 (i.e., the variations of the useful signal have been levelled out completely) and a noise process η_k with time-variant power $N_0/|\ c_k\ |^2$, where $|\ c_k\ |^2 = (\sum_{l=1}^{L}\ |\ c_{l;k}\ |^2)$ is the power of the composite path process. Both the diversity transmission system and its equivalent are depicted in Figure 13-1. Obviously, the effect of diversity manifests itself in the statistics of the equivalent noise power and thus the composite path process power $|\ c_k\ |^2$. Realizations of $|\ c_k\ |^2$ illustrating the averaging effect of diversity are shown in Figure 13-2 for L=1, 2, 4, 8, and 16 Rayleigh fading diversity branches $c_{l,k}$ with Jakes Doppler spectrum and relative Doppler frequency λ'_D=0.01.

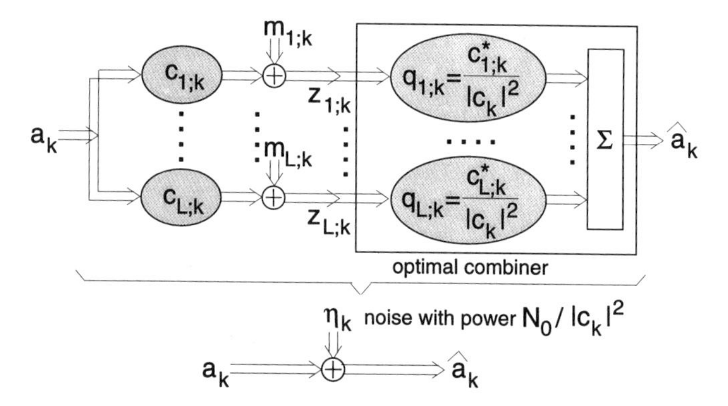

Figure 13-1 Lth-Order Diversity Optimal Combiner and Equivalent Diversity Transmission Model

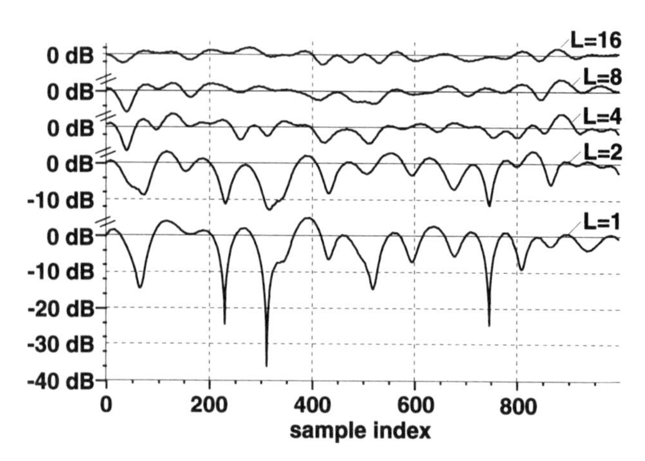

Figure 13-2 Realization of Lth-Order Diversity Equivalent Channel Trajectory $|\, c_k \,|^2$

Assuming independent Rayleigh fading of the diversity branches, the composite path weight power $|\, c_k \,|^2$ and hence the instantaneous SNR per symbol $\gamma_{s;k} = |\, c_k \,|^2 / N_0$ behind the combiner follows a χ^2 distribution with $2L$ orders of freedom [5]. The pdf $p(r)$ of the random variable $r = \gamma_{s;k}/\overline{\gamma}_s$, being a measure of the variability of the SNR about its average $\overline{\gamma}_s$, and also the probability $P(r < R)$, being a measure of the probability of deep residual fades (behind the optimal combiner) below a small threshold $R \ll 1$, are then given by

$$p(r) = \frac{L^L}{(L-1)!}\, r^{L-1}\, e^{-L\cdot r} \qquad (r \geq 0)$$

$$\begin{aligned} P(r < R) &= \frac{\Gamma(L,0) - \Gamma(L,R\,L)}{(L-1)!} = \frac{\int_0^{RL} x^{L-1} e^{-x}\, dx}{(L-1)!} \\ &\approx \frac{L^{L-1}}{(L-1)!}\, R^L \qquad (0 \leq R \ll 1) \end{aligned} \tag{13-3}$$

respectively, with $\Gamma(a,z) = \int_z^{\infty} x^{a-1} e^{-x}\, dx$ the incomplete gamma function. The pdf $p(r)$ and probability $P(r < R)$ are illustrated in Figures 13-3 and 13-4,

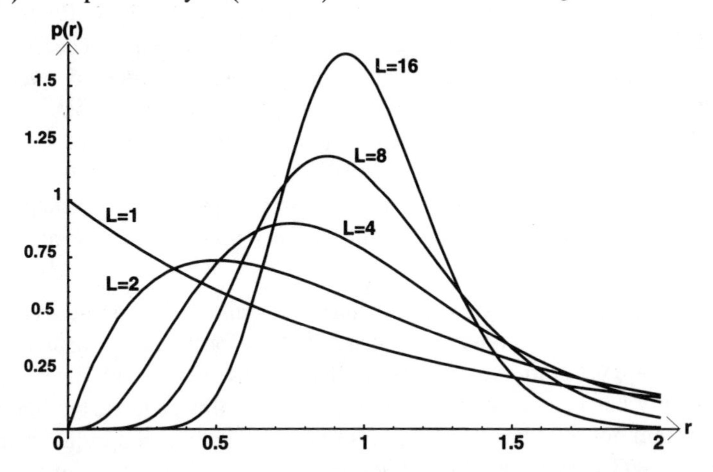

Figure 13-3 Probability Density of Normalized SNR for Lth-Order Diversity

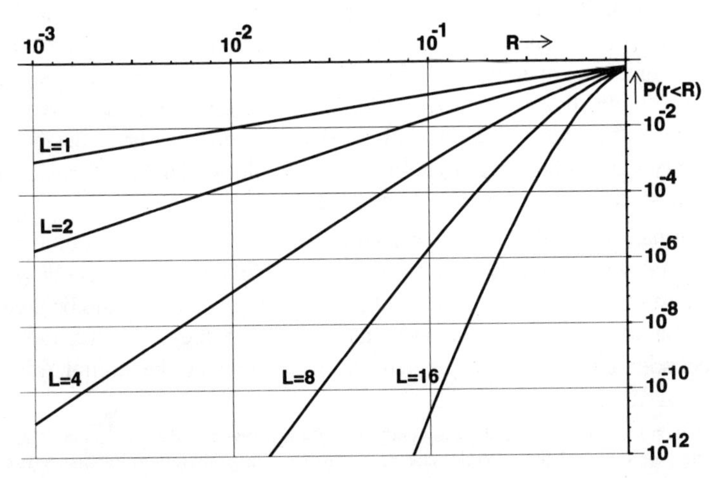

Figure 13-4 Probability of Normalized SNR Being below a Threshold R for Lth-Order Diversity

respectively, for L= 1, 2, 4, 8 and 16. The SNR variations are seen to become smaller with rising L until, in the limiting case of $(L \rightarrow \infty)$-dim. diversity, the fading channel in fact approaches the stationary AWGN channel [$p(r)$ reduces to a Dirac pulse at $r = 1$], which is exactly the goal that the inner transmission system is supposed to strive for.

From eq. (13-3) and Figure 13-4, the probability $P(r < R)$ of deep residual fades behind the optimal combiner is seen to be essentially proportional to the Lth power of the threshold R. For example, with a threshold of $R = 0.1$ (instantaneous SNR below average by 10 dB or more), we have $P(r < 0.1) = 0.095$ without diversity ($L = 1$), $P(r < 0.1) = 0.0175$ with dual diversity ($L = 2$), $P(r < 0.1) = 7.76 \times 10^{-4}$ with $L = 4$ and $P(r < 0.1) = 2.05 \times 10^{-6}$ with L=8. Deep fades thus occur with much smaller probability already for low orders of diversity (L= 2 or 4), which is easily achievable, e.g., by appropriate interleaved channel coding. However, some mild SNR fluctuations remain even for high orders of diversity (see also Figure 13-2); for instance, with a threshold of $R = 0.5$ [instantaneous SNR below average by 3 dB or more; here the approximation (third row) of eq. (13-3) is no more valid], we have $P(r < 0.5)$ =0.39, 0.26, 0.143, and 0.051 for L =1, 2, 4, and 8, respectively. In summary, however, these results demonstrate that providing for and making use of diversity is very effective in making a channel *look* stationary Gaussian.

Following Shannon's argument, the inner transmission system may be taken to encompass all transceiver components that help transform a *fading* channel into an "equivalent", almost *stationary Gaussian* channel (as seen by the outer receiver). Since diversity has been shown to play a key role in this context, the inner transmission system not only has to take care of synchronization and prefiltering (as in the case of AWGN channels), but, in addition, it has to provide for and make use of *diversity*. More specifically, the inner transmit system should provide for diversity, e.g., via explicit tranmit antenna diversity and/or appropriate channel coding, and the inner receiver has to exploit as many diversity mechanisms as possible, e.g., explicit antenna diversity (if available) via optimal combining, polarization diversity, multipath diversity via *equalization* (if the channel is selective), and/or time diversity via *channel decoding*, given that the code has been designed for a large diversity gain rather than a coding gain[7]. A block diagram of such an inner transmission system is depicted in Figure 13-5.

Unfortunately, the diversity-like effect resulting from channel coding is often difficult to analyze in practice, especially in the case of concatenated coding, so that it is really a matter of taste whether the channel coding system should be considered a part of the inner transmission system or not. In the case that the channel encoder/decoder pair is well separated[8] from the rest of the digital transceiver,

---

[7] A diversity gain (on fading channels) essentially calls for a large effective code length (ECL), while a large coding gain (on nonfading AWGN channels) calls for a large minimum Euclidean distance [6].

[8] For instance, the decoder is well separated if it receives a soft decision and possibly some additional channel state information from the inner receiver. The opposite is true if there is feedback from the decoder back into the inner transmission system.

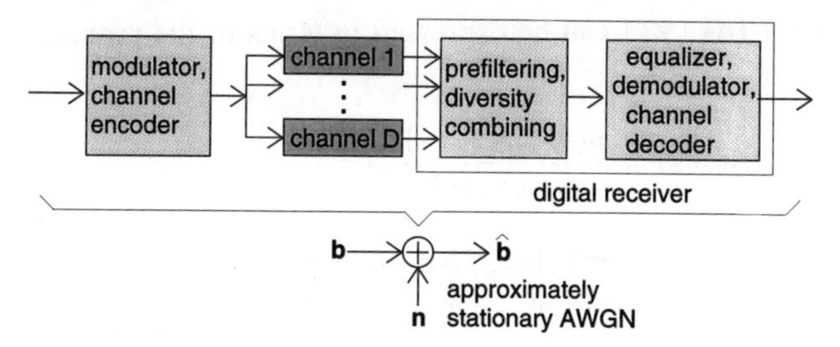

Figure 13-5 Inner Transmission System for Fading Channels, High-Order Diversity

it should be viewed as part of the outer transmission system, as we shall do in the sequel. The tasks that remain to be performed by the inner transmission system are still formidable and include modulation, filtering, channel accessing, equalization, diversity combining, demodulation, and synchronization.

Main Points

Generally, the inner transmission system should serve as a good – i.e., Gaussian – channel for the outer transmission system. On fading channels, the inner system should therefore not only perform synchronization and preprocessing of the received signal, but also provide for and make use of signal diversity. By virtue of the averaging effect of high-order diversity, signal fading is effectively mitigated so that the inner system approaches a stationary Gaussian channel.

13.2 Inner Receiver for Flat Fading Channels

The inner receiver for flat fading includes a *preprocessing* unit (pulse matched filtering, decimation, frequency offset correction) which generates the T-spaced signal z_k (or diversity signals $z_{d;k}$), and a *detection* unit which attempts to extract the information on symbols a_k from z_k. The clue to deriving the detector structure is the *recursive* formulation of the optimal decision rule. This derivation – which is due to Häb [7] – is found to yield not only expressions for decision metrics and metric increments, but also the synchronizer structure necessary for non-data-aided (NDA) channel estimation discussed in Section 14.1.

Setting $\mathbf{r} = \mathbf{z}$ in eq. (12-15), the optimal decision rule reads

$$\hat{\mathbf{a}} = \arg\max_{\mathbf{a}} \; P(\mathbf{a}|\mathbf{z}) \tag{13-4}$$

Denoting the data and received sequences up to time k by $\mathbf{a}_k$ and $\mathbf{z}_k$, respectively, the probability $P(\mathbf{a}_k \mid \mathbf{z}_k)$ can be expressed in terms of the probability $P(\mathbf{a}_{k-1} \mid \mathbf{z}_{k-1})$:

$$\begin{aligned} P(\mathbf{a}_k|\mathbf{z}_k) &= \frac{p(\mathbf{a}_k, \mathbf{z}_{k-1}, z_k)}{p(\mathbf{z}_{k-1}, z_k)} \\ &= \frac{p(z_k|\mathbf{z}_{k-1}, \mathbf{a}_k)\, P(\mathbf{a}_k, \mathbf{z}_{k-1})}{p(z_k|\mathbf{z}_{k-1})\, p(\mathbf{z}_{k-1})} \\ &= \frac{P(\mathbf{a}_k|\mathbf{z}_{k-1})\, p(z_k|\mathbf{z}_{k-1}, \mathbf{a}_k)}{p(z_k|\mathbf{z}_{k-1})} \\ &\propto P(a_k|\mathbf{a}_{k-1})\, p(z_k|\mathbf{z}_{k-1}, \mathbf{a}_k)\, P(\mathbf{a}_{k-1}|\mathbf{z}_{k-1}) \end{aligned} \tag{13-5}$$

where the denominator $p(z_k \mid \mathbf{z}_{k-1})$ has been omitted since it does not depend on the data $\mathbf{a}$.

The first term $P(a_k \mid \mathbf{a}_{k-1})$ reflects the structure of the message and the memory of the modulated symbols. If a_k is a known (training) symbol, then $P(a_k \mid \mathbf{a}_{k-1}) = 1$ for that symbol and 0 for all other symbols. If $\mathbf{a}$ is an encoded message, $P(a_k \mid \mathbf{a}_{k-1})$ is dependent on the particular code; most often all allowed symbols are equally likely so that, if there are Q such symbols a_k, we have $P(a_k \mid \mathbf{a}_{k-1}) = 1/Q$ if a_k is an allowed symbol and zero otherwise. In the case of memoryless uncoded M-QAM or M-PSK transmission, we have $P(a_k \mid \mathbf{a}_{k-1}) = 1/M$ for all a_k. Assuming that only allowed (coded or uncoded) equally probable data sequences are "tested" in the decision unit, the term $P(a_k \mid \mathbf{a}_{k-1})$ may also be omitted.

The second term $p(z_k \mid \mathbf{z}_{k-1}, \mathbf{a}_k)$ is the pdf pertaining to the *one-step prediction* $\hat{z}_{k|k-1}$ of the received sample z_k, given the "past" received signal $\mathbf{z}_{k-1}$ and the data hypothesis $\mathbf{a}_k$ up to the "present". Assuming that all random quantities are Gaussian, this pdf is determined by the conditional mean and variance

$$\begin{aligned} \hat{z}_{k|k-1} &= E[z_k|\mathbf{z}_{k-1}, \mathbf{a}_k] \\ \sigma^2_{z;k|k-1} &= E\Big[\big|z_k - \hat{z}_{k|k-1}\big|^2 \,\big|\, \mathbf{z}_{k-1}, \mathbf{a}_k\Big] \end{aligned} \tag{13-6}$$

respectively. Observing the transmission model $z_k = c_k a_k + m_k$, the predicted channel output $\hat{z}_{k|k-1}$ and its covariance $\sigma^2_{z;k|k-1}$ can equivalently be expressed in terms of the channel prediction $\hat{c}_{k|k-1}$ and its error covariance $\sigma^2_{c;k|k-1}$, respectively:

$$\begin{aligned} \hat{z}_{k|k-1} &= a_k\, E[c_k|\, \mathbf{z}_{k-1}, \mathbf{a}_{k-1}] &&= a_k\, \hat{c}_{k|k-1} \\ \sigma^2_{z;k|k-1} &= E\Big[\big|a_k\,(c_k - \hat{c}_{k|k-1}) + m_k\big|^2 \,\big|\, \mathbf{z}_{k-1}, \mathbf{a}_{k-1}\Big] &&= p_k\, \sigma^2_{c;k|k-1} + N_0 \end{aligned} \tag{13-7}$$

Recursion eq. (13-5) therefore becomes

$$
\begin{aligned}
P(\mathbf{a}_k|\mathbf{z}_k) &\propto p(z_k|\mathbf{z}_{k-1},\mathbf{a}_k)\,P(\mathbf{a}_{k-1}|\mathbf{z}_{k-1}) \\
&= \left(\frac{1}{\pi\,\sigma^2_{z;k|k-1}}\exp\left\{-\frac{\left|z_k-\hat{z}_{k|k-1}\right|^2}{\sigma^2_{z;k|k-1}}\right\}\right)P(\mathbf{a}_{k-1}|\mathbf{z}_{k-1}) \\
&= \left(\frac{1}{\pi\left(p_k\,\sigma^2_{c;k|k-1}+N_0\right)}\exp\left\{-\frac{\left|z_k-a_k\hat{c}_{k|k-1}\right|^2}{p_k\,\sigma^2_{c;k|k-1}+N_0}\right\}\right)P(\mathbf{a}_{k-1}|\mathbf{z}_{k-1})
\end{aligned}
\tag{13-8}
$$

Taking the logarithm and inverting the sign then yields the recursion for the decision metric (to be minimized):

$$
m_{D;k}(\mathbf{a}_k) = \underbrace{\left(\frac{\left|z_k-a_k\hat{c}_{k|k-1}\right|^2}{p_k\sigma^2_{c;k|k-1}+N_0}+\ln\left[\pi\left(p_k\,\sigma^2_{c;k|k-1}+N_0\right)\right]\right)}_{\Delta m_{D;k}(\mathbf{a}_k)} + m_{D;k-1}(\mathbf{a}_{k-1})
\tag{13-9}
$$

The metric increment $\Delta m_{D;k}(\mathbf{a}_k)$ is seen to depend not only on the "new" symbol a_k and its energy p_k, but also involves the "online" computation of the NDA channel estimate $\hat{c}_{k|k-1}$ and its error covariance $\sigma^2_{c;k|k-1}$. The increment $\Delta m_{D;k}(\mathbf{a}_k)$ is not memoryless even in the case of memoryless uncoded modulation since the information contained in the entire (hypothetical) past symbol sequence $\mathbf{a}_{k-1}$ is absorbed in the respective channel estimate $\hat{c}_{k|k-1}$. In the case that known training symbols $\mathbf{a}_T$ are available, the NDA estimate $\hat{c}_{k|k-1}$ may be replaced by the data-aided (DA) estimate $\hat{c}_k$ obtained from $\mathbf{a}_T$ (Section 14.2). Then $\hat{c}_k$ and its error covariance $\sigma^2_{c;k}$ are to be inserted into eq. (13-9) in lieu of $\hat{c}_{k|k-1}$ and $\sigma^2_{c;k|k-1}$, respectively. In the following, let us denote the channel estimate in general (NDA, DA, or any other kind) and its error covariance by $\hat{c}_k$ and $\sigma^2_{c;k}$, respectively.

Channel estimation, in particular NDA 1-step prediction, yields estimates $\hat{c}_k$ which are clearly suboptimal with respect to the smoothed estimates of Chapter 12. Nevertheless, the metric $m_D(\mathbf{a}) = m_{D;k=N-1}(\mathbf{a}_{k=N-1})$ at the end of the message remains to be optimal for detection; it just has been computed in a recursive manner. Like in Chapter 12, it needs, in principle, to be evaluated for each and every symbol hypothesis $\mathbf{a}$.

The expression for the decision metric eq. (13-9) may be cast into a simpler form as follows. Assuming steady-state channel estimation, the error covariance $\sigma^2_c = \sigma^2_{c;k}$ is the same for all k. Normalizing the MAP estimation error covariance σ^2_c to the error covariance $N_0/p_k = 1/(p_k\overline{\gamma}_s)$ in the ML channel estimate $\hat{c}_{S;k} = (a_k^*/p_k)z_k$ [eqs. (14-6) and (14-7)],

$$
\begin{aligned}
r_{c;k} &= \frac{\text{error cov. in } \hat{c}_k}{\text{error cov. in } \hat{c}_{S;k}} &&= \frac{p_k}{N_0}\,\sigma^2_c \\
\overline{r}_c &= \frac{1}{N_0}\,\sigma^2_c &&= \overline{\gamma}_s\,\sigma^2_c
\end{aligned}
\tag{13-10}
$$

the metric increment of eq. (13-9) reads

$$\begin{aligned}\Delta m_{D;k}(a_k) &= \frac{|z_k - a_k \hat{c}_k|^2}{(1+p_k\overline{r}_c)N_0} + \ln\left[\pi(1+p_k\overline{r}_c)N_0\right] \\ &\propto \frac{|z_k - a_k \hat{c}_k|^2}{1+p_k\overline{r}_c} + \frac{1}{\overline{\gamma}_s}\ln\left[1+p_k\overline{r}_c\right]\end{aligned} \tag{13-11}$$

The second term can safely be neglected since it is much smaller than the first; for phase-modulated symbols or in the limiting case of perfect sync ($\overline{r}_c \to 0$) it is independent of a_k anyway. For instance, with 16-QAM, $\overline{\gamma}_s$ =20 dB and $\overline{r}_c$ =1, the second term assumes values between 0.002 and 0.01, whereas the first term is in the order of 0 if a_k is correct, and in the order of 0.14 to 0.34 otherwise. Furthermore, the denominator of the first term is usually not much dependent on p_k and thus can be omitted also, leading to the simplified metric increment

$$\Delta m_{D;k}(a_k) \approx \frac{|z_k - a_k \hat{c}_k|^2}{1+p_k\overline{r}_c} \simeq |z_k - a_k \hat{c}_k|^2 \tag{13-12}$$

which is seen to have collapsed to the squared Euclidean distance between the estimated channel output $\hat{z}_k = a_k \hat{c}_k$ and the actual received signal z_k.

The decision metric of eq. (13-12) calls for the inner receiver to generate the signal pair $(z_k, \hat{c}_k)$. As shown below, the inner receiver may equivalently generate the signal pairs $(\hat{a}_k, |\hat{c}_k|^2)$ or $(\hat{a}_k, \gamma_{s;k})$, with soft symbol decisions:

$$\hat{a}_k = \frac{z_k}{\hat{c}_k} = \frac{\hat{c}_k^*}{|\hat{c}_k|^2} z_k \approx \frac{\hat{c}_k^*}{|c_k|^2} z_k = \frac{\hat{c}_k^*}{E_k} z_k \tag{13-13}$$

obtained by compensating for the fading, and CSI given either by the (estimated) instantaneous channel energy $|\hat{c}_k|^2 \approx |c_k|^2 = E_k$ or the (estimated) instantaneous SNR $|\hat{c}_k|^2/N_0 \approx |c_k|^2/N_0 = E_k/N_0 = \gamma_{s;k}$:

$$\begin{aligned}\Delta m_{D;k}(a_k) &\simeq |z_k - a_k \hat{c}_k|^2 = |\hat{c}_k|^2 \left| \underbrace{z_k/\hat{c}_k}_{\hat{a}_k} - a_k \right|^2 = |\hat{c}_k|^2 \, |\hat{a}_k - a_k|^2 \\ &\approx |c_k|^2 \, |\hat{a}_k - a_k|^2 \propto \gamma_{s;k} \, |\hat{a}_k - a_k|^2\end{aligned} \tag{13-14}$$

Hence, this form of the (simplified) metric increment reduces to the squared Euclidean distance between the soft decision $\hat{a}_k$ and the trial symbol a_k, weighted by the instantaneous energy or SNR of the channel.

In the case of diversity reception (D received signals $z_{d;k}$ and channel estimates $\hat{c}_{d;k}$), the decision metric can be computed in the same manner. However, the soft decisions $\hat{a}_k$ are now formed by diversity combination

$$\hat{a}_k = \sum_{d=1}^{D} q_{d;k} \, z_{d;k} \tag{13-15}$$

with weights $q_{d;k}$, and the CSI is now given either by the (estimated) instantaneous combined channel energy $(\sum_{d=1}^{D} \mid \hat{c}_{d;k} \mid^2) \approx (\sum_{d=1}^{D} \mid c_{d;k} \mid^2) = E_k$ or the (estimated) instantaneous combined SNR $(\sum_{d=1}^{D} \mid \hat{c}_{d;k} \mid^2)/N_0 \approx (\sum_{d=1}^{D} \mid c_{d;k} \mid^2)/N_0 = E_k/N_0 = \gamma_{s;k}$. The combiner weights can be optimized by inspection of the residual error $\hat{a}_k - a_k$, assuming that the trial symbol a_k has actually been transmitted:

$$\hat{a}_k - a_k = \left[\left(\sum_{d=1}^{D} q_{d;k}\, c_{d;k} \right) - 1 \right] a_k + \left(\sum_{d=1}^{D} q_{d;k}\, m_{d;k} \right) \tag{13-16}$$

The minimization [8] of the combined noise power of $\eta_k = (\sum_{d=1}^{D} q_{d;k} m_{d;k})$ [second term of eq. (13-16)] subject to the constraint $(\sum_{d=1}^{D} q_{d;k} c_{d;k}) = 1$ [then the first term of eq. (13-16) is zero] yields optimal combiner weights:

$$q_{d;k} = \frac{c^*_{d;k}}{E_k} = \frac{c^*_{d;k}}{\sum_{d=1}^{D} |c_{d;k}|^2} \tag{13-17}$$

Strictly speaking, eq. (13-16) – and thus the optimization – is correct only for the true channel $c_{d;k}$, but $c_{d;k}$ may safely be replaced by its estimate in eq. (13-17) as long as $\hat{c}_{d;k}$ is reasonably accurate.

A block diagram of the detection path of an inner receiver including preprocessing and fading correction units is shown in Figure 13-6 (upper part: no diversity, center part: Dth-order diversity combining). The fading correction unit is a very simple equalizer that attempts to "undo" both the amplitude and the phase variations of the fading process c_k by a rotate-and-scale operation: the MF output signal(s) $z_{[d];k}$ are phase-aligned via multiplication by c^*_k (no diversity) or $c^*_{d;k}$ (diversity) and scaled via division by $E_k = \mid c_k \mid^2$ [no diversity, eq. (13-13)] or $E_k = (\sum_{d=1}^{D} \mid c_{d;k} \mid^2)$ [diversity, eqs. (13-15) and (13-17)]. In the absence of diversity, fading correction may likewise be performed implicitly by scaling and rotating the QAM or PSK decision grid. If c_k (or its estimate) is very small or exactly zero, it suffices to suppress the rotate-and-scale operation and output an erasure flag.

Using optimal combiner weights [eq. (13-17)], the minimum combined noise power results to be $N_0/(\sum_{d=1}^{D} \mid c_{d;k} \mid^2) = N_0/E_k = 1/\gamma_{s;k}$. The entire discrete-equivalent inner system (bottom of Figure 13-6) is therefore characterized by AWGN $\eta_k = (\sum_{d=1}^{D} q_{d;k} m_{d;k})$ with time-variant power $\sigma^2_{\eta;k} = N_0/E_k = 1/\gamma_{s;k}$ or time-variant SNR (per symbol) $\gamma_{s;k} = E_k/N_0 = E_k \overline{\gamma}_s$.

As opposed to the nonfading channel case, the sequence of soft decisions $\hat{a}_k$ [eqs. (13-1), (13-13), and (13-15), and Figure 13-6] alone is not sufficient for near-optimal detection. Rather, a sufficient statistic for detection – to be generated by the inner and transferred to the outer receiver – is given by the signal pair

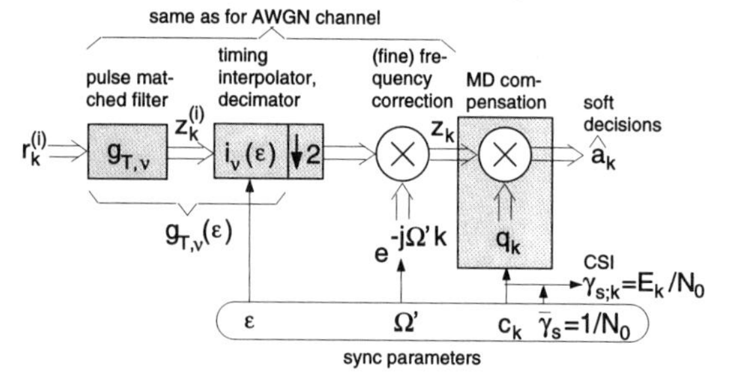

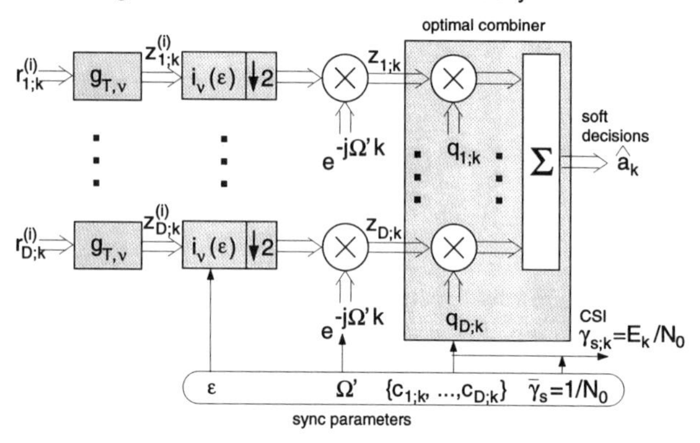

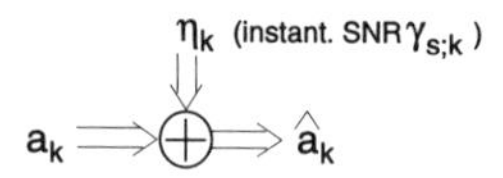

Figure 13-6 Inner Receiver for Flat Fading Channels without and with Diversity

$(z_k, \hat{c}_k)$ or, if MD estimation and compensation are reasonably accurate, the signal pairs $(\hat{a}_k, \hat{c}_k)$ or $(\hat{a}_k, \gamma_{s;k})$.

From this discussion of the inner receiver's *detection* path it is once again apparent that its *synchronization* path must strive for generating up-to-date and accurate estimates of relative frequency Ω', timing ε, and the MD c_k (MDs $c_{d;k}$ for diversity reception). Since frequency and timing estimation have been discussed thoroughly in the previous chapters, we shall concentrate on MD estimation, keeping in mind that the MD is often taken to absorb a stationary oscillator phase and/or a very small frequency shift since it is immaterial whether these effects stem from oscillator imperfections or the physical channel.

Main Points

- Exact and simplified expressions for the decision metric and metric increment for flat fading channels have been derived by means of the recursive formulation of the decision rule. The near-optimal metric increment can be cast into the form

$$\begin{aligned} \Delta m_{D;k}(a_k) \simeq \left|z_k - a_k \hat{c}_k\right|^2 &= \left|\hat{c}_k\right|^2 \left|\hat{a}_k - a_k\right|^2 \\ \approx \left|c_k\right|^2 \left|\hat{a}_k - a_k\right|^2 &\propto \gamma_{s;k} \left|\hat{a}_k - a_k\right|^2 \end{aligned} \tag{13-18}$$

  [eqs. (13-12) and (13-14)] with soft symbol decisions $\hat{a}_k$ [eqs. (13-13) and (13-15)]. The inner receiver should therefore deliver one of the signal pairs $(z_k, \hat{c}_k)$, $(\hat{a}_k, \hat{c}_k)$ or $(\hat{a}_k, \gamma_{s;k})$.
- Synchronized detection according to eq. (13-18) necessitates an explicit flat fading channel estimate $\hat{c}_k$. This estimate may be generated by one-step prediction $\hat{c}_{k|k-1}$ in the case of online NDA synchronization (Section 14.1), or by estimation from training symbols in the case of DA synchronization (Section 14.2).

13.3 Inner Receiver for Selective Fading Channels

The inner receiver for selective fading has to process the $(T/2)$-spaced received signal $\mathbf{r} = \mathbf{r}^{(0)} \oplus \mathbf{r}^{(1)}$, where $\mathbf{r}^{(i)} = \mathbf{H}^{(i)} \cdot \mathbf{a} + \mathbf{n}^{(i)}$ [eq. (12-10), zero frequency offset]. As in the flat fading case, this receiver input signal can be directly used for synchronized detection (next subsection). For many channels of practical interest, however, it is more advantageous to apply appropriate preprocessing such that $\mathbf{r}$ is transformed into another signal $\mathbf{v}$ which remains to be a sufficient statistic but is much better suited for *reduced-complexity* detection. For this purpose, the finite-length *whitening matched filter* (WMF) will turn out to be a particularly effective preprocessing device.

13.3.1 Recursive Computation of the Decision Metric

Since part of this material resembles that of the previous section, we shall sketch the derivation of the recursive decision metric for selective fading channels only briefly. Again, the starting point is the optimal decision rule of eq. (12-15):

$$\hat{\mathbf{a}} = \arg \max_{\mathbf{a}} P(\mathbf{a}|\mathbf{r}) \tag{13-19}$$

Denoting the data and received sequences up to time k by $\mathbf{a}_k$ and $\mathbf{r}_k = \mathbf{r}_k^{(0)} \oplus \mathbf{r}_k^{(1)}$, respectively, and omitting terms that are not dependent on $\mathbf{a}$, the probability

$P(\mathbf{a}_k \mid \mathbf{r}_k)$ can be expressed in terms of $P(\mathbf{a}_{k-1} \mid \mathbf{r}_{k-1})$:

$$P(\mathbf{a}_k|\mathbf{r}_k) \propto p(r_k|\mathbf{r}_{k-1}, \mathbf{a}_k)\, P(\mathbf{a}_{k-1}|\mathbf{r}_{k-1}) \tag{13-20}$$

[see eq. (13-5)], where $p(r_k \mid \mathbf{r}_{k-1}, \mathbf{a}_k)$ is the pdf related to the one-step prediction $\hat{r}_{k|k-1}$ of the tuple $r_k = (r_k^{(0)}, r_k^{(1)})$ of partial received samples, given the past received signal $\mathbf{r}_{k-1} = (\mathbf{r}_{k-1}^{(0)}, \mathbf{r}_{k-1}^{(1)})$ and the data hypothesis $\mathbf{a}_k$ up to the present.

In order to simplify the metric computation (and later also channel estimation; Chapter 15), the partial received signals

$$r_k^{(i)} = \sum_{l=0}^{L} h_{l;k}^{(i)}\, a_{k-l} + n_k^{(i)} \tag{13-21}$$

[$i = 0, 1$, eq. (12-1), finite channel memory, zero frequency shift, no diversity] are assumed to be essentially uncorrelated, even though this is in fact not true since (i) pulse shaping makes the $T/2$-spaced channel taps correlated (Chapter 1), and (ii) the $T/2$-spaced noise samples are in general also correlated due to receiver prefiltering. As noted already in Chapter 12 [remarks following eq. (12-74)], intertap correlations $\rho_{\nu,\nu+\mu} = E\{h_{\nu;k} h^*_{\nu+\mu;k}\}$ ($\mu \neq 0$) are difficult to estimate in practice and are therefore neglected so that, for the purpose of receiver design, the channel autocorrelation matrix $\mathbf{R}_h$ may be taken to be diagonal. Hence, the pdf $p(r_k^{(0)}, r_k^{(1)} \mid \mathbf{r}_{k-1}^{(0)}, \mathbf{r}_{k-1}^{(1)}, \mathbf{a}_k)$ reduces to the product $\Pi_{i=0}^{1}\, p(r_k^{(i)} \mid \mathbf{r}_{k-1}^{(i)}, \mathbf{a}_k)$. Also, the pdf $p(r_k^{(i)} \mid \mathbf{r}_{k-1}^{(i)}, \mathbf{a}_k)$ is Gaussian with mean $\hat{r}_{k|k-1}^{(i)}$ and covariance $\sigma_{r^{(i)}}^2$ so that, after taking the logarithm and inverting the sign, the metric and metric increment, respectively, become

$$\begin{aligned} m_{D;k}(\mathbf{a}_k) &= \Delta m_{D;k}(\mathbf{a}_k) + m_{D;k-1}(\mathbf{a}_{k-1}) \\ \Delta m_{D;k}(\mathbf{a}_k) &= \sum_{i=0}^{1} -\ln\left[p\left(r_k^{(i)}|\mathbf{r}_{k-1}^{(i)}, \mathbf{a}_k\right)\right] \\ &= \sum_{i=0}^{1}\left(\frac{\left|r_k^{(i)} - \hat{r}_{k|k-1}^{(i)}\right|^2}{\sigma_{r^{(i)}}^2} + \ln\left[\pi\, \sigma_{r^{(i)}}^2\right]\right) \end{aligned} \tag{13-22}$$

As this result has been derived by neglecting inter-tap correlations $\rho_{\nu,\nu+\mu}$ for $\mu \neq 0$ the metric of eq. (13-22) is truly optimal only if the random channel tap processes are indeed uncorrelated, implying that the channel transfer function is in general nonzero over the entire bandwidth $1/T_s = 2/T$ represented by the digital transmission model. Most often, however, the channel is bandlimited to $B = (1+\alpha)(1/T) < 2/T$ by virtue of the (sampled) pulse shaping filter $g_{T,n}^{(i)}$ (Section 11.2). Hence, the metric eq. (13-22) will in practice be suboptimal in the sense that it neglects the SNR variations over frequency and thus does not suppress noise beyond the signal bandwidth or in deep channel notches. Truly optimal reception calls for channel matched filtering discussed in Section 13.3.3.

Considering the transmission model of eq. (13-21), the predicted channel output $\hat{r}^{(i)}_{k|k-1}$ and its error covariance $\sigma^2_{r^{(i)}}$ are linked with the predicted channel tap estimates $\hat{h}^{(i)}_{l;k|k-1}$ and their error covariances $\sigma^2_{h^{(i)}_l}$ through the relations

$$
\begin{aligned}
\hat{r}^{(i)}_{k|k-1} &= E\left[r^{(i)}_k | \mathbf{r}^{(i)}_{k-1}, \mathbf{a}_k\right] &= \sum_{l=0}^{L} \hat{h}^{(i)}_{l;k|k-1}\, a_{k-l} \\
\sigma^2_{r^{(i)}} &= E\left[\left|r^{(i)}_k - \hat{r}^{(i)}_{k|k-1}\right|^2 | \mathbf{r}^{(i)}_{k-1}, \mathbf{a}_k\right] &= \sum_{l=0}^{L} p_{k-l}\, \sigma^2_{h^{(i)}_l} + N_0
\end{aligned}
\tag{13-23}
$$

Defining the average tap error covariance ratio as $\overline{r}^{(i)}_l = \sigma^2_{h^{(i)}_l}/N_0 = \sigma^2_{h^{(i)}_l}\overline{\gamma}_s$, the metric increment thus obtained

$$
\Delta m_{D;k}(\mathbf{a}_k) = \sum_{i=0}^{1} \left(\frac{\left| r^{(i)}_k - \sum_{l=0}^{L} \hat{h}^{(i)}_{l;k|k-1} a_{k-l} \right|^2}{\left(1 + \sum_{l=0}^{L} p_{k-l} \cdot \overline{r}^{(i)}_l \right) N_0} + \ln\left[\pi N_0 \left(1 + \sum_{l=0}^{L} p_{k-l} \overline{r}^{(i)}_l \right)\right] \right)
\tag{13-24}
$$

depends not only on the present symbol a_k but also on the symbols $a_{k-1}, \ldots, a_{k-L}$ residing in the channel tap delay line (channel "memory"). Since coping with (historically: attempting to undo) the channel memory is referred to as *equalization* [9] and the channel can be viewed as a finite-state machine (FSM), these symbols may also be termed *equalizer state* $s_k \leftrightarrow \{a_{k-1}, \ldots, a_{k-L}\}$.

Motivated by the same arguments as in the previous subection (metric increment for flat fading reception), both the second term $\ln[\ldots]$ and the denominator of the first term, $(1 + \sum_{l=0}^{L} \cdots)\, N_0$, may be dropped so that the metric increment reduces to the familiar Euclidean branch metric for ML sequence detection [10, 11]:

$$
\Delta m_{D;k}(\mathbf{a}_k) \simeq \sum_{i=0}^{1} \left| r^{(i)}_k - \sum_{l=0}^{L} \hat{h}^{(i)}_{l;k|k-1}\, a_{k-l} \right|^2
\tag{13-25}
$$

with the only difference that the actual channel impulse response (CIR) $h^{(i)}_{l;k}$ has been replaced by its one-step predictor estimate $\hat{h}^{(i)}_{l;k|k-1}$ (online NDA channel estimation) or, in general, the estimate $\hat{h}^{(i)}_{l;k}$ obtained by any kind of channel estimation. For M-PSK or in the limiting case of perfect sync ($\overline{r}^{(i)}_l \to 0$), the simplified metric eq. (13-25) is equivalent to the original metric since then the term

$(1 + \sum_{l=0}^{L} \cdots)$ of eq. (13-24) is independent of the particular symbol hypothesis $\{a_k, s_k\} \leftrightarrow \{a_k, a_{k-1}, \ldots, a_{k-L}\}$.

In the case of diversity reception (D received signals $r_{d;k}$ and channel estimates $\hat{h}_{d,l;k}^{(i)}$), it is in general not possible to achieve near-to-optimal co-phasing of signals by means of a memoryless combiner operating directly on the received signals $r_{d;k}$. Hence, it is more appropriate to combine the individual metric increments $\Delta m_{D,d;k}(\mathbf{a}_k)$ ($d = 1, \ldots, D$). By a lengthy but straightforward analysis, the optimal combiner weights are found to be equal for all diversity branches, so that the combined metric increment is obtained by simply adding up all individual branch metric increments:

$$\Delta m_{D;k}(\mathbf{a}_k) = \sum_{d=1}^{D} \Delta m_{D,d;k}(\mathbf{a}_k) \quad = \sum_{d=1}^{D} \sum_{i=0}^{1} \left| r_{d;k}^{(i)} - \sum_{l=0}^{L} \hat{h}_{d,l;k}^{(i)} a_{k-l} \right|^2 \tag{13-26}$$

13.3.2 Maximum-Likelihood Sequence Detection

As discussed above, the inner receiver should generate soft symbol decisions $\hat{a}_k$ – along with channel state information – so that as much information as possible is passed on to the outer receiver. In systems employing channel coding, however, the encoded symbols are most often spread in time by means of *interleaving* in order to provide for implicit time diversity. Then the code memory (uninterleaved time scale) and the channel memory (interleaved time scale) are decoupled which, in general, precludes recursive-type symbol decoding. Therefore, equalization of symbols a_k (index k: interleaved time scale) is usually performed independently from the process of decoding the symbols a_i (index i: noninterleaved or deinterleaved time scale). An important exception discussed in Section 15.2 is the method of combined equalization and decoding (CED) [12, 13] made possible by coordinate interleaving.

In the common case that equalization and decoding are viewed as two tasks being well separated by the deinterleaving device, there should be little or no feedback from the decoder back into the equalizer. A very powerful technique avoiding any such feedback is the symbol-by-symbol *maximum a posteriori (MAP) equalizer* [14] which generates the set of probability estimates (metrics) $\{P(a_k \mid \mathbf{r})\}$ for each possible channel symbol a_k, given the entire received signal $\mathbf{r}$. These probability metrics incorporate the channel state information and can be regarded as soft symbol decisions indicating the likelihood that a particular symbol a_k has been sent. Using the deinterleaved sets $\{P(a_i \mid \mathbf{r})\}$ for decoding yields near-optimal performance , but the forward and backward recursions necessary for generating the sets $\{P(a_k \mid \mathbf{r})\}$ are extremely complex.

A simpler, yet suboptimal technique is the so-called *soft-output Viterbi algorithm (SOVA) equalizer* [15] which delivers the most likely sequence $\hat{\mathbf{a}}$ of hard symbol decisions $\hat{a}_k$ (ML decision), along with a sequence of (estimated) soft probability metrics P_k indicating the reliability of these symbol decisions. The decoder then makes a final decision based on the deinterleaved symbols $\hat{a}_i$ and their reliabilities P_i. Since the basic SOVA equalizer does not have or use any knowledge of the code, its deinterleaved output sequences are not necessarily allowed code sequences. For this reason, more advanced SOVA algorithms introduce some degree of feedback from the decoder to the equalizer. For instance, the *generalized Viterbi algorithm* (GVA) [16] delivers not only the (single) ML symbol sequence but a list of several most likely sequences; the decoder then searches that list until an allowed code sequence is found. Searching for the most likely allowed code sequence may also be performed in an iterative manner, switching back and forth between the equalizer and the decoder.

The most straightforward method of equalization and decoding consists in a simple *concatenation* of equalizer, deinterleaver, and decoder without any feedback. A conventional equalizer [linear equalizer (LE) with or without noise prediction [17] or decision-feedback equalizer (DFE) [18]] generates hard or (preferably) soft decisions $\hat{a}_k$ to be delivered to the outer receiver. Due to noise enhancement and coloration (LE) or unreliable hard decisions at low SNR (DFE), these decisions are, in general, much less reliable than those of the MAP or SOVA algorithms.

For *uncoded* transmission, ML detection – also known as ML sequence estimation (MLSE), ML sequence detection (MLSD), or Viterbi equalization (VE) – is the optimal equalization algorithm [10]. In the case of *coded* transmission with interleaving, ML detection remains to be an integral part of many important equalization algorithms, most notably the SOVA and GVA, but also the LE with noise prediction [19], the DFE – which can be interpreted as the simplest reduced-complexity variant of the ML detector [20] – and even the MAP algorithm, in particular its popular near-optimal variant with maximum rule (MR-MAP) [21, 22]. For this reason, we shall now discuss in some detail several optimal and reduced-complexity ML sequence detection algorithms and their implications on parameter synchronization.

For sequence lengths N exceeding the channel memory length L, ML sequence detection is best performed *recursively* via the well-known *Viterbi algorithm* (VA) [10] based on Bellman's optimality principle of dynamic programming [23]. The Viterbi algorithm takes advantage of the finite channel memory $L < N$ in a way that the computational effort for performing an exhaustive search over all Q^N candidate sequences $\mathbf{a}$ is reduced to an effort not exceeding $\propto Q^L \ll Q^N$ at any given time instant k. The quantity Q, denoting the number of allowed candidate symbols a_k given the previous symbols $\mathbf{a}_{k-1}$, is less than or equal to the cardinality M of the M-PSK or M-QAM symbol alphabet.

Each received sample $r_k^{(i)}$ or $r_{d;k}^{(i)}$ [eq. (13-21)], as well as the metric increment $\Delta m_{D;k}(\mathbf{a}_k)$ of eq. (13-25) or (13-26), do not depend on the entire sequence $\mathbf{a}_k$ but only on the most recent symbol a_k and the equalizer state $s_k \leftrightarrow \{a_{k-1}, \ldots, a_{k-L}\}$. For this reason, the minimization of the decision metric may be performed in a recursive manner as follows:

$$\begin{aligned}
\hat{\mathbf{a}} &= \arg\min_{\mathbf{a}} m_D(\mathbf{a}) \\
&= \arg\min_{\mathbf{a}} m_{D;N+L-1}(\mathbf{a}_{N+L-1}) \quad = \arg\min_{\mathbf{a}} \left(\sum_{k=0}^{N+L-1} \Delta m_{D;k}(a_k, s_k) \right) \\
&= \arg\min_{a_{N-1}} \underbrace{\left[\min_{a_{N-2}} \underbrace{\left[\dots \left(\min_{a_1} \underbrace{\left(\min_{a_0} \underbrace{\left(\sum_{k=0}^{L} \Delta m_{D;k} \right)}_{\text{dep. on } a_0, s_{L+1}} + \Delta m_{D;L+1} \right)}_{\text{dependent on } a_1 \text{ and } s_{L+2}} \right) \dots \right]}_{\text{dependent on } a_{N-2} \text{ and } s_{N+L-1}} + \Delta m_{D;N+L-1} \right]}_{\text{dependent on } a_{N-1} \text{ and } s_{N+L}}
\end{aligned} \tag{13-27}$$

The Viterbi algorithm thus starts with accumulating the first $L+1$ metric increments (branch metrics) $\Delta m_{D;0}(a_0, s_0), \dots, \Delta m_{D;L}(a_L, s_L)$ during time instants $k = 0, \dots, L$. The resulting path metric $m_{D;L}(\mathbf{a}_L) = (\sum_{k=0}^{L} \Delta m_{D;k}(a_k, s_k))$ at time $k = L$ depends on the set of symbols $\{a_0, a_1, \dots, a_L\} \leftrightarrow \{a_0, s_{L+1}\}$ and therefore has to be formed for all Q^{L+1} possible sets $\{a_0, s_{L+1}\} = \{s_L, a_L\}$. Since the sum contains all metric increments which are dependent on the first symbol a_0, the first minimization step (iteration $k = L \to L+1$) can be carried out with respect to a_0, yielding the Q^L possible symbol sets $\{a_1, \dots, a_L\} \leftrightarrow s_{L+1}$, along with the "survivor" sequence $\mathbf{a}_L$ (including the symbol a_0) associated with each "new" state $\{a_1, \dots, a_L\} \leftrightarrow s_{L+1}$.

The second minimization step (iteration $k = L+1 \to L+2$) then starts with augmenting each of the Q^L states $\{a_1, \dots, a_L\} \leftrightarrow s_{L+1}$ and survivor sequences $\mathbf{a}_L$ by Q possible new symbols a_{L+1}, giving a total of Q^{L+1} "extended" states $\{a_1, \dots, a_L, a_{L+1}\} \leftrightarrow \{s_{L+1}, a_{L+1}\} = \{a_1, s_{L+2}\}$ and survivors $\mathbf{a}_{L+1}$. The path metric $m_{D;L}(\mathbf{a}_L)$ associated with an "old" state s_{L+1} is likewise extended to $m_{D;L}(\mathbf{a}_L) + \Delta m_{D;L+1}$, where the symbols $\{a_1, \dots, a_{L+1}\}$ needed to compute the branch metric $\Delta m_{D;L+1}(a_1, \dots, a_{L+1})$ are those of the extended state $\{s_{L+1}, a_{L+1}\} = \{a_1, s_{L+2}\}$ of a particular state transition $s_{L+1} \to s_{L+2}$. There are Q extended states merging in a particular new state s_{L+2}, and these differ only in the symbol a_1. The second minimization is now carried out with respect to a_1, yielding the best of the "paths" ending in the new state s_{L+2}, along with the associated survivor $\mathbf{a}_{L+1}$ and path metric $m_{D;L+1}(\mathbf{a}_{L+1}) = \min\{m_{D;L}(\mathbf{a}_L) + \Delta m_{D;L+1}\}$.

The procedure of iteratively minimizing the ML decision metric via the Viterbi algorithm is illustrated in Figure 13-7 for the simple example of uncoded BPSK

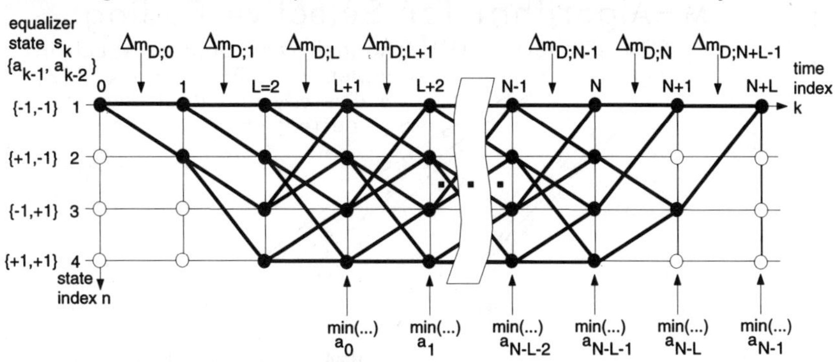

Figure 13-7 The Viterbi Algorithm for ML Sequence Detection on Selective Channels

transmission (Q=2) and a channel memory length L=2 [i.e., the (partial) channel has three taps $\{h^{(i)}_{0;k}, h^{(i)}_{1;k}, h^{(i)}_{2;k}\}$] so that there are Q^L=4 states and Q^{L+1}=8 extended states. In the *trellis* (state transition diagram) of Figure 13-7, the states and possible state transitions are shown for each time instant k. During ordinary operation of the VA (iterations $k = L \to L+1, \ldots, N-1 \to N$), there are Q=2 branches originating from each old state, and also Q=2 branches merging into each new state. Of these merging paths, the VA selects and keeps the best one. During the startup phase (iterations $k = 0 \to 1, \ldots, L-1 \to L$) while the algorithm just accumulates the branch metrics $\Delta m_{D;k}$, the trellis expands until all Q^L=4 states can be reached. During the final phase (iterations $k = N \to N+1, \ldots, N+L-1 \to N+L$), the algorithm just performs survivor selection by metric minimization until a single state s_{N+L} remains, here $s_{N+2} \leftrightarrow \{a_{N+1}, a_N\}$. The detected message $\hat{\mathbf{a}}$ is then given by the (first N symbols of the) survivor sequence $\mathbf{a}_{N+L-1}$ associated with that final state.

In Figure 13-7, the initial state $s_0 \leftrightarrow \{a_{-1}, a_{-2}\}$ has been arbitrarily set to $\{-1,-1\}$ but can as well be any other state if the message is preceded by a preamble. Likewise, the final state $s_{N+L} = s_{N+2} \leftrightarrow \{a_{N+1}, a_N\}$, which also has been set to $\{-1,-1\}$ in Figure 13-7, can be any other sequence if the message is succeeded by a postamble. Notice also that neither the preamble nor the postamble symbols need to be of the same symbol alphabet as the message; they just need to be known for metric computation. Hence, ML symbol detection does not impose any constraint on the design of pre- and postambles or any other training segments within the message (Section 15.2).

13.3.3 Reduced-Complexity ML Sequence Detection

When the channel memory L is large, optimal ML sequence detection searching the entire trellis by means of the Viterbi algorithm quickly becomes unfeasible since the number Q^L of equalizer states rises exponentially with L. Therefore,

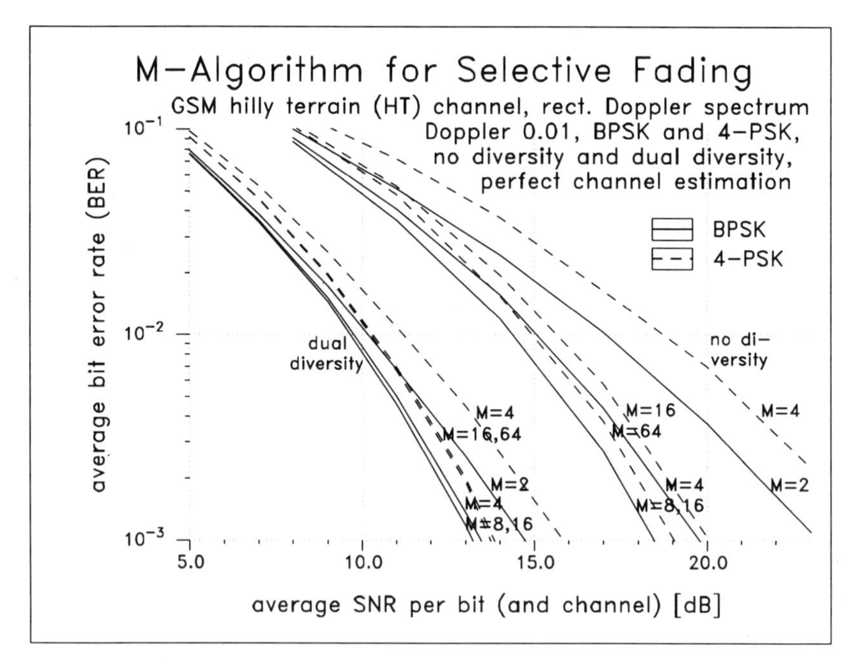

Figure 13-8 Performance of M-Algorithm Equalizer Working on the Received Signal

reduced-complexity variants of the VA have been devised which search only a small subtrellis and thus retain a small number of survivors, say $M \ll Q^L$, and drop all other sequences (*M algorithm* [24]). In the simplest case when $M = 1$, the M algorithm reduces to the operation of the backward filter of the very popular decision-feedback equalizer (DFE) [18].

However, the M algorithm with small $M \ll Q^L$ is often found to be ineffective when working directly on the received signal $\mathbf{r}$ [25, 26, 27], even in the case of simple uncoded transmission and ideal channel estimation. As an example, consider the simulated bit error curves displayed in Figure 13-8 for BPSK and 4-PSK transmission over the GSM hilly terrain (GSM-HT) channel (Section 11.3.2). Despite of the small PSK signal alphabets and the mild intersymbol interference (ISI) spanning at most L=4 symbol intervals, the number M of equalizer states to be considered by the M algorithm must be quite large, especially when diversity is not available. If the "convergence" of BER curves with rising M is taken as a measure of (near-) optimality, M should be 8 (no diversity) or 4 (diversity) with 2-PSK, and as large as 64 (no diversity) or 16 (diversity) with 4-PSK. Similar results have also been obtained for a two-ray fading channel with equal average path gains, delay τ'=2, and L=4. Since the full Viterbi algorithm would have to consider Q^L=16 (2-PSK) and 256 (4-PSK) states, the reduction in complexity achieved by the M algorithm working directly on the received signal $\mathbf{r}$ is far from the desired goal $M \ll Q^L$.

In addition to the suboptimality mentioned in Section 13.3.1, the dominant mechanism responsible for this insatisfactory behavior is the fact that, on fading channels, the first arriving multipath ray may undergo deep fading. During such a deep fade, the "first" taps $\{h_{0;k}^{(i)}, h_{1;k}^{(i)}, \ldots\}$ of the channel impulse response are virtually zero. With zero $h_{0;k}^{(i)}$, the term $h_{0;k}^{(i)} a_k$ is also zero so that the branch metrics $\Delta m_{D;k}(\mathbf{a}_k) = \Delta m_{D;k}(a_k, s_k)$ [eq. (13-25)] are no more dependent on the latest symbol a_k. Thus all Q extended metrics $m_{D;k-1} + \Delta m_{D;k}(a_k, s_k)$ of paths that have originated from a particular predecessor state s_k are equal. When the first two taps $h_{0;k}^{(i)}, h_{1;k}^{(i)}$ are zero, we have $h_{0;k}^{(i)} a_k + h_{1;k}^{(i)} a_{k-1}$=0 so that all Q^2 extended metrics of paths that have originated from a particular state s_{k-1} (two time steps ago) are equal, and so forth. Therefore, the number of contenders having virtually the same path metric may quickly rise when the first multipath ray undergoes a deep fade. This is no problem with full Viterbi processing, but with reduced-complexity sequence detection where only few contenders are kept after each iteration, the correct path is very likely to be dropped from the survivor list.

There are two basic methods of resolving this dilemma. The first method consists in having the sequence detection procedure still operate directly on $\mathbf{r}$ but augmenting it with some *adaptive control* unit which continually monitors the actual CIR $\{h_{0;k}^{(i)}, h_{1;k}^{(i)}, \ldots\}$ and adjusts the ML sequence detector accordingly such that the detrimental mechanism explained above is avoided. For instance, the set of extended equalizer states $\{a_k, s_k\} \leftrightarrow \{a_k, a_{k-1}, \ldots, a_{k-L}\}$ considered for sequence detection may be truncated to $\{a_{k-F}, \ldots, a_{k-L}\}$ where a_{k-F} is the symbol weighted by the first *nonzero* channel coefficient $h_{F;k}^{(i)}$ above a certain threshold [26]. Because the position F is unknown and may vary, this "precursor control" – the first channel taps $\{a_k, \ldots, a_{k-(F-1)}\}$ are temporarily ignored – necessitates a considerable control effort. Also, when $h_{F;k}^{(i)}$ is barely above the threshold, the correct survivor is more likely to be dropped than if $h_{F;k}^{(i)}$ were a strong tap. On the other hand, precursor taps $\{a_k, \ldots, a_{k-(F-1)}\}$ barely below the threshold lead to an irreducible noiselike effect in the detection process.

The second basic method whereby frequent droppings of the correct survivor can be counteracted consists in *preprocessing* the received signal $r_k^{(i)}$ by a prefilter $\mathbf{u}_k$ and a decimator to symbol rate, thus transforming the $(T/2)$-spaced received signal $r_k = r_k^{(0)} \oplus r_k^{(1)}$ into a T-spaced signal v_k. Denoting the resulting *equivalent* T-spaced transmission system – accounting for the cascade of physical channel, pulse shaping filter, prefilter $\mathbf{u}_k$ and decimator – by $\mathbf{f}_k = \{\ldots, f_{-1;k}, f_{0;k}, f_{1;k}, \ldots, f_{L;k}, f_{L+1;k}, \ldots\}$, the equalizer faces the signal

$$v_k = \sum_l f_{l;k}\, a_{k-l} + \eta_k \tag{13-28}$$

where η_k is the decimated noise behind the filter $\mathbf{u}_k$.

We have seen that the shape of the impulse response $\mathbf{h}_k$ or $\mathbf{f}_k$ "seen" by the equalizer is most critical to the performance of reduced-complexity ML sequence detection. From the discussion above it follows that $\mathbf{f}_k$ should have the following desired properties:

- The length of the equivalent response $\mathbf{f}_k$ should not exceed that of $\mathbf{h}_k$ so that the number of equalizer states does not increase. Hence, $\mathbf{f}_k$ must be *finite-length* (FIR) with no more than $L+1$ nonzero coefficients.
- Allowing noncausal preprocessing filters, the FIR response $\mathbf{f}_k$ can always be made *causal* so that the desired equivalent channel becomes $\mathbf{f}_k = \{f_{0;k}, f_{1;k}, \ldots, f_{L;k}\}$.
- The taps of $\mathbf{f}_k$ should be maximally *concentrated near zero delay*, i.e., the first coefficients, in particular $f_{0;k}$, must be as strong as possible with respect to the postcursors of $\mathbf{f}_k$. Then variations in the symbol a_k (now weighted by the strong tap $f_{0;k}$) immediately translate into large variations in the branch metrics $\Delta m_{D;k}(a_k, s_k)$ and thus the extended branch metrics. Trellis pruning to $M \ll Q^L$ survivors then affects only weak postcursors, so that the correct survivor is dropped with much smaller probability. The branch metrics are computed like in eq. (13-25), now with $r_k^{(i)}$ and taps $h_{l;k}^{(i)}$ replaced by the decimated prefilter output v_k and the equivalent channel taps $f_{l;k}$, respectively:

$$\Delta m_{D;k}(\mathbf{a}_k) = \left| v_k - \left(\sum_{l=0}^{L} f_{l;k}\, a_{k-l} \right) \right|^2 \tag{13-29}$$

- Preprocessing should aim at (near-) optimal *noise suppression*, or should at least avoid noise enhancement.
- Finally, preprocessing should not destroy the *whiteness* property of the noise.

Obviously, the first three objectives can be satisfied by choosing the desired $\mathbf{f}_k$ to be $\mathbf{f}_k = \{1, 0, \ldots, 0\}$, which calls for perfect channel equalization by means of a zero-forcing linear equalizer $\mathbf{u}_k$. Optimal equalization based on the LE output v_k is still possible, but the noise enhancement and coloration have to be counteracted by means of a noise predicting Q^L-state Viterbi algorithm. Reducing the complexity by disregarding the noise correlation, i.e., performing symbol-by-symbol processing of v_k without noise prediction, is far from optimal. For these reasons, maximizing the SNR and preserving the whiteness property of the noise are very desirable constraints to be imposed on the choice of $\mathbf{u}_k$ and $\mathbf{f}_k$. As a consequence, the preprocessing device (filter $\mathbf{u}_k$ plus decimator) should have an *allpass* response so that η_k remains white (assuming that the noise $n_k^{(i)}$ before prefiltering is white) and the SNR is maximized or at least not compromized. On the other hand, the filter $\mathbf{u}_k$ must remain to be *implementable*, i.e., of finite length with a certain number of coefficients, say $2U+1$ $(T/2)$-spaced taps spanning U symbol intervals, where the tap weights can be adapted to the changing channel conditions.

The structure and coefficients of a suitable preprocessing device complying with the above objectives can be devised by combining two important elements of communication theory: (i) the *canonical receiver structure* for ML sequence estimation in the presence of ISI, originally derived by Forney [3], and (ii) the structure of a finite-length *decision-feedback equalizer* [18]. In Forney's receiver, the ML sequence detector — or, in the case of coded transmission with interleaving, any other equalizer such as MAP or SOVA — is preceded by a *whitening matched filter* (WMF).

Before explaining why the WMF is the optimal prefilter and how to obtain its coefficients, let us establish the relationship between the WMF and matched filtering discussed before. This is more than a theoretical subtlety since the WMF is often implemented as the cascade of the $T/2$-spaced *channel matched filter* (MF) $\mathbf{m}_k$, a decimator to symbol rate $1/T$, and a T-spaced *whitening filter* (WF) $\mathbf{w}_k$. In Chapters 4 and 1, the filter matched to the (partial T-spaced) transmitting pulse shape $g_{T,n}^{(i)}(\varepsilon)$ [eq. (11-18)] or the [$(T/2)$-spaced] pulse $g_{T,\nu}(\varepsilon)$, followed by a decimator to symbol rate $1/T$, was established to be the optimum prefilter in the case of AWGN or flat fading channels without ISI, in the sense that the decimated MF output is a sufficient statistic for symbol-by-symbol detection at rate $1/T$. In the presence of ISI, the decimated output of the $(T/2)$-spaced MF $\mathbf{m}_k$ matched to the channel $\mathbf{h}_k$ (pulse shape convolved with the physical channel impulse response) is also an optimal sufficient statistic for detection at rate $1/T$, but the MF does not remove the ISI. On the contrary, the MF spreads the ISI over $2L$ (L noncausal and L causal) symbol intervals [eq. (13-31)]. The T-spaced WF $\mathbf{w}_k$ following the decimator then attempts to cancel the noncausal precursor ISI – hence, the WF is sometimes also termed *precursor equalizer* [20]. The resulting structure of the canonical inner receiver is visualized in the upper part of Figure 13-9.

The T-spaced equivalent channel $\mathbf{f}_k$ seen by the equalizer incorporates the transmit rate converter $1/T \rightarrow 2/T$, CIR $\mathbf{h}_k$, channel MF $\mathbf{m}_k$, decimator $2/T \rightarrow 1/T$, and the WF $\mathbf{w}_k$. As discussed above, the MF, decimator, and WF can be merged into the $(T/2)$-spaced WMF $\mathbf{u}_k$ followed by a decimator. Both alternatives are also visualized in the lower left part of Figure 13-10.

In the case of Dth-order diversity reception (central part of Figure 13-9), the processing for each of the D prefiltering branches is the same as above, with the channel MF $\mathbf{m}_{d;k}$ and WF $\mathbf{w}_{d;k}$ now matched to the respective diversity channel $\mathbf{h}_{d;k}$. The dth equivalent T-spaced channel seen by the optimal combiner is again denoted by $\mathbf{f}_{d;k}$.

13.3.4 Adjustment of Inner Receiver Components for Selective Fading Channels

In the course of discussing the adjustment of time-variant receiver component parameters – also termed *parameter synchronization* – it will soon become apparent why the WMF (or MF+WF) is the optimal prefilter for reduced-complexity ML sequence detection. Disregarding frequency shifts Ω', a sufficient statistic for parameter synchronization (Figure 13-9) is given by the channel impulse response(s)

Selective Fading Inner Receiver without Diversity

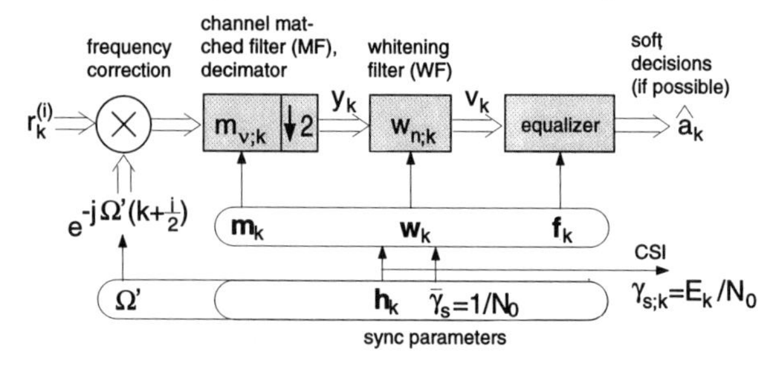

Selective Fading Inner Receiver with D-th order Diversity

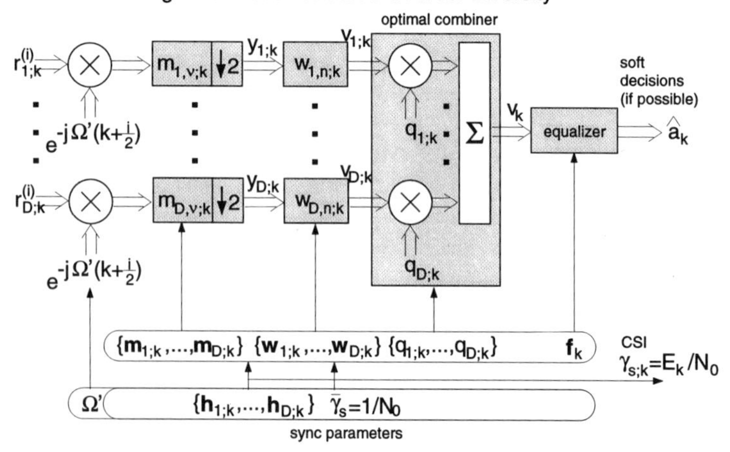

Discrete-Equivalent Transmission System (without Equalizer; Correct Sync)

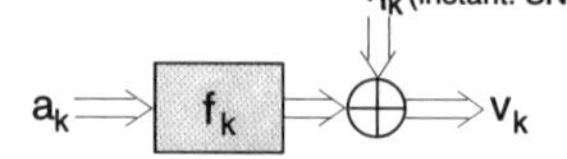

Figure 13-9 Inner Receiver for Selective Fading Channels without and with Diversity

$\mathbf{h}_{(d);k}$, and – less importantly – the (combined) noise power or average SNR $\overline{\gamma}_s$ per symbol. In other words, the CIR(s) and the noise level can be viewed as the "system state" from which all other sync parameters can be derived by a mathematical mapping. Hence, channel estimation is the most important synchronization task; it is discussed thoroughly in Chapter 14.

In the following, the tasks to be performed for parameter synchronization based on the knowledge of the channel(s)[9] will be discussed, viz. the mappings from the CIR $\mathbf{h}_k$ onto the MF $\mathbf{m}_k$, WF $\mathbf{w}_k$ (the WMF $\mathbf{u}_k$ is then the convolution

[9] The diversity channel index d is dropped here.

between the MF and the WF), the combiner parameters $q_{d;k}$, as well as the equivalent channel $\mathbf{f}_k$.

Under the assumption of quasi-stationarity of the channel within the duration L of the causal channel impulse response ($2[L+1]$ $T/2$-spaced taps $h_{\nu;k}$, $\nu = 0, \ldots, 2L+1$), the normalizing *channel matched filter* $\mathbf{m}_k$ is anticausal with $T/2$-spaced coefficients:

$$m_{\nu;k} = \frac{1}{E_k} h^*_{-\nu;k} \qquad (\nu = -[2L+1], \ldots, 0) \tag{13-30}$$

with $E_k = \|\mathbf{h}_k\|^2 = \sum_{\nu=0}^{2L+1} | h_{\nu;k} |^2$ the instantaneous channel energy. The equivalent channel between the transmit rate converter $1/T \rightarrow 2/T$ and the decimated MF output

$$\begin{aligned} y_k &= \sum_{\nu=-(2L+1)}^{0} m_{\nu;k}\, r_{k-\nu/2} \\ &= \frac{1}{E_k}\left(\sum_{\nu=0}^{2L+1} h^*_{\nu;k}\, r_{k+\nu/2}\right) = \frac{1}{E_k}\left(\sum_{i=0}^{1}\sum_{n=0}^{L} h^{(i)*}_{n;k}\, r^{(i)}_{k+n}\right) \\ &= \sum_{n=-L}^{L} x_{n;k}\, a_{k-n} + n_{c;k} \end{aligned} \tag{13-31}$$

(Figure 13-9) is then given by the T-spaced Hermitian symmetric deterministic channel autocorrelation

$$x_{n;k} = \begin{cases} \dfrac{1}{E_k}\left(\displaystyle\sum_{\nu=0}^{2[L-n]+1} h_{\nu+2n;k}\, h^*_{\nu;k}\right) = \dfrac{1}{E_k}\left(\displaystyle\sum_{i=0}^{1}\sum_{j=0}^{L-n} h^{(i)}_{n+j;k}\, h^{(i)*}_{j;k}\right) & (n=1,\ldots,L) \\ 1 & (n=0) \\ x^*_{-n;k} & (n=-L,\ldots,-1) \end{cases} \tag{13-32}$$

The normalizing MF effectuates both digital amplitude gain control (center tap $x_{0;k}$ is unity) and optimal noise suppression. Since (perfect) matched filtering does not entail a loss of information, the MF output signal y_k is a sufficient statistic just like the received signal $r_k^{(i)}$. However, the ISI now spreads over $2L$ (L causal and L noncausal) symbols, and the filtered noise $n_{c;k}$ is now colored with time-variant power $N_0/E_k = 1/\gamma_{s;k}$ and autocorrelation $R_{n_c}(n) = (N_0/E_k) x_{n;k}$.

Therefore, further processing by means of the *whitening filter* $\mathbf{w}_k$ is necessary, considering the objectives that the equivalent channel $\mathbf{f}_k = \mathbf{x}_k * \mathbf{w}_k$ must satisfy. In Forney's canonical receiver [3], the WF is specified in terms of its transfer function $\mathbf{w}_k \circ\!\!-\!\!\bullet W_k(z)$ as follows. A spectral factorization of the channel autocorrelation $X_k(z)$ yields the cascade of filters $F_k^*(1/z^*)$ and $F_k(z)$, i.e., $\mathbf{x}_k \circ\!\!-\!\!\bullet X_k(z) = F_k^*(1/z^*) F_k(z)$. If $F_k(z)$ is chosen to be causal and *minimum-phase*, it is exactly this transfer function $F_k(z) \bullet\!\!-\!\!\circ \mathbf{f}_k$ which is identified as the equivalent channel with the desired properties, viz. (i) causal, (ii) finite memory L, and (iii) the energy

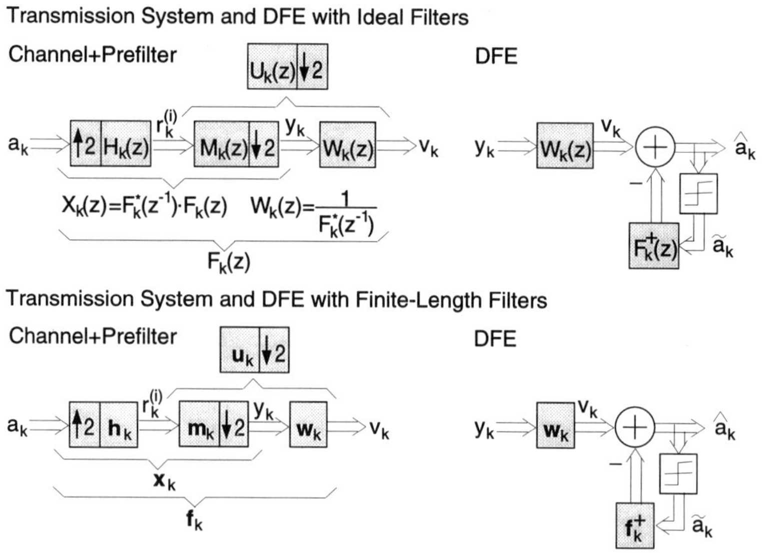

Figure 13-10 Ideal and Finite-Length Prefiltering and Decision-Feedback Equalization

of $\mathbf{f}_k$ is maximally concentrated near zero delay, thanks to the minimum-phase property of $F_k(z)$. A transfer function $F_k(z)$ is (loosely) minimum phase if all L zeros of $F_k(z)$ lie inside (or on) the unit circle, which implies that the energy of the first coefficients $\{f_{0;k}, f_{1;k}, \ldots\}$ is as large as possible among all impulse responses with the same magnitude of the transfer function, or equivalently, that the postcursor ISI $\{f_{1;k}, f_{2;k} \ldots\}$ is minimized with respect to the center tap $f_{0;k}$.

As a consequence, the ideal whitening filter transfer function $W_k(z)$ is given by the inverse of the second part $F_k^*(1/z^*)$ of the spectral factorization. Since $F_k(z)$ is (loosely) minimum-phase and stable, all zeros and poles of $F_k(z)$ lie inside (or on) the unit circle so that all zeros and poles of $F_k^*(1/z^*)$ lie outside (or on) the unit circle. Hence, the inverse filter $W_k(z) = 1/F_k^*(1/z^*)$ is maximum-phase and, in order to be stable, the WF $W_k(z)$ must be anticausal. The ideal transmission and prefiltering system in the z-domain is visualized in the upper left part of Figure 13-10.

Unfortunately, the canonical WF transfer function $W_k(z) = 1/F_k^*(1/z^*)$ generally translates into an *infinite-length* anticausal filter response $\mathbf{w}_k \circ\!\!-\!\!\bullet W_k(z)$. Observing the finite-length constraint ($W+1$ coefficients spanning W symbol intervals), the WF response $\mathbf{w}_k$ cannot be easily determined in the z-domain. This is where the structure of a decision-feedback equalizer comes into play. The (infinite-length) anticausal DFE *feedforward* (FF) filter, optimized according to the zero-forcing (ZF) criterion, is known to be identical to the ideal canonical WF $W_k(z)$, and the causal DFE *feedback* (FB) filter is identical to the strictly causal part $F_k^+(z)$ of the equivalent channel $F_k(z)$ [11] (see upper right side of

Figure 13-10). These correspondences originally established in the z-domain can now be transferred to *finite-length* time-domain prefiltering [28]. Hence, the WF coefficients $\mathbf{w}_k$ are given by those of the length $(W+1)$ DFE-FF filter, and the strictly causal equivalent channel coefficients $\mathbf{f}_k^+ = \{f_{1;k}, \ldots, f_{L;k}\}$ are identical to those of the DFE-FB filter. These correspondences are visualized in the lower part of Figure 13-10.

The DFE-FF and FB filter coefficients can be computed from the channel autocorrelation $\mathbf{x}_k$ and the noise power N_0/E_k behind the matched filter as follows. The DFE forms a symbol estimate $\hat{a}_k$ by filtering the MF output y_k via $\mathbf{w}_k$ — this operation cancels most of the precursor ISI — and subtracting from the DFE-FF filter output v_k an estimate of the remaining postcursor ISI formed via $\mathbf{f}_k^+$ and past symbol decisions:

$$\hat{a}_k = \underbrace{\left(\sum_{n=-W}^{0} w_{n;k}\, y_{k-n}\right)}_{v_k = \mathbf{w}_k^T \cdot \mathbf{y}_k} - \underbrace{\left(\sum_{n=1}^{L} f_{n;k}\, \tilde{a}_{k-n}\right)}_{\mathbf{f}_k^{+T} \cdot \tilde{\mathbf{a}}_k} \tag{13-33}$$

(see also lower right side of Figure 13-10), where $\mathbf{y}_k$ and $\tilde{\mathbf{a}}_k$ denote the MF output samples in the DFE-FF delay line and the past symbol decisions in the DFE-FB delay line, respectively.

For a given prefilter length W, the DFE coefficients are now determined so as to minimize the error $\hat{a}_k - a_k$ in the symbol estimates (correct decisions $\tilde{a}_k = a_k$ assumed). Since a finite-length WF $\mathbf{w}_k$ cannot cancel noncausal "precursor" ISI completely, the DFE filter coefficients should be optimized according to the mean-square-error (MSE) criterion whereby the composite noise power resulting from residual ISI and additive noise is minimized. The derivation of the optimal FF (=WF $\mathbf{w}_k$) and FB (= $\mathbf{f}_k^+$) coefficients via the orthogonality theorem [29] yields

Coefficients of noncausal WF $\mathbf{w}_k$:

$$\underbrace{\begin{pmatrix} \psi_{-W,-W} & \cdots & \psi_{-W,0} \\ \vdots & \ddots & \vdots \\ \psi_{0,-W} & \cdots & \psi_{0,0} \end{pmatrix}_k}_{\mathbf{\Psi}_k} \underbrace{\begin{pmatrix} w_{-W} \\ \vdots \\ w_0 \end{pmatrix}_k}_{\mathbf{w}_k} = \underbrace{\begin{pmatrix} x_{-W} \\ \vdots \\ x_0 \end{pmatrix}_k}_{\mathbf{x}_k^+}$$

with elements of Hermitian matrix $\mathbf{\Psi}_k$: (13-34)

$$\psi_{i,j;k} = \left(\sum_{l=-L}^{-j} x_l\, x^*_{l+[j-i]}\right)_k + \frac{1}{\gamma_{s;k}}\, x_{i-j;k} \qquad (i,j = -W, \ldots, 0)$$

Coefficients of causal equivalent channel $\mathbf{f}_k$:

$$f_{n;k} = \sum_{l=n-L}^{0} w_l\, x_{n-l} \qquad (n = 0, \ldots, L)$$

where $\mathbf{x}_k^+ = \{x_0 \; x_1 \; \dots \; x_L \; 0 \; \dots \; 0\}$ denotes the channel autocorrelation vector padded with zeros so that its dimension is augmented to $[W+1] \geq L$.

For the purpose of performance evaluation, the *entire* equivalent channel $\mathbf{f}_k = \mathbf{x}_k * \mathbf{w}_k$ is of interest, particularly its noncausal part $\mathbf{f}_k^- = \{\dots, f_{-2;k}, f_{-1;k}\}$ being responsible for residual precursor ISI. The response $\mathbf{f}_k$ including $\mathbf{f}_k^-$ and the minimum MSE associated with finite-length prefiltering are given by

$$f_{n;k} = (\mathbf{x}_k * \mathbf{w}_k)_n = \begin{cases} f_{n;k}^- = \sum\limits_{l=n-L}^{0} w_l \, x_{n-l} & (n = -[W+L], \dots, -1) \\ f_{n;k} = \sum\limits_{l=n-L}^{0} w_l \, x_{n-l} & (n = 0, \dots, L) \end{cases}$$

$$\zeta_{\min;k} = E\left[|\hat{a}_k - a_k|^2\right] \quad = 1 - \left(\sum_{l=-W}^{0} w_l \, x_{-l} \right)_k \tag{13-35}$$

For quasi-perfect prefiltering where the impact of residual precursors $\mathbf{f}_k^-$ can be neglected, the equivalent transmission model in front of the equalizer (ML sequence detector or its variants) shown in the bottom part of Figure 13-9 [see also eq. (13-29)] reduces to

$$\begin{aligned} v_k &= \sum_{l=0}^{L} f_{l;k} \, a_{k-l} \quad + \eta_k \\ E\left[|\eta_k|^2\right] &= ||\mathbf{f}_k||^2 \frac{N_0}{E_k} \quad = ||\mathbf{f}_k||^2 \frac{1}{\gamma_{s;k}} \\ \gamma_{s;k} &= \frac{||\mathbf{f}_k||^2}{E\left[|\eta_k|^2\right]} \quad = \frac{E_k}{N_0} \end{aligned} \tag{13-36}$$

In the case of diversity reception (center part of Figure 13-9), quasi-optimal diversity combining may be performed by weighting the WF outputs $v_{d;k}$ of the diversity branches according to the instantaneous channel energies $E_{d;k}$ [30]. The combined equivalent channel $\mathbf{f}_k$ is then given by the weighted sum of the individual channels $\mathbf{f}_{d;k}$,

$$\begin{aligned} q_{d;k} &= \frac{E_{d;k}}{E_d} \qquad (d = 1, \dots, D) \qquad \text{with} \quad E_d = \sum_{d=1}^{D} E_{d;k} \\ \mathbf{f}_k &= \sum_{d=1}^{D} q_{d;k} \, \mathbf{f}_{d;k} \end{aligned} \tag{13-37}$$

For quasi-perfect prefiltering, the equivalent diversity transmission model in front of the equalizer (see again bottom part of Figure 13-9) reduces to

$$
\begin{aligned}
v_k &= \sum_{l=0}^{L} f_{l;k}\, a_{k-l} \quad + \eta_k \\
E\left[|\eta_k|^2\right] &= \left(\sum_{d=1}^{D} |q_{d;k}|^2\, ||\mathbf{f}_{d;k}||^2\, \frac{N_0}{E_{d;k}}\right) \quad = \frac{1}{\gamma_{s;k}}\left(\sum_{d=1}^{D} q_{d;k}\, ||\mathbf{f}_{d;k}||^2\right) \\
\gamma_{s;k} &= \frac{\left(\sum_{d=1}^{D} q_{d;k}\, ||\mathbf{f}_{d;k}||^2\right)}{E\left[|\eta_k|^2\right]} \quad = \frac{\left(\sum_{d=1}^{D} E_{d;k}\right)}{N_0} \quad = \frac{E_k}{N_0}
\end{aligned}
\tag{13-38}
$$

Hence, the equivalent transmission model seen by the equalizer is essentially the same as for nondiversity reception, except that the equivalent channel is the weighted superposition of all diversity channels and the instantaneous SNR $\gamma_{s;k}$ is determined by the combined channel energies.

Concluding this subsection, it has been shown how the components of the inner receiver for selective fading channels can be obtained from the channel impulse response(s) and the noise level. In particular, the MF coefficients $\mathbf{m}_k$ are given by the mapping $\mathbf{h}_k \rightarrow \mathbf{m}_k$ [eq. (13-30)], the WF coefficients $\mathbf{w}_k$ by the mappings $\mathbf{h}_k \rightarrow \mathbf{x}_k$ [eq. (13-32)] and $\mathbf{x}_k \rightarrow \mathbf{w}_k$ [eq. (13-34)] using $\gamma_{s;k}$ and the parameter W, and the equivalent channel $\mathbf{f}_k$ is given by the mapping $\mathbf{x}_k, \mathbf{w}_k \rightarrow \mathbf{f}_k$ [eq. (13-34)]. If needed, the noncausal part $\mathbf{f}_k^-$ of the equivalent channel and the MMSE $\zeta_{\text{min};k}$ can also be obtained from $\mathbf{x}_k, \mathbf{w}_k$ [eq. (13-35)].

For diversity reception, these same mappings apply to each of the D diversity branches, viz. $\mathbf{h}_{d;k} \rightarrow \mathbf{m}_{d;k}$, $\mathbf{h}_{d;k} \rightarrow \mathbf{x}_{d;k}$, $\mathbf{x}_{d;k} \rightarrow \mathbf{w}_{d;k}$, and $\mathbf{x}_{d;k}, \mathbf{w}_{d;k} \rightarrow \mathbf{f}_{d;k}$. In addition, the combiner weights $p_{d;k}$ and the combined equivalent channel $\mathbf{f}_k$ are given by the mappings $\{E_{d;k}\} \rightarrow \{p_{d;k}\}$ and $\{\mathbf{f}_{d;k}, p_{d;k}\} \rightarrow \mathbf{f}_k$ [eq. (13-37)], respectively.

13.3.5 Example: Inner Receiver Prefiltering for Selective Fading Channels

To illustrate the effect of prefiltering, consider the simple example of a single (no diversity) two-ray channel:

$$
h_{\nu;k} = \begin{cases} 1/\sqrt{1+a^2} & (\nu = 0) \\ a/\sqrt{1+a^2} \quad (= a\, h_{\nu=0;k}) & (\nu = 2L) \\ 0 & (\text{else}) \end{cases}
\tag{13-39}
$$

with energy $||\mathbf{h}_k||^2$=1, memory length L, and two nonzero weights $h_{\nu;k}$, where $h_{\nu=2L;k} = a h_{\nu=0;k}$. Notice that this example is artificial in the sense that two isolated spikes in the $T/2$-spaced CIR do not represent a band-limited channel (see Chapter 1). This, however, is immaterial since the three-tap respective channel autocorrelation $\mathbf{x}_k$ could as well have been produced by a band-limited channel in cascade with its respective matched filter. Behind the normalizing MF [eq. (13-30)], the T-spaced channel correlation [eq. (13-32)] has three nonzero taps $x_{0;k}=1$ and $x_{\mp L;k} = a/(1+a^2)$.

Let us assume a whitening filter with span W, where $W = ZL$ is a multiple of the channel length L. Since only two channel taps and three autocorrelation coefficients are nonzero, the $(W+1)$-dim. system of equations for the mapping $\mathbf{x}_k \to \mathbf{w}_k$ [eq. (13-34)] collapses to the $(Z+1)$-dim. system of equations:

Coefficients of $T-$spaced noncausal WF $\mathbf{w}_k$:

$$\underbrace{\begin{pmatrix} c & e & f & & \\ e & d & \ddots & \ddots & \\ f & \ddots & \ddots & \ddots & f \\ & \ddots & \ddots & \ddots & e \\ & & f & e & d \end{pmatrix}_k}_{\mathbf{\Psi}_k} \underbrace{\begin{pmatrix} w_0 \\ w_{-L} \\ \vdots \\ \vdots \\ w_{-ZL} \end{pmatrix}_k}_{\mathbf{w}_k} = \underbrace{\begin{pmatrix} 1 \\ x_L \\ \\ \\ \end{pmatrix}_k}_{\mathbf{x}_k^+} \tag{13-40}$$

with elements of vector $\mathbf{x}_k^+$ and matrix $\mathbf{\Psi}_k$:

$$x_L = h_{\nu=0}\, h_{\nu=2L} \quad = \frac{a}{1+a^2}$$

$$c = 1 + x_L^2 + \frac{1}{\gamma_s}, \qquad d = 1 + 2\,x_L^2 + \frac{1}{\gamma_s},$$

$$e = 2\,x_L + x_L\,\frac{1}{\gamma_s}, \qquad f = x_L^2$$

(for convenience, time indices k have been dropped). The $(Z+1)$ nonzero WF coefficients $\mathbf{w}_k$ can now be used to compute the $(Z+2)$ nonzero WMF (cascade of MF and WF) coefficients $\mathbf{u}_k$, $(Z+1)$ noncausal equivalent channel taps $\mathbf{f}_k^-$, two causal taps $\mathbf{f}_k$, and the minimum MSE:

Coefficients of $(T/2)$-spaced whitening matched filter $\mathbf{u}_k$:

$$u_n = \begin{cases} h_{\nu=0}\, w_0 & (n=0) \\ h_{\nu=0}\, w_n + h_{\nu=2L}\, w_{n+2L} & (n = -2L, \dots, -2ZL) \\ h_{\nu=2L}\, w_{-2ZL} & (n = -2[Z+1]L) \\ 0 & (\text{else}) \end{cases}$$

Coefficients of T-spaced equivalent channel $\mathbf{f}_k^-, \mathbf{f}_k$:

$$f_n^- = \begin{cases} w_n + x_L(w_{n-L} + w_{n+L}) & (n = -L, -2L, \dots, -[Z-1]L) \\ w_{-ZL} + x_L\, w_{-[Z-1]L} & (n = -ZL) \\ x_L\, w_{-ZL} & (n = -[Z+1]L) \\ 0 & (\text{else}) \end{cases} \tag{13-41}$$

$$f_n = \begin{cases} w_0 + x_L\, w_{-L} & (n=0) \\ x_L\, w_0 & (n=L) \\ 0 & (\text{else}) \end{cases}$$

Minimum mean-square-error $\zeta_{\min;k}$:

$$\zeta_{\min} = 1 - w_0 - x_L\, w_{-L}$$

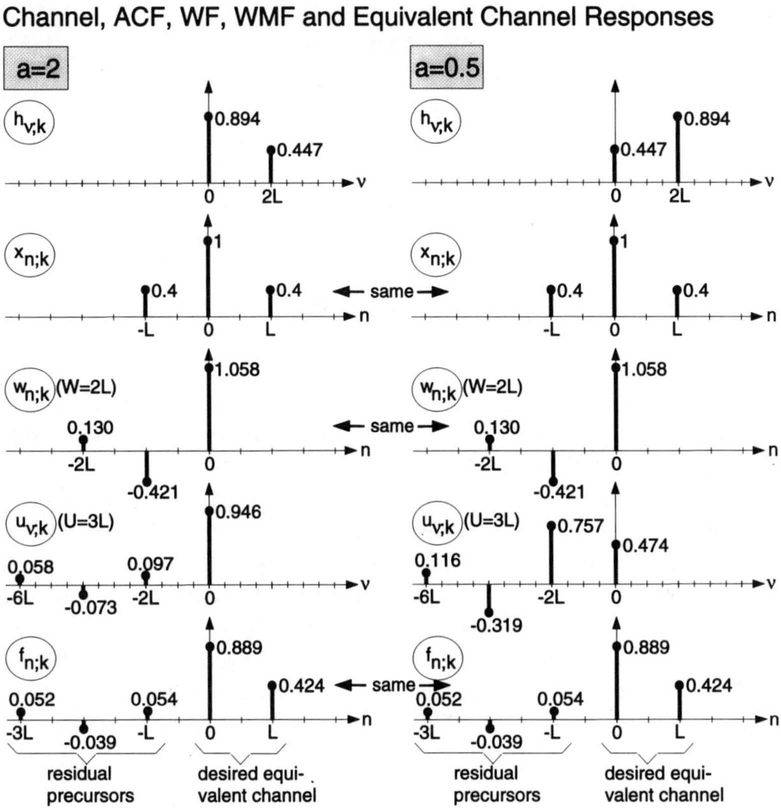

Figure 13-11 Numerical Example of Finite-Length Prefiltering for Parameter a=2 and 0.5

Some interesting conclusions can be drawn from a closer investigation of eqs. (13-40) and (13-41). In Figure 13-11, the coefficients $h_{\nu;k}$ (channel), $x_{n;k}$ (channel acf), $w_{n;k}$ (WF), $u_{\nu;k}$ (WMF), and $f_{n;k}$ (equivalent channel) are shown for the numerical examples of $a = 2$ (first channel tap $h_{\nu=0;k}$ is twice as large as second tap; left side of Figure 13-11) and $a = 0.5$ (second channel tap $h_{\nu=L;k}$ is twice as large as first tap; right side of the figure). In this example, the instantaneous SNR has been chosen as $\gamma_{s;k}$=10 dB, and Z has been set to 2 so that the WF spanning $W = ZL = 2L$ symbol intervals has $Z+1$=3 nonzero T-spaced taps, and the WMF spanning $U = 3L$ symbol intervals has Z+2=4 nonzero $(T/2)$-spaced taps.

First of all, one observes that the channel autocorrelation $\mathbf{x}_k$, WF $\mathbf{w}_k$, and the equivalent channel $\mathbf{f}_k$ are independent of whether a is 2 or 0.5. The same applies to the MMSE $\zeta_{\min}$= –9.56 dB [eq. (13-41)], which is only 0.44 dB worse than with ideal (infinite-length) prefiltering. So it doesn't matter whether the first or second channel tap is larger; prefiltering in both cases yields an equivalent channel $\mathbf{f}_k$ that has the desired properties.

When $a > 1$ (first channel tap $h_{\nu=0;k}$ larger than the second tap $h_{\nu=2L;k}$;

left-hand side of Figure 13-11), the WF attempts to "undo" the effect of the MF, i.e., all noncausal precursors of the WMF $\mathbf{u}_k$ vanish as $W, U \to \infty$ and SNR $\gamma_{s;k} \to \infty$, and the equivalent channel approaches $\mathbf{f}_k \propto \mathbf{h}_k$. This is because the CIR $\mathbf{h}_k = \{h_{\nu=0;k}\ 0\ \ldots\ 0\ h_{\nu=2L;k}\}$ has already the desired properties (minimum phase) so that, if the channel were known to satisfy $a > 1$ at all times, there would be no need for prefiltering at all. Finite-length prefiltering then unnecessarily produces precursor ISI and thus may become counterproductive if W is too small. This adverse effect of prefiltering is worst when both channel taps are equal ($a = 1$).

On the other hand, when $a < 1$ (first channel tap $h_{\nu=0;k}$ smaller than the second tap $h_{\nu=2L;k}$; right-hand side of Figure 13-11), the WMF $\mathbf{u}_k$ approaches an allpass filter response as $W, U \to \infty$ and $\gamma_{s;k} \to \infty$, shifting the phase such that the equivalent channel approaches $\mathbf{f}_k \propto \{h_{\nu=2L;k}\ 0\ \ldots\ 0\ h_{\nu=0;k}\}$. Hence, the stronger second channel tap $h_{\nu=2L;k}$ has essentially been "moved" to the first position $f_{0;k}$ as desired.

In order to explore the MMSE performance of prefiltering, Figure 13-12 displays MMSE curves for the prefilter cascade (MF + finite-length WF) versus the WF length W. Parameters are the SNR per symbol ($\gamma_{s;k} \to \infty$ and $\gamma_{s;k}$ =20 dB) and the ratio $a = (h_{\nu=2L;k}/h_{\nu=0;k})$ between the second and first channel taps. From Figure 13-12, a coarse estimate of the WF filter length necessary for reliable reception can be deduced. Given a certain expected channel ISI profile – thus L and a rough estimate of factor a – as well as a certain SNR threshold (average SNR $\overline{\gamma}_s$) necessary for reliable transmission – this SNR threshold again

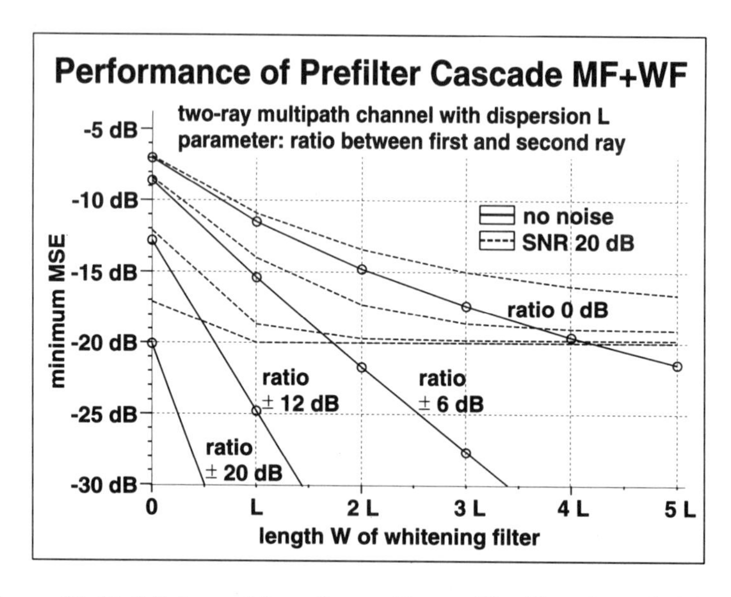

Figure 13-12 Minimum Mean-Square Error of Prefilter Cascade MF+WF

depends on the symbol constellation and the power of channel coding – the WF length should be chosen such that the MMSE resulting from precursor ISI alone remains well below the noise level. If this is accomplished, the symbol estimates $\hat{a}_k$ corresponding to the correct survivor are dominated by additive noise only. If, for example, equal-gain channels occur frequently (a =0 dB) and the (average) SNR is 20 dB, Figure 13-12 reveals that the WF should span no less than 4 or 5 channel lengths. If, on the other hand, the expected ISI is milder, say $a = \mp 6$ dB, the WF needs to span only one (SNR 10 dB) or two (SNR 20 dB) channel lengths. These figures, by the way, have proven to be useful guiderails in the design and implementation of modems for mobile and high frequency (HF) communications [31, 32]. At any rate, the choice of the whitening filter length W is a very typical trade-off between performance and complexity.

The beneficial effect of prefiltering on the BER performance is demonstrated in Figure 13-13 for the GSM-HT channel with Doppler 0.001 and 0.01. In the SNR region of interest (below 20 dB), WF filter lengths of $W = L$ =4 (2- and 4-PSK), $W = 2L$ =8 (8-PSK), and $W = 4L$ =16 (16-PSK) have been found (by simulation) to be sufficient for the mildly selective GSM-HT channel. In the case of severe selectivity ("GSM-2–ray" channel with equal average gains, τ' =2 and L =4), WF filter lengths of $W = 2L$ =8 (2-PSK) and $W = 4L$ =16 (4-, 8-, 16-PSK) are found to be necessary for near-optimal performance. These figures are close to those suggested by the guiderails established above.

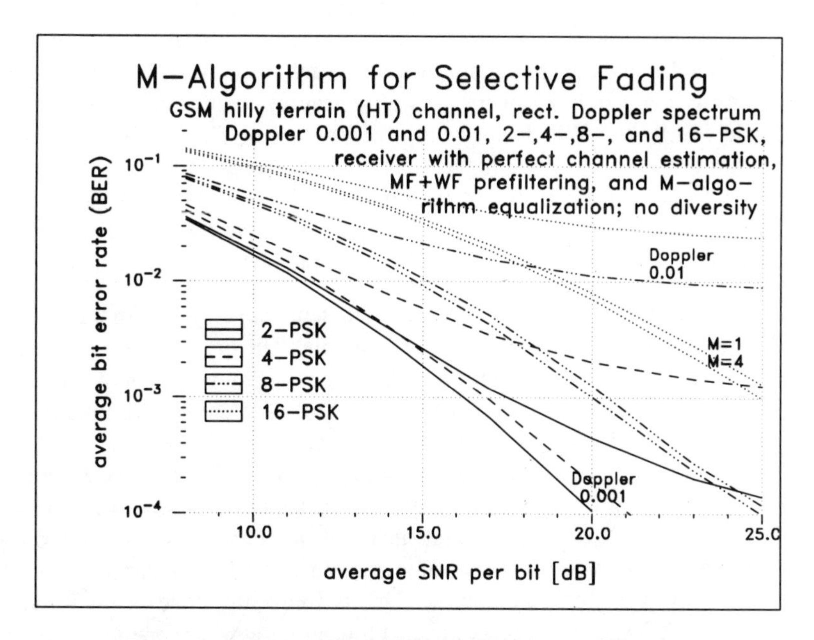

Figure 13-13 Performance of M-Algorithm Equalizer Working on MF+WF Prefilter Output

Comparing the BER results of Figure 13-13 with those of Figure 13-8 above (M algorithm equalizer working directly on the received signal), prefiltering results in the BER curves "converging" much faster with rising M. In fact, the simple DFE (M =1) almost always yields quasi-optimal performance when the channel is only mildly selective. Selecting an M larger than 1 (M =4 in Figure 13-13) leads to a slight improvement only in the case of denser symbol constellations (8- and 16-PSK) and low Doppler. One further observes that the BER results of Figure 13-13 are 2 to 3 dB better than those obtained without prefiltering. As mentioned in Section 13.3.1, this can be attributed to the matched filter attempting to minimize the noise power in its output signal $\mathbf{v}$, while, without prefiltering, all the noise in the received signal $\mathbf{r}$ – including some noise outside of the signal bandwidth – contributes to the distortion in the decision metric. In summary, the use of prefiltering significantly improves on the BER performance. It does necessitate more complex parameter synchronization but greatly reduces the computational burden on the equalizer side. The only drawback of prefiltering is the somewhat increased sensitivity against fading: at Doppler λ'_D =0.01, an irreducible BER level of about 2×10^{-4} (2-PSK) and 2×10^{-3} (4-PSK) is observed from Figure 13-13. Notice, however, that λ'_D =0.01 (assuming $1/T$ =135 kBd) corresponds to a vehicle speed of 1500 km/h (900 MHz) or 750 km/h (1800 MHz), which is much larger than can be tolerated by the GSM system.

Of course, there are many variations on the theme of receiver prefiltering. For example, the MF and WF filters may be merged into a single $(T/2)$-spaced WMF $\mathbf{u}_k$ whose coefficients are computed directly from the channel $\mathbf{h}_k$ via a system of equations similar to eq. (13-34). Thus, a degradation is avoided in case the channel is already minimum phase. In that case, the WMF reduces to a simple memoryless weighting operation (all noncausal coefficients zero), whereas the cascade of MF and WF does so only for $W \to \infty$. This modest advantage, however, has to be paid for by a higher computational effort. Given a certain span W (WF) or $U = W+L$ (WMF) of the prefilter, $2U+1 = 2(W+L)+1$ $(T/2)$-spaced WMF coefficients must be adjusted, a figure more than twice as large as the number $W+1$ of T-spaced WF coefficients computed via eq. (13-34).

Since the computation of prefiltering coefficients via a system of equations is costly, reduced-complexity algorithms are desirable. The per-symbol effort can be drastically reduced by performing the mapping $\mathbf{h}_k \longrightarrow (\mathbf{m}, \mathbf{w}, \mathbf{f})_k$ only once every I symbol intervals. The resulting parameter sets $(\mathbf{m}, \mathbf{w}, \mathbf{f})_K$ valid for time instants $k = KI,\ K = 0, 1, 2, \ldots$ may then be interpolated linearly so as to obtain the parameters valid for intermediate time indices k [33]. The parameter I naturally depends on the fading dynamics and the frame structure of symbol transmission. If the fading is very slow so that the channel does not change significantly during I symbol intervals, one may use the same parameter set for the entire block of I symbols until the next set is computed. This is frequently done in time-division multiple access (TDMA)-like burst or packet transmission where one set of receiver parameters is computed from a pre- or midamble and used for the entire packet.

Recalling the fact that we are primarily interested in large "first" coefficients of the equivalent channel, in particular $f_{0;k}$, another interesting strategy for reduced-complexity prefiltering consists in the partitioning of the instantaneous CIR $\mathbf{h}_k$ into two parts, viz. unwanted precursors or "early echos" $\mathbf{h}_{E;k}$ preceding a strong "center tap", and postcursors or "late echos" $\mathbf{h}_{L;k}$ following the center tap. Let L_E and L_L denote the precursor and postcursor lengths, respectively, with $L_E + L_L = L$ the total CIR length. Then the prefilter can be adjusted so as to cancel the precursor ISI $\mathbf{h}_{E;k}$ only; the corresponding mapping $\mathbf{h}_{E;k} \longrightarrow \mathbf{w}_{E;k}$ is of reduced complexity thanks to the smaller filter length W_E (depending on L_E instead of L) of $\mathbf{w}_{E;k}$ and thus the smaller dimension of the system of equations to be solved. However, the position of the center tap needs to be continually monitored, and the noise in the resulting equivalent channel is colored since $\mathbf{w}_{E;k}$ is no true whitening filter.

13.3.6 Main Points

- Exact and simplified expressions for the metric and metric increment for synchronized detection on selective fading channels have been derived, again by means of the recursive formulation of the decision rule. The near-optimal metric increment for ML sequence detection (MLSD) based on the received signal $\mathbf{r}$ is given by

$$\Delta m_{D;k}(\mathbf{a}_k) \simeq \sum_{i=0}^{1} \left| r_k^{(i)} - \sum_{l=0}^{L} \hat{h}_{l;k}^{(i)} \, a_{k-l} \right|^2 \tag{13-42}$$

  [eq. (13-25)].
- The effort associated with an exhaustive search over all Q^N candidate sequences can be reduced by applying the Viterbi algorithm (VA) to ML sequence detection. The VA searches a trellis with Q^L equalizer states in a recursive manner.
- Reduced-complexity ML sequence detection by means of searching a subset of $M \ll Q^L$ equalizer states (M algorithm) often performs poorly on fading channels when the metric increment [eq. (13-42)] is based on the received signal $\mathbf{r}$. An effective remedy is the use of preprocessing the received signal and performing ML detection on the prefilter output v_k:

$$\Delta m_{D;k}(\mathbf{a}_k) = \left| v_k - \left(\sum_{l=0}^{L} f_{l;k} \, a_{k-l} \right) \right|^2 \tag{13-43}$$

  [eq. (13-29)], where $\mathbf{f}_k$ is the equivalent transmission system seen by the ML detector, i.e., the cascade of channel $\mathbf{h}_k$, prefilter $\mathbf{u}_k$, and decimator to symbol rate.
- The optimal preprocessing device prior to the reduced-complexity MLSD (and most other types of equalizers) must transform any channel response $\mathbf{h}_k$ into

an equivalent channel response $\mathbf{f}_k$ which is causal, of finite length L, and minimum-phase so that the first coefficients $\{f_{0;k}, f_{1;k}, \ldots\}$ are as large as possible. All of this is accomplished by the ideal whitening matched filter (WMF) with transfer function $U_k(z)$ being the cascade of the matched filter (MF) $M_k(z)$, decimator, and the whitening filter (WF) whose transfer function $W_k(z) = 1/F_k^*(1/z^*)$ is given in terms of a spectral factorization.

- The optimal finite-length approximation $\mathbf{w}_k$ to the ideal WF – the WMF $\mathbf{u}_k$ is again given by the cascade of MF $\mathbf{m}_k$ [eq. (13-30)], decimator, and WF $\mathbf{w}_k$ – is identical to the feedforward filter of a decision-feedback equalizer (DFE) optimized according to the MMSE criterion. The causal portion of the equivalent channel $\mathbf{f}_k$ is identical to the feedback filter of the DFE. The coefficients of both the WF $\mathbf{w}_k$ and the equivalent channel $\mathbf{f}_k$ can be determined from the channel response $\mathbf{h}_k$ and the SNR $\gamma_{s;k}$ through a systems of equations [eqs. (13-32) and (13-34)].

13.4 Spread Spectrum Communication

In digital radio communications, spread-sectrum (SS) techniques, in particular, code division multiple access (CDMA), is becoming increasingly important. This section describes how SS communication relates to the baseband model discussed so far. We assume that the reader is familiar with the principles of CDMA.

13.4.1 Modulator and Demodulator Structure

The basic building block of a spread spectrum QPSK modulator is shown in Figure 13-14.

The same binary input symbol a_n is modulated by two PN (pseudo-noise) sequences $d_\nu^{(I)}$ (in-phase) and $d_\nu^{(Q)}$ (quadrature-phase), respectively. The spreaded signal is subsequently modulated onto a sinusoidal carrier. The reason for using both quadrature channels is to avoid any dependence on the carrier phase, see Chapter 2.3.2 of [34]. Notice that a BPSK modulator uses only one of the two signals in Figure 13-14. Introducing the complex PN sequence

$$d_\nu = d_\nu^{(I)} + j d_\nu^{(Q)} \tag{13-44}$$

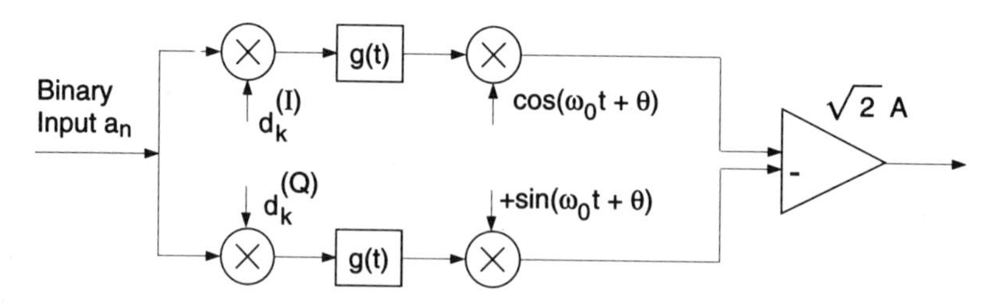

Figure 13-14 Spread-Spectrum QPSK Modulator

with independent PN sequences $d_\nu^{(I)}$ and $d_\nu^{(Q)}$, the transmitted signal can be written as

$$s(t) = \sqrt{2}A \operatorname{Re}\left\{ \sum_n a_n \sum_{\nu=0}^{N_c-1} d_\nu \, g(t-nT-\nu T_c-\varepsilon_c T_c)\, e^{j(\omega_0 t+\theta)} \right\} \tag{13-45}$$

with T_c the chip duration, ε_c a fraction of T_c, T the symbol duration, and N_c the number of chips per symbol. There are two extreme cases for the pulse shaping filter, viz. (i) the pulse $g(t)$ is a constant rectangular pulse of duration T_c, and (ii) $G(\omega)$ is strictly band-limited to $B = 1/2\ T_c$. In both cases successive chip outputs are uncorrelated.

The receiver performs the corresponding operations shown in Figure 13-15. These operations include downconversion, chip matched filtering with the conjugate complex of the transmit filter $g(t)$, and sampling at the chip intervals. Subsequently, the signal is despreaded. This is accomplished by correlating the received samples with a synchronized replica of the PN code. The result is a baseband signal. It can be readily verified that the squared magnitude P_n is independent of the carrier phase. It is used for noncoherent detection and, as will be shown shortly, also for the synchronization of the PN sequence.

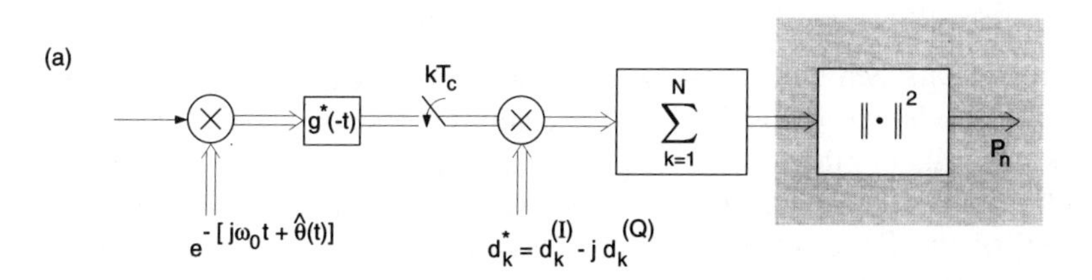

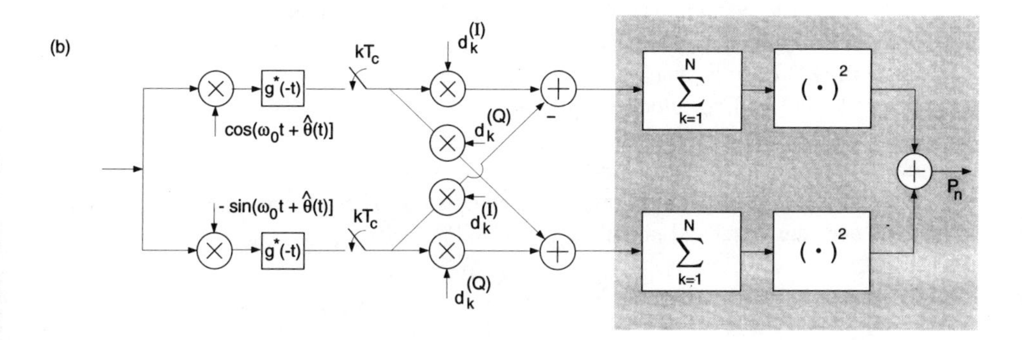

Figure 13-15 Spread Spectrum Demodulator
(a) Complex Representation, (b) Real and Imaginary Representation

In a multipath environment, the correlator output is weighted by the magnitude of the (cluster of) multipath ray(s) whose delays are in alignment with the replica of the PN code. If the chip duration T_c is much shorter than the relative delays between multipaths, then (by virtue of the cross correlation properties of PN sequences) these multipath rays can, in principle, be resolved by using a number $D > 1$ of cross correlators per symbol, each matched (local replica of the PN sequence synchronized) to the particular multipath delay. By virtue of the uncorrelated-scattering assumption on multipath propagation, the D correlator output signals so obtained can be viewed as having undergone independently fading "channels" which, given that the symbol duration is much longer than the largest multipath delay, behave like flat fading diversity channels. The D correlator output signals may then be combined. Such a scheme (sometimes termed RAKE receiver) is equivalent to the inner transmission model for flat fading channels with Dth-order diversity as shown in Figure 13-6. The diversity branches, however, are now unequally weighted according to the average powers of the individual multipath rays or ray clusters. Given that only a few of these resolvable multipaths are of significant strength, most of the multipath diversity gain inherent in selective channels can be recovered. Hence, a spread-spectrum system in principle faces the same performance limit (viz., matched filter bound) as a serial transmission system occupying the same bandwidth.

13.4.2 Synchronization of Pseudorandom Signals

The process of synchronizing the locally generated replica(s) of the transmitted PN sequence ("code synchronization") is accomplished in two steps. The first step consists in bringing the local spreading signal into coarse alignment with the (delayed) transmitted code sequence. In principle, this requires a search over the entire period of the PN sequence. The ML estimator can readily be shown to be a device which correlates the incoming samples with the local replica $d_{\nu+M}$ and searches for the delay M which maximizes the magnitude-squared correlator output P_n (see Figure 13-15b). Once the signal is acquired within $\pm T_c/2$, the fine alignment and tracking of the fractional delay ε_c is accomplished by means of an error feedback system. The error signal is obtained by computing an early (late) version delayed by $T_c/2$ $(-T_c/2)$ of the variable P_n, see the following example. Once the PN code is synchronized, symbol timing recovery is completed since the PN code and symbol clock are locked.

Example: Time Tracking Loop

Assume that the acquisition process is completed. Then, for an unmodulated pilot sequence, the baseband signal of eq. (13-45) reads

$$U(t) = \sum_{\nu} d_\nu(\varepsilon_c)\, g(t-\nu T_c-\varepsilon_c T_c)\, e^{j\theta} \tag{13-46}$$

Following exactly the steps of Section 5.3 we leave it as an exercise to the reader to verify that the low SNR approximation of the log-likelihood function for (θ, ε_c) is independent of θ and given by

$$L(\varepsilon_c) = \sum_{\nu} \underbrace{|d_\nu(\varepsilon_c)|^2}_{1} \; |z(\nu T_c + \varepsilon_c T_c)|^2 \tag{13-47}$$

where $z(t)$ is the output of the analog chip matched filter $g^*(-t)$ taken at $\nu T_c = \varepsilon_c T_c$. The error signal is obtained by computing $L_+(\varepsilon_c)$ and $L_-(\varepsilon_c)$, respectively:

$$e(e_c) = L_+(\varepsilon_c) - L_-(\varepsilon_c) \tag{13-48}$$

where

$$L_\pm(\varepsilon_c) = \sum_{\nu} |z(\nu T_c \pm T_c/2 + \varepsilon_c T_c)|^2 \tag{13-49}$$

The error feedback system is called a delay-locked loop (DLL) in the literature.

A crucial issue in cellular multiuser CDMA communications is to ensure that the power at the base station received from each user (uplink) is nearly equal. This requires elaborate power control algorithms which are discussed in Chapter 4.7 of [34].

13.5 Bibliographical Notes

Material on the diversity-like effect of coding and soft decision decoding is found in [35, 36] and [37, 3, 4, 38]. Methods of combined equalization and decoding based on coordinate interleaving are presented in [19, 13, 12, 39]. References [10, 14, 40, 41, 42] give details on MAP equalization. Publications on the Viterbi algorithm and its applications include [23] on Bellman's principle of optimality, [15, 40, 43] on soft-output Viterbi equalization, and [16, 44] on the generalized Viterbi algorithm. Details on decision-feedback equalization and the M algorithm are found in [45, 46, 20, 18] and [24, 25, 26, 27], respectively.

Optimal receiver structures for ISI channels are discussed in Forney's papers [3, 10] and a number of standard textbooks, e.g., [20, 5]. Methods of adjusting receiver components, founded in estimation and adaptive filtering theory [29, 8], are detailed in [30, 5]. Applications include modem design for mobile and HF communications [33, 31, 32], and many more.

The initial application of spread spectrum was in military and guidance systems. The books by Simon, Omura, Scholtz, and Levitt [47] and Holmes [48] provide a comprehensive discussion of the subject with many references to the pioneers of this technology. Spread spectrum multiple access communication (CDMA) has a number of unique properties which make it attractive for commercial applications. A detailed discussion of CDMA is found in the highly recommended book by A. J. Viterbi [34]; in particular, synchronization is treated in Chapter 3.

Bibliography

[1] A. J. Viterbi, "Wireless Digital Communication: A View Based on Three Lessons Learned," *IEEE Communic. Magaz.*, pp. 33–36, Sept. 1991.

[2] H. Meyr and R. Subramanian, "Advanced Digital Receiver Principles and Technologies for PCS," *IEEE Communic. Magaz.*, pp. 68–78, Jan. 1995.

[3] G. D. Forney, "Maximum-Likelihood Sequence Estimation of Digital Sequences in the Presence of Intersymbol Interference," *IEEE Trans. Inform. Theory*, vol. 18, no. 3, pp. 363–378, May 1972.

[4] J. Hagenauer and E. Lutz, "Forward Error Correction Coding for Fading Compensation in Mobile Satellite Channels," *IEEE J. Sel. Areas Commun.*, vol. 5, no. 2, pp. 215–225, Feb. 1987.

[5] J. G. Proakis, *Digital Communications*. New York: McGraw-Hill, 1989.

[6] D. Divsalar and M. K. Simon, "The Design of Trellis Coded MPSK for Fading Channels: Performance Criteria," *IEEE Trans. Commun.*, vol. 36, no. 9, pp. 1004–1011, Sept. 1988.

[7] R. Häb and H. Meyr, "A Systematic Approach to Carrier Recovery and Detection of Digitally Phase-Modulated Signals on Fading Channels," *IEEE Trans. Commun.*, vol. 37, no. 7, pp. 748–754, July 1989.

[8] S. Haykin, *Adaptive Filter Theory*. Englewood Cliffs, NJ: Prentice-Hall, 1986.

[9] S. U. H. Qureshi, "Adaptive Equalization," *IEEE Proc.*, vol. 73, pp. 1349–1387, Sept. 1985.

[10] G. D. Forney, "The Viterbi Algorithm," *IEEE Proc.*, vol. 61, no. 3, pp. 268–278, Mar. 1973.

[11] E. A. Lee and D. G. Messerschmitt, *Digital Communication*. Boston: Kluwer Academic, 1994.

[12] R. Mehlan and H. Meyr, "Combined Equalization/Decoding of Trellis Coded Modulation on Frequency Selective Fading Channels," in *Coded Modulation and Bandwidth-Efficient Transmission*, (Tirrenia, Italy), pp. 341–352, Sept. 1991.

[13] S. A. Fechtel and H. Meyr, "Combined Equalization, Decoding and Antenna Diversity Combining for Mobile/Personal Digital Radio Transmission Using Feedforward Synchronization," in *Proc. VTC*, (Secaucus, NJ), pp. 633–636, May 1993.

[14] L. R. Bahl *et al.*, "Optimal Decoding of Linear Codes for Minimizing Symbol Error Rate," *IEEE Trans. Inform. Theory*, vol. 20, pp. 284–287, Mar. 1974.

[15] J. Hagenauer and P. Höher, "A Viterbi Algorithm with Soft-Decision Outputs and its Applications," in *Proc. Globecom*, (Dallas, TX), pp. 1680–1686, Nov. 1989.

[16] T. Hashimoto, "A List-Type Reduced-Constraint Generalization of the Viterbi Algorithm," *IEEE Trans. Inform. Theory*, vol. 33, no. 6, pp. 866–876, Nov. 1987.

[17] H. K. Thapar, "Real-Time Application of Trellis Coding to High-Speed Voiceband Data Transmission," *IEEE J. Sel. Areas Commun.*, vol. 2, no. 5, pp. 648–658, Sept. 1984.

[18] J. Salz, "Optimum Mean-square Decision Feedback Equalization," *Bell Syst. Tech. J.*, vol. 52, pp. 1341–1373, Oct. 1973.

[19] M. V. Eyuboglu, "Detection of Coded Modulation Signals on Linear, Severely Distorted Channels Using Decision Feedback Noise Prediction with Interleaving," *IEEE Trans. Commun.*, vol. 36, no. 4, pp. 401–409, Apr. 1988.

[20] G. Lee and M. P. Fitz, "Bayesian Techniques for Equalization of Rapidly Fading Frequency Selective Channels," in *Proc. VTC*, (Stockholm, Sweden), pp. 287–291, May 1994.

[21] H. Dawid and H. Meyr, "Real-Time Algorithms and VLSI Architectures for Soft Output MAP Convolutional Decoding," in *Proc. PIMRC*, (Toronto, Canada), Sept. 1995.

[22] P. Robertson, E. Villebrun, and P. Höher, "A Comparison of Optimal and Sub-Optimal MAP Decoding Algorithms Operating in the Log Domain," in *Proc. ICC*, (Seattle, WA), June 1995.

[23] R. E. Bellman and S. E. Dreyfus, *Applied Dynamic Programming*. Princeton, NJ: Princeton University Press, 1962.

[24] J. B. Anderson and S. Mohan, "Sequential Coding Algorithms: A Survey and Cost Analysis," *IEEE Trans. Commun.*, vol. 32, no. 2, pp. 169–176, Feb. 1984.

[25] A. Baier and G. Heinrich, "Performance of M-Algorithm MLSE Equalizers in Frequency-Selective Fading Mobile Radio Channels," in *Proc. ICC*, (Boston, MA), pp. 281–284, 1989.

[26] A. P. Clark and S. M. Asghar, "Detection of Digital Signals Transmitted over a Known Time-Varying Channel," *IEE Proc.*, vol. 128, no. 3, pp. 167–174, June 1981.

[27] W. Sauer and W. Rupprecht, "Ein aufwandsgünstiges suboptimales Detektionsverfahren für stark verzerrte und gestörte Datensignale," *ntz Archiv*, vol. 9, pp. 155–169, 1987.

[28] J. M. Cioffi, P. H. Algoet, and P. S. Chow, "Combined Equalization and Coding with Finite-Length Decision Feedback Equalization," in *Proc. Globecom*, (San Diego, CA), pp. 1664–1668, 1990.

[29] B. D. O. Anderson and J. B. Moore, *Optimal Filtering*. Englewood Cliffs, NJ: Prentice-Hall, 1979.

[30] S. A. Fechtel, *Verfahren und Algorithmen der robusten Synchronisation für die Datenübertragung über dispersive Schwundkanäle (in German)*. Wissenschaftsverlag Mainz, Aachen, 1993. Ph.D. Thesis, Aachen Univ. Tech. (RWTH).

[31] S. A. Fechtel and H. Meyr, "A New Mobile Digital Radio Transceiver Concept

Using Low-Complexity Combined Equalization / Trellis Decoding and a Near-Optimal Receiver Sync Strategy," in *Proc. PIMRC*, (Boston, MA), pp. 382–386, Oct. 1992.

[32] S. A. Fechtel and H. Meyr, "Parallel Block Detection, Synchronization and Hypothesis Decision in Digital Receivers for Communications over Frequency-Selective Fading Radio Channels," in *Proc. Globecom/CTMC*, (San Francisco, CA), pp. 95–99, Dec. 1994.

[33] S. A. Fechtel and H. Meyr, "Optimal Parametric Feedforward Estimation of Frequency-Selective Fading Radio Channels," *IEEE Trans. Commun.*, vol. 42, no. 2/3/4, pp. 1639–1650, Feb./Mar./Apr. 1994.

[34] A. J. Viterbi, *CDMA. Principles of Spread Spectrum Communication.* New York: Addison-Wesley, 1995.

[35] S. A. Fechtel and H. Meyr, "Matched Filter Bound for Trellis-Coded Transmission over Frequency-Selective Fading Channels with Diversity," *European Trans. Telecomm.*, vol. 4, no. 3, pp. 109–120, May-June 1993.

[36] M. L. Moher and J. H. Lodge, "TCMP: A Modulation and Coding Strategy for Rician Fading Channels," *IEEE J. Sel. Areas Commun.*, vol. 7, no. 9, pp. 1347–1355, Dec. 1989.

[37] S. A. Fechtel and H. Meyr, "An Investigation of Near-Optimal Receiver Structures Using the M-Algorithm for Equalization of Dispersive Fading Channels," in *Proc. EUSIPCO*, (Brussels, Belgium), pp. 1603–1606, Aug. 1992.

[38] G. Ungerboeck, G., "Adaptive Maximum-Likelihood Receiver for Carrier-Modulated Data Transmission Systems," *IEEE Commun. Magazine*, vol. 22, no. 5, pp. 624–636, May 1974.

[39] K. Zhou, J. G. Proakis, and F. Ling, "Decision-Feedback Equalization of Time-dispersive Channels with Coded Modulation," *IEEE Trans. Commun.*, vol. 38, no. 1, pp. 18–24, Jan. 1990.

[40] P. Höher, "Detection of Uncoded and Trellis-coded PSK-Signals on Frequency-Selective Fading Channels," in *Proc. ITG Fachbericht 107*, (Nürnberg, Germany, VDE Verlag), pp. 225–232, Apr. 1989.

[41] P. Höher, *Kohärenter Empfang trelliscodierter PSK-Signale auf frequenzselektiven Mobilfunkkanälen – Entzerrung, Decodierung und Kanalparameterschätzung (in German).* PhD thesis, Kaiserslautern University of Technology, 1990.

[42] W. Koch, "Combined Design of Equalizer and Channel Decoder for Digital Mobile Radio Receivers," in *Proc. ITG Fachbericht 107*, (Nürnberg, Germany, VDE Verlag), pp. 263–270, Apr. 1989.

[43] P. Höher, "TCM on Frequency-Selective Fading Channels: A Comparison of Soft-Output Probabilistic Equalizers," in *Proc. Globecom*, (San Diego, CA), pp. 376–381, Dec. 1990.

[44] N. Seshadri and C. E. W. Sundberg, "Generalized Viterbi Algorithms for Error Detection with Convolutional Codes," in *Proc. Globecom*, (Dallas, TX), pp. 1534–1537, Nov. 1989.

[45] C. A. Belfiore and J. H. Park, "Decision Feedback Equalization," *IEEE Proc.*, vol. 67, pp. 1143–1156, Aug. 1979.

[46] J. W. M. Bergmans, S. A. Rajput, and F. A. M. van de Laar, "On the Use of Decision Feedback for Simplifying the Viterbi Detector," *Philips J. Res.*, vol. 42, no. 4, pp. 399–428, 1987.

[47] M. K. Simon, J. K. Omura, R. A. Scholtz, and B. K. Levitt, *Spread Spectrum Communications*. Rockville, MA: Computer Science Press, 1985.

[48] J. K. Holmes, *Coherent Spread Spectrum Systems*. Malabar, FL: Robert E. Krieger Publishing Company, 1990.

Chapter 14 Parameter Synchronization for Flat Fading Channels

In Chapters 14 and 15, important aspects of digital receiver synchronization for fading channels are discussed. The two sections of Chapter 14 concentrate on linear sync parameter estimation for the flat fading case, i.e., flat fading channel estimation. The derivation of many of the algorithms presented here draws from the ideas and results of Chapter 12. However, while the optimal algorithms of Chapter 12 are, in most situations of interest, far too complex to realize, we shall now turn our attention to reduced-complexity yet close-to-optimal synchronizer structures and algorithms that can actually be implemented using today's DSP or ASIC technology.

14.1 Non-Data-Aided (NDA) Flat Fading Channel Estimation and Detection

In this section, data detection on flat fading channels without the aid of known training symbols is investigated in more detail. Following again the concept of synchronized detection (Chapter 12), we consider estimation-detection type of receivers with "online" one-step channel prediction (Chapter 13) which are simplified in a systematic way so as to arrive at realizable receivers. In the next two sections, optimal and near-optimal methods for NDA one-step channel prediction are discussed in detail. Finally, suboptimal but simple decision-directed (DD) channel estimation and symbol detection is investigated.

14.1.1 Optimal One-Step Channel Prediction

In Section 13.2, the one-step predictor estimate $\hat{c}_{k|k-1}$ of the flat fading channel gain c_k has been identified as the sync parameter necessary for metric computation [eq. (13-12)] in NDA synchronized detection. The optimal channel predictor estimate is given by the conditional expected value

$$\hat{c}_{k|k-1} = E[c_k|\mathbf{z}_{k-1}, \mathbf{a}_{k-1}] \tag{14-1}$$

given the "past" observation $\mathbf{z}_{k-1}$ and symbol sequence $\mathbf{a}_{k-1}$ [1]. The transmission model for $\mathbf{z}_{k-1}$ is given by the vector/matrix equation (12-48) which is now truncated at index $k-1$, i.e., $\mathbf{z}_{k-1} = \mathbf{A}_{k-1} \cdot \mathbf{c}_{k-1} + \mathbf{m}_{k-1}$. As in Section 12.2.2, the desired quantity c_k and the observation $\mathbf{z}_{k-1}$ are understood to be jointly Gaussian dynamic parameters. Furthermore, as motivated by the discussion in Section 12.2.3, c_k can, for the purpose of channel estimation, safely be assumed to be zero-mean even if a LOS path is present. Then the optimal channel predictor and its MMSE are given by

$$\begin{aligned}\hat{c}_{k|k-1} &= \mu_{c_k|\mathbf{z}_{k-1}} = \mathbf{\Sigma}_{c_k,\mathbf{z}_{k-1}} \cdot \mathbf{\Sigma}_{\mathbf{z}_{k-1}}^{-1} \cdot \mathbf{z}_{k-1} \\ \sigma^2_{c;k|k-1} &= \mathbf{\Sigma}_{c_k|\mathbf{z}_{k-1}} = \mathbf{\Sigma}_{c_k} - \mathbf{\Sigma}_{c_k,\mathbf{z}_{k-1}} \cdot \mathbf{\Sigma}_{\mathbf{z}_{k-1}}^{-1} \cdot \mathbf{\Sigma}_{c_k,\mathbf{z}_{k-1}}^{H}\end{aligned} \tag{14-2}$$

[see [1] and eqs. (12-28) and (12-30)]. Similar to eq. (12-29), the matrices $\mathbf{\Sigma}_{c_k}$ (scalar), $\mathbf{\Sigma}_{c_k,\mathbf{z}_{k-1}}$ $(1 \times k)$, and $\mathbf{\Sigma}_{\mathbf{z}_{k-1}}$ $(k \times k)$ evaluate as

$$\begin{aligned}\mathbf{\Sigma}_{c_k} &= E\left[|c_k|^2\right] &&= 1 \\ \mathbf{\Sigma}_{c_k,\mathbf{z}_{k-1}} &= E\left[c_k \cdot \mathbf{z}_{k-1}^H\right] &&= \mathbf{r}_{D;k-1}^H \, \mathbf{A}_{k-1}^H \\ \mathbf{\Sigma}_{\mathbf{z}_{k-1}} &= E\left[\mathbf{z}_{k-1} \cdot \mathbf{z}_{k-1}^H\right] &&= \mathbf{A}_{k-1} \, \mathbf{R}_{D;k-1} \, \mathbf{A}_{k-1}^H + \mathbf{R}_{m;k-1}\end{aligned} \tag{14-3}$$

where $\mathbf{R}_{D;k-1}$ denotes the truncated channel autocorrelation matrix of eq. (12-50) [dynamic part only, i.e., $\alpha_d(m) = \alpha(m)$ and $\rho_d = 1$], $\mathbf{r}_{D;k-1} = (\alpha(k) \, \ldots \, \alpha(2) \; \alpha(1))^T$ the vector of channel autocorrelation samples, and $\mathbf{R}_{m;k-1}$ the truncated noise covariance matrix. Inserting $\mathbf{\Sigma}_{c_k}$, $\mathbf{\Sigma}_{c_k,\mathbf{z}_{k-1}}$, and $\mathbf{\Sigma}_{\mathbf{z}_{k-1}}$ into eq. (14-2) yields

$$\begin{aligned}\hat{c}_{k|k-1} &= \mathbf{r}_{D;k-1}^H \mathbf{A}_{k-1}^H \cdot \left(\mathbf{A}_{k-1}\mathbf{R}_{D;k-1}\mathbf{A}_{k-1}^H + \mathbf{R}_{m;k-1}\right)^{-1} \cdot \mathbf{z}_{k-1} \\ \sigma^2_{c;k|k-1} &= 1 - \mathbf{r}_{D;k-1}^H \mathbf{A}_{k-1}^H \cdot \left(\mathbf{A}_{k-1}\mathbf{R}_{D;k-1}\mathbf{A}_{k-1}^H + \mathbf{R}_{m;k-1}\right)^{-1} \cdot \mathbf{A}_{k-1}\mathbf{r}_{D;k-1}\end{aligned} \tag{14-4}$$

which is of the same form as eq. (12-31) except for the zero mean. By invoking the matrix inversion lemma eq. (12-39) twice (similarly as in Section 12.2.6), eq. (14-4) can be reformulated to

$$\begin{aligned}\hat{c}_{k|k-1} &= \mathbf{r}_{D;k-1}^H \cdot \left(\mathbf{R}_{D;k-1} + \left[\mathbf{A}_{k-1}^H \mathbf{R}_{m;k-1}^{-1} \mathbf{A}_{k-1}\right]^{-1}\right)^{-1} \\ &\quad \cdot \left(\mathbf{A}_{k-1}^H \mathbf{R}_{m;k-1}^{-1} \mathbf{A}_{k-1}\right)^{-1} \cdot \mathbf{A}_{k-1}^H \mathbf{R}_{m;k-1}^{-1} \cdot \mathbf{z}_{k-1} \\ \sigma^2_{c;k|k-1} &= 1 - \mathbf{r}_{D;k-1}^H \cdot \left(\mathbf{R}_{D;k-1} + \left[\mathbf{A}_{k-1}^H \mathbf{R}_{m;k-1}^{-1} \mathbf{A}_{k-1}\right]^{-1}\right)^{-1} \cdot \mathbf{r}_{D;k-1}\end{aligned} \tag{14-5}$$

The term $\left(\mathbf{A}_{k-1}^H \mathbf{R}_{m;k-1}^{-1} \mathbf{A}_{k-1}\right)^{-1} \cdot \mathbf{A}_{k-1}^H \mathbf{R}_{m;k-1}^{-1} \cdot \mathbf{z}_{k-1}$ has the same form as eq. (12-27) and is therefore identified as the *ML estimate* $\hat{\mathbf{c}}_{S;k-1}(\mathbf{a}_{k-1})$ of the past channel trajectory up to time $k-1$. Hence, optimal prediction is again separable into ML estimation from the observation $\mathbf{z}_{k-1}$ and the data hypothesis $\mathbf{a}_{k-1}$, followed by

MAP prediction from the ML estimate, making use of the channel parameters:

ML channel estimation:

$$\hat{\mathbf{c}}_{S;k-1}(\mathbf{a}_{k-1}) = \left(\mathbf{A}_{k-1}^H \mathbf{R}_{m;k-1}^{-1} \mathbf{A}_{k-1}\right)^{-1} \mathbf{A}_{k-1}^H \mathbf{R}_{m;k-1}^{-1} \cdot \mathbf{z}_{k-1}$$

$$\mathbf{\Sigma}_{S;k-1}(\mathbf{a}_{k-1}) = \left(\mathbf{A}_{k-1}^H \mathbf{R}_{m;k-1}^{-1} \mathbf{A}_{k-1}\right)^{-1}$$

MAP channel prediction from ML estimate: (14-6)

$$\hat{c}_{k|k-1} = \mathbf{r}_{D;k-1}^H \left(\mathbf{R}_{D;k-1} + \left[\mathbf{A}_{k-1}^H \mathbf{R}_{m;k-1}^{-1} \mathbf{A}_{k-1}\right]^{-1}\right)^{-1} \cdot \hat{\mathbf{c}}_{S;k-1}(\mathbf{a}_{k-1})$$

$$\sigma_{c;k|k-1}^2 = 1 - \mathbf{r}_{D;k-1}^H \cdot \left(\mathbf{R}_{D;k-1} + \left[\mathbf{A}_{k-1}^H \mathbf{R}_{m;k-1}^{-1} \mathbf{A}_{k-1}\right]^{-1}\right)^{-1} \cdot \mathbf{r}_{D;k-1}$$

If the noise is AWGN with power N_0, eq. (14-6) can be further simplified. Since the MAP estimator and its error covariance are not much dependent on the noise power (Figures 12-2 and 12-3), the term $\left[\mathbf{A}_{k-1}^H \mathbf{R}_{m;k-1}^{-1} \mathbf{A}_{k-1}\right]^{-1} = N_0 \cdot \mathbf{P}_{k-1}^{-1}(\mathbf{a}_{k-1})$ may be safely simplified to $N_0\mathbf{I}$ even if there are amplitude variations; i.e., for the purpose of channel prediction from the ML estimate, p_k is set to 1 as if amplitude variations in M-QAM symbols were not present. Then eq. (14-6) boils down to [see also eq. (12-53)]:

ML channel estimation:

$$\hat{\mathbf{c}}_{S;k-1}(\mathbf{a}_{k-1}) = \mathbf{P}_{k-1}^{-1}(\mathbf{a}_{k-1}) \cdot \mathbf{A}_{k-1}^H \cdot \mathbf{z}_{k-1}$$

$$\mathbf{\Sigma}_{S;k-1}(\mathbf{a}_{k-1}) = N_0 \cdot \mathbf{P}_{k-1}^{-1}(\mathbf{a}_{k-1})$$

MAP channel prediction from ML estimate: (14-7)

$$\hat{c}_{k|k-1} = \underbrace{\mathbf{r}_{D;k-1}^H (\mathbf{R}_{D;k-1} + N_0 \mathbf{I})^{-1}}_{\mathbf{w}_{k-1}^H} \cdot \hat{\mathbf{c}}_{S;k-1}(\mathbf{a}_{k-1})$$

$$\sigma_{c;k|k-1}^2 = 1 - \mathbf{r}_{D;k-1}^H \cdot (\mathbf{R}_{D;k-1} + N_0 \mathbf{I})^{-1} \cdot \mathbf{r}_{D;k-1}$$

MAP channel prediction from $\hat{\mathbf{c}}_{S;k-1}(\mathbf{a}_{k-1})$ and the (nominal) prediction error covariance $\sigma_{c;k|k-1}^2$ remain to be dependent on the channel parameters (λ'_D, N_0), but have become entirely independent of the data $\mathbf{a}_{k-1}$. Thus, a single set of real-valued Wiener predictor weights $\mathbf{w}_{k-1}$ can be precomputed and used for all data hypotheses $\mathbf{a}_{k-1}$.

14.1.2 Reduced-Complexity One-Step Channel Prediction

Up to this point, the optimal decision metric can be computed recursively [eq. (13-9)], but optimal channel prediction still needs to be performed nonrecursively by means of finite-impulse-response (FIR)-type filters $\mathbf{w}_{k-1}$ [eq. (14-7)]. These filters are causal – as opposed to the smoothing filters of Chapter 12 – but dependent on the time index k, regarding both the filter length ($=k-1$) and the tap weights.

In theory, $N-1$ sets of predictor filters $\mathbf{w}_0, \ldots, \mathbf{w}_{N-2}$ need to be precomputed, stored, and used for prediction. Hence, this procedure is still not simple enough for implementation.

Reduced-complexity one-step prediction may be accomplished by either *finite-length* FIR- or *recursive* infinite-impulse-response (IIR)-type filtering. The former method is an obvious modification to optimal Wiener prediction in that the number of filter coefficients is fixed at an arbitrary number ν; the predictor therefore reduces to a single, time-invariant ν-tap FIR filter $\mathbf{w} = (w_0\ w_1\ \ldots\ w_{\nu-1})^T$. Its tap weights and error covariance are obtained from eq. (14-7) with time index set to $k = \nu$. During the transmission startup phase when only $k < \nu - 1$ samples of the ML estimate $\hat{\mathbf{c}}_{S;k-1}(\mathbf{a}_{k-1})$ are to be processed, one may either use shorter k-dim. optimal filters $\mathbf{w}_{k-1}$, or otherwise simply the k last taps of the length-ν filter $\mathbf{w}$. Performance results on both optimal and length-ν Wiener prediction are presented at the end of this section.

The second solution to reduced-complexity prediction consists in recursive IIR-type filtering. Then both metric computation and channel prediction are performed recursively. If the dynamics of the fading process $\{c_k\}$ can be cast into a (here: stationary) state-space *Gauss-Markov* process model

$$\begin{aligned} \mathbf{x}_{k+1} &= \mathbf{F} \cdot \mathbf{x}_k + \mathbf{G} \cdot w_k \\ c_k &= \mathbf{H}^H \cdot \mathbf{x}_k \end{aligned} \tag{14-8}$$

(system state $\mathbf{x}_k$, system noise w_k, system matrix $\mathbf{F}$, input matrix $\mathbf{G}$, output matrix $\mathbf{H}^H$), the optimal IIR-type predictor is given by the *Kalman filter* (KF) [2]

$$\begin{aligned} &\text{State update recursion:} \\ &\hat{\mathbf{x}}_{k|k-1} = \mathbf{F} \cdot \left(\hat{\mathbf{x}}_{k-1|k-2} + \mathbf{K}_{k-1} \cdot \left[\hat{c}_{S;k-1} - \hat{c}_{k-1|k-2}\right]\right) \\ &\text{Predictor estimate:} \\ &\hat{c}_{k|k-1} = \mathbf{H}^H \cdot \hat{\mathbf{x}}_{k|k-1} \end{aligned} \tag{14-9}$$

where both the (time-variant) Kalman gain $\mathbf{K}_{k-1}$ and the state error covariance $\mathbf{\Sigma}_{x;k|k-1}$ can be precomputed via the Kalman Riccati equations [1]. If the time-variant noise power (N_0/p_k) in the ML estimate $\hat{c}_{S;k-1}(a_{k-1})$ is replaced by its average N_0, these computations yield a Kalman predictor which again is entirely independent of the data hypothesis.

The block diagram of the Gauss-Markov process model for $\{c_k\}$, followed by the modulator (multiplication by "true" symbols $a_{k-1}^{(0)}$), the AWGN m_{k-1}, ML channel estimator (multiplication by a_{k-1}^*/p_{k-1}), and the Kalman channel predictor, is shown in Figure 14-1.

Unfortunately, the actual channel dynamics, in particular a strictly band-limited fading process $\{c_k\}$ [eq. (12-49)], can only be represented with sufficient accuracy by a high-order Gauss-Markov process model. Then, however, the complexity of the corresponding Kalman predictor quickly rises beyond that of

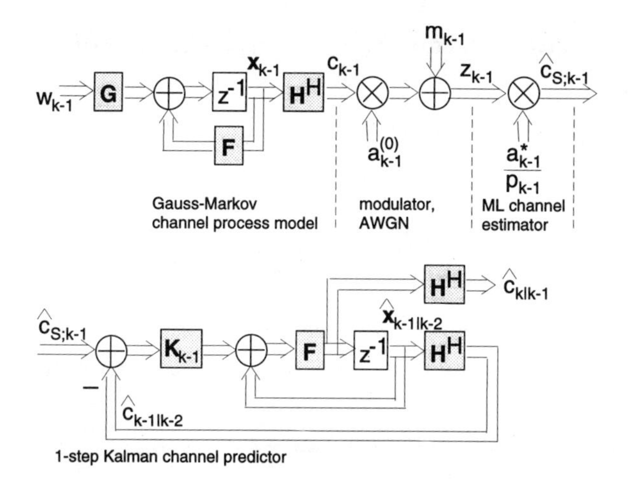

Figure 14-1 Gauss-Markov Flat Fading Channel Model and One-Step Kalman Predictor

the length-ν FIR Wiener predictor. For this reason, Kalman filtering based on high-order process modeling is not pursued further.

In order to keep channel prediction as simple as possible, let us constrain the Kalman filter to be a *first-order* stationary IIR filter with scalar gain $\mathbf{K}_{k-1} = K$, unity system matrix $\mathbf{F} = 1$, state estimate $\hat{\mathbf{x}}_{k|k-1} = \hat{c}_{k|k-1}$, and output matrix $\mathbf{H}^H = 1$ (Figure 14-1). The recursive channel predictor then simplifies to

$$\begin{aligned} \hat{c}_{k|k-1} &= \hat{c}_{k-1|k-2} + K\left(\hat{c}_{S;k-1} - \hat{c}_{k-1|k-2}\right) \\ &= (1-K)\,\hat{c}_{k-1|k-2} + K\,\hat{c}_{S;k-1} \end{aligned} \tag{14-10}$$

This algorithm is identical with the well-known least-mean-square (LMS) adaptive filter [3] with gain factor K, so that this filter may also be termed *LMS-Kalman* predictor. Its z-transform and 3-dB cutoff frequency are given by

$$\begin{aligned} H(z) &= \frac{K}{z - [1-K]} \\ \lambda'_{c,\,3\,\mathrm{dB}} &= \frac{1}{2\pi}\cos^{-1}\left[\frac{\left(1+[1-K]^2\right) - \sqrt{2}K^2}{2(1-K)}\right] \end{aligned} \tag{14-11}$$

respectively.

Unfortunately, the gain K (and also the Kalman filter error covariance) cannot be determined by means of ordinary Kalman filter theory (Riccati equations [1]) because of the mismatch between the (high-order) process model and the (first-order) prediction filter. Thanks to the simple form of eq. (14-10), however, the gain K may be optimized directly by means of a prediction error analysis. Using eq. (14-10), the prediction error $\Delta_k = c_k - \hat{c}_{k|k-1}$ can be expressed as

$$\begin{aligned}\Delta_k &= c_k - \hat{c}_{k|k-1} \\ &= c_k - \left[\hat{c}_{k-1|k-2} + K\left(\hat{c}_{S;k-1} - \hat{c}_{k-1|k-2}\right)\right] \\ &= \underbrace{(c_k - c_{k-1})}_{\delta_k} + \underbrace{(c_{k-1} - \hat{c}_{k-1})}_{\Delta_{k-1}} - K\left[\underbrace{\frac{a^*_{k-1}}{p_{k-1}}\left(a^{(0)}_{k-1}c_{k-1} + m_{k-1}\right)}_{\hat{c}_{S;k-1}} - \hat{c}_{k-1}\right]\end{aligned} \tag{14-12}$$

where $\delta_k = c_k - c_{k-1}$ is the channel trajectory increment from time step $k-1$ to k. Assuming correct symbol hypotheses $a_{k-1} = a^{(0)}_{k-1}$, the channel error recursion reduces to

$$\Delta_k = (1-K)\,\Delta_{k-1} + \delta_k - K\,\tilde{m}_{k-1} \tag{14-13}$$

where the noise $\tilde{m}_{k-1} = (a^*_{k-1}/p_{k-1})m_{k-1}$ has time-variant power N_0/p_{k-1}. This recursion may be split into two independent recursions for the "lag" error $\Delta^{(L)}_k$ and the "noise" error $\Delta^{(N)}_k$, driven by fading δ_k and noise $\tilde{m}_{k-1}$, respectively, whose solutions are given by

$$\begin{aligned}\Delta^{(L)}_k &= (1-K)\,\Delta^{(L)}_{k-1} + \delta_k &&= \left(\sum_{l=0}^{k}(1-K)^l\,\delta_k\right) \\ \Delta^{(N)}_k &= (1-K)\,\Delta^{(N)}_{k-1} - K\,\tilde{m}_{k-1} &&= -K\left(\sum_{l=0}^{k-1}(1-K)^l\,\tilde{m}_{k-1-l}\right)\end{aligned} \tag{14-14}$$

Assuming white noise $\tilde{m}_{k-1}$ and setting the time-variant noise power N_0/p_{k-1} to its average N_0, the average noise error in the channel predictor estimate is easily evaluated as

$$\left(\sigma^2_{c;k|k-1}\right)^{(N)} = E\left[\left|\Delta^{(N)}_k\right|^2\right] = \frac{K}{2-K}N_0 \underset{(K\ll 1)}{\approx} \frac{K}{2}N_0 \tag{14-15}$$

The lengthy but straightforward analysis of the average lag error yields the steady-state ($k \to \infty$) result:

$$\begin{aligned}\left(\sigma^2_{c;k|k-1}\right)^{(L)} &= E\left[\left|\Delta^{(L)}_k\right|^2\right] \\ &= E\left[\sum_{i=0}^{k\to\infty}\sum_{j=0}^{k\to\infty}(1-K)^i(1-K)^j\,\delta_{k-i}\,\delta^*_{k-j}\right] \\ &= \frac{2K}{(2-K)(1-K)}\left(\sum_{m=1}^{\infty}(1-K)^m\,\bar{\alpha}(m)\right)\end{aligned} \tag{14-16}$$

with $\overline{\alpha}(m) = 1 - \alpha(m)$ the complementary channel autocorrelation function.

In order to facilitate the optimization of K, a simple functional approximation to $(\sigma^2_{c;k|k-1})^{(L)}$ would be of great value. Considering the special case of Jakes and rectangular lowpass Rayleigh Doppler spectra, a Taylor series analysis of the corresponding $\overline{\alpha}(m)$ [eq. (12-49)] about $m = 0$ yields $\overline{\alpha}(m) \approx \pi^2(\lambda'_D)^2 m^2$, valid for both Jakes and rectangular Doppler spectra as long as $(\lambda'_D m) \leq 0.2$ (less than 5 percent error). Using this parabolic approximation in eq. (14-16) for all $m = 0, \ldots, \infty$ yields $(\sigma^2_{c;k|k-1})^{(L)} \approx 2\pi^2(\lambda'_D)^2/K^2$. Since the summation has been carried out over some m for which the approximation $(\lambda'_D m) \leq 0.2$ is not valid, this result should be used with caution. Nevertheless, one may surmise that the approximation would be of the form

$$\left(\sigma^2_{c;k|k-1}\right)^{(L)} \approx C_L \left(\frac{\lambda'_D}{K}\right)^2 \tag{14-17}$$

Then the expression for the total error covariance becomes

$$\begin{aligned} \left(\sigma^2_{c;k|k-1}\right) &= \left(\sigma^2_{c;k|k-1}\right)^{(L)} + \left(\sigma^2_{c;k|k-1}\right)^{(N)} \\ &\approx C_L \left(\frac{\lambda'_D}{K}\right)^2 + \frac{K}{2-K} N_0 \end{aligned} \tag{14-18}$$

In essence, a wider prediction filter bandwidth (K larger) reduces the lag error but increases the noise error, and vice versa. Assuming for the moment that this approximation is correct, the optimum gain $K^{(\text{opt})}$ and the minimum error covariance $\left(\sigma^2_c\right)^{(\min)} = \left(\sigma^2_{c;k|k-1}\right)^{(\min)}$ or, equivalently, the minimum ratio $r_c^{(\min)} = \left(\sigma^2_c\right)^{(\min)}/N_0$ between ML and MAP error covariances, are found by setting the derivative with respect to K to zero. Setting $C_L = 12$ (see below) and using $K/(2-K) \approx K/2$ ($K \ll 1$) then leads to the result

$$\begin{aligned} K^{(\text{opt})} &\approx 3.6 \sqrt[3]{\frac{\left(\lambda'_D\right)^2}{N_0}} \\ r_c^{(\min)} = \frac{\left(\sigma^2_c\right)^{(\min)}}{N_0} &\approx 2.7 \sqrt[3]{\frac{\left(\lambda'_D\right)^2}{N_0}} = \frac{3}{4} K^{(\text{opt})} \end{aligned} \tag{14-19}$$

Using this $K^{(\text{opt})}$, the 3-dB cutoff frequency λ'_c of the optimized filter $H^{(\text{opt})}(z)$ [eq. (14-11)] turns out to be much larger than the Doppler frequency λ'_D, and the filter response $H^{(\text{opt})}\left(e^{j2\pi\lambda'}\right)$ remains close to unity (0.98,...,1) in the passband region $|\lambda'| \leq \lambda'_D$. Hence, predictor performance is sensitive to passband distortions whereas the stopband filter response is of minor importance. It therefore does not make sense to employ higher-order filters of the form $H(z) = K^M/(z - [1-K])^M$; the corresponding error analysis yields reduced noise (sharper cutoff) but a larger lag error due to increased passband distortion.

Of course, the validity of the approximative result [eq. (14-19)] has to be checked. A numerical evaluation shows that eq. (14-17) is indeed valid as long as K is larger than some minimum $K_{\min}\left(\lambda'_D\right)$. Within that range, C_L is found

Table 14-1 Optimal Kalman Gains and Minimum Error Covariance of LMS-Kalman Flat Fading Channel Predictor

	$K^{(\text{opt})}$			$r_c^{(\min)}$		
$\lambda_D' \ \backslash \ N_0$	0.1	0.01	0.001	0.1	0.01	0.001
1×10^{-3}	0.08	0.17	0.36	0.058	0.13	0.27
2.5×10^{-3}	0.14	0.31	0.66	0.11	0.23	0.50
5×10^{-3}	0.23	0.49		0.17	0.37	
1×10^{-2}	0.36	0.78		0.27	0.58	
2×10^{-2}	0.57			0.43		
3×10^{-2}	0.75			0.56		

to be only weakly dependent on K or λ_D', so that fixing C_L at 12 is correct (accuracy $\pm 10\%$) for $K_{\min}\left(\lambda_D' = 0.001\right) \approx 0.02$, $K_{\min}\left(\lambda_D' = 0.01\right) \approx 0.1$, and $K_{\min}\left(\lambda_D' = 0.03\right) \approx 0.22$. Examining eq. (14-19), one finds that $K^{(\text{opt})} > K_{\min}$ is always satisfied for relevant noise powers $N_0 < 1$.

However, K must satisfy not only $K > K_{\min}$ but also $K \ll 1$ so that the filter "memory" remains sufficiently large. From Table 14-1 it is seen that the optimization according to eq. (14-19) calls for very large K ($K > 1$ left blank in Table 14-1) in the case of large Doppler λ_D' and/or small $N_0 = 1/\overline{\gamma}_s$. Hence, an optimal trade-off between noise and lag error [eq. (14-19)] is only achievable for small Doppler $\lambda_D' \leq 3 \times 10^{-2}$ (average SNR per symbol $\overline{\gamma}_s = 10$ dB), $\lambda_D' \leq 1 \times 10^{-2}$ ($\overline{\gamma}_s = 20$ dB), and $\lambda_D' \leq 3 \times 10^{-3}$ ($\overline{\gamma}_s = 30$ dB). When the fading is fast, the lag error becomes dominant and the gain K must be fixed at some value < 1.

Wrapping up the discussion of channel prediction, let us compare the performance of optimal Wiener FIR prediction, its finite-length variant, and first-order IIR prediction using the LMS-Kalman algorithm. Figure 14-2 displays the respective optimal ratios $r_c^{(\min)} = \left(\sigma_c^2\right)^{(\min)} / N_0$ for Doppler frequencies between 10^{-3} and 10^{-1} and SNRs (per symbol) of $\overline{\gamma}_s = 10$, 20, and 30 dB.

Optimal Wiener prediction is seen to be only slightly dependent on the noise power – this has already been observed in Chapter 12 where $r_c^{(\min)}$ is almost independent of the noise power – and it yields a gain with respect to memoryless ML estimation ($r_c^{(\min)}$ below 0 dB) for Doppler frequencies up to about $\lambda_D' \approx 5 \times 10^{-2}$. Beyond that, the predictor estimate becomes worse than the ML estimate; the lag error increasingly dominates so that $r_c^{(\min)}$ rises faster with λ_D'.

Compared with infinite-length prediction, the length-10 Wiener prediction gain

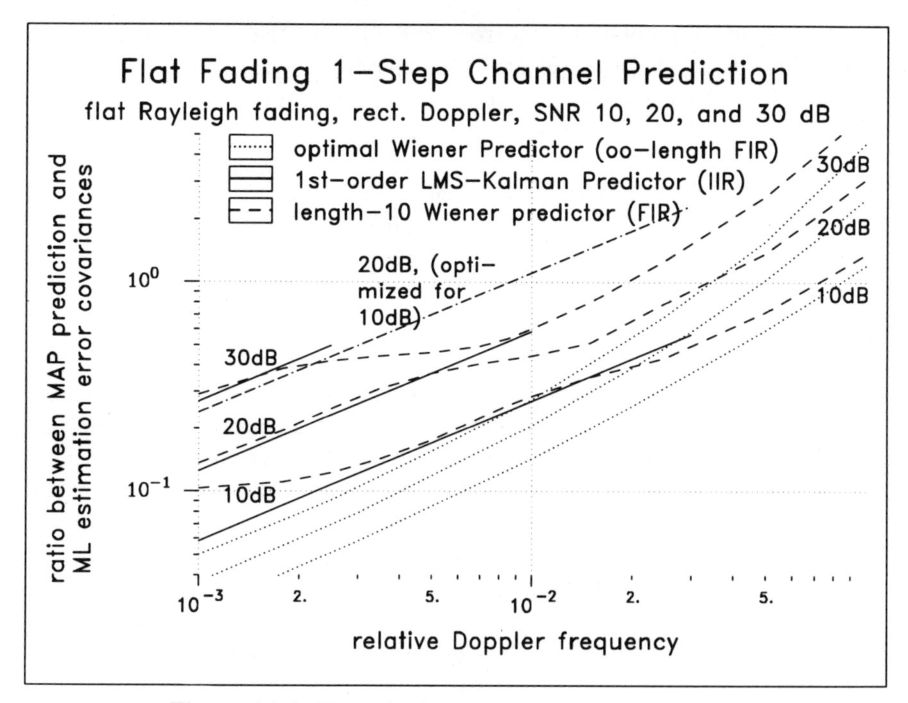

Figure 14-2 Error Performance of Wiener FIR and LMS-Kalman IIR Flat Fading Channel Predictors

factor experiences a bottoming effect for very small Doppler, simply because the noise reduction factor of a length-ν averaging filter is limited to $1/\nu$. On the other hand, the FIR predictor performs quite well in the critical region of large Doppler, which is particularly useful since LMS-Kalman prediction cannot be applied in that region. Furthermore, the predictor is tolerant against a mismatch in the SNR assumption; the filter matched to $\overline{\gamma}_s$ =10 dB but used when the actual SNR is $\overline{\gamma}_s$ =20 dB leads to a certain loss (see Figure 14-2), which, however, is quite moderate.

Finally, the simple LMS-Kalman IIR predictor with optimized gain performs well in the region where it can be applied. At $\lambda'_D \approx 5 \times 10^{-3}$ the performance is almost the same as with length-10 FIR filtering ($\overline{\gamma}_s \leq$20 dB), and at very low Doppler, the LMS-Kalman algorithm has advantages thanks to its simplicity and good noise averaging capability. However, IIR-type channel estimation is more popular when it comes to selective channels where the fading is usually slower (Section 15.1).

14.1.3 Decision-Directed (DD) Flat Fading Channel Estimation and Detection

Whether the decision metric $m_D(\mathbf{a})$ [eq. (13-9)] is computed recursively or not, an optimal decision can only be made after the entire message has been processed. Fortunately, the recursive form of the metric $m_{D;k}(\mathbf{a}_k) = \Delta m_{D;k}(\mathbf{a}_k) + m_{D;k-1}(\mathbf{a}_{k-1})$ [eq. (13-9)] allows for systematic simplifications. In particular, the receiver can perform premature symbol detection before the end of the message. For instance, at time instant k, one may decide on the symbol a_{k-S} – and thus the sequence $\mathbf{a}_{k-S}$ – corresponding to the sequence $\mathbf{a}_k$ with the best metric $m_{D;k}(\mathbf{a}_k)$ at time instant k. Then the first $k-S+1$ symbols $\{a_0, \ldots, a_{k-S}\}$ of $\mathbf{a}_k$ can be taken to be known, and only Q^S (instead of Q^N) distinct partial sequences $\{a_{k-S+1}, \ldots, a_k\}$ $(S \geq 1)$ and their metrics $m_{D;k}(\mathbf{a}_k)$ need to be processed. This procedure would be quasi-optimal if S were in the order of the channel memory length; choosing $S \approx 1/\lambda'_D$, however, remains to be far from being implementable.

The simplest suboptimal receiver is obtained when *all* previously transmitted symbols $\mathbf{a}_{k-1}$ are taken to be known ($S = 1$), which is most often accomplished by DD online detection [2]. However, online DD detection is, in general, unable to incorporate deinterleaving so that symbol detection must be separated from deinterleaving and decoding. Hence, detection is performed suboptimally on a symbol-by-symbol basis as if uncoded M-QAM or M-PSK symbols were transmitted. Since only one sequence $\mathbf{a}_{k-1} = \tilde{\mathbf{a}}_{k-1}$ and its metric $m_{D;k-1}(\tilde{\mathbf{a}}_{k-1})$ has "survived", the decision on symbol a_k is based solely on the metric increment $\Delta m_{D;k}(a_k)$ [eq. (13-9)], which, in turn, depends on the *single* channel predictor estimate $\hat{c}_{k|k-1}$.

The performance of the receiver with decision-directed one-step prediction can be assessed by assuming that the past decisions $\tilde{\mathbf{a}}_{k-1}$ are correct. Hence, the BER curves analytically derived here have the character of best-case lower bounds; the effect of error propagation resulting from incorrect decisions has to be determined by simulation. Using the simplified metric of eq. (13-12), the average pairwise symbol error event probability can be expressed as

$$\begin{aligned} P_e &= P\left[\Delta m_{D;k}\left(a_k^{(1)}\right) < \Delta m_{D;k}\left(a_k^{(0)}\right)\right] \\ &= P\left[|Y_1|^2 - |Y_0|^2 < 0\right] \end{aligned} \tag{14-20}$$

with

$$\begin{aligned} Y_0 &= z_k - a_k^{(0)}\,\hat{c}_{k|k-1} \\ Y_1 &= z_k - a_k^{(1)}\,\hat{c}_{k|k-1} \end{aligned} \tag{14-21}$$

where $a_k^{(1)} \neq a_k^{(0)}$ is an incorrect and $a_k^{(0)}$ the correct symbol hypothesis. Since Y_0 and Y_1 are complex Gaussian random variables, the error event probability is given by [2, 4, 5, 6]

$$\begin{aligned} P_e &= 0.5\,(1-\mu) \\ \mu &= \frac{R_{11} - R_{00}}{\sqrt{(R_{00} + R_{11})^2 - 4R_{01}R_{10}}} \\ R_{ij} &= E\left[Y_i Y_j^*\right] \end{aligned} \tag{14-22}$$

Assuming Rayleigh fading, the expected values R_{00}, R_{11}, R_{01}, R_{10} are evaluated as

$$\begin{aligned} R_{00} &= E\left[|Y_0|^2\right] = (1+p_0\overline{r}_c)N_0 \\ R_{11} &= E\left[|Y_1|^2\right] = (1-\overline{r}_c N_0)d^2 + (1+p_0\overline{r}_c)N_0 \\ R_{01} &= E[Y_0 Y_1^*] = (1+p_0\overline{r}_c)N_0 \\ R_{10} &= E[Y_1 Y_0^*] = R_{01} \end{aligned} \tag{14-23}$$

with squared distance $d^2 = \mid a_k^{(1)} - a_k^{(0)} \mid^2$ and symbol energy $p_0 = p_k^{(0)}$. Inserting these quantities into eq. (14-22) yields

$$\mu = \frac{1}{\sqrt{1 + 4\dfrac{1+p_0\overline{r}_c}{1-\overline{r}_c/\overline{\gamma}_s}\dfrac{1}{d^2}\dfrac{1}{\overline{\gamma}_s}}} \tag{14-24}$$

In the important special case of M-PSK we have $p_0 = p_1 = 1$ and $\overline{r}_c = r_c = \sigma^2_{c;k|k-1}\overline{\gamma}_s$. Considering only error events with minimal distance $d^2 = d_M^2 = 2\,(1-\cos[2\pi/M])$ (PSK symbol constellation), μ reduces to

$$\mu_{\text{MPSK}} = \frac{1}{\sqrt{1 + \dfrac{2}{1-\cos[2\pi/M]}\left(\dfrac{1+r_c}{1-r_c/\overline{\gamma}_s}\right)\dfrac{1}{\overline{\gamma}_s}}} \tag{14-25}$$

Assuming that a symbol error entails only one bit error (Gray coding), and denoting by $n_b = \text{ld}[M]$ the number of bits per symbol, by N_b the number of neighboring symbols with minimal distance (1 for 2-PSK and 2 otherwise), and by $\overline{\gamma}_b = \overline{\gamma}_s/n_b$ the average SNR per bit, the average bit error probability can be approximated (lower bounded) by

$$\begin{aligned} P_{b;\text{MPSK}} &\approx \frac{N_b}{2n_b}(1-\mu_{\text{MPSK}}) \\ &= \frac{N_b}{2n_b}\left(1 - \frac{1}{\sqrt{1+\dfrac{2}{n_b(1-\cos[2\pi/M])}\left(\dfrac{1+r_c}{1-r_c/[n_b\overline{\gamma}_b]}\right)\dfrac{1}{\overline{\gamma}_b}}}\right) \end{aligned} \tag{14-26}$$

For Dth-order diversity with independent diversity branches, the approximate BER is given by [2, 7]

$$P_{b;\left(\substack{\text{MPSK}\\ D\ \text{diversity}}\right)} \approx \frac{N_b}{2n_b}\left(1 - \mu_{\text{MPSK}}\sum_{d=0}^{D-1}\binom{2d}{d}\left(\frac{1-\mu^2_{\text{MPSK}}}{4}\right)^d\right) \tag{14-27}$$

Letting $r_c \rightarrow 0$ in μ_{MPSK} yields the well-known result for coherent M-PSK detection with perfect synchronization. Hence, the term $\eta = (1 - r_c/[n_b\overline{\gamma}_b])/(1 + r_c) < 1$ is identified as the SNR degradation factor resulting from imperfect one-step channel prediction.

For a particular prediction algorithm, Doppler frequency λ'_D and SNR $\overline{\gamma}_b$, the ratio r_c may be extracted from Figure 14-2 or computed from eqs. (14-7) (optimal or finite-length Wiener) or (14-19) (LMS-Kalman) and inserted into eq. (14-26) or (14-27) to obtain the best-case BER curves for decision-directed sync on flat fading channels. These results, along with reference curves for perfect sync, have been generated for 2-, 4-, 8- and 16-PSK modulation, Wiener channel prediction via optimal (∞ length) as well as length-10 FIR filtering, and for no diversity (D=1) as well as dual diversity (D=2). As an example, Figure 14-3 shows resulting BER curves for Doppler λ'_D=0.005 and no diversity. The other results are concisely summarized in Table 14-2. The BER performance of 4-PSK is only very slightly worse than that of 2-PSK so that the respective curves are indistinguishable in Figure 14-3. When the fading is relatively slow (λ'_D=0.005), optimal prediction requires a small extra SNR of 0.3–0.7 dB compared with perfect sync (both without and with diversity), and length-10 Wiener (as well as LMS-Kalman) prediction costs another 0.4–1.9 dB. As expected, the higher PSK constellations are somewhat more sensitive against imperfect channel estimation. When the fading is fast (λ'_D=0.05; see Table 14-2), the minimum extra SNR needed

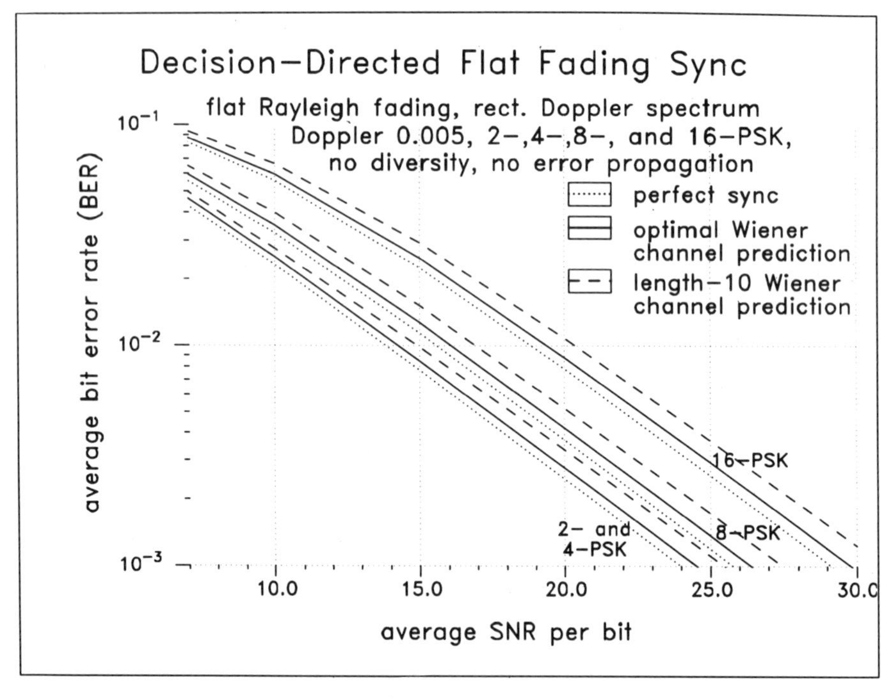

Figure 14-3 BER Performance of Receiver with DD Flat Fading Channel Prediction, Doppler 0.005

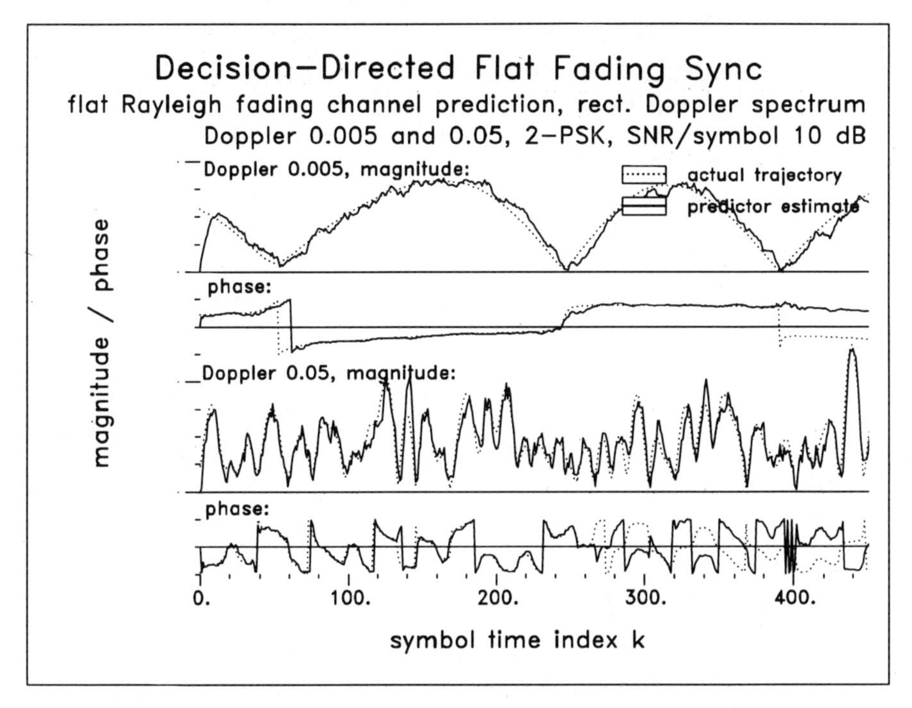

Figure 14-4 Flat Fading Channel Estimate of DD Predictor, Doppler 0.005 and 0.05, SNR/Symbol 10 dB

for optimal prediction rises considerably to 2.4–4 dB @ BER=10^{-2} and 2.6–5.3 dB @ BER=10^{-3}. On the other hand, the additional SNR required for length-10 Wiener prediction, ranging between 0.4 and 1.9 dB, is moderate for both slow and fast fading.

The analytical BER results [eqs. (14-26) and (14-27)] have been obtained under idealistic assumptions. In reality, not only minimum-distance error events (the receiver decides on a neighboring symbol, and only one bit error is made thanks to Gray coding), but also error events with larger distance (then more than one bit error may ensue) contribute to the BER. Also, the bootstrap mechanism of alternating between detection and synchronization (symbol decisions depend on the predictor channel estimate, and vice versa) may give rise to error propagation. Isolated decision errors are usually leveled out by the predictor memory, while error bursts may lead to phase slips which cannot be resolved by a coherent symbol-by-symbol detector. This effect is visualized in Figure 14-4 where the magnitudes and phases (2π normalized to unity) of the channel predictor estimates are displayed for the first 450 iterations, along with the magnitudes and phases of the actual channel. One observes that the magnitudes are tracked well, even at low average SNR (10 dB), for both slow (λ'_D=0.005) and fast (λ'_D=0.05) fading. Phase sync, however, is lost after some tens or hundreds of iterations. In Figure 14-4, the first such cycle slip (2-PSK: $\pm\pi$) occurs after about 400 (λ'_D=0.005) and 260 (λ'_D=0.05) iterations. Figure 14-4 also reveals that these cycle slips correspond

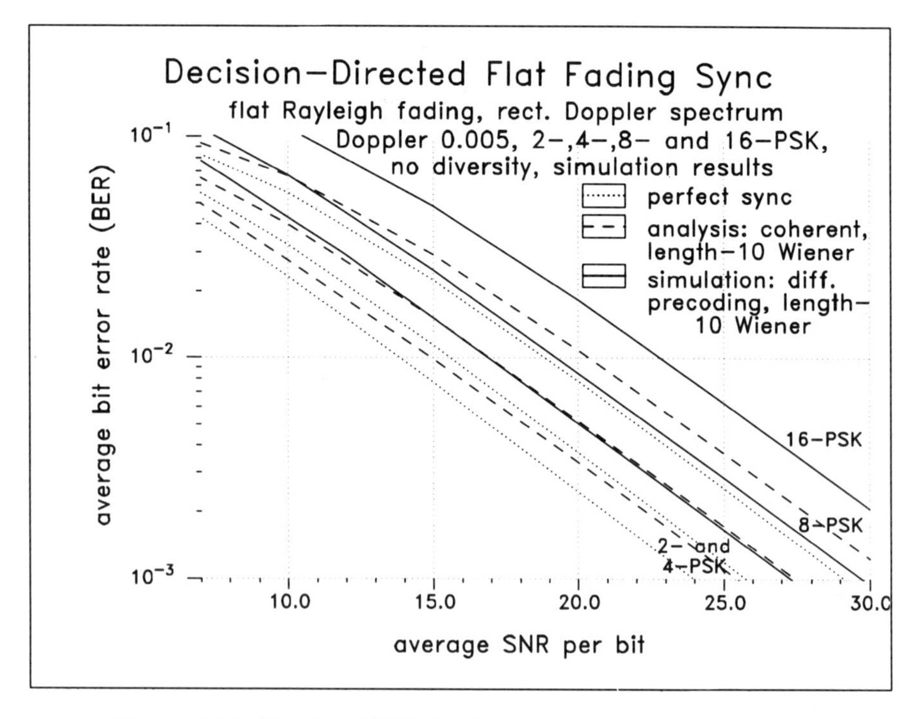

Figure 14-5 Simulated BER Performance of Receiver with DD Flat Fading Channel Prediction, Doppler 0.005

with very deep fades where the symbol decisions are likely to be incorrect. In order to avoid catastrophic error propagation resulting from the loss of an absolute phase reference, coherent detection of long symbol sequences must be aided by some kind of differential precoding.

The combined detrimental effect of all these mechanisms on the BER performance is best assessed by simulation. At each iteration $k-1 \rightarrow k$, upon reading the new received (diversity) signal sample(s) $z_{d;k}$, the simulation module performs (diversity) channel prediction $\hat{\mathbf{c}}_{Sd;k-1}, \mathbf{w} \rightarrow \hat{c}_{d;k|k-1}$ from the (old) ML channel estimate vector(s) $\hat{\mathbf{c}}_{Sd;k-1}$, computation of combiner weights $\hat{c}_{d;k|k-1} \rightarrow \hat{q}_{d;k}$, diversity combination $z_{d;k}, \hat{q}_{d;k} \rightarrow \hat{a}_k$, detection $\hat{a}_k \rightarrow \tilde{a}_k$, and decision-directed ML channel estimation $z_{d;k}, \tilde{a}_k \rightarrow \hat{c}_{Sd;k}$ needed for the next iteration.

Figures 14-5 and 14-6 display simulation results of differentially precoded 2-, 4-, 8- and 16-PSK detection (without diversity) based on length-10 Wiener channel prediction for Doppler frequencies λ'_D=0.005 and 0.05, respectively. Corresponding simulations have also been performed for dual diversity with maximum ratio combining (see Section 13.1 and Figure 13-6); the results are included in Table 14-2. The table lists both analytically derived and simulated SNR levels (per bit, per channel) necessary to achieve bit error rates of 10^{-2} and 10^{-3}. The simulation runs were terminated after 10^7 symbols or 10^5 symbol errors, whichever occurred first. While phase slips occurred frequently (Figure 14-4), catastrophic

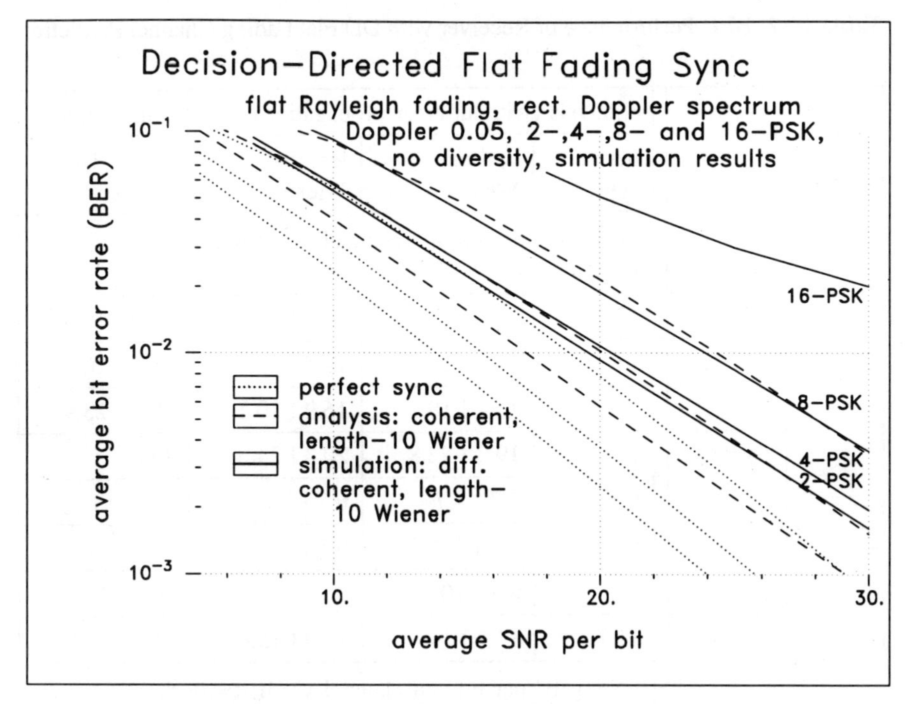

Figure 14-6 Simulated BER Performance of Receiver with DD Flat Fading Channel Prediction, Doppler 0.05

error propagation was never observed in any of the simulation runs, this being due to the constant-modulus property of PSK symbols which have been matched filtered and sampled at the correct timing instants. In the lower SNR regions, one observes losses of about 2–4 dB (no diversity) and 1–2 dB (dual diversity) with respect to the analytically derived BER curves (length-10 Wiener prediction). Without diversity, however, the curves begin to flatten out at high SNR when the fading is fast (λ'_D=0.05, Figure 14-6); for 16-PSK, the error floor even exceeds 10^{-2}, whereas with dual diversity such a bottoming effect is observed only at BER levels below 10^{-4}. The use of diversity is therefore not only effective in averaging out deep fades but also in mitigating considerably the effects of imperfect channel estimation.

In summary, the decision-aided receiver is a good candidate for suboptimal detection of differentially precoded (but otherwise uncoded) M-PSK transmission with alphabet sizes up to M=8 or even 16. Without diversity, the very simple 2-, 4- and 8-PSK receiver is robust against up to 5 percent Doppler. When the fading is slow or diversity is available, even 16-PSK is a viable option. In order to aid the decoder in the case that coding with interleaving is employed, the receiver can be made to output channel state information, viz. the channel predictor estimate $\hat{c}_{k|k-1}$ or its power $|\,\hat{c}_{k|k-1}\,|^2$, and/or soft symbol decisions given by the symbol estimates $\hat{a}_k$ in front of the slicer.

Table 14-2 BER Performance of Receiver with DD Flat Fading Channel Prediction

		Analysis (without error propagation)			Simulation
		perfect sync	optimal Wiener	length-10 Wiener	length-10 Wiener
		SNR [dB] per bit and channel @ BER=10^{-2}, Doppler $\lambda'_D = 0.005 \vert 0.05$			
$D = 1$	2-PSK	13.9	14.3 | 16.6	15.0 | 17.1	17.0 | 19.8
	4-PSK	13.9	14.3 | 16.6	15.0 | 17.1	17.0 | 20.3
	8-PSK	15.6	16.1 | 19.1	16.9 | 20.1	19.3 | 23.9
	16-PSK	18.9	19.3 | 22.8	20.3 | 24.3	22.8
$D = 2$	2-PSK	5.3	5.7 | 7.7	6.1 | 8.1	7.8 | 9.1
	4-PSK	5.3	5.7 | 7.7	6.1 | 8.1	7.8 | 9.3
	8-PSK	8.1	8.4 | 10.7	8.9 | 11.2	10.6 | 12.2
	16-PSK	11.7	12.2 | 14.8	12.9 | 15.6	14.8 | 17.0
		SNR [dB] per bit and channel @ BER=10^{-3}, Doppler $\lambda'_D = 0.005 \vert 0.05$			
$D = 1$	2-PSK	24.0	24.5 | 27.8	25.5 | 29.1	27.2 | 32.7
	4-PSK	24.0	24.5 | 27.8	25.5 | 29.1	27.2 | 34.0
	8-PSK	25.8	26.4 | 30.4	27.6 | 32.2	29.7 | 38.0
	16-PSK	29.1	29.8 | 34.4	30.8 | 36.3	33.3
$D = 2$	2-PSK	11.1	11.5 | 13.7	12.0 | 14.2	13.5 | 15.3
	4-PSK	11.1	11.5 | 13.7	12.0 | 14.2	13.5 | 15.7
	8-PSK	13.8	14.2 | 17.1	15.0 | 18.0	16.5 | 19.0
	16-PSK	17.7	18.2 | 21.4	19.2 | 22.8	20.6 | 24.2

14.1.4 Main Points

- Optimal MAP one-step flat fading channel prediction as part of the NDA on-line detection process (Chapter 13) can again be performed via ML estimation:

$$\begin{aligned}
&\text{ML channel estimation:}\\
&\hat{\mathbf{c}}_{S;k-1}(\mathbf{a}_{k-1}) = \mathbf{P}_{k-1}^{-1}(\mathbf{a}_{k-1}) \cdot \mathbf{A}_{k-1}^{H} \cdot \mathbf{z}_{k-1}\\
&\text{MAP channel prediction from ML estimate:}\\
&\hat{c}_{k|k-1} = \mathbf{w}_{k-1}^{H} \cdot \hat{\mathbf{c}}_{S;k-1}(\mathbf{a}_{k-1})
\end{aligned} \tag{14-28}$$

  [eq. (14-7)] where $\mathbf{w}_{k-1}$ is the length-k optimal Wiener filter.

- Channel estimator complexity is reduced by truncating the Wiener filter to a fixed-length FIR filter $\mathbf{w}$ or by adopting an IIR Kalman filtering approach. Simplifying the Kalman filter to first order leads to the LMS-Kalman algorithm

$$\hat{c}_{k|k-1} = (1-K)\,\hat{c}_{k-1|k-2} + K\,\hat{c}_{S;k-1} \tag{14-29}$$

[eq. (14-10)]. Through minimizing the total error covariance $\sigma^2_{c;k|k-1}$, an approximate expression for the optimal gain factor K as a function of λ'_D and N_0 has been derived [eq. (14-19)]. The LMS-Kalman algorithm is best suited to slow fading (up to about $\lambda'_D \approx 5\times10^{-3}$) while the FIR Wiener filter performs quite well in the critical region of large Doppler above 10^{-2} where LMS-Kalman is no more applicable.
- The decision-directed (DD) receiver for online detection and synchronization features just a single NDA channel estimator being fed by symbol-by-symbol decisions. The BER performance of DD reception has been assessed analytically (best-case lower bound) and by simulation. In the case of slow fading (λ'_D=0.005), the extra SNR required for (both LMS-Kalman and length-10 Wiener) DD estimation is moderate. The use of antenna diversity not only reduces this SNR loss (here from 3–4 dB to 2.5–3 dB), but also mitigates the effects of faster fading. Differentially precoded M-PSK detection with LMS-Wiener DD channel prediction is feasible also for fast fading (λ'_D=0.05) and symbol constellations up to 8-PSK (no diversity) and 16-PSK (dual diversity).

14.2 Data-Aided (DA) Flat Fading Channel Estimation and Detection

In this section, linear coherent *data-aided* (DA) detection and flat fading channel estimation is investigated where known *training* symbols are multiplexed into the unknown information-bearing data symbol stream. The received signal is then demultiplexed into "training" and "data" signal streams. If channel estimation is based on training signals *only*, estimation and detection – which have hithereto been viewed as tasks to be performed jointly – become well-separated tasks. As opposed to the feedback-type DD receiver, the DA receiver is of an entirely *feedforward* nature; channel estimation does not rely on past data decisions, nor can the decision process be disturbed by errors in the channel estimate other than those caused by additive noise and possibly aliasing.

Feedforward DA reception using training or pilot symbols multiplexed into the data stream in a TDMA (time division multiple access) -like fashion has been proposed independently by Aghamohammadi and Meyr [8] ("smoothed synchronization"), Cavers [9] ("pilot symbol aided modulation", PSAM) and Moher and Lodge [10] ("trellis-coded modulation with pilot", TCMP). It has been shown that this method outperforms the more traditional pilot tone assisted modulation (PTAM) where the fading process is estimated continually via a pilot tone (sometimes two tones) embedded in the information-bearing signal such that data and pilot are orthogonal (similar to frequency division multiple access, FDMA). Using an orthogonal FDMA-like pilot tone necessitates fairly complex in-band filtering for tone separation, consumes extra bandwidth (considering realizable filters), is sensitive against frequency shifts, and it aggravates adjacent channel interference (ACI) if the tone is placed at the edge of the useful signal spectrum. Also, the composite transmitted signal exhibits a larger peak-to-average power ratio thus placing more stringent requirements on the linearity of the transmitter amplifier.

Naturally, training symbol insertion has to be paid for by a slightly reduced power and bandwidth efficiency. On the other hand, there are a number of favorable consequences associated with DA channel estimation and detection. First of all, DA reception is of relatively low complexity since channel estimation (carrier synchronization) and detection are totally decoupled. Due to the phase ambiguity being resolved by training, fully coherent demodulation can be maintained at any time, even during and following deep fades. Catastrophic error propagation is circumvented, as long as the positions of training symbols in the data stream are known or have been detected correctly ("frame sync"). Also, amplitude-sensitive multilevel symbol constellations (M-QAM) can be employed and demodulated as easily as M-PSK since the fading compensation unit or diversity combiner of the inner receiver (Figure 13-6) acts as an inherent amplitude gain control (AGC). As discussed in Section 13.1, providing for and making use of diversity techniques requires the synchronizer to operate at conditions where the noise power is comparably high as on nonfading AWGN channels. Whereas DD channel estimation is based on unreliable detection at low SNR, synchronized DA diversity reception still functions in these lower SNR regions since DA channel estimation deteriorates gracefully (and not catastrophically) with increasing noise. Interleaved channel coding – being a particularly interesting form of diversity – is thus made possible also. Moreover, there is no need for the inner receiver to generate hard decisions at all; transferring soft symbol decisions $\hat{a}_k$ along with the CSI $\gamma_{s;k}$ (Figure 13-6) through the outer receiver's deinterleaving device preserves all relevant information needed for hard detection at the end of the inner/outer receiver chain.

Further advantages of data-aided reception include its applicability to multiuser TDMA-based channel access since DA estimation exhibits a robust behavior — no significant transient effects — near the ends of short messages (this is proven in Section 14.2). As opposed to DD reception, channel estimation via training can make use of information contained in "future" samples (smoothed sync, see Chapter 12). Furthermore, DA reception can – if necessary – be further improved by iterative detection and estimation: in the first pass, tentative symbol decisions are generated (most often delayed) based on pure DA channel estimation; in the second pass, tentative decisions (or a subset of reliable decisions) may be used for improved channel estimation as if they were training symbols [11].

All receivers based on the flat fading assumption are subject to serious degradation in the case that this assumption is violated. Naturally, DA reception also shares this high sensitivity to ISI resulting from delay spreads. As a rule of thumb, the error floor varies as the square of rms delay spread τ'_D. For example, with uncoded BPSK or QPSK DA reception and bandwidth expansion factor α=0.2, rms delay spreads τ'_D of 0.01 and 0.1 lead to BER floors in the order of 10^{-4} and 10^{-2}, respectively. A wide bandwidth expansion factor of α=1 reduces the BER floors by about one order of magnitude [12, 13].

14.2.1 DA Flat Fading Channel Estimation

Essentially, all kinds of optimal channel estimation discussed in this book are variantions on the same basic theme, viz. optimal linear estimation of a desired quantity (channel) via an intermediate ML estimate of a related quantity (channel samples or a subset thereof), based on appropriate subsets of received samples $\mathbf{z}$ and transmitted symbols $\mathbf{a}$, respectively. For instance, the optimal receiver for synchronized detection (Chapter 12) performs ML estimation $\hat{\theta}_S(\mathbf{a}) = \hat{\mathbf{c}}_S(\mathbf{a}) = \mathbf{P}^{-1}(\mathbf{a}) \cdot \mathbf{A}^H \cdot \mathbf{z}$ attempting to generate a modulation-free channel trajectory, followed by MAP estimation $\hat{\theta}_D(\mathbf{a}) = \hat{\mathbf{c}}(\mathbf{a}) = E[\mathbf{c}|\mathbf{z}, \mathbf{a}] = \mathbf{N}(\mathbf{a}) \cdot \hat{\mathbf{c}}_S(\mathbf{a})$ [$\boldsymbol{\mu}(\mathbf{a})$ set to zero] from the ML estimate, attempting to suppress as much noise as possible by smoothing [eq. (12-53)], making use of the entire observation $\mathbf{z}$ and the entire symbol sequence hypothesis $\mathbf{a}$. Similarly, the optimal DD receiver (previous subsection) performs ML estimation $\hat{\mathbf{c}}_{S;k-1}(\mathbf{a}_{k-1}) = \mathbf{P}_{k-1}^{-1}(\mathbf{a}_{k-1}) \cdot \mathbf{A}_{k-1}^H \cdot \mathbf{z}_{k-1}$ followed by one-step prediction $\hat{c}_{k|k-1}(\mathbf{a}_{k-1}) = E[c_k|\mathbf{z}_{k-1}, \mathbf{a}_{k-1}] = \mathbf{w}_{k-1}^H \cdot \hat{\mathbf{c}}_{S;k-1}(\mathbf{a}_{k-1})$ [eq. (14-7)], making use of past observations $\mathbf{z}_{k-1}$ and symbols $\mathbf{a}_{k-1}$ only.

By the same token, the optimal DA receiver performs ML and MAP channel estimation based on the subset of $\overline{N}$ training symbols

$$\mathbf{a}_T = \begin{pmatrix} a_{k_0} & a_{k_1} & a_{k_2} & \dots & a_{k_{\overline{N}-1}} \end{pmatrix}^T \tag{14-30}$$

located at positions $k = k_0, k_1, \dots, k_{\overline{N}-1}$ within the length-N message $\mathbf{a}$, and the associated "punctured" observation

$$\mathbf{z}_T = \mathbf{A}_T \cdot \mathbf{c}_T + \mathbf{m}_T \tag{14-31}$$

with diagonal symbol matrix $\mathbf{A}_T = \text{diag}\left\{a_{k_0}, a_{k_1}, \dots, a_{k_{\overline{N}-1}}\right\}$, and punctured channel $\mathbf{c}_T$ and noise $\mathbf{m}_T$ vectors. Note that appropriately chosen training symbols $\mathbf{a}_T$ can also be used for purposes other than channel estimation, viz. frame sync, estimation of small frequency offsets Ω' [eq. (12-8)], and synchronization of other receiver components, e.g., deinterleaver and decoder.

Analogous to joint detection and estimation [eq. (12-28)] and one-step prediction [eq. (14-2)], the optimal DA channel estimator and its MMSE are given by [1][10]

$$\begin{aligned} \hat{c}_k &= \mu_{c_k|\mathbf{z}_T} = \boldsymbol{\Sigma}_{c_k,\mathbf{z}_T} \cdot \boldsymbol{\Sigma}_{\mathbf{z}_T}^{-1} \cdot \mathbf{z}_T \\ \sigma_{c;k}^2 &= \boldsymbol{\Sigma}_{c_k|\mathbf{z}_T} = \boldsymbol{\Sigma}_{c_k} - \boldsymbol{\Sigma}_{c_k,\mathbf{z}_T} \cdot \boldsymbol{\Sigma}_{\mathbf{z}_T}^{-1} \cdot \boldsymbol{\Sigma}_{c_k,\mathbf{z}_T}^H \end{aligned} \tag{14-32}$$

[10] Again, the channel is assumed to be zero-mean Gaussian even if a LOS path is present (see Section 12.2.3) so that only the dynamic part is considered for channel estimator design, i.e., $\alpha_d(m) = \alpha(m)$ and $\rho_d = 1$.

The quantities $\boldsymbol{\Sigma}_{c_k}$ (scalar), $\boldsymbol{\Sigma}_{c_k,\mathbf{z}_T}$ $(1 \times \overline{N})$, and $\boldsymbol{\Sigma}_{\mathbf{z}_T}$ $(\overline{N} \times \overline{N})$ evaluate as

$$\begin{aligned}
\boldsymbol{\Sigma}_{c_k} &= E\left[|c_k|^2\right] &&= 1 \\
\boldsymbol{\Sigma}_{c_k,\mathbf{z}_T} &= E\left[c_k \cdot \mathbf{z}_T^H\right] &&= \mathbf{r}_{D;T}^H \mathbf{A}_T^H \\
\boldsymbol{\Sigma}_{\mathbf{z}_T} &= E\left[\mathbf{z}_T \cdot \mathbf{z}_T^H\right] &&= \mathbf{A}_T \mathbf{R}_{D;T} \mathbf{A}_T^H + \mathbf{R}_{m;T}
\end{aligned} \tag{14-33}$$

where

$$\mathbf{R}_{D,T} = \begin{pmatrix} \alpha(k_0 - k_0) & \alpha(k_0 - k_1) & \dots & \alpha\left(k_0 - k_{\overline{N}-1}\right) \\ \alpha(k_1 - k_0) & \alpha(k_1 - k_1) & \dots & \alpha\left(k_1 - k_{\overline{N}-1}\right) \\ \vdots & \vdots & \ddots & \vdots \\ \alpha\left(k_{\overline{N}-1} - k_0\right) & \alpha\left(k_{\overline{N}-1} - k_1\right) & \dots & \alpha\left(k_{\overline{N}-1} - k_{\overline{N}-1}\right) \end{pmatrix}$$

$$\mathbf{r}_{D,T;k} = \left(\alpha(k - k_0) \quad \alpha(k - k_1) \quad \dots \quad \alpha\left(k - k_{\overline{N}-1}\right)\right)^T \tag{14-34}$$

are the "punctured" channel autocorrelation matrix [eq. (12-49)] and the vector of channel autocorrelation samples with regard to the particular position k of the desired channel sample c_k, respectively, and $\mathbf{R}_{m,T}$ the punctured noise covariance matrix. Inserting $\boldsymbol{\Sigma}_{c_k}$, $\boldsymbol{\Sigma}_{c_k,\mathbf{z}_T}$, and $\boldsymbol{\Sigma}_{\mathbf{z}_T}$ into eq. (14-32) and applying the matrix inversion lemma [eq. (12-39)] twice yields the result [analogous to eq. (14-6)]:

ML channel estimation:

$$\begin{aligned}
\hat{\mathbf{c}}_{S,T}(\mathbf{a}_T) &= \left(\mathbf{A}_T^H \mathbf{R}_{m,T}^{-1} \mathbf{A}_T\right)^{-1} \mathbf{A}_T^H \mathbf{R}_{m,T}^{-1} \cdot \mathbf{z}_T \\
\boldsymbol{\Sigma}_{S,T}(\mathbf{a}_T) &= \left(\mathbf{A}_T^H \mathbf{R}_{m,T}^{-1} \mathbf{A}_T\right)^{-1}
\end{aligned}$$

MAP channel estimation from ML estimate:

$$\begin{aligned}
\hat{c}_k &= \mathbf{r}_{D,T;k}^H \left(\mathbf{R}_{D,T} + \left[\mathbf{A}_T^H \mathbf{R}_{m,T}^{-1} \mathbf{A}_T\right]^{-1}\right)^{-1} \cdot \hat{\mathbf{c}}_{S,T}(\mathbf{a}_T) \\
\sigma_{c;k}^2 &= 1 - \mathbf{r}_{D,T;k}^H \cdot \left(\mathbf{R}_{D,T} + \left[\mathbf{A}_T^H \mathbf{R}_{m,T}^{-1} \mathbf{A}_T\right]^{-1}\right)^{-1} \cdot \mathbf{r}_{D,T;k}
\end{aligned} \tag{14-35}$$

If the noise is AWGN with power N_0, eq. (14-35) can be further simplified via replacing the term $\left[\mathbf{A}_T^H \mathbf{R}_{m,T}^{-1} \mathbf{A}_T\right]^{-1} = N_0 \cdot \mathbf{P}_T^{-1}(\mathbf{a}_T)$ by $N_0\mathbf{I}$ (see previous section), so that eq. (14-35) reduces to

ML channel estimation:

$$\begin{aligned}
\hat{\mathbf{c}}_{S,T}(\mathbf{a}_T) &= \mathbf{P}_T^{-1}(\mathbf{a}_T) \cdot \mathbf{A}_T^H \cdot \mathbf{z}_T \\
\boldsymbol{\Sigma}_{S,T}(\mathbf{a}_T) &= N_0 \cdot \mathbf{P}_T^{-1}(\mathbf{a}_T)
\end{aligned}$$

MAP channel estimation from ML estimate:

$$\begin{aligned}
\hat{c}_k &= \underbrace{\mathbf{r}_{D,T;k}^H \cdot (\mathbf{R}_{D,T} + N_0\mathbf{I})^{-1}}_{\mathbf{w}_k^H} \cdot \hat{\mathbf{c}}_{S,T}(\mathbf{a}_T) \\
\sigma_{c;k}^2 &= 1 - \mathbf{r}_{D,T;k}^H \cdot \underbrace{(\mathbf{R}_{D,T} + N_0\mathbf{I})^{-1} \cdot \mathbf{r}_{D,T;k}}_{\mathbf{w}_k}
\end{aligned} \tag{14-36}$$

optimal flat fading channel estimation (Wiener smoothing)
for joint detection and estimation

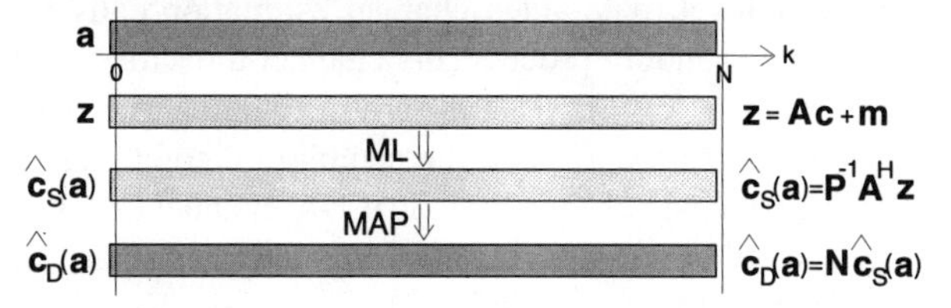

decision-directed (DD) optimal 1-step flat fading channel prediction

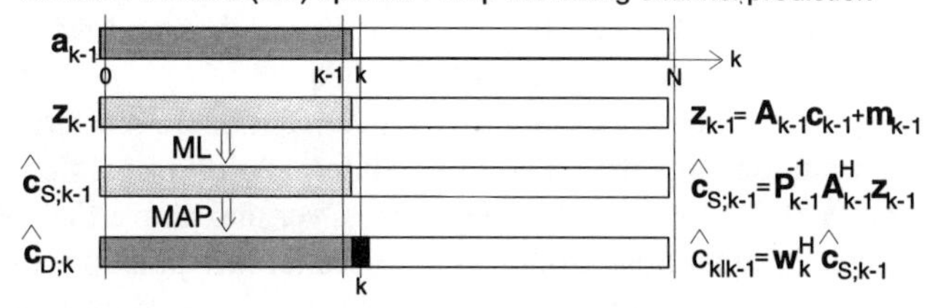

data-aided (DA) optimal flat fading channel estimation

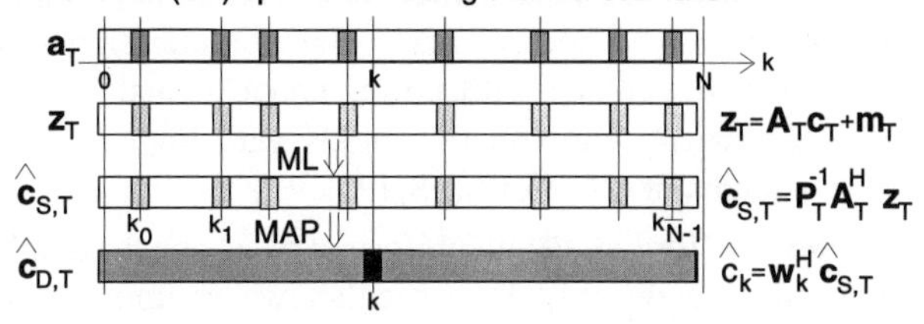

Figure 14-7 Optimal Flat Fading Channel Estimation: Wiener Smoothing (for Joint Detection), DD Estimation, and DA Estimation

[see also eqs. (12-53) and (14-7)]. Again, ML estimation depends on the particular choice of training symbols $\mathbf{a}_T$, while MAP estimation from $\hat{\mathbf{c}}_{S,T}(\mathbf{a}_T)$ as well as the (nominal) error covariance $\sigma^2_{c;k}$ depend on the channel parameters (λ'_D, N_0) and the positions k_0, k_1, ..., $k_{\overline{N}-1}$ of the training symbols relative to the position k of the desired channel estimate $\hat{c}_k$. In theory, a distinct set of (real-valued) Wiener coefficients $\mathbf{w}_k$ needs to be computed for each instant k.

The scenarios of all three types of optimal flat fading channel estimation, viz. Wiener smoothing using *all* symbols (joint detection and estimation), optimal one-step channel prediction using past symbols (DD receiver), and Wiener smoothing using training symbols (DA receiver), are illustrated in Figure 14-7. Depending on the position k of the desired channel sample, DA estimation is seen to encompass the tasks of optimal channel *interpolation* (if there are both past and future training instants k_K relative to k), *extrapolation* (if there are only past *or* future training instants k_K), and *filtering* (if k corresponds to one of the k_K).

14.2.2 Uniform DA Channel Sampling and Estimation

On a more abstract level, data-aided channel estimation calls for the reconstruction of a band-limited random process (the channel trajectory $\mathbf{c}$) from its own noisy samples (ML estimate $\hat{\mathbf{c}}_{S,T}$). Hence, DA estimation is intimately linked with concepts of *sampling theory*. In fact, the sampling theorem has been and remains to be the theoretical basis of most practical DA channel estimation schemes where the training symbols are spaced F symbol intervals apart such that the channel is sampled at or above the Nyquist rate $2\lambda'_D$, i.e., $F \leq 1/(2\lambda'_D)$. Conversely, aspects of sampling and signal reconstruction can be viewed from the perspective of *estimation theory*. In that vein, linear Wiener MMSE estimation – in particular, Wiener smoothing according to eq. (14-36) – constitutes a more general mathematical framework within which conventional reconstruction of uniformly sampled random signals via $\sin(x)/x$ interpolation (regardless of whether the signal is noisy or not) is just a special case. Over and above that, Wiener estimation effectuates optimal noise suppression, optimal interpolation, extrapolation, and signal reconstruction from unequally spaced samples. In all of these scenarios, estimator performance (noise and/or aliasing) remains completely analyzable via eq. (14-36). This also extends to intercellular cochannel interference (CCI) which can be treated as if it were AWGN with power equal to that of the interfering signal(s) [14]. The Gaussian approximation holds with high accuracy for both uncoded and coded systems at all relevant SNR levels.

Let us now study in further detail uniform DA channel sampling and estimation under *best-case* conditions. In particular, we assume *quasi-continuous* transmission (message length much larger than the channel coherence time, i.e., $N \gg 1/\lambda'_D$), *steady-state* Wiener filtering (time index k far from either end of the message, i.e., $k \gg 1/\lambda'_D$ and $N-k \gg 1/\lambda'_D$), and channel *oversampling* through uniformly-spaced training symbols positioned at time instants $k_K = KF$ $(K = 0, \ldots, \overline{N}-1,\ F \leq 1/(2\lambda'_D))$. With uniform sampling and steady-state receiver processing, only F sets of interpolation filter coefficients $\mathbf{w}_{k'}$ [eq. (14-36)] need to be precomputed, one for each relative ("intrablock") position $k' = 0, \ldots, F-1$[11].

With infinite-length uniform filtering, the set of coefficients $\mathbf{w}_{k'}$ $(k' = 0, \ldots, F-1)$ can easily be determined as follows. The sequence $\hat{c}_{S,T;k} = \hat{c}_{S;K}$ of ML channel probes, sampled at rate $1/FT$ from a channel trajectory with Doppler frequency λ_D, has the same statistical properties as a sequence $\hat{c}_{S;k}$ of channel probes, sampled at symbol rate $1/T$ from a channel trajectory with F times larger Doppler frequency $F \cdot \lambda_D$. Making use of the approximations discussed in Chapter 12, the filter coefficients are therefore given by those of the "rectangular" Wiener filter for optimal smoothing [eq. (12-63)], with λ'_D replaced by $F\lambda'_D$ and index

[11] The *relative* position k' of a symbol (or channel sample) with respect to the previous training symbol (at position $k = k_K$) corresponds to the *absolute* position $k = k_K + k'$.

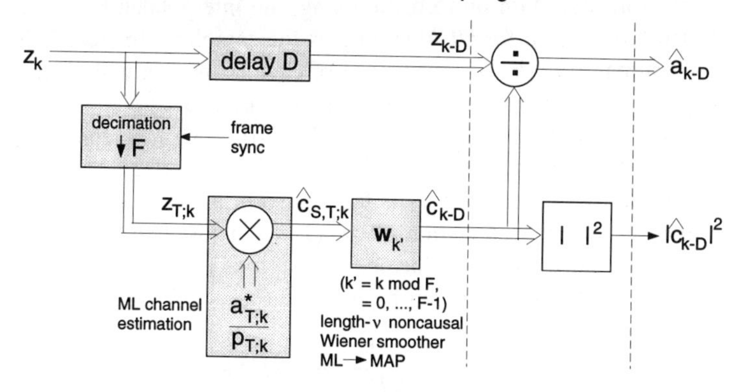

Figure 14-8 Flat Fading DA Receiver for Continuous Data Transmission (Uniform Sampling)

k replaced by $n - k'/F$:

$$w_{n;k'} = \frac{1}{1 + 2F\lambda'_D/\overline{\gamma}_s} 2F\lambda'_D \, \mathrm{si}\left[2\pi F\lambda'_D\left(n - \frac{k'}{F}\right)\right] \tag{14-37}$$

Naturally, the filters $\mathbf{w}_{k'}$ must be truncated to some finite length, say ν. With symmetric filter length truncation ($\nu/2$ "causal" and $\nu/2$ "noncausal" coefficients), quasi-optimal performance results if the filter spans $(\nu/2)F \gg 1/\lambda'_D$ symbol intervals to either side of the present block. Such a receiver with uniform channel sampling is shown in Figure 14-8. Since uniform DA channel estimation from training samples $z_{T;k}$ using symmetric filters $\mathbf{w}_{k'}$ introduces a delay of $D = (\nu/2)F$ symbol intervals. the data samples z_k must also be delayed by D. In the figure, the inner receiver outputs the (delayed) soft symbol decisions $\hat{a}_k$ and the CSI $\mid \hat{c}_k \mid^2$), but it may equivalently produce the tuple $(\hat{a}_k, \gamma_{s;k})$ as in Figure 13-6 or the signal pair $(z_k, \hat{c}_k)$ [see eq. (13-14)] as needed for the computation of the decision metric increment $\Delta m_{D;k}(a_k)$ according to eq. (13-12); the metric is identical to that of DD reception with the DD predictor estimate $\hat{c}_{k|k-1}$ replaced by the DA estimate $\hat{c}_k$. The receiver of Figure 14-8 readily extends to the case of diversity reception where DA channel estimation must be performed in each diversity branch prior to combining (Figure 13-6).

The error analysis of uniform DA channel sampling and estimation closely follows that of DD channel estimation or optimal channel smoothing. Of course, ML channel estimation $\hat{c}_{S,T;k} = (a^*_{T;k}/p_{T;k})z_{T;k}$ [eq. (14-36)] yields the same error covariance $\sigma^2_{S,T;k} = 1/(\overline{\gamma}_s p_{T;k})$ as in eqs. (14-7) (DD) and (12-53) (optimal smoothing). The performance $\sigma^2_{c;k}$ of MAP estimation according to eq. (14-37) is assessed as follows. With uniform sampling, the interpolated estimates $\hat{c}_k$ bear the same error covariance for all k as long as the sampling theorem holds. Since the optimal estimator works on a sequence of channel probes with F times larger Doppler than the original channel, the error covariance $\sigma^2_{c;k} = \sigma^2_c$ is identical to that of optimal Wiener smoothing [eq. (12-53)], with λ'_D replaced by $F\lambda'_D$. However, since the approximation $r_c(\lambda'_D, \overline{\gamma}_s) \approx r_c(\lambda'_D) \approx 2F\lambda'_D$ [eq. (12-62)] neglecting the impact of the SNR $\overline{\gamma}_s$ is too coarse in the case of smaller effective sampling margins, the approximation of eq. (12-61) is used so that the expression for the error covariance reduction factor r_c – the total error covariance, assuming PSK training symbols, is then given by $\sigma^2_c = r_c/\overline{\gamma}_s$ – reduces to

$$r_c = \frac{1}{1 + 2F\lambda'_D/\overline{\gamma}_s}\, 2F\lambda'_D \tag{14-38}$$

As expected, the error performance of uniform DA channel estimation (every Fth symbol is used for training) is worse by a factor of $\approx F$ than that of optimal smoothing where *all* symbols are used as if they were training symbols.

The total SNR loss factor $\eta = \eta_F \eta_c$ of DA reception depends on the quality of channel estimation and on the percentage $\eta_F = (F-1)/F$ of training symbols inserted into the symbol stream. The SNR reduction factor η_c resulting from imperfect channel estimation – originally derived for symbol-by-symbol QAM or PSK detection but, thanks to symbol randomization through interleaving, also valid for interleaved coded modulation – is obtained as follows. Since the performance analysis of the previous section [eqs. (13-12) through (14-27)] does not depend on the particular method of channel estimation, the term $(1 - r_c/[n_b\overline{\gamma}_b])/(1 + r_c)$, which has been identified as the (average) SNR degradation factor resulting from DD channel prediction, is also valid for DA channel estimation, with σ^2_c now given by eq. (14-38). Hence, the average BER for DA reception on flat fading channels is well approximated by eqs. (14-26) (no diversity) or (14-27) (diversity), with SNR degradation factor:

$$\begin{aligned}
&\text{Total SNR loss :}\\
\eta &= \eta_F\,\eta_c\\
&\text{SNR loss due to training symbol insertion :}\\
\eta_F &= \frac{F-1}{F}\\
&\text{SNR loss due to imperfect channel estimation :}\\
\eta_c &\simeq \frac{1 - r_c/\overline{\gamma}_s}{1 + r_c} = \frac{1}{1 + (1 + 1/\overline{\gamma}_s)\,2F\lambda'_D}
\end{aligned} \tag{14-39}$$

and $\overline{\gamma}_s = n_b\overline{\gamma}_b$ the (average) SNR per symbol.

Obviously, there is a trade-off between η_F and η_c; a higher percentage of training symbols improves on the quality of channel estimation but reduces the power (and bandwidth) efficiency, and vice versa. Figure 14-9 displays SNR loss

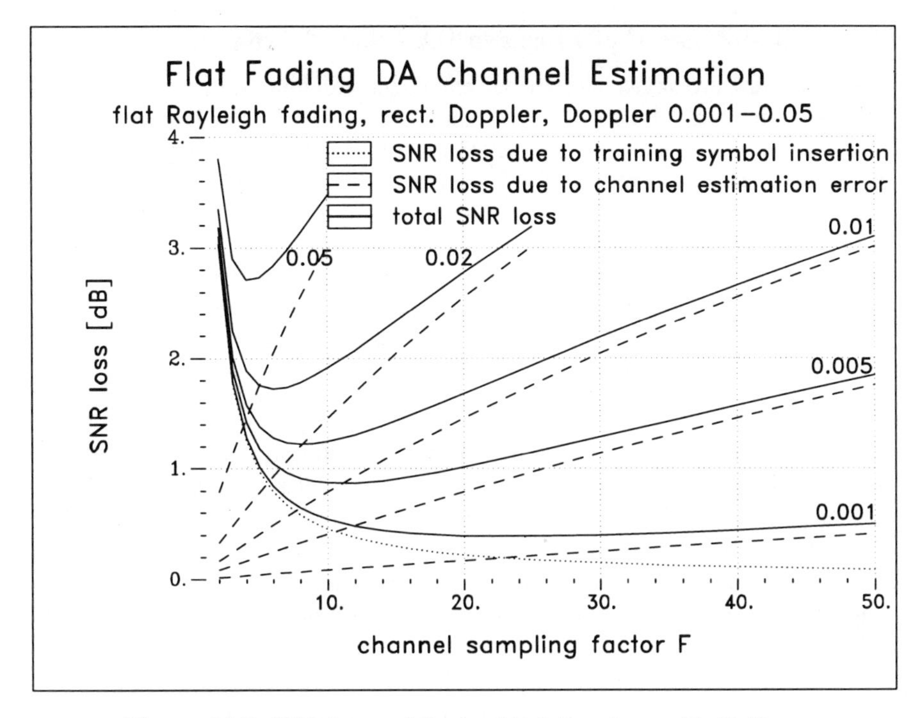

Figure 14-9 SNR Loss of Optimal DA Receiver with Uniform Channel Sampling for Flat Fading Channels

factors $10 \log_{10}(1/\eta)$ (in decibels) for Doppler frequencies $\lambda'_D = 10^{-3}, 5 \times 10^{-3}$, 0.01, 0.02 and 0.05, versus the channel sampling factor (block length) F. Since η_c is not much dependent on the SNR $\overline{\gamma}_s$, the curves are given for the limiting case $\overline{\gamma}_s \to \infty$. The estimation loss $10 \log_{10}(1/\eta_c)$ is seen to rise almost linearly with F until it reaches 3 dB at $F_{\max} = 1/\left(2\lambda'_D\right)$ (Nyquist rate) where the total SNR loss is dominated by η_c (see, e.g., λ'_D=0.01, F=50). Depending on λ'_D, the total loss has a distinct minimum. By setting the derivative of η [eq. (14-39)] with respect to F to zero, one obtains the associated optimal sampling factor

$$F_{\text{opt}} \approx \text{int}\left[1 + \frac{1}{\sqrt{2\,\lambda'_D}}\right] = \text{int}\left[1 + \sqrt{F_{\max}}\right] \tag{14-40}$$

with int$[x]$ the nearest integer to x. This optimal frame length is usually much smaller than the maximum sampling factor $F_{\max}$ determined by the Nyquist rate, which suggests that DA flat fading channel estimation should be designed such that there is a considerable oversampling margin. Apart from reducing the total SNR loss, selecting smaller values of F provides for a safety margin against larger Doppler (or frequency offsets) and may also facilitate other tasks such as frame sync or online estimation of unknown statistical channel parameters.

14.2.3 Aliasing and End Effects in DA Channel Estimation

Let us now investigate aliasing and end effects resulting from various kinds of "nonideal sampling" of a flat fading channel. Nonideal sampling is often dictated by real-world constraints such as channel access, receiver complexity, or latency. For instance, limited complexity and latency both call for finite-length channel smoothing. Also, most multiuser cellular radio systems feature some kind of TDMA channel access so that the receiver must be able to perform synchronization based on a single message burst. This is even more important in the case of spontaneous packet transmission where frequency and frame synchronization may be required in addition to timing sync and channel estimation. Some examples of typical message formats are shown in Figure 14-10. The message formats studied here are the following:

a. *Finite-length* DA channel estimation using ν uniformly spaced channel samples with sampling factor $F = F_{\text{opt}} = 11 \mid 4$ (Doppler $\lambda'_D = 0.005 \mid 0.05$)
b. DA channel estimation near the *start* (first four blocks) of a long message with uniform channel sampling (sampling factor $F = F_{\text{opt}}$) and no preamble
c. DA channel estimation near the start (first four blocks) of a long message with uniform sampling, preceded by a short *preamble* block of length $P = 5 \mid 2$

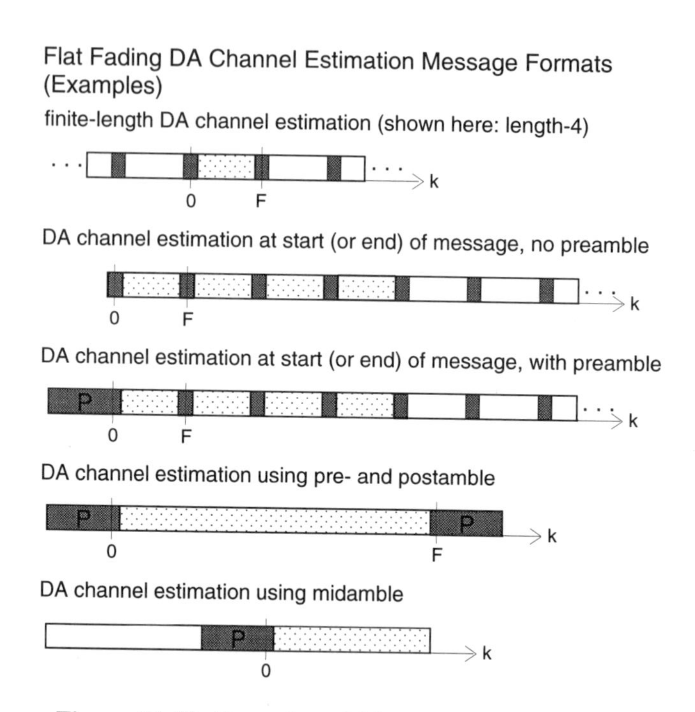

Figure 14-10 Examples of Message Formats for DA Flat Fading Channel Estimation

d. DA channel estimation (interpolation) within the data block of length $F_{\max}/2 - 1 = 49 \mid 4$ embedded in two short *pre-* and *postamble* blocks, both of length $P = 5 \mid 2$
e. DA channel estimation (extrapolation) within the data block following a short *midamble* of length $P = 5 \mid 2$

The error covariance factors $r_{c;k} = \sigma^2_{c;k}\overline{\gamma}_s$ evaluated via eq. (14-7) are now dependent not only on $(\lambda'_D, \overline{\gamma}_s)$ but also on the sample index k. The result corresponding to the frame formats (a) through (e) listed above are shown in Figures 14-11 through 14-14.

From Figure 14-11 (λ'_D=0.005, $F = F_{\text{opt}}$= 11) one observes that finite-length symmetric DA filtering (case a) will not significantly compromise estimator performance even in the case of short filters with only ν= 6 or 4 coefficients. Very simple length-2 filters exhibit a noticeable – yet still tolerable – degradation which is due to the limited degree of noise averaging. The same is also true for fast Doppler (λ'_D=0.05, $F = F_{\text{opt}}$= 4), except for the somewhat larger estimation error ($r_{c;k}$ between 0.35 and 1; results not shown). At any rate, the use of finite-length Wiener filtering with spacing $F = F_{\text{opt}}$ is seen to yield significant improvements ($r_{c;k}$ usually well below 1) with respect to ML estimation alone, especially when the SNR is low.

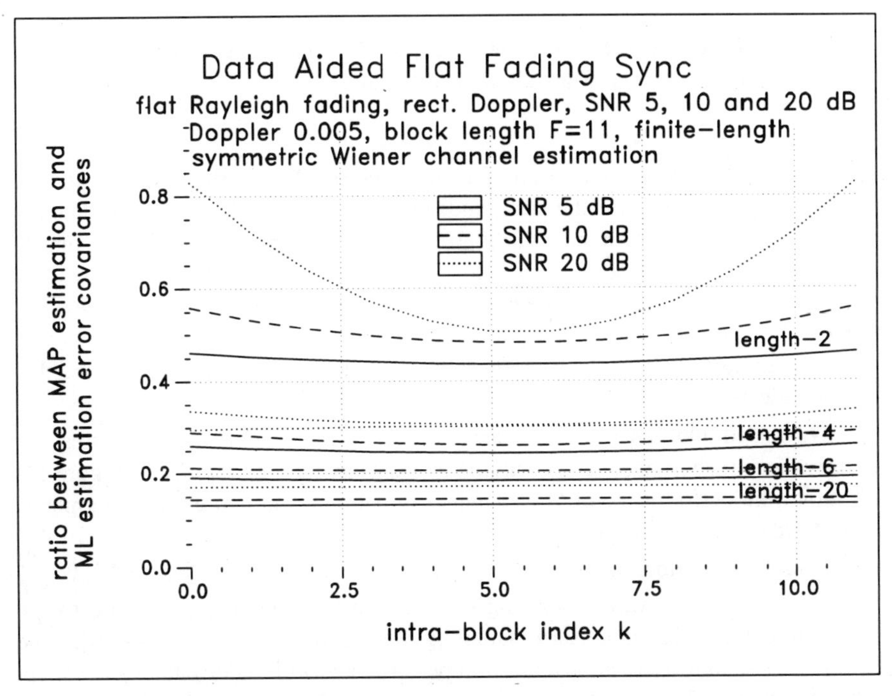

Figure 14-11 Performance of Finite-Length Symmetric DA Flat Fading Channel Estimation

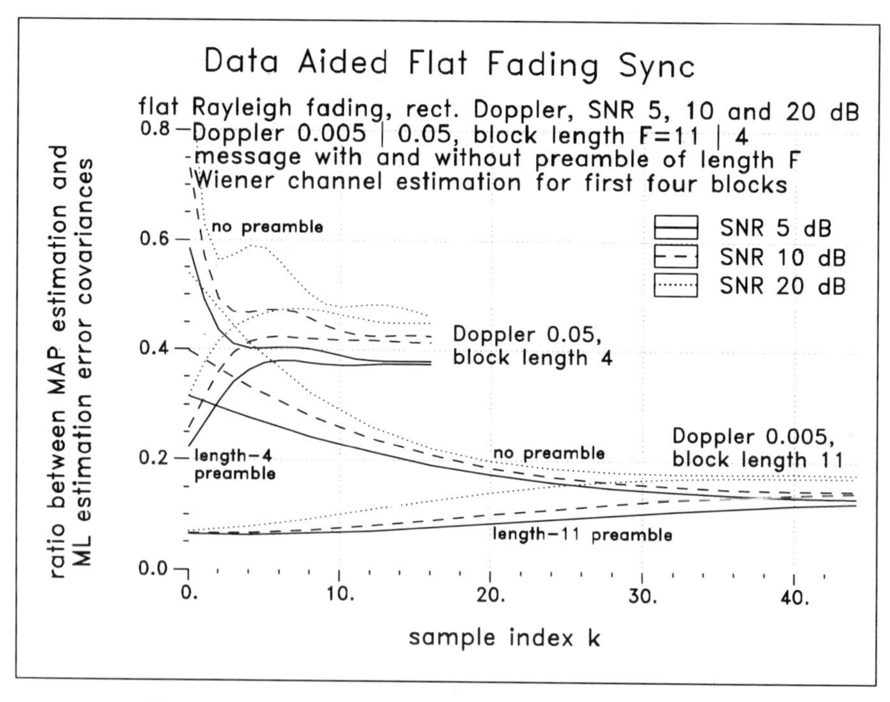

Figure 14-12 Performance of DA Flat Fading Channel Estimation at Start (End) of Message

Figure 14-12 shows the results for unsymmetric DA filtering (cases b and c), most often performed near the start (end) of a message, or when a severe latency constraint precludes any noncausal filtering. Without a preamble (case b), some – again tolerable – degradation results for the first two or three data blocks ($r_{c;k}$ is shown for the first four data blocks of length $F = F_{\text{opt}} = 11 \mid 4$). If necessary, this can be compensated by using a preamble (case c); Figure 14-12 reveals that a short preamble (length $11 \mid 4$) suffices to more than offset any degradations due to end effects.

Advanced multiuser channel access with TDMA component requires the transmission of short data bursts comprising very few training segments. Figure 14-13 shows the performance of DA channel estimation using two training segments, one at either end of the message, viz. a pre- and a postamble (case d). In this example, the data block is of length $F_{\max}/2 - 1 = 49 \mid 4$ where $F_{\max} = 1/\left(2\lambda_D'\right)$ $= 100 \mid 10$ is the maximum spacing between training symbols according to the sampling theorem [as if uniform sampling were applied; see eq. (14-40)]. The estimator is seen to yield almost uniform performance within the data block, except when the SNR is large. In that case aliasing effects dominate in the central region of the message, and the interpolation error can exceed the error in the ML estimate ($r_{c;k}$ larger than 1), whereas the estimation error near the training segments at either end of the message is dominated by the noise averaging capability of the training

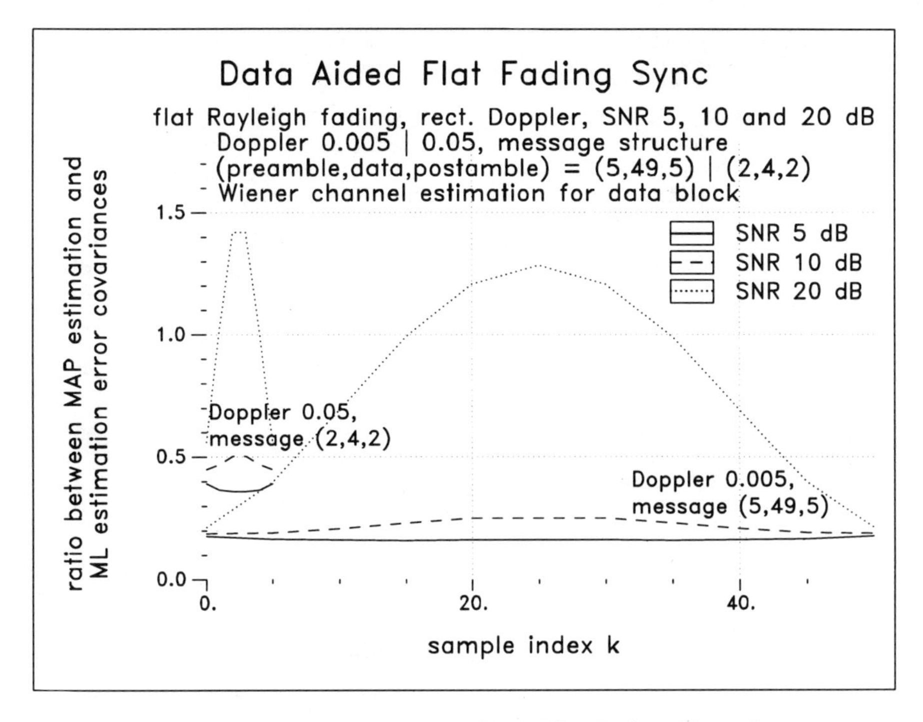

Figure 14-13 Performance of DA Flat Fading Channel Estimation in Between Pre- and Postamble

segments. When the SNR is small, on the other hand, the noise may completely mask any aliasing effects even in the center of the message (Figure 14-13, results for 5 and 10 dB).

As a general rule of thumb, one may state that a message format should be designed such that aliasing effects can only become the dominant source of error at SNR levels considerably higher than the nominal point (or region) of operation. In the examples studied so far, this has always been the case, at least when considering average SNRs below 20 dB. If a message burst comprises only a single training segment (midamble, case e), the degree of aliasing heavily depends on the channel dynamics relative to the length of the data blocks preceding and following the midamble. As shown by Figure 14-14, the error in the extrapolated DA channel estimate quickly rises with increasing distance from the midamble. As a consequence, using a midamble is only advisable if the data blocks are much shorter than the channel coherence time. In this example, the estimation error remains below or is comparable to that of ML channel estimation at 10 dB SNR ($r_{c;k}$ smaller than or in the order of 1) for data blocks not longer than about (1/7)$\times 1/(2\lambda'_D) \approx 14 \mid 1.4$. Hence, if fading is a problem, it is advisable to augment the message by a postamble (case d) since then the data blocks are allowed to be much longer (here: data block length 49 | 4).

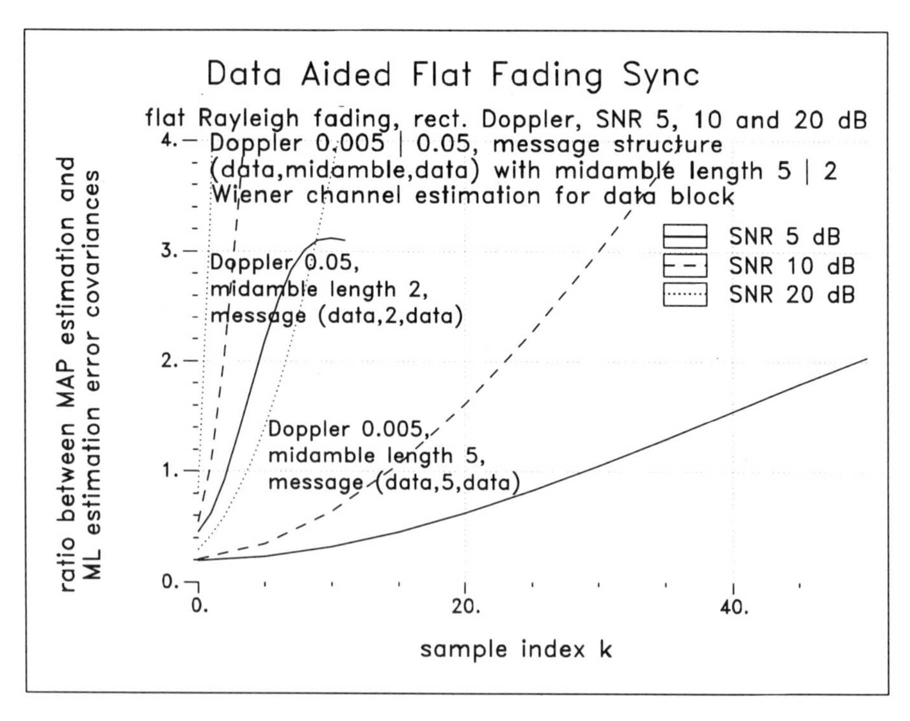

Figure 14-14 Performance of DA Flat Fading Channel Estimation to Either Side of Midamble

14.2.4 Example: DA Receiver for Uncoded M-PSK

Let us now examine a DA receiver design example for uncoded M-PSK transmission. Again, the two design Doppler frequencies of 0.005 and 0.05 are considered. The corresponding uniform sampling factors F are chosen optimally [eq. (14-40)] as 11 and 4, respectively. With infinite-length optimal filtering [eq. (14-37)], the error covariance factors r_c [eq. (14-38)] are in the range 0.106,...,0.11 ($\overline{\gamma}_s$= 5,...,30 dB) for (λ'_D=0.005, F=11), and 0.36,...,0.4 ($\overline{\gamma}_s$= 5,...,30 dB) for (λ'_D=0.05, F=4). The corresponding total SNR loss η [eq. (14-39)] evaluates as 1.02,...,0.87 dB (λ'_D=0.005, F=11, η_F=10/11) and 3.3,...,2.7 dB (λ'_D=0.05, F=4, η_F=3/4), respectively.

A good compromise between performance and complexity results if the length ν of the symmetric DA estimation filter (Figure 14-8) is chosen such that the filter spans the main lobe of the channel autocorrelation function: $(\nu/2)F \approx 1/(2\lambda'_D)$ $\rightarrow \nu \approx 1/(F\lambda'_D)$. With ν=21 (λ'_D=0.005, F=11) and 7 (λ'_D=0.05, F=4), the error covariance factors r_c [eq. (14-36)] evaluate as 0.14,...,0.19 and 0.42,...,0.70, respectively; these figures are not much worse than those associated with infinite-length filtering (see above).

Like with DD estimation, these error covariance factors translate into bit error curves via eqs. (14-26) (no diversity) or (14-27) (diversity). Figures 14-15 (λ'_D=0.005, F=11) and 14-16 (λ'_D=0.05, F=4) display BER curves for 2-, 4-, 8- and 16-PSK without diversity. In addition, Table 14-3 summarizes these results – along with those obtained by simulation – for no diversity as well as dual diversity with maximum ratio combining. With length-ν DA filtering and no diversity, the SNR loss is observed to be only slightly larger than with optimal infinite-length filtering; the additional degradation is in the order of 0.1–0.3 dB (λ'_D=0.005, F=11, ν=21) and between 0.3 and 1 dB (λ'_D=0.05, F=4, ν=7) without diversity, and 0.1–0.3 dB and 0.2–0.8 dB with dual diversity, respectively. Again, the use of diversity tends to mitigate the effects of imperfect channel estimation.

The performance of the DA receiver (Figure 14-8) and its extension to diversity reception (Figure 13-6) was also verified by simulation. When it comes to receiver implementation, the dependence of the filter coefficient sets $\mathbf{w}_{k'}$ on the parameters λ'_D, k', and $\overline{\gamma}_s$ is a nuisance that needs some attention. Assuming that the block length F is tailored to a certain (maximum) Doppler λ'_D, the filters $\mathbf{w}_{k'}$ can likewise be designed for this particular Doppler frequency λ'_D so that the two parameters k' and $\overline{\gamma}_s$ remain. Since the block length $F \approx F_{\text{opt}}$ is considerably smaller than the maximum allowable F_{max} [eq. (14-40)], the channel process is oversampled by a factor of 9 (λ'_D=0.005, F_{max}=100, F=11) and 2.5 (λ'_D=0.05, F_{max}=10, F=4) for this example. Therefore, Wiener filtering needs not be performed for each and every $k' = 0, \ldots, F-1$ but for very few intermediate "pole" positions only, say k'=0 (λ'_D=0.005, F=11) and k'=0 and 2 (λ'_D=0.05, F=4). For k' in between these pole positions, the estimated trajectory is reconstructed by simple linear interpolation.

With infinite-length filtering, the remaining parameter $\overline{\gamma}_s = 1/N_0$ affects only the overall gain factor $1/(1+2F\lambda'_D/\overline{\gamma}_s)$ [eq. (14-37)] so that the computation of $\mathbf{w}_{k'}(\overline{\gamma}_s)$ is very simple in that case. On the other hand, finding a simple functional approximation to the length-ν filter coefficients

$$\mathbf{w}_{k'}(\overline{\gamma}_s) = \Bigg(\underbrace{\mathbf{R}_{D,T} + \frac{1}{\overline{\gamma}_s}\mathbf{I}}_{\mathbf{\Phi}}\Bigg)^{-1} \cdot \mathbf{r}_{D,T;k'} \tag{14-41}$$

[eq. (14-36)] is less obvious. As shall be shown below, the νth-order matrix operations of eqs. (14-37) and (14-41) can be circumvented by an eigensystem synthesis approach where the matrix $\mathbf{\Phi}$ is expressed as

$$\begin{aligned}\mathbf{\Phi}(\overline{\gamma}_s) &= \mathbf{R}_{D,T} + \frac{1}{\overline{\gamma}_s}\mathbf{I} = \mathbf{U}\mathbf{\Lambda}\mathbf{U}^H = \sum_{i=0}^{\nu-1} \lambda_i \cdot \mathbf{u}_i\mathbf{u}_i^H \\ &= \mathbf{U}\left(\mathbf{\Lambda}_{D,T} + \frac{1}{\overline{\gamma}_s}\mathbf{I}\right)\mathbf{U}^H = \mathbf{U}\mathbf{\Lambda}_{D,T}\mathbf{U}^H + \mathbf{U}\left(\frac{1}{\overline{\gamma}_s}\mathbf{I}\right)\mathbf{U}^H\end{aligned} \tag{14-42}$$

with eigenvector matrix $\mathbf{U} = (\mathbf{u}_0 \ \ldots \mathbf{u}_{\nu-1})$ satisfying $\mathbf{U}^H\mathbf{U} = \mathbf{U}\mathbf{U}^H = \mathbf{I}$, $\mathbf{\Lambda} = \text{diag}\{\lambda_0, \ldots, \lambda_{\nu-1}\}$ the (diagonal) eigenvalue matrix of $\mathbf{\Phi}$, and $\mathbf{\Lambda}_{D,T} = \text{diag}\{\lambda_{D,T;0}, \ldots, \lambda_{D,T;\nu-1}\}$ the eigenvalue matrix of $\mathbf{R}_{D,T}$. The eigenvalues λ_i of $\mathbf{\Phi}$ are therefore given by $\lambda_i = \lambda_{D,T;i} + 1/\overline{\gamma}_s$, and the eigenvectors $\mathbf{u}_i$ of $\mathbf{\Phi}$ are the same as those of $\mathbf{R}_{D,T}$ and thus independent of the SNR $\overline{\gamma}_s$. Hence, the weight vector $\mathbf{w}_{k'}(\overline{\gamma}_s)$ can be written as

$$\begin{aligned} \mathbf{w}_{k'}(\overline{\gamma}_s) &= \mathbf{\Phi}^{-1}(\overline{\gamma}_s) \cdot \mathbf{r}_{D,T;k'} = \mathbf{U}\mathbf{\Lambda}^{-1}\mathbf{U}^H \cdot \mathbf{r}_{D,T;k'} \\ &= \left(\sum_{i=0}^{\nu-1} \frac{1}{\lambda_{D,T;i} + 1/\overline{\gamma}_s} \cdot \mathbf{u}_i\mathbf{u}_i^H \right) \cdot \mathbf{r}_{D,T;k'} \\ &= \sum_{i=0}^{\nu-1} \frac{1}{\lambda_{D,T;i} + 1/\overline{\gamma}_s} \cdot \underbrace{\left([\mathbf{u}_i\mathbf{u}_i^H] \cdot \mathbf{r}_{D,T;k'} \right)}_{\mathbf{w}_{i;k'}} \end{aligned} \tag{14-43}$$

Since the partial weigth vectors $\mathbf{w}_{i;k'}$ are independent of $\overline{\gamma}_s$, they can be precomputed and stored, together with the set of eigenvalues $\{\lambda_{D,T;i}\}$. During online receiver operation, the SNR-dependent weight vector $\mathbf{w}_{k'}(\overline{\gamma}_s)$ can now be easily computed by superposition.

When performing the eigensystem analysis of eq. (14-42), one finds that only a small number $\overline{\nu} < \nu$ of eigenvalues $\{\lambda_{D,T;0}, \ldots, \lambda_{D,T;\overline{\nu}-1}\}$ and respective eigenvectors $\{\mathbf{u}_0, \ldots, \mathbf{u}_{\overline{\nu}-1}\}$ are significant. Reformulating the inverse of $\mathbf{\Phi}$ as

$$\begin{aligned} \mathbf{\Phi}^{-1} &= \sum_{i=0}^{\nu-1} \frac{1}{\lambda_{D,T;i} + 1/\overline{\gamma}_s} \cdot \mathbf{u}_i\mathbf{u}_i^H \\ &\simeq \left(\sum_{i=0}^{\overline{\nu}-1} \frac{1}{\lambda_{D,T;i} + 1/\overline{\gamma}_s} \cdot \mathbf{u}_i\mathbf{u}_i^H \right) + \left(\sum_{i=\overline{\nu}}^{\nu-1} \overline{\gamma}_s \cdot \mathbf{u}_i\mathbf{u}_i^H \right) \\ &= \left(\sum_{i=0}^{\overline{\nu}-1} \frac{1}{\lambda_{D,T;i} + 1/\overline{\gamma}_s} \cdot \mathbf{u}_i\mathbf{u}_i^H \right) + \overline{\gamma}_s \underbrace{\left(\sum_{i=0}^{\nu-1} \mathbf{u}_i\mathbf{u}_i^H \right)}_{=\mathbf{U}\mathbf{U}^H=\mathbf{I}} - \overline{\gamma}_s \left(\sum_{i=0}^{\overline{\nu}-1} \mathbf{u}_i\mathbf{u}_i^H \right) \\ &= \left(\sum_{i=0}^{\overline{\nu}-1} \left[\frac{1}{\lambda_{D,T;i} + 1/\overline{\gamma}_s} - \overline{\gamma}_s \right] \cdot \mathbf{u}_i\mathbf{u}_i^H \right) + \overline{\gamma}_s \mathbf{I} \\ &= \overline{\gamma}_s \cdot \left(\mathbf{I} - \left[\sum_{i=0}^{\overline{\nu}-1} \frac{\lambda_{D,T;i}}{\lambda_{D,T;i} + 1/\overline{\gamma}_s} \cdot \mathbf{u}_i\mathbf{u}_i^H \right] \right) \end{aligned} \tag{14-44}$$

leads to the simplified weight vector synthesis

$$\mathbf{w}_{k'}(\overline{\gamma}_s) = \overline{\gamma}_s \cdot \left(\mathbf{r}_{D,T;k'} - \sum_{i=0}^{\overline{\nu}-1} \frac{\lambda_{D,T;i}}{\lambda_{D,T;i} + 1/\overline{\gamma}_s} \cdot \underbrace{\left([\mathbf{u}_i\mathbf{u}_i^H] \cdot \mathbf{r}_{D,T;k'} \right)}_{\mathbf{w}_{i;k'}} \right) \tag{14-45}$$

Hence, only $\overline{\nu}$ partial weight vectors $\{\mathbf{w}_{0;k'}, \ldots, \mathbf{w}_{\overline{\nu}-1;k'}\}$ and eigenvalues $\{\lambda_{D,T;0}, \ldots, \lambda_{D,T;\overline{\nu}-1}\}$ need to be precomputed and stored for online SNR-dependent weight vector generation.

If the filter $\mathbf{w}_{k'}$ exhibits even symmetry about the pole position k' of interest[12], $\overline{\nu}$ is even smaller; for the DA receiver example discussed here, only $\overline{\nu}$=3 symmetric partial weight vectors $\mathbf{w}_{i;k'}$ and respective eigenvalues remain. This motivates the choice of ν=21 for (λ'_D=0.005, F=11, k'=0), ν=7 for (λ'_D=0.05, F=4, k'=0), and ν=6 for (λ'_D=0.05, F=4, $k' = F/2$=2).

The results obtained by simulation, making use of linear interpolation between pole positions and the weight vector generation procedure just described, are shown in Figures 14-15 and 14-16 (no diversity) along with the analytically derived BER curves. Again, Table 14-3 summarizes these results for 2-,4-,8- and 16-PSK DA reception without and with dual diversity.

Compared with DD reception, one observes a better agreement between analysis and simulation. For 2- and 4-PSK (no diversity) and all signal alphabets from 2- to 16-PSK (dual diversity, not shown here), the simulated BER curves are indistinguishable from their analytically obtained counterparts. Only in the case of no diversity and dense symbol constellations (8- and 16-PSK) the simulated BER

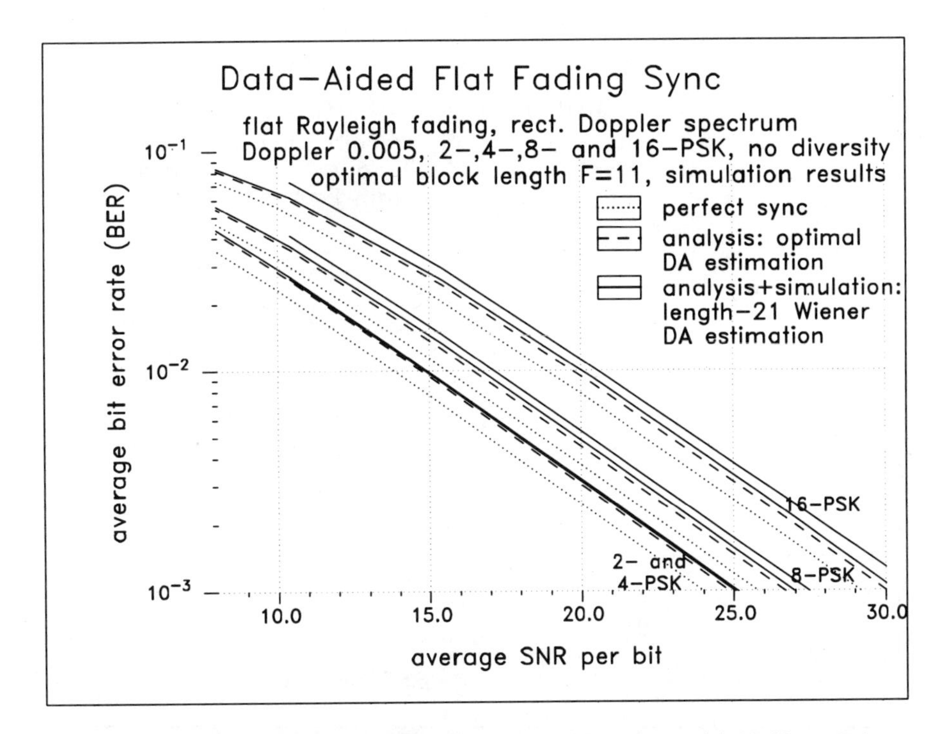

Figure 14-15 Simulated BER Performance of Receiver with Uniform DA Flat Fading Channel Estimation, Doppler 0.005, F=11

[12] The filter $\mathbf{w}_{k'}$ is symmetric for pole position k'=0 and ν odd, or $k' = F/2$ and ν even.

is seen to be slightly larger (Figures 14-15 and 14-16), which can be attributed to the fact that symbol errors occurring at low instantaneous SNR levels may entail more than a single bit error.

According to Tables 14-2 (DD) and 14-3 (DA), DA reception is superior in terms of power efficiency (simulated BER) for all scenarios investigated here, even when considering the power that has to be spent for training. However, the insertion of training symbols also costs bandwidth which has to be taken into account for a fair comparison. Figure 14-17 compares DD and DA reception in terms of both power and bandwidth efficiency, based on the SNR (per bit, per channel) needed for a (simulated) BER of 0.01. For low Doppler (λ'_D=0.005), DA reception remains to be superior; here the reduction in bandwidth efficiency (factor 10/11) is more than offset by the good power efficiency. This is also true for high Doppler (0.05) and no diversity, in spite of the compromised bandwidth efficiency (factor 3/4), since with DD reception denser symbol constellations are more sensitive against fading. DD reception can be superior only when the DA training overhead is large (high Doppler) and dual diversity is available. In this case, however, DA reception can also be made more bandwidth efficient by using less frequent training, e.g., $F = F_{\max}$=10 (λ'_D=0.05), and performing two-stage detection and channel estimation [11].

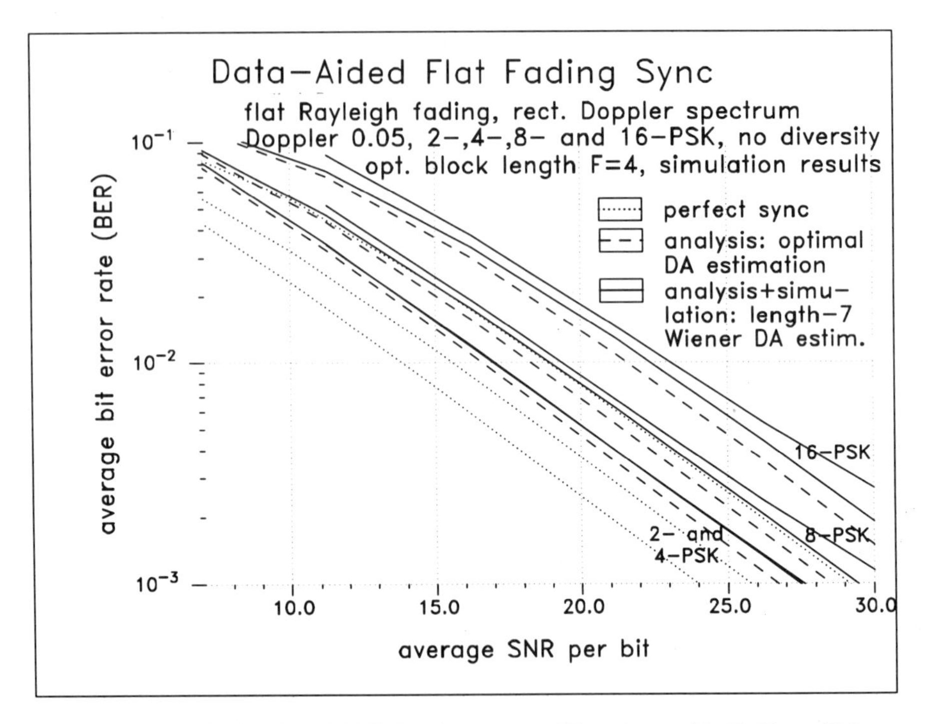

Figure 14-16 Simulated BER Performance of Receiver with Uniform DA Flat Fading Channel Estimation, Doppler 0.05, F=4

Table 14-3 BER Performance of Receiver with Uniform DA Flat Fading Channel Estimation

		Analysis			Simulation
		perfect sync	optimal uniform DA	length-21 I 7 uniform DA	length-21 I 7 uniform DA
		Doppler $\lambda'_D = 0.005 \vert 0.05$, optimal block length (sampling factor) F=11 I 4			
		SNR [dB] per bit and channel @ BER=10^{-2}			
$D=1$	2-PSK	13.9	14.8 I 16.7	14.9 I 17.0	14.9 I 17.0
	4-PSK	13.9	14.8 I 16.7	14.9 I 17.0	14.9 I 17.0
	8-PSK	15.6	16.5 I 18.3	16.6 I 18.9	17.0 I 19.2
	16-PSK	18.9	19.8 I 21.6	19.9 I 22.4	20.4 I 23.0
$D=2$	2-PSK	5.3	6.3 I 8.3	6.4 I 8.5	6.4 I 8.5
	4-PSK	5.3	6.3 I 8.3	6.4 I 8.5	6.4 I 8.5
	8-PSK	8.1	9.0 I 10.8	9.2 I 11.3	9.2 I 11.3
	16-PSK	11.7	12.6 I 14.4	12.8 I 14.9	12.8 I 14.9
		SNR [dB] per bit and channel @ BER=10^{-3}			
$D=1$	2-PSK	24.0	24.9 I 26.7	25.2 I 27.5	25.2 I 27.6
	4-PSK	24.0	24.9 I 26.7	25.2 I 27.5	25.2 I 27.6
	8-PSK	25.8	26.7 I 28.5	27.0 I 29.5	27.4 I 30.6
	16-PSK	29.1	30.0 I 31.8	30.3 I 32.8	31.0
$D=2$	2-PSK	11.1	12.0 I 13.9	12.1 I 14.2	12.1 I 14.2
	4-PSK	11.1	12.0 I 13.9	12.1 I 14.2	12.1 I 14.2
	8-PSK	13.8	14.7 I 16.5	14.9 I 17.1	14.9 I 17.1
	16-PSK	17.7	18.6 I 20.4	18.9 I 21.2	18.9 I 21.2

14.2.5 Example: DA Receiver for Trellis-Coded Modulation

In the case of uncoded transmission, both DD and DA reception are viable options. With interleaved channel coding and high noise/interference, however, DD channel estimation is more critical, and feedforward DA receiver processing is highly desirable in order to avoid error propagation resulting from unreliable tentative symbol decisions. A channel-coded transmission system featuring DA

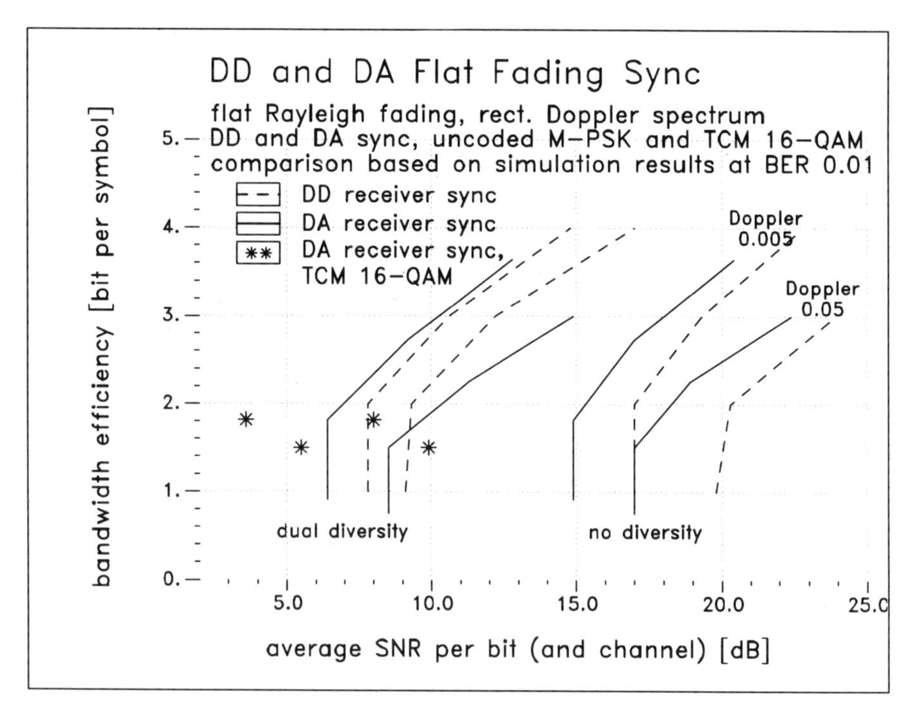

Figure 14-17 Performance Comparison of DD and DA Receivers: Power and Bandwidth Efficiency

reception (inner receiver) and interleaved trellis-coded modulation (TCM, outer receiver) is shown in Figure 14-18.

The inner receiver performs uniform DA channel estimation and delivers fading-corrected soft decisions $\hat{a}_k$ and channel state information (CSI) $\gamma_{s;k}$ (Figure 13-6). The outer transmission system comprises trellis encoding, modulation (symbol mapping), block interleaving/deinterleaving, and Viterbi decoding using the CSI. The input bits b_i (index i: noninterleaved time scale) are trellis-encoded and mapped onto PSK or QAM data symbols which are written *row-wise* into the block interleaver matrix. After adding training symbols for DA channel estimation, the channel symbols a_k (index k: interleaved time scale) are read out *column-wise* for transmission. In the outer receiver, both the soft decisions $\hat{a}_k$ and the CSI $\gamma_{s;k}$ are deinterleaved and used for Viterbi decoding.

The interleaver depth is assumed to be a multiple $I \times F$ of the frame length F. In the case of continuous transmission without frequency hopping (FH), I should be sufficiently large so as to effectuate near-ideal interleaving. In this example, $I = F$ is chosen so that the interleaver depth of $F^2 = F_{\text{opt}}^2 \approx F_{\max} = 1/\left(2\lambda_D'\right)$ [eq. (14-40)] approximately equals the (one-sided) width of the main lobe of the channel autocorrelation function. Then the channel samples pertaining to

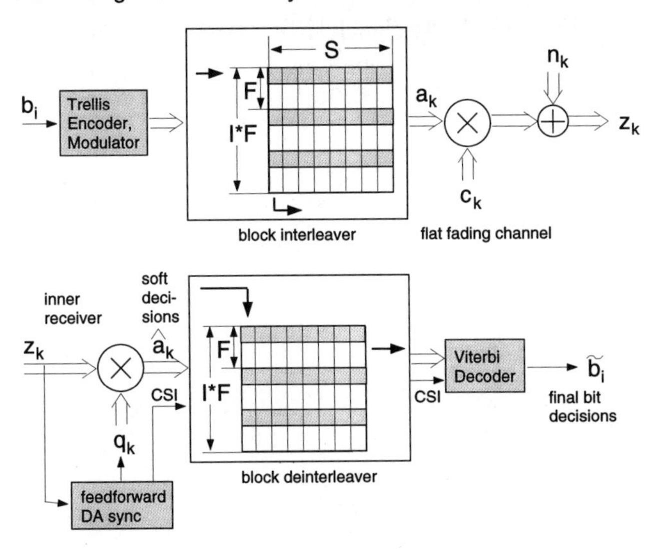

Figure 14-18 Transmission System with DA Reception and TCM Channel Coding

neighboring deinterleaved symbols, i.e., along a row in the deinterleaver matrix, are sufficiently uncorrelated. The interleaver length S, which must be large with respect to the code constraint length, is chosen here to be equal to the survivor depth D_S (here $S = D_S$ =25) of the Viterbi decoder. This choice is particularly advantageous in the context of combined decoding and equalization discussed in Section 15.2.

It is important to note that this scheme may also be combined with TDMA channel access and/or frequency hopping (FH). This is made possible by DA reception being also applicable to short packets or bursts (Section 14.2.3). For instance, one interleaver column of length $I \times F$, possibly enhanced by some additional preamble or training symbols, may constitute one FH-TDMA burst. This not only aids near-ideal interleaving at a reduced latency ($I \times F$ can be small), but has also a number of other advantages in multiuser environments, e.g., interferer and frequency diversity effects [15].

For the use on fading channels – especially on flat Rayleigh channels without antenna diversity where deep fades occur frequently – the codes should be selected for maximum *effective code length* (ECL) rather than minimal distance [16]. The ECL is the minimum number of channel symbols with nonzero Euclidean distance in an error event path and can therefore be interpreted as the order of time diversity provided by the code. Using interleaved coded modulation designed for large ECL, combined with FH-TDMA channel access and feedforward DA receiver synchronization, constitutes a very effective anti-fading and anti-interference technique [17].

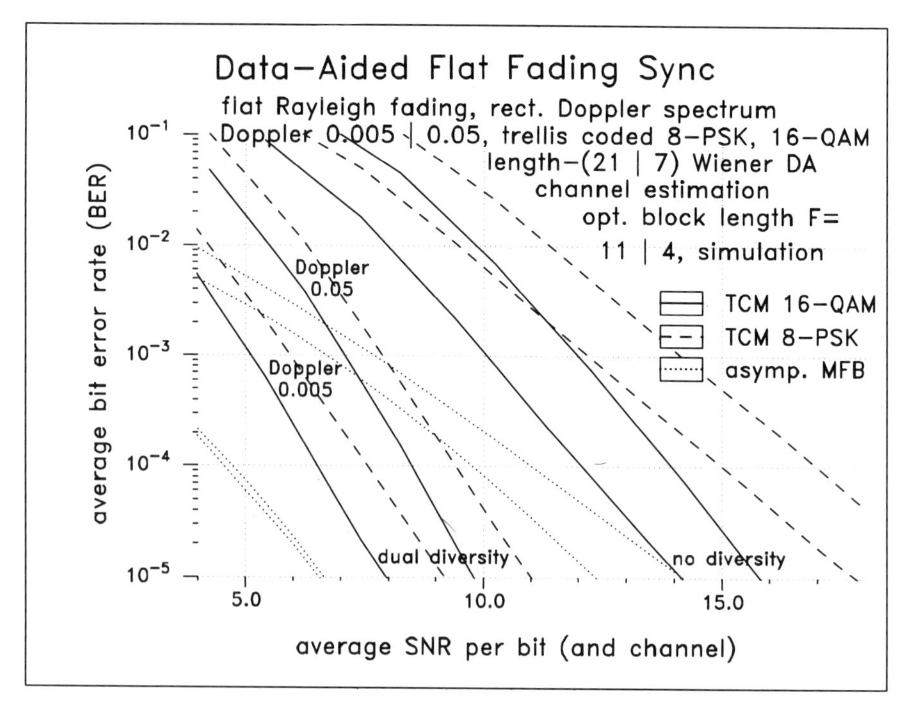

Figure 14-19 Simulated BER Performance of Transmission System with DA Flat Fading Channel Estimation and TCM Channel Coding

Two trellis codes with large ECL have been selected for this example. Both codes convey two information bits per symbol and are therefore comparable (in terms of bandwidth efficiency) with uncoded 4-PSK. The first code ("8-PSK") is a rate-2/3, 16-state 8-PSK Ungerboeck code [18] with asymptotic coding gain (AWGN channel) 4.1 dB and ECL 3. The second code ("16-QAM") by Moher and Lodge [10] is a 2× rate-1/2, 2×16 state, 2×4-PAM (16-QAM with independently encoded I and Q rails) code with asymptotic coding gain 3.4 dB and ECL 5.

Figure 14-19 shows the BER simulation results for 8-PSK and 16-QAM trellis-coded transmission over flat Rayleigh channels with $\lambda'_D = (0.005 \mid 0.05)$ and DA reception with uniform sampling and $F = (11 \mid 4)$, along with the asymptotic matched filter lower bounds (MFB) for these codes [17, 19]. The SNRs needed to achieve BERs of 10^{-2} and 10^{-3} extracted from Figure 14-19 are listed in Table 14-4. As expected, the 16-QAM code is most effective especially when antenna diversity is not available. Thanks to the large ECL of 5, the BER decays as fast as one decade per 2 dB SNR, and a BER of 10^{-2} is achievable at (8.0 | 9.9) dB SNR per bit ($\lambda'_D = (0.005 \mid 0.05)$), which is superior to the 8-PSK code by about 1.5 dB. With dual diversity, the 16-QAM code remains to be superior to the 8-PSK code (by about 0.7 dB), and a BER of 10^{-2} is achievable already at (3.6 | 5.5) dB SNR per bit and channel. Compared with uncoded 4-PSK – see also Figure 14-17 where

Table 14-4 Performance of Transmission System with DA Flat Fading Channel Estimation and TCM Channel Coding

	SNR [dB] per bit and channel @ BER=10^{-2}		SNR [dB] per bit and channel @ BER=10^{-3}	
	Doppler $\lambda'_D = 0.005 \vert 0.05$, optimal block length (sampling factor) F=11 | 4, length-21 | 7 uniform DA reception			
	8-PSK code	16-QAM code	8-PSK code	16-QAM code
$D = 1$	9.5 | 11.5	8.0 | 9.9	12.2 | 14.2	10.1 | 12.0
$D = 2$	4.3 | 6.3	3.6 | 5.5	6.0 | 7.9	5.1 | 7.1

the power and bandwidth efficiencies of uncoded M-PSK and TCM 16-QAM are visualized – the 16-QAM code yields gains as large as 7 dB without diversity and still about 3 dB with dual diversity. Hence, at the expense of complexity and some latency, well-designed trellis-coded modulation combined with DA channel estimation can improve on the power efficiency and thus the system capacity by a considerable margin without compromising the bandwidth efficiency.

14.2.6 Main Points

- Feedforward data-aided (DA) flat fading channel estimation based on a set of training symbols $\mathbf{a}_T$ has a number of advantages, resulting from the fact that channel estimation and data detection have become well-separated tasks. Optimal MAP or MMSE estimation from the received training sections $\mathbf{z}_T$ can be performed via ML estimation:

$$\begin{aligned} &\text{ML channel estimation:} \\ &\hat{\mathbf{c}}_{S,T}(\mathbf{a}_T) = \mathbf{P}_T^{-1}(\mathbf{a}_T) \cdot \mathbf{A}_T^H \cdot \mathbf{z}_T \\ &\text{MAP channel estimation from ML estimate:} \\ &\hat{c}_k = \mathbf{w}_k^H \cdot \hat{\mathbf{c}}_{S,T}(\mathbf{a}_T) \end{aligned} \tag{14-46}$$

  [eq. (14-36)] with Wiener filter $\mathbf{w}_k$. Estimator performance now depends not only on the channel parameters (λ'_D, N_0) but also the time instant k relative to the positions k_K of training symbols. Both the power allocated to training and the estimation error contribute to the total SNR degradation.
- In the important case of uniform DA channel sampling (sampling factor F), quasi-continuous transmission and steady-state estimation, a set of F Wiener filters $\mathbf{w}_{k'}$ suffices for optimal channel interpolation. As long as F is chosen such that the channel is sampled above the Nyquist rate, the estimator

performance is the same for all instants k. Best estimator performance results if the channel is well oversampled [optimal F_{opt} of eq. (14-40)].

- Aliasing and end effects caused by "nonideal" message formats or finite-length filtering entail some performance degradation, which, however, is remarkably small in many relevant cases, e.g., near the start or end of a message, within very short message blocks, or when a truncated interpolation filter $\mathbf{w}_k$ is used.
- A receiver design example has been studied featuring coherent M-PSK detection, uniform DA channel sampling, linear interpolation between selected pole positions, and eigenvalue-based synthesis of Wiener filter vectors $\mathbf{w}_{k'}$. Simulation results agree well with the analysis, and the extra SNR to be spent for DA channel estimation lies in the order of 1–1.5 dB (λ'_D =0.005) and 3–4 dB (λ'_D =0.05). Compared with DD reception, this is superior in terms of power efficiency and, except for dual diversity and fast fading, also in terms of bandwidth efficiency. DA reception is particularly advantageous when combined with interleaved channel coding. For example, using trellis codes with large effective code length yields additional gains in the order of 7 dB (no diversity) and 3 dB (dual diversity).

14.3 Bibliographical Notes

For fundamentals of estimation theory, the reader is once again referred to the textbooks [1, 3, 20, 21, 22]. IIR-type NDA channel prediction based on Kalman filtering is detailed in [2, 23, 24], and material for further readings on DA flat channel estimation using FDMA-like pilot tones and TDMA-like pilot samples is found in [10, 25, 26] and [8, 9, 10, 12, 13, 27], respectively.

For the application of trellis-coded modulation to fading channels and criteria for code selection, the reader is referred to [10, 16, 17, 28, 29, 30, 31, 32]. Publications [18] and [10] detail on the codes by Ungerböck and Moher/Lodge, respectively, and references [17, 19] present results on the matched filter bound.

Bibliography

[1] B. D. O. Anderson and J. B. Moore, *Optimal Filtering*. Englewood Cliffs, NJ: Prentice-Hall, 1979.

[2] R. Häb and H. Meyr, "A Systematic Approach to Carrier Recovery and Detection of Digitally Phase-Modulated Signals on Fading Channels," *IEEE Trans. Commun.*, vol. 37, no. 7, pp. 748–754, July 1989.

[3] S. Haykin, *Adaptive Filter Theory*. Englewood Cliffs, NJ: Prentice-Hall, 1986.

[4] P. Y. Kam and C. H. Teh, "An Adaptive Receiver with Memory for Slowly Fading Channels," *IEEE Trans. Commun.*, vol. 32, pp. 654–659, June 1984.

[5] R. Maciejko, "Digital Modulation in Rayleigh Fading in the Presence of Cochannel Interference and Noise," *IEEE Trans. Commun.*, vol. 29, pp. 1379–1386, Sept. 1981.

[6] S. Stein, "Unified Analysis of Certain Coherent and Noncoherent Binary Communication Systems," *IEEE Trans. Inform. Theory*, vol. 10, pp. 43–51, Jan. 1964.

[7] J. G. Proakis, *Digital Communications*. New York: McGraw-Hill, 1989.

[8] A. Aghamohammadi and H. Meyr, "A New Method for Phase Synchronization and Automatic Gain Control of Linearly-Modulated Signals on Frequency-Flat Fading Channels," *IEEE Trans. Commun.*, vol. 39, pp. 25–29, Jan. 1991.

[9] J. K. Cavers, "An Analysis of Pilot Symbol Assisted QPSK for Digital Mobile Communications," in *Proc. Globecom*, (San Diego, CA), pp. 928–933, Dec. 1990.

[10] M. L. Moher and J. H. Lodge, "TCMP: A Modulation and Coding Strategy for Rician Fading Channels," *IEEE J. Sel. Areas Commun.*, vol. 7, no. 9, pp. 1347–1355, Dec. 1989.

[11] J. P. Seymour and M. P. Fitz, "Two-Stage Carrier Synchronization Techniques for Nonselective Fading," *IEEE Trans. Vehic. Technol.*, vol. 44, no. 1, pp. 103–110, Feb. 1995.

[12] J. K. Cavers, "Pilot Symbol Assisted Modulation in Fading and Delay Spread," in *Proc. VTC*, (Stockholm, Sweden), pp. 13–16, June 1994.

[13] J. K. Cavers, "Pilot Symbol Assisted Modulation and Differential Detection in Fading and Delay Spread," *IEEE Trans. Commun.*, vol. 43, no. 7, pp. 2206–2212, July 1995.

[14] J. K. Cavers and J. Varaldi, "Cochannel Interference and Pilot Symbol Assisted Modulation," *IEEE Trans. Vehicular Technol.*, vol. 42, no. 4, pp. 407–413, Sept. 1993.

[15] D. D. Falconer, F. Adachi, and B. Gudmundson, "Time Division Multiple Access Methods for Wireless Personal Communications," *IEEE Commun. Magazine*, vol. 43, no. 1, pp. 50–57, Jan. 1995.

[16] D. Divsalar and M. K. Simon, "The Design of Trellis Coded MPSK for Fading Channels: Set Partitioning for Optimum Code Design," *IEEE Trans. Commun.*, vol. 36, no. 9, pp. 1013–1021, Sept. 1988.

[17] S. A. Fechtel and H. Meyr, "Matched Filter Bound for Trellis-Coded Transmission over Frequency-Selective Fading Channels with Diversity," *European Trans. Telecomm.*, vol. 4, no. 3, pp. 109–120, May-June 1993.

[18] G. Ungerboeck, "Channel Coding with Multilevel/Phase Signals," *IEEE Trans. Inform. Theory*, vol. 28, pp. 55–67, Jan. 1982.

[19] M. V. Clark, L. J. Greenstein, W. K. Kennedy, and Shafi, M., "Matched Filter Performance Bounds for Diversity Combining Receivers in Digital Mobile Radio," in *Proc. Globecom*, (Phoenix, AZ), pp. 1125–1129, Dec. 1991.

[20] J. L. Melsa and D. L. Cohn, *Decision and Estimation Theory*. New York: McGraw-Hill, 1978.

[21] L. L. Scharf, *Statistical Signal Processing*. New York: Addison-Wesley, 1991.

[22] H. L. van Trees, *Detection, Estimation, and Modulation Theory – Part I*. New York: Wiley, 1968.

[23] A. Aghamohammadi, H. Meyr, and G. Ascheid, "Adaptive Synchronization and Channel Parameter Estimation Using an Extended Kalman Filter," *IEEE Trans. Commun.*, vol. 37, no. 11, pp. 1212–1219, Nov. 1989.

[24] P. Höher, "An Adaptive Channel Estimator for Frequency-Selective Fading Channels," in *Aachener Symposium fuer Signaltheorie*, (Aachen, Germany), pp. 168–173, Sept. 1990.

[25] J. K. Cavers, "An Analysis of Pilot Symbol Assisted Modulation for Rayleigh Fading Channels," *IEEE Trans. Vehicular Technol.*, vol. 40, no. 4, pp. 686–693, Nov. 1991.

[26] J. K. Cavers and M. Liao, "A Comparison of Pilot Tone and Pilot Symbol Techniques for Digital Mobile Communication," in *Proc. Globecom*, (Orlando, FL), pp. 915–921, Dec. 1992.

[27] S. Gurunathan and K. Feher, "Pilot Symbol Aided QPRS for Digital Land Mobile Applications," in *Proc. ICC*, (Chicago, IL), pp. 760–764, June 1992.

[28] K. Boulle, G. Femenias, and R. Agusti, "An Overview of Trellis Coded Modulation Research in COST 231," in *Proc. PIMRC*, (The Hague, Netherlands), pp. 105–109, Sept. 1994.

[29] D. Divsalar and M. K. Simon, "The Design of Trellis Coded MPSK for Fading Channels: Performance Criteria," *IEEE Trans. Commun.*, vol. 36, no. 9, pp. 1004–1011, Sept. 1988.

[30] C. Schlegel and D. J. Costello, "Bandwidth Efficient Coding for Fading Channels: Code Construction and Performance Analysis," *IEEE J. Sel. Areas Commun.*, vol. 7, no. 9, pp. 1356–1368, Dec. 1989.

[31] B. Vucetic and J. Nicolas, "Construction of M-PSK Trellis Codes and Performance Analysis over Fading Channels," in *Proc. ICC*, (Singapore), pp. 614–618, April 1990.

[32] B. Vucetic and J. Du, "Performance Bounds and New 16-QAM Trellis Codes for Fading Channels," in *Coded Modulation and Bandwidth-Efficient Transmission*, (Tirrenia, Italy), pp. 353–363, Sept. 1991.

Chapter 15 Parameter Synchronization for Selective Fading Channels

In the two sections of this chapter, we shall concentrate on linear sync parameter estimation for the selective fading case, i.e., frequency-selective fading channel estimation. Again, many of the sync algorithms presented here draw from the ideas and results of Chapter 12, with emphasis on reduced-complexity yet close-to-optimal synchronizer structures and algorithms.

15.1 Non-Data-Aided (NDA) Selective Fading Channel Estimation and Detection

In this section, detection and selective fading channel estimation without the use of known data is discussed. Analogous to NDA flat fading channel estimation, the one-step selective channel predictor estimate $\hat{h}^{(i)}_{l;k|k-1}$ has been identified as the sync parameter necessary for metric computation [eq. (13-25)] in NDA synchronized detection (Section 13.3). Algorithms for generating such an estimate are studied in Section 15.1.1, and differentially coherent detection using decision-directed (DD) selective channel estimation is discussed in Section 15.1.2.

15.1.1 LMS-Kalman and LMS-Wiener One-Step Channel Prediction

As with flat fading channel estimation, the selective channel predictor estimate $\hat{\mathbf{h}}_{k|k-1} = \hat{\mathbf{h}}^{(0)}_{k|k-1} \oplus \hat{\mathbf{h}}^{(1)}_{k|k-1}$ can, in principle, be generated either via nonrecursive Wiener estimation or recursive Kalman-type algorithms. From Section 12.2.4 we know that true ML estimation of the modulation-free channel tap trajectories $h_{\nu;k}$ ($\nu = 0, 1, \ldots, 2L{+}1$) is not possible, so optimal Wiener estimation must resort to a direct computation of the MAP predictor estimate [eq. (12-71)] or to a partitioning of the signal $\mathbf{r}_{k-1}$ and symbols $\mathbf{a}_{k-1}$ into several subblocks so that ML snapshot acquisition becomes feasible [eq. (12-83)]. Both methods can easily be adapted to one-step prediction but remain to be too complex to be actually implemented. Nevertheless, we shall see that an FIR-type Wiener tap estimator can easily be incorporated into the LMS-Kalman recursive channel estimator structure.

Again, the LMS-Kalman algorithm for (partial[13]) selective channel estimation

[13] In this subsection, partial channel indices i are dropped.

is based on the Gauss-Markov model [1] of eq. (14-8), now with state vector $\mathbf{x}_k = \left(\mathbf{x}_{0;k}^T \ldots \mathbf{x}_{L;k}^T\right)^T$ comprising $L+1$ substates $\mathbf{x}_{n;k}$, one for each (partial) channel tap $h_{n;k}$. Assuming uncorrelated taps $h_{n;k}$, eq. (14-8) falls apart into $L+1$ independent Gauss-Markov models for each of the tap processes. The resulting Kalman filter thus exhibits $L+1$ respective filtering branches ("subfilters"). Once again, by constraining these to be first-order stationary IIR subfilters, one arrives at the LMS-Kalman algorithm [analogous to eq. (14-10)] [2] whose filtering recursions are given by:

$$\begin{aligned}
&\text{Model output}: \\
&\hat{r}_{k-1|k-2} = \sum_{n=0}^{L} a_{k-1-n}\, \hat{h}_{n;k-1|k-2} \\
&\text{Pseudo} - \text{ML tap estimates } (n = 0, \ldots, L): \\
&\hat{h}_{S,n;k-1} = \hat{h}_{n;k-1|k-2} + \frac{a_{k-1-n}^*}{p_{k-1-n}} \left(r_{k-1} - \hat{r}_{k-1|k-2}\right) \\
&\text{Channel tap update recursions } (n = 0, \ldots, L): \\
&\hat{h}_{n;k|k-1} = (1 - K_n)\, \hat{h}_{n;k-1|k-2} + K_n\, \hat{h}_{S,n;k-1}
\end{aligned} \tag{15-1}$$

The structure of the LMS-Kalman selective one-step (partial) channel predictor is visualized in Figure 15-1.

This LMS-Kalman estimator structure is similar to that of the well-known LMS or *stochastic-gradient* (SG) algorithm [3], with the important difference that the (often heuristically chosen) *global* LMS-gain K has been replaced by the set of systematically derived *individual* gain factors $\{K_n\}$[14] tailored to the statistics

[14] These gain factors may even be time-variant. For example, using gains $K_{n;k}$ whose initial values are larger than their steady-state settings can improve the convergence behavior in the case that the LMS-Kalman algorithm is used not only for channel *tracking* but also channel *acquisition*, most often in a training-aided fashion.

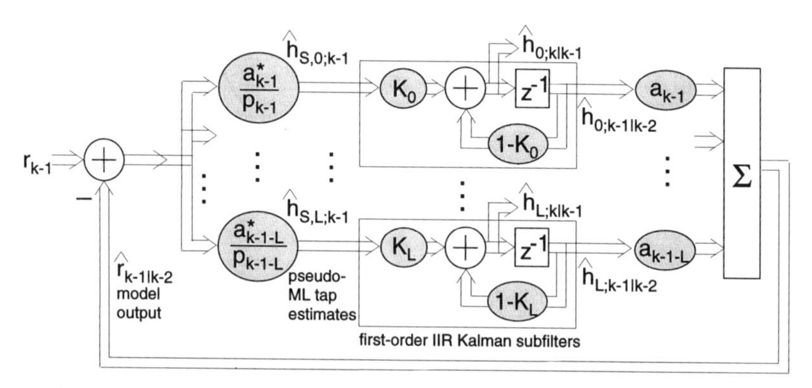

Figure 15-1 LMS-Kalman One-Step Selective (Partial) Channel Predictor

of each channel tap process $h_{n;k}$. The LMS-Kalman algorithm ranges in between the simple LMS and the full Kalman algorithms, in terms of both performance and complexity. Interestingly, it can be shown that the LMS-Kalman is actually *superior* to the more complex *recursive least-squares* (RLS) algorithm [3], provided that the data sequence $\{a_k\}$ is white.

The channel update recursions [third part of eq. (15-1)] are seen to be of the same form as the update recursion for flat fading channels [eq. (14-10)], where the true ML estimate $\hat{c}_{S;k-1}$ has been replaced by a "pseudo-ML" tap estimate $\hat{h}_{S,n;k-1}$. Expanding $\hat{h}_{S,n;k-1}$ [second part of eq. (15-1)] yields

$$\begin{aligned}\hat{h}_{S,n;k-1} = \; & \hat{h}_{n;k-1|k-2} \quad + \quad \Delta_{n;k-1|k-2} \\ & + \underbrace{\left(\sum_{l=0,\; l\neq n}^{L} \frac{a^*_{k-1-n}\, a_{k-1-l}}{p_{k-1-n}} \Delta_{l;k-1|k-2} \right) + \frac{a^*_{k-1-n}}{p_{k-1-n}}\, n_{k-1}}_{\text{total disturbance } \tilde{n}_{k-1}}\end{aligned} \tag{15-2}$$

One observes that (i) the pseudo-ML estimate is driven toward the actual tap weight $h_{n;k-1}$ by virtue of the prediction error $\Delta_{n;k-1|k-2} = h_{n;k-1} - \hat{h}_{n;k-1|k-2}$ in the nth tap and (ii) that $\hat{h}_{S,n;k-1}$ is not disturbed by the other taps themselves but – apart from the thermal noise – a term involving the prediction errors $\Delta_{l;k-1|k-2}$ in the other taps. Under the vital assumption of white data $\{a_k\}$, the latter disturbance is approximately white Gaussian with average power $\sum_{l=0,l\neq n}^{L} \sigma^2_{l;k|k-1} = N_0 \left(\sum_{l=0,l\neq n}^{L} \overline{r}_l\right)$, so that the total disturbance $\tilde{n}_{k-1}$ [last two terms of eq. (15-2)] is approximately AWGN with average power:

$$\overline{N}_{0,n} = N_0 \left(1 + \sum_{l=0,\; l\neq n}^{L} \overline{r}_l \right) \tag{15-3}$$

where $\overline{r}_n = \sigma^2_{n;k|k-1}/N_0$ are the average tap error covariance reduction factors. Since $\hat{h}_{S,n;k-1}$ is disturbed by noise whose power $\overline{N}_{0,n}$ depends on the prediction errors in the other taps, the tap update recursions [third part of eq. (15-1)] are not entirely independent, even if the tap processes $h_{n;k}$ are perfectly uncorrelated.

As with flat fading channel estimation, the weights K_n and error covariance factors $\overline{r}_n$ cannot be determined by means of the Kalman filter Riccati equations. A prediction error analysis completely analogous to that of Section 14.1.2 yields the prediction error recursion [eq. (14-13)]

$$\Delta_{n;k} = (1 - K_n)\, \Delta_{n;k-1} + \delta_{n;k} - K_n\, \tilde{n}_{k-1} \tag{15-4}$$

with $\delta_{n;k} = h_{n;k} - h_{n;k-1}$ the tap trajectory increment due to fading. Like eq. (14-14), eq. (15-4) can be split into independent lag and noise error recursions.

Solving these leads to approximative expressions for the tap lag and noise error covariances

$$\begin{aligned} \left(\sigma^2_{n;k|k-1}\right)^{(L)} &\approx C_L\,\rho_n \left(\frac{\lambda'_{D,n}}{K_n}\right)^2 \\ \left(\sigma^2_{n;k|k-1}\right)^{(N)} &= \frac{K_n}{2-K_n}\,\overline{N}_{0,n} \underset{(K_n \ll 1)}{\approx} \frac{K_n}{2}\,\overline{N}_{0,n} \end{aligned} \tag{15-5}$$

respectively [similar to eq. (14-18)]. Setting C_L to 12 yields less than 10 percent approximation error as long as the gain factor K_n satisfies $K_{\min}\left(\lambda'_D\right) \le K_n < 1$, with lower limit $K_{\min}\left(\lambda'_D = 0.001\right) \approx 0.02$, $K_{\min}\left(\lambda'_D = 0.01\right) \approx 0.1$, and $K_{\min}\left(\lambda'_D = 0.03\right) \approx 0.22$. Making the common assumption that all tap Doppler frequencies are equal ($\lambda'_{D,n} = \lambda'_D$), the optimal gains K_n and minimal error covariance factors $\overline{r}_n = \overline{r}_n^{(L)} + \overline{r}_n^{(N)}$ can then be approximated in closed form by

$$\begin{aligned} K_n^{(\text{opt})} &\approx 3.6\,\sqrt[3]{\rho_n \frac{\left(\lambda'_D\right)^2}{\overline{N}_{0,n}}} = 3.6\,\sqrt[3]{\underbrace{\left(\frac{1}{1+\sum\limits_{l=0,\ l\neq n}^{L} \overline{r}_l}\right)}_{\eta_n} \underbrace{\frac{\rho_n}{N_0}}_{\rho'_n} \left(\lambda'_D\right)^2} \\ \overline{r}_n^{(\min)} &\approx 2.7\,\sqrt[3]{\rho_n \frac{\left(\lambda'_D\right)^2}{\overline{N}_{0,n}}} = 0.75\,K_n^{(\text{opt})} \end{aligned} \tag{15-6}$$

[see also eq. (14-19)], where $\rho'_n = \rho_n/N_0$ is the tap SNR and the quantity $\eta_n \le 1$ reflects the impact of the other taps (perfect sync: $\eta_n = 1$). The condition $K_n \ge K_{\min}\left(\lambda'_D\right)$ is found to be satisfied for tap SNRs $\rho'_n = \rho_n/N_0$ larger than about 0.25. If ρ'_n is considerably smaller, $h_{n;k}$ is a weak tap deeply buried in noise which can thus be set to zero, as well as the respective gain K_n. In that case, the tap error covariance $\sigma^2_{n;k|k-1} = \rho_n$ (covariance factor $\overline{r}_n = \rho'_n \ll 1$) remains small.

Since the gains $K_n^{(\text{opt})}$ and factors $\overline{r}_n^{(\min)}$ are interrelated through the set of noise powers $\{\overline{N}_{0,n}\}$ [eq. (15-3)], the sets $\left\{K_n^{(\text{opt})}\right\}$ and $\left\{\overline{r}_n^{(\min)}\right\}$ $(n = 0, \ldots, L)$ should be computed iteratively as follows. Starting with $\overline{N}_{0,n}$ set to $N_0 = 1/\overline{\gamma}_s$, i.e., neglecting the impact of errors in the other taps as if channel estimation were perfect, eqs. (15-6) (update of $\left\{K_n^{(\text{opt})}\right\}$, $\left\{\overline{r}_n^{(\min)}\right\}$) and (15-3) (update of $\{\overline{N}_{0,n}\}$) are applied alternately. This iterative procedure converges very fast, and the final $\left\{K_n^{(\text{opt})}\right\}$, $\left\{\overline{r}_n^{(\min)}\right\}$ are not much different from their initial settings.

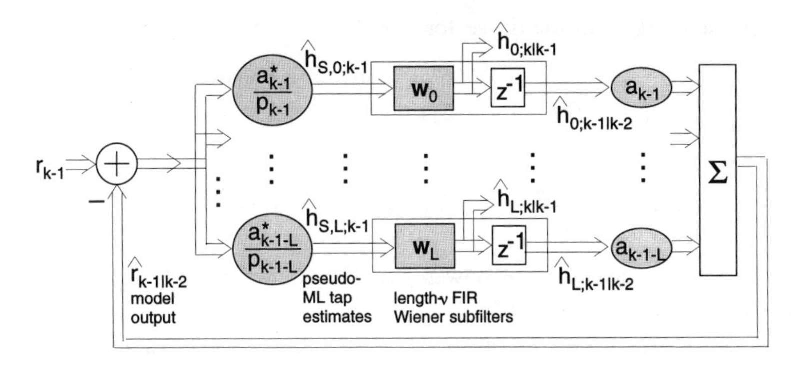

Figure 15-2 LMS-Wiener One-Step Selective (Partial) Channel Predictor

A straightforward modification to the LMS-Kalman algorithm, termed here *LMS-Wiener* algorithm, can be devised by replacing the $L+1$ first-order IIR-type Kalman subfiltering branches [third part of eq. (15-1), Figure 15-1] by length-ν FIR-type Wiener subfilters $\mathbf{w}_n$, driven by the respective pseudo-ML tap estimates $\hat{h}_{S,n;k-1}$. In comparison with the Wiener estimator for flat fading channels [MAP channel prediction from ML estimate, third part of eq. (14-7)], this LMS-Wiener structure visualized in Figure 15-2 remains to be recursive, which is a prerequisite for the generation of pseudo-ML tap estimates $\hat{h}_{S,n;k-1}$ in lieu of the – nonexisting – true ML estimates.

While the first two parts of eq. (15-1) (model output, pseudo-ML estimation) remain unaltered, the third part of eq. (15-1) (channel tap update) is replaced by

$$\hat{h}_{n;k|k-1} = \mathbf{w}_n^H \cdot \hat{\mathbf{h}}_{S,n;k-1} = \sum_{l=0}^{\nu-1} w_{n,l}\, \hat{h}_{S,n;k-\nu+l} \tag{15-7}$$

$(n=0,\ldots,L)$ with $\mathbf{w}_n = (w_{n,0} \ldots w_{n,\nu-1})^T$ the real-valued FIR Wiener tap filter coefficients and $\hat{\mathbf{h}}_{S,n;k-1} = \left(\hat{h}_{S,n;k-\nu} \ldots \hat{h}_{S,n;k-1}\right)^T$ the samples of pseudo-ML tap estimates in the filter memory. An analysis completely analogous to that of Section 14.1.2 [eq. (14-7)] yields the optimal tap filters $\mathbf{w}_n$ and associated error covariance factors $\overline{r}_n$:

$$\begin{aligned} \mathbf{w}_n &= \left[\mathrm{A}_D + \frac{1}{\eta_n \rho_n'}\mathbf{I}\right]^{-1} \boldsymbol{\alpha}_D \\ \overline{r}_n &= \rho_n' \left(1 - \boldsymbol{\alpha}_D^H \underbrace{\left[\mathrm{A}_D + \frac{1}{\eta_n \rho_n'}\mathbf{I}\right]^{-1} \boldsymbol{\alpha}_D}_{\mathbf{w}_n}\right) \end{aligned} \tag{15-8}$$

with tap autocorrelation matrix/vector

$$\begin{aligned} \mathrm{A}_D &= \frac{\mathbf{R}_{D,n}}{\rho_n} &&= \begin{pmatrix} \alpha(0) & \dots & \alpha(\nu-1) \\ \vdots & \ddots & \vdots \\ \alpha(\nu-1) & \dots & \alpha(0) \end{pmatrix} \\ \boldsymbol{\alpha}_D &= \frac{\mathbf{r}_{D,n}}{\rho_n} &&= \begin{pmatrix} \alpha(\nu) & \dots & \alpha(1) \end{pmatrix}^T \end{aligned} \tag{15-9}$$

normalized to the average tap power ρ_n, where the tap SNR ρ_n' and quantity $\eta_n \leq 1$ are defined by eq. (15-6).

As discussed in Section 14.2.4 (DA flat fading channel estimation), online computation of weight vectors $\mathbf{w}_n$ can be substantially simplified via eigensystem synthesis. Since the expression for $\mathbf{w}_n$ [eq. (15-8)] is of the same form as that of eq. (14-41), the result of eq. (14-45) holds also here, where $\mathbf{R}_{D,T}$, $\mathbf{r}_{D,T;k'}$, and the average SNR $\overline{\gamma}_s$ have to be replaced by A_D, $\boldsymbol{\alpha}_D$, and the weighted average tap SNR $\eta_n \rho_n'$, respectively:

$$\mathbf{w}_n\left(\rho_n', \eta_n\right) = \left(\eta_n \rho_n'\right)\left(\boldsymbol{\alpha}_D - \sum_{i=0}^{\overline{\nu}-1} \frac{\lambda_{D,i}}{\lambda_{D,i} + \frac{1}{\left(\eta_n \cdot \rho_n'\right)}} \underbrace{\left(\left[\mathbf{u}_i \mathbf{u}_i^H\right] \boldsymbol{\alpha}_D\right)}_{\mathbf{w}_{(i)}}\right) \tag{15-10}$$

with $\{\lambda_{D,0}, \ldots, \lambda_{D,\overline{\nu}-1}\}$ and $\{\mathbf{u}_0, \ldots, \mathbf{u}_{\overline{\nu}-1}\}$ the sets of the $\overline{\nu}$ largest eigenvalues and respective eigenvectors of matrix A_D, and $\{\mathbf{w}_{(0)}, \ldots, \mathbf{w}_{(\overline{\nu}-1)}\}$ the set of partial weight vectors. Both $\{\lambda_{D,0}, \ldots, \lambda_{D,\overline{\nu}-1}\}$ and $\{\mathbf{w}_{(0)}, \ldots, \mathbf{w}_{(\overline{\nu}-1)}\}$, although dependent on the Doppler frequency λ_D', are independent of the parameters ρ_n', η_n and can thus be computed offline and stored. Given the tap SNR profile $\{\rho_n'\}$ (or an estimate thereof), online computation of the corresponding set of optimal weights $\{\mathbf{w}_n\}$ can then be performed iteratively; each iteration comprises an update of weights [eq. (15-10)], error covariance factors $\{\overline{r}_n = \rho_n'(1 - \boldsymbol{\alpha}_D^H \mathbf{w}_n)\}$ [eq. (15-8)], and quantities $\{\eta_n = 1/(1 + \sum_{l=0, l\neq n}^{L} \overline{r}_l)\}$ [see eq. (15-6)], where the η_n are initially set to unity.

Figure 15-3 displays the error covariance factors $\overline{r}_n$ associated with LMS-Kalman and LMS-Wiener one-step prediction of the GSM hilly-terrain (GSM-HT) channel (Section 11.3.2). The error covariances of all $2(L+1) = 10$ significant tap estimates $\hat{h}_{\nu,k|k-1}$ – both partial channels have been merged into the $(T/2)$-spaced channel impulse response – have been obtained analytically [LMS-Kalman: eq. (15-6), LMS-Wiener: eq. (15-8)] for Doppler frequencies $\lambda_D' = 0.001$, 0.01, and 0.1, and 10 dB SNR per symbol. The error covariance profiles $\{\overline{r}_\nu\}$ resemble somewhat the average tap power profile $\{\rho_\nu\}$ of the GSM-HT channel, however with less pronounced peaks and dips. In the case of very large Doppler $\lambda_D' = 0.1$, the approximation eq. (15-6) for both the Kalman gains K_ν and error covariance factors $\overline{r}_\nu$ (upper solid curve in Figure 15-3) is in fact no more valid; the largest Kalman gain would have to be in the order of 1 which is clearly impossible. The dotted (reference) curves in Figure 15-3 are associated with near-optimal

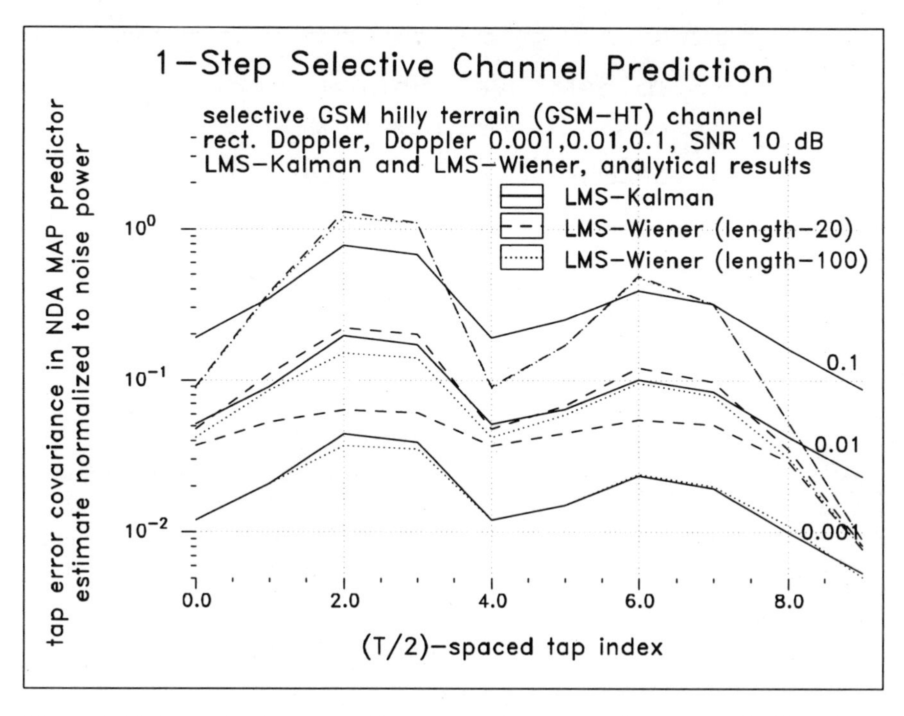

Figure 15-3 Error Performance (Set of Error Covariance Factors) of LMS-Kalman and LMS-Wiener One-Step Selective Channel Prediction, GSM-HT Channel

length-100 LMS-Wiener prediction, while the dashed curves refer to the (more realistic) length-20 LMS-Wiener algorithm. In this example, between $\overline{\nu}$=2 and 4 eigenvalues and eigenvectors are needed for eigensystem synthesis of the LMS-Wiener weight vectors $\mathbf{w}_\nu$. From Figure 15-3, the simple IIR-type LMS-Kalman can be expected to perform comparably well as the more complex LMS-Wiener; it may even outperform (in theory) the length-20 LMS-Wiener when the fading is rather slow (λ'_D =0.001).

Simulation results on MSE factors $\zeta = \sum_\nu \overline{r}_\nu$ vs. the Doppler frequency λ'_D are shown in Figure 15-4, again for the GSM-HT channel. In the case of low Doppler and/or low SNR, the LMS-Kalman is seen to yield simulated MSEs which are somewhat larger than predicted by the analysis. At low SNR (10 dB), optimal Kalman gain factors K_ν according to eq. (15-6) can be used for Doppler up to about 0.03, while at high SNR (30 dB), optimal Kalman gain factors are only applicable to Doppler up to about 0.003. Of course, at high SNR the LMS-Kalman algorithm also works when λ'_D is larger than 0.003, but then the gains K_ν and error covariance factors $\overline{r}_n$ cannot be computed via eq. (15-6); instead, the largest gains of the set $\{K_\nu\}$ must be clipped so that they remain well below 1.

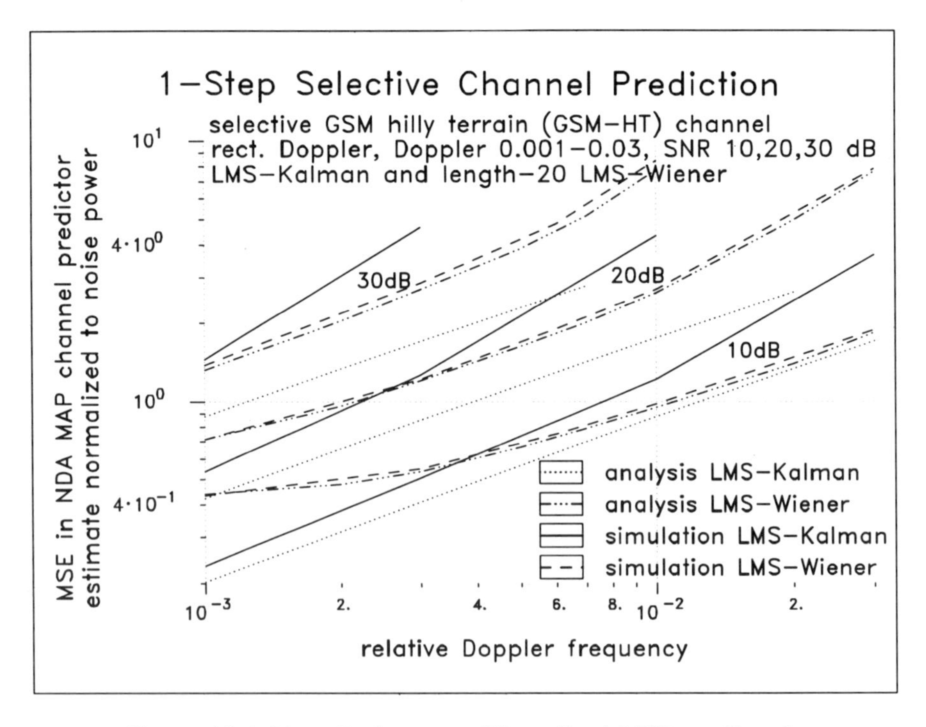

Figure 15-4 Error Performance (Normalized MSE vs. Doppler Frequency) of LMS-Kalman and LMS-Wiener One-Step Selective Channel Prediction, GSM-HT Channel

While the LMS-Kalman simulation results of Figure 15-4 differ considerably from their analytical counterparts at high Doppler and/or high SNR, the LMS-Wiener performance is predicted very well by the analysis, even at very high Doppler. As a consequence, the length-20 LMS-Wiener algorithm – which, according to the analysis, should have been inferior to the LMS-Kalman – in fact outperforms the LMS-Kalman algorithm already at moderate Doppler from about 0.004, 0.003, and 0.001 at 10, 20, and 30 dB SNR per symbol, respectively. When the fading is slow, on the other hand, the LMS-Kalman retains its advantage – both in terms of performance and complexity – and thus remains to be the prediction algorithm of choice.

15.1.2 Decision-Directed (DD) Selective Fading Channel Estimation and Detection

Analogous to NDA online flat fading channel estimation and detection (Section 14.1), a one–step (partial) channel predictor estimate $\hat{\mathbf{h}}^{(i)}_{k|k-1}(\mathbf{a}_{k-1})$ as well as an associated set of sync parameters must, in theory, be computed for each and every symbol hypothesis $\mathbf{a}_{k-1}$. By virtue of whitening matched prefiltering (WMF, previous chapter), only very few – in all of the simulation examples below, between M=1 and 4 – contending symbol hypotheses need to be kept by the equalizer for quasi-optimal detection. The number of NDA parameter sync units working on the corresponding survivors is reduced accordingly, i.e., only M channel estimates (and parameter sets derived from the channel estimates) remain to be generated. Estimating the channel — or any other parameters of interest — in such a survivor-dependent fashion is a rather general concept sometimes referred to as *per-survivor-processing* (PSP) [4].

Recall that, in uncoded systems with proper prefiltering, a single survivor (M=1) already suffices to achieve near-optimal BER performance (Section 13.3.5). In other words, the decision-feedback equalizer (DFE) symbol decisions $\tilde{\mathbf{a}}_{k-1}$ are sufficiently reliable to be used for NDA channel estimation. Hence, the NDA receiver can be further simplified by retaining just a single, now DD online parameter synchronization unit, regardless of the number M of survivors actually kept for detection. If there is more than one survivor (M algorithm), the DD synchronizer uses the symbols (tentative symbol "decisions") $\mathbf{a}_{k-1}$ associated with the best survivor having the smallest metric $m_{D;k-1}(\mathbf{a}_{k-1})$ at time instant $k-1$.

The block diagram of a complete digital receiver for differentially encoded M-PSK detection is shown in Figure 15-5. This receiver, which has been implemented to generate all of the simulation results reported below, features a detection path comprising matched filtering, decimation to symbol rate, whitening filtering, M algorithm equalization, and (not shown in the figure) differential decoding in order to resolve phase slips. The parameter synchronization path includes DD channel estimation, parameter computation from the channel estimate, and a unit referred to as *timing slip control*. As explained below, this unit is necessary to prevent

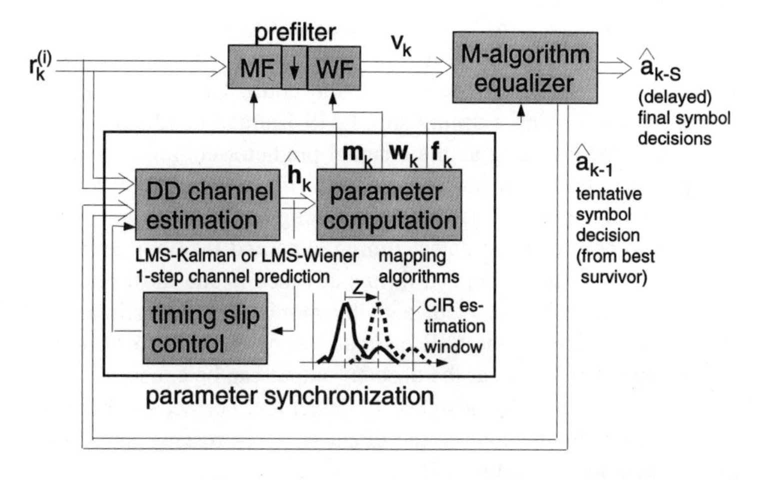

Figure 15-5 M-PSK Receiver with DD Selective Fading Channel Estimation

the channel impulse response (CIR) estimate from shifting out of the estimation window.

In receivers with DD selective channel estimation, timing slips are possible because the pair of unknowns, i.e., the channel $\mathbf{h}$ and symbol sequence $\mathbf{a}$, is not unique. For instance, a noise-free sample $s_k^{(i)}$ of the (partial) observation $\mathbf{r}^{(i)} = \mathbf{s}^{(i)} + \mathbf{n}^{(i)}$ may have been produced not only by the (partial) CIR $\mathbf{h}_k^{(i)} = (h_{0;k}^{(i)} \; \dots \; h_{L;k}^{(i)})^T$ convolved with the symbol sequence $(a_k \; \dots \; a_{k-L})^T$ (as has been assumed so far), but also by the *shifted* (partial) CIR $(h_{-Z;k}^{(i)} \; \dots \; h_{L-Z;k}^{(i)})^T$ convolved with the *time-shifted* symbol sequence $(a_{k+Z} \; \dots \; a_{k+Z-L})^T$:

$$
\begin{aligned}
s_k^{(i)} &= (a_k \; \dots \; a_{k-L}) \begin{pmatrix} h_{0;k} \\ \vdots \\ h_{L;k} \end{pmatrix}^{(i)} = \sum_{n=0}^{L} h_{n;k}^{(i)} \, a_{k-n} \\
&= \sum_{n=Z}^{L+Z} h_{n-Z;k}^{(i)} \, a_{k+Z-n} \underset{\begin{pmatrix}\text{if CIR remains in} \\ \text{estimation window}\end{pmatrix}}{=} \sum_{n=0}^{L} h_{n-Z;k}^{(i)} \, a_{k+Z-n} \\
&= (a_{k+Z} \; \dots \; a_{k+Z-L}) \begin{pmatrix} h_{-Z;k} \\ \vdots \\ h_{L-Z;k} \end{pmatrix}^{(i)}
\end{aligned}
\tag{15-11}
$$

For positive Z (timing advance), the shifted CIR $(h_{-Z;k}^{(i)} \dots h_{L-Z;k}^{(i)})^T$ remains in the estimation window as long as the "tail" section of the original CIR $(h_{L-Z+1;k}^{(i)} \; \dots \; h_{L;k}^{(i)})^T$ is zero (notice that this is not the case in the example shown in Figure 15-5), and for negative Z (timing delay), the same is true as long as the "head" section $(h_{0;k}^{(i)} \dots \; h_{-Z-1;k}^{(i)})^T$ is zero. At any rate, timing slips are not desirable since (i) the CIR may shift out of the estimation window when either the head or the tail section of the CIR or the entire CIR (deep fade) are temporarily small, (ii) following a timing slip, LMS-Kalman or LMS-Wiener channel prediction is perturbed as long as the channel predictor continues to work on the unshifted tap power profile or its estimate, and (iii) an undetected timing shift in the detected symbol sequence may result in a long error burst. For these reasons, some kind of slip control must detect timing slips and take corrective action. A timing slip, along with the associated timing advance Z and most probable instant of occurrence k_Z, may be detected, e.g., by monitoring the estimate of the tap power profile (derived from the CIR estimate) or with the help of short training blocks. Also, long error bursts at the decoder output can be counteracted by means of coding schemes specifically designed to detect and correct symbol shift errors. Examples of such codes combating the effect of zero-crossing jitter are found in the magnetic recording literature [5].

Let us now study the BER performance of the DD receiver shown in Figure

15-5 for some typical examples. We consider the GSM-HT channel with mild ISI (Section 11.3.2), and a GSM two-ray channel with strong ISI (two rays of equal average gain and mutual delay τ'=2). Both channels span L=4 symbol intervals and thus bear $2(L+1)$=10 significant taps. Parameter synchronization is based on DD channel estimation featuring either LMS-Kalman or length-20 LMS-Wiener one-step prediction; parameter computation from the ideal channel has also been simulated for reference. As test simulations have shown, choosing the whitening filter length as W =4, 8, 12, and 16 for 2-, 4-, 8-, and 16-PSK, respectively, and setting the number of survivors in the M algorithm equalizer to M=4 (the best survivor is used for DD channel estimation) yields the best trade-off between performance and complexity, in the sense that additional gains obtained by using larger W or M are very small. Sets of BER results have been generated for Doppler frequencies of $\lambda'_D = 3\times 10^{-4}$, 10^{-3}, and 3×10^{-3}, which – again assuming a symbol rate of $(1/T)$=135 kBd – correspond to vehicle speeds of about 50, 150, and 500 km/h (900 MHz) or 25, 75, and 250 km/h (1800 MHz), respectively.

Figure 15-6 shows BER curves for differentially encoded 2-, 4-, 8- and 16-PSK, slow Doppler $\lambda'_D = 3\times 10^{-4}$ and no diversity reception; Figure 15-7 displays the same for dual diversity reception. As expected (Section 15.1.1), the LMS-

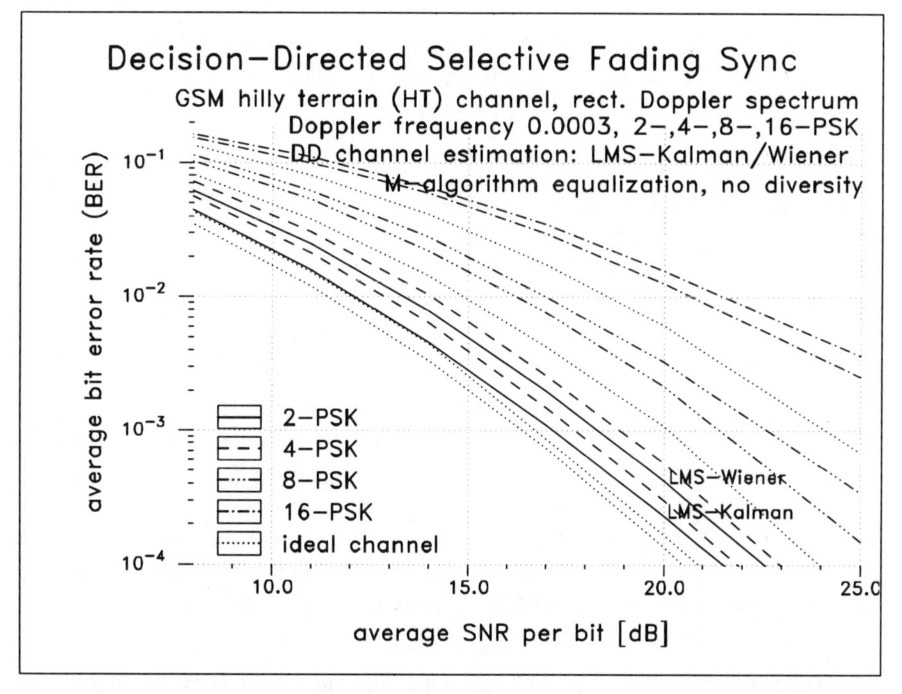

Figure 15-6 BER Performance of Transmission System with DD Selective Fading Channel Estimation and Differentially Encoded M-PSK Detection without Diversity; GSM-HT Channel, Doppler 0.0003

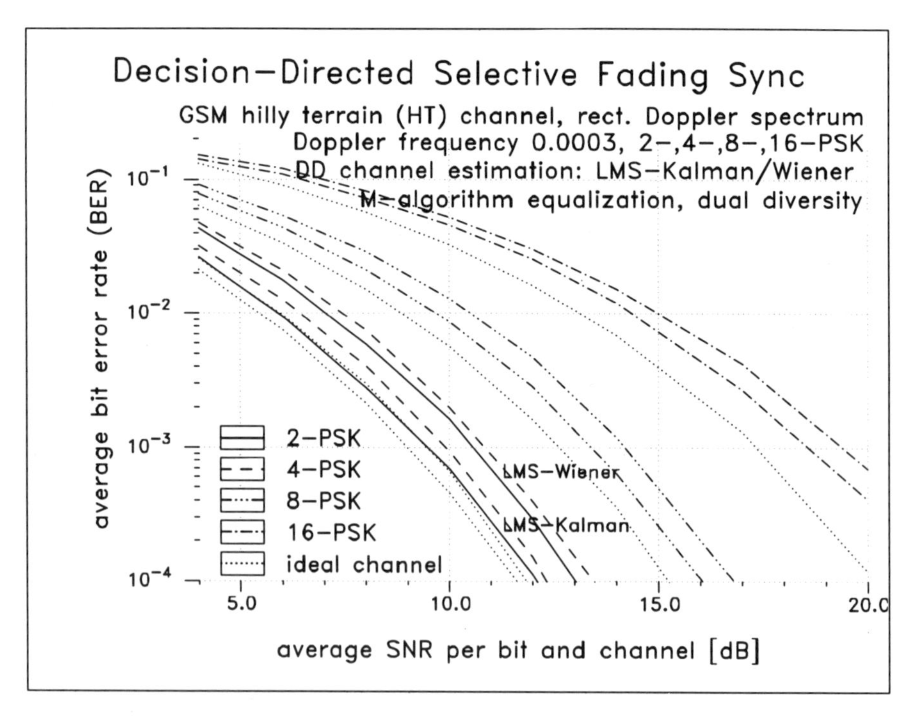

Figure 15-7 BER Performance of Transmission System with DD Selective Fading Channel Estimation and Differentially Encoded M-PSK Detection with Dual Diversity; GSM-HT Channel, Doppler 0.0003

Kalman algorithm for steady-state channel tracking has clear advantages at these slow fading rates. Without diversity, DD parameter sync using LMS-Kalman channel prediction is seen to cost less than 1 dB (2-, 4-PSK), 1.5 dB (8-PSK), and 2 dB (16-PSK, BER $\approx 10^{-2}$) with respect to parameter sync based on the ideal channel reference. With dual diversity, the DD sync loss is further reduced to about 0.5 dB (2-, 4-PSK), 0.8 dB (8-PSK), and 1.2 dB (16-PSK).

As exemplified by Figure 15-8 (GSM-HT channel, Doppler $\lambda'_D = 3 \times 10^{-3}$, no diversity), the DD sync loss as well as the error floor grow rapidly as the fading rate becomes large (here: ten times as large as in Figure 15-6). Without diversity, a sync loss of 4 dB or more entails. The 8-PSK error floor lies barely below 10^{-2}, and denser symbol alphabets such as 16-PSK are no more feasible at all (thus not shown in Figure 15-8). As conjectured from Section 15.1.1, the LMS-Wiener algorithm now outperforms the LMS-Kalman, particularly at high SNR where the error floor is reduced by a substantial margin.

The simulation results, as well as the *matched-filter bound* (MFB) on the BER performance [6, 7], have been summarized in Tables 15-1 and 15-2 for the GSM-HT and GSM two-ray channels, respectively. The tables show the minimum SNR

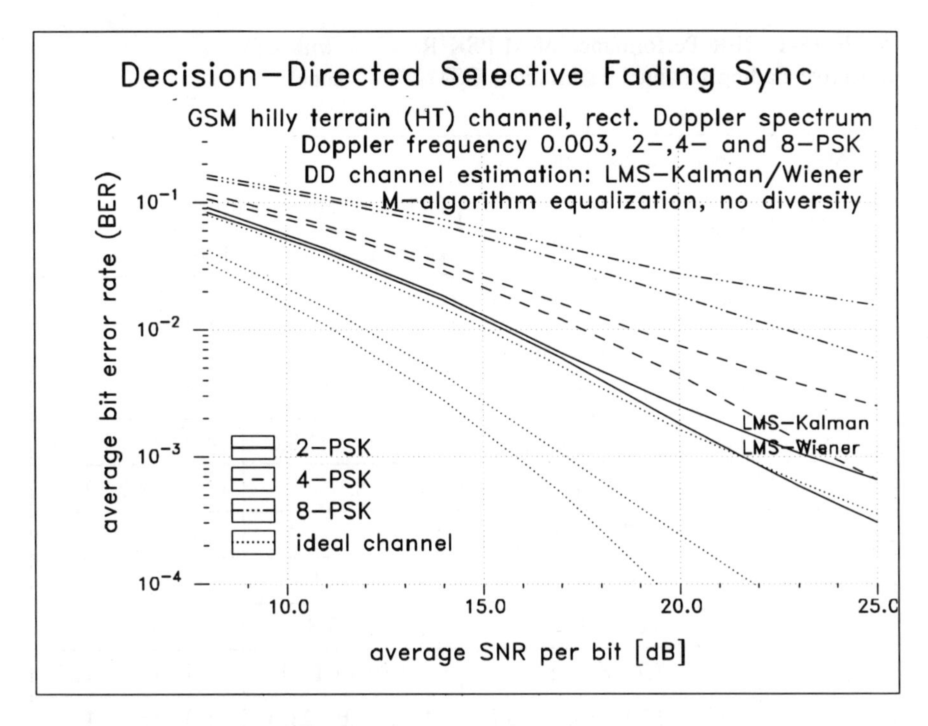

Figure 15-8 BER Performance of Transmission System with DD Selective Fading Channel Estimation and Differentially Encoded M-PSK Detection without Diversity; GSM-HT Channel, Doppler 0.003

per bit and per channel necessary to achieve BER levels of 10^{-2} and 10^{-3}; blank spots in the list denote SNR values above 30 dB or infinity, i.e., the BER level of interest (10^{-2} or 10^{-3}) cannot be attained at all.

First of all, the best simulation results, i.e., those based on the ideal channel, are seen to be inferior to the matched filter bound by a margin of 2.5–3.5 dB (no diversity) and 1.5–3.0 dB (dual diversity). This, however, is to be expected since the derivation of the MFB relies on a number of idealistic assumptions, viz. perfect prefiltering, equalization, synchronization, and coherent detection. Moreover, the MFB is a tight lower bound only asymptotically for very high SNR or very low BER.

While the results based on the ideal channel do not show much sensitivity against fading up to $\lambda_D' = 3\times10^{-3}$ and even 10^{-2} (2- and 4-PSK, Figure 13-13), fading has a noticeable effect on DD reception already at $\lambda_D' = 10^{-3}$. LMS-Kalman channel prediction is superior at low Doppler up to $\lambda_D' \approx 10^{-3}$–$2\times10^{-3}$ but rapidly deteriorates at larger Doppler. On the other hand, the length-20 LMS-Wiener algorithm yields more uniform performance and is superior at larger Doppler starting at $\lambda_D' \approx 2\times10^{-3}$ in this example.

Table 15-1 BER Performance of M-PSK Receiver with DD Selective Fading Channel Estimation, GSM-HT Channel

GSM-HT channel (mild ISI)		Anal.	Simulation		
		MFB	ideal channel	DD parameter synchronization	
				LMS-Kalman	LMS-Wiener
		Doppler λ'_D =0.0003 I 0.001 I 0.003			
D	PSK	SNR [dB] per bit and channel @ BER=10^{-2}			
1	2	8.8	11.2 I 11.2 I 11.3	11.8 I 13.0 I 15.7	13.2 I 13.7 I 15.5
	4	8.8	12.0 I 12.0 I 12.1	12.7 I 14.3 I 18.6	14.0 I 14.7 I 17.4
	8	11.3	14.8 I 14.8 I 15.0	16.3 I 18.1	17.1 I 18.7 I 22.6
	16	15.1	18.7 I 18.9	20.8	21.6
2	2	3.7	5.5 I 5.5 I 5.5	5.9 I 6.6 I 8.0	7.0 I 7.2 I 7.8
	4	3.7	5.8 I 5.8 I 5.8	6.4 I 7.0 I 8.8	7.5 I 7.6 I 8.6
	8	6.3	8.7 I 8.7 I 8.8	9.7 I 10.4 I 13.0	10.5 I 10.7 I 12.7
	16	10.3	13.1 I 13.1 I 13.2	14.3 I 15.6 I 21.0	15.0 I 15.8 I 18.8
		SNR [dB] per bit and channel @ BER=10^{-3}			
1	2	13.6	16.3 I 16.3 I 16.3	17.2 I 18.4 I 23.3	18.4 I 19.0 I 21.3
	4	13.6	16.9 I 16.9 I 17.0	17.8 I 20.5	19.0 I 20.7 I 23.8
	8	16.4	20.0 I 20.0 I 21.6	21.5 I 25.6	22.8 I 26.3
	16	20.5	24.2 I 25.0	28.0	30.0
2	2	7.5	9.1 I 9.2 I 9.2	9.4 I 10.3 I 11.8	10.6 I 10.9 I 11.6
	4	7.5	9.5 I 9.6 I 9.6	10.0 I 10.7 I 13.1	10.8 I 11.2 I 12.7
	8	10.5	12.6 I 12.7 I 13.0	13.5 I 14.5 I 19.1	14.2 I 15.0 I 17.6
	16	14.7	17.2 I 17.3 I 17.8	18.6 I 21.3	19.4 I 21.3

Table 15-2 BER Performance of M-PSK Receiver with DD Selective Fading Channel Estimation, GSM Two-Ray Channel

GSM two-ray channel (strong ISI)		Anal.	Simulation		
		MFB	ideal channel	DD parameter synchronization	
				LMS-Kalman	LMS-Wiener
		Doppler λ'_D =0.0003 ǀ 0.001 ǀ 0.003			
D	PSK	SNR [dB] per bit and channel @ BER=10^{-2}			
1	2	8.3	11.0 ǀ 11.0 ǀ 11.0	12.0 ǀ 13.0 ǀ 16.7	12.6 ǀ 13.0 ǀ 16.0
	4	8.3	11.6 ǀ 11.6 ǀ 12.0	12.6 ǀ 14.0 ǀ 19.5	13.4 ǀ 14.2 ǀ 18.2
	8	10.8	14.3 ǀ 14.3 ǀ 15.1	16.0 ǀ 18.0	16.5 ǀ 18.2 ǀ 24.2
	16	14.7	18.2 ǀ 18.5	20.5	21.0
2	2	3.2	5.2 ǀ 5.2 ǀ 5.2	5.5 ǀ 6.3 ǀ 7.9	6.5 ǀ 6.7 ǀ 7.4
	4	3.2	5.5 ǀ 5.5 ǀ 5.5	6.2 ǀ 6.8 ǀ 8.8	6.9 ǀ 7.1 ǀ 8.3
	8	5.9	8.7 ǀ 8.7 ǀ 8.8	9.6 ǀ 10.5 ǀ 13.6	10.2 ǀ 10.5 ǀ 13.3
	16	9.9	13.0 ǀ 13.0 ǀ 13.2	14.3 ǀ 16.2	14.9 ǀ 16.4 ǀ 20.7
		SNR [dB] per bit and channel @ BER=10^{-3}			
1	2	14.1	16.7 ǀ 16.7 ǀ 16.7	17.7 ǀ 20.0 ǀ 26.5	19.2 ǀ 20.0 ǀ 24.0
	4	14.1	17.2 ǀ 17.2 ǀ 17.7	18.8 ǀ 20.8	20.5 ǀ 21.6 ǀ 26.5
	8	16.7	20.3 ǀ 20.3 ǀ 22.6	24.0 ǀ 26.5	25.0 ǀ 26.8
	16	20.6	24.0 ǀ 25.0	28.0	29.0
2	2	7.0	8.7 ǀ 8.8 ǀ 8.8	9.2 ǀ 10.1 ǀ 11.7	10.1 ǀ 10.5 ǀ 11.1
	4	7.0	9.2 ǀ 9.2 ǀ 9.3	9.8 ǀ 10.7 ǀ 13.7	10.6 ǀ 10.8 ǀ 13.1
	8	10.0	12.6 ǀ 12.6 ǀ 12.7	13.7 ǀ 14.4 ǀ 21.2	14.3 ǀ 14.5 ǀ 18.5
	16	14.1	17.1 ǀ 17.2 ǀ 18.4	18.7 ǀ 21.3	19.4 ǀ 21.7

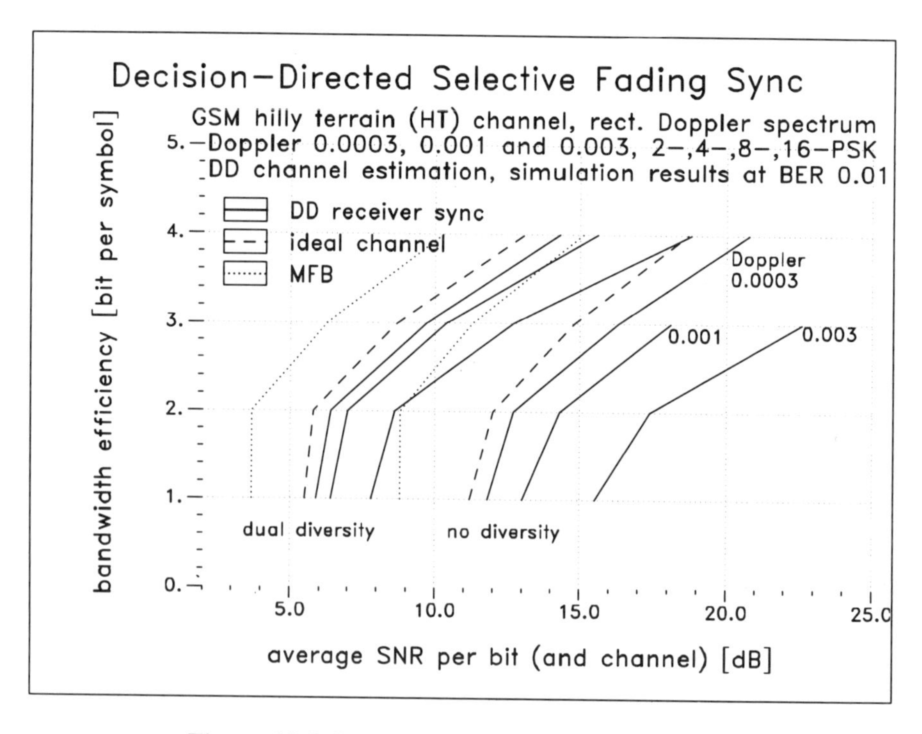

Figure 15-9 Power and Bandwidth Efficiency of DD Receiver for GSM-HT Channel

The performance of DD reception in terms of both power and bandwidth efficiency is visualized in Figures 15-9 and 15-10 for the GSM-HT and GSM two-ray channel, respectively. Like in Section 14.2.4, the SNR needed for a (simulated) BER of 0.01 – see Tables 15-1 and 15-2 – is taken as the measure of power efficiency. In the absence of training, the bandwidth efficiency is 1, 2, 3, and 4 bits per symbol for 2-, 4-, 8-, and 16-PSK.

In comparison with parameter sync based on the ideal channel (dotted curves), DD reception (solid curves) without diversity is seen to cost about 0.7–2.5 dB (2- to 16-PSK), 1.8–3.5 dB (2- to 8-PSK) and from 4 dB for the set of Doppler frequencies $\lambda'_D = 3\times10^{-4} \mid 10^{-3} \mid 3\times10^{-3}$. With dual diversity, the SNR loss figures are 0.5–1.2 dB, 1–3 dB, and from 2.3 dB (2- to 16-PSK) for the same Doppler $\lambda'_D = 3\times10^{-4} \mid 10^{-3} \mid 3\times10^{-3}$. Again, antenna diversity does not only improve detection but also helps mitigating the effect of fading – 16-PSK thus becomes feasible in this example – and reduces the extra SNR needed for DD parameter sync.

In general, the amount of ISI present on the channel may also heavily affect receiver performance. The two channels chosen in this example have very different *ISI profiles* [GSM-HT: mild ISI (fast decaying profile); GSM two-ray: strong ISI (strong postcursors)]. However, the degree of *multipath diversity* inherent in the

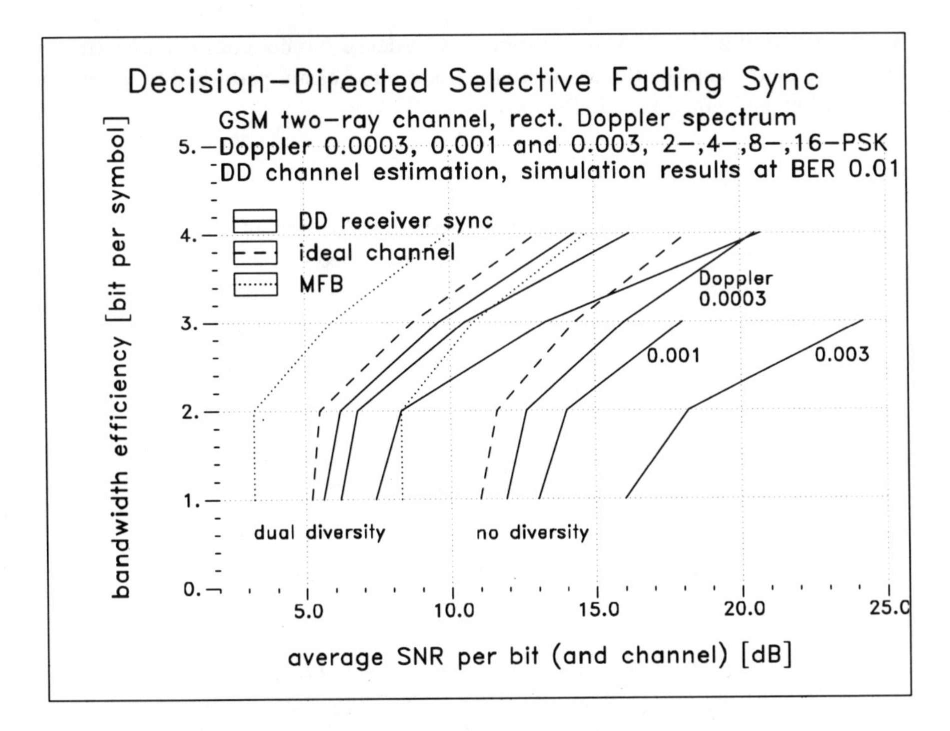

Figure 15-10 Power and Bandwidth Efficiency of DD Receiver for GSM Two-Ray Channel

channels is almost the same. This is because the matched-filter bound – reflecting the degree of multipath diversity in a band-limited selective fading channel – differs by 0.5 dB or less in this example. To further illustrate this point, consider the GSM two-ray channel whose delay τ'=2 is large so that the rays remain resolvable after transmitter pulse shaping (see Figure 11-4) and also receiver prefiltering. Since these equal-average-gain multipath rays are assumed to fade independently (WSSUS scattering model), this ISI channel is, from the viewpoint of diversity, equivalent to a dual diversity channel, provided that perfect equalization can be accomplished. Since the MFBs of the GSM-HT and GSM two-ray channels are almost identical, the GSM-HT channel can also be regarded as nearly equivalent to a dual diversity channel.

Comparing Figures 15-9 and 15-10, receiver performance is seen to be not much affected by the ISI profile (mild vs. severe ISI) as long as the fading remains slow or moderate (up to $\lambda'_D = 10^{-3}$); the DD sync loss margins are very much the same for both channels. In other words, the receiver is capable of recovering most of the inherent multipath diversity gain thanks to near-optimal prefiltering and M algorithm equalization. However, 8- and 16-PSK transmission over the GSM two-ray channel is seen to be significantly more sensitive against faster fading ($\lambda'_D = 3\times10^{-3}$). In the presence of strong ISI, many nonzero filter taps are involved in prefiltering and equalization so that dense symbol constellations (more precisely, their associated decision metrics) are more easily perturbed by errors in the prefilter and equalizer coefficients.

15.1.3 Main Points

- One-step selective fading channel prediction as part of the NDA online detection process (Chapter 13) can be accomplished by means of an LMS-like adaptive algorithm which involves the computation of model output $\hat{r}_{k-1|k-2}$, pseudo-ML tap estimates $\hat{h}_{S,n;k-1}$, and the channel tap update algorithm yielding the tap estimates $\hat{h}_{n;k|k-1}$ [eq. (15-1)]. The latter are obtained from the pseudo-ML estimates via the recursive LMS-Kalman or FIR-type LMS-Wiener algorithm:

$$
\begin{aligned}
&\text{LMS}-\text{Kalman algorithm } (n=0,\ldots,L):\\
&\hat{h}_{n;k|k-1} = (1-K_n)\,\hat{h}_{n;k-1|k-2} + K_n\,\hat{h}_{S,n;k-1}\\
&\text{LMS}-\text{Wiener algorithm } (n=0,\ldots,L):\\
&\hat{h}_{n;k|k-1} = \mathbf{w}_n^H \cdot \hat{\mathbf{h}}_{S,n;k-1}
\end{aligned}
\tag{15-12}
$$

  [eqs. (15-1) and (15-7)], where the parameter sets $\{K_n\}$, $\{\mathbf{w}_n\}$ are functions of λ'_D and the tap SNR profile $\{\rho'_n\}$. The LMS-Kalman algorithm is best suited to slow fading, while the LMS-Wiener has advantages at larger Doppler.
- The DD receiver features just a single NDA parameter synchronizer which performs DD channel estimation, parameter computation, and timing slip control (Figure 15-5). For slow fading, the SNR loss incurred by DD parameter sync is moderate. Antenna diversity further reduces this extra SNR (from 1–2 dB to 0.5–1.2 dB in the examples reported) and also mitigates the detrimental effects of faster fading. By proper prefiltering and equalization, most of the inherent multipath diversity gain can be recovered, and the DD receiver performance is largely independent of the shape of the ISI profile. In the presence of strong ISI, however, dense symbol constellations are more sensitive against fading.

15.2 Data-Aided (DA) Selective Fading Channel Estimation and Detection

In this section, the idea of *data-aided* (DA) parameter synchronization and detection is extended to selective fading channel estimation [8, 9]. As in the flat fading case (Section 14.2), known training data are multiplexed into the information-bearing symbol stream. Because of ISI, however, the received signal cannot be cleanly partitioned into "training" and "data" signal streams; some of the received samples depend on both training and random data. In order to support true data-aided sync, the received signal must contain samples which depend on training *only*. Hence, isolated training symbols are no more appropriate but have to be replaced by *segments* of consecutive training symbols. Transmission systems incorporating such DA *dispersive* channel estimation schemes — thereby separating the tasks of parameter sync and data detection — inherit all the favorable properties that have been identified in the context of DA *flat* fading channel estimation (Section 14.2), including the applicability of TDMA channel access using short bursts, fully coherent detection, amplitude-sensitive modulation formats, near-optimal diversity combining, interleaved channel coding, and soft-decision equalization and decoding.

Following the idea of two-step channel estimation (Section 12.2.2), it is desirable to perform MAP DA channel estimation via ML estimation from individual training segments. In Section 12.2.4 it was shown that such intermediate ML estimates — termed channel *snapshots* — can indeed be generated, provided that (i) the training blocks have some minimum length and (ii) the training symbols are suitably chosen. The topic of ML snapshot acquisition, in particular, the optimal choice of training symbols and the performance of ML snapshot acquisition under fading conditions, is further explored in Section 15.2.2.

In the final two sections we validate and discuss the performance of systems employing DA parameter sync. BER results are presented for the two examples of (i) continuous uncoded M-PSK transmission and M-algorithm equalization (Section 15.2.3) and (ii) TDMA channel access using interleaved trellis coded modulation and combined equalization and decoding (Section 15.2.4). The chapter concludes with a comparison between the potentials of DA and DD parameter synchronization in terms of power and bandwidth efficiency.

15.2.1 DA Selective Fading Channel Estimation

As pointed out in Section 12.2, MAP channel estimation via ML estimation can be accomplished by partitioning the received signal $\mathbf{r}$ into a number of blocks $\mathbf{r}_K$ of length R_K in such a way that (partial[15]) "block" ML estimates, or ML snapshots:

$$\hat{\boldsymbol{\theta}}_{S,K}(\mathbf{a}_K) = \hat{\mathbf{h}}_K(\mathbf{a}_K) = \underbrace{\left[\mathbf{A}_K^H \mathbf{R}_{n,K}^{-1} \mathbf{A}_K\right]^{-1}}_{\boldsymbol{\Sigma}_{S,K}(\mathbf{a}_K)} \cdot \left[\mathbf{A}_K^H \mathbf{R}_{n,K}^{-1}\, \mathbf{r}_K\right]$$
$$\boldsymbol{\Sigma}_{S,K}(\mathbf{a}_K) = \left[\mathbf{A}_K^H \mathbf{R}_{n,K}^{-1} \mathbf{A}_K\right]^{-1} \tag{15-13}$$

[eq. (12-83)] exist for particular data hypotheses $\mathbf{a}_K$. This, of course, requires that the fading is slow enough so that the actual CIR process $\{\mathbf{h}_k\}$ is approximately constant within the Kth block $\mathbf{r}_K$, i.e., $\mathbf{h}_k \approx \mathbf{h}_{k_K}$ for time instants k =

[15] Again, partial channel indices i are suppressed here.

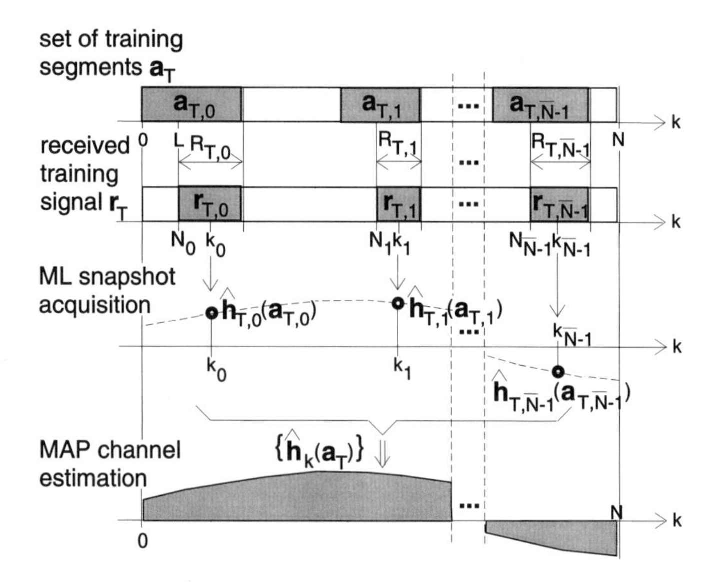

Figure 15-11 MAP DA Selective Channel Estimation via ML Snapshot Acquisition

$N_K, \ldots, N_K + R_{T,K} - 1$ (Figure 15-11), so that the simplified block transmission model $\mathbf{r}_K = \mathbf{A}_K \cdot \mathbf{h}_{k_K} + \mathbf{n}_K$ [eq. (12-76)] holds.

This same principle of MAP selective channel estimation via ML snapshot acquisition can be applied to DA selective channel estimation. The overlapping blocks of unknown data symbols $\mathbf{a}_K$ and corresponding received samples $\mathbf{r}_K$ (Figures 12-6 and 12-7) are now replaced by a number $\overline{N}$ of nonoverlapping training sequences $\mathbf{a}_{T,K}$ and corresponding received blocks $\mathbf{r}_{T,K}$, respectively. The block transmission model for training thus reads

$$\underbrace{\begin{pmatrix} r_{N_K} \\ \vdots \\ r_{N_K+R_{T,K}-1} \end{pmatrix}}_{\mathbf{r}_{T,K}} = \underbrace{\begin{pmatrix} a_{N_K} & \cdots & a_{N_K-L} \\ \vdots & \ddots & \vdots \\ a_{N_K+R_{T,K}-1} & \cdots & a_{N_K+R_{T,K}-1-L} \end{pmatrix}}_{\mathbf{A}_{T,K}} \underbrace{\begin{pmatrix} h_{0;k_K} \\ \vdots \\ h_{L;k_K} \end{pmatrix}}_{\mathbf{h}_{k_K}} + \mathbf{n}_{T,K} \tag{15-14}$$

so that, using eq. (12-83), the ML snapshot acquisition rule and corresponding error covariance are reformulated as

$$\begin{aligned} \hat{\boldsymbol{\theta}}_{S,T,K}(\mathbf{a}_{T,K}) &= \hat{\mathbf{h}}_{T,K}(\mathbf{a}_{T,K}) \\ &= \underbrace{\left[\mathbf{A}_{T,K}^H \mathbf{R}_{n,T,K}^{-1} \mathbf{A}_{T,K}\right]^{-1}}_{\boldsymbol{\Sigma}_{S,T,K}(\mathbf{a}_{T,K})} \cdot \left[\mathbf{A}_{T,K}^H \mathbf{R}_{n,T,K}^{-1}\, \mathbf{r}_{T,K}\right] \\ \boldsymbol{\Sigma}_{S,T,K}(\mathbf{a}_{T,K}) &= \left[\mathbf{A}_{T,K}^H \mathbf{R}_{n,T,K}^{-1} \mathbf{A}_{T,K}\right]^{-1} \end{aligned} \tag{15-15}$$

respectively. The final MAP channel estimates — obtained by filtering the ML snapshots $\hat{\mathbf{h}}_{T,K}(\mathbf{a}_{T,K})$ — and their error covariances for a particular time instant k are then given by

$$
\begin{aligned}
\hat{\theta}_{D,T,k}(\mathbf{a}_T) &= \hat{\mathbf{h}}_k(\mathbf{a}_T) \\
&= \left(\mathbf{R}_h(k-k_0) \vdots \dots \vdots \mathbf{R}_h\left(k-k_{\overline{N}-1}\right) \right) \\
&\cdot \left[\begin{pmatrix} \mathbf{R}_h(k_0-k_0) & \dots & \mathbf{R}_h\left(k_0-k_{\overline{N}-1}\right) \\ \vdots & \ddots & \vdots \\ \mathbf{R}_h\left(k_{\overline{N}-1}-k_0\right) & \dots & \mathbf{R}_h\left(k_{\overline{N}-1}-k_{\overline{N}-1}\right) \end{pmatrix} + \begin{pmatrix} \mathbf{\Sigma}_{S,T,0}(\mathbf{a}_{T,0}) & & \\ & \ddots & \\ & & \mathbf{\Sigma}_{S,T,\overline{N}-1}\left(\mathbf{a}_{T,\overline{N}-1}\right) \end{pmatrix} \right]^{-1} \\
&\cdot \underbrace{\begin{pmatrix} \hat{\mathbf{h}}_{T,0}(\mathbf{a}_{T,0}) \\ \vdots \\ \hat{\mathbf{h}}_{T,\overline{N}-1}\left(\mathbf{a}_{T,\overline{N}-1}\right) \end{pmatrix}}_{\hat{\theta}_{S,T}(\mathbf{a}_T)} \\
\mathbf{\Sigma}_{D,T,k}(\mathbf{a}_T) &= \mathbf{R}_h(0) - \left(\mathbf{R}_h(k-k_0) \vdots \dots \vdots \mathbf{R}_h\left(k-k_{\overline{N}-1}\right) \right) \\
&\cdot \left[\begin{pmatrix} \mathbf{R}_h(k_0-k_0) & \dots & \mathbf{R}_h\left(k_0-k_{\overline{N}-1}\right) \\ \vdots & \ddots & \vdots \\ \mathbf{R}_h\left(k_{\overline{N}-1}-k_0\right) & \dots & \mathbf{R}_h\left(k_{\overline{N}-1}-k_{\overline{N}-1}\right) \end{pmatrix} + \begin{pmatrix} \mathbf{\Sigma}_{S,T,0}(\mathbf{a}_{T,0}) & & \\ & \ddots & \\ & & \mathbf{\Sigma}_{S,T,\overline{N}-1}\left(\mathbf{a}_{T,\overline{N}-1}\right) \end{pmatrix} \right]^{-1} \\
&\cdot \begin{pmatrix} \mathbf{R}_h^H(k-k_0) \\ \vdots \\ \mathbf{R}_h^H\left(k-k_{\overline{N}-1}\right) \end{pmatrix}
\end{aligned}
\tag{15-16}
$$

[see eqs. (12-85) and (12-82)]. The procedure of performing MAP DA selective channel estimation via ML snapshot acquisition is visualized in Figure 15-11.

As in the case of NDA selective channel estimation (Section 15.1), estimator complexity can be reduced to a reasonable level by making a number of simplifying assumptions. By taking the tap processes $h_{n;k}$ to be mutually independent with equal tap Doppler spectra, the tap cross-correlation matrix simplifies to $\mathbf{R}_h(m) = \mathbf{R}_h \cdot \alpha(m) = \text{diag}\{\rho_0, \dots, \rho_L\}\, \alpha(m)$ [eqs. (11-72) and (12-74)] for each of the partial channels. Making the white noise assumption, i.e., $\mathbf{R}_{n,T,K} = N_0 \cdot \mathbf{I} = (1/\overline{\gamma}_s) \cdot \mathbf{I}$, and selecting the training sequences $\mathbf{a}_{T,K}$ such that the matrix products $\mathbf{A}_{T,K}^H \cdot \mathbf{A}_{T,K}$ become diagonal, i.e., $\left(\mathbf{A}_{T,K}^H \cdot \mathbf{A}_{T,K}\right) = R_{T,K} \cdot \mathbf{I}$ (this is accomplished by using *perfect* polyphase sequences, see Section 15.2.2), the ML snapshot acquisition algorithm [eq. (15-15)] simplifies to:

$$\text{ML channel estimation (snapshot acquisition)}:\quad \begin{aligned} \hat{\mathbf{h}}_{T,K}(\mathbf{a}_{T,K}) &= \frac{1}{R_{T,K}} \cdot \mathbf{A}_{T,K}^{H} \cdot \mathbf{r}_{T,K} \\ \mathbf{\Sigma}_{S,T,K}(\mathbf{a}_{T,K}) &= \frac{N_0}{R_{T,K}} \cdot \mathbf{I} \;=\; \frac{1}{(R_{T,K} \cdot \overline{\gamma}_s)} \cdot \mathbf{I} \end{aligned} \tag{15-17}$$

Moreover, the MAP DA estimation rule and the associated error covariance [eq. (15-16)] fall apart into $(L+1)$ *independent* expressions for each individual (partial) channel tap estimate $\hat{h}_{n;k}$. For the purpose of computing the weight vector $\mathbf{w}_{T,n;k}$ for MAP estimation from ML snapshots, the received training sections can be taken to be of equal length $R_{T,K} = R_T$ (otherwise, just take the average length), so that the MAP DA tap estimator and its (average) error covariance can be written as:

MAP DA channel tap estimation from ML estimate:

$$\begin{aligned} \hat{h}_{n;k}(\mathbf{a}_T) &= \underbrace{\mathbf{r}_{D,T,n;k}^{H} \cdot \left(\mathbf{R}_{D,T,n} + \frac{N_0}{R_T} \cdot \mathbf{I}\right)^{-1}}_{\mathbf{w}_{T,n;k}^{H}} \cdot \underbrace{\begin{pmatrix} \hat{h}_{n,T,0}(\mathbf{a}_{T,0}) \\ \vdots \\ \hat{h}_{n,T,\overline{N}-1}\left(\mathbf{a}_{T,\overline{N}-1}\right) \end{pmatrix}}_{\hat{\mathbf{h}}_{S,T,n}} \\ &= \mathbf{w}_{T,n;k}^{H} \cdot \hat{\mathbf{h}}_{S,T,n} \\ \sigma_{n;k}^{2} &\simeq 1 - \mathbf{r}_{D,T,n;k}^{H} \cdot \underbrace{\left(\mathbf{R}_{D,T,n} + \frac{N_0}{R_T} \cdot \mathbf{I}\right)^{-1} \cdot \mathbf{r}_{D,T,n;k}}_{\mathbf{w}_{T,n;k}} \end{aligned} \tag{15-18}$$

These expressions are similar to those for flat fading channel estimation [eq. (14-36)] and selective LMS-Wiener tap prediction [eq. (15-7)]. Analogous to eq. (15-9), the tap autocorrelation matrix and cross-correlation vector are given by

$$\begin{aligned} \mathbf{R}_{D,T,n} = \rho_n \, \mathbf{A}_{D,T} &= \rho_n \begin{pmatrix} \alpha(k_0 - k_0) & \cdots & \alpha\left(k_0 - k_{\overline{N}-1}\right) \\ \vdots & \ddots & \vdots \\ \alpha\left(k_{\overline{N}-1} - k_0\right) & \cdots & \alpha\left(k_{\overline{N}-1} - k_{\overline{N}-1}\right) \end{pmatrix} \\ \mathbf{r}_{D,T,n;k} = \rho_n \, \boldsymbol{\alpha}_{D,T;k} &= \rho_n \left(\alpha(k - k_0) \;\; \cdots \;\; \alpha\left(k - k_{\overline{N}-1}\right)\right)^{T} \end{aligned} \tag{15-19}$$

so that the optimal Wiener tap filters $\mathbf{w}_{T,n;k}$ and associated (average) error covariance factors $\overline{r}_{n;k}$ [analogous to eq. (15-8)] can also be expressed as

$$\begin{aligned} \mathbf{w}_{T,n;k} &= \left[\mathrm{A}_{D,T} + \frac{1}{R_T\, \rho_n'} \cdot \mathbf{I} \right]^{-1} \cdot \boldsymbol{\alpha}_{D,T;k} \\ \overline{r}_{n;k} \simeq \frac{\sigma^2_{n;k}}{N_0} &= \rho_n' \left(1 - \boldsymbol{\alpha}^H_{D,T;k} \cdot \underbrace{\left[\mathrm{A}_{D,T} + \frac{1}{R_T \rho_n'} \cdot \mathbf{I} \right]^{-1} \cdot \boldsymbol{\alpha}_{D,T;k}}_{\mathbf{w}_{T,n;k}} \right) \end{aligned} \tag{15-20}$$

The Wiener tap filters $\mathbf{w}_{T,n;k}$ and covariance factors $\overline{r}_{n;k}$ are seen to depend on the tap SNR $\rho_n' = \rho_n / N_0$ [eq. (15-6)], the received training length R_T, the set of training reference positions k_K (incorporated – together with the Doppler – in $\mathrm{A}_{D,T}$ and $\boldsymbol{\alpha}_{D,T;k}$), and the particular position k of the estimate (through $\boldsymbol{\alpha}_{D,T;k}$).

Just like in the flat fading case (Section 14.2), DA MAP selective channel tap estimation from ML snapshots is intimately linked with *sampling* theory. Wiener estimation effectuates (close-to-) optimal noise suppression and tap interpolation, extrapolation, or filtering. Depending upon the set of reference positions k_K relative to the positions k for which estimates are to be generated, the channel tap trajectories are reconstructed from a —possibly small — number of noisy samples which may or may not be uniformly spaced in time. In all of these scenarios, the tap estimation error performance remains to be completely analyzable through eq. (15-20).

15.2.2 Maximum-Likelihood DA Snapshot Acquisition

In the previous subsection, the use of *perfect* polyphase training sequences has been found to have two most advantageous consequences: (i) the channel estimator is greatly simplified, and (ii) MAP channel estimation from the acquired ML estimate can be performed *independently* for each individual (partial) channel tap [eq. (15-18)] without loss in optimality. The question remains of whether perfect polyphase training sequences also yield optimal ML estimates. As proven below, this is indeed the case.

Under the white noise assumption, the ML snapshot acquisition rule and corresponding error covariance [eq. (15-15)] become

$$\begin{aligned} \hat{\mathbf{h}}_{T,K}(\mathbf{a}_{T,K}) &= \left[\mathbf{A}^H_{T,K} \cdot \mathbf{A}_{T,K}\right]^{-1} \cdot \left[\mathbf{A}^H_{T,K} \cdot \mathbf{r}_{T,K}\right] \\ \boldsymbol{\Sigma}_{S,T,K}(\mathbf{a}_{T,K}) &= N_0 \left[\mathbf{A}^H_{T,K} \cdot \mathbf{A}_{T,K}\right]^{-1} \end{aligned} \tag{15-21}$$

Now optimal acquisition calls for minimizing the $(L+1)\times(L+1)$-dim. error covariance matrix $\boldsymbol{\Sigma}_{S,T,K}(\mathbf{a}_{T,K})$ (more precisely, its trace) and thus the inverse deterministic training autocorrelation matrix $(\mathbf{A}_{T,K}^{H} \cdot \mathbf{A}_{T,K})^{-1}$. When the training symbols are taken from a PSK alphabet ($| a_k | = 1$), all main diagonal elements of $(\mathbf{A}_{T,K}^{H} \cdot \mathbf{A}_{T,K})$ equal $R_{T,K}$ so that the trace – and thus the sum of eigenvalues $(\sum_{n=0}^{L} \lambda_n)$ – of $(\mathbf{A}_{T,K}^{H} \cdot \mathbf{A}_{T,K})$ evaluate as $(L+1)R_{T,K}$. Noting that the inverse $(\mathbf{A}_{T,K}^{H} \cdot \mathbf{A}_{T,K})^{-1}$ has eigenvalues $(1/\lambda_n)$ $(n = 0, \ldots, L)$ [10] and performing minimization of $\mathrm{tr}\left\{(\mathbf{A}_{T,K}^{H} \cdot \mathbf{A}_{T,K})^{-1}\right\} = (\sum_{n=0}^{L}(1/\lambda_n))$ under the constraint $(\sum_{n=0}^{L} \lambda_n) = (L+1)R_{T,K}$ via the method of Lagrange factors yields the optimality condition

$$(\mathbf{A}_{T,K}^{H} \cdot \mathbf{A}_{T,K}) = R_{T,K} \cdot \mathbf{I} \tag{15-22}$$

which leads to the ML channel acquisition rule and associated error covariance of eq. (15-17) above.

Training sequences thus should have ideal (or close-to-ideal) deterministic autocorrelation, i.e., largest possible integration gain $R_{T,K}$ and zero sidelobes. Polyphase sequences satisfying this conditions are called *CAZAC* (constant amplitude, zero autocorrelation) or *perfect* sequences [11]. Such sequences exist for arbitrary training lengths $(R_T + L) \geq (2L+1)$ [received training lengths $R_T \geq (L+1)$] and can be easily constructed from a *periodic* sequence with period length $P_T \geq (L+1)$. As an example, some M_T-PSK CAZAC sequences, which have been generated using known construction methods, are listed in Table 15-3 for period lengths $P_T = 2, \ldots, 8$, and 16.

In summary, perfect polyphase sequences should be used as training sequences wherever possible, since they yield both (quasi-)optimal ML/MAP estimation performance and lowest estimator complexity.

As mentioned above, all results on DA selective channel estimation established here are based on the snapshot assumption, i.e., the fading must be slow

Table 15-3 Some CAZAC (perfect) M_T-PSK Training Sequences

Period Length $P_T \geq (L+1)$	M_T-PSK symbol alphabet	Training Symbols (one period) $a_k = \exp\{j(2\pi/M_T)\, l_k\}$; training symbol labels (one period) l_k:
2	4-PSK	{ 0, 1 }
3	3-PSK	{ 0, 1, 0 }
4	2-PSK	{ 0, 0, 1, 0 }
5	5-PSK	{ 0, 1, 3, 1, 0 }
6	12-PSK	{ 0, 1, 4, 9, 4, 1 }
7	7-PSK	{ 0, 1, 3, 6, 3, 1, 0 }
8	4-PSK	{ 0, 0, 1, 2, 0, 2, 1, 0 }
16	4-PSK	{ 0, 0, 0, 0, 1, 2, 3, 0, 2, 0, 2, 0, 3, 2, 1, 0 }

relative to the training block length $(R_{T,K}+L)$ (Figure 15-11) so that the snapshots $\hat{\mathbf{h}}_{T,K}(\mathbf{a}_{T,K})$ are not "blurred" by fading effects. Hence, it is vital to check whether the snapshot assumption is satisfied for all channel conditions that can be covered by DA channel estimation.

To this end, let us investigate the effect of time variations in the CIR trajectory $\mathbf{h}_k$ within the training block ($k = N_K, \ldots, N_K+R_{T,K}-1$); Figure 15-11) [12, 13]. The signal transmission model, including channel fading [eq. (12-1)], can be expressed as:

$$
\begin{aligned}
&\text{(partial) received signal:}\\
r_k &= \underbrace{\left(a_k \; a_{k-1} \; \ldots \; a_{k-L} \right)}_{\mathbf{a}_k^T} \cdot \begin{pmatrix} h_{0;k} \\ \vdots \\ h_{L;k} \end{pmatrix} \quad + n_k \\
&= \mathbf{a}_k^T \cdot \mathbf{h}_k \quad + n_k \\
&= \mathbf{a}_k^T \cdot (\mathbf{h}_{k_K} + \mathbf{d}_k) \quad + n_k \\
&= \mathbf{a}_k^T \cdot \mathbf{h}_{k_K} + \underbrace{\mathbf{a}_k^T \cdot \mathbf{d}_k}_{e_k} \quad + n_k \\
&\text{(partial) received training vector:}\\
\mathbf{r}_{T,K} &= \mathbf{A}_{T,K} \cdot \mathbf{h}_{k_K} \; + \mathbf{e}_{T,K} \; + \mathbf{n}_{T,K}
\end{aligned}
\tag{15-23}
$$

with $\mathbf{d}_k = \mathbf{h}_k - \mathbf{h}_{k_K}$ the difference in CIR vectors with respect to the reference position k_K, and $e_k = \mathbf{a}_k^T \cdot \mathbf{d}_k$, and $\mathbf{e}_{T,K}$ the extra observation noise due to fading within the Kth training block. Performing optimal CAZAC ML snapshot acquisition on $\mathbf{r}_{T,K}$ [eq. (15-17)] then yields

$$
\begin{aligned}
\hat{\mathbf{h}}_{T,K}(\mathbf{a}_{T,K}) &= \frac{1}{R_{T,K}} \cdot \mathbf{A}_{T,K}^H \cdot \mathbf{r}_{T,K} \\
&= \underbrace{\frac{1}{R_{T,K}} \cdot \mathbf{A}_{T,K}^H \mathbf{A}_{T,K} \mathbf{h}_{k_K}}_{\mathbf{h}_{k_K}} + \underbrace{\frac{1}{R_{T,K}} \cdot \mathbf{A}_{T,K}^H \mathbf{e}_{T,K}}_{\Delta \mathbf{h}_{T,K}^{(F)}} + \underbrace{\frac{1}{R_{T,K}} \cdot \mathbf{A}_{T,K}^H \mathbf{n}_{T,K}}_{\Delta \mathbf{h}_{T,K}^{(N)}} \\
&= \mathbf{h}_{k_K} \; + \Delta \mathbf{h}_{T,K}^{(F)} \; + \Delta \mathbf{h}_{T,K}^{(N)}
\end{aligned}
\tag{15-24}
$$

with $\Delta \mathbf{h}_{T,K}^{(F)}$ and $\Delta \mathbf{h}_{T,K}^{(N)}$ the tap error vectors due to fading and noise, respectively. While the *noise* error covariance of each tap is given by $\sigma_{n,K}^{(N)^2} = N_0/R_{T,K} = 1/(R_{T,K}\overline{\gamma}_s)$ [main diagonal elements of $\mathbf{\Sigma}_{S,T,K}^{(N)}$, eq. (15-17)], we are now interested in the *fading* error covariance $\sigma_{n,K}^{(F)^2} = E[| \Delta h_{n,T,K}^{(F)} |^2]$, with tap estimation error:

$$
\begin{aligned}
\Delta h_{n,T,K}^{(F)} &= \left(\Delta \mathbf{h}_{T,K}^{(F)}\right)_n \\
&= \left(\frac{1}{R_{T,K}} \cdot \underbrace{\begin{pmatrix} a_{N_K}^* & \cdots & a_{N_K+R_{T,K}-1}^* \\ \vdots & & \vdots \\ a_{N_K-L}^* & \cdots & a_{N_K+R_{T,K}-1-L}^* \end{pmatrix}}_{\mathbf{A}_{T,K}^H} \cdot \begin{pmatrix} e_{N_K} \\ \vdots \\ e_{N_K+R_{T,K}-1} \end{pmatrix} \right)_n \\
&= \frac{1}{R_{T,K}} \begin{pmatrix} a_{N_K-n}^* & \cdots & a_{N_K+R_{T,K}-1-n}^* \end{pmatrix} \begin{pmatrix} \sum\limits_{l=0}^{L} a_{N_K-l} \cdot d_{l;N_K} \\ \vdots \\ \sum\limits_{l=0}^{L} a_{N_K+R_{T,K}-1-l} \cdot d_{l;N_K+R_{T,K}-1} \end{pmatrix} \\
&= \frac{1}{R_{T,K}} \left(\sum_{k=N_K}^{N_K+R_{T,K}-1} \sum_{l=0}^{L} \left(a_{k-n}^* a_{k-l}\right) d_{l;k} \right)
\end{aligned}
\tag{15-25}
$$

so that

$$
\begin{aligned}
\sigma_{n,K}^{(F)^2} &= E\left[\Delta h_{n,T,K}^{(F)} \, \Delta h_{n,T,K}^{(F)^*}\right] \\
&= \frac{1}{R_{T,K}^2} \left(\sum_{i=N_K}^{N_K+R_{T,K}-1} \sum_{j=N_K}^{N_K+R_{T,K}-1} \sum_{l=0}^{L} \sum_{m=0}^{L} \left(a_{i-n}^* a_{i-l}\right) \left(a_{j-n}^* a_{j-m}\right)^* E\left[d_{l;i} d_{m;j}^*\right] \right)
\end{aligned}
\tag{15-26}
$$

Assuming independently fading (partial) channel taps, we have

$$
\begin{aligned}
E\left[d_{l;i} d_{m;j}^*\right] &\simeq E\left[d_{l;i} d_{l;j}^*\right] \delta_{l,m} \\
&= E\left[\left(h_{l;i} - h_{l;k_K}\right)\left(h_{l;j} - h_{l;k_K}\right)^*\right] \delta_{l,m} \\
&= \rho_l \left[(1 + \alpha(i-j)) - (\alpha(i-k_K) + \alpha(j-k_K))\right] \delta_{l,m}
\end{aligned}
\tag{15-27}
$$

with $\alpha(m)$ the (normalized) channel autocorrelation function [eq. (12-49)]. Using the parabolic approximation to $\alpha(m)$, i.e., $\alpha(m) \approx 1 - \pi^2 (\lambda_D')^2 m^2$ (valid for both Jakes and rectangular Doppler spectra as long as $\mid \lambda_D' m \mid \leq 0.2$, see Section 14.1) and setting the reference index k_K to the center of the received block $\mathbf{r}_{T,K}$, i.e., $k_K = N_K + (R_{T,K} - 1)/2$ (the fading error covariance is thus minimized), the expected value simplifies to

$$E\left[d_{l;i}d^*_{m;j}\right] \simeq \rho_l\,\pi^2\left(\lambda'_D\right)^2\left[(i-k_K)^2+(j-k_K)^2-(i-j)^2\right]\delta_{l,m}$$
$$=\rho_l\,2\pi^2\left(\lambda'_D\right)^2\left[\left(i-\underbrace{\left\lfloor N_K+\frac{R_{T,K}-1}{2}\right\rfloor}_{k_K}\right)\left(j-\underbrace{\left\lfloor N_K+\frac{R_{T,K}-1}{2}\right\rfloor}_{k_K}\right)\right]\delta_{l,m} \qquad (15\text{-}28)$$

Substituting $i \leftarrow (i-k_K)$ and $j \leftarrow (j-k_K)$ finally yields the tap error covariance for the nth tap of Kth snapshot $\hat{\mathbf{h}}_{T,K}(\mathbf{a}_{T,K})$ due to fading:

Snapshot tap error covariance due to fading:

$$\sigma^{(F)^2}_{n,K} = 2\pi^2\left(\frac{\lambda'_D}{R_{T,K}}\right)^2\left(\sum_{l=0}^{L}\rho_l\,|X_{n,l}|^2\right)$$

with quantities:

$$X_{n,l} = \sum_{i=-(R_{T,K}-1)/2}^{(R_{T,K}-1)/2}(i)\left(a^*_{N_K+\left(i+\frac{R_{T,K}-1}{2}\right)-n}\;a_{N_K+\left(i+\frac{R_{T,K}-1}{2}\right)-l}\right) \qquad (15\text{-}29)$$

where index i is integer when $R_{T,K}$ is odd, and half-integer otherwise. As shown in Figure 15-12 for the example: GSM-HT channel (L=4), received training length R_T=16, 4-PSK CAZAC training sequence with period length P_T=8 from Table 15-3 (see example below), the analysis [eq. (15-29)] agrees well with simulation results.

The total fading MSE $\zeta^{(F)}_{T,K}$ in the snapshots $\hat{\mathbf{h}}_{T,K} = \hat{\mathbf{h}}^{(0)}_{T,K}\oplus\hat{\mathbf{h}}^{(1)}_{T,K}$ – taking into account the contributions from both partial channels – is then given by

$$\zeta^{(F)}_{T,K} = \sum_{i=0}^{1}\sum_{n=0}^{L}\sigma^{(F,i)^2}_{n,K}$$
$$= 2\pi^2\left(\frac{\lambda'_D}{R_{T,K}}\right)^2\underbrace{\left(\sum_{i=0}^{1}\sum_{l=0}^{L}\rho^{(i)}_l\left[\sum_{n=0}^{L}|X_{n,l}|^2\right]\right)}_{C(\boldsymbol{\rho},\mathbf{a}_{T,K})} \qquad (15\text{-}30)$$

As expected, the fading MSE depends on the Doppler, the training length $R_{T,K}$, the channel tap energy profile $\boldsymbol{\rho}$, and the particular training sequence $\mathbf{a}_{T,K}$.

The question remains under which conditions the snapshot assumption holds. Given some SNR of operation $\overline{\gamma}_s$, the fading effects must not dominate the total

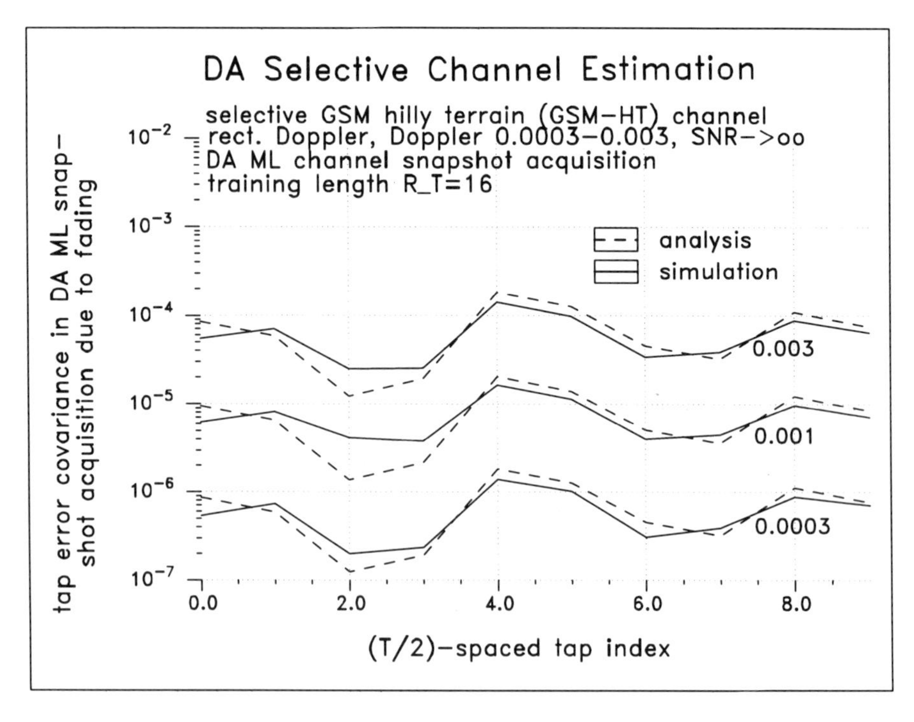

Figure 15-12 Acquisition Tap Error Covariance Due to Fading, GSM-HT Channel

MSE $\zeta_{T,K} = \zeta_{T,K}^{(F)} + \zeta_{T,K}^{(N)}$, i.e.,

$$\zeta_{T,K}^{(F)} = 2\pi^2 \left(\frac{\lambda_D'}{R_{T,K}} \right) C(\boldsymbol{\rho}, \mathbf{a}_{T,K}) \overset{!}{<} \zeta_{T,K}^{(N)} = \sum_{i=0}^{1} \sum_{n=0}^{L} \sigma_{n,K}^{(N,i)^2} = \frac{2(L+1)}{R_{T,K}} \frac{1}{\overline{\gamma}_s}$$

$$\Rightarrow \quad \lambda_D' \overset{!}{<} \frac{1}{\pi} \sqrt{\frac{R_{T,K}(L+1)}{C(\boldsymbol{\rho}, \mathbf{a}_{T,K})}} \frac{1}{\sqrt{\overline{\gamma}_s}} \tag{15-31}$$

where the total noise MSE $\zeta_{T,K}^{(N)}$ is simply given by $\sigma_{n,K}^{(N,i)^2} = 1/(R_{T,K}\overline{\gamma}_s)$ [eq. (15-17)] multiplied by the total number of taps.

A simple example may help to illustrate this condition on the validity of the snapshot assumption. Consider the two training scenarios:

1. GSM-HT channel (L=4), received training length R_T=16, training block length $R_T + L$=20 = 2.5 periods of 4-PSK CAZAC training sequence with period length P_T=8 from Table 15-3,
2. GSM-HT channel (L=4), received training length R_T=8, training block length $R_T + L$=12 = 1.5 periods of the same 4-PSK CAZAC training sequence.

The first scenario corresponds to the preamble designed for continuous transmission of uncoded M-PSK (Section 15.2.3), and the second to both pre- and postamble used for TDMA transmission of interleaved TCM (Section 15.2.4).

Numerical evaluation yields $C(\boldsymbol{\rho}, \mathbf{a}_{T,K}) \approx 1077$ and 270 for the two scenarios so that the snapshot assumption is valid as long as $\lambda_D^{'} < 0.08675/\sqrt{\overline{\gamma}_s}$ and $\lambda_D^{'} < 0.1225/\sqrt{\overline{\gamma}_s}$ [or $\overline{\gamma}_s < \left(0.08675\,\lambda_D^{'}\right)^2$ and $\overline{\gamma}_s < \left(0.1225\,\lambda_D^{'}\right)^2$], respectively. Considering a transmission frame length of $F\ =156$ (leading to a reasonable training overhead of about 13–15 percent), snapshots are generated at rate $1/F$ so that the sampling theorem (snapshot interpolation) allows for a maximum Doppler of $\lambda_D^{\prime} \approx 0.003$. The snapshot assumption is then satisfied for SNRs up to about 29 and 32 dB, respectively. Hence, in cellular mobile environments where SNRs (SIRs) of interest range between 5 and 20 dB, the snapshot assumption certainly holds. The fading-induced error in the snapshots may have a significant effect on very dense symbol constellations (e.g., 64- or 256-QAM), but by far the dominant source of error under such conditions will be the disturbance caused by other receiver components, in particular, the WMF prefilter (Chapter 13).

15.2.3 Example: DA Receiver for Uncoded M-PSK

In this first example, consider the *continuous* transmission of noninterleaved uncoded M-PSK information symbols. The digital receiver *detection path* includes matched and whitening filters, optimal diversity combiner (if necessary), and an M-algorithm equalizer with M=4 equalizer states for near-optimal reduced-complexity ML detection. The receiver *sync path* comprises DA channel estimation (described below), followed by the adjustment of receiver components (MF, WF, combiner, equivalent channel $\mathbf{f}_k$) according to the algorithms discussed in Section 13.3.4. The DA receiver block diagram without diversity – the extension to diversity reception is straightforward, see receiver structure of Figure 13-9 above – is shown in Figure 15-13.

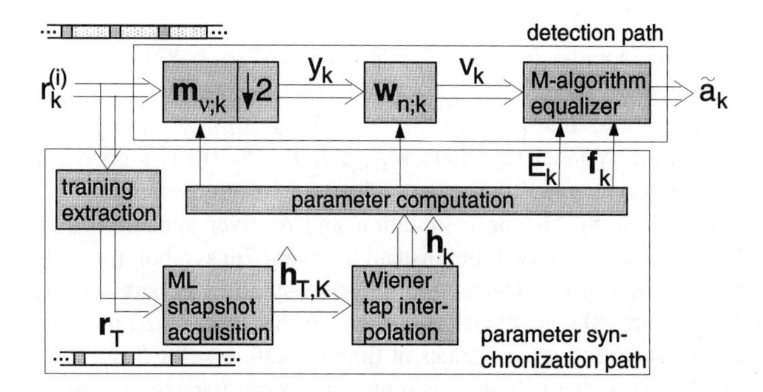

Figure 15-13 DA Selective Fading Receiver for Continuous Data Transmission

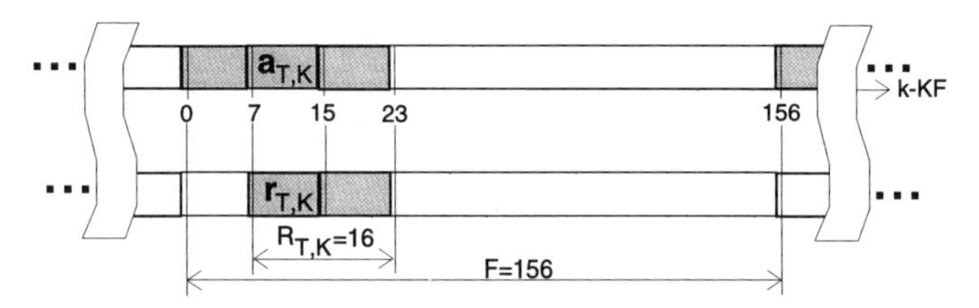

Figure 15-14 Frame Format for Continuous Data Transmission and DA Reception

The data transmission format used here is illustrated in Figure 15-14. In the *flat* fading case (Section 14.2), noise suppression relies on smoothing alone due to the large average noise power N_0 in the ML estimate $\hat{\mathbf{c}}_{S,T}(\mathbf{a}_T)$ [eq. (14-36)]. For this reason, optimal training symbol insertion rates $1/F$ were found to be substantially larger than the minimum sampling rate [eq. (14-40)]. In the *selective* fading case, on the other hand, large channel oversampling margins are not necessary since ML snapshot acquisition already effectuates good noise suppression [average noise power $N_0/R_{T,K}$ in $\hat{\mathbf{h}}_{T,K}(\mathbf{a}_{T,K})$, eq. (15-17)] by virtue of the integration gain R_T. Therefore, training sequence insertion rates $1/F$ should be as small as possible. Hence, the frame length has been fixed at F=156 for all Doppler up to the maximum (see above) of λ'_D ≈0.003.

The training sequence selected here (perfect 4-PSK, length 23, received length R_T=16, period length P_T=8; see Table 15-3 and Figure 15-14) allows for a maximum dispersion of L=7, thus leaving some margin for frame sync when GSM-like channels (L=4) are considered. Besides, this kind of training sequence also allows for frame and frequency sync (or monitoring) since each of the received training sections $\mathbf{r}_{T,K}$ comprises *two* periodic (disregarding noise and fading effects) blocks of received samples. With 23 training and 133 data symbols per frame, training efficiency is $\eta_F \approx 0.85$ so that training insertion entails a penalty of about 0.7 dB.

Considering Doppler frequencies in the range $\lambda'_D \approx$ 0.0003–0.003, the fading autocorrelation main lobe spans up to about 1500 symbols (10 frames) in either direction. Therefore, interpolating between 20 uniformly spaced snapshots by means of length-20 Wiener tap filters $\mathbf{w}_{T,n;k}$ [eq. (15-20)] is a good compromise between performance and filtering complexity. Receiver complexity is further reduced by performing channel estimation and receiver synchronization not for each and every symbol interval, but instead for every 7th symbol interval within the data portion of the frame. Hence, with 133 data symbols in between two training segments, there are 20 sync cycles per frame so that a set of 20 tap filters $\mathbf{w}_{T,n;k}$ is needed for generating tap estimates at time instants $k = 23, 30, \ldots, 156$ (Figure 15-14). In between these "pole positions", the sync parameters (MF, WF, and equalizer coefficients, Figure 15-13) are generated by simple linear interpolation.

Since, in this example, Wiener weight computation according to eq. (15-20) would involve solving a 20-dimensional matrix equation whenever the average tap SNRs ρ'_n or their estimates change, an eigensystem synthesis of the tap estimation weight vectors $\mathbf{w}_{T,n;k}$ (discussed in Sections 14.2.4 and 15.1.1) is desirable. Recognizing that eq. (15-20) is of the same form as eqs. (14-41) and (15-8), the results of eqs. (14-45) (DA flat channel estimation) and (15-10) (NDA LMS-Wiener selective channel prediction) can easily be adapted to DA selective channel estimation to yield

$$\mathbf{w}_{T,n;k}\left(R_T,\rho_n'\right) = \left(R_T\rho_n'\right)\left(\boldsymbol{\alpha}_{D,T;k} - \sum_{i=0}^{\overline{\nu}-1} \frac{\lambda_{D,T,i}}{\lambda_{D,T,i}+\frac{1}{R_T\rho_n'}} \underbrace{\left(\left[\mathbf{u}_i\cdot\mathbf{u}_i^H\right]\boldsymbol{\alpha}_{D,T;k}\right)}_{\mathbf{w}_{(i;k)}}\right) \tag{15-32}$$

with $\{\lambda_{D,T,0},\ldots,\lambda_{D,T,\overline{\nu}-1}\}$ and $\{\mathbf{u}_0,\ldots,\mathbf{u}_{\overline{\nu}-1}\}$ the sets of the $\overline{\nu}$ largest eigenvalues and eigenvectors of matrix $\mathrm{A}_{D,T}$ [eq. (15-19)], and $\{\mathbf{w}_{(0;k)},\ldots,\mathbf{w}_{(\overline{\nu}-1;k)}\}$ the set of partial weight vectors that – dependent on the Doppler λ'_D and k but independent of ρ'_n – can be precomputed and stored. Weight vector computation can then be performed online (e.g., once per frame), based on the current estimate of ρ'_n which itself can be continually updated using the ML channel snapshots $\hat{h}_{n,T,K}$.

Unfortunately, weight vector synthesis according to eq. (15-32) may call for storing many partial weight vectors $\mathbf{w}_{(i;k)}$, especially when there are many nonzero eigenvalues $\lambda_{D,T,i}$ and a large set of positions k for which to generate weight vectors and channel estimates. For instance, considering Doppler frequencies of λ'_D =0.0003, 0.001, and 0.003 ($\approx$ the Nyquist limit), one finds that there are 6, 12, and 20 nonzero eigenvalues, respectively. Channel estimation at 20 positions k within the data block then calls for storing 120, 240, and 400 distinct partial weight vectors $\mathbf{w}_{(i;k)}$. This large storage requirement can be traded against a more computationally complex weight generation algorithm by expressing eq. (15-32) in the form

$$\mathbf{w}_{T,n;k}\left(R_T,\rho_n'\right) = \underbrace{\left(R_T\rho_n'\right)\left(\mathbf{I} + \sum_{i=0}^{\overline{\nu}-1} \frac{\lambda_{D,T,i}}{\lambda_{D,T,i}+\frac{1}{R_T\rho_n'}}\left[\mathbf{u}_i\cdot\mathbf{u}_i^H\right]\right)}_{\boldsymbol{\Gamma}\left(R_T,\rho_n'\right)}\cdot\boldsymbol{\alpha}_{DTk} \tag{15-33}$$

Now the matrix $\boldsymbol{\Gamma}(R_T,\rho'_n)$ needs to be constructed online using the stored eigenvalues $\lambda_{D,T,i}$ and eigenvectors $\mathbf{u}_i$, but this algorithm is still simpler than solving for a high-dimensional systems of equations [eq. (15-20)].

When approaching the Nyquist limit (here at Doppler λ'_D $\approx$0.003), all 20 eigenvalues $\lambda_{D,T,i}$ are found to be nonzero. Fortunately, eigenvalue synthesis is still viable since one of the eigenvalues is in fact a multiple eigenvalue $\lambda_{D,T,\mathrm{mult}}$ with multiplicity $\overline{\nu}_{\mathrm{mult}}$ (for λ'_D =0.003 we have $\overline{\nu}_{\mathrm{mult}}$=5). Reformulating eq. (15-33) for this case yields, after some elementary operations, an appropriate expression for matrix $\mathbf{\Gamma}(R_T, \rho_n')$:

$$\mathbf{\Gamma}\left(R_T, \rho_n'\right) = \frac{1}{c_{\text{mult}}}\mathbf{I} + \left(\sum_{i=0}^{\nu-\overline{\nu}_{\text{mult}}-1} \frac{\lambda_{D,T,\text{mult}} - \lambda_{D,T,i}}{c_{\text{mult}}\, c_i} \cdot \left[\mathbf{u}_i \cdot \mathbf{u}_i^H\right] \right) \tag{15-34}$$
$$\text{with} \qquad c_{\text{mult}} = \lambda_{D,T,\text{mult}} + \frac{1}{R_T \rho_n'} \qquad c_i = \lambda_{D,T,i} + \frac{1}{R_T \rho_n'}$$

to be used in lieu of eq. (15-33) when $\mathbf{A}_{D,T}$ has multiple eigenvalues.

As mentioned in Section 14.2.2, steady-state uniform DA channel estimation yields the same average error covariance – here tap error covariance factors $\overline{r}_{n;k}$ of eq. (15-20) – for all positions k as long as the sampling theorem holds. Figure 15-15 displays covariance factors $\overline{r}_n$ obtained by analysis [eq. (15-20)] and simulation (averaging over 10,000 frames and all 20 intraframe positions k) for the GSM-HT channel. The total tap error covariance reduction relative to the noise level is brought about by (i) the integration gain from snapshot acquisition (here $R_T = 16$ or 12 dB) and (ii) the Wiener filtering gain. This extra gain is large for slow fading and weak taps, but theoretical gains are small for stronger taps with Doppler near the Nyquist limit ($\lambda_D' \approx 0.003$), and simulation results even reveal a loss relative

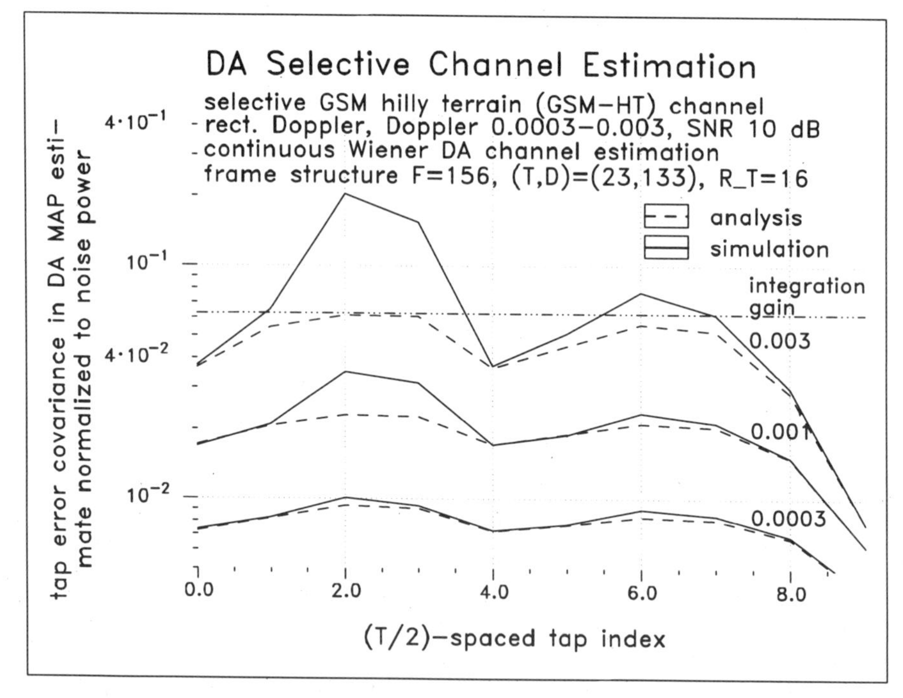

Figure 15-15 Tap Error Covariance of DA Selective Fading Channel Estimation; Continuous Transmission, GSM-HT Channel, SNR 10 dB

to the error performance of snapshot acquisition. This also translates into the MSE factors $\zeta = \sum_i \sum_n \overline{r}_n^{(i)}$ (both partial channels) vs. Doppler λ'_D shown in Figure 15-16. The formidable deviations between analysis and simulation at large SNR and larger Doppler can be attributed to irreducible effects, in particular, fading errors in the acquired snapshots (Section 15.2.2) but also inaccuracies in the tap fading process simulation (the simulated Doppler spectrum is not exactly band-limited due to 8th-order Butterworth filtering).

The following figures illustrate the BER performance of the inner DA receiver (Figure 15-13) for uncoded M-PSK continuous transmission, DA parameter synchronization (WF lengths of W=4, 8, 12, and 16 for 2-, 4-, 8-, and 16-PSK, respectively), and M-algorithm equalization with 4 equalizer states. Again we consider the GSM-HT and GSM two-ray channels (L =4), and Doppler frequencies of λ'_D = 0.0003, 0.001, and 0.003.

Figure 15-17 shows BER curves for slow Doppler λ'_D= 0.0003, GSM-HT channel, and no diversity reception; Figure 15-18 displays the same for dual diversity reception. It is important to note that the 0.7-dB training insertion loss has been taken into account not only in the results for DA parameter sync but also in the curves for the "ideal channel". This is because the presence of training

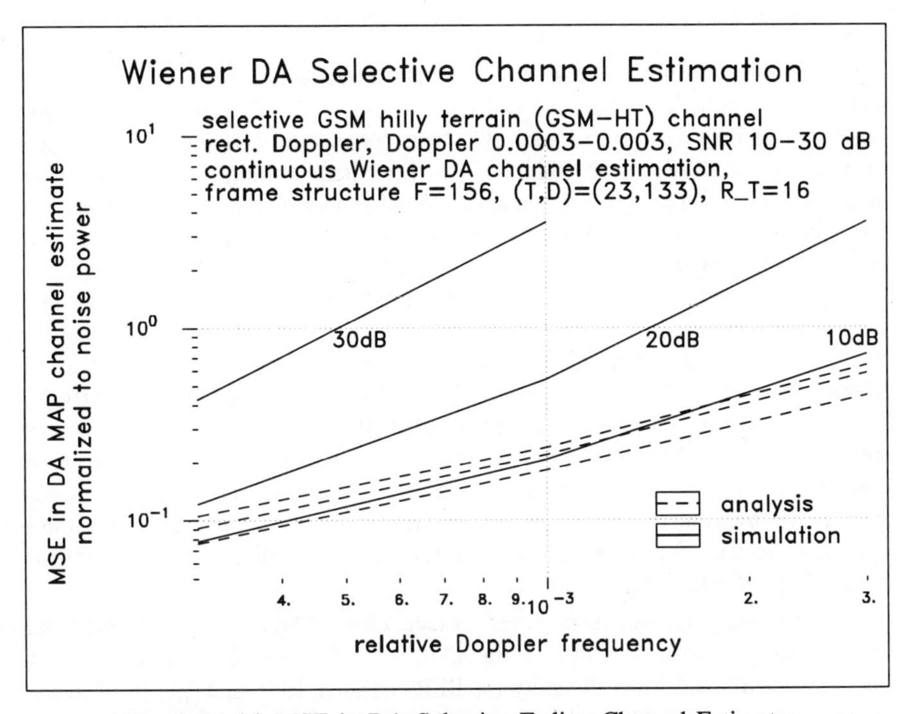

Figure 15-16 MSE in DA Selective Fading Channel Estimate; Continuous Transmission, GSM-HT Channel

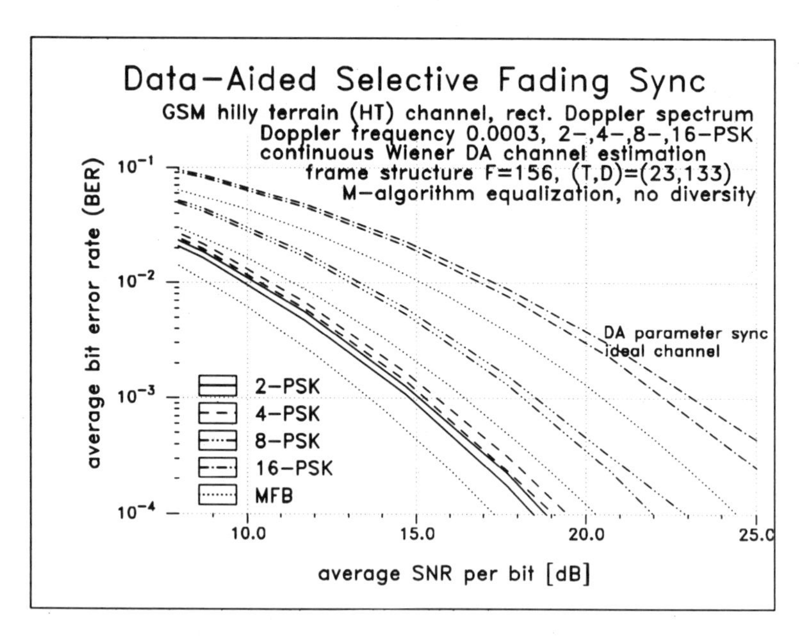

Figure 15-17 BER Performance of Transmission System with DA Selective Fading Channel Estimation and M-PSK Detection without Diversity; GSM-HT Channel, Doppler 0.0003

segments does not only enable DA sync but also effectively aids in the equalization and detection process. While the DD receiver (Figure 15-5) continually operates on unknown data and is thus subject to error propagation during and following deep fades, the DA receiver (Figure 15-13) periodically reinitializes the equalizer state by virtue of the training symbols. In other words, the equalizer trellis of quasi-infinite extension (DD receiver) is replaced by a succession of finite-length equalizer trellises (Figure 13-7), thus limiting the extent of error bursts. A clear performance advantage of about 1.5–2 dB (no diversity) and 0.7–1.5 dB (dual diversity) is observed by comparing Figures 15-6 and 15-7 (dotted curves labeled ideal channel, no equalizer reinitialization) and Figures 15-17 and 15-18 (curves labeled ideal channel, with training-aided equalizer reinitialization), respectively. In fact, by virtue of training-aided detection, the BER curves approach the matched filter lower bound (MFB, dotted curves in Figures 15-17 and 15-18) by a margin of 1–2 dB (no diversity) and 1–1.5 dB (dual diversity). Considering that this includes the 0.7-dB training insertion loss, the reduced-complexity detection algorithms (prefilter, equalizer) are proven to perform near optimally, leaving very little potential for further improvement.

Compared to receiver sync based on ideal channel knowledge, DA parameter sync is seen to cost very little extra SNR, viz. 0.2–0.8 dB without diversity and 0.2–0.4 dB with dual diversity (at BER between 10^{-2} and 10^{-3}) when the fading is slow (λ'_D=0.0003). As shown in Figure 15-19, however, fast fading with Doppler near the Nyquist limit (here λ'_D=0.003) leads to considerable bottoming

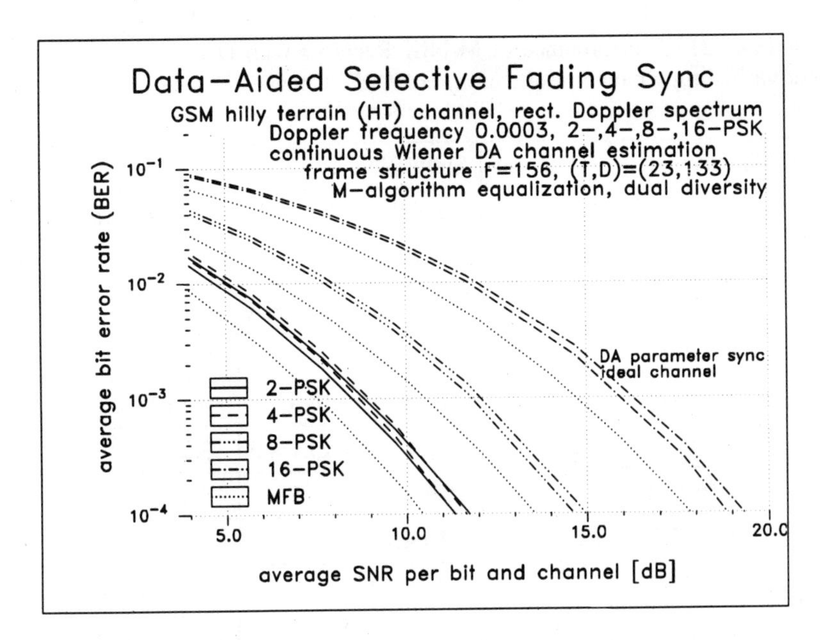

Figure 15-18 BER Performance of Transmission System with DA Selective Fading Channel Estimation and M-PSK Detection with Dual Diversity; GSM-HT Channel, Doppler 0.0003

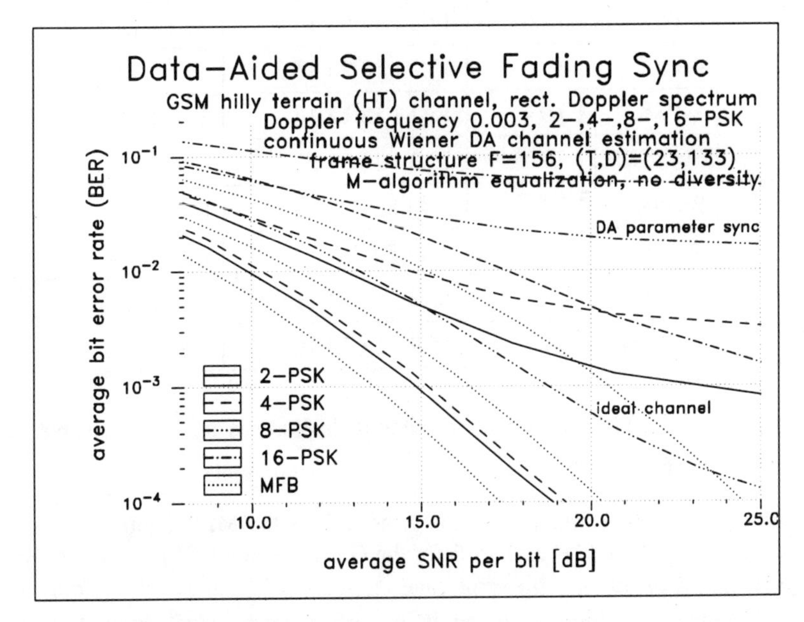

Figure 15-19 BER Performance of Transmission System with DA Selective Fading Channel Estimation and M-PSK Detection without Diversity; GSM-HT Channel, Doppler 0.003

Table 15-4 BER Performance of M-PSK Receiver with DA Selective Fading Channel Estimation, GSM-HT Channel

GSM-HT channel (mild ISI)		Analysis	Simulation	
		MFB	ideal channel	DA parameter sync
		Doppler λ'_D =0.0003 I 0.001 I 0.003		
		SNR [dB] per bit and channel @ BER=10^{-2}		
$D = 1$	2-PSK	8.8	9.8 I 9.8 I 9.8	10.2 I 10.6 I 12.6
	4-PSK	8.8	10.4 I 10.5 I 10.5	10.6 I 11.1 I 14.7
	8-PSK	11.3	13.0 I 13.1 I 13.1	13.2 I 14.1
	16-PSK	15.1	16.9 I 17.0 I 17.5	17.3 I 20.0
$D = 2$	2-PSK	3.7	4.7 I 4.7 I 4.8	5.1 I 5.5 I 6.5
	4-PSK	3.7	5.0 I 5.0 I 5.0	5.3 I 5.6 I 6.9
	8-PSK	6.3	7.7 I 7.7 I 7.7	7.9 I 8.3 I 11.3
	16-PSK	10.3	11.7 I 11.8 I 11.9	11.9 I 12.8
		SNR [dB] per bit and channel @ BER=10^{-3}		
$D = 1$	2-PSK	13.6	14.8 I 14.8 I 14.8	15.1 I 15.7 I 22.8
	4-PSK	13.6	15.2 I 15.2 I 15.3	15.6 I 16.4
	8-PSK	16.4	18.1 I 18.1 I 18.7	18.5 I 21.3
	16-PSK	20.5	22.2 I 22.7 I 27.2	23.0
$D = 2$	2-PSK	7.5	8.4 I 8.4 I 8.5	8.7 I 9.3 I 10.9
	4-PSK	7.5	8.6 I 8.6 I 8.7	8.9 I 9.5 I 12.0
	8-PSK	10.5	11.8 I 11.8 I 11.9	12.0 I 12.8
	16-PSK	14.7	15.9 I 16.2 I 16.7	16.2 I 19.2

effects; 8- and 16-PSK transmission without diversity is no more feasible under such conditions.

Once again, the simulation results – as well as the MFB figures – have been summarized in Tables 15-4 and 15-5 for the GSM-HT and GSM two-ray channels, respectively. Considering bit error rates between 10^{-2} and 10^{-3}, fading up to λ'_D=0.001 has little impact on the BER performance. Dual diversity tends to mitigate irreducible effects near the Nyquist limit (4- and 8-PSK transmission become feasible at λ'_D=0.003), but beyond this limit the receiver performance rapidly deteriorates.

Table 15-5 BER Performance of M-PSK Receiver with DA Selective Fading Channel Estimation, GSM Two-Ray Channel

GSM Two-ray channel (strong ISI)		Analysis	Simulation	
		MFB	ideal channel	DD parameter sync
		Doppler λ'_D =0.0003 I 0.001 I 0.003		
		SNR [dB] per bit and channel @ BER=10^{-2}		
$D = 1$	2-PSK	8.3	9.4 I 9.4 I 9.5	9.8 I 10.3 I 12.0
	4-PSK	8.3	9.9 I 9.9 I 10.0	10.2 I 10.7 I 13.3
	8-PSK	10.8	12.5 I 12.5 I 12.8	12.8 I 13.8
	16-PSK	14.7	16.4 I 16.6 I 17.3	16.8 I 19.9
$D = 2$	2-PSK	3.2	4.3 I 4.3 I 4.3	4.6 I 4.9 I 6.0
	4-PSK	3.2	4.5 I 4.5 I 4.5	4.8 I 5.1 I 6.3
	8-PSK	5.9	7.4 I 7.4 I 7.5	7.6 I 8.1 I 10.6
	16-PSK	9.9	11.5 I 11.7 I 11.9	11.8 I 12.7
		SNR [dB] per bit and channel @ BER=10^{-3}		
$D = 1$	2-PSK	14.1	15.0 I 15.0 I 15.1	15.4 I 16.3
	4-PSK	14.1	15.5 I 15.5 I 15.6	15.9 I 17.2
	8-PSK	16.7	18.0 I 18.5 I 19.7	18.6 I 23.0
	16-PSK	20.6	23.9 I 24.5	25.0
$D = 2$	2-PSK	7.0	8.0 I 8.0 I 8.1	8.2 I 8.7 I 10.3
	4-PSK	7.0	8.2 I 8.2 I 8.3	8.4 I 8.9 I 11.3
	8-PSK	10.0	11.3 I 11.3 I 11.5	11.5 I 12.3
	16-PSK	14.1	16.0 I 16.3 I 17.2	16.3 I 19.6

Results for the two-ray channel (Table 15-5) are very similar, indicating that the prefilter and equalizer can cope with more severe ISI. The BER approaches the MFB by a margin of 1–1.7 dB (slow fading, ideal channel). For denser symbol constellations, more pronounced bottoming effects are perceived (irreducible BER between 10^{-3} and 10^{-4}), but in the normal operating range of inner receivers (BER $10^{-2} - 10^{-3}$) the performance of 16-PSK is almost the same as for channels with mild ISI. Again, when the fading is slow, DA parameter sync costs very little extra SNR (no diversity: 0.3–0.6 dB, dual diversity: 0.2–0.3 dB). The receiver gradually deteriorates with faster fading, but, of course, fails for Doppler beyond the Nyquist limit.

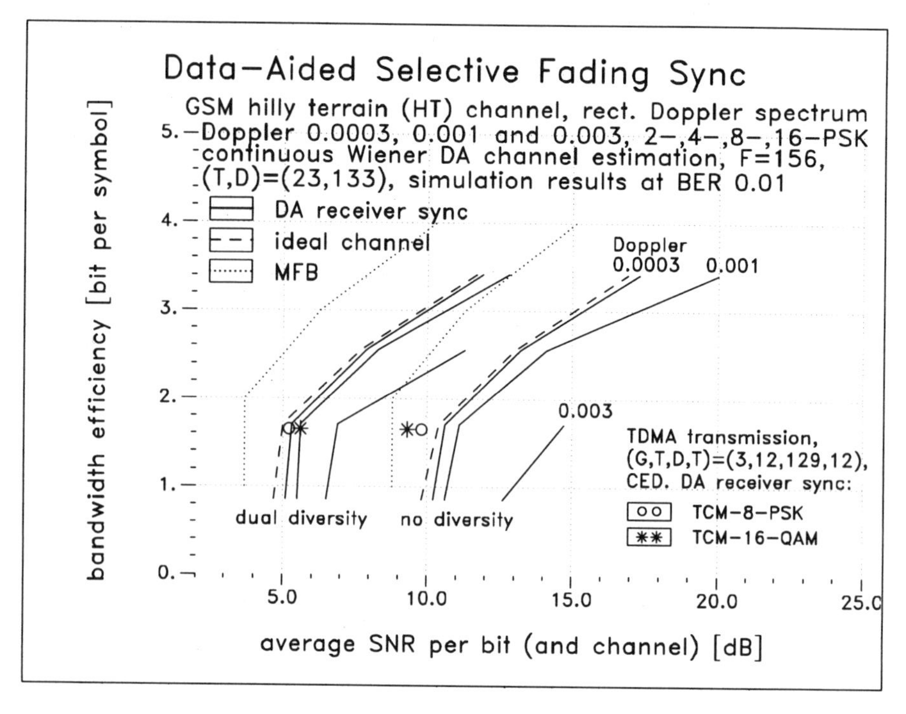

Figure 15-20 Power and Bandwidth Efficiency of DA Receiver for GSM-HT Channel

DA reception performance in terms of power and bandwidth efficiency is shown in Figures 15-20 and 15-21 for the GSM-HT and two-ray channels, respectively. The power efficiency is again given by the SNR needed to achieve a BER of 0.01, and, with 15 percent training, the bandwidth efficiency is 0.85, 1.7, 2.55, and 3.4 data bits per symbol (2-, 4-, 8-, and 16-PSK). The *power* efficiency of DA reception (Figures 15-20 and 15-21) is definitely superior to that of DD reception (Figures 15-9 and 15-10) as long as the Doppler remains below the Nyquist limit. Even considering the compromized bandwidth efficiency, DA reception remains superior for 2- and 4-PSK. With denser symbol constellations – thus higher training insertion loss in terms of bandwidth efficiency – DD reception performs comparably well or sometimes slightly better. In this example, the Nyquist limit λ'_D=0.003 is the boundary beyond which both DA and DD receivers tend to break down; the DA receiver does so quickly while the DD receiver fails more gracefully (8-PSK transmission is still feasible at λ'_D=0.003). However, if robustness against *fading* is of highest concern, the DA receiver can easily be made more robust at the expense of an increased percentage of training. For instance, with 50 percent training and shortest possible training length $2L+1$=9 [thus reducing the frame length F from 156 (Figure 15-14) to 18], the Nyquist limit could be increased by a factor of more than 8. If, on the other hand, *bandwidth* efficiency is of high concern, there are many options of deliberately undersampling the channel fading process and combining the advantages of DA and DD reception. For instance,

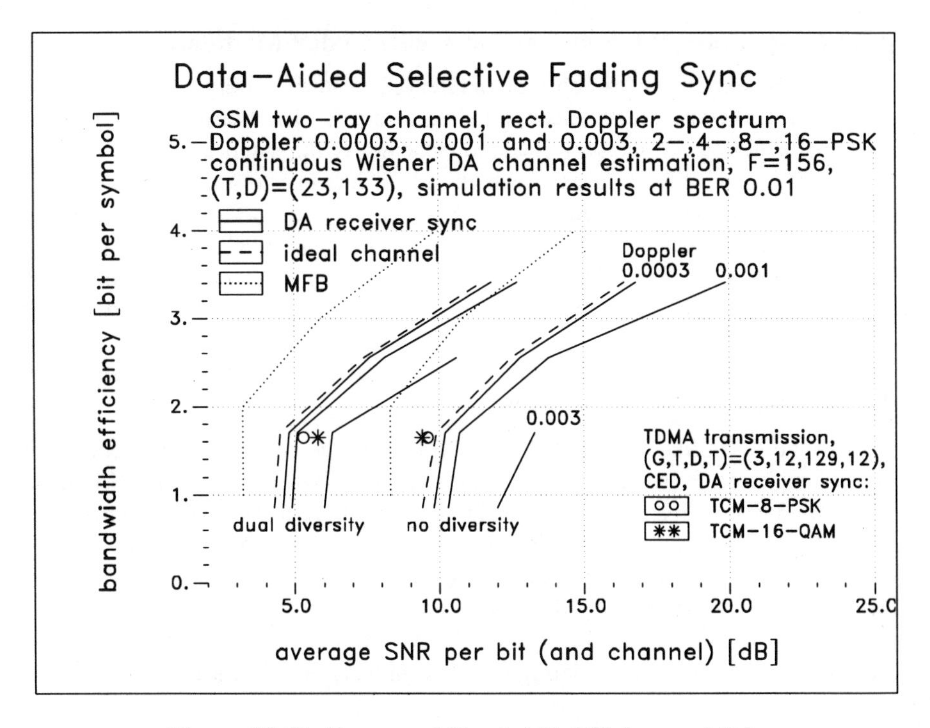

Figure 15-21 Power and Bandwidth Efficiency of DA Receiver for GSM Two-Ray Channel

the portions of a longer data block near a training segment may be detected using reliable DA channel extrapolator estimates, while the center section of the data block – where aliasing precludes the use of DA sync – is decoded after switching to DD detection, possibly both in forward and backward directions. If necessary, this process may be repeated iteratively. Since (hard or soft) symbol decisions near training segments tend to be more reliable, increasing the bandwidth efficiency in this manner may also affect outer transmission system design.

Returning to Figures 15-20 and 15-21, the results also show that most of the inherent multipath diversity gain – manifesting itself in the matched filter bound and being almost the same for both the GSM-HT and two-ray channels – can be recovered by equalization, regardless of the ISI profile (mild vs. severe ISI). As a consequence, the presence of ISI can be viewed as an asset rather than a nuisance. For instance, when comparing the results for selective fading DA reception (Figures 15-20 and 15-21) with those for DA flat fading reception (solid curves in Figure 14-17), a gain of about 3–5 dB (no diversity) and 1–2 dB (dual diversity) is observed (at BER 0.01). This gain, of course, has to be paid for by a lower tolerance against Doppler and a significantly increased signal processing complexity. Also, the presence of multipath (depends on the terrain, range, etc.) and thus diversity-like gains cannot be guaranteed. Hence, *worst-case* outer transmission design should be based on the inner receiver performance for *flat* fading.

15.2.4 Example: DA Receiver for Trellis-Coded Modulation

In this section, an example of interleaved channel coding and decoding is discussed in the context of selective fading channels. The transmitter part of the system, including the channel encoding scheme, is the same as that of Section 14.2.5. Although block codes match the block interleaver structure particularly well, we have selected again the 8-PSK (ECL 3) and 16-PSK (ECL 5) trellis codes by Ungerboeck [14] and Moher/Lodge [15], respectively. The presence of ISI, however, calls for different kinds of data format, receiver prefiltering (MF, WF), DA selective fading sync parameter estimation, and, most importantly, a suitable way of performing equalization and decoding. The block diagram of the system and the TDMA transmission frame format considered here are shown in Figures 15-22 and 15-23, respectively.

In the GSM-like frequency hopped (FH) TDMA channel access scheme of this example, each user occupies one time slot per frame, where a TDMA frame accommodates 8 time slots. FH-TDMA is particularly attractive in interference-limited cellular environments for a number of reasons. *Frequency diversity* is available by virtue of both frequency hopping and some frequency selectivity in each channel; in fact, FH-TDMA can be interpreted as a CDMA system with the code given by the hopping sequence [16]. As long as the combined effect of ISI and differential propagation delays of users (uplink) does not exceed a couple of

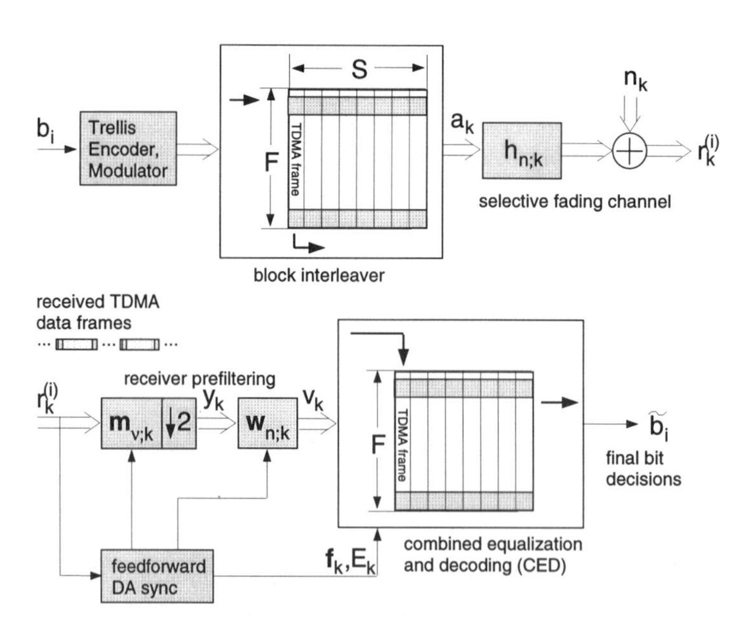

Figure 15-22 DA Selective Fading Receiver for TDMA Transmission with Interleaved Channel Coding

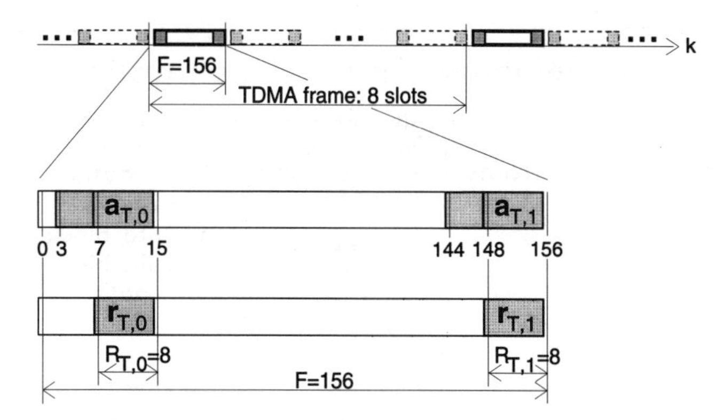

Figure 15-23 Frame Format for TDMA Transmission and DA Reception

symbol durations, *orthogonality* between users of the same cell is easily maintained by inserting a small guard time between consecutive bursts, thus avoiding the need for multiuser detection. Combining interleaved channel coding with sophisticated *dynamic channel allocation* (DCA) techniques which allocate channels according to the instantaneous interference situation, advanced FH-TDMA systems feature a high degree of immunity against interference from neighboring cells – thus high capacity – thanks to an *interferer diversity* or averaging effect.

As explained in Section 14.2.5, TCM-encoded symbols are written row-wise into the block interleaver matrix. Each interleaver column, extended by a pre- and postamble, then forms a TDMA burst to be transmitted over the channel. In this example, each burst of length F=156 (Figure 15-23) comprises a guard time of 3 symbol durations, a data block of length 129, and length-12 pre- and postambles ($\mathbf{a}_{T,0}$, $\mathbf{a}_{T,1}$) of 4-PSK CAZAC sequences with period length P_T=8 (Table 15-3). DA receiver sync starts afresh for each individual burst, using the two received training blocks ($\mathbf{r}_{T,0}$, $\mathbf{r}_{T,1}$) for ML acquisition of snapshots ($\hat{\mathbf{h}}_{T,0}$, $\hat{\mathbf{h}}_{T,1}$) at reference positions $(k_0, k_1) = (11, 151)$. From these two snapshots, channel impulse responses $\hat{\mathbf{h}}_k$ (Wiener estimation) and corresponding receiver parameters (MF, WF, equivalent channel coefficients) are computed for 20 "pole" positions k=15, 22, 29,. . ., 148. Digital prefiltering (WF, MF) is then performed on the burst just received, thus eliminating the need for storing prefilter coefficients. However, equivalent channel vectors $\mathbf{f}_k$ and CSI samples E_k or $\gamma_{s;k}$ (here 20 per burst) need to be stored while filling the deinterleaver matrix with (decimated) WF output samples v_k.

As discussed in Section 14.2.3, the error covariance of DA channel estimation using only two snapshots cannot be expected to be equal for all intraburst positions k. Therefore, Figure 15-24 displays the analytically obtained tap error covariance factors $\overline{r}_{n;k}$ [eq. (15-20), again GSM-HT channel] for the margins (k=15 and 143) as well as the center ($k \approx 80$) of the data block. Close to the block margins (dashed lines), the estimator performs reasonably well also for larger Doppler

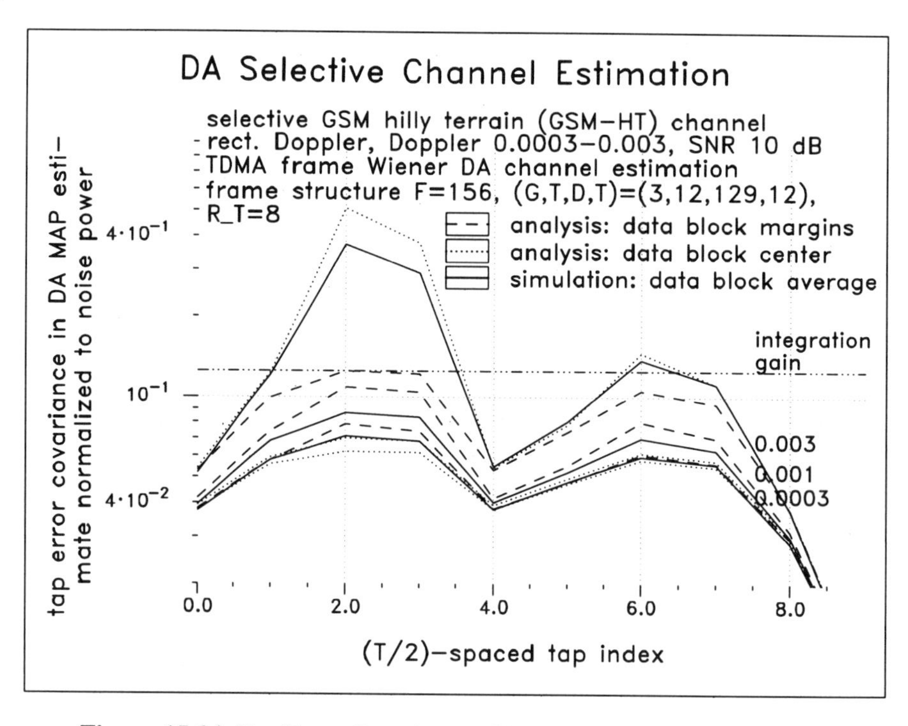

Figure 15-24 Tap Error Covariance of DA Selective Fading Channel Estimation; TDMA Transmission, GSM-HT Channel, SNR 10 dB

near the Nyquist limit.[16] In the center (dotted lines), estimator performance for slowly changing taps ($\lambda'_D \leq 0.001$) is actually better than near the margins, while the error covariance is dominated by aliasing near the Nyquist limit (λ'_D=0.003). Simulation results obtained by averaging over all tap errors within the data block (solid lines) and also results on the corresponding MSE factors ζ shown in Figure 15-25 corroborate these findings: near the block margins, estimator performance is almost independent of the Doppler; in the center, the same is true up to some "critical" Doppler frequency which depends on the SNR of operation. Beyond this critical Doppler, the error is dominated by aliasing and rapidly increases with λ'_D.

When the channel is *flat*, the outer transmission system (channel coding and decoding) is cleanly separable from the inner system; only soft decisions and the CSI are to be transferred from the inner to the outer system (Figure 14-18). When the channel is *selective*, this is no more the case, at least when close-to-optimal performance is of concern. Of course, one may force the decoder and equalizer (typically LE or DFE) to be independent units being well separated by the deinterleaving device, but this is clearly suboptimal. The simple concatenated

[16] Strictly speaking, a Nyquist limit does not exist for finite-length sampling. Here, the term Nyquist limit is taken to refer to a hypothetical system which periodically transmits pre- and postambles spaced apart by $\approx (k_1 - k_0)$=140 symbol intervals.

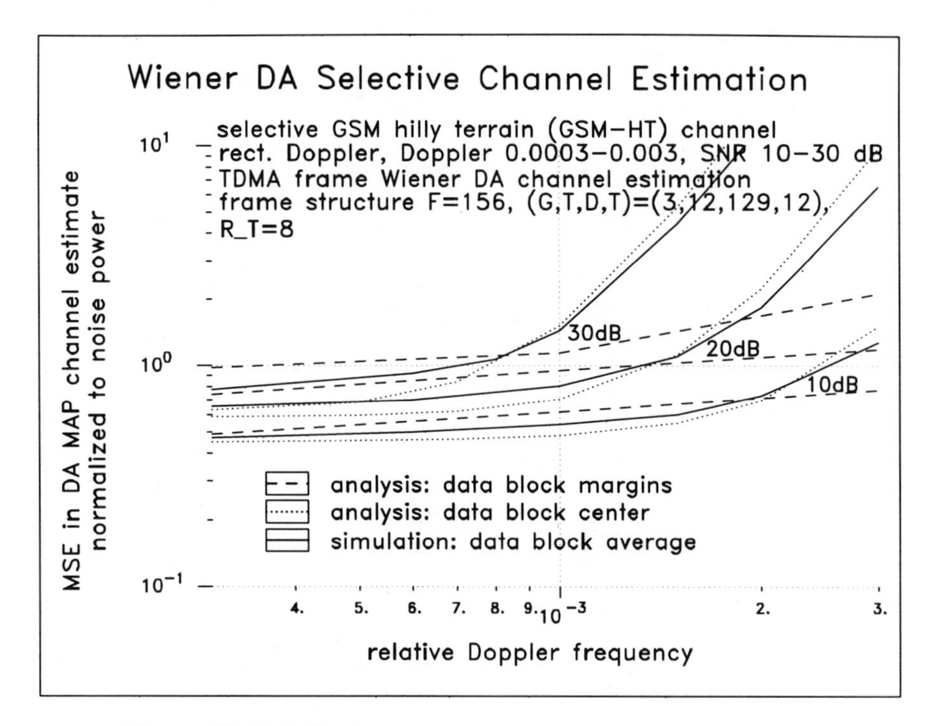

Figure 15-25 MSE in DA Selective Fading Channel Estimate; TDMA Transmission, GSM-HT Channel

scheme can be improved by means of soft-output equalization techniques such as MAP and SOVA (for a concise introduction see Section 13.3.2). Recognizing that tasks which are interrelated in some way – such as equalization and decoding – should generally be performed *jointly*, we shall, in this example, resort to a class of algorithms termed *combined equalization and decoding* (CED).

The basic idea of CED is to perform equalization, deinterleaving, and decoding in such a way that – exploiting the power of channel coding – delayed *final* (thus more reliable) symbol decisions are generated and used for equalization in a decision-feedback manner. The first such scheme by Eyuboglu [17] features periodic interleaving and, using delayed symbol decisions, noise-predictive linear equalization, but a fraction of noise samples still remains unpredictable. Zhou et al. [18] and Wang [19] propose the use of reference symbols and deinterleaving on a vector-by-vector basis, respectively, such that unpredictable noise samples are avoided. In this example, we consider a variant of CED which has been proposed by Mehlan and Meyr [20] and developed by the author [21]. This technique – which has meanwhile been studied for a variety of applications [22, 23, 24] – yields good performance at relatively low complexity.

While the data block length of a TDMA burst (or a multiple thereof, see Figure 14-18) determines the interleaver depth, the basic form of the CED calls for the *interleaver length* S being equal to the *survivor depth* D_S of the Viterbi decoder

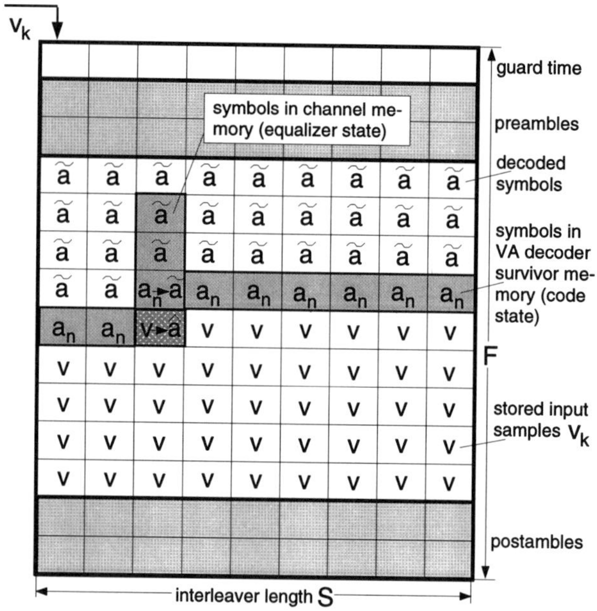

Figure 15-26 Combined Equalization and Decoding (CED) for TDMA Transmission with Interleaved Channel Coding

(VD). The deinterleaver matrix and the process of equalization and decoding are illustrated in Figure 15-26. As soon as $S = D_S$ TDMA bursts have been received and the matrix is filled up with samples v_k, the decoding process begins in the upper left corner with the first data-dependent sample v_k (k: time scale of column-wise channel transmission). Each iteration $(i-1) \rightarrow i$ (i: time scale of row-wise channel decoding) starts with a DFE-like ISI cancellation step where ISI postcursors are subtracted from v_k:

$$\begin{aligned} \hat{a}_k &= \underbrace{\left(a_k + \sum_{l=1}^{L} f_{l;k} a_{k-l} + \eta_k \right)}_{v_k} - \underbrace{\left(\sum_{l=1}^{L} f_{l;k} \tilde{a}_{k-l} \right)}_{\text{ISI estimate}} \\ &= a_k + \eta_k \qquad \text{(if decisions correct, i.e., } \tilde{a}_{k-l} = a_{k-l}) \end{aligned} \tag{15-35}$$

This kind of ISI cancellation thus uses final symbol decisions $\tilde{a}_{k-l}$ residing (in the deinterleaver matrix) *above* the current sample v_k. As long as the $\tilde{a}_{k-l}$ are correct, the ISI-cancelled sample $\hat{a}_k$ is just a noisy replica of the transmitted symbol a_k so that $\hat{a}_k$ can, along with the CSI (E_k or $\gamma_{s;k}$), be passed on to the Viterbi decoder.

Using the ISI-cancelled *soft* symbol decision $\hat{a}_i$ (= $\hat{a}_k$ relabeled in terms of channel decoding time scale i) and the CSI, the VD then updates the decision metric increment

$$\Delta m_{D;i}(a_i) \propto \gamma_{s;i} \left| \hat{a}_i - a_i \right|^2 \tag{15-36}$$

[eq. (13-18)] for each allowed code symbol a_i as if the channel were nonselective. Following survivor extension [extended survivors $\mathbf{a}_i$ with decision metrics $m_{D;i}(\mathbf{a}_i)$; eq. (13-9)] and survivor selection (keeping the best of merging survivor paths), the VD makes a final decision on the oldest symbol in the survivor memory. This decision $\tilde{a}_{i-S-1}$ happens to be the channel symbol decision $\tilde{a}_{k-1}$ residing "above" the next sample v_k to be ISI-cancelled in the following iteration.

A couple of important comments are in order here. First, notice that DA receiver sync ideally complements the CED concept. The training symbols serve not only for parameter synchronization but also for initialization of the ISI cancellation process (preamble) and the termination of error events produced by error propagation (postamble). Moreover, receiver prefiltering according to Section 13.3.4 effectuates optimal suppression of detrimental effects caused by ISI cancellation: thanks to minimum postcursor ISI (the first tap $f_{0;k} \approx 1$ of $\mathbf{f}_k$ is maximized relative to the postcursors $f_{1;k}, \ldots, f_{L;k}$), both the noise in the soft decisions $\hat{a}_k$ and the disturbance produced by using incorrect hard decisions $\tilde{a}_{k-1}, \ldots, \tilde{a}_{k-L}$ in the ISI cancellation process is minimized.

Another comment concerns the existence of a so-called *super state*. When using convolutional channel coding on ISI channels without interleaving (this makes sense when the ISI channel is nonfading), the code and channel states can be lumped together into one super state representing the "memory" of both the code and the ISI channel [25]. Uncoordinated interleaving in general forces the code and channel memories to be separated in an irregular fashion. In the case of schemes which use coordinated interleaving (such as the CED), on the other hand, a super state exists in the sense that, for each sample v_k in the deinterleaver matrix to be processed, both the code state $\mathbf{a}_i$ (symbols in code memory along rows) and the equalizer state $\mathbf{a}_k$ (symbols in channel memory along columns) contain only *causal* symbols which the decoder has already processed when arriving at v_k. In Figure 15-26, the positions to the left of v_k and the rows above v_k constitute the "causal region". Both the code memory and the channel memory extend backwards in processing direction (horizontally to the left and vertically upward, respectively). Hence, merging the states $\mathbf{a}_i$ and $\mathbf{a}_k$ into a super state is possible just as in the noninterleaved case, thus allowing for a large variety of reduced-complexity algorithms to be used for searching the super trellis [24, 26].

Some final remarks concern variations on the theme of CED which follow from the existence of a super state. While the basic CED considers all code states (VD) and just a single equalizer state (past final decisions), an M-algorithm CED (M-CED) [27, 28] may be devised which keeps all code states and M equalizer states per code state in the survivor memory, thus increasing the number of survivors by a factor of M. In the example discussed here, keeping M=4 (TCM 8-PSK) and 2 (TCM 16-QAM) equalizer states per code state has been found to suffice for near-optimal performance; the gain with respect to the basic CED is in the order of 1 dB. If the *latency* is critical, the interleaver size can be reduced by retarding the final decoder decisions until Z rows in the deinterleaver matrix have been processed, i.e., $D_S = ZS$ (instead of $D_S = S$) so that, using the

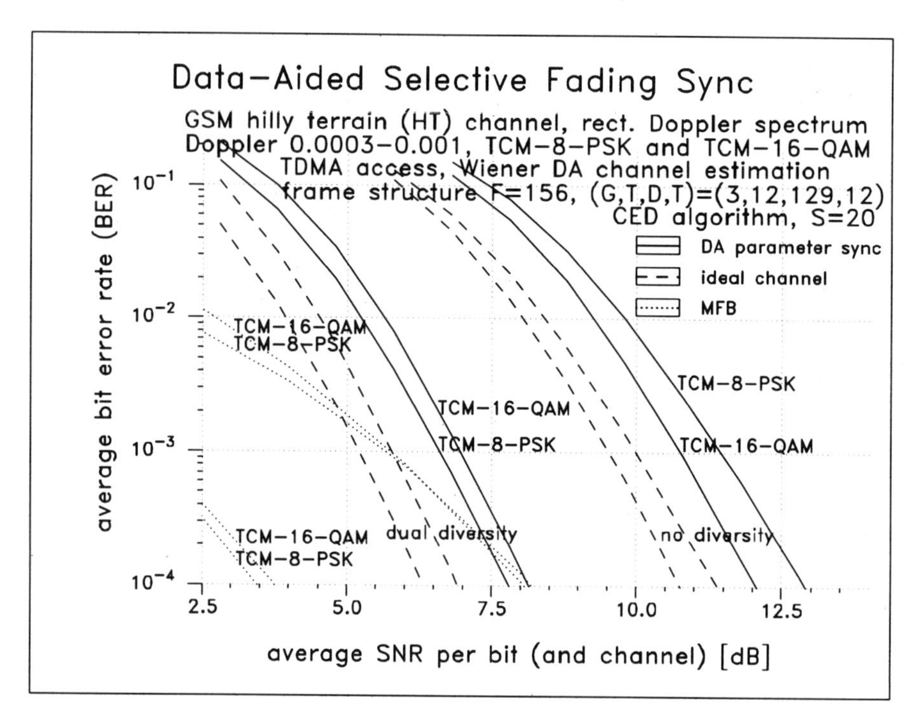

Figure 15-27 BER Performance of TDMA Transmission System with DA Selective Fading Channel Estimation and TCM Channel Coding; GSM-HT Channel, Doppler 0.0003–0.001

same survivor depth D_S, the interleaver size can be reduced by a factor of Z.[17] Then, however, ISI cancellation must work on symbols $a_{k-l}, \ldots, a_{k-Z}$ "above" v_k which the basic CED has not yet decided upon. Considering a particular code state (M=1) or super state (M-CED with M >1), however, the survivor history $\mathbf{a}_i$ and thus also the equalizer state $\mathbf{a}_k$ are known. Hence, *per-survivor processing* (PSP) [4] can be applied to ISI cancellation, leading to the M-CED with PSP [28]. As a final remark, firm CED decisions or symbols corresponding to the survivor history can also be used for channel estimation and parameter synchronization purposes thus performing DD receiver sync, possibly in a PSP manner [24].

The BER performance of a TDMA transmission system featuring DA receiver sync, interleaved TCM channel coding, and basic CED with survivor depth D_S=20 is shown in Figures 15-27 (GSM-HT) and 15-28 (GSM two-ray). The BER results, which are summarized in Table 15-6, include 0.8-dB training insertion loss for both ideal channel and DA receiver sync.

[17] However, in order to maintain sufficient interleaving, the interleaver length S must not be smaller than the code ECL; the symbols in one column tend to fade simultaneously since they belong to the same burst.

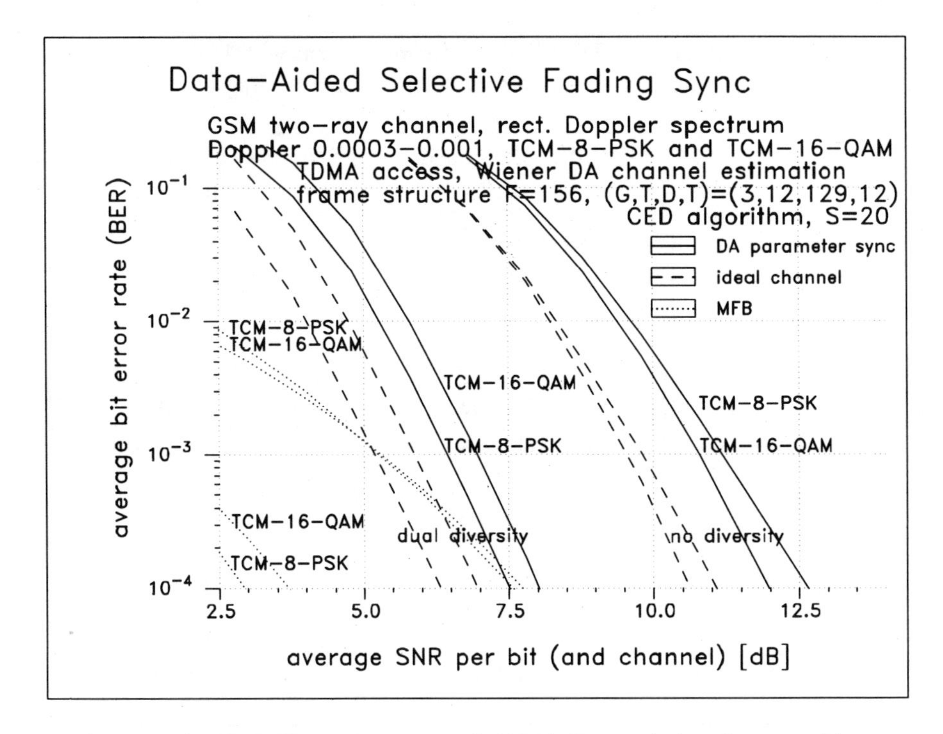

Figure 15-28 BER Performance of TDMA Transmission System with DA Selective Fading Channel Estimation and TCM Channel Coding; GSM Two-Ray Channel, Doppler 0.0003–0.001

With channel coding, the distance between the simulated BER curves and the asymptotic matched filter bound is observed to be quite large (5–8 dB at BER 10^{-2}, still 4–5 dB at BER 10^{-4}), indicating that the asymptotic MFB curves are too optimistic in relevant BER regions. Concerning the choice of codes, TCM 8-PSK is superior in the presence of other diversity mechanisms (D=2); here the larger coding gain (4.1 dB under AWGN conditions) dominates the performance. On the other hand, TCM 16-QAM is superior when little or no other diversity mechanisms are present (D=1); here the code diversity by virtue of the larger ECL (=5) more than offsets the lower AWGN coding gain (3.4 dB). Regarding the fading, the relatively sharp cutoff in the channel estimation MSE (Figure 15-25) translates into a likewise sharp BER cutoff (not shown here) near the Nyquist limit; dual diversity somewhat mitigates this effect. As expected, the intraburst BER profile experiences strong aliasing effects in the center of the burst, but no catastrophic error propagation has been observed in the simulations. The total extra SNR needed for DA receiver sync ranges between 1 and 1.5 dB throughout.

Results on power and bandwidth efficiency for the CED have been included in Figures 15-20 (GSM-HT) and 15-21 (GSM two-ray) above. In the high-BER region (10^{-2}) and without diversity, the kind of channel coding considered here

Table 15-6 Performance of TDMA Transmission System with DA Selective Fading Channel Estimation and TCM Channel Coding

		SNR [dB] per bit and channel @ BER=10^{-2}		SNR [dB] per bit and channel @ BER=10^{-3}	
		Doppler $\lambda'_D = 0.0003 - 0.001 \vert 0.003$, TDMA frame structure (G,T,D,T) = (3,12,129,12), DA reception			
		8-PSK code	16-QAM code	8-PSK code	16-QAM code
$D = 1$	GSM-HT	9.8	9.3	11.4	10.7
	GSM 2-ray	9.6	9.4	11.1	10.8
$D = 2$	GSM-HT	5.2 | 7.8	5.6 | 7.4	6.6 | 10.8	7.0 | 9.8
	GSM 2-ray	5.3 | 7.8	5.8 | 7.7	6.5 | 9.8	7.0 | 9.9

features only a small gain (about 1 dB) with respect to uncoded 4-PSK. With dual diversity, no gain or even a loss (about 1 dB) is observed. On the other hand, gains in the order of 4–5 dB (D=1) and 1–2 dB (D=2) are obtained at lower target BER (10^{-3}).

15.2.5 Main Points

- Feedforward data-aided selective fading channel parameter estimation by multiplexing segments of consecutive training symbols into the data stream has a number of inherent advantages resulting from the separation of detection and synchronization tasks. Under certain conditions, MAP DA channel estimation can again be separated into ML estimation of snapshots and MAP estimation from the intermediate ML estimates.
- ML snapshot acquisition requires blocks $\mathbf{r}_{T,K}$ (length $R_{T,K}$) of received samples which are dependent on the respective training sequences $\mathbf{a}_{T,K}$ but independent of random data. The use of perfect training sequences makes ML acquisition both optimal and simple:

$$\hat{\mathbf{h}}_{T,K}(\mathbf{a}_{T,K}) = \frac{1}{R_{T,K}} \, \mathbf{A}^H_{T,K} \cdot \mathbf{r}_{T,K} \tag{15-37}$$

[eq. (15-17), white noise assumption]. Furthermore, subsequent MAP channel estimation from ML snapshots can then be performed independently for each channel tap by Wiener filtering the sequence of ML tap estimates $\hat{\mathbf{h}}_{S,T,n} = (\hat{h}_{n,T,0} \ldots \hat{h}_{n,T,\overline{N}-1})^T$:

$$\hat{h}_{n;k}(\mathbf{a}_T) = \mathbf{w}^H_{T,n;k} \cdot \hat{\mathbf{h}}_{S,T,n} \tag{15-38}$$

[eq. (15-18)].

- Polyphase sequences having the CAZAC (constant amplitude, zero autocorrelation) property are perfect training sequences for optimal ML estimation. They exist for any sequence length and are easily constructed. For most (underspread) channels of interest, the snapshot assumption holds, i.e., the fading does not dominate the ML estimation error.
- In DA selective channel estimation, much of the noise suppression is brought about by the integration gain inherent in ML acquisition so that additional gains by virtue of MAP estimation from snapshots don't need to be large. Hence, providing for large oversampling margins is not necessary, but training insertion rates must remain larger than the Nyquist rate. Near-optimal MAP estimation is accomplished by FIR Wiener filters spanning the fading coherence time. For a particular channel tap SNR (estimate), optimal filter coefficients can be generated via eigensystem synthesis. Alternatively, suboptimal ordinary interpolation filters may be applied.
- The DA inner receiver components (MF, WF, diversity combiner, equalizer) are adjusted by means of DA channel estimation and parameter computation. Complexity is reduced by performing only few sync cycles per frame. Below the Nyquist limit, the SNR loss incurred by DA parameter sync is small (between 0.2 and 0.8 dB in the examples reported), and, for continuous uncoded PSK transmission, the BER curves approach the matched filter bound by a margin of 1–2.5 dB, thus leaving very little room for further improvement. Despite the bandwidth efficiency being compromised by training insertion, the clearly better power efficiency makes DA reception superior to DD reception in terms of total power/bandwidth efficiency.
- In advanced FH-TDMA systems transferring short channel-encoded signal bursts, DA reception ideally complements combined equalization and decoding (CED) techniques since the training segments are used for both DA parameter sync and CED initialization/termination. For a specifically chosen burst and block interleaving format, there exists a super state encompassing both code and equalizer memories. Hence, near-optimal and/or latency-reduced variants of the CED can be devised. The extra SNR needed for DA parameter sync is moderate (1–1.5 dB in the example). Depending on the existence of other diversity mechanisms, channel codes with larger AWGN gain or with larger ECL are preferable. At high target BER (10^{-2}), channel coding is of little or no use, while substantial coding gains are achieved at moderate BER levels in the order of 10^{-3} or below.

15.3 Bibliographical Notes

The simple LMS and the LMS-Kalman adaptive algorithms for NDA selective channel estimation are discussed in [3, 29, 30] and [2, 31, 32], respectively. Information on the more complex RLS algorithm is found in the textbooks [3, 33]. The concept of per-survivor processing (PSP), being applicable to DD channel estimation as well as combined equalization and decoding (CED), is explained in [24, 34, 4].

References [8, 35, 9, 12] give details on DA selective fading channel estimation. Perfect training sequences for optimal ML DA channel acquisition and the construction of perfect polyphase sequences are discussed in [36, 37, 38, 12, 11, 39] and [40, 38, 41, 42], respectively. The effect of fading on DA channel acquisition is analyzed in [12, 13]. Methods of applying perfect training sequences also to frame and frequency sync are discussed in [25, 43, 44].

References [17, 19, 18] introduce the idea of combined equalization and decoding. Details on the CED algorithm featuring block deinterleaving and training-aided initialization are found in [45, 21, 22, 23, 24, 20], and variants of the CED algorithm (M–CED and CED with PSP) are studied in [27, 28]. For further information on the concept of super states and on searching the respective super trellis, see references [25, 46, 47, 48, 49] and [26, 24].

Bibliography

[1] B. D. O. Anderson and J. B. Moore, *Optimal Filtering*. Englewood Cliffs, NJ: Prentice-Hall, 1979.

[2] S. A. Fechtel and H. Meyr, "Near-Optimal Tracking of Time-Varying Digital Radio Channels Using a-priori Statistical Channel Information," in *Coded Modulation and Bandwidth-Efficient Transmission*, (Tirrenia, Italy), pp. 367–377, Sept. 1991.

[3] S. Haykin, *Adaptive Filter Theory*. Englewood Cliffs, NJ: Prentice-Hall, 1986.

[4] R. Raheli, A. Polydoros, and C. K. Tzou, "Per-Survivor Processing: A General Approach to MLSE in Uncertain Environments," *IEEE Trans. Commun.*, vol. 43, pp. 354–364, Feb./Mar./Apr. 1995.

[5] JSAC *IEEE J. Sel. Areas Commun.*, vol. 10, no. 1, Jan. 1992.

[6] M. V. Clark, L. J. Greenstein, W. K. Kennedy, and Shafi, M., "Matched Filter Performance Bounds for Diversity Combining Receivers in Digital Mobile Radio," in *Proc. Globecom*, (Phoenix, AZ), pp. 1125–1129, Dec. 1991.

[7] S. A. Fechtel and H. Meyr, "Matched Filter Bound for Trellis-Coded Transmission over Frequency-Selective Fading Channels with Diversity," *European Trans. Telecomm.*, vol. 4, no. 3, pp. 109–120, May-June 1993.

[8] N. W. K. Lo, D. D. Falconer, and A. U. H. Sheikh, "Channel Interpolation for Digital Mobile Radio Communications," in *Proc. ICC*, (Denver, CO), pp. 773–777, June 1991.

[9] S. A. Fechtel and H. Meyr, "An Investigation of Channel Estimation and Equalization Techniques for Moderately Rapid Fading HF-Channels," in *Proc. ICC*, (Denver, CO), pp. 768–772, June 1991.

[10] B. Noble and J. W. Daniel, *Applied Linear Algebra.* Englewood Cliffs, NJ: Prentice-Hall, 1977.

[11] H. D. Lüke, *Korrelationssignale.* Berlin: Springer-Verlag, 1992.

[12] S. A. Fechtel and H. Meyr, "Optimal Parametric Feedforward Estimation of Frequency-Selective Fading Radio Channels," *IEEE Trans. Commun.*, vol. 42, no. 2/3/4, pp. 1639–1650, Feb./Mar./Apr. 1994.

[13] R. A. Ziegler and J. M. Cioffi, "Estimation of Time-Varying Digital Mobile Radio Channels," in *Proc. Globecom*, (Orlando, FL), pp. 1130–1134, Dec. 1992.

[14] G. Ungerboeck, "Channel Coding with Multilevel/Phase Signals," *IEEE Trans. Inform. Theory*, vol. 28, pp. 55–67, Jan. 1982.

[15] M. L. Moher and J. H. Lodge, "TCMP: A Modulation and Coding Strategy for Rician Fading Channels," *IEEE J. Sel. Areas Commun.*, vol. 7, no. 9, pp. 1347–1355, Dec. 1989.

[16] D. D. Falconer, F. Adachi, and B. Gudmundson, "Time Division Multiple Access Methods for Wireless Personal Communications," *IEEE Commun. Magazine*, vol. 43, no. 1, pp. 50–57, Jan. 1995.

[17] M. V. Eyuboglu, "Detection of Coded Modulation Signals on Linear, Severely Distorted Channels Using Decision Feedback Noise Prediction with Interleaving," *IEEE Trans. Commun.*, vol. 36, no. 4, pp. 401–409, Apr. 1988.

[18] K. Zhou, J. G. Proakis, and F. Ling, "Decision-Feedback Equalization of Time-dispersive Channels with Coded Modulation," *IEEE Trans. Commun.*, vol. 38, no. 1, pp. 18–24, Jan. 1990.

[19] T. Wang and C. L. Wang, "Improved Adaptive Decision-Feedback Equalization with Interleaving for Coded Modulation Systems," in *Proc. IEEE Globecom'94*, (San Francisco, CA), pp. 6–10, Nov. 1994.

[20] R. Mehlan and H. Meyr, "Combined Equalization/Decoding of Trellis Coded Modulation on Frequency Selective Fading Channels," in *Coded Modulation and Bandwidth-Efficient Transmission*, (Tirrenia, Italy), pp. 341–352, Sept. 1991.

[21] S. A. Fechtel and H. Meyr, "Combined Equalization, Decoding and Antenna Diversity Combining for Mobile/Personal Digital Radio Transmission Using Feedforward Synchronization," in *Proc. VTC*, (Secaucus, NJ), pp. 633–636, May 1993.

[22] G. Femenias, A. Gelonch, and J. L. Galvez, "Joint Antenna Diversity and Combined DFE/Decoding of Trellis Coded Modulation on Frequency-Selective Fading Mobile Radio Channels," in *Proc. VTC*, (Stockholm, Sweden), pp. 952–956, June 1994.

[23] A. Gusmao and N. Esteves, "On Mobile Broadband System Specifications and Radio Transmission Performance." RACE II MBS Report R 2067, Apr. 1994.

[24] E. Katz and G. Stüber, "Sequential Sequence Estimation for Trellis-Coded Modulation on Multipath Fading ISI Channels," *IEEE Trans. Commun.*, vol. 43, no. 12, pp. 2882–2885, Dec. 1995.

[25] P. R. Chevillat, D. Maiwald, and G. Ungerboeck, "Rapid Training of a Voiceband Data Modem Receiver Employing an Equalizer with Fractional T-spaced Coefficients," *IEEE Trans. Commun.*, vol. 35, pp. 869–876, Sept. 1987.

[26] J. B. Anderson and S. Mohan, "Sequential Coding Algorithms: A Survey and Cost Analysis," *IEEE Trans. Commun.*, vol. 32, no. 2, pp. 169–176, Feb. 1984.

[27] S. A. Fechtel, "Equalization for Coded Modulation on Fading Channels," in *Proc. IEEE Workshop on Synchronization and Equalization*, (Gent, Belgium), pp. 22–28, May 1995.

[28] Fechtel, S. A. and Meyr, H., "M-Algorithm Combined Equalization and Decoding for Mobile/Personal Communications," in *Proc. ICC*, (Seattle, WA), pp. 1818–1822, June 1995.

[29] M. L. Honig and D. G. Messerschmitt,, *Adaptive Filters*. Boston: Kluwer Academic, 1984.

[30] G. Ungerboeck, "Theory on the Speed of Convergence in Adaptive Equalizers for Digital Communication," *IBM J. Res. Develop.*, pp. 546–555, Nov. 1972.

[31] P. Höher, "An Adaptive Channel Estimator for Frequency-Selective Fading Channels," in *Aachener Symposium fuer Signaltheorie*, (Aachen, Germany), pp. 168–173, Sept. 1990.

[32] P. Höher, *Kohärenter Empfang trelliscodierter PSK-Signale auf frequenzselektiven Mobilfunkkanälen – Entzerrung, Decodierung und Kanalparameterschätzung (in German)*. PhD thesis, Kaiserslautern University of Technology, 1990.

[33] L. Ljung and T. Söderström, *Theory and Practice of Recursive Identification*. MIT Press, 1986.

[34] R. Raheli, A. Polydoros, and C. K. Tzou, "Per-Survivor Processing," *Digital Signal Processing*, vol. 3, pp. 175–187, July 1993.

[35] N. W. K. Lo, D. D. Falconer, and A. U. H. Sheikh, "Adaptive Equalization and Diversity Combining for Mobile Radio Using Interpolated Channel Estimates," *IEEE Trans. Vehicular Technol.*, vol. 40, no. 3, pp. 636–645, Aug. 1991.

[36] W. O. Alltop, "Complex Sequences with Low Periodic Correlations," *IEEE Trans. Inform. Theory*, vol. 26, pp. 350–354, May 1980.

[37] R. W. Chang and E. Y. Ho, "On Fast Start-up Data Communication Systems Using Pseudo-random Training Sequences," *Bell Syst. Tech. J.*, vol. 51, pp. 2013–2027, Nov. 1972.

[38] A. P. Clark, Z. C. Zhu, and J. K. Joshi, "Fast Start-up Channel Estimation," *IEE Proc.*, vol. 131, pp. 375–382, July 1984.

[39] A. Milewski, "Periodic Sequences with Optimal Properties for Channel Estimation and Fast Start-up Equalization," *IBM J. Res. Develop.*, vol. 27, pp. 426–431, Sept. 1983.

[40] D. C. Chu, "Polyphase Codes with Good Periodic Correlation Properties," *IRE Trans. Inform. Theory*, vol. 18, pp. 531–532, July 1972.

[41] R. L. Frank and S. A. Zadoff, "Phase Shift Pulse Codes with Good Periodic Correlation Properties," *IRE Trans. Inform. Theory*, vol. 8, pp. 381–382, Oct. 1962.

[42] R. C. Heimiller, "Phase Shift Pulse Codes with Good Periodic Correlation Properties," *IRE Trans. Inform. Theory*, vol. 6, pp. 254–257, Oct. 1961.

[43] S. A. Fechtel and H. Meyr, "Fast Frame Synchronization, Frequency Offset Estimation and Channel Acquisition for Spontaneous Transmission over Unknown Frequency-Selective Radio Channels," in *Proc. PIMRC*, (Yokohama, Japan), pp. 229–233, Sept. 1993.

[44] S. A. Fechtel and H. Meyr, "Improved Frame Synchronization for Spontaneous Packet Transmission over Frequency-Selective Radio Channels," in *Proc. PIMRC*, (The Hague, Netherlands), pp. 353–357, Sept. 1994.

[45] S. A. Fechtel and H. Meyr, "A New Mobile Digital Radio Transceiver Concept Using Low-Complexity Combined Equalization / Trellis Decoding and a Near-Optimal Receiver Sync Strategy," in *Proc. PIMRC*, (Boston, MA), pp. 382–386, Oct. 1992.

[46] A. Duel and C. Heegard, "Delayed Decision Feedback Sequence Estimation," *IEEE Trans. Commun.*, vol. 37, no. 5, pp. 428–436, May 1989.

[47] M. V. Eyuboglu and S. U. H. Qureshi, "Reduced-State Sequence Estimation with Set Partitioning and Decision Feedback," *IEEE Trans. Commun.*, vol. 36, no. 1, pp. 13–20, Jan. 1988.

[48] M. V. Eyuboglu, S. U. H. Qureshi, and M. P. Chen, "Reduced-state Sequence Estimation for Trellis-coded Modulation on Intersymbol Interference Channels," in *Proc. Globecom*, (Hollywood, FL), pp. 878–882, Nov. 1988.

[49] M. V. Eyuboglu and S. U. H. Qureshi, "Reduced-state Sequence Estimation for Coded Modulation on Intersymbol Interference Channels," *IEEE J. Sel. Areas Commun.*, vol. 7, no. 6, pp. 989–995, Aug. 1989.

Index